KANGAROO MEAT

Australian frontiersmen in remote areas boil or roast the legs of the kangaroo, which look something like ham, and which are especially good when spiked with bacon. Kangaroo tail is used as the base for a rich and heavy soup, which is considered quite a delicacy and is sold commercially. Fearing that the animal might become extinct, in 1973 the Australian Parliament passed a law banning the sale to other countries of live kangaroos or of kangaroo hides or meat.

KARELL DIET

A special diet for the early stage of victims of congestive heart failure or myocardial infarction. It is a low-calorie fluid diet consisting of only 27 oz (*800 ml*) of milk per day. The milk is given every four hours in 6½ oz (*200 ml*) servings. No other food and little additional fluid is offered for 2 to 3 days.

KCAL

The abbreviation for kilocalorie, a measure of food energy. (Also see KILOCALORIE.)

KEI-APPLE (UMKOKOLA) *Dovyalis caffra; Aberia caffra*

The fruit of a shrub (of the family *Bixaceae*) that is native to Africa, and is used locally to make jam.

KELP (SEAWEED)

One of the names by which seaweed is known. (Also see SEAWEED.)

KEMPNER DIET

Used in the treatment of high blood pressure, some kidney disorders, cirrhosis of the liver, and pregnancy toxemia, the Kempner diet is a rice-fruit diet. As such it is low in fat, protein, and sodium. Specifically, it is comprised of 10.5 oz (*300 g*) of raw rice cooked by boiling, or by steaming without milk, fat, or salt. With this, liberal amounts of canned or fresh fruit may be eaten. Daily fluid intake consists of 1 qt (*700 to 1,000 ml*) of fruit juice, but no additional water.

KERATIN

A sulfur-containing protein which is the primary component of epidermis, hair, wool, hoof, horn, and the organic matrix of the teeth. (Also see PROTEIN[S].)

KERATINIZATION

A condition occurring in a severe vitamin A deficiency, in which the epithelial cells either slough off or become dry and flattened, then gradually harden and form rough horny scales. The process may occur in the cornea, the respiratory tract, the genitourinary tract, or the skin. (Also see VITAMIN A, section on "Deficiency Symptoms.")

KERNEL

The whole grain of a cereal. The meats of nuts and drupes (single-stoned fruits). (Also see CEREAL GRAINS.)

KETO–

A prefix denoting the presence of the carbonyl (CO) group.

KETO-ACID

The amino acid residue after deamination. The glycogenic keto-acids are used to form carbohydrates; the ketogenic keto-acids are used to form fats.

KETOGENESIS

The formation of ketones from fatty acids and some amino acids.

KETOGENIC

Capable of being converted into ketone bodies. The fatty acids and certain amino acids are the ketogenic substances in metabolism.

KETOGENIC AMINO ACID

The amino acids leucine, isoleucine, phenylalanine, tryptophan, and tyrosine are capable of undergoing a metabolic conversion to acetoacetate, a ketone body. Thus, they are said to be ketogenic. (Also see AMINO ACID[S]; KETOSIS; and METABOLISM.)

KETOGENIC DIET

Diets high in fat and low in carbohydrate are termed ketogenic since they cause the accumulation of ketone bodies in the tissues. These diets eliminate carbohydrate sources such as breads, cereals, fruits, desserts, sweets, and sugar-containing beverages while foods high in fat such as butter, cream, bacon, mayonnaise, and salad dressing are eaten in generous amounts. Ketones are produced from the breakdown (oxidation) of fats. When the ratio of fatty acids to available glucose in the diet exceeds 2:1, ketosis occurs.

A ketogenic diet is considered monotonous and unpalatable. In some cases a ketogenic diet may be prescribed to control epilepsy, if drugs prove ineffective. (Also see KETOSIS; METABOLISM; and MODIFIED DIETS.)

KETONE

Any compound containing a ketone (CO) grouping.

KETONE BODIES

Acetoacetic acid, acetone, and beta-hydroxybutyric acid.

KETONIC RANCIDITY

Decomposition of fats and oils results in the production of ketones. These ketones give fats or oils an off flavor or odor. The decomposition occurs slowly and spontaneously, however, certain molds of *Aspergillus* and *Penicillium* species may also be responsible.

KETOSIS

Shifting the metabolic machinery of the body to excessive utilization of fats instead of carbohydrates or a balance of fats and carbohydrates results in the buildup of ketone bodies—acetoacetate, beta-hydroxybutyrate, and acetone—in the blood and their appearance in the urine. This condition is referred to as ketosis, and outwardly noted by the sweetish, acetone odor of the breath. Three circumstances can cause ketosis: (1) high dietary intake of fat but low carbohydrate intake as in ketogenic diets; (2) diminished carbohydrate breakdown and high mobilization of fats as in starvation; or (3) disorders in carbohydrate metabolism as in diabetes mellitus. Unless ketosis goes unchecked and results in acidosis, it is a normal metabolic adjustment.

(Also see CARBOHYDRATE[S], section on "Regulation of Fat Metabolism"; FATS AND OTHER LIPIDS, section on "Recommended Fat Intake"; and METABOLISM.)

KIBBLED

Coarsely ground grain or meal.

KIDNEY DISEASES

Due to their importance, any disorder in the function or structure of the kidneys can have serious consequences, and often require some dietary adjustments. Glomerulonephritis (nephritis), nephrosis, nephroptosis, nephrosclerosis, renal (kidney) failure, kidney stones (nephrolithiasis), pyelitis, and pyelonephritis are all specific diseases which may in general be called kidney diseases.

Before there are any dietary adjustments, the extent of the kidney disease must be evaluated. It may be necessary to restrict sodium and potassium intake, and control the levels of protein and fluid intake. If protein intake is controlled, then it is important that sufficient high caloric but nonprotein food be eaten to meet the body's demand for energy, and to prevent the breakdown of the body's own protein for energy. Those proteins that are eaten should be proteins of high biological value—containing a large number of essential amino acids—like eggs, meat, milk, and milk products.

(Also see AMINO ACID[S]; GLOMERULONEPHRITIS; KIDNEY STONES; MODIFIED DIETS; NEPHROSCLEROSIS; NEPHROSIS; and PROTEIN[S].)

KIDNEYS

They are paired, bean-shaped organs located in the abdominal cavity at the level of the lower ribs next to the back with one on each side of the spine. The kidneys form urine and control the volume and the composition of the blood (extracellular fluid). Each day, the kidneys (1) filter about 50 gal *(190 liter)* of fluid, and (2) remove excesses of substances such as urea and uric acid while conserving needed substances such as water and glucose. Without their cleansing-filtering function, death results in 8 to 14 days. Kidneys also regulate red blood cell production, adrenal secretion of aldosterone, blood pressure, and calcium metabolism.

(Also see CALCIUM, section on "Excretion of Calcium"; GLOMERULONEPHRITIS; HEMODIALYSIS; and WATER BALANCE.)

KIDNEY STONES (NEPHROLITHIASIS; RENAL CALCULI; URINARY CALCULI; UROLITHIASIS)

Hard concentrations primarily composed of calcium oxalate or calcium phosphate which form within the kidney. They vary in size from very small gravel to very large stones. Apparently, a variety of factors may contribute to the formation of kidney stones, and to some extent they may be controlled by diet. Some common contributing factors are (1) a high phosphorus–low calcium diet; (2) a high potassium intake; (3) vitamin A deficiency; (4) high animal protein (meat, fish, poultry) intake; (5) renal infection; (6) urinary tract obstruction; (7) chronic dehydration; (8) long periods of immobilization; (9) hypercalcemia; or (10) heredity. Generally, a kidney stone goes unnoticed until it starts to move out of the kidney and down the ureter. When this occurs, there is excruciating pain until the stone is passed or removed surgically or nonsurgically.

Note: Currently, a German-built machine which uses high-energy shock waves to break up kidney stones from outside the body is being tested. The nonsurgical procedure is called extracorporeal shock wave lithotripsy, or stone exploding.

If at all possible the cause and type of kidney stones should be determined since a person forming a kidney stone once is apt to form them repeatedly.

Dietary modification depends on the type of stones formed. The formation of calcium phosphate stones may be lessened by a calcium- and phosphorus-restricted diet, and by aluminum gel which diminishes the absorption of phosphorus. Modification of the urine pH also prevents the formation of kidney stones. An acid urine helps prevent stones of calcium and magnesium phosphate and carbonates, while an alkaline urine helps prevent oxalate and uric acid stones. The diet should support therapy with alkalinizing or acidifying agents by supplying acid-producing or alkaline-producing foods. Also the formation of uric acid stones may be lessened by a purine-restricted diet.

Most importantly, a liberal intake of fluid—3 to 4 qt *(3 to 4 l)* or more depending upon the environment—is essential to prevent the formation of a concentrated urine which allows the stone-forming salts to precipitate.

(Also see CALCIUM, section headed "Calcium Related Diseases"; MODIFIED DIETS; and URIC ACID KIDNEY STONES.)

KILOCALORIE (kcal)

The amount of energy as heat required to raise the temperature of 1 kilogram of water 1°C (from 14.5 to 15.5°C). It is equivalent to 1,000 calories. In human nutrition, it may also be referred to as a kilogram calorie, k-calorie, kcal, or a *large Calorie*. The last designation is spelled with a capital C to distinguish it from the *small calorie*. However, in some of the literature, the word *calorie* is used, even though it is technically incorrect.

(Also see CALORIE.)

KILOGRAM

A metric measure of weight which is equal to 1000 g or about 2.2 lb.

(Also see WEIGHTS AND MEASURES.)

KILOJOULE (KJ)

A metric unit of energy, equivalent to 0.239 kcal.

KINETIC ENERGY (MECHANICAL ENERGY)

A body, an object, or a molecule in motion which is capable of performing work at once. It possesses kinetic energy, energy of motion.

KJELDAHL

Relating to a method of determining the amount of nitrogen in an organic compound. The quantity of nitrogen measured is then multiplied by 6.25 to calculate the protein content of the food or compound analyzed. The method was developed by the Danish chemist, J. G. C. Kjeldahl, in 1883.

(Also see ANALYSIS OF FOOD.)

KOHLRABI *Brassica oleracea,* variety *caulorapo*

The name of this vegetable, which belongs to the mustard family *(Cruciferae),* literally means "cabbage (kohl)-turnip (rabi)." It has both the wild cabbage *(B. oleracea)* and the wild turnip *(B. campestris,* species *rapifera)* as ancestors. Kohlrabi is unlike other cabbage vegetables in that it is comprised of stem tissue, while the rest of these vegetables are buds (sprouts), flowers, and leaves. Fig. K-1 shows a typical kohlrabi.

Fig. K-1. Kohlrabi, a vegetable derived from the wild cabbage and the wild turnip. (Courtesy, Field Museum of Natural History, Chicago, Ill.)

ORIGIN AND HISTORY. It is not known for certain when the kohlrabi was first developed, because a vegetable described in the writings of the Roman botanist, Pliny, (1st century, A.D.) had very similar characteristics. However, it is not mentioned in any other writings until after the Middle Ages, when it was a popular vegetable in the central and eastern parts of Europe. Since then, the utilization of kohlrabi has declined to the extent that it is difficult to find in vegetable markets. One of the reasons for the decline in popularity of the vegetable may have been the production of large, tough bulbs the size of softballs, rather than the more tender bulbs that are usually smaller than baseballs.

PRODUCTION. There is little commercial production of the kohlrabi in the United States; hence, statistics on the size of the crop are not available. But per capita consumption is low.

Kohlrabi seeds may be planted as soon as the soil can be cultivated in the spring. The vegetable may also be planted in the summer for harvesting in the fall. A rich soil is needed for high-quality kohlrabi. Therefore, it is often desirable to fertilize with nitrogen, phosphorus, and potassium.

The bulblike vegetables may be harvested when they are between 2 and 3 in. *(5 and 8 cm)* in diameter. It is not wise to let them grow any larger because the edible portion becomes fibrous and tough when it reaches full maturity. The edible portion is the bulbous enlargement of the stem, which like all stems, becomes very woody as it ages.

PROCESSING. It appears that most, or all, of the kohlrabi produced in the United States is marketed fresh, and that very little, if any, is processed.

SELECTION AND PREPARATION. Best quality kohlrabi is from 2 to 3 in. *(5 to 8 cm)* in diameter, and has fresh green tops and a tender rind. Bulbs that are overly large (over 3 in. in diameter), and that have blemishes or growth cracks, are likely to be of poor quality.

Raw kohlrabi may be sliced and added to relish trays and salads. The vegetable is usually cooked by steaming, although the people in the Central European countries may stuff and bake it after scooping out some of the center. Cooked kohlrabi is enhanced by a well-seasoned cheese sauce or cream sauce.

CAUTION: Kohlrabi and other vegetables of the mustard family *(Cruciferae)* contain small amounts of goiter-causing (goitrogenic) substances that may interfere with the utilization of iodine by the thyroid gland. In the quantities normally consumed, this is not a concern. Besides, this effect may be counteracted by the consumption of ample amounts of dietary iodine, a mineral that is abundantly present in iodized salt, ocean fish, seafood, and seaweeds such as kelp.

NUTRITIONAL VALUE. The nutrient composition of kohlrabi is given in Food Composition Table F-21.

Some noteworthy observations regarding the nutrient composition of kohlrabi follow:

1. Kohlrabi is high in water content (over 90%) and low in calories (only about 40 kcal per cup *[240 ml]).*
2. The levels of most nutrients are lower in kohlrabi than

in the other cabbage vegetables because it is comprises of tem tissue, whereas, the other vegetables are buds (sprouts), flowers, and leaves. However, kohlrabi is a good source of potassium and vitamin C.

3. Kohlrabi is most comparable to the potato and the other tuberous vegetables that are enlarged stems of plants. On this basis, kohlrabi has about the same amount of protein (2%) as the potato, but contains less then half as many calories, and more than three times as much vitamin C.

(Also see VEGETABLE[S], Table V-2 Vegetables of the World.)

KOJI

A yeast or other starter prepared in Japan from rice inoculated with the spores of the mold *Asperigillus oryzae* and permitted to develop a mycelium. Koji is used for making miso from soybeans, a popular fermented food in Japan.

KORSAKOFF'S SYNDROME (PSYCHOSIS)

A mental disease involving disintegration of the personality or escape from reality. Typical behavior includes confused thinking, making up stories to fill in gaps of memory (confabulation), irresponsibility, and the inability to learn new things. Eventually a stupefied state may result. These effects may be commonly found in chronic alcoholics, semistarved persons such as prisoners of war, and persons suffering from beriberi. Most evidence seems to indicate that the Korsakoff's Syndrome is due to a long-standing thiamin deficiency. Vitamin therapy restores responsiveness and alertness, but some irreversible structural damage to the nervous system may have occurred. Often the early symptoms of a thiamin deficiency—Wernicke's Syndrome—are noted prior to the mental disorders, and the physical and mental symptoms are together called the *Wernicke-Korsakoff Syndrome*.

(Also see ALCOHOLISM; and BERIBERI.)

KUBAN

A type of fermented milk.

KWASHIORKOR

Kwashiorkor is most common in the tropical areas of the world, where it afflicts children from weaning to age 4.

Kwashiorkor results from a severe protein deficiency, and is characterized by changes in pigmentation of the skin and hair, edema, skin lesions, anemia, and apathy. Kwashiorkor differs from marasmus in terms of the greater role of protein

Fig. K-2. Kwashiorkor, a protein deficiency disease. Note characteristic bloated belly. Other usual symptoms are: stunted growth, diarrhea, brittle hair with an abnormal reddish color, and retarded mentality. (Courtesy, FAO, Rome, Italy)

deficiency in causing it.

The recommended treatment: Providing 1 to 2 g of high-quality protein per pound of body weight, along with sufficient energy. Cautious replacement of mineral salts may also be needed. In case of severe diarrhea, the protein intake should be temporarily restricted to 0.5 g per pound of body weight.

Prevention consists in feeding children an adequate amount of protein (particularly after weaning) from such sources as fish, legumes, meats, milk, and nuts. (Milk supplies 1 g of protein per oz [30 ml].)

(Also see MALNUTRITION, PROTEIN-ENERGY; and PROTEIN(S), section headed "Protein Malnutrition.")

LABELS, FOOD

On November 8, 1990, President Bush signed into law *The Nutrition Labeling and Education Act of 1990*, the most significant revision of the U.S. food labeling law since 1938. The new labeling law requires that the majority of food products under FDA authority carry nutrition labeling, regulates the terms used on labels to describe a food's nutrient content, authorizes and regulates food health claims, and establishes national uniformity by federal preemption for many areas of food labeling other than safety warning statements. In 1992, FDA regulatory authority encompassed about 75% of all food consumed in this country. The remaining foods—mainly meat, poultry, and egg products—are largely under the U.S. Department of Agriculture regulatory authority.

The Nutrition Labeling and Education Act of 1990 (Public Law 101–538–Nov. 8, 1990) amends the Federal Food, Drug, and Cosmetic Act to require the following changes in food labels:

• **Nutrition labeling**—Nutrition labeling is required for all retail food products, including fresh produce, fish, and seafood. **NOTE WELL:** Subsequently President Bush ruled that the law would apply only to packaged foods.

• **Health claims on food**—The Nutrition Labeling and Education Act permits health claims on foods, provided (1) that such claims are based on scientific evidence (including evidence from well-designed experiments), and (2) that there is significant agreement among experts qualified by scientific training and experience.

• **Nutrient content claims**—The FDA is required to define the following nutrient terms as they relate to nutrient content: *free, reduced, low, less, light* or *lite,* and *high.*

Further, the Act prohibits a label claim for the absence of a nutrient (e.g., "sodium-free," "no cholesterol") where the nutrient is not normally present in the food.

• **Implementation of food labeling law**—On November 8, 1990, President George Bush signed into law The Nutrition Labeling and Education Act. However, implementation of the law was delayed due to disagreement between the U.S. Secretary of Agriculture and the U.S. Secretary of Health and Human Services relative to certain aspects of food labeling. Finally, on December 2, 1992, President Bush resolved the dispute between the two cabinet secretaries, made some changes in the application of the law, announced that the regulations will be final following their publication in the Federal Register, and ruled that all labels must be on packaged foods by May 1994.

Fig. L-1 shows the format and content of a new food label similar to the type that will be standard on all packaged foods in the future. The label on a carton of yogurt will look just like the label on a can of corn.

Pertinent new food labeling rules which were revealed on December 2, 1992, at the time of implementing the food labeling law, follow.

Nutrition facts

| Serving size | ½ cup (114 g) |
| Servings per container | 4 |

Amount per serving

| Calories 260 | Calories from fat 120 |

		% Daily value*
Total fat	13 g	**20%**
Saturated fat	5 g	25%
Cholesterol	30 mg	**10%**
Sodium	660 mg	**28%**
Total carbohydrate	31 g	**11%**
Sugars	5 g	
Dietary fiber	0 g	0%
Protein	5 g	

Vitamin A 4% • Vitamin C 2% • Calcium 15% • Iron 4%

*Percents (%) of a Daily Value are based on a 2,000-calorie diet. Your Daily Values may vary higher or lower depending on your calorie needs:

Nutrient		2,000 calories	2,500 calories
Total fat	Less than	65 g	80 g
Saturated fat	Less than	20 g	25 g
Cholesterol	Less than	300 mg	300 mg
Sodium	Less than	2,400 mg	2,400 mg
Total carbohydrate		300 g	375 g
Fiber		25 g	30 g

1 g fat = 9 calories • 1 g carbohydrates = 4 calories • 1 g protein = 4 calories

Fig. L-1. A new food label.

1. **All packaged foods must be labeled.** The regulations apply to all packaged foods, but not to fresh meat, poultry, fish, and produce.

2. **Percent Daily Value will be used.** Instead of Recommended Daily Allowance (RDA), the new labels will refer to percent of *Daily Value.*

3. **The nutritional value of a food can be compared with daily dietary needs.** The new labels will provide consumers with an opportunity to compare the nutritional value of a food with the daily dietary needs. Thus, the label will show how much fat is in a product, and how much of the daily quotient of fat the consumer will get from one serving of that product. Also, the label provides consumers a way in which to compare the nutrients found in their food with their total daily dietary needs, based on recommended diets of both 2,000 calories and 2,500 calories. Thus, from the label, consumers will be able to compare the percentage of the *Daily Value* of each nutrient contained in a particular product with a chart that shows the total amount of each nutrient they should be eating.

4. **Serving sizes and descriptive terms will be standardized.** Thus, the serving size in canned soups will be the same, no matter what the brand. Likewise, descriptive phrases like "low-fat" and "light" will be standardized.

5. **Labels must be rewritten and printed.** An estimated 257,000 labels must be rewritten and printed.

6. **Complete nutritional information will be required.** If manufacturers do not report complete nutritional information on their product labels, they will be required to have their products tested.

7. **Calorie information will be given.** The new labels will include information on how many calories are in a gram of fat, in a gram of carbohydrate, and in a gram of protein.

8. **Amounts of essential nutrients, along with the requirements, must be listed.** Every label must list total grams or milligrams, along with required amounts, of fat, saturated fat, cholesterol, sodium, total carbohydrates (sugar and dietary fiber), protein, and essential vitamins and minerals.

• **Basic label and other information to be continued**—In addition to requiring changes in nutrition labeling as detailed above, the following basic information will be continued on food nutrition labels: (1) the name of the product, (2) the net contents or net weight, and (3) the name and place of business of the manufacturer, packer, or distributor. Also, label information may include the following:

• **Grades**—Some food products carry a grade on the label, such as "U.S. Grade A." Grades are set by the U.S. Department of Agriculture, based on the quality levels inherent in a product—its taste, texture, and appearance. U.S. Department of Agriculture grades are not based on nutritional content.

Milk and milk products in most states carry a "Grade A" label. This grade is based on FDA recommended sanitary standards, for the production and processing of milk and milk products, which are regulated by the states.

• **Open dating**—To help consumers obtain food that is fresh and wholesome, many manufacturers date their product. Open dating, as this practice is often called, is not regulated by FDA. Four kinds of open dating are commonly used:

1. Pack date is the day the food was manufactured or processed or packaged. In other words, it tells how old the food is at the time of purchase. The importance of this information to consumers depends on how quickly the particular food normally spoils. Most canned and packaged foods have a long shelf life when stored under dry, cool conditions.

2. Pull or sell date is the last date the product should be sold, assuming it has been stored and handled properly. The pull date allows for some storage time in the home refrigerator. Cold cuts, ice cream, milk, and refrigerated fresh dough products are examples of foods with pull dates.

3. Expiration date is the last date the food should be eaten or used. Baby formula and yeast are examples of products that may carry expiration dates.

4. Freshness date is similar to the expiration date but may allow for normal home storage. Some bakery products that have a freshness date are sold at a reduced price for a short time after the expiration date.

• **Code dating**—Many companies use code dating on products that have a long *shelf life*. This is usually for the company's information, rather than for the consumer's benefit. The code gives the manufacturer and the store precise information about where and when the product was packaged, so if a recall should be required for any reason the product can be identified quickly and withdrawn from the market.

• **Universal product code**—Many food labels now include a small block of parallel lines of various widths, with accompanying numbers. This is the Universal Product Code (UPC). The code on a label is unique to that product. Some stores are equipped with computerized checkout equipment that can read the code and automatically ring up the sale. The UPC, when used in conjunction with a computer, also can function as an automated inventory system. The computer can tell management how much of a specific item is on hand, how fast it is being sold, and when and how much to order.

• **Symbols**—Food labels can possess a variety of symbols. The letter "R" on a label signifies that the trademark used on the label is registered with the U.S. Patent Office. The letter "C" indicates that the literary and artistic content of the label is protected against infringement under the copyright laws of the United States. Copies of such labels have been filed with the Copyright Office of the Library of Congress. The symbol which consists of the letter "U" inside the letter "O", is one whose use is authorized by the Union of Orthodox Jewish Congregations of America, more familiarly known as the Orthodox Union, for use of foods which comply with Jewish dietary laws. The symbol which consists of the letter "K" inside the letter "O" is used to indicate that the food is *kosher*, that is, it complies with the Jewish dietary laws, and its processing has been under the direction of a rabbi. None of these symbols is required by, nor under the authority of, any of the laws enforced by the FDA.

(Also see FOOD AND DRUG ADMINISTRATION; and U.S. DEPARTMENT OF AGRICULTURE.)

LABILE

Unstable. Easily destroyed.

LABILE PROTEIN

The reserve protein available in most tissues.

LABRADOR TEA

This term may refer to (1) any of a group of low evergreen shrubs growing in bogs and swamps of the arctic and subarctic regions, or (2) tea made from the leaves of these plants. Labrador tea was popular with miners, mountainmen, and American colonists during the Revolution.

LACTALBUMIN

It is one of the proteins of milk, sometimes called a *whey protein*. During the making of cheese, lactalbumin is not precipitated with the protein casein; and thus, it appears in the whey. One half, or more, of the protein in human milk is lactalbumin; the rest is casein.

(Also see MILK AND MILK PRODUCTS; and PROTEIN[S].)

LACTASE

An enzyme in intestinal juice which acts on lactose to produce glucose and galactose.

(Also see DIGESTION AND ABSORPTION.)

LACTATION

That process of milk formation common to all mammals for the purpose of nurturing their young. Milk formation occurs in the mammary gland following preparation during pregnancy, and it is initiated near the time of birth. Lactating mothers require more food and more nutritious food than non-lactating females.

(Also see BREAST FEEDING; INFANT DIET AND NUTRITION; and MILK AND MILK PRODUCTS.)

LACTIC ACID (C₃H₆O₃)

During strenuous exercise lactic acid is produced in the muscles of the body. Also, this important biochemical forms during some food processes. The fermentation of pickles, sauerkraut, cocoa, tobacco, and silage produces lactic acid. Lactic acid formed by *lactobacillus* bacteria in the fermentation of milk sugar (lactose) is responsible for the sour milk flavor, and for the formation of cottage cheese and yogurt. Furthermore, it is a food additive employed as an acidulant and antimicrobial in cheeses, beverages, and frozen desserts. In addition to the above, lactic acid has some nonfood uses in the tanning, plastic, and textile industries. Commercially, lactic acid is produced by the fermentation of whey, cornstarch, potatoes, or molasses.

(Also see METABOLISM; and MILK AND MILK PRODUCTS.)

LACTOBACILLUS ACIDOPHILUS

A lactic-acid forming organism, which occurs as rods sometimes united in short chains; found in intestinal contents of young infants.

LACTOFLAVIN

An obsolete name for riboflavin (vitamin B-2) which described the origin of the isolate in the early days of riboflavin research. Hence, riboflavin isolated from milk was called lactoflavin.

(Also see RIBOFLAVIN.)

LACTOSE (MILK SUGAR)

A disaccharide found in milk having the formula $C_{12}H_{22}O_{11}$. It hydrolyzes to glucose and galactose. Commonly known as milk sugar.

(Also see CARBOHYDRATE[S], and MILK AND MILK PRODUCTS, section headed "Carbohydrates in Milk.")

LAETRILE (VITAMIN B-17, AMYGDALIN, NITRILOSIDES)

Contents	Page
History	601
Chemistry, Metabolism, Properties	601
Measurement	602
Functions	602
Deficiency Symptoms	602
Dosage	602
Toxicity	602
Sources of Laetrile	602
Summary	602

NOTE WELL: Laetrile (amygdalin, nitrilosides) has no known value for humans. The authors present this section relative to the controversial compound for informational purposes only. Further, it is listed with the vitaminlike substances because it is sometimes erroneously designated as vitamin B-17. Both the pros and cons relative to Laetrile are given, then the reader may make a judgment.

Laetrile, or amygdalin, is a natural substance obtained from apricot pits, which its advocates claim to have cancer preventive and controlling effects. Dr. Ernest Krebs, Sr., who was the first to use Laetrile therapeutically in this country, considered it to be an essential vitamin and called it vitamin B-17.

In the United States, Laetrile therapy is not approved by the Food and Drug Administration for treatment of cancer.

In a report before the American Society for Clinical Oncology (the society of cancer specialists) May 1, 1981, Dr. Charles G. Moertel of Mayo Clinic reported on the Laetrile treatment of 156 patients, all with cancers that either had not responded or were not likely to respond to other treatments. Nine months after the beginning of the study in July 1980, 102 of the patients were dead, and the other 54 had seriously *progressive cancer,* which did not respond to the Laetrile treatment. The results: Laetrile was not effective; the results were about the same as would be expected had the doctors given the patients either placebos (dummy pills, with no effectiveness), or no treatment at all.

NOTE WELL: This study pertained to the use of Laetrile as a treatment of patients "with cancers that either had not responded or were not likely to respond to other treatments"; and not as a preventive of cancer, or as a treatment of early cancers.

The following statement explains the position of the Soviet Union relative to Laetrile:

"The preparation has never been tested in the Soviet Union. And in our opinion, today there are no grounds to show any interest in it. Why? Because the Soviet medical profession feels that this chemical is a useless preparation. Laetrile does not possess those wonderful properties which are ascribed to it. Moreover, being toxic, Laetrile may actually harm a patient's health.

Soviet specialists in chemotherapy subscribe to the competent opinion of the American Cancer Society and the World Health Organization, who have conducted thorough tests of Laetrile and convincingly proved its quackish nature."[1]

Despite the above, the advocates of Laetrile advance the following arguments: There is usually a considerable time lag between the scientific validation of a medical treatment and its acceptance. Moreover, the bias against nutrition is frequently deep-seated when it comes to cancer. Then, they make the following statements in support of the use of Laetrile in the prevention and treatment of cancer:

1. It is harmless when not taken in excessive amounts all at once.

2. It is low cost.

3. The ancient Chinese got pretty good results using the same treatment (apricot kernels) for cancer.

4. Several distinguished doctors feel that it gives good results.

5. It is manufactured and used legally in about 20 countries throughout the world, including Belgium, Germany, Italy, Mexico, and the Philippines.

HISTORY. Laetrile was first extracted from apricot pits in the early 1900s, and first used as a therapy for cancer in the United States in the 1920s.

CHEMISTRY, METABOLISM, PROPERTIES.

• **Chemistry**—Laetrile or amygdalin is a nitriloside, a simple chemical compound consisting of two molecules of sugar (glucose), one molecule of benzaldehyde, and one molecule of hydrogen cyanide (HCN) (see Fig. L-2).

---

[1]Garbin, Avgust, Professor, Doctor of Science (Medicine), Deputy General, Director of the Oncological Research Center of the USSR Academy of Medical Sciences, "What is the Soviet Attitude Toward Laetrile?," *Soviet Life*, November 1980, p. 30.

Fig. L-2. Structural formula of Laetrile ($C_{20}H_{27}NO_{11}$).

There are some 20 different nitrilosides occurring in at least 1,200 different plants, many of which were or are used as food, historically or at present. The natural cyanide in such foods, including apricot kernels or bitter almonds, is locked in a sugar molecule, and is released slowly in the digestive tract of man.

• **Metabolism**—Oral doses of Laetrile pass from the stomach into the small intestine, where the substance is acted upon by enzymes. The enzymes break down Laetrile into four components, which are then absorbed into the lymph and portal systems and circulated throughout the body.

• **Properties**—White crystals; bitter taste; soluble in water and alcohol; insoluble in ether.

MEASUREMENT.   Laetrile is measured in grams.

FUNCTIONS.   Nitrilosides, including apricot kernels and bitter almonds, supply the body a low, but steady, level of HCN. Man and other mammals have an enzyme, rhodanase, which converts the cyanide to thiocyanate.

There are at least three theories as to how the HCN interferes with tumor growth, but all theories relative to the action of nitrilosides recognize the steady low-level supply of HCN as the active agent. The simplest theory postulates that cancer cells do not have rhodanase; instead, they are surrounded by another enzyme, beta-glucuronidase, which releases the bound cyanide from the Laetrile at the site of the malignancy. So, Laetrile is believed to attack and destroy only the malignant cells.

The advocates claim that Laetrile is effective in preventing tumor cells from getting a foothold. In support of this thesis, the following epidemiological data has been cited by Dr. James Cason, Professor of Chemistry, University of California, Berkeley, an authority on carcinogenic hydrocarbons:[2]

1.   In a 1958 publication, famed African physician Dr. Albert Schweitzer, stated that for *several decades* his hospital at Lamberene in Gabon did not see a single case of cancer among the cassava-eating tribes. This was attributed to the fact that the natives in the Lamberene area obtain 80 to 90% of their calories from cassava, the tubers of which contain about 0.5% of a nitriloside.

2.   Studies by the Loma Linda Hospital and the U.S.C. Medical School on the incidence of cancer in the Los Angeles basin showed that the Seventh-day Adventists had

only one-third the incidence of cancer suffered by the rest of the population. For religious reasons, the Adventists' diet is heavily vegetarian; hence, rich in nitrilosides. It is estimated that the Adventists consumed 6 to 8 mg of nitrilosides per person daily, in comparison with an estimated average daily consumption of less than 1 mg per capita by the U.S. population.

3.   Among the Hunzakuts (natives of the Himalayan Kingdom of Hunza), it has been estimated that the average daily intake of nitrilosides per capita is more than 100 mg. In addition to being Moslems, a local twist to the religion calls for eating apricot kernels; with the priests seeing that the brethren keep that part of the faith. Moreover, a young woman is not regarded as properly marriageable unless she has at least seven apricot trees in her dowry. After doing a study of morbidity of the Hunzakuts covering 100 years, the World Health Organization reported that not a single death from cancer was recorded.

NOTE WELL: The three reports cited above pertain to the use of Laetrile as a preventive of cancer, and not as a treatment of cancer.

DEFICIENCY SYMPTOMS.   The advocates claim that prolonged deficiency of Laetrile, or amygdalin, may lead to lowered resistance to malignancies.

DOSAGE.   The usual dosage is 0.25 to 1.0 g taken at meals. Cumulative daily amounts of more than 3.0 g are sometimes taken, *but more than 1.0 g should never be taken at any one time.*

According to the advocates, 5 to 30 apricot kernels eaten through the day may be effective as a preventive of cancer, but they should not be eaten all at one time.

• **Laetrile intake in average U.S. diet**—It is estimated that the average daily consumption of nitrilosides by the U.S. population is less than 1 mg per person.

TOXICITY.   Toxicity levels of Laetrile haven't been established, but one should exercise extreme caution in order to avoid ingesting excessive amounts. *More than 1.0 g should not be taken at any one time.*

SOURCES OF LAETRILE.   A concentration of 2 to 3% Laetrile is found in the whole kernels of most fruits, including apricots, apples, cherries, peaches, plums, and nectarines.

SUMMARY.   Regardless of the validity of the claims and counter claims pertaining to Laetrile, it is noteworthy that more and more medical authorities are accepting the nutrition approach in the prevention and treatment of cancer. Although the use of such nutrients as Laetrile, vitamin A (carotene), folic acid, and vitamin C, along with a general supplemental program of minerals and vitamins, is controversial as a cancer treatment, perhaps even more important in the long run is the apparent success of nutrients of many kinds in strengthening the body's immune system—as preventives.

There is evidence that a person's immune capability is enhanced by top physical and mental condition; in turn, physical and mental condition depend heavily on diet. Together, they impart the *fighting spirit* and the *will to live,* terms that are often heard in the medical profession.

(Also see VITAMIN[S], Table V-5.)

---

[2]Cason, James, "Ascorbic Acid, Amygdalin, and Carcinoma," *The Vortex,* June 1978, pp. 9-23.

FOOD PRODUCTION

Plate 1. Wheat—the staff of life. It provides more nourishment for more people throughout the world than any other food. (Courtesy, USDA)

Plate 4. Adzuki beans, green and flowering. This legume ranks next to soybeans in the Orient. (Courtesy, University of Minnesota, St. Paul)

Plate 2. Summer squash. This shows a Yellow Crookneck Squash ready for harvest, when they are tender and before the rind has hardened. (Courtesy, USDA)

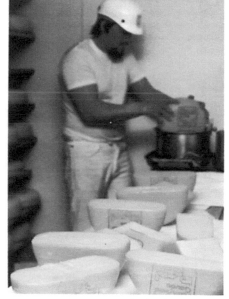

Plate 5. Cheese. This milk product is manufactured by (a) exposing milk to specific bacterial fermentation, and/or (b) treating with enzymes. (Courtesy, University of Minnesota, St. Paul)

Plate 3. Pineapple. This shows ripe Hawaiian pineapple being hand-harvested and placed on a conveyor belt built into a long boom. (Courtesy, Castle & Cook, Inc.—Dole Pineapple)

Plate 6. Turkeys. This shows Broad-Breasted White turkeys on the range. (Courtesy, USDA)

FOOD MARKETING

Plate 7. Cabbage. This shows several types of cabbage in a market, where it is sold by bundle and priced by weight. (Courtesy, University of Hawaii)

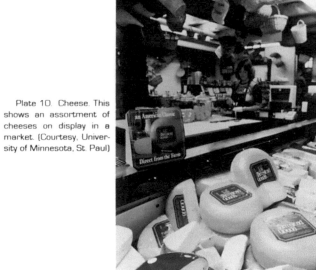

Plate 10. Cheese. This shows an assortment of cheeses on display in a market. (Courtesy, University of Minnesota, St. Paul)

Plate 8. Cucumbers. Five long, slender cucumbers are shown on display in a market. (Courtesy, USDA)

Plate 11. Meat. Customer being assisted at the supermarket. (Courtesy, National Live Stock & Meat Board, Chicago, Ill.)

Plate 9. Plums. This shows plums on display in a market. (Courtesy, USDA)

Plate 12. Vegetables and fruits. An assortment of vegetables and fruits on display in a market. (Courtesy, USDA)

BREAD, CEREAL, RICE, & PASTA GROUP

Plate 13. Bread—made from whole wheat. The Bread, Cereal, Rice, & Pasta Group forms the broad base of the *Food Guide Pyramid*, with 6-11 daily servings of this group recommended. (Courtesy, Washington State University, Pullman)

Plate 16. Bran muffins, made from the outer coat of wheat, are high in fiber (10 to 14%) and a good source of minerals and vitamins. (Courtesy, USDA)

Plate 14. Wheat cereal grain products. Whole cereal products supply most of the essential nutrients. (Courtesy, USDA)

Plate 17. Rice. There are 25 species and thousands of varieties of rice. (Courtesy, USA Rice Council, Houston, Tex.)

Plate 15. A bowl of cooked cereal decorated with a face made of assorted fruits, for a preschooler. (Courtesy, USDA)

Plate 18. Pasta (macaroni), pieces of dough that have been formed into various shapes, then dried. Pasta contains sufficient calories and protein to be a major dietary source of these essentials. (Courtesy, USDA)

VEGETABLE GROUP

Plate 19. Carrots, the richest source of vitamin A among the common vegetables. The *Food Guide Pyramid* calls for 3-5 daily servings of the Vegetable Group. (Courtesy, USDA)

Plate 20. Tomatoes. Fresh tomatoes are high in water content (about 94%), low in calories, and a good source of vitamins A and C. (Courtesy, Shell Chemical Company, Houston, Tex.)

Plate 21. Onions. Dry mature onions are high in water content (89%) and low in calories. They are distinguished by their pungency when raw and the appetizing flavor that they impart when cooked with various other foods. (Courtesy, USDA)

Plate 22. Green beans. They have higher nutritive value than most vegetables; they are a good source of protein, calcium, iron, and potassium. (Courtesy, USDA)

Plate 23. Chef's salad. The calories in such a salad may be minimized by using lowfat dressing, lowfat dairy products, and/or lean meat, poultry, or fish. (Courtesy, USDA)

Plate 24. Corn-on-the-cob, a rich source of carbohydrates and fats, featured with the main meal. (Courtesy, USDA)

FRUIT GROUP

Plate 25. Grapes, one of the richest sources of chromium and a good source of potassium. The *Food Guide Pyramid* calls for 2-4 servings of the Fruit Group daily. (Courtesy, California Table Grape Commission)

Plate 28. Blueberries. They are a good source of fiber, iron, and bioflavonoids, and fair sources of potassium and vitamin C. (Courtesy, USDA)

Plate 26. Peach, a good source of potassium and an excellent source of vitamin A. (Courtesy, National Film Board of Canada)

Plate 29. Strawberries, a good source of iron and potassium, and an excellent source of vitamin C. (Courtesy, USDA)

Plate 27. Oranges. Note that ripe oranges and blossoms are on the same tree at the same time. Oranges are a good source of potassium and an excellent source of vitamin C. (Courtesy, USDA)

Plate 30. Pineapple juice, a refreshing beverage, and a fair source of vitamins A and C. (Courtesy, USDA)

MEAT, POULTRY, FISH, DRY BEANS, EGGS, & NUTS GROUP

Plate 31. Beef burger, an excellent source of high-quality protein, iron, and the B-complex vitamins. Vitamin B-12 is found in all foods of animal origin, but not in plants. The *Food Guide Pyramid* calls for 2-3 daily servings of the Meat, Poultry, Fish, Dry Beans, Eggs, & Nuts Group. (Courtesy, National Live Stock and Meat Board, Chicago, Ill.)

Plate 34. Eggs and milk. Eggs contain an abundance of proteins, minerals, and vitamins. Also, egg protein is highly digestible and of high quality, having the highest biological value of any food. (Courtesy, USDA)

Plate 32. Roasted turkey, a Thanksgiving and Christmas tradition. Poultry meat is high in quality of protein and a rich source of all the amino acids. (Courtesy, Poultry International)

Plate 35. Peanut butter sandwich. Peanut butter is an excellent source of protein, contains no cholesterol, and the fat is mostly unsaturated. (Courtesy, Peanut Advisory Board, Atlanta, Ga.)

Plate 36. *Below:* Meat, Poultry, Fish, Dry beans, Eggs, and Nuts Group—foods that are good, and good for you. (Courtesy, USDA)

Plate 33. A variety of dishes made with beans and ground beef. Meat may be extended and lowered in cost by the use of beans. (Courtesy, USDA)

MILK, YOGURT, & CHEESE GROUP

Plate 40. Cottage cheese-fruit parfait—cottage cheese layered with fruits. The most important cheese produced from skimmed milk is cottage cheese. (Courtesy, USDA)

Plate 37. Milk, which Hippocrates, the father of medicine, described as "the most nearly perfect food." This must be so, for all newborn mammals rely almost entirely on their mother's milk for food. The *Food Guide Pyramid* calls for 2-3 servings daily of the Milk, Yogurt, & Cheese Group. (Courtesy, National Dairy Board, Arlington, Va.)

Plate 38. Yogurt, frozen. The increased consumption of yogurt in the U.S.A. in recent years is largely attributed to fruit-flavored and low-calorie yogurts. (Courtesy, American Dairy Assn., Rosemont, Ill.)

Plate 41. Pizza. This is an Italian word meaning pie, which a pizza resembles. Typically, pizzas contain mozzarella and parmesan cheeses, sausages, and other good things. (Courtesy, American Dairy Assn., Rosemont, Ill.)

Plate 42. *Below:* Assorted cheeses. There are more than 2,000 different named cheeses. Cheese is made by (a) exposing milk to specific bacterial fermentation, or (b) treating with enzymes, or both methods, to coagulate some of the proteins. More than 30% of all the milk used in manufactured dairy products in the U.S.A. is processed into cheese (exclusive of cottage cheese). (Courtesy, USDA)

Plate 39. *Below:* Ice cream. It consists primarily of milk fat, nonfat milk solids, sugar, flavoring, and stabilizers. (Courtesy, American Dairy Assn., Rosemont, Ill.)

FATS, OILS, & SWEETS GROUP

Plate 43. Salad dressings, many of which are high in fats and oils. *Caution:* The 1989 Recommended Dietary Allowances calls for maintaining total fat intake at or below 30% of the total caloric intake and saturated fatty acid intake of less than 10%. So, less fats and oils should be used in salad dressings.

Plate 46. Sugar being poured from a spoon. *Caution:* The new *Food Guide Pyramid* admonishes that Fats, Oils, and Sweets should be *used sparingly.* (Courtesy, USDA)

Plate 44. Doughnuts are generally high in oil because they are fried in deep oil. Thus, those who eat doughnuts will need to restrict the consumption of other foods high in fats/oils in order to restrict the total fat intake at or below the 30% of the total caloric intake as called for by the 1989 RDA. (Courtesy, American Soybean Assn., St. Louis, Mo.)

Plate 47. Candy. This shows candy on display. The story of candy is the story of sugar, because basically candymaking is simply the boiling of sugar. (Photo by A. H. Ensminger)

Plate 45. Sugarcane, source of 60 to 62% of the world's refined sugar. (Courtesy, USDA)

Plate 48. Coffee cake. Generally, coffee cake is low in fat, but high in sugar. So, remember the admonition of the *Food Guide Pyramid: Use sparingly.*

FOOD PRODUCTION

Plate 49. Buckwheat, in full bloom. Buckwheat is used for about the same purposes as the cereal grains, but it is not a cereal; it is related to rhubarb. (Courtesy, University of Minnesota, St. Paul)

Plate 52. Milking. This shows dairy cows being milked by machine in a milking parlor. (Courtesy, University of Maryland, College Park)

Plate 50. Amaranth. It is grown for the leaves as a green vegetable and for the seed as a grain. (Courtesy, University of Minnesota, St. Paul)

Plate 53. *Below:* Cowpeas. This shows a field of cowpeas in south-eastern U.S.A. The plants have a trailing and bushy vine, but do not climb. (Courtesy, University of Minnesota, St. Paul)

Plate 51. *Below:* Cherries. This shows a cluster of the fruit ready for harvest. (Courtesy, USDA)

Plate 54. Layers in cages. More than 90% of U.S. layers are in cages or on wire. (Courtesy, USDA)

FOOD MARKETING

Plate 55. Celery. This shows harvested celery being packed in the field for shipment to market. (Courtesy, USDA)

Plate 58. Cheese, bread, and wine—an attractive display. (Courtesy, USDA)

Plate 56. Brussel sprouts in a supermarket. (Courtesy, USDA)

Plate 59. *Below:* Meat, showing customer selecting from a meat display in a supermarket. (Courtesy, USDA)

Plate 57. *Below:* Strawberries, fully ripe and glistening in a market. Strawberries are the world's leading berry-type fruit. (Courtesy, USDA)

Plate 60. *Below:* Vegetables and fruits—an attractive display. (Courtesy, University of Florida, Gainesville)

BREAD, CEREAL, RICE, & PASTA GROUP

Plate 61. Whole wheat bread—a member of the Bread, Cereal, Rice, & Pasta Group, of which 6-11 daily servings are recommended in the *Food Guide Pyramid.* (Photo by A. H. Ensminger)

Plate 64. Rice, of which 7 kinds are displayed. Of the thousands of varieties, only about 25 are grown commercially in the U.S.A. (Courtesy, USA Rice Council, Houston, Tex.)

Plate 62. Corn flakes. This ready-to-eat breakfast cereal is manufactured from deskinned, degerminated kernels of soaked dried corn by flavoring, rolling, and toasting. (Courtesy, USDA)

Plate 65. A pita bread sandwich. (Courtesy, USDA)

Plate 66. *Below:* Pasta (macaroni). This shows pasta used to extend meat—a high-quality protein. (Courtesy, National Live Stock and Meat Board, Chicago, Ill)

Plate 63. Muffins with turkey bacon. (Courtesy, Oscar Mayer Food Corp., Madison, Wisc.)

VEGETABLE GROUP

Plate 67. Onions. This vegetable is used primarily for its flavor and appetite appeal. The *Food Guide Pyramid* calls for 3-5 daily servings of the Vegetable Group. (Courtesy, USDA)

Plate 70. Red beans and rice. A mixture of 50% of each beans and rice has a protein quality that approaches meat, milk, and other animal proteins. (Courtesy, USDA)

Plate 68. Tomatoes, a member of the nightshade family. Ripe (red) tomatoes are low in calories and a good source of vitamins A and C. (Courtesy, Rutgers University, New Brunswick, N.J.)

Plate 71. Peas in a pod. Peas are usually picked when immature. They are rich in protein, an excellent source of vitamin A, and a good source of vitamin C. (Courtesy, USDA)

Plate 69. Spinach salad, with turkey and bacon. Spinach has a high water content (90 to 93%), is low in calories, and is an excellent source of magnesium, iron, potassium, and Vitamin A. (Courtesy, Oscar Mayer Food Corp., Madison, Wisc.)

Plate 72. *Below:* Corn-on-the-cob. Three ears of sweet corn-on-the-cob contain about 240 Calories (kcal). Yellow corn is also a fair source of vitamin A. (Courtesy, USDA)

FRUIT GROUP

Plate 73. Grapes, the leading fruit crop of the world. Fresh table grapes are one of the richest sources of chromium and a good source of potassium. The *Food Guide Pyramid* calls for 2-4 servings of the Fruit Group daily. (Courtesy, California Table Grape Commission)

Plate 76. Strawberries, a good source of iron and potassium, and an excellent source of vitamin C. (Courtesy, National Film Board of Canada)

Plate 74. Peaches. Fresh peaches contain 89% water. Additionally, each 100 g (about 3 1/2 oz or I medium-sized peach) contain 38 Calories (kcal) of energy, 202 mg of potassium, 1,330 IU of vitamin A, and only I mg of sodium. (Courtesy, USDA)

Plate 77. *Below:* Fruits for snacks. Fruits provide fiber, few calories, no cholesterol, and little sodium or fat. (Courtesy, North Dakota State University, Fargo)

Plate 75. Oranges ready for harvest. Fresh oranges are low in calories (47 kcal per 100 g), but they are good sources of fiber, pectin, potassium, vitamin C, inositol, and bioflavonoids. (Courtesy, Sunkist, Van Nuys, Calif.)

Plate 78. Lemonade, a refreshing beverage and a good source of vitamin C. (Courtesy, Sunkist, Van Nuys, Calif.)

Plate 79. Crown roast of lamb—an elegant dish. Lamb is an excellent source of high-quality protein, a rich source of phosphorus and iron, and an excellent source of vitamins B-12, B-6, biotin, niacin, pantothenic acid, and thiamin. The *Food Guide Pyramid* calls for 2-3 daily servings of the Meat, Poultry, Fish, Dry Beans, Eggs, & Nuts Group. (Courtesy, National Live Stock and Meat Board, Chicago, Ill.)

Plate 82. Garbanzo beans (Chickpeas). Beans may be used (a) to upgrade the protein quantity and quality of diets based mainly on cereals and starchy foods, and (b) to extend and lower the cost of animal proteins. (Courtesy, University of Minnesota, St. Paul)

Plate 80. Roasted turkey. Poultry meat is a good source of phosphorus, iron, copper, and zinc; a rich source of vitamins B-12 and B-6; and a fair source of vitamins A, biotin, niacin, pantothenic acid, riboflavin, and thiamin. (Courtesy, Canadian Marketing Agency, Ontario, Canada)

Plate 83. Bacon and eggs, the traditional American breakfast. Bacon contains all the essential amino acids, is rich in iron, and is high in the B-vitamins. Eggs contain an abundance of proteins, minerals, and vitamins. (Courtesy, USDA)

Plate 81. Seafood dinner. Fish and seafood are high in the essential amino acids, and rich in minerals and vitamins. Also, the fatty acids present are polyunsaturated; hence, they can play a major role in low-cholesterol diets. (Courtesy, U.S. Department of Interior)

Plate 84. Almonds, which come in more forms than any other nut. Nuts are high in energy and protein; a good source of magnesium, zinc, and copper; and a good source of the B-vitamins. (Courtesy, Almond Board of California, Sacramento)

MILK, YOGURT, & CHEESE GROUP

Plate 85. Milk, the major components of which are carbohydrates, fat, and protein. Milk is also a valuable source of certain minerals and vitamins. The *Food Guide Pyramid* calls for 2-3 servings daily of the Milk, Yogurt, & Cheese Group. (Courtesy, American Dairy Assn., Rosemont, Ill.)

Plate 88. Sour cream. Today, most sour cream is cultured, much the same way as buttermilk. It is a uniformly textured, smooth product which is widely used for flavoring. (Courtesy, American Dairy Assn., Rosemont, Ill.)

Plate 89. Pizzas. This Italian creation is a flat leavened bread topped with such good things as mozzarella and parmesan cheeses, sausages, anchovies, tomatoes, mushrooms, olives, herbs and spices. (Courtesy, USDA)

Plate 86. Yogurt, a fermented milk product prepared from lowfat milk, skim milk, or whole milk. (Courtesy, National Dairy Board, Arlington, Va.)

Plate 90. *Below:* Assorted cheeses. Milk can be, and is, processed into many different varieties of cheese. Some are made from whole milk, others from milk that has had part of the fat removed, and still others from skimmed milk. American types of cheese (Cheddar, Colby, Washed curd, Stirred curd, Monterey, and Jack) make up to 60% of the nation's cheese output. (Courtesy, American Dairy Assn., Rosemont, Ill.)

Plate 87. *Below:* Ice cream. Of all the frozen dairy desserts, ice cream contains the most milk solids (about 16 to 24%). (Courtesy, Carnation, Los Angeles, Calif.)

FATS, OILS, & SWEETS GROUP

Plate 91. Cooking oil pouring out of a glass bottle. **Caution:** The new *Food Guide Pyramid*, released by the USDA in 1992, lists fats, oils, and sweets at the apex of the pyramid, where they are accorded the least space, along with the admonition to *use sparingly*. (Courtesy, USDA)

Plate 94. The Candy Store. An assortment of candy and chocolate covered nuts and raisins. **Note well:** The new *Food Guide Pyramid* admonishes that sweets should be *used sparingly*. (Photo by A. H. Ensminger)

Plate 92. Peanut products. The nuts and skin yield about 47.5% oil. Peanuts supply about $\frac{1}{5}$ of the world's edible oil production. Peanut oil contains 76 to 82% unsaturated fatty acids. (Courtesy, Peanut Advisory Board, Atlanta, Ga.)

Plate 95. Raisin pie, which is moderately low in sugar, but moderately high in fat because of the crust. The calories in raisins are derived primarily from carbohydrates.

Plate 93. Sugarcane—a coarse grass of tropical and semitropical climates from which refined sugar is derived. (Courtesy, USDA)

Plate 96. Yogurt cheese tart. Luscious desserts need not be high in fat. This creamy dessert made with lowfat yogurt has only 6 g of fat per serving. (Courtesy, National Dairy Board, Arlington, Va.)

LAMB AND MUTTON

Contents | Page
Qualities in Lamb Desired by Consumers 603
Federal Grades of Lamb and Mutton 603
Lamb Cuts and How to Cook Them 603
U.S. Lamb Production 603
Per Capita Lamb and Mutton Consumption 603
Lamb and Mutton as Foods 603

Fig. L-3. Live sheep and other animals sent to Persia and India with an Expeditionary Force in 1670. Until meat preservation was developed, live animals had to be taken along by the explorers. (Courtesy, the Bettmann Archive, Inc., New York, N.Y.)

Lamb is the flesh of young sheep; mutton is the flesh of mature sheep. Lamb differs from mutton in both tenderness and flavor.

QUALITIES IN LAMB DESIRED BY CONSUMERS.
Consumers desire the following qualities in lamb:

1. Palatability.
2. Attractiveness.
3. Maximum muscling; minimum fat.
4. Small cuts.
5. Ease of preparation.
6. Tenderness.
7. Repeatability.

FEDERAL GRADES OF LAMB AND MUTTON. The
quality grades of lambs, yearlings, and mutton may be defined as a measure of their degree of excellence based on conformation and quality, or eating characteristics of the meat; and the yield, or percent of retail cuts, is based on the thickness of fat over the ribeye. **Note:** 0.25 in. of fat over the ribeye is considered the line between desirable and somewhat overfat lambs.

The current federal grades, which became effective July 6, 1992, on a voluntary basis, call for both quality and yield grading after removal of most of the kidney and pelvic fat, although up to 1.0% of the carcass weight in kidney and pelvic fat is allowed in carcasses.

Federally graded meats are so stamped (with an edible vegetable dye) that the grade will appear on the retail cuts as well as on the carcass and wholesale cuts. The federal quality grades of lambs, yearlings, and mutton are:

Lambs and Yearlings	Mutton
Prime	Choice
Choice	Good
Good	Utility
Utility	Cull

In addition to the quality grades given, there are the following yield grades of lamb, yearling, and mutton carcasses: Yield Grade 1, Yield Grade 2, Yield Grade 3, Yield Grade 4, and Yield Grade 5. The new standards call for both quality and yield grades. Also, the new grades apply to both (1) lamb, yearling, and mutton carcasses; and (2) slaughter (live) lambs, yearlings, and mutton sheep.

The word *quality* infers superiority. It follows that quality grades indicate the relative superiority of carcasses or cuts in palatability characteristics. In turn, palatability is associated with tenderness, juiciness, and flavor. The major quality-indicating characteristics are (1) color and texture of bone, which indicates the age of the animal; (2) firmness; (3) texture; and (4) marbling—the intermixing of fat among the muscle fibers.

Yield grades indicate the *quantity* of meat—the amount of retail, consumer-ready, ready-to-cook, or edible meat that a carcass contains. Yield grade is not to be confused with dressing percent. Dressing percent refers to the amount of carcass from a live animal, whereas yield grade refers to the amount of edible product from the carcass.

The grades of lambs, yearlings, and mutton are based on separate evaluation of (1) the palatability indicating characteristics of the lean—the *quality grade;* and (2) the indicated percent of trimmed, boneless major retail cuts to be derived from the carcass—the *yield grade.*

LAMB CUTS AND HOW TO COOK THEM. Fig. L-5
illustrates the retail cuts of lamb, and gives the recommended method or methods of cooking each.

U.S. LAMB PRODUCTION. In 1990, 362,000,000 lb
(165,000,000 kg) of lamb and mutton were produced in the United States, which represented only 0.9% of the nation's total meat production. On January 1, 1991, there were 11,200,000 sheep and lambs in the United States.

PER CAPITA LAMB AND MUTTON CONSUMP-
TION. In 1990, the U.S. per capita consumption of lamb and mutton (and most of it was lamb) was a mere 1.1 lb.

LAMB AND MUTTON AS FOODS.

Fig. L-4. Roast leg of lamb. (Courtesy, National Live Stock and Meat Board, Chicago, Ill.)

Lamb
· RETAIL CUTS ·
WHERE THEY COME FROM
HOW TO COOK THEM

LEG *LOIN*
RIB *FORESHANK & BREAST*
SHOULDER

LEG

Whole Leg
Roast

Short Cut Leg, Sirloin Off
Roast

Shank Portion Roast
Roast

Center Leg Roast
Roast

Center Slice
Broil, Panbroil, Panfry

American-Style Roast
Roast

Frenched-Style Roast
Roast

Boneless Leg Roast
Roast, Broil if butterflied

Hind Shank
Braise, Cook in Liquid

Sirloin Chop
Broil, Panbroil, Panfry, Braise

Boneless Sirloin Roast
Roast

LOIN

Loin Roast
Roast

Loin Chop
Broil, Panbroil, Panfry

Double Loin Chop
Broil, Panbroil, Panfry

FORESHANK & BREAST

Shank
Braise, Cook in Liquid

Spareribs
Braise, Broil, Roast

Boneless Rolled Breast
Roast, Braise

Riblets
Braise, Cook in Liquid, Broil

RIB

Rib Roast
Roast

Rib Chop
Broil, Panbroil, Panfry, Roast

Frenched Rib Chop
Broil, Panbroil, Panfry

Crown Roast
Roast

SHOULDER

Square-Cut Shoulder, Whole
Roast, Braise

Pre-Sliced Shoulder
Roast, Braise

Boneless Shoulder Roast
Roast, Braise

Neck Slice
Braise, Cook in Liquid

Blade Chop
Braise, Broil, Panbroil, Panfry

Arm Chop
Braise, Broil, Panbroil, Panfry

OTHER CUTS

Lamb for Stew
Braise, Cook in Liquid

Cubes for Kabobs
Broil, Braise

Ground Lamb
Broil, Panbroil, Roast (Bake)

THIS CHART APPROVED BY
NATIONAL LIVE STOCK & MEAT BOARD

Fig. L-5. The retail cuts of lamb: where they come from, and how to cook them. (Courtesy, National Live Stock and Meat Board, Chicago, Ill.)

Lamb has long been noted for its delicate flavor and tenderness. In biblical times, frequent reference was made to the desirability of lamb meat. Today, it may be featured for gourmet dining at home and in hotels and restaurants.

Lamb is important in the diet, not for its appetite appeal alone, but for the essential food nutrients that it contains. It is an excellent source of high-quality protein for muscle building and body maintenance. In most cuts of lamb, the fat may be easily separated from the lean; hence, the calories can be adjusted to desired levels. Lamb is one of the best sources of iron, needed for hemoglobin formation; and it is rich in phosphorus, needed for bones and teeth. It is also an excellent source of vitamin B-12, vitamin B-6, biotin, niacin, pantothenic acid, and thiamin. Also, lamb muscle is easily digested, so it is included in the diet of both the young and aged.

In the past, lamb was often considered to be a seasonal meat—spring lamb. Today, it is available the year-round, primarily due to freezer storage and imported lamb. Since lamb will keep frozen for long periods (12 months), large volumes are now marketed as portion-cut, frozen, and ready-to-cook.

LARD

Lard is the fat rendered (melted out) from fresh, fatty pork tissue.

The proportion of live market weight made into lard varies with the type, weight, and finish of the hogs, and the relative price of lard and the cuts of meat. A market weight hog grading U.S. No. 1 yields about 8 lb of lard per 100 lb (3.6 kg/45 kg) liveweight, whereas the same weight hog grading U.S. No. 3 yields about 16 lb per cwt (7.3 kg/45 kg).

It is noteworthy, too, that lard consumption has steadily declined since 1950. The U.S. per capita consumption of lard in 1950 was 12.6 lb; in 1990, it was only 2.2 lb.

The three most important sources of lard obtained from a hog carcass are: (1) leaf fat, (2) fat trimmings, and (3) fat backs and plates. About 75% of the fat back is rendered for lard, the other 25% is marketed fresh, frozen, or cured.

LARDING

A method of adding fat to lean meat so that it does not dry out during cooking. Narrow strips of fat (as bacon fat) are inserted into the meat with a special larding needle—a large needle with a hollow split end. The strips of fat are called lardoons.

LARK (MEADOWLARK)

Singing and game birds of which there are numerous species, found mostly in Europe, Asia, and northern Africa. The lark has a very delicate flesh and is greatly esteemed by gastronomes.

LATHYRISM

A disease due to the toxicant found in the Lathyrus seed (*Lathyrus sativus*). The toxin produces an irreversible, gradual weakness, followed by paralysis of both legs (paraplegia). Outbreaks occur in Asia and North Africa mainly during years of poor wheat crops. In these areas, it is common practice to plant Lathyrus with the wheat. If rainfall is adequate, the wheat overgrows the Lathyrus. However, in a year when the rains fail, mainly Lathyrus seed is harvested and it becomes the main dietary energy source; and the symptoms

of Lathyrism begin appearing. The disease can affect horses and cattle, too.

(Also see Table L-2, Legumes of the World, Lathyrus Pea.)

LAXATIVE

A food or drug that will induce bowel movements and relieve constipation. Bran and other fibrous foods will soften the stool and encourage more rapid movement through the digestive tract. Numerous laxative drugs are sold, most of which act by increasing the water content of the stool, or by increasing the contraction of the bowel, or by lubricating the lining of the bowel. Persons seeking relief from constipation should exercise care in the choice and use of a laxative; and, when in doubt, consult a doctor.

(Also see CONSTIPATION.)

LEAD (Pb)

Lead poisoning has dogged man for centuries. It is even speculated that it may have caused the decline and fall of the Roman Empire. In recent years, lead poisoning has been reduced significantly with the use of lead-free paints.

(Also see MINERAL[S], section headed ''Contamination of Drinking Water, Food, or Air with Toxic Minerals''; and POISONS, Table P-2, Some Potentially Poisonous [Toxic] Substances.)

LEAVEN

To make light by aerating (adding air) in any of three different methods: (1) by the fermentation of yeast, (2) by the action of baking powder, or (3) by beating air into egg whites.

(Also see BREADS AND BAKING.)

LEBEN

• A type of fermented milk in Egypt.

• A sour milk delicacy of biblical times, made from goat milk.

LEBKUCHEN

A Christmas cookie made with honey, brown sugar, almonds, candied fruit peel, and spices.

LECITHIN (PHOSPHATIDYL CHOLINE)

A versatile phospholipid found in all living organisms. It is a mixture of the diglycerides of the fatty acids stearic, palmitic, and oleic combined with the choline ester of phosphoric acid. The body is capable of synthesizing lecithin. In addition, lecithin is found in a wide variety of foods. *At this time, there is no evidence that lecithin has any nutritional significance.*

The commercial source of lecithin is predominately soybeans. It is an FDA approved food additive employed as a stabilizer and emulsifier in margarine, dressings, chocolate, frozen desserts, and baked goods. Furthermore, lecithin is also used in such products as paints, soaps, printing inks, and cosmetics, to name only a few.

(Also see ADDITIVES, Table A-3; CHOLINE, section on sources of Choline; and SOYBEAN, Table S-12, Soybean Products and Uses For Human Food.)

LEGUMES

Contents	Page
History	606
World and U.S. Production	606
Preparation and Serving of Legumes	607
Nutritional Value	609
Protein Quantity and Quality	609
Legumes of the World	610

This very large family of plants contains about 13,000 species, most of which are distinguished by their seed-bearing pods. These species occur in such diverse forms as short, erect broad-leafed plants; climbing vines; and various types of trees. Legumes rank second only to cereals in supplying calories and protein for the world's population. They supply about the same number of calories per unit weight as the cereals, but they contain from 2 to 4 times more protein. Also, the amino acid patterns of legumes complement those of cereal grains so that combinations of the two foods provide dietary protein that is used much more efficiently by the human body than that from either food alone.

One or more types of edible legumes may be found almost everywhere in the world where the soil is arable and sufficient water is available, because these plants are endowed with certain characteristics which favor their growth and propagation. First, the roots of many legumes harbor bacteria which convert nitrogen gas from the air into nitrogen compounds that may be utilized by the plant. This process is called *nitrogen fixation*. Hence, these plants may grow on soils in which the fixed nitrogen has been exhausted by the growth of other plants such as cereal grains. Also, legumes release their nitrogen compounds upon decaying, so that grains may again grow well on soils that were once depleted. Second, wild legumes have pods that burst open when mature, and scatter their seeds in all directions.

HISTORY. Historians express wonderment at the apparent wisdom of primitive peoples who produced and utilized legumes and other foods in ways which have only recently received the endorsement of professional agriculturists and nutritionists. Therefore, the history of legumes merits special consideration.

It appears that the first cultivation of legumes occurred in Southeast Asia, rather than in the Middle East, as was originally thought. The recently discovered Spirit Cave near the border between Burma and Thailand contained seeds of beans, peas, and other plants that had been there since about 9750 B.C. (as estimated by radiocarbon dating), and which closely resembled the seeds of today's cultivated plants.

The next oldest sites of agriculture, which date to about 8000 B.C., are those in the Middle East, in the region commonly referred to as the Fertile Crescent, a broad arc of land that curved northward and eastward from the Mediterranean coast of what is now Israel to the Zagros Mountains near the border between Iraq and Iran. This region was probably the place of origin of chickpeas, fava beans, and lentils.

The long dependence of China and India on almost wholly vegetarian diets led to (1) the development of soybean-based imitations of dairy products by the Chinese, and (2) the utilization of legumes for food in India; in a manner and to a degree that was unequaled elsewhere in the world.

Apparently, the first cultivation of legumes in the New World occurred around 4000 B.C. in both Peru and Mexico, but the practice gradually spread to other parts of North and South America. Hence, there was much similarity between the American Indian practices observed by the Spanish when they arrived in Latin America and those noted by the English settlers of New England. The Indians of both places planted beans among rows of corn so that the corn stalks might provide convenient supports for the climbing legume vines. This practice eliminated much of the necessity for weeding the crops.

The English colonists also learned from the Indians how to make succotash (a mixture of corn and beans). Boston baked beans were derived from the Indian method of soaking beans until they swelled, then baking them with deer fat and onions in clay pots surrounded by hot stones. The Pilgrims baked their beans on Saturday nights in order to avoid the performance of servile work on Sundays. Later, the custom changed as the strict observance of the sabbath abated, and baked beans were served on Saturday nights in New England.

During the great westward expansion in the 19th century, the settlers of the southwestern United States adopted many of the Mexican bean dishes such as frijoles (refried beans) and chili. It is believed that the original chili dish was invented by Mexican nuns.

At the present time, food technologists around the world are engaged in testing many new food products made from legumes in the hope that this approach might help to combat protein deficiency. Also, agricultural scientists are conducting experiments designed to (1) improve the per acre yields of legumes, and (2) increase the amount of nitrogen fixed by these plants.

WORLD AND U.S. PRODUCTION. The world production of major legumes totals about 186,503,000 metric tons annually. Fig. L-7 shows the leading legumes of the world, and the production of each. Fig. L-8 shows the contribution of each of the major legume crops to world production. It is noteworthy that soybeans account for 58% of the world's legumes.

Fig. L-6. Soybeans, the leading legume of the world. (Courtesy, USDA)

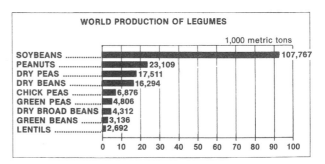

WORLD PRODUCTION OF LEGUMES

1,000 metric tons

SOYBEANS	107,767
PEANUTS	23,109
DRY PEAS	17,511
DRY BEANS	16,294
CHICK PEAS	6,876
GREEN PEAS	4,806
DRY BROAD BEANS	4,312
GREEN BEANS	3,136
LENTILS	2,692

0 10 20 30 40 50 60 70 80 90 100

Fig. L-7. The leading legume crops of the world, and the annual production of each. Based on data from *FAO Production Yearbook 1990*, FAO/UN, Rome, Italy, Vol. 44. **Note well:** Annual production fluctuates as a result of weather and profitability of the crop.

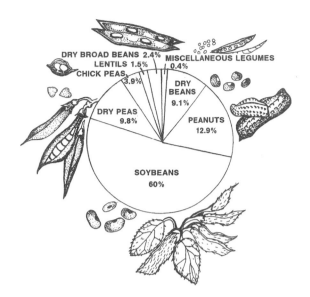

Fig. L-8. Contributions of the U.S. major legume crops to world production. (Based on data from *FAO Production Yearbook 1990*, FAO/UN, Rome, Italy, Vol. 44)

The U.S. production of major legumes totals about 60.4 million short tons annually. Soybeans and peanuts accounted for 92% and 3%, respectively, of the nation's legumes; hence, combined, these two legumes supplied an astounding 95% of U.S. legumes.

U.S PRODUCTION OF LEGUMES

million short tons

SOYBEANS	55.7
PEANUTS	1.8
DRY BEANS	1.6
GREEN BEANS	0.8
GREEN PEAS ...	0.5

0 10 20 30 40 50 60 70 80

Fig. I-9. The leading legume crops of the U.S., and the annual production of each. Based on data from *Agricultural Statistics 1991*, USDA. **Note well:** Annual production fluctuates as a result of weather and profitability of the crop.

PREPARATION AND SERVING OF LEGUMES.

Today, many American homemakers work either part time or full time and have little time to boil beans for two or more hours, unless they wish to use a crock pot and start them cooking before they leave for work in the morning. However, long cooking uses a lot of electricity or gas, which are becoming more expensive. Therefore, the U.S. Department of Agriculture and various university departments have conducted research to find methods of simplifying the cooking and serving of legumes. Their findings have led to the suggestions which follow:

1. Soaking dry beans cuts cooking time considerably. However, the traditional overnight soaking is not needed if the beans are placed in water, heated to boiling and the boiling maintained for at least 2 minutes, then set aside to soak for at least an hour. The initial boiling rapidly softens the seed coats so that water may penetrate more readily.

2. If hard water is used in cooking, it will take much longer for the legumes to cook because the calcium and magnesium in hard water combine with pectin in the peas or beans and render them tough. The remedy is to add not more than ⅛ to ¼ teaspoon (0.6 to 1.3 ml) of baking soda per cup of legumes to the cooking water. Also, such acid ingredients as tomato sauce and vinegar should not be added until the peas or beans have become tender.

3. Most legumes can be made tender by pressure cooking for 10 to 15 minutes, if the pressure is allowed to fall without allowing the steam to escape. Pressure cooking has not usually been recommended because certain legumes generate considerable foam which may plug the vent on the cooker. However, the addition of about a teaspoon or so of oil or melted fat reduces foaming greatly.

4. Whenever feasible, at least small amounts (1 oz or more) of cheese, eggs, fish, meat, milk (1 glass per meal), or poultry should be served with cooked legumes since ½ cup (120 ml) or more of the latter are required to furnish an amount of protein equivalent to only 1 oz (28 g) of meat. (Adults require a minimum daily protein intake which is equivalent to that in 8 oz of meat.) Most legumes go well with most animal foods when appropriate seasoning is used.

5. Strict vegetarians who eat no animal products (vegans) need to keep track of their protein intake so as to be certain that they get enough, because most plant foods are low in protein. (However, legume flours are high in protein.) Therefore, vegans should eat from 1½ to 2 times as much legumes per meal as a person who eats some animal foods. For example, ¾ cup (180 ml) of cooked legumes furnishes about 11 to 13 g of protein. This would need to be supplemented with at least 1½ cups (360 ml) of cooked cereal product or 3 to 4 slices of bread to insure that the meal contained sufficient protein. Also, the breads used by vegans should either (1) contain plenty of nuts and/or soy flour; or (2) be eaten with peanut butter or mashed cooked bean spreads (homemade or commercial bean dips are suitable).

• **Sprouting legume seeds**—Using sprouted seed adds considerably to the dietary intake of vitamins with few added calories because much of the food energy stored in a seed is used up during the growth of the embryonic plant. Also, extra vitamins are synthesized by the seed when the embryo starts to grow. Therefore, it is noteworthy that a dormant seed usually starts sprouting when it has absorbed sufficient water to allow a stepping up of its metabolism. Directions for sprouting seeds follow.

1. Purchase seeds that are guaranteed to sprout. Most health food stores now carry alfalfa seeds for sprouting. Some also sell viable mung bean seeds and soybean seeds. It is not wise to use seeds intended for planting as food because they may have been treated with one or more chemical agents used to prevent fungus diseases.

2. Use only unbroken seeds, and soak them overnight in an amount of water at least four times as great as the seeds. After soaking, drain off water through a strainer. Place seeds in a special sprouter, or in a wide-mouthed jar or a tray. Cover with wet paper toweling. Rinse seeds in a strainer three or four times a day, and replace paper towels. Sprouts should be sufficiently developed for use after 3 to 4 days.

3. Fresh sprouts may be used in salads, sandwiches, soups, stews, casseroles, and baked goods. Generally, seeds yield six or more times their original weight in sprouts. Sprouts have only about ½ the solids content of cooked beans or peas. Hence, one has to eat from 1 to 1½ cups

(*360 ml*) of sprouts to obtain the amount of protein furnished by about ½ cup (*120 ml*) of cooked dried beans. However, the sprouts furnish fewer calories, but more vitamins per gram of protein.

CAUTION: Deleterious substances of various kinds are present in some legumes eaten by people; among them, the trypsin inhibitor, and the following toxins: cyanogenetic glucosides, saponins, alkaloids, goitrogenic factors, hemagglutinins (substances which agglutinate red cells and destroy them) and the unknown toxic factor which produces lathyrism. Further research on many of these substances is needed. And consumption of certain legumes should be limited to nontoxic species or to those where toxicity can be eliminated or reduced to safe limits by appropriate preparation and cooling. But present knowledge suggests that most of these substances are present only in raw grain and are eliminated by ordinary methods of preparation; e.g., adequate soaking and boiling. There is no existing evidence, for example, that poisoning by cyanogens has ever occurred in human beings as a result of eating cooked legumes, or that goitrogenic substances in legumes have caused goiter. Two toxic substances in certain legumes are, however, of special importance; namely, the factor in lathyrus pea (*Lathrus sativus*) which causes the disease lathyrism, and the hemolytic factor in broad bean (*Vicia faba*) associated with the disease favism.

Consumers should be aware of the potentially harmful substances present in certain legumes so that there will be proper preparation. To the latter end, the major antinutritional and toxic factors in certain legumes are presented in Table L-1.

TABLE L-1
ANTINUTRITIONAL AND/OR TOXIC FACTORS WHICH MAY BE PRESENT IN CERTAIN LEGUMES

Type of Factor(s)	Effect of Factor(s)	Method(s) of Counteracting Undesirable Factor(s)	Legumes Containing the Factor(s)
Antivitamin factors: Antivitamin A	Interfere with the actions of certain vitamins. Lipoxidase oxidizes and destroys carotene (provitamin A).	Roasting at 212°F (*100°C*) or higher.	Soybeans.
Antivitamin B-12	Increases requirement for vitamin B-12.	Heating.	Soybeans.
Antivitamin D	Causes rickets unless extra vitamin D is provided.	Autoclaving (pressure cooking).	Soybeans.
Antivitamin E	Damage to the liver and muscles.	Heat treatment (only partially effective).	Alfalfa, Common beans (*Phaseolus vulgaris*), Peas (*Pisum sativum*).
Cyanide-releasing glucosides	An enzyme in legumes causes the release of hydrocyanic acid (a cellular toxin) from the otherwise inert glucoside. The toxic factor may also be released by an enzyme in *E. coli* a normal inhabitant of the human intestine.	Soaking and cooking the beans releases much of the toxic factor. The *E. coli* population of the human intestine may be reduced by encouraging the growth of *Lactobacilli* species, which are present in fermented dairy products (sour milk, yogurt, etc.).	All legumes contain at least small amounts of these factors. However, certain varieties of lima beans (*Phaseolus lunatus*) may contain much larger amounts.
Favism factor	Causes the breakdown of red blood cells in susceptible individuals only (some of the people with ancestors from the Mediterranean, Taiwan, and various parts of Asia). Other people are *not* affected.	Avoidance of fava beans by people who are susceptible to this type of hemolytic anemia. Some of the more sensitive people are even affected by inhalation of the pollen from the bean flower.	Fava beans (*Vicia faba*)
Gas-generating carbohydrates	Certain indigestible carbohydrates are acted upon by gas-producing bacteria in the lower intestine.	Much of the indigestible carbohydrates may be broken down by (1) soaking and cooking (and discarding the water after each step); (2) acid treatment, followed by incubation; or (3) fermentation and sprouting.	Many species of mature dry legume seeds, but *not* peanuts. The immature (green) seeds contain much lower amounts.
Goitrogens	Interfere with the utilization of iodine by the thyroid gland. As a result, the thyroid gland may become enlarged (forms a goiter).	Heat treatment is partially effective. However it is a good idea to provide extra dietary iodine when large amounts of peanuts and/or soybeans are consumed.	Peanuts and Soybeans.
Inhibitors of trypsin	The inhibitor(s) binds with the digestive enzyme trypsin. As a result, the pancreas may become overworked as it attempts to produce additional enzyme. Normally, the secretion of adequate trypsin acts as a feedback to prevent excessive secretion of pancreatic enzymes.	Heat treatment (soaking, followed by steaming, pressure cooking, extrusion cooking); fermentation; or sprouting.	All legumes contain trypsin inhibitors to some extent.
Lathyrogenic neurotoxins	Consumption of the lathyrogenic legumes in large quantities (¼ or more of the total diet) for long periods (several months) results in severe neurological disorders.	Dehulling, then soaking overnight and steaming or roasting.	Lathyrus pea (*L. sativus*), which is grown mainly in India. Common vetch (*Vicia sativa*) may also be lathyrogenic.
Metal binders	These factors bind certain essential minerals (such as copper, iron, manganese, and zinc) so that they are not well utilized by the body.	Heat treatment.	Soybeans, Peas (*Pisum sativum*)
Red blood cell clumping agents (*hemagglutinins*)	The agents cause the red blood cells to clump together.	Soaking followed by heat treatment.	Occurs in all legumes to some extent.

It may be seen from Table L-1 that various types of heat treatment inactivate most of the antinutritional and toxic factors in legumes. However, heating has no effect on the favism-producing factor in fava beans. Hence, people troubled with favism should avoid eating the offending beans. Those who tolerate gassiness (flatulence) poorly would be better off eating only immature (green) beans or mature beans which have been specially treated.

The most effective detoxifying treatment for the home preparation of beans is soaking until the legumes begin to swell, discarding the water and covering with fresh water, then cooking them until tender. Also, some researchers have found that pressure cooking prior to grinding them into flour or meal inactivates most of the potentially harmful substances. Finally, it is noteworthy that various types of fermentation, which have been utilized by Asian peoples for centuries, are also means of detoxifying legumes.

NUTRITIONAL VALUE.

Seeds such as those of the common food legumes are good sources of certain nutrients, because they are storehouses of the carbohydrates, fats, proteins, minerals, and vitamins needed to sustain the embryonic plant until it has developed sufficiently to take care of its own needs. The nutritive composition of common legumes is given in Food Composition Table F-21.

Some noteworthy observations regarding the nutrient composition of legumes follow:

1. The items lowest in calories are those with the highest water content. Hence, weight watchers may eat their fill of cooked snap beans (92.4% water and 25 kcal per 100 g), but they should curtail their consumption of roasted, salted peanuts (1.6% water and 585 kcal per 100 g). However, it is noteworthy that the high caloric value of the latter item is also due to a high fat content, since defatted peanut flour is much lower in energy content (371 kcal per 100 g).

2. A 100 g portion (approximately ½ cup) of most types of cooked mature dry beans supplies between 100 and 130 Calories (kcal) and an average of about 7 g of protein, which is about the same as the amount of protein in 1 oz (28 g) of cooked lean meat. However, the quality of protein in legumes is somewhat lower than that of meat and other animal proteins.

3. Sprouted seeds supply more protein per calorie than unsprouted seeds because much of the stored carbohydrate is metabolized for energy during sprouting, whereas little protein is used for this purpose. For example, cooked sprouted soybeans contain 13.9 g of protein per 100 kcal, whereas cooked unsprouted soybeans contain only 8.4 g of protein per 100 kcal.

4. Defatted peanut flour and defatted soybean flour each contain almost 50% protein. Hence, small amounts of these products make major protein contributions when added to cereal flours, which contain only about 8 to 15% protein.

5. The various soybean flours are good sources of calcium and iron, although the beany flavor sometimes limits the amounts which may be used in certain foods.

6. All of the legumes are good to excellent sources of iron, potassium, thiamin, riboflavin, and niacin.

7. Immature cowpea (black-eyed pea) pods and seeds are the best legume sources of vitamins A and C, which are lacking in the mature dry seeds.

8. Cooked pigeon peas are high in fiber, which might be beneficial to people bothered with chronic constipation and certain other disorders. Some researchers have reported that the consumption of fibrous legumes lowers blood cholesterol significantly, even when substantial amounts of animal fats are consumed with the legumes.

Additional information on certain nutritional and antinutritional aspects of legumes are given in the sections which follow.

Protein Quantity and Quality.

There is much current worldwide interest in the value of legumes as major sources of dietary protein because (1) the prices of meat and other animal products are rising rapidly, and (2) certain densely populated developing countries lack sufficient land to raise large numbers of animals for meat. Therefore, certain facts regarding protein quantity and quality of legumes are noteworthy:

1. Cooked legumes furnish less protein per gram, and per calorie than low and medium fat types of cheeses, eggs, fish, meats, and poultry. For example, 3½ oz (100 g) of cooked common beans supplying 118 Calories (kcal) are required to provide the protein in 1 oz (28 g) of cooked lean meat (55 kcal) or medium-fat meat (78 kcal). Also, the protein quality of the legumes is lower than that of the animal foods.

2. The proteins in legumes are moderately deficient in the sulfur-containing amino acids methionine and cystine. However, the importance of this problem has been exaggerated somewhat, because tests of protein quality are usually conducted with rats, which have a higher requirement for these amino acids than people. The feeding of a little extra protein from legumes usually insures that adequate amounts of the deficient amino acids are provided.

3. Mixtures of legumes and cereals have a protein quality which comes close to that of meat, milk, and other animal proteins. The highest protein quality is usually achieved in mixtures comprised of 50% legume protein and 50% cereal protein because the amino acid patterns of the two types of foods complement each other. Some examples of food combinations utilizing this principle are corn tortillas and refried beans, baked beans and brown bread, peanut butter sandwiches, and macaroni products fortified with soy protein.

4. Cooked legumes contain from 2 to 4 times the protein of most cooked cereals. For example, a cup of cooked dried legumes (approximately 200 g) supplies from 10 to 20 g of protein, whereas a cup of cooked cereal (150 g - 250 g) provides from 3 to 7 g of protein. Also, more calories (about 50% more) are supplied by the legumes.

5. Legumes may be used to upgrade the protein quantity and quality of diets based mainly on cereals and/or starchy foods such as bananas, cassava, and sweet potatoes. The latter foods are the mainstays of people in the developing tropical regions. For example, a protein-enriched flour or meal may contain about 15% defatted peanut or soybean flour or meal, and about 85% of cereal product made from corn, rice, or wheat. Highly starchy diets may be upgraded considerably by mixing 20% defatted legume flour with 80% starchy food. For example, one of the first attempts to rehabilitate African children suffering from protein-calorie malnutrition utilized a mixture of soybean flour and bananas.

6. Animal protein foods may be extended by the use of legumes. Some approximate equations of protein value follow:

a. 1 frankfurter + ½ cup cooked beans = 2 frankfurters

b. 10% legume flour + 5% skim milk powder + 85% cereal flour = 10% skim milk powder + 90% cereal flour

Both examples show that the requirement for expensive animal protein may be halved by the judicious use of legume products. However, it should be noted that the substitution of legumes for animal foods may raise the caloric value of the diet.

LEGUMES OF THE WORLD.

About a dozen species account for the greater part of the world's legume production. Additionally, many luguminous food plants with only minor roles in commercial food production at the present time have the potential to become important food crops. Also, most of these lesser known species thrive in subtropical and tropical areas where most of the world's impoverished people live. Table L-2 gives pertinent information relative to the legumes of the world. Additionally, the most important legumes are accorded narrative status, alphabetically, in this book.

NOTE WELL: Where production figures are given in Table L-2, Legumes of the World, in column 2 under "Importance," unless otherwise stated, they are based on 1990 data. Most world figures are from *FAO Production Yearbook 1990*, FAO/UN, Rome, Italy, Vol. 44. Most U.S. figures are from *Agricultural Statistics 1991*. Annual production fluctuates as a result of weather and profitability of the crop.

Although beans are often classed as vegetables, especially those that are used as immature green pods (snap beans, or green beans), all beans are legumes. So, in order to avoid double listing, the species of beans commonly used in the United States are listed and summarized in Table L-2, Legumes of the World.

Fig. L-10. Cowpeas (black-eyed peas) used in an entree. Mixtures of legumes and cereals have a protein quality close to that of animal proteins. (Courtesy, California Dry Bean Advisory Board, Dinuba, Calif.)

TABLE L-2
LEGUMES OF THE WORLD

Popular and Scientific Name(s); Origin and History	Importance; Principal Areas; Growing Conditions	Processing; Preparation; Uses	Nutritional Value; Caution[1,2]
Adzuki Bean[3] *Phaseolus angularis* **Origin and History:** Originated in China and Japan, where it has been cultivated for centuries. (Also see BEAN[S]; BEANS, COMMON; and ZEN MACROBIOTIC DIETS.)	**Importance:** Ranks next to soybeans among the Chinese and Japanese. Recently, became known in the U.S. as a staple food in the Zen Macrobiotic Diets. **Principal Areas:** China, Japan, and Korea. **Growing Conditions:** Temperate climate with hot humid summers.	**Processing:** Usually, the beans are picked when mature. They may be left whole or pounded into a meal. **Preparation:** Whole beans boiled, then mashed; or used as a flour. **Uses:** Vegetable dish. Bean flour is used for cakes and various desserts and sweets, and as a substitute for milk, called *kokoh*.	**Nutritional Value:** 100 g of uncooked dried beans supply 324 kcal, 21.1 g protein, 1.0 g fat, 59.5 g carbohydrate, 3.9 g fiber, 82 mg calcium, and 6.4 mg iron. Adzuki beans are low in calcium, high in phosphorus, rich in iron and potassium, and almost totally lacking in vitamins A and C.
Alfalfa (Lucerne) *Medicago saliva* **Origin and History:** Alfalfa was seeded for livestock forage in southwestern Asia long before recorded history. The Persians took it to Greece when they invaded the country in 490 B.C. From Greece, it was taken to Italy in the first century A.D. from whence it spread to other parts of Europe.	**Importance:** The most valuable hay plant of the U.S., with an estimated annual production of 83 million short tons. **Principal Areas:** U.S., Argentina, Europe, and Asia. **Growing Conditions:** Temperate climate.	**Processing:** Mature seeds are sprouted. A flour made from the dried leaves, and a protein concentrate from the juice of the fresh leaves. **Preparation:** Sprouts may be cooked or left raw. **Uses:** Sprouts, in salads, and sandwiches. Flour may be added in small amounts to various cereal products. Protein concentrate is used mainly in livestock feeds.	**Nutritional Value:** 100 g of sprouted seeds contain 41 kcal, 5.1 g protein, 0.6 g fat, 1.7 g fiber, 28 mg calcium, and 1.4 mg iron. *CAUTION:* Usually, only the sprouts are eaten by people. Sprouting counteracts the potentially harmful agents.
Bambarra Groundnut[3] *Voandzeia subterranea* **Origin and History:** Native to tropical Africa. Taken to other parts of Africa, Brazil, and the Orient in the 17th century.	**Importance:** Its use in certain regions of Africa has been replaced to a large extent by the peanut. **Principal Areas:** District of Bambarra in Mali, Zambia, and Madagascar. (All are in Africa.) **Growing Conditions:** Tropical climate.	**Processing:** This plant, like peanuts, must be unearthed for harvesting of the seeds. The hard seeds require soaking and/or breaking up into pieces prior to cooking. **Preparation:** Dry seeds are roasted. Seeds in pods are boiled. **Uses:** Vegetable dish, source of calories and protein.	**Nutritional Value:** 100 g of uncooked dried seeds contain 370 kcal, 16.0 g protein, 6.0 g fat, 65.0 g carbohydrates, 4.8 g fiber, 62 mg calcium, and 12.2 mg iron.
Bean(s) (See BEAN[S].)			
Bean, common *Phaseolus vulgaris* (See BEAN, COMMON.)			
Broad Bean (Fava Bean) *Vicia faba* **Origin and History:** Believed to have originated in northern Africa and in the eastern Mediterranean region. The Chinese used broad beans as food 5,000 years ago. It has been cultivated in various parts of southern and northern Europe (particularly the British Isles) since the Iron Age (1000 B.C.). (Also see BEAN[S]; and BEAN, COMMON.)	**Importance:** Worldwide, about 4.3 million metric tons of dry broad beans are produced annually. Broad bean is one of the leading nonoilseed legumes grown in the temperate areas of the world. However, it is a minor crop in the U.S.A. **Principal Areas:** China, Ethiopia, Egypt, and Italy. **Growing Conditions:** Thrives under temperatures ranging from 50°F *(10°C)* to 86°F *(30°C)*, but is also planted in the British Isles in the autumn or early winter, for harvesting in June or in the late summer.	**Processing:** Immature pods and seeds may be picked, or mature seeds may be harvested, shelled, and dried. **Preparation:** May be boiled, cooked in casseroles, or steamed. Immature pods may be cooked whole or sliced. **Uses:** Vegetable dish, and in casseroles, soups, and stews.	**Nutritional Value:** Mature, uncooked, dried beans are rich in calories (328 kcal per 100 g), proteins (25%), and carbohydrates (56.9%), but low in fat (1.2%). Immature beans and cooked dried beans have a high water content and much lower levels of these nutrients. However, the immature beans are a fair source of vitamin C. *CAUTION:* Contains a hemolytic anemia factor[4] which triggers favism, but which affects *only* people who have an inherited susceptibility. Other potentially harmful constituents are rendered safe by soaking and cooking, sprouting, or fermentation.
Carob (Locust Bean, St. John's Bread) *Ceratona siliqua* **Origin and History:** Originated in the eastern Mediterranean region and introduced into tropical areas around the world.	**Importance:** Appears to be increasing in importance because the seeds yield a valuable gum, and the pods provide a substitute for expensive cocoa. **Principal Areas:** Subtropics and tropics. **Growing Conditions:** Thrives in hot, dry climate.	**Processing:** First, the pods with seeds are dried in the sun. Then, the pods are ground to a powder. Seeds are crushed, roasted, and boiled in water to extract the gum. **Preparation:** The pod powder may be mixed in hot or cold water or milk to make a drink. **Uses:** Powdered pods are a substitute for cocoa. The seed gum is a food additive.	**Nutritional Value:** 100 g of the dried pod powder supply 380 kcal, 3.8 g protein, 0.2 g fat, 90.6 g carbohydrate, 5.4 g fiber, and 290 mg of calcium per 100 g. *CAUTION:* The pods are not known to contain any harmful substances, and the seed gum is generally safe in the amounts used. Seeds are rarely prepared and eaten like other legumes.
Chickpea (Garbanzo Bean)[3] *Cicer arietinum* **Origin and History:** Most likely originated in the Near East. From there its cultivation spread eastward to India, and westward throughout the countries of north Africa and southern Europe. (Also see BEAN[S]; and BEAN, COMMON.)	**Importance:** Worldwide about 6.9 million metric tons are produced annually; and chickpeas rank fifth among the legume crops of the world. **Principal Areas:** Over 70% of the crop is produced in India. Most of the remainder is grown in Turkey, Pakistan, and Mexico. **Growing Conditions:** Grows best in areas where the temperature ranges from 36°F *(2°C)* to 86°F *(30°C)*. Hence, it may be grown during the warm season in northern India, and during the cool season in the dry tropics.	**Processing:** Dehulling, then drying. In India, the dried seeds are sometimes ground into flour. **Preparation:** Boil, fry, or roast. **Uses:** Snack, vegetable dish, and in soups, salads, and stews. In India, the flour is used to make various confections.	**Nutritional Value:** Similar to the various beans in nutritional value. Hence, chickpeas are a good source of calories and protein (100 g, or about ½ cup of cooked beans supplies about 10 g of protein). Also, they are a good source of phosphorus and iron.

Footnotes at end of table

(Continued)

TABLE L-2 *(Continued)*

Popular and Scientific Name(s); Origin and History	Importance; Principal Areas; Growing Conditions	Processing; Preparation; Uses	Nutritional Value; Caution[1],[2]
Cowpea (Black-eyed Pea)[3],[4] *Vigna sinensis; V. unguiculata* **Origin and History:** Originated in central Africa. Brought to the West Indies and to the southeastern U.S. by African slaves. Its cultivation also spread northward to the Mediterranean and eastward to China and India.	**Importance:** A valuable crop in the subtropical and tropical areas of the world. **Principal Areas:** Africa, India, China, West Indies, and southeastern U.S. **Growing Conditions:** 68°F (20°C) to 95°F (35°C). Hence, the growing season is not long enough in northern U.S. to produce mature beans. However, the immature pods and seeds may be produced in cool, temperate climates.	**Processing:** May be picked early for immature pods. Mature beans are dried, canned, and frozen. In Africa, the dried seed may be ground before cooking. **Preparation:** Mature dried beans may be boiled or baked. Immature pods and seeds may be chopped and stir-fried with meats and other foods. **Uses:** Vegetable dish, with or without cooked pork; and an ingredient of casseroles, salads, soups, and stews.	**Nutritional Value:** 3½ oz (100 g, or about ⅔ cup) of the cooked, mature beans supply 76 kcal and 5 g protein. Compared with cooked, mature beans, the cooked immature seeds contain more calories (108 kcal per 100 g), protein (8%), iron (50% more), and vitamins A and C.
Fenugreek *Trigonella foenum-graceum* **Origin and History:** Originated in the Mediterranean region.	**Importance:** Widely used spice. **Principal Areas:** Southern Europe, north Africa, Egypt, and India. **Growing Conditions:** Subtropical climate.	**Processing:** Dried seeds are often ground into flour. **Preparation:** Pods may be cooked. **Uses:** Pods as a vegetable. Powdered seeds as a condiment.	**Nutritional Value:** 100 g of uncooked dried seeds contain 323 kcal, 23 g protein, 6.4 g fat, and 58.4 g carbohydrate, 10.1 mg fiber, 176 mg calcium, and 34 mg iron.
Guar (Cluster Bean) *Cyanopsis tetragonoloba* **Origin and History:** Originated in India. (Also see GUMS.)	**Importance:** Source of a valuable gum. **Principal Areas:** Southeast Asia, India, Indonesia, Burma, and the U.S. **Growing Conditions:** Subtropical climate.	**Processing:** In Asia, immature pods and seeds are picked. In the U.S., the mature seeds are harvested and the gum is extracted. **Preparation:** In Asia, pods are cooked for eating. **Uses:** As a food and livestock feed in Asia. As a source of a food additive in the U.S.	**Nutritional Value:** Data not available. Most likely the composition of the immature pods and seeds is similar to that of other legumes. *CAUTION:* The mature seeds[4] are used almost exclusively for extraction of the gum, which is a safe food additive. However, the residue from extraction requires detoxification.
Hyacinth Bean (Lablab Bean)[3] *Dolichos lablab* **Origin and History:** Originated in India. Cultivated for many centuries as a food plant, and lately as an ornamental plant. (Also see BEAN[S]; and BEAN, COMMON.)	**Importance:** Its drought resistance makes it valuable in the developing countries, but it is little known in the developed countries. **Principal Areas:** Southern and eastern Asia and parts of Africa. It is an important crop of India. **Growing Conditions:** Dry, tropical climate.	**Processing:** Picked when immature or mature for human consumption. In India, the mature beans are dried, and sometimes split. **Preparation:** The whole beans or the split beans are cooked. The immature pods and seeds may be prepared as a green vegetable. **Uses:** The mature beans are used primarily as cooked whole beans or cooked split seeds. The immature pods and seeds may be used like snap beans. Because of its purple flowers and attractive purplish-red pods, the hyacinth bean is sometimes used for ornamental purposes.	**Nutritional Value:** 100 g of uncooked, dried whole mature seeds supply 338 kcal, 22.2 g protein, 1.5 g fat, 61 g carbohydrate, 6.9 g fiber, 73 mg calcium, and 5.1 mg iron. The seeds are low in calcium (they contain only ⅛ as much calcium as phosphorus); and the protein quality is inferior to animal proteins (the protein is low in the amino acids methionine and cystine).
Jack Bean (Horse Bean) *Canavalia ensiformis* **Origin and History:** Native to Mexico, Central America, and the West Indies. Known for 5,000 years in Mexico. (Also see BEAN[S]; and BEAN, COMMON.)	**Importance:** A minor crop grown mainly for green manure or forage. However, the dried seeds and the immature pods and seeds are eaten at times of food scarcity. **Principal Areas:** Tropical regions around the world, including the southern U.S. **Growing Conditions:** Dry, tropical climate.	**Processing:** Immature pods or mature beans may be harvested. May also be plowed under as green manure. **Preparation:** Seeds require prolonged soaking and boiling with several changes of water to soften. Immature pods are cooked in water like snap beans. **Uses:** Human food, cattle fodder, green manure. During times of food scarcity, both the immature pods and seeds and the mature seeds are used for human food.	**Nutritional Value:** 100 g of uncooked dried beans contain 348 kcal, 21.0 g protein, 3.2 g fat, 61.0 g carbohydrate, 7.6 g fiber, 134 mg calcium, and 8.6 mg iron. The dried mature beans are deficient in calcium and in the amino acids cystine and methionine. The immature beans are similar in nutrient composition to snap beans, but they are poorer sources of vitamins A and C. *CAUTION:* They may contain more antinutritional factors than occur in most legumes. Hence, prolonged soaking and thorough cooking[5] are required.
Lathyrus Pea (Grass Pea) *Lathyrus sativus* **Origin and History:** Native to southern Europe and western Asia. Excessive consumption of the seeds, which occurs in time of famine, can cause a paralyzing disease known as lathyrism. (Also see LATHYRISM.)	**Importance:** An emergency food used by the impoverished and during famines because it grows well on poor soils and is drought resistant. **Principal Areas:** India, southern Europe, and South America. **Growing Conditions:** Semitropical climate.	**Processing:** Dehusked seeds may be soaked in water, dried in the sun, then ground into flour. Sometimes the seeds are steamed and/or roasted to remove a highly toxic constituent which causes the disease lathyrism. **Preparation:** Whole seeds cooked as other legumes. Flour made into thin, unleavened cakes (chapatis) and baked. **Uses:** Human food, livestock feed, green manure.	**Nutritional Value:** 100 g of uncooked, dried whole seeds supply 348 kcal, 27.4 g protein, 1.1 g fat, 59.8 g carbohydrate, 7.3 g fiber, 127 mg calcium, and 10.0 mg iron. *CAUTION:* Contains lathyrogens, in addition to factors found in other legumes. Toxicity is counteracted by overnight soaking and thorough cooking.[5] It is *not* sufficient to bake uncooked flour products.
Lentil *Lens esculenta* (See LENTIL.)			
Lima Bean (Butter Bean) *Phaseolus lunatus* (See LIMA BEAN.)			

Footnotes at end of table

TABLE L-2 *(Continued)*

Popular and Scientific Name(s); Origin and History	Importance; Principal Areas; Growing Conditions	Processing; Preparation; Uses	Nutritional Value; Caution[1,2]
Lupines *Lupinus* genus **Origin and History:** There are about 200 species of lupines, most of which are native to the U.S.	**Importance:** Mainly as an emergency food because the species grows in poor soils. **Principal Areas:** Southern Europe, northern Africa, and North and South America. **Growing Conditions:** Temperate to subtropical climate, depending upon species.	**Processing:** Seeds may be treated as most other legumes, or they may be roasted and ground to make a coffee substitute. **Preparation:** Seeds are soaked in water, then boiled. **Uses:** Vegetable dish, coffee substitute, livestock feed, green manure.	**Nutritional Value:** 100 g of uncooked dried seeds furnish 407 kcal, 44.3 g protein, 16.5 g fat, 28.2 g carbohydrate, 7.1 g fiber, 90 mg calcium, and 6.3 mg iron. *CAUTION:* Some varieties contain a toxic alkaloid which is removed by soaking in water. It might also be a good idea to rinse the beans after thorough cooking,[5] and to discard the cooking water.
Mesquite[3] *Prosopis* genus **Origin and History:** Most species probably originated in the tropical areas of the Americas, where they were long used by the Indians.	**Importance:** These wild plants are usually grazed by livestock, but occasionally serve as human food. Some species grow on poor soils and withstand severe droughts. **Principal Areas:** South, Central, and North America; Africa; and other subtropical areas. **Growing Conditions:** Subtropical to tropical climate.	**Processing:** Depending upon the species, either the pods and/or the seeds may be used. Sometimes, the dried seeds may be ground into a flour. **Preparation:** Pod and/or seeds may be boiled or steamed. **Uses:** Human food, livestock feed.	**Nutritional Value:** 100 g of uncooked dried seeds contain 347 kcal, 15.4 g protein, 1.6 g fat, 75.5 g carbohydrate, 7.3 g fiber, and 421 mg of calcium. *CAUTION:* The pods and pod pulp contain large amounts of indigestible carbohydrates (fiber) which may interfere with the utilization of various nutrients.
Mung Bean (Golden Gram; Green Gram)[3] *Phaseolus aureus* **Origin and History:** Probably originated in India. (Also see BEAN(S); and BEAN, COMMON.)	**Importance:** Important food in India and Pakistan. Used mainly for sprouting in China and U.S. **Principal Areas:** India and Pakistan. **Growing Conditions:** Does well in temperatures between 68°F *(20°C)* and 113°F *(45°C)*.	**Processing:** Picked when fully mature. In India and Pakistan, the beans are dried, debranned, and ground into flour, whereas in China and the U.S., the beans are usually allowed to sprout for about 4 days. **Preparation:** Boil, or sprout for eating raw or cooked. **Uses:** Cooked beans as a vegetable dish. Sprouts in salads, soups, sandwiches, etc. In Chinese cooking, the sprouts may be stir-fried with green onions and seasoned with soy sauce.	**Nutritional Value:** 3½ oz *(100 g)* of the raw, sprouted beans supply 53 kcal, 4.3 g of protein, and 16 mg of vitamin C.
Pea, Field *Pisum arvense* **Origin and History:** Thought to have originated in Central Asia and Europe, with possibly secondary developments in the Near East and north Africa, as evidenced by 9,000 year-old buried seeds found in archeological sites.	**Importance:** Less important for human food than garden peas (*Pisum sativum*), except in Africa, where it has long been used as human food. **Principal Areas:** Africa, Asia, and central and northern Europe. **Growing Conditions:** Grows best at temperatures between 50°F *(10°C)* and 86°F *(30°C)*. However, they may also be grown in some parts of Africa and Europe as a cold season crop. Field peas are more hardy than garden peas.	**Processing:** Shelling, followed by drying. Entire plant may be plowed under as green manure, or used as forage. **Preparation:** Boil or bake, with or without meat. Field peas do not become soft like garden peas. **Uses:** Vegetable dish, livestock feed.	**Nutritional Value:** The nutrient composition of field peas is given in Food Composition Table F-21. A 3½ oz *(100 g)* serving (about ½ cup) of cooked field peas provides about 103 calories (kcal) and 7 g of protein. This is about double the caloric value and approximately equal to the protein value of 1 oz *(28 g)* of cooked lean meat. Field peas contain less than half as much calcium as phosphorus. Peas are a good source of iron and potassium. *CAUTION:* Antivitamin E factor (the diet should contain ample vitamin E and selenium). Metal-binding factor(s). Pressure cooking[4] is more certain to eliminate this problem than boiling at normal pressure.
Pea, Garden (English pea, Green pea) *Pisum sativum* (See PEA, GARDEN.)			
Peanut (Groundnut) *Archis hypogaea* (See PEANUT.)			
Pigeon Pea[3,4] *Cajanus cajan* **Origin and History:** Probably native to Africa; now widely grown in tropical and subtropical countries, paticularly India, equatorial Africa, the East Indies, and the West Indies. African slaves took it to the Caribbean.	**Importance:** A valuable drought-resistant tropical food crop. However, the tightly adhering seed coat has an acrid taste which reduces its acceptability in certain areas. **Principal Areas:** India, Africa, Burma, Dominicam Republic, Malawi. **Growing Conditions:** Tropical climate that is moderately dry.	**Processing:** Either immature pods with seeds or mature seeds may be harvested. The latter are canned in the West Indies. In India, the seed coats are removed before preparation. **Preparation:** Cooked as other legumes. **Uses:** Human food, forage plant.	**Nutritional Value:** 100 g of uncooked dried beans furnish 342 kcal, 20.4 g protein, 1.4 g fat, 63.7 g carbohydrate, 7.0 mg fiber, 129 mg calcium, and 8.0 mg iron.
Scarlet Runner Bean[3,4] *Phaseolus coccineus* **Origin and History:** Originated in Central America or Mexico. The Indians in Mexico used it as food 9,000 years ago. (Also see BEAN(S); and BEAN, COMMON.)	**Importance:** Popular in U.K., Asia, and Africa, for its immature pods and seeds. Mature seeds used mainly in Central America. **Principal Areas:** Temperate climate areas of Europe, Asia, Africa, and Central America (grown in the Highlands). Commercial production is limited to Great Britain. **Growing Conditions:** Needs a mild climate.	**Processing:** Immature pods and seeds are picked and cut into string beans. May be sold fresh, cooked and canned, or frozen. **Preparation:** Cook briefly in water and serve with butter or margarine. Toss with hot oil and chopped garlic (Italian style). Dried seeds are prepared like kidney beans. **Uses:** Dried seeds are used like dried kidney beans. Immature pods and seeds are used like snap beans. Also, as an ornamental plant.	**Nutritional Value:** The immature pods and beans contain over 90% water. Hence, the nutrient levels are much lower than those for mature beans. The immature pods and seeds are fair to good sources of vitamins A and C. The dried seeds are deficient in the amino acids cystine and methionine, and deficient in calcium.

Footnotes at end of table

TABLE L-2 *(Continued)*

Popular and Scientific Name(s); Origin and History	Importance; Principal Areas; Growing Conditions	Processing; Preparation; Uses	Nutritional Value; Caution[1,2]
Soybean *Glycine max* (See SOYBEAN.)			
Tamarind Tree *Tamarindus indica* **Origin and History:** Native to eastern tropical Africa; spread to India long ago.	**Importance:** For a very long time, it was of limited local importance, but it may be increasing due to exporting of such products as tamarind juice. **Principal Areas:** India, Arabic countries, East Indies, and countries in the Caribbean. **Growing Conditions:** Tropical climate.	**Processing:** Pulp from ripened pods is made into juice, fruit paste, and other products. Seeds may be roasted or boiled, dehulled, and ground into a meal. **Preparation:** The pulp is usually mixed with other fruit products, because its strong flavor and acidity accentuate fruity tastes. **Uses:** Consumed fresh where produced, and exported in preserved forms (juices, pastes, etc.).	**Nutritional Value:** 100 g of dried pod pulp furnish 270 kcal, 5.0 g protein, 0.6 g fat, 70.7 g carbohydrate, 166 mg calcium, and 2.2 mg iron. *CAUTION:* The seeds should be soaked overnight and cooked thoroughly[5] like other mature legume seeds.
Tepary Bean[3] *Phaseolus acutifolius* **Origin and History:** Probably originated in Central Mexico, where it was cultivated by the Mexican Indians 5,000 years ago. It grows wild in Mexico and Arizona. (Also see BEAN[S]; and BEAN, COMMON.)	**Importance:** Not grown commercially, but still cultivated by various Indian tribes in U.S. and Mexico. It grows in areas of little rainfall. Less gas-forming than other beans. **Principal Areas:** Western Mexico and southwestern U.S. **Growing Conditions:** Hot, dry climate in tropical areas.	**Processing:** Shelling, drying. **Preparation:** Cook as other legumes. However, tepary beans absorb more water than other beans during soaking and require longer cooking time. **Uses:** Vegetable dish, in mixed meat and bean dishes.	**Nutritional Value:** 100 g of uncooked dried beans contain 353 kcal, 19.3 g protein, 1.2 g fat, 67.8 g carbohydrate, and 4.8 g fiber. *CAUTION:* Tepary beans contain less gas-forming carbohydrates than most other mature beans.
Yard-Long Bean; Asparagus Bean *Vigna unguiculata; Vigna sesquipedalis; Dolichos sesquipedalis* **Origin and History:** Thought to have originated somewhere in tropical Asia. However, the wild ancestors of cowpeas came from Central Africa over 5,000 years ago. (Also see BEAN[S]; and BEAN, COMMON.)	**Importance:** Mainly used as a food where grown. Green pods are much liked in Asia and the Americas. **Principal Areas:** Tropical parts of Asia. **Growing Conditions:** Semitropical climate.	**Processing:** Much of the crop is sold as a fresh vegetable, but some is also canned and frozen. **Preparation:** Brief cooking in water. **Uses:** Served in the same ways as snap beans.	**Nutritional Value:** The immature pods and seeds have a high water content (88.3%). Hence, a 100 g portion supplies only 37 kcal, 3.0 g protein, 0.2 g fat, 7.9 g carbohydrate, 44 mg calcium, and 0.7 mg iron. *CAUTION:* The mature seeds[4] should be treated the same as other mature beans.

[1]Refers to substances other than gas-generating carbohydrates, which are present in almost all types of mature legume seeds other than peanuts. The seeds are rendered less gassy by sprouting or fermentation.
[2]Additional details on specific antinutritional and/or toxic factors are given in the section of this article, "CAUTION," and Table L-1.
[3]Much of the gassiness of these beans may be alleviated by (1) soaking and cooking (and by discarding the water after each step); (2) acid treatment, followed by incubation; or (3) fermentation and sprouting.
[4]Immature pods with seeds do not appear to contain the antinutritional factors found in the mature seeds of the same species. Hence, the green pods need only be cooked briefly.
[5]Time may be saved by pressure cooking. However, oil or melted fat should always be added to beans before cooking in a pressure cooker so that foaming is reduced. Foaming may result in the plugging of the safety valve and a buliding up of pressure which may cause the counterweight to be blown off.

LEMON *Citrus limon*

Fig. L-11. Lemons, a tangy-flavored fruit that is used to enhance the flavors of many beverages and foods. (Courtesy, USDA)

A small, yellow, oval-shaped citrus fruit. Although it is too sour to be eaten like other citrus fruits, it imparts an excellent flavor to a wide variety of beverages and other preparations. The lemon is a member of the rue family (*Rutaceae*).

ORIGIN AND HISTORY. The lemon tree is native to southeastern Asia. It was intoduced in the Mediterranean area about l000 A.D. From the Mediterranean countries, it was taken to Europe, thence to the United States.

WORLD AND U.S. PRODUCTION. Worldwide, about 6.6 million metric tons of lemons and limes are produced annually; and the leading lemon-producing countries, by rank, are: United States, Mexico, Italy, Spain, India, Brazil, Iran, Argentina, Turkey, and Egypt.[3]

About 722,000 short tons of lemons are produced annually in the United States, with all the commercial production in California and Arizona. California produces 78% of the U.S. lemon crop.[4]

[3]Data from *FAO Production Yearbook 1990*, FAO/UN, Rome, Italy, Vol. 44, p. 163, Table 71. **Note well:** Annual production fluctuates as a result of weather and profitability of the crop.

[4]Data from *Agricultural Statistics 1991*, USDA, p. 189, Table 278. **Note well:** Annual production fluctuates as a result of weather and profitability of the crop.

PROCESSING. More than half of the U.S. lemon crop is processed. A major portion of the processed lemons are converted into lemon juice and related products such as frozen lemon juice concentrates, lemonade, frozen lemonade concentrates, and lemon flavored soft drinks. The peel, pulp, and seeds are used to make lemon oil, lemon essence, pectin, bioflavonoids (vitaminlike substances), and cattle feed ingredients.

SELECTION, PREPARATION, AND USES. Fresh lemons that have a fine-textured skin and are heavy for their size are generally of better quality than those that are coarse-skinned and light in weight. Deep-yellow-colored lemons are usually relatively mature and are not so acid as those of the lighter or greenish-yellow color; they are also generally thinner skinned and may have a relatively larger proportion of juice but they are not so desirable since lemons are wanted for their acid flavor.

Fresh lemons are available year long. Also they may be stored for much longer than other major citrus fruits. Lastly, lemon juice, peel, and pulp are strongly flavored so that small amounts of each of these items may be used to enhance relatively large amounts of food. Therefore, it would be worthwhile for most homemakers to use the fruit more often in their preparations. Some suggestions follow:

1. Lemon juice may be used instead of vinegar in salad dressings and sauces.

2. Squirt a little lemon juice on the cut surfaces of fresh fruits and vegetables in order to prevent them from darkening.

3. Mixtures of herbs and lemon juice may be substituted for oily, salty, and/or sugary dressings when certain types of restricted diets must be followed.

4. Lemon juice, peel, and pulp may be used to make homemade items such as candied peel, ice cream, ices, pie fillings, puddings, punches, and sherbets.

NUTRITIONAL VALUE. The nutrient compositions of various forms of lemons and lemon products are given in Food Composition Table F-21.

Some noteworthy observations regarding the nutrient composition of these items follow:

1. Fresh lemons are very low in calories (27 kcal per 100 g), and they are a good source of potassium, vitamin C, and bioflavonoids.

2. Bottled, canned, and fresh lemon juice are approximately equal to fresh lemons, except that the bioflavonoid levels may be much lower in the juices than in the fruit because most of the latter nutrient is found in the peel and the membranes of the fruit.

3. Frozen lemon juice concentrate has four times the caloric and nutrient levels of bottled, canned, or fresh lemon juice. Hence, it may make a significant nutritional contribution when added undiluted to drinks, salad dressings, and sauces. However, additional caloric or noncaloric sweetener may be needed because the pure juice concentrate is very tart.

4. Frozen lemonade concentrate contains considerable added sweetener and is 68% higher in calories (it contains 195 kcal per 100 g [3½oz]) than unsweetened frozen lemon juice concentrate. Furthermore, the lemonade concentrate contains much lower levels of potassium and vitamin C.

5. Candied lemon peel is high in calories (316 kcal per 100 g) and is low in almost all nutrients other than carbohydrates and fiber. However, it may contain significant amounts of bioflavonoids.

6. Lemon-flavored carbonated drinks consist primarily of water, sugar, and imitation or natural flavoring(s).

(Also see BIOFLAVONOIDS; CITRUS FRUITS; POTASSIUM; and VITAMIN C.)

LENHARTZ DIET

Initially, patients suffering from a bleeding peptic ulcer were sometimes offered this type of diet. It is mainly fluid; it consists of raw eggs or milk and vegetable purees fed frequently throughout the day.

(Also see MODIFIED DIETS; SIPPY DIET; and ULCERS, PEPTIC.)

LENTIL *Lens esculenta*

This appetizing legume is one of the oldest cultivated plants in the world. Fig. L-12 features both the plant and a closeup of the seed.

Fig. L-12. The lentil, an ancient food.

ORIGIN AND HISTORY. The lentil is indigenous to southwestern Asia, where it was cultivated 8,000 years ago. It was introduced into Greece and Egypt before Biblical times, and it was taken as far east as China. According to the Bible, Esau traded his inheritance to Jacob for some lentils.

WORLD AND U.S. PRODUCTION. Lentils rank ninth among the legume crops of the world. Worldwide, about 2.7 million metric tons of lentils are produced annually; and the leading producing countries of the world, by rank, are: Turkey, India, Canada, Bangladesh, and China. Turkey produces one-third of the world's crop of lentils.[5]

The United States produces about 40,000 metric tons of lentils annually, mostly in Washington and Idaho.[6]

[5]Data from *FAO Production Yearbook 1990*, FAO/UN, Rome, Italy, Vol. 44, p. 106, Table 36. **Note well:** Annual production fluctuates as a result of weather and profitability of the crop.

[6]*Ibid.*

PROCESSING. Lentils may be marketed as whole dried seeds, split dried seeds, or as an ingredient of canned soups. Occasionally, the legume is ground into a flour and combined with cereal flour to make a protein supplement for children in the developing countries of North Africa and the Middle East.

SELECTION, PREPARATION, AND USES. Lentils should be mature, plump, free from foreign matter, and dry.

Lentils may be boiled or stewed.

Cooked lentils go well with most other vegetables in casseroles, soups, and stews. Also, they may be used as substitutes for part or all of the more expensive animal foods such as cheeses, eggs, fish, meats, and poultry.

NUTRITIONAL VALUE. The nutrient composition of lentils is given in Food Composition Table F-21.

Some noteworthy observations regarding the nutrient compositon of lentils follow:

1. A 3½ oz (*100 g*) serving (about ½ cup) of cooked lentils is equivalent in protein value to 1 oz of cooked lean meat, but the legume supplies about twice as many calories.

2. Lentils are a good source of iron, but they contain less than one-fourth as much calcium as phosphorus. Therefore, foods richer in calcium and lower in phosphorus (such as dairy products and green leafy vegetables) should be consumed to improve the dietary calcium to phosphorus ratio.

3. The substitution of lentils for animal protein foods on an equal protein basis usually results in an increase in the number of calories, whereas the replacement of cereal grains by lentils results in an increase in the amount of protein supplied per calorie.

4. The protein in lentils is quite deficient in the amino acids methionine and cystine, which are supplied amply by the protein in cereal grains. Furthermore, the grains are short of lysine, an amino acid that is present more abundantly in lentils. Therefore, the amino acid patterns of lentils and grains complement each other so that mixtures of the two foods contain higher quality protein than either one alone.

Protein Quantity and Quality. Many of the people in India are vegetarians, so they rely heavily upon grains, lentils, and other legumes to supply much of their protein needs. (Milk and other dairy products may also be utilized, but the supply of these foods is limited.) Hence, the grams of protein per 100 Calories (kcal) provided by lentils compared to other selected foods are noteworthy:

Food	Grams of Protein per 100 Calories (kcal) (g)
Lentils	7.4
Cottage cheese	13.2
Lean meat	12.7
Eggs	8.6
Sweet corn	3.9
Brown rice	2.1

LETHARGY

Drowsiness and lack of energy.

LETTUCE *Lactuca sativa*

This vegetable is the major salad crop of the United States. Lettuce belongs to the Sunflower family (*Compositae*). Fig. L-13 shows a typical head of lettuce.

Fig. L-13. A head of the "New York" variety of lettuce. (Courtesy, J. C. Allen & Son, Inc., West Lafayette, Ind.)

ORIGIN AND HISTORY. Lettuce is thought to have originated in Asia Minor from wild lettuce. It was popular with pre-Christian people, having appeared on the royal tables of Persian Kings as long ago as 550 B.C.

WORLD AND U.S. PRODUCTION. Although lettuce is grown over a wide area, it is of lesser importance worldwide than many other vegetables.

Fig. L-14. Field of mature Iceberg lettuce, ready for harvest. (Courtesy, California Iceberg Lettuce Commission, Monterey, Calif.)

Lettuce ranks fourth among the leading vegetable crops of the United States.

About 3.3 metric tons of lettuce are produced annually in the United States, 76% of which is produced in California.

The leading lettuce-producing states of the nation, by rank, are: California, Arizona, Florida, and Colorado.[7]

PROCESSING. Lettuce is processed to a very limited extent, since it cannot be frozen or heated without becom-

Fig. L-15. Harvested lettuce field—wrapped in a cello wrapper being packed in cardboard boxes, following which, it well be picked up and delivered to a centrally located vacuum-cooling unit. (Courtesy, California Iceberg Lettuce Commission, Monterey, Calif.)

ing limp and losing much of its appeal. However, the demand for ready-to-use shredded lettuce by many food service establishments has led to the production of prepacked shredded lettuce. This product is made from the freshly harvested vegetable by coring, shredding, cooling, and washing. The shredded lettuce is then packed in plastic bags that are shipped in styrofoam containers to prevent damage during transit.

SELECTION, PREPARATION, AND USES. Consumers are more likely to get the best buy if they know something about (1) the types of lettuce marketed in their locality, and (2) the indicators of quality.

Most of the varieties of lettuce sold in the United States fall into one of the types shown in Fig. L-16.

Good quality lettuce should be clean, crisp, and tender. The heads of crisp-head lettuce should be fairly firm to firm.

The enjoyment of lettuce dishes may be enhanced by keeping lettuce as fresh as possible until it is served.

7Data from *Agricultural Statistics 1991*, USDA, p. 155, Table 219. **Note well:** Annual production fluctuates as a result of weather and profitability of the crop.

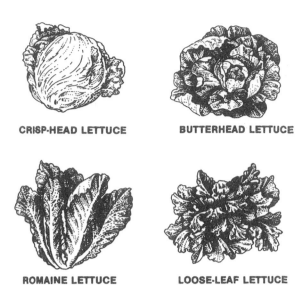

CRISP-HEAD LETTUCE   BUTTERHEAD LETTUCE

ROMAINE LETTUCE   LOOSE-LEAF LETTUCE

Fig. L-16. The four main types of lettuce grown in the United States.

Maximum meal pleasure and nutritional benefits may be obtained by utilizing the principles that (1) lettuce is a low-calorie food which goes well with cheese, eggs, fish and seafood, grains and cereal products, legumes, meats, nuts, poultry, and other vegetables; and (2) the larger the quantity of lettuce used, the more likely it will be that hunger will be satisfied and the digestive system stimulated to optimal functioning by the abundant water and gentle bulk provided by this vegetable. Fig. L-17 shows an appetizing and nutritious lettuce dish.

Fig. L-17. An appetizing and nutritious lettuce salad. (Courtesy, Iceberg Lettuce Commission, Monterey, Calif.)

Some appetizing ways of serving lettuce are: (1) as a salad, or as an ingredient of a salad (lettuce is by far the most commonly used salad green in the United States); (2) as a pocket bread sandwich; (3) as broiled prawns and cherry tomatoes on a lettuce raft; (4) as stir-fried; (5) as a wilted salad in which a hot dressing is poured over the raw vegetable, and (6) as cream of lettuce and minestrone soups.

NUTRITIONAL VALUE. The nutrient composition of the major types of lettuce are given in Food Composition Table F-21.

Some noteworthy observations regarding the nutrient composition of lettuce follow:

1. All types of lettuce have a high water content (94 to 96%) and are low in calories (13 to 18 kcal per 100 g). Therefore, a 1-lb (454 g) head of iceberg lettuce furnishes only about 56 Calories (kcal). Hence, people trying to cut their caloric intake in order to lose weight may assuage their hunger pangs by eating large quantities of this vegetable.

2. The darker the green color of lettuce, the richer it is likely to be in nutrients. For example, the butterhead, loose-leaf, and romaine types contain significantly more iron and vitamin A than the crisp-head type. Also, it is noteworthy that romaine lettuce contains about 10 times as much folic acid as the other types. Although each of the four types is only a fair source of vitamin C, lettuce may make a significant contribution of this vitamin to the diet if sufficiently large quantities are consumed.

LEUCINE

One of the essential amino acids.
(Also see AMINO ACID[S].)

LEUKEMIA

A malignant condition, a blood cancer, wherein there is uncontrolled production of abnormal white blood cells, leukocytes. These are produced in the bone marrow, spleen, and lymph tissues. Blood concentrations of leukocytes may increase a hundredfold. There are several forms of the disease, the exact causative factors of which are unknown. However, excessive radiation, certain chemicals, heredity, and hormonal abnormalities have been implicated. New techniques employed to combat leukemia have greatly increased a patient's chances of complete remission. During treatment, a supportive diet should be given which is higher than normal in energy and nutrient content and which is acceptable and appetizing.
(Also see CANCER.)

LEVULOSE

It is another name for fructose, fruit sugar. It is levoratatory (turning to the left) to polarized light; hence, the name levulose. Levulose, which is much sweeter than cane sugar, is found in honey, ripe fruits, and some vegetables.
(Also see CARBOHYDRATE[S]; and FRUCTOSE.)

LICORICE

Licorice is a herb, the root of which produces a flavoring that is used to flavor candy, chewing gum, and soft drinks.
Caution: Licorice raises the blood pressure of some people dangerously high, due to the retention of sodium.

LIEBERKUHN'S CRYPT

Tubular glands located in the mucous membrane of the small intestine, named after Johann Lieberkuhn, German anatomist, who first described them. Now they are simply called intestinal glands. They secrete intestinal juices.
(Also see DIGESTION AND ABSORPTION.)

LIFE EXPECTANCY AND NUTRITION

Contents	Page
Well-Known Studies of Health and Longevity	.618
Experiments with Laboratory Animals	.618
Observations of Human Populations	.619
Practical Applications	.620
Selecting Foods	.620
Physical Activity	.620

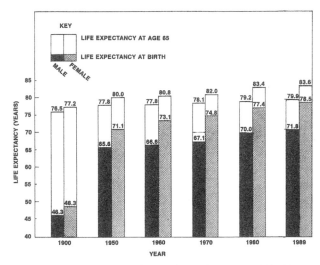

Fig. L-18. Life expectancies in the United States from 1900 to 1989. (Source: *Health, United States 1979,* U.S. Dept. HEW, p. 90, Table 9; and *Statistical Abstract of the United States 1991,* p. 73)

The great increase in life expectancy at birth has been due mainly to the control of many of the infectious diseases that formerly killed many infants and children. Therefore, researchers are now trying to identify the factors that may increase the life expectancies of older people by substantial amounts.

WELL-KNOWN STUDIES OF HEALTH AND LONGEVITY. Certain studies are referred to frequently in the popular and scientific literature on longevity. Hence, details of some of the better known observations follow.

Experiments with Laboratory Animals. In the early 1930s, Dr. Clive McKay of Cornell University succeeded in doubling the life expectancy of laboratory rats by drastically restricting their diets from after weaning to death. Furthermore, the incidence of tumors in the restricted rats was much lower than normal. However, the rats given the stringent diets failed to grow and develop at the normal rate and had an increaseed susceptbility to diseases and death early in their lifespan. These drawbacks make it unlikely that similar measures will be applied deliberately to people, although it has been suggested that very rigorous conditions of life in certain parts of the world have similar effects on human populations. That is, the more vulnerable people die young, while those with unusually strong constitutions live to ripe old ages.

Observations of Human Populations.   The longest-lived peoples of the world are believed to be those that live in the geographic areas depicted in Fig. L-19. Descriptions of the people shown in Fig. L-19 follow:

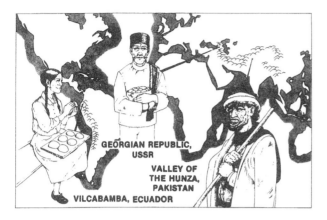

Fig. L-19. The locations where some people of apparently great longevity are found.

- **Ecuadorians of Vilcabamba**—These people live in a village that is located at an elevation of 4,600 ft (*1,400 m*) in the Andes mountains of southern Ecuador. They are reputed to live as long as 125 years, although this belief is now being questioned due to the fact that many people take the names of their parents and other close relatives. Hence, it is difficult to determine whether the names listed in the baptismal records of the local church are those for people that are now alive, or those that are deceased. Nevertheless, the older people in this village appear to be sufficiently vigorous and healthy to engage in a primitive type of farming. Furthermore, they have to contend with a mountainous terrain that makes getting about a test of physical fitness. Their caloric intake is low and they eat mainly vegetables because animal foods are scarce in their region. Most Vilcabambans drink moderate amounts of a rumlike alcoholic beverage called zuhmir, and they smoke.

- **Georgians of the Soviet Union**—Gerontologists have long thought that these people had the longest lives known to man, since some were once thought to be at least 150 years old. However, the Russian gerontologist Zhores Medvedev believes that the advanced ages may be gross exaggerations which started in the early 1900s as a means of avoiding military service under the Czars. Even if the reports of the advanced ages prove to be false, the fitness of the people and the scarcity of degenerative diseases in this region provide ample testimony to the benefits of a varied diet and strenuous exercise. The Georgians eat dairy products, meats, plenty of vegetables, and some sweets such as halvah. Many also drink and smoke. Apparently, the potential ill effects of the latter vices are offset by the heavy work on the farms.

- **Hunzas of Pakistan**—There are few American enthusiasts for health foods who have not heard of these people in the Karakorum Mountains (the western end of the Himalayas) in the northeastern part of Pakistan. Now, it appears that the reports of exceptional health and longevity of these people have been less than accurate, since a Japanese medical team found many cases of cancer, heart disease, and tuber-

culosis, and little evidence to confirm exceptional longevity.[8] Furthermore the death rates for the younger generations were quite high, since 30% of the children died before age 10, and 10% of the survivors died before 40. Therefore, it appears that the sprightly Hunza oldsters observed by the seekers of the secrets of longevity represented the hardiest of the group, who were endowed with exceptional ability to adapt to primitive conditions.

Even if it should be impossible to extend the human life-span much beyond 100 years, it would still be a great achievement to cut the incidences of the major degenerative diseases drastically, so that more people might live healthier and longer lives. Hence, the groups of people who have lower-than-normal rates of these diseases are noteworthy.

- **Bulgarians**—In the early 1900s, the Russian microbiologist Elie Metchnikoff promoted the consumption of the yogurt bacteria *Lactobacillus bulgaricus* because he observed that the yogurt-consuming Bulgarians appeared to have healthier and more long-lived people than the other groups he studied. Furthermore, he theorized that the *Lactobacillus* organisms prevented the growth of undesirable toxin-producing species of bacteria in the intestine. Many other scientists of that era scoffed at Metchnikoff's ideas, but recent studies have yielded evidence that supports some of his concepts.[9] For example, the feeding of *Lactobacilli* has been beneficial to (1) infants afflicted with diarrhea induced by pathogenic microorganisms, and (2) people suffering from ammonia toxicity due to liver failure. In the latter case, intestinal bacteria convert urea that is present in digestive secretions to ammonia which is then absorbed into the blood. Normally, the liver converts excess ammonia to urea, but in liver failure the ammonia accumulates in toxic amounts. *Lactobacilli* reduce the buildup of ammonia.

- **Longshoremen in San Francisco**—A 22-year study (from 1951 to 1972) of 3,686 longshoremen revealed that those who performed heavy work (that requiring an average of 1,876 Calories [kcal] per day above the BMR) had a much lower risk of a fatal heart attack than those who performed light work which required an expenditure of about 1,066 Calories (kcal) per day over the BMR.[10]

- **Masai tribesmen of Africa**—The males of this tribe, who consume liberal amounts of cow's blood, milk, and meat, do not develop heart disease as they grow older even though atherosclerotic plaques form in their coronary arteries, because the exercise of walking about 11 miles (*18 km*) per day results in compensatory enlargement of the arteries.[11]

- **Mormons**—Various studies of these people have shown them to have a cancer death rate that is about ⅓ lower than the national average, and from ⅓ to ½ fewer deaths from

[8]Jarvis, W. T., "The Myth of the Healthy Savage," *Nutrition Today*, March-April 1981, pp. 14-22.

[9]"Interrelationships of Diet, Gut Microflora, Nutrition, and Health," *Dairy Council Digest*, July-August 1976, pp. 19-24.

[10]Paffenbarger, R.S., Jr., *et al.*, "Work-Energy Level, Personal Characteristics, and Fatal Heart Attack: A Birth-Cohort Effect," *American Journal of Epidemiology*, Vol. 105, 1977, pp. 200-213.

[11]Mann, G. V., Testimony in the hearings of the U.S. Senate Select Committee on Nutrition and Human Needs, March 24, 1977.

heart attacks.[12] Furthermore, many elderly Mormons stay active in businesses, farming, and church affairs well into their 80s and 90s. Their superior health and longevity is thought to result from factors such as (1) abstinence from alcoholic beverages, coffee and tea, habit-forming drugs, and smoking; (2) consumption of liberal amounts of whole grains, fruits and vegetables, and only moderate amounts of meat; (3) adequate physical activity and sufficient rest and recreation; (4) family unity; (5) extensive participation of the laity in church affairs; and (6) striving for high levels of occupational achievement.

• **Seventh-day Adventists**—Adventists men live an average of 6 years longer than other American men, and the women live 3 years longer than their counterparts.[13] Furthermore, the death rates of cancer, heart disease, and stroke for Adventists are only about 50 to 60% of that of other Americans. It appears that the reduced susceptibility of Adventists to degenerative diseases results from (1) abstinence from alcoholic drinks, caffeine-containing beverages, drugs, and smoking; (2) consumption of strict vegetarian (vegan) diets or lacto-ovo-vegetarian diets (those which contain milk and eggs, but no meat); (3) exercising on a regular basis; and (4) participation in the activities of their church.

PRACTICAL APPLICATIONS. The observations of the various groups of people who live full lives here and abroad provide some lessons regarding the types of health practices that promote longevity. First of all, the healthiest people usually eat a wide variety of fresh, minimally-processed foods such as meats and fish, dairy products, fruits and vegetables, and whole grain breads and cereals; and they may limit their consumption of potentially harmful items such as strong alcoholic drinks, caffeine-containing beverages, and refined fats, starches, and sugars. Furthermore, the quantities of foods consumed are sufficient to promote health and vitality, but not enough to bring on life-threatening obesity. Finally, the winners in the game of survival have harmonious relationships with their families and neighbors, and engage in moderate to vigorous levels of physical activity on a regular basis.

SELECTING FOODS. The food groups which comprise the framework of the daily food plan are: (1) bread, cereal, rice, and pasta group; (2) vegetable group; (3) fruit group; (4) milk, yogurt, and cheese group; (5) meat, poultry, fish, dry beans, eggs, and nuts group; and (6) fats, oils, and sweets. The minimum quantities indicated in the food groups form a foundation for an adequate diet, safeguarding the quantity and proportion of minerals, vitamins, and other nutrients. Many people will use more than the minimum number of servings suggested in each food group; and everyone will add some sugars, fats, and oils during food preparation and at the table. Perhaps a third to a half of the day's calories will come from such additions to the minimal number of servings. However, people who are overweight would be wise to limit their food selections to the basic foods in each group, and to restrict drastically their intake of sugars, fats, and oils.

PHYSICAL ACTIVITY. Regular physical activity helps to utilize dietary calories and prevent obesity, and imparts a good feeling. It is also useful in preventing degenerative diseases and for rehabilitating people who contract them. For example, patients with heart disease have had their health and physical capacity improved by physical training.[14] Also, moderate to strenuous exercise performed on a regular basis helps to retard the loss of bone calcium that occurs with aging. The exercise is most effective in conjunction with adequate amounts of dietary calcium, phosphorus, and vitamin D.

CAUTION: It is wise to consult a doctor before starting strenuous activities that have not been performed recently.

LIGHT OFF-FLAVOR

In some foods, exposure to a light source creates an off-flavor. Light provides the energy for a chemical change which creates the undesirable odor. These chemical changes due to light are known to occur in fats, milk, and beer, and, depending upon the product, they may have a special name such as light-struck flavor in milk and sunlight off-flavor or skunky in beer.

LIGNIN

A practically indigestible compound which along with cellulose is a major component of the cell wall of certain plant foods, such as very mature vegetables.
(Also see FIBER.)

LIMA BEAN (BUTTER BEAN) *Phaseolus lunatus*

This bean is believed to have been named after Lima, Peru, the area where it was first cultivated. It belongs to the family *Leguminosae*. The beans may be picked while still immature and green for use as *baby* limas, or when fully mature. Fig. L-20 shows the beans at various stages of growth.

Fig. L-20. Lima beans, flowers and seeds.

ORIGIN AND HISTORY. The lima bean is native to Peru, where there is archeological evidence that it was first cultivated about 3800 B.C. It was spread throughout the Americas and the West Indies by the Indians; and it was taken to Europe by the Spanish explorers.

---

[12]Davidson, B., "What Can We Learn About Health from the Mormons?", *Family Circle*, July 1976.

[13]Walton, L. R., et al., *How You Can Live Six Extra Years*, Woodbridge Press Publishing Company, Santa Barbara, Calif., 1981, pp. 4-7.

[14]Redwood, D. R., et al., "Circulatory and Symptomatic Effects of Physical Training in Patients with Coronary-Artery Disease and Angina Pectoris," *The New England Journal of Medicine*, Vol. 286, May 4, 1972, pp. 959-965.

WORLD AND U.S. PRODUCTION. Lima beans are grown in temperate and subtropical areas throughout the world, although production figures are not available.

About 140,000 short tons are produced annually in the United States, largely in the coastal area of California.

PROCESSING. Only a very small part of the green lima beans grown in the United States is marketed fresh; most of the crop is processed. About 30% of the processed beans is canned and 70% is frozen.

Dried lima beans store well without refrigeration, as long as they are kept under dry conditions and protected from pests. Therefore, they are usually stored after harvesting and either marketed or sent to processors at the most profitable time. Most of the dried beans which are processed are cooked and canned, but small amounts are also ground into flour.

SELECTION, PREPARATION, AND USES. Fresh green limas are often marketed in their pods. Hence, the suggestions which follow are noteworthy.

The pods of good quality unshelled lima beans are well filled, clean, bright, fresh, and dark green in color.

Shelled lima beans should be plump, have tender skins, and be green or greenish-white in color.

Green limas may be prepared and served in the same ways as green peas—in casseroles, salads, soups, and stews. Usually, they are made tender by 25 to 30 minutes of cooking.

Dry limas have a mild flavor which allows them to mix well with many other foods. Also, their ample protein content makes them nutritionally valuable for extending more expensive animal protein foods such as cheeses, eggs, fish, meats, and poultry.

• **Antinutritional and toxic factors**—Many foods contain components which may be harmful under certain circumstances, especially if they have not been improved through breeding, and/or if they are not processed properly; and lima beans are no exception.

Some varieties of lima beans, such as those that are native to the Caribbean region—and which are usually colored, contain harmful levels of cyanide-releasing glucosides. In such lima seeds, an enzyme in the legume causes the release of hydrocyanic acid (a cellular toxin) from the otherwise inert glucoside. The toxic factor may also be released by the enzyme in *E. coli* bacteria, a normal inhabitant of the human intestine. Fortunately, soaking and cooking the beans releases much of the toxic principle. Also, the *E. coli* population of the human intestine may be reduced by encouraging the growth of *Lactobacilli* species, which are present in fermented dairy products (sour milk, yogurt, etc.). Nevertheless, travelers in the Caribbean region should be wary of eating colored limas.

The varieties of limas grown in the United States contain only negligible amounts of cyanide. Besides, U.S. law prohibits the marketing of lima beans that contain harmful amounts of this toxic factor.

Lima beans also contain inhibitors of trypsin, which bind the digestive enzyme trypsin; and red blood cell clumping agents (hemagglutinins), which cause the red blood cells to clump together. Soaking followed by cooking (steaming, pressure cooking, extrusion cooking), fermentation, or sprouting greatly reduce the effects of trypsin inhibitors and blood cell clumping agents.

Fig. L-21. Golden limas and lamb. This casserole dish features large limas, lean lamb stew meat, celery, small white onions, carrots, fresh mushrooms, and a variety of delectable seasonings (butter or oil, garlic salt, chicken soup stock, onion salt, turmeric, white pepper, and a dash of brandy or Cognac). (Courtesy, The California Dry Bean Advisory Board, Dinuba, Calif.)

• **Gas-producing substances**—Lima beans, in common with most other mature beans, also contain gas-generating carbohydrates—certain indigestible carbohydrates which are acted upon by gas-producing bacteria in the lower intestine. Much of the indigestible carbohydrate may be broken down by incubation of the raw beans in a mildly acid solution for 1 to 2 days. However, people who are greatly troubled with gassiness (flatulence) are admonished not to eat lima beans.

NUTRITIONAL VALUE. The nutrient composition of lima beans is given in Food Composition Table F-21.

Some noteworthy observations regarding the nutritive value of lima beans follow:

1. Cooked mature limas furnish considerably more calories, protein, carbohydrate, phosphorus, iron, and potassium than cooked immature beans (*green limas*). However, the latter are much higher in vitamins A and C. Both types of limas are low in fat.

2. One cup (*240 ml*) of cooked mature beans contains about the same amount of protein (*about 14 g*) as 2 oz (*57 g*) of lean meat, but the beans supply over three times as many calories (360 kcal vs 110 kcal).

3. Lima beans, like most other legume seeds, contain much less calcium than phosphorus. Hence, they should be eaten with foods such as milk and green leafy vegetables which have more favorable calcium to phosphorus ratios. An excessively low ratio of dietary calcium to phosphorus will result in poor utilization of calcium by the body.

4. Lima bean flour is a concentrated source of both calories and protein.

Fig. L-22. California potluck salad. An appetizing dish made from large limas, lettuce, celery, green pepper and pimiento strips, mayonnaise, pickle relish, mustard, dried dill weed, and fresh ground black pepper. (Courtesy, The California Dry Bean Advisory Board, Dinuba, Calif.)

Protein Quantity And Quality. Various types of beans are being used all over the world as protein supplements because many people consume only small amounts of animal protein. Lima beans are more acceptable in mixed dishes than many other beans because limas have a milder taste and a lighter color. Therefore, certain facts concerning the quantity and quality of protein in lima beans are noteworthy:

1. Cooked limas furnish less protein per gram of food, and per calorie than low and medium fat types of cheeses, eggs, fish, meats, milk, and poultry. For example, 3½ oz (100 g) of cooked green lima beans supplying 133 Calories (kcal) are required to provide the amount of protein (about 8 g) in 1¾ oz (50 g) of cottage cheese (45 kcal) or 1 large egg (82 kcal). It is noteworthy, too, that only 3 oz (85 g) of cooked mature limas are equivalent in calorie and protein values to 3½ oz (100 g) of cooked green limas.

2. Measure for measure, cooked lima beans contain from 2 to 3 times the protein of most cooked cereals. Similarly, lima bean flour provides twice as much protein per calorie as wheat flour.

3. The protein in lima beans is moderately deficient in the sulfur-containing amino acids methionine and cystine, but it contains ample amounts of lysine, which is deficient in cereal grains. Furthermore, the grains are good sources of methionine and cystine. Therefore, mixtures of the beans with cereal products have a higher quality of protein than either food alone. A good example of this type of mixture is succotash, a combination of lima beans and corn.

4. Bread made from 70% wheat flour and 30% lima bean flour contains about 40% more protein than bread made from wheat flour alone.

(Also see FLOURS, Table F-13, Special Flours.)

5. Expensive animal protein foods may be extended by replacing approximately half of the animal protein with an equivalent amount of protein from lima beans, without any significant reduction in the utilization of the dietary protein by the body. Some approximate *equations* of protein value follow:

  a. 1 frankfurter + ½ cup cooked beans = 2 frankfurters

  b. 10% lima bean flour + 5% skim milk powder + 85% cereal flour = 10% skim milk powder + 90% cereal flour.

However, the substitution of lima beans for protein rich animal foods usually results in the consumption of considerably more calories.

(Also see BEAN[S]; and BEAN, COMMON.)

LIME *Citrus aurantifolia*

A rounded fruit that is pointed at both ends and greener than the lemon. The juice of this fruit is used in various refreshing beverages. Limes, like the other citrus fruits, belong to the rue family (*Rutaceae*).

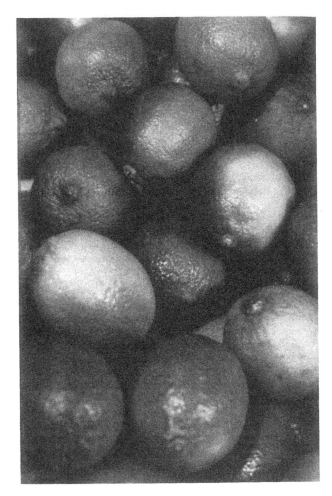

Fig. L-23. Limes, famed for protecting early-day sailors against scurvy. Because of this, British sailors were called limeys. (Courtesy, USDA)

ORIGIN AND HISTORY. The lime tree originated in India. The Arab traders first took the fruit to Arabia sometime before 900 A.D.; and about the same time, they introduced it to Egypt. Limes were taken to Spain in the 13th century. Columbus took lime seeds to Haiti during his second voyage to the New World in 1493.

WORLD AND U.S. PRODUCTION. FAO, United Nations, reports combined production figures for lemons and limes, without any breakdown between the two fruits. (See LEMONS.) However, it is estimated that annual world production of limes is about 360,000 metric tons. Most of the limes are grown in the warmer countries, especially in Mexico, Egypt, the West Indies, and in Florida.

The U.S. production of limes commercially is limited to Florida, where about 72,000 short tons are produced annually.[15]

SELECTION, PREPARATION, AND USES. Fresh limes that are green in color and heavy for their size are the most desirable. Many adult Americans have used alcoholic drinks made with lime juice. It is commonly used in the *coolers* served in hot weather. However, there are other good uses for fresh limes and lime juice. For example, the whole fruit makes excellent marmalade after slicing and removing the seeds. Also, the freshly squeezed juice imparts a superb flavor to fish and seafood dishes, fruits and fruit juices, meats, pie fillings, puddings, salad dressings, and sauces.

NUTRITIONAL VALUE. The nutrient compositions of various forms of limes are given in Food Composition Table F-21.

Some noteworthy observations regarding the nutrient composition of limes follow:

1. Fresh limes are low in calories (28 kcal per 100 g); and good sources of potassium, vitamin C, and bioflavonoids.

2. Fresh lime juice is similar to fresh limes in caloric and nutrient levels, except that the juice is likely to be much lower in bioflavonoids, which are found mainly in the peel and the membranes within the fruit.

3. Reconstituted frozen limeade concentrate contains about 50% more calories, but is much lower in nutrients than fresh lime juice.

4. Bottled lime juice contains only about two-thirds as much vitamin C as fresh lime juice.

(Also see BIOFLAVONOIDS; CITRUS FRUITS; POTASSIUM; and VITAMIN C.)

LIMITING AMINO ACID

The essential amino acid of a protein which shows the greatest percentage deficit in comparison with the amino acids contained in the same quantity of another protein selected as a standard.

(Also see AMINO ACID[S]; and PROTEIN.)

LIMONIN

The chemical responsible for the bitter principle of lemons. It is contained in the white spongy inner part of the rind. Too much limonin released into the juice during extraction produces bitter juice.

LINGONBERRY (COWBERRY)
Vacinnium vitis-idaea

The fruit of a vine (of the family *Ericaceae*) that grows wild throughout the cool, temperate regions of North America and Europe. Lingonberries, which are much esteemed in the Scandinavian countries, closely resemble their nearest

relative, the cranberry, and are about the same size. They are too tart to be eaten raw, but they make excellent jams and jellies.

LINOLEIC ACID

An 18-carbon unsaturated fatty acid having two double bonds. It is one of the essential fatty acids.
(Also see FATS AND OTHER LIPIDS.)

LINOLENIC ACID

An 18-carbon unsaturated fatty acid having three double bonds.
(Also see FATS AND OTHER LIPIDS.)

LIPASE

A fat-splitting enzyme, present in gastric juice and pancreatic juice. It acts on fats to produce fatty acids and glycerol.
(Also see DIGESTION AND ABSORPTION.)

LIP, CLEFT (HARELIP)

Usually infants born with a cleft palate (roof of the mouth) also have a cleft lip. These are congenital defects characterized by a gap in the palate and a gap in the upper lip. The condition may be bilateral—occurring on both sides of the face under each nostril. Infants with a cleft palate and/or lip have difficulty sucking, drinking, eating and swallowing. Hence, they require some special feeding techniques, depending upon the degree of the defect. Fortunately new plastic surgery methods have been developed and may be employed as early as the third month of life. Nevertheless, afflicted infants must build up their reserves for surgery; hence, the following suggestions will help to cope with feeding problems:

1. Sucking difficulties can be overcome by using a nipple with an enlarged opening, a medicine dropper, or a 10-20 cc syringe with a 2-in. (5 cm) piece of rubber tubing.

2. The tendency to choke should be countered by offering liquids slowly and in small amounts.

3. Frequent burpings are necessary as large amounts of air may be swallowed.

4. Irritating food such as spicy and acid foods should be avoided.

5. Foods such as peanut butter, nuts, leafy vegetables, and creamed dishes may adhere to the palate.

6. Pureed foods may be given from a bottle with a large nipple opening if diluted with milk, fruit juice, or broth.

7. Extra time should be allowed for feeding, and possibly five or six small meals may be better than three larger ones.

8. Supplementing the diet with the proper amounts of vitamins will insure optimal health and growth.

LIPIDS

An all-embracing term referring to any compound that is soluble in chloroform, benzene, petroleum, or ether. Included are fats, oils, waxes, sterols, and complex compounds such as phospholipids and sphingolipids. There are three basic types of lipids—simple lipids, compound lipids, and derived lipids. When fatty acids are esterified with alcohols, simple lipids result. If compounds such as choline or serine are esterified to alcohols in addition to fatty acids, compound lipids result. The third type of lipid, derived lipids, result from the hydrolysis of simple and compound lipids. The sterols and fatty acids are derived lipids.

---

[15]Data from *Agricultural Statistics 1991*, USDA, p. 177, Table 254. **Note well:** Annual production fluctuates as a result of weather and profitability of the crop.

(Also see FATS AND OTHER LIPIDS; and OILS, VEGETABLE.)

LIPOGENESIS

The formation of fat.

LIPOIC ACID

Lipoic acid is a fat-soluble, sulfur-containing substance. It is not a true vitamin because it can be synthesized in the body and is not necessary in the diet of animals. However, it functions in the same manner as many of the B-complex vitamins.

HISTORY.   The continuing study of thiamin as a coenzyme in carbohydrate metabolism revealed that this metabolic system required other coenzyme factors in addition to thiamin. In work with lactic acid bacteria, Reed discovered, in 1951, that one of these factors is a fat-soluble acid which he named lipoic acid (after the Greek *lipos* for fat).

FUNCTIONS.   Lipoic acid functions as a coenzyme. It is essential, together with the thiamin-containing enzyme, pyrophosphatase (TPP), for reactions in carbohydrate metabolism which convert pyruvic acid to acetyl-coenzyme A. Lipoic acid, which has two sulfur bonds of high-energy potential, combines with TPP to reduce pyruvate to active acetate, thereby sending it into the final energy cycle. It joins the intermediary products of protein and fat metabolism in the Krebs cycle in the reactions involved in producing energy from these nutrients. A metal ion (magnesium or calcium) is involved in this oxidative decarboxylation, along with lipoic acid and four vitamins: thiamin, pantothenic acid, niacin, and riboflavin. Thus, this underscores the concept of the interdependent relationships among the vitamins.

RECOMMENDED DIETARY ALLOWANCE OF LIPOIC ACID.   Because the body can synthesize the needed lipoic acid, no dietary requirement for humans or animals has been established.

SOURCES OF LIPOIC ACID.   Lipoic acid is found in many foods. Yeast and liver are rich sources.
(Also see VITAMIN[S], Table V-5.)

LIPOLYSIS

The hydrolysis of fats by enzymes, acids, alkalis, or other means to yield glycerol and fatty acids.

LIPOLYTIC RANCIDITY

Food spoilage that is caused by the enzymatic breakdown of fats by lipases. Lipases are produced by some microorganisms, and are present in tissues. Heating destroys the lipases, preventing this type of spoilage.
(Also see FATS AND OTHER LIPIDS; and RANCIDITY.)

LIPOPROTEIN

A lipid-protein complex that is water soluble. Hence, it is involved in the transport of lipids in the blood. Four types of lipoprotein circulate in the blood, all of which contain the lipids, triglycerides, cholesterol, and phospholipid in varying proportions. The four types of lipoprotein are (1) chylomicrons—lowest density; (2) very low density, VLDL; (3) low density, LDL; and (4) high density, HDL.

(Also see CHOLESTEROL; DIGESTION AND ABSORPTION, section on "Lipids"; and FATS AND OTHER LIPIDS, section headed "Other Lipids".)

LIPOTROPIC (FACTOR)

A substance that prevents accumulation of fat in the liver. Choline is probably the most important of the lipotropic factors. So, any substance capable of contributing methyl groups for choline synthesis is lipotropic.
(Also see CHOLINE, section headed "Functions.")

LIQUID DIETS

This term covers a wide variety of dietary preparations that range from clear liquids which provide little in the way of nourishment to nearly complete diets in liquid form. For example, an infant formula is one type of liquid diet, whereas tube feedings are another.

Fig. L-24. A hospital patient being tube fed a liquid diet. This type of feeding may be a lifesaver for people who are unable to eat in the normal way.

The use of liquid diets is growing rapidly, as evidenced by the various types of canned products that may be found in supermarkets and pharmacies. Some liquid diets are used for therapeutic purposes. Others are intended as replacements for, or supplements to, regular diets. Some department stores now have vending machines which dispense a liquid dietary formula in paper cups so that employees and shoppers may gulp their lunch down quickly and spend most of their lunch hour shopping.

TYPES OF LIQUID DIETS.   The major types of liquid diets are described in Table L-3.

NOTE: Liquid diets should be used for therapeutic purposes only after consulting a doctor. *This caution also applies to liquid reducing diets.*

TABLE L-3
MAJOR TYPES OF LIQUID DIETS

Diet	Composition		Uses	Comments
	Foods[1] Included	Foods Excluded		
Clear Liquid	Only clear liquids such as tea; coffee; carbonated beverages; cereal beverages (coffee substitutes); flavored gelatin; ices made from strained fruit juices; fruit juices and drinks without pulp; fat-free fish, meat, and poultry broths; consomme; bouillon; sugar; clear syrups and honey.	All others.	To maintain the body's water balance when other foods and beverages cannot be taken by mouth, or when it is necessary to clear the colon of residue (in preparation for x rays; prior to, and following, surgery on the digestive track; during diarrheal disease; or as an initial oral feeding during the recovery of a critically ill patient).	This diet should be used for only a few days because it is deficient in almost all of the essential nutrients. Elemental diets serve most of the same purposes and are much more suitable for longterm use.
Elemental (semi-synthetic liquid diet) (Also see ELEMENTAL DIETS.)	Formula powders that are mixed with water and/or carbonated or noncarbonated fruit drinks that are free of fibrous residues, or bouillon. Essential nutrients are provided in the following forms: Carbohydrates as sugars and other well absorbed, rapidly utilized saccharides. Fat (usually only 0 to 10% of the diet) as a vegetable oil. (1) Amino acids as chemically pure crystalline powders; or (2) A protein hydrolysate such as digested casein; or (3) Egg albumin. Mineral salts and vitamins in chemically pure forms.	All others, unless allowed by a doctor, or a dietitian.	To promote ease of digestion and absorption and minimize food residue in the colon, while providing nearly all of the required nutrients, prior to and following surgery on the digestive tract; during diarrheal disease, infections, or inflammatory conditions; for the rehabilitation of severly injured and/or malnourished patients; minimizing nausea due to chemotherapy and/or radiation treatments; or as an initial oral feeding during the recovery of a critically ill patient. May also be used in tube feeding.	None of the products now available contain the essential trace minerals chromium and selenium, and only some brands contain vitamin K. Therefore, these nutrients must be provided by other means of nourishment for more than a few weeks. Patients have been maintained in good condition on elemental diets for as long as a year or more.

Full liquid diet (General liquid diet)
(See MODIFIED DIETS, Table M-28, Modified Diets—General liquid diet [Full liquid diet] for the composition and uses of Full liquid diet.)

Diet	Foods[1] Included	Foods Excluded	Uses	Comments
Infant formulas (Also see BABY FOODS; BREAST FEEDING; and INFANT DIET AND NUTRITION.)	Commercially produced infant formulas based mainly on milk or a milk substitute (usually soy protein isolates or pureed beef or lamb) that may contain added vegetable oil, sweetener, mineral salts, and/or vitamins. Homemade formulas prepared from diluted cow's milk with or without an added sweetener.	Homogenized cow's milk is not considered suitable for infants younger than 6 months, but evaporated milk is usually tolerated well.	In lieu of breast feeding young infants, and for older infants who do not tolerate cow's milk well. May be used by older children and adults who require liquid diets, when the latter is not available. Special modified types of infant formulas are available for children who have intolerances to certain foods and nutrients, but should be used only after consulting a doctor.	Some formulas furnish all the nutrients required by infants, whereas others should be supplemented by mineral and vitamin preparations. Infants aged 6 months and older should receive other foods besides a formula in order to develop good eating habits.
Special dietary products	Liquid or powdered (for mixing with water, milk, or other suitable liquid) dietary preparations that are designed to meet the special needs of patients placed on certain modified diets. Most of the products contain the appropriate types of carbohydrates, fats, amino acids or proteins, mineral salts and vitamins; but some furnish only supplementary nutrients for meeting unusual requirements.	Some products may contain little or no allergens, cholesterol, electrolytes, lactose, milk protein, protein in general, saturated fat, sodium, or other substances that are deemed to aggravate the patient's condition.	As a substitute for a normal diet that may be unsuitable in composition for the patient. Provision of calories and/or nutrients in addition to those in the normal diet where patients cannot or will not consume foods that supply the essential nutrients.	The use of these products alone may become very expensive and monotonous for the patient, who may benefit more by consuming homemade items prepared from common foods by modified recipes (when it is feasible for a lay person to prepare them).
Supplemental feedings	Dietary formulations designed to provide extra calories, proteins, and other nutrients in easily digested and readily absorbed forms when the patient is already receiving some, but not all, of his or her nutrient requirements from another source. Most products contain sugars, powdered milk or a substitute for milk protein, vegetable oils or their derivatives, added mineral salts and vitamins, and flavorings such as vanilla and chocolate.	Nutrients that are poorly digested or slowly absorbed, such as fiber, and long chain fatty acids.	Supplementation of the normal or therapeutic diets of growing children athletes, underweight adults, people recovering from debilitating illnesses, injuries, burns, surgery, gastrointestinal disorders, anorexia; and various other conditions in which nutritional needs are likely to be in excess of the nourishment provided by the unsupplemented diets.	Some of the products in this category are called *nourishments* after the traditional dietary practice of providing certain patients with extra feedings between meals. In the past, these feedings often consisted of items such as cream, milk, milk drinks, puddings, custards, and other high-calorie foods.

Footnotes at end of table

(Continued)

TABLE L-3 *(Continued)*

Diet	Composition		Uses	Comments
	Foods[1] Included	**Foods Excluded**		
Tube feedings	Mixtures of liquids, dissolved powders, and/or pureed solid foods of sufficiently fluid consistencies to pass through feeding tubes without clogging them. Some typical ingredients of the common types of feedings are: commercial formulas; whole milk, evaporated milk, strained buttermilk, and/or ice cream; strained pureed meats, hard cooked yolk, fruits, and/or vegetables; fruit juices; sugar, syrup, and/or honey; added mineral salts and vitamins; and flavorings such as vanilla and chocolate.	Foods and nutrients that are poorly digested or slowly absorbed, such as fiber and long chain fatty acids. Substances that are likely to settle out and clog the tube are also excluded, such as pieces of foods, nuts, seeds, chocolate chips, milk curds, solidified fats, and powders that are difficult to dissolve.	Nutritional support of people who are unable to chew and/or swallow ordinary foods due to surgery on the face, oral cavity, neck, or throat; fractures of the jaw; esophageal strictures; injury, inflammation, or surgery affecting one or more parts of the upper digestive tract; partial paralysis, or unconsciousness. Sometimes, tube feedings are used for force feeding people who refuse to eat and become emaciated, such as those with anorexia nervosa or protesters on a *hunger strike.*[2] The tube for feeding may be inserted (1) up through the nose and down to the stomach (nasogastric feeding), (2) through the nose and down to the duodenum (nasoduodenal feeding), (3) into an opening on the abdomen leading to the stomach (gastrostomy feeding), or (4) into an opening on the abdomen leading to the jejunum (jejunostomy feeding).	Care has to be taken that tube feedings are not regurgitated and aspirated into the lungs. This may be prevented by giving the initial feeding at only ½ strength, keeping the rate of feeding slow; and by allowing the patient to rest in a sitting position after an intermittent feeding of a large volume of formula. It is noteworthy that some tube feeding formulas have a high concentration of dissolved materials (solutes) and may cause dehydration and/or diarrhea unless some additional water is given. Detailed instructions for tube feedings may be obtained from the manufacturers of the formulas and from nursing manuals.

[1]Many liquid diets provide nutrients in highly purified forms rather than in the forms of ordinary foods.

[2]Long-term deliberate or unavoidable abstention from the consumption of adequate food may result in a loss of appetite and other abnormalities that often necessitate forced feeding to initiate recovery of the normal eating and digestive functions.

Fig. L-25. Liquid diets can be monotonous, so different tumblers and frills can *zip* them up. (Courtesy, American Dairy Assn., Chicago, Ill.)

SUMMARY. Providing most of a patient's nutrient needs by a liquid diet may be done successfully over long periods of time if this means of nourishment is carried out according to the recommendations of a doctor and/or a dietitian.
(Also see ELEMENTAL DIETS; and MODIFIED DIETS.)

LITHIASIS

The formation of calculi (stones) of any kind (as in the urinary tract and gallbladder).

LIVETINS

A collective term for the principal water-soluble proteins that are found in egg yolk—an albumin, a glycoprotein, and a globulin.
(Also see EGG.)

LOFENALAC

This term is the registered trademark of the commercially prepared protein formula sold by Mead Johnson & Company of Evansville, Indiana. Ninety-five percent of the amino acid phenylalanine is removed from Lofenalac. It provides the basis for dietary management of phenylketonuria (PKU), an inborn error of metabolism. Most proteins contain 4 to 6% phenylalanine and would be detrimental to an infant with PKU. Lofenalac is used to meet the infant's protein and energy needs while the requirement for phenylalanine, an essential amino acid, is just satisfied—no excess—with foods whose phenylalanine content is known.
(Also see INBORN ERRORS OF METABOLISM; and PHENYLKETONURIA.)

LOGANBERRY *Rubus loganbaccus*

A reddish-purple type of blackberry (fruit of the *Rosaceae* family) that was discovered growing in the garden of Judge Logan of Santa Cruz, California in 1881.

It is believed to have originated as a result of a natural crossbreeding (hybridization) between a western wild blackberry and a domesticated variety of red raspberry. The loganberry is grown mainly in the central coast region of California, the Willamette Valley of Oregon, and the southwestern part of British Columbia around Vancouver.

Only about 2% of the U.S. crop of loganberries is marketed fresh. The rest is canned, frozen, and made into jam, jelly, juice, and wine. It is noteworthy that the sharp tartness of the fresh berries is improved by cooking.

The nutrient compositions of various forms of loganberries are given in Food Composition Table F-21.

Some noteworthy observations regarding the nutrient composition of loganberries follow:

1. The raw fruit is moderately high in calories (62 kcal per 100 g) and carbohydrates (15%). It is an excellent source of fiber, a good source of potassium, iron, and vitamin C, and only a fair to poor source of vitamin A.

2. Canned loganberries contain about two-thirds of the nutrients of the fresh fruit, except that much of the vitamin C is destroyed during processing. Also, the products containing syrups may be rather high in caloric content.

LOGARITHMIC PHASE (EXPONENTIAL PHASE; LOG PHASE)

When pertaining to bacterial cultures in an adequate

medium, it is the period of most rapid growth when numbers increase in geometric progression. On a rich medium bacteria can double in numbers every 20 minutes.

LONGAN *Euphoria longana; Nephelium longana*

The fruit of a tree (of the *Sapindaceae* family) that is native to the Asian tropics, and is now grown in some of the other tropical regions of the world. Longans are round, yellowish-brown fruits that contain a white, juicy pulp.

The nutrient compositions of raw and dried longans are given in Food Composition Table F-21.

Some noteworthy observations regarding the nutrient composition of longans follow:

1. The raw fruit is moderately high in calories (62 kcal per 100 g) and carbohydrates (14.9%). It is a good source of iron, but a poor source of vitamin C.

2. Dried longans contain about 5 times the nutrient levels of the fresh fruit. The vitamin C content is apparently retained completely during drying. Hence, the dried fruit is suitable for a nutritious high-calorie (286 kcal per 100 g) snack that contains almost 5% protein, 2% fiber, and is an excellent source of phosphorus and potassium, and a good source of vitamin C.

LONGEVITY

Long-lived.

LOW DENSITY LIPOPROTEIN (LDL)

A lipoprotein with a density of 1.006 to 1.063; composed of about 21% protein; high in cholesterol.
(Also see BETA-LIPOPROTEINS.)

LUCERNE (ALFALFA) *Medicago sativa*

The plant known as alfalfa (botanically *Medicago sativa*) in the United States is called lucerne in many parts of the world.
(See ALFALFA.)

LUPUS ERYTHEMATOSUS

Recent research indicates that systemic lupus erythematosus is primarily a disease of the immune system in which antigen-antibody complexes cause tissue damage. An early manifestation of the disease is the erosive inflammation of the skin typically in the form of a *butterfly rash* over the nose and cheeks. This rash gave the disease its original name—*lupus erythematosus* or *red wolf*. Besides the rash, other symptoms of lupus include an intermittent fever, severe joint pain, and periods of extreme fatigue. It afflicts about one woman in 500 in the United States, and about one-tenth as many men. Lupus affects not only the skin, but, at different times in different individuals, it may affect the joints, the blood vessels, the heart, the lungs, and the brain—and most significantly the kidneys. The disease probably arises from an interplay of genetics, drugs, hormones and viruses. Methods for early diagnosis have led to improved control of lupus, thereby preventing irreversible tissue damage, particularly to the kidneys.

Part of the treatment for lupus may involve cortisone or other steroids to control the inflammation; hence, a sodium-restricted diet may be necessary. Should the kidneys become involved, dietary changes similar to those for kidney diseases may be recommended.

(Also see MODIFIED DIETS.)

LYCOPENE

A chemical similar to carotene, but it has no vitamin A activity. It is the principal red pigment present in tomatoes, watermelon, pink grapefruit, paprika, rose hips, and palm oil.

LYMPH

The slightly yellow, transparent fluid occupying the lymphatic channels of the body. It is derived from the fluid between the cells—interstitial fluid. The chemical makeup of lymph is similar to that of blood plasma. Following a meal, some of the fat from the diet is absorbed into the lymph in the form of chylomicrons.

(Also see BODY FLUIDS; and DIGESTION AND ABSORPTION.)

LYMPHATIC SYSTEM

All the vessels and structures that carry lymph from the tissues to the blood.

LYOPHILIZATION (FREEZE DRYING)

A process of rapid freezing followed by drying under a vacuum. The ice sublimes off as water vapor without melting. Freeze drying results in the least damage to foods of all the commercial processes for drying. It can be applied to all foods, both raw and cooked, which can withstand freezing. Freeze drying yields a product that is dried without shrink, and which has a highly porous structure favoring rapid rehydration. After rehydration, freeze-dried foods are often indistinguishable from their commercially frozen counterparts.

(Also see PRESERVATION OF FOOD.)

LYSERGIC ACID

A chemical derived from the hydrolysis (breakdown) of the alkaloids of the fungus ergot. It is thought to be the active component of ergot. Lysergic acid can produce a state resembling psychosis. Also, its diethylamide derivative, LSD, is a potent hallucinogenic drug.

(Also see POISONS, Table P-2, Some Potentially Poisonous [Toxic] Substances—"Ergot.")

LYSINE

Contents	Page
Involvement in the Body	628
Sources	628
Plant vs Animal	629
Other Lysine Sources	629
Factors Affecting Availability	629
Lysine Supplementation	629
Lysine and Herpes Infections	629

Lysine is an essential amino acid. The body requires lysine but cannot synthesize it so it must be provided in the diet. High quality protein foods provide lysine. However, lysine is the limiting amino acid in most cereals—foods for much of the world. It is present in cereal grains, but at low levels. Inadequate lysine in the diet may create a negative nitrogen balance—protein being broken down and excreted. Chemically, lysine is unique among the essential amino acids since it possesses two amino groups (NH_2). This can further affect the availability of lysine when it is in short supply. All other essential amino acids possess one amino group. Since lysine, in the form of lysine monohydrochlorine, can be commercially produced at a reasonable cost, much consideration has been given to the supplementation of cereals with lysine in an effort to improve the nutrition of people who depend largely upon cereal grains.

INVOLVEMENT IN THE BODY. The requirement of lysine for growth is quite high, since tissue proteins contain a high proportion of lysine. Infants require 45 mg/lb (*99 mg/kg*) of body weight per day, and growing children require 20 mg/lb (*44 mg/kg*) of body weight per day, while adults require only 5.5 mg/lb (*12 mg/kg*) of body weight per day. Rats fed diets deficient in lysine grow very little while the addition of lysine to the same diet significantly increases growth. Since protein synthesis—growth of tissue—is an all-or-none phenomenon, it requires that all amino acids necessary for synthesis be present. Otherwise, synthesis will not proceed; hence, growth will be limited. The protein in most cereal grains is poorly utilized by infants and children due to its low lysine content.

SOURCES. Lysine may be obtained from both plant and animal dietary proteins. However, in terms of lysine content, plant proteins and animal proteins are not created equal. In general, plant proteins contain less lysine than animal proteins.

Fig. L-26. Lysine, an essential amino acid, is generally more abundant in animal proteins than in plant proteins. Beef is an especially rich source. (Courtesy, National Livestock & Meat Board, Chicago, Ill.)

Plant vs Animal.   Protein is a part of all tissue—plant or animal. The seeds of many plants supply protein, hence lysine, to mankind; for example, wheat, corn, barley, lentils, peas, rice, and soybeans. Likewise, all animal foods—except pure fats—are good protein sources and, therefore, provide lysine. Table L-4 shows the comparative lysine content of some common protein sources.

TABLE L-4
LYSINE CONTENT OF SOME COMMON FOODS[1]

Food	Lysine	
	(mg/100 g)[2]	(mg/g protein)[3]
Beef steak, cooked	2,999	98
Chicken, cooked	1,830	86
Cheese, cheddar	2,072	83
Pork chop, cooked	2,044	83
Cod, cooked	2,421	83
Sardines, canned	1,552	83
Lamb leg, roasted	2,081	82
Milk, cow's (2% fat)	264	80
Soybeans, cooked	759	69
Eggs	820	68
Lentils, cooked	476	61
Peas, green, cooked	254	47
Peanuts, roasted	1,176	45
Rice, brown, cooked	99	40
Oatmeal, dry	521	37
Wheat, parboiled (bulgur)	217	35
Barley, pearled light, uncooked	279	34
Corn meal, cooked	29	26

[1]Data from Protein and Amino Acid Content of Selected Foods, Table P-16.

[2]One hundred grams is approximately equal to 3½ oz.

[3]Two grams per kilogram of body weight per day of a protein containing 51 mg of lysine per gram of protein would meet the lysine needs of the infant.

Cereal grains—barley, corn, oats, rice, and wheat—are low in lysine as Table L-4 shows, whereas the lysine content of legumes—lentils, peas, and soybeans—is better. The lysine content of animal products is highest.

It is a recognized fact that lysine is the limiting amino acid—the essential amino acid most deficient—of the cereal grain proteins. Since cereal grains provide much of the world's food, there have been attempts to increase their lysine content. Plant breeding programs have successfully increased the lysine content of barley, sorghum, and corn as shown in Table L-5. This brings the protein quality of the cereal grains closer to that of animal products.

TABLE L-5
LYSINE CONTENT OF CEREAL GRAINS

Grain	Lysine	
	(mg/100 g)	(mg/g protein)
Barley	435	34
High-lysine barley	1,089	56
Corn	290	29
High-lysine corn	491	45
Sorghum	297	27
High-lysine sorghum	611	33

These improved grains have been tested in animal and human trials with good results. However, farmers are reluc-tant to grow the hybrid varieties because they yield 10 to 15% less grain per acre than the more common varieties.

(Also see CORN, section headed "High-Lysine Corn [Opaque 2, or O$_2$].")

Other Lysine Sources.   Lysine itself may be produced by two processes: (1) the fermentation of carbohydrate materials; and (2) the Toray process, which converts cyclohexene, a by-product of nylon production, to lysine. Following purification procedures, lysine is available as a pure white powder, often as lysine monohydrochlorine. Currently, the cost of producing pure lysine is about $1.50 to $2.00 per pound ($3.30 to $4.40/kg).

FACTORS AFFECTING AVAILABILITY.   Beside lysine being in short supply in cereal grains, the biological availability—availability to the body once ingested—may be adversely affected. As pointed out, lysine has an extra amino group (NH_2). This amino group can react with other compounds, primarily the aldehyde group of reducing sugars such as glucose or lactose. The reaction yields an amino-sugar complex that is no longer available for use in the body. Digestive enzymes cannot split the amino-sugar complex; hence, it is not absorbed. This reaction—the Maillard reaction—occurs due to extensive heating or prolonged storage. In protein sources such as the cereal grain products where lysine is already in short supply, the reaction can further reduce the lysine content of the diet.

LYSINE SUPPLEMENTATION.   Since lysine is the limiting amino acid of cereal grains which make up the diets of a large portion of the world, some scientists propose the enrichment of these grains with pure lysine derived from processes described above. This practice would be similar to that in the United States wherein thiamin, niacin, riboflavin, and iron are added to many cereal products. Indeed, there have been experimental trials with animals and humans demonstrating the beneficial effects of adding lysine. Still, in countries where diets are extremely poor, more may be gained by increasing the food supply in general and providing other protein sources such as legumes which complement the protein of cereal grains. However, some countries have tried lysine supplemented foods. The Japanese school lunch programs fortified bread with lysine and vitamins. Government-controlled bakeries in India produced bread fortified with lysine, vitamins, and minerals. In Guatemala, Incaparina, a new product to increase the quality and quantity of protein, is fortified with lysine. In the United States, Pillsbury test marketed a flour containing added vitamins, minerals, and lysine.

Lysine enrichment is a debatable issue, depending on a variety of factors, primarily the continued availability of major sources of calories and protein. Supplementation is of little value if there is not enough food. But lysine supplementation is a potential tool.

(Also see AMINO ACID[S]; and PROTEIN[S].)

LYSINE AND HERPES INFECTIONS.   The Herpes simplex viruses are responsible for cold sores, fever blisters, and genital herpes, while a close relative, Herpes zoster, is responsible for chickenpox, shingles, and *infectious mono*. In 1979, the Lilly Research Laboratories, a division of Eli Lilly and Company of Indianapolis, released research results showing that supplementary lysine speeded recovery and suppressed recurrences of herpes infections.

Tablets of lysine as the hydrochloride may be purchased at many general health food stores. Each tablet contains about 300 mg of lysine. A monograph from Lilly Research laboratories recommends the following:

1. When a clearly established herpes infection is present, two lysine tablets should be taken two to four times daily until the infection has cleared.

2. Those individuals who can predict an episode of herpes due to its association with such things as menstruation, exposure to sunlight, eating nuts, or any stressful situation, may take suppressive amounts of lysine—one, two, three, or four tablets daily depending upon the individual. Some individuals may prevent further infection with one tablet daily, while others may require two tablets twice a day.

3. Attacks may be completely aborted if lysine is taken in a dose of two tablets, three times per day for 3 to 5 days as soon as the signs of an attack are noted—a stinging and burning sensation followed by a little reddish nodule.

4. Some individuals can prevent herpes by avoiding known sources of another amino acid—arginine. These sources include: peanut butter, cashews, pecans, almonds, and chocolate. Tissue culture studies have shown that arginine enhances herpes growth.

Supplemental lysine does not produce any undesirable effect nor is it toxic. It is a normal constituent of protein. However, some individuals may not receive adequate amounts, and the needs from individual to individual may vary widely. Hence, to suppress herpes the dose of lysine must be varied.

Some recent work indicates that the continued use of lysine may antagonize (counteract) arginine, another amino acid. Also, drugs and vaccines raise the hope of suppressing the herpes virus. Thus, those who are persistently plagued with herpes are advised to confer with their physician.

LYSOSOMES

Structures of cell cytoplasm that contain digestive enzymes.

LYSOZYME

An enzyme that digests some gram-positive bacteria and certain high molecular weight carbohydrates; present in saliva, tears, and egg white.

Fig. L-27. Balancing the lysine! Corn and other cereals are low in lysine, while chicken and other animal products are high in lysine. (Courtesy, USDA)

MACARONI AND NOODLE PRODUCTS

Contents   Page

History 631
Production 632
  Commercial Products 632
    Ingredients 632
    Manufacturing Procedures 632
    Types of Macaroni and Noodles 633
Nutritive Values of Macaroni and Noodle Products 633
  Enrichment 634
  Fortification 634
Preparation and Serving of Pasta 634

These items, which are also called pasta, are pieces of dough that have been formed into various shapes, then dried. In the United States, macaroni products may contain egg as an optional ingredient; but products labeled as noodles must contain a certain minimum amount of egg. The names *macaroni*, *spaghetti*, and *vermicelli* are applied to macaroni products that are cord-shaped and have certain diameters. *Macaroni* has the largest diameter of the macaroni products, while *vermicelli* has the smallest diameter.

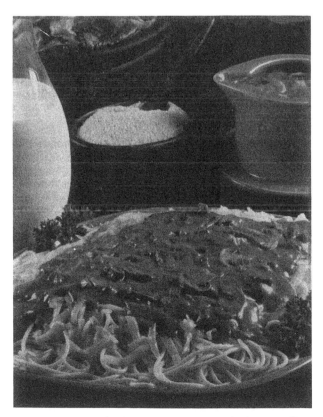

Fig. M-1. Spaghetti, a favorite pasta. (Courtesy, United Dairy Assoc., Rosemont, Ill.)

Macaroni and noodle products are important foods for many peoples around the world because they (1) are easy to make with simple equipment from ingredients that are available in most places, (2) keep well without refrigeration, (3) are bland flavored and mix well with most other foods, and (4) contain sufficient calories and protein to be major dietary sources of these nutrients.

HISTORY. Evidence suggests that forms of pasta were used by the ancient Etruscans (people who settled the upper western coast of the Italian peninsula around 700 B.C.). Some historians suspect that macaroni and noodle types of pasta were introduced into Italy during the Middle Ages by the Asian slaves who cooked for the affluent Italians, since threadlike noodles were then made in various places across Asia—from China to Arabia.

Fig. M-2. The macaroni seller, in Southern Italy. (After a colored lithograph by Muller. Courtesy, The Bettman Archive, Inc., New York, N.Y.)

Whatever the origin of pasta, the southern Italians became the masters of its production in Europe, because the northern Italians ate rice in their soups and with their sauces. Also, the warm, dry climate of southern Italy favored the drying of freshly made pasta in the sun. Finally, the southern Italians preferred their pasta firm (*aldente,* which means that it requires chewing), whereas the northern Italians cooked it until it was much more tender.

Pasta appears to have spread throughout Europe from Italy, although the peoples in the northern countries apparently preferred flat noodles to the rounded types (macaroni and spaghetti). It is believed that Thomas Jefferson brought pasta to America from France, where he had served as U.S. Ambassador. Later, immigrants from northern Europe brought their own noodle dishes. Still later, Italian immigrants came to America, and for a long time thereafter imported their pasta from Italy.

It is noteworthy that the production of good pasta (that which holds together in boiling water) requires hard wheats with a high gluten content. The Italians obtained most of their wheat from the U.S.S.R. until the Crimean War (1854-1856) cut off their supplies. Shortly thereafter, hard wheats were produced in Canada and the United States. Hence, North America soon became the major source of durum wheat for the pasta industry around the world. However, there was not much production of macaroni or spaghetti in the United States until World War I cut off the supply from Italy.

PRODUCTION.   The consumption of macaroni and noodle products in the United States has grown steadily since World War II, even though the total consumption of all types of wheat products declined. It is noteworthy that the production of macaroni utilizes most of the durum wheat grown in North Dakota. However, good macaroni may also be made from weaker flours if certain optional ingredients such as eggs are added.

Commercial Products.   The first machine powered pasta presses were developed around the mid-1800s. Prior thereto, most macaroni was produced in small shops which utilized hand operated equipment. Today, large machines make it possible to turn out large batches of these products. Details of commercial production follow.

INGREDIENTS.   The items which follow are the ingredients specified in the U.S. standards of identity for macaroni and noodle products:[1]

• **Semolina, durum flour, farina, flour**—One or more of these wheat products constitute(s) the major ingredient of pasta. Semolina is the coarsely milled endosperm of durum wheat, while durum flour is more finely milled. Farina is similar to semolina, but it is obtained from other than durum wheats, whereas flour is the more finely milled endosperm of nondurum wheats. Each of these four fractions may contain only minimal amounts of branny particles. Generally, pasta makers prefer to use semolina and/or farina since products made from these meals hold up better in cooking than those made from the flours.

• **Water**—This is the usual liquid for making a dough, although it may be replaced completely by fluid milk in certain products. Care is taken that the water is pure and free from undesirable flavors.

• **Eggs**—Egg whites (fresh, frozen, or dried) are optional ingredients for macaroni products, in which they help to strengthen the dough. On the other hand, noodles must contain at least 5½% egg solids from either whole egg or egg yolk. The latter is preferred for the color it imparts to noodles.

• **Disodium phosphate**—Small amounts (0.5 to 1.0%) of this optional additive make pasta cook more quickly.

• **Seasonings and spices**—Onions, celery, garlic, bay leaf, and salt are optional ingredients for macaroni and noodle products. When used, they must be listed on the label.

• **Gum gluten**—Sometimes, this product, which is made from wheat flour by washing out the starch, is added in small amounts to strengthen the dough.

• **Glyceryl monostearate**—Less than 2% of this additive (it is an emulsifying agent) may be used to prevent the formation of clumps of dough.

• **Minerals and vitamins**—Products labeled enriched must contain prescribed amounts of iron, thiamin, and riboflavin, which may be (1) added as chemically pure compounds, or (2) supplied by optional ingredients such as dried yeast, dried torula yeast, partly defatted wheat germ, enriched farina, or enriched flour. Certain specified amounts of calcium and vitamin D are optional.

• **Protein supplements**—The sources of extra protein are usually corn flour and soybean flour or meal, although other ingredients such as nonfat dry milk, fish flour, and derivatives of oilseeds (usually peanut, safflower seed, or sunflower seed) also may serve this purpose. These supplements raise both the quantity and quality of protein so that the macaroni and noodle products may be substituted for part of the dietary animal protein.

• **Milk products**—Whole fluid milk may replace completely the water used in making a dough, or the milk ingredient may be concentrated milk, evaporated milk, dried milk, a mixture of butter and skim milk, concentrated skim milk, evaporated skim milk, or nonfat dry milk. In the case of Milk Macaroni Products, the milk ingredient(s) must supply an amount of milk solids equal to 3.8% of the finished product. Also, the weight of nonfat milk solids cannot be more than 2.275 times the weight of milk fat. This regulation insures uniformity by making all products equivalent to those made with whole milk. It is not applicable to items designated as "Macaroni products made with nonfat milk."

• **Carrageenan or its salts**—This optional additive, which is a gum, is used to strengthen products containing greater than average amounts of nonfat dry milk, because the latter ingredient weakens doughs.

• **Vegetable products**—Colorful macaroni and noodle products are produced by the addition of red tomato, artichoke, beet, carrot, parsley, or spinach derivatives. The vegetable ingredient may be fresh, canned, dried, or in the form of a paste or a puree. However, the pasta product must contain at least 3% vegetable solids. Spinach is the most commonly used vegetable ingredient. It is noteworthy that 3% vegetable solids corresponds to a content of about 30% fresh vegetable matter since most fresh vegetables contain about 90% water.

MANUFACTURING PROCEDURES.   There must be careful control of each manufacturing stage in order to ensure that the product holds up well under (1) handling between the factory and the consumer, and (2) boiling in water. Therefore, the major processes that are utilized in manufacturing of pasta are noteworthy:

1. **Milling of the durum wheat into semolina.**   This milling procedure differs from that used to make flour in that a granular product is desired, with a minimum amount of flour. Hence, certain aspects of the flour milling procedure are altered in the production of semolina. The durum wheat is moistened before milling to toughen the outer layers of the wheat kernels so that they may be removed readily from

---

[1] Code of Federal Regulations, Title 21, Revised 4/1/78, Part 139, Sections 139.110 through 139.160.

the inner portion (endosperm) which yields the semolina. Then, the wheat is broken into coarse particles by corrugated rolls. Other rolls then crush the grain further and scrape the branny material from the pieces of endosperm. Finally, the particles of endosperm and bran are separated by sifting, and by a stream of air which lifts away the smaller flakes of bran.

2. **Preparation of the pasta dough.** The wheat ingredient may be semolina, durum flour, farina, nondurum flour, or various combinations of these items. About 31 parts of water (by weight) are mixed with about 69 parts of meal and/or flour to make a dough. Then, the dough is mixed well and a vacuum is applied to remove any air that might have been mixed into the dough, because air bubbles weaken the pasta.

3. **Shaping the products.** Macaroni products are given their shapes by a process called extrusion, in which the dough is forced through openings in a die made from bronze or Teflon. The shaped raw pasta which emerges from the die is cut to the proper length by either a rotating knife (in the case of short products), or by other means.

Fig. M-3. Two shapes of macaroni (Pasta)-shells and rings. (Courtesy, Hershey Foods Corp. Hershey, Pa.)

4. **Drying.** The newly shaped pieces of pasta are carefully dried under strictly controlled conditions to bring the moisture content down to between 12 and 13%.

5. **Packaging.** Most long, cordlike, or tubular macaroni products are packaged in boxes that protect the strands against breakage. However, noodle products are usually packed in plastic bags.

TYPES OF MACARONI AND NOODLES. Many new products have been developed recently by food technologists

for use in various parts of the world. However, it could be very confusing for the American consumer if each manufacturer were allowed to make products and label them in an arbitrary manner, since the ingredients and nutritive values might vary widely. Hence, the U.S. Food and Drug Administration, with the cooperation of American pasta manufacturers, has developed names and specifications (standards of identity) for each of the more common macaroni and noodle products. These follow:

1. Macaroni products.
2. Enriched macaroni products.
3. Enriched macaroni products with fortified protein.
4. Milk macaroni products.
5. Nonfat milk macaroni products.
6. Enriched nonfat milk macaroni products.
7. Vegetable macaroni products.
8. Enriched vegetable macaroni products.
9. Whole wheat macaroni products.
10. Wheat and soy macaroni products.
11. Noodle products.
12. Enriched noodle products.
13. Vegetable noodle products.
14. Enriched vegetable noodle products.
15. Wheat and soy noodle products.

NUTRITIVE VALUES OF MACARONI AND NOODLE PRODUCTS. People with low incomes have long relied on pasta dishes to supply most of their dietary calories and other nutrients. Furthermore, these dishes are served regularly in the school lunch programs around the United States. Therefore, it is noteworthy that the nutritive values of these foods vary widely, depending upon the type of macaroni or noodle product and the other items that might be served with them. The nutrient compositions of macaroni, noodles, pastinas, spaghetti, and some of the most common dishes prepared with these products are given in Food Composition Table F-21.

Some of the more noteworthy observations regarding the products listed in Food Composition Table F-21 are as follows:

1. A cup of a plain, tender cooked macaroni product contains about 5 oz (140 g) of food, 155 Calories (kcal), and 4.8 g of protein, whereas a cup of noodles supplies 5⅔ oz (160 g) of food, 200 Calories (kcal), and 6.6 g of protein. The superior nutritive values of the noodles are due mainly to the egg content.

2. Enriched macaroni and noodles supply fair amounts of iron, thiamin, riboflavin, and niacin, but only negligible amounts of calcium, phosphorus, vitamin A, and vitamin C.

3. The addition of cheese to macaroni provides considerable extra calories, protein, calcium, phosphorus, and vitamin A. Similarly, the addition of tomato sauce plus cheese or meat to spaghetti enhances the nutritive value of a pasta dish.

4. Chow mein noodles are a concentrated source of calories and protein because they contain almost no water. However, the ratio of protein to calories (2.7 g per 100 kcal) is less than those for either macaroni (3.1 g per 100 kcal) or noodles (3.3 g per 100 kcal), because the frying of the Chinese noodles raises the fat content without increasing the protein.

5. Another important observation regarding macaroni and noodles, which is not apparent from the data presented in Food Composition Table F-21, is that the protein in these products is deficient in the amino acid lysine, which is critical for the growth of infants and children. This defect is offset partly by the high quality of the protein in the eggs that are

added to noodles.

There is a great need for supplementing the nutritive contributions of pasta products when they are used as staple foods for infants and children. (Tiny egg noodles called pastina are used as a baby food in Italy.) Therefore, the current enrichment and fortification practices that are utilized in macaroni and noodle products merit some discussion.

Enrichment. The most common practice is to add the specified amounts of iron, thiamin, riboflavin, and niacin in chemically pure form. However, the restoration of iron and three vitamins replaces only a few of the two dozen or more essential nutrients that are removed in substantial amounts during the conversion of whole wheat kernels into semolina or flour.

Better, but more expensive, means of enrichment are permitted by the standards of identity for enriched macaroni products and enriched noodle products.[2] It consists of providing the specified amounts of iron and the B vitamins through the addition of the highly nutritive ingredients dried yeast, dried torula yeast, and/or partly defatted wheat germ. The use of these ingredients adds extra protein, minerals, and B vitamins that are not provided by the usual means of enrichment.

The federal standards of identity also provide for the optional calcium and vitamin D enrichment of both macaroni and noodles. These nutrients are rarely added, although they would be very beneficial for people who drink little milk or who receive limited amounts of sunshine.

Fig. M-3a. Pasta salad. (Courtesy, Castle & Cooke, Inc.)

Fortification. The low quality of the wheat protein in macaroni and noodles is of considerable concern to nutritionists, because these foods are popular with children around the world. This concern led to the development of new products by food companies, which in turn led to the establishment of a standard of identity for "Enriched macaroni products with fortified protein."[3]

These products must contain at least 20% protein with a quality at least 95% that of casein (a major protein in cow's milk). The supplemental protein may be supplied by flours or meals made from nonwheat cereals or oilseeds such as soybeans, peanuts, sunflower seeds, and cottonseeds.

Similar protein fortified products are specified under the standards of identity for "Wheat and soy macaroni products" and for "Wheat and soy noodle products."[4] The quantities of protein in these items are at least 50% greater than those in similar items made from wheat flour alone. They are often sold in health food stores. However, products labeled "imitation soy macaroni" and "imitation soy noodles" may not contain as much protein because the word "imitation" denotes products which do not conform to a federal standard of identity.

Finally, "Nonfat milk macaroni products" contain between 12 and 25% nonfat milk solids.[5] These items contain from 36 to 75% more protein than is present in similar products made from wheat flour alone. A cup *(240 ml)* of cooked macaroni would also furnish from 89 to 175 mg of calcium and from 131 to 197 mg of phosphorus. Unfortunately, this type of fortified product does not appear to be available at the present time.

PREPARATION AND SERVING OF PASTA. Macaroni and noodle dishes may be utilized as main courses, appetizing side dishes, or as substitutes for starchy vegetables such as corn, potatoes, or rice.

MACEDOINE

A French word meaning a mixture of fruits or vegetables which are chopped and mixed with a dressing, or put into a gelatin. A macedoine can be served as a salad or as a dessert.

MACROBIOTICS, ZEN

This term encompasses both a diet and a philosophy of life. Zen refers to meditation, and macrobiotic suggests a tendency to prolong life.

(Also see ADZUKI BEAN, section on "Zen Macrobiotic Diets"; and ZEN MACROBIOTIC DIET.)

MACROCYTE

An exceptionally large red blood cell occurring chiefly in anemias.

MACRO (OR MAJOR) MINERALS

The major minerals—calcium, phosphorus, sodium, chlorine, magnesium, potassium, and sulfur.
(Also see MINERAL[S].)

MACRONUTRIENT

These are nutrients that are present in the body and required by the body in amounts ranging from a few tenths of a gram to one or more grams. Macronutrients include fat, water, protein, calcium, phosphorus, sodium, chlorine, magnesium, potassium, and sulfur.
(Also see MINERAL[S].)

MAGMA

• A suspension that is comprised of finely divided insoluble or nearly insoluble material in water; for example, magnesia magma (milk of magnesia).

---

[2]*Ibid.,* Sections 139.115 and 139.155.
[3]*Ibid.,* Section 139.117

[4]*Ibid.,* Sections 139.140 and 139.180
[5]*Ibid.,* Section 139.121.

• A crude organic mixture that is in the form of a paste; for example, the mixture of sugar syrup and sugar crystals produced during the refining process.

MAGNESIUM (Mg)

Contents   Page

History...635
Absorption, Metabolism, Excretion....................635
Functions of Magnesium...........................635
Deficiency Symptoms..............................635
Interrelationships................................635
Recommended Daily Allowance of Magnesium........635
Toxicity...636
Sources of Magnesium............................636

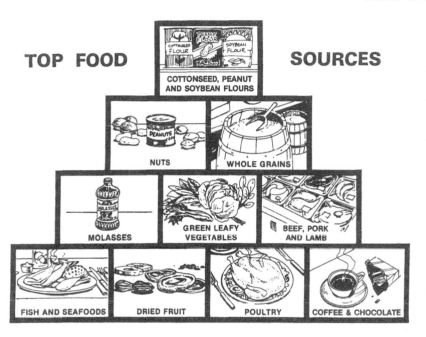

Fig. M 1. Top food sources of magnesium.

Magnesium is an essential mineral that accounts for about 0.05% of the body's total weight, or about 20 to 30 g. Nearly 60% of this is located in the bones in the form of phosphates and carbonates, while 28% of it is found in the soft tissues and 2% in the body fluids. The highest concentration in the soft tissues is in the liver and muscles. The red blood cells also contain magnesium. Blood serum contains 1 to 3 mg of magnesium per 100 ml, most of which is bound to proteins; the rest is in the form of ions.

HISTORY. Centuries ago, the ancient Romans claimed that *magnesia alba* (white magnesium salts from the district of Magnesia in Greece, from which the element was eventually named) cured many ailments. But, it was not until 1808 that Sir Humphrey Davy, the British chemist, announced that he had isolated the element, magnesium.

In 1926, LeRoy, in France, using mice, first proved that magnesium is an essential nutrient for animals. Subsequently, McCollum and co-workers described several magnesium deficiency signs in rats and dogs, including magnesium tetany, a form of convulsions in which the nerves and muscles are affected. Indications that magnesium is required by man followed during the period 1933-1944.

ABSORPTION, METABOLISM, EXCRETION. From 30 to 50% of the average daily intake of magnesium is absorbed in the small intestine. Almost all of the magnesium in the feces represents unabsorbed dietary magnesium. Its absorption is interfered with by high intake of calcium, phosphate, oxalic acid (spinach, rhubarb), phytate (whole grain cereals), and poorly digested fats (long chain saturated fatty acids). Its absorption is enhanced by protein, lactose (milk sugar), vitamin D, growth hormone, and antibiotics.

Magnesium is generally reabsorbed in the kidneys, thus minimizing the loss of body reserves.

The main route of excretion of magnesium is the urine. Aldosterone, a hormone secreted by the adrenal gland, helps regulate the rate of magnesium excretion through the kidneys. Losses tend to increase with the consumption of alcohol or the use of diuretics.

FUNCTIONS OF MAGNESIUM. Magnesium is involved in many functions. It is a constituent of bones and teeth; it is an essential element of cellular metabolism, often as an activator of enzymes involved in phosphorylated compounds and of the high energy phosphate transfer of ADP and ATP; and it is involved in activating certain peptidases in protein digestion. Magnesium relaxes nerve impulses and muscle contraction; functioning antagonistically to calcium which is stimulatory.

DEFICIENCY SYMPTOMS. A deficiency of magnesium is characterized by (1) muscle spasms (tremor, twitching) and rapid heartbeat; (2) confusion, hallucinations, and disorientation; and (3) lack of appetite, listlessness, nausea, and vomiting.

Magnesium deficiency has been observed in alcoholics, in severe kidney disease, in acute diarrhea, and in kwashiorkor.

Some scientists have cautiously theorized a magnesium-cancer link. In support of this thinking, it is noteworthy that fewer cases of leukemia (a form of blood cancer characterized by abnormal increase of white blood cells) have been reported in Poland in areas where magnesium is plentiful in the soil and drinking water than in areas where it is scarce. At this time the magnesium-cancer link theory is without adequate proof, but it merits further research.

INTERRELATIONSHIPS. When magnesium intake is extremely low, calcium is sometimes deposited in soft tissues forming calcified lesions.

An excess of magnesium upsets calcium and phosphorus metabolism. Also, with diets adequate in magnesium and other nutrients, increasing the calcium and/or phosphorus results in magnesium deficiency.

Magnesium activates many enzyme systems, particularly those concerned with transferring phosphate from ATP and ADP. Magnesium is also capable of inactivating certain enzymes, and is known to be a component of at least one enzyme.

Overuse of such substances as *milk of magnesia* (magnesium hydroxide, an antacid and laxative) or *Epsom salts* (magnesium sulfate, a laxative and tonic) may lead to deficiencies of other minerals, or even to toxicity.

RECOMMENDED DAILY ALLOWANCE OF MAGNESIUM. The Food and Nutrition Board (FNB) of

Allowances of magnesium are given in the section on MINERAL[S], Table M-25, Mineral Table.

• **Magnesium intake in average U.S. diet**—The U.S. Department of Agriculture reported that foods available for consumption in the United States in 1988 provided an average of 330 mg of magnesium per person per day, and that the leading sources were as follows: dairy products, 18.5%; flour and cereal products, 18.5%; meat, poultry, and fish, 15.4%; legumes, including dry beans and peas, nuts, soya flour, and grits, 12.7%; and vegetables, including tomatoes, 16.1%.

Surveys indicate that the magnesium content of the average American diet is about 120 mg per 1,000 Calories (kcal). Thus, a person consuming a varied diet of about 2,000 Calories (kcal) will get 240 mg of magnesium. However, a woman following diets limiting intake to 1,000 to 1,500 Calories (kcal) will likely have magnesium intakes below the recommended daily allowance.

TOXICITY. Magnesium toxicity may occur when the kidneys are unable to get rid of a large overload; characterized by slowed breathing, coma, and sometimes death. It is noteworthy that magnesium salts taken by pregnant mothers may affect their newborn babies.

SOURCES OF MAGNESIUM. Magnesium is relatively widespread in foods.

Human milk contains approximately 3 mg of magnesium per 100 g, while cow's milk contains about 13 mg/100 g. Processing grains causes them to lose most of their magnesium. Likewise, refined sugar, alcohol, fats, and oils do not contain magnesium. Excessive cooking of foods in large amounts of water can cause losses.

Groupings by rank of common food sources of magnesium are given in the section on MINERAL[S], Table M-25, Mineral Table.

For additional sources and more precise values of magnesium, see Food Composition Table F-21 of this book.

(Also see MINERAL[S], Table M-25, Mineral Table.)

Fig. M-5. Soybean products, made from soybean flour—a top source of magnesium. (Courtesy, Worthington Foods, Worthington, Ohio)

MAILLARD REACTION

Nonenzymatic browning that takes place upon heating or prolonged storage of food is caused by this reaction. The responsible chemical reaction occurs between a sugar and a free amino group (NH_2) of an amino acid. Once the reaction occurs the amino acid is no longer biologically available, thus reducing the nutritive value of the food. Most Maillard reactions take place with the essential amino acid lysine, since it has a free amino group not involved in a peptide bond. Dr. L. C. Maillard, who first described the reaction, died in 1936 without ever receiving recognition for the importance of his pioneering work.

(Also see LYSINE.)

MALABSORPTION SYNDROME

This term is used to describe a number of symptoms and signs indicating a defect which limits or prevents absorption of one or more essential nutrient from the small intestine. Individuals suffering from a malabsorption syndrome display in varying degrees the following clinical manifestations: (1) diarrhea; (2) steatorrhea (fatty diarrhea) due to impaired fat absorption; (3) progressive weight loss and muscle wasting; (4) abdominal distention; and (5) evidence of vitamin and mineral deficiencies, such as macrocytic anemia due to inadequate folic acid and vitamin B-12 absorption, iron-deficiency anemia, hypocalcemic (low calcium) tetany, and inflammation of the tongue, mouth, and skin—glossitis, stomatitis, and dermatitis, respectively. The laboratory findings characteristic of malabsorption consist of decreases in the blood concentrations of electrolytes, albumin, and carotene, and increases in fecal fat and nitrogen (protein).

The causes of malabsorption syndrome are diverse, but they fall into seven general categories: (1) heart and blood vessel disorders, (2) endocrinal disorders, (3) inadequate absorptive surface, (4) inadequate digestion, (5) lymphatic obstruction, (6) defects in the absorption surface, and (7) reduced bile salts.

Diagnosis of a malabsorption syndrome is based on the results of absorption tests such as xylose absorption, fat absorption balance study, the Schilling test for vitamin B-12 absorption, and the folic acid test. Also gastrointestinal x ray studies, small intestine biopsy, prothrombin time, and serum levels of vitamin A are useful diagnostic tools.

Treatment is directed toward (1) eliminating, insofar as possible, the abnormality causing the malabsorption, (2) vitamin and mineral supplements, and (3) dietary modifications.

Dietary modification will differ due to the cause of the malabsorption syndrome. In general, the diet should be high in calories and protein, but some disorders require the elimination of certain carbohydrates, proteins, or amino acids. Persistent diarrhea may be dealt with by a soft or fiber-restricted diet. Furthermore, a modification of fat intake is beneficial; for example, the common practice of incorporating medium chain triglycerides (8- and 10-carbon chains) into the diet. Medium chain triglycerides substitution can reduce the steatorrhea and the losses of calcium, sodium, and potassium observed in many malabsorption syndromes.

MALAISE

A feeling of illness or depression.

MALARIA

Initially, there was some confusion over the cause-and-effect of this disease; hence, the name *malaria* which literally means *bad air*. Swamps do produce bad air but, more importantly, they also produce *Anopheles* mosquitoes which carry the malaria parasites. In the cycle of this parasite, man serves as an intermediate host, and the mosquito serves as a vector. Transmission of the disease requires at least two mosquito bites—one to pick up the malaria parasite, and another to infect the victim. Fig.M-6 illustrates the malaria cycle.

There are four types of *Plasmodium* which are responsible for three distinct forms of malaria. Each form of malaria can be distinguished by the length of time between the recurrent malaria attacks.

• **Tertian malaria**—Merozoites are released every other day—every 48 hours.

• **Quartan malaria**—Attacks occur every fourth day, or 72 hours apart; hence, the name quartan.

• **Falciparum malaria**—In this form of malaria there is no characteristic length of time between attacks since the mero-

zoites multiply in asynchronous cycles.

It is not possible to prevent infection, but while residing in an area where there is a risk of contracting malaria, drugs can be taken that suppress the symptoms.

Prevention can only come about through public health programs that eliminate the *Anopheles* mosquito.

From a nutritional standpoint, malaria is important for two reasons. First, malaria survives in areas where the mosquito and the infected human population remain above a critical density for each. This situation often occurs in time of famine when people congregate, following which rains come producing breeding grounds for mosquitoes. Malaria then spreads rapidly among the weakened survivors of the famine. Second, malaria attacks, chronic malaria, or repeated infections of malaria are taxing on the nutritional reserves of the body. Good nutrition is essential to recovery. A high calorie, high protein, high fluid intake with vitamin supplementation is recommended.

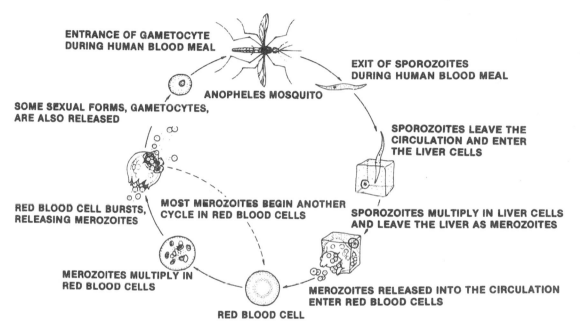

Fig. M-6. The malaria cycle.

MALAY APPLE (POMERAC) *Eugenia malaccensis*

The fruit of an evergreen tree (of a family *Myrtaceae*) that is native to Malaysia and is now grown throughout the tropics. Malay apples are reddish-pink, oblong or pear-shaped fruit that are about 3 in. (8 cm) long. They have a white flesh that surrounds the single large seed. The rather bland fruit may be eaten fresh or made into jam or jelly.

Malay apples have a high water content (91%) and are low in calories (32 kcal per 100 g) and carbohydrates (8%). They are a fair source of vitamin C.

MALFORMATION

Any abnormal development in the physical and physiological makeup of the body.

MALIGNANT

Cancerous growth, as distinguished from a benign growth. (Also see CANCER.)

MALNUTRITION

Contents   Page
Malnutrition Around the World......................638
Causes of Malnutrition.............................639
Consequences of Malnutrition......................639
Common Types of Malnutrition.....................639
Energy-Protein Malnutrition.......................639
Deficiencies of Minerals or Vitamins.................639
Disorders of Nutrient Utilization....................640
Signs of Malnutrition...............................640
Indicators of Malnutrition in Communities............640
Signs of Malnutrition in Individuals..................640
Treatment and Prevention of Malnutrition............641

This term refers to an impairment of health resulting from a failure of the diet, or from a failure of the physiologic processes of the body itself, to provide to the tissues the correct proportions of nutrients. Thus, malnutrition may involve nutrient deficiencies, excesses of nutrients such as carbohydrates, or a combination of deficiencies and excesses.

(Also see DEFICIENCY DISEASES; and DISEASES.)

Disproportions or imbalances between the dietary supply and the tissue requirement for nutrients are responsible for severe deficiency diseases, like beriberi. Populations relying on dietary staples, such as commercially milled white rice, which is high in carbohydrates but low in the thiamin required for its metabolism, are very susceptible to beriberi. Other groups eating a similar diet may avoid the disease by using rice which has been parboiled prior to milling; parboiling rice drives the thiamin from the outside of the grain into the inner parts. As a result, less thiamin is lost when the outer layers of rice are removed during milling.

(Also see BERIBERI; and RICE, section headed "Nutritional Losses.")

Another example of a potential imbalance is the substitution of heat-processed polyunsaturated fats for saturated fats without the addition of vitamin E to the diet. The natural sources of these fats are plant seeds which usually contain ample amounts of vitamin E. But the vitamin E is easily destroyed in the extracted oils by exposure to heat, light, and oxygen. Although severe deficiencies of vitamin E are rare, marginal deficiencies may, over a long period, be responsible for cellular damage and premature senility. Both disorders may be caused by the peroxidation of polyunsaturated fatty acids, which is prevented by vitamin E.

(Also see VITAMIN E.)

Finally, there is evidence that high intake of dietary protein, such as 100 to 150 g/day, results in increased calcium excretion in the urine, thereby raising the requirement for dietary calcium to replace the extra loss from the body. Failure to provide the additional calcium in the diet may lead to the removal of calcium from bone, since the body maintains the level of the mineral in the blood by taking it from bone.
(Also see CALCIUM; and OSTEOPOROSIS.)

MALNUTRITION AROUND THE WORLD. Many adverse physical conditions are brought about by prolonged periods of malnutrition, especially the absence of sufficient quantities of critical compounds in the diet. Rickets, scurvy, beriberi, pellagra, vitamin A deficiency (a major cause of blindness in many countries), anemia, iodine deficiency (which often leads to endemic goiter and cretinism), kwashiorkor, and marasmus are but a few of the many classical examples of nutritional deficiency diseases. When malnutrition begins at an early age and persists over a substantial period of time, a variety of defects may develop involving bone and body structures and mental condition, all of which may be irreversible. Unfortunately, in the developing countries, the basic problem is one of gross dietary inadequacies which may be compounded by specific vitamin or protein insufficiency.

The most malnutrition is found in the developing countries where 70% of the people in the world live, but where only 40% of the world's food is produced. Fig. M-7 shows the countries where the greatest deficits of protein and calories are found.

In Latin America, where protein-poor cereal grains are the imperfect staff of life, 82 out of every 1,000 children die before their first birthday; and another 12 die before they reach the age of 4. Even the survivors may envy the dead. Often brain damaged, they become the adults who are most in need of help and least able to help themselves.

Fig. M-7. World geography of calories and proteins. (Courtesy, *The New York Times*)

In the Far East and Africa, 25 to 30% of the population is estimated to suffer from significant undernutrition. FAO estimates that (1) altogether in the developing world (exclusive of the Asian centrally planned economies for which insufficient information is available), malnutrition affects around 460 million people—that's nearly twice the population of the United States; and (2) one-half of the young children in the developing countries suffer in varying degrees from inadequate nutrition.[6]

According to the World Health Organization (WHO), 10 million children under the age of 5 are now chronically and severely malnourished, and 90 million more may be moderately affected. While undernourished children may remain alive, they are extremely vulnerable to minor infectious diseases. WHO figures also show that of all the deaths in the poor countries, more than half occur among children under 5, and that the vast majority of these deaths, perhaps as many as 75%, are due to malnutrition complicated by infection.

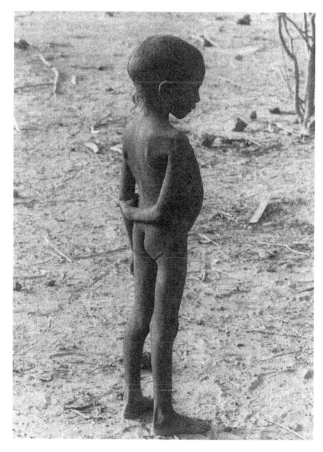

Fig. M-8. A malnourished youngster in the Sahel. Five years of drought caused human suffering and loss of cattle in the six countries along the southern edge of the Sahara. The Sahelian Zone includes Upper Volta, Mauritania, Niger, Senegal, Mali and Chad. (Courtesy, Agency for International Development)

CAUSES OF MALNUTRITION. Poor health due to malnourishment may be the result of one or more factors contributing to poor diets or reduced utilization of nutrients of the body. A listing of some of the more important causes of malnutrition follows:

1. Failure to cultivate available arable land.
2. Catastrophic events, including droughts, floods, earthquakes, wars, and civil disorders.
3. The population explosion.
4. Bad weather.
5. Inefficient food production due to lack of knowledge, skills, seeds, improved animals, feeds, fertilizer, pesticides, farming equipment, and irrigation.
6. Shortages of grains (a) forcing low-income people to resort to starchy plants, such as bananas, sweet potatoes, and cassava, for their staple foods, and (b) creating nutritional problems due to the low content of protein and other nutrients in these high-carbohydrate foods.
7. Lack of money.
8. High food costs.
9. Poor sanitation.
10. Poor nutritional knowledge and practices.
11. Disorders which hamper the utilization of nutrients, including uncontrolled diabetes, hyperthyroidism, hyperparathyroidism, diarrhea, vomiting, parasites, wasting diseases, infections, fevers, inborn errors of metabolism, and toxicants in food.
12. Medical treatments which interfere with nutritional processes; among them, the administration of antibiotics, laxatives, enemas, hormones, anticonvulsants, diuretics, and gastrointestinal surgery, and the severe restriction of diet.

(Also see DEFICIENCY DISEASES.)

CONSEQUENCES OF MALNUTRITION. The cost to the individual and to the society of illness, disability, and death from malnutrition far outweigh the expense of adequate nutrition. Some of these consequences are:

1. Impairment of physical and mental development.
2. Increased susceptibility to infectious diseases.
3. Increased susceptibility to metabolic disorders.
4. Reduced work capacity.

COMMON TYPES OF MALNUTRITION. About one billion cases of malnutrition around the world result mainly from lack of sufficient food.[7] However, hundreds of millions of people suffer from mild to severe nutritional deficiency diseases such as anemia, beriberi, blindness due to vitamin A deficiency (xerophthalmia), goiter, marasmus, pellagra, protein-energy malnutrition, rickets (children) or osteomalacia (adults), and scurvy.

Some of the more prevalent types of deficiencies are briefly described in the paragraphs which follow.

Energy-Protein Malnutrition. An insufficient intake of energy and/or protein needed to maintain the body functions—for activity, for growth, and for resistance to infectious diseases—may be manifested, in keeping with the severity of the deficiency, from the slight impairment of growth or thinness seen in mildly undernourished children to the gross alterations shown by persons suffering from kwashiorkor (protein deficiency) or marasmus (energy deficiency).

(Also see MALNUTRITION, PROTEIN-ENERGY; and STARVATION.)

Deficiencies of Minerals or Vitamins. The lack of sufficient quantities of critical vitamins or minerals may be the cause of many adverse conditions, even when the dietary

[6]The World Food Situation and Prospects to 1985, Foreign Ag. Econ. Report No. 98, Economic Research Service, USDA, p. 50.

[7]McLaren, D. S., "Hunger/Malnutrition: Some Misconceptions," The Professional Nutritionist, Fall 1978, p. 13.

amounts of protein and energy are adequate.

(Also see ANEMIA; BERIBERI; BLINDNESS DUE TO VITAMIN A DEFICIENCY; DEFICIENCY DISEASES; GOITER; PELLAGRA; RICKETS; and SCURVY.)

Disorders of Nutrient Utilization.

Malnutrition may result even when the diet contains adequate amounts of all of the nutrients which are known to be required. In these cases, disease may be caused by a failure of the body to utilize fully certain nutrients, by accumulation of excess amounts of nutrients and/or their metabolites, or by diet-induced changes of the normal environment in and around tissues so that these areas develop increased vulnerability to pathogenic agents and stress factors. However, there may also be nonnutritional causes of some of these disorders, since it has been difficult to explain why a given diet results in disease for some, and fairly good health for others.

(Also see HEART DISEASE; DENTAL CARIES; DIABETES MELLITUS; INBORN ERRORS OF METABOLISM; MALAB-SORPTION SYNDROME; OBESITY; OSTEOPOROSIS.)

SIGNS OF MALNUTRITION.

There are many signs which indicate moderate to severe malnutrition. Also, a number of laboratory tests are available for detection of tissue pathologies, metabolic imbalances, and deficiency diseases. It would be worthwhile, however, to be able to diagnose early stages of malnutrition, since prompt correction of dietary imbalances will, in the long run, be much less costly than later treatment of the more severe stages.

The incidence of malnutrition may be estimated for communities by examining hospital and autopsy records, and by interviewing the parents of deceased children. Also, some individual cases may be diagnosed by their features. Each of these types of assessment will be discussed under the items which follow.

Indicators of Malnutrition in Communities.

Although cases of malnutrition may be found among all socioeconomic classes, one is most likely to find this condition in areas where there are serious problems concerning the following factors: the production and distribution of food, unemployment and low incomes, environmental sanitation, public health and other community services, and unusual food preferences.

A survey of the living conditions in a community might, therefore, save time and money in assessing the nutritional status of its residents. The following types of data might be collected in such a survey: (1) population from latest census; (2) total food consumption by the population (which may be roughly estimated from figures showing local production and shipments of food into the area); (3) foods and nutrients available per capita, which may be calculated from (1) and (2) and a food composition table; (4) morbidity and mortality statistics, plus estimates of school attendance and absenteeism in local industry; (5) infant feeding practices, such as breast feeding, use of formulas, baby foods, table foods, etc.; (6) food preferences and customs of the ethnic and religious groups in the community; (7) employment and income status; (8) water quality and sanitary facilities for disposal of garbage and human and animal waste; (9) food, health, and financial services provided by public and voluntary agencies; and (10) the professional training and number of workers in the service agencies.

Analyses of the data should provide indications of the types of malnutrition and other health problems which are most likely to be present.

Signs of Malnutrition in Individuals.

Assessment of the nutritional status of individuals by laboratory tests is expensive, so it is necessary to have some screening procedures for identifying persons who are most likely to suffer from malnutrition. Such screenings may include medical histories of individuals and their families; heights, weights, and other growth data for children; dietary histories; and observations of the face, tongue, mouth, eyes, skin, muscular and skeletal development, and weight in proportion to height.

Also, the following are valuable adjuncts in screening: measurements of hemoglobin in finger prick samples of blood, testing of urine with pretreated tapes for sugar and ketone bodies, and examination for the presence of sediment in the urine. In adolescents, the stage of sexual development in relation to age may be determined by physicians or nurses, since maturation may be delayed in malnutrition.

Criteria for the diagnosis of specific nutritional disorders are given in this book under the separate entries for each disease. Physical signs often associated with malnutrition are shown in Table M-1.

(Also see DEFICIENCY DISEASES.)

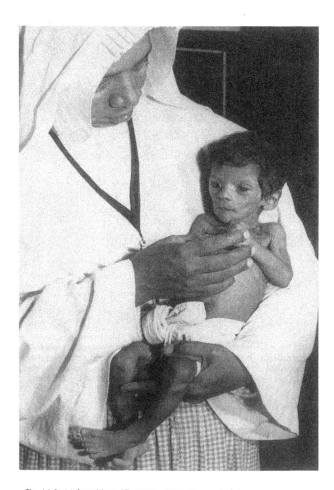

Fig. M-9. Malnutrition. (Courtesy, FAO, Rome, Italy)

TABLE M-1
PHYSICAL SIGNS WHICH ARE OFTEN ASSOCIATED WITH MALNUTRITION[1]

Area of Body Examined	Normal Appearance	Common Signs of Malnutrition	Differential Diagnoses (other causes of clinical signs)
Appearance in general	Normal weight for age, sex, and height. Alert and emotionally stable. No areas of edema.	Significant overweight or underweight. Apathetic or hyper-irritable. Loss or slowness of ankle or knee reflexes. Pitting edema.	Nonnutritional nervous or metabolic disorders. Endocrine diseases.
Eyes	Shiny and free from area of abnormal pigmentation, opacity, vascularization, and dryness.	Paleness, dryness, redness, or pigmentation of membranes (conjunctiva). Foamy patches on conjunctiva (Bitot's spots). Dullness, softness, or vascularization of the cornea. Redness or fissures on eyelids.	Exposure to environmental or chemical irritants. Tissue changes accompanying aging.
Face	Skin is clear and uniform in color, free of all except minor blemishes. Free of swollen or lumpy areas.	Skin is lighter (depigmentation) and darker (over cheeks and under eyes) areas. Greasy scales around nose and lips. Swollen or lumpy areas.	Poor hygiene. Addison's disease (moonface).
Glands	No swollen areas on face or neck.	Swelling of parotids (enlarged "jowls") or thyroid (front of neck near its base).	Mumps. Inflammation, tumor, or hyperfunction of the thyroid.
Gums	Red, free from bleeding or swelling.	Receded. *Spongy* and bleeding. Swelling of the gingiva.	Medication. Periodontal disease. Poor oral hygiene.
Hair	Shiny, firmly attached to scalp (not easily plucked without pain to patient).	Dullness, may be brittle and easily plucked without pain. Sometimes lighter in color than normal (depigmentation may be bandlike when hair is held up to a source of light).	Endocrine disorders.
Lips	Smooth, not chapped, cracked, or swollen.	Swollen, red, or cracked (cheilosis). Angular fissures and/or scars at the corners of the mouth.	Herpes (blisters). Exposure to strong sunshine, dry or cold climates.
Muscles	Muscles are firm and of normal size.	Wasting and flabbiness of muscle. Bleeding into muscle.	Wasting diseases. Trauma.
Nails	Firm, pink.	Spoon-shaped nails. Brittle and ridged nails.	Cardiopulmonary disease.
Organs, internal	Normal heart rate and rhythm. Normal blood pressure. Internal organs cannot be palpated (except liver in children).	Racing heartbeat (over 100 beats per minute). Abnormal rhythm of heart. High blood pressure. Palpable enlargement of liver or spleen.	Rheumatic fever. Nonnutritional diseases of the heart, liver, spleen, and kidneys.
Skeleton	Bones have normal sizes and shapes.	Softening, swelling, or distorted shapes of bones and joints.	Nonnutritional connective tissue disorders.
Skin	Smooth, free of rashes, swellings, and discoloration.	Roughness (follicular hyperkeratosis), dryness or flakiness. Irregular pigmentation, black and blue marks, *crazy pavement* lesions. Symmetrical, reddened lesions (like sunburn). Looseness of skin (lack of subcutaneous fat).	Secondary syphilis. Poor or improper hygiene (excessive washing with soap). Environmental irritants. Trauma. Anticoagulant therapy.
Teeth	Enamel is unbroken and unspotted. None or a few small cavities.	Caries. Mottled or darkened areas of enamel.	Developmental abnormalities. Stains from foods or cigarettes.
Tongue	Normal size of papillae (no atrophy or hypertrophy). Color is uniform and deep red. Sense of taste is without impairment.	Atrophy of papillae (the tongue is smooth) or hypertrophy of papillae. Irregularly shaped and distributed white patches. Swollen, scarlet, magenta (purple colored), or raw tongue.	Dietary irritants. Colors or dyes from food. Nonnutritional anemias. Antibiotics. Uremia. Malignancy.

[1]Adapted by the authors from *American Journal of Public Health*, Vol. 63, November 1973 Supplement, p. 19, Table 1.

TREATMENT AND PREVENTION OF MALNUTRITION.
The first step in the correction of a dietary imbalance is the use of a modified or supplemented diet, or when necessary, medical correction of a disorder which limits the utilization of nutrients.

(Also see MINERAL[S]; MINERAL SUPPLEMENTS; MODIFIED DIETS; NUTRITIONAL SUPPLEMENTS; and VITAMIN[S].)

Sometimes it is necessary to administer nutrients by tube feeding to hospitalized patients who can't eat, or by injection to persons who require a supplement such as vitamin B-12, which is stored in the body and needs only to be supplied once a month.

Overtreatment of malnutrition by administration of excessive amounts of nutrients in a short period of time may be dangerous, since the victim usually has a reduced ability to detoxify and excrete excess materials. The treatment may, therefore, produce effects worse than the disease. All treatments require supervision by health professionals who are able to recognize the clinical signs of deterioration in the patient.

It may be more expeditious at the community, state, or national levels to enrich and fortify foods of low nutrient quality with amino acids, minerals, and vitamins, than to try to supply foods which naturally contain the optimal quantities of nutrients. This is particularly applicable to situations where foods, which are rich in the desired nutrients, are rejected by the population to be fed. For example, there is a high rejection rate for broccoli, carrots, and other vegetable sources of vitamin A in the U.S. School Lunch Program. Therefore, food technologists have proposed the development of a snack food, like a potato chip, which would contain spinach powder and be fortified with minerals and vitamins.

Prevention of malnutrition requires correction of the factors leading to a poor diet. This calls for education of the consumer; instituting new patterns of food selection, preparation, and consumption; and insuring that supplies of the appropriate foods are available.

(Also see DEFICIENCY DISEASES; and ENRICHMENT [Fortification, Nutrification, Restoration].)

MALNUTRITION, PROTEIN-ENERGY

Contents   *Page*
Dietary Protein and Energy Supplies Around the World .. 642
Types of Protein-Energy Malnutrition 642
   Kwashiorkor 642
   Marasmus 643
   Marasmic Kwashiorkor 643
   Nutritional Dwarfing or Growth Failure 643
   Protein-Energy Malnutrition (PEM) in Adults 643
Signs of Protein-Energy Malnutrition 643
   Indicators of Protein-Energy Malnutrition
      in Communities 643
   Signs of Protein-Energy Malnutrition
      in Individuals 643
Treatment and Prevention of
      Protein-Energy Malnutrition 645
   Emergency Care of Severely Ill Patients 645
   Treatment of Protein Depletion in Hospital Patients
      (With Trauma, Infection, Burns, or After Surgery) 646
   Correction of Mild to Moderate
      Protein-Energy Malnutrition 646
   Prevention of Protein-Energy Malnutrition 646
   Recent Developments in the Treatment and
      Prevention of Protein-Energy Malnutrition 646

One of the most serious nutritional problems in the developing countries of the world is a group of disorders associated with inadequate or unbalanced intakes of protein and energy. It is often convenient to consider these disorders as various types of protein-energy malnutrition (PEM).

DIETARY PROTEIN AND ENERGY SUPPLIES AROUND THE WORLD. Although protein-energy malnutrition (PEM) may be found even in the developed countries, by far the greatest number of cases occur in the developing countries where food shortages limit the average consumption of protein and energy. Therefore, rough indicators of the incidence of PEM are the per capita deficits in protein and energy supplies of various countries. This means that there is the likelihood of there being a significant number of cases of PEM in countries where the per capita supplies of protein and/or energy fall below the recommended daily requirements defined by the Joint FAO/WHO Expert Committee on Energy and Protein Requirements.[8]

An assessment of the world situation, per capita protein and calories, by FAO/WHO, is summarized in Table M-2. It is noteworthy that, of the 127 countries included in Table M-2, 33 are classed as developed and 94 are classed as developing. Table M-2 also shows the per capita proteins and calories per day of countries in 2 groups: (1) developed countries, and (2) developing countries.

Additional facts pertinent to world per capita protein and energy in supplies follow:

1. In general, countries that are low in calories are also

TABLE M-2
PER CAPITA PROTEINS AND CALORIES PER DAY IN DEVELOPED COUNTRIES AND DEVELOPING COUNTRIES[1]

Year	Per Capita Protein/Day			Per Capita Calories/Day		
	World	Developed Countries	Developing Countries	World	Developed Countries	Developing Countries
	(g)	(g)	(g)	(kcal)	(kcal)	(kcal)
1970	69.0	96.4	57.4	2,480	3,150	2,200
1985	72.6	100.0	63.3	2,610	3,220	2,400

[1]United Nations data.

low in protein supplies; and countries that are low in total protein are also low in animal protein (which generally has a higher biological value than protein from plant sources).

2. A total of 65 countries—over half of all countries studied—fail to meet the recommended daily requirement of 2,385 Calories (kcal) per person.

3. All but two of the 127 countries studied meet or exceed the recommended daily requirement of 38.7 g of protein per person. But it should be noted that, on a worldwide basis, much protein is diverted to meet dietary energy deficits; it is estimated that 11% of the energy comes from protein. Remember, too, that many of the countries that are low in total protein depend to a large extent on protein from plant sources which is required in greater amounts than animal protein.

4. The developed countries average over 3,150 Calories (kcal) per capita per day in their energy supplies (23% above requirement), whereas the developing countries average only 2,200 Calories (kcal) per capita per day (see Table M-2). Similarly, the protein supply is over 96 g per capita per day in the developed countries vs less than 58 g in the developing regions; and much of the latter is vegetable protein rather than animal protein, and is diverted in an attempt to meet the energy deficits (see Table M-2).

5. Despite an apparently adequate supply of energy and protein, a country may have a considerable number of people who fail to have their requirements met due to the un-equal distribution of food between the different socioeconomic classes. There may also be disproportionate distributions of food within households, with the male head receiving well over his needs while the wife and children receive less than their needs.[9]

TYPES OF PROTEIN-ENERGY MALNUTRITION. Depending upon the nature and extent of the deficiency, there is at one extreme *Kwashiorkor*, or severe deficiency of protein; and at the other, *marasmus*, or severe deficiency of energy. In between are the combination disorders such as *marasmic kwashiorkor* and *nutritional dwarfing*. More than likely, deficiencies of certain minerals and vitamins accompany all of these disorders, except that the problems due to severe deficiencies of protein or energy overshadow the other problems. A discussion of the characteristics of each type of PEM follows.

Kwashiorkor. Infants fed low-protein, starchy foods (such as bananas, yams, and cassava) after weaning may develop kwashiorkor. This imbalanced diet, which has a subnormal protein-to-calorie ratio for infants, prevents some

---

[8]Per capita requirements per day of 2,385 Calories (kcal) and 38.7 g of protein (as defined by the FAO/WHO Expert Committee in April 1971). From *Nutrition Newsletter*, Vol. 11, No. 4, FAO of the United Nations, October-December 1973, p. 4., Table 1.

[9]den Hartog, A. P., "Unequal Distribution of Food Within the Household," *Nutrition Newsletter*, Vol. 10, No. 4, FAO of the United Nations, October-December 1972, p. 8.

of the adaptive mechanisms of the body from operating the way that they do in the case of starvation or marasmus.

In kwashiorkor there is less breakdown of protein and release of amino acids from muscle than in marasmus since provision of adequate energy in the diet reduces the adrenal cortical response to starvation and, consequently, the flow of amino acids from muscle to viscera. The release of amino acids from muscle is, therefore, not sufficient to meet the needs of the internal organs.

Particularly critical is the loss of tissue cells from the gastrointestinal lining which leads to various types of malabsorption. Also, the severe shortage of protein results in a reduction in synthesis of digestive enzymes, normally secreted from the pancreas or present in the intestinal wall. Thus, there is likely to be diarrhea which may cause excessive loss of water or dehydration, loss of mineral salts, and an electrolyte imbalance.

There may also be glucose intolerance, and, in some cases, elevated levels of plasma growth hormones, although the particular role of this hormone in malnutrition is not fully understood. Also, there may be reductions in the secretions of the thyroid gland which could lead to failure of the body to maintain its temperature.

Liver function is greatly reduced due to lack of protein for the synthesis of enzymes. Also, there is likely to be fat accumulation in the liver due to its inability to package fat with protein for transport in the blood. Low blood levels of albumin, along with anemia, result from the scarcity of amino acids for protein synthesis; and edema is a consequence of the depletion of plasma proteins and electrolytes.

The production of antibodies is likely to be greatly reduced and the victim, therefore, will be prone to infectious diseases. In chronic cases, there is likely to be delayed eruption of teeth, poor enamel with many caries, and pale gums and mucous membranes due to anemia.

Marasmus. Starvation in adults has a counterpart in children called marasmus. It is a severe deficiency of energy usually found in infants and young children who are not getting enough food or who have suffered from bouts of infection and diarrhea which have depleted body reserves. Marasmus in infants and children is more severe than starvation in adults, who can survive for a long period (1 to 3 months) on their tissue stores of nutrients. An infant, or child, has very high requirements for nutrients in proportion to his size. His rate of metabolism is much higher than that of adults, and he has additional requirements for a rapid rate of growth.

Physiological adaptations in marasmus involve the breakdown of body tissues for energy which is stimulated by catabolic hormones, such as those from the adrenal cortex, and the reduction or cessation of protein synthesis and growth in many tissues. The specific effects of the adrenal cortical hormones are the breakdown of proteins to amino acids, and conversion of amino acids to glucose for use by the brain and nervous system.

Survival during extreme calorie deprivation is moderately enhanced by a large reduction in the utilization of nutrients for energy which is brought about by a diminished secretion of thyroid hormones and a reduced response of cells to insulin. Eventually, there may be dehydration and electrolyte imbalances due to depletion of sodium, potassium, magnesium, and chloride ions, which may in turn lead to disorders of the brain, heart, kidneys, and nerves. Therefore, these conditions can rapidly lead to death if not promptly corrected.

Marasmic Kwashiorkor. This condition is found in many situations where children are fed diluted gruels after weaning. It is characterized by a combination of disorders which accompany both marasmus and kwashiorkor. Today, the tendency is to use the generalized term, protein-energy malnutrition, to describe this condition.

Nutritional Dwarfing or Growth Failure. Nutritional dwarfing, or growth failure, while not always a severe handicap, is nonetheless an indicator of underlying protein-energy malnutrition, and may be the beginning of the development of kwashiorkor or marasmus.

Protein-Energy Malnutrition (PEM) in Adults. These disorders may occur in adults under the following circumstances: chronic illness or alcoholism; unemployment or retirement—which means living on very limited resources; or hospitalized and maintained on intravenous glucose solutions without any other nutrients. The latter situation is particularly dangerous in the case of postsurgical patients who need extra amounts of protein for the healing of their tissues. These patients may show signs like those of kwashiorkor since calories are provided without protein.

SIGNS OF PROTEIN-ENERGY MALNUTRITION.
Every effort should be made to detect PEM in its early stages, since the cost of treating a severe case of this disorder in a hospital may easily be several times that of the treatment of a mild case on an outpatient basis.[10]

Indicators of Protein-Energy Malnutrition in Communities. Even in the developed countries resources for health surveillance and follow-up care are limited. Therefore, it is necessary to identify communities where these services might yield maximum benefits. Regional health departments usually collect data which is helpful in locating areas where the needs for public health services are greatest. Statistics which suggest the possibility of finding PEM in a community are increased mortality and morbidity of preschool children, a high incidence of infectious diseases, and significant numbers of children who are below standard weights and heights for their ages. Unfortunately, it is not a common practice to obtain on a regular basis the heights and weights of preschool and school age children—not even in developed countries such as the United States.

Signs of Protein-Energy Malnutrition in Individuals. The extreme forms of PEM (kwashiorkor and marasmus) are not often found in the developed countries, unless there are provocative factors such as parental neglect of children, untreated diseases such as those that affect the gastrointestinal tract, and improper infant feeding. For example, a case of kwashiorkor was identified by a foreign-trained physician in Bronx, New York.[11] The circumstances of the case follow: An attending physician in a clinic told the mother of a child suspected of having an allergy to cow's milk to feed her child soy milk and rice. The mother misunderstood the physician and fed her child only the rice, and the child became very ill. Another physician who was foreign trained, later saw the child in the hospital and recognized kwashiorkor (by the depigmented hair). Fortunately, the child was brought

[10]Beghin, I. D., "Nutritional Rehabilitation Centers in Latin America: A Critical Assessment," *American Journal of Clinical Nutrition,* Vol. 23, 1970, p. 1412.

[11]Taitz, L. S., and L. Finberg, "Kwashiorkor in the Bronx," *American Journal of Diseases of Children,* Vol. 112, 1966, p. 76.

back to health. Thus, it is helpful for all health professionals to be able to recognize severe PEM. Some of the features of kwashiorkor and marasmus are given in Table M-3.

TABLE M-3
FEATURES OF KWASHIORKOR AND MARASMUS

Features[1]	Kwashiorkor	Marasmus
Albumin concentration in plasma[1]	Markedly low	Normal or slightly low
Anemia	Moderate	Moderate
Appetite	Depressed	Normal to increased
Dehydration	Sometimes	Often
Edema[1]	Often on the legs and feet	Sometimes
Electrolyte metabolism	Potassium depletion	Near to normal
Endocrine function:		
Adrenal cortex (glucocorticoids)	Normal to decreased secretion	Increased secretion
Pancreas (insulin)	Decreased secretion, impaired glucose tolerance	Normal to low secretion
Pituitary (growth hormone)	Increased secretion	Increased secretion
Thyroid	Decreased secretion	Decreased secretion
Essential/nonessential amino acids (blood)[1]	Subnormal ratio	Normal ratio
Gastrointestinal function	Decreased secretions, diarrhea	Atrophy of digestive tract
Growth failure[1]	Moderate	Severe
Hair	Depigmented in spots	Easily plucked
Liver[1]	Fatty (may be palpated)	Atrophied
Mouth	Smooth tongue	Normal
Muscle wasting[1]	Masked by edema	Noticeable
Psychological	Apathetic, miserable	Alert (early stages)
Skin[1]	Depigmentation, rashes	Dry and baggy
Weight (for age and height)[1]	Moderately subnormal	Severely subnormal

[1]Features which are commonly used to distinguish between marasmus and kwashiorkor.

It is also worthwhile to be able to distinguish between the types and degrees of PEM since in most cases the treatment should be directed towards correction of the causative factors and their resulting consequences. Distinguishing characteristics are discussed under the items which follow:

• **Kwashiorkor**—The specific features which distinguish this disorder from marasmus are (1) a significantly subnormal albumin concentration in plasma, (2) swollen parotid glands (just under and in front of the ears), (3) a depressed ratio of essential to nonessential amino acids in the blood plasma, (4) fatty liver (which often may be palpated), and (5) a moderate deficit in weight for height and age (the weight is usually 80% or more of normal).

Fig. M-10 shows a typical case of kwashiorkor.

• **Marasmus**—In contrast to kwashiorkor, this condition is distinguished by (1) severe growth failure where weight may be only 60 to 80% of normal, (2) noticeable muscle wasting, (3) dry and baggy skin (edema is not usually apparent), and (4) normal values for both the plasma albumin and the ratio of essential to nonessential amino acids.

Fig. M-10. An African boy suffering from kwashiorkor. Note the misery reflected in his expression and the characteristic signs of swollen parotid glands and distended abdomen, flabby thigh muscles, flaky and roughened skin (on his right thigh). (Courtesy, FAO, Rome, Italy)

• **Marasmic kwashiorkor**—Patients who show features of both marasmus and kwashiorkor are usually given the general diagnosis of protein-energy malnutrition.

• **Nutritional dwarfing or growth failure**—One of the first effects of serious malnutrition in infants and children is growth retardation. Thus, it may often be possible to diagnose PEM in children who show none of the abnormal features other than growth deficits. New standard charts of growth curves are available for making comparisons.[12] These revised charts are now based on data from all contemporary U.S. children of the same age and sex, which is in contrast to the older standards based upon data from white, middle-class children living in Boston or Iowa 30 to 40 years ago.

Subnormal values for height and weight at a given age are considered to be those which are more than two standard deviations below the mean values (or below the curve representing the third percentile of values). Growth records should be kept over a period of time, since the child's placement on the chart for a single measurement is not as indicative of nutritional status as is the growth velocity, or increments of growth which occur between several measurements. For example, a child that was initially placed well up on a growth chart may subsequently experience protein-energy malnutrition and, as a result, drop down on the chart but still not fall below the curve representing the third per-

[12]National Center for Health Statistics, *NCHS Growth Charts*, 1976, Health Resources Administration, Monthly Vital Statistics Report, Vol. 25, No. 3, June 1976 Supplement.

Fig. M-11. *Molok, a 4-month-old Iranian girl suffering from marasmus. Left:* When admitted to the demonstration foundling home in Teheran, she was 24 in. *(61 cm)* tall and weighed 7 lb *(3.2 kg). Right:* Following 10 months in the foundling home, Molok was 31 in. *(79 cm)* tall and weighed 16 lb *(7.3 kg).* She was restored to good health by a diet consisting of UNICEF milk powder, orange juice, green vegetables, potatoes, dried apricots, beef, liver, lentils, and bread. (Courtesy, FAO, Rome, Italy)

centile. When, however, the pattern of measurements is carefully examined, it can be seen that there has been a marked departure from the normal progress of growth.

• **Protein-energy malnutrition in adults**—Many of the signs observed in children suffering from PEM will be found in adult patients who have protein and/or energy deficiencies. However, in the case of adults, the ratio of weight to height has been found to be less sensitive than other indicators of PEM.[13] Better indicators of the severity of protein deficiency are serum albumin, triceps skinfold, and muscle circumference. These three measurements are distinctly subnormal in moderate to severe protein deficiencies.

TREATMENT AND PREVENTION OF PROTEIN-ENERGY MALNUTRITION.
Depending upon the severity of the disorder, the process of rehabilitation may require one or more levels of treatment. The main therapeutic levels follow.

Emergency Care of Severely Ill Patients.
The severely malnourished person has usually adapted over a period of weeks or even months to greatly reduced amounts of food and drink. Therefore, a great deal of care has to be exercised during rehabilitation in order to avoid overloading the patient's imbalanced physiologic functions. Death has frequently resulted from overzealous administration of food, fluids, and mineral salts. The priorities for rehabilitation follow:

• **Correction of dehydration of electrolyte imbalances**—Intravenous administrations of fluid may be harmful to a patient with impaired kidney function resulting from malnutrition. This therapy should be undertaken only when it is certain that there is dehydration (it is not usually present in the malnourished unless there has been diarrhea and/or vomit-

ing). Signs of dehydration are dryness inside the mouth, and a rapid and weak pulse as a result of shrinkage of blood volume due to water loss, which, in turn, causes cardian insufficiency. If there is no vomiting, oral administration of an electrolyte solution for rehydration is preferable to intravenous administration. Half-strength Darrow's solution with 2.5% dextrose is suitable for either mode of administration (32 ml of solution per pound of body weight over a period of 24 hours). This procedure should be stopped when the pulse slows to normal.

Repletion with magesium should *not* be undertaken until rehydration has been accomplished. Dangerously high levels of this element may accumulate when there is dehydration. Magnesium should not be given intravenously, because it may cause a dangerous drop in blood pressure. Instead, it should be given either orally (as magnesium hydroxide) or intramuscularly (as 50% solution of magnesium sulfate).

The malnourished patient usually has an excess of body sodium, so intravenous sodium bicarbonate should be given only when there is severe acidosis.

• **Feeding by nasogastric tube**—Acutely ill infants and children may be very irritable, making it difficult to feed them by spoon or bottle. Also, the feeding of large quantities of food in a short period of time may aggravate diarrhea. Therefore, some rehabilitation centers usually begin the refeeding procedure with administration of milk-oil formula[14] by nasogastric drip. The compositon of the formula is as follows (percent by weight): dried skim milk, 9.3%; vegetable oil (medium-chain triglycerides, a special dietetic fat derived from coconut oil, have also been used successfully in rehabilitation), 6.8%; magnesium hydroxide, 0.12%; potassium chloride, 0.4%; and water, 83.4%. One-half cup of this formula provides approximately 4 g of protein and

---

[13]Blackburn, G. L., and B. Bistrian, "Report from Boston," *Nutrition Today*, Vol. 9, No. 3, May/June 1974, p. 30.

[14]Robson, J. R. K., and C. de Joya-Agregado, "The Operation and Function of Malward in the Philippines," *The Journal of Tropical Pediatrics and Environmental Child Health*, March 1973, p. 43, Table IV.

115 Calories (kcal). The formula is usually diluted at the beginning of rehabilitation and fed at a rate of 0.5, 1.0, or 1.5 g of protein per pound (*1.1, 2.2, or 2.3 g/kg*) of body weight per day. The lower concentrations are used in the case of diarrhea, but the amounts of magnesium and potassium salts should be proportionately increased. Following 1 to 3 days the nasogastric feeding may be replaced with bottle, spoon, or cup feeding of the full-strength mixture at the rate of 2 g protein/pound/day.

There may initially be a loss of weight due to correction of edema by protein repletion. The excess water in the tissues is gradually excreted. Once weight gain is steady, the dried skim milk may be replaced by dried whole milk, or by an equivalent quantity of evaporated or pasteurized milk, provided suitable sanitary storage facilities are available for these products.

Treatment of Protein Depletion in Hospital Patients (With Trauma, Infection, Burns, or After Surgery).
This has been most successfully accomplished by either (1) feeding meat, or (2) intravenous infusion of amino acids (compared to the lesser effect of intravenous glucose alone, or a combination of glucose and smaller amounts of amino acids).[15]

The rationale for not feeding patients who have had gastrointestinal surgery has been to prevent irritation or opening up of the surgical wounds. However, there are now available nutrient supplement powders which contain no indigestible residue, and which may be cautiously given to patients soon after surgery. These powders, generally called elemental diets, are marketed by several pharmaceutical companies.

Correction of Mild to Moderate Protein-Energy Malnutrition.
The rehabilitation procedure in developing countries usually consists of cautiously feeding nonfat dry milk (dietary fat is restricted at the start of treatment to avoid aggravation of diarrhea) reconstituted with boiled water (to avoid bacterial contamination) until other foods are tolerated. Supplements of vitamins and minerals are given along with the food to prevent or correct deficiencies. Vitamin A deficiency frequently accompanies kwashiorkor, but may not be noticed until the protein deficiency has been corrected. As soon as possible, part of the milk should be replaced with low-cost, nutritious foods which are readily available in the local community. Sometimes the milk may be mixed with foods of high-energy content, such as butter, banana, cassava, beans, and mixtures of grains.

Part of the rehabilitation procedure should involve teaching the patient (or the mothers in the cases of infants and children) the principles of selecting a diet containing the correct proportions of nutrients. This is not achieved easily where food supplies are scarce, or cultural practices limit the use of certain foods.

Prevention of Protein-Energy Malnutrition.
A high priority should be given to the identification of children who have been weaned early, or are being breast fed by mothers who cannot produce adequate amounts of milk. They should be fed a supplemental formula. When feeding formulas under primitive or unsanitary conditions, every effort must be made to prevent contamination of the food given to the baby. Lack of proper preparation of formulas is the most common cause of diarrhea; spoilage of the milk formula is another.

The diets of infants after weaning should be carefully planned to include a high-quality protein source, since infants and children require suitable patterns of amino acids in their diets. This means that if grains are a major dietary staple, there should be supplementation with animal proteins—such as milk, eggs, fish, or if these are unavailable, with legumes such as soybeans or peanuts (which may be in the form of a powder). None of the commonly used grains—corn, rice, wheat, and barley—are adequate as the major source of protein for growing children, unless the grain preparations have been supplemented with lysine and/or other deficient amino acids.

Adults are not as likely to suffer from protein deficiencies as are children because (1) their requirements for protein, in proportion to their body size, are much less; and (2) they do not require high-quality proteins—for example, adults have been adequately nourished when wheat flour constituted the sole source of protein. Lysine is not required by adults at as high a level as it is for growing infants and children.

Many cultures believe that males require the most animal protein. So, they tend to feed women less meat than men, and to feed children largely on cereals. These priorities should be reversed by feeding young children the highest quality protein, including animal protein whenever possible; by giving women more meat than men, in order to offset menstrual losses of blood and iron; and, when necessary, by providing mainly vegetarian diets for men. However, some animal proteins, such as egg and milk, are needed by men in order to provide vitamin B-12.

RECENT DEVELOPMENTS IN THE TREATMENT AND PREVENTION OF PROTEIN-ENERGY MALNUTRITION.
A variety of new high-protein food products has been developed by large food concerns for distribution in developing countries. Many of these products are currently being test marketed in Latin American countries. For example, the protein quality of tortilla flour in Mexico has been enhanced by the mixing of soy flour with the traditional corn flour.

Another approach to meeting the protein needs of growing populations is the development of new, improved varieties of grains. Plant geneticists have recently produced high-lysine corn and both high-protein and high-yielding strains of wheat, rice, corn, barley, and sorghum; and they have crossed wheat and rye to obtain a nutritious new hybrid grain called *triticale*. Collectively, these developments, which are known as the *Green Revolution*, have made these much-used cereals more nearly the staff of life for growing children.

(Also see DEFICIENCY DISEASES; ENRICHMENT [Fortification, Nutrification, Restoration]; GREEN REVOLUTION; HUNGER, WORLD; POPULATION, WORLD; PROTEIN; and WORLD FOOD.)

MALT

Contents   Page
History of Malting...............................647
Malt Derivatives..................................647
Nutritive Values of Malt Products....................647
Commercial Uses of Malt and Its Derivatives...........647
Beverages and Foods.............................647
Feeds..647
Using Malt Extracts in Home Recipes.................647

The term commonly applied to grains of barley which have been allowed to germinate under controlled condi-

---

[15]"Protein Sparing Produced by Protein and Amino Acids," *Nutrition Reviews*, Vol. 34, 1976, p. 174.

tions, then stripped of their sprouts and dried to prevent further sprouting or spoiling. However, the term malt may also refer to (1) the process of making malt; (2) germinated grains other than barley, such as wheat and rye; or (3) a water extract of the malted grain which may be either in the form of a dried powder or a concentrated syrup.

HISTORY OF MALTING.

It is known that the Chaldeans and Egyptians brewed beer on a regular basis as early as 5000 B.C. However, the ancient Egyptians became the first commercial malsters in 1300 B.C., when they produced malt for export by baking ground, malted barley into cakes which could be transported without spoilage. Beer was brewed from these malted cakes by soaking them in water until fermentation occurred.

The Greeks acquired these skills from the Egyptians around 700 B.C. About 300 years later, beer became a major beverage of the Roman legions when they learned about brewing from the Greeks. Shortly thereafter, all parts of the Roman Empire malted barley and brewed beer.

Although most of the malt which is produced today is used by brewers and distillers to make alcoholic beverages, some tasty, nonalcoholic foods are also made from malted grain.

MALT DERIVATIVES.

Malt retains most of its enzyme activity. Also, it contains maltose and other sugars which result from the breakdown of about 18% of the starch in the original grain. These properties lend themselves well to the production of the items which follow:

1. Beers.
2. Malt vinegar.
3. Distilled malt vinegar.
4. Whiskeys.
5. Malt flour.
6. Caramel malts.
7. Chocolate malts.
8. Malt extracts.
9. Malt bread.
10. Malted milk powder.

NUTRITIVE VALUES OF MALT PRODUCTS.

Table M-4 compares the nutrient content of the original barley grain, malt, and two of the products derived from it.

TABLE M-4
NUTRIENT COMPOSITION OF BARLEY AND MALT

	Unhulled Barley Grain	Dried Malt	Malt Extract Dried[1]	Malted Milk Powder[1]
	◄――――――(per 100 g portion)――――――►			
Food energy .... kcal	274	368	367	410
Carbohydrates .... g	66.6	77.4	89.2	72.9
Fat .... g	1.7	1.9	Trace	8.3
Protein .... g	12.2	13.7	6.0	13.1
Fiber .... g	5.3	3.3	Trace	.6
Calcium .... mg	40	60	48	266
Phosphorus .... mg	330	460	294	380
Sodium .... mg	30	80	80	440
Magnesium .... mg	140	180	140	93
Potassium .... mg	400	430	230	758
Iron .... mg	8	4	9	1
Niacin .... mg	8.5	5.7	9.9	.3
Riboflavin .... mg	.2	.3	.4	.5
Thiamin .... mg	.4	.4	.4	.3

[1]Data from Food Composition Table F-21.

Table M-4 shows that malt contains higher levels of energy and protein, but less fiber, than the grain from which it was derived. In general, the grain is rendered more digestible by malting as a result of enzymatic activity generated in this process.

Malt extract is composed mainly of carbohydrates in the form of sugars, since much of the protein in malt is insoluble, and, therefore, is not extracted by water. Also, there is only a negligible content of fiber in the extract. Thus, the malt extract is not as complete a food as the malted grain.

Malted milk powder is the most nutritious of the products listed in Table M-4 since the whole milk used in its preparation contributes extra energy, fat, protein, calcium, phosphorus, sodium, and potassium. It may be mixed with milk and used as a nearly complete food for convalescents, except that supplemental vitamins, minerals, and fiber should be provided if it is the exclusive food for a long period of time.

Additional nutritional information may be found on malt and malt products in Food Composition Table F-21.

COMMERCIAL USES OF MALT AND ITS DERIVATIVES.

Although malt products are almost exclusively consumed by man in the form of beverages and foods, producers of these items recover and sell the by-products from their processes for use as ingredients in livestock feeds. A brief summary of some of the commercial uses of malt and its derivatives follows.

Beverages and Foods.

The manufacturers of these products use malt because of (1) its enzymatic activity which helps to make foods more digestible, or more readily fermentable; (2) its content of sugars, such as maltose; and (3) its flavor. Examples of these applications follow:

1. Diastatic malts.
2. Malt-flavored beverages and foods.
3. Malt vinegars.

Feeds.

Even though malt has a higher nutritive value for feeding livestock than unmalted barley grain (see Table M-4), the higher cost of producing the malt makes such a use unprofitable. But, the by-products which are not consumed by man make good ingredients for animal feeds. Among them, diastatic malts, malt hulls, and sprouts, miller's screenings (from the production of malt flour), brewers' grains, brewers' yeast, distillers' grains, and distillers' solubles. These by-products are combined with various other ingredients to make up rations for both nonruminants like chickens and hogs, and for ruminants such as sheep and cattle.

USING MALT EXTRACTS IN HOME RECIPES.

Diastatic malts are sold mainly by wholesalers to bakers, brewers, and distillers. Therefore, consumers are not likely to be able to purchase such products in retail stores.

Malt extracts are available in powdered and syrup form, but they are rather expensive since they are sold mainly in drug stores as laxatives and tonics. However, a slightly less expensive diastatic malt is sold by some health food stores.

An inexpensive, hop-flavored, nondiastatic malt syrup is sold in some supermarkets. This syrup might be used to flavor beverages where a bit of bitterness (from the hops) is desired. It makes a good yeast food for bread baking because of its sugar content.

(Also see BEERS AND BREWING, section headed "Malting.")

MALTASE

An enzyme which acts on maltose to produce glucose. Salivary amylase is present in saliva, and intestinal maltase is present in intestinal juice.

(Also see DIGESTION AND ABSORPTION.)

MALTHUS

An English clergyman whose full name was Thomas Robert Malthus. In 1798, he prophesied that world population grows faster than man's ability to increase food production. He stated, "The power of population is infinitely greater than the power in the earth to provide subsistence for man." For almost 200 years, we proved Malthus wrong because, as the population increased, new land was brought under cultivation; and machinery, chemicals, new crops and varieties, and irrigation were added to step up the yields. Now, science has given us the miracle of better health and longer life; and world population is increasing at the rate of about 240,000 people a day, or 87.5 million per year; and it is predicted that world population will be 6.2 billion by the year 2000, and be doubled by the year 2045. To match population growth and feed people adequately, world food production needs to increase at an average rate of about 2.5% per year.

(Also see HUNGER, WORLD; POPULATION, WORLD; and WORLD FOOD.)

MALTOL

This substance is used both for its own flavor and for enhancing flavors of sweet, fruit-containing foods. It has a fragrant caramellike odor and a bittersweet taste.

FORMATION AND OCCURRENCE OF MALTOL. Chemists have demonstrated that maltol may be formed (1) by heating maltose at 375°F *(191°C)* for 1 hour (note the similarities of these conditions to those in baking), or (2) by heating mixtures of sugars, such as maltose and lactose, with amino acids, such as glycine (the latter procedure is known as nonenzymatic browning reaction of the Maillard type).

Maltol is found in roasted materials which have a moderate to high carbohydrate content, such as bread crusts, cocoa beans, cellulose, cereals, chicory, coffee beans, diastatic flour doughs (where some of the starch has been converted by enzyme action to maltose), malt products, soft woods, and soybeans. It is also found in heated products which contain moderate amounts of both sugars and amino acids, such as condensed and dried milks, dried whey, and soy sauce. Apparently heating is not always required for the production of maltol, since it also occurs in larch bark and the dry needles of cone-bearing evergreen trees.

USES OF MALTOL. When maltol flavor is desired in a food, it is often produced by careful attention to the conditions under which food is processed. For example, the roasting of cereal grains results in the production of maltol and other flavors which are similar to those in roasted cocoa and coffee. Thus, the distinctive flavors and aromas of many roasted foods are due to the different mixtures of the individual substances produced by the browning reactions.

The enhancement of fruit flavors in sweetened foods requires more precise control of the amounts of maltol in the food product than can be achieved by processing, since the flavor of maltol itself should not be evident in these foods. Thus, pure maltol is used in trace amounts ranging from 5 to 350 ppm (from about a teaspoon to 12 oz per ton of food) in products such as cakes, fruit drinks, gelatin desserts, ice cream, jams, soft drinks, and sweet rolls which have fruit flavors (strawberry, raspberry, pineapple, black cherry, and orange). Maltol also enhances the sensation of sweetness in sweet, fruit-containing foods, but it is less effective for this purpose than for enhancing fruit flavor.

ETHYL MALTOL. The strength of maltol as an enhancer of fruit flavors and as a sweetener is increased by a factor of from 4 to 6 when the methyl group (CH_3) in the molecule is replaced by an ethyl group (CH_2CH_3). Also, water solutions of ethyl maltol at room temperature are 9 times more volatile, and, therefore, more aromatic than those of maltol. The use of ethyl maltol as an additive in foods allows the sugar content to be reduced by 15%.

MALTOSE (MALT SUGAR)

A disaccharide with the formula $C_{12}H_{22}O_{11}$; obtained from the partial hydrolysis of starch; yields 2 molecules of glucose on further hydrolysis.

MAMEY (MAMMEE APPLE) *Mammea americana*

The fruit of an evergreen tree (of the family *Guttiferae*) that is native to the West Indies, Central America, and South America.

The fruit is pear-shaped and from 3 to 6 in. *(7 to 14 cm)* in diameter with brown thick skin and an orange-colored flesh. It is eaten fresh, stewed and sweetened, or made into jam or jelly. Also, a liqueur is made from the flowers and a seasoning agent is made from the seeds of the fruit.

The fruit is fairly low in calories (59 kcal per 100 g) and carbohydrates (12%). It is a fair source of vitamin C.

MANATEE (MANATI; SEA COW)

A formerly common American herbivorous sea mammal of the West Indies and neighboring mainland coasts from Florida to Yucatan, now rare because of excessive killing. The manatee is about 10 ft long, nearly black, thick-skinned, and almost free from hair. The flesh of the manatee, which tastes something like pork, is much esteemed in the West Indies.

MANGANESE (Mn)

Contents

Contents	Page
History	649
Absorption, Metabolism, Excretion	649
Functions of Manganese	649
Deficiency Symptoms	649
Interrelationships	650
Recommended Daily Allowance of Manganese	650
Toxicity	650
Sources of Manganese	650

In industry, manganese is a metallic element used chiefly as an alloy in steel to give it toughness. In nutrition, it is an essential element for many animal species. Manganese is an activator of several enzyme systems involved in protein and energy metabolism and in the formation of mucopolysaccharides. The human body contains 12 to 20 mg of manganese. So, an essential function of the element must be assumed to exist in man.

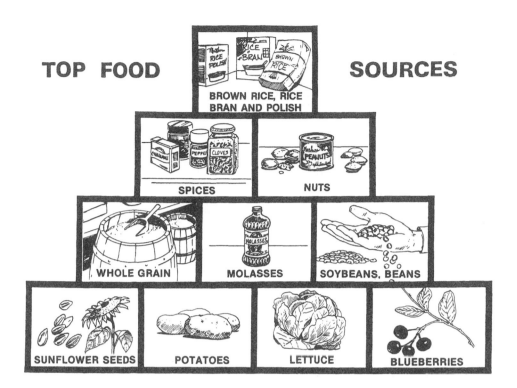

TOP FOOD SOURCES

BROWN RICE, RICE BRAN AND POLISH

SPICES   NUTS

WHOLE GRAIN   MOLASSES   SOYBEANS, BEANS

SUNFLOWER SEEDS   POTATOES   LETTUCE   BLUEBERRIES

Fig. M-12. Top food sources of manganese.

HISTORY. Manganese was first recognized as an element in 1774 by the famous Swedish chemist Carl W. Scheele. It was isolated in the same year by his co-worker, Johann G. Ghan. The name *manganese* is a corrupted form of the Latin word for a form of magnetic stone, *magnesia*.

About 95% of the world's annual production of manganese is used by the steel industry. Also, manganese is essential for plant growth, and it is found in trace amounts in higher animals, where it activates many of the enzymes involved in metabolic processes.

In 1931, University of Wisconsin researchers reported that manganese was a dietary essential for growth of rats. Later, it was shown to be essential for poultry (a deficiency results in slipped tendon; see Fig. M-13), swine, guinea pigs, cattle, and other animals. Undoubtedly, manganese is an essential nutrient for man, even though the deficiency symptoms have never been determined with certainty.

ABSORPTION, METABOLISM, EXCRETION. Manganese is rather poorly absorbed, primarily in the small intestine. In the average diet about 45% of the ingested manganese is absorbed, and 55% is excreted in the feces. Absorption can be depressed when excessive amounts of calcium, phosphorus, or iron are consumed.

Following absorption, manganese is loosely bound to a protein and transported as transmanganin. The bones, and to a lesser extent the liver, muscles, and skin, serve as storage sites.

Manganese is mainly eliminated from the body in the feces as a constituent of bile, but much of this is again reabsorbed, indicating an effective body conservation. Very little manganese is excreted in the urine.

The concentration of manganese in the various body tissues is quite stable under normal conditions, a phenomenon attributed to well controlled excretion rather than regulated absorption.

FUNCTIONS OF MANGANESE. Manganese is involved in the formation of bone and the growth of other connective tissues; in blood clotting; in insulin action; in cholesterol synthesis; and as an activator of various enzymes in the metabolism of carbohydrates, fats, proteins, and nucleic acids (DNA and RNA).

DEFICIENCY SYMPTOMS. Signs of deficiency in many animal species include poor reproductive performance, growth retardation, congenital malformations in the offspring, abnormal formation of bone and cartilage, and impaired glucose tolerance.

Fig. M-13. A bone disease known as perosis, or slipped tendon, in a chicken due to manganese deficiency. (Courtesy, Department of Poultry Science, Cornell University)

The only confirmed deficiency of manganese in man was in connection with a vitamin K deficiency, where administration of the vitamin did not correct the abnormality in blood clotting until supplemental manganese was provided.

Analyses of hair and blood samples for manganese content indicate that subclinical deficiencies of the mineral might aggravate such disorders as growth impairments, bone abnormalities, diabeticlike carbohydrate metabolism, lack of muscle coordination in the newborn, and abnormal metabolism of lipids (fatty acids, choline, and cholesterol).

In 1963, researchers at the University of California—Davis reported, in the *American Journal of Physiology,* on a study of convulsions in rats; they found that manganese levels helped determine the susceptibility to convulsions of rats subjected to electroshocks and a convulsive drug. In 1976, Yukio Tanaka, a clinical chemist at St. Mary's Hospital in Montreal, reported, before the American Chemical Society in Chicago, that about 30% of all children who have epileptic convulsions have low levels of manganese in their blood. These leads merit further study.

INTERRELATIONSHIPS.
Manganese interacts with other nutrients. Excess calcium and phosphorus interfere with the absorption of manganese; the functions of manganese, copper, zinc, and iron may be interchangeable in certain enzyme systems; and manganese and vitamin K work together in the promotion of blood clotting.

RECOMMENDED DAILY ALLOWANCE OF MANGANESE.
Extreme dietary habits can result in manganese intakes outside the limit suggested as safe, but the consumption of a varied diet, balanced with regard to bulk nutrients, can be relied on to furnish adequate and safe amounts.

Estimated safe and adequate intakes of manganese are given in the section on MINERAL(S), Table M-25, Mineral Table.

• **Manganese intake in average U.S. diet**—In the United States, the normal daily intake of manganese varies from 2 to 9 mg/day in adults, depending on the composition of the diet.

TOXICITY.
Toxicity in man as a consequence of dietary intake has not been observed. However, it has occurred in workers exposed to high concentrations of manganese dust in the air (such as in mining ores rich in manganese, and in the production of dry-cell batteries where manganese dioxide is used). The excess accumulates in the liver and central nervous system. The symptoms resemble those found in Parkinson's and Wilson's diseases.

SOURCES OF MANGANESE.
The manganese content of plants is dependent on soil content. It is noteworthy, however, that plants grown on alkaline soils may be abnormally low in manganese.

• **Rich sources**—Rice (brown), rice bran and polish, spices, walnuts, wheat bran, wheat germ.

• **Good sources**—Blackstrap molasses, blueberries, lettuce, lima beans (dry), navy beans (dry), peanuts, potatoes, soybean flour, soybeans (dry), sunflower seeds, torula yeast, wheat flour, whole grains (barley, oats, sorghum, wheat).

• **Fair sources**—Brewers' yeast, liver, most fruits and vegetables, orange pekoe tea, white enriched bread.

• **Negligible sources**—Fats and oils, fish, eggs, meats, milk, poultry, sugar.

• **Supplemental sources**—Alfalfa leaf meal, dried kelp, manganese gluconate.

For additional sources and more precise values of manganese, see Table M-5, Some Food Sources of Manganese, which follows:

TABLE M-5
SOME FOOD SOURCES OF MANGANESE

Food	Manganese	Food	Manganese
	(mg/100 g)		(mg/100 g)
Rice bran	34.7	Barley, whole grain	1.2
Cloves	30.3	Wheat flour	1.0
Rice polish	17.1	Lettuce	1.0
Ginger	17.8	Potato, raw	1.0
Walnuts	15.2	Blueberries, raw	1.0
Wheat germ	13.3	Spinach, raw	.8
Wheat bran	11.0	Beet, common red, raw	.7
Rice, whole grain	9.6	Yeast, brewers'	.6
Molasses, blackstrap	4.3	Corn, whole grain	.5
Wheat, whole grain	3.7	Bananas, raw	.5
Oats, whole grain	3.7	Carrot, raw	.4
Alfalfa leaf meal (powder), dehydrated	3.6	Turnip, raw	.4
		Cherries, raw	.4
Soybean flour	3.2	Green beans	.3
Soybeans, dry	3.0	Apple, raw	.3
Peanuts	2.5	Bread, enriched white	.3
Sunflower seeds	2.3	Sweet potato	.3
Beans, navy, dry	2.1	Cabbage, raw	.3
Beans, lima, dry	1.6	Strawberries, raw	.3
Sorghum, whole grain	1.6	Liver	.2
Yeast, torula	1.3	Orange pekoe tea	.2
		Rutabaga, raw	.1

MANGO *Mangifera indica*

The mango is a tropical evergreen tree of the cashew family, *Anacardiaceae,* which produces the mango fruit—a slightly sour juicy oval fruit with a thick yellowish-red rind.

Fig. M-14. A fruiting mango tree. (Courtesy, USDA)

Although not very popular in the United States, it is an important fruit crop of tropical countries. Mangoes have a spicy taste. Centuries of selection have produced mangoes free of fiber and offensive flavors.

ORIGIN AND HISTORY. The mango tree originated in India, from the wild trees of the area. From here, it spread throughout the tropical regions of the world. Settlers brought it to America in the 1700s.

WORLD AND U.S. PRODUCTION. Worldwide, mangoes are the seventh leading fruit crop. About 15.7 million metric tons are produced annually; and the leading countries, by rank, are: India, Mexico, Pakistan, Thailand, China, Indonesia, Brazil, Philippines, Haiti, and Zaire.[16]

In the tropical countries, mangoes are more important than peaches and apples.

No production figures on mangoes for the United States are available. But it will grow in California, Florida, and Hawaii; and there is a limited amount of commercial production and sale in Florida and Hawaii.

PROCESSING. Virtually all of the mango crop is marketed fresh. However, small amounts are canned.

SELECTION, PREPARATION, AND USES. Mangoes should be nearly ripe, or ripe, firm, and free from bruises. Ripe fruits are eaten raw as a dessert, fruit dish, or used in juice, jams, jellies, or preserves. Small amounts are canned. Unripe mangoes are used in pickles and chutneys—a delicacy of India. Some are sun dried, and seasoned with turmeric to produce *amchur*, which may be ground and used in soups and chutneys.

NUTRITIONAL VALUE. Raw mangoes are about 82% water and contain 66 Calories (kcal) of energy per 100 g (about 3½ oz). Unripe fruits contain starch which changes to sugars during ripening. Also, mangoes are a good source of potassium, and an excellent source of vitamin A (4,800 IU/ 100 g), and a fair source of vitamin C (35mg/100 g). More complete information regarding the nutritional value of mangoes is presented in Food Composition Table F-21.

MANGO MELON *Cucumis melo*

This is a fairly uncommon variety (*Chito*) of muskmelon with fruit as small as lemons. The fruits are also called melon apples, orange melons, and vegetable oranges. They are used to make preserves.

MANGROVE *Rhizophera mangle*

The fruit of a tree (of the family *Rhizophoraceae*) which grows along the Atlantic coast from South Carolina to the American tropics. Mangrove fruit is sweet and edible if it is picked before it starts to develop roots. It is one of the rare fruits that sprouts while on the tree.

MANNA

The food provided by God to the Israelites during their 40 years of wandering in the wilderness as they left Egypt for Canaan—the Promised Land. Manna fell from the sky every day except the Sabbath, for which an extra portion was gathered the previous day. (Exodus 16; Numbers 11)

Some historians say that manna was a gluey sugar from the tamarisk tree.

MANNITOL ($C_6H_{14}O_6$)

Mannitol is a sugar alcohol. It tastes about 70% as sweet as sucrose—table sugar; hence, it is used as a sweetener in foods.

(Also see SWEETENING AGENTS.)

MANNOSE ($C_6H_{12}O_6$)

This is a hexose monosaccharide—a 6-carbon sugar. Mannose is not found free in foods, but it is derived from ivory nuts, orchid tubers, pine trees, yeast, molds, and bacteria. Although unimportant to human nutrition, it is a component of some glycoproteins and mucoproteins of the body. Its chemical structure is very similar to glucose.

(Also see CARBOHYDRATE[S]; and PROTEIN[S], Table P-11, "Classification Of Some Common Proteins.").

MAPLE SYRUP

A product unique to North America, made by boiling the sap of the sugar maple (*Acer saccharum*). Collection of the sap commences in the spring of the year, when warm days begin to follow cool nights, causing the sap of the sugar maple to flow. During the winter, some of the starch that the tree made the previous summer and stored in its roots is converted to sugar, primarily sucrose, and carried in the first sap of the spring. Spring sap contains 4 to 10% sugar. Collected sap is boiled to concentrate the sugar and produce the characteristic flavor. Maple syrup is esteemed for its sweet taste and *maple* flavor. Interestingly, the maple flavor of the syrup is not present in the sap, but develops during the boiling.

Although the United States produces more than one million gallons (*3.8 million liter*) of maple syrup yearly, this production is dwarfed by the Canadian province of Quebec which produces about 20 million gallons (*76 million liter*) of maple syrup yearly. Maple syrup produced in the United States is for both home and commercial use.

Fig. M-15. Fresh maple syrup pie is the first sign of spring. (Courtesy, Agriculture Canada, Ottawa, Ontario, Canada)

[16]Based on data from *FAO Production Yearbook 1990*, FAO/UN, Rome, Italy, Vol. 44, p. 167–168, Table 73. **Note well:** Annual production fluctuates as a result of weather and profitability of the crop.

Production of maple syrup begins with the spring tapping of maple trees. Although the sugar maple (*Acer saccharum*) is the major source of sap, the black maple (*Acer nigrum*) and the red maple (*Acer rubrum*) are also syrup sources. All of these types of maple trees are native to the eastern half of the North American continent. Their sap is harvested during a 4 to 6 week period before buds on the tree begin to open, usually between January and April.

Each day the accumulation of sap is collected from the pails, poured into a large tank, and hauled by sled or wagon to the sugarhouse.

PROCESSING.   Processing occurs in the sugarhouse. It is here that the sap is strained and then placed in shallow pans—evaporators—over wood, oil, or gas fires. As the sap boils, the water evaporates. When the sugar concentration reaches 66.5%, it is drawn off, filtered, and bottled as maple syrup. During the boiling and evaporation, the characteristic maple flavor and color develop. Depending on the sugar content of the sap, 30 to 50 gal (*115 to 190 liter*) of sap are required to make one gallon (*3.8 liter*) of maple syrup. Maple sugar is produced by further boiling and evaporation of most of the water. One gallon of syrup yields about 8 lb of sugar. An old-fashioned treat enjoyed by those making maple sugar is called Jack wax—a taffylike confection formed by pouring the hot syrup onto the snow. The process of forming maple sugar is referred to as sugaring off. Often sugaring off gatherings are held at which people sample the maple syrup and sugar.

NUTRITIONAL VALUE AND USES.   During colonial times, maple syrup and sugar were important and abundant, but nowadays these products are rather a treat and a luxury.

The composition of maple syrup is listed in Food Composition Table F-21 of this book. Nutritionally, maple syrup is primarily a source of energy; 1 Tbsp (*15 ml*) of maple syrup contains about 50 Calories (kcal) of energy, along with some calcium and potassium.

No doubt, maple syrup is esteemed more for its flavor than its nutritional value. Maple syrup on pancakes, French toast, or ice cream, is a real treat. The flavor can also be enjoyed in glazed hams, glazed carrots, and baked beans—all made with maple syrup. A delicious hot toast spread can be made by boiling two cups of syrup in a pot to 230°F (*110°C*), placing the hot pan in ice water without stirring, cooling to tepid but pliant state, and then beating until creamy.

U.S. Grade A Light syrup has a subtle flavor, while Grade A Medium and Dark syrups have a more robust flavor for cooking; besides, they cost less.

(Also see SUGAR, section headed "Other Sources of Sugar.")

MARASMUS

From the Greek *marasmos,* meaning a *dying away.* A progressive wasting and emaciation.

This disease is due to chronic protein-calorie malnutrition. It is found in infants and young children in many areas of the world, particularly in the developing and overpopulated countries, where protein-rich foods, especially those of animal origin, are practically unavailable to the poorer segments of the population. This is because (1) protein itself is deficient, and (2) total food consumption, and hence energy intake, is so inadequate that the protein eaten is not spared to function as an essential nutrient.

Physiological adaptations in marasmus involve the break-

Fig. M-16. Marasmus, due to lack of energy foods. *Left:* A 2 year-old girl, weighing 11½ lbs. *Right:* Same little girl after 10 months of treatment. (Courtesy, FAO, Rome, Italy)

down of body tissues for energy which is stimulated by catabolic hormones from the adrenal cortex. Moreover, there is a reduction or cessation of protein synthesis and growth in many tissues. Outward signs and symptoms include wasting of muscle, loss of subcutaneous fat, dry baggy skin, general appearance of old age, low body weight (often less than 60% of weight expected for age and height), large sunken eyes, diarrhea, subnormal body temperature, and malabsorption. Survival during marasmus is moderately enhanced by a large reduction in the utilization of nutrients for energy which is brought about by a diminished secretion of the thyroid hormones, and a reduced response of the cells to insulin. Eventually, there may be dehydration and electrolyte imbalances due to depletion of sodium, potassium, magnesium, and chloride ions, which may in turn lead to disorders of the brain, heart, kidneys, and nerves. These conditions can rapidly lead to death if not promptly corrected. Treatment of marasmus consists of feeding an easily digested diet, such as milk and oil, which supplies at least 1.0 g of protein and 50 to 60 kcal per lb of body weight (2.2 g of protein, 110 to 132 kcal per kg of body weight).

Also see MALNUTRITION, PROTEIN-ENERGY.)

MARGARINE (OLEOMARGARINE)

This is an economical, nondairy product resembling butter in appearance, form, composition—and taste. Margarine contains about 80% fat as does butter, but the fat is generally entirely of plant origin—corn, soybean, safflower, and/or cottonseed, although a small amount of animal fat (not butterfat) is sometimes used by processors. Like many other products, margarine was developed in response to a need and an incentive.

ORIGIN AND HISTORY.   Napoleon III, the ruler of France from 1852 to 1870, appealed to his people for an economical and nutritious butter substitute. Near the end of his rule, a chemist, Hippolyte Mege-Mouries, developed margarine, for which he was awarded a prize by the emperor. Mouries used oil obtained by pressing beef fat, oleo oil, salt, milk, and annatto for coloring. These ingredients were partially emulsified, then chilled rapidly with ice water to separate the solidified fatty granules from the excess water of the milk. This formed a solid emulsion which could be kneaded to the finished margarine. Even today, Mouries' method is the basis of margarine production. This first margarine had a pearly appearance and was called oleomargarine—a name derived from the oleo oil, and from the Greek word *margaron,* meaning pearl. In the United States, the name oleomargarine was required by law until 1952. Subsequently, the terms margarine and oleomargarine have been used interchangeably, but the term margarine is used most frequently.

Today, Americans consume much more margarine than butter, as Table M-6 shows. Most of the increased consumption of margarine can be explained by the price difference between butter and margarine; butter costs about 1¾ times more than margarine.

TABLE M-6
PER CAPITA CONSUMPTION OF BUTTER AND MARGARINE IN THE UNITED STATES SINCE 1950[1]

Year	Butter		Margarine	
	(lb)	(kg)	(lb)	(kg)
1950	10.7	4.9	6.1	2.8
1955	9.0	4.1	8.2	3.7
1960	7.5	3.4	9.4	4.3
1965	6.4	2.9	9.9	4.5
1970	5.3	2.4	11.0	5.0
1975	4.8	2.2	11.2	5.1
1980	4.5	2.0	11.4	5.2
1985	4.8	2.2	10.7	4.9
1990	4.3	2.0	10.9	5.0

[1]Data from *Agricultural Statistics,* USDA, 1965, p. 150, Table 220; 1980, p. 148, Table 208; and 1991, p. 139, Table 194.

PRODUCTION. Several types of plant oils may be used in the production of margarine. The most common is soybean oil, but corn, cottonseed, palm, peanut, and safflower oils may also be used by processors. Some of these oils may be hardened by hydrogenation. Also, some processors may employ a small amount of animal fat. Regardless of the oil source, the U.S. government stipulates that margarine must contain at least 80% fat.

Fig. M-17. About 80% of all the margarines in the U.S. market contain soybean oil. (Courtesy, American Soybean Assoc., St. Louis, Mo.)

In 1990, about 2.8 billion lb *(1.3 billion kg)* of margarine were produced in the United States. Although most people consider margarine a household item, about 10 to 15% of the total production is used by the baking industry.

NUTRITIONAL VALUE. With the exception of the origin of the fat, the nutrient composition of margarine is very nearly the same as that of butter, as Table M-7 reveals.

TABLE M-7
COMPARISON OF THE NUTRIENT COMPOSITION OF BUTTER AND MARGARINE[1]

Nutrient		Butter	Margarine
		◄─────(per 100 g)─────►	
Food energy	kcal	717.0	718.7
Protein	g	0.9	0.9
Fats	g	81.1	80.5
Calcium	mg	24.0	29.9
Phosphorus	mg	23.0	22.9
Sodium	mg	826.0	943.4
Magnesium	mg	2.0	2.6
Potassium	mg	26.0	42.4
Iron	mg	0.2	—
Zinc	mg	0.1	—
Copper	mg	0.4	—
Vitamin A	IU	3,058.0	3,307.0[2]
Vitamin E	mg	1.6	15.0[3]
Thiamin	mcg	1.6	10.0
Riboflavin	mcg	34.0	37.0
Niacin	mcg	42.0	23.0
Vitamin B-6	mcg	3.0	9.0
Folic acid	mcg	3.0	1.2
Vitamin B-12	mcg	—	0.1

[1]Values from Food Composition Table F-21.

[2]Vitamin A must be added to yield a finished margarine with not less than 15,000 IU per pound (0.45 kg).

[3]Varies depending on the type of oil used.

The major difference between margarine and butter—the fat—is expanded in Table M-8.

TABLE M-8
COMPARISON OF THE FATS OF BUTTER AND MARGARINE[1]

Fat	Butter	Margarine
	◄─────(g/100 g)─────►	
Total fat	81.1	81.1
Animal fat	81.1	—
Plant fat	—	81.1
Saturated fat	50.5	14.8
Polyunsaturated fat	3.0	—
Oleic acid	20.4	41.4
Linoleic acid	1.8	22.2
Cholesterol	.2	.0

[1]Values from Table F-4, Fats and Fatty Acids in Selected Foods.

Table M-8 indicates a difference that has been exploited as a selling point—no cholesterol in margarine. Furthermore, margarine contains significant amounts of the essential fatty acid, linoleic acid.

Despite attempts to stymie the growth of the margarine industry, it has become a success and is established as a high-quality, wholesome, nutritious food, with worldwide acceptance.

(Also see FATS AND OTHER LIPIDS; MILK AND MILK PRODUCTS; and OILS, VEGETABLE.)

MARGARINE, KOSHER

Made only from vegetable fats, and fortified only with carotene (derived from vegetable sources).

MARINADE

A mixture of liquids with herbs and spices which is used to steep meats and fish before cooking. By using pineapple juice, the marinade can be used to tenderize meats. The marinades containing oil and seasonings add oil and flavor to the fish or meat being cooked.

MARMITE

• A large metal or pottery soup kettle with a lid. Small individual ones are called *petite marmite*.

• Soup that is cooked in a marmite.

• A yeast product used for flavoring soups and meats; also as a spread or beverage.

MARRON

Chestnuts preserved in syrup flavored with vanilla.

MATÉ (BRAZILIAN TEA, PARAGUAY TEA, YERBA MATÉ)

This is a stimulating drink prepared from the dried leaves of maté or ilex plant (*Ilex paraguayensis*) of South America. It is prepared in much the same manner as tea. Boiling water is poured over maté leaves or powder and allowed to steep for about 10 minutes. Then the drink is strained and taken either hot or cold, with or without sugar. Maté is enjoyed throughout most of South America. It contains caffeine.

MATOKE *Musa paradisiaca* **var.** *sapientum*

The East African name for the local varieties of unripe bananas that are used mainly for cooking.
(Also see BANANA; and PLANTAINS.)

MATRIX

The ground work in which something is enclosed or embedded; for example, protein forms the bone matrix into which mineral salts are deposited.

MAY APPLE (INDIAN APPLE; MAYFLOWER) *Podophyllum peltatum*

The small, yellow fruit of a wild herb (of the *Podophyllaceae* family) which is native to eastern North America from southern Canada to the Gulf of Mexico.
It is good fresh or when made into fruit drinks, jam, jelly, marmalade, pies, and wine. It is noteworthy that the unripe fruit, leaves, and roots of the plant are poisonous.

MAYONNAISE

This is an oil-in-water type of semisolid emulsion with egg yolk acting as the emulsifying agent. It consists of vegetable oil, egg yolk or whole egg, vinegar, lemon and/or lime juice, with one or more of the following: salt, a sweetener, mustard, paprika or other spice, monosodium glutamate, and other food seasonings. Vegetable oil comprises 65 to 80% of the composition. Thus, mayonnaise is high in calories. Each 3½ oz (*100 g*) contains 650 to 700 Calories (kcal). Vinegar and salt act as preservatives. A number of methods for manufacturing mayonnaise are available, but all of them make use of high-speed beating or dispersing equipment in some stage of the process. Manufacturers employ machines and systems which allow the semicontinuous and continuous production of mayonnaise. Mayonnaise may be packaged in glass or plastic wide-mouth jars, tubes, sachets, or individual portion cups. It is used on salads, fish, vegetables, other foods, and in a variety of recipes.

MAYPOP *Passiflora incarnata*

The fruit of a wild species of passion fruit (family *Passifloraceae*) that grows in the southern United States. Maypop berries are yellow-colored, egg-shaped, and about 2 in. (5 cm) long. They may be eaten raw or made into jam, jelly, pies, and sauces.
The raw fruit is moderately high in calories (111 kcal per 100 g) and carbohydrates (21%). It is also high in fiber and a good source of iron.

MEAL

• A food ingredient having a particle size somewhat larger than flour.

• Mixtures of cereal foods, in which all of the ingredients are usually ground.
(Also see CEREAL GRAINS.)

MEAL PLANNING

Many factors enter into planning meals. Perhaps the most important one is nutritive value; but other things such as appearance, palatability, cost, and food preferences must be considered at the same time.

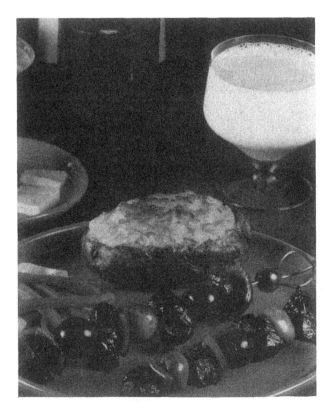

Fig. M-18. Meal planning should begin by using the food groups as a pattern. (Courtesy, American Dairy Assoc., Rosemont, Ill.)

The procedure should begin by using the food groups as a pattern. Choose the entree from the protein group, then add the milk group, follow with the vegetable and the fruit, and finish up with the cereal group. It is quite simple if every meal is planned this way.

There are some well-known food combinations which have stood the test of time, thus when combining foods it is best not to try unorthodox combinations. The public has demanded that restaurant meals be well balanced; thus, when in doubt, their menus can be used as patterns.

Table M-9 Meal Planning Guide, can be used for the purpose of planning the day's meals. It gives the number of servings of each food group, for the different ages. Thus, one menu can take care of every age, merely by varying the number of servings from each food group.

TABLE M-9
MEAL PLANNING GUIDE

Food Group and Serving Size[1]			Servings Per Person Per Day											
				Children			Adolescents		Adults (20–50 yrs)				Persons Older than 50 Yrs	
			Infants under 1 yr	1–3 yrs	4–7 yrs	8–11 yrs	12–19 yrs M	12–19 yrs F	M	Non-Preg. F	Preg-nant F	Lac-tating F	M	F
Low-Energy Foods	Medium-Energy Foods	High-Energy Foods												
MEATS, FISH OR SEAFOOD, EGGS, BEANS, PEAS, OR NUTS—each serving provides 12 to 20 g of protein			½	1	2	2	2	2	2	2	3	3	2	2
(50–150 kcal per serving)	*(200–300 kcal per serving)*	*(350–500 kcal per serving)*												
MILK PRODUCTS—each serving provides 250 to 300 mg of calcium and 8 g of protein			1½–2	3	3	3	4	4	3	3	4	4	3	3
(about 100 kcal per serving)	*(150–200 kcal per serving)*	*(300–500 kcal per serving)*												
BREADS AND CEREALS—should be whole-grain or enriched. Each serving provides 70 kcal and 2 g of protein			2	3	4	4	6	6	4	4	6	6	4	4
DARK-GREEN AND DEEP-YELLOW FRUITS AND VEGE-TABLES—each serving provides 3,000 or more IU of vitamin A			⅓	½	½	½	1	1	1	1	1½	1½	1	1
(less than 100 kcal per serving)		*(100–200 kcal per serving)*												
CITRUS FRUITS OR THEIR EQUIVALENTS—each serving provides 30 or more mg of vitamin C and less than 100 kcal			1	1	1	1	1	1	1	1	2	2	1	1
OTHER FRUITS AND VEGETABLES—sources of carbohydrates, fiber, minerals, and vitamins			½	1	3	3	3	3	3	3	4	4	4	4
(less than 100 kcal per serving)	*(100–200 kcal per serving)*	*(200–350 kcal per serving)*												
ESSENTIAL FATS—each serving provides approximately 1 g of essential fatty acids and 40–70 kcal			1	1½	2	2½	3	2½	3	2	2½	2½	2½	2
SUPPLEMENTAL SOURCES OF ENERGY—to bring the daily dietary energy up to the total allowance														
(Less than 100 kcal per serving)	*(150–300 kcal per serving)*	*(350–500 kcal per serving)*												
ENERGY ALLOWANCES (kcal per person per day)			50/lb	1,300	1,800	2,400	3,000	2,250	2,700	2,000	2,300	2,500	2,400	1,800

[1]Secure the foods according to the calories required for serving, from the Food Composition Table F-21.

Here is how to use Table M-9: Let us suppose that it is necessary to plan a daily meal pattern for a 25-year-old single, moderately active female, who is not overweight.

Step 1—Look down the column headed "Adults (20-50 years)" and subheaded "nonpregnant F." Note that the essential food groups and the number of servings are as follows:

Meat, Fish, or Seafood, etc. — 2
Milk Products — 3
Breads and Cereals — 4
Dark-Green Vegetables — 1
Citrus Fruits — 1
Other Fruits and Vegetables — 3
Essential Fats — 2

Optional items to make up a total of 2,000 Calories (kcal).

Step 2—Select items from the essential groups. Take care to balance their distribution between the meals and snacks, but avoid exceeding the total allowance for energy—2,000 Calories (kcal) per day. However, if the subject gains weight on 2,000 Calories (kcal) per day, then it will be necessary to choose servings from the *Low-Energy Foods*. On the other hand, if she wishes to gain weight, then choose servings from the *High-Energy Foods*. Optional items may be added from the group *Supplemental Sources of Energy* in order to bring the total energy up to the amount allowed. However, it would be better to obtain the supplemental sources of energy from among the essential groups. A typical day's selection might be as follows:

MEAL	FOOD	AMOUNT	CALORIES
Breakfast	Grapefruit	½	100
	Poached egg	1	74
	Whole wheat toast	2 slices	109
	w/butter	2 tsp	70
	Coffee w/evaporated milk	¼ c	50
	or Milk, 2%	1 c	122
Morning Snack	Yogurt, flavored	1 c	231
Lunch	Sandwich made w/		
	Cooked ham	2 oz	112
	Swiss cheese	½ oz	51
	Sliced tomato	4 slices	22
	Lettuce leaf		
	Mayonnaise	2 tsp	36
	Whole wheat bread	2 slices	109
	Milk, nonfat	1 c	83
Dinner	Roast beef	3 oz	300
	Tomato catsup	1 Tbsp	16
	Baked potato	1	93
	Mixed vegetables	½ c	64
	Biscuits made w/whole		
	wheat flour	2	204
	Butter	2 tsp	70
	Jam	1 Tbsp	54
Bedtime Snack	Apple	1	84
	Total		1932-2004

Step 3—After writing down the menu plan for the day, check the number of servings of each of the essential food groups and the total energy supplied as given in Table M-9.

Although Table M-9 is a useful guide in planning the day's diet, always remember the Golden Rule of Eating, which is—"The greater the variety of foods in the diet, the greater the chances of getting everything that the body requires."

(Also see FOOD GROUPS.)

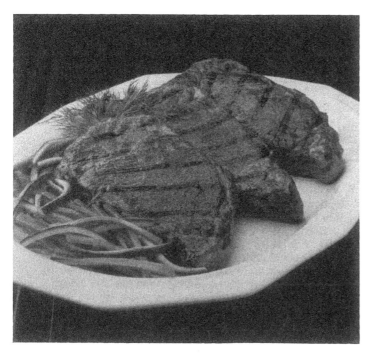

Fig. M-19. Rib eye steaks. People eat steaks because they're good, and good for them. (Courtesy, National Live Stock & Meat Board, Chicago, Ill.)

MEAT(S)

Contents   Page

History of Meat..657
From Whence Meat Comes..........................657
Meat Packing...657
  Meat Tenderizing..................................657
  Meat Curing...658
  Federal and State Meat Inspection.................659
  Federal Grades of Meats...........................659
  Packer Brand Names...............................659
  Kosher Meats...................................... 660
  By-Products 660
Kinds of Meat..661
Meat Preservation....................................661
Fabricated; Packaged Meat..........................663
Qualities in Meats Desired by Consumers.............663
Meat Buying..664
  What Determines the Price of Meat?................664
  Where the Consumer's Food Dollar Goes............664
  Income—Proportion Spent for Food;
    Food—Kinds Bought..........................666
  Selecting the Meat Market..........................666
  Selecting the Cut of Meat..........................666
  Watching for Specials.............................666
  Deciding How Much to Buy.........................666
Meat Cooking..666
Meat Carving..668
U.S. Meat Production.................................668
U.S. Per Capita Meat Consumption...................668
Eating Habit Trends..................................669
World Meat Production and Consumption.............669
Nutritive Composition of Meats.......................670
  Energy 670
  Carbohydrate 670
  Fats...671
  Protein...671
  Minerals..671
    Macrominerals 671
    Microminerals................................ 672
  Vitamins..673
    Fat-Soluble Vitamins 673
    Water-Soluble Vitamins 673
  Summary 674
Meat Facts and Myths............................... 674

For centuries, animals have made a major contribution to human welfare as sources of food, pharmaceuticals, clothing, transportation, power, soil fertility, fuel, and pleasure. The most important of these is food—meat and other animal products. This section contains information on the value of animal foods to humans, with particular emphasis on the meat and meat by-product supply.

Meat is the edible flesh, organs, and glands of animals used for food. Meat by-products include all products, both edible and inedible, other than the carcass meat. In a broad sense, the term meat may include the flesh of poultry, fish and other seafoods, and game. In this section, however, the discussion of meat will largely be limited to red meat and to by-products therewith—to beef, lamb, and pork. Poultry, fish and seafoods, game, and milk and milk products are treated in separate sections of this book.

That Americans enjoy eating meat is attested by the fact that, in 1990, the average American ate 112.3 lb of beef, veal, pork, and lamb. Of this total, beef accounted for 64 lb, pork for 46.3 lb, veal for 0.9 lb, and lamb for 1.1 lb.

Not only are meat and animal by-products good, but they are good for you. Foods from animals are the most nutritionally complete known. Today, they (meat, poultry, fish, and milk) supply to the U.S. diet an average of 95% of the vitamin B-12, 79% of the calcium (mostly from milk and milk products), 55% of the riboflavin, 63% of the protein, 64% of the phosphorus, 51% of the vitamin B-6, 46% of the niacin, 36% of the vitamin A, 33% of the thiamin, 25% of the iron (in a readily absorbed form), and 34% of the magnesium. Additionally, animal proteins provide the nine essential amino acids in the proportions needed for humans; hence, they are *high quality* proteins.

HISTORY OF MEAT.

The use of meat by man antedates recorded history. Primitive man recognized that a meat-rich diet was far more concentrated than leaves, shoots, and fruits. Thus, the adoption of the meat-eating habit allowed him, if he was lucky in his kill, more time for other pursuits and skills, such as devising new tools and learning the beginnings of pictorial art.

Until less than 10,000 years ago, man obtained his food by hunting, scavenging, or gathering from the animal and vegetable kingdoms in his environment. It is probable that taming, and then domestication, of animals occurred without people being aware of what was happening. Certainly, gatherers and hunters—the people who first domesticated animals—could not forsee any use for them other than those they knew already—for meat and skins. Only later, after a change from a nomadic to a more settled life-style, were animals used for such things as milk, wool, motive power, war, sport, and prestige.

It is recorded that pork was used as a food in Egypt, probably at certain feasts, as early as 3400 B.C., and in China beginning in 2900 B.C. Old Testament Laws, written about 1100 B.C., dictated what meats were considered wholesome and sanitary. In 1 Corinthians, 15:39, written about 54 A.D., the Bible tells us—"All flesh is not the same flesh, but there is one kind of flesh of men, another flesh of beasts, another of fishes, and another of birds."

FROM WHENCE MEAT COMES.

Animal agriculture is worldwide. Most of the livestock and poultry, along with 58% of the agricultural land, are found in Asia, Africa, and South America. Yet, North America, Europe, and Oceania, where scientific methods of production are most widely applied, produces 58% of the world's beef and veal, 62% of the milk, 41% of the lamb and mutton, 53% of the wool, 55% of the pork, 60% of the poultry meat, and 48% of the eggs.

Consumers—all—need to know from whence their meat comes; they need to know that meat on the table is the ultimate objective in producing cattle, sheep, and swine. Consumers need to know that only farmers produce foods—that neither governments nor supermarkets produce foods; and they need to know that people do those things that are most profitable for them—and that farmers are people.

HOW MEAT REACHES THE TABLE

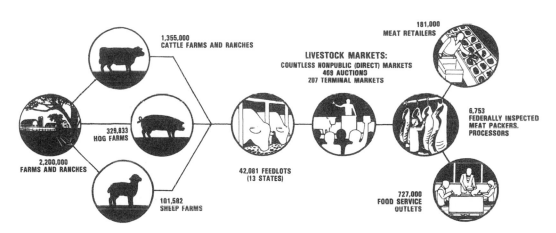

Fig. M-20. From whence meat comes. Back of the meat on the table are 2,200,000 farms, of which 1,355,000 are cattle farms and ranches, 329,833 are hog farms, and 101,582 are sheep farms. All statistics are from the USDA, and for 1988, except number of cattle farms, number of hog farms, and number of sheep farms are from 1982 Agricultural Census.

MEAT PACKING.

Meat packing is the business of slaughtering cattle, hogs, and sheep, and of preparing the meat for transportation and sale. It is an important industry in many countries; the United States ranks first in meat packing. In 1986, 6,753 U.S. establishments slaughtered 37.6 million cattle, 3.5 million calves, 5.8 million sheep and lambs, and 80 million hogs; and produced 39.3 billion pounds of red meat. Also, the meat packing industry employed 143,000 people.

Today, the name *packing* is a misnomer, for the barrelled or pickled pork from whence the name originated is only a memory. But modern meat slaughtering and processing developed slowly, and the original name was so well established that there has been no attempt to rename the industry.

Meat Tenderizing.

Consumers want tenderness, as well as flavor, in the meat they buy. The common methods of tenderizing are:

• **Aging, Natural (low temperature)**—Following slaughter, dressed carcasses are rapidly chilled in a room at 32°F *(0°C)*, at a relative humidity of about 85% in order to reduce moisture loss. The chilled side or cuts may be held for aging.

Aging is accomplished by holding the beef carcass or cuts for 1 to 6 weeks at a temperature of 38 to 40°F (3 to 4°C) and a relative humidity of 70 to 90%, with a gradual flow of air to provide a fresh atmosphere. These conditions permit selective enzymatic and bacterial action; the enzymes aid in the breakdown of various tissues, and the bacteria provide the nutlike flavor characteristic of aged beef. Today, very little beef is aged more than a week, because aging in carcass or wholesale cuts has become too costly in terms of shrinkage and cooler space.

Lamb carcasses are not aged.

• **Vegetable enzymes**—Three enzymes of vegetable origin which dissolve or degrade collagen and elastin are:

1. *Papain*, secured from the tropical American tree, *Carica papaya* (Also see PAPAYA.)

2. *Bromelin*, secured from the juice of the pineapple (Also see PINEAPPLE.)

3. *Ficin*, secured from figs (Also see FIG.)

The *tenderizing* action of enzymes does not occur until the meat is heated during cooking.

• **Mechanical tenderization (blade tenderization)**—Mechanical tenderizers, consisting of a machine that presses many sharp blades or needles through roasts and steaks, are being used by some segments of the meat trade. It is chiefly used on primal cuts (rib and loin) from lower quality grades and on the less tender cuts (chuck, round, shank) of higher grade carcasses. These machines have gained wide acceptance in the hotel, restaurant, and institutional trade in recent years. Research indicates that mechanical tenderization improves tenderness by 20 to 50%, without increasing cooking losses.

• **Electrical stimulation**—Electrical stimulation is a method of tenderizing meat with electricity. The passing of a current through the carcass causes the muscle fibers to loosen. Before rigor mortis sets in, the carcass is shocked until all the energy in the muscle that causes the fibers to contract is used up.

Electrical tenderizing is effective on beef, lamb, and goat carcasses, but not pork.

Meat Curing. In this country, meat curing is largely confined to pork, primarily because of the keeping qualities

and palatability of cured pork products. Considerable beef is corned or dried, and some lamb and veal are cured, but none of these is of such magnitude as cured pork.

Meat is cured with salt, sugar, and certain curing adjuncts (ascorbate, erythorbate, etc.); with sodium nitrite or sodium nitrate (the latter is used only in certain products); and with smoke. Each of the most common curing additives will be discussed:

• **Salt**—Sodium chloride is added to processed meats for preservative and palatability reasons. The addition of salt makes it possible to distribute certain perishable meats through what are often lengthy and complicated distribution systems. Also, in products such as wieners, bologna, and canned ham, fat solubilizes myosin, the major meat protein, and causes the meat particles to hold together; the meat can then be sliced without falling apart.

(Also see SALT.)

• **Sodium nitrite (NaNO₂), sodium nitrate (NaNO₃), potassium nitrate or saltpeter (KNO₃)**—Nitrite and nitrate contribute to (1) the prevention of *Clostridium botulinum* spores in or on meat (*Clostridium botulinum* bacteria produce the botulin toxin that causes botulism, the most deadly form of food poisoning); (2) the development of the characteristic flavor and pink color of cured meats; (3) the prevention of *warmed-over flavor* in reheated products; and (4) the prevention of rancidity. Of all the effects of nitrite and nitrate, the antibotulinial effect is by far the most important. Nitrite is the essential ingredient for curing; the effectiveness of nitrate results from change of part of the nitrate to nitrite by reactions that occur during processing.

No entirely satisfactory substitute is known for nitrite in meat curing.

• **Smoking**—Smoking produces the distinctive smoked-meat flavor which consumers demand in certain meats.

Fig. M-21. Hot dogs (Frankfurters, or wieners), the most popular sausage in America, processed by a large meat packing company. (Courtesy, University of Tennesse, Knoxville, Tenn.)

Fig. M-22. The Smoke House, in which meats were smoked in colonial days. This method of preserving meats is very old. Modern meat packers still smoke many of their products. (Courtesy, Swift and Company)

• **Phosphate (PO₄)**—Phosphate is added to increase the water-binding capacity of meat; its use increases yields up to 10%. Also, it enhances juiciness of the cooked product. Federal regulations restrict the amount of phosphates to: (1) not to exceed 0.5% in the finished product, and (2) not more than 5% in the pickle solution based on 10% pumping pickle. The addition of phosphate must be declared on the label.

Federal and State Meat Inspection.
Most foods from animals are inspected by government officials to assure wholesomeness, proper labeling, and freedom from disease and adulteration. Federal inspection of animal products is the responsibility of the Food Safety and Quality Service of the U.S. Department of Agriculture. Imported meat and poultry are produced under USDA supervision in foreign countries where they originate, and samples are inspected by USDA upon arrival in the United States.

If meat passes federal inspection, an inspection stamp (Fig. M-23) is affixed to the wholesale cuts or package to assure the retailer and consumer that the products have been inspected. Products that do not pass inspection are denatured and rendered unusable for human consumption.

THIS IS THE STAMP USED ON MEAT CARCASSES. NOTE THAT IT ALSO BEARS THE OFFICIAL NUMBER ASSIGNED TO THE ESTABLISHMENT---IN THIS CASE NUMBER 38. THE STAMP IS USED ONLY ON THE MAJOR CUTS OF THE CARCASS; HENCE, IT MAY NOT APPEAR ON THE ROAST OR STEAK THE CONSUMER BUYS.

THIS MARK IS FOUND ON EVERY PREPACKAGED PROCESSED MEAT PRODUCT---SOUPS TO SPREADS--- THAT HAS BEEN FEDERALLY INSPECTED.

Fig. M-23. Federal inspection marks used on animal products.

Federal Grades of Meats.
The grade of meat may be defined as a measure of its degree of excellence based on quality, or eating characteristics of the meat, and the yield, or total proportion, of primal cuts. Naturally, the attributes upon which the grades are based vary between species. Nevertheless, it is intended that the specifications for each grade shall be sufficiently definite to make for uniform grades throughout the country and from season to season, and that on-hook grades shall be correlated with on-foot grades.

Fig. M-24. Federal grader shown rolling (grading or stamping) beef with an edible vegetable dye. (Courtesy, *Livestock Breeder Journal*, Macon, Ga.)

Government grading, unlike meat inspection, is not compulsory. Official graders are subject to the call of anyone who wishes their services, (packer, wholesaler, or retailer) with a charge per hour being made.

Consumers, especially the housewife who buys most of the meat, should know the federal grades of meats, because (1) in these days of self-service, prepackaged meats there is less opportunity to secure the counsel and advice of the meat cutter when making purchases, and (2) the average consumer is not the best judge of the quality of the various kinds of meats on display in the meat counter.

Federally graded meats are so stamped (with an edible vegetable dye) that the grade will appear on the retail cuts as well as on the carcass and wholesale cuts. These are summarized in Table M-10.

TABLE M–10
QUALITY GRADES OF MEATS BY CLASSES[1]

Beef[2]	Veal	Mutton and Lamb[2]	Pork
1. Prime[3]	1. Prime	1. Prime[4]	1. U.S. No. 1
2. Choice	2. Choice	2. Choice	2. U.S. No. 2
3. Select	3. Select	3. Select	3. U.S. No. 3
4. Standard	4. Standard	4. Utility	4. U.S. No. 4
5. Commercial	5. Utility	5. Cull[5]	5. U.S. Utility
6. Utility			
7. Cutter			
8. Canner			

[1]In rolling meat, the letters U.S. precede each federal grade name. This is important as only government-graded meat can be so marked. For convenience, however, the letters U.S. are not used in this table or in the discussion which follows.
[2]In addition to the quality grades given herein, there are the following yield grades of beef and lamb (and mutton) carcasses: Yield Grade 1, Yield Grade 2, Yield Grade 3, Yield Grade 4, and Yield Grade 5.
[3]Cow beef is not eligible for the Prime grade.
[4]Limited to lamb and yearling carcasses.
[5]Limited to mutton carcasses.

As would be expected, in order to make the top grade in the respective classes, the carcass or cut must possess a very high degree of the attributes upon which grades are based. The lower grades of meats are deficient in one or more of these grade-determining factors. Because each grade is determined on the basis of a composite evaluation of all factors, a carcass or cut may possess some characteristics that are common to another grade. It must also be recognized that all of the wholesale cuts produced from a carcass are not necessarily of the same grade as the carcass from which they are secured.

NOTE WELL: Federal grades of meats change from time to time, reflecting consumer preferences, technological developments in meat processing and marketing, and political expediency.

• **Proportion of U.S. meat federally graded**—In 1988, 93.6% of all lamb and mutton was quality graded, 56.4% of the beef was quality graded, but only negligible amounts of pork were quality graded.

Packer Brand Names.
Practically all packers identify their higher grades of meats with alluring private brands so that the consumer as well as the retailer can recognize the quality of a particular cut.

A meat packer's reputation depends upon consistent standards of quality for all meats that carry his brand names. The brand names are also effectively used in advertising campaigns.

Kosher Meats.[17]—Meat for the Jewish trade—known as kosher meat—is slaughtered, washed, and salted according to ancient Biblical laws, called *Kashruth,* dating back to the days of Moses, more than 3,000 years ago. The Hebrew religion holds that God issued these instructions directly to Moses, who, in turn, transmitted them to the Jewish people while they were wandering in the wilderness near Mount Sinai.

The Hebrew word *kosher* means fit or proper, and this is the guiding principle in the handling of meats for the Jewish trade. Also, only those classes of animals considered clean—those that both chew the cud and have cloven hooves—are used. Thus, cattle, sheep, and goats—but not hogs—are koshered (Deuteronomy 14:4–5 and Leviticus 11:1–8).

Poultry is also koshered. Rabbinical law dictates that only cold water shall be used; hot or warm water cannot be used in defeathering or at any time in the processing of poultry products. Also, following picking and cleaning, kosher poultry must be washed, hung up to drip, hand-salted internally and externally, and rinsed; with constant inspection at all phases.

Both forequarters and hindquarters of kosher-slaughtered cattle, sheep, and goats may be used by Orthodox Jews. However, the Jewish trade usually confines itself to the forequarters. The hindquarters (that portion of beef carcass below the twelfth rib) are generally sold as nonkosher for the following reasons:

1. The Sinew of Jacob (*the sinew that shrank,* now known as the sciatic nerve), which is found in the hindquarters only, must be removed by reason of the Biblical story of Jacob's struggle with the Angel, in the course of which Jacob's thigh was injured and he was made to limp.

Actually, the sciatic nerve consists of two nerves; an inner long one located near the hip bone which spreads throughout the thigh, and an outer short one which lies near the flesh. Removal of the sinew (sciatic nerve) is very difficult.

The Biblical law of the sciatic nerve applies to cattle, sheep, and goats, but it does not apply to birds because they have no spoon-shaped hip (no hollow thigh).

2. The very considerable quantity of forbidden fat (Heleb) found in the hindquarters, especially around the loins, flanks, and kidneys, must be removed; and this is difficult and costly. Forbidden fat refers to fat (tallow) (a) that is not intermingled (marbled) with the flesh of the animal, but forms a separate solid layer; and (b) that is encrusted by a membrane which can be easily peeled off.

The Biblical law of the forbidden fat applies to cattle, sheep, and goats, but not birds and nondomesticated animals.

3. The blood vessels must be removed, because the consumption of blood is forbidden; and such removal is especially difficult in the hindquarters.

NOTE WELL: Forbidden fat and blood (the blood vessels) must be removed from both fore and hindquarters, but such removal is more difficult in the hindquarters than in the forequarters. However, the Sinew of Jacob (the sciatic nerve), which must also be removed, is found in the hindquarters only.

Because of the difficulties in processing, meat from the hindquarters is not eaten by Orthodox Jews in many countries, including England. However, the consumption of the hindquarters is permitted by the Rabbinic authorities where there is a special hardship involved in obtaining alternative supplies of meat; thus, in Israel the sinews are removed and the hindquarters are eaten.

Because the forequarters do not contain such choice cuts as the hinds, the kosher trade attempts to secure the best possible fores; thus, this trade is for high grade slaughter animals.

Kosher meat must be sold by the packer or the retailer within 72 hours after slaughter, or it must be washed (a treatment known as *begiss,* meaning to wash) and reinstated by a representative of the synagogue every subsequent 72 hours. At the expiration of 216 hours after the time of slaughter (after begissing three times), however, it is declared *trafeh,* meaning forbidden food, and is automatically rejected for kosher trade. It is then sold in the regular meat channels. Because of these regulations, kosher meat is moved out very soon after slaughter.

Kosher sausage and prepared meats are made from kosher meats which are soaked in water ½ hour, sprinkled with salt, allowed to stand for an hour, and washed thoroughly. This makes them kosher indefinitely.

The Jewish law also provides that before kosher meat is cooked, it must be soaked in water for ½ hour. After soaking, the meat is placed on a perforated board in order to drain off the excess moisture. It is then sprinkled liberally with salt. One hour later, it is thoroughly washed. Such meat is then considered to remain kosher as long as it is fresh and wholesome.

Meats and fowl are the only food items which require ritual slaughter, washing, and salting before they are rendered kosher.

As would be expected, the volume of kosher meat is greatest in those Eastern Seaboard cities where the Jewish population is most concentrated. New York City alone uses about one-fourth of all the beef koshered in the United States.

While only about half of the total of more than six million U.S. Jewish population is orthodox, most members of the faith are heavy users of kosher meats.

By-Products. By-products include everything of value produced on the killing floor other than the dressed carcass; and they are classified as edible or inedible. The edible by-products include blood, brains, casings, fats, gelatin, hearts, kidneys, livers, oxtails, sweetbreads, tongues, and tripe. The inedible by-products include animal feeds, bone meal, bone products, brushes, cosmetics, feathers, fertilizer, glue, glycerin, hides and skins, lanolin, ligatures, lubricants, neat's foot oil, pluck (lungs, etc.), soap, and wool.

In addition to the edible (food) and nonfood (inedible) products, certain chemical substances useful as human drugs or pharmaceuticals are obtained as by-products. Among such drugs are ACTH, cholesterol, estrogen, epinephrine, heparin, insulin, rennet, thrombin, TSH, and thyroid extracts—all valuable pharmaceuticals which are routinely recovered from meat animals in the United States.

---

[17]Authoritative information relative to kosher meats was secured from: (1) Grunfeld, Dayton Dr. I., *The Jewish Dietary Laws,* The Soncino Press, London/Jerusalem/New York, 1975; (2) a personal letter to M. E. Ensminger dated December 22, 1981, from Rabbi Menachem Genack, Rabbinic Coordinator, Union of Orthodox Jewish Congregations of America, New York, N.Y.; and (3) John Ensminger, Lawyer, New York, N.Y.

Fig. M-25. Some of the items for which by-products are used—items which contribute to the convenience, enjoyment, and health of people in all walks of life. (Courtesy, American Meat Institute, Washington, D.C.)

1. Bone for bone china
2. Horn and bone handles for carving sets
3. Hides and skins for leather goods
4. Rennet for cheese making
5. Gelatin for marshmallows, photographic film, printers' rollers
6. Stearin for making chewing gum and candies
7. Glycerin for explosives used in mining and blasting
8. Lanolin for cosmetics
9. Chemicals for tires that run cooler
10. Binders for asphalt paving
11. Medicines such as various hormones and glandular extracts, insulin, pepsin, epinephrine, ACTH, cortisone; and surgical sutures
12. Drumheads and violin strings
13. Animal fats for soaps
14. Wool for clothing
15. Camel's hair (actually from cattle ears for artists' brushes)
16. Cutting oils and other special industrial lubricants
17. Bone charcoal for high-grade steel such as ball bearings
18. Special glues for marine plywoods, paper, matches, window shades
19. Curled hair for upholstery. Leather for covering fine furniture
20. High-protein livestock feeds

KINDS OF MEAT.
The meat of cattle, hogs, sheep, and goats is known by several different names. These follow:

• **Beef**—This is the flesh of adult cattle. Good beef has white fat and bright cherry-red lean meat.

• **Veal**—This is the flesh of calves, usually 2 to 12 weeks old. Older calves are usually sold as *calves*. Veal is pink in color and has a fine texture.

• **Mutton**—This is the flesh of mature sheep. It is darker colored and stronger flavored than lamb.

• **Lamb**—This is the flesh of young sheep. The meat of a sheep becomes mutton when the animal is about a year old. Lamb has a light-pink color and white fat, and a much milder flavor than mutton.

• **Pork**—This is the flesh of hogs. It is known as pork no matter how old the animal from which it comes. The eating quality of pork does not change much with the animal's age.

• **Chevon (goat meat)**—This is goat meat. The meat from young goats (kids) is delicious. Chevon from older goats is likely to possess a strong flavor.

• **Variety meat**—The various organs of animals are called variety meats. The heart, liver, brains, kidneys, tongue, cheek meat, tail, feet, sweetbreads (thymus and pancreatic glands), and tripe (pickled rumen of cattle and sheep) are sold over the counter as variety meats or fancy meats.

• **Game meat**—The flesh of wild animals or birds used as food.

MEAT PRESERVATION.
Methods of preserving meat were well established before the dawn of recorded history. Meat was dried by the aborigines; smoking and salting techniques were well established long before Homer's time (about 1000 B.C.); preparation and spicing of some kinds of sausages were common practices in Europe and in the Mediterranean countries before the time of the Caesars.

Fundamentally, meat preservation is a matter of controlling putrefactive bacterial action. Various methods of preserving meats have been practiced through the ages, the most common of which (listed alphabetically), are: (1) acidity, (2) air exclusion, (3) canning, (4) drying, (5) freeze-drying, (6) radiation, (7) refrigeration and freezing, (8) salting, (9) dry sausage, (10) smoking, (11) spicing, and (12) sugar.

• **Acidity**—Commonly referred to as pickling or souring, an acid condition is unfavorable to bacteria. This method of preservation is used for pigs' feet, certain sausages, sauerkraut, pickles, and milk.

• **Air exclusion (vacuum packed)**—Vacuum packaging, or packaging with inert gas (e.g., N_2 or CO_2), are examples of the exclusion of air method. Elimination of oxygen from the package is the primary objective in vacuum and gas packaging. It extends the shelf life of fresh meats.

• **Canning**—Canning is the temporary increasing of the heat content of a product that has been enclosed in an airtight container for the purpose of preventing the spoilage of the product being canned by the inactivation of particular organisms. Two levels of heat may be applied in canning meat:

Fig. M-26. Canning meat, a method of meat preservation. (Courtesy, *Meat Magazine*)

1. **Pasteurization.** A low level of heat in the process of pasteurization, which is designed to kill all pathogenic organisms, but not all spoilage organisms. Since all spoilage organisms have not been destroyed, the product must still be stored under refrigeration. Most canned hams are pasteurized.

2. Sterilization. A high temperature in the process of sterilization. In this process, the canned meat is processed at a sufficiently high temperature and held there long enough to kill bacteria that cause spoilage, food poisoning, or infection. This calls for using a steam-pressure canner. By holding steam under pressure, one can get a temperature of 240°F (115.5°C) or more. Many luncheon meats are processed in this manner.

When canning, checking the sea level is important. More pressure in the canner is required at high altitudes than at low altitudes. A rule of thumb is: add 1 lb pressure for each 2,000 ft above sea level. Processing time depends on the container size.

Fig. M-27. Indians drying and smoking venison. (Photo from a watercolor original by Ernest Smith. Owned by the Rochester Museum of Arts and Sciences)

Today, a limited amount of beef is cured and dried, in which form it is known as *beef jerky*. It has a tough, chewy consistency and may be smoked, unsmoked, or air- or oven-dried.

• **Freeze-drying**—This process involves the dehydration of meat by the freeze-drying process, which consists of subjecting frozen pieces of meat to heat under vacuum. Products retain their shape and form and do not shrink as with other methods of dehydration. Freeze-dried meat and poultry have a moisture content of 2% or less and will not support the growth of most microbes.

• **Radiation**—This is the newest development in meat preservation. Nonionizing and ionizing radiation are capable of deactivating microorganisms.

Ultraviolet light is an example of nonionizing radiation; however, it does not sterilize beneath the surface.

Ionizing radiation, in the form of soft x rays or gamma rays, is capable of destroying organisms in canned, packaged, or exposed meat products without raising the temperature appreciably (hence, it is known as *cold sterilization*). Some researchers report that radiation may result in slight odor, flavor, and color changes. Also, it is noted that small doses of radiation do not completely inactivate all the proteolytic enzymes in meat, with the result that there is danger of these enzymes causing serious changes in the product during storage. However, much research is in progress; hence, these deficiencies may soon be overcome.

In 1981, the control of the Mediterranean fruit fly in California caused the Food and Drug Administration to ease the ban on treating food with gamma rays. However, each food processed by radiation requires FDA's approval. Also, labels on foods that have been irradiated must so indicate.

Consumer reaction to radiation of foods will be watched closely. It has great potential, provided people are not squeamish about the *dangers*.

(Also see RADIATION PRESERVATION OF FOOD; and ULTRAVIOLET LIGHT.)

• **Refrigeration and freezing**—Meat is perishable, so proper care is essential to maintaining its keeping qualities.

Meat may be refrigerated for short-time storage, at a temperature of 36° to 40°F (2° - 4°C); or it may be frozen for longtime storage, at a temperature of 0°F or lower. Table M-11 gives recommended times for refrigerator and freezer storage. Additional pointers relative to refrigeration and freezing follow:

TABLE M-11
STORAGE TIME FOR REFRIGERATED AND FROZEN MEATS[1]
(Maximum storage time recommendations for fresh, cooked, and processed meat)

Meat	Refrigerator (36° to 40°F)[2]	Freezer (at 0°F or lower)
Beef (fresh)	2 to 4 days	6 to 12 months
Veal (fresh)	2 to 4 days	6 to 9 months
Pork (fresh)	2 to 4 days	3 to 6 months
Lamb (fresh)	2 to 4 days	6 to 9 months
Ground beef, veal and lamb	1 to 2 days	3 to 4 months
Ground pork	1 to 2 days	1 to 3 months
Variety meats	1 to 2 days	3 to 4 months
Luncheon meats	1 week	not recommended
Sausage, fresh pork	1 week	60 days
Sausage, smoked	3 to 7 days	—
Sausage, dry and semidry (unsliced)	2 to 3 weeks	—
Frankfurters	4 to 5 days	1 month
Bacon	5 to 7 days	1 month
Smoked ham, whole	1 week	60 days
Ham slices	3 to 4 days	60 days
Beef, corned	1 week	2 weeks
Leftover cooked meat	4 to 5 days	2 to 3 months
Frozen Combination Food		
Meat pies (cooked)	—	3 months
Swiss steak (cooked)	—	3 months
Stews (cooked)	—	3 to 4 months
Prepared meat dinners	—	2 to 6 months

[1]*Lessons on Meat*, published by the National Live Stock and Meat Board, Chicago, Ill., p. 60.
[2]The range in time reflects recommendations for maximum storage time from several authorities. For top quality, fresh meats should be used in 2 or 3 days, ground meat and variety meats should be used in 24 hours.

1. Storage of meat. Fresh meat which is not to be refrozen should be stored in the coldest part of the refrigerator or, when available, in the compartment designed for meat storage. Additional guides for storage of meats follow:

a. **Fresh meat, prepackaged by the meat retailer (self-service).** It may be stored in the refrigerator in the original wrapping for no more than 2 days, or it may be frozen without rewrapping and stored in the freezer 1 to 2 weeks. For longer freezer storage, the original package should be overwrapped with special freezer material.

b. **Fresh meat, not prepackaged.** It should be removed from the market wrapping paper, wrapped loosely in waxed paper or aluminum foil, and refrigerated for no more than 2 days.

c. **Variety meats and ground or chopped meats.** They are more perishable than other meats; so, they should be cooked in 1 or 2 days if not frozen.

d. **Cured, cured and smoked meats, sausages, and ready-to-serve meats.** These should be left in their original wrapping and stored in the refrigerator.

e. **Canned hams, picnics and other perishable canned meats.** These should be stored in the refrigerator unless the directions on the can read to the contrary. These meats should not be frozen.

f. **Frozen meat.** Meat that is frozen at the time of purchase should be placed in the freezer soon after purchase unless it is to be defrosted for immediate cooking. The temperature of the freezer should not go above 0°F (-18°C). Packages should be dated as they are put into the freezer, and the storage time should be limited as recommended in Table M-11.

g. **Cooked meats.** Leftover cooked meats should be cooled within 1 to 2 hours after cooking, then covered or wrapped to prevent drying, and stored in the refrigerator. Meats cooked in liquid for future serving should be cooled uncovered for 1 to 2 hours, then covered and stored in the refrigerator.

2. **Freezing of meat.** Most meats may be frozen satisfactorily if properly wrapped, frozen quickly, and kept at 0°F or below. The following guides will help ensure good quality in frozen meats:

a. **Freeze meat while it is fresh and in top condition.** Properly done, freezing will maintain quality, but what comes out of the freezer will be no better than what goes in.

b. **Select proper wrapping material.** Choose a moisture-vapor-proof wrap so that air will be sealed out and moisture locked in. When air penetrates the package, moisture is drawn from the surface of the meat and the condition known as *freezer burn* develops. Several good freezer wraps are on the market.

c. **Prepare meat for freezing before wrapping.** Trim off excess fat and remove bones when practical; wrap in family-sized packages; place double thickness of freezer wrap between chops, patties, or individual pieces of meat, for easier separation during thawing.

d. **Label properly.** This should include the kind and cut of meat enclosed, the number of possible servings (including weight if convenient), and the date of packaging.

e. **Freeze immediately.** Start freezing immediately following wrapping and labeling.

f. **Keep freezer temperature 0°F (–18°C) or lower.** Higher temperatures and fluctuations above that temperature impair quality.

g. **Check freezer, Table M-11, for maximum storage times.** Frozen meat will be of best quality if not held longer than indicated.

h. **Avoid refreezing if possible.** Refreezing results in loss of juices when rethawed, thereby affecting juiciness and flavor.

• **Salting**—Dry salting of meats has been practiced since the 5th century B.C., and possibly longer. However, it was not until the latter part of the 18th century that salt curing of meat was done on a scientific basis.

The presence of salt has an effect on osmotic pressure. In addition, the action of chloride on microorganisms is detrimental.

Currently, salt curing is limited to a few specialized products, such as hams and bacon. Although there is no specified salt content for products labeled *salt cured,* most cured meats are acceptable at levels of 2.5 to 3% salt.

• **Dry sausages**—Dry sausages are often called summer sausages because of their keeping qualities. They can be held in a fairly cool room for a very long period of time. Dry salami is an example of a dry sausage.

• **Smoking**—This method of preserving meats is very old. The drying and smoking of meats was known to the Egyptians as well as to the ancient Sumerian civilization which preceded them. The advantage to this type of preservation lay in the fact that the smoke overcame objectionable flavors that developed if the drying did not proceed at a rapid rate. The favorite wood smoke in colonial America was produced by hickory or oak, although the Indians also used sage and various aromatic seeds and plants.

Smoking permits formaldehyde and phenolic compounds to accumulate on the surface of the meat. This, along with the surface drying, prevents microbial growth.

• **Spicing**—Many of the essential oils and other substances found in spices are effective bacteriostatic agents *provided* they are used at high levels. Mustard oil and allicin (allythiosulfinic allyl ester) in garlic are examples of effective bacteriostatic substances. Contrary to historical and popular belief, in the concentrations generally employed in flavoring agents, spices cannot be depended upon to have any preservative action. Rather, it appears that they gained early popularity as a means of masking undesirable flavors resulting from microbial spoilage and other causes.

• **Sugar**—A high concentration of sugar exerts a high osmotic pressure and causes water to be withdrawn from microorganisms. However, the sugar concentrations normally employed in meat curing are far short of those needed to contribute to any preservative action. Mincemeat, of which only a minor proportion is meat, is an example of a meat-containing food which is preserved by a combination of sugar, acetic acid, and spices.

FABRICATED; PACKAGED MEAT. With the advent of the refrigerator car, meat was shipped in exposed halves, quarters, or wholesale cuts, and divided into retail cuts in the back rooms of meat markets. But this traditional procedure leaves much to be desired from the standpoints of efficiency, sanitation, shrink, spoilage, and discoloration. To improve this situation, more and more packers are fabricating and packaging (boxing) meat in their plants, thereby freeing the back rooms of 200,000 supermarkets.

After chilling, the carcass is subjected to a disassembly process, in which it is fabricated or broken into counter-ready cuts; vacuum-sealed; moved into storage by an automated system; loaded into refrigerated trailers; and shipped to retailers across the nation.

QUALITIES IN MEATS DESIRED BY CONSUMERS.
Consumers desire the following qualities in meats:
1. Palatability
2. Attractiveness; eye appeal
3. Minimum amount of fat
4. Tenderness
5. Small cuts
6. Ease of preparation
7. Repeatability

If these qualities are not met by meats, other products will. Recognition of this fact is important, for competition is keen for space on the shelves of a modern retail food outlet.

MEAT BUYING.   Meat buying is important because, (1) meat prices change, (2) one-fifth of the disposable personal income spent on food goes for red meats, and (3) buying food for a family may be a more intricate affair than its preparation at home. Hence, meat buying merits well informed buyers.

What Determines the Price of Meat?   During those periods when meat is high in price, especially the choicest cuts, there is a tendency on the part of the consumer to blame either or all of the following: (1) the farmer or rancher, (2) the packer, (3) the meat retailer, or (4) the government. Such criticisms, which often have a way of becoming quite vicious, are not justified. Actually, meat prices are determined by the law of supply and demand; that is, the price of meat is largely dependent upon what the consumers as a group are able and willing to pay for the available supply.

• **The available supply of meat**—Because the vast majority of meats are marketed on a fresh basis rather than cured, and because meat is a perishable product, the supply of this food is dependent upon the number and weight of cattle, sheep, and hogs available for slaughter at a given time. In turn, the number of market animals is largely governed by the relative profitability of livestock enterprises in comparison with other agricultural pursuits. That is to say, farmers and ranchers—like any other good businessmen—generally do those things that are most profitable to them. Thus, a short supply of market animals at any given time usually reflects the unfavorable and unprofitable production factors that existed some months earlier and which caused curtailment of breeding and feeding operations.

• **The demand for meat**—The demand for meat is primarily determined by buying power and competition from other products. Stated in simple terms, demand is determined by the spending of money available and the competitive bidding of millions of housewives who are the chief home purchasers of meats. On a nationwide basis, a high buying power and great demand for meats exist when most people are employed and wages are high.

But the novice may wonder why these choice cuts are so scarce, even though people are able and willing to pay a premium for them. The answer is simple. Nature does not make many choice cuts or top grades, regardless of price. Moreover, a hog is born with two hams only, a lamb with two hind legs; and only two loins (a right and a left one) can be obtained from each carcass upon slaughter. In addition, not all weight on foot can be cut into meat. For example, the average steer weighing 1,050 lb *(477 kg)* on foot will only yield 448.8 lb *(204 kg)* of retail cuts (the balance consists of hide, internal organs, etc., see Fig. M-29, next page). Secondly, this 448.8 lb will cut out about 34.8 lb *(15.8 kg)* of porterhouse, T-bone, and top loin steaks. The balance of the cuts are equally wholesome and nutritious; and, although there are other steaks, many of the cuts are better adapted for use as roasts, stews, and soup bones. To make bad matters worse, not all cattle are of a quality suitable for the production of steaks. For example, the meat from most worn-out dairy animals and thin cattle of beef breeding is not sold over the block. Also, if the moneyed buyer insists on buying only the top grade of meat—namely U.S. Prime or its equivalent—it must be remembered that only a small proportion of slaughter cattle produce carcasses of this top grade. To be sure, the lower grades are equally wholesome, but they are simply graded down because the carcass is somewhat deficient in conformation, finish, and/or quality.

Thus, when the national income is exceedingly high, there is a demand for the choicest but limited cuts of meat from the very top but limited grades. This is certain to make for high prices, for the supply of such cuts is limited, but the demand is great. Under these conditions, if prices did not move up to balance the supply with demand, there would be a marked shortage of the desired cuts at the retail counter.

Where the Consumer's Food Dollar Goes.   In recognition of the importance of food to the nation's economy and the welfare of its people, it is important to know (1) what percentage of the disposable income is spent for food, and (2) where the consumer's food dollar goes—the proportion of it that goes to the farmer, and the proportion that goes to the middleman.

Fig. M-28 reveals that of each food dollar, the farmer's share ranged from 7¢ (for white bread) to 62¢ (for eggs). For choice beef, the farmer's share was 52¢ in 1987; the rest—48¢—went for processing and retailing. This means that nearly one-half of today's meat dollar goes for meat packing and meat retailing.

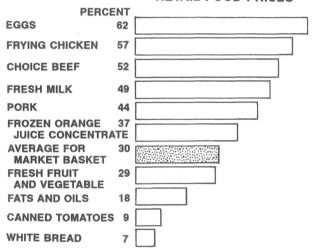

FARM VALUE SHARE OF RETAIL FOOD PRICES

PERCENT

EGGS	62
FRYING CHICKEN	57
CHOICE BEEF	52
FRESH MILK	49
PORK	44
FROZEN ORANGE JUICE CONCENTRATE	37
AVERAGE FOR MARKET BASKET	30
FRESH FRUIT AND VEGETABLE	29
FATS AND OILS	18
CANNED TOMATOES	9
WHITE BREAD	7

Fig. M-28. Farm share of retail food prices. (From: *1989 Handbook of Agricultural Charts*, Agricultural Handbook No. 684, USDA, p. 61)

The farmer's share of the retail price of eggs, meat products, poultry, and beef is relatively high because processing is relatively simple, and transportation costs are low due to the concentrated nature of the products. On the other hand, the farmer's share of bakery and cereal products is low due to the high processing and container costs, and their bulky nature and costly transportation.

Among the reasons why farm prices of foods fail to go up, thereby giving the farmer a greater share of the consumer's food dollar, are: (1) rapid technological advances on the farm, making it possible for the farmer to stay in business despite small margins; (2) overproduction; and (3) the relative ease with which cost pressures within the marketing system can be passed backward rather than forward.

Over the years, processing and marketing costs have increased primarily because consumers have demanded, and gotten, more and more processing and packaging—more built-in services. For example, few consumers are interested in buying a live hog—or even a whole carcass. Instead, they want a pound of pork chops—all trimmed, packaged, and ready for cooking. Likewise, few housewives are interested in buying flour and baking bread. But consumers need to be reminded that, fine as these services are, they cost money—

A STEER IS NOT ALL STEAK!

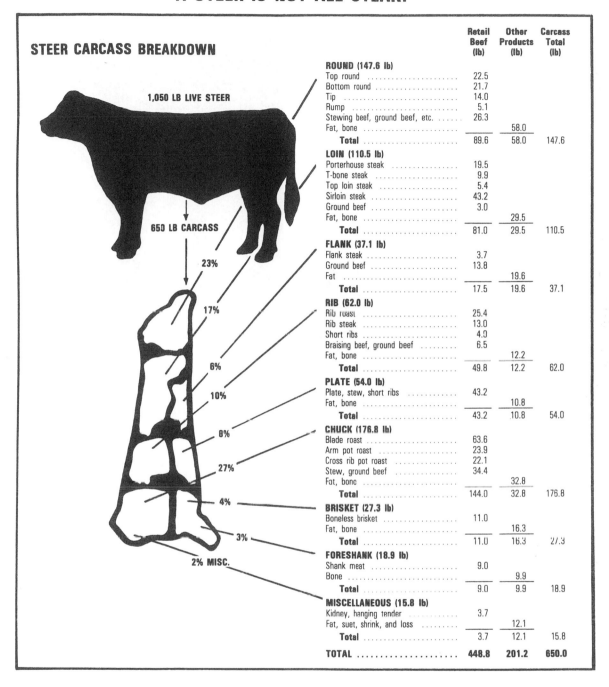

STEER CARCASS BREAKDOWN

1,050 LB LIVE STEER

650 LB CARCASS

23%
17%
6%
10%
8%
27%
4%
3%
2% MISC.

	Retail Beef (lb)	Other Products (lb)	Carcass Total (lb)
ROUND (147.6 lb)			
Top round	22.5		
Bottom round	21.7		
Tip	14.0		
Rump	5.1		
Stewing beef, ground beef, etc.	26.3		
Fat, bone		58.0	
Total	89.6	58.0	147.6
LOIN (110.5 lb)			
Porterhouse steak	19.5		
T-bone steak	9.9		
Top loin steak	5.4		
Sirloin steak	43.2		
Ground beef	3.0		
Fat, bone		29.5	
Total	81.0	29.5	110.5
FLANK (37.1 lb)			
Flank steak	3.7		
Ground beef	13.8		
Fat		19.6	
Total	17.5	19.6	37.1
RIB (62.0 lb)			
Rib roast	25.4		
Rib steak	13.0		
Short ribs	4.9		
Braising beef, ground beef	6.5		
Fat, bone		12.2	
Total	49.8	12.2	62.0
PLATE (54.0 lb)			
Plate, stew, short ribs	43.2		
Fat, bone		10.8	
Total	43.2	10.8	54.0
CHUCK (176.8 lb)			
Blade roast	63.6		
Arm pot roast	23.9		
Cross rib pot roast	22.1		
Stew, ground beef	34.4		
Fat, bone		32.8	
Total	144.0	32.8	176.8
BRISKET (27.3 lb)			
Boneless brisket	11.0		
Fat, bone		16.3	
Total	11.0	16.3	27.3
FORESHANK (18.9 lb)			
Shank meat	9.0		
Bone		9.9	
Total	9.0	9.9	18.9
MISCELLANEOUS (15.8 lb)			
Kidney, hanging tender	3.7		
Fat, suet, shrink, and loss		12.1	
Total	3.7	12.1	15.8
TOTAL	**448.8**	**201.2**	**650.0**

Fig. M-29. Cattle are not all beef, and beef is not all steak! This shows the approximate (a) percentage yield of carcass in relation to the weight of the animal on foot, and (b) the yield of differnet retail cuts. Note that a 1,050-lb live steer produces appoximately a 650-lb carcass, and ends up with only 448.9 lb of retail beef. Note, too, the small amount of steaks. (Source: Adapted by the authors from *Meat Facts*, published by the American Meat Institute, Washington, D.C. Data derived from USDA and industry figures.)

hence, they should be expected to pay for them. Without realizing it, American consumers have 18.2 million people working for them in the food and fiber sector after the products leave the nation's farms and ranches. Of these 18.2 million workers, 1.3 million are employed in food processing, 2.7 million in manufacturing, 6.6 million in transportation, trade and retailing, 3.7 million in eating establishments, and 3.9 million in other related work (1987 figures, latest available). They're the people who make it possible for the housewife to choose between quick-frozen, dry-frozen, quick-cooking, ready-to-heat, ready-to-eat, and many other conveniences. Hand in hand with this transition, and accentuating the demand for convenience foods, more women work outside the home; the proportion of the nation's labor force made up of women rose from 28% in 1947 to 52.2% in 1987. All this is fine, but it must be realized that these people engaged in food preparation must be paid, for they want to eat, too.

Income—Proportion Spent for Food; Food—Kinds Bought. U.S. consumers are among the most favored people in the world in terms of food costs and the variety of food products available. In 1988, only 14.5% of the U.S. disposable income was spent for food; 18.8% of this was spent for red meats, with 9.5% of the red meat share spent for beef and 1.8% of it spent for pork.

Food takes about ⅕ of the income in most other developed countries, and ½ or more of the income in most of the developing countries.

(Also see INCOME, PROPORTION SPENT FOR FOOD.)

Selecting the Meat Market.

Fig. M-30. The way it used to be done. Early-day butcher shop, from a painting by W. S. Mount. There were neither wholesale cuts nor grades. (Courtesy, The Bettmann Archive)

Today's family may need help in meat buying. Meat is available in many places, in both supermarkets and meat specialty stores. Also, there is a selection from different species; and meat comes in many forms—fresh, cured, cured and smoked, frozen, freeze-dried, canned, and ready-to-serve. Then, there is the matter of deciding on which of the popular cuts to buy and serve on different occasions. All this leads most consumers to shop around until they find a retailer who provides the quality of meat and customer services desired, then remain loyal to that food store.

Selecting the Cut of Meat. The open refrigerated cases in a modern supermarket or meat specialty store present a sea of appetizing meats and meat products. Generally, meats are displayed wrapped in polyvinyl chloride (PVC) film that is sealed, on an affixed label on which is stamped the net weight, total price, price per pound, and the name of the cut.

The arrangement for displaying meats is rather standard. They are segregated as to kind (beef, veal, pork, lamb, poultry) and type of cut (steak, roast, ground beef, etc.). Smoked meats have separate display space as do fish, liver and other specialty items. Usually a clerk behind the display cases adds a personal touch and furnishes a source of information.

As the consumer pushes the cart along the refrigerated cases, selection is based primarily on preference, satiety, and nutritional value, along with the following practical considerations:

Fig. M-31. Consumer preference determines sales. (Courtesy, University of Tennessee, Knoxville, Tenn.)

1. The kind of meat that is best for the use planned. Skill in identifying cuts of meat and in cooking each of them properly is imperative.

2. The best buy on the basis of cost per serving. Except for such items as chopped meat and stew, the price per pound includes bone and fat, which are not eaten. Sometimes cuts like spareribs seem relatively inexpensive, but the amount of edible meat is small.

(Also see BEEF AND VEAL, section headed "Beef Cuts and How to Cook Them"; PORK, section headed "Pork Cuts and How to Cook Them"; and LAMB AND MUTTON, section headed "Lamb Cuts and How to Cook Them.")

Watching for Specials. Most meat retailers feature a good many meat specials. Experienced homemakers watch for these announcements and take advantage of them. When a home freezer is available, it may be possible to effect additional savings by buying extra large amounts of meat during special sales, then storing them.

Deciding How Much to Buy. Each meat shopper needs to be able to estimate how much meat, fish, or poultry to buy. Consideration should be given to the appetites and preferences of the members of the family. For example, active men and teen-agers generally have hearty appetites, and one child may like pork chops while another prefers hamburgers.

Other factors that should be considered in determining how much meat to buy are:

1. Time available for preparation.
2. The practicality of buying a particular cut for more than one meal.
3. Storage facilities.

MEAT COOKING—Every grade and cut of meat can be made tender and palatable provided it is cooked by the proper method. Also, it is important that meat be cooked at low temperature, usually between 300° and 350°F (*149 and 177°C*). At this temperature, it cooks slowly; and, as a result, it is juicier, shrinks less, and is better flavored than when cooked at high temperatures.

The method used in meat cookery depends on the nature of the cut to which it is applied. The common methods are:

1. **Dry-heat cooking**—Dry-heat cooking is used in preparing the more tender cuts; those that contain little connective tissue. The common methods of cooking by dry-heat are: (a) roasting, (b) broiling, and (c) panbroiling (see Fig. M-32).

Fig. M-32. Common methods of meat cookery. *Dry-heat cooking:* A, roasting; B, broiling; and C, panbroiling. *Moist-heat cooking:* D, braising; and E, cooking in liquid.

a. **How to roast:**
(1) Season with salt and pepper, if desired.
(2) Place fat side up on rack in open roasting pan.
(3) Insert thermometer.
(4) Roast in oven at 300° to 350°F (*149 to 177°C*).
(5) Do not add water, nor cover, nor baste.
(6) Roast until the meat thermometer registers rare, medium, or well-done, as desired.

For best results, a meat thermometer should be used to test the doneness of roasts (and also for thick steaks and chops). It takes the guess work out of cooking. Allowing a certain number of minutes to the pound is not always accurate; for example, rolled roasts take longer to cook than ones with bones.

The thermometer is inserted into the cut of meat so that the bulb reaches the center of the largest muscle, and so that it is not in contact with fat or bone. Naturally, frozen roasts need to be partially thawed before the thermometer is inserted, or a metal skewer or ice pick will have to be employed in order to make a hole in frozen meat.

As the oven heat penetrates, the temperature at the center of the meat gradually rises and is registered on the thermometer. The meat can be cooked as desired—rare, medium, or well-done, except for pork which should always be cooked well-done (160° to 170°F [*71° to 77°C*] for cured pork; 185° [*85°C*] for fresh pork).

b. **How to broil:**
(1) Set the oven regulator for broiling.
(2) Place meat on the rack of the broiling pan and cook 2 to 5 inches from heat.
(3) Broil until the top of meat is brown.
(4) Season with salt and pepper.
(5) Turn the meat and brown the other side.
(6) Season and serve at once.

Fig. M-33. Beef Kabobs may be broiled or barbecued, forms of dry-heat cooking. (Courtesy, National Live Stock & Meat Board, Chicago, Ill.)

Fig. M-34. Barbecued steaks, a form of dry-heat cooking. (Courtesy, USDA)

c. **How to panbroil:**
(1) Place meat in a heavy, uncovered frying pan. Cook slowly.
(2) Do not add fat or water.
(3) Turn at intervals to ensure even cooking.
(4) As fat accumulates, pour it off.
(5) Brown meat on both sides.
(6) Do not overcook. Season.

2. **Moist-heat cooking**—Moist-heat cooking is generally used in preparing the less tender cuts, those containing more connective tissues that require moist heat to soften them and make them tender. In this type of cooking, the meat is surrounded by hot liquid or by steam. The common methods of moist-heat cooking are: (a) braising, and (b) cooking in liquid (see Fig. M-32).

a. **How to braise:**

(1) Brown the meat on all sides in a small amount of hot fat in a heavy utensil.

(2) Season with salt and pepper.

(3) Add small amount of liquid, if necessary.

(4) Cover tightly.

(5) Cook at simmering temperature, without boiling until tender.

b. **How to cook in liquid (large cuts and stews):**

(1) Brown on all sides in hot fat, if desired.

(2) Season with salt and pepper, if desired.

(3) Cover with water and cover kettle tightly.

(4) Cook slowly (simmer but not boil) until done.

(5) Add vegetables just long enough before serving to be cooked, if desired.

MEAT CARVING. *Carving is the art of cutting up meat, poultry, or game to serve at the table.*

Carving should be a proud accomplishment rather than drudgery. To master carving for the best presentation and maximum yield, the following simple rules should be observed:

Fig. M-35. Roast beef on the table. (Courtesy, National Live Stock and Meat Board, Chicago, Ill.)

1. **Have proper tools.** Most carving needs can be met with a standard carving set and a steak set.

2. **Use sharp knife.** Always use a sharp carving knife (and never sharpen it at the table).

3. **Stand up if you prefer.** It is perfectly proper for the carver to stand if he prefers to do so.

4. **Know bone structure.** The carver should know something about anatomy; otherwise, the bones will get in his way.

5. **Know direction of muscle fiber and cut across the grain.** Except for steaks, meat should always be cut across the grain to avoid long meat fibers giving a stringy texture to the slices.

6. **Have a plan.** Start with a plan, cut with a plan, and make neat slices.

7. **Have ample platter or carving board.** The carving platter or board should be of ample size; large enough to accommodate slices.

8. **Have elbow room.** The carver must have plenty of elbow room.

9. **Use fork to hold cut.** The fork should be used for its intended purpose—to hold the cut; and not to dull the knife.

10. **Cut large and even slices.** The slices should be as large and even as possible.

11. **Be at ease.** The carver should always appear at ease. If the carver is nervous, the guests will be nervous, too.

U.S. MEAT PRODUCTION. U.S. production of red meat (total and by kinds) in 1990 follows:

	Million pounds
Total all meat	38,785
Beef	22,743
Pork, excluding lard	15,353
Veal	327
Lamb and mutton	362

U.S. PER CAPITA MEAT CONSUMPTION. Although comprising only 5.0% of the world's population, the people of the United States consume 10.5% of the total world production of meat. The amount of meat consumed in this country varies from year to year (see Fig. M-36).

Fig. M-36. Per capita meat consumption in the United Sates, by kind of meat. As noted, the amount of meat consumed in this country varies from year to year. In recent years, the average American has consumed more beef than any other kind of meat. (Based on USDA figures)

In 1990 the average per capita red meat consumption was 163.2 lb, with distribution by types of meat as shown in Fig. M-36.

For the most part, meat consumption in this country is on a domestic basis, with only limited amounts being either imported or exported. Although cured meats furnish somewhat of a reserve supply—with more meats going into cure during times of meat surpluses—meat consumption generally is up when livestock production is high. Also, when good crops are produced and feed prices are favorable, market animals are fed to heavier weights. On the other hand, when feed-livestock ratios are unfavorable, breeding operations are curtailed, and animals are marketed at lighter weights. But during the latter periods, numbers are liquidated, thus tending to keep the meat supply fairly stable.

EATING HABIT TRENDS.
Figs. M-37 and M-38, along with other data from the same source, clearly point up the following U.S. trends in eating habits in per capita consumption of meat and other foods by groups.

● The consumption of crop products is increasing, while the consumption of animal products is decreasing.

● Consumption of poultry and fish is on the increase; consumption of eggs continues to decrease; consumption of meat and dairy products has leveled off.

● Consumption of fruits stays about the same, vegetables show a slight increase, and cereal products continue to rise.

● Total consumption of sugars and sweeteners is rising again after a flat period during the 1970s. The consumption of sugar has dropped because it has been replaced by corn sweeteners, which are used primarily in soft drinks.

● Consumption of coffee, tea, and cocoa has declined rather sharply.

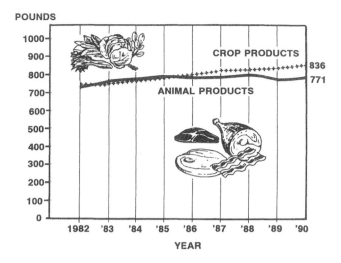

Fig. M-38. Per capita consumption of animal products vs crop products. (*Agricultural Statistics 1991*, USDA, p. 479, Table 672)

● Consumption of vegetable oils has increased, while consumption of animal fats has fallen.

The selection of food is based on taste preference, relative prices, dietary guidelines issued by government agencies and scientific groups, and myths.

WORLD MEAT PRODUCTION AND CONSUMPTION.
In general, meat production and consumption are highest in those countries which have extensive grasslands, temperate climates, well-developed livestock industries, and sparse populations. In many of the older and more densely populated regions of the world, insufficient grain is produced to support the human population even when consumed directly. This lessens the possibility of keeping animals, except for consuming forages and other humanly inedible feeds. Certainly, when it is a choice between the luxury of meat and animal by-products or starvation, people will elect to accept a lower standard of living and go on a grain diet. In addition to the available meat supply, food habits and religious restrictions affect the kind and amount of meat produced and consumed.

Table M-12 shows the per capita consumption, by type, of the leading red meat-eating countries of the world.

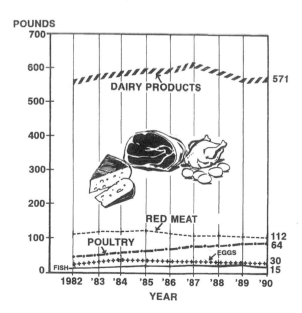

Fig. M-37. Per capita consumption of red meat, fish, poultry, eggs, and dairy products. (*Agricultural Statistics 1991*, USDA, p. 479, Table 672)

TABLE M–12
LEADING RED MEAT-EATING COUNTRIES OF THE WORLD[1]

Country	Beef and Veal Quantity[2]	Country	Pork Quantity[2]
	(lb)		(lb)
1. Argentina	143.3	1. Hungary	168.7
2. Uruguay	127.0	2. Denmark	147.7
3. U.S.A.	97.2	3. Czechoslovakia ...	130.3
4. Australia	88.9	4. Austria	116.0
5. Canada	86.2	5. Germany	115.8

[1]*Statistical Abstract of the United States 1991*, U.S. Dept. of Commerce, p. 843, Table 1452.
[2]Pounds per capita. To convert to kilograms divide by 2.2.

As shown in Table M-12, in 1990 the United States with a per capita consumption of 97.2 lb *(44.2 kg)* ranked third in consumption of beef and veal, edging out the Australians who averaged 88.9 lb *(40.4 kg)*. But the United States was exceeded in per capita beef and veal consumption by Argentina, with 143.3 lb *(65.1 kg)*; and Uruguay, with 127 lb *(57.7 kg)*.

In per capita pork consumption, the five leading countries in 1990 were: Hungary, 168.7 lb *(76.7 kg)*; Denmark, 147.7 lb *(67 kg)*; Czechoslovakia, 130.3 lb *(59 kg)*; Austria, 116 lb *(52.7 kg)*; and Germany, 115.8 lb *(52.6 kg)*. The United States ranked twenty-fourth in per capita pork consumption.

The per capita red meat consumption of the leading meat-eating countries changes from time to time. Thus, from 1965 to 1968, New Zealand was the world's largest per capita consumer of red meat, followed closely by Australia. In 1969, Australia took the lead. In 1989, the leading countries in red meat consumption by rank were: East Germany, Uruguay, Czechoslovakia, New Zealand, Australia, and the United States.

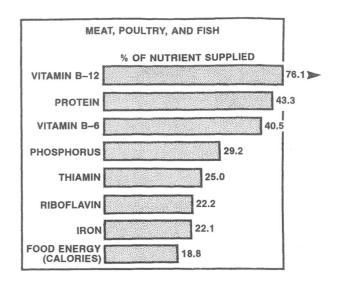

Fig. M-40. Contribution of meat, poultry, and fish combined, to the dietary intake in the United States. (*Agricultural Statistics 1991*, USDA, pp. 474-479)

NUTRITIVE COMPOSITION OF MEATS.

Meat is far more than just a very tempting and delicious food. From a nutritional standpoint, it contains the essentials of an adequate diet. This is important, for how we live and how long we live are determined in large part by our diet.

Meat is an excellent source of high-quality protein, of certain minerals, especially iron, and of the B-complex vitamins. It supplies phosphorus and nutrients which contribute significantly to the dietary balance of meals, and it is easily digested.

Although the RDAs are believed to be more than adequate to meet the requirements of most consumers, it is recognized that individuals vary greatly in their requirements. So, Fig. M-40 presents data showing the contribution of meat, poultry, and fish combined, toward the total dietary intake of each of the nutrients.

As shown in Fig. M-40, consumers obtain large proportions of their vitamin B-12, protein, Vitamin B-6, and iron from meat, poultry, and fish. Additionally, meat, poultry, and fish are fair sources of phosphorus, food energy, riboflavin, and thiamin.

The nutritive qualities of meats are detailed in the sections that follow.

Energy. The energy value of meat is largely dependent upon the amount of fat that it contains.

Today, meat-poultry-fish combined supply 18.8% of the dietary energy intake in the United States. Thus, meat is not a major contributor to excess energy intake.

(Also see CALORIC [ENERGY] EXPENDITURE; and ENERGY UTILIZATION BY THE BODY.)

Carbohydrate. Although the carbohydrates provide a major source of energy for man, they are found only in very limited amounts in meats and other animal products.

About half the carbohydrate is distributed through the muscles and in the bloodstream; the other half is stored in the liver in the form of glycogen (animal starch), where it constitutes 3 to 7% of the weight of that organ. Yet, most animal products contain little, if any, carbohydrate, for the reason that when an animal is slaughtered, the glycogen stored in the liver and muscles is rapidly broken down to lactic acid and pyruvic acid. Oysters and scallops contain some glycogen, but the amount is not significant to the diet. Milk, which contains the carbohydrate lactose or milk sugar, is the only animal food of importance as a carbohydrate source.

Only 0.4 g of carbohydrate comes from red meat daily, with most of it derived from the glycogen and reducing tissues naturally present in the tissues. Neither meat nor meat-poultry-fish make any significant contribution to total carbohydrate consumption; in each case they account for only 0.1%. A small amount of carbohydrate in the diet also

Fig. M-39. Rib eye steak. (Courtesy, National Live Stock and Meat Board, Chicago, Ill.)

comes from sugars and other carbohydrates added to sausages. The low level of carbohydrate in meat helps to keep its caloric contribution to a minimum. On the other hand, there is evidence that complex carbohydrates, i.e., fiber, contribute to bowel movement and the elimination of toxic substances from the digestive tract. Thus, it is recommended that fruit and vegetables be included in the diet to provide fiber.

(Also see CARBOHYDRATE[S].)

Fats. Energy is supplied by animal fats, which are highly digestible. Fats also supply needed fatty acids, transport fat-soluble vitamins (A, D, E, K), provide protection and insulation to the human body, and add palatability to lean meat. Highly unsaturated fats are soft and oily (as in soft pork). Generally, animal fats contain higher levels of saturated fat than vegetable oils.

The amount and composition of fats in the animal body are highly variable, depending upon the species, diet, and maturity. The species differences in the composition of fats are particularly noticeable between ruminants and nonruminants; ruminants (through rumen microorganisms) may alter the composition of the dietary fats, whereas nonruminants tend to deposit lipids in the form found in the diet. Australian researchers have developed a procedure for protecting dietary fats from the action of the rumen organisms by encapsulation of the fat, thus permitting alteration of the characteristics of fatty acid composition in both meat and milk.

The fats of the diet serve as a source of the essential fatty acids—arachidonic, linoleic, and linolenic acid, but only linoleic acid is required preformed in the diet. Although meats provide fat, the vegetable oils are generally richer sources of the essential fatty acids than animals fat; hence, no deficiency of fatty acids is likely to develop in a meatless diet unless there is impaired lipid absorption.

There is no RDA for fat consumption since there is seldom a deficiency. Rather, the main problem from fat consumption is its contribution to obesity.

Meat-poultry-fish combined contribute 53.3 g of fat per capita per day and account for 31.7% of the total consumption.

(Also see FATS AND OTHER LIPIDS.)

Protein. The need for protein in the diet is actually a need for amino acids, of which nine are essential for humans. (Ten are essential for animals; they require arginine, which is not required by humans.) But all proteins are not created equal! Some proteins of certain foods are low or completely devoid of certain essential amino acids. Hence, meeting the protein requirement demands quality as well as quantity. It is noteworthy that all the essential amino acids are present in muscle meats, heart, liver, and kidney. However, there is considerable variation in the protein content of meat, depending on the degree of fatness and the water content; the total protein content of animal bodies ranges from about 10% in very fat, mature animals to 20% in thin, young animals.

Because animal proteins are in short supply in the developing countries, it has been estimated that 20 to 30% of the children in these countries suffer severe protein-calorie malnutrition, and that it may contribute to as much as 50% of the mortality for this group. Kwashiorkor and marasmus are common manifestations of protein and protein-calorie deficiencies in infants and children. Also, the ability of the human body to produce antibodies (substances that attack specific foreign bodies) is dependent upon an adequate supply of amino acids in the diet. (The antibody molecule is actually a molecule of globulin—a class of protein.)

Fig. M-40 shows that meat-poultry-fish supply 43.3% of the dietary protein intake in the United States.

In spite of the high level of high-quality protein in the average American diet, there is evidence that some protein deficiencies occur in young children shortly after weaning. In these cases, it appears that protein intake is low due either to their economic status or a strict vegetarian diet. Regardless of the reason for the deficiency, inclusion of small amounts of meat or other animal protein would alleviate all signs of protein deficiency.

(Also see PROTEIN[S].)

Minerals. Meat is a rich source of several minerals, but it is an especially rich source of phosphorus and iron.

In the discussion that follows, minerals are divided into macrominerals or microminerals, based on the relative amounts needed in the diet.

(Also see MINERAL[S].)

MACROMINERALS. The macrominerals, or major minerals, are those that are needed in greatest abundance.

• **Calcium**—Meat is a poor source of calcium. However, it is noteworthy that mechanically deboned meat, which contains some bone, contains up to 0.75% calcium, and that it can make a considerable contribution to the calcium of the diet of man.

(Also see CALCIUM.)

• **Phosphorus**—Although meat is a poor source of calcium, it is usually a good source of phosphorus. Meat-poultry-fish combined contribute 56.2% of the RDA.

Meat Is A Good Source of Phosphorus

Low phosphorus diet
(no meat)
Wt. 168 grams

Same diet plus meat
Wt. 225 grams

Fig. M-41. This shows that meat is a rich source of phosphorus. Minerals are necessary in order to build and maintain the body skeleton and tissues and regulate body functions. Although meat is a good source of several minerals, it is especially rich as a source of phosphorus and iron. (Courtesy, National Live Stock and Meat Board and Rush Medical College)

Meat-poultry-fish combined provide 29.2% of the phosphorus intake in the United States (Fig. M-40).

(Also see PHOSPHORUS.)

• **Sodium**—Most of the sodium in the diet is in the form of salt (sodium chloride—NaCl). Deficiencies of sodium are rare because nearly all foods contain some sodium. However, excess dietary salt over a long period of time is believed to lead to high blood pressure in susceptible people. Also, excess salt may cause edema. So, a prudent diet should limit excess salt intake; this would call for limiting the intake of cured meats.

(Also see SODIUM.)

• **Chlorine**—Chlorine is provided by table salt (sodium chloride—NaCl) and by the foods that contain salt. Persons suffering from diseases of the heart, kidney, or liver, whose sodium intake is severely restricted by deleting table salt and such salty foods as cured meats, may need an alternative source of chloride; a number of chloride-containing salt substitutes are available for this purpose.

(Also see CHLORINE or CHLORIDE.)

• **Magnesium**—Meat, poultry, and fish are low in magnesium. Most of the magnesium in meat is found in the muscle and liver.

(Also see MAGNESIUM.)

• **Potassium**—Potassium is the third most abundant element in the body, after calcium and phosphorus; and it is present in twice the concentration of sodium.

Potassium is widely distributed in foods. Meat, poultry, fish, many fruits, whole grains, and vegetables are good sources. (Although meat, poultry, and fish are good sources of potassium, these items may be restricted when dietary sodium is restricted.)

(Also see POTASSIUM.)

• **Sulfur**—Meeting the sulfur needs of the body seems to be primarily a matter of providing sulfur for the sulfur-containing amino acids: methionine, cystine, and cysteine.

Inorganic sulfur is poorly utilized by man. Thus, meat, poultry, and fish, which are good sources of sulfur, serve an important function in supplying the body with its sulfur requirements.

(Also see SULFUR.)

MICROMINERALS. The microminerals, or trace or minor minerals, are those that are needed in least amounts.

• **Chromium**—Chromium, which has a number of important functions in the human body, is widely distributed in animal tissues and in animal products. In the section on Chromium, Table C-16, of this book, the chromium content of the following foods of animal origin, in mcg/100 g is reported: dried liver, 170; egg, 52; cheese, 51; fresh liver, 50; and beef, 32.

(Also see CHROMIUM.)

• **Cobalt**—This mineral, which is an integral part of vitamin B-12—an essential factor in the formation of red blood cells, must be ingested in the form of vitamin B-12 in order to be of value to man; hence, a table showing the cobalt content of human foods serves no useful purpose. The organ meats (liver, kidney) are excellent sources of vitamin B-12 (hence, of cobalt). The vitamin B-12 content, in mcg/100 g, of some rich animal food sources follows: beef liver, 111; clams, 98; lamb kidneys, 63; turkey liver, 48; and calf kidney, 25.

(Also see COBALT.)

• **Copper**—Organ meats (liver, kidney, brains) and shell fish are rich sources of copper.

(Also see COPPER.)

• **Fluorine**—Fluorine, which is necessary for sound bones and teeth, is widely, but unevenly, distributed in nature. Seafoods and tea are the richest dietary sources.

Because fluorine tends to concentrate in the bones and teeth and is toxic if used in excess, there has been some concern about fluorine levels in deboned meat. Yet, there seems to be little basis for such apprehension unless the bones come from animals suffering from fluorosis.

(Also see FLUORINE.)

• **Iodine**—Most of the iodine need of humans is met by iodized salt. Saltwater fish and most shellfish are rich sources of iodine. Dairy products and eggs may be good sources if the producing animals have access to iodine-enriched rations. Red meats and poultry are poor sources of iodine.

(Also see IODINE.)

• **Iron**—Liver is an excellent source of easily assimilated iron and is prescribed in the diet of anemia sufferers. Muscle meat, poultry, and fish are good sources of iron, but they contain less than half as much iron as liver.

There are two forms of food iron—heme (organic) and nonheme (inorganic). Of the two, heme is absorbed from food more efficiently than inorganic iron and is independent of vitamin C or iron binding chelating agents. Although the proportion of heme iron in animal tissue varies, it amounts to about ⅓ of the total iron in all animal tissues—including meat, liver, poultry, and fish. The remaining ⅔ of the iron in animal tissues and all the iron of vegetable products are treated as nonheme iron.

Meat also enhances the absorption of iron, as does vitamin C; it improves iron absorption by supplying what is referred to as the *meat factor*. It has been demonstrated that beef, veal, lamb, pork, liver, chicken, and fish all increase iron absorption by two to fourfold, while milk, cheese, and egg do not increase iron absorption.

Fig. M-40 shows that meat-poultry-fish combined provide 22.1% of the dietary iron intake in the United States.

There are four situations in life in which the iron requirement is increased (according to *Recommended Dietary Allowances, NRC-National Academy of Sciences,* 1980, p. 186): (1) in infancy; (2) during rapid growth and adolescence; (3) during the female reproductive period; and (4) in pregnancy. During these four stages in life, the iron status could be greatly improved by including meat, poultry, and fish in the diet on a regular basis.

(Also see IRON.)

• **Manganese**—Only a trace of manganese is found in muscle meat. Liver is a fair source.

(Also see MANGANESE.)

• **Molybdenum**—The organ meats (liver, kidney) are a good source of molybdenum.

(Also see MOLYBDENUM.)

• **Selenium**—Meat, poultry, and fish all contain small amounts of selenium, but liver and kidney are the best sources. The selenium content of animal products depends upon the content of the element in livestock feeds; for example, the Ohio Station reported that the addition of 0.1 ppm selenium to the animal's diet increased the selenium content of beef liver by 72%.[18]

[18]Moxon, A. L. and D. L. Palmquist, "Selenium Content of Foods Grown or Sold in Ohio," *Ohio Report,* Ohio State University, Columbus, Ohio, January-February 1980, pp. 13-14, Table 1.

Fortunately, foods eaten in the United States are generally varied in nature and origin; so, there does not appear to be any deficiency of selenium in man. However, a deficiency of selenium has been reported in a localized area in China where only locally produced foods are consumed.
(Also see SELENIUM.)

• **Silicon**—Meat, poultry, and fish all contribute silicon to the diet in appreciable quantities. However, the best sources of silicon are the fibrous parts of whole grains, followed by various organ meats (liver, kidney, brain), and connective tissues.

Because of the abundance of silicon in all but the most highly refined foods, it is of little concern in the human diet.
(Also see SILICON.)

• **Zinc**—Meat, poultry, fish, eggs, and dairy products are good sources of zinc. The animal proteins are generally much better sources of dietary zinc than plants because the phytate present in many plant sources complexes the zinc and makes it unavailable.

Since meat, poultry, fish, eggs, and dairy products are good sources of zinc, they should be included regularly in the diet. This recommendation is further reinforced by the fact that zinc deficiency is virtually unknown on a diet containing animal products.
(Also see ZINC.)

Vitamins. Many phenomena of vitamin nutrition are related to solubility—vitamins are soluble in either fat or water. This is a convenient way in which to discuss vitamins in meats; hence, they are grouped and treated as fat-soluble vitamins or water-soluble vitamins.
(Also see VITAMIN[S].)

FAT-SOLUBLE VITAMINS. The fat-soluble vitamins are stored in appreciable quantities in the animal body, whereas the water-soluble vitamins are not. Any of the fat-soluble vitamins can be stored wherever fat is deposited; and the greater the intake of the vitamin, the greater the storage. It follows that most meats, poultry, and fish are good sources of the fat-soluble vitamins.

• **Vitamin A**—Both vitamin A and the carotenes are found in the animal body associated with fats (lipids).

Cod and other fish liver oils are extremely high in vitamin A, sufficiently so that they are used as supplemental sources. Livers of all kinds are rich food sources of vitamin A, exceeding raw carrots.
(Also see VITAMIN A.)

• **Vitamin D**—Fish liver oils (from cod, halibut, or swordfish) are very high in vitamin D, sufficiently so that they are used as supplemental sources. Fatty fish and fish roe are rich food sources, while liver, egg yolk, cream, butter, and cheese are fair sources. Muscle meats and unfortified milk are negligible sources.
(Also see VITAMIN D.)

• **Vitamin E**—Meat and animal products are good to fair sources of vitamin E (the tocopherols); plant materials are much richer sources, especially vegetable oils, some grains, nuts, and green leafy vegetables.

The amount of vitamin E in animal tissues and products is influenced by the dietary consumption of tocopherols by the animal. Good animal sources are: beef and organ meats, butter, eggs, and seafoods. Cheese and chicken are fair sources.
(Also see VITAMIN E.)

• **Vitamin K**—Certain animal foods are good to fair sources of vitamin K, although they are far outranked by green tea, turnip greens, and broccoli. Beef liver is a rich source of vitamin K; bacon, cheese, butter, and pork liver are good sources; and beef fat, ham, eggs, pork tenderloin, ground beef, and chicken liver are fair sources.
(Also see VITAMIN K.)

WATER-SOLUBLE VITAMINS. The water-soluble vitamins are not stored in the animal body to any appreciable extent.

• **Biotin**—Biotin is widely distributed in foods of animal origin, with liver and kidney being rich sources, followed by eggs, sardines and salmon, cheese, chicken, oysters, and pork.
(Also see BIOTIN.)

• **Choline**—All animal tissues contain some choline, although it is more abundant in fatty than in lean tissues. Rich sources are: eggs, liver, and dried buttermilk.
(Also see CHOLINE.)

• **Folacin (folic acid)**—Folacin is widely distributed in animal foods, with liver and kidney being particularly rich sources. Eggs and fish are good sources, while cheese, cod, and halibut are fair sources. Chicken, milk, and most muscle meats are poor sources of folacin.
(Also see FOLACIN [FOLIC ACID].)

• **Niacin (nicotinic acid; nicotinamide)**—Generally speaking, niacin is found in animal tissue as nicotinamide. Animal foods are excellent dietary sources, with the richest sources being liver, kidney, lean meats, poultry, fish, and rabbit. Although low in niacin content, milk, cheese, and eggs are good sources because (1) of their tryptophan content (which may yield niacin), and (2) of their niacin being in available form.

The role of niacin-rich animal foods in eliminating pellagra, which was once a serious scourge in the United States, is generally recognized.
(Also see NIACIN.)

• **Pantothenic acid (vitamin B-3)**—Pantothenic acid is widely distributed in animal foods, with organ meats (liver, kidney, and heart) being particularly rich sources. Salmon, blue cheese, eggs, and lobster are good sources, while lean muscle and chicken are only fair sources.
(Also see PANTOTHENIC ACID.)

• **Riboflavin**—The organ meats (liver, kidney, heart) are rich sources of riboflavin; lean meat (beef, pork, lamb), cheese, eggs, and bacon are good sources; while chicken and fish are only fair sources.

Meat-poultry-fish combined supply 22.2% of the per capita daily consumption of riboflavin (Fig. M-40), and 31.3% of the RDA of riboflavin.
(Also see RIBOFLAVIN.)

• **Thiamin (vitamin B-1)**—Some thiamin is found in a large variety of animal products, but it is abundant in few.

Lean pork (fresh cured) is a rich source; kidney is a good source; while egg yolk, poultry, beef liver, luncheon meat, and fish are only fair sources. The thiamin content of pork

MEAT IS A GOOD SOURCE OF THIAMIN (VITAMIN B-1)

RAT NO. 899 RECEIVED ADEQUATE DIET EXCEPT FOR THIAMIN
RAT NO. 1805 RECEIVED SAME DIET + 2% DRIED PORK HAM WHICH IS RICH IN THIAMIN

Fig. M-42. Meat is a good source of thiamin (vitamin B-1). (Studies by the University of Wisconsin; supported by National Live Stock and Meat Board)

can be greatly altered by the thiamin level of the feed. Curing of meat causes only small losses.

Meat-poultry-fish combined contribute approximately 25% of the total thiamin intake (Fig. M-40).

(Also see THIAMIN.)

• **Vitamin B-6 (pyridoxine; pyridoxal; pyridoxamine)**—In animal tissues, vitamin B-6 occurs mainly as pyridoxal and pyridoxamine. Vitamin B-6 is widely distributed in foods of animal origin. Liver, kidney, lean meat, and poultry are good sources; eggs are a fair source; and fat, cheese, and milk are negligible sources.

Meat-poultry-fish combined contribute 40.5% of the total vitamin B-6 intake (Fig. M-40).

(Also see VITAMIN B-6.)

• **Vitamin B-12 (cobalamins)**—Vitamin B-12 is found in all foods of animal origin. Plants cannot manufacture vitamin B-12; hence, except for trace amounts absorbed from the soil (because of soil bacteria, soil is a good source of B-12) by the growing plant, very little is found in plant foods.

Liver and other organ meats—kidney, heart—are rich sources of vitamin B-12; muscle meats, fish, shellfish, eggs, and cheese are good sources; and milk, poultry, and yogurt are fair sources.

Fig. M-40 shows meat-poultry-fish contribute 76.1% of the consumption of vitamin B-12.

(Also see VITAMIN B-12.)

• **Vitamin C (ascorbic acid)**—Vitamin C is found only in small amounts in animal tissues, with the largest amounts being localized in the adrenal glands. Vitamin C is added to cured meats at a level of 550 ppm to reduce nitrosamine formation.

Summary. The unique contributions of meat, poultry, and fish to the nutritional needs of man have been presented in this section. In summary form, the major nutritive

contribution of foods of animal origin are:

1. They are an excellent source of protein, from the essential amino acids.

2. They are rich sources of phosphorus, iron, copper, and zinc.

3. They are the major dietary source of vitamin B-12 and vitamin B-6, and they supply appreciable amounts of vitamin A, biotin, niacin, pantothenic acid, riboflavin, and thiamin.

MEAT FACTS AND MYTHS. Much has been spoken and written linking the consumption of meat to certain health related problems in humans, including heart disease, cancer, high blood pressure (hypertension), and harmful residues. A summary of four incorrect statements, along with the correct facts, follows:

• **Fact or Myth: Meat fats cause coronary heart disease**—Coronary heart disease (CHD) is the leading cause of death in the United States, accounting for one-third of all deaths, according to *The Surgeon General's Report on Nutrition and Health* (1988, p. 4, Table 2). The major form of CHD results from atherosclerosis, a condition characterized by fatty deposits in the coronary arteries. These deposits are rich in cholesterol, a complex fatlike substance. Also, in general, serum cholesterol levels are relatively high among individuals with atherosclerosis.

Fact: Three major factors are associated with the risk of coronary heart disease; namely, elevated blood pressure, cigarette smoking, and elevated serum cholesterol. A serum cholesterol level in excess of 280 mg/dl is considered a major risk for heart disease. The concentration of cholesterol in the blood is strongly affected by the degree of saturation of the dietary fat.

Professional groups suggest that many people would benefit (1) if the amount of fat in the diet were reduced from the present level of 37% to less than 30% of the total calories; (2) if the amount of dietary cholesterol were restricted to less than 300 mg per day; (3) if polyunsaturated fats were substituted for some of the saturated fat in the diet so that the distribution among polyunsaturated, monounsaturated, and saturated fatty acids would be about equal; (4) if saturated fat were limited to 10% of the total caloric intake; and (5) if caloric intake were adjusted to maintain desired body weight.

The major dietary sources of fat in the American diet are meat, poultry, fish, dairy products, and fats and oils. Animal products tend to be higher in both total and saturated fats than most plant sources. Also, dietary cholesterol is found only in foods of animal origin.

The intake and types of fatty acids that made up an average American diet in 1956 is shown in Table M-13.

Since the 1956 report by the U.S. Department of Agriculture (Table M-13), health and dietary professionals have urged the American people to (1) reduce the consumption of total fat, especially saturated fat and cholesterol; (2) increase the intake of fruits, vegetables, and whole grain products and cereals; and (3) increase the consumption of fish, poultry prepared without skin, lean meats, and low-fat dairy products. Dietary changes have been made since 1956, but further improvements will lessen the risk of coronary heart disease.

(Also see HEART DISEASE.)

• **Fact or Myth: Meat causes bowel cancer**—This question has been prompted by the following reports: (1) that the age-adjusted incidence of colon cancer has been found to increase with the per capita consumption of meat in countries; (2) that, in a study done in Hawaii, the incidence of colon cancer in persons of Japanese ancestry was found to be greater among those who ate Western-style meals, especially those who ate beef; and (3) that an examination of (a) international food consumption patterns, and (b) food consumption survey data from the United States showed that a higher incidence of colon cancer occurred in areas with greater beef consumption.

Fact: A direct cause-effect relationship between diet and cancer has not been established. Such studies as the three cited provide valuable leads for researchers who are trying to determine the cause of a certain disease such as colon cancer, but they do not establish the cause. The reason is that the factor measured and found associated with the incidence of colon cancer or other condition is not the only difference among the population groups studied, and the factor measured in the study may be only associated in some way with the real cause.

(Also see CANCER.)

• **Fact or Myth: Meat causes high blood pressure (hypertension)**— Some have implicated meat as a cause of high blood pressure.

Fact: There is no evidence that meat per se has any major effect on high blood pressure. However, consumption of cured meat containing large amounts of salt should be minimized as should the amount of salt used as a condiment on meat and other foods. Other dietary factors that contribute to high blood pressure include obesity and excessive intake of alcohol.

(Also see HIGH BLOOD PRESSURE [HYPERTENSION].)

• **Fact or Myth: Meat contains harmful residues**—Do meats contain harmful toxic metals, pesticides, insecticides, animal drugs and additives?

Fact: If one pushed the argument of how safe is "safe" far enough, it would be necessary to forbid breast feeding as a source of food, because, from time to time, human milk has been found to contain DDT, antibiotics, thiobromine, caffeine, nicotine, and selenium.

The FDA requires drug withdrawal times on some of them, in order to protect consumers from residues. Additionally, federal agencies (USDA and/or FDA), as well as certain state and local regulatory groups, conduct continuous surveillance, sampling programs, and analyses of meats and other food products on their content of drugs and additives.

(Also see FOOD MYTHS AND MISINFORMATION; and POISONS.)

TABLE M-13
AVERAGE CONSUMPTION, IN GRAMS PER PERSON PER DAY, OF FATTY ACIDS FROM FOODS THAT MADE UP AN AVERAGE AMERICAN DIET AS REPORTED BY THE U.S. DEPARTMENT OF AGRICULTURE IN 1956[1]

Commodity	Total Saturated Acids	Oleic Acid	Other Monounsaturated Acids	Linoleic Acid	Other Polyunsaturated Acids
Dairy products (milk, cheese, butter)	28.2	7.5	3.0	0.8	0.4
Fats, oils	10.0	18.9	0.4	9.5	0.4
Flour, cereal products	0.7	1.5	—	2.2	0.2
Bakery products, purchased	2.7	3.7	—	2.4	0.3
Meats, poultry, eggs	14.1	17.0	0.8	3.8	0.6
Sugar, sweets	1.1	1.0	—	0.4	—
Potatoes	0.4	0.3	—	0.3	0.2
Fruits, vegetables	0.3	0.6	—	0.4	0.1
Miscellaneous foods	0.9	2.2	—	0.7	—
Totals	58.4	52.7	4.2	20.5	2.2
% of Total	42.3	38.2	3.0	14.9	1.6

[1]From: *Diet and Health*, Council for Agricultural Science and Technology, Report No. 111, March 1987, p. 25. The per capita fat consumption represented 96.4% of the fat consumed.

MEAT ANALOGS

Food material usually prepared from vegetable protein to resemble specific meats in texture, color, and flavor.
(Also see SOYBEANS; and VEGETARIAN DIETS.)

MEAT EXTRACT

This is a boiled down and concentrated extract of beef, veal, poultry, or game. Meat extracts should be regarded as condiments which impart the same flavor as the meat stock from which they were obtained.

MEAT FACTOR

A factor in meat which enhances the absorption of iron in the body. It has been demonstrated that beef, veal, pork, liver, chicken, and fish all increase iron absorption by two- to fourfold, while milk, cheese, and egg do not increase iron absorption.

MECHANICALLY EXTRACTED

A method of extracting the fat content from oilseeds by the application of heat and mechanical pressure. The hydraulic and expeller processes are both methods of mechanical extraction.
(Also see OILS, VEGETABLE.)

MEDICINAL PLANTS

Contents   Page
History...676
Plant Remedies to Modern Medicine..................677
Language of Plant Remedies.........................677
Common Medicinal Plants..........................678
Words of Caution................................678

Fig. M-43. A bouquet of familiar medicinal herbs.

Through the ages various plants have been tried as medicine to see which ones helped cure certain ailments of mankind. Thus, by trial and error, people came to use thousands of plants as remedies for their ills. For example, many American Indian tribes used willow bark to treat pain, but how the Indians came to choose willow bark is not known. Willow bark contains salicin, a forerunner of aspirin. While not all plants actually cured or relieved ailments, many did. Prescribing and dispensing medicinal plants, was the pharmaceutical industry of the past. Then, as science advanced, those plants which demonstrated positive effects on certain ailments were extracted, chemically analyzed, and their active chemicals were discovered. Thus, some traditional medicinal plants became stepping stones for the development of modern day drugs as active chemicals were isolated from plants or synthesized in laboratories. Although drugs used in medicine today are composed of specific chemicals extracted or derived from plants or synthesized by other methods, the direct use of plants for medicine still survives. There are endless books available describing the medicinal plants, their combinations, and conditions which they prevent, relieve, or cure. Sometimes it is called folk medicine, or herbal medicine. Indeed, whole businesses have grown up around the uses of herbs or medicinal plants. This article deals specifically with plants used directly for medicinal purposes.

HISTORY. Almost every civilization has a history of medicinal plants. Perhaps, one of the oldest civilizations to record their use of plants for medicine is China. Books describing the use of plants as medicine are sometimes called herbals, and one of the earliest known herbals was supposedly written by the Chinese emperor Shen-Nung about 2700 B.C. About the time of the Chin and Han dynasties (202 B.C. to 220 A.D.), the first Chinese book on pharmacology was written. In it, 365 medicinal substances were recorded, with notes on the collection and preparation of the drugs. During the Tang dynasty, the government established a pharmacological institute which was called the Herbary. A total of 850 medicinal plants were grown on its 50 acres of fertile land. The cultivation, collection, and preparation of drugs were taught as basic subjects in this institute. In his *Magnum Opus*, Li Shih-chen (1518 to 1593 A.D.), the pharmacologist, recorded 1,892 drugs. In China today, there is an attempt to mesh traditional Chinese medical practices with Western medical practices. Chinese traditional medical prescriptions are the roots, stems, bark, leaves, flowers, and fruits of shrubs and trees. A small part comes from birds, animals, insects, fish, and minerals. About 70% of the prescriptions made out in the hospitals and in the rural areas are for herbal medicine.

In India, the oldest sacred writings of the Hindus, the four Vedas, refer to many healing plants.

Ancient Egyptian doctors practiced a combination of herbal medicine and faith healing. Carvings on tomb and temple walls indicate that people used plants for medicine around 3000 B.C., while a document written about 1500 B.C. described more than 800 remedies.

Hippocrates, the Father of Medicine, and other Greeks probably copied some of their plant lore from the Egyptians. About 370 B.C., Theophrastus, a pupil of Plato, wrote *An Enquiry Into Plants*, which contained a section on the medicinal properties of plants.

Dioscorides, a Greek surgeon in the Roman army, recorded the combined Egyptian, Greek, and Roman knowledge on the properties of plants. Much of the work of Dioscorides remained the basis of European medical practice for nearly 1500 years.

During the Middle Ages, Moslem pharmacists prepared herbal medicines for the disease-plagued Europeans. Men paid fantastic prices for these remedies since the Arabs controlled the drug and spice trade.

With the invention of printing, all knowledge became more widespread, and some of the first works printed following the *Bible* were books dealing with medicinal plants. In 1471, *De Agricultura* by Peter Crescentius, was published, containing complete, woodcut illustrated references to the useful and medicinal plants of southern Europe. More and more publications followed, and in 1597 *Gerard's Herbal* was published in England. Therefore, explorers of the New World must have been well versed in medicinal plants.

In the New World, Spanish explorers found the Incas and Aztecs using medicinal plants. The British and French explorers found the American Indians using still other plant remedies. As these plants were brought back to Europe, they were added to the European medicine chest and studied.

As the New World was settled, the herbal knowledge of the English soon meshed with that of the Indians who depended upon their medicine men. Benjamin Rush, a Philadelphia physician, investigated and wrote about the Indian cures. Soon, other compilers began to publish guides to Indian medicines during the early 1800s.

In the late 18th century and early 19th century, medicinal plants were primarily available to those who collected them in the wild. However, members of the Church of the United Society of Believers—the Shakers—were the first to mass produce herbs. By 1857, a Shaker settlement in New Lebanon, New York produced tons of medicinal plants which were shipped to every state and to England and Australia.

Gradually, as curiosity increased, scientists studied these medicinal plants handed down through the ages. They sought to discover the chemicals responsible for the healing properties of medicinal plants.

But herbal medicine still lives in this day and age of antibiotics, antibodies, vaccines, and synthetic chemicals! In fact, there seems to have been a revitalization of herbal medicine, with numerous new books printed and old books reproduced on the subject, and with increased sales of herbs and mixtures of herbs to cure, relieve, or prevent almost every ailment.

PLANT REMEDIES TO MODERN MEDICINE.

As scientists studied the medicinal plants, they isolated the active chemicals, which could then be put into liquids or pills. Furthermore, once the active chemicals were discovered, the hope was that these chemicals could be synthesized in the laboratory, thus eliminating the need for collecting and extracting tons of plants to meet medicinal needs. Discovery of some active chemicals did lead to their laboratory synthesis, while some drugs still must come from plants. Examples of some important chemicals derived from plants include (1) the pain killer morphine from the opium poppy, (2) the tranquilizer reserpine from the snakeroot plant (rauwolfia), (3) the anesthetic cocaine from the coca plant, (4) the muscle relaxant and deadly poison curare from the curare vine, (5) the first malaria treatment, quinine from the bark of the cinchona tree, (6) the heart stimulant digitalis from the foxglove plant, and (7) the pupil-dilating drug atropine from the deadly nightshade. Thus, many medicinal plants handed down through the ages possess a chemical reason for their action. No doubt other important drugs will be discovered in plants as the search continues with more sophisticated techniques. Drug companies and government agencies are screening and testing thousands of plants yearly in an effort to find new and more effective cures for the ailments of mankind.

LANGUAGE OF PLANT REMEDIES.

Along with the knowledge of the use of medicinal plants, a particular language has resulted which describes the method of use and/or the ailments for which plant remedies are employed. Some of the more important words commonly used follow:

- **Alteratives**—These are medicines that produce a gradual change for the better, and restore normal body function.

- **Anodyne**—Agents which relieve pain are called anodynes.

- **Astringent**—Substances which contract tissues and check the discharge of blood or mucus are termed astringents.

- **Carminative**—This is a substance that acts to reduce flatulence (gas).

- **Cathartics**—These are plant substances which relieve constipation by stimulating the secretions of the intestines.

- **Decoctions**—These are medicines made by simmering plant parts in water.

- **Demulcents**—These plant medicines are soothing to the intestinal tract and usually of an oily or mucilaginous nature.

- **Diaphoretic**—This type of substance has the ability to produce sweating.

- **Emollients**—These act similar to demulcents, but their soothing action is for the skin rather than the intestines.

- **Emmenagogues**—Plant remedies with this action are said to promote menstrual discharge.

- **Expectorants**—These are remedies which aid the patient in bringing up and spitting out excessive secretions of phlegm (mucus) from the lungs and windpipe.

- **Febrifuges**—These are the same as antipyretics—agents which help dissipate a fever.

- **Infusion**—This term applies to soaking of plant parts in water to extract their virtues. Infusions provide a quick and simple method for removing the medicinal principle from dried plant parts.

- **Nervines**—These substances are said to calm, soothe, and relax tensions caused by nervous excitement, strain, or fatigue.

- **Ointments and linaments**—The major difference between these is largely their consistency. An ointment is a semisolid in a fatty material, while a linament is a liquid or semiliquid preparation in a base of alcohol or oil.

- **Stimulants**—These are plant medicines which temporarily increase mental or physical activity.

- **Stomachics**—Plants with this property stimulate stomach secretions.

• **Tincture**—This refers to a solution of plant substances in alcohol. Tinctures contain those oils, resins, or waxes not soluble in water.

• **Tonic**—A tonic is said to be an agent which restores and invigorates the system and stimulates the appetite.

• **Vulneraries**—This word is often found in old herbals. It refers to medicines useful in healing wounds.

COMMON MEDICINAL PLANTS. Table M-14 groups some of the medicinal plants according to the language of herbal medicine.

TABLE M-14
SOME MEDICINAL PLANTS GROUPED ACCORDING TO THEIR REPORTED EFFECTS AND USES

Alterative	Anodyne	Appetite Stimulation	Astringent	Calmative
agrimony	birch bark	alfalfa	agrimony	catnip
black cohosh root	hops	anise	bayberry	chamomile
blue flag root	white willow bark	chamomile	blackberry	fennel seeds
burdock root	wintergreen	celery	valerian	linden flowers
dandelion		dandelion	witch hazel	
echinacea		ginseng		
ginseng		juniper		
goldenseal		mint		
red clover flowers		parsley		
sarsaparilla root		rosemary		
		sweet cicely		
		Virginia snakeroot		
		winter savory		
		wormwood		

Carminative	Cathartic	Demulcent	Diaphoretic	Diuretic
anise seed	barberry	borage	angelica root	alfalfa
capsicum	buckthorn bark	comfrey	borage	angelica
cardamom	chicory	marshmallow	catnip	bearberry leaves
catnip	colcynth		chamomile	buchu leaves
cumin	dandelion		ginger root	celery
fennel seed	senna		hyssop	chicory
ginger root			pennyroyal	cleaver's herb
goldenrod			senega root	corn silk
lovage root			serpentaria root	dandelion
nutmeg				elecampane root
peppermint				goldenrod
spearmint				horehound
valerian root				horse tail grass
				juniper berries
				parsley root
				wild carrot

Expectorant	Febrifuge	Nervine	Stimulant	Tonic
acacia	angelica	catnip	angelica	barberry root and bark
angelica	balm	chamomile	bayberry leaves	cascarilla bark
colt's foot	birch bark	hops	capsicum fruit	celery seed
garlic	borage	linden flower	cardamom	dandelion root
horehound	dandelion	passion flower	mayweed	gentian root
licorice root	eucalyptus	skullcap	Paraguay tea	ginseng
senega root	lobelia	valerian root	sarsaparilla root	goldenseal
	meadowsweet	yarrow	tansy	hops
	pennyroyal		vervain	mugwort
	senna		wintergreen	wormwood
	willow bark			

WORDS OF CAUTION. Herbal medicine, folk medicine, or whatever it is called, is back in vogue. It is appealing because most remedies are handy, inexpensive, and *natural.* Moreover, folk medicine gives people the feeling that they are taking care of themselves. With this return to folk medicine, there are numerous purveyors; hence, people should be wary. Many generalized and meaningless claims are employed to describe the action of certain plant remedies. For example, such claims include the following: purify the blood; strengthen the glands; establish a normal balance; strengthen the nerves; balance the female organism; encourage rejuvenation; and ease nerve pain. These claims could describe any number of symptoms or cures—or none at all. Moreover, some plant remedies are still recommended on the basis of the *doctrine of signatures,* which held that plant remedies could be identified by their resemblance to the afflicted part of the body. Names like liverwort and heartsease recall this belief.

As long as herbal medicine is not poisonous and does not replace traditional, professional medicine in serious ailments, no harm is likely. However, the individual subscribing to herbal medicine should be certain of the identity of the plants used, since some are poisonous and others may have serious side effects. Overall, herbal medicine is safe, fun, and some of it probably even helps ease minor ailments.

(Also see POISONOUS PLANTS; and WILD EDIBLE PLANTS.)

MEDIUM CHAIN TRIGLYCERIDES (MCT)

This is a special dietary product made from coconut oil by (1) steam and pressure hydrolysis of the oil into free fatty acids and glycerol, (2) separation of the resulting hydrolysate into medium-chain and long-chain fatty acids, and (3) recombination of the medium-chain fatty acids with glycerol to form MCT oil. It is composed of about ¾ caprylic acid, a saturated fatty acid containing 8 carbon atoms; and about ¼ capric acid, a saturated fatty acid containing 10 carbon atoms. The special nutritive values of this product are as follows:

1. It is much more readily digested, absorbed, and metabolized than either animal fats or vegetable oils which contain mainly long-chain triglycerides. The enzyme, pancreatic lipase, can easily break down MCT. Hence, MCT oil is valuable in the dietary treatment of fatty diarrhea and other digestive disorders in which the absorption of fat is impaired —malabsorption.

2. Medium-chain triglycerides, unlike other saturated fats, do *not* contribute to a rise in blood cholesterol. Therefore, they are used to treat some forms of hyperlipoproteinemia.

At the present time, MCT oil is used almost exclusively for special dietary formulations in the United States, but it is also available in various nonmedical consumer products in Europe.

(Also see HYPERLIPOPROTEINEMIAS; and MALABSORPTION SYNDROME.)

MEGADOSE

A very large dose; for example, taking 20 to 100 times the recommended allowance of vitamin C.

MEGAJOULE (MJ)

A metric unit of energy, equivalent to 240 kcal.

MEGALOBLAST

A large embryonic type of red blood cell with a large nucleus; present in the blood in cases of pernicious anemia, vitamin B-12 deficiency, and/or folacin deficiency.

MELANIN

Any of the various dark brown or black pigments of the skin, hair, and certain other tissues; derived from tyrosine metabolism.

MELBA TOAST

This is very thin bread which is toasted or baked until crisp and well browned. Thus, rusks, croutons, and melba toast all have one thing in common—all of them are made from bread which is baked until dry and crisp.

MELON(S)

The term *melon* is applied to the fruit of several closely related plants of the cucurbit family (Cucurbitaceae)—a family whose members include cucumbers, pumpkins, squashes, watermelons, and gourds.

Melons grow on plants that are either climbing or trailing vines with round, pointed or folded leaves, and small, yellow flowers. Melons are oblong to round. The surface of the melon is a skin called a rind which may be smooth, wrinkled, warty, netted, or ridged, and colored tan and yellow to light or dark green. The edible flesh—the pulp—may be green, white, yellow, pink, or red.

Melons include the summer melons, muskmelons, or cantaloupes; the winter melons, casaba, honeydew, crenshaw, and Persian melon; and watermelons. Descriptions of these melons follow:

• **Cantaloupe**—This is a variety of muskmelon, sometimes it is even called muskmelon. The netted-rind of the cantaloupe is yellow-green while the pulp is orange. Cantaloupes range in size from 4 to 7 in. *(10 to 17.5 cm)* in diameter. Although there are several varieties of cantaloupes which are grown in different climatic regions of the United States, California is by far the largest producer of the fruit. Usually, cantaloupes require about 3 months or so to reach maturity. Most of the crop is sold fresh, but some is canned or frozen in the form of melon balls.

• **Casaba**—The casaba is a golden-yellow, wrinkled-rind, round variety of muskmelon which has a white-colored flesh. Casabas range from 6 to 8 in. *(15 to 20 cm)* in diameter. This fruit requires a hot, dry climate and reaches maturity in about 3½ months. Most of the U.S. production is grown on irrigated fields in the southwestern and western states. All of the crop is marketed fresh right after harvesting or after a brief period of storage.

• **Honeydew**—This variety of melon has a white, smooth rind and light-green pulp. It is about 7 in. *(17.5 cm)* in diameter and 8 in. *(20 cm)* long. Honeydew melons grow best in hot, dry climates and require about 4 months to reach maturity. They are called winter melons because they ripen late in the season and are marketed during the winter. Most of the U.S. production comes from irrigated fields in the southwestern and western states. They are usually used fresh and as canned or frozen melon balls packed in syrup.

• **Crenshaw**—Crenshaw melons are oblong, with a yellow-tan, smooth rind. Their pulp is salmon-orange in color. Crenshaws are usually about 6 in. *(15 cm)* in diameter and about 7 in. *(17.5 cm)* long. These melons require a hot, dry climate and reach maturity in about 4 months. Most of the U.S. production is grown in irrigated fields in the southwestern and western states. All of this crop is marketed fresh.

• **Persian**—These melons are round with a dark green netted, rind and orange pulp. There are small Persian melons that are 4 to 5 in. *(10 to 12.5 cm)* in diameter and large types that are 7 to 10 in *(17.5 to 25 cm)* in diameter. Persian melons require a hot, dry climate and reach maturity in about 4 months. Most of the U.S. crop is grown in the southwestern states.

• **Watermelon**—All the melons listed above belong to the genus and species *Cucumis melo*, but the scientific name for watermelon is *Citrullus vulgaris*. Depending upon the variety, watermelons weight 5 to 85 lb *(2.3 to 38.3 kg)* and vary in shape from round to oval to oblong-cylindrical. The rind may be very light to very dark-green with stripes or mottling, while the pulp is red, pink, orange, yellow, or white. Watermelons are grown in tropical, semitropical, and temperate climates. In the United States commercial watermelon production occurs primarily in the southern states.

Nutritionally, melons are all quite similar containing only about 30 calories (kcal) per 100 g and being good sources of potassium, vitamin C, and vitamin A when the flesh is deep orange. Melons are all about 90% water.

(Also see CANTALOUPE; and WATERMELON.)

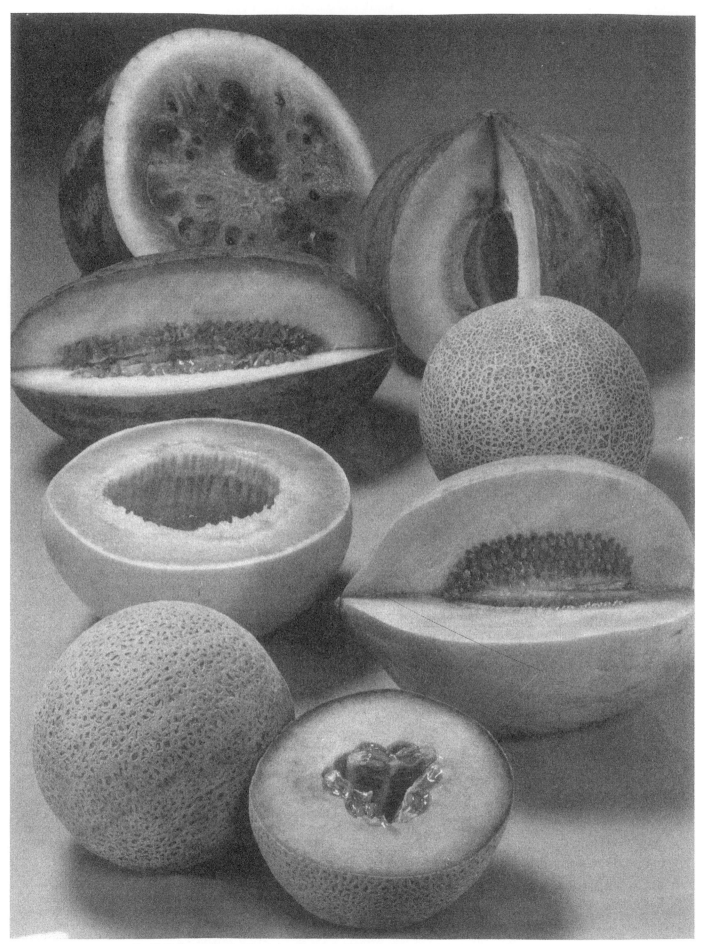

Fig. M-44. You can't get melancholy with melons. (Courtesy, United Fresh Fruit & Vegetable Assoc., Alexandria, Va.)

MELON PEAR (PEPINO) *Solanum muricatum*

The yellow and purple streaked fruit of a shrub (of the family *Solanaceae*) that is native to Peru, but is now grown in various parts of the American tropics. It is between 4 and 6 in. (*10 and 15 cm*) long, and has an aromatic pulp that is used to make drinks, jams, and jellies.

The pulp of the fruit is high in water content (92%) and low in calories (32 kcal per 100 g) and carbohydrates (6%). It is a good source of vitamin C.

MENIERE'S SYNDROME

Most common in the fifth decade of life, Meniere's syndrome denotes symptoms resulting from a disturbance of the inner ear due to fluid accumulation. The disease, whose symptoms were first collectively described by Prosper Meniere in 1861, is characterized by the sensation of whirling around in space—vertigo—fluctuating incidents of deafness, high pitch ringing or buzzing in the ear, nausea, vomiting, and headache. Many inner ear disorders can be traced to a variety of causes. However, the cause of Meniere's syndrome is difficult to pinpoint and it is rarely traceable to a disease or infection. Treatment, therefore, may proceed on a trial-and-error basis. Bed rest is an effective treatment since the sufferer can usually find a position in which vertigo is minimal. Other forms of treatment consist of diuretics and low salt diets to relieve the fluid accumulation in the inner ear and the use of sedatives and tranquil-izers. A preparation such as Dramamine may be given to relieve the vertigo. Should the condition become so annoying that normal activities are impossible, then surgery is often the only cure.

Some sufferers of Meniere's syndrome report obtaining relief from taking niacin (both oral and injections simultaneously) and other B complex vitamins, and from adding yogurt or buttermilk to the diet. Also, a diet low in (1) saturated fats, (2) carbohydrates (especially sugar), and (3) salt has been found to be helpful.

(Also see MODIFIED DIETS.)

MENTAL DEVELOPMENT

Mental development is a complex, on-going process influenced by a multitude of factors. Furthermore, it is difficult to assess and compare due to various social, racial, and ethnic backgrounds. In general, mental development may be defined as the increased ability to function. It involves motor skills, social traits, behavioral traits and learning skills. Normal mental development is evidenced by the gradual development of certain skills—roll over, sit, creep, walk, coordination of eyes and hands, formation of words, formation of whole thoughts, and ability to relate to people. The overall desired outcome of mental development is the attainment of an individual's full genetic potential.

In humans, it is difficult to say whether nutrition *per se* contributes to mental development since the conditions of malnutrition and deprived social and environmental conditions coexist. However some animal models have been designed to separate the effects of nutrition and environment on mental development. Therefore, animal models and observations of human populations combine to form the following salient points in regards to nutrition and mental development:

1. The brain and spinal cord continue rapid growth during infancy.
2. Malnutrition during the early life of animals and humans alters the number of brain cells, brain cell size, and/or other biochemical parameters depending upon the time of onset and the duration of the malnutrition.
3. Malnourished animals demonstrate impaired learning abilities.
4. Teachers have observed that hungry, undernourished children are apathetic and lethargic.
5. Children suffering from kwashiorkor are apathetic, listless, withdrawn, and seldom resist examination or wander off when left alone.
6. Children who recover from marasmus have a reduced head circumference—brain size—even after five years of rehabilitation.
7. Numerous studies suggest that infants subjected to acute malnutrition during the first years of life develop irreversible gaps in mental development, and thus never attain their full genetic potential.

Nevertheless, no amount of food can compensate for the emotional and social deprivation, resulting primarily from the lack of fulfilling infant-parent interaction.

(Also see MALNUTRITION; and STARVATION.)

MENTAL ILLNESS

This is a catchall term referring to any alteration in the normal function of the mind, ranging from periods of brief depression to severe, long lasting, disabling personality disorders. Rather broadly, mental illnesses can be classified as (1) neuroses, (2) psychophysiologic disorders, (3) personality disorders, (4) psychoses, and (5) transient situational disturbances. It can be characterized by lack of control over emotions and actions plus anxiety and even the stimulation of various physical disorders; for example, a peptic ulcer. Mental illness may be inherited and/or induced by environmental factors—chemical, bacterial, viral, nutritional, psychological, and social. Indeed there are probably as many or more forms of mental illness as there are physical illnesses. It should also be noted that no sharp line exists between mental health and mental illness because every human is unique. What is normal or abnormal may be merely a matter of degree and the manner in which a person's behavior is interpreted by family, friends, and physicians. However, it is well documented that there are certain nutritional disorders which cause some modifications in normal nervous system function—mental illness. The following examples demonstrate that nutritional states modify nervous system function:

1. **Vitamins and minerals.** Dietary deficiencies of thiamin, riboflavin, niacin, pantothenic acid, vitamin B-6, and vitamin B-12 will modify sensory functions, motor ability, and personality. Deficiencies of minerals such as sodium and magnesium, and toxic intakes of minerals such as lead and mercury, produce mild to severe forms of mental disorders—hyperactivity, learning difficulties, hallucinations, confusion, and giddiness.

(Also see MINERAL[S]; and VITAMIN[S].)

2. **Hypoglycemia.** When hypoglycemia (low blood sugar) develops over a short period of time, the symptoms include rapid pulse, trembling, hunger, and a slight amount of mental confusion. However, hypoglycemia which has developed over many hours—meal skipping—is characterized by headache, depression, blurred vision, incoherent speech, and considerable mental confusion.

(Also see HYPOGLYCEMIA.)

3. **Starvation.** Short periods of starvation—4 days of water only—in healthy individuals decreases coordination and ability to concentrate. Semistarvation over long periods of time, such as existed in World War II concentration camps, decreased the ability to concentrate, lessened inter-

ests, decreased the ability of sustained mental effort, and created irritability, apathy, and sullenness.

(Also see STARVATION.)

4. **Beriberi.** Beriberi is a disease due to thiamin deficiency. Behavioral changes which have been noted include mental depression, loss of interest, diminution of the sense of humor, loss of ability to concentrate, and loss of patience with others. Rapid reversal of these symptoms depends upon restoration of thiamin to the diet.

(Also see BERIBERI.)

5. **Alcoholism.** Alcoholics make up about 10% of the first admissions to psychiatric wards. They possess some of the same symptoms noted in thiamin deficiency, and indeed respond to thiamin therapy. Often Korsakoff's syndrome—confusion, confabulation, fear, impaired learning, delirium—is associated with alcoholism. Vitamin therapy helps to restore the individual's attentiveness, alertness, and responsiveness.

(Also see ALCOHOLISM.)

(Also see ANOREXIA NERVOSA; DEFICIENCY DISEASES; INBORN ERRORS OF METABOLISM; MENTAL DEVELOPMENT; MENTAL RETARDATION; and STRESS.)

MENTAL RETARDATION

Various degrees of mental retardation occur, but all are characterized by faulty development or set-back in the mental processes which impairs an individual's ability to cope with and adapt to the demands of society. In the United States, there are about 6 million mentally retarded individuals. It is generally recognized that there are three classes of mental retardation: (1) high grade or educable; (2) middle grade or trainable; and (3) low grade. A physical injury to the brain; some childhood diseases; an infection such as rubella, syphilis, or meningitis during pregnancy; a genetic defect such as Down's syndrome; and inborn errors of metabolism such as phenylketonuria and galactosemia without dietary adjustments, can all cause mental retardation. Nutritionally, a low iodine intake by the mother during pregnancy may result in cretinism in the baby, and alcohol consumption during pregnancy may contribute to varying degrees of mental retardation in the infant. A poor diet by the mother during pregnancy, and/or a poor diet by the infant may contribute to abnormal mental development. In fact, a large number of cases of mild mental retardation, which are educable, are not related to any known physical or organic cause. In these cases, genetic, environmental, nutritional, or psychological factors, or a combination of factors, may be the cause. Depending upon the degree, a mentally retarded person can present some obvious feeding problems. Feeding may be messy and slow, and thus require frequent small feedings in an effort to provide adequate food. Mentally retarded individuals have food likes and dislikes, and, despite their mental differences, their nutritional requirements are like those of everyone else.

(Also see ALCOHOLISM; INBORN ERRORS OF METABOLISM; IODINE; and MENTAL DEVELOPMENT.)

MERCURY (Hg)

The element mercury is discharged into air and water from industrial operations and is used in herbicide and fungicide treatments. Although known cases of mercury poisoning have been limited, there is widespread concern over environmental pollution by this element.

(Also see MINERAL[S]; section on "Contamination of Drinking Water, Foods, or Air with Toxic Minerals"; and POISONS.)

METABOLIC POOL

The nutrients available at any given time for the metabolic activities of the body; e.g., the amino acid pool, the calcium pool.

METABOLIC RATE

Essentially all of the energy released from foods becomes heat. Therefore, the metabolic rate is a measure of the rate of heat production by the body. It is expressed in Calories (kcal) and can be determined by direct calorimetry which measures the heat given off by a person engaged in various activities. However, indirect calorimetry is more frequently used. This method determines the heat produced by measuring the oxygen consumed, since it is known that on the average for each liter of oxygen burned by the metabolic fires of the body 4.825 Calories (kcal) of energy are released. When a normal person is in a quiet resting state the metabolic rate is about 60 to 80 Calories (kcal) per hour. However, a variety of factors influence the metabolic rate. Many of these factors are interrelated and some are more important than others. Among the factors that affect the metabolic rate are: exercise, recent ingestion of food, environmental temperature, body surface area, sex, pregnancy, age, emotional state, state of nutrition, body temperature, and hormones.

(Also see CALORIC [ENERGY] EXPENDITURE, section headed "Special Circumstances that Affect Energy Requirements"; CALORIMETRY; and METABOLISM, section on "Factors Controlling Metabolism.")

METABOLIC REACTIONS

The chemical changes that occur in the body; some are synthetic (the formation of new compounds), others are degradative (the breaking down of compounds).

METABOLIC WATER

Water formed within the cells of the body during the combustion (breakdown) of nutrients for the release of energy. On the average 13 ml of metabolic water are formed for every 100 kcal of metabolizable energy in the typical diet of a human. It contributes to the daily intake of water, and must be accounted for in the maintenance of water balance. Table M-15 illustrates the oxygen requirement and the production of carbon dioxide, water, and energy for carbohydrates, proteins, and fats.

(Also see METABOLISM; and WATER BALANCE.)

TABLE M-15
OXYGEN REQUIREMENTS AND CARBON DIOXIDE, WATER, AND ENERGY PRODUCTION FOR THE BREAKDOWN OF EACH 100 GRAMS OF CARBOHYDRATE, PROTEIN, AND FAT

Nutrient	Oxygen Consumed	Carbon Dioxide Produced	Water Produced	Energy Released
	(liter)	(liter)	(ml)	(kcal)
Carbohydrate	82.9	82.9	60	410
Protein	96.6	78.2	40	410
Fat	201.9	142.7	110	930

METABOLISM

Contents   Page

Cells—Functional Units............................683
Nutrients For Metabolism.........................684
  Carbohydrates...................................684
    Catabolism....................................684
    Anabolism.....................................686
  Fats...686
    Catabolism....................................686
    Anabolism.....................................687
  Proteins..687
    Catabolism....................................687
    Anabolism.....................................688
  Minerals And Vitamins...........................688
  Water...689
Excretion..689
Factors Controlling Metabolism.....................689
Inborn Errors Of Metabolism........................690
Summary..690

PROTEINS CARBOHYDRATES TRIGLYCERIDES

Fig. M-45. Metabolism of foods for the production of energy—the number one need.

Beefsteak, lettuce, tomatoes, and beans never get to the cells and tissues of our bodies, including our brains. But the nutrient chemicals—more than 40 of them, including amino acids, minerals, vitamins—do reach the body cells and tissues and are essential to their life. These are the ABCs of modern nutrition.

Once foods are eaten and digested, the nutrients are absorbed into the blood and distributed to the cells of the body. Still more chemical changes are required before the nutrients can be put to work in the body, by transforming them into energy or structural material. Thus, nutrients—carbohydrates, fats, proteins, minerals, vitamins, and water—are subjected to various chemical reactions. These chemical reactions occur on the cellular and subcellular level. The sum of all of these chemical reactions is termed metabolism. It has two phases: (1) catabolism, and (2) anabolism:

1. **Catabolism.** Catabolism is the oxidation—burning—of nutrients, liberating energy which is used to fulfill the body's immediate demands. When reactions liberate energy they are termed *exergonic reactions*.

2. **Anabolism.** Anabolism is the process by which nutrient molecules are used as building blocks for the synthesis of complex molecules. Anabolic reactions are endergonic—they require the input of energy into the system.

Basic to understanding the process of metabolism, is the realization that all of the reactions are catalyzed by enzymes —protein molecules which speed up biochemical reactions without being used up in the reactions. Furthermore, most all of the biochemical reactions of the metabolic processes are reversible. This is an important concept.

Life is a conquest for energy. Energy is required for practically all life processes—for the action of the heart, maintenance of blood pressure and muscle tone, transmission of nerve impulses, ion transport across membranes, reabsorption in the kidneys, protein and fat synthesis, the secretion of milk, and muscular contraction. A deficiency of energy is manifested by slow or stunted growth, body tissue losses, and/or weakness.

The diet contains carbohydrates, fats, and proteins. Although each of these has specific functions in maintaining a normal body, all of them can be used to provide the number one requirement—energy. Hence, the following discussion of their metabolism centers first, on their catabolism, and second, on their anabolism. Although they are not used for energy, minerals, vitamins, and water are essential to the metabolic processes.

In many ways, metabolism is similar to a factory. Like any manufacturing process, there are waste products with which to be dealt. Since the waste products of metabolism are toxic to the body at high levels, the body has developed reliable means for the removal of these wastes. Also, like any manufacturing process, there must be controls. Various factors control the rate and the direction of metabolism—anabolism or catabolism. Occasionally, a faulty piece of equipment hinders a manufacturing process. The body, too, can inherit faulty equipment for performing the metabolic processes.

CELLS—FUNCTIONAL UNITS. Every living thing is composed of cells and cell products. In an organism as complex as a human—with trillions of cells—the proper function of the body processes is dependent upon each individual cell carrying out its proper metabolic processes for the organ or tissue in which it is located. Just as the body has organs, the cells of the body contain organelles (little organs) which are involved in the metabolic processes.

A typical cell with the structures involved in metabolic processes is shown in Fig. M-46.

Fig. M-46. A typical cell showing some of the structural components most cells have in common, though cells of the body are variable in structure and function.

NUTRIENTS FOR METABOLISM. From the standpoint of supplying the normal energy needs, carbohydrates are by far the most important, with fats second, and proteins third. Carbohydrates are usually more abundant and cheaper, and most of them are easily digested and absorbed. Moreover, excess carbohydrates can be transformed into body fat or into amino acids for protein synthesis.

Carbohydrates. The carbohydrates are organic compounds composed of carbon (C), hydrogen (H), and oxygen (O). They are formed in the plant by photosynthesis as follows: $6CO_2 + 6H_2O$ + energy from sun = $C_6H_{12}O_6$ (glucose) + $6O_2$. Carbohydrates form the woody framework of plants as well as the chief reserve food stored in seeds, roots, and tubers. As far as man is concerned, starch and sugars, primarily 6-carbon sugars, are the carbohydrates available for his use. Starch is composed of long chains of glucose which the digestive processes split into individual glucose molecules. Fructose (fruit sugar), glucose (grape sugar), sucrose (cane and beet sugar), and lactose (milk sugar) are the important dietary sugars. Other carbohydrates such as cellulose, hemicellulose, and pectin are not digestible; hence, they are unavailable for the metabolic processes. Sometimes these carbohydrates are called fiber or roughage. Ultimately, after being digested and absorbed, carbohydrates are converted to glucose.

No appreciable amount of carbohydrate is found in the body at any one time, but the blood supply of glucose is held rather constant at about 70 to 100 mg/100 ml. However, this small quantity of glucose in the blood, which is constantly replenished by changing the glycogen of the liver back to glucose, serves as the chief source of fuel with which to maintain the body temperature and to furnish the energy needed for all body processes. Maintenance of the blood glucose level is so important that at times amino acids and part of the fat molecule (glycerol) may be used to make glucose. The storage of glycogen (so-called animal starch) in the liver amounts to 3 to 7% of the weight of that organ. Glycogen is also stored in the muscles, but the latter is not available to raise blood glucose.

(Also see CARBOHYDRATE[S]; FIBER; HYPOGLYCEMIA; and STARCH.)

CATABOLISM. The primary use for carbohydrates is to produce energy. In the presence of oxygen, glucose will produce a maximum of 686 kcal/mole (*180.2 g*), as seen in the following equation:

$$C_6H_{12}O_6 + 6 O_2 \rightarrow 6 CO_2 + 6 H_2O + 686 \text{ kcal/mole (heat)}$$

The above reaction is a combustion or burning reaction. Hence, the expression of *burning off calories* can be understood.

No physiological system can approach 100% efficiency in the production of energy from the catabolism of a nutrient. Likewise, the body cannot produce 686 kcal from one mole (*180.2 g*) of glucose. Nevertheless, a great deal of energy is produced from the catabolism of glucose by a series of enzymatic reactions which permit the orderly transfer of energy from glucose to energy-rich compounds:

• **ATP—energy currency**—For metabolism to permit the orderly transfer of energy from foods to processes requiring energy, there must be a common carrier of energy— something cells can use to exchange goods for services. The

energy currency—the primary mechanism by which energy is captured and stored—is a compound called adenosine triphosphate, often abbreviated ATP. Just as pennies are small units of a dollar, ATP is a small unit of energy.

ATP is a compound formed from the purine adenine, the 5-carbon sugar ribose, and three phosphate molecules as shown in Fig. M-47. Two of the phosphates are joined by high energy bonds.

Fig. M-47. Structure of adenosine triphosphate.

When ATP is formed from ADP (adenosine diphosphate) by the addition of another phosphate, 8 kcal/mole must be put into the system—like a deposit in a bank account. The ATP then acts as a high-energy storage compound for the energy produced in the catabolism of various nutrients. When energy is required for body functions, ATP is broken down to ADP by releasing one phosphate and 8 kcal/mole of energy.

Releasing all of the energy contained in carbohydrates (fats and proteins, too) in one step would result in lost and wasted heat energy. Instead, energy is released in steps. The body utilizes a series of biological oxidations to form ATP. Oxidation refers to the loss of electrons (or hydrogen) by a compound. Conversely, reduction refers to the acceptance of electrons (or hydrogen) by a compound. Chemical energy is released as the lost electrons flow down energy gradients of an elevated state to the next lower hydrogen (electron) acceptor or carrier. Several coenzymes act as hydrogen (H^+) acceptors, many of which involve vitamins; for example, niacin and riboflavin. Compounds such as nicotinamide adenine dinucleotide (NAD^+), nicotinamide adenine dinucleotide phosphate ($NADP^+$), and flavin adenine dinucleotide (FAD) can accept and transfer hydrogen (or electrons) through a cytochrome system known as the electron transport system. Cytochromes are pigmented proteins similar to hemoglobin. In the electron transport system, electrons cascade from one intermediate carrier molecule—cytochrome—to the next lower in line, like water over a small waterfall. For every two electrons that tumble down the cytochrome system, enough energy is released to form three ATP molecules by adding a high energy phosphate to ADP. The ultimate hydrogen acceptor in the system is oxygen; hence, water (H_2O) is formed. This is metabolic water. Production of ATP through this system is known as oxidative phosphorylation. Oxygen is required. It should be noted that when oxidation is initiated and NAD^+ carries the hydrogen ($NADH + H^+$) there is a net production of three ATP molecules. Only two ATPs are yielded, however, when hydrogen is transferred by FAD ($FADH_2$).

Fig. M-48 illustrates the cytochrome system, and the formation of ATP as hydrogen drops to different energy levels.

Fig. M-48. The formation of adenosine triphosphate (ATP) from adenosine diphosphate as hydrogen tumbles down a chain of hydrogen carriers—the electron transport system. NAD + stands for nicotinamide adenine dinucleotide, and FAD stands for flavin adenine dinucleotide—both hydrogen carriers similar to the cytochromes.

• **Glycolysis**—Before glucose can be used by the cells it must first enter from the blood. Entry into the cells depends upon the hormone insulin from the pancreas. Without insulin, glucose stays in the blood and creates the condition known as diabetes mellitus. Once inside the cell, the initial steps in the catabolism of glucose begin with a process called glycolysis—meaning glucose destruction. This pathway of chemical changes is referred to as the Embden-Meyerhof or glycolytic pathway. It is a series of reactions that are oxygen independent—no free or molecular oxygen needs to be involved; hence, these reactions are said to be anaerobic. The first reaction of glucose is that of providing activation energy—energy to get glucose *over a hump* before proceeding *downhill*. This is accomplished by a reaction in which ATP donates a high energy phosphate yielding a new compound called glucose-6-phosphate.

Next, glucose atoms are slightly rearranged to yield fructose-6-phosphate. Then another phosphate is added by a donation from ATP. The compound is now called fructose-1, 6-phosphate. Another enzyme splits this 6-carbon compound into two 3-carbon compounds each containing a phosphate. One compound is glyceraldehyde-3-phosphate and the other is dihydroxyacetone phosphate. Dihydroxyacetone phosphate can be converted to glyceraldehyde-3-phosphate which receives another phosphate after two hydrogen atoms are removed via NADH + H +. When oxygen is present (aerobic), this hydrogen enters the electron transport system and yields three ATP. Next, the compound with two phosphates—1, 3, diphosphoglyceric acid—loses a phosphate thus forming ATP from ADP and converting the compound to 3-phosphoglyceric acid. Eventually, the other phosphate is lost to convert ADP to another ATP and pyruvic acid is formed. Since each glucose provides two 3-carbon compounds for glycolysis, the net gain in ATP to this point is two ATP. Two ATP were used to add phosphate to the 6-carbon compounds while two were gained for each 3-carbon compound converted to pyruvic acid.

Lactic acid is produced only under anaerobic conditions such as exercise, when oxygen cannot be delivered fast enough by the blood. Since the conversion of pyruvic acid to lactic acid requires the input of two hydrogen atoms, the NADH + H + generated earlier in glycolysis can be reconverted to NAD + thus allowing glycolysis to continue. Normally, NADH + H + would go to the electron transport system, but this requires oxygen. Lactic acid is a dead end. In order for the body to use it, it must be reconverted to pyruvic acid, and in some cases glucose.

The following items are important features of glycolysis:
1.   Two ATP are netted for each glucose.

2.   The system allows for the anaerobic breakdown of glucose.
3.   It provides a system whereby energy from glucose stored as glycogen can be made readily available in muscle even when oxygen is in short supply during exercise.
4.   Intermediates for the synthesis of other nutrients, such as glycerol and amino acids, are provided.
5.   In the step whereby glyceraldehyde-3-phosphate is converted to 1, 3-diphosphoglyceric acid, 2 moles of NADH + H + are produced which can subsequently be oxidized in the electron transport system under aerobic conditions to produce 6 additional ATP.

Through more chemical reactions, more energy can be derived from glucose which has reached the point of pyruvic acid. These additional reactions take place in a cellular structure appropriately called the powerhouse of the cell. These are the mitochondria.

• **Mitochondria**—These are watermelon-shaped microscopic structures within the cells of the body. The number and distribution of mitochondria within a cell depends upon the energy requirement of the cell. Those cells with large energy requirements possess a large number of mitochondria. The mitochondria contain the electron transport system. Moreover, they contain the enzymes necessary for the further metabolism of pyruvic acid, and other chemical transformations. It is their unique internal structure of cristae—a series of baffles—that allows for very efficient production of ATP.

Fig. M-49. Microscopic view of a mitochondrion—powerhouse of the cell.

Pyruvic acid enters the mitochondria of the cell and begins another series of reactions variously referred to as the Krebs cycle, the tricarboxylic acid (TCA) cycle, or the citric acid cycle. First, pyruvic acid is converted to acetyl coenzyme A (acetyl CoA) by a loss of a carbon atom as carbon dioxide (CO_2) and two hydrogen atoms. This conversion yields 3 ATP, since the hydrogen atoms enter the electron transport system via NADH + H +. Acetyl CoA is then condensed with oxaloacetic acid to form citric acid, thus initiating the Krebs cycle. As the cycle progresses, more electrons (hydrogens) are transferred to different coenzymes, which, in turn, enter the electron transport chain producing ATP. One complete turn of the cycle (1) produces 12 ATP and the loss of two carbons as carbon dioxides (CO_2) from citrate, and (2) the restoration of oxaloacetic acid; completing the cycle. Since 2 pyruvic acid molecules can be formed from 1 glucose molecule, 24 ATP are produced via the Krebs cycle. Thus, 38 ATP are produced from 1 molecule of glucose: 8 from glycolysis, 6 in the conversion of pyruvic acid to acetyl CoA, and 24 from the Krebs cycle. Each pass of the 2-carbon acetyl CoA through the Krebs cycle results in the formation of CO_2 (carbon dioxide) and H_2O (water), the end products of catabolism.

• **Energy efficiency**—One ATP molecule will yield about 8 Calories (kcal). Since 38 ATP are derived from the complete oxidation of glucose, 304 Calories (kcal) of energy are formed (38 × 8 = 304). Under ideal conditions, the oxidation of glucose yields 686 Calories (kcal). Hence, the human body is 44% efficient in converting glucose to energy (304 ÷ 686 × 100 = 44). Diesel engines have a maximum efficiency of about 40%, while many gasoline engines only convert about 25% of their fuel energy into work. The human body burns its fuel quite efficiently.

• **Hexosemonophosphate pathway (Pentose shunt)**—Not all glucose enters the glycolytic pathway. A small amount continually enters another pathway known as the pentose shunt or hexosemonophosphate (HMP) pathway. The functions of this pathway are twofold: (1) to provide some energy in the form of NADPH, another hydrogen carrier, which will function in the formation of fatty acids; and (2) to provide the 5-carbon sugar ribose for use in the nucleic acid DNA and RNA, or nucleotide coenzymes such as ATP. Products of this pathway eventually re-enter the glycolytic pathway.

ANABOLISM. Two processes involved in carbohydrate metabolism may be considered anabolic in that they require energy: (1) storage of glucose as glycogen; and (2) the formation of glucose from noncarbohydrate sources, or gluconeogenesis.

• **Glycogen**—Upon absorption into the bloodstream, glucose travels to the liver where it is incorporated into glycogen. Glycogen is composed of numerous glucose molecules which are all chemically joined. Starch in plants is also composed of numerous glucose molecules joined together; hence, glycogen is sometimes referred to as animal starch. Each time a glucose molecule is linked with the glycogen already in the liver, two high-energy phosphate bonds are required. Then, as glucose is needed to maintain the blood glucose levels, glycogen is broken down to release glucose.

Some glycogen is manufactured and stored in the muscles. However, glucose derived from this glycogen is available only for use by the muscle and not the general circulation. Lactic acid may leave the muscle and be reconverted to glucose in the liver and released into the blood. This process requires six ATPs for each glucose molecule, and is referred to as the lactic acid cycle or Cori cycle. A husband and wife team, Carl and Gerty Cori, described it; and they were jointly awarded the Nobel Prize in 1947.

• **Gluconeogenesis**—Gluconeogenesis is the formation of glucose from nutrients other than carbohydrates. This pathway is extremely important when there is an insufficiency of dietary carbohydrates. The liver is the main site for gluconeogenesis. Amino acids are the primary precursors in gluconeogenesis. However, amino acids must first have their amino group (NH_2) removed by a process called deamination, following which the remaining carbon skeleton can enter the Krebs cycle and, essentially, be worked backwards through glycolysis to glucose. All of the amino acids except leucine are glycogenic—capable of forming glucose. Production of glucose by this means is possible since most reactions catalyzed by enzymes are reversible. There is, however, one exception. Pyruvic acid in the mitochondria of cells cannot be directly converted to phosphoenolpyruvic acid. Rather, it is transformed into oxaloacetic acid and malic acid. Malic acid is then able to cross the mitochondrial membrane into the cytoplasm where it is reconverted to oxaloacetic acid. Oxaloacetic acid is then converted to phosphoenolpyruvic acid due to the presence of a specific enzyme in the cytoplasm. Once this is accomplished, the process closely resembles the reverse of the glycolytic pathway. The whole process requires an expenditure of energy.

Table M-16 indicates where each amino acid fits into the metabolic scheme following deamination.

TABLE M-16
GLUCONEOGENESIS FROM AMINO ACIDS

Amino Acid	Compounds Formed Following Deamination	Metabolic Fate[1]
Alanine	Pyruvic acid	Glycogenic
Arginine	Alpha-ketoglutaric acid	Glycogenic
Aspartate	Fumaric acid	Glycogenic
Cysteine	Pyruvic acid	Glycogenic
Glutamate	Alpha-ketoglutaric acid	Glycogenic
Glycine	Pyruvic acid	Glycogenic
Histidine	Alpha-ketoglutaric acid, beta-ketoglutaric acid	Glycogenic
Proline	Alpha-ketoglutaric acid	Glycogenic
Hydroxyproline	Alpha-ketoglutaric acid	Glycogenic
Methionine	Succinyl CoA	Glycogenic
Serine	Pyruvic acid	Glycogenic
Threonine	Succinyl CoA	Glycogenic
Valine	Succinyl CoA	Glycogenic
Isoleucine	Acetyl CoA, succinyl CoA	Glycogenic and ketogenic
Lysine	Alpha-ketoglutaric acid, acetyl CoA, acetoacetyl CoA	Glycogenic and ketogenic
Phenylalanine	Fumaric acid, acetyl CoA	Glycogenic and ketogenic
Tryptophan	Succinyl CoA, acetyl CoA	Glycogenic and ketogenic
Leucine	Acetyl CoA, acetoacetic acid	Ketogenic

[1]Glycogenic forms glucose while ketogenic forms ketone bodies—acetoacetic acid, beta-hydroxybutyric acid, and acetone.

Fats. Fats are compounds composed of carbon, hydrogen, and oxygen—much like carbohydrates. However, because of the larger portion of carbon and hydrogen in their makeup, fats provide about 2.25 times as much energy as do the carbohydrates—more hydrogen atoms for the electron transport system. Dietary fats are primarily compounds composed of three fatty acid molecules linked to one glycerol molecule—a combination called triglycerides.

(Also see FATS AND OTHER LIPIDS.)

CATABOLISM. When fats are hydrolyzed (water added), glycerol and fatty acids are released and subsequently catabolized separately. Glycerol is converted to glycerol-3-phosphate, then to dihydroxyacetone phosphate. Dihydroxyacetone phosphate can be readily converted to 3-phosphoglyceraldehyde which enters the last part of the glycolytic pathway and eventually enters into the Krebs cycle. Thus, it yields an amount of energy similar to that of one-half of the glucose molecule passing through glycolysis and the Krebs cycle.

The catabolism of fatty acids occurs in the mitochondria through a systematic process called beta-oxidation, whereby 2-carbon fragments are successively chopped from the fatty acid molecule to form acetyl CoA which then enters the Krebs cycle. The term lipolysis indicates mobilization of fats and their oxidation. Carnitine, a vitaminlike substance, facilitates the transport of fatty acids across the mitochondrial membrane. The first step in the beta-oxidation of a fatty acid is the addition of coenzyme A (CoA) to the end of the molecule. This requires energy, but not nearly as much energy as is subsequently generated. For example, the 16-carbon fatty acid palmitic acid can be oxidized seven times. Each time a 2-carbon fragment—acetyl CoA—is released from a fatty acid, 5 ATP are formed. However, 1 ATP is consumed in activating the fatty acid with CoA. Then, each acetyl CoA enters the Krebs cycle and produces 12 ATP. The energy balance sheet is as follows:

			ATP
Palmitic acid	→	Palmitoyl CoA............	− 1
Palmitoyl CoA	→	8 Acetyl CoA (7 × 5 ATP).	35
8 Acetyl CoA	→	H_2O + 18 CO_2 (8 × 12 ATP)	96
	Net		130

Each mole of ATP contains about 8 Calories (kcal). Therefore, 1 mole of palmitic acid yields 1,040 Calories (kcal), which demonstrates that fats are high-energy compounds. Comparing this value to that obtained by burning fat in a laboratory bomb calorimeter, the efficiency of the biological burning of palmitic acid is about 42%—not bad for any machine.

Acetyl CoA is a common point—a crossroad— for many of the biochemical reactions of the body. It can be used for building other substances including (1) new fatty acids and cholesterol, (2) the formation of the ketone body, acetoacetic acid, and (3) the nerve transmitter substance, acetylcholine. It is formed during the catabolism of fatty acids, glucose, and amino acids; hence, excesses of these nutrients can be stored as fat. As already pointed out, acetyl CoA can enter the Krebs cycle. However, entry into this cycle is dependent upon continued availability of carbohydrates.

• **Ketosis**—Pyruvate, a product of carbohydrate catabolism, is needed to produce oxaloacetic acid. Oxaloacetic acid must then condense with acetyl CoA in order to produce citric acid. If there is insufficient oxaloacetic acid in the cell to keep the Krebs cycle functioning efficiently, the acetyl CoA is converted to acetoacetic acid, beta-hydroxybutyric acid, and acetone. These ketone bodies accumulate in the blood; and if the condition goes unchecked, acidosis will occur, often resulting in coma and death. Ketosis can be a complication of diabetes mellitus and starvation.

(Also see DIABETES MELLITUS; and STARVATION.)

ANABOLISM. Saturated fatty acids and monounsaturated fatty acids are rapidly and abundantly formed from acetyl CoA. Hence, any nutrient capable of yielding acetyl CoA during the metabolic processes can potentially contribute to fatty acid synthesis—a process called lipogenesis. While a limited amount of fat synthesis occurs in the mitochondria of the cell, most synthesis takes place outside the mitochondria via enzymes in the cytoplasm when supplemented with energy (ATP), carbon dioxide (CO_2), hydrogen, and the mineral magnesium (Mg^+). The 2-carbon compound, acetyl CoA, is the primary building block. It is first carboxylated (organic acid group COOH added) to form malonyl CoA, and then it is transferred to a carrier protein—acyl-carrier protein (ACP). Malonyl-ACP then combines with acetyl CoA via acetyl ACP to form acetoacetyl-ACP which then has hydrogen added by NADPH generated in the pen-

tose shunt to convert to butyryl-ACP. More and more acetyl CoA and hydrogen are added to this molecule until the chain is 16 carbons long—palmitic acid. All fatty acids formed as described above are then transported back into the mitochondria where three fatty acids are joined—esterified—with one glycerol molecule, thus forming triglycerides. The glycerol is derived from the glycolytic pathway. Fatty acids may also be elon-gated in the mitochondria.

Acetyl CoA may also be used to synthesize cholesterol in many tissues, but the liver is the major site of synthesis. (Also see CHOLESTEROL; and FATS AND OTHER LIPIDS.)

Proteins. Proteins are complex organic compounds made up chiefly of amino acids, which are present in characteristic proportions for each specific protein. This nutrient always contains carbon, hydrogen, oxygen, and nitrogen; and, in addition, it usually contains sulfur and frequently phosphorus. Proteins are essential in all plant and animal life as components of the active protoplasm of each living cell. The need for protein in the diet is actually a need for its constituents—the amino acids. For more than a century, the amino acids have been studied and recognized as important nutrients.

The dietary proteins are broken down into amino acids during digestion. They are then absorbed and distributed by the bloodstream to the body cells, which rebuild these amino acids into body proteins. Although the primary use of amino acids in the body is synthesis, amino acids can be catabolized as a source of energy.

CATABOLISM. Amino acids from the proteins are the primary sources of carbon skeletons utilized in gluconeogenesis—a process previously explained. In order that amino acids may enter energy metabolic pathways, they must first have their amino (NH_2) group removed—a process called oxidative deamination. This forms ammonia and a carbon skeleton that can be used for energy.

Following deamination, the carbon skeletons, depending upon their structure, can enter the Krebs cycle as pyruvic acid, alpha-ketoglutaric acid, succinic acid, fumaric acid, or oxaloacetic acid and be completely oxidized to carbon dioxide and water yielding the production of ATP (See Table M-16). Transamination—the transfer of an amino group (NH_2) to synthesize new amino acids—also forms carbon skeletons which can enter the Krebs cycle.

Most deaminated amino acids can be used to synthesize glucose by working backwards through glycolysis. This process is called gluconeogenesis, and amino acids capable of this are called gluconeogenic or glycogenic. Alanine is an important gluconeogenic amino acid. Alanine is formed in the muscles by transamination. When it is deaminated in the liver, it becomes pyruvic acid which the liver converts to glucose. Leucine is the only amino acid incapable of gluconeogenesis.

• **Urea**—The fate of the amino group (NH_2) split off from amino acids, presents a special metabolic problem. Fortunately, the body has developed the urea cycle for handling this substance. Most of the NH_2 is converted to urea in the liver for excretion by the kidneys. To facilitate elimination, 1 mole of ammonia (NH_3) combines with 1 mole of carbon dioxide (CO_2), another metabolic waste product. Phosphate is then added to this compound to produce carbamyl phosphate. Carbamyl phosphate then combines with ornithine to form citrulline—an intermediate in the urea cycle. The amino acid, aspartartic acid, contributes another amino group (NH_2), and citrulline is then converted to the amino acid arginine. Urea splits off from arginine forming ornithine, and the cycle is completed. The kidneys remove urea from the blood and excrete it in the urine.

$$H_2N-\underset{\underset{O}{\|}}{C}-NH_2$$

Fig. M-50. The structure of urea—the means of eliminating amino groups from amino acids.

(Also see UREA CYCLE.)

ANABOLISM. The various amino acids are systematically joined together to form peptides and proteins. The amino portion (NH_2) of one amino acid will combine with the carboxyl (COOH) portion of another amino acid releasing water (H_2O) and forming a peptide linkage—like railroad cars joining to form a train. When numerous of these junctions occur, the resulting molecule is called a protein. The proteins of the body are the primary constituents of many structural and protective tissues—bones, ligaments, hair, fingernails, skin, organs, and muscles. Perhaps, most importantly enzymes are proteins, and they are responsible for most all of the catabolic and anabolic reactions.

Protein synthesis is not a random process whereby a number of amino acids are joined together; rather, it is a detailed predetermined procedure. Within the cell, DNA contains the coded information concerning the amino acid sequences of the various proteins to be synthesized in the cell. When DNA is decoded, amino acids are linked to form a specific protein which has its own particular physiological function.

Protein synthesis involves a series of reactions which are specific for *each* protein. A four step outline of this procedure follows:

1. Messenger RNA (ribonucleic acid) transcribes the sequence *message* from DNA to form a template (a pattern or guide) for protein synthesis. The sequences of the nucleotides in the DNA are the keys to the sequence pattern forming this template. This is possible because the nucleotides always pair off in the following manner: adenine (A) and thymine (T); adenine (A) and uracil (U); and guanine (G) and cytosine (C). Triplets of the purine (adenine and guanine) and pyrimidine (cytosine, thymine, and uracil) bases in the DNA form *codons* which correspond to specific amino acids and are signals which control protein synthesis. For example, the codon AGG (adenine, guanine, and guanine) signals the incorporation of arginine.

2. A specific transfer RNA (tRNA) combines with each respective amino acid to form an aminoacyl complex. This reaction requires the expenditure of energy. There is at least one specific tRNA for most of the 22 amino acids.

3. The initiation of protein synthesis occurs when the ribosome of the rough endoplasmic reticulum (site of protein synthesis) recognizes a codon specific for initiation. This first amino acid will be the amino (NH_2) terminal end of the protein. This amino acid is formulated to prevent the amino group of the amino acid from being incorporated in a peptide bond. Upon completion of the protein, the N-formyl group is cleaved from the protein.

4. Once synthesis has been initiated, the protein is elongated through a series of successive additions of amino acids as determined by the messenger RNA template. Each new amino acid is linked to the next by the formation of a peptide bond. Eventually the procedure will be terminated when a codon specific for terminating protein synthesis is reached. Thereupon, the protein splits off from the ribosomes.

Each sequence of amino acids is a different protein. Hence, different proteins are able to accomplish different functions in the body. With 22 amino acids, the different arrangements possible are endless, yielding a variety of proteins. For example, egg albumin, a small protein, contains approximately 288 amino acids. Thus, if one assumes that there are about 20 different amino acids in the albumin molecule, then mathematical calculations show that the possible arrangements of this number of amino acids are in excess of 10^{300}. (For comparison, one million is equal to 10^6.)

Since the DNA of each cell carries the master plan for forming proteins, genetic mutations often are manifested by disorders in protein metabolism. Many metabolic defects—inborn errors of metabolism—are the result of a missing or modified enzyme or other protein. The misplacement or omission of just one amino acid in a protein molecule can result in a nonfunctional protein. Therefore, when proteins are formed three requirements must be met: (1) the proper amino acids, (2) the proper number of amino acids, and (3) the proper order of the amino acid in the chain forming the protein. Meeting these requirements allows the formation of proteins which are very specific and which gives tissues their unique form, function and character.

(Also see INBORN ERRORS OF METABOLISM; and NUCLEIC ACIDS.)

Minerals And Vitamins. The metabolism of minerals and vitamins does not produce energy. However, minerals and vitamins are involved in many of the reactions in the body which comprise metabolism. The following table indicates some of the areas where minerals and vitamins are directly involved in metabolism. More information may be gained about minerals and vitamins under their separate entries.

(Also see MINERAL[S]; and VITAMIN[S].)

TABLE M-17
MINERAL AND VITAMIN FUNCTIONS IN METABOLISM

Mineral or Vitamin	Function
Minerals:	
Calcium	Cell wall permeability; enzyme activation; hormone secretion.
Phosphorus	Energy utilization (ATP formation); amino acid metabolism; protein formation; enzyme systems; formation of some fats (phospholipids).
Sodium	Acid-base balance of the body; absorption of sugars.
Chlorine	Acid-base balance; stomach acid; enzyme activitation.
Magnesium	Enzyme activation.
Potassium	Acid-base balance; enzyme reactions; nutrient transfer.
Sulfur	Component of the vitamins biotin and thiamin; component of coenzyme A (CoA).
Chromium	Component of the Glucose Transfer Factor; enzyme activation; stimulation of fatty acid synthesis.
Copper	Constituent of several enzyme systems.
Iodine	Constituent of thyroid hormones which act to regulate metabolism.
Iron	Component of cytochromes; component of hemoglobin which brings oxygen to the tissues.

(Continued)

TABLE M-17 *(Continued)*

Mineral or Vitamin	Function
Manganese	Enzyme activator.
Molybdenum	Component of enzyme systems.
Selenium	Enzyme component.
Zinc	Component of enzyme systems.
Vitamins:	
A	Synthesis of protein, some hormones, and glycogen.
D	Regulates calcium and phosphorus.
E	Regulator of DNA synthesis, and coenzyme Q, a hydrogen carrier.
Biotin	Coenzyme in decarboxylation (removal of carbon dioxide) and carboxylation (addition of carbon dioxide) reactions of carbohydrate, fat, and protein metabolism; for example, pyruvic acid to oxaloacetic acid, and acetyl CoA to malonyl CoA; formation of purines; formation of urea; deamination of amino acids.
Choline	Transport and metabolism of fats; source of methyl groups (CH_3).
Folacin	Formation of coenzymes for synthesis of DNA and RNA; amino acid synthesis.
Niacin	Constituent of the two coenzymes nicotinamide adenine dinucleotide (NAD) and nicotinamide adenine dinucleotide phosphate (NADP) which transport hydrogen during energy formation.
Pantothenic acid	Form part of coenzyme A (CoA) and the acyl carrier protein (ACP).
Riboflavin	Forms coenzymes like flavin adenine dinucleotide (FAD) which is responsible for transport of hydrogen ions.
Thiamin	A coenzyme in energy metabolism via decarboxylation of pyruvic acid and formation of acetyl CoA.
Vitamin B-6	As a coenzyme for: transamination, decarboxylation, deamination, transsulfuration, absorption of amino acids, conversion of glycogen to glucose, and fatty acid conversions.
Vitamin B-12	Coenzyme in protein synthesis.
Vitamin C	Makes iron available for hemoglobin synthesis, metabolism of the amino acid tyrosine and tryptophan.

Water. Water is a vital nutrient. It is the solvent wherein the metabolic reactions of the body take place. Also, as a solvent, water carries (1) the nutrients which are subjected to cellular metabolism, and (2) the waste products of metabolism. Also, it serves to disperse the heat generated by the metabolic reactions. In many of the metabolic reactions water is either added or subtracted. Subtracted water is termed metabolic water. The addition of water is termed hydrolysis.

(Also see WATER; and WATER AND ELECTROLYTES.)

EXCRETION. Products not used and products formed during metabolism must be continually removed from the body. The three major routes of removal are: (1) feces, (2) urine, and (3) lungs. The feces serve to remove materials that cannot be digested or absorbed. Urea, produced from the metabolism of amino acids is excreted by the kidneys into the urine, along with excess water, mineral salts, and some other compounds formed during metabolism. Carbon dioxide produced during metabolism is excreted from the body via the lungs. For every 2,100 Calories (kcal), about 327 liters of carbon dioxide are produced in the body.

FACTORS CONTROLLING METABOLISM. The rate at which the body utilizes nutrients, and the direction of metabolism—anabolism or catabolism, are controlled by a variety of factors, but primarily the nervous system and endocrine systems. Many of these factors are interrelated and some are more important than others. Generally, metabolism is discussed in terms of energy, since energy to drive the machinery of the body is the first concern. The metabolic rate is the rate at which the body uses energy—calories—packaged in ATP. Some factors affecting metabolism follow:

• **Exercise**—Perhaps, muscular activity is the most powerful stimulus for increasing the metabolic rate. Short bursts of strenuous exercise can increase the metabolic rate to 40 times that of the resting state.

• **Recent ingestion of food**—Following a meal, the metabolic rate rises. This increase is dependent upon the type of food eaten. Carbohydrates and fats elevate the metabolic rate 4 to 15%, whereas proteins increase the rate 30 to 60%. This phenomenon is known as the specific dynamic action of foods (SDA).

• **Environmental temperature**—Increases or decreases in the environmental temperature increase the metabolic rate.

• **Body surface area**—Body surface area is a function of height and weight. A person weighing 200 lb (91 kg) has about 30% more surface area than a person weighing 100 lb (45.5 kg), when both are the same height. Hence, the larger person's metabolic rate is about 30% greater, not twice as great as is the weight.

• **Sex**—Females have a slightly lower metabolic rate than males throughout life.

• **Pregnancy and lactation**—The metabolic rate rises during pregnancy and lactation.

• **Age**—Metabolic rate gradually declines with age. During growth, energy is needed for the anabolic processes.

• **Emotional state**—Various emotional states—fear, anger, shock—stimulate the release of the hormones epinephrine and norepinephrine which can increase the metabolic rate 100% on a short term basis.

• **State of nutrition**—During prolonged undernutrition the body adapts by decreasing the metabolic rate by as much as 50%.

• **Body temperature**—For every one degree increase in body temperature during a fever, the metabolic rate increases 7%.

• **Hormones**—Many of the above changes in metabolism are ultimately due to some hormonal change. Most of the hormones direct the metabolic processes. The storage and release of glucose is hormone mediated. Protein synthesis is hormone mediated, as is the release of fatty acids from fats. The thyroid hormones can, on a long-term basis, raise the metabolic rate 100%. Conversely, undersecretion can cause the metabolic rate to fall 50% below normal. Other hormones control water and minerals, and still others control the digestive processes.

(Also see CALORIC [ENERGY] EXPENDITURE; ENDOCRINE GLANDS; and ENERGY UTILIZATION BY THE BODY.)

INBORN ERRORS OF METABOLISM. At times the metabolism of the nutrients cannot proceed normally due to some defect in the genetic information that exists at birth or shortly thereafter. These defects can affect the metabolism of carbohydrates, proteins, and fats; hence, they are referred to as inborn errors of metabolism. Often they are due to production of a nonfunctional enzyme or complete lack of an enzyme involved in the metabolic scheme. Since enzymes are protein, their production relies upon correct genetic information. Many of these inborn errors have serious consequences, but fortunately most are rare. Familiar examples of errors in carbohydrate metabolism include lactose intolerance and galactosemia. Familiar examples of errors in protein metabolism include albinism, maple syrup urine disease, and phenylketonuria. The hyperlipoproteinemias are familiar examples of inborn errors of fat metabolism.

(Also see INBORN ERRORS OF METABOLISM.)

SUMMARY. Although metabolism is discussed in parts —pathways, nutrients, and cycles—it is an interrelated, continuous process—the sum of all chemical changes in the body. While some organs have more important roles, metabolic processes occur in every cell of the body—from those in the big toe to brain cells. Energy production in the form of ATP is the first concern of metabolism, and water (H_2O), carbon dioxide (CO_2), and urea are the major waste products of metabolism.

METABOLITE

Any substance produced by metabolism or by a metabolic process.

METALLOENZYME

An enzyme of which a metal (ion) is an integral part of its active structure. Examples: carbonic anhydrase (zinc) and cytochromes (iron, copper).

METHIONINE

An essential amino acid which contains sulfur. Furthermore, methionine participates in a biochemical reaction in the body called transmethylation by donating a methyl group (CH_3).

(Also see AMINO ACID[S].)

METHYL ALCOHOL (WOOD ALCOHOL; CH₃OH METHANOL)

The simplest of the alcohols, but it is highly toxic when ingested. As little as 20 ml may cause (1) permanent blindness as a result of the formaldehyde formed by the metabolism of the methyl alcohol damaging the retina of the eye, or (2) death due to severe acidosis. Poisoning due to methyl alcohol is almost always a case of ingestion as a substitute for ethyl alcohol—the alcohol of liquors. Rather ironically, part of the treatment for methyl alcohol poisoning consists of administering ethyl alcohol. Aside from its adverse effects when ingested, methyl alcohol is an important industrial chemical utilized for numerous processes and products.

METHYLATED SPIRITS

Ethyl alcohol to which methyl alcohol has been added to render it unfit for consumption as a beverage. Often called denatured alcohol. The added methyl alcohol makes it highly poisonous.

(Also see METHYL ALCOHOL.)

MEULENGRACHT DIET

A diet for peptic ulcer patients first introduced by the Danish Physician Meulengracht in 1935. It is more liberal than the *Sippy Diet*, which was initiated in 1915 by the American Physician, Bertram Sippy. The Sippy Diet holds rather rigidly to a program of milk and cream feeding with slow additions of single soft food items over a prolonged period of time. To begin with, the Meulengracht Diet consists of milk, eggs, pureed fruits and vegetables, custard, ice cream, gelatin, plain pudding, crackers, bread and butter. Then, after 2 days, ground or minced meats and broiled, baked, or creamed fish are included. Taboo foods are coffee, tea, cocoa, soft drinks, alcoholic beverages, spices, nuts, and pastries.

(Also see MODIFIED DIETS; and ULCERS, PEPTIC.)

MEVALONIC ACID (C₆H₁₂O₄)

In the human body, mevalonic acid occurs as a precursor in the manufacture of cholesterol. In bacterial cultures, mevalonic acid can replace acetate as a growth promoting factor for *Lactobacillus acidophilus*, a bacterium producing lactic acid.

MICELLAR BILE-FAT COMPLEX

A micelle is a particle formed by an aggregate of molecules —a microscopic unit of protoplasm. In the micellar bile-fat complex, the particle is formed by the combination of bile salts with fat substances (fatty acids and glycerides) to facilitate the absorption of fat across the intestinal mucosa.

MICELLE

A microscopic particle of lipids and bile salts.

MICROCYTE

A small red blood cell.

MICROGRAM

A metric measure of weight which is equal to one millionth of a gram or one thousandth of a milligram. One ounce of a substance weighs over 28 million micrograms.

MICRONUTRIENTS

These are nutrients that are present in the body, and required by the body in minute quantities, ranging from millionths of a gram (microgram) to thousandths of a gram (milligram). Examples are vitamin B-12, pantothenic acid, chromium, cobalt, copper, fluorine, iodine, iron, manganese, molybdenum, selenium, silicon, and zinc. Their minuteness in no way diminishes their importance to human nutrition— many are known to be absolutely essential.

(Also see MINERAL[S]; TRACE ELEMENTS; and VITAMIN[S].)

MICROVILLI

Minute (visible only through an electron microscope) surface projections that cover the edge of each intestinal villus, called the brush border. The microvilli add a tremendous surface area for absorption.

MICROWAVE COOKING

Although microwave cooking has been available since the 1950s, it took 20 years for the microwave oven to become a common item on the market.

The magnetron tube is a vacuum tube which can convert electricity into electromagnetic energy radiation. In microwave cooking, food is cooked by the heat generated in the food itself.

The frequencies for microwave heating come under the rules of the Federal Communications Commission, which has designated four frequencies—915, 2450, 5800, and 22,125 mc/sec. The majority of the microwave ovens on the market today use 2450 megacycles.

It requires skill and experience to use the microwave oven successfully. Although it cooks foods in ½ to ⅓ the time of an electric oven, there are the following problems:

1. The microwave oven does not brown the food.
2. Frequently, microwaves do not penetrate a roast uniformly, with the result that there may be variations in doneness of meat. Thus, unless pork cooked in a microwave oven is subjected to a temperature of 170°F (77°C) *throughout* to destroy any organisms that might be present, *Trichinella spiralis* (the parasite that causes trichinosis) and food poisoning bacteria may not be destroyed.

(Also see FOOD—BUYING, PREPARING, COOKING, AND SERVING, section headed "Methods and Media of Cooking".)

MILK-ALKALI SYNDROME

A condition usually associated with sufferers of peptic ulcers wherein the sufferer consumes large amounts of milk and a readily absorbed alkali over a period of years resulting in hypercalcemia—high levels of calcium in the blood. Symptoms consist of vomiting, gastrointestinal bleeding, and increased blood pressure. Furthermore, kidney stones are common and calcium may be deposited in other soft tissues. The use of nonabsorbable antacids in peptic ulcer therapy seems to have lessened the prevalence of the milk alkali syndrome.

(Also see ANTACIDS; and HYPERCALCEMIA.)

MILK AND MILK PRODUCTS

Contents   Page
History and Development of the Dairy Industry 692
Milk from Various Species 692
  Goat Milk 692
Milk From Cow to Table 693
Nutrients of Milk 694
  Carbohydrates in Milk 694
  Milk Fat 695
  Protein in Milk 695
  Minerals in Milk 696
  Vitamins in Milk 696
Milk as a Food 697
Health Aspects of Milk Consumption 697
  Health Benefits from Milk 698
  Health Problems from Milk 698
    Milk Allergy (Milk Sensitivity) 698
    Lactose Intolerance 698
    Milk Intolerance 698
    Galactose Disease (Galactosemia) 699
    Milk Anemia 699
    After Gastric Surgery 699
    Milk Myths 699

Contents (cont.)   Page
U.S. Production of Milk 699
  Dairy Producers 699
    Milking and Handling Milk 699
    Hauling Milk 700
Processing Milk 700
  Dairy Processing Plants 701
  Kinds and Uses of Milk and Milk Products 701
    Fluid Milk 701
    Special Milks 702
    Evaporated and Condensed Milk 702
    Dry Milk 703
    Cultured Milk and Other Products 703
    Specialty Milks 704
    Cream 704
    Ice Cream and Frozen Desserts 705
    Butter 705
    Cheese 705
    Caseinates (Coffee Whiteners; Whipped Toppings) 706
    Whey 706
Marketing and Distributing Milk 706
  Milk Supply 706
  Milk Demand 707
    Factors Influencing Demand 707
  Federal Milk Marketing Orders 707
  State Milk Controls 707
  Other Regulatory Programs 707
    Sanitary Regulations 707
    Standards and Grades 708
      Grade A Milk 708
      Manufacture Grade Milk 708
  Market Distribution of Fluid Milk 708
  Dairy Imports and Exports 708
  Per Capita Consumption of Milk and Milk Products 709
  Care of Milk in the Home 709

Fig. M-51. Foster mother of the human race. (Courtesy, Holstein-Friesian Assn. of America, Brattleboro, Vt.)

Milk is the fluid normally secreted by female mammals for the nourishment of their young.

Milk products include a diverse group of foods made from milk, including butter, cheese, dried nonfat milk, condensed and evaporated milk, yogurt, etc.

In different parts of the world, milk from various species of animals is used for food. In the United States, however, the cow furnishes virtually all of the available market milk. Therefore, unless otherwise stated, the terms *milk* and *milk products* as used in this section refer to cow's milk.

HISTORY AND DEVELOPMENT OF THE DAIRY INDUSTRY.

Fig. M-52. Dairy scene from ancient Egypt, from bas-relief found in the tomb of Princess Kewitt. Note man milking cow while calf is fastened to left foreleg. (Courtesy, The Bettmann Archive, Inc., New York, N.Y.)

Long before recorded history, the first food provided for mankind was from women's mammary glands. In remote times, when nature failed to bless the newborn child with a lactating mother, the baby suckled another mother, or suckled an animal, or died.

In due time, man found that milk was good—and good for him—with the result that he began domesticating milk-producing animals. However, under natural conditions, wild animals produce only enough milk for their offspring. So, man began selecting them for higher production for his own use. For the most part, this included the cow, the buffalo, and the goat—although the ewe, the mare, the sow, and other mammals have been used for producing milk for human consumption in different parts of the world. The importance of the cow in milk production is attested to by her well-earned designation as *the foster mother of the human race.*

As the cow population expanded throughout the world and as milk production became a specialized phase of agriculture, methods were developed for concentrating and preserving milk foods for future use, barter, and/or international trade. Dairying progressed from an art to a science—drawing upon genetics, chemistry, physics, bacteriology, heat, refrigeration, engineering, business, and other branches of knowledge, in order to improve dairy commodities, create new ones, and profitably distribute them in both domestic and world commerce.

Hand in hand with improvements and expansion in production and processing, recognition of the nutritional importance of milk and milk products increased. Following World War II, milk and milk foods were established as one of the groups of foods contributing the essentials of an adequate diet—first as one division of the Seven Food Groups, later in the Four Food Groups, and presently in the Six Food Groups (Food Guide Pyramid).

MILK FROM VARIOUS SPECIES. Most of the world's milk is produced by cows. However, 8.6%—slightly less than one-twelfth—of the global milk supply comes from buffaloes, goats, and sheep. Yet, in 16 Asian countries for which data are available, only 51.5% of the milk is from cows, while water buffaloes, goats, and sheep produce 11.1, 15.2, and 22.2%, respectively. Mares are also used for producing milk for human consumption in different parts of the world.

A comparison of milks from various species, including humans, is given in Table M-18.

Goat Milk. Goats have played an important role in supplying milk for man since prehistoric times, and they still do in many areas. Approximately 400 million goats, including milk, mohair and meat goats, exist throughout the world, most of them in India, China, Turkey, Ethiopia, Iran, and Brazil. Although goats supply less than 3% of the worldwide milk supply for man, it is believed that more people consume goat milk than cow's milk, even though the per capita consumption may be small. This is because the highly populated countries of Asia and Africa have 70% of the world's goats.

Dairy goats commonly produce 3 to 4 qt *(2.9 to 3.8 liter)* of milk (6 to 8 lb) daily, although some yield 20 lb *(9.1 kg)* or more daily at the peak of lactation. Goat milk resembles cow's milk in composition (see Table M-18). However, the fat globules of goat milk are smaller and tend to remain suspended (little cream rises to the top); hence, homogenization is unnecessary. Also, goat milk is whiter in color than cow's milk, because the goat is essentially 100% efficient in converting carotene to vitamin A; it follows that butter made from goat milk is white, also. Goat milk produced under sanitary conditions is sweet, tasty, and free of off-flavors. Goat milk should be produced, handled, and processed much like cow's milk. The most important dairy goat products are milk and cheese, although goat milk can be manufactured into all the products made from cow's milk.

Since the time of Hippocrates, physicians have recommended goat milk for infants and invalids because it is easily digested. Also, those allergic to cow milk can usually drink goat milk without ill effects; and goat milk is widely used by persons afflicted with stomach ulcers.

Fig. M-53. Saanen doe, Laurelwood Acres Merna, owned by Laurelwood Acres. Ripon. Calif. (Courtesy, *Dairy Goat Journal*)

TABLE M-18
COMPOSITION OF MILKS FROM DIFFERENT SPECIES[1]
(Amount in 100 g edible portion. Dashes denote lack of reliable data for a constituent believed to be present in measurable amounts.)

Nutrient	Species				
	Cow	Human	Buffalo	Goat	Sheep
Water .... (g)	87.99	87.50	83.39	87.03	80.70
Food energy .... (kcal)	61	70	97	69	108
.... (kJ)	257	291	404	288	451
Carbohydrate, total .... (g)	4.66	6.89	5.18	4.45	5.36
Fat .... (g)	3.34	4.38	6.89	4.14	7.00
Cholesterol .... (mg)	14	14	19	11	—
Protein (N × 6.38) .... (g)	3.29	1.03	3.75	3.56	5.98
Fiber .... (g)	0	0	0	0	0
Ash .... (g)	.72	.20	.79	.82	.96
Macrominerals					
Calcium .... (mg)	119	32	169	134	193
Phosphorus .... (mg)	93	14	117	111	158
Sodium .... (mg)	49	17	52	50	44
Magnesium .... (mg)	13	3	31	14	18
Potassium .... (mg)	152	51	178	204	136
Microminerals					
Iron .... (mg)	.05	.03	.12	.05	.10
Zinc .... (mg)	.38	.17	.22	.30	—
Fat-soluble vitamins					
Vitamin A .... (RE)	31	64	53	56	42
.... (IU)	126	241	178	185	147
Water-soluble vitamins					
Folacin .... (mcg)	5	5	6	1	—
Niacin .... (mg)	.084	.177	.091	.277	.417
Pantothenic acid .... (mg)	.314	.223	.192	.310	.407
Riboflavin .... (mg)	.162	.036	.135	.138	.355
Thiamin .... (mg)	.038	.014	.052	.048	.065
Vitamin B-6 .... (mg)	.042	.011	.023	.046	—
Vitamin B-12 .... (mcg)	.357	.045	.363	.065	.711
Vitamin C (ascorbic acid) .... (mg)	.94	5.00	2.25	1.29	4.16

[1]*Newer Knowledge of Milk and Other Fluid Dairy Products*, National Dairy Council, Rosemont, Ill., 1979, p. 44, Table D.

MILK FROM COW TO TABLE. Fig. M-54 illustrates how milk gets from the cow to the table.

Fig. M-54. Steps and channels for getting milk from cows to table.

In summary form, the stages in getting milk from the cow to the table follow:

1. Milk is secreted by the mammary glands of a cow, beginning with the birth of a calf and continuing for about 305 days.

2. Milk is removed through the teats, generally by machine milking, although hand milking is still done.

3. Milk is cooled, usually in bulk tanks, to at least 50°F (preferably 40°F) as soon after milking as possible in order to inhibit bacterial growth.

4. Insulated trucks usually pick up milk from dairy farms daily and transport it to processing plants.

5. Processors (a) process and package fluid milk, or (b) manufacture milk into various products.

6. Milk and milk products are distributed through wholesale and retail channels to consumers.

NUTRIENTS OF MILK.[19]   Although fluid whole milk is a liquid food (88% water), it contains an average of 12% solids (total solids include all the constituents of milk except water) and 8.6% solids-not-fat (includes carbohydrate, protein, minerals, and water-soluble vitamins).

More than 100 components have been identified in milk. Two glasses (1 pt) of whole milk provides approximately 23 g of carbohydrate (total), 16.3 g of fat, 16 g of protein, 3.5 g of minerals, plus fat-soluble and water-soluble vitamins. But the composition of milk varies in response to physiological factors (inherited or genetic) and environmental factors. Variations in composition occur among breeds, between milkings, and between milk taken from different sections of the udder. Also, the composition of milk is affected by the feed, the environment temperature, the season, and the age of the cow.

Milk plays an important role in the diet of the average American, as is shown in Table M-19 and Fig. M-55. The major nutrients provided by dairy products in the American diet are food energy (9.6%), fat (11.7%), protein (19.5%), calcium (75.1%), phosphorus (34.3%), vitamin A (15.8%), and riboflavin (32.6%).

TABLE M-19
FOOD NUTRIENTS: U.S. PERCENTAGE OF TOTAL CONTRIBUTED BY LIVESTOCK AND POULTRY PRODUCTS, 1988[1]

	Food Energy	Carbo-hydrates	Fat	Pro-tein	Cal-cium	Phos-phorus	Mag-nesium	Iron	Vitamin A Value	Niacin	Ribo-flavin	Thiamin	Vitamin B-6	Vitamin B-12
							(%)							
Dairy products, excluding butter	9.6	5.3	11.7	19.5	75.1	34.3	18.5	2.4	15.8	1.6	32.6	7.5	10.2	18.5
Meat, fish, and poultry ...	18.8	0.1	31.7	43.3	3.9	29.2	15.4	22.1	20.6	44.8	22.2	25.0	40.5	76.1
Eggs	1.4	0.1	2.0	4.0	1.9	3.9	1.0	2.9	4.0	0.1	7.2	0.9	2.2	3.7
Total	29.8	5.5	45.4	66.8	80.9	67.4	34.9	27.4	40.4	46.5	62.0	33.4	52.9	98.3

[1]*Agricultural Statistics 1991*, p. 478.

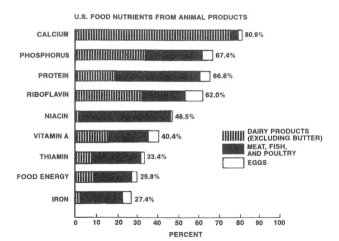

Fig. M-55. Percentage of food nutrients contributed by animal products of the total supply in the U.S.

The major components of milk—carbohydrate, fat, and protein—possess a wide range of nutritional and functional properties which lend themselves to application in a variety of food formulations. Milk is also a valuable source of certain minerals and vitamins.

Carbohydrates In Milk.   Lactose is, by far, the most abundant sugar in milk, constituting about 4.9% of cow's milk and 7.0% of human milk. This disaccharide, consisting of glucose and galactose, is only one-sixth as sweet as sucrose (table sugar) and is less soluble in water. In the small intestine, lactose is cleaved to form its constituent monosaccharides. Glucose is readily absorbed, while galactose is slowly absorbed, thereby serving as a growth promotant for intestinal bacteria that synthesize the vitamins biotin, folacin, and riboflavin in the intestine. Lactose is also converted to lactic acid by intestinal bacteria. When lactic acid is produced in the intestines, growth of pathogenic bacteria is inhibited.

Lactose has been shown to facilitate the absorption of calcium. Also, there are reports that lactose enhances the absorption of barium, magnesium, phosphorus, radium, strontium, and zinc.

Glucose and galactose are the products of lactose hydrolysis by the enzyme lactase. Some individuals are unable to metabolize lactose because of reduced lactase levels, a condition termed lactose intolerance. Preparations of the enzyme lactase are now being marketed for commercial or consumer addition to milk. The resulting milk is sweeter tasting because the majority of the lactose is hydrolyzed to glucose and galactose.

Lactose has many food uses; among them, the following: as an anticaking ingredient in dry mixtures such as nondairy coffee whiteners; as a flavor enhancer in dusting powders to flavor potato chips, barbecue sauces, salad dressings, pie fillings, and puddings; in candies and confectionery products; and in the bakery industry.

(Also see subsequent section headed "Lactose Intolerance.")

---

[19]In the preparation of this section, the authors adapted material from the following publications by the National Dairy Council: *Newer Knowledge of Milk and Other Fluid Dairy Products*, and *Dairy Council Digest*, Vol. 48, No. 5; published by National Dairy Council, Rosemont, Ill.

Milk Fat. Milk fat, sometimes called butterfat, contributes about 48% of the calories of whole milk. Fat also gives milk its flavor.

Milk fat is highly emulsified, which facilitates its digestion. Since fats must be liquid or emulsified at body temperature to be digested and absorbed, and since milk fat has a melting point lower than body temperature, it is efficiently utilized—particularly by the young and the elderly.

Fat is the most variable constituent of milk, the amount being affected primarily by such factors as breed of cow and environmental temperature (a given cow will produce higher fat milk in cool weather than in hot weather). Minor factors affecting the amount of milk fat include the type and amount of feed and the stage of lactation. Feed also influences the fatty acid composition of the fat produced; for example, cows consuming succulent feed or pasture produce much more oleic acid than cows receiving dry roughage.

Milk fat consists of 12.5% glycerol and 87.5% fatty acids, by weight. The fatty acids combine with glycerol in units of three fatty acid molecules per glycerol molecule to form triglycerides which make up 95 to 96% of the total milk fat.

Milk fat contains a greater number of different fatty acids than any other food fat; approximately 500 different fatty acids and fatty acid derivatives have been identified in milk. Milk fat contains a relatively high proportion of short-chain fatty acids. These short-chain fatty acids (which are virtually absent in most vegetable oils) are water-soluble, are absorbed through the intestinal wall without being resynthesized to glycerides, and are transported in the portal vein directly to the liver where they are immediately converted to forms utilizable for energy. For this reason, they serve as a quick source of energy, which may be important—especially early in life.

Generally, the fatty acids are approximately 66% saturated (devoid of double bonds between the carbon atoms), 30% monounsaturated, and 4% polyunsaturated. The saturated fatty acids present in largest amounts in milk fat are myristic, palmitic, and stearic. The chief unsaturated (double-bond linkage between the carbon atoms) fatty acids are oleic, linoleic, and linolenic. Arachidonic acid, which has four double bonds, is present in trace amounts. Arachidonic acid and its precursor, linoleic acid, are essential fatty acids—they are not synthesized in the human body, or they are not synthesized fast enough; consequently, they must be supplied in the diet.

Milk fat is available in the food industry as conventional butter, anhydrous milk fat, or anhydrous butter oil.

Besides butter, fluid whole milk, cheeses, and ice cream are the main dairy foods in which milk fat is consumed in the United States. Milk fat is also used in the candy industry, in the baking industry; and, in the form of butter, in soups, and sauces for frozen and canned vegetables.

Cholesterol is the principal sterol in milk. An 8-oz (1 cup, or *240 ml*) serving of whole fluid milk (3.34% fat) contains 33 mg of cholesterol, whereas a similar serving of fluid skim milk (0.1% fat) contains 4 mg of cholesterol. In order to limit the consumption of cholesterol to 300 mg per day, as is recommended by most health professionals, low-fat dairy products may be used.

Protein In Milk. Fluid whole milk contains an average of 3.3% protein, mainly casein and whey proteins. Protein accounts for about 38% of the total solids-not-fat content of milk and for about 22% of the calories of whole milk. Dairy products (excluding butter) contributed 19.5% of the total protein in the U.S. diet in 1990.

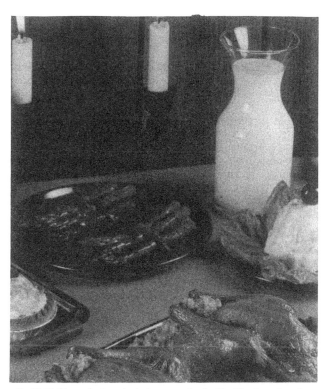

Fig. M-56. Dinner for two, includes, a carafe of milk. (Courtesy, American Dairy Assn., Rosemont, Ill.)

The proteins of milk are of very high quality, having a biological value of 85 as compared to 50 to 65 for the cereal proteins. They contain, in varying amounts, all of the amino acids required by man. Except for the sulfur-containing amino acids (methionine and cystine), 1 pt (*470 ml*) of milk supplies the recommended daily allowance of all the essential amino acids. Additionally, the protein to calorie ratio is very favorable in milk, assuring that the consumer is not ingesting *empty* calories.

It is also noteworthy that milk is rich in the amino acid tryptophan, a precursor of niacin; hence, milk is an excellent source of niacin equivalents. (One niacin equivalent is defined as 1 mg of niacin or 60 mg of tryptophan.) It is noteworthy, too, that the immunoglobulins in milk are a heterogeneous group of antibody proteins that serve as a source of passive immunity for the newborn, and that colostrum contains large quantities of globulins. A discussion of milk proteins follows.

• **Casein**—Comprising about 82% of the total protein content, casein is found only in milk, in which it is the principal protein. It gives milk its white color. Casein alone has a somewhat lower biological value than the whey proteins because of the limited quantity of the sulfur-containing amino acids. Commercial casein is made by either of two techniques: (1) precipitation by acid, or (2) coagulation by rennin. This procedure is followed by washing and drying the casein curd to a granular powder. The caseins and caseinates (sodium caseinate and calcium caseinate) are concentrated forms of proteins which have exceptional water-binding, fat-emulsifying and whipping capabilities.

Casein has been utilized as the main source of protein in the manufacture of meat analogs and as a protein supplement in some meat products. Also, it has found application in baked goods and as an ingredient of coffee whitener and whipped toppings.

Sodium caseinates are used in the meat industry as water-binding agents, as well as in the preparation of toppings, cream substitutes, and numerous desserts as whip-imparting agents.

Calcium caseinate, while significantly less soluble than the sodium salt, is used in various specialized dietary foods where low sodium intake is essential.

• **Whey protein**—A by-product of cheese and casein manufacture, whey is the liquid that remains after the removal of casein and fat from milk.

Like casein, whey protein is a complex mixture of a large number of protein fractions, with beta-lactoglobulin and alpha-lactalbumin being the principal ones of commercial interest. The whey proteins have a relative surplus of the sulfur-containing amino acids.

Recent advances have made the production of undenatured whey protein concentrates (WPC) a commercial reality. Among the uses made of WPC in food products are: in the fortification of acid fruit drinks and carbonated beverages; in cake flour; in dessert topping; and as a binder in the production of textured vegetable protein products. Also, the excellent binding, emulsifying, and gelling properties of WPC have been utilized in the formulation of meat products, soups, and sauces. In addition to the contributions that whey proteins make toward improved organoleptic characteristics in other foods, their nutritional contributions should not be underestimated.

• **Co-precipitates**—As the name implies, co-precipitates contain the total milk proteins that are precipitated from fresh milk. They are unique in that they contain both caseins and whey proteins, thus making them of high nutritional value. The low content of lactose also makes these coprecipitates useful in formulations where the use of dry milk powder may be unacceptable. Co-precipitates have found uses in baking, in dessert and confectionery products, and in meat analogs.

Minerals In Milk.   Minerals are generally classified as macrominerals or microminerals, according to the amounts needed in the diet. These two groupings will be used in the discussion that follows:

• **Macrominerals**—Minerals that are needed at levels of 100 mg or more daily are generally classified as macrominerals.

1. **Calcium.** Of all the foods commonly eaten in the American diet, milk is one of the best and most commonly consumed sources of calcium. Dairy products (exclusive of butter) contributed 75.1% of the total calcium in the U.S. diet in 1990.

It is difficult to meet the RDA for calcium without including milk or milk products in the diet. Two 8-oz glasses (1 pt or *480 ml*) of milk daily will provide about 73% of the adult RDA for calcium. An additional allowance of 400 mg of calcium per day (a total of 1.2 g) during pregnancy and lactation will provide for the needs of the mother, the growing fetus, and the infant; a quart *(950 ml)* of milk will meet approximately 97% of this calcium allowance. For children 1 to 10 years of age, 800 mg of calcium daily are recommended, an allowance which is amply met by three 8-oz glasses of milk. During periods of rapid growth in preadolescence and puberty, 1.2 g of calcium per day are recommended, an amount supplied by 1 qt of milk.

2. **Phosphorus.** Dairy products (exclusive of butter) contributed 34.4% of the total phosphorus in the U.S. diet in 1990.

The RDA for phosphorus is the same as that for calcium after 1 year of age, thus two glasses *(480 ml)* of milk will supply the same percentage of the RDA for phosphorus as that for calcium.

3. **Magnesium.** Dairy products (exclusive of butter) contributed 18.5% of the total magnesium in the U.S. diet in 1990.

The RDA for magnesium is about 350 and 280 mg for adult males and females, respectively. Milk contains about 13 mg per 100 g; so, two 8-oz glasses (1 pt) furnish about 19% of the magnesium RDA for the adult male and 20.5% for the adult female.

• **Microminerals**—These are the minerals which are needed at levels of a few milligrams daily. Of the 103 known microminerals, 17 exhibit biological function in man and animals. The contribution which milk makes to the essential trace element intake is greatest with respect to zinc and possibly iodine.

The microminerals in milk are derived from the cow's feedstuffs, water, and environment. The concentrations are highly variable due to stage of lactation, season, milk yield, amount of the trace element in the cow's diet, handling of the milk following pasteurization, and storage conditions.

1. **Iodine.** Iodine is present in milk, but in variable amounts. As sea water is rich in iodine, milk from coastal agricultural areas contains more iodine than milk from inland areas. The iodine content of milk is highly dependent upon the composition of the animal's feed; it can be increased by as much as 200 times by adding iodine to the feed or by giving iodized salt blocks. Whole milk normally contains 13 to 37 mcg iodine per 100 g. So, two 8-oz glasses *(480 ml)* of milk supply about 25 to 69% of the iodine RDA for an adult man, 25 to 50 years of age.

2. **Iron.** Milk is a poor source of iron. The iron content of milk ranges from 10 to 90 mcg per 100 g. Feeding supplemental iron to cows does not increase the iron content of milk. However, milk is a very satisfactory food for iron fortification.

Two 8-oz glasses (1 pt) of milk provide about 1% of the iron recommended for women of childbearing age.

3. **Zinc.** This mineral occurs in many foods, including milk. The zinc content of milk varies from 0.3 to 0.6 mg per 100 g of milk, with an average of 0.38 mg. It may be slightly increased by feeding a zinc supplement to the cow. About 88% of zinc in milk is associated with casein; thus, the level of zinc in milk is mostly related to the milk protein content. It follows that the zinc content of lowfat and skim milk is virtually identical to that of whole milk.

Two 8-oz glasses (1 pt) of milk provide about 14% of the zinc RDA for adults.

(Also see MINERAL[S].)

Vitamins in Milk.   For convenience, the vitamins in milk are grouped and discussed as fat-soluble vitamins and water-soluble vitamins in the discussion that follows:

• **Fat-soluble vitamins**—Vitamins A, D, E, and K are associated with the fat component of milk. Their quantities in milk, with the exception of vitamin K are dependent upon the dietary intake of the cow.

1. Vitamin A and its precursors, the carotenoids (principally beta-carotene), are present in high, but variable, quantities in milk. The carotenoids, which give milk its characteristic creamy color, constitute from 11 to 50% of the total vitamin A activity of milk. The carotenoid intake of the cow varies seasonally, generally being highest when the animals are on pasture.

One hundred grams of whole milk contain about 126 IU of vitamin A (31 RE); so, 1 qt *(960 ml)*, of whole milk will supply about 30% of the vitamin A recommended dietary allowance (RDA) of 5,000 IU (1,000 RE) for the adult male.

2. **Vitamin D.** Vitamin D (ergocalciferol and cholecalciferol), the ricket-preventing vitamin, is present in low concentrations in unfortified milk. However, approximately 98% of the fluid milk marketed in the United States is fortified with vitamin D to obtain standardized amounts of 400 IU (10 mcg cholecalciferol) per quart, the recommended intake for most persons.

The addition of vitamin D to fluid whole milk—thereby resulting in a widely used food providing calcium, phosphorus, and vitamin D—has been credited with the virtual eradication of rickets among children in the United States. The vitamin D fortification of milk to provide a concentration of 400 IU per quart has been endorsed by the Food and Nutrition Board, National Research Council–National Academy of Sciences; the American Academy of Pediatrics; and the American Medical Association.

3. **Vitamin E.** Vitamin E (principally alpha-tocopherol) is present in low concentrations in milk.

4. **Vitamin K.** Vitamin K is found only in trace amounts in milk. Also, there is some indication that part of the vitamin K of milk is destroyed by pasteurization.

• **Water-soluble vitamins**—All of the vitamins known to be essential for man have been detected in milk. The water-soluble vitamins are in the nonfat portion. The content of water-soluble vitamins in milk is relatively constant and not easily influenced by the vitamin content of the cow's ration.

1. **Biotin.** Milk is a fairly good source of biotin, generally providing about 3 mcg per 100 g. Two glasses of milk would provide between 7 and 23% of the average daily dietary intake of this vitamin. Less than 10% of biotin is destroyed by pasteurization.

2. **Folate (folic acid).** According to recent analyses, milk contains an average of 5 mcg of folate per 100 g. Two glasses *(480 ml)* of milk would supply about 12% of the 200 mcg RDA. Folate of milk is decreased by heat, but this can be minimized by excluding oxygen.

3. **Niacin.** The average niacin content of milk is only 0.08 mg per 100 g; so, two glasses of milk make little contribution (2 to 3%) to the 15 mg niacin RDA for adults. Nevertheless, milk and milk products are among the most effective pellagra-preventive foods because of (a) the complete availability of niacin in milk; and (b) the presence of the amino acid tryptophan (about 46 mg tryptophan per 100 g milk) in milk protein, which can be used for the synthesis of niacin in the body. (A dietary intake of 60 mg tryptophan is equivalent to 1 mg niacin.) Thus, the niacin value of milk is considerably greater than is reflected by its niacin concentration. Also, it is noteworthy that pasteurization does not destroy the niacin content of milk.

4. **Pantothenic acid (vitamin B-3).** Milk is a good source of pantothenic acid; it averages about 0.31 mg of pantothenic acid per 100 g. Pasteurization does not destroy the pantothenic acid in milk.

5. **Riboflavin (vitamin B-2).** Milk is an important source of this vitamin; dairy products (excluding butter) contributed 32.6% of the total riboflavin in the U.S. diet in 1990. The average riboflavin content of fluid whole milk is about 0.16 mg per 100 g. Two glasses *(480 ml)* of milk would supply between 45 and 60% of the 1.3 to 1.7 mg RDA for adults. Although riboflavin is not sensitive to heat, it can be destroyed by exposure to light. Riboflavin losses due to light can be minimized by protecting milk from exposure to strong light through utilizing opaque containers, transporting milk in special trucks, and storing milk in darkened refrigerators.

6. **Thiamin (vitamin B-1).** Significant quantities of this vitamin are found in milk—an average of 0.04 mg per 100 g. Since the RDA for thiamin is between 1.1 and 1.5 mg for the adult, two glasses of milk per day would supply about 13 to 18% of the RDA. Pasteurization results in the loss of about 10% of the thiamin.

7. **Vitamin B-6 (pyridoxine, pyridoxal, pyridoxamine).** An average of about 0.04 mg of vitamin B-6 are found in 100 g of milk; but there is considerable variation in the content of this vitamin in milk. Two glasses of milk furnish about 10% of the 2 mg RDA for adults. Pasteurization does not affect the vitamin B-6 content of milk.

8. **Vitamin B-12.** Milk is a good source of vitamin B-12; it contributed 18.5% of the total vitamin B-12 in the U.S. diet in 1990. Milk averages about 0.36 mcg of vitamin B-12 per 100 g. Two glasses of milk would furnish about 88% of the 2 mcg RDA for most adults. Pasteurization causes only minor destruction of the vitamin B-12 in milk.

9. **Vitamin C (ascorbic acid).** Fresh raw milk contains a small amount of ascorbic acid (about 0.94 mg per 100 g of milk), but processing and exposure to light and heat reduce the amount.

10. **Other vitamins.** Milk also contains choline, myoinositol, and para-aminobenzoic acid, the quantitative requirements of which have not been established.

(Also see VITAMIN[S].)

MILK AS A FOOD. Milk is often referred to as "the most nearly perfect food." It is consumed as a liquid and in the form of such products as butter, cheese, ice cream, and numerous other foods. Milk contains all the nutrients (food elements) needed for growth and good health.

The milk food group makes important contributions to the diet. Milk and milk products are the leading source of calcium and a good source of phosphorus. Also, they provide high-quality protein, the fat-soluble vitamins—A, D (when fortified), E, and K, and all the water-soluble vitamins. Lowfat or skim-milk products fortified with vitamins A and D have essentially the same nutrients as whole milk products, but with fewer calories.

No food has a wider acceptability or offers a greater variety of uses than milk. It is the first food of newborn babies, whether they are breast-fed or bottle-fed; and milk and milk products are important throughout life. For this reason, milk and dairy foods comprise a separate group in The Six Food Groups (Food Guide Pyramid).

(Also see FOOD GROUPS.)

Milk also fills a special role in meeting the nutrient needs of each of the following groups:

1. Pregnant and lactating women. (See PREGNANCY AND LACTATION NUTRITION.)

2. Infants. (See INFANT DIET AND NUTRITION.)

3. Children and adolescents. (See CHILDHOOD AND ADOLESCENT NUTRITION.)

4. Adults. (See ADULT NUTRITION.)

5. Elderly. (See GERONTOLOGY AND GERIATRIC NUTRITION.)

The recommended nutrient allowances (RDA) for each of the above groups are given in the section on Nutrients: Requirements, Allowances, Functions, Sources, Table N-3.

HEALTH ASPECTS OF MILK CONSUMPTION. Numerous advantages, along with some disadvantages, should be considered when incorporating milk in a nutrition program.

Health Benefits from Milk. The health, strength and vitality of people is dependent upon their getting an adequate supply of the right nutrients and foods. Milk, the most nearly perfect food, contributes richly to these needs. It bridges the gap all the way from the dependent fetus to independent old age.

The health imparting qualities of milk are indicated by the fact that daily consumption of a quart *(950 ml)* of cow's milk furnishes an average adult approximately all the fat, calcium, phosphorus, and riboflavin; ½ the protein; ⅓ of the vitamin A, ascorbic acid, and thiamin; ¼ of the calories; and, with the exception of iron, copper, manganese, and magnesium, all the minerals needed daily. These and other nutritional and health benefits of milk are detailed under two earlier headings of this section entitled "Nutrients of Milk" and "Milk as a Food"; hence, the reader is referred thereto.

Additional health benefits of milk are evidenced by its contribution to longevity, tranquilizing effect, ulcer therapy, symbolic meaning, and perhaps, lessening stomach cancer.

Health Problems from Milk. Several metabolic disorders are found in certain individuals when milk is introduced into the diet; among them, milk allergy, lactose intolerance, milk intolerance, galactose disease, milk anemia, or following gastric surgery.

MILK ALLERGY (MILK SENSITIVITY). Cow's milk is probably the most common food allergen in the United States, estimates of the incidence of which range from 0.3% to 3%. It tends to run in families; and an infant who has an allergy to milk may also be allergic to other foods, such as eggs.

When allergy to cow's milk exists, the following dietary alternatives are suggested:

1. Substitute goat's milk.

2. Change the form of the milk; that is, try boiled, powdered, acidulated, or evaporated milk.

3. Eliminate cow's milk and milk products from the diet, and substitute formulas in which the protein is derived from meat or soybeans.

(Also see ALLERGIES.)

LACTOSE INTOLERANCE. During digestion, nature ordained that lactose (milk sugar) be hydrolyzed (split) into its component monosaccharides, glucose and galactose, by the enzyme lactase. Sometimes this physiological process does not occur, and lactose intolerance results.

Lactose intolerance is the malabsorption of the milk sugar lactose due to a decrease in, or absence of, the enzyme lactase, which results in characteristic clinical symptoms (abdominal pain, diarrhea, bloating, flatulence). Three types of lactose intolerance are known:

1. Congenital lactose intolerance (often regarded as an inborn error of metabolism), a relatively rare condition, caused by an absence of lactase from birth.

2. Secondary lactose intolerance caused by damage by viruses, bacteria, allergens, etc. to the outer cell layer of the intestinal epithelium, possibly the brush border or the glycocalyx which surrounds the cell membrane, where the enzyme is confined.

3. Primary lactose intolerance due to an apparently normal development decrease in lactase activity.

The latter type—primary lactose intolerance—involving a normal decrease in the enzyme lactase after early childhood, is the most common type; hence, the discussion that follows pertains primarily thereto.

Normally, the enzyme lactase reaches maximum activity soon after birth, and its activity remains high throughout infancy (the suckling period). But, by late childhood, it decreases to a very low level in all but the Caucasian race.

Because of low intestinal lactase activity in a high proportion of non-Caucasian adults, their consumption of appreciable quantities of lactose can create considerable problems. In such cases, only a portion of the lactose is hydrolyzed; the excess passes down into the lower regions of the gastrointestinal tract, where it undergoes bacterial fermentation. As a consequence, large quantities of fluids are drawn into the lumen of the gut due to the osmotic effect of the uncleaved lactose molecule and its fermented by-products. A reaction follows, generally characterized by abdominal bloating, gaseousness, cramps, flatulence, and watery diarrhea. Initial symptoms are usually observed 30 to 90 minutes after the administration of the disaccharide; diarrhea usually occurs within 2 hours; and the symptoms disappear within 2 to 6 hours after the intake of lactose.

Lactose intolerance is of practical concern in developing countries, as well as in school lunch programs, where milk is commonly used in attempts to correct or avoid serious nutritional problems.

This metabolic disorder interests the anthropologist as well as the physiologist because certain populations throughout the world show extremely high rates of lactose intolerance. Less than 15% of Scandinavians and those of Western and Northern European extraction exhibit this affliction, while 60 to 80% of Greek Cypriots, American Indians, Arabs, Ashkenazi Jews, Mexican-Americans, and African Americans, and 90% of African Bantus and Orientals have been reported to be lactose intolerant. An estimated 30 to 50 million Americans are lactose intolerant. In general, populations in which dairying is traditional seem to tolerate lactose, whereas people from nondairying parts of the world do not; and the inability to utilize lactose in the latter groups persist in successive generations in spite of migration. The question posed by anthropologists is whether the intolerance to lactose evolved due to the inadequate consumption of milk over a period of generations or whether the low consumption of milk was a result of low gut lactase activity originally. That is, did these people once have the ability to digest lactose and eventually lose it, or have they always had this lactase insufficiency?

Fortunately, many people having lactose intolerance can digest most cheeses and other fermented dairy products, such as yogurt. Fortunately, too, lactase is now on the market. By adding it to milk, lactose is changed into two simple sugars, glucose and galactose. The resulting milk is four times sweeter than regular milk, but is otherwise unchanged. Another alternative is lactose-reduced lowfat milk currently available that contains 70% less lactose.

MILK INTOLERANCE. Milk intolerance is not the same as lactose intolerance. Rather, it refers to the development of significant symptoms similar to those described for lactose intolerance following the consumption of usual amounts of milk or milk-containing products. However, many persons with lactose intolerance (lactase deficiency) can consume milk in 1-cup quantities at meals without discomfort, and others can tolerate smaller quantities.

Suitable alternatives for milk-intolerant individuals should provide the same nutrients as does milk. Lactose hydrolyzed milk has been shown to be a suitable alternative in most cases. In addition, other dairy foods such as cheese and other fermented dairy products are well tolerated by the truly milk-intolerant individual.

On the basis of present evidence, it would appear inappropriate to discourage supplemental milk feeding programs targeted at children on the basis of primary lactose intolerance. Also, the use of milk should not be discouraged in feeding malnourished children—except when they have severe diarrhea.

GALACTOSE DISEASE (GALACTOSEMIA).

Galactose disease, an inborn error of carbohydrate metabolism, inherited as an autosomal recessive trait, is caused by insufficient levels of the enzyme galactose-1-phosphate uridyl transferase, sometimes abbreviated P-Gal-transferase, which is needed in the liver for the conversion of galactose to glucose. Galactose is derived from the hydrolysis of lactose (milk sugar) in the intestine. (The enzyme lactase splits lactose into glucose and galactose.)

Where this genetic defect exists, galactose disease becomes apparent within a few days after birth by such symptoms as loss of appetite, vomiting, occasional diarrhea, drowsiness, jaundice, puffiness of the face, edema of the lower extremities, and weight loss. The spleen and liver enlarge. Mental retardation becomes evident very early in the course of the disease, and cataracts develop within the first year.

Dietary treatment consists of early diagnosis and the complete exclusion from the diet of milk, the only food that supplies lactose, along with all foods that contain milk. As a rule, the substitution of a nonmilk formula (usually a meatbase or a soy-base formula, supplemented with calcium gluconate or chloride, iron, and vitamins) leads to rapid improvement. All the symptoms disappear except the mental retardation which has already occurred and is not reversible.

Complete elimination of galactose is necessary for the young child, but breads and other prepared foods containing milk are usually permitted when the child enters school. Milk must be permanently excluded from the diet, however.

(Also see INBORN ERRORS OF METABOLISM, Table I-2, Inborn Errors of Carbohydrate Metabolism.)

MILK ANEMIA.

Milk anemia is the condition that results when infants and children drink milk over an extended period without iron supplementation by an iron salt or an iron-rich food(s).

During growth, the demand for positive iron balance is imperative. At birth, a newborn infant has about 3 to 6 months' supply of iron, which was stored in the liver during fetal development. Since milk does not supply iron, supplementary iron-rich foods must be provided to prevent the classic milk anemia of young children fed cow's milk only after 6 months of age.

Iron is also needed for continued growth during the toddler stage (1 to 3 years). Sometimes excessive milk intake during this period, a habit carried over from infancy, may exclude some solid foods from the diet. As a result, the child may be lacking iron and develop a milk anemia.

AFTER GASTRIC SURGERY.

The patient whose gastric or duodenal ulcer does not respond to medical treatment is the most common candidate for gastric surgery.

Immediately after gastric surgery, milk may or may not be included in the small frequent feedings, for the reason that the contents of the stomach may pass into the small intestine before it is in proper solution and cause distention of the jejunum. When this happens, the patient experiences nausea, cramps, diarrhea, light-headedness, and extreme weakness. This occurs 15 to 30 minutes after meals and is known as the *dumping syndrome.*

Also, some, but not all, individuals develop lactose intolerance following gastric surgery; therefore, some physicians regularly exclude milk after such surgery. However, other physicians exclude milk only after the patient has experienced distention and diarrhea that are relieved by the exclusion of milk.

Milk Myths: Some myths and incorrect statements related to health concerns appear to have acted as barriers to milk consumption; among them, the following:

Milk is toxic to nonwhite people.
Milk causes coronary heart disease.
Milk causes cancer.
Milk causes iron deficiency anemia.
Milk causes urinary calculi.
Milk causes acne.

U.S. PRODUCTION OF MILK.

The United States ranks fourth in number of milk cows and third in production per cow, among the leading dairy countries of the world. The former U.S.S.R. holds a commanding lead in the number of milk cows; and Israel and Japan rank first and second, respectively, in average milk production per cow.

Milk is produced in every state of the Union. However, the greatest concentration of dairy cows is found in those areas with the densest human populations. This is as one would expect from the standpoint of the demand for, and the marketing of, fresh milk—a highly perishable product.

Fig. M-57. Cows convert the photosynthetic energy derived from solar energy and stored in grass into milk for humans. (Courtesy, Union Pacific Railroad, Omaha, Nebr.)

Dairy Producers. It is impossible to describe the *average* American dairy farm, for the dairyman can choose to tailor his operation in an infinite number of ways. Herd size, breeds, milking equipment, frequency of milking, and intensity of production all enter into the makeup of the dairy farm.

MILKING AND HANDLING MILK.

Milk is secreted by the mammary glands of mammals. In the United States, cows furnish virtually all market milk.

Lactation begins at freshening, when a calf is born, and usually lasts 305 days. Milk is removed through the teats (the elongated nipples). Milking is carried out by either of two methods: (1) hand milking, or (2) machine milking.

The composition of milk depends mainly on the breed

of cows. The flavor and quality of milk vary according to how the cows are cared for, what they are fed, and how the milk is handled. Milk must be kept cool and clean from the moment it comes from the cows.

Milk possesses two characteristics which make it ideal for the development of bacteria: (1) It is a well-balanced food in which bacteria thrive; and (2) the temperature, as it comes from the cow, is ideal for bacterial growth. For these reasons, milk must be cooled to at least 50°F (preferably 40°F, or 4°C) as soon as possible in order to inhibit bacterial growth.

Fig. M-58. Cows being machine milked in a milking parlor. Note that the milk is piped directly into a refrigerated storage tank; without exposure—all very sanitary. (Courtesy, USDA)

Quality milk can be produced only when dairymen pay special attention to a number of factors; among them, herd health, clean cows, care of the milking equipment, cooling and storage of milk, and proper transportation of milk to market.

HAULING MILK.

Fig. M-59. The bulk system of handling milk. This over-the-road transport is capable of moving 6,000 gal (22,800 liter) of milk. The insulated stainless steel tank holds milk within a few degrees of its original temperature. (Courtesy, Dairymen, Inc., Lexington, Ky.)

Practically all milk is now (1) placed in mechanically cooled bulk tanks on farms, and (2) transported to processing plants in bulk tank trucks with capacities ranging from 15,000 to 50,000 lb (6,803 to 22,676 kg).

Because ownership changes at loading, this places responsibility on the truck operator to evaluate odor, flavor, and appearance of the milk, and to measure and sample the milk accurately and properly.

Fig. M-60. Seventeenth century dairy manufacturing plant. Note (1) division of labor among women workers; and (2) buttermaking, using two different types of churns: a barrellike churn with paddles inside (elevated in back of room and to the left), and a churn with a stirring stick called a *dasher* (right). (Courtesy, The Bettmann Archive, New York, N.Y.)

PROCESSING MILK. Upon arrival at the processing plant, trained laboratory technicians make many tests to ensure the quality and purity of milk, including temperature, flavor, odor, and appearance. They reject milk that does not meet the required high standards for quality.

Also milk is tested to determine its fat content; and different batches may be blended, in a process called standardization, to adjust the fat content.

Today, most milk is homogenized, pasteurized, and fortified (at least with vitamin D), the details of which follow:

• **Homogenization**—Most whole milk is homogenized to prevent formation of a cream layer. Homogenization is accomplished by passing hot milk under high pressure through a narrow constriction with a specially designed pump. In the process, fat globules are reduced in size so that they are no longer acted upon by force of gravity and remain evenly dispersed throughout the milk. The keeping quality and nutritive value of homogenized and nonhomogenized milk are similar. The main **advantages** of homogenizing are: (1) no separation of the cream—the product retains uniform consistency; and (2) a softer curd forms in the stomach, which aids digestion. The main **disadvantages**: (1) The protein is more readily coagulated by heat or acid so that care must be taken to avoid curdling; and (2) increased susceptibility of off-flavor induced by sun or fluorescent light.

• **Pasteurization**—*Pasteurization is the heating of raw milk in properly approved and operated equipment at a sufficiently high temperature for a specified length of time to destroy pathogenic bacteria.* Pasteurization is required by law for all Grade A fluid milk and milk products moved in interstate commerce for retail sale. This process was developed by and named after Louis Pasteur, the French scientist who in the early 1860s demonstrated that wine and beer could be preserved by heating above 135°F (57.2°C).

In addition to extending the shelf life of milk, pasteurization destroys pathogens that might be conveyed from infected cows to man.

Various time-temperature relationships are utilized in the pasteurization process. The first three that follow are most commonly used by milk processors. The fourth—sterilized or aseptic—milk is a relatively new development.

1. **Low temperature pasteurization.** Milk is heated to a minimum of 145°F (63°C) for at least 30 minutes.

2. **High temperature pasteurization.** Milk is heated to a minimum of 160.7°F (71.5°C) for at least 15 seconds, following which it is immediately cooled.

3. **Ultrapasteuriztion.** Milk is heated to 191.3°F (88.5°C) for 1 second. This method is common in Europe and is widely used for cream and eggnog in the United States.

4. **Sterilized milk (aseptic milk).** This refers to ultra-pasteurized milk products which are packaged aseptically. They can be stored at room temperature for several weeks.

• **Fortification**—Fortified milks are those that contain added minerals, vitamins, and/or milk solids-not-fat. Details follow:

1. **Fortification with vitamin D.** 400 IU of vitamin D per quart (950 ml) are added to most milk and lowfat milk. This has been common practice for many years. It has contributed significantly to a reduction in the incidence of rickets in infants. Milk provides the necessary calcium and phosphorus, but vitamin D must also be present for normal calcification of bones and teeth.

2. **Fortification of skim milk with vitamin A.** 2,000 to 4,000 IU of vitamin A per quart may be added to skim milk in most states, in addition to vitamin D.

3. **Multivitamin-mineral milk.** In addition to vitamins A and D, multivitamin-mineral milk typically has the following vitamins and minerals added per quart: thiamin (1 mg), riboflavin (2 mg), niacin (10 mg), iron (10 mg), and iodine (0.1 mg).

4. **Protein fortified lowfat and skim milks.** Federal standards for lowfat and skim milks provide for a protein fortified product containing a minimum of 10% milk-derived nonfat solids, with the stipulation that the product must be labeled *protein fortified* or *fortified with protein.* Fortification of milk with nonfat solids is desirable for two reasons: (1) The flavor of most milk is improved, and (2) the nutritive value is increased.

NOTE WELL: The Food and Drug Administration requires that fortified foods or foods that are labeled or advertised with a claim to their nutrition be labeled to show content, in terms of percent Recommended Daily Allowance, of protein, vitamin A, vitamin C, thiamin, riboflavin, niacin, calcium, and iron. Also, serving size and quantities per serving must be stated for calories, fat, protein, and carbohydrate.

State laws vary relative to fortification. In those states permitting such fortification, the levels are regulated by law.

Dairy Processing Plants. Dairy plants process fluid milk and/or manufacture dairy products from milk. Some of them produce several kinds of milk and milk products, others specialize—for example, they make cheese only, or they may condense and evaporate milk.

Fluid milk processors are located in and near population centers, where most fluid milk is consumed. The Great Lakes and Midwest regions of the United States are the primary centers for manufacturing grade milk. Thus, with the exception of ice cream production, the processing of manufactured dairy products—butter, nonfat dry milk, cheese, evaporated and condensed milk, and other products of minor importance—is concentrated near these areas of production. Other major production areas—California, New York, and Pennsylvania—produce substantial quantities of manufactured dairy products from excess Grade A milk.

Kinds And Uses Of Milk And Milk Products. Table M-20 shows the quantity of milk going into different uses, and Fig. M-61 shows the relative importance of each use.

TABLE M-20
HOW THE U.S. MILK SUPPLY IS USED[1]

Product	Milk Equivalent	
	(mil lb)	(mil kg)
Fluid milk and cream sales (25.8 billion quart)	55,370	25,168
Cheese	47,368	21,531
Creamery butter	25,043	11,383
Frozen dairy products[2]	12,307	5,594
Used on farms where produced	2,048	931
Evaporated and condensed milk	1,926	875
Other uses	4,413	2,006

[1]*1991 Milk Facts*, Milk Industry Foundation, Washington, D.C., p. 29.
[2]Plus 2,014 mil lb of milk equivalent in other manufactured dairy products used in production of frozen dairy products.

Fig. M-61. The uses of milk, in percentages. (From *1991 Milk Facts*, Milk Industry Foundation, p. 28)

Each of the branches of the dairy processing segment and each of the products will be listed and described briefly in the sections that follow.

FLUID MILK. This type of processing requires only a minimum of product alteration. Milk is usually homogenized, pasteurized, and fortified with vitamin D. The common forms of fluid milk follow:

• **Whole milk**—*Raw whole milk is defined as the lacteal secretion, practically free from colostrum, obtained by milking one or more cows.* Federal regulations require that whole milk as marketed contain not less than 3.25% milkfat and not less than 8.25% milk solids-not-fat. (Solids-not-fat include carbohydrate, protein, minerals, and water-soluble vitamins, whereas the total solids of milk includes all the constituents of milk except water.)

Various state standards provide that milkfat minimums may vary from 3.0 to 3.8% and solids-not-fat minimums from

8.0 to 8.7%. At the milk plant, the milk from different farms is pooled and *standardized* by removing excess milkfat or by adding cream, concentrated milk, dry whole milk, skim milk, concentrated skim milk, or nonfat dry milk to meet or exceed the minimum legal standards.

Whole fluid milk shipped in interstate commerce is usually pasteurized or ultrapasteurized; it must be pasteurized or ultrapasteurized before being sold for beverage use. Where milk is ultrapasteurized, it must be so indicated on the label.

The addition of vitamin A and/or vitamin D to fresh fluid milk is optional. If added, vitamin A must be present at a level not less than 2,000 IU per quart (*950 ml*) and vitamin D must be present at a level of 400 IU per quart. These amounts must be stated on the label. Milk is an excellent food for vitamin D fortification as it contains the proportion of calcium and phosphorus desirable for normal calcification of bones and teeth. Either vitamin D_2 or D_3 can be added to milk. Although vitamin D fortification of milk is optional, estimates indicate that approximately 98% of all homogenized milk contains vitamin D.

Fig. M-62. Fluid milk can be used in many forms and in many ways. This shows milk served with kabobs in outdoor cooking. (Courtesy, United Dairy Industry Assn., Rosemont, Ill.)

● **Lowfat milk**—Lowfat milk is milk from which sufficient milkfat has been removed to produce a food having one of the following milkfat contents: 0.5, 1.0, 1.5, or 2.0%; along with not less than 8.25% milk solids-not-fat. In 1987, for the first time ever, low-fat fluid milk consumption in the United States exceeded whole fluid milk consumption.

The addition of vitamin A to lowfat milk to a level of 2,000 IU per quart is mandatory. The addition of vitamin D is optional, but if added it must be present at a level of 400 IU per quart. If nonfat milk solids are added to reach the 10% solids-not-fat level, the product must be labeled *protein fortified* or *fortified with protein*.

● **Skim or nonfat milk**—This is milk from which as much fat as possible has been removed. But the milk solids-not-fat must be a minimum of 8.25%. The provisions for the addi-

tions of vitamins A and D and nonfat milk solids to skim or nonfat milk are identical to those described for lowfat milk.

● **Chocolate milk; other flavored milks**—Chocolate whole, lowfat, and skim milk are flavored with a chocolate syrup, cocoa, or a chocolate powder to give a final chocolate solids concentration of 1 to 1.5%. Additionally, a nutritive sweetener (about 5 to 7% sucrose) is added. Chocolate lowfat and skim milk must contain not less than 2,000 IU of vitamin A per quart. If vitamin D is added, it must be present in not less than 400 IU per quart.

Numerous other flavorings may also be added to milk; among them, strawberry, cherry, raspberry, pineapple, apple, orange, and banana. But chocolate is the most popular flavoring of milk.

SPECIAL MILKS. Certain fluid milks are sold under special labels. Among them are:

1. **Certified milk.** This is milk that is produced under special sanitary conditions prescribed by the American Association of Medical Commissioners. It is sold at a higher price than ordinary milk.

2. **Golden Guernsey milk.** Golden Guernsey milk is produced by owners of purebred Guernsey herds who comply with the regulations of the American Guernsey Cattle Club. Such milk is sold under the trade name "Golden Guernsey," at a premium price.

3. **All-Jersey milk.** This is produced by registered Jersey herds whose owners comply with the regulations of the American Jersey Cattle Club. It is sold at a premium price under the trademark of "All Jersey."

EVAPORATED AND CONDENSED MILK. The introduction of condensed milk in 1853 (patented in 1856) made available dairy products that expanded markets at home and abroad and significantly enhanced growth of the U.S. dairy industry. Today, candy manufacturers, bakers, and ice cream processors are large users of concentrated milk.

The primary products within this category are evaporated and concentrated (condensed) milk packed in cans for consumer use, and concentrated whole and skim milk shipped in bulk.

● **Evaporated milk**—Evaporated milk is made by preheating to stabilize proteins, concentrating in vacuum pans at 122°

Fig. M-63. Evaporated milk. (Courtesy, Carnation Co., Los Angeles, Calif.)

to 131°F (50° to 55°C) to remove about 60% of the water, homogenizing (which is mandatory), standardizing to the required percentage of components, adding vitamins (vitamin D must be added to provide 400 IU per reconstituted quart), canning, and stabilizing (sealed in the container, then heat treated [240° to 245°F, or 115.5° to 118.5°C, for 15 minutes] to sterilize).

Federal standards require that the milkfat and total milk solids content of evaporated milk be not less than 7.5 and 25.5%, respectively. Evaporated skim milk is concentrated, fortified skim milk containing not less than 20% milk solids.

In recent years, per capita consumption of evaporated milk has declined, but bulk shipments of unsweetened concentrated milk to food and ice cream processors have increased.

• **Concentrated or condensed milk**—This is similar to evaporated milk because it is made by the partial removal of water from fluid milk; hence, the milkfat and total milk solids must not be less than 7.5 and 25.5%, respectively. Unlike evaporated milk, however, concentrated milk is usually not subjected to further heat treatment to prevent spoilage. The product is perishable and spoilage may occur rapidly at temperatures above 44.6°F (7°C).

• **Sweetened condensed milk**—This product is made by the removal of approximately 60% of the water from whole milk and the addition of a suitable sweetener such as sucrose. The amount of the sweetener (40 to 45% of the condensed milk) used is sufficient to prevent spoilage. Federal standards of identity require that sweetened condensed milk contain not less than 8.5% of milkfat and not less than 28% of total milk solids.

DRY MILK. Perishability and bulk of fluid milk have been a major drawback to transporting milk over long distances. However, the development of the dry milk industry made it possible to export milk abroad. Also, dry milk may be transported over relatively long distances domestically (1) to consumers who desire a less costly source of milk, and (2) to processors of other dairy products in areas where milk for manufacturing may not be readily available. The leading states producing dry milk are California, Wisconsin, and Minnesota. A discussion of each of the common dry milks follows:

• **Nonfat dry milk**—This product is prepared by removing water from pasteurized skim milk. Federal standards of identity require that it contain not more than 5% by weight of moisture and not more than 1.5% by weight of milkfat unless indicated. Except for small losses of ascorbic acid, thiamin, vitamin B-12, and biotin, the processing has no appreciable effect on the nutritive value of the milk. Due to its low moisture content, it can be kept for long periods.

• **Nonfat dry milk fortified with vitamins A and D**—This product is similar to nonfat dry milk except that it contains 2,000 IU of vitamin A and 400 IU of vitamin D per quart (950 ml) when reconstituted according to label directions. Almost all nonfat dry milk sold retail is fortified with vitamins A and D.

• **Dry whole milk**—This product is made from pasteurized whole milk from which water has been removed. Except for losses in ascorbic acid (20% for spray dried), vitamin B-6 (30% for spray dried), and thiamin (10% for spray dried, and 20% for roller dried), the processing has no appreciable

effect on the nutritive value of the product. The major deterioration of dry whole milk is oxidative changes in the fat. Stability of the product can be increased by special packaging, such as gas-pack (removal of air and replacement with inert gas such as nitrogen) or vacuum-pack containers.

CULTURED MILK AND OTHER PRODUCTS. Cultured milks are fluid products that result from the souring of milk or its products by bacteria that produce lactic acid, whereas acidified milks are obtained by the addition of food-grade acids to produce an acidity of not less than 0.20%, expressed as lactic acid. The word *cultured* is used because pure bacterial cultures are employed in commercial manufacture. It is proper also to use the words *fermented* or *sour* because lactic acid, which imparts sourness, is produced by fermentation of milk sugar (lactose). Kefir, koumiss, acidophilus milk, cultured buttermilk, sour cream, and yogurt are examples of lactic acid fermentations.

Sales of cultured products are increasing as consumers recognize their nutritional value and appreciate their flavors. Cultured products are also a practical means of introducing large numbers of particular organisms into the intestinal tract.

The body, flavor, and aroma developed in the finished product vary with the type of culture and milk, the concentration of milk solids-not-fat, the fermentation process, and the temperature. The vitamin concentration may be altered; the variation of specific vitamins depends on both the types of culture and the length of ripening. A discussion of each of the common cultured milk products follows:

• **Kefir and koumiss**—These are the two major types of cultured milks which have been subject to both acid and alcoholic fermentation.

Kefir, made from the milk of mares, goats, or cows, is popular in countries of southwestern Asia.

Koumiss, made from unheated mare's milk, is the typical fermented milk of the U.S.S.R. and western Asia.

• **Acidophilus milk**—Pasteurized milk, usually lowfat or skim, cultured with *Lactobacillus acidophilus* and incubated at 100.4°F (38°C) for at least 18 hours until a soft curd forms, is called acidophilus milk.

Today, there is increasing demand for health foods; and, frequently, acidophilus milk is promoted as such a food. Although much research is still needed, *L. acidophilus* is believed to be a normal bacterial component of the gut microflora that helps to maintain the balance of microorganisms in the intestinal tract. Cultures of products containing *L. acidophilus* are sometimes used to regenerate intestinal flora after antibiotic treatment or other conditions that upset the microfloral balance in the gut.

• **Cultured buttermilk; Bulgarian buttermilk**—The fluid remaining after churning cream to make butter is called buttermilk. Today, this product is used primarily by the baking industry; and most buttermilk for beverage purposes is a cultured product. Most of the cultured buttermilk marketed in the United States is made from fresh pasteurized skim or lowfat milk with added nonfat dry milk solids, cultured with *Streptococcus lactis*. However, cultured buttermilk can also be made from fluid whole milk, concentrated fluid forms, or reconstituted nonfat dry milk.

Bulgarian buttermilk is made with *Lactobacillus bulgaricus*.

• **Sour cream or cultured sour cream**—Pasteurized, homogenized cream cultured with *Streptococcus lactis* at 71.6°F (22°C) until the acidity is at least 0.5%, calculated as lactic acid, results in an acid gel product known as cultured sour cream. Federal standards of identity specify that cultured sour cream contain not less than 18% milkfat. If nutritive sweeteners or bulky flavoring ingredients are added, the product must not contain less than 14.4% milkfat.

Cultured sour cream is a uniformly textured, smooth product that is widely used for flavoring.

• **Acidified sour cream**—This product results from the souring of pasteurized cream with safe and suitable acidifiers, with or without the addition of lactic acid producing bacteria. Federal standards of identity call for not less than 18% milkfat and a titratable acidity of not less than 0.5% calculated as lactic acid. In the event nutritive sweeteners or bulky flavoring ingredients are added, the product may not contain less than 14.4% milkfat.

• **Sour half-and-half or cultured half-and-half**—The federal standards of identity state that this is the product of pasteurized half-and-half containing not less than 10.5% and not more than 18% milkfat and containing lactic acid producing bacteria.

• **Acidified sour half-and-half**—This product results from the souring of half-and-half with safe and suitable acidifiers, with or without the addition of lactic acid producing bacteria.

• **Sour cream dressing and sour half-and-half dressing**—The word dressing denotes a product in which other dairy ingredients such as butter have been substituted for cream and/or milk. Sour cream dressing is similar to sour cream and contains not less than 18% milkfat or not less than 14.4% milkfat if nutritive sweeteners or bulky ingredients are added. Sour half-and-half dressing is made in semblance of sour half-and-half and contains not less than 10.5% milkfat or not less than 8.4% if nutritive sweeteners or bulky ingredients are added.

• **Yogurt**—This product can be manufactured from fresh whole milk, lowfat milk, or skim milk. In the United States, yogurt usually is made from a mixture of fresh, partially skimmed milk and nonfat dry milk; and fermentation is generally accomplished by a one-to-one mixed culture of *Lactobacillus bulgaricus* and *Streptococcus thermophilus*, in a symbiotic relationship, with each microorganism metabolizing milk products which are subsequently used by the other.

The milk is pasteurized, homogenized, inoculated, incubated at 107.6° to 114.8°F (*42° to 46°C*) until the desired stage of acidity and flavor is reached, and chilled to 44.6°F (*7°C*) or lower to halt further fermentation. The usual milkfat levels of current products are: yogurt, at least 3.25% milkfat; lowfat yogurt, 0.5 to 2% milkfat; and nonfat yogurt, less than 0.5%. Milk solids-not-fat range from 9 to 16%, the higher amount reflecting the addition of solids. Today, three main types of yogurt are produced:
1. Flavored, containing no fruit.
2. Flavored, containing fruit.
3. Unflavored yogurt.

The rise in popularity of yogurt in the United States in recent years has been phenomenal. This increased consumption is largely attributed to development of fruit-flavored yogurts and the promotion of low-calorie, but highly nutritional, yogurts as diet foods.

• **Eggnog**—Eggnog is a mixture of milk, cream, sugar, milk solids, eggs, stabilizers, and spices that is pasteurized, homogenized, cooled, and packaged. Eggnog may refer to a product with 6.0 to 8.0% milkfat, 1.0% egg yolk solids, and 0.5% stabilizer. Also, eggnog-flavored milk is available; it contains a minimum milkfat level of 3.25%, minimum egg yolk solids of 0.5%, and a maximum stabilizer level of 0.5%.

SPECIALTY MILKS. Several specialty milks are on the market, including certified milk, low sodium milk, imitation milk, and filled milk. A discussion of each of these follows:

• **Certified milk**—The first certified milk was produced in 1893, in Essex County, New Jersey, before dairy sanitation had achieved the excellence that now exists in producing Grade A milk. Certified milk refers to raw or pasteurized milk produced and handled by dairies that operate according to the rules and sanitary regulations stated in the book *Methods and Standards for the Production of Certified Milk* issued and revised periodically by the American Association of Medical Milk Commissioners, Inc. Today, the quality of Grade A milk is so high that it has largely replaced certified milk—and at a lower cost, so certified milk is produced in only a few localities. The production of certified milk, conducted under the auspices of a Medical Milk Commission, involves the veterinary examination of cows, the sanitary inspection of the dairy farm and equipment, and the medical examination of employees who handle the milk.

• **Low sodium milk**—Ninety-five percent or more of the sodium that occurs naturally in milk can be removed by ion-exchange. Thus, the sodium content of whole milk generally can be reduced from a normal amount of about 49 mg to about 2.5 mg per 100 g of milk. Fresh whole milk is passed through an ion-exchange resin to replace the sodium in milk with potassium, following which the milk is pasteurized and homogenized.

Low sodium milk permits the inclusion of milk and milk-containing foods that might otherwise be limited in therapeutic diets because of their sodium content.

• **Imitation milk**—Imitation milks purport to substitute for and resemble milk. These products usually contain water, corn syrup solids, sugar, vegetable fat (coconut, soybean, cottonseed), and protein from soybean, fish, sodium caseinate, or other sources. Although imitation fluid milks do not contain dairy products as such, they may contain derivatives of milk such as casein, salts of casein, milk proteins other than casein, whey, and lactose. Sometimes vitamins A and/or D are added. Ingredient composition, and hence nutrient composition, vary widely. The American Academy of Pediatrics considers imitation milk products inappropriate for feeding infants and young children.

• **Filled milk**—These are milk products (milk, lowfat milk, half-and-half or cream, whether or not condensed, evaporated, concentrated, powdered, dried, or desiccated) from which all or part of the milkfat has been removed and to which any fat or oil other than milkfat has been added. The American Academy of Pediatrics does not recommend the use of filled milks for feeding infants and small children.

CREAM. Cream is the liquid milk product, high in fat, separated from milk, which may have been adjusted by the addition of milk, concentrated milk, dry whole milk, skim milk, concentrated skim milk, or nonfat dry milk. Federal standards of identity require that cream contain not less than 18% milkfat. The several cream products on the market are:

• **Half-and-half**—This is a mixture of milk and cream containing not less than 10.5% nor more than 18% milkfat. It is pasteurized or ultrapasteurized, and it may be homogenized.

• **Light cream (coffee cream, table cream)**—This product contains not less than 18% nor more than 30% milkfat. It is pasteurized or ultrapasteurized, and it may be homogenized.

• **Light whipping cream (whipping cream)**—This cream contains not less than 30% nor more than 36% milkfat. It is pasteurized, or ultrapasteurized, and it may be homogenized.

• **Heavy cream (heavy whipping cream)**—This cream contains not less than 36% milkfat. It is pasteurized or ultrapasteurized, and it may be homogenized.

ICE CREAM AND FROZEN DESSERTS.

Fig. M-64. Homemade ice cream. (Courtesy, Carnation, Los Angeles, Calif.)

Today, the ice cream manufacturing industry is one of the largest in the realm of dairy processing. U.S. production of frozen dairy products increased from 936 million gallon (3.5 billion liter) in 1960 to 1.3 billion gallons (4.9 billion liter) in 1988.

Several products are classified under the category of frozen desserts. Some of these follow:

1. **Ice cream.** Of all the frozen dairy desserts, ice cream contains the most milk solids (about 16 to 24%). It consists primarily of milk fat, nonfat milk solids, sugar, flavoring, and stabilizers.

2. **Custards and French ice cream.** These products closely resemble ice cream except they must contain at least 1.4% egg yolk solids, additionally.

3. **Ice milk.** Ice milk contains less fat and less total solids than ice cream; hence, it contains fewer calories. It must contain at least 2%, but not more than 7%, milkfat.

4. **Mellorine.** Mellorine is essentially the same product as ice cream except that vegetable fat is used in place of milkfat.

5. **Sherbet.** Sherbet is a sweet, tart-flavored frozen dairy product that is low in total milk solids (about 3 to 5%). It can be made either by mixing water with ice cream or by

manufacturing from the raw ingredients. Since its melting point is lower than ice cream, it is softer than ice cream when the two products are stored at the same temperature.

BUTTER. Butter is made from cream. As marketed, it contains about 80 to 82% butterfat, 14 to 16% water, 0 to 4% salt, and 0.1 to 1.0% curd. *Sweet cream butter* is made from pasteurized sweet cream to which no starter has been added. *Ripened cream butter* is made by starter ripened cream. *Sweet butter* contains no salt. *Unsalted butter* is butter to which no salt has been added. *Salted butter* is butter to which salt has been added.

Butter is classed as Grade AA, A, B, or C based on flavor, body, color, and salt. Butter which does not meet the standards of Grade C cannot carry a USDA label.

Table M-21 shows the per capita consumption of butter and margarine for selected years from 1910 to 1990. Per capita consumption of margarine surpassed butter in 1957. In 1990, the per capita consumption of butter was 4.4 lb, as compared to 10.9 lb of margarine. Low prices and the controversy about cholesterol and heart disease are the primary reasons for the decrease in both butter and margarine consumption.

TABLE M–21
PER CAPITA CONSUMPTION OF BUTTER AND MARGARINE[1]

Year	Per Capita Consumption			
	Butter		Margarine	
	(lb)	(kg)	(lb)	(kg)
1910	18.3	8.3	1.6	0.7
1920	14.9	6.8	3.4	1.5
1930	17.6	8.0	2.6	1.2
1940	17.0	7.7	2.4	1.1
1950	10.7	4.9	6.1	2.8
1960	7.5	3.4	0.4	4.3
1970	5.3	2.4	11.0	5.0
1980	4.5	2.0	11.4	5.2
1990	4.4	2.0	10.9	4.5

[1]*Dairy Statistics Through 1960*, Stat. Bull No. 303, Supp. for 1963–64, Economic Research Service, USDA; *National Food Situation*, Economic Research Service, USDA, February 1966; *Agricultural Statistics 1980*, USDA, p. 148 and *1991*, p. 479.

CHEESE. Cheese may be defined as the fresh or matured product obtained by draining after coagulation of milk, cream, skimmed or partly skimmed milk, buttermilk, or a combination of some or all of these products.

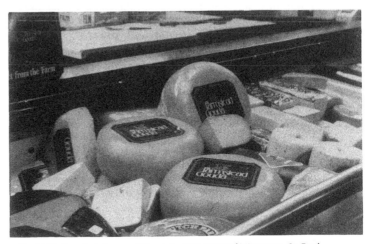

Fig. M-65. Assorted cheeses. (Courtesy, University of Minnesota, St. Paul, Minn.)

Cheese is manufactured by: (1) exposing milk to specific bacterial fermentation, and/or (2) treating with enzymes; to coagulate some of the protein. The first step in cheese making is to separate the casein and milk solids from the water in milk. The coagulated milk is then cut into small pieces and the water separated from the solids by a series of drainings and pressings. The characteristic flavor of the cheese is created by bacteria growth and subsequent acid production in the cheese during the cheese-making process and also by the bacteria and mold development during a period of curing.

Milk can be, and is, processed into many different varieties of cheese. In fact, there are over 2,000 different named cheeses. Some are made from whole milk, others from milk that has had part of the fat removed, and still others from skim milk. (The most important variety produced from skim milk is cottage cheese.) In 1990, American types of cheese accounted for 50% of the U.S. total cheese production, of which 96% was American Cheddar and 2.2% was other American types (Colby, washed curd, stirred curd, Monterey). Other important types of cheese are Italian (mostly soft varieties), Swiss, Muenster, cream, blue, and Neufchatel.

Fig. M-66 shows the kinds of cheese produced in 1990, exclusive of cottage cheese.

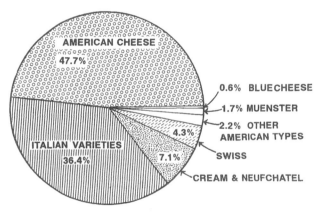

Fig. M-66. Relative importance of kinds of cheeses produced in the United States. (Source: USDA)

The U.S. per capita consumption of cheese has tripled since 1960, increasing from 8.3 lb in 1960 to 24.7 lb in 1990.

Hard cheese, such as Cheddar, is one of the most concentrated of common foods; 100 g (about 3.5 oz) supplies about 36% of the protein, 80% of the calcium, and 34% of the fat in the recommended daily allowance. Cheese is also a good source of some minerals and vitamins.

Cheese is a high-protein food, and the protein (principally casein) is partially digested. It is rich in the essential amino acids, calcium, phosphorus, certain other minerals, and vitamins, and has a high caloric value. Only a trace of the lactose present in milk remains in the cheese.

Cottage cheese (as made from skim milk) is low in fat and high in protein, but has only about one-eighth of the calcium content of American (Cheddar) cheese. Cream cheese differs from other soft cheeses in having more fat and less protein.

One ounce (about 28 g) of Cheddar-type cheese provides about the same nutritive value as a glass of milk. It is a particularly good food for children.

• **Processed cheese**—This product is formed by grinding, heating, and mixing hard-type cheeses. This type of cheese serves as an outlet to salvage defective cheeses as well as provide a uniform, widely marketable product.

• **Nutritive value of cheeses**—The composition and nutritive value of a number of varieties of cheeses is given in Food Composition Table F-21, in the section on "Milk and Products."

CASEINATES (COFFEE WHITENERS; WHIPPED TOPPINGS). *Caseinates are salts of casein.* They are classified as food chemicals derived from milk and are commonly used in (1) nondairy coffee whiteners (coffee whiteners are made from sodium caseinate) and (2) whipped toppings.

To produce caseinates, enough alkali, commonly called calcium hydroxide or sodium hydroxide, is added to acid coagulated casein to reach a pH of 6.7. The resulting suspension is pasteurized and spray dried.

WHEY. *Whey is the watery part of milk separated from the curd.* Whey is available as dried whey, condensed whey, dried whey solubles, condensed whey solubles, dried hydrolyzed whey, condensed hydrolyzed whey, condensed whey product, dried whey product, and condensed cultured whey.

Whey provides food processors with an inexpensive product that can be used as a source of lactose, milk solids, milk proteins, or total solids. Additionally, whey, which is relatively high in digestibility and nutritive value, is commonly used as a livestock feed.

MARKETING AND DISTRIBUTING MILK. Marketing milk is that all-important end of the line; it's that which gives point and purpose to all that has gone before.

Satisfactory milk marketing requires one basic ingredient—quality milk; and this begins with production on the farm.

The difference in price between Grade A milk and the lower grades is substantial. But it goes beyond this; quality influences consumer demand.

In our present system, the vast majority of the marketing of milk and dairy products is handled by specialists, usually under a myriad of complex regulations and controls. Both successful milk producers and enlightened consumers should be familiar with milk markets, regulations and controls, and pricing systems, along with the factors affecting them.

Milk Supply. In 1990, a total of 148 billion pounds *(67 billion kilograms)* of milk were marketed by U.S. dairy farmers for $20 billion delivered to plants. Of this amount, 37.3% was consumed in fluid form and 62.7% was processed into manufactured dairy products.

Fluid milk is retailed as pasteurized milk, homogenized milk, fortified milk (usually vitamin D), skim milk, and flavored milk. Of the milk used for manufacturing in 1990, 31.9% was used for making cheese (exclusive of creamed

Fig. M-67. How it used to be done. Milkman transferring milk from can into customer's own pitcher, using a quart measure. (Courtesy, The Bettmann Archive, Inc., New York, N.Y.)

cottage cheese) and 16.9% was used for making butter.

Future milk supply will depend primarily on the profitability of dairy enterprises, for, like all other businesses, dairies are owned by people—and all people like to make a profit. In turn, profitability is determined by milk prices and cost of production. Also, dairy farmers will consider alternative farm and off-farm opportunities.

Milk Demand. There have been dramatic changes in the per capita consumption of milk and manufactured dairy products in recent years. Fig. M-68 clearly shows these changes.

PERCENT CHANGE IN PER CAPITA SALES 1977–87

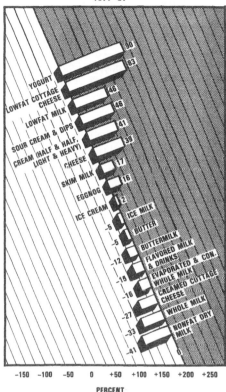

Fig. M-68. Change in U.S. per capita consumption of milk and dairy products 1977 to 1987, based on sales. Bars to the left of the vertical line represent decreases; bars to the right represent increases. (Source: *1988 Milk Facts*, Milk Industry Foundation, Washington, D.C., p. 16)

Because of increasing consumer infatuation with weight watching, low calorie diets, and cholesterol levels, it appears that present trends will continue—that there will be further increases in per capita consumption of low-fat and high-solids nonfat dairy products, and further decreases in per capita consumption of high-fat products.

Contrary to popular opinion, the United States ranks low among nations in per capita consumption of dairy products. Of the 18 major countries in per capita consumption of fluid milk, the United States ranks fifteenth in per capita consumption of fluid milk, last in consumption of butter, seventh in consumption of cheese, and twelfth in consumption of dry milk. Thus, based on (1) nutritive qualities, and (2) per capita consumption in those countries that exceed us, it would appear that there is a place and a need for increased consumption of milk and dairy products in the United States.

FACTORS INFLUENCING DEMAND. The demand for milk and milk products depends on a number of factors, chief of which are those which follow:

1. Price.
2. Availability and price of substitute products.
3. Consumer income.
4. Population growth.
5. Changes in consumer preference.
6. Age and sex.
7. Carbonated beverages.
8. Type and size of container.
9. Promotion of milk and dairy products.

Federal Milk Marketing Orders. Federal Milk Marketing Orders are established and administered by the Secretary of Agriculture under acts of Congress passed in 1933 and 1937. They are legal instruments, and they are very complex. However, stated in simple terms, they are designed to stabilize the marketing of fluid milk and to assist farmers in negotiating with distributors for the sale of their milk.

Federal Milk Orders (1) regulate only Grade A farm milk prices—they do not regulate Grade B milk prices, and (2) establish minimum, not maximum, farm milk prices. Prices paid to farmers are controlled, but there is no direct control of retail prices.

Federal orders are not concerned with sanitary regulations. These are administered by state and local health authorities.

State Milk Controls. Several of the states have milk control programs that also include some authority over milk prices.

In setting minimum farm prices, state control agencies often operate in a manner similar to Federal Milk Orders. They price only Grade A milk. State Orders price milk only within the particular state whereas Federal Orders can cover several states. Federal Milk Orders establish only farm milk prices, while State Orders establish prices at various stages along the way—farm, wholesale, retail, as well as regulating trade practices.

Some states set maximum prices at either wholesale or retail, or at both levels. Others set both minimum and maximum wholesale and retail milk prices. Still others require a minimum markup by retailers, while others require that milk prices be reported to the state. Most states specify Class I prices in their Orders.

All state control agencies have authority to require distributors to be licensed and to inspect and investigate operations of milk dealers and audit their records.

Other Regulatory Programs. Because of the essential nature of milk, plus the fact that it is easily contaminated and a favorable medium for bacterial growth, it is inevitable that numerous regulatory programs have evolved around it—federal, state, and local—some having been designed to control prices and assure a reasonably uniform flow of milk, and others for sanitary reasons.

SANITARY REGULATIONS. The sanitation of milk and dairy products is assured by the enforcement of sanitary regulations by federal, state, and local authorities.

All major cities and states have sanitary regulations governing the production, transportation, processing, and delivery of milk. Unfortunately, from area to area, there are a bewildering number of different regulations, with the result that milk going to more than one city market is often subjected to duplication and confusion in inspection. Also, sanitary and health regulations have sometimes been used as barriers to keep milk out of a certain area for competitive reasons.

In 1923, the U.S. Public Health Service (USPHS) established an Office of Milk Investigations; and, in 1924, the USPHS published its first Grade A pasteurized milk ordinance. Subsequently, this regulation has been revised several times.

Producers are issued permits allowing them to ship Grade A milk. The permit is revoked if either the bacteria count of raw milk exceeds 100,000 per milliliter or the cooling temperature exceeds 40°F in 3 of the last 5 samples.

The standard plate count of Grade A pasteurized milk may not exceed 20,000 per milliliter nor the coliform count 10 per milliliter in three of the last five samples or the processor's permit will be revoked.

A sample must be taken from each bulk tank at each farm every time the milk is collected.

The Food and Drug Administration (FDA) is charged with inspecting dairy products and processing plants for contamination and adulteration.

STANDARDS AND GRADES. Milk standards are established on the state and local levels and are patterned after the Pasteurized Milk Ordinance (PMO) formulated by the Food and Drug Administration (FDA). Essentially, the PMO is a set of recommendations for voluntary adoption by state and local regulatory agencies. Legal responsibility for the provision of quality milk rests mostly with state and local governments whose requirements are in some instances more stringent than the guidelines of the PMO.

The common grades are Grade A, manufacturing grade, or reject. In some areas, Grades A, B, and C are used.

Details relative to *Grade A Milk* and *Manufacture Grade Milk* follow, with a separate section devoted to each. For more specific regulations concerning the grading of milk in any particular area, the reader is advised to contact his local dairy extension agent.

Grade A Milk. *Grade A milk is milk that meets certain quality standards and is produced, processed, and handled in a stipulated manner in approved facilities and equipment.*

Grade A milk is produced and processed in accordance with the quality standards of the Pasteurized Milk Ordinance (PMO) formulated (and revised from time to time) by the Food and Drug Administration (FDA) for voluntary adoption by state and local milk regulatory agencies and the dairy industry. Federal specifications for the procurement of Grade A milk and milk products are based on the PMO, as are sanitary regulations for Grade A milk and milk products entering interstate commerce.

The Pasteurized Milk Ordinance is a very effective instrument for protecting milk quality. Strict compliance therewith is imposed so that the milk will be of high uniform quality. The sanitary dairy practices outlined by the PMO include:

1. Inspection and sanitary control of farms and milk plants.

2. Examination and testing of herds for the elimination of bovine diseases related to public health.

3. Regular instruction on desirable sanitary practices for persons engaged in production, processing, and distribution of milk.

4. Proper pasteurization or ultrapasteurization of milk.

5. Laboratory examination of milk.

6. Monitoring of milk supplies by federal, state, and local health officials to protect against unintentional chemical, physical, and microbiological adulterants.

Most rigid control is placed on the production and processing of Grade A market milk which is sold to consumers in its fluid state. Grade A pasteurized milk is obtained from dairy farms that conform with the sanitary requirements of the PMO or its equivalent as enforced by state or local authorities. The milk must be obtained from cows tested and found free of disease and disease-producing organisms. The raw milk is cooled immediately to the specified legal temperature and maintained at no higher temperature from the completion of milking until processing at a dairy plant that also conforms with state and local sanitation requirements.

At the processing plant, the milk is analyzed to be sure it contains not more than a specified bacterial count and is pasteurized according to the specified time-temperature relationships. Public health authorities advocate pasteurization since it is the only practical commercial process that destroys all disease-producing organisms that may be present in fluid milk. After processing, the milk is cooled again to the specified legal temperature and maintained at no higher temperature until sold. Today, practically all fluid market milk sold to consumers is Grade A pasteurized milk.

In order to become a producer of Grade A milk, the milking and milk storage facilities of the dairy must first pass the careful inspection of the state inspector. If approved by the inspector, the dairy is granted a permit for the shipment of Grade A milk. Once a permit is issued, the dairy operation is periodically inspected by a local health department official to ensure that adequate sanitation practices are maintained. The permit may be revoked if the bacteria count is too high, if the cooling temperature exceeds 40°F, and/or if the producer is found in repeated violation of other requirements of the ordinance.

Manufacture Grade Milk. As with Grade A milk, each state adopts and enforces its own regulations for the production, processing, and handling of Manufacture Grade milk, which is produced for, and processed into, ice cream, cheeses, etc. There is much less uniformity in the regulations governing Manufacture Grade milk than in Grade A milk, and the standards are not so high (although the trend is higher and more rigid).

Dairy farms producing Manufacture Grade milk must also conform to the sanitary requirements of state and local authorities, and the milk must come from cows tested and found free of disease and disease-producing organisms.

Processing plants manufacturing dairy products are subjected to general sanitation requirements similar to those for plants processing Grade A milk. Additionally, there are supplemental and specific requirements for plants that process dry milk products, butter, cheeses, and evaporated, condensed, or sterilized milk products.

Market Distribution of Fluid Milk. Once milk is processed, it must be marketed rather rapidly because of its highly perishable nature. When a consumer has a bad experience with milk, it is difficult to get him or her back into the habit of drinking milk. Hence, it is imperative that fluid milk be of high quality and possess desirable taste and packaging.

Milk is distributed in a number of ways: (1) home delivery, (2) retail stores, and (3) vending machines and self-service.

Dairy Imports and Exports. In order to protect the dairy industry from depressed prices caused by other countries dumping dairy products onto American markets, import quotas have been established. A number of dairy products, including several types of cheeses, butter, malted milk, butter oil, and dried milk are covered by specific import quotas. Although not formally restricted, certain other dairy products may be limited by agreement between the United States and the exporting country. Total imports on a milk equivalent basis have been very small. In 1988, imports of dairy products amounted to 2.4 million pounds *(1.1 mil kg)* when total milk production was more than 148 billion pounds *(67 bil kg)*. Hence, imports of dairy products were equivalent to only 0.002% of the total U.S. market milk.

As long as domestic prices are above world prices and world supplies are ample, exporting countries will look to the United States as a possible market. The four countries exporting the largest amount of dairy products are Italy, New Zealand, Denmark, and The Netherlands. As a result, import pressure will persist; yet, it is expected that imports of many commodities will continue to be limited by quotas.

As with imports, exports of dairy products by the United States are rather small. They fluctuate from year to year, but never amount to more than about 1% of the total milk supply.

Per Capita Milk Consumption of Milk and Milk Products. Milk is an important constituent of the American diet in terms of per capita consumption (Table M-22).

In 1990, the total milk equivalent consumption of all dairy products was 570. 6 lb *(259.4 kg)*, a 13% decline from 1960. But the growth in human population offset per capita decline, with the result that total consumption increased. From Table M-22, several trends are apparent: Americans have become weight conscious in recent years leading to a decreased per capita consumption of fluid whole milk and cream and an increased per capita consumption of low-fat milk. Per capita fluid whole milk consumption declined 67% from the period of 1960 to 1990, while per capita low-fat milk consumption increased 313%. Of the manufactured dairy products, per capita consumption of butter declined 41%, and evaporated milk 42%, in the period 1960 to 1990. Per capita total cheese consumption increased 34% in this same period. Per capita consumption of ice cream declined due to the increased popularity of frozen yogurt.

TABLE M-22
DAIRY PRODUCTS: U.S. PER CAPITA CIVILIAN CONSUMPTION, SELECTED YEARS 1960–88[1]

	Fluid Milk Products										Manufactured													
Year	Fluid Whole Milk		Cream		Low-fat Milk		Total Product		Whole Milk Equiv. of Butterfat		Butter		Total Cheese		Cottage Cheese		Ice Cream		Evap. & Cond. Milk		Nonfat Dry Milk		Yogurt		All Milk Equivalent	
	(lb)	(kg)	(lb)	(kg)	(lb)	(kg)	(lb)	(kg)	(lb)	(kg)	(lb)	(kg)	(lb)	(kg)	(lb)	(kg)	(lb)	(kg)	(lb)	(kg)	(lb)	(kg)	(lb)	(kg)	(lb)	(kg)
1960	276	125	9.1	4.1	23.8	10.8	309	140	309	140	7.5	3.4	8.3	3.8	4.8	2.2	18.3	8.3	13.7	6.2	6.2	2.8	0.3	0.1	653	296
1965	264	120	7.6	3.4	34.0	15.4	306	139	294	133	6.4	2.9	9.6	4.4	4.7	2.1	18.5	8.4	10.7	4.9	5.6	2.5	0.3	0.1	620	281
1970	229	104	5.6	2.5	57.5	26.1	292	132	260	118	5.3	2.4	11.5	5.2	5.1	2.3	17.7	8.0	7.1	3.2	5.4	2.4	0.9	0.4	561	254
1975	195	88	5.9	2.7	84.7	38.4	286	130	244	111	4.8	2.2	14.5	6.6	4.7	2.1	18.6	8.4	5.3	2.4	3.3	1.5	2.1	1.0	546	248
1980	148	67	5.7	2.6	97.8	44.5	251	114	231	105	4.1	1.9	17.1	7.8	4.6	2.1	17.6	8.0	7.2	3.3	3.0	1.3	2.5	1.1	554	252
1985	120	55	6.0	2.7	107.0	48.6	239	109	186	85	4.9	2.2	22.6	10.3	4.1	1.9	18.0	8.2	7.9	3.6	2.2	1.0	4.1	1.9	593	270
1988	100	45	6.0	2.7	132.0	60.0	238	108	184	84	4.4	2.0	23.6	10.7	3.8	1.7	17.8	8.1	7.9	3.6	2.7	1.2	4.7	2.1	585	266

[1]*1989 Dairy Producer Highlights*, National Milk Producers Federation, p. 17.

Fig. M-69. Milk and milk products in a modern retail store. (Photo by A. H. Ensminger)

Care of Milk in the Home.[20] The consumer also has responsibility for protecting the quality of milk. Proper handling of dairy products and open dating are designed to assure consumers of dairy products with a good shelf life, which is the length of time after processing that the product will retain its quality.

Open dating is the sometimes mandatory, often voluntary, inclusion of a date on milk containers to indicate when they should be withdrawn from retail sale. It is used by industry to reflect the age of individual packages. It does not indicate the shelf life of products. Generally, depending upon storage conditions and care in the home, a product will remain fresh and usable for a few days beyond this *pull date*, or *sell-by date*. Regulation of open dating varies among states and municipalities.

[20]This section and Table M-23 were adapted by the authors from *Newer Knowledge of Milk and Other Fluid Dairy Products*, National Dairy Council, Rosemont, Ill.

Dairy products are highly perishable, therefore, it is recommended that consumers observe the following practices to preserve quality:

1.   Use proper containers to protect milk from exposure to sunlight, bright daylight, and strong fluorescent light to prevent the development of off-flavor and a reduction in riboflavin, ascorbic acid, and vitamin B-6 content.

2.   Store milk at refrigerated temperatures 44.6°F (7°C) or below as soon as possible after purchase.

3.   Keep milk containers closed to prevent absorption of other food flavors in the refrigerator. An absorbed flavor alters the taste, but the milk is still safe.

4.   Use milk in the order purchased.

5.   Serve milk cold.

6.   Return milk container to the refrigerator immediately to prevent bacterial growth. Temperatures above 44.6°F (7°C) for fluid and cultured milk products for even a few minutes reduces shelf life. Never return unused milk to the original container.

7.   Keep canned milk in a cool, dry place. Once opened it should be transferred to a clean opaque container and refrigerated.

8.   Store dry milk in a cool, dry place and reseal the container after opening. Humidity causes dry milk to lump and may affect flavor and color changes. If such changes occur, the milk should not be consumed. Once reconstituted, dry milk should be treated like any other fluid milk: covered and stored in the refrigerator.

Guidelines on the storage life of various milk products at specific temperatures are shown in Table M-23. In this table, storage life refers to the approximate length of time after processing, not after purchasing, that the product will retain its quality.

TABLE M-23
APPROXIMATE STORAGE LIFE OF MILK PRODUCTS AT SPECIFIC TEMPERATURES
(Storage life refers to the length of time after processing—not after purchasing—that the product will retain its quality.)

Product	Approximate Storage Life at Specific Temperatures		
	(Time)	(°F)	(°C)
Fresh fluid milk (whole, lowfat, skim, chocolate, and unfermented acidophilus)	8 to 20 days below ..	39.2	4
Sterilized whole milk	4 months at	69.8	21
	12 months at	39.2	4
Frozen whole milk	12 months at	-2.2	-23
Evaporated milk	1 month at	89.6	32
	12 to 24 months at ..	69.8	21
	24 months at	39.2	4
Concentrated milk	2 or more weeks at ..	34.7	1.5
Concentrated frozen milk	6 months at	-6.8	-26
Sweetened condensed milk	3 months at	89.6	32
	9 to 24 months at ...	69.8	21
	24 months at	39.2	4
Nonfat dry milk, extra grade .... (in moisture-proof pack)	6 months at	89.6	32
	16 to 24 months at ..	69.8	21
	24 months at	39.2	4
Dry whole milk, extra grade .... (gas pack: maximum oxygen 2%)	6 months at	89.6	32
	12 months at	69.8	21
	24 months at	39.2	4
Buttermilk	2 to 3 weeks at	39.2	4
Sour cream	3 to 4 weeks at	39.2	4
Yogurt	3 to 6 weeks at	39.2	4
Eggnog	1 to 2 weeks at	39.2	4
Ultrapasteurized cream	6 to 8 weeks at	39.2	4

Three to five days may elapse between processing and purchasing, so the shelf life from the point of purchasing would be correspondingly less.

Various dairy products such as fluid and concentrated milk may be preserved by freezing. Milk has a lower freezing point than water due to dissolved constituents such as lactose and salts. The average freezing point of milk is 31°F (-0.54°C).

A very low temperature is needed to freeze milk completely. About 75% of the milk will freeze at 14°F (-10°C). A temperature below 7.6°F (-18°C) is recommended for frozen storage.

Milk can be frozen and thawed in the refrigerator; however, it is not recommended. The quality of milk is impaired by freezing due to protein destabilization and settling of some milk solids. A watery wheylike liquid may collect on the top and curd particles may also appear. Mixing returns the product to its normal dispersion and subsequent separation does not occur. However, thawed milk may be susceptible to the development of an oxidized flavor. The homogenization of milk overcomes the problem of demulsification of fat, a result of freezing milk.

Cream can be frozen, as evidenced by the successful use of frozen cream by the ice cream industry for several decades. However, particles of fat are evident when frozen cream is thawed. Homogenization of high-fat products such as cream has little effect in preventing demulsification of fat which occurs during freezing. The addition of sugar and very rapid freezing tend to retard the fat coalescence. Generally, the home freezing of cream produces an unsatisfactory product. Similarly, sour cream does not freeze well, although some dishes prepared with sour cream can be frozen without adverse effects.

Freezing buttermilk is not advocated due to the separation of the watery portion from the solids. If inadvertently frozen, buttermilk can be thawed in the refrigerator, gently mixed and used in cooked products.

Freezing per se of milk, cream, and other dairy products does not influence their nutritional properties. However, ascorbic acid and thiamin levels in dairy products may decrease during frozen storage.

MILK, FERMENTED

Milk is fermented with selected bacteria, which convert some of the milk sugar, lactose, into lactic acid. In selected cases, some alcohol is formed additionally.

(Also see MILK AND MILK PRODUCTS, section headed "Cultured Milk and Other Products.")

MILK FREEZING POINT TEST

This test serves as an indicator of adulteration, especially the addition of water. Most milk samples freeze within the range of 31.01° to 31.05°F (-0.530° to -0.550°C). The addition of 1% of water to milk will raise the freezing point slightly more than 0.011°F (0.006°C). Thermistor cryoscopes are sensitive to changes in temperature of 0.002°F (0.001°C).

MILK, FROZEN OR FRESH FROZEN

This is milk that has been pasteurized, treated with an ultrasonic vibrator at 5 million cycles per second for 5 minutes, then frozen to 10°F (-12.2°C). It will keep for a year; when thawed, it is indistinguishable from the original milk.

MILK SHAKE

Milk and a flavored syrup sometimes with added ice cream either shaken up and down with a hand shaker or blended in an electric mixer.

MILK, TURBIDITY TEST

This is a test to distinguish sterilized milk from pasteurized milk. During sterilization, all the albumin is precipitated. So, in the test the filtrate from an ammonium sulfate precipitation should remain clear on heating, indicating that no albumin was present in solution and the milk had therefore been sterilized.

MILLET *Panicum miliaceum*

Millet is a rapid-growing, warm weather cereal grass of the family, *Gramineae*, genus *Panicum* and species *miliaceum*, whose small grains are used primarily for human food.

Millet provides the major part of the diet of millions of poor people living on the poor, dry lands of India, Africa, China, Russia, and elsewhere.

Fig. M-70. Millet the chief cereal grain of millions of people. (Courtesy, USDA)

ORIGIN AND HISTORY. Although millet has been grown in Asia and North Africa since prehistoric times, little is known of its origin. It probably came from Eastern and Central Asia. Millet was important in Europe during the Middle Ages, before corn and potatoes were known there. But it is of minor importance in Europe today.

WORLD AND U.S. PRODUCTION. The annual production of millet totals 30 million metric tons, grown on some 94 million acres *(38 million hectares)*, more than 87% of which is produced in Asia and Africa. India alone produces 39% of the world's millet, followed by China and Nigeria, each produces 13% of the global total.[21]

Millet has never been of major agricultural importance in the United States. It is grown primarily as a substitute or emergency crop in the Great Plains and southeastern states.

NUTRITIONAL VALUE. Food Composition Table F-21 provides nutritional information regarding millet.

Some salient points concerning the nutritional value of millet follow:

1. Millet is high in starch—about 70%; hence, it serves as an energy food.

2. Like other grains, millet is low or lacking in calcium, vitamins A, D, C, and B-12.

3. The protein content varies greatly among various types, ranging from 5 to 20%, with an average of 10 to 12%.

MILLET PRODUCTS AND USES. Globally, about 85% of the millet crop is used for food. Millions of the poor in India, Africa, China, and parts of Eurasia depend on millet for 70% or more of their food calories.

Millet grain does not contain any gluten; hence, it is unsuited for making leavened breads. Millet meal and flour are used primarily for making flat breads and griddle-type cakes, and in boiled gruels. Ground millet is used in puddings, steamed meals, and deep-fried doughs, and is mixed with pulses (peas, beans, and lentils), vegetables, milk, cheese or dates. Whole grains are eaten with soups and stews or they are popped, roasted, sprouted, or malted.

Table M-24 presents, in summary form, the use of millet (1) in fortified foods, based on millet, for developing countries, and (2) in fermented beverages.

(Also see CEREAL GRAINS; and FLOURS, Table F-16, Special Flours.)

[21]Data from *FAO Production Yearbook 1990*, FAO/UN, Rome, Italy, Vol. 44, p. 83, Table 23. **Note well:** Annual production fluctuates as a result of weather and profitability of the crop.

TABLE M-24
MILLET PRODUCTS AND USES FOR HUMAN FOOD

Product	Description	Uses	Comments
FORTIFIED FOOD, BASED ON MILLET, FOR DEVELOPING COUNTRIES			
Ailment de Sevrage	Mixture of millet flour, peanut flour, skim milk powder, sugar, calcium, and vitamins A and D. It contains 20% protein.	High-protein infant food	Only the Senegal version of this food contains millet. The Algerian version contains wheat.
FERMENTED BEVERAGES			
Millet beers	Brewed beverages made from finger millet, a variety of millet, with or without corn (maize) and sorghum.	Alcoholic beverages	The beers are made by (1) sprouting millet by immersion in a stream, (2) drying and grinding the sprouted grain, (3) preparing a yeasty mixture from part of the sprouted grain by exposure to wild yeasts in the air, or on the surface of the brewing vessel, (4) mixing the freshly prepared yeast with a batch of sprouted grain, and pouring the mixture into gourds where fermentation occurs, (5) diluting the crude beer paste with hot water, and (6) allowing sediment to settle out prior to drinking.

MILLIEQUIVALENT (mEq)

An expression of concentration of a substance per liter of solution calculated by dividing the concentration in mg percent by the molecular weight.

MILLIGRAM

Metric unit of weight equal to 1/1,000 gram.

MINCE

To mince is to chop very finely. Sometimes the food is put through a food grinder to get it as fine as desired. In England, mince refers to ground beef or hamburger. Therefore, in England, rather than going for a hamburger, you would go for a mince.

MINCEMEAT

A special mixture used in pies, coffee cakes, and cookies. It is made with raisins, citron, apple, sugar and spices and usually contains some cooked beef and suet, which has been put through the food chopper along with the apple. The whole mixture is then simmered together. It can be used immediately, packed in sterilized jars, or put in the freezer.

MINERAL(S)

Contents

	Page
Historical Background	712
Mineral Table	713
Causes of Mineral Deficiencies and/or Toxicities	719
Agricultural Practices Which May Be Harmful	719
Contamination of Drinking Water, Foods, or Air with Toxic Minerals	720
Dietary and Medical Treatments	722
Drugs Which May Affect the Utilization of Minerals by the Body	723
Food Additives	723
Food Processing	724
Metabolic Disorders Which May Alter Mineral Metabolism	725
Mineral Composition of the Soil	726
Naturally Occurring Mineral Antagonists in Foods	727
Overconsumption of Alcohol	728
Poor Food Choices	728
Procedures Used in the Preparation of Meals	728
Stress and Trauma	728
Diagnosis of Mineral Deficiencies	729
Dietary Histories	729
Signs or Symptoms of Mineral Deficiencies	729
Laboratory Tests for Mineral Deficiencies	729
Prevention and Treatment of Mineral Deficiencies and/or Toxicities	730
Selection of Foods Rich in Essential Minerals	730
Production and Preparation of Foods to Maximize Their Mineral Values	730
Beneficial Practices in Agricultural Production	730
Food Processing Procedures Which Raise the Mineral Values of Foods	732
Meal Preparation Practices Which Conserve Essential Minerals	732
Enrichment and Fortification of Foods with Minerals	733
Special Dietary Products	734
Specialty Foods and Highly Potent Nutritional Supplements	734
Specialty Foods	734
Highly Potent Nutritional Supplements	735
Mineral Supplements	736

In nutrition, the term *mineral* denotes certain chemical elements which are found in the ash that remains after a food or a body tissue is burned. Some of these elements are essential to the proper functioning of the body—hence, they must be regularly supplied by the diet. Other elements are not known to be essential, yet they may get into the body by various means.

Each of the essential elements is considered to be either a macromineral or a micromineral, depending upon the quantity which is required in the diet.

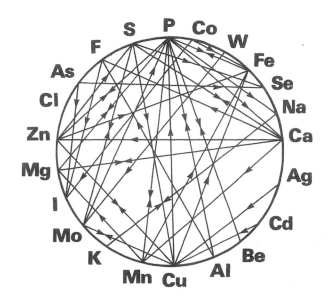

Fig. M-71. Mineral interaction chart, showing the interrelationship of minerals. The importance of such relationships is evidenced by the following:

1. A great excess of dietary calcium and/or phosphorus interferes with the absorption of both minerals, and increases the excretion of the lesser mineral.
2. Excess magnesium upsets calcium metabolism.
3. Excess zinc interferes with copper metabolism.
4. Copper is required for the proper utilization of iron, but excess copper can markedly depress iron absorption.

And there are others! The maze of connecting lines in this figure shows the relation of each mineral to other minerals.

• **Macrominerals**—These elements—which include calcium, phosphorus, sodium, chlorine, magnesium, potassium, and sulfur—are required in amounts ranging from a few tenths of a gram to one or more grams per day.

• **Microminerals (Trace elements)**—These elements are also known as *trace* elements because they are required in minute quantities, ranging from millionths of a gram (microgram) to thousandths of a gram (milligram) per day. As of this writing, the trace elements known to be essential for humans are chromium, cobalt, copper, fluorine, iodine, iron, manganese, molybdenum, selenium, silicon, and zinc. It seems likely that arsenic, vanadium, tin, and nickel may soon be added to this list.

HISTORICAL BACKGROUND. It is noteworthy that certain ancient peoples recognized and treated mineral deficiencies, although they did not understand the bases of their treatments. For example, a Chinese document dated about 3000 B.C. described goiter and recommended that the afflicted people eat seaweed and burnt sponge, which are good sources of the trace element iodine. Another deficiency disease, anemia, was treated in ancient Greece (around the time of Hippocrates, or about 4th century, B.C.) by giving the patient iron-containing water in which heated swords had been quenched. However, the effects of such treatments were often unpredictable, because there were no means of identifying or measuring the quantity of the active ingredients in the various medicinal substances.

The breakthroughs in our understanding of the functions of minerals in the body did not come for many centuries because laboratories for research were not developed until the Renaissance; although the medieval alchemists appear to have invented some of the techniques and tools of chemistry in their futile efforts to change base metals into gold.

The great French chemist Lavoisier, who is credited with being the founder of the science of nutrition, predicted in 1799 that such elements as sodium and potassium would soon be discovered because he believed that they were present in certain mineral compounds which were then known as *earths*. Sure enough, within a few years the British chemist Davy discovered not only sodium and potassium, but also calcium, sulfur, magnesium, and chlorine. Davy's discoveries were so highly regarded by the French Academy that in 1806 they awarded him their new Volta medal, even though France and England were at war when the award was presented.

The pace of progress quickened as the noted Swedish chemist Berzelius added to our knowledge of minerals by (1) reporting his analysis of the calcium and phosphorus content of bone in 1801, and (2) concluding in 1838 that the iron in hemoglobin made it possible for the blood to absorb much oxygen. Similar contributions were made by the French chemist Boussingault, who (1) noted in 1822 that South American villagers who used salt containing iodine were protected from goiter which affected those who used plain salt; and (2) showed by means of animal feeding trials the necessity of providing dietary calcium and iron. Calcium studies were also conducted by Chossat, a physician in Switzerland, who won a prize in 1840 for his demonstration that the addition of calcium carbonate to a diet of wheat and water improved bone growth in pigeons.

Another half century elapsed before the value of iodine became widely accepted; thanks to (1) the discovery by the German biochemist Baumann in 1895 that the thyroid gland contained iodine, and (2) the demonstrations by the American medical scientist Marine and his co-workers between 1907 and 1918 that the administration of minute amounts of iodine prevented goiter in animals and in school children.

By the end of the 19th century only about 1/3 of the minerals now accepted as essential were known to be required in the diet. This state of affairs existed in spite of the many nutritional investigations which were conducted during that century because the need for vitamins had not yet been discovered. Hence, it was difficult to differentiate between the deficiency diseases due to lack of minerals and those due to lack of vitamins.

The first two decades of the 20th century were marked by a flurry of discoveries concerning the roles of various vitamins. Then, various minerals were found to be required by animals and man, as indicated by the dates of the discoveries which follow: phosphorus, 1918; copper, 1925; magnesium, manganese, and molybdenum, 1931; zinc, 1934; and cobalt, 1935. However, there was often uncertainty as to the metabolic roles of various minerals, even though they were known to be essential. For example, it was not until 1948 that it was established that cobalt functions as a component of vitamin B-12. Hence, the two decades between the 1930s and the 1950s were marked by feverish activity aimed at learning how each of the newly discovered essential elements acted in the body.

The most recent chapter in this story began when Schwarz, a medical scientist who had emigrated to the United States from Germany, and his co-workers discovered the essentiality of selenium in 1957, and of chromium in 1959. They then developed isolation equipment for shielding their animals and their ultrapure diets from contamination by minute amounts of elements in the environment. Their painstaking work paid off because in 1972 they were able to show that fluorine and silicon were also essential. It is ironic that only a short time ago each of these four trace elements was considered to be only an unwanted toxic contaminant of foods, water, and air.

Much work remains to be done in the areas of (1) testing whether other elements might be essential, (2) defining the limits of safe and toxic doses for those already known, and (3) determining how the various elements interact with each other in the human body.

MINERAL TABLE.   The general functions of minerals are as follows:

1. Give rigidity and strength to the skeletal structure.
2. Serve as constituents of the organic compounds, such as protein and lipid, which make up the muscles, organs, blood cells, and other soft tissues of the body.
3. Activate enzyme systems.
4. Control fluid balance—osmotic pressure and excretion.
5. Regulate acid-base balance.
6. Exert characteristic effects on the irritability of muscles and nerves.
7. Act together with hormones, vitamins, and other metabolic regulators.

In addition to the general functions in which several minerals may be involved, each essential mineral has at least one specific role. Those which are present in large amounts, like calcium and phosphorus, are likely to have multiple and diverse roles. However, trace elements that are present at barely detectable levels, such as chromium and cobalt, appear to have highly specific functions centered around the organic molecules to which they are bound.

A summary of individual mineral functions, deficiency and toxicity symptoms, recommended daily allowance, and sources is given in Table M-25, Mineral Table. (Calcium, phosphorus, sodium, and chlorine are listed in this order, and first, under macrominerals, then the rest of the minerals are listed alphabetically under their respective categories—as macromineral or micromineral.)

Additionally, a separate presentation relative to each mineral appears alphabetically in this book under its respective name.

TABLE M-25
MINERAL TABLE

Functions	Deficiencies and Toxicity Symptoms	Recommended Daily Allowance			Sources	Comments
		Category	Age (Yrs)	Mineral		

MACROMINERALS:
Calcium (Ca)

Functions

The primary function of calcium is to build the bones and teeth, and to maintain the bones..

Other functions are:
1. Blood clotting.
2. Muscle contraction and relaxation, especially the heartbeat.
3. Nerve transmission.
4. Cell wall permeability.
5. Enzyme activation.
6. Secretion of a number of hormones and hormone-releasing factors.
(Also see CALCIUM.)

Deficiencies and Toxicity Symptoms

Deficiency symptoms—The most dramatic deficiency symptoms are manifested in the bones and teeth of the young, evidenced in—
1. Stunting of growth.
2. Poor quality bones and teeth.
3. Malformation of bones—rickets.
The clinical manifestations of calcium related diseases are—
1. Rickets in children.
2. Osteomalacia, the adult counterpart of rickets.
3. Osteoporosis, a condition of too little bone, resulting when bone resorption exceeds bone formation.
4. Hypercalcemia, characterized by high serum calcium.
5. Tetany, characterized by muscle spasms and muscle pain.
6. Kidney stones.

Toxicity—Normally, the small intestine prevents excess calcium from being absorbed. However, a breakdown of this control may raise the level of calcium in the blood and lead to calcification of the kidneys and other internal organs.

High calcium intake may cause excess secretion of calcitonin and such bone abnormalities as osteopetrosis (dense bone).

High calcium intakes have also been reported to cause kidney stones.

Recommended Daily Allowance — Calcium[1] (mg)

Category	Age (Yrs)	Mineral
Infants	0–0.5	400
	0.5–1.0	600
Children	1–3	800
	4–6	800
	7–10	800
Males	11–14	1,200
	15–18	1,200
	19–24	1,200
	25–50	800
	51+	800
Females	11–14	1,200
	15–18	1,200
	19–24	1,200
	25–50	800
	51+	800
Pregnant		1,200
Lactating		1,200

[1]Recommended Dietary Allowances, 10th ed., NRC–National Academy of Sciences, 1989, p. 285.

Sources

Rich sources—Cheeses, wheat-soy flour, blackstrap molasses.

Good sources—Almonds, Brazil nuts, caviar, cottonseed flour, dried figs, fish with soft edible bones, green leafy vegetables, hazel nuts, ice cream, milk, oysters, sour cream, soybean flour, yogurt.

Fair sources—Beans, bread, broccoli, cabbage, clams, cottage cheese crab, dehydrated apricots, dehydrated peaches, eggs, legumes, lettuce, lobster, okra, olives, oranges, parsnips, peanut butter, prunes, raisins, rhubarb, spinach, Swiss chard, turnips, wheat germ.

Negligible sources—Asparagus, beef, beets, Brussels sprouts, carrots cauliflower, corn and cornmeal and other grains and cereals, cucumbers, fats and oils, juices, kohlrabi, most fish, most fresh fruits, mushrooms, peas, pickles, popcorn, pork, potatoes, poultry, pumpkin, radishes, roe, squash, sugar, tomatoes, tuna.

Supplemental sources—Bone meal, calcium, carbonate, calcium gluconate, calcium lactate, dicalcium phosphate, dolomite, kelp. If calcium is taken as a pill, it should be in the form of some soluble salt, such as calcium lactate.

Comments

Calcium is the most abundant mineral in the body. It makes up about 2.0% of the body weight, or about 40% of the total mineral present; 99% of it is in the bones and teeth.

Normally, only 20 to 30% of the calcium in the average diet is absorbed from the intestinal tract and taken into the bloodstream. But growing children and pregnant-lactating women utilize calcium most efficiently—they absorb 40% or more of the calcium in their diets.

Calcium is involved in a number of relationships, which are detailed in the Calcium section of this book.

Generally, nutritionists recommend a calcium-phosphorus ratio of 1.5:1 in infancy, decreasing to 1:1 at 1 year of age and remaining at 1:1 throughout the rest of life; although they consider ratios between 2:1 and 1:2 as satisfactory.

Phosphorus (P)

Functions

Essential for bone formation and maintenance.

Important in the development of teeth.

Essential for normal milk secretion.

Important in building muscle tissue.

As a component of nucleic acids (RNA and DNA), which are important in genetic transmission and control of cellular metabolism.

Maintenance of osmotic and acid-base balance.

Important in many metabolic functions, especially—
1. Energy utilization.
2. Phospholipid formation.
3. Amino acid metabolism; protein formation.
4. Enzyme systems.
(Also see PHOSPHORUS.)

Deficiencies and Toxicity Symptoms

Deficiency symptoms—General weakness, loss of appetite, muscle weakness, bone pain, demineralization of bone, and loss of calcium.

Severe and prolonged deficiencies of phosphorus may be manifested by rickets, osteomalacia, and other phosphorus related diseases.

Toxicity—There is no known phosphorus toxicity per se. However, excess phosphate consumption may cause hypocalcemia (a deficiency of calcium in the blood).

Recommended Daily Allowance — Phosphorus[1] (mg)

Category	Age (Yrs)	Mineral
Infants	0–0.5	400
	0.5–1.0	600
Children	1–3	800
	4–6	800
	7–10	800
Males	11–14	1,200
	15–18	1,200
	19–24	1,200
	25–50	800
	51+	800
Females	11–14	1,200
	15–18	1,200
	19–24	1,200
	25–50	800
	51+	800
Pregnant		1,200
Lactating		1,200

[1]Recommended Dietary Allowances, 10th ed., NRC–National Academy of Sciences, 1989, p. 285.

Sources

Rich sources—Cocoa powder, cottonseed flour, fish flour, peanut flour, pumpkin and squash seeds, rice bran, rice polish, soybean flour, sunflower seeds, wheat bran.

Good sources—Beef, cheeses, fish and seafood, lamb, liver, nuts, peanut butter, pork, poultry, whole grain flours.

Fair sources—Breads and cereals, cottage cheese, dehydrated fruits, eggs, ice cream, kidneys, milk, molasses, most vegetables, mushrooms, sausages and luncheon meats, white flour, yogurt.

Negligible sources—Fats and oils, juices and beverages, raw and canned fruits, some vegetables (lettuce, celery, carrots, tomatoes), sugar.

Supplemental sources—Ammonium phosphate, bone meal, calcium phosphate, dicalcium phosphate, lecithin, monosodium phosphate, wheat germ (toasted), yeast (brewers', torula).

Comments

Phosphorus comprises about ¼ of the total mineral matter in the body.

Eighty percent of the phosphorus is in the bones and teeth in inorganic combination with calcium.

Normally, 70% of the ingested phosphorus is absorbed.

Generally, nutritionists recommend a calcium-phosphorus ratio of 1.5:1 in infancy, decreasing to 1:1 at 1 year of age, and remaining at 1:1 throughout the rest of life; although they consider ratios between 2:1 and 1:2 as satisfactory.

Sodium (Na)

Functions

Helps to maintain the balance of water, acids, and bases in the fluid outside the cells.

As a constituent of pancreatic juice, bile, sweat, and tears.

Associated with muscle contraction and nerve functions.

Plays a specific role in the absorption of carbohydrates.
(Also see SODIUM.)

Deficiencies and Toxicity Symptoms

Deficiency symptoms—Reduced growth, loss of body weight due to loss of water, reduced milk production of lactating mothers, muscle cramps, nausea, diarrhea, and headache.

Excess perspiration and salt depletion may be accompanied by heat exhaustion.

Toxicity—Salt may be toxic (1) when a high intake is accompanied by a restriction of water, (2) when the body is adapted to a chronic low salt diet, or (3) when it is fed to infants.

Recommended Daily Allowance — Sodium[1] (mg)

Category	Age (Yrs)	Mineral
Infants	0–0.5	120
	0.5–1.0	200
Children and adolescents	1	225
	2–5	300
	6–9	400
	10–18	500
Adults	18+	500

[1]**Note well:** The sodium values are "Estimated Minimum Requirements," from Recommended Dietary Allowances, 10th ed., NRC–National Academy of Sciences, 1989, p. 253, Table 11–1.

Sources

Generally speaking, the need is for low sodium diets, which calls for avoiding high sodium foods. Groupings of common food sources of sodium in descending order follow:

Rich sources—Anchovy paste, bacon, bologna, bran cereal, butter and margarine, Canadian bacon, corned beef, cornflakes cereal, cucumber pickles, cured ham, dehydrated cod, dried squid, frankfurters, green olives, luncheon meats, oat cereal, parmesan cheese, pasteurized process

Comments

Most of the salt in the diet is in the form of sodium chloride (NaCl).

Excess dietary salt, over a long period of time, is believed to lead to high blood pressure in susceptible people. Also, excess sodium may cause edema, evidenced by swelling of such areas as the legs and face. The simplest way to reduce sodium intake is to eliminate the use of table salt.

Deficiencies of sodium may occur from strict vegetarian diets without salt, or when there has been heavy,

(Continued)

TABLE M-25 *(Continued)*

Functions	Deficiencies and Toxicity Symptoms	Recommended Daily Allowance			Sources	Comments
		Category	Age (Yrs)	Mineral		
Sodium (Na) *(Continued)*	or others whose kidneys cannot excrete the excess in the urine.				cheese, potato chips, pretzels, sausages, seaweed, shrimp, smoked herring, soda crackers, soy sauce, tomato	prolonged sweating, diarrhea, vomiting, or adrenal cortical insufficiency. In such cases, extra salt should be taken.

catsup, wheat flakes cereal.

Good sources—Breads and cookies, canned vegetables, cheeses, condensed soups, many prepared foods and mixes, peanut butter, pickled foods, salad dressings, seafoods.

Fair sources—Beef, desserts and sweets, eggs, lamb, milk, pork, poultry, salt-loving vegetables (spinach, beets, chard, celery, carrots), some fish (cod, haddock, salmon, tuna) depending upon preparation, veal, yogurt.

Negligible sources—Lard, legumes, most fresh fruits and vegetables, rye flour, shortening, soybean flour, sugar, vegetable oils, wheat bran, wheat flour.

Supplemental sources—Salt (sodium chloride), baking soda, monosodium glutamate (MSG), sodium-containing baking powder.

Functions	Deficiencies and Toxicity Symptoms	Category	Age (Yrs)	Mineral	Sources	Comments
Chlorine or chloride (Cl) Plays a major role in the regulation of osmotic pressure, water balance, and acid-base balance. Required for the production of hydrochloric acid of the stomach; this acid is necessary for the proper absorption of vitamin B-12 and iron, for the activation of the enzyme that breaks down starch, and for suppressing the growth of microorganisms that enter the stomach with food and drink. (Also see CHLORINE or CHLORIDE.)	**Deficiency symptoms**—Severe deficiencies may result in alkalosis (an excess of alkali in the blood), characterized by slow and shallow breathing, listlessness, muscle cramps, loss of appetite, and occasionally, by convulsions. Deficiencies of chloride may develop from prolonged and severe vomiting, diarrhea, pumping of the stomach, injudicious use of diuretic drugs, or strict vegetarian diets without salt. **Toxicity**—An excess of chlorine ions is unlikely when the kidneys are functioning properly.	*Chlorine*[1] (mg) Infants Children and adolescents Adults	 0–0.5 0.5–1.0 1 2–5 6–9 10–18 18+	 180 300 350 500 600 750 750	Table salt (sodium chloride) and foods that contain salt. Persons whose sodium intake is severely restricted (owing to diseases of the heart, kidney, or liver) may need an alternative source of chloride; a number of chloride-containing salt substitutes are available for this purpose.	Chlorine is an essential mineral, with a special function in forming the hydrochloric acid (HCl) present in the gastric juice.

[1]**Note well:** The chloride values are "Estimated Minimum Requirements," from *Recommended Dietary Allowances*, 10th ed., NRC–National Academy of Sciences, 1989, p. 253, Table 11–1.

Functions	Deficiencies and Toxicity Symptoms	Category	Age (Yrs)	Mineral	Sources	Comments
Magnesium (Mg) Constituent of bones and teeth. Essential element of cellular metabolism, often as an activator of enzymes involved in phosphorylated compounds and of high energy phosphate transfer of ADP and ATP. Involved in activating certain peptidases in protein digestion. Relaxes nerve impulses, functioning antagonistically to calcium which is stimulatory. (Also see MAGNESIUM.)	**Deficiency symptoms**—A deficiency of magnesium is characterized by (1) muscle spasms (tremor, twitching) and rapid heartbeat; (2) confusion, hallucinations, and disorientation; and (3) lack of appetite, listlessness, nausea, and vomiting. **Toxicity**—Magnesium toxicity is characterized by slowed breathing, coma, and sometimes death.	*Magnesium*[1] (mg) Infants Children Males Females Pregnant Lactating	 0–0.5 0.5–1.0 1–3 4–6 7–10 11–14 15–18 19–24 25–50 51+ 11–14 15–18 19–24 25–50 51+	 40 60 80 120 170 270 400 350 350 350 280 300 280 280 280 320 355	Groupings by rank of common food sources of magnesium follow: **Rich sources**—Coffee (instant), cocoa powder, cottonseed flour, peanut flour, sesame seeds, soybean flour; spices, wheat bran, wheat germ. **Good sources**—Blackstrap molasses, nuts, peanut butter, whole grains (oats, barley, wheat, buckwheat), whole wheat flour, yeast. **Fair sources**—Avocados, bananas, beef and veal, breads, cheese, chicken, corn, cornmeal, dates, dehydrated fruit, fish and seafoods, lamb, liver, olives, pork, raspberries, rice, turkey, most green leafy vegetables. **Negligible sources**—Cabbage, egg plant, eggs, fats and oils, ice cream, lettuce, milk, most fruits, mushrooms, rhubarb, rutabagas, sausages and luncheon meats, sugar, tomatoes. **Supplemental sources**—Dolomite, magnesium gluconate, magnesium oxide, wheat germ.	Overuse of such substances as "milk of magnesia" (magnesium hydroxide, an antacid and laxative) or "Epsom salts" (magnesium sulfate, a laxative and tonic) may lead to deficiencies of other minerals or even to toxicity.

[1]*Recommended Dietary Allowances*, 10th ed., NRC–National Academy of Sciences, 1989, p. 285

Functions	Deficiencies and Toxicity Symptoms	Category	Age (Yrs)	Mineral	Sources	Comments
Potassium (K) Involved in the maintenance of proper acid-base balance and the transfer of nutrients in and out of individual cells. Relaxes the heart muscle—action opposite to that of calcium, which is stimulatory. Required for the secretion of insulin by the pancreas, in enzyme reactions involving the phosphorylation of creatine, in carbohydrate metabolism, in protein synthesis. (Also see POTASSIUM.)	**Deficiency symptoms**—Potassium deficiency may cause rapid and irregular heartbeats, and abnormal electrocardiograms; muscle weakness, irritability, and occasionally paralysis; and nausea, vomiting, diarrhea, and swollen abdomen. Extreme and prolonged deficiency of potassium may cause hypokalemia, culminating in the heart muscles stopping. **Toxicity**—Acute toxicity from potassium (known as hyperpotassemia or hyperkalemia) will result from sudden increases of potassium to levels of about 18 g per day for an adult,	*Potassium*[1] (mg) Infants Children and adolescents Adults	 0–0.5 0.5–1.0 1 2–5 6–9 10–18 18+	 500 700 1,000 1,400 1,600 2,000 2,000	**Rich sources**—Dehydrated fruits, molasses, potato flour, rice bran, seaweed, soybean flour, spices, sunflower seeds, wheat bran. **Good sources**—Avocado, beef, dates, guavas, most raw vegetables, nectarine, nuts, pork, poultry, sardines, veal. **Fair sources**—Breads, cereals, cheeses, cooked or canned vegetables, eggs, fruit juices, milk, raw, cooked, or canned fruits, shellfish, whole wheat flour, wine, yogurt. **Negligible sources**—Cooked rice, cornmeal, fats and oils, honey, olives,	Potassium is the third most abundant element in the body, after calcium and phosphorus; and it is present in twice the concentration of sodium. Sources of potassium other than foods, such as potassium chloride, should be taken only on the advice of a physician or nutritionist.

[1]**Note well:** The potassium values are "Estimated Minimum Requirements," from *Recommended Dietary Allowances*, 10th ed., NRC–National Academy of Sciences, 1989, p. 253, Table 11–1.

TABLE M-25 *(Continued)*

Functions	Deficiencies and Toxicity Symptoms	Recommended Daily Allowance			Sources	Comments
		Category	Age (Yrs)	Mineral		

Potassium (K) *(Continued)*

	provided the kidneys are not functioning properly and there is not an immediate and sharply increased loss of potassium from the body. The condition may prove fatal due to cardiac arrest.				sugar. **Supplemental sources**—Potassium gluconate, potassium chloride, seaweed, yeast (brewers', torula), wheat germ.	

Sulfur (S)

Functions	Deficiencies and Toxicity Symptoms	Category	Age (Yrs)	Mineral	Sources	Comments
As a component of the sulfur-containing amino acids methionine, cystine, and cysteine. As a component of biotin, sulfur is important in fat metabolism. As a component of thiamin and insulin, it is important in carbohydrate metabolism.	**Deficiency symptoms**—Primarily retarded growth, because of sulfur's association with protein synthesis. **Toxicity**—Except in rare, inborn errors of metabolism where the utilization of the sulfur-containing amino acids is abnormal, excess organic sulfur intake is essentially nonexistent. However, inorganic sulfur can be dangerous if ingested in large amounts.			There are no recommended allowances for sulfur because it is assumed that the sulfur requirements are met when the methionine and cystine intakes are adequate.	For the most part, the sulfur needs of the body are met from organic complexes, notably the amino acids of proteins, rather than from inorganic sources. Good food sources are: cheese, grains and grain products, eggs, fish, legumes, meat, nuts, and poultry.	Sulfur is found in every cell in the body and is essential for life itself. Approximately 10% of the mineral content of the body is sulfur.

As a component of coenzyme A, it is important in energy metabolism.
As a component of certain complex carbohydrates, it is important in various connective tissues.
As a component of insulin and glutathione, it regulates energy metabolism.
As a converter of toxic substances to nontoxic forms, it rids the body (via excretion) of such toxic substances as phenols and cresols.
(Also see SULFUR.)

MICROMINERALS OR TRACE ELEMENTS:
Chromium (Cr)

Functions	Deficiencies and Toxicity Symptoms	Category	Age (Yrs)	Chromium[1,2] (mcg)	Sources	Comments
Component of the Glucose Tolerance Factor (GTF), which enhances the effect of insulin. Activator of certain enzymes, most of which are involved in the production of energy from carbohydrates, fats, and proteins. Stabilizer of nucleic acids (DNA and RNA). Stimulation of synthesis of fatty acids and cholesterol in the liver. (Also see CHROMIUM.)	**Deficiency symptoms**—Impaired glucose tolerance, which may be accompanied by high blood sugar and the spilling of sugar in the urine; seen especially in older persons, in maturity-onset diabetes, and in infants with protein-calorie malnutrition. Disturbance in lipid and protein metabolism. (Also see DIABETES.) **Toxicity**—Chromium is seldom toxic because (1) only small amounts	Infants	0–0.5 0.5–1.0	10–40 20–60	Groupings by rank of some common food sources of chromium follow: **Rich sources**—Blackstrap molasses, cheese, eggs, liver. **Good sources**—Apple peel, banana, beef, beer, bread, brown sugar, butter or margarine, chicken, cornflakes, cornmeal, flour, oysters, potatoes, vegetable oils, wheat bran, whole wheat.	Further studies are needed to determine chromium's role in metabolsim and its nutritional significance. Chromium levels in the body decline with aging. Hence, it is suspected that some of the cases of adult-onset diabetes are due to lack of GTF-chromium.
		Children and adolescents	1–3 4–6 7–10 11+	20–80 30–120 50–200 50–200		
		Adults		50–200		

[1]**Note well:** The chromium values are "Estimated Minimum Requirements."
[2]From *Recommended Dietary Allowances*, 10th ed., NRC–National Academy of Sciences, 1989, p. 284.

Fair sources—Carrots, green beans, oranges, spinach, strawberries.
Negligible sources—Milk, most fruits and vegetables, sugar.
Supplemental sources—Brewers' yeast, dried liver.
Note well: The content and/or availability of chromium in foods may be affected by the following:
1. The chromium content of the soil.
2. The processing of grain.
3. The refining of molasses.
4. The type of cooking utensil.
5. Fermentation.
6. The alcohol-extractable fraction.
(Also see GLUCOSE TOLERANCE FACTOR.)

are present in most foods, (2) the body utilizes it poorly, and (3) there is a wide margin of safety between helpful and harmful doses.
Note well: Excesses of inorganic chromium are much more toxic than similar amounts of GTF-chromium.

Cobalt (Co)

Functions	Deficiencies and Toxicity Symptoms	Category	Age (Yrs)	Mineral	Sources	Comments
The only known function of cobalt is that of an integral part of vitamin B-12, an essential factor in the formation of red blood cells. (Also see COBALT; and VITAMIN B-12.)	A cobalt deficiency as such has never been produced in humans. The signs and symptoms that are sometimes attributed to cobalt deficiency are actually due to lack of vitamin B-12, characterized by pernicious anemia, poor growth, and occasionally neurological disorders. (Also see ANEMIA, PERNICIOUS.)			There is no known human requirement for cobalt, except for that contained in vitamin B-12.	Cobalt is present in many foods. However, the element must be ingested in the form of vitamin B-12 in order to be of value to man; hence, a table showing the cobalt content of human foods serves no useful purpose. Instead, the need is for rich sources of vitamin B-12—liver, kidneys, fish, poultry, eggs, and tempeh.	Cobalt is an essential constituent of vitamin B-12 and must be ingested in the form of the vitamin molecule inasmuch as humans synthesize little of the vitamin. (A small amount of vitamin B-12 is synthesized in the human colon by *E. coli*, but absorption is very limited.

Copper (Cu)

Functions	Deficiencies and Toxicity Symptoms	Category	Age (Yrs)	Copper[1] (mg)	Sources	Comments
Facilitating the absorption of iron from the intestinal tract and releasing it from storage in the liver and the reticuloendothelial system. Essential for the formation of hemoglobin, although it is not a part of hemoglobin as such. Constituent of several enzyme systems. Development and maintenance of the vascular and skeletal structures (blood vessels, tendons, bones).	**Deficiency symptoms**—Deficiency is most apt to occur in malnourished children and in premature infants fed exclusively on modified cow's milk and in infants breast fed for an extended period of time. Deficiency leads to a variety of abnormalities, including anemia, skeletal defects, demyelination and degeneration of the nervous system, defects in pigmentation and structure of the hair, reproductive failure, and pronounced cardiovascular lesions. **Toxicity**—Copper is relatively non-	Infants	0–0.5 0.5–1.0	0.4–0.6 0.6–0.7	Groupings by rank of common food sources of copper follow: **Rich sources**—Black pepper, blackstrap molasses, Brazil nuts, cocoa, liver, oysters (raw). **Good sources**—Lobster, nuts and seeds, olives (green), soybean flour, wheat bran, wheat germ (toasted). **Fair sources**—Avocado, banana,	Most cases of copper poisoning result from drinking water or beverages tht have been stored in copper tanks and/or that pass through copper pipes. Dietary excesses of calcium, iron, cadmium, zinc, lead, silver, and molybdenum plus sulfur reduce the utilization of copper.
		Children and adolescents	1–3 4–6 7–10 11+	0.7–1.0 1.0–1.5 1.0–2.0 1.5–2.5		
		Adults		1.5–3.0		

[1]**Note well:** The copper values are "Estimated Minimum Requirements," from *Recommended Dietary Allowances*, 10th ed., NRC–National Academy of Sciences, 1989, p. 284.

beans, beef, breads and cereals, butter, Cheddar cheese, coconut, dried fruits, eggs, fish, granola, green vegetables (turnip greens, collards, spinach), lamb, peanut butter, pork, poultry, turnip, yams.
Negligible sources—Fats and oils, milk, and most milk products (ice

(Continued)

TABLE M-25 *(Continued)*

Functions	Deficiencies and Toxicity Symptoms	Recommended Daily Allowance			Sources	Comments
		Category	Age (Yrs)	Mineral		
Copper (Cu) *(Continued)* Structure and functioning of the central nervous system. Required for normal pigmentation of the hair. Component of important copper-containing proteins. Reproduction (fertility). (Also see COPPER.)	toxic to monogastric species, including man. The recommended copper intake for adults is in the range of 2 to 3 mg/day. Daily intakes of more than 20 to 30 mg over extended periods would be expected to be unsafe.				cream, cottage cheese), other fruits and vegetables than listed above, sugar. **Supplemental sources**—Alfalfa leaf meal, brewers' yeast, copper carbonate, copper sulfate. **Note well:** Copper sulfate or copper carbonate should only be taken on the advice of a physician or nutritionist.	
Fluorine (F) Constitutes .02 to .05% of the bones and teeth. Necessary for sound bones and teeth. Assists in the prevention of dental caries. (Also see FLUORINE.)	**Deficiency symptoms**—Excess dental caries. Also, there is indication that a deficiency of fluorine results in more osteoporosis in the aged. However, excesses of fluorine are of more concern than deficiencies. **Toxicity**—Deformed teeth and bones, and softening, mottling, and irregular wear of the teeth.	Infants Children and adolescents Adults	0–0.5 0.5–1.0 1–3 4–6 7–10 11+	*Fluorine[1]* (mg) 0.1–0.5 0.2–1.0 0.5–1.5 1.0–2.5 1.5–2.5 1.5–2.5 1.5–4.0	Fluorine is found in many foods, but seafoods and dry tea are the richest food sources. A few rich sources follow: **Source** / **Fluorine (F)** (ppm) Dried seaweed 326.0 Tea 32.0 Mackerel 19.0 Sardines 11.0 Salmon 6.8 Shrimp 4.5	Large amounts of dietary calcium, aluminum, and fat will lower the absorption of fluorine. Fluoridation of water supplies (1 ppm) is the simplest and most effective method of providing added protection against dental caries. Fluoridation of water supplies to bring the concentration of fluoride to 1 ppm.
			[1]Note well: The fluoride values are "Estimated Minimum Requirements," from *Recommended Dietary Allowances*, 10th ed., NRC–National Academy of Sciences, 1989, p. 284.			
Iodine (I) The sole function of iodine is making the iodine-containing hormones secreted by the thyroid gland, which regulate the rate of oxidation within the cells; and in so doing influence physical and mental growth, the functioning of the nervous and muscle tissues, circulatory acitivity, and the metabolism of all nutrients. (Also see IODINE.)	**Deficiency symptoms**—Iodine deficiency is characterized by goiter (an enlargement of the thyroid gland at the base of the neck), coarse hair, obesity, mental fatigue, and high blood cholesterol. Iodine-deficient mothers may give birth to infants with a type of dwarfism known as cretinism, a disorder characterized by malfunctioning of the thyroid gland, goiter, mental retardation, and stunted growth. A similar disorder of the thyroid gland, known as myxedema, may develop in adults. **Toxicity**—Long-term intake of large excesses of iodine may disturb the utilization of iodine by the thyroid gland and result in goiter. (Also see ENDOCRINE GLANDS; and GOITER.)	Infants Children Males Females Pregnant Lactating	0–0.5 0.5–1.0 1–3 4–6 7–10 11–14 15–18 19–24 25–50 51+ 11–14 15–18 19–24 25–50 51+	*Iodine[1]* (mcg) 40 50 70 90 120 150 150 150 150 150 150 150 150 150 150 175 200	Among natural foods the best sources of iodine are kelp, seafoods, and vegetables grown on iodine-rich soils. Dairy products and eggs may be good sources if the producing animals have access to iodine-enriched rations. Most cereal grains, legumes, roots, and fruits have low iodine content. Of the various methods for assuring an adequate iodine intake, iodized salt has thus far proved to be most successful, and therefore the most widely adopted method. In the U.S., iodination is on a voluntary basis, nevertheless slightly more than half of the table salt consumed is iodized. Stabilized iodized salt contains 0.01% potassium iodide (0.0076% I), or 76 mcg of iodine per gram. Iodine may also be provided in bread. But the practice of using iodates in bread making appears to be on the decline.	The enlargement of the thyroid gland (goiter) is nature's way of attempting to make sufficient thyroxine under conditions where a deficiency exists. Certain foods (especially plants of the cabbage family) contain goitrogens, which interfere with the use of thyroxine and may produce goiter. Fortunately, goitrogenic action is prevented by cooking. For intakes of iodine, man is dependent upon food, soil, and water. The iodine content of food varies widely, depending chiefly on (1) the iodine content of the soil, (2) the iodine content of the animal feeds (to which iodized salt is routinely added in most countries), and (3) the use of iodized salt in food processing operations. Iodine in drinking and cooking water varies widely in different regions; in some areas, such as near oceans, it is high enough to meet the daily requirement.
			[1]Recommended Dietary Allowances, 10th ed., NRC–National Academy of Sciences, 1989, p. 285.			
Iron (Fe) Iron (heme) combines with protein (globin) to make hemoglobin, the iron-containing compound in red blood cells; so, iron is involved in transporting oxygen. Iron is also a component of enzymes which are involved in energy metabolism. (Also see IRON.)	**Deficiency symptoms**—Iron-deficiency (nutritional) anemia, the symptoms of which are: paleness of skin and mucous membranes, fatigue, dizziness, sensitivity to cold, shortness of breath, rapid heartbeats, and tingling of the fingers and toes. **Toxicity**—Approximately 2,000 cases of iron poisoning occur each year in the U.S., mainly in young children who ingest the medical iron supplements of their parents. The lethal dose of ferrous sulfate for a 2-year-old is about 3 g; for an adult it's between 200 and 250 mg/kg of body weight. (Also see ANEMIA.)	Infants Children Males Females	0–0.5 0.5–1.0 1–3 4–6 7–10 11–14 15–18 19–24 25–50 51+ 11–14 15–18 19–24 25–50 51+	*Iron[1]* (mg) 6 10 10 10 10 12 12 10 10 10 15 15 15 15 10	Groupings by rank of common sources of iron follow: **Rich sources**—Beef kidneys, blackstrap molasses, caviar, chicken giblets, cocoa powder, fish flour, liver, orange pekoe tea, oysters, potato flour, rice polish, soybean flour, spices, sunflower seed flour, wheat bran, wheat germ, wheat-soy blend flour. **Good sources**—Beef, brown sugar, clams, dried fruits, egg yolk, heart, light or medium molasses, lima beans (cooked), nuts, pork, pork and lamb kidneys. **Fair sources**—Asparagus, beans, chicken, dandelion greens, enriched bread, enriched cereals, enriched cornmeal, enriched flour, enriched rice,	About 70% of the iron is present in the hemoglobin, the pigment of the red blood cells. The other 30% is present as a reserve store in the liver, spleen, and bone marrow. An excess of iron in the diet can tie up phosphorus in an insoluble iron-phosphate complex, thereby creating a deficiency of phosphorus. Babies are not born with sufficient iron to meet their needs beyond 6 months. Milk and milk products are poor sources of iron. About 20% of the iron in the average U.S. diet comes from fortified products. Ferrous sulfate is the supplemental

TABLE M-25 (Continued)

Functions	Deficiencies and Toxicity Symptoms	Recommended Daily Allowance — Category	Age (Yrs)	Mineral[1]	Sources	Comments
		Pregnant[2]		30		
		Lactating		15		

[1] *Recommended Dietary Allowances*, 10th ed., NRC–National Academy of Sciences, 1989, p. 285.

[2] The increased requirement during pregnancy cannot be met by the iron content of habitual American diets nor by the existing iron stores of many women; therefore, the use of 30 to 60 mg of supplemental iron is recommended. Iron needs during lactation are not substantially different from those of nonpregnant women, but continued supplementation of the mother for 2 to 3 months after parturition is advisable in order to replenish stores depleted by pregnancy.

Sources (top right): fish, lamb, lentils, mustard greens, peanuts, peas, sausages and luncheon meats, spinach, Swiss chard, turkey, turnip greens, whole eggs.

Negligible sources—Cheese, fats and oils, fresh and canned fruits, fruit juices and beverages, ice cream, milk, most fresh and canned vegetables, sour cream, sugar, yogurt.

Supplemental sources—Dried liver, ferrous gluconate, ferrous succinate, ferrous sulfate, iron furmarate, iron peptonate, seaweed, yeast.

Comments: iron source of choice.

Manganese (Mn)

Functions: Formation of bone and the growth of other connective tissues. Blood clotting. Insulin action. Cholesterol synthesis. Activator of various enzymes in the metabolism of carbohydrates, fats, proteins, and nucleic acids (DNA and RNA). (Also see MANGANESE.)

Deficiency symptoms—The only confirmed deficiency of manganese in man was in connection with a vitamin K deficiency, where administration of the vitamin did not correct the abnormality in blood clotting until supplemental manganese was provided.

Analysis of hair and blood samples for manganese content indicate that subclinical deficiencies of the mineral might aggravate such disorders as growth impairments, bone abnormalities, diabeticlike carbohydrate metabolism, lack of muscle coordination of the newborn, and abnormal metabolism of lipids (fatty acids, choline, and cholesterol).

Toxicity—Toxicity in man as a consequence of dietary intake has not been observed. However, it has occurred in workers (miners, and others) exposed to high concentrations of manganese dust in the air. The symptoms resemble those found in Parkinson's and Wilson's diseases.

Category	Age (Yrs)	Manganese[1] (mg)
Infants	0–0.5	0.3–0.6
	0.5–1.0	0.6–1.0
Children and adolescents	1–3	1.0–1.5
	4–6	1.5–2.0
	7–10	2.0–3.0
	11+	2.0–5.0
Adults		2.0–5.0

[1] **Note well:** The manganese values are "Estimated Minimum Requirements," from *Recommended Dietary Allowances*, 10th ed., NRC–National Academy of Sciences, 1989, p. 284.

Sources: Groupings by rank of common food sources of manganese follow:

Rich sources—Rice (brown), rice bran and polish, spices, walnuts, wheat bran, wheat germ.

Good sources—Blackstrap molasses, blueberries, lettuce, lima beans (dry), navy beans (dry), peanuts, potatoes, soybean flour, soybeans (dry), sunflower seeds, torula yeast, wheat flour, whole grains, (barley, oats, sorghum, wheat).

Fair sources—Brewers' yeast, liver, most fruits and vegetables, orange pekoe tea, white enriched bread.

Negligible sources—Fats and oils, eggs, fish, meats, milk, poultry, sugar.

Supplemental sources—Alfalfa leaf meal, dried kelp, manganese gluconate.

Comments: In average diets, only about 45% of the ingested manganese is absorbed. The manganese content of plants is dependent on soil content. It is noteworthy, however, that plants grown on alkaline soils may be abnormally low in manganese.

Molybdenum (Mo)

Functions: As a component of three different enzyme systems which are involved in the metabolism of carbohydrates, fats, proteins, sulfur-containing amino acids, nucleic acids (DNA and RNA), and iron. As a component of the enamel of teeth, where it appears to prevent or reduce the incidence of dental caries, although this function has not yet been proven conclusively. (Also see MOLYBDENUM.)

Deficiency symptoms—Naturally occurring deficiency in man is not known, *unless* utilization of the mineral is interfered with by excesses of copper and/or sulfate.

Molybdenum-deficient animals are especially susceptible to the toxic effects of bisulfite, characterized by breathing difficulties and neurological disorders.

Severe molybdenum toxicity in animals (molybdenosis), particularly cattle, occurs throughout the world wherever pastures are grown on high-molybdenum soils. The symptoms include diarrhea, loss of weight, decreased production, fading of hair color, and other symptoms of copper deficiency.

Toxicity—In the U.S.S.R., molybdenum toxicity in man is reported to have caused a high incidence of goutlike syndrome associated with elevated blood levels of molybdenum, uric acid, and xanthine oxidase.

Category	Age (Yrs)	Molybdenum[1] (mcg)
Infants	0–0.5	15–30
	0.5–1.0	20–40
Children and adolescents	1–3	25–50
	4–6	30–75
	7–10	50–150
	11+	75–250
Adults		75–250

[1] **Note well:** The molybdenum values are "Estimated Minimum Requirements," from *Recommended Dietary Allowances*, 10th ed., NRC–National Academy of Sciences, 1989, p. 284.

Sources: The concentration of molybdenum in food varies considerably, depending on the soil in which it is grown.

Most of the dietary molybdenum intake is derived from organ meats, whole grains, leafy vegetables, legumes, and yeast.

Comments: The utilization of molybdenum is reduced by excess copper, sulfate, and tungsten.

In cattle, a relationship exists between molybdenum, copper, and sulfur. Excess molybdenum will cause copper deficiency. However, when the sulfate content of the diet is increased, the symptoms of toxicity are avoided inasmuch as the excretion of molybdenum is increased.

Selenium (Se)

Functions: Component of the enzyme glutathione peroxidase, the metabolic role of which is to protect against oxidation of polyunsaturated fatty acids and resultant tissue damage. Protecting tissues from certain poisonous substances, such as arsenic, cadmium, and mercury. Interrelation with vitamin E—they spare each other, and with the sulfur-containing amino acids. (Also see SELENIUM.)

Deficiency symptoms—There are no clear-cut deficiencies of selenium, because this mineral is so closely related to vitamin E that it is difficult to distinguish deficiency due to selenium alone.

The selenium content of the blood, hair, and fingernails may be used as rough indicators of the selenium content of the rest of the body.

Toxicity—Poisonous effects of selenium are manifested by (1) abnormalities of the hair, nails, and skin; (2) garlic odor on the breath; (3) intensification of selenium toxicity by arsenic or mercury; (4) higher than normal rates of dental caries.

Category	Age (Yrs)	Selenium[1] (mcg)
Infants	0.0–0.5	10
	0.5–1.0	15
Children	1–3	20
	4–6	20
	7–10	30
Males	11–14	40
	15–18	50
	19–24	70
	25–50	70
	51+	70
Females	11–14	45
	15–18	50
	19–24	55
	25–50	55
	51+	55
Pregnant		65
Lactating		75

[1] **Note well:** The selenium values are "Estimated Minimum Requirements," from *Recommended Dietary Allowances*, 10th ed., NRC–National Academy of Sciences, 1989, p. 285.

Sources: The selenium content of plant and animal products is affected by the soil and animal feed, respectively. Nevertheless, such data are useful as they show which foods are likely to be good sources of the mineral. Groupings by rank of some common food sources of selenium follow:

Rich sources—Brazil nuts, butter, fish flour, lobster, smelt.

Good sources—Beer, blackstrap molasses, cider vinegar, clams, crab, eggs, lamb, mushrooms, oysters, pork kidneys, spices (garlic, cinnamon, chili powder, nutmeg), Swiss chard, turnips, wheat bran, whole grains (wheat, barley, rye, oats).

Fair sources—Cabbage, carrots, cheese, corn, grape juice, most nuts, orange juice, whole milk.

Negligible sources—Fruits and sugar.

Supplemental sources—Wheat germ, yeast (brewers' and torula).

Note well: Many drug stores and health food stores now carry special types of yeast which were grown on media rich in selenium; hence, they contain much greater amounts of this mineral than ordinary yeast. Tablets made from high-selenium yeast are also available.

Comments: The high selenium areas are in the Great Plains and the Rocky Mountain States—especially in parts of the Dakotas and Wyoming.

The functions of selenium are closely related to those of vitamin E and the sulfur-containing amino acids.

Selenium-rich supplements should be taken only on the advice of a physician or dietitian.

TABLE M-25 *(Continued)*

Functions	Deficiencies and Toxicity Symptoms	Recommended Daily Allowance			Sources	Comments
		Category	Age (Yrs)	Mineral		
Silicon (Si) Necessary for normal growth and skeletal development of the chick and the rat. These findings suggest that it may also have essential functions in man. (Also see SILICON.)	**Deficiency symptoms**—Deficiencies in chicks and rats are characterized by growth retardation and skeletal alterations and deformities, especially of the skull. Deficiencies of silicon have not been produced in man. **Toxicity**—Silicon does not appear to be toxic in the levels usually found in foods.	A human requirement for silicon has not been established. The silicon intake in the average U.S. diet has been estimated at approximately 1 g per day.			The best sources of silicon are the fibrous parts of whole grains, followed by various organ meats (liver, lungs, kidneys, and brain) and connective tissues. Much of the silicon in whole grains is lost when they are milled into highly refined products.	A common nondietary form of silicon toxicity is a fibrosis of the lungs known as silicosis, due to inhalation of airborne silicon oxide dust.
Zinc (Zn) Needed for normal skin, bones, and hair. As a component of several different enzyme systems which are involved in digestion and respiration. Required for the transfer of carbon dioxide in red blood cells; for proper calcification of bones; for the synthesis and metabolism of proteins and nucleic acids; for the development and functioning of reproductive organs; for wound and burn healing; for the functioning of insulin; and for normal taste acuity. (Also see ZINC.)	**Deficiency symptoms**—Loss of appetite, stunted growth in children, skin changes, small sex glands in boys, loss of taste sensitivity, lightened pigment in hair, white spots on the fingernails, and delayed healing of wounds. In the Middle East, pronounced zinc deficiency in man has resulted in hypogonadism and dwarfism. In pregnant animals, experimental zinc deficiency has resulted in malformation and behavioral disturbances in offspring. **Toxicity**—Ingestion of excess soluble salts may cause nausea, vomiting, purging.	Infants Children Males Females Pregnant Lactating	0–0.5 0.5–1.0 1–3 4–6 7–10 11–14 15–18 19–24 25–50 51+ 11–14 15–18 19–24 25–50 51+	Zinc[1] (mg) 5 5 10 10 10 15 15 15 15 15 12 12 12 12 12 15 19	Groupings by rank of common food sources of zinc follows: **Rich sources**—Beef, liver, oysters, spices, wheat bran. **Good sources**—Cheddar cheese, crab, granola, lamb, peanut butter, peanuts, popcorn, pork, poultry. **Fair sources**—Beans, clams, eggs, fish, sausages and luncheon meats, turnip greens, wheat cereals, whole grain products (wheat, rye, oats, rice, barley). **Negligible sources**—Beverages, fats and oils, fruits and vegetables, milk, sugar, white bread. **Supplemental sources**—Wheat germ, yeast (torula), zinc carbonate, zinc gluconate, zinc sulfate. (Zinc carbonate or zinc sulfate are commonly used where zinc supplementation is necessary.)	The biological availability of zinc in different foods varies widely; meats and seafoods are much better sources of available zinc than vegetables. Zinc availability is adversely affected by phytates (found in whole grains and beans), high calcium, oxalates (in rhubarb and spinach), high fiber, copper (from drinking water conveyed in copper piping), and EDTA (an additive used in certain canned foods).

[1]*Recommended Dietary Allowances,* 10th ed., NRC–National Academy of Sciences, 1989, p. 284.

CAUSES OF MINERAL DEFICIENCIES AND/OR TOXICITIES.

Recently, there has been much speculation as to whether many people in the United States might have subclinical deficiencies of certain minerals. It has also been speculated that mild, but chronic, toxicities, due to excesses of certain nutrients, might contribute to the development of such disorders as atherosclerosis, diabetes, heart failure, high blood pressure, and kidney stones. These concerns prompt conjecture that our distant ancestors were also prone to similar diet-related problems. Hence, it is revealing to review certain aspects of man's early existence and to compare dietary factors of the past with those of the present.

Our knowledge of prehistoric diets has been derived mainly from archaeological studies of charred food remains and dried fecal material found in caves and at the sites of ancient camps. Also, there have been many studies of the aborigines who are still engaged in stone age practices in such parts of the world as Africa, Australia, New Guinea, and the Philippines.

Some causes of mineral deficiencies and or toxicities are listed and discussed in the sections that follow.

Agricultural Practices Which May Be Harmful.
Plants use the energy of the sun, gases from the air, and water and minerals from the soil, to make the various substances which are essential for their life processes and for those of animals and man. Hence, the mineral composition of the soil is one of the major factors affecting the nutritive composition of the foods we eat. Agricultural practices affect the utilization of soil minerals by plants in ways which may, in turn, affect animals and people that eat the plants. However, the reader should note that the discussions which follow are not intended to be condemnations of certain practices; rather, they are designed to point out some pitfalls of carrying them to extremes. Consider:

• **Application of sewage sludge to soils**—Although organic wastes may be applied to the soil in order to improve its chemical and physical characteristics, sewage sludge may contain such toxic elements as cadmium, lead, and mercury—particularly when the sludge comes from highly industrialized areas.

• **Close grazing of seleniferous ranges**—This practice forces the grazing animals to eat plants which may contain extremely high levels of selenium. Some of these plants would not be eaten under conditions of light or moderate grazing.

• **Depletion of soil minerals by intensive single cropping**—Growing the same crop on the same area year after year may deplete the soil of certain minerals because they are removed more rapidly than they are replaced by the commonly used fertilizers.

• **Excessive use of copper-containing sprays**—Plants are sprayed with copper compounds in order to prevent certain fungus diseases. However, overspraying may cause the buildup of toxic levels of copper in the plants and/or the soil, which may interfere with the utilization of other minerals.

• **Feeding livestock unsupplemented feeds and forages which have been grown on mineral-deficient soils**—It is well known that products obtained from animals given such feedstuffs are likely to reflect the mineral deficiencies of the soils on which the feeds and/or forages were grown.

• **Overfertilization of soils with certain minerals**—The overapplication of such minerals as nitrogen, potash, potassium, lime, and sulfur to soils may (1) kill soil microorganisms which convert minerals to forms which are readily picked up by plants, and (2) change the chemical composition of the soil so that certain minerals form insoluble compounds which are not available to plants.

(Also see DEFICIENCY DISEASES, section headed "Eating Patterns Which Lead to Dietary Deficiencies.")

Contamination of Drinking Water, Foods, or Air with Toxic Minerals. Toxic chemical elements—which include some of the essential trace minerals—may get into foods or drinking water from the environment, piping, commercial processing equipment, packaging, or utensils used in the home. In certain cases, such contamination might even be beneficial, if it results in the ingestion of essential minerals which are not supplied in adequate amounts by the foods or the drinking water. However, the margins between beneficial or harmless amounts and toxic levels are rather narrow for certain elements. Furthermore, excesses of certain elements may produce deficiencies of others that they counteract. Hence, it is important to consider how potentially toxic elements get into water, foods, or air so that appropriate measures might be taken to control the amounts which are ingested or inhaled. Consider:

• **Aluminum**—It is not certain whether it is safe to cook such highly acid foods as rhubarb and tomatoes in aluminum pots and pans, because aluminum is dissolved by acid solutions. Abnormally large intakes of aluminum have irritated the digestive tract. Also, unusual conditions have sometimes permitted the absorption of sufficient aluminum from antacids to cause brain damage.[22] Even if the amounts of aluminum leached from aluminum utensils by acid foods are not toxic, there is the possibility that insoluble complexes might be formed between the aluminum and some of the trace minerals, so that the absorption of these nutrients is blocked.

(Also see ANTACIDS; and POISONS, Table P-2, Some Potentially Poisonous [Toxic] Substances.)

• **Arsenic**—Although the public has long thought arsenic to be synonymous with poison, it may be less toxic than the essential trace element selenium, which it counteracts within the body. Also, recent studies on rats suggested that arsenic may be essential.

Minute amounts of arsenic are widely distributed in the common foods, in quantities that are more likely to be beneficial than toxic. However, poisoning may result from foods contaminated with excessive amounts of arsenic-containing sprays used as insecticides and weed killers.

(Also see POISONS, Table P-2, Some Potentially Poisonous [Toxic] Substances.)

• **Cadmium**—In the early 1970s there were many cases of cadmium toxicity in the Jinzu River basin near Toyama, Japan, which is about 160 miles northwest of Tokyo. The disease that resulted was known as itai-itai, or *ouch-ouch* disease, because of the pains the victims had in their bones. Eventually, the poisonings were traced to the chronic consumption of rice and soybeans containing about 3 ppm (parts per million) of cadmium, a nonessential trace mineral which came from the wastes of nearby mines and smelters. Therefore, it is noteworthy that cadmium levels of around 1 ppm have been found in soybeans grown in soil fertilized by sewage sludge, and that oysters taken from waters contaminated by industrial wastes contained about 3 ppm of cadmium.

Other sources of potential environmental contamination by cadmium are cigarette smoke, electroplating processes, paint pigments, the cadmium-nickel type of automobile storage battery, certain phosphate fertilizers, and some of the older types of galvanized water tanks.

While severe cadmium toxicity like that which occurred near Toyama has not been found elsewhere, it is suspected that mild to moderate types of chronic cadmium toxicity may cause disorders of the kidneys leading to high blood pressure. However, the milder forms of cadmium poisoning may be counteracted by such essential minerals as calcium, copper, iron, manganese, selenium, and zinc. Therefore, a few scientists believe that high ratios of cadmium to zinc in the diet and in the various tissues of the body are better indicators of potential cadmium toxicities than the dietary and tissue levels of this toxicant alone.

• **Chlorine**—This element may be ingested as essential chloride ions from inorganic salts, or as toxic organic compounds such as chlorinated hydrocarbons, which may cause liver damage, or even cancer.[23] The toxic chlorine compounds are formed when lake or river water containing nonchlorinated hydrocarbons is chlorinated prior to its use as drinking water. Chlorinated hydrocarbons have been found to be present at various levels in many of the drinking water supplies in the United States, but information is lacking as to the levels which pose either short-term or long-term hazards.

(Also see CHLORINE OR CHLORIDE.)

• **Copper**—Several groups of scientists have recently warned about the possibility of copper poisoning by acid and/or soft drinking water which has been conveyed through copper piping. The blue stains on the surface of sinks where the water drips from the faucet are an indication that drinking water may contain too much copper. About twice as much copper may be present in hot tap water as in cold tap water. Infants are particularly susceptible to the toxic effects of copper in tap water because (1) their water needs are high relative to their size, so they ingest more water in proportion to their body weights than adults; and (2) their diets are usually limited in variety, which means

---

[22]"Possible Aluminum Intoxication," *Nutrition Reviews,* Vol 34, 1976, pp. 166-167.

[23]Ibrahim, M.A., and R. F. Christman, "Drinking Water and Carcinogenesis: The Dilemmas," *American Journal of Public Health,* Vol. 67, 1977, p. 719.

that they may not receive much of the other mineral elements which counteract the effects of copper.

In view of the disorders associated with copper toxicity, it is heartening to note that many of these conditions may be counteracted by increasing the dietary levels of iron, molybdenum, sulfur, zinc, and vitamin C. However, in cases of severe poisoning, it is necessary to administer special metal-binding (chelating) agents which draw copper out of the body.

(Also see POISONS, Table P-2, Some Potentially Poisonous [Toxic] Substances.)

• **Fluorine**—This essential element, like chlorine, is beneficial when it is provided in ionic form (inorganic fluorides) at levels up to 1 ppm in drinking water. The major benefit obtained from such fluoridated water is the increased resistance to tooth decay which occurs when it is consumed by children whose teeth are developing. Higher levels of fluoride (3 to 10 ppm), such as may occur in certain naturally fluoridated waters, may cause mottling of the teeth in some of the children who drink it, although this does not occur if the initial exposure to excessive fluoride takes place after the teeth are fully developed. Furthermore, growing children appear to have greater resistance than adults to the other toxic effects of fluoride, like the development of brittleness in bones and the inhibition of certain enzymes, because growing bones apparently take up fairly large amounts of fluoride without harm.

In addition to fluoridated water, some other sources of significant amounts of fluoride are (1) the environments around aluminum refining plants; and (2) the consumption of tea, crude sea salt, bone meal, or fish (particularly, fish bones). However, the effects of excess fluoride may be reduced somewhat by diets rich in calcium and magnesium.

(Also see POISONS, Table P-2, Some Potentially Poisonous [Toxic] Substances.)

• **Iodine**—Only the radioactive isotope of this element—Iodine 131—is likely to be an environmental hazard, because it emits dangerous radiation. It is formed during the explosions of nuclear weapons; and it may get into foods.

When the contaminated foods are eaten, the radioactive iodine is utilized by the body as if it were normal, nonradioactive iodine. Most of it is picked up by the thyroid gland, which may be injured by the radioactivity that is emitted.

(Also see RADIOACTIVE FALLOUT.)

• **Iron**—The worldwide concern over iron-deficiency anemia has often obscured the fact that excesses of dietary iron might sometimes lead to such disorders as toxic accumulations of the mineral in the liver, spleen, pancreas, heart muscle, and kidneys. This overloading of the body with iron, called hemochromatosis, is most likely to occur in adult males who consume liberal amounts of alcohol. Females are less likely to have such an accumulation of iron because they regularly lose significant amounts of the mineral during their menstrual flow, or as a result of pregnancy and childbirth.

Moderate excesses of iron, which do *not* have direct toxic effects, may produce nutritional deficiencies by interfering with the absorption of copper, manganese, and zinc, and by destroying vitamins C and E.

Some of the sources of iron which may lead to overaccumulation of the mineral in the body are (1) acid foods which have been cooked in iron pots, (2) certain ciders and wines, and (3) occasionally, drinking water. Fortunately, diets containing abundant quantities of phosphorus, copper, manganese, and zinc tend to protect against excesses of iron, because these minerals interfere with its absorption.

• **Lead**—Apparently, lead poisoning has dogged man for a longer time than most people might suspect, because toxic levels of this nonessential mineral were found in the disinterred bones of ancient Romans. In fact, the use of lead for household utensils and wine containers by the Romans is suspected of having been responsible for widespread lead poisoning which, it is conjectured, may have led to weakness and infertility, and the eventual decline of the empire.

Today, there is great concern over acute lead poisoning in young children (ages 1 to 6) who live in urban slums where they may eat chips of lead-containing paints, peeled off from painted wood. The effects of such poisoning may be anemia, hyperactivity, learning difficulties, mental retardation, or even death.[24] There is also concern for older children and adults who might develop chronic lead poisoning from (1) exposure to such sources of the toxic mineral as automobile exhaust fumes, cigarette smoke, fumes from lead smelters, or smoke from coal fires; or (2) consumption of beer held in lead-containing pewter mugs, fruits from orchards sprayed with lead arsenate, fruits and vegetables grown near heavily traveled highways.

Children and pregnant women appear to be most susceptible to lead poisoning because (1) they are likely to have deficiencies of calcium and iron, minerals which protect against lead toxicity; (2) rapidly growing children absorb more lead and excrete less of this metal than other people, and (3) there is a rapid transfer of lead from the blood of pregnant women, through the placenta, to the fetus.

(Also see POISONS, Table P-2, Some Potentially Poisonous [Toxic] Substances.)

• **Mercury**—It is common knowledge that mercury fumes may be responsible for the mental deterioration encountered by some workers in the felt hat industry (often called *mad hatters*). Yet, widespread concern over environmental pollution by the mineral did not develop until the late 1950s when an epidemic of mercury poisonings occurred in Japan.

The mercury poisonings in Japan—which were characterized by losses of balance, hearing, speech, and vision; and by mental disorders—came to be known as "Minamata disease," because they first occurred in people who lived around Minamata Bay, about 50 miles (80 km) southeast of Nagasaki. However, another rash of poisonings occurred around the Agano River in Nugata, a region about 150 miles (240 km) north of Tokyo. Eventually, investigators identified more than 700 cases of mercury toxicity at Minamata, and more than 500 cases at Nugata.[25] The poisonings were traced to the consumption of large amounts of fish and shellfish which contained methylmercury, an organic compound of the element that is much more toxic than its inorganic compounds. It appears that the chain of events which led to the buildup of methylmercury in the fish was as follows:

1. Certain industries discharged their inorganic mercury wastes into Minamata Bay and the Agano River.

---

[24]King, B. G., A. F. Schaplowsky, and E. B. McCabe, "Occupational Health and Child Lead Poisoning: Mutual Interest and Special Problems," *American Journal of Public Health*, Vol. 62, 1972, pp. 1056-1058.

[25]*An Assessment of Mercury in the Environment*, National Academy of Sciences, 1978, p. 90.

2.   The bay and the river contained abundant amounts of microscopic organisms which converted inorganic mercury into methylmercury. It also appears likely that the profuse growth of these organisms was promoted by the dumping of organic wastes into these waters.

3.   The minute organisms which methylated and stored the mercury were eaten by larger forms of life, which in turn stored the methylmercury. Hence, the poisonous compound eventually became concentrated in the bodies of fish and shellfish.

Another recent outbreak of mercury poisonings occurred in Iraq, where more than 6,000 people were hospitalized after eating bread baked from wheat seed that had been treated with a methylmercury fungicide.[26] The poisonous seed grain had not been intended for human consumption; rather, it had been sent to Iraq for planting. Such fungicide-treated seed does not produce poisonous plants, because the methylmercury is detoxified in the soil.

Around the world, investigations were triggered by the reports of mercury poisonings. Only a few people were found to have unmistakable signs of mercury toxicity, even though some had abnormally high levels of the metal in their tissues. Recently, it has been found that the essential trace mineral selenium sometimes protects people against the toxic effects of mercury. Hence, a food which contains both mercury and selenium may not be as dangerous as one which contains mercury alone. Furthermore, the selenium in one food might also provide some protection against mercury in other foods.[27]

(Also see POISONS, Table P-2, Some Potentially Poisonous [Toxic] Substances.)

• **Selenium**—Although it is suspected that people living in areas with high levels of selenium in the soil may occasionally suffer from such effects of selenium toxicity as dental caries and dermatitis; poisoned livestock commonly develop severe deformities; or they may even die from eating excessive amounts of this essential mineral. It is not that people are immune to severe poisoning from selenium; rather, the grains eaten by people are relatively low in selenium compared to the forage plants eaten by grazing animals. Furthermore, high selenium soils alone will not yield wheat containing abnormally large amounts of the mineral unless it is in a form which is readily available to the growing wheat. Usually, this requires that *selenium converter plants,* such as *Astragulus racemosus,* be present to convert soil selenium to a form which is absorbable by wheat.

In the event that conditions in high-selenium areas should be conducive to the poisoning of people, it is noteworthy that selenium toxicity may be counteracted by arsenic and copper, which are less toxic.

(Also see SELENIUM.)

• **Strontium**—This nonessential element, which is normally nontoxic, has an artificial radioactive isotope—strontium 90—that is produced in man-made nuclear reactions, such as those occurring during the explosion of nuclear weapons or during the generation of nuclear energy. Strontium 90 is toxic because it is absorbed by the body and deposited in the bones, where it continuously emits radiation which may cause such disorders as bone cancer and leukemia.

Strontium is chemically similar to calcium, and is likely to be mixed with this element in foods such as milk, vege-

tables, and cereals. Hence, it follows that the foods which are most frequently fed to infants and young children are those most likely to be contaminated with radioactive strontium. However, diets high in noncontaminated calcium tend to reduce the absorption of strontium 90.

In the long run, the best protection against the buildup of radioactive strontium in the environment is the cessation of nuclear weapons tests conducted in the atmosphere.

(Also see RADIOACTIVE FALLOUT.)

• **Zinc**—It is not likely that people will receive toxic excesses of zinc from eating ordinary, unsupplemented diets unless some of the foods were stored in zinc-coated (galvanized) containers, because most of the common foods contain only small fractions of the levels of zinc which may be ingested safely. However, zinc toxicity due to contamination of food by galvanized containers or similar means may be accompanied by nausea, vomiting, stomachache, diarrhea, and fever.

Dietary and Medical Treatments.   Various types of therapies may eventually lead to deficiencies or toxic excesses of certain essential minerals because they alter the ways in which the body absorbs, utilizes, retains, or excretes these nutrients. Often, countermeasures may be used to offset the undesirable effects of treatments, while retaining the desirable effects. Hence, the typical treatments and effects which follow are noteworthy:

• **Blood transfusions**—Sometimes, blood is given regularly, and over a period of time, to people with various diseases characterized by short-lived red cells. The danger of this practice is that it may lead to the overaccumulation of iron, since each pint of blood contains over 200 mg of the mineral, most of which is likely to be retained within the body.

• **Dialysis of patients with kidney failure**—Dialysis procedures, which are used to remove wastes from the blood of people whose kidney function is impaired, also remove essential minerals. Although certain procedures are designed to prevent the withdrawal of such major minerals as calcium, magnesium, potassium, and sodium, there is still the likelihood that various bloodborne trace elements will be reduced to deficient levels.

• **Dietary modifications**—Restriction or elimination of certain foods in order to produce diets which are low in carbohydrates, fats, or protein may lead to mineral deficiencies, unless great care is taken to provide other sources of the deleted elements. For example, it was found that some typical hospital diets tended to be deficient in copper, iron, magnesium, and manganese.[28]

Even more likely to lead to mineral deficiencies are the *crash diets* which cause rapid loss of weight primarily through loss of water, because there also may be excessive losses of minerals in the urine.

• **Intravenous administration of glucose and/or other nutrients**—Long-term maintenance of hospitalized patients on intravenous solutions may accentuate the effects of mild to moderate deficiencies of chromium, manganese, and zinc, because (1) each of these essential trace elements is involved, somehow, in the actions of insulin, and (2) increases in blood sugar add to the workload of the insulin-secreting system. Also, it is noteworthy that the solutions

[26]*Ibid.,* p. 91.
[27]"Mercury Toxicity Reduced by Selenium," *Nutrition Reviews,* Vol. 31, 1973, p. 25.

[28]Gormican, A., "Inorganic Elements in Foods Used in Hospital Menus," *Journal of the American Dietetic Association,* Vol. 56, 1970, p. 403.

commonly used for total parenteral nutrition may or may not contain manganese and zinc, and that most do not contain chromium.

• **Pumping the stomach**—Normally, the minerals which are abundantly present in the digestive juices secreted by the stomach are almost totally absorbed in the intestine so that only negligible amounts of these essential substances are lost in the stool. However, the pumping of gastric fluid from the stomach results in the removal of large quantities of chloride and potassium ions, so that deficiencies of these elements may result, unless measures are taken to replace them.

• **Tube feeding**—When concentrated liquid formulas are fed too rapidly to sick people by stomach tube, there may be diarrhea, dehydration, and losses of both the minerals which are present in the formulas and those which are secreted in the digestive juices. The causes of these undesirable effects are (1) reduction in the secretion of the digestive juices which would normally dilute the formula, due to the illness and the unnatural means of feeding; and (2) the drawing of water from the intestinal wall by the formula, so that a strong laxative effect is produced.

Drugs Which May Affect the Utilization of Minerals by the Body.
Certain commonly used drugs affect the utilization of both essential and nonessential minerals by the body. Furthermore, many of these medicines may be used to treat chronic conditions over long periods of time. Therefore, brief descriptions of the effects of some of the most frequently used drugs are presented, so that appropriate measures may be taken to prevent mineral deficiencies or toxicities:

• **Antacids**—The nonabsorbable types of these compounds, such as aluminum hydroxide, may form insoluble complexes in the intestine, which reduce the absorption of phosphate, fluoride, and other essential minerals. Also, it recently became apparent that certain patients with kidney diseases who had been put on diets low in phosphate and who were given regular dialysis treatments absorbed considerable amounts of aluminum from antacids, because they developed toxic deposits of the metal in the brain and the bones.[29]

When consumed in excess, the absorbable antacids, which yield such absorbable ions as calcium, magnesium, sodium, and bicarbonate, may lead to toxic excesses of these ions in the body.

(Also see ANTACIDS.)

• **Antibiotics**—The tetracycline types of these drugs may form insoluble, unabsorbable complexes with such essential minerals as calcium, iron, and magnesium.

• **Anticonvulsants**—Epileptic patients given these drugs for the prevention of seizures have sometimes developed softening and distortion of their bones which may lead to rickets in children, or to softening of the bones (osteomalacia) in adults.[30] Apparently, the drugs produce their harmful effects on the bones by provoking increased rates of destruction of vitamin D and its active metabolites, so that the absorption and utilization of dietary calcium is reduced. Sometimes, their effects have been overcome by providing

extra vitamin D and/or additional exposure to sunlight.

• **Blood-cholesterol-lowering agents**—It is suspected that the types of these drugs which bind with cholesterol and bile salts in the intestine may similarly tie up calcium and other minerals so that their absorption is reduced.

• **Diuretics (water pills)**—Usually, these drugs are given to people with water accumulation in their tissues and/or high blood pressure. Some of them may promote excessive urinary losses of potassium and other essential minerals.

• **Hormones**—Although the administration of adrenocorticotropic hormone (ACTH) and cortisone has often provided great relief for those who suffer from allergies and from inflammatory diseases like arthritis, the long-term use of these hormones has sometimes caused loss of bone minerals to the extent that weakening of the bones (osteoporosis) and/or collapse of the spine have occurred.

• **Laxatives**—Habitual use of these drugs may result in mineral deficiencies because they reduce the intestinal absorption of (1) dietary minerals, and (2) minerals secreted in the various digestive juices.

• **Mineral supplements**—The consumption of excessive amounts of certain minerals in the form of pills or tonics may lead to the deficiencies of other minerals, because there are diverse interactions between the various essential elements. For example, high intakes of iron may interfere with the absorption of copper, phosphorus, and zinc.

• **Oral contraceptives (birth control pills)**—The full details are still uncertain, but it appears that these drugs alter the metabolic processes which involve calcium, magnesium, and phosphorus.

Food Additives.
Many natural and synthetic substances are added to foods to improve their color, odor, texture, and taste. The U.S. Food and Drug Administration has established strict regulations governing food additives so that there is little likelihood of any direct toxic effects resulting from the proper use of such ingredients. However, some of these substances affect the utilization of certain minerals. Such effects may not be important to people who eat a little of everything, but they may be responsible in part for mineral deficiencies or toxicities in people whose diets are narrowly limited by choice, economic circumstances, or other factors. Therefore, some effects of various food additives on mineral metabolism follow:

• **Aluminum compounds**—The amounts of aluminum that most people consume in the form of food additives—mainly present in such items as baking powder, pickles, and processed cheeses—are probably less than that ingested by people who regularly take aluminum-containing antacids. For example, 3½ oz (100 g) of one brand of processed American cheese was found to contain 70 mg of aluminum,[31] which is about the amount present in a single tablet of a commonly used antacid. Nevertheless, even small amounts of aluminum may form nonabsorbable complexes with essential trace elements such as iron.

• **Bran**—This product of flour milling is sometimes added to breads and cereals to increase their fiber content.

[29]Alfrey, A. C., et al., "The Dialysis Encephalopathy Sydrome," The New England Journal of Medicine, Vol. 294, 1976, p. 184.

[30]"Anticonvulsant Drugs and Calcium Metabolism," Nutrition Reviews, Vol. 33, 1975, pp. 221-222.

[31]Gormican, A., "Inorganic Elements in Foods Used in Hosptial Menus," Journal of the American Dietetic Association, Vol. 56, 1970, p. 399, Table 1.

However, it is rich in phytates, phosphorus compounds that bind many of the essential minerals so that they are not readily absorbed. It is noteworthy that the leavening of breads with yeast overcomes much of the effects of phytates.[32]

• **Chelating agents**—These additives are used to bind metallic mineral elements like copper, iron, and zinc, because the unbound forms of the minerals may promote the deterioration and/or discoloration of foods.

The effects of added chelating agents may be beneficial, unfavorable, or uncertain with respect to mineral nutrition, because (1) foods may contain naturally occurring chelating agents which interact with the food additives; (2) some agents interfere with mineral absorption, while others enhance it; (3) sometimes, agents which enhance the absorption of metals also increase their urinary excretion; and (4) the overall mineral content of the diet often determines which of the various metals will be bound to the chelating agent(s), and which will remain free.

• **Gums**—These water-soluble mucilaginous materials may be added to a wide variety of processed foods to produce clarification (by binding with clouding agents), gelling, stabilization (prevention of the separation of various components), and thickening. Many of the gums used in foods are neither digested nor absorbed by man; hence, they may tie up essential minerals so that they are poorly absorbed. For example, sodium alginate—a seaweed derivative which is sometimes added to beers, cheeses, ice creams, salad dressings, sausages, whipped toppings, and wines—forms insoluble, nonabsorbable complexes with iron and copper.

• **Iodates**—Some nutritionists have expressed concern lest the use of these compounds as dough conditioners in the baking industry may lead to excessive consumption of iodine. Hence, it is noteworthy that the iodate fortification of bread for the prevention of goiter in Tasmania was accompanied by an increased incidence of iodine-induced overactivity of the thyroid gland (hyperthyroidism) in goiter-prone people.[33]

• **Iron**—In 1970, the American Bakers Association and the Millers' National Federation asked the U.S. Food and Drug Administration to triple the amount of iron permitted in the enrichment of flour and bread because there was widespread concern over iron-deficiency anemia in women and children.[34] However, it was strongly opposed by some of the leading blood specialists on the grounds that it might lead to toxic accumulations of iron in the tissues of susceptible people.

Finally, in late 1977, the FDA decided that there were too many unresolved questions regarding the safety of adding the proposed levels of iron to breads and flour, so they rejected the proposal by the bakers and the millers.

There is no evidence that the current amounts of iron which are added to foods pose any hazards.

• **Phosphates**—It has been estimated that the widespread use of phosphates as food additives may make the calcium to phosphorus ratio in the American diet as low as 1:4. This mineral imbalance may lead to poor absorption and utilization of dietary calcium, and perhaps even to certain bone disorders.

Sometimes, these effects may be prevented by extra dietary vitamin D, or by exposure to sunlight. However, newborn infants, whose kidneys do not excrete excess phosphate as well as those of older infants and children, may develop high blood levels of phosphate and have seizures when they are fed evaporated milk containing phosphate additives.[35] The seizures are attributed to the fact that cow's milk contains almost four times the calcium and over six times the phosphorus present in human breast milk.[36] Furthermore, extra phosphate is often added to evaporated milk to lengthen its shelf life. Hence, newborn babies fed evaporated milk (diluted with an equal volume of water) may receive between seven and eight times the phosphorus they would get from the same amount of human breast milk. An elevation of the blood level of phosphate causes a corresponding drop in the level of ionized calcium. The direct cause of the milk-induced seizures is the lack of sufficient ionized calcium in the blood.

Fortunately, such seizures have become much less of a problem in the United States, because (1) the use of commercial infant formulas which contain much less phosphate have essentially replaced the use of evaporated milk for feeding newborn infants; and (2) pediatricians generally advise mothers who neither breast feed nor use commercial formulas either to dilute fresh cow's milk or to use extra water in diluting evaporated milk.

Another problem which may result from high levels of dietary phosphate is the formation of nonabsorbable complexes with essential trace elements such as iron.

• **Sodium compounds**—It is suspected that high intakes of sodium may be one of the factors which lead to the development of high blood pressure in susceptible people. Natural, unprocessed foods contain only small amounts of sodium compared to most commercially processed foods. Furthermore, in addition to being present in foods in the form of common salt (sodium chloride), sodium may also be present in such additives as sodium bicarbonate, monosodium glutamate, and various types of phosphates. Therefore, the consumer who is on a low-sodium diet must read carefully the labels of food products because items which are low in salt may contain liberal amounts of other sodium compounds.

(Also see SODIUM.)

Food Processing. The development of food processing has enabled man to modify such undesirable qualities of natural foods as bulkiness, toughness, susceptibility to spoilage, and in some cases, even toxicity. However, certain processes may lessen the mineral values of foods, while others may enhance them. The favorable effects are given in the section of this article headed, "Production and Preparation of Foods to Maximize Their Mineral Values." The details which follow cover some of the detrimental effects:

• **Canning fruits and vegetables**—Some of the minerals in canned fruits and vegetables escape into the packing fluid,

---

[32]"Zinc Availability in Leavened and Unleavened Bread," *Nutrition Reviews*, Vol. 33, 1975, pp. 18-19.

[33]"Iodine Fortification and Thyrotoxicosis," *Nutrition Reviews*, Vol. 28, 1970, pp. 212-213.

[34]"Anatomy of a Decision,"*Nutrition Today*, Vol. 13, January/February, 1978, pp.6-7.

[35]Filer, L., Jr., "Excessive Intakes and Imbalance of Vitamins and Minerals," *Nutrients in Proposed Foods: Vitamins-Minerals*, (Papers from a symposium sponsored by the American Medical Association), Publishing Sciences Group, Inc., Acton, Mass., 1974, p. 27.

[36]Robson, et al., *Malnutrition: Its Causation and Control*, Gordon and Breach, Science Publishers, Inc., 1972, Vol. 1, p. 125, Table 12.

which may be discarded. The amount of mineral loss depends upon such factors as the type of processing, degree of acidity of the packing medium, and the length of storage.

• **Cheese making**—During this process, milk is clotted by either acid or rennet (an enzyme-containing substance derived from animal stomachs) so that it separates into cheese and whey. Usually, much more calcium and phosphorus are lost in the whey from acid-clotted items like cottage cheese than from the rennet-clotted cheeses like Cheddar and Swiss.

• **Milling grains**—The greater part of the mineral content of grains is lost during milling because these elements are usually concentrated in the outer layers of the seed. Hence, such by-products of milling as bran, hominy feed, and rice polishings are excellent sources of many essential minerals. Usually, these by-products are fed to livestock, although they are sometimes added to special breads and cereals.

Fig. M-72. Milling wheat to produce white flour for the making of bread removes much of the mineral content because the minerals are concentrated in the outer layers of the seed. (Courtesy, University of Minnesota, St. Paul, Minn.)

• **Refining sugar**—Most of the minerals that are present in raw sugar made from either sugarcane or sugar beets are (1) removed as the sugar is refined, and (2) end up in the crude molasses, which is a by-product. Cane molasses may be marketed for human consumption as dark molasses, or *blackstrap* molasses, but because beet molasses is bitter, it is sold as a feed ingredient.

Metabolic Disorders Which May Alter Mineral Metabolism. Various disorders of metabolism may interfere with the ways in which the body normally absorbs, utilizes, and excretes essential minerals, so that deficiencies and/or toxicities may result. Sometimes, the disorders produce conditions so precarious that the maintenance of the proper blood levels of sodium, potassium, calcium, and magnesium may become a matter of life or death. Brief

descriptions of some of these medical problems are given so that the reader may realize the importance of consulting a doctor before taking mineral supplements or making drastic dietary changes, since people differ widely in their susceptibility to these metabolic disorders. Metabolic disorders which may alter mineral metabolism follow:

• **Acidosis**—This condition is characterized by an excess of acid in the blood which may result from such diverse causes as diabetes; diarrhea; diets which are high in fat and protein, but low in carbohydrates; fever; kidney diseases; lung diseases; severe stress; starvation; or trauma. If prolonged, acidosis may lead to (1) excessive urinary loss of calcium and demineralization of the bones; (2) depletion of water, sodium, and potassium; and (3) retention of excessive amounts of phosphate and sulfate ions. The blood levels of potassium may be normal, even when the cells are low in the mineral. Hence, the feeding of injured, sick, or starved people must be done cautiously so as not to aggravate the imbalances in the body's mineral salts.
(Also see ACID-BASE BALANCE.)

• **Addison's disease**—Victims of this disorder have abnormally low secretions of adrenal cortical hormones, so they are unable to retain sufficient amounts of sodium and water in their bodies to counter such stresses as dehydration and low-sodium diets. However, they may retain excessive amounts of potassium in their blood, which may cause cardiac arrest.
(Also see ADDISON'S DISEASE; and ENDOCRINE GLANDS.)

• **Alkalosis**—The main causes of alkalosis, which is an excess of alkali in the blood, are: (1) the consumption of large amounts of such absorbable antacids as sodium bicarbonate, (2) the loss of large amounts of stomach acid through vomiting, (3) the depletion of the body's supply of chloride ions through vomiting or diarrhea, and (4) excessively deep breathing—due to nervousness or overdoses of aspirin—which eliminates too much carbon dioxide. Alkalosis may be accompanied by the development of a potassium deficiency, due to an excessive loss of the mineral in the urine; and a reduction in the blood level of calcium ions, so that muscle spasms may occur.
(Also see ACID-BASE BALANCE.)

• **Congestive heart failure**—Abnormalities such as the accumulation of fluid in the tissues, and an overtaxing of the heart muscle, make the patient with heart failure extra susceptible to the toxic effects of imbalances between the blood and tissue levels of such ions as sodium, potassium, calcium, and magnesium. Extra magnesium and potassium may be given to prevent irregular heartbeats when digitalis is used as a medication. However, too much potassium may cause cardiac arrest, if the kidneys are not able to excrete excesses of this mineral in the urine.
(Also see HEART DISEASE.)

• **Cushing's syndrome**—This disorder is characterized by oversecretion of adrenal cortical hormones which cause the breakdown and loss of protein from the bones and muscles. Potassium is lost along with muscle protein, and the bone minerals—mainly calcium and phosphorus—are lost when the protein structure of bone breaks down. People with Cushing's syndrome often have an accumulation of salt and water in their tissues, a condition which gives rise to a rounded body, *moon-face,* and high blood pressure.

(Also see CUSHING'S SYNDROME; and ENDOCRINE GLANDS.)

• **Diabetes**—Uncontrolled diabetes is frequently accompanied by acidosis and ketosis—conditions characterized by excesses of acid and ketones in the blood—which may lead to depletion of bone calcium and of muscle potassium. It is noteworthy that the treatment of diabetes with insulin results in the rapid uptake of bloodborne potassium by the tissues. Hence, there may be a dangerous drop in the blood level of this mineral, unless extra amounts are given.

(Also see DIABETES MELLITUS.)

• **Goiter**—People with goiter may be overly susceptible to the toxic effects of excess iodine because their thyroid glands have become extra efficient in the utilization of this mineral. Therefore, they may develop oversecretion of thyroid hormones (hyperthyroidism) when given extra iodine. This toxic condition may sometimes be accompanied by the loss of calcium from the bones, and of potassium from the muscles.

(Also see ENDOCRINE GLANDS; GOITER; and IODINE.)

• **Hemolytic anemia**—Several types of anemias are characterized by an overly rapid breakdown of red blood cells, which may be caused by genetic, nutritional, and/or toxic factors. Administration of extra iron or blood transfusions to people with these disorders may be dangerous because the iron which is released during the destruction of red cells may accumulate in various tissues where it may cause damage.

(Also see ANEMIA.)

• **High blood pressure**—Sometimes, this condition results from a failure of the kidneys to excrete excess sodium, which promotes the accumulation of water in the body. Certain people appear to be overly susceptible to the effects of only moderate excesses of dietary sodium, so they should restrict their salt intake in order to avoid high blood pressure.

(Also see HIGH BLOOD PRESSURE.)

• **Ketosis**—The most common causes of this condition—which is characterized by an excess of ketones in the blood—are (1) diabetes; (2) diets high in fat and protein, but low in carbohydrate; (3) fevers; and (4) starvation. A mild ketosis in normally healthy people is usually not dangerous, unless it occurs regularly over a long period of time. Then, it may lead to such problems as (1) excessive urinary loss of sodium and water; (2) acidosis which provokes the loss of calcium from bone, and potassium from muscle; and (3) the accumulation of uric acid (a waste product of protein metabolism) in the blood, and sometimes in the soft tissues where it causes damage and pain (the latter disorder is commonly called gout). Uric acid buildup is usually treated with alkalizers to prevent the formation of kidney stones. However, the alkalizers may cause other alterations in mineral metabolism.

It is noteworthy that two men who lived for 1 year on a ketogenic diet consisting only of beef, veal, lamb, pork, and chicken lost about 300 mg of calcium from their bodies each day.[37]

(Also see DIABETES MELLITUS; MODIFIED DIETS; and STARVATION.)

---

[37]Randall, H. T., "Water, Electrolytes, and Acid-Base Balance," *Modern Nutrition in Health and Disease,* 5th ed., edited by R. S. Goodhart and M. E. Shils, Lea & Febiger, Philadelphia, Pa., 1973, p. 360.

• **Kidney diseases**—These disorders are usually accompanied by excessive loss of certain minerals in the urine and/or excessive retention of others, depending upon the parts of the kidney which are diseased. Kidney function may also be impaired whenever the blood supply to the organs is reduced, such as occurs in heart failure and shock. Furthermore, certain kidney diseases impair the metabolism of vitamin D, so that the utilization of calcium and phosphorus may be greatly reduced.

(Also see KETOSIS; and KIDNEY DISEASES.)

• **Lack of stomach acid**—Older people may have a deficiency of stomach acid. This may contribute to mineral malnutrition because acid is needed to offset the alkalinity of bile and pancreatic juice which tends to reduce the absorption of certain minerals.

• **Oversecretion of parathyroid hormone (PTH)**—This condition may be due to (1) a tumorous growth of the parathyroid glands, or (2) enlargement of the glands due to prolonged stimulation by a chronically low blood level of calcium ions. The hormone raises the blood calcium by causing its withdrawal from the bones. Hence, the chronic oversecretion of PTH may lead to such troubles as demineralization of bones, excessive excretion of phosphorus so that muscle weaknesses result, and deposition of calcium in tissues such as the kidneys.

(Also see CALCIUM; ENDOCRINE GLANDS; and PHOSPHORUS.)

• **Rickets**—This disease of infancy and childhood, which is usually due to a deficiency of vitamin D, is characterized by poor utilization of dietary calcium and phosphorus for the mineralization of bones. A similar condition, which may occur in adults, is called *softening of the bones* or osteomalacia.

(Also see CALCIUM; PHOSPHORUS; RICKETS; and VITAMIN D.)

• **Starvation**—The extreme deprivation of food is increasingly being used as a means to achieve a rapid loss in body weight, even though there may be such harmful effects as severe depletion of water, sodium, and potassium. These effects have often led to serious disorders of the heart, and even death—in cases of both voluntary and involuntary starvation. Furthermore, there may be abnormal shifts of mineral salts and water between the tissues and the blood, so it may be difficult to decide how to rehabilitate starved people without aggravating their water and salt imbalances.

(Also see STARVATION.)

• **Vitamin D poisoning**—Overdoses of this vitamin may raise the blood calcium so high that the mineral forms harmful deposits in many of the soft tissues in the body. If this toxicity is prolonged, there may be demineralization of the bones since the high blood calcium is maintained at the expense of the bones. A similar, but milder type of poisoning may result from the overexposure of fair-skinned people to sunlight or other forms of ultraviolet light, such as sunlamps.

(Also see VITAMIN D.)

Mineral Composition of the Soil. There has been continuing controversy over how the mineral composition of the soil might affect the nutrient levels of the plants grown on it. The U.S. Food and Drug Administration has long maintained that (1) mineral deficiencies in soils may lead to reduced yields of plants rather than to plants containing sub-

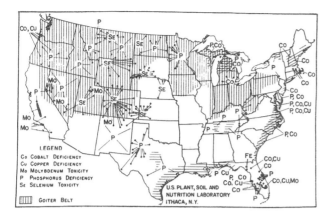

Fig. M-73. Known areas in the U.S. where mineral nutritional diseases of animals occur. The dots indicate approximate locations where troubles occur. The lines not terminating in dots indicate a generalized area or areas where specific locations have not been reported.

People need to be aware of any soil mineral dificiencies or of certain minerals that may cause toxicities because there is a direct and most important relationship between soil minerals and plant composition. (Courtesy, USDA)

normal levels of minerals; and (2) most people eat a wide variety of foods from different areas of the United States, so that variations in the mineral contents of foods tend to average out. On the other hand, certain proponents of organic farming may go so far as to assert that various human nutritional deficiencies may be due to the consumption of products grown on mineral-deficient soils. It is difficult to evaluate the information presented in support of each side of the controversy, because comparisons have often been made between different nutrients in different crops which have been grown on different soils.

Present research, together with practical observation, points to the fact that the mere evaluation of crop yields in terms of tons of forage or bushels of grain produced per acre is not enough. Neither does a standard food analysis (a proximate analysis) tell the whole story. Rather, there is a direct and most important relationship between the fertility of the soil and the composition of the plant.

Fig. M-74. Mineral deficient soils may produce mineral deficient plants, which can cause mineral deficient people. (Courtesy, USDA)

The land surface of the earth is covered with many kinds of soil. Some of them naturally contain an abundant supply of most of the elements needed by both plants and animals. Other soils may have an abundant supply of most required elements and yet be deficient in one or more essential. For example, southwestern United States is known as a phosphorus-deficient area; northwestern United States and the Great Lakes region are iodine-deficient areas, and southeastern United States is a cobalt-deficient area.

At every step in the chain from soils to man, the essential mineral elements interact with other elements, and these interactions may profoundly affect the availability of essential elements or the amount of the essential element required for normal growth or metabolic function. For example, a high level of sulfate in the soil may depress the uptake of selenium by plants, and cause the people that eat the plants to suffer from selenium deficiency. The availability of zinc may be depressed if the diet is high in calcium. These and other interactions must be considered in assessing whether a given soil will supply plants with needed nutrients, and, in turn, whether plants will supply the people that consume them with needed nutrients.

Thus, the transfer of essential nutrient elements from soils to plants, thence to man, is a complicated process.

Naturally Occurring Mineral Antagonists in Foods. Although there has been much recent praise of the virtues of natural, unprocessed foods, certain naturally occurring substances present in foods may interfere with the absorption and utilization of minerals by the body. Sometimes, these interfering substances or antagonists may contribute to the development of mineral deficiencies, when diets are barely adequate with respect to these nutrients. Hence, it is important to identify the most common types of naturally occurring mineral antagonists, such as—

• **Fiber**—Plant foods that contain large amounts of indigestible carbohydrate (fiber) tend to stimulate more rapid movements of the intestine, so that the absorption of minerals and other nutrients may be reduced. Another way in which fiber may reduce mineral absorption is by enveloping certain mineral elements so that they pass into the stool with the indigestible matter.

• **Goitrogens**—Certain plants contain substances called goitrogens, which interfere with the utilization of iodine by the thyroid gland, so that the gland becomes enlarged. An enlarged thyroid gland is called a goiter. The most common food sources of goitrogens are Brussels sprouts, cabbage, kohlrabi, rutabaga, and turnip. Even the milk of cows which have been fed substantial amounts of these plants may be goitrogenic. Also, it is noteworthy that soybeans contain a goitrogen, which may be rendered less toxic by heating.[38]

(Also see DEFICIENCY DISEASES, section headed "Major Dietary Deficiency Diseases"; and GOITER.)

• **Interacting minerals**—Excesses of various essential or nonessential minerals may reduce the absorption and/or the activity in the body of other minerals. Such interferences are most likely to happen when large excesses of special mineral supplements are taken. However, they may occur to a lesser degree when (1) the diet lacks variety, and disproportionate amounts of phosphate are obtained from foods like fish, meats, nuts, and poultry; and (2) most of the

[38]Wills, J.H., Jr., "Goitrogens in Foods," *Toxicants Occurring Naturally in Foods*, National Academy of Sciences, 1966, p. 8.

foods have been produced in a particular area where the environment is heavily contaminated with such toxic elements as arsenic, cadmium, copper, lead, mercury, or selenium.

(Also see Table M-25 for additional details on mineral interrelationships.)

• **Naturally occurring chelating agents**—Many naturally occurring substances in foods—such as amino acids and other organic acids like citric, oxalic, salicylic, and tartaric—may bind with (chelate) metallic mineral elements in the digestive tract. The effects of such chelating agents may be either beneficial or detrimental, depending upon whether mineral absorption is enhanced or reduced. Also, the effects of chelating agents may be counteracted by dietary excesses of metal which are chemically similar, because such metals may compete for binding to the chelator. Therefore, it is not known at the present time whether these agents may have significant effects on mineral nutrition when ordinary, mixed diets are consumed.

• **Oxalates**—Nutritionists have long been concerned about whether oxalates, which are present in such foods as rhubarb and spinach, might bind with sufficient amounts of calcium and/or iron so as to produce deficiencies of these minerals. However, experiments have shown that such effects are not likely to occur in man unless (1) the mineral intake is marginal and (2) unusually large amounts of these foods are eaten, along with other sources of oxalates like almonds, beet greens, cashew nuts, cocoa, or tea.[39]

• **Phytates**—These poorly utilized compounds of phosphorus—which are present in the outer layers of grains—bind with such minerals as calcium, iron, and zinc and interfere with their absorption. Hence, mineral deficiencies may occur in people whose diets are composed mainly of whole grain breads. For example, diets based upon these breads were found to be the cause of combined zinc and protein deficiencies which led to stunting of growth and impaired sexual development of males in rural Iran.[40]

The interference of phytates with calcium absorption may be counteracted by supplemental vitamin D and/or exposure of the skin to ultraviolet light from the sun, or from an artificial source such as a sunlamp.[41]

Overconsumption of Alcohol.
Alcoholics may suffer from deficiencies of minerals such as magnesium, potassium, and zinc, provoked in part by erratic eating habits and in part by an increased urinary loss of these minerals due to the effects of alcohol.

Alcoholism may also increase the toxicity of certain dietary minerals by causing alterations in the ways in which the body absorbs, utilizes, and excretes these substances. For example, iron deposits in organs such as the liver, pancreas, and spleen may sometimes result from the enhancement of iron absorption by alcohol. Similarly, a moderate to heavy consumption of alcohol appears to have contributed to the cobalt poisoning of certain beer drinkers, because the amounts of cobalt ingested by these people were less than the doses of the mineral which have been used medicinally without such ill effects.[42]

Poor Food Choices.
Although a wide variety of foods rich in minerals is available in most supermarkets, people may eat only a few of them for reasons such as (1) lack of money to buy certain items, (2) dependence upon eating most meals away from home, (3) dislike of many types of nutritious foods, and (4) lack of knowledge of and/or the facilities for the preparation of various items. Even the consumption of a wide variety of foods may not guarantee the achievement of adequate mineral nutrition if most of the foods selected are high in fats and carbohydrates, but low in minerals.

Procedures Used in the Preparation of Meals.
The preparation of meals may involve procedures which result in the loss of essential minerals. These losses may be great enough to make the difference between adequate and inadequate mineral nutrition. Hence, the food preparation procedures that follow are noteworthy:

1. The thawing of frozen fish, meat, and poultry results in the loss of some of the juices, but data on their mineral content are scarce. Usually, the amount of fluid which is lost increases as the amount of cut surface increases. Thus, small cubes of beef for stewing will have greater losses than large pieces of meat. It may not always be safe to save the thaw juices from the food, because there may be considerable bacterial growth in the fluids during the period of thawing.

2. Large amounts of peelings and trimmings may be removed from fruits and vegetables during meal preparation. Often, the material which is removed and discarded may be richer in certain minerals than the part which is retained for eating. For example, the outer green leaves of cabbage may contain from 1.5 to 3 times as much iron as the inner bleached leaves.

3. Soaking vegetables in water prior to cooking tends to leach out some of the water-soluble minerals.

4. Cooking fruits and vegetables in lots of water, then throwing away the cooking water, is a sure way to lose plenty of minerals. For example, in one study it was found that the boiling of cabbage led to the following losses of minerals: calcium, 72%; magnesium, 76%; phosphorus, 60%; and iron, 67%.[43] Boiling results in much greater losses of minerals than pressure cooking, baking, frying, steaming, or roasting. The greater the amount of cut surface, the greater the mineral losses. Hence, French cut green beans are subject to greater losses during cooking than cross cut beans.

Stress and Trauma.
Many types of stresses and traumas (the term used for sudden shocks, injuries, or wounds to the body) may alter various aspects of mineral metabolism. The reactions of the body to various types of stresses and traumas are similar because (1) such stresses as the deprivation of food or water, extremes of heat or cold, or emotional upsets may damage the body tissues in ways which are similar to the effects of trauma; and (2) most traumas provoke substantial increases in the secretion of stress hormones by the pituitary and adrenal glands. The prolonged secretion of abnormally high levels of stress hormones may lead to excessive losses of potassium, phosphorus, calcium, magnesium, and zinc in the urine. Usually, such conditions are accompanied by greater than normal

[39]Fassett, D. W., "Oxalates," *Toxicants Occurring Naturally in Foods*, National Academy of Sciences, 1966, pp. 259-264.

[40]Prasad, A. S., "Nutritional Aspects of Zinc," *Dietetic Currents*, Vol. 4, September/October 1977, p. 27.

[41]"Phytates and Rickets," *Nutrition Reviews*, Vol. 31, 1973, pp. 238-239.

[42]"Synergism of Cobalt and Ethanol," *Nutrition Reviews*, Vol. 29, 1971, pp. 43-45.

[43]Lachance, P. A., "Effects of Food Preparation Procedures on Nutrient Retention with Emphasis upon Food Service Practices," *Nutritional Evaluation of Food Processing*, 2nd ed., edited by R. S. Harris and E. Karmas, The Avi Publishing Company, Inc., Westport, Conn., 1975, p. 476.

retention of sodium and water, so that fluid sometimes accumulates in tissues and the blood pressure may be raised.

Bad burns produce effects somewhat different from those due to other stresses and traumas because substantial amounts of body fluid and mineral salts may be lost. Hence, burn patients are often very thirsty and drink large amounts of water, which can be dangerous because there may be excessive dilution of the body fluids and the salts which remain. A severe deficiency of sodium might occur under such conditions. Therefore, considerable care must be taken to replenish the body with the proper mixture of mineral salts and water. It is noteworthy that attempts to replace the large losses of calcium and phosphorus rapidly have sometimes led to kidney stones, because the body was unable to use all of the minerals which were provided, so the excesses were filtered through the kidneys.

Another factor which causes increased mineral loss in the urine after an injury is the confinement of the patient to a bed and/or wheel chair. It seems that a certain amount of physical activity is required for the maintenance of optimal bone mineralization.

DIAGNOSIS OF MINERAL DEFICIENCIES. There are various degrees of mineral deficiencies, ranging from a mild depletion of the body's stores of the mineral(s), to severe clinical disorders which may result in disability or even death. Hence, it may be possible to improve health and increase longevity if mineral malnutrition is detected early by means such as those which follow:

Dietary Histories. Dietary patterns which might lead to mineral deficiencies may sometimes be discovered by (1) asking people to list the types and amounts of food that they usually eat; (2) calculating their mineral intakes with the help of food composition tables (see Food Composition Table F-21); and (3) comparing their mineral intakes with the mineral allowances which have been set by the Food and Nutrition Board of the National Research Council. These dietary histories may also reveal the presence of conditions that are known to interfere with the absorption and/or the utilization of certain minerals.

Table M-25, Mineral Table, gives the mineral allowances recommended by the Food and Nutrition Board, National Academy of Sciences. However, official allowances are given for only seven essential elements, while estimated safe and adequate ranges of intakes are given for five, and minimum requirements are given for three.

The Recommended Daily Allowances in Table M-25, Mineral Table, may be used for purposes of comparing the official mineral allowances with the actual mineral intakes of people who may be eating inadequate diets. However, a person who consumes a diet which lacks sufficient amounts of certain minerals may *not* necessarily have deficiencies of these minerals because (1) individual needs for minerals vary widely; (2) other dietary and nondietary factors affect the ways in which minerals are utilized; and (3) there may be stores of the minerals in that individual's body, which may be used to offset inadequacies of the diet. Nevertheless, the chances are that a markedly inadequate diet will eventually lead to nutritional deficiencies.

Calculating individual mineral intakes for large groups of people is an expensive and time-consuming procedure. Hence, a practical shortcut is to try to identify the most likely candidates for mineral deficiencies by scanning individual patterns of food selection. Some guidelines for identifying diets which may lead to certain mineral deficiencies follow:

1. **Calcium deficiency** may result from diets containing only small amounts of dairy products and/or green, leafy vegetables.
2. **Phosphorus deficiency** is likely only when the diet is low in protein foods.
3. **Iodine deficiency** usually occurs in areas where the soil is low in the mineral, unless iodized salt is used or there are such unsuspected sources of dietary iodine as (a) breads containing iodates as additives or (b) milk from cows which have been fed supplemental iodine, and/or which have had their udders washed with disinfectant solutions that contain iodine compounds.
4. **Iron deficiency** may often be due to lack of sufficient meat in the diet, because the forms of iron which are present in blood and muscle are utilized about three times as well as those in cereals, legumes, and vegetables; and about twice as well as those in fish.
5. **Magnesium deficiency** is not thought to be common in the United States, except among people whose diets consist mainly of highly milled grains and cereal products, and which are lacking in whole grains, leafy green vegetables, nuts, legumes, fish, and meats.
6. **Zinc deficiency** may occur in a mild form when the diet lacks meat, poultry, or fish.

Signs or Symptoms of Mineral Deficiencies. Only a few clinical signs, such as a goiter which may result from an iodine deficiency, are specific indicators of certain mineral deficiencies. Other signs, like stunted growth, may be due to deficiencies of one or more minerals, vitamins, carbohydrates, fats, and/or proteins. (See Table M-25, Mineral Table, for a more complete coverage of the symptoms associated with various mineral deficiencies.) Hence, it is not wise to rely upon signs or symptoms for the early diagnosis of mineral deficiencies. Some other types of diagnostic measures are discussed in the next section.

Laboratory Tests for Mineral Deficiencies. Tests for iron-deficiency anemia are widely used all over the world; but most of the tests for other mineral deficiencies are limited mainly to research studies, because there is a lack of information regarding normal values for such groups as infants, children, adults, and pregnant or nursing mothers. Nevertheless, a growing number of commercial laboratories offer analyses of hair and/or urine directly to customers. These laboratories may also sell mineral and vitamin supplements to the people who use their services.

It is not wise for lay people to attempt to interpret laboratory tests for mineral deficiencies without the aid of a doctor, because many factors other than the diet may raise or lower the levels of minerals in the body liquids and the tissues. Therefore, some of the better-known types of diagnostic tests are covered briefly, so that the reader may be knowledgable when discussing them with his or her physician. These follow:

• **Blood levels of various minerals**—Mild to moderate deficiencies of minerals may not be detected by analysis of the blood, because in mild deficiencies blood levels of essential elements are maintained at the expense of tissue supplies. Hence, subnormal blood values are likely to mean that the mineral reserves have been depleted. Unfortunately, there are only a few convenient procedures for measuring minerals in the tissues.

• **Bone marrow biopsies**—This procedure involves the sampling (by means of a needle) of a small amount of bone marrow for its iron content. It gives an indication of the amount of iron which is stored in the skeleton. However, it is used mainly in research studies.

• **Hair levels of minerals**—Hair is one of the last tissues of the body to be nourished and one of the first to show the effects of poor nutrition. Hence, there is much current interest in the use of hair as a test medium for various mineral deficiencies and/or toxicities. New analytical methods are now available for the rapid analysis of hair for more than a dozen minerals, so certain laboratories offer this service to the public. Some of the merits and demerits of these tests follow:

1. Abnormally high amounts of arsenic, lead, and mercury in hair usually indicate internal poisoning by these toxic elements.

2. Subnormal levels of such trace elements as chromium, manganese, and zinc are usually correlated with deficiencies of these essential minerals.

3. Sometimes, the minerals in hair originate from environmental contamination—from hair dyes, shampoos, and swimming pool water, rather than from the diet.

4. Standards for the mineral content of hair should take into consideration the color of the hair because certain minerals are constituents of hair pigments. For example, blonde hair has lower ratios of zinc to copper than dark shades of hair.[44] Also, red hair contains more iron than other colors, because iron is present in the red pigment. Also, age and sex can affect hair analyses.

5. The part of the body from which the hair sample is taken influences mineral content.

6. The rate of hair growth, which is influenced by several factors such as disease and season of the year, influences the concentration of minerals.

7. Genetic constitution, gender, and drugs may influence mineral content.

So, the results of hair analysis may be both inaccurate and misleading.

Many of the laboratories that analyze hair recommend supplements to correct deficiencies. It's no small coincidence, of course, that they just happen to sell the supplement that they are recommending.

(Also see HAIR ANALYSIS.)

• **Urinalysis**—Mineral deficiencies are likely to result in reductions in the amounts of the elements which are excreted in the urine, as the body attempts to conserve its dwindling supply of these essential nutrients. The trouble with urinalysis as a test for mineral nutrition status is that other factors, such as dehydration or stress, may raise or lower the rate of urinary excretion of minerals; hence, the effects of nutritional status alone cannot be readily measured.

• **X rays of bone**—X rays are used to diagnose demineralization of bones—which is usually due to deficiencies of calcium, phosphorus, and/or vitamin D—and the deposition of lead in the bones whch occurs in lead poisoning.

Often, tests such as those described are used in conjunction with tests of other metabolic functions, because the metabolic activities of certain minerals are not always correlated with the levels of the elements which are present in body fluids and tissues. For example, hair may be analyzed for its chromium content in order to determine whether a diabeticlike response to a glucose tolerance test is due to

a lack of chromium in the form of the glucose tolerance factor (GTF).

More details on the detection of specific mineral deficiencies will be found in the articles dealing with each of the individual minerals.

PREVENTION AND TREATMENT OF MINERAL DEFICIENCIES AND/OR TOXICITIES. The prevention and treatment of the various mineral deficiencies require the cautious use of mineral-rich foods or special supplements because (1) there is a narrow range between the safe levels and the toxic levels of such trace elements as iron, copper, fluorine, and selenium; (2) excesses of certain minerals are likely to reduce the absorption and utilization of others; and (3) large excesses of most minerals—except for potassium, sodium, and chloride—are not well absorbed, but are excreted in the stool. Therefore, lay persons should *not* dose themselves with *shotgun* amounts of mineral supplements; rather, they should limit themselves to more prudent measures, such as those which follow:

Selection of Foods Rich in Essential Minerals. Foods which comprise the most vital tissues of plants and animals are likely to be the richest in essential minerals, because mineral elements participate in the important life functions of these organisms. Hence, the animal products highest in mineral content are blood, bones, milk, and the vital organs, whereas leaves and seeds contain the highest levels of minerals found in the plant foods. Therefore, a mineral-rich diet may be assured by including a wide variety of such foods as dairy products; fish and other seafoods; meats and poultry (hearts, kidneys, and livers are much richer in minerals than the muscle meats); beans and peas; molasses; nuts; herbs; spices; tea; raw, unpeeled vegetables; and whole grain breads and cereal products.

Generally, foods which are only minimally refined contain the most minerals, because these elements are often present in the parts which are removed during processing. However, certain types of food processing may favorably alter the forms of the minerals present in foods so that these nutrients are better absorbed and utilized by the body. For example, yeast fermentation during the leavening of whole grain bread doughs is known to increase the availability of zinc and other minerals.

Production and Preparation of Foods to Maximize Their Mineral Values. Sometimes, animals and plants which have been produced *naturally*—with a minimum of interference from man—are not as nutritious as they might be. Correction of certain soil deficiencies and problems may enhance the mineral value of plants, and, likewise, the supplementation of natural animal feeds may improve the nutritive value of meat, milk, and eggs. Also, the procedures used in processing foods and in preparing meals may contribute to better mineral nutrition. Therefore, it is worth noting how the mineral values of various foods may be maximized through careful attention to certain details of food production and preparation.

BENEFICIAL PRACTICES IN AGRICULTURAL PRODUCTION. The profitable production of both animal and plant foods depends partly on the minerals which are provided by the farmers as supplements to the feedstuffs for livestock and as fertilizers for the soil. Although large excesses of certain elements will interfere with the utilization of other elements, it seems that small excesses—beyond the minimal amounts required for good production—may raise the levels

[44]Eads, E. A., and C. E. Lambdin, "A Survey of Trace Metals in Human Hair," *Environmental Research*, Vol. 6, 1973, p. 251.

of minerals present in the food products. However, this concept is highly controversial, so some of the evidence favoring liberal mineral supplementation for animals and plants follow:

1. Feeding livestock extra amounts of certain minerals like copper, iodine, manganese, selenium, and zinc tends

to raise the levels of these elements in meat (including poultry) and eggs. But the amounts of mineral supplements which are required are considerably greater than the gains in mineral content of products achieved; hence, the method is inefficient.

2. Practices which encourage the growth of certain soil

SOME FOODS RICH IN ESSENTIAL MINERALS

Fig. M-75. Barbecued steaks. (Courtesy, Agriculture Canada, Ottawa, Ontario)

Fig. M-77. Octopus—a seafood. (Courtesy, Department of Food Science and Human Nutrition, University of Hawaii)

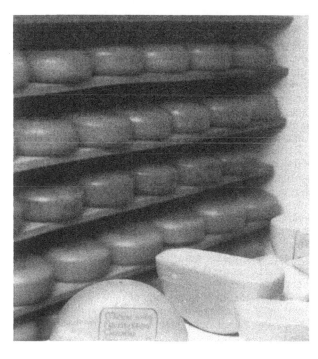

Fig. M-76. Cheeses. (Courtesy, University of Minnesota, St. Paul, Minn.)

Fig. M-78. Whole grain breads. (Courtesy, University of Minnesota, St. Paul, Minn.)

microorganisms may also enhance the uptake of minerals by plants because the microorganisms convert insoluble mineral compounds to more soluble forms, which are more readily utilized by plants. The growth of beneficial microorganisms in the soil is favored by (a) organic matter which may be added as compost or manure; (b) the proper balance between acidity and alkalinity; and (c) minerals which are required by the microorganisms for growth and reproduction.

3. The improvement of mineral-depleted soils by balanced mineral fertilization has often resulted in both better overall crop yields and higher levels of certain minerals in individual plants. However, there are limits of the extent to which the mineral composition of plants may be improved by fertilization, because the utilization of minerals by plants is controlled mainly by genetic factors.

FOOD PROCESSING PROCEDURES WHICH RAISE THE MINERAL VALUES OF FOODS.

It is well known that certain types of food processing are detrimental to nutritive values. On the other hand, mention is seldom made of the types of food processing which either raise the level(s) of one or more minerals, or which alter foods in ways that make their minerals more available. Details follow:

1. The preparation of purees by homogenizing vegetables in water makes the iron in these items more available, because it breaks down the fibrous cell walls which enclose minerals and other nutrients.

2. Yeast fermentation which occurs during the leavening of whole grain breads breaks down the phytates (poorly utilized phosphorus compounds in the bran of the wheat grain) that interfere with mineral absorption.[45] Hence, there is better utilization of the minerals in such breads when yeast is used instead of baking powder.

3. The production of undistilled alcoholic beverages (beers and wines) from grains, grapes, and other plant materials appears to render minerals like iron more available for absorption. Alcoholic iron tonics have long been used to treat anemia. Also, alcoholic fermentation by yeast may convert the inorganic chromium which is present in various fermentable materials to the glucose tolerance factor (GTF)—an organically complexed form of chromium that augments the effects of insulin. (GTF has been found in beer and in yeast which has been grown on chromium-containing nutrient broths. Hence, it seems likely that yeast might similarly convert the chromium that is abundantly present in grape juice into GTF.)

(Also see BEERS AND BREWING; GLUCOSE TOLERANCE FACTOR; and WINE.)

4. Cheese making may raise or lower the amount of calcium in 100-Calorie (kcal) portions of the product compared to the original milk because variable amounts of calcium, phosphorus, protein, and lactose are lost in the whey, depending upon whether the milk is clotted with rennet, or with acid alone. Table M-26 shows how the contents of calories, calcium, and phosphorus vary between milk and several types of popular cheeses.

Dieters seeking to cut down their caloric intake while obtaining most of their calcium requirements from dairy products can use a few ounces of Swiss and/or Cheddar cheeses instead of cupfuls of whole milk or cottage cheese. However, each unit of calcium from Cheddar or Swiss cheese costs about 50% more than an equivalent amount from fresh whole milk.

[45]"Zinc Availability in Leavened and Unleavened Bread," *Nutrition Reviews*, Vol. 33, 1975, pp. 18-19.

TABLE M-26
SELECTED COMPONENTS OF MILK AND FOUR TYPES OF CHEESES[1]

Component	Milk[2]	Cheddar Cheese	Swiss Cheese	Cottage Cheese
Calories .... (kcal/100 g)	62	402	372	90
Calcium .... (mg/100 g)	119	718	961	68
Phosphorus .... (mg/100 g)	89	510	605	150
Calcium in 100 kcal portion of food .... (mg)	192	179	258	76
Calcium to phosphorus ratio	1.33:1	1.41:1	1.59:1	.45:1
Amount of product needed to supply 800 mg calcium .... (oz)	23.7	3.9	2.9	41.5
.... (g)	*672*	*111*	*83*	*1,176*
Amount of energy which must be consumed to obtain 800 mg calcium .... (kcal)	417	446	309	1,058

[1]The food composition data were obtained from Food Composition Table F-21 of this book.
[2]Milk which contains 3.5% fat.

5. The extraction of fat from soybean products prior to production of flour may raise the ratio of the mineral content (as represented by ash content) to the caloric content by more than 60%. This means that the consumer gets more minerals, but fewer calories, when defatted soybean products are selected.

6. Ascorbic acid (vitamin C) is a beneficial food additive because it converts oxidized iron to the reduced form, which is better absorbed. Also, the addition of sulfites (substances used to prevent discoloration and/or microbial growth) to foods may indirectly promote better utilization of iron, because these additives preserve ascorbic acid against oxidation.

7. The use of limewater (calcium hydroxide solution) in the production of tortillas (flat, thin cakes made of corn meal) adds about 100 mg of calcium per 100 g of product.

8. Heat treatment of soybeans improves the utilization of copper, iron, manganese, and zinc from this food because it inactivates a metal-binding constituent.

MEAL PREPARATION PRACTICES WHICH CONSERVE ESSENTIAL MINERALS.

It may not have been too important for prehistoric man to minimize the loss of mineral nutrients during cooking because the lack of laborsaving devices made it necessary to eat large amounts of food so as to provide for great expenditures of energy. Today, people expend much less energy and eat considerably less food; so, it is now necessary to use better methods of meal preparation. Some guidelines for conserving minerals follow:

1. Soaking spinach and rhubarb in water removes some of the oxalic acid from these items so that dietary minerals are better utilized. Oxalic acid forms insoluble, unabsorbable compounds with calcium, iron, and other minerals.

2. Cleaning fruits and vegetables with a stiff brush is preferable to peeling them, because the outer layers of many items are much richer in minerals than the inner parts. Hence, the mineral losses due to peeling may be disproportionately high for the amount of material that is removed.

3. It is better to purchase smaller amounts of fresh, leafy vegetables more frequently than to store large amounts in the refrigerator for long periods of time, because cutting down on storage time minimizes wilting and the need to discard the wilted outer leaves, which are the richest in minerals.

4. The loss of minerals during pressure cooking or steam cooking of vegetables is significantly less than that during boiling. Other ways of cooking which conserve minerals are baking and microwave heating. When other conditions are

the same, mineral losses may also be minimized by cooking vegetables in large pieces rather than small pieces, and by keeping the cooking time as short as possible.

5. There are considerable amounts of minerals in the packing fluids used in most canned foods, so the fluids should be saved and used whenever feasible.

6. The thawing of frozen foods results in losses of minerals with the juices which escape during thawing. Freezing food ruptures cell walls so that the cell contents escape readily during thawing. Although the thaw juices might be used in meal preparation, it may be dangerous to do so if there has been considerable time for the growth of microorganisms during thawing. Hence, it may be desirable to cook frozen foods without preliminary thawing so that the juices which escape may be used safely.

Microwave heating may be preferable to other ways of cooking unthawed frozen foods because when food is penetrated by the microwaves cooking proceeds uniformly throughout, whereas in other ways of cooking the heat acts mainly on the outside of the food, which may become overcooked, while the interior is undercooked.

ENRICHMENT AND FORTIFICATION OF FOODS WITH MINERALS.
Extra minerals may be added to foods so as (1) to restore those lost during processing (such restoration is called *enrichment*), or (2) to make certain popular foods richer in various nutrients than is natural for these items (this process is called *fortification*). For example, the level at which iron is added to white flour and other refined cereal products constitutes enrichment or restoration of the iron lost during milling, whereas the addition of iron to milk-based infant formulas is fortification, because milk is normally low in iron.

From such early beginnings as the iodine fortification of salt in the 1920s, and the enrichment of flour and bread with iron and vitamins in the 1930s, the addition of essential minerals has been extended to both traditional foods and new items which have been recently concocted by food technologists. Therefore, it is important to be knowledgeable relative to the following items that may be enriched or fortified with minerals:

• **Breakfast cereals**—Many of these items are made from highly milled grains, so they are often fortified with vitamins and iron, and sometimes with calcium. However, milling removes large amounts of other minerals like magnesium, chromium, manganese, copper, zinc, selenium, and molybdenum, which are not usually added during enrichment or fortification, unless the manufacturer does so indirectly by adding wheat germ, bran, or rice polish to his products. (The by-products of milling removed from the whole grains are rich sources of the minerals.) Hence, the consumer receives better overall mineral nutrition with whole grain cereals than with those made from refined grains which have been enriched or fortified with one or two minerals.

• **Complete diet formulas**—There are various types of powders and liquids, which usually contain skim milk or other sources of protein, that are fortified with minerals and vitamins so that they may be taken instead of regular meals. Nevertheless, they are not likely to contain adequate amounts of such newly discovered trace minerals as chromium, selenium, and silicon, because these elements are not usually added to the products. Hence, the formulas are best used as replacements for only one or two meals per day, so that the missing minerals may be obtained from regular foods.

• **Flour, grains, and bakery goods**—Usually, iron is the only mineral used to enrich these products when they are made from milled grains. It may be wise to avoid these items and select whole wheat or bran breads and muffins, and brown rice, unless the rest of the diet makes up for the mineral deficiencies in the highly refined products.

• **Imitation foods or analogs**—High prices for such staple foods as fruit juices, milk, ice cream, butter, meats, and poultry, along with the concern over the cholesterol and fat content of certain of these foods, have stimulated the development of imitation products or analogs. These products are made from lower-cost materials, like soybean derivatives, and they are usually fortified with minerals, vitamins, and amino acids so as to be nutritionally equivalent to the items which they are designed to replace in the diet.

Analogs are generally wholesome and nutritious for most healthy people. However, those who are prone to heart disorders and/or high blood pressure should note that (1) some of the soybean-based imitations of meats and poultry contain much more sodium than the authentic animal products, and (2) artificial fruitlike drinks may contain little or none of the potassium present in real fruit juices, which are often prescribed as sources of the mineral for therapeutic diets. Hence, there is a need for more extensive nutritional labeling on the packages of imitation foods.

• **Infant cereals**—These products generally provide better mineral nutrition than the breakfast cereals designed for older children and adults, because they are often fortified with calcium, iron, copper, and other minerals. Of course, whole grain cereals are also good, but the fiber they contain may have too much of a laxative effect on infants.

• **Infant formulas**—The manufacturers of these products have usually designed them to be nutritionally equivalent to human breast milk, perhaps with the following notable exceptions:

1. The calcium and phosphorus levels in the formulas may be considerably higher than those in human milk, but they are usually somewhat lower than those in fresh or evaporated cow's milk. Newborn infants may have seizures if they are fed excessive amounts of phosphorus. Hence, the formulas are safer than either undiluted, fresh cow's milk or evaporated milk diluted with only an equal volume of water.

2. The iron levels may be unusually high in certain formulas, because it is believed that the very low levels of this mineral in both human and cow's milk may sometimes be responsible for the development of anemia in infants and children.

• **Instant breakfast powders**—These powders, which are usually designed to be mixed with a glass of milk, are fortified with minerals and vitamins so as to supply at least one-fourth of the daily nutrient requirement for an adult. Probably they supply better balanced mineral nutrition than most people regularly receive from ordinary types of breakfast foods. However, they may not be suitable for people prone to constipation, unless the low fiber content of the formulas is offset by eating pulpy fruits or drinking unstrained juices.

• **Milk**—Minerals are not usually added to ordinary cow's milk, but vitamin D is usually added, so that the calcium and phosphorus which are present in the milk will be well utilized.

• *Nutrified* cakes, cookies, and food bars—The new term *nutrification,* which is derived from the words nutrient and fortification, was coined by food scientists to denote the providing of nutrients through fabricated, fortified foods such as snack cakes, cookies, food bars, and *chips* laced with minerals, vitamins, and amino acids. However, some of these products are too high in fats and/or sugars to be used by sedentary adults, who might easily eat too much of these concentrated sources of calories.

• Salt—Slightly more than half of the salt sold at the retail level is fortified with iodine for the purpose of preventing iodine-deficiency goiter.

Special Dietary Products. The use of special foods and mineral supplements may be justified when it appears that the diet is deficient in certain essential elements. Ordinary diets may not suffice to meet mineral needs under circumstances such as (1) pregnancy; (2) breast feeding of infants; (3) illnesses which depress the appetite; (4) types of malnutrition that require rapid restoration of the body's supply of certain minerals (*these conditions require the attention of physicians who are proficient in nutritional therapy*); (5) sharp restrictions in caloric intake in order to achieve rapid reduction in body weight; or (6) shortages of wholesome, nutritious foods.

The authors do *not* wish to suggest that any of the conditions listed are to be treated with special dietary products; rather, they wish to point out that certain products may be used to fill the gaps in mineral nutrition which may result from inadequate diets. Furthermore, the nutritional assessment of diets and the selection of products to provide minerals may be hazardous for a lay person to undertake without the assistance of a doctor or a dietitian. Unfortunately, this type of assistance is not always readily available.

Hence, some practical guidelines are presented with the hope that they may help to counteract the current tendency for many people to select nutritional products on the basis of questionable advertising claims.

SPECIALTY FOODS AND HIGHLY POTENT NUTRITIONAL SUPPLEMENTS. It is fortunate that most consumers are not able to purchase pure mineral salts and to mix them into their foods, because without special measuring equipment and experience in preparing such mixtures, it is very easy to add toxic excesses of trace elements like iodine and selenium to foods. However, even some of the products that are sold by health food stores may have undesirable, or even toxic effects, because they are very rich in certain minerals. The potentially hazardous products may be considered as *highly potent nutritional supplements,* in order to distinguish them from the less potent specialty foods, which are usually safe in the amounts consumed. Table M-27 shows the approximate amounts of these minerals that are provided by various specialty foods and nutritional supplements.

It may be seen from Table M-27 that it is safer to use specialty foods as sources of minerals than it is to use highly potent nutritional supplements because the former contain considerably lower levels of minerals than the latter, so the chance of receiving dangerous excesses of any of the elements is much less. For example, 3½ oz (*100 g*) of wheat germ contain about half of the daily allowance for copper, whereas an equal amount of liver powder contains seven times as much of the mineral. Furthermore, the contents of protein, vitamins, and other nutrients in the different products vary considerably. Hence, it is desirable to use a variety of these items, rather than a single one, to meet specific mineral needs. Some guidelines for using the various products follow.

Specialty Foods. The following items are good to excellent sources of energy, minerals, protein, and vitamins; so,

TABLE M-27
AMOUNTS OF SELECTED MINERALS IN SOME SPECIALTY FOODS AND NUTRITIONAL SUPPLEMENTS[1]

Product	Macrominerals					Microminerals					
	Calcium (Ca)	Phosphorus (P)	Potassium (K)	Sodium (Na)	Magnesium (Mg)	Iron (Fe)	Zinc (Zn)	Copper (Cu)	Selenium (Se)	Iodine (I)	Chromium (Cr)
	(mg/100 g)	(mg/100 g)	(mg/100 g)	(mg/100 g)	(mg/100 g)	(mg/100 g)	(mg/100 g)	(mg/100 g)	(mcg/100 g)	(mcg/100 g)	(mcg/100 g)
Specialty Food:											
Alfalfa (leaf) powder	1,640	230	2,070	60	350	36.0	1.6	1.0	28	—	—
Almonds, dried	234	504	773	4	293	4.7	3.1	1.2	—	—	—
Bran	119	1,276	1,121	9	598	11.0	10.4	1.3	63	6	40
Flaxseeds	271	462	1,460	30	400	9.0	—	—	—	—	—
Nonfat milk, dried, powder	1,308	1,016	1,794	532	144	.3	4.1	.5	—	—	12
Peanuts, roasted w/skins ..	72	407	701	5	175	2.2	3.0	.3	—	—	—
Rice polishings	69	1,106	714	100	650	16.1	2.6	.7	—	7	—
Sesame seeds	131	776	407	40	347	7.8	10.3	1.6	—	—	—
Soybean flour (low fat) ...	263	634	1,859	1	289	9.1	2.0	1.6	—	—	—
Sunflower seeds	120	837	920	30	38	7.1	4.6	1.8	—	—	—
Wheat germ, toasted	47	1,084	947	2	365	8.9	15.4	1.3	111	—	25
Nutritional Supplement:											
Blackstrap molasses	684	84	2,927	96	209	16.1	2.2	6.0	—	158	115
Bone meal	29,820	12,490	180	5,530	320	85.0	12.6	1.1	—	—	—
Brewers' yeast	210	1,753	1,894	121	231	17.3	3.9	3.3	125	—	118
Dolomite	22,080	40	360	—	9,870	76.0	—	—	—	—	—
Kelp	1,904	240	5,273	3,007	213	89.6	3.5	.6	43	62,400	Trace
Liver, dried, powder	40	830	930	450	—	20.0	—	9.0	180	—	170
Torula yeast	424	1,713	2,046	15	165	19.3	1.9	1.4	132	—	—

[1]A dash indicates a lack of authoritative data, *not* that the mineral is absent.

they may be used alone as foods, or they may be mixed with other items:

• **Alfalfa (leaf) powder or tablets**—Because of its grasslike taste, alfalfa powder should be mixed with other foods like baked goods (not more than a tablespoon per cup of flour), thick vegetable soups, or *green sauces* for pasta. It is noteworthy that the flavor of alfalfa blends well with mint or spices. Alfalfa is also available in the form of tablets that contain no other ingredient. Tablets avoid flavor problems because they may be swallowed whole. Alfalfa powder is an excellent source of calcium, iron, and magnesium.

• **Almonds, dried**—Although these nuts cost about twice as much as peanuts, they may be worth the price because (1) they are tasty without salt, and (2) their mineral content is higher than peanuts (three times the calcium, one and one-half times the magnesium, and two times the iron). Many people eat them plain, but they are also good in toppings, coatings (such as fish almondine), or in pastries and other desserts. Almonds, like other nuts, are also good sources of protein, polyunsaturated fats, and vitamins.

• **Bran**—This product has about nine times the overall mineral content of white flour. It may be (1) used alone as a cereal (a little goes a long way, because it soaks up liquid and is, therefore, very filling); (2) mixed with other cereals; (3) used as a breading for baked, broiled, or fried foods; or (4) mixed with chopped fish, meat, and poultry as a filler which aids in the retention of their juices. However, bran may have a laxative effect on some people, varying according to the amount eaten and individual susceptibilities. So, it should be added cautiously to the diet, in small amounts, until the level of tolerance is known. Also, plenty of liquid should be taken with bran, or it may cause irritation in the digestive tract. Bran is an excellent source of many of the essential minerals.

• **Flaxseeds**—These seeds, when ground, have long been used as the basis of various *natural* laxative preparations. They contain a mucilaginous material which thickens the porridge that is formed when they are cooked in water with other grains. Their bland taste makes it convenient to add them to various cereal products, where their ample supply of calcium, iron, and magnesium adds to the mineral values.

• **Nonfat milk powder**—This product, when mixed with water, yields a drink which is equal in mineral value to whole milk, but which is lower in calories. However, even greater mineral nutrition may be promoted by adding the dry powder to various foods, to which it contributes calcium and magnesium. For example, it blends with the other ingredients in items such as cooked cereals, doughs, fudge-type candies, puddings, coffee, cocoa, and tea. Furthermore, milk sugar—lactose, which is abundantly present in nonfat milk powder, appears to promote the absorption of calcium and other essential elements.

• **Peanuts**—Peanuts and peanut butter are often used as meat substitutes, because of their protein content. However, they contain much more calcium, chromium, copper, magnesium, and calories than do muscle meats. They may be eaten alone, or in various combinations with other foods.

• **Rice polishings**—This product, like wheat bran, contains the minerals removed from whole grains (brown rice) during milling. It is an excellent source of iron, magnesium, and the vitamin B complex. The bland taste makes it a good addi-tive for fortifying baked items, breading, and breakfast cereals.

• **Sesame seeds**—These seeds are often sprinkled over breads, cookies, crackers, and rolls. They are higher than muscle meats in content of several minerals and calories. Sesame seed butter (which resembles peanut butter) is sold in many health food stores.

• **Soy flour**—Usually soybeans have had most of their fat removed by crushing and extraction prior to being ground into flour. Hence, soy flour contains more minerals, but fewer calories, than most nuts and seeds. It is a good source of calcium and magnesium, as well as protein and certain vitamins. However, the use of this product is limited by (1) the beany flavor; and (2) its content of certain carbohydrates, which ferment in the digestive tract to form gas. Some ways of overcoming these disadvantages are: (1) adding soy flour to highly spiced foods like gingerbread cookies and ground sausage meat, so that the beany taste is disguised; and (2) using small amounts of soy flour at first, until the limit of one's tolerance to the gas-producing effects is determined.

• **Sunflower seeds**—Although many Americans formerly considered this chewy product to be best suited for feeding birds, it is now sold as a snack food. The seeds keep well without refrigeration, because of their protective outer covering. Their content of copper and iron, along with protein and vitamins, makes them a much better snack item than popcorn, potato chips, and sweets. Some health food stores carry sunflower seed butter and sunflower seed meal; but these items sometimes have a bitter taste, which may be due to rancidity that develops when the seed is crushed and the oil exposed to the air.

• **Wheat germ**—This by-product of wheat milling has about the same overall mineral content as bran, but it is better digested and it provides more protein, calories, and vitamins. Also, it ranks close to the vegetable oils as a source of vitamin E. It may be used alone as a breakfast cereal, or it may be added to various cereal products. (Also see BREAKFAST CEREALS; and WHEAT.)

Highly Potent Nutritional Supplements. The following products should be used with caution so as to avoid over-doses of certain minerals or other potentially harmful substances:

• **Blackstrap molasses**—About two heaping tablespoons (40 g total) per day of this thick, sugary fluid will furnish liberal amounts of calcium, magnesium, chromium, copper, iodine, iron, and zinc. However, many people may find its licoricelike taste too strong for their liking when they attempt to take it alone, like a tonic. Also, it may be a strong laxative for some people. Hence, it might be mixed cautiously into other foods where its flavor may be either diluted or masked by other strong flavors. For example, it may be added to items like (1) a glass of milk, because the addition of 1 teaspoon (5 ml) of molasses per 1 or 2 teaspoons of carob powder makes a drink that is more nutritious than chocolate milk; (2) spice cakes or gingerbread cookies, which may also contain added soy flour; or (3) pumpernickel or other dark breads leavened with yeast, because the yeast may convert the inorganic chromium from the molasses into the glucose tolerance factor, an organic form of the mineral, which acts along with insulin in promoting the metabolism of carbohydrates.

(Also see MOLASSES, section headed "Food.")

• **Bone meal powder or tablets**—A little of this item goes a long way. A teaspoon of the powder provides over 1,000 mg of calcium, which appears to be more than enough for people other than pregnant or nursing mothers. Tablets may be more convenient than powder for most people, who usually lack scales for accurately measuring small quantities of powder. However, if the powder is to be mixed into such items as cereals, doughs, gravies, and sauces—*not* more than ¼ teaspoon per person per serving should be used. Careful trials should be conducted on a small scale, before adding the powder to recipes, because bone meal imparts a slightly gritty texture to foods.

• **Brewers' yeast powder or tablets**—This special product has long been used to prevent or treat malnutrition by providing extra protein and the vitamin B complex. However, it is also one of the richest sources of both chromium and selenium—minerals that are likely to be supplied inadequately by ordinary diets. Furthermore, the chromium is present in an organic form (glucose tolerance factor) which is much more beneficial to the body than the inorganic forms of the mineral.

Brewers' yeast is also rich in nucleic acids, which are substances that may cause certain susceptible people to develop high blood levels of uric acid (hyperuricemia) and/or gout. These disorders tend to run in certain families, so it may be wise to inquire whether any close relatives have had these disorders, before taking this supplement on a regular basis.

A suitable daily portion of the dried yeast—for people who are *not* prone to either hyperuricemia or gout—is 1 or 2 tablespoons (*15 or 30 ml*). It may be mixed in vegetable drinks like tomato juice, cooked cereals, flavored instant breakfast drinks, gravies, meat loaves, or various sauces. If there is a choice of product, one that is *debittered* should be selected. Tablets are also available for those who prefer a more convenient form of the product.

• **Dolomite tablets**—Usually, each tablet supplies about 130 mg of calcium, and about 60 mg of magnesium. Four such tablets daily should be an adequate supplement for most people, except for pregnant or nursing mothers, who may need as many as six tablets to fill the gap between intake and recommended allowances of these minerals. However, excesses of calcium and magnesium may interfere with the utilization of other essential minerals.

• **Kelp tablets**—Just about every brand of these tablets contains an amount of iodine equal to the U.S. RDAs for adults (150 mcg per day) in each tablet. It is not wise to exceed this amount because people who have tendencies to develop goiter are extra susceptible to iodine toxicity. Those who have suffered from dietary iodine deficiency may have thyroid glands which have adapted to low intakes of the mineral by becoming very efficient in the production of thyroid hormones. Thus, their glands may overproduce these hormones when the dietary level of iodine is raised.

CAUTION: Due to its high content of iodine, it is very easy to get an overdose of iodine from only small amounts of kelp powder.

(Also see ENDOCRINE GLANDS; and GOITER.)

• **Liver, dried, powder**—This excellent source of chromium, copper, iron, and selenium may also be rich in cholesterol and fat, unless it has been defatted. Also, the defatted types of liver powder have a more pleasant taste. However,

this supplement should *not* be used by people who might have tendencies to develop high blood levels of uric acid or gout because it is rich in nucleic acids, which are converted into uric acid in the body.

Two teaspoons (*10 ml*) of liver powder per day provide plenty of extra minerals for most people, when it is used on a regular basis. It may be added to gravies, meat loaves, or various sauces, provided sufficient seasonings are used to disguise the liver flavor.

MINERAL SUPPLEMENTS. Sometimes, it may be more convenient and economical to provide supplementary minerals in pill form than to attempt to meet all of one's mineral needs through ordinary foods and/or specialty products. However, mineral pills or tablets have certain disadvantages in that (1) people may neglect to take their pills regularly; and (2) small pills which contain large amounts of certain minerals may lead to overdosage by people who are impatient to obtain the expected beneficial effects, and to poisoning of small children who might be tempted to sample the pills. Nevertheless, there are circumstances under which it is necessary to take mineral pills in order to prevent nutritional deficiencies. Hence, some suggestions for using these products follow:

1. Whenever feasible, consult with a doctor or a dietitian before attempting to correct dietary inadequacies by means of special supplements.

2. Buy mineral pills which are labeled in accordance with the U.S. RDAs proposed by the Food and Drug Administration.

3. Use mineral supplements according to the directions on their lablels.

4. Keep all types of pills out of the reach of infants and children.

5. Report any unusual signs or symptoms, such as changes in bowel habits, to a physician.

(Also see NUTRIENTS: REQUIREMENTS, ALLOWANCES, FUNCTIONS, SOURCES; NUTRITIONAL SUPPLEMENTS; and U.S. RECOMMENDED DAILY ALLOWANCES.)

MISCIBLE

Capable of being mixed easily with other substances.

MITOCHONDRIA

Minute spheres, rods, or filaments in the cytoplasm. Mitochondria are the sites of numerous biochemical reactions including amino acid and fatty acid catabolism, the oxidative reactions of the Krebs cycle, respiratory electron transport, and oxidative phosphorylation. As a result of these reactions, mitochondria are the major producers of the high energy compound adenosine triphosphate (ATP) in aerobically grown cells.

MODIFIED DIETS

Contents	Page
History	737
Common Types of Dietary Modifications	737
Use of Exchange Lists in Planning Diets	746
Special Dietetic Foods	748
Summary	748

This term refers to modifications of the ordinary American diet which are made for the purpose of providing nourishment when the consumption of a normal diet is inadvisable. (Strictly speaking, the term diet alone refers to whatever pattern of foods is consumed regularly, whether it conforms to a normal pattern or not.)

There is widespread current interest in various types of modified diets because many people are interested in increasing their productive years while minimizing the amounts spent on medical care. Also, an ever increasing number of nutritional supplements and special dietary products are marketed each year.

HISTORY. During most of human history, food patterns for most people have been based mainly upon whatever edible items could be obtained readily from nearby fields, forests, oceans, rivers, and streams, although some prehistoric peoples traveled long distances while hunting and gathering their foods. However, it appears that each of the preagricultural societies had some type of lore that attributed curative powers to certain special animal foods and druglike preparations from plants. For example, it was widely believed that eating the eyes of animals was the proper treatment for disorders of vision. (This idea was confirmed as valid in the 20th century, when it was found that eyes and other rich sources of vitamin A helped to cure visual problems due to deficiencies of this nutrient.)

The rise of agriculture made it possible for large numbers of people to dwell together in complex societies, but it also brought new health and nutritional problems because (1) infectious diseases spread rapidly through densely populated regions, and (2) the need to provide for large populations resulted in fewer types of foods being produced and consumed. The latter condition increased the likelihood of nutritional deficiencies because there were no means of identifying specific deficiencies and the nutrients that prevented them until a more scientific approach to nutrition was initiated by the great French chemist Antoine Lavoisier in the latter part of the 18th century. Nevertheless, the actual practice of dietetics began much earlier.

Dietetic concepts in the ancient civilizations of Babylon, China, Egypt, and India were based on various combinations of religious beliefs, superstitions, taboos, and trial-and-error observations. Even the last-named basis of judgment was clouded frequently by illness due to a contagious disease, contaminated water, or spoiled food. (The long-standing bias of some Greek and Roman physicians against the eating of fruit is believed to have been due to the fact that fruits were usually abundant in warm weather when diarrheal diseases were rampant because the conditions favored the rapid growth of microorganisms.) Another common type of prescientific concept was that (1) all matter was constituted of varying proportions of the elements air, earth, fire, and water (the Hindus believed that there were three elements, whereas the Chinese postulated five); and (2) the characteristic temperament of the patient (choleric, melancholic, phlegmatic, or sanguine) was due to the excess of one of the elements. Hence, it was considered desirable for an ill person to eat foods which counteracted the excess rather than those which augmented it. For example, a person with a fiery (choleric) temperament was usually told to eat only *cool* foods. In this case, coolness or hotness of foods referred to attributes other than temperature, which were deduced by rather convoluted reasoning. It is noteworthy that beliefs in cool and hot foods are still held by many poorly educated groups of people around the world.

Two major developments which contributed to the beginnings of modern dietetics in the 18th century were (1) the experiments on metabolism by Lavoisier whose career was literally cut short by the guillotine, and (2) founding of voluntary hospitals in England by private philanthropy. (The latter development helped to fill the gap created by the closing of the many hospitals operated by Catholic religious orders that occurred during the Reformation.) At first, the diet which was given to all patients was based mainly on beef, beer, and bread. Later, the diet was modified for (1) convalescent patients who were given *half diets* consisting of smaller quantities of meat that were augmented with light puddings, soups, broths, and vegetables; and (2) febrile patients who were fed *low diets* of gruels, milk, porridges, puddings, and small portions of cheese and/or meat. Use of these diets and variations of them persisted until the 20th century, because the nutritional science of the 19th century was dominated by the belief that adequate amounts of carbohydrates, fats, proteins, and minerals were all that were required to maintain good health. The need to prevent scurvy and other nutritional deficiencies by the provision of fruits and vegetables was not common knowledge until the latter part of the 19th century. Also, these items spoiled readily.

The major factors which accelerated the growth of modern dietetics in the first few decades of the 20th century were (1) the discoveries of the requirements for certain amino acids, trace minerals, and vitamins; (2) the founding of the American Dietetic Association in 1917 (this organization has contributed greatly to the standardization of the education and training of dietitians and the types of diets used for treating various disorders); and (3) the widespread recognition of the need for dietary modification in the treatment of certain conditions. (The designation *dietitian* originated in 1899 at the meeting of the National Home Economics Association at Lake Placid, New York.) It is noteworthy that the dramatic recoveries of some critically ill patients which resulted from the appropriate diet therapy led to a certain amount of unscrupulous promotion of nutritional supplements. These abuses inspired leading dietitians, doctors, and public health professionals to conduct vigorous educational campaigns against food faddism and nutritional hucksterism. The campaigns also emphasized that balancing diets with respect to all of the essential nutrients was much more important than relying on unusual dietary modifications or large quantities of a few minerals and vitamins to bring about the desired improvement of health.

Today, there are many types of modified diets that are used regularly in the treatment of a wide variety of abnormal conditions. However, there is a need for more accurate record keeping and monitoring of the patient's daily food intake so that the efficacy of these modifications may be judged with more certainty.

COMMON TYPES OF DIETARY MODIFICATIONS. Two of the most important principles of dietary modification are (1) only necessary changes should be made in order to minimize the difficulty the patient will have in following the dietary prescription, and (2) every effort should be made to ensure that a modified diet is an adequate source of all essential nutrients, or if this is not feasible, appropriate nutritional supplementation should be recommended. The most common dietary modifications are given in Table M-28.

NOTE: This table lists only diets comprised of common foods and beverages that are taken orally. Special formulas and special means of administering nutrients are covered in the following articles: ELEMENTAL DIETS; INTRAVENOUS (PARENTERAL) NUTRITION, SUPPLEMENTARY; LIQUID DIETS; and TOTAL PARENTERAL (INTRAVENOUS) NUTRITION.

Diet	Composition		Uses	Comments
	Foods Included	**Foods Excluded**		
General diet	The regular diet consumed by most Americans.	None	Patients who do not require any dietary modifications.	This is also called the house or full hospital diet and varies somewhat from one institution to another.
I. Modifications in texture and consistency				
General diet, pureed	Same as general diet.	None	Toothless patients.	
General diet, liquid (Also see LIQUID DIETS.)	Finely homogenized and strained fish, meat, or poultry (visible fat should be removed before homogenization).	All other forms of fish, meat, or poultry.	For patients who are unable to chew and/or swallow foods due to circumstances such as surgery on the face, oral cavity, neck, or throat; fractures of the jaw; esophageal strictures; partial paralysis; or unconsciousness. As nutritional supplements for people who find it difficult to eat sufficient amounts of ordinary food. By athletes prior to, and/or during sporting events because liquids leave the stomach more rapidly than solid foods. Weight reduction when a low-calorie liquid diet is used at one or more daily meals in place of solid foods which contain more calories.	The daily allotment is usually divided into six feedings. Milk, milk drinks, and strained fruit or vegetable juices may be served between meals. Underweight patients may need supplemental feedings (commonly called "nourishments") that are rich in calories from foods such as melted butter or margarine (added to hot liquids), or honey, sugar, or syrups (added to fruit juices). Liquid diets are usually deficient in certain nutrients. Hence, the doctor or dietitian may recommend certain mineral and vitamin supplements if the diet is to be used for more than a week or so.
	Any type or kind of milk or cream.	None		
	Strained cottage cheese.	All other forms of cheese.		
	Eggs in cooked foods such as puddings and soft custards.	All other forms of eggs.		
	Strained fruit juices, pureed and strained fruits.	Berries with small seeds.		
	Strained vegetable juices, pureed and strained mild flavored vegetables such as asparagus, beans (green and wax), beets, carrots, peas, potatoes, spinach, squash, sweet potatoes, and tomatoes.	Strong flavored vegetables.		
	Cereal gruels made from enriched refined cooked cereals.	All prepared or dry cereals, and whole grain cooked cereals. All forms of bread.		
	Moderate amounts of butter, margarine, oil, or cream.	All other forms of fat.		
	Cream or broth-type soups made with pureed vegetables; strained fish, meats, or poultry; broth, bouillon, or consomme.	Highly seasoned soups or those containing pieces of vegetables, meats, or poultry.		
	Sugar, honey, syrups, jelly, and plain sugar candy.	Jams and marmalades.		
	Plain ice cream, sherbets, ices, puddings, junket preparations, soft custard, tapioca, and plain dessert gels.	Desserts that contain nuts, fruit, coconut, or other solids.		
	Carbonated beverages, cocoa, coffee, tea, and coffee substitutes.	All other beverages.		
	Salt, mild spices, and vanilla and other mild flavorings in moderate amounts.	All other types of condiments, flavorings, seasonings, and spices.		
Soft diet	Fresh, canned, or frozen fish; beef, lamb, liver, lean roast pork or ham, crisp bacon; or poultry meat. (These items should be cooked by baking, broiling, panbroiling, or roasting.)	Fried or deep fat fried fish, meats, or poultry; fish, meats, or poultry with small bones, gristle, or skin; fatty pork or ham, franks, corned beef; spiced or smoked items.	An intermediate diet that is given after a full liquid diet, and before a regular diet. Patients who are bedridden and likely to suffer from gas and other digestive problems when coarser foods are consumed. When difficulties in chewing and/or swallowing are present.	The compositions of soft diets vary somewhat between different hospitals. The diet may not furnish enough iron for women unless there is a liberal use of meats, enriched breads, and strained legumes. (The latter may be made into dips, soups, or spreads for bread.) Soft diets can be used for long periods of time, but they should not be used longer than necessary or the patient may lose his or her tolerance for coarser foods.
	Any type or kind of milk or cream.	None		
	Cottage and other mild-flavored varieties of cheeses.	Sharp or strong-flavored cheeses.		
	Eggs that are baked, hard or soft cooked, poached, or scrambled without fat (cooked in a nonstick fry pan).	Fried eggs or scrambled eggs cooked in fat.		
	All fruit juices; cooked and canned apples (without skin), applesauce, apricots, cherries, peaches, pears, plums, and prunes; and raw, ripe banana, grapefruit, and orange sections, ripe peeled peaches, and ripe peeled pears.	All fruits with small seeds or tough skins, unless these parts are removed completely. Dates, figs, and raisins.		
	All vegetable juices; cooked and canned asparagus, beets, carrots, eggplant, green beans, peas, pumpkin, spinach, squash, and tomatoes; corn and lima or navy beans that have been passed through a sieve; and raw lettuce.	Cooked broccoli, Brussels sprouts, cabbage, cauliflower, corn, cucumbers, kale, onions, parsnips, and whole lima or navy beans. All raw vegetables except lettuce.		
	Day old or toasted white bread; salted, soda, or graham crackers; refined cereals such as cornmeal, cream of wheat, oatmeal, or rice; dry cereals such as cornflakes.	All whole grain breads and cereals except oatmeal. Hot breads.		
	Sweet or white potatoes (without skins); macaroni, noodles, or spaghetti made with white flour; and white rice.	Spicy dressings or sauces on potatoes, pasta, or rice.		

(Continued)

Diet	Composition		Uses	Comments
	Foods Included	**Foods Excluded**		
Soft diet *(Continued)*	Butter or margarine, cooking oils, cream, crisp bacon, mild-flavored salad dressings.	Fried foods and gravies.		
	Canned or homemade broths or soups made from items in this column.	Spicy soups or those containing large amounts of fat.		
	Sugar, syrups, honey, and clear jellies in moderation.	Jams and preserves which contain nuts, seeds, or skins.		
	Cakes (angel food, mild chocolate, plain white, sponge, yellow); cookies; custards; fruit whips; gelatin desserts with allowed fruits; ice cream or sherbet without fruit pulp or nuts; and puddings (bread, butterscotch, chocolate, rice, tapioca, vanilla).	Pastries; pies; excessively sweet desserts; or those containing dates, fruits with small seeds, or raisins.		
	Carbonated beverages, coffee and coffee substitutes, and tea in moderation.	Alcoholic beverages, unless ordered by the doctor.		
	Salt, lemon juice, vanilla, cocoa, cinnamon, allspice, and ripe olives.	Spicy or fibrous condiments.		
Soft diet, pureed	Same foods as in the soft diet.	Same foods as in the soft diet.	Patients who have considerable difficulty in chewing and/or swallowing.	All solid foods are chopped fine, strained, or blended.
Bland diet	All of the foods in the soft diet, except those items excluded in the next column.	Those excluded from the soft diet *plus* chocolate, cocoa, coffee, decaffeinated coffee, cola beverages, tea; tomatoes, tomato juice, and citrus juices; and excessively fatty foods.	In some cases of chronic digestive disorders such as esophagitis, hiatus hernia, ulcers, colitis, diverticulitis, etc.	There is now some doubt as to whether the traditional types of bland diets promote healing any faster than regular diets which are free of strong irritants such as mustard and pepper. Pureed vegetables may also be used to make bland dressings and sauces. A vitamin C supplement is advisable.
Bland diet, low fiber	All the foods listed for the soft diet, except those items excluded in the next column.	Those excluded from the soft diet *plus* chocolate, coffee, decaffeinated coffee, cola beverages, tea; all fruit juices and fruits; all vegetables other than those pureed, strained, and prepared in cream soups; excessively fatty foods; and all spices other than salt.	In some cases of chronic digestive disorders such as esophagitis, hiatus hernia, ulcers, colitis, diverticulitis, etc.	The foods are usually distributed between 6 small meals per day. Additional feedings of milk and/or cream may be given between meals. A vitamin C supplement should be given. There is no reason for continuing this diet after the acute inflammatory stage of the disorder has subsided.
Milk and cream diet	A 1:1 mixture of whole milk and cream; or *coffee cream* with 12.5% butterfat; or if calories are to be reduced, only whole milk or skim milk.	All others	Acute stages of gastrointestinal disorders such as high gastric acidity, gastritis, ulcers, etc.	Generally, about 4 oz (*120 ml*) of the mixture are given every hour to keep some food in the stomach at all times. Mineral and vitamin supplements are needed.
Clear liquid diet (Also see LIQUID DIETS.)	Fat-free broth, bouillon, carbonated beverages, coffee, decaffeinated coffee, flavored gelatin, honey, popsicles, strained fruit juices, sugar, syrups (clear), and tea.	All others	Zero residue feeding used for (1) preparation for a barium enema and/or bowel surgery, (2) after surgery, (3) diarrhea, and (4) very weakened patients.	Supplies only a few calories (from the sugars in the allowed items) and some electrolytes (mainly sodium, chloride and potassium).
Tea and fat-free broth diet	Tea and fat-free broth (the doctor may allow the use of sugar).	All others	First feeding after surgery on the digestive tract.	Provides mainly water and little in the way of nutrients.
II. Modifications of caloric content				
High-calorie diet (Also see UNDERWEIGHT.)	The regular American diet with an emphasis on eating greater-than-usual quantities of carbohydrate-rich items (breads, desserts, pasta products, and sweets); fat-rich items (bacon, butter, cream, margarine, mayonnaise, oils, peanut butter, salad dressings, and sour cream); and protein-rich items (cheeses, custards, eggs, eggnogs, fish, meats, poultry, and undiluted evaporated milk).	None, but low-calorie foods should be replaced in part by calorie-rich foods.	Providing the energy needs of (1) hypermetabolic patients (in cancer, hyperthyroidism, severe burns, and trauma), (2) underweight people, (3) victims of malnutrition and malabsorption syndromes, and (4) very active people such as athletes, explorers, laborers, miners, and soldiers.	Many of the people who have high energy needs also have high protein requirements. Hence, extra protein sources should be given. The metabolic status of the patient should be considered in determining whether the extra calories should be supplied by carbohydrates, fats, and/or protein.

Diet	Composition		Uses	Comments
	Foods Included	**Foods Excluded**		
Low-calorie diet (Also see OBESITY.)	Low-calorie items that are high in protein, fiber, minerals, and vitamins, such as eggs, fish, lean meats, poultry, low-fat cheeses and milk products, low-carbohydrate fruits and vegetables, and whole grain and bran types of breads and cereals. (The last named group of foods is not much lower in calories than those made from refined flour, but it contains very filling items that reduce the desire to overeat.)	Those which supply mainly calories, but little else in the way of nutrients (alcoholic beverages, cakes, cookies, desserts, highly refined fatty, starchy, and sugary foods, and many of the common snack foods).	Weight reduction for overweight people. (Loss of weight may also help to correct other undesirable conditions such as high blood pressure, an impaired utilization of sugars, and elevations in the blood levels of cholesterol and triglycerides.)	Drastic restrictions in dietary calories may lead to great reductions in certain energy-rich foods that are good sources of minerals and vitamins. Therefore, nutritional supplements may be needed in these cases. Vigorous exercise may offset some of the need for dietary restrictions. People attempting to lose considerable weight should be examined by a doctor before going on a very low calorie diet.

III. Modifications in the intakes of specific nutrients

Diet	Foods Included	Foods Excluded	Uses	Comments
Carbohydrate restricted diet (Also see DUMPING SYNDROME; HYPO-GLYCEMIA; and OBESITY.)	All types of cheeses, eggs, meats, and poultry; milk products without added sweetener; nuts; and peanut butter.	Sweetened milk products.	Treatment of the dumping syndrome in patients who have had all or part of their stomach removed. Correction of elevated blood levels of triglycerides when used with concurrent restrictions of alcohol, cholesterol, and saturated fats. Stabilization of the blood sugar in people with a tendency to low blood sugar (hypoglycemia). Weight reduction of patients who are obese due to an oversecretion of insulin in response to dietary sugars. Management of some types of juvenile diabetes in which it is difficult to adjust the dosages of insulin, or in the control of adult onset diabetes.	Some patients may also require the restriction of all milk products after they have had stomach surgery. This diet is high in cholesterol, saturated fats, and fats in general and should not be used without further modifications for patients with high blood levels of lipids (hyperlipoproteinemias). The high protein content of this diet may be dangerous for people with certain kidney disorders. (Also see the ketogenic diet which is outlined in this table).
	Unsweetened fruit juices and fruits (limited to 3 exchanges a day).	Sweetened fruit juices and fruits.		
	Low-carbohydrate vegetables (limited to 2 exchanges per day).	Vegetables prepared with sweetener.		
	Unsweetened breads, cereals, pasta, and starchy vegetables (limited to 5 exchanges per day).	Breads and cereals with sugar, dates, or raisins (includes granola types).		
	Bacon, butter or margarine, cream (unsweetened), fats, French dressing, mayonnaise, shortenings, and vegetable oils.	Sweetened imitation cream toppings.		
	Soups made with allowed foods (starchy items limited as above).	Gravies thickened with cornstarch or flour.		
	Custards, gelatin desserts, and junket puddings made without a natural sweetener (an artificial sweetener may be used).	All desserts made with natural sweeteners.		
	Artificially sweetened carbonated beverages, coffee, herb teas, and tea (only artificial sweetener may be used in coffee and teas).	Alcoholic beverages, naturally sweetened carbonated beverages, sweetened coffee, imitation coffee creams, and sweetened ice tea.		
	Unsweetened condiments and spices.	Honey, jams, jellies, marmalade, molasses, sugar, and syrups.		
Carbohydrate, fiber, and fluid restricted diet	All of the foods listed for the soft diet, except the excluded items listed in the next column. Fluid intake should be restricted to 4 to 8 oz (120 to 240 ml) taken at each of 6 small meals per day. (The patient may be allowed to sip water or suck on cracked ice between meals.) Strained fruit juices should be diluted 1:1 with water. Only unsweetened cooked or canned fruits are allowed. The only allowed desserts are gelatin or baked custard made without added sugar (artificial sweeteners may be used).	Those excluded from the soft diet *plus* dried fruits, sweetened cooked or canned fruits, sweet potatoes, graham crackers, all types of sweetened desserts, all forms of sugar and sweets, all foods preserved with sugar or salt, and salt that is added at the table. (Some salt may be used in cooking.)	After gastrointestinal surgery when it is necessary to reduce the volume of fluids and foods in the stomach and intestines.	The amounts of food fluid served at each of the 6 daily meals should be increased cautiously in accordance with the patient's tolerance for food. Most nutrient requirements will be met if a variety of the allowed foods are consumed, and adequate amounts of food are eaten. However, a vitamin C supplement should be provided to compensate for the small amounts of fruits.
Fat restricted diet (Also see HYPERLIPO-PROTEINEMIAS.)	Canned fish (water pack), fresh or frozen fish (skinless and unbreaded), lean meats, and poultry prepared by baking, or broiling (the pieces may be wrapped in foil to retain the juices), panbroiling (in a nonstick frying pan without added fat), and roasting.	Fish canned in oil, fresh or frozen salmon, fish cakes and frozen fish sticks, bacon, ham, pork, duck, goose, sandwich meats and sausages, canned and frozen meat dishes.	Patients who show signs of an impairment in the clearing of bloodborne fats such as chylomicrons and triglycerides from the blood within a normal period after a fat-containing meal. The major signs of this condition (Type I Hyperlipoproteinemia) are (1) a creamy upper layer that forms on a sample of blood plasma left overnight in a refrigerator, (2) abnormal pain after the consumption of fat, and (3) visible deposits of fat under the skin (eruptive xanthomas).	This diet imposes a drastic restriction on the dietary fats which form chylomicrons. However, the fat-soluble vitamins (vitamins A, D, E, and K) are absorbed and transported in the blood along with these fats. Hence, deficiencies of these vitamins may result from the diet unless supplements containing water-soluble forms of the vitamins are provided. Sometimes, patients are allowed to use specially prepared fats called medium chain triglycer-
	Evaporated skim milk, fat-free buttermilk, nonfat dry milk, and skim milk.	Chocolate milk, cultured buttermilk, whole milk, and yogurt.		
	Fat-free cottage cheese, pot cheese, and other skim-milk cheeses.	All other cheeses		
	Medium to hard cooked, poached, or scrambled egg (in a nonstick pan without added fat). Limit of 1 whole egg per day, but whites are not limited.	Eggs cooked with added fat or omelettes or souffles with fatty foods.		

(Continued)

Diet	Composition		Uses	Comments
	Foods Included	**Foods Excluded**		
Fat restricted diet *(Continued)*	All fruits and their juices, except avocado and coconut.	Avocado and coconut.	Other uses are in disorders of the biliary tract and/or the pancreas in which dietary fats are digested poorly and likely to cause a fatty diarrhea (steatorrhea).	ides (MCT) because they do not accumulate in the blood.
	All canned, fresh, or frozen vegetables prepared without a cream sauce, an oil, a fat, or a fatty food (such as a rich cheese).	Creamed vegetables or those prepared with a fat or an oil.		
	Breads and other baked goods made with little or no fat.	All other breads and baked goods.		
	All cooked and dry cereals (served with skim milk).	None		
	Baked, boiled, or mashed (with skim milk) potatoes served without any fat; hominy (corn), pasta, or rice without a fatty sauce; and unbuttered popcorn.	Escalloped, creamed, or fried potatoes; potato chips, ovenbrowned potatoes; buttered popcorn; and egg noodles.		
	Lowfat desserts such as cake (angel food), cookies (arrowroot, ladyfingers, and vanilla wafers), crackers (graham), fruit whips made with egg whites or gelatin, gelatin desserts, meringues, puddings (cornstarch, junket, rice, and tapioca) made with skim milk and egg whites, and water ices.	All cakes, cookies, ice cream, pastries, puddings, and other desserts made with eggs, egg yolks, fats, and whole milk.		
	Hard candy, fondant, jelly beans, honey, jams, and other lowfat sweets.	Candy made with fats or nuts, chocolate candy or chocolate syrup.		
	Carbonated beverages, coffee, grain beverages, and tea.	Alcoholic beverages, chocolate drinks.		
	Condiments, pickles, spices, white sauce made with skim milk.	Fat-containing gravies and sauces.		
Cholesterol and saturated fat restricted diet (Also see HEART DISEASE; and HYPERLIPO-PROTEIN-EMIAS.)	A modified regular diet that emphasizes foods containing little or no cholesterol and/or saturated fat (typical items are breads, cereals, eggs whites, fish, fruits, lean muscle meats, legumes [beans and peas], lowfat cheeses [such as cottage], margarines made from liquid vegetable oils other than coconut, nonfat milk products, oils, peanut butter [nonhydrogenated], poultry breast meats, salad dressings made from vegetable oils, and vegetables). Also, it is usually recommended that polyunsaturated vegetable oil products replace most of the eliminated saturated fats.	The foods that follow are *not* excluded, but should be eaten only in limited amounts: bacon, brains, butter, coconut oil, cream, creamy or fatty cheeses, chicken fat, egg yolks, kidneys, lard, liver, margarines made from animal fats, mayonnaise, meat fats, rich ice creams, sandwich meats and sausages, shellfish, sour cream, and whole milk products.	To prevent or treat certain hyperlipoproteinemias (high blood levels of cholesterol and/or triglycerides) in patients who (1) have a family history of early cardiovascular and/or cerebrovascular diseases (tendencies to have heart attacks and/or strokes), (2) are overweight and have high blood fats, or (3) have an abnormally rapid clotting of the blood.	The replacement of dietary saturated fats with polyunsaturated vegetable oils is considered by some doctors to be potentially hazardous because it may (1) increase the risk of cancer, (2) aggravate gallbladder disease, and (3) produce a deficiency of vitamin E, unless supplements of the latter are taken. It is not certain whether this diet is of any benefit to people other than those who tend to accumulate fats in the blood. Hyperlipemic patients should be given a glucose tolerance test before being put on this diet (to rule out the possibility that their condition is induced by dietary carbohydrates).
High fiber diet (Also see FIBER.)	A modification of the diet consumed by most Americans, with an emphasis on consuming greater-than-usual amounts of milk products, fruits, vegetables, whole-grain and bran-type breads and cereals, and nuts.	None, but some or all of the highly refined carbohydrate foods should be replaced by their unrefined equivalents.	Stimulation of larger, more frequent bowel movements to correct or prevent chronic constipation and its consequences (diverticulitis, hemorrhoids, and narrowing of the colon diameter).	Dietary fiber should be increased gradually for people accustomed to low fiber diets in order to prevent flatulence and looseness of the bowels. Should *not* be used when the bowel is inflamed.
Fiber-restricted diet	All the foods for the soft diet, except the excluded items listed in the next column. Milk intake should be restricted to 2 cups per day.	Those excluded from the soft diet *plus* all fruit and vegetable products other than strained fruit and vegetable juices and cooked white potatoes without skin; popcorn and potato chips, and prune juice (even when strained it has a laxative effect).	When gastrointestinal passages such as the esophagus or the intestines are narrowed by strictures and there is danger of blockage. Severe inflammation of the bowel in diverticulitis, infectious enterocolitis, or ulcerative colitis. As an intermediate diet that is used after a full liquid diet, and before a soft or regular diet.	A low fiber diet should not be used any longer than necessary because it produces small infrequent stools. This condition may encourage (1) excessive constriction of the colon, (2) narrowing of its diameter, (3) herniation of the colonic muscles and the production of pouches (diverticula), and (4) general aggravation of bowel problems.

(Continued)

Diet	Composition		Uses	Comments
	Foods Included	**Foods Excluded**		
Nonfiber, milk-free diet	All of the foods listed for the soft diet, except for the items listed in the next column.	Those excluded from the soft diet *plus* all milk products and cheeses, all fruit products other than strained fruit juices, all vegetable products other than strained tomato juice, potatoes, and all spices and condiments other than salt.	After intestinal surgery or trauma, when it is necessary to reduce bowel movements to a minimum.	This diet is usually given for only a few days until healing occurs because small, infrequent stools may result in (1) excessive con striction of the colon, (2) narrowing of its diameter, and (3) general aggravation of bowel problems. A calcium supplement should be given.
High protein, high calorie diet	The normal American dietary pattern supplemented by extra servings of protein-rich foods such as cheeses, eggs, fish, meats, milk products, peanut butter, poultry, and special proprietary formulas when necessary.	None, although certain filling low protein foods such as fruits and vegetables may be replaced in part by high protein foods.	Patients recovering from burns, cancer therapy, injuries, hepatitis, malnutrition, surgery, and ulcerative colitis and /or other inflammatory conditions of the gastrointestinal tract, and underweight.	High protein diets often increase the urinary losses of calcium so that a calcium supplement may be advisable. Should not be used when protein metabolism or nitrogen excretion is impaired.
Controlled protein, potassium, sodium, and water diet	A daily food pattern that provides 30 g of protein and about 800 mg each of potassium and sodium follows: 1½ oz *(28 g)* of unsalted eggs, fish, liver or muscle meat, or poultry.	Brains; kidneys; salted eggs; fish, meats, or poultry; sandwich meats; and sausages.	To provide as much nourishment as possible to patients with advanced kidney disease while preventing the accumulation in the body of excessive amounts of nitrogenous wastes and minerals salts. However, patients who are being treated regularly with hemodialysis or peritoneal dialysis may usually be given more liberal diets.	This type of diet should not be used until it is absolutely necessary, because if given prematurely, the patient may tire of it by the time it is needed the most. The diet may not supply sufficient calories for all patients and may need to be supplemented with intravenous feeding. Also, a mineral and vitamin supplement is required when a patient is given the diet for more than a few days.
	½ cup *(120 ml)* of unsalted cottage cheese or ½ oz *(14 g)* of unsalted American cheese.	All other forms of cheese and milk products.		
	½ cup *(120 ml)* of a canned, fresh, or frozen non-citrus-type of fruit.	Citrus fruits and dried fruits.		
	1 cup *(240 ml)* of cranberry juice or reconstituted frozen lemonade or limeade.	All other fruit juices.		
	1 cup *(240 ml)* of salt-free or unsalted cooked or raw cabbage, canned carrots, canned green beans, canned peas, canned wax beans, fresh cucumber, or lettuce.	All vegetables prepared with salt, and all others not listed.		
	3 slices of enriched white bread or 5 saltine crackers.	All other breads and baked goods.		
	½ cup *(120 ml)* of cooked cream of wheat, farina, or hominy grits served without milk.	All other cereals.		
	1 cup *(240 ml)* of cooked enriched pasta or enriched white rice that has been prepared without salt.	All forms of sweet and white potatoes, potato chips, and brown rice.		
	3 tsp *(15 ml)* of butter, cream cheese, margarine, mayonnaise, or salad dressing; and unrestricted amounts of corn oil.	Cream gravies and all other fatty foods.		
	No soups.	All types of soups.		
	1 tsp *(5 ml)* of jam or jelly; moderate amounts of gum drops, hard candy, marshmallows, and sugar syrup (colorless-type).	All other types of sugars and sweets.		
	Moderate amounts of a homemade gelatin dessert prepared with sugar and artificial flavoring; dietetic gelatin; or lime ice.	Commercial gelatin desserts and all other types of desserts.		
	2 cups *(480 ml)* of ginger ale, or root beer.	All other beverages.		
	Unlimited amounts of cinnamon, mace, nutmeg, peppermint extract, vanilla, vinegar, and white pepper.	All other condiments, flavorings, herbs, and spices.		
Purine restricted diet	Cheeses; eggs; fishes and meats, except for those listed in the next column as excluded; and seafoods.	Brains, kidneys, liver, and sweetbreads; broths, extracts, gravies, sauces, and soups made from meats; fried meats; and anchovies, cavier, roe, and sardines.	Patients with high blood levels of uric acid (hyperuricemia), gout, and gouty arthritis.	A high fat diet may aggravate hyperuricemia. Hence, the amounts of dietary fats should be moderately restricted. The blood level of uric acid is also raised by the heavy consumption of alcoholic beverages and fasting. Vegetarian diets that are rich in microbically-fermented products, seeds, and yeast or yeast derivatives may furnish more
	All milk products, except that high fat products should be used in moderation.	None		
	All fruits and fruit juices.	None		
	Vegetables and vegetable juices as desired, except that cooked dried beans and peas should be limited to ½ cup *(120 ml)* per day.	None		

Diet	Composition		Uses	Comments
	Foods Included	**Foods Excluded**		
Purine restricted diet *(Continued)*	Most breads and cereals.	Wheat germ.		purines (uric-acid forming substances) per calorie than diets which contain fish, meats, and poultry.
	Most forms of sweet and white potatoes; and all pasta and rice products.	Fried potatoes and potato chips.		
	All types of fats when used in moderation.	None		
	Desserts that contain only small to moderate amounts of fats, such as custards made with skim milk, gelatin desserts, lowfat cakes, cookies, ice cream, and puddings.	High fat items such as rich cakes, cookies, ice cream, mince meat, pastries, and whipped cream or imitation whipped toppings.		
	Carbonated beverages and cereal beverages (coffee substitutes).	Alcoholic beverages, chocolate, cocoa, coffee, and tea.		
	Condiments, herbs, nuts, olives, peanut butter, pickles, popcorn, relishes, salt, spices, vinegar, and white sauce.	Fatty broths, extracts, gravies, sauces, and soups; and yeast.		
Ketogenic diet	This diet is very low in carbohydrates, high in fat content, and contains moderate amounts of protein. It is characterized by (1) the use of large and carefully calculated amounts of bacon, butter, cheese, cream, eggs, fish, meats, and poultry; (2) minimal amounts of low carbohydrate fruits and vegetables; and (3) exclusion of foods of moderate to high carbohydrate contents (such as those listed in the next column).	Beverages that contain sugar, breads and cereals, desserts other than small amounts of fruits with plenty of whipped cream, fruits and vegetables of moderate to high carbohydrate contents, milk products other than butter and cream, and all forms of sweets.	Prevention of seizures in epileptic children by inducing and maintaining a state of ketosis (the presence of ketone bodies in the blood and urine). Sometimes, this type of diet is used to induce a rapid loss of body weight in patients who find it difficult to reduce by other means. (The weight loss is largely made up of lost fluid.)	Some patients on this diet are prone to abdominal cramps, nausea, and vomiting. Small, frequent feedings are tolerated better than a few large meals per day. A special fat preparation called medium chain triglycerides is often given to patients on these diets.
Gluten-free diet	All types of cheeses, eggs, fish, meats, and poultry that are *not* breaded, creamed, extended, or served with a gravy made from barley, oats, rye, or wheat; sandwich meats and sausages if labeled as "pure meat."	Cheese, egg, fish, meat or poultry mixtures such as chili, croquettes, meat loaf, and canned products that are likely to contain flour.	The life-long diet of patients who have an intolerance to foods that contain the cereal protein gluten.	Sometimes, adults who have no previous history of gluten sensitivity have developed this condition after surgery on the digestive tract. (It has been suggested that any condition which allows incompletely digested gluten-containing proteins to be absorbed may lead to the development of a sensitivity.)
	Milk products that are not likely to contain cereal ingredients.	Chocolate and malted milk products that contain cereals or flours.		
	All fruit juices and fruits.	None		
	All vegetable juices, vegetables, except those listed in the next column as excluded.	Items prepared with bread, bread crumbs, or wheat flour.		
	Baked goods, breads and cereals made from corn, potato, rice, soybean, and/or tapioca flours or meals only.	Products containing barley, oat, rye, or wheat flour or meal.		
	Sweet or white potatoes other than those prepared by breading or flouring.	Macaroni, noodles, and spaghetti.		
	All types of fat except those listed in the next column as excluded.	All commercial salad dressings except pure mayonnaise.		
	Clear broths, consomme, and soups that contain no barley, oats, rye, or wheat; and cream soups that are thickened only with cornstarch or potato flour.	All canned soups other than clear broth; and all creamed soups that are thickened with the excluded grain products.		
	Gluten-free desserts such as custards, fruit products made without flour, gelatin desserts, homemade ice cream, ices, and sherbets, and junket, rice and other puddings thickened only with cornstarch, potato flour, or tapioca.	Cakes, cookies, ice cream containing a stabilizer made from wheat, ice cream cones, commercial baking mixes other than those labeled "gluten-free," and commercial puddings.		
	Corn syrup, honey, jams, and jellies, molasses, and brown and white sugars.	Commercial candies that contain the excluded grain products.		
Lactose-free diet	All types of eggs, fish, meats, nuts, and poultry without added lactose or milk.	Creamed or breaded preparations, sandwich meats and sausages unless labeled "all meat," and omelets and souffles made with cheeses or milk.	For people who cannot digest milk sugar (lactose) and suffer from unpleasant symptoms when lactose-containing foods are consumed.	It is believed that a large percentage of the world's people may have mild to severe symptoms of lactose intolerance (abdominal cramps,
	Nondairy creamers and similar lactose-free substitutes for milk.	All forms of milks, cheeses, ice creams, malted milks, etc.		

(Continued)

Diet	Composition		Uses	Comments
	Foods Included	**Foods Excluded**		
Lactose-free diet *(Continued)*	All fruits and vegetables and their juices that have been prepared without cheeses, cream sauces, lactose, or other milk products.	Fruit and vegetable products prepared with lactose or a milk product (such as breaded, buttered, or creamed items, corn curls, frozen french fries, and instant potatoes.)	In many cases the intolerance is congenital or hereditary, but it may also occur spontaneously in people who showed no prior indications of the condition. Usually, acquired lactose intolerance occurs as a result of injury to the intestinal lining in celiac disease (gluten intolerance), gastointestinal milk protein allergy, irritable bowel syndrome, kwashiorkor, malnutrition, regional enteritis (Crohn's disease), ulcerative colitis, viral enteritis; and after some types of gastrointestinal surgery.	diarrhea, distention, gas, and malabsorption of nutrients) after the consumption of fairly small amounts of milk (as little as ¼ cup or *60 ml*. However, some types of fermented milks contain little or no unfermented lactose and do not produce the symptoms. (Many of the brands of yogurt sold in the U.S. contain added nonfermented skim milk solids.) Patients should be taught to scrutinize food labels for the presence of butter, casein, cheese, cream, lactose, milk, or whey. A commercial enzyme preparation that digests lactose in milk to harmless products is now available.
	Breads, cereals, and pasta made without lactose or milk.	Breads and other baked goods made from prepared mixes that contain lactose or milk products, cereals, and pasta made from milk.		
	Dressings, margarines, and shortenings made without any milk products; bacon; meats fats; nut butters that are lactose-free; oils, and certain brands of nondairy creamers and whipped toppings that contain no milk derivatives.	Dressings, margarines, certain nondairy creamers, peanut butter, and whipped toppings made with butter, cheese, cream lactose, or other milk products.		
	Broths, consomme, cream soups, and soups without milk products.	Items containing lactose or milk.		
	Candies, cakes, cookies, pies, and puddings made without milk.	Dessert items that contain milk.		
	Alcoholic and carbonated beverages, coffee, milk substitutes, tea.	Hot chocolate and some types of cocoa.		
	Condiments, jams, jellies, marmalades, spices, and sweeteners.	Artificial sweeteners that contain lactose.		
Calcium restricted diet	Most common foods, except those that are rich in calcium and are listed in the excluded list in the next column. However, meats are limited to 6 oz *(170 g)* per day, eggs to 1 per day, nonexcluded vegetables to 1 cup *(240 ml)* per day, and potatoes (prepared without milk) to 1 cup *(240 ml)* per day.	All milk products (except butter); canned mackerel, salmon, sardines, shellfish; foods that contain milk; baking powders and products made from them; green leafy vegetables; and molasses.	When the blood calcium of a patient is excessively high, and in certain tests of calcium metabolism. (Usually, only a few days of dietary calcium restriction will bring the blood calcium down to normal.)	This diet is *no* longer used to prevent the formation of calcium-containing kidney stones because keeping the urine acidic (by feeding cranberry juice or other means) works better.
Phosphorus restricted diet	3 oz *(85 g)* of cod, crab, haddock, lobster, and oysters; beef, bologna, corned beef, frankfurters, ham, heart, pork, tongue; and chicken, and duck.	Other fish and shellfish; bacon, brains, lamb, liver, turkey, and veal.	In hyperphosphatemia (excessively high blood levels of phosphate that occurs in hyperparathyroidism) and certain kidney diseases. Sometimes, a diet that is low in both phosphorus and calcium is given to prevent or treat calcium phosphate kidney stones.	When a rapid reduction in blood phosphate is essential to the patient's health, the diet is used in conjunction with the oral dosage of aluminum hydroxide antacids that bind with phosphate and prevent its absorption.
	½ cup *(120 ml)* of heavy cream (35% butterfat).	All other milk products		
	No cheeses	All cheeses		
	No eggs	All forms of eggs		
	3 cups *(720 ml)* of any canned, fresh, or frozen fruit or juice other than those listed as excluded in the next column.	Avocado, banana, blackberries; and all dried fruits.		
	1 ½ cups *(360 ml)* beets (roots are permitted, but not the leaves), cabbage, carrots, chicory, cucumbers, green bean, green peppers, lettuce, sauerkraut, spinach, squash, and tomatoes.	All dried beans and peas, and beet greens, broccoli, Brussels sprouts, and corn.		
	5 slices of enriched white bread or the equivalent in crackers made from enriched white flour.	Breads made from rye, or whole or cracked wheat, and all items made with baking powder.		
	½ cup *(120 ml)* of a dry or a cooked cereal made from refined grain.	All cereals made from whole or cracked (but not fully refined) grains.		
	½ cup *(120 ml)* of sweet or white potato; pasta made from refined white flour and white rice.	No milk products except the cream allowance may be used in dressings or sauces.		
	8 tsp *(120 ml)* of butter, margarine.	All other types of fats and salad dressings.		
	3 tsp *(15 g)* of white sugar, or 3 Tbsp *(45 ml)* of honey, jam, or jelly; and /or hard candy, drops, fondant, and mints as desired.	Candies that contain chocolate, cream or milk, dried fruits, or nuts; and all other desserts.		
	Unlimited coffee and tea	All other beverages		
	Unlimited olives and pickles	All other condiments and snacks		
High-potassium diet	A regular diet that emphasizes the consumption of potassium-rich foods such as those which follow: milk, citrus fruits and juices, apricots, bananas, cantaloupe, dates, figs, prunes, raisins, cooked dry beans and peas, green leafy vegetables, tomatoes and tomato juice, whole-grain and bran-type breads and cereals, nuts, catsup, chocolate, cocoa, herbs, spices, and sauces.	None. However, it would be wise to substitute the high-potassium foods in each food group (such as fruits and vegetables) for the low-potassium items in each group.	Patients who have become potassium depleted as a result of diarrhea, malnutrition, stresses, surgery, and treatment with certain diuretics.	When there is only mild to moderate potassium depletion it is generally safer to supply it in foods than as pure potassium salts which may cause gastrointestinal irritation. Potassium-containing salt substitutes should be used cautiously, and only after a doctor has authorized their use.

Diet	Composition		Uses	Comments
	Foods Included	**Foods Excluded**		
Sodium restricted diet	The plan for a moderately restricted diet (800 to 1,000 mg of sodium per day) follows: 10 oz (*284 g*) of unsalted canned, fresh, or frozen fish, meats, or poultry; except the types listed as excluded in the next column. (Liver may be served only once every 2 weeks.)	All types of fish; cured meats such as bacon and ham; and poultry that contains added salt. (Also see Sodium, Table S-6.)	Patients with (1) fluid and sodium retention that is associated with congestive heart failure, liver disease such as cirrhosis, and certain types of kidney diseases; and (2) high blood pressure (hypertension).	Careful adherence to a moderately low salt diet may reduce or eliminate the need for some people to take diuretics and other medications. Many low salt foods are now available in health food stores, large supermarkets, and pharmacies. (A pharmacist can prepare a low sodium baking powder if a commercial product is not available.)
	3 cups (*720 ml*) of skim milk, unsalted buttermilk, and/or whole milk; or the equivalent in evaporated or nonfat dry milk.	Condensed milk, cultured buttermilk, and yogurt; all milk drinks containing chocolate syrup, ice cream, or a malt product.		
	1 oz (*28 g*) of unsalted American cheese, unsalted cottage cheese, or unsalted cream cheese.	All other cheeses		
	2 eggs, boiled, poached, scrambled, or fried in unsalted butter.	None		
	1½ cups (*360 ml*) of canned, fresh, or frozen fruit juice(s) and fruits, one of which should be citrus.	Candied fruits, dried figs, or raisins; tomato juice; all fruit drinks, juices, and other products to which sodium coloring, flavoring, or preservatives have been added.		
	1½ cups (*360 ml*) of salt-free canned, fresh, and frozen vegetables except those listed as excluded in the next column; dried lima beans, lentils, soybeans, and split peas, prepared without salt, salted pork, or a salty sauce.	Canned vegetables and vegetable juices that contain added salt; and frozen vegetables processed with salt.		
	6 slices of unsalted yeast bread, quick breads made with low sodium baking powder, and unsalted crackers. (Salt-containing yeast may be permitted, but not salt sticks.)	Products prepared with salt and/or baking soda that contains sodium.		
	½ cup (*120 ml*) of an unsalted, slow-cooking cereal such as barley, buckwheat groats (kasha), corn-meal, and hominy grits; unsalted dry cereals such as puffed rice and puffed wheat and some brands of shredded wheat (unsalted corn flakes are available in some health food stores).	Quick-cooking and enriched hot cereals; and dry cereals that contain salt.		
	1 cup (*240 ml*) of pasta, potatoes, and/or rice cooked without salt.	Potato chips and prepared potato products.		
	6 tsp (*30 ml*) of unsalted butter, cream, margarine, and homemade salt-free mayonnaise and salad dressings (commercial salt-free products are sold in some health food stores); cooking oils, salt-free shortenings and gravies.	Bacon, pork fat, salted butter and mayonnaise and salad dressings unless labeled "salt-free".		
	1 cup (*240 ml*) of unsalted broth, consomme, cream soup (made with part of the day's allowances for butter and milk), and other soups made from allowed foods listed in this column.	All canned, dehydrated, and frozen broths, bouillon cubes, consomme, extracts and soups that contain salt.		
	Moderate amounts of hard candy, honey, jams and jellies made without sodium benzoate, maple syrup, and brown and white types of sugar.	Chocolate syrups, commercial soft candies or syrups, molasses, and artificial sweeteners that contain sodium.		
	½ cup (*120 ml*) of unsalted desserts such as custards (made with part of the day's allowances for eggs and milk), fruit ices, gelatin desserts made with fresh fruit juices, ice cream and sherbets (commercial types must be deducted from the day's milk allowance), and salt-free puddings (cornstarch, rice, and tapioca).	All commercial cakes, cookies, gelatin desserts, chocolate junket desserts, pies, and puddings unless labeled "salt-free"; and all other desserts made with baking powder, soda, or salt.		
	Unlimited cocoa (milk deducted from day's allowance), coffee, decaffeinated coffee, fruit drinks without sodium benzoate or an artificial sweetener. Limit of 1 cup of carbonated drink.	Commercial chocolate drinks, diet soft drinks, ginger ale, and instant cocoa.		
	All seasonings, spices, and unsalted condiments except those excluded; unsalted nuts and popcorn; and low sodium catsup.	Salted condiments, seasonings, snacks, and spices; celery seed and dried celery leaves.		

Use of Exchange Lists in Planning Diets. Table M-28 shows that many types of modified diets are based upon the selection of certain allowed foods and the exclusion of others. However, it is often necessary to control the composition of diets quantitatively in regard to the levels of protein, carbohydrates, fats, and calories. In the early days of prescribing diets for diabetics this was done by stipulating the weights of the foods that were allowed. The preparation of meals for these patients was a tedious task for homemakers, who had to control their recipes carefully and weigh each portion of food that was served. This laborious task was simplified greatly in 1950, when the food exchange lists prepared by a joint committee of the American Dietetic Association, American Diabetes Association, and the U.S. Public Health Service were published. A revised version of this system is outlined in Table M-29.

TABLE M-29
FOOD EXCHANGE LISTS GROUPED ACCORDING TO CALORIC CONTENT AND SIZE OF PORTION[1]

Name and Description of Exchange	Low-Calorie Items	Medium-Calorie Items	High-Calorie Items
Meat and protein-rich meat substitutes 1 oz (*28 g*) of cooked fish, meat, or poultry; or the protein equivalent in cheese, eggs, cooked dried beans or peas, peanut butter, or a vegetable protein product.	***Lean meats and other protein-rich foods:*** 1 exchange furnishes about 7 g of protein, 3 g of fat, and 55 kcal. Baby beef, chipped beef, chuck, flank steak, tenderloin, plate ribs, plate skirt steak, round (bottom, top), all cuts rump, spare ribs, tripe 1 oz (*28 g*) Leg, rib, sirloin, loin, shank, shoulder of lamb .... 1 oz (*28 g*) Leg (whole rump, center shank) of pork, smoked ham (center slices) .... 1 oz (*28 g*) Leg, loin, rib, shank, shoulder, cutlets of veal .... 1 oz (*28 g*) Meat without skin of chicken, turkey, Cornish hen, guinea hen, pheasant .... 1 oz (*28 g*) Any fresh or frozen fish .... 1 oz (*28 g*) Canned crab, lobster, mackerel, salmon, and tuna .... 1 oz (*28 g*) Clams, oysters, scallops, shrimp 5 or 1 oz (*28 g*) Sardines, drained .... 3 Cheeses containing less than 5% fat .. 1 oz (*28 g*) Cottage cheese, dry and 2% fat ... ¼ c (*60 ml*) Cooked dried peas and beans (omit 1 bread exchange) .... ½ c (*120 ml*)	***Medium fat meats and other protein-rich foods:*** 1 exchange furnishes about 7 g of protein, 5 g of fat, and 75 kcal (*omit ½ fat exchange*).[2] Canned beef, corned beef, ground (15% fat), ground round, rib eye .... 1 oz (*28 g*) Loin of pork (all cuts), tenderloin, shoulder arm shoulder blade (Boston butt), Canadian bacon, boiled ham, loin, picnic ham, shoulder 1 oz (*28 g*) Liver, heart, kidney, sweetbreads ... 1 oz (*28 g*) Cheese (farmer's, mozzarella, Neufchatel, ricotta) .... 1 oz (*28 g*) Cheese, Parmesan .... 3 Tbsp (*45 ml*) Egg .... 1 Peanut butter (omit 2 fat exchanges) . 2 Tbsp (*30 ml*)	***High fat meats and other protein-rich foods:*** 1 exchange furnishes about 7 g of protein, 8 g of fat, and 100 kcal (*omit 1 fat exchange*).[2] Brisket of beef, corned beef (brisket), ground beef (more than 20% fat), ground chuck, hamburger, rib roasts, club and rib steaks ..... 1 oz (*28 g*) Breast of lamb .... 1 oz (*28 g*) Ground pork loin (back ribs), spare ribs, country-style ham, deviled ham .... 1 oz (*28 g*) Breast of veal .... 1 oz (*28 g*) Capon, duck, goose .... 1 oz (*28 g*) Cheddar cheese and similar types .... 1 oz (*28 g*) Cold cuts—4½" (*113 mm*) diameter x ⅛" (*3 mm*) thick .... 1 slice Frankfurter 1
Milk and milk products 1 c (*240 ml*) or the equivalent in evaporated or nonfat dry milk.	***Nonfat items:*** 1 exchange furnishes about 8 g of protein, 12 g of carbohydrate, a trace of fat, and 80 kcal. Fluid nonfat or skim milk ..... 1 c (*240 ml*) Buttermilk made from skim milk ... 1 c (*240 ml*) Plain (unflavored) yogurt made with skim milk .... 1 c (*240 ml*) Canned evaporated skim milk .... ½ c (*120 ml*) Nonfat dry (powdered skim) milk .. ⅓ c (*80 ml*)	***Low fat items:*** 1 exchange furnishes about 8 g of protein, 12 g of carbohydrate, 5 g of fat, and 125 kcal (*omit 1 fat exchange*). Lowfat (2% fat) fluid milk .... 1 c (*240 ml*)	***Medium fat items:*** 1 exchange furnishes about 8 g of protein, 12 g of carbohydrate, 8.5 g of fat, and 160 kcal (*omit 2 fat exchanges*). Buttermilk made from whole milk .. 1 c (*240 ml*) Canned evaporated whole milk .... ½ c (*120 ml*) Fluid whole milk .... 1 c (*240 ml*) Yogurt made from whole milk .. 1 c (*240 ml*)
Fruits and fruit juices Each of the designated portions furnishes 10 g of carbohydrate and 40 kcal.	***Large portions:*** 4½ to 7 oz (*130 to 200 g*) Cantaloupe, 6" (*152 mm*) diameter .... ¼ Grapefruit, small .... ½ Honeydew melon 7" (*178 mm*) diameter .. ⅛ Raspberries .... ⅔ c (*160 ml*) Strawberries .... ¾ c (*180 ml*) Watermelon .... 1 c (*240 ml*)	***Medium portions:*** 2¼ to 4½ oz (*65 to 129 g*) Apple juice or apple cider .... ⅓ c (*80 ml*) Apple, 2" (*51 mm*) diameter .... 1 Applesauce, unsweetened .... ½ c (*120 ml*) Apricots, fresh, medium .... 2 Apricot nectar .... ⅓ c (*80 ml*) Blackberries .... ½ c (*120 ml*) Blueberries .... ½ c (*120 ml*) Fruit cocktail .... ½ c (*120 ml*) Grapefruit juice .... ½ c (*120 ml*) Grapes .... 12 Mango, small .... ½ Nectarine, small .... 1 Orange, small .... 1 Orange-apricot nectar .... ⅓ c (*80 ml*) Orange juice .... ½ c (*120 ml*) Papaya, medium .... ⅓ Peach, medium 1	***Small portions:*** ½ to 2¼ oz (*15 to 64 g*) Apricots, dried .... 4 halves Banana, small .... ½ Dates .... 2 Fig, dried, small .... 1 Fig, fresh, large .... 1 Grape juice .... ¼ c (*60 ml*) Prunes, dried, medium .... 2 Prune juice .... ¼ c (*60 ml*) Raisins .... 2 Tbsp (*30 ml*) Peach nectar .... ⅓ c (*80 ml*) Pear, small .... 1 Pear nectar .... ⅓ c (*80 ml*) Pineapple .... ½ c (*120 ml*) Pineapple juice .... ⅓ c (*80 ml*) Plums, medium .... 2 Tangerine, medium 1

Footnotes at end of table

(Continued)

TABLE M-29 *(Continued)*

Name and Description of Exchange	Low-Calorie Items	Medium-Calorie Items	High-Calorie Items
Vegetables and vegetable juices (The caloric contents of these items are dependent upon the carbohydrate contents. Hence, the low-calorie items are also low-carbohydrate items, and vice versa.)	***Free items to be used as desired when consumed raw:*** (If cooked, 1 c [*240 ml*] may be used without counting it, but the second cup of the cooked item should be counted as furnishing 2 g of protein, 5 g of carbohydrates, and 25 kcal.) Celery, Chicory, Chinese cabbage, Cucumbers, Endive, Escarole, Lettuce, Watercress. Asparagus, Bean Sprouts, Beets, Broccoli, Brussels sprouts, Cabbage, Carrots, Cauliflower, Eggplant, Green pepper	***Counted items:*** Each ½ c of the cooked or raw items (or their corresponding juices) that follow are to be counted as furnishing 2 g of protein, 5 g of carbohydrates, and 25 kcal. Greens: Beet, Chard, Collard, Dandelion, Kale, Mustard, Spinach, Turnip, Mushrooms, Okra, Onions, Radishes. Rhubarb, Rutabaga, Sauerkraut, String beans, green or wax, Summer squash, Tomatoes, Tomato juice, Turnips, Vegetable juice cocktail, Zucchini	***Starchy vegetables:*** 1 exchange furnishes 2 g of protein, 15 g of carbohydrates, and 70 kcal. Baked beans, no pork (canned) ..... ¼ c (*60 ml*) Cooked dried beans, lentils, peas .... ½ c (*120 ml*) Corn kernels, regular or cream-style .. ⅓ c (*80 ml*) Lima beans .... ½ c (*120 ml*) Parsnips .... ⅔ c (*160 ml*) Peas, green (canned or frozen) .... ½ c (*120 ml*) Potato, white, small .... 1 Pumpkin .... ¾ c (*180 ml*) Winter squash, acorn or butternut ... ½ c (*120 ml*) Yam or sweet potato .... ¼ c (*60 ml*)
Breads and cereals	***Low-fat items:*** 1 exchange furnishes 2 g of protein, 15 g of carbohydrate, and 70 kcal. Bread (French, Italian, pumpernickel, raisin, rye, white, whole wheat) .... 1 slice Bread crumbs, dry .... 3 Tbsp (*45 ml*) English muffin, small .... ½ Frankfurter roll .... ½ Hamburger bun .... ½ Pancake 5" (*127 mm*) diameter .... 1 Roll, plain, small .... 1 Tortilla 6" (*152 mm*) diameter .... 1 Waffle 5" (*127 mm*) square .... 1 Cooked cereals .... ½ c (*120 ml*) Ready-to-eat cereals: Bran flakes .... ½ c (*120 ml*) Others (unsweetened) .... ¾ c (*180 ml*) Puffed grains .... 1 c (*240 ml*) Pastas, cooked .... ½ c (*120 ml*) Flour .... 2½ Tbsp (*38 ml*) Wheat germ .... ¼ c (*60 ml*)	***Medium-fat items:*** 1 exchange furnishes 2 g of protein, 15 g of carbohydrate, 5 g of fat, and 115 kcal (*omit 1 fat exchange*). Biscuit, 2" (*51 mm*) diameter .... 1 Corn muffin, 2" (*51 mm*) diameter .... 1 Crackers, round butter-type .... 5 Muffin, plain, small .... 1 Potatoes, french fried, 2 to 3½" (*51 to 89 mm*) long .... 8 Crackers: Arrowroot .... 3 Graham 2½" (*64 mm*) square .... 2 Matzoh 4" x 6" (*102 x 152 mm*) ...... ½ Oyster .... 20 Pretzels 3⅛" (*80 mm*) long x ⅛" (*3 mm*) diameter .... 25 Rye wafers 2" (*51 mm*) x 3½" (*89 mm*) .... 3 Saltines .... 6 Soda 2½" (*64 mm*) square .... 4 Sponge cake, plain, 1½" (*38 mm*) cube .. 1 Vanilla wafers 5	***High-fat items:*** 1 exchange furnishes 2 g of protein, 15 g carbohydrate, 10 g of fat, and 160 kcal (*omit 2 fat exchanges*). Corn or potato chips .... 15 Ice cream .... ½ c (*120 ml*)
Fats Each of the designated portions furnishes 5 g of fat and 45 kcal.	***Least concentrated sources:*** 2 to 3 Tbsp (*30 to 45 g*). Avocado 4" (*102 mm*) diameter .... ⅛ Cream, light, sweet or sour ...... 2 Tbsp (*30 ml*) Half and half .... 3 Tbsp (*45 ml*) Olives, green, medium 3	***Intermediate sources:*** 1 to 2 Tbsp (*15 to 30 g*). Cream, heavy .... 1 Tbsp (*15 ml*) Cream cheese .... 1 Tbsp (*15 ml*) French dressing .... 1 Tbsp (*15 ml*)	***Most concentrated sources:*** 1 to 3 tsp (*5 to 15 g*). Bacon, crisp .... 1 strip Butter, lard, margarine, mayonnaise, meat fat (melted), oil, shortening ...... 1 tsp (*5 ml*) Nuts, shelled, small .... 6 Salad dressing, mayonnaise type .... 2 tsp (*10 ml*)

[1]Adapted by the authors from *Exchange Lists for Meal Planning*, The American Dietetic Association and the American Diabetes Association, in cooperation with the National Institute of Arthritis, Metabolism and Digestive Diseases, and the National Heart, Blood and Lung Institute, Public Health Service, U.S. Dept. HEW, 1976.

[2]Most diets are based upon the use of lean meats and nonfat milk products. Hence, the use of items richer in fat must be compensated for by the deduction of fat exchanges from the total allowed.

The use of the food exchange lists presented in Table M-29 is illustrated by the sample calculations shown in Table M-30 and the typical menu in Table M-31.

The bases of the calculations shown in Table M-30 and the typical daily menu in Table M-31 are as follows:

1. A 48-year-old businessman who is 6 ft, 2 in. (*188 cm*) tall, is of medium build, and weighs 198 lb (*90 kg*), has been diagnosed recently as an adult onset diabetic who does not require insulin injections but would benefit from a reduction of his body weight to one that is more desirable for his age, height, and condition.

2. The ideal weight of the patient is 176 lb, or 80 kg (from a standard life insurance table) and the caloric intake for a sedentary person at that weight should be 176 lb x 13 Calories (kcal) per pound (*28.6 kcal/kg*), or 2,288 kcal per day. (This allowance may be rounded to 2,300 kcal per day.)

3. It is decided that the weight reduction of the patient should be gradual (about 1 lb per week). Hence, his dietary prescription is for 1,800 kcal per day, distributed as follows: 20% from proteins, 45% from carbohydrates, and 35% from fats. (A caloric deficit of 500 kcal per day equals a weekly deficit of 3,500 kcal, or the amount needed to lose 1 lb [*0.45 kg*] of body weight.) The daily allowances (in grams) of proteins, carbohydrates, and fats are determined by dividing the caloric factors for nutrients (proteins, 4 kcal/g; carbohydrates, 4 kcal/g; and fats, 9 kcal/g) into the kcal contributed by each nutrient.

TABLE M-30
CALCULATION OF FOOD EXCHANGES FOR A PRESCRIBED DIET

Prescription: 1,800 kcal 20% protein (*90 g*), 45% carbohydrate (*202 g*), 35% fat (*70 g*)

Exchange	No. of Exchanges	Protein, g	Carbohydrate, g	Fat, g	Calories (kcal)
Meat, lean	3	(× 7) = 21	—	(× 3) = 9	(× 55) = 165
Meat, medium fat	3	(× 7) = 21	—	(× 5) = 15	(× 75) = 225
Running subtotal		(42)		(24)	(390)
Milk, lowfat	2	(× 8) = 16	(× 12) = 24	(× 5) = 10	(× 125) = 250
Milk, medium fat	1	(× 8) = 8	(× 12) = 12	(× 8.5) = 8.5	(× 160) = 160
Running subtotal		(66)	(36)	(42.5)	(800)
Fruit	3	—	(× 10) = 30	—	(× 40) = 120
Running subtotal		(66)	(66)	(42.5)	(920)
Vegetables, free items	unlimited	—	—	—	—
Vegetables, counted	2	(× 2) = 4	(× 5) = 10	—	(× 25) = 50
Vegetables, starchy	2	(× 2) = 4	(× 15) = 30	—	(× 70) = 140
Running subtotal		(74)	(106)	(42.5)	(1,110)
Bread, low fat	3	(× 2) = 6	(× 15) = 45	—	(× 70) = 210
Bread, medium fat	2	(× 2) = 4	(× 15) = 30	(× 5) = 10	(× 115) = 230
Bread, high fat	1	(× 2) = 2	(× 15) = 15	(× 10) = 10	(× 160) = 160
Running subtotal		(86)	(196)	(62.5)	(1,710)
Fats	2	—	—	(× 5) = 10	(× 45) = 90
Total		86	196	72.5	1,800
× calorie factor		× 4	× 4	× 9	
Calories		344	+ 784	+ 652.5 =	1,781

$$\text{Proteins} = \frac{1{,}800 \text{ kcal} \times 20\%}{4 \text{ kcal/g}} = 90 \text{ g}$$

$$\text{Carbohydrates} = \frac{1{,}800 \text{ kcal} \times 45\%}{4 \text{ kcal/g}} = 202 \text{ g}$$

$$\text{Fats} = \frac{1{,}800 \text{ kcal} \times 35\%}{9 \text{ kcal/g}} = 70 \text{ g}$$

4. The allowed food exchanges are worked out by (1) considering the meat and milk items to have the highest priorities, and (2) keeping a running count of the dietary content of proteins, carbohydrates, fats, and calories as the various exchanges are added to the diet. This calculation is simplified by the tabular format of Table M-30 in which the values shown in parentheses represent the nutrient values of the exchanges listed in the column on the left margin. The values shown after the equal signs were obtained by multiplying the nutrient values for the exchanges by the number of exchanges selected. It is noteworthy that the patient was allowed some medium fat exchanges for meat, milk, and bread products because he has to eat his lunch in a fast food restaurant.

5. After the bread exchanges were added to the diet, the running subtotal showed that only two separate fat exchanges could be added. Then, a check of the dietary totals of proteins, carbohydrates, fats, and calories showed that they were reasonably close to those in the dietary prescription.

6. The menu shows how the allowed exchanges may be used in meals. Although it may not always be possible to adhere perfectly to the plan, lapses on certain days should be noted carefully in terms of the allowances that were exceeded and they should be compensated for in the days that follow the deviations. For example, extra fat taken on one day should be compensated for by an equivalent deduction the next day, such as the substitution of a lowfat exchange for an allowed medium fat exchange.

Special Dietetic Foods. Many types of special dietary products are now available for people who must limit their consumption of substances present in ordinary foods. For example, there are artificially sweetened jams, jellies, salad dressings, sauces, and soft drinks. The term *dietetic* does not always mean that the product contains an artificial sweetener. Rather, it may mean that considerably less than the normal amounts of natural sweetener were used in the recipe. It may also be used to designate lowfat varieties of items such as salad dressings. Also, both dry and liquid forms of low-cholesterol egg products are sold in most supermarkets. However, it is best for anyone who has been placed on a modified diet to consult his or her doctor or dietitian before using any special dietetic foods because use of the new product may result in a significant alteration in the nutritional value of the prescribed diet.

SUMMARY. The proper types of modified diets may be very useful in the treatment of people with various disorders or undesirable conditions such as overweight or underweight. Nevertheless, the combined expertise of a doctor and a dietitian are usually needed to determine the dietary patterns most suitable for individual patients.

A TYPICAL DAILY 1,800-CALORIE MENU THAT WAS PLANNED FROM FOOD EXCHANGE LISTS[1]

Type of Food Exchange	No. of Exchanges	Typical Foods in Designated Exchange
Breakfast:		
Meat, medium fat . .	1	An egg, a 1 oz-slice of boiled ham, Canadian bacon, or 1 lamb kidney (cooked without fat, unless 1 of the allowed fat exchanges is used).
Milk, low fat (2%) . .	1	Taken as a beverage, in coffee or tea, and/or on cereal.
Fruit	1	Glass (4 oz, or *120 ml*) of fruit juice, or a serving of a fruit.
Bread, low fat	2	Bread and/or cereal.
Unrestricted items . .	—	Coffee, cereal beverage (coffee substitute), dietetic (artificially sweetened) soft drink, herb tea, or tea without a natural sweetener (unless ⅓ bread exchange is deducted for each tsp of sweetener used during the day).
Midmorning:		
Unrestricted items . .	—	Coffee, cereal beverage, dietetic soft drink, herb tea, etc. (See Breakfast.)
Bread, if *not* used at breakfast	1	Bread, cake, cookies, or crackers.
Milk, if *not* used at breakfast	1	Part of the allowed medium fat milk exchange (whole or evaporated milk) may be used in the unrestricted beverage.
Lunch:		
Meat, medium fat . .	2	2 slices (2 oz, or *56 g*) of broiled or roasted (not fried) meat.
Fruit	1	Glass (4 oz, or *120 ml*) of fruit juice, or a serving of a fruit.
Vegetable, free	—	Celery, cucumber, lettuce, etc. as desired.
Vegetable, counted .	1	Beets, carrots, green pepper, mushrooms, onions, tomatoes, etc.
Vegetable, starchy . .	1	Baked beans (without pork), cooked dried beans or peas, corn, green peas, sweet or white potato.
Bread, medium fat .	2	Allowance is intended to cover items such as a small serving of french fried potatoes, biscuits, or corn muffins. If 1 or more low fat items are used instead, 1 or more extra fat exchanges may be allowed.
Fat	1	Dressing for salad or vegetable, or spread for bread.
Unrestricted items . .	—	Coffee, cereal beverage, dietetic soft drink, herb tea, etc. (See Breakfast.)
Midafternoon:		
Unrestricted items . .	—	Coffee, cereal beverage, dietetic soft drink, herb tea, etc. (See Breakfast.)
Bread, if *not* used at breakfast or lunch .	1 to 2	Bread, cake, cookies, or crackers.
Milk, if *not* used as allowed at another meal	—	(See the note next to milk under Midmorning.)
Supper:		
Meat, lean	3	Fish, meat, poultry, or a meat substitute.
Fruit	1	Glass (4 oz, or *120 ml*) of fruit juice, or a serving of a fruit.
Vegetable, free	—	Celery, cucumber, lettuce, etc. as desired.
Vegetable, counted .	1	Beets, carrots, green pepper, mushrooms, onions, tomatoes, etc.
Vegetable, starchy . .	1	Baked beans (without pork), cooked dried beans or peas, corn, green peas, sweet or white potato.
Bread, low fat	1	Bread, roll, tortilla, etc.
Bread, high fat	1	Items such as 1 serving of ice cream or potato chips (if low or medium-fat bread is used instead, an extra fat exchange may be allowed).
Fat	1	Dressing for salad or vegetable, or spread for bread.
Unrestricted items . .	—	Coffee, cereal beverage, dietetic soft drink, herb tea, etc. (See Breakfast.)
Bedtime:		
Milk, low fat (2%) . .	1	
Bread, if the exchanges allowed at meals during the day are not all used	1 to 3	Bread, cake, cereal, cookies, or crackers.

[1]See Table M-30 and the accompanying text for the calculations used to determine the numbers of allowed food exchanges.

MOIETY

Any equal part.

MOISTURE

A term used to indicate the water contained in foods—expressed as a percentage.

MOISTURE-FREE (M-F, oven-dry, 100% dry matter)

This refers to any substance that has been dried in an oven at 221 °F (*105 °C*) until all the moisture has been removed. (Also see ANALYSIS OF FOODS.)

MOLASSES

Molasses is the thick brownish syrup by-product of the manufacture of cane or beet sugar from which part of the crystallizable sugar has been removed. However, molasses is also the by-product of several other industries. Citrus molasses is produced from the juice of citrus wastes. Wood molasses is a by-product of the manufacture of paper, fiberboard, and pure cellulose from wood; it is an extract from the more soluble carbohydrates and minerals of the wood material. Starch molasses, Hydrol, is a by-product of the manufacture of dextrose (glucose) from starch derived from corn or grain sorghums in which the starch is hydrolyzed by use of enzymes and/or acid. Cane molasses and beet molasses are, by far, the most extensively used types of molasses. The different types of molasses are available in both liquid and dehydrated forms.

Molasses contains sugar, which forms the basis of determining its quality. The sugar content is expressed as *Brix*. Brix is determined by measuring the specific gravity of molasses. After the specific gravity has been obtained, the value is applied to a conversion table from which the level of sucrose (or degrees Brix) can be determined. As sugar content increases, degrees Brix likewise increases. Since molasses also contains lipids, protein, inorganic salts, waxes, gums, and other material, the Brix classification can often be misleading, because each of these contaminants has an influence on the specific gravity of the solution. However, degrees Brix does give a relatively accurate indication of the sugar content of molasses and is, therefore, a good means of determining quality.

PRODUCTION. Molasses may be made by the open kettle method or by the vacuum pan method. In the open kettle method, the cane juice is boiled in a large open pan. Large sugar factories generally use the vacuum pan method, in which large, covered vacuum pans are used.

USES. Only a small portion of the molasses available in the United States is consumed by humans. Some molasses is used in industries for the production of yeast and organic fermentation chemicals. But the major utilization of molasses in the United States is in animal feeds, as blackstrap molasses—the molasses left after several boilings.

Food. The baking industry is the largest consumer of edible molasses. It imparts a pleasant flavor to breads, cakes, and cookies. Additionally, molasses serves as a humectant—helping to maintain the freshness of baked goods, and enhances the flavor of chocolate in baked goods or drinks.

In home cooking, molasses is employed for the same purposes as in the baking industry. Also, molasses is often incorporated in baked beans, and in glazes for sweet potatoes and meats such as hams. The flavor of toffees and caramel may be rounded out with molasses. Some table syrups contain molasses, and some individuals sweeten foods with molasses, while others mix a spoonful or two with hot water for a hot beverage. Many recipes have been developed which take advantage of the pleasant flavor and smell of molasses.

Generally, molasses is eaten in such small amounts it does not make an important contribution to the diet. The nutritional value of molasses is given in Food Composition Table F-21.

Molasses is (1) a good source of some minerals—especially calcium, magnesium, potassium, and iron; and (2) a concentrated energy source. Blackstrap molasses contains several of the minerals at higher levels than light or medium molasses. Vitamins are present only in low levels, or nonexistent.

Dark molasses (blackstrap molasses) is a fair, but variable and unreliable, source of iron. Today, there is less iron in molasses that is processed in stainless steel or aluminum vessels than in molasses processed in old-fashioned iron vats and pipes. Besides, molasses is generally eaten infrequently and/or in small servings. Some opinions to the contrary, molasses has no miraculous curative properties; rather, it should be eaten for reason of taste.

Rum is produced from the fermentation of molasses. After distillation, it is aged from 5 to 7 years. Rum gets its color from the oaken cask in which it is aged and from caramel added before aging. Some light rums are rapidly fermented with cultured yeast and aged from 1 to 4 years.

(Also see DISTILLED LIQUORS.)

MOLD INHIBITORS

Substances such as sodium and calcium propionate which are added to foods to prevent the growth of mold. Often, substances with this action are referred to as antimicrobials.

(Also see ADDITIVES, FOODS.)

MOLDS (FUNGI)

Fungi which are distinguished by the formation of mycelium (a network of filaments or threads), or by spore masses.

(Also see POISONS, Table P-2, Some Potentially Poisonous [Toxic] Substances—Mycotoxins; and SPOILAGE OF FOOD.)

MOLECULE

A chemical combination of two or more atoms.

MOLYBDENUM (Mo)

Contents   Page
History...750
Absorption, Metabolism, Excretion...................750
Functions of Molybdenum.........................750
Deficiency Symptoms.............................750
Interrelationships.................................750
Recommended Daily Allowance of Molybdenum........ 750
Toxicity... 750
Sources of Molybdenum............................ 751

Although the essential role of molybdenum in plants is well known, the essentiality of this element in man is less well established. Evidence that molybdenum is an essential trace element is based on the facts that it is part of the molecular structure of two enzymes, xanthine oxidase (involved in the oxidation of xanthine to uric acid) and aldehyde oxidase (involved in the oxidation of aldehydes

to carboxylic acids), and that diets low in molybdenum adversely affect growth in small animals. There is no evidence, however, that a low molybdenum intake produces deficiency signs and symptoms in man.

The adult human body contains only about 9 mg of molybdenum.

HISTORY. In 1778, Karl Scheele of Sweden recognized molybdenite as a distinct ore of a new element. Then, in 1782, P. J. Hjelm obtained the metal by reducing the oxide with carbon and called it molybdenum. The name molybdenum is derived from the Greek *molybdos*, meaning lead.

Molybdenum was long known as essential for the growth of all higher plants. Then, in 1953 it was found in the essential enzyme, xanthine dehydrogenase.

ABSORPTION, METABOLISM, EXCRETION. Molybdenum is readily absorbed as molybdate in the small intestine, although some absorption occurs throughout the intestinal tract. There is little retention of this element except in the liver, adrenals, kidneys, and bones. Molybdenum is excreted rapidly in the urine, and in limited amounts via the bile and feces.

FUNCTIONS OF MOLYBDENUM. Molybdenum is a component of three different enzyme systems which are involved in the metabolism of carbohydrates, fats, proteins, sulfur-containing amino acids, nucleic acids (DNA and RNA), and iron. Also, it is found in the enamel of teeth, where it appears to prevent or reduce the incidence of dental caries, although this function has not yet been proven conclusively.

DEFICIENCY SYMPTOMS. Naturally occurring deficiency of molybdenum in human subjects is not known, *unless* utilization of the mineral is interfered with by excesses of copper and/or sulfate.

Molybdenum-deficient animals are especially susceptible to the toxic effects of bisulfite, which is both a food additive and a product of the metabolism of sulfur-containing amino acids. Bisulfite toxicity is characterized by breathing difficulties and neurological disorders.

INTERRELATIONSHIPS. The utilization of molybdenum is reduced by excess copper, sulfate, and tungsten (an environmental contaminant). Also, even a moderate excess of molybdenum causes significant urinary loss of copper; and toxicity due to excess copper is counter-acted by molybdenum.

RECOMMENDED DAILY ALLOWANCE OF MOLYBDENUM. Man's requirement for molybdenum is so low that it is easily furnished by the diet.

The Food and Nutrition Board of the National Research Council (FNB-NRC) recommends that the molybdenum intake of an adult be within the range of 75 to 250 mcg/day.

Estimated safe and adequate intakes of molybdenum are given in the section on MINERAL(S), Table M-25, Mineral Table.

• **Molybdenum intake in average U.S. diet**—The daily intake from a mixed diet in the United States has been estimated at about 180 mcg/day.

TOXICITY. The greatest concern about molybdenum

is its toxicity. Adverse effects of high molybdenum concentrations in the environment have been reported in the human population living in a province of the U.S.S.R.[46] According to the U.S.S.R. report, an excessive dietary intake of 10 to 15 mg/day of molybdenum caused a high incidence of a goutlike syndrome associated with elevated blood levels of molybdenum, uric acid, and xanthine oxidase.

Potentially toxic levels of molybdenum are contained in some soils and herbage in parts of Florida, California, and Manitoba.

SOURCES OF MOLYBDENUM. The concentration of molybdenum in food varies considerably, depending on the soil in which it was grown. Most of the dietary molybdenum intake is derived from organ meats, whole grains, wheat germ, legumes, leafy vegetables, and yeast. Table M-32 gives the molybdenum content of some common foods. The molybdenum requirement should be met by most diets. Supplements of additional molybdenum are not recommended.

(Also see MINERAL[S], Table M-25.)

TABLE M-32
MOLYBDENUM CONTENT OF SOME COMMON FOODS

Food	Molybdenum
	(mg/100 g)
Lima beans, dry	.323
Wheat germ	.210
Liver	.150
Green beans	.067
Eggs	.050
Whole wheat flour	.048
Poultry	.040
White flours	.025
Spinach	.025
Cabbage	.017
Potato	.016
Cantaloupe	.016
Apricot	.011
Carrot	.010
Banana	.003
Milk	.003
Lettuce	.002
Celery	.002

MONOGLYCERIDE

An ester of glycerol with one fatty acid.

MONOPHAGIA

The habit of eating, or the desire to eat, only one type of food.
(Also see APPETITE.)

MONOSACCHARIDE

Any one of several simple, nonhydrolyzable sugars. Glucose, fructose, galactose, arabinose, xylose, and ribose are examples.
(Also see CARBOHYDRATE[S]; and SUGARS.)

MONOUNSATURATED

Having one double bond, as in a fatty acid; e.g., oleic acid.

MONOVALENT

Having a valence (the power of an atom to combine with another atom) of one. Example: sodium (Na +) has a valence of one; it combines with chloride (Cl-) to form sodium chloride (NaCl).

MORBIDITY

A state of sickness.

MORTADELLA

A dry or semidry sausage made of chopped pork and pork fat; seasoned with red pepper and garlic; stuffed in large casings; and cooked and smoked.

MOTILITY

The power of spontaneous movement.

MUCIN

A glycoprotein secreted by the secretory cells of the digestive tract. It functions to (1) facilitate the smooth movement of food through the digestive tract, and (2) provide a protective coating to the gastric and duodenal mucosa against the corrosive action of hydrochloric acid.
(Also see DIGESTION AND ABSORPTION; and ULCERS, PEPTIC.)

MUCOPOLYSACCHARIDE (GLYCOSAMINO-GLYCAN)

A group of polysaccharides which contains hexosamine (as glucosamine), which may or may not be combined with protein, and which, dispersed in water, form many of the mucins—thick gelatinous material that cements cells together and lubricates joints and bursas.

MUCOPROTEIN

Substances containing a polypeptide chain and disaccharides, found in mucous secretions of the digestive glands.

MUCOSA (MUCOUS MEMBRANE)

A membrane rich in mucous glands which lines the gastrointestinal, respiratory, and genitourinary tracts.

MUCOUS MEMBRANE

A membrane lining the cavities and canals of the body, kept moist by mucus.

MUCUS

A slimy liquid secreted by the mucous glands and membranes.

MULTIPLE SCLEROSIS (MS)

A strange and baffling disease of the nervous system of unknown origin whose victims are between the ages of 20 and 40. Some evidence suggests that it is caused by a slow-acting virus to which the body's immune system does not respond. Nerve tracts of the brain and spinal cord develop areas of faulty nerve transmission due to the development of areas of demyelination—loss of myelin, the insulation of nerves—which is thought to be the result of some alteration in lipid metabolism. These areas occur at random and may come and go. Symptoms signaling the onset of multiple sclerosis may be so minor that they are overlooked; for exam-

[46]Kovalskiy, V. V., and G. A. Yarovaya, Molybdenum-infiltrated Biogeochemical Provinces, *Agrokhimiya*, Vol. 8, 1966, pp. 68-91.

ple, blurred or double vision. Once the initial symptoms disappear they may never recur, and a period of months or years may pass before other symptoms appear such as weakness of certain muscles, unusual tiring of a limb, minor interference with walking, muscle stiffness, dizziness, loss of bladder control and disturbances in the senses of touch, pain, and heat. Each symptom may appear and then disappear only to be followed by another. The disease continues to progress, and eventually the victim becomes crippled and even bedridden. However, the progression and remission of multiple sclerosis is different for every victim. Maintenance of good physical health, resistance to infection, and good mental health play a role in the severity of the symptoms. To this end a nutritionally adequate and appealing diet simply prepared is recommended. As the disease progresses, a pureed or liquefied diet may be necessary since victims have difficulty swallowing.

(Also see MODIFIED DIETS.)

MUSCLES

All physical functions of the body involve muscular contractions, which, in turn, require nutrients derived from the diet. Muscles conduct nervous impulses, contract in response to nervous impulses, and modify their activity according to the degree they are stretched. In the body, there are three types of muscles:

1. **Skeletal muscles.** These are attached to the 206 bones that make up the skeleton. Their coordinated contraction and relaxation results in the movement of body parts. Skeletal muscles are voluntarily controlled. They are also classified as striated muscles due to their microscopic appearance.

2. **Smooth muscles.** Most of the internal organs of the body contain smooth muscles. They are responsible for contractions of such organs as the intestines, blood vessels, and uterus. For the most part, smooth muscles are involuntarily controlled.

3. **Cardiac muscle.** Obviously, it is found only in the heart. Continually contracting and relaxing with a built-in rhythmicity until death, it pumps about 8,000 gallons (30,400 liters) of blood through 12,000 miles (19,200 km) of blood vessels each day.

Muscular contraction is a complicated process involving the actin and myosin filaments of the cell; requiring chemical energy and oxygen. It forms heat and a force capable of performing work.

(Also see ENERGY UTILIZATION OF THE BODY, section on "Muscle Contraction";METABOLISM; and MUSCULAR WORK.)

MUSCOVADO

Unrefined raw sugar obtained from the juice of sugarcane by evaporation and drawing off the molasses.

MUSCULAR DYSTROPHY (MD)

Any of the hereditary diseases of the skeletal muscles characterized by a progressive shrinking or wasting of the muscles. At the onset only certain muscles seem affected, but ultimately all muscles become involved. The exact cause is not known. The most common form of muscular dystrophy occurs almost entirely in males since its inheritance is sex linked—passed from mother to son on the X chromosome. There is no effective treatment. Physical therapy may delay the progression of the disease. Dietary adjustments due to limited mobility and feeding difficulties are necessary

as the disease advances.

(Also see MODIFIED DIETS.)

MUSCULAR WORK

Energy is the ability to do work. Mechanical work is accomplished when a force acts to move an object some distance. Muscles require chemical energy derived from the diet to contract and produce a force capable of performing work. Contraction of the heart produces a force which propels the blood through the blood vessels. Contracting limb muscles are capable of exerting a force to move an object from the floor to a table. Contracting limb muscles work to move the body—exercise. The amount of energy required is in direct relation to the intensity of the muscular work.

(Also see CALORIC [ENERGY] EXPENDITURE.)

MUSH

This is a dish made by boiling cornmeal in water, which is eaten as a hot cereal; or it can be put into a mold until cold, then sliced and fried in a little butter.

MUSHROOMS, CULTIVATED
Basidiomycetes, Ascomycetes

These much esteemed foods are the fleshy, fruiting bodies of higher fungi that grow in soil rich in organic matter, or on living trees or dead wood. From a strict botanical point of view, the term mushroom designates certain club fungi *(Basidiomycetes)*. However, many people extend the term to cover cup fungi *(Ascomycetes)* such as truffles.

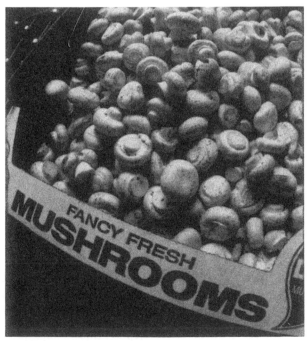

Fig. M-79. Mushrooms marketed in bulk. (Courtesy, Castle and Cooke, Inc.)

MAJOR TYPES. There are about 38,000 known species of mushrooms. Of course, it is impossible in this article to list or describe all the kinds of mushrooms. However, brief descriptions of the five most commonly cultivated mushrooms follow.

• **Common Cultivated Mushroom** (*Agaricus bisporus*)—This is the type that is best known to Americans. It is light colored and very mild flavored.

Fig. M-80. The common Cultivated Mushroom (*Agaricus bisporus*).

• **Oyster Mushroom** (*Pleurotus ostreatus*)—The color and shape of this type are like those of an oyster. It is grown mainly in the Orient and Europe. The canned oyster mushrooms sold in the United States are usually imported from Taiwan.

Fig. M-81. The Oyster Mushroom (*Pleurotus ostreatus*).

• **Padi Straw Mushroom** (*Volvariella volvaceae*)—Most of this crop is produced in the Orient, where it is grown on rice (paddy) straw. These mushrooms are usually sold in the dried form.

Fig. M-82. The Padi Straw Mushroom (*Volvariella volvaceae*).

• **Shiitake** (*Lentinus edodes*)—This type is grown mainly in China and Japan, and is exported in canned, dried, and pickled forms.

Fig. M-83. Shiitake (*Lentinus edodes*), a mushroom which grows on hardwood logs.

• **Truffles** (*Tuber melanosporum; tuber magnatum*)—These are by far the most sought after and expensive of the mushrooms. Actually, they are cup fungi (*Ascomycetes*), rather than mushrooms, and they grow underground in association with the roots of oak and beech trees. Most of the world's supply of this delicacy is produced in France and Italy. Terfezios are fungi that resemble truffles closely and are found mainly in the desert areas of North Africa and the Middle East.

Fig. M-84. Truffles (*Tuber melanosporum*), a gourmet fungus that is highly esteemed for its fragrant aroma.

NUTRITIONAL VALUE. The nutrient compositions of various forms of mushrooms are given in Food Composition Table F-21.

Mushrooms are high in water content (90%) and low in calories (28 to 35 kcal per 100 g). They contain about 20% more protein than potatoes, but they furnish less than half as many calories. Furthermore, they are very low in calcium, vitamin A, and vitamin C; moderately low in thiamin and riboflavin; and a good source of phosphorus, potassium, and niacin. However, many of the nutrients that are lacking in mushrooms may be provided by green leafy vegetables and milk products, which go well with the fungi.

(Also see VEGETABLE(S), Table V-2, Vegetables of the World.)

MUSKRAT Ondatra zibethicus

A small rodent found throughout the United States and Canada, and in parts of Europe, living in the holes in the banks of ponds and streams or in dome-shaped houses of rushes and mud. The muskrat is about the size of a small cat, has a long tail and webbed hind feet, is brown in color, and possesses small glands that give off a musky odor. Muskrat fur is sold as *Hudson seal*. Muskrat meat is tasty and is sold as *marsh rabbit*.

MUSTARD SEED Sinapis alba; Brassica spp

Mustard, which is made from the seeds of certain plants of the mustard family (*Cruciferae*), is the leading spice in worldwide usage. The four species of mustard seeds used for this purpose are: (1) white mustard (*Sinapis alba*), (2) brown mustard (*Brassica juncea*), (3) black mustard (*Brassica nigra*), and (4) Ethiopian mustard (*Brassica carinata*). It is noteworthy that brown mustard, which is also called Indian mustard, may also be grown for its green leaves rather than for its seeds.

PROCESSING. Mustard seed is usually marketed as a dry powder, or as a preparation with a pastelike consistency. Both of these forms are made by grinding the dry seed into a powder, then removing the seed hulls by milling, screening, and sifting. The dried mustard powder has no pungent odor because the ''hot'' components are present in a chemically inactive form. However, wetting the powder brings about an enzymatic action which liberates the pungent components.

Most prepared mustard is made by mixing the dried mustard powder with salt, vinegar, and other spices. Also, mustard is utilized in other products such as mayonnaise, salad dressings, and pickles.

SELECTION, PREPARATION, AND USES. Mustard powder, which is available in most supermarkets, may be added directly to casseroles (such as baked beans and pork), soups, stews, and other mixed dishes. However, it should be used cautiously, since the enzymatic action which releases the pungent components from the wetted powder requires a bit of time in which to develop the maximum flavor intensity. It is noteworthy that mixing the powder with very hot water, or with moderately strong acid such as lemon juice or vinegar, stabilizes the flavor intensity by stopping the enzyme action.

Prepared mustard may be used to spice up deviled eggs, fish, meats, and poultry. It also enhances the flavor of dressings and sauces.

NUTRITIONAL VALUE. Mustard seed is rich in nutrients (469 kcal per 100 g, 25% protein, and 29% fat), but it is not likely to contribute much to the diet, nutritionally, because the acrid components limit the amounts which may be consumed safely. (Mustard oil blisters the skin.) Nevertheless, the small amounts normally used are helpful in enhancing the taste of food and in stimulating the appetite. The nutrient composition of mustard powder is given in Food Composition Table F-21.

MUTAGENS

Any of a number of chemical compounds capable of inducing mutations (gene changes) in DNA and in living cells. The alkyl mustards, as well as dimethyl sulfate, diethyl sulfate, and ethylmethane sulfonate, comprise a group of alkylating agents, reacting with the nitrogen atoms of guanine (a purine base), a constituent of both RNA and DNA. This reaction affects the guanine molecule in such a way as ultimately to induce a mutation in DNA. Nitrous oxide can deaminate (remove the amino group from) both guanine and cytosine (a pyrimidine base).

Caffeine appears to be a weak mutagen in some nonmammalian systems. But its significance as a mutagen in humans is unknown. It is noteworthy, however, that studies of the caffeine intake of pregnant women have shown no association between caffeine and birth defects.
(Also see NUCLEIC ACIDS.)

MYASTHENIA GRAVIS

An autoimmune disease—failure of the body's immune system to recognize self—in which the body forms antibodies against its own muscle cell membranes. The antibodies attack the muscle cell membranes and destroy their responsiveness to nerve impulses. Nerve impulses fail to reach the muscle and contraction fails. Muscles unable to contract are paralyzed.

The cause of the disease is not known, but it is associated with an enlarged thymus. Myasthenia gravis is characterized by various groups of muscles which become easily fatigued and weakened. Respiratory, neck, tongue, trunk and shoulder muscles are affected. Some drugs are capable of improving muscular functions. Due to the muscles involved, there are some feeding difficulties as swallowing, and coughing; and grasping abilities are impaired.

MYELIN

The white, fatlike substance which forms a sheath around certain nerve fibers.

MYO–

Prefix meaning muscle.

MYOCARDIUM

The heart muscle.

MYOFIBRILS

The hundreds of small contractile units found in each muscle fiber. Myofibrils are composed of the muscle protein filaments, myosin and actin.

MYOGEN

A muscle protein, an albumin, that is found in the cytoplasm of the muscle cell. It is not a protein of the actin and myosin filaments.

MYOGLOBIN

A protein similar to hemoglobin in structure and function, found only in the muscles. Myoglobin receives oxygen from the blood and stores it for use by the muscle cells. It is responsible for the red color of meat.
(Also see HEMOGLOBIN.)

MYOSIN

One of two muscle cell proteins which play an important role in the contraction and elasticity of muscles. The other protein is actin.
(Also see MUSCLES.)

MYSORE FLOUR

This flour was developed in India and used for large-scale feeding trials as a partial substitute for cereals. It is a mixture of 75% tapioca flour and 25% peanut flour.

NAILS (FINGERNAILS)

Nails may be a sign of good health or of poor health. In good health, the nails are firm and pink. In poor health, the nails are rigid and brittle. In chronic iron deficiency anemia, the nails may be spoon-shaped (*koilonychia*). Severe protein deficiency may result in transverse white bands in the nails, occurring symmetrically on both hands. In other forms of malnutrition, the nails may be brittle, thickened, or lined on the surface either transversely or longitudinally; but these changes may also be seen in well-nourished people.

(Also see DEFICIENCY DISEASES, Table D-2, Minor Dietary Deficiency Disorders—Nails; and HEALTH, Sections headed "Signs of Good Health" and "Signs of Ill Health.")

NAPHTHOQUINONE

A derivative of quinone. Some of these derivatives have vitamin K activity.

NARANJILLA *Solanum quitoense*

The fruit of a very large herbaceous plant (of the family *Solanaceae*) that is 6 to 10 ft (2 to 3 m) tall and a native of the northern Andes.

Naranjilla fruits are orange-colored, fuzzy tomatolike fruits that have a sour green pulp. They are used mainly to make jam, jelly, juice, and pies.

The pulp of the fruit has a high water (92%) and a low content of calories (28 kcal per 100 g) and carbohydrates (7%). It is a good source of vitamin C.

NATAL PLUM *Carissa grandiflora*

Fruit of a shrub (of the family *Apocynaceae*) that is native to South Africa. Natal plums are pear-shaped and are up to 2 in. (5 cm) long. They have a reddish skin and reddish pulp with a white milky latex. The ripe fruits are used for making jellies and sauces. They may also be eaten fresh or stewed.

Natal plums are moderately high in calories (68 kcal per 10 g) and carbohydrates (16%). They are a good source of iron and ascorbic acid.

NATIONAL FLOUR

Presently, this is the name given to 85% extraction wheat flour in the United Kingdom.

NATIONAL RESEARCH COUNCIL (NRC)

A division of the National Academy of Sciences established in 1916 to promote the effective utilization of scientific and technical resources. Periodically, this private, nonprofit organization of scientists publishes bulletins giving nutrient requirements and allowances for man and animals, copies of which are available on a charge basis through the National Academy of Sciences, National Research Council, 2101 Constitution Avenue, N.W., Washington, D.C. 20418.

NATURAL FOODS

Natural foods are those that are grown naturally and subjected to little or no processing. But there is no official definition of a natural food. The Food and Drug Administration's policy on food labeling only prohibits a manufacturer from calling an entire food natural if it contains artificial colors or flavors or any synthetic ingredients.

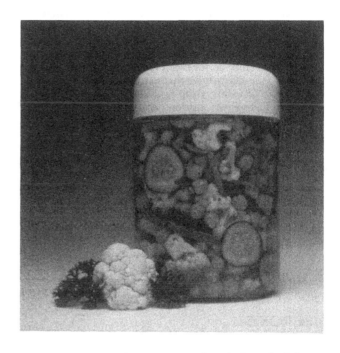

Fig. N-1. Natural foods—foods grown naturally and subjected to little or no processing. (Courtesy, USDA)

Natural food enthusiasts stress nature and the return to a more primitive way of life. Some of these advocates go so far as to suggest that something valuable has been removed from the traditional foods sold in the supermarket; that they are counterfeit, prefabricated, worthless, or devitalized. It is inferred that processing destroys the nutritive value of the food. Such notions have created doubts in the minds of some Americans about the integrity, purity, and nutritive content of the nation's food supply. It is true that modern food refining removes some valuable nutrients from such foods as white flour and wild rice. However, these foods, and many other processed foods, are enriched with minerals and vitamins to make them nutritious.

Usually, natural foods are grown with organic fertilizer and without chemical sprays in a garden, on a tree, or a farm, and just washed, possibly hulled and cracked, or cut before eating raw or cooked. And they contain no chemical additives, such as preservatives, emulsifiers, and antioxidants. Such foods include fresh fruits and vegetables, nuts, seeds, and stone-ground cereals. Honey, brown sugar, and fertile eggs are other foods which natural food enthusiasts consider to have superior power over their actual nutritive content.

The back-to-nature movement popularized granola, a *natural* food that had been around for years. Granola is a heavy, chewy, dry cereal made from such *healthy* ingredients as whole grains, nuts, seeds, raisins, and honey, and often (though not always) without chemical preservatives. The granola labels proclaim *all natural*. To its credit, granola does contain more protein, fiber, vitamins, and minerals than most popular cereals. However, its nutritious aspects may be offset by lots of fats (especially highly saturated coconut oil), and sugars (after all, honey is sugar, natural or not), and calories—four to six times as many calories as the same volume of more traditional cereals.

Some natural food advocates use raw milk, despite the hazard of brucellosis (undulant fever) and tuberculosis—disease-producing organisms that are destroyed by pasteurization. Others consider fish, chicken, and animal flesh as natural foods so long as the animals are grown without the use of commercial feeds, hormones, and antibiotics. Still, others imagine that natural vitamins are superior to synthetic vitamins; despite the fact that a vitamin is a vitamin whatever its source, and that a vitamin has a chemical formula and functions in a certain manner in the body whether it is natural or synthetic.

But the natural food movement has made for nutritional awareness and interest. In this respect, it has succeeded where mothers have failed. Many young people are now eating vegetables and a greater variety of other foods. Also, they are turning away from highly refined foods like candy, soft drinks, and other *empty calories*.

(Also see FOOD MYTHS AND MISINFORMATION; HEALTH FOODS; and ORGANICALLY GROWN FOODS.)

NAUSEA

Sickness at the stomach associated with an urge to vomit.

NECROSIS

Death of a part of the cells making up a living tissue.

NECTAR

The two major nutritional meanings of this term are as follows:

• A syrupy liquid produced by special glands of flowers that is collected by bees and converted by them into honey. The unique flavors of the honeys derived from the different nectars are due mainly to the essential oils and other aromatic substances produced by the flowers.
(Also see HONEY.)

• A pulpy fruit drink made from various combinations of ingredients such as water, fruit puree, fruit pulp, fruit juice, sweetener(s), citric acid, and vitamin C. The minimum amounts of fruit ingredient for each of the types of nectar are specified in the Standards Of Identity which have been established by the FDA. Some of the fruit nectars sold in the United States are apricot, guava, mango, papaya, peach, and pear.
(Also see the articles on the individual fruits for details regarding the various nectars.)

NEONATE

A newborn baby.

NEOPLASM (TUMOR)

A new and abnormal growth—a tumor, which serves no physiological purpose.

NEPHRITIS

Inflammation of the kidneys.
(Also see DISEASES.)

NEPHRON

The structural and functional unit of the kidneys consisting of a tuft of capillaries known as the glomerulus attached to the renal tubule. Urine is formed by filtration of blood in the glomerulus and by the selective reabsorption and secretion of solutes by cells that comprise the walls of the renal tubules. There are aproximately 1 million nephrons in each kidney.
(Also see GLOMERULONEPHRITIS [Nephritis].)

NEPHROSCLEROSIS

Hardening and narrowing of the arteries of the kidneys, a condition usually associated with high blood pressure and arteriosclerosis. Nephrosclerosis causes degeneration of the renal tubules, and fibrosis of the glomeruli, the points where filtration occurs. Its incidence increases with age but it often produces no outward symptoms. It may, however, account for the decrease in kidney reserve noted in the elderly. During its late stages some sodium and protein restrictions may be necessary.

NEPHROSIS

A degenerative kidney disease occurring without signs of inflammation. It may occur as a consequence of acute glomerulonephritis, lupus erythematosus, an allergic reaction, diabetes mellitus, or mercury poisoning. In children it appears for no apparent reason. Clinically, the patient exhibits massive losses of protein in the urine—proteinuria; low blood protein—hypoproteinemia; and water in the tissues of the body—edema. Treatment consists of replacing protein losses with a high protein diet, and controlling the edema with a restricted sodium intake. Possibly diuretics or steroids may also be involved in the treatment. Successful therapy is a long-term process. Full recovery seldom requires less than 2 years.
(Also see MODIFIED DIETS; GLOMERULONEPHRITIS; and LUPUS ERYTHEMATOSUS.)

NEROLI OIL

This is a fragrant essential oil obtained from certain flowers, especially the sour orange, which is used chiefly in perfumes, but which is also used in flavoring foods.
(Also see SEVILLE ORANGE.)

NERVOUS SYSTEM

The entire nerve apparatus of the body consisting of the brain and spinal cord, nerves, ganglia, and parts of the receptor organs that receive and interpret stimuli and transmit impulses to the effector organs.

NERVOUS SYSTEM DISORDERS

These may encompass everything from numbness in a limb to quadraplegia to mental illness, which in itself includes a multitude of disorders.

A variety of interrelationships exist between nutrition and nervous disorders. Among the nutritional factors which may cause or contribute to certain nervous system disorders are: energy, protein, lipid, minerals, and vitamins. Also many toxins have detrimental effects on the nervous system; for example, pesticides, ergot, mycotoxins, polybrominated biphenyls (PBBs), polychlorinated biphenyls (PCBs), neurotoxins, hallucinogens, and cholinesterase inhibitors.

Nutrition and nervous system disorders are further related since nervous disorders often require special dietary adjustments, and may present some feeding difficulties.

(Also see MODIFIED DIETS; MENTAL DEVELOPMENT; and MENTAL ILLNESS.)

NEURITIC

Of, relating to, or affected by neuritis.

NEURITIS

Inflammation of the peripheral nerves—the nerves which link the brain and spinal cord with the muscles, skin, organs, and other parts of the body.

NEUROPATHY

Disease of the nervous system, especially when involving degenerative changes.

NEUTROPENIA

The presence of neutrophile cells (white blood cells which do not stain readily) in abnormally small numbers in the peripheral bloodstream.

NIACIN (NICOTINIC ACID; NICOTINAMIDE)[1]

Contents	Page
History	758
Chemistry, Metabolism, Properties	758
Measurement/Assay	759
Functions	759
Deficiency Symptoms	760
Recommended Daily Allowance of Niacin	760
Toxicity	761
Niacin Losses During Processing, Cooking, and Storage	761
Sources of Niacin	761

[1]Niacin has, along the way, been known as vitamin P-P, pellagra-preventive vitamin, vitamin G, vitamin B-3, vitamin B-4, and vitamin B-5.

NIACIN

IT'S ESSENTIAL FOR HUMANS TOO!

NIACIN MADE THE DIFFERENCE! LEFT: CHICK ON NIACIN-DEFICIENT DIET. NOTE POOR GROWTH AND ABNORMAL FEATHERING. RIGHT: CHICK THAT RECEIVED PLENTY OF NIACIN. (COURTESY, UNIVERSITY OF WISCONSIN)

TOP FOOD SOURCES

LIVER AND KIDNEY | LEAN MEAT, (BEEF, PORK, LAMB) | POULTRY | FISH | RABBIT

MUSHROOMS | NUTS | MILK AND CHEESE | EGGS | ENRICHED CEREAL PRODUCTS

Fig. N-2. Niacin made the difference! *Left*: Chick on niacin-deficient diet. Note poor growth and abnormal feathering. *Right*: Chick that received plenty of niacin. (Courtesy, H.R. Bird, Department of Poultry Science, University of Wisconsin) Note, too, top ten food sources of niacin.

Niacin, a member of the B-complex, is a collective term which includes nicotinic acid and nicotinamide, both natural forms of the vitamin with equal niacin activity. In the body, they are active as nicotinamide adenine dinucleotide (NAD) and nicotinamide adenine dinucleotide phosphate (NADP) and serve as coenzymes, often in partnership with thiamin and riboflavin coenzymes, to produce energy within the cells, precisely when needed and in the amount necessary.

The discovery of the role of niacin as a vitamin of the B group was the result of man's age-old struggle against pellagra. The disease was first described by the physician Gaspar Casal in Spain in 1730, soon after the introduction of corn (maize) into Europe; and it was given the name pellagra (*pelle,* for skin; and *agra,* for sour) by physician Francesco Frapoli in Italy in 1771.

Pellagra spread with the spread and cultivation of corn. In the 19th century, it was common in almost all of the European and African countries bordering on the Mediterranean sea; and it later spread to other African countries. Pellagra had long been present in both North and South America, but it reached epidemic proportions in southern United States after the Civil War, which left poverty in its wake, as a result of which many of the poor subsisted almost entirely on corn. Outbreaks of the disease were so widespread and severe that most physicians considered the cause to be either an infectious agent or a toxic substance present in spoiled corn.

HISTORY. Nicotinic acid was first discovered and named in 1867, when Huber, a German chemist, prepared it from the nicotine of tobacco. But for the next 70 years it remained idle on the chemist's shelves because no one thought of it, even remotely, as a cure for pellagra. In the meantime, thousands of people died from the disease.

Subsequent to 1867, nicotinic acid was *rediscovered* many times as a compound present in foods and tissues. In 1912, Funk, in England, isolated nicotinic acid from rice polishings while attempting to isolate the antiberiberi vitamin; and, that same year, Suzuki, in Japan, isolated nicotinic acid from rice bran. But both researchers lost interest in the acid when they found it ineffective in curing beriberi.

In the early 1900s, pellagra reached epidemic proportions in southern United States, where the diet was based primarily on corn, which is extremely low in both available niacin and in tryptophan. In 1915, 10,000 people died of the disease; and, in 1917-18, there were 200,000 cases of pellagra in this country.

In 1914, the U.S. Public Health Service dispatched a team under the direction of Dr. Joseph Goldberger, a physician-researcher, to study the cause of, and hopefully find a cure for, pellagra. In a series of studies initiated in 1914 and continuing throughout the 1920s, Goldberger proved that the disease was caused by a dietary deficiency, and not an infection or toxin.

In 1925, Goldberger and Tanner classified common foods on the basis of their effectiveness in preventing or curing human pellagra. They studied the effects of supplementing a pellagra-producing diet with certain foods. From these results, (1) they concluded that there is a specific dietary factor of unknown nature present in certain foods, which is involved in preventing and curing pellagra; and (2) they rated foods on their content of this factor as follows: *abundant,* yeast; *good,* lean meat and milk; *fair,* peas and beans (and other vegetables).

In 1926, Goldberger and Wheeler showed that pellagra in man and blacktongue in dogs were similar. Goldberger and his associates observed that in towns where many people suffered from pellagra, a large percentage of the dogs had blacktongue, thereby noting the similarity between pellagra in man and blacktongue in dogs. Further, they confirmed that dogs with blacktongue could be cured by yeast. Goldberger and Wheeler designated this preventive and curative factor, present in certain foods, as P-P (pellagra-preventive); others designated it vitamin G for Goldberger.

In 1935, von Euler, Albers, and Schlenck studied the preparation of cozymase, the coenzyme which is necessary for the alcoholic fermentation of glucose by apozymase, shown later to be diphosphopyridine nucleotide (DPN). On hydrolysis, cozymase yielded nictotinic acid. This was the first evidence that nicotinic acid (in the form of its amide) formed a part of the structure of an enzyme, and placed it among the organic compounds of great importance in biological chemistry.

In 1936, two German scientists, Warburg and Christian, showed that nicotinamide was an essential component of the hydrogen transport system in the form of nicotinamide adenine dinucleotide (NAD).

In 1937, Dr. Conrad Elvehjem, and co-workers, at the University of Wisconsin, discovered that niacin (as either nicotinic acid or nicotinic acid amide, which he isolated from liver) cured blacktongue in dogs, a condition recognized as similar to pellaga in man. Shortly thereafter, several investigators found that niacin was effective in the prevention and treatment of pellagra in humans. Soon, the vitamin became recognized as a dietary essential for man, monkeys, pigs, chickens, and other species.

In 1945, Willard Krehl and his associates at the University of Wisconsin finally solved another mystery in the story of pellagra prevention when they discovered that tryptophan is a precursor of niacin, thereby explaining two things: (1) why milk, which is low in niacin but high in tryptophan, will prevent or cure pellagra; and (2) why, in earlier concepts, protein deficiency was often related to pellagra—for without protein, there could be no tryptophan (the precursor of niacin). Corn is low in tryptophan whereas meat contains both tryptophan and niacin.

In 1971, the name *niacin* was adopted by the American Institute of Nutrition and international agencies for all forms of the vitamin.

It is noteworthy that recent findings indicate that most persons suffering from pellagra have multiple deficiencies—that certain symptoms formerly associated with the disease are not relieved until thiamin and riboflavin are supplied along with niacin.

Today, pellagra is rare in the United States. Even in Latin America and Mexico, where many people eat large amounts of corn, pellagra is seldom seen. This is because of their common practice of soaking the corn in lime, which makes the niacin present in the corn in the bound form (niacytin) more available to the body. This probably explains why Mexicans who eat tortillas are relatively free of pellagra. In making tortillas, the pre-Columbian civilizations of Mexico (Aztec, Mayan, Toltec) devised a procedure to treat corn flour with lime water (alkali) before cooking in order to improve the plastic properties of the dough; and, presumably unbeknown to them, the lime water treatment also freed the niacin from the niacytin and made it fully available to the body tissues.

Africa is the only continent in which pellagra is still a public health problem.

CHEMISTRY, METABOLISM, PROPERTIES. The chemistry, metabolism, and properties of nicotinic acid and nicotinamide follow:

• **Chemistry.**—The structure of nicotinic acid and nicotinamide are shown in Fig. N-3.

Fig. N-3. The formulas of nicotinic acid and nicotinamide reveal that the compounds are derivatives of pyrimidine.

• **Metabolism**—Niacin is readily absorbed from the small intestine into the portal blood circulation and taken to the liver. There it is converted to the coenzyme nicotinamide adenine dinucleotide (NAD). Also, some NAD is synthesized in the liver from tryptophan. NAD formed in the liver is broken down, releasing nicotinamide, which is excreted into the general circulation. This nicotinamide and the niacin that was not metabolized in the liver are carried in the blood to other body tissues, where they are utilized for the synthesis of niacin-containing coenzymes.

Niacin is found in the body tissues largely as part of two important coenzymes, nicotinamide adenine dinucleotide (NAD) and nicotinamide adenine dinucleotide phosphate (NADP); together, NAD and NADP are known as the pyridine nucleotides. The structure of NAD is given in Fig. N-4.

Fig. N-4. Structure of NAD.

NAD is composed of nicotinamide, adenine, two molecules of ribose, and two molecules of phosphate. NADP is similar in structure except it contains three phosphate groups.

Little niacin is stored in the body. Most of the excess is methylated and excreted in the urine, principally as N-methylnicotinamide and N-methyl pyridine (in about equal quantities). Also, small amounts of nicotinic acid and niacinamide are excreted in the urine. With a low niacin intake, there is a low level of metabolite excretion in any form.

• **Properties**—Nicotinic acid appears as colorless needle-like crystals with a bitter taste, whereas nicotinamide is a white powder when crystallized. Both are soluble in water (with the amide being more soluble than the acid form) and are not destroyed by acid, alkali, light, oxidation, or heat.

Nicotinic acid is easily converted to nicotinamide in the body. In large amounts, nicotinic acid acts as a mild vasodilator (as a mild dilator of blood vessels), causing flushing of the face, increased skin temperature, and dizziness. Since nicotinamide does not cause these unpleasant reactions, its use is preferred in therapeutic preparations.

MEASUREMENT/ASSAY. Niacin in foods and niacin requirements are expressed in milligrams of the pure chemical substance.

Chemical and microbiological methods for niacin assay are now generally used rather than animal assays.

The biological vitamin activity of new compounds can be assayed by the curative dog test (blacktongue disease) or by growth test with chicks and rats.

FUNCTIONS. The principal role of niacin is as a constituent of two important hydrogen-transferring coenzymes in the body: nicotinamide adenine dinucleotide (NAD) and nicotinamide adenine dinucleotide phosphate (NADP). These coenzymes function in many important enzyme systems that are necessary for cell respiration. They are involved in the release of energy from carbohydrates, fats, and protein. Along with the thiamin- and riboflavin-containing coenzymes, they serve as hydrogen acceptors and donors in a series of oxidation-reduction reactions that bring about the release of energy (see Fig. N-5).

Fig. N-5. Hydrogen acceptor function of nicotinamide containing coenzymes. R = adenine dinucleotide (= NAD): = adenine dinucleotide phosphate (= NADP).

Also, NAD and NADP are involved in the synthesis of fatty acids, protein, and DNA. For many of these processes to proceed normally, other B-complex vitamins, including vitamin B-6, pantothenic acid, and biotin, are required.

Niacin also has other functions. It is thought to have a specific effect on growth. Also, there are reports that nicotinic acid (but not nicotinamide) reduces the levels of cholesterol; and that niacin in large doses is slightly beneficial in protecting to some degree against recurrent nonfatal myocardial infarction. However, because of possible undesirable effects, ingestion of large amounts (therapeutic doses) of niacin should be under the direction of a physician.

DEFICIENCY SYMPTOMS.   In man, a deficiency of niacin results in pellagra, which generations of medical students have remembered as the disease of the three "Ds"—dermatitis, diarrhea, and dementia (insanity).

The typical features of pellagra are: dermatitis, particularly of areas of skin which are exposed to light or injury; inflammation of mucous membranes including the entire gastrointestinal tract, which results in a red, swollen, sore tongue and mouth, diarrhea, and rectal irritation; and psychic changes, such as irritability, anxiety, depression, and, in advanced cases, delirium, hallucinations, confusion, disorientation, and stupor.

Dogs develop a characteristic black tongue and lesions in the mouth, along with a skin rash, bloody diarrhea, and wasting; followed by eventual death.

(Also see PELLAGRA.)

Fig. N-6. Niacin made the difference! *Top*: A dog that had been fed a diet extremely low in niacin. *Bottom*: The same dog after having been fed meat—a good source of niacin—for only 2 weeks. (Courtesy, University of Wisconsin)

RECOMMENDED DAILY ALLOWANCE OF NIACIN.   Estimation of niacin requirements are complicated (1) by the fact that some tryptophan is converted to niacin in man, (2) by the paucity of people of different ages receiving diets varying in niacin and tryptophan content, and (3) by the possible unavailablity of niacin in some foods (such as corn).

The Food and Nutrition Board (FNB) of the National Research Council (NRC) recommended daily allowances of niacin are given in the section on VITAMIN(S), Table V-5, Vitamin Table, Niacin. Allowances for niacin are commonly related (1) to energy expenditure, based on the essential role of niacin in energy formation—the involvement of the coenzymes NAD and NADP in the functions of respiratory enzymes; and (2) to protein intake, because (a) a diet that fur-

nishes the recommended allowances of protein usually also provides enough niacin through the conversion of tryptophan to niacin, and (b) protein-rich foods are generally, except for milk, rich in preformed niacin. Hence, the section on VITAMIN(S), Table V-5, Vitamin Table, Niacin, gives niacin equivalent allowances in relation to both calories and protein.

Also in this same table, the recommended dietary allowances of niacin are presented as niacin equivalents (NE), recognizing that the contribution from tryptophan may be variable and unpredictable but may represent a substantial portion of the niacin activity of the diet. In estimating the amount of niacin available from foods, the average value of 60 mg of tryptophan should be considered equivalent to 1 mg of niacin. (See Fig. N-8.)

As with the other B-complex vitamins, the niacin requirements are increased whenever metabolism is accelerated as by fever or by the stress of injury or surgery.

Additional pertinent information relative to Table V-5, Vitamin Table, Niacin, follows:

• **Recommended allowance for infants, children, and adolescents**—Human milk contains approximately 0.17 mg of niacin and 22 mg of tryptophan per 100 ml or 70 Calories (kcal). Milk from a well-nourished mother appears to be adequate to meet the niacin needs of the infant. Therefore, the niacin allowance recommended for infants up to 6 months of age is 7.7 niacin equivalents per 1,000 Calories (kcal), about two-thirds of which will ordinarily come from tryptophan. The niacin allowance for children over 6 months of age and for adolescents is 7.1 niacin equivalents per 1,000 Calories (kcal).

• **Recommended allowance for adults**—The allowance recommended for adults, expressed as niacin equivalents, is 6.6 niacin equivalents per 1,000 Calories (kcal) and not less than 13 niacin equivalents at caloric intakes of less than 2,000 Calories (kcal). This amount provides an allowance for the differences in the contributions from tryptophan and the availability of niacin in diets.

• **Recommended allowances for pregnancy and lactation**—The recommended allowance provides an increase of 2 niacin equivalents daily during pregnancy, based on the recommended increase in energy intake of 300 Calories (kcal) daily.

With a recommended increase of 500 Calories (kcal) to support lactation, an additional intake of 3.3 niacin equivalents would be indicated; hence, a total additional intake of 5 niacin equivalents per day is recommended during lactation.

• **Pharmacological intakes of niacin**—Niacin in large doses has been found to be slightly beneficial in protecting to some degree against recurrent nonfatal myocardial infarction. However, ingestion of large amounts of nicotinic acid (3 g or more daily), *but not of the amide,* may produce vascular dilation, or *flushing,* along with other harmful side effects. So, it is recommended that great care and caution be exercised if this vitamin is to be used for the treatment of individuals with coronary heart disease. Megadoses of niacin should be taken under the supervision of a physician.

• **Niacin intake in average U.S. diet**—Average diets in the United States for women ages 19 to 50 supply 700 mg of tryptophan daily, and for men 19 to 50, 1,100 mg. The corresponding values for preformed niacin are 16 and 24 mg, respectively. Thus, the calculated intakes of total NEs are 27 mg for women and 41 mg for men. Proteins of animal origin (milk, eggs, and meat) contain approximately 1.4% tryptophan; most vegetable proteins contain about 1% tryptophan, whereas corn products contain only 0.6%. Some foodstuffs, such as corn, contain niacin-containing compounds from which the niacin may not be completely available. It has been estimated that the enrichment of cereal products adds about 20% more niacin to the food supply than would be provided if these products were not enriched.

The U.S. Department of Agriculture reports that there are sufficient available food sources in the United States to provide an average consumption of 26 mg of niacin per person per day; with 45% of the total contributed by meat, poultry, and fish, and 31% contributed by flour and cereal products.

TOXICITY.   Only large doses of niacin, sometimes given to an individual with a mental illness, are known to be toxic. However, the ingestion of large amounts of nicotinic acid (2 to 3 g per day) may result in vascular dilation, or *flushing* of the skin, itching, liver damage, elevated blood glucose, elevated blood enzymes, and/or peptic ulcer. So, high doses, which are sometimes prescribed for cardiovascular diseases and other clinical symptoms, should only be taken on the advice of a physician.

NIACIN LOSSES DURING PROCESSING, COOKING, AND STORAGE.   Niacin is the most stable of the B-complex vitamins. It can withstand reasonable periods of heating, cooking, and storage with little loss.

Canning, dehydration, or freezing result in little destruction of the vitamin.

Because niacin is water-soluble, some of it may be lost in cooking, but in a mixed diet usually such losses do not amount to more than 15 to 25%. Using a small amount of cooking water will minimize this loss.

Storage results in little loss. In a study of the niacin content of potatoes stored for 6 months at 40°F (4.4°C), only small losses of niacin were observed.

SOURCES OF NIACIN.   Generally speaking, niacin is found in animal tissues as nicotinamide and in plant tissues as nicotinic acid; both forms of which are of equal niacin activity and commercially available. For pharmaceutical use,

Fig. N-7. Poultry, a rich source of niacin. (Courtesy, National Turkey Federation, Reston, Va.)

nicotinamide is usually used; for food nutrification (fortification), nicotinic acid is usually used.

As is true of other B vitamins, the niacin content of foods varies widely. A grouping and ranking of foods on the basis of normal niacin content is given in the section on VITAMIN(S), Table V-5, Vitamin Table, Niacin.

Food Composition Table F-21 gives the niacin content of foods. Proteins and Amino Acids in Selected Foods, Table P-16, gives the tryptophan content on a limited number of foods. In the absence of information on the tryptophan content of foods, the following thumb rules for estimating tryptophan content may be used: Proteins of animal origin (milk, eggs, and meat) contain about 1.4% tryptophan, those of vegetable origin about 1.0%, and corn products about 0.6%. An average mixed diet in the United States provides about 1% of protein as tryptophan. Thus, a diet supplying 60 g of protein contains about 600 mg of tryptophan, which will yield about 10 mg of niacin (on the average, 1 mg of niacin is derived from each 60 mg of dietary tryptophan). (See Fig. N-8.)

EXAMPLE OF CALCULATING NIACIN EQUIVALENTS

Fig. N-8. How to calculate the niacin equivalents in a quart of milk. A quart of milk containing 449 mg of tryptophan will form 7.5 mg of niacin (449 ÷ 60) from tryptophan. So, this is added to the 1.7 mg of niacin found as such in milk, bringing the total niacin equivalents of milk to 9.2.

Niacin may be present in foods in a *bound* form (i.e., niacytin in corn) which is not absorbable. This is particularly true of corn, wheat, oats, rice, and rye. Yet, pellagra occurs less frequently than one might expect in Mexico, where corn consumption is high. Born of centuries of experience, it is the custom in Mexico to treat corn with limewater before making tortillas, thereby liberating the nicotinic acid. Likewise, the Hopi Indians of Arizona roast sweetcorn in hot ashes, another traditional practice that liberates the nicotinic acid. But the ways of food preparation in Africa do not have this effect.

Coffee is a good source of niacin (a dark roast provides about 3 mg of niacin per cup). In certain areas of the world where the diet of the people is low in niacin and tryptophan, their high consumption of coffee may explain their low incidence of pellagra.

When considering sources of niacin, it should be noted that niacin can be, and is, synthesized by the intestinal flora. However, the amount produced is only of minor importance in the human. By contrast, as with thiamin and riboflavin, ruminants (cattle, sheep, etc.) have no dietary requirements for niacin because of bacterial synthesis in the rumen.

Niacin was one of the original vitamins, along with thiamin and riboflavin, first added to white flour in the United States in 1941; and later to other processed cereal products. Nicotinic acid is the form usually added to foods.

As with most other vitamins, the most inexpensive source of niacin is the synthetic source. Nicotinamide (niacinamide), the form usually taken as a vitamin supplement, is available in most food stores and pharmacies at a very reasonable cost.

NOTE: When estimating the total amount of niacin supplied in the diet, the tryptophan content of the foods should be considered, also. Note, too, that most foods that are rich in animal protein are also rich in tryptophan.

(Also see VITAMIN[S], Table V-5.)

NIACINAMIDE

The biologically active form of niacin occurring in the tissues.

NIACIN EQUIVALENT (NE)

Because tryptophan is a precursor of niacin and thus an additional source, dietary requirements for niacin are usually given in terms of total niacin or niacin equivalents. Niacin equivalent is the total niacin available from the diet including (1) preformed niacin, plus (2) niacin derived from the metabolism of tryptophan (60 mg tryptophan = 1 mg niacin).

(Also see NIACIN; and VITAMIN(S).)

NICKEL (Ni)

This element, first discovered in 1751 by Cronstedt and named nickel after "Old Nick," a demon, is the same nickel that the U.S. five-cent piece is named after.

In 1970, F.H. Nielsen, of the U.S. Department of Agriculture, reported that chickens deficient in nickel (Ni) developed slightly enlarged hocks, thickened legs, bright orange leg color (instead of pale yellow-brown), a dermatitis, and a less friable liver; deficiency symptoms which were corrected by adding nickel at a level of 3 to 5 mg per kg of diet. Subsequently, deficiencies have also been reported in rats, pigs, and goats, and nickel has been found in a serum protein (called nickeloplasmin) in rabbits and humans.

NICOTINAMIDE

The amide of nicotinic acid. It has niacin (one of the B vitamins) activity as a constituent of two coenzymes.

(Also see NIACIN; and VITAMIN[S].)

NICOTINAMIDE ADENINE DINUCLEO-TIDE (NAD)

A coenzyme formed by the chemical combination of nicotinamide, adenine, ribose and phosphate. Its formation requires the vitamin niacin. In the body it is employed as a hydrogen (electron) acceptor during the oxidation (breakdown) of foods to form energy.

(Also see METABOLISM; and NIACIN.)

NICOTINAMIDE ADENINE DINUCLEO-TIDE PHOSPHATE (NADP)

A coenzyme of niacin with three high-energy phosphate bonds which facilitates oxidation within the cells.

NICOTINIC ACID

Another name for niacin.
(Also see NIACIN; and VITAMIN[S].)

NISIN

A naturally-occurring antibiotic, sometimes found in milk. Many countries use nisin as a food preservative. However, the direct addition of antibiotic to food is not permitted in the United States.

(Also see ANTIBIOTICS.)

NITRATES AND NITRITES

Contents   Page
  Occurrence and Exposure...........................762
    Naturally Occurring...............................762
    Food Additives...................................762
    Other Sources...................................763
  Dangers of Nitrates and Nitrites....................763
    Toxicity 763
    The Cancer Question............................763
    Exposure to Nitrosamines.......................763
    The FDA and the Delaney Clause.................763

Nitrate refers to the chemical union of one nitrogen (N) and three oxygen (O) atoms, or NO_3, while nitrite refers to the chemical union of one nitrogen (N) and two oxygen (O) atoms, or NO_2. Of prime concern in foods are sodium nitrate (Chile saltpeter), potassium nitrate (saltpeter), sodium nitrite, and potassium nitrite.

OCCURRENCE AND EXPOSURE. Nitrates and nitrites are common chemicals in our environment whether they come from *natural* or *unnatural* sources.

Naturally Occurring. Most green vegetables contain nitrates. The level of nitrates in vegetables depends on (1) species, (2) variety, (3) plant part, (4) stage of plant maturity, (5) soil condition such as deficiencies of potassium, phosphorus, and calcium or excesses of soil nitrogen, and (6) environmental factors such as drought, high temperature, time of day, and shade. Regardless of the variation of nitrate content in plants, vegetables are the major source of nitrate ingestion as Table N-1 shows.

TABLE N-1
ESTIMATED AVERAGE DAILY INGESTION OF NITRATE AND NITRITE PER PERSON IN THE UNITED STATES[1]

Source	Nitrate (NO_3-)	Nitrite (NO_2-)
	(mg)	(mg)
Vegetables	86.1	.20
Cured meats	9.4	2.38
Bread	2.0	.02
Fruits, juices	1.4	.00
Water	.7	.00
Milk and products	.2	.00
Total	99.8	2.60
Saliva[2]	30.0	8.62

[1]*Nitrates: An Environmental Assessment*, 1978, National Academy of Sciences, p. 437, Table 9.1.
[2]Not included in the total since the amount of nitrite produced by bacteria in the mouth depends directly upon the amount of nitrate ingested.

Those vegetables which are most apt to contain high levels of nitrates include beets, spinach, radishes, and lettuce. Despite the shift from manure to chemical fertilizers over the years, the overall average concentration of nitrate in plants has remained unchanged. Other natural sources of nitrate are negligible.

Nitrite occurrence in foods is minimal. Interestingly, a naturally occurring source of nitrites appears to be the saliva.

Food Additives. Both nitrates and nitrites are used as food additives, mainly in meat and meat products, according to the guidelines presented in Table N-2. Their use in meat has been the subject of much publicity, though their use

TABLE N-2
FEDERAL NITRATE AND NITRITE ALLOWANCES IN MEAT

Meat Preparation	Level Allowed	
	Sodium or Potassium Nitrate	Sodium or Potassium Nitrite
Finished product ...	200 ppm or 91 mg/lb (maximum)	200 ppm or 91 mg/lb (maximum)
Dry cure	3.5 oz/100 lb or 991 mg/lb	1.0 oz/100 lb or 283 mg/lb
Chopped meat	2.75 oz/100 lb or 778 mg/lb	0.25 oz/100 lb or 71 mg/lb

to cure meat is lost in antiquity. The role of nitrates in meats is not clear, though it is believed that they provide a reservoir source of nitrite since microorganisms convert nitrate to nitrite. It is the nitrite which decomposes to nitric oxide, NO, and reacts with heme pigments to form nitrosomyoglobin giving meats their red color. Furthermore, taste panel studies on bacon, ham, hot dogs, and other products have demonstrated that a definite preference is shown for the taste of those products containing nitrites. In addition, nitrites retard rancidity, but more importantly, nitrites inhibit microbial growth, especially *Clostridium botulinum*. Hence, cured meats provide a source of ingested nitrates and nitrites. However, as a source of nitrate, cured meats are minor compared to vegetables. The major dietary source of nitrites is cured meats, but this is small when compared to that produced by the bacteria in the mouth and swallowed with saliva. Currently, there is no other protection from botulism as effective as nitrites. So, for those wishing to eliminate nitrites, possible carcinogens, there is a Catch 22—eliminate the nitrites and increase botulism poisoning, a proven danger.

NOTE: Since 1990, 90% of the cured meat samples have contained less than 50 ppm nitrate; only 0.1% have contained more than 200 ppm.

Other Sources. Aside from very unusual circumstances, other sources of exposure to nitrates and nitrites are relatively minor. Nitrate concentrations in groundwater used for drinking range from several hundred micrograms per liter (1.06 qt) to a few milligrams per liter. Nitrates are generally higher in groundwater than surface water since plants remove nitrates from surface water.

DANGERS OF NITRATES AND NITRITES. Possibly these chemicals may present a hazard to man through two routes. First, under certain circumstances nitrates and nitrites can be directly toxic. Second, nitrates and nitrites contribute to the formation of cancer causing nitrosamine.

Toxicity. Our knowledge of the toxic effects of nitrates and nitrites is derived from its long use in medicine, accidental ingestion, and ingestion by animals. Overall, poisoning by nitrates is uncommon. An accidental ingestion of 8 to 15 g causes severe gastroenteritis, blood in the urine and stool, weakness, collapse, and possibly death. Fortunately, nitrate is rapidly excreted from the adult body in the urine, and the formation of methemoglobin generally is not part of the toxic action of nitrates.

Almost all cases of nitrate-induced methemoglobinemia in the United States have resulted from the ingestion of infant formula made with water from a private well containing an extremely high nitrate level. Overall, it is comforting to note that several hundred million pounds of beets and spinach—nitrate-containing vegetables—are eaten yearly without injury.

The Cancer Question. Without doubt, the greatest concern of people is the involvement of nitrates and nitrites in directly causing cancer, or in indirectly producing compounds known as nitrosamines. Since nitrosamines are definitely accepted as carcinogens in test animals, a majority of the furor around nitrates and nitrites stems from this fact.

It cannot be stated that any human cancer has been positively attributed to nitrosamines. However, some nitrosamines have caused cancer in every laboratory animal species tested.

EXPOSURE TO NITROSAMINES. When nitrosamines are mentioned, foods are the first items which come to mind. However, there are numerous other sources of nitrosamines. Furthermore, people are exposed to such nitrosamines as cosmetics, lotions, and shampoos containing N-nitrosodiethanolamine (NDELA), (the latter is carcinogenic in the rat). Other preformed nitrosamines have been found in tobacco and tobacco smoke. Hence, human exposure can result from breathing or eating preformed nitrosamines, or by applying them to the skin. Foods are not the only source of nitrosamines.

THE FDA AND THE DELANEY CLAUSE. To ban or not to ban the use of nitrates and nitrites in foods is the question. In the summer of 1978, this "fire" received more fuel when a study conducted for the FDA by Dr. Paul Newberne of Massachusetts Institute of Technology (MIT) reported that nitrite alone fed to rats increased the incidence of cancers of the lymphatic system. Immediately, and before the study was properly reviewed, the USDA and the FDA announced they would soon ban nitrites. Tempers flared and pork producers lost money due to the implication of bacon containing a cancer-causing substance. The MIT study has now been reviewed by an independent group, and the research has been shown to be in error. For the time being, the FDA and USDA have backed down from their earlier stand to ban nitrite; they now say that the evidence is insufficient to initiate any action to remove nitrite from foods. However, it is noteworthy that the U.S. Supreme Court has cleared the way for the USDA to approve no-nitrate labels in processed meats, should they wish to do so.

On an individual basis, after carefully considering the issue, it is the old question of benefit versus risk. The risk of botulism in cured meats in the absence of nitrite is both real and dangerous, while the risk of cancer from low levels of nitrosamines and/or nitrites remains uncertain. Thus, the risk of botulism is considered greater than the risk of nitrates, so their use is allowed. Furthermore, no acceptable alternative is as effective as nitrite in preventing botulism. Nevertheless, nitrite should be reduced in all products to the extent protection against botulism is not compromised.

(Also see ADDITIVES; and MEAT[S], section on "Meat Curing.")

NITROGEN (N)

A chemical element essential to life. All plant and animal tissues contain nitrogen. Animals and humans get it from protein foods; plants get it from the soil; and some bacteria get it directly from the air. Nitrogen forms about 80% of the air.

(Also see METABOLISM, section on "Proteins"; and PROTEIN[S].)

NITROGEN BALANCE (Nitrogen Equilibrium)

The nitrogen in the food intake minus the nitrogen in the feces, minus the nitrogen in the urine is the nitrogen balance. Normal adults are in nitrogen balance—intake equals output. Nitrogen is obtained from the proteins we eat.

• **Positive nitrogen balance**—When nitrogen intake exceeds nitrogen output a positive nitrogen balance exists. Such a condition is present in the following physiological states: pregnancy, lactation, recovery from a severe illness, growth, and following the administration of an anabolic steroid such as testosterone. A positive nitrogen balance indicates that new tissue is being built.

• **Negative nitrogen balance (Nitrogen deficit)**—This occurs when nitrogen intake is less than output. Starvation, diabetes mellitus, fever, surgery, burns or shock can all result in a negative nitrogen balance. This is an undesirable state since body protein is being broken down faster than it is being built up.

(Also see PROTEIN[S], section on "Quantity and Quality"; and BIOLOGICAL VALUE [BV] OF PROTEINS.)

NITROGEN-FREE EXTRACT (NFE)

It consists principally of the available carbohydrates—sugars, starches, pentoses, and nonnitrogenous organic acids in any given food. The percentage is determined by subtracting the sum of the percentages of moisture, crude protein, crude fat, crude fiber, and ash from 100. This fraction represents a catchall for the organic compounds for which there is no specific analysis when performing a proximate food analysis.

(Also see ANALYSIS OF FOODS.)

NITROGEN, METABOLIC

That nitrogen which is lost in the urine and feces due to the metabolic processes of the body and not due to that derived from the diet. Metabolic nitrogen consists of digestive enzymes, cells from the lining of the gastrointestinal tract and bacteria.

(Also see NITROGEN BALANCE.)

NITROSAMINES

A whole family of chemical compounds formed when chemicals containing nitrogen dioxide (NO_2)—so called nitrites—react with amine (NH_2) groups of other chemicals. Many nitrosamines are potent carcinogens when tested in animals. Because there are numerous chemicals capable of reacting with nitrite, nitrosamines have been found in the air, water, tobacco smoke, cured meats, cosmetics, pesticides, tanneries, alcoholic beverages, and tire manufacturing plants. It is even possible that they are formed in our body, though this process is not clearly demonstrated. Thus, human exposure may result from several routes. However, very few data are available to make estimates of levels of exposure by each route. Some epidemiological studies have associated increased incidence of human cancer with the presence of nitrosamines in the diet. Direct evidence that nitrosamines are carcinogenic for humans is lacking.

(Also see NITRATES AND NITRITES.)

NITROSOMYOGLOBIN

The chemical responsible for the red color of cured meat. It is formed by the decomposition of nitrites to nitric oxide (NO) which reacts with the myoglobin of the muscle.

(Also see MEATS, section headed "Meat Curing"; MYOGLOBIN; and NITRATES AND NITRITES.)

NITROUS OXIDE (N_2O)

To most people it is better known as laughing gas, an inhalation anesthetic and analgesic. However, in the food industry it is approved by the FDA as a propellant and aerating agent in certain sprayed foods canned under pressure such as whipped cream or sprayed vegetable fats.

NOCTURIA

Excessive urination at night.

NONESTERIFIED FATTY ACIDS (NEFA)

Those fatty acids which are freed from triglycerides, and released into the blood. Often they are called free fatty acids (FFA).

(Also see FREE FATTY ACIDS; and TRIGLYCERIDES.)

NONHEME

Iron that is not a part of the hemoglobin molecule; a designation for iron in foods of plant origin.

(Also see IRON, section headed "Absorption, Metabolism, Excretion.")

NONPAREILS

Another name for colored sugar crystals used in decorating candies, cakes, or cookies.

NONPROTEIN NITROGEN (NPN)

Nitrogen which comes from other than a protein source but which under certain circumstances may be used by man in the building of body protein. NPN sources include compounds like urea and salts of ammonia. Persons on low protein diets make better use of NPN than those fed adequate or high levels of protein.

(Also see UREA.)

NONVEGAN

A nonvegetarian; a person who includes animal proteins in the diet.

NOURISH

To furnish or sustain with food or other substances necessary for life and growth.

NUCLEIC ACIDS

Contents	Page
Synthesis	765
Digestion, Absorption, and Metabolism	766
Functions	766
Required Intake	766
Sources	766
Genetic Engineering	766

Nucleic acids were so named because they were originally isolated from cell nuclei. They are the carriers and mediators of genetic information, of which there are two types: *deoxy-*

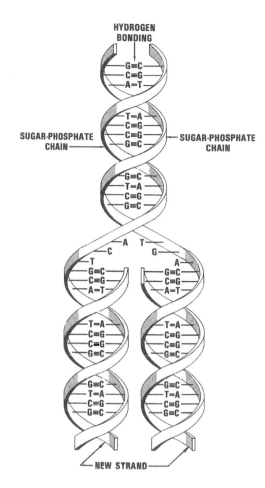

HYDROGEN
BONDING

SUGAR-PHOSPHATE
CHAIN

SUGAR-PHOSPHATE
CHAIN

NEW STRAND

Fig. N-9. The spiral structure of deoxyribonucleic acid, or DNA—the basic building block of life on earth. It's a double helix (a double spiral structure), with the sugar (deoxyribose)-phosphate (phosphoric acid) *backbone* represented by the two spiral ribbons. Connecting the *backbone* are four nitrogenous bases (a base is the nonacid part of a salt): adenine (A) paired with thymine (T), and guanine (G) paired with cytosine (C); with the parallel spiral ribbons held together by hydrogen bonding between these base pairs. Adenine and guanine are purines, while thymine and cytosine are pyrimidines.

ribonucleic acid (DNA) and *ribonucleic acid (RNA)*. The two types of nucleic acids, DNA and RNA, differ in that DNA has one less oxygen molecule (carbon atom number 2) in its component sugar ribose, and is a double strand.

Every cell of the body contains the same amount of DNA with the exception of the sperm and egg cells. It is the DNA of the chromosomes in the nuclei of cells that carries the coded master plans for all of the inherited characteristics—size, shape, and orderly development from conception to birth to death. DNA is different for each species, and even for each individual within a species. These differences consist of minor rearrangements of sequences among the nitrogenous bases, which constitute a code containing all the information on the heritable characteristics of cells, tissues, organs, and individuals.

The messages carried by DNA are put into action in the cells by the other nucleic acid, RNA. To do this, DNA serves as a template (as the pattern or guide) for the formation of RNA. The genetic message is coded by the sequence of purine and pyrimidine bases attached to the *backbone* of the DNA structure—a long chain of the sugar deoxyribose and

phosphoric acid. Purine bases in DNA include adenine and guanine, while pyrimidine bases include cytosine and thymine. One molecule of DNA may contain 500 million bases. The *backbone* of RNA is also a sugar, the sugar ribose, plus phosphoric acid. However, in RNA the pyrimidine base thymine is replaced by uracil, another pyrimidine. RNA molecules are considerably smaller than DNA, containing from less than a hundred to hundreds of bases—not millions.
(Also see METABOLISM, section headed "Proteins.")

SYNTHESIS. Each time a cell divides, a new DNA is made, while within most cells RNA is continually being synthesized and broken down into its components. When the purines and pyrimidines are combined with the sugars ribose or deoxyribose, the resultant compound is referred to as a *nucleoside*. *Nucleotides* are nucleosides that are esterified (acid and alcohol combination) with phosphoric acid. When a series of nucleotides are joined together, nucleic acids are formed. The secret of the code is a process called base pairing.

PYRIMIDINES

CYTOSINE   THYMINE   URACIL

PURINES

ADENINE   GUANINE

SUGARS

DEOXYRIBOSE   RIBOSE

ACID

PHOSPHORIC ACID

Fig. N-10. Components of nucleic acids.

During the formation of new DNA or RNA, the same bases always pair off—guanine with cytosine, adenine with thymine, and adenine with uracil in RNA. This ensures that the code is duplicated time and time again. During synthesis, the double helix splits, pulling the base pairs apart. New bases line up with the proper partner and form another sugar-phosphoric acid *backbone* (see Fig. N-9).

Purines, pyrimidines, ribose, and deoxyribose may be synthesized in the body from other compounds, or they may be recycled. Phosphorus, for the formation of phosphoric acid, is a required dietary mineral, and it is employed in a variety of other compounds besides DNA and RNA.

DIGESTION, ABSORPTION, AND METABOLISM.
The enzymes from the pancreas, called nucleases, split the nucleic acids into nucleotides. Then the nucleotides are split into nucleosides and phosphoric acid by other enzymes in the intestine. Finally, the nucleosides are split into their (1) constituent sugars (deoxyribose or ribose), (2) purine bases (adenine and guanine), and (3) pyrimidine bases (cytosine, thymine, and uracil). These bases are then absorbed into the blood via active transport. The sugars are absorbed slowly, and, as far as is known, there is no specific transport mechanism.

Once in the body, the purine and pyrimidine bases may be reused. They are not only used for DNA and RNA but also as components of a variety of coenzymes. If not used, pyrimidines are catabolized to beta-alanine, beta-aminoisobutyric acid, carbon dioxide (CO_2), and ammonia (NH_3), while the purines are converted to uric acid. Ammonia, uric acid, and beta-aminoisobutyric acid are excreted in the urine.

Ribose can be metabolized via the pentose shunt or hexosemonophosphate pathway—the same metabolic pathway that produces ribose.

(Also see METABOLISM.)

FUNCTIONS.
DNA is the component of the chromosomes that carries the blueprint for a species—the heritable characteristics of each cell in the body and its descendants. It functions in the egg and sperm cells to pass the blueprint along from parent to offspring. Messages are relayed from DNA in the nucleus of cells to the cytoplasm by RNA, whereupon the sequences of amino acids in protein synthesis are dictated by the order of the purine and pyrimidine bases transcribed from the DNA. Each cell of the body contains a full genetic blueprint, but only small parts of the genetic message are normally transcribed by RNA for cells to fulfill their role in whatever tissue they are located.

REQUIRED INTAKE.
There is no required level of intake of nucleic acids *per se*. A nutritious, well-balanced diet provides the precursors necessary for the synthesis of the purines, pyrimidines, and ribose sugar in the body. The only direct requirement is for the mineral phosphorus. Since it is a component of DNA and RNA, it is in every cell of the body. Children and adults require 800 to 1,200 mg daily.

SOURCES.
Since DNA and RNA are components of all cells, any food in which the cells are concentrated is a rich source of nucleic acids. Organ meats such as liver, kidneys, and pancreas are rich sources. Since these meats are rich sources of nucleic acids, it follows that they are rich sources of purines; hence, sufferers of gout are advised to avoid them. Other meats, poultry, and fish, the embryo or germ of grains and legumes, and the growing parts of young plants are good sources of nucleic acids. Butter and other fats, cheese, eggs, fruits, milk, nuts, starch, sugar, and vegetables are low in nucleic acids.

(Also see ARTHRITIS, section headed "Gout [Gouty Arthritis].")

GENETIC ENGINEERING.
The recent development of gene-splicing (also known as recombinant DNA) ushered in a new era of genetic engineering—with all its promise and possible peril. The scientific community is bitterly divided about the unknown risks of "tinkering with life." Proponents of research in DNA are convinced that it can help point the way to new scientific horizons—of understanding and perhaps curing cancer and such inherited diseases as diabetes and hemophilia; of a vastly improved knowledge of the genetics of all plants and animals, including eventually humans. The outcome being the potential of creating new or improved animals and plants, and correcting errors in human genetics. On May 23, 1977, scientists at the University of California-San Francisco reported a major breakthrough as a result of altering genes—turning ordinary bacteria into factories capable of producing insulin, a valuable hormone previously extracted at slaughter from pigs, sheep, and cattle, so essential to the survival of 1.8 million diabetics. The feat opened the door to further genetic engineering or splicing. Already, this genetic wizardry has been used in transplanting into bacteria (and recently into yeast cells) genes responsible for many critical biochemicals in addition to insulin.

Also, the advocates of DNA research argue, recombinant DNA techniques are of enormous help to scientists in mapping the positions of genes and learning their fundamental nature. They also point out that man has been intervening in the natural order for centuries—by breeding animals and hybrid plants, and more recently by the use of antibiotics. On the other hand, the opponents of tinkering with DNA raise the specter (1) of reengineered creatures proving dangerous, and ravaging the earth, and (2) of moral responsibility in removing nature's *evolutionary barrier* between species that do not mate.

Genetic manipulation to create new forms of life make biologists custodians of a great power. Despite different schools of thought, scare headlines, and political hearings, molecular biologists will continue recombinant DNA studies, with reasonable restraints, and work ceaselessly away at making the world a better place in which to live.

NUCLEOLUS

The nucleus usually contains a distinct body—the *nucleolus*, which is rich in RNA.

NUCLEOPROTEIN

A compound of one or more proteins and nucleic acid found in the nuclei of cells; found in large amounts in glandular tissue.

NUCLEOTIDES

Hydrolytic products of nucleic acid; they contain a sugar-phosphate component and a purine or pyrimidine base.

NUCLEUS

That part of the cell that contains the chromosomes; it's the control center of the cell, for both chemical reactions and reproduction; it contains large quantities of DNA.

NUTRIENTS: REQUIREMENTS, ALLOWANCES, FUNCTIONS, SOURCES

Nutrients are the chemical substances found in food that can be used, and are necessary for the maintenance, growth, and health of people. There are more than 40 nutrients,

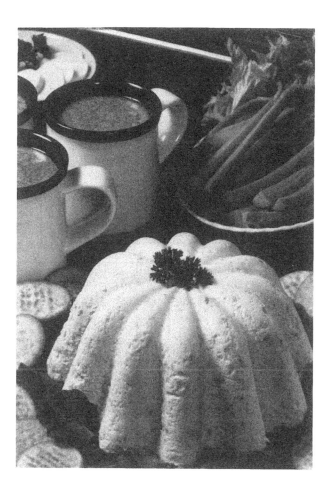

Fig. N-11. Soup in mugs, crackers with cheese spread, and celery sticks provide many nutrients for a luncheon. (Courtesy, Carnation Co., Los Angeles, Calif.)

including minerals, vitamins, and the amino acids from protein. Specific amounts of the nutrients are required, but their intake in the diet is expressed as allowances. Each nutrient from the various food sources reaches the cells of the body where it performs essential functions.

• **Requirements**—Nutrient requirements are the amount of nutrients necessary to meet an individual's minimum needs, without margins of safety, for maintenance, growth, pregnancy, and lactation. To meet nutrient requirements, individuals must receive sufficient food to furnish the necessary quantity of energy, protein, minerals, and vitamins. However, good nutrition is more than just meeting minimal needs.

• **Allowances**—To ensure good nutrition, allowances for the daily intake of nutrients are determined. These allowances provide reasonable margins of safety because of variations in such things as food composition, environment, stress, and individuality. Moreover, the nutritive needs vary according to age, sex, pregnancy, and lactation. Allowances must be determined; otherwise, imbalances may result, excesses may cause toxicity and needless expense, or deficiency diseases may develop. Recommended daily allowances of nutrients for humans are based on the *Recommended Dietary Allowances* compiled by the Committee on Dietary Allowances and the Food and Nutrition Board, National Research Council. The most recent edition

of the *Recommended Dietary Allowances* is the tenth revised edition, 1989, published by the National Research Council, National Academy of Sciences. (See Table N-3.)

• **Functions**—Nutrients perform a variety of specific functions in the body, from providing energy, to becoming tissue components, to acting as cofactors in enzymatic reactions. Altogether, the functions of these nutrients ensure good health as reflected in the general appearance, weight, posture, muscles, nerves, digestive system, heart, lungs, skin, hair, teeth, eyes, nails, and skeleton.

• **Sources**—However, nutrients are more than just chemicals. They are the components of foods, but nutrient distribution in foods is not equal. Some foods may contain virtually no nutrients, while others are good sources of several nutrients. Since there are thousands of foods from which to select, knowledge of foods providing good nutrition becomes very important.

Table N-3 provides summary information relative to the recommended daily allowance of each nutrient, its function in the body, and some of the best foods for obtaining the nutrient. More information about specific nutrients and their food sources may be obtained from individual articles on each nutrient. Also, Food Composition Table F-21 of this book lists complete nutrient composition data for a large number of foods.

(Also see RDA [Recommended Dietary Allowances]; and U.S. Recommended Daily Allowances [U.S.RDA].)

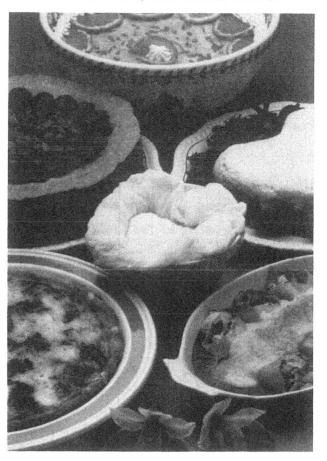

Fig. N-12. Nutrients are more than just chemicals, they are the components of foods. Shown: a variety of good foods. (Courtesy, National Film Board of Canada)

Nutrient	Recommended Daily Dietary Allowances By Age Group																
	Infants		Children			Males					Females						
Years	0–0.5	0.5–1	1–3	4–6	7–10	11–14	15–18	19–24	25–50	50+	11–14	15–18	19–24	25–50	50+	Pregnant	Lactating
Calories																	
REE ....kcal	320	500	740	950	1130	1440	1760	1780	1800	1530	1310	1370	1350	1380	1280	+300	+300
Energy allowance ....kcal/kg	108	98	102	90	70	55	45	40	37	30	47	40	38	36	30		
Total ....kcal	650	850	1300	1800	2000	2500	3000	2900	2900	2300	2200	2200	2200	2200	1900	2500	2700
Protein ....g	13	14	16	24	28	45	59	58	63	63	46	44	46	50	50	60	65
MACROMINERALS																	
Calcium ....mg	400	600	800	800	800	1200	1200	1200	800	800	1200	1200	1200	800	800	1200	1200
Phosphorus ....mg	300	500	800	800	800	1200	1200	1200	800	800	1200	1200	1200	800	800	1200	1200
Sodium[2] ....mg	120	200	225	300	400	500	500	500	500	500	500	500	500	500	500	500	500
Chloride ....mg	180	300	350	500	600	750	750	750	750	750	750	750	750	750	750	750	750
Magnesium ....mg	40	60	80	120	170	270	400	350	350	350	280	300	280	280	280	320	355
Potassium ....mg	500	700	1000	1400	1600	2000	2000	2000	2000	2000	2000	2000	2000	2000	2000	2000	2000
MICROMINERALS																	
Chromium[2] ....mcg	10–40	20–60	20–80	30–120	50–200	50–200	50–200	50–200	50–200	50–200	50–200	50–200	50–200	50–200	50–200	50–200	50–200
Copper[2] ....mg	.4–.6	.6–.7	.7–1.0	1.0–1.5	1.0–2.0	1.5–2.5	1.5–2.5	1.5–3.0	1.5–3.0	1.5–3.0	1.5–2.5	1.5–2.5	1.5–3.0	1.5–3.0	1.5–3.0	1.5–3.0	1.5–3.0
Fluoride[2] ....mg	.1–.5	.2–1.0	.5–1.5	1.0–2.5	1.5–2.5	1.5–2.5	1.5–2.5	1.5–4.0	1.5–4.0	1.5–4.0	1.5–2.5	1.5–2.5	1.5–4.0	1.5–4.0	1.5–4.0	1.5–4.0	1.5–4.0
Iodine ....mcg	40	50	70	90	120	150	150	150	150	150	150	150	150	150	150	175	200
Iron ....mg	6	10	10	10	12	12	10	10	10	15	15	15	15	15	10	30[3]	15[3]
Manganese[2] ....mg	.3–.6	.6–1	1–1.5	1.5–2	2–3	2–5	2–5	2–5	2–5	2–5	2–5	2–5	2–5	2–5	2–5	2–5	2–5
Molybdenum[2] ......mcg	15–30	20–40	25–50	30–75	50–150	75–250	75–250	75–250	75–250	75–250	75–250	75–250	75–250	75–250	75–250	75–250	75–250
Selenium ....mcg	10	15	20	20	30	40	50	70	70	70	45	50	55	55	55	65	75
Zinc ....mg	5	5	10	10	10	15	15	15	15	15	12	12	12	12	12	15	19
FAT-SOLUBLE VITAMINS																	
Vitamin A ....mcg RE	375	375	400	500	700	1000	1000	1000	1000	1000	800	800	800	800	800	1300	1200
Vitamin D ....mcg	7.5	10	10	10	10	10	10	10	5	5	10	10	10	5	5	10	10
Vitamin E ....mg alpha TE	3	4	6	7	7	10	10	10	10	10	8	8	8	8	8	10	12
Vitamin K ....mcg	5	10	15	20	30	45	65	70	80	80	45	55	60	65	65	65	65

Footnotes at end of table.

N–3
FUNCTIONS, SOURCES[1]

Nutrient	Major Function(s) in the Body	Best Food Source
Calories		
Carbohydrates	Supply food energy; in certain cases, fiber and bulk	Bread and cereal products made from whole grains, starchy vegetables, fruits, and sugars.
Fats	Provide food energy and the essential fatty acid; carrier of fat-soluble vitamins; body structure; and regulatory functions.	Vegetable oils, butter, whole milk and cream, margarine, lard, shortening, salad and cooking oils, nuts, and meat fat
Protein	Formation of new tissues; maintenance, regulatory functions; and provide energy.	Meats, fish, poultry, eggs, cheese, dried beans, soybeans, peas, lentils, nuts, and milk.
MACROMINERALS		
Calcium	Builds and maintains bones and teeth; blood clotting, muscles contraction and relaxation, cell wall permeability; enzyme activation; and nerve transmission.	Milk and milk products, most nuts, dried figs, fish with soft edible bones, and green leafy vegetables.
Phosphorus	Bone formation and maintenance, development of teeth, building muscles, genetic transmission and control of cellular metabolism, maintenance of osmotic and acid-base balance, and important in many metabolic functions.	Milk and milk products, rice bran, rice polish, wheat bran, sunflower seeds, beef, lamb, pork, poultry, liver, fish and seafoods, nuts, and whole grains.
Sodium	Helps maintain balance of water, acids, and bases in fluids outside of cells; a constituent of pancreatic juice, bile, sweat, and tears; associated with muscle contraction and nerve functions; involved in absorption of carbohydrates	Table salt, and most fresh or processed foods other than fresh fruits and vegetables. **NOTE WELL:** Generally, the need is for low-sodium diets, which calls for avoiding high-sodium foods.
Chlorine	Helps regulate osmotic pressure, water balance and acid-base balance; necessary for stomach acid formation and vitamin B-12 and iron absorption.	Table salt and foods containing table salt.
Magnesium	Constituent of bones and teeth; essential element of cellular metabolism; involved in protein metabolism; relaxes nerve impulses.	Nuts, sesame seeds, spices, wheat bran, wheat germ, whole grains, and molasses.
Potassium	Involved in acid-base balance and transfer of nutrients in and out of individual cells; relaxes heart muscle; required for secretion of insulin, phosphorylation of creatin, carbohydrate metabolism, and protein synthesis.	Fruits, molasses, rice bran, sunflower seeds, wheat bran, beef, most raw vegetables, nuts, pork, poultry, and sardines.
MICROMINERALS		
Chromium [2]	Component of the Glucose Tolerance Factor (GTF) which enhances the effect of insulin; enzyme activator; stabilizer of nucleic acids.	Apple peel, banana, beef, beer, blackstrap molasses, bread, brown sugar, butter or margarine, cheese, eggs, flour and whole wheat, liver, oysters, and potatoes.
Copper [2]	Facilitates iron absorption and release from storage; necessary for hemoglobin formation; hair pigmentation, component of important proteins.	Black pepper, blackstrap molasses, cocoa, green olives, liver, nuts and seeds, oysters, soybean flour, and wheat bran and germ.
Fluoride [2]	Necessary for sound bones and teeth.	Dry tea, fluoridated water, and seafoods.
Iodine	Essential component of hormones secreted by the thyroid gland.	Iodized table salt, kelp, seafoods, and vegetables grown on iodine-rich soils.
Iron	Iron (heme) combines with protein (globin) to make hemoglobin for red blood cells. Also, iron is a component of enzymes which are involved in energy metabolism.	Organ meats (liver, kidney, heart) blackstrap molasses, oysters, rice polish, wheat bran, beef, brown sugar, egg yolk, lima beans, nuts, pork, and iron-fortified products.
Manganese [2]	Formation of bone and growth of connective tissues; blood clotting; insulin action; enzyme activation.	Brown rice, legumes, rice bran and polish, spices, and whole grains.
Molybdenum [2]	Component of enzymes which metabolize carbohydrates, fats, proteins, and nucleic acids; component of tooth enamel.	Leafy vegetables, legumes, organ meats (liver, kidney, heart), whole grains, and yeast.
Selenium	Component of the enzyme glutathione peroxidase; interrelated with vitamin E as an antioxidant.	Beer, blackstrap molasses, Brazil nuts, butter, clams, crab, egg, lamb, lobster, mushrooms, oysters, smelt, Swiss chard, turnips, and whole grains
Zinc	Needed for normal skin, bones, and hair; component of many enzymes of energy metabolism and of protein synthesis.	Beef, lamb, liver, oysters, peanuts, pork, poultry, spices, and wheat bran.
FAT-SOLUBLE VITAMINS		
Vitamin A	Necessary for vision in dim light, the growth and repair of certain tissues, and resistance to infection.	Dark-green leafy vegetables, yellow vegetables, yellow fruits, crab, halibut, oysters, salmon, swordfish, butter, cheese, egg yolk, margarine (fortified), and whole milk.
Vitamin D	Essential for the absorption and utilization of calcium in the building of strong bones and teeth.	Fatty fish, liver, egg yolk, cream, butter, cheese, milk fortified with vitamin D. Exposure of the skin to sunlight results in the conversion of provitamin D (converted in the body to vitamin D).
Vitamin E	Serves along with selenium as an antioxidant; protects fatty acids from oxidative destruction.	Wheat germ, salad and cooking oils, margarine, nuts, sunflower seed kernels, beef and organ meats, butter, eggs, green leafy vegetables, oatmeal, and seafoods.
Vitamin K	Necessary for the clotting of the blood.	Green leafy vegetables, green tea, beef liver. Produced by microorganisms in the intestine.

(Continued)

Nutrient	Recommended Daily Dietary Allowances By Age Group																
	Infants		Children			Males					Females						
Years	0–0.5	0.5–1	1–3	4–6	7–10	11–14	15–18	19–24	25–50	50+	11–14	15–18	19–24	25–50	50+	Pregnant	Lactating
WATER-SOLUBLE VITAMINS																	
Biotin [2] ....mcg	10	15	20	25	30	30–100	30–100	30–100	30–100	30–100	30–100	30–100	30–100	30–100	30–100	30–100	30–100
Folate ....mcg	25	35	50	75	100	150	200	200	200	200	150	180	180	180	180	400	280
Niacin ....mg NE	5	6	9	12	13	17	20	19	19	15	15	15	15	15	13	17	20
Pantothenic acid [2] ...mg	2	3	3	3–4	4–5	4–7	4–7	4–7	4–7	4–7	4–7	4–7	4–7	4–7	4–7	4–7	4–7
Riboflavin (vitamin B-2)........mg	.4	.5	.8	1.0	1.2	1.5	1.8	1.7	1.7	1.4	1.3	1.3	1.3	1.3	1.2	1.6	1.8
Thiamin (vitamin B-1) mg	.3	.4	.7	.9	1.0	1.3	1.5	1.5	1.5	1.2	1.1	1.1	1.1	1.1	1.0	1.5	1.6
Vitamin B-6 (pyridoxine) ....mg	.3	.6	1.0	1.1	1.4	1.7	2.0	2.0	2.0	2.0	1.4	1.5	1.6	1.6	1.6	2.2	2.1
Vitamin B-12 (cobalamins) ....mcg	.3	.5	.7	1.0	1.4	2	2	2	2	2	2	2	2	2	2	2.2	2.6
Vitamin C (ascorbic acid) .....mg	30	35	40	45	45	50	60	60	60	60	50	60	60	60	60	70	95

[1]Compiled from *Recommended Dietary Allowances*, 10th ed., National Academy of Sciences, 1989.

[2]These figures are given in the form of ranges of recommended intakes, since there is insufficient information to determine allowances.

[3]The increased requirement during pregnancy cannot be met by the iron content of habitual American diets nor by the existing iron stores of many women; therefore the use of 30–60 mg of supplemental iron is recommended. Iron needs during lactation are not substantially different from those of nonpregnant women, but continued supplementation of the mother for 2 to 3 months after parturition (birth) is advisable in order to replenish stores depleted by pregnancy.

NUTRITION

Nutrition can be defined as the science of food and its nutrients and their relation to health.

NUTRITIONAL DEFICIENCY DISEASES

Those disorders in normal structure and function of the body resulting from dietary shortages of one or more essential nutrients. They may be prevented or cured by the administration of the missing nutrient(s), except when there is irreparable damage to vital tissues of the body.

(Also see DEFICIENCY DISEASES.)

NUTRITIONAL STATUS

An evaluation or assessment of how well the needs of the body for essential nutrients are being met.

NUTRITIONAL SUPPLEMENTS

A nutritional supplement is any food(s) or nutrient(s), or *a mixture of both, used to improve the nutritional value of the diet.* Usually, a nutritional supplement consists of minerals, vitamins, and unidentified factor source(s), although it may include protein, one or more amino acids, and other substances.

Clearly, the above definition states that a nutritional supplement is "used to improve." It follows that if the diet is complete and balanced, and not in need of improvement, no supplementation is necessary. Certainly, nutritional supplementation is unnecessary for anyone who always eats a balanced diet consisting of such foods as fresh fruit, fresh vegetables, an animal protein (meat, milk, and/or eggs), whole grain breads and cereals, unroasted nuts, and honey—all produced on unleached, mineral-rich soils; the type of good eating that grandmother served up. But how many people eat such foods regularly today? Also, and most important, there is a wide difference between (1) the amount of a nutrient(s) required to prevent deficiency symptoms, and (2) the amount required for buoyant good health.

N–3 *(Continued)*

Nutrient	Major Function(s) in the Body	Best Food Source
WATER-SOLUBLE VITAMINS		
Biotin [2]	Essential for the metabolism of carbohydrates, fats, and proteins.	Cheese (processed), kidney, liver, most vegetables, eggs, nuts, sardines, salmon, and wheat bran. Also, considerable biotin is synthesized by the microorganisms in the digestive tract.
Folate	Important in metabolism, in cell division and reproduction, and in the formation of heme—the iron-containing protein in hemoglobin.	Liver and kidneys, beans, beets, eggs, fish, green leafy vegetables, nuts, oranges, and whole wheat products.
Niacin	Constituent of coenzymes that function in cell respiration and in the release of energy from carbohydrates, fats, and protein; involved in synthesis of fatty acids, protein, and DNA.	Liver, kidney, lean meats, poultry, fish, rabbit, cornflakes (enriched), nuts, milk, cheese, and eggs.
Pantothenic Acid [2]	Functions as part of two enzymes important in metabolism, nerve impulses, hemoglobin synthesis, synthesis of steroids, maintenance of normal blood sugar, and formation of antibodies.	Organ meats (liver, kidney, heart), wheat bran, rice bran, rice polish, nuts, salmon, eggs, brown rice, and sunflower seeds.
Riboflavin (vitamin B-2)	Essential for metabolism of amino acids, fatty acids, and carbohydrates, accompanied by release of energy; and necessary for formation of niacin from the amino acid tryptophan.	Organ meats (liver, kidney, heart), cheese, eggs, lean meat (beef, pork, lamb), enriched breads, turnip greens, wheat bran, and bacon.
Thiamin (vitamin B-1)	Energy metabolism—without thiamin there could be no energy; needed for conversion of glucose to fats, healthy nerves, normal appetite, muscle tone, and good mental attitude.	Rice bran, wheat germ, rice polish, lean pork, sunflower seeds, nuts, wheat bran, kidney, enriched breads, rye bread, whole wheat bread, and soybean sprouts.
Vitamin B-6 (pyridoxine)	In its coenzyme forms, involved in a number of physiologic functions, particularly (1) protein metabolism, (2) carbohydrate and fat metabolism, and (3) central nervous system disturbances.	Rice polish, rice bran, wheat bran, sunflower seeds, bananas, corn, fish, kidney, liver, lean meat, nuts, poultry, brown rice, and whole grains.
Vitamin B-12 (cobalamins)	Needed for red blood cell production, healthy nerves, and metabolism.	Organ meats, muscle meats, fish, shellfish, egg, cheese. Most plant foods contain little or none of this vitamin.
Vitamin C (ascorbic acid)	Formation and maintenance of collagen; involved in metabolism of amino acids tyrosine and tryptophan, absorption and movement of iron, metabolism of fats and lipids and cholesterol control; and makes for sound teeth and bones, and strong capillary walls and blood vessels.	Citrus fruits, guavas, peppers (green or hot), green leafy vegetables, cantaloupe, papaya, strawberries, and tomatoes.

NUTRITION EDUCATION

Contents

	Page
Nutrition Educators	772
Government Agencies	772
Educational Institutions	772
Professional Organizations	772
Voluntary Agencies	772
Food Industries	772
Nutrition Education Channels	773
Nutrition Education Sources	773
Summary	773

Nutrition education is concerned with (1) imparting sound knowledge of how food selection influences the health and well being of the individual, and (2) motivating the individual to use that information in daily living. Nutrition is, amongst other things, a behavioral science. It follows that nutrition education needs to be a practical program with a positive approach based on nutrition science and food composition, along with an understanding of human behavior. Additionally, cognizance should be taken that American consumers want freedom of choice in what they believe and in what they consume. This challenges nutrition educators to present scientifically based nutrition information in such a way that its authenticity is recognized and acted upon.

More people in the United States are malnourished because of nutritional ignorance and misinformation than because of poverty. The explanation is simple: In human folly, facts have never stood in the way of myths and misinformation. Faddism is costly, too. Americans spend more than $10 billion each year on magic to lose weight—more than is spent on research to discover the major causes of diseases that kill. Only a massive program of nutrition education can hope to have any overall effect in raising the quality of diets.

In the developing countries of the world where population growth is resulting in hunger and malnutrition, food production is the first requisite. Nutrition education has little effect where people have neither the food nor the money with which to buy food to quiet their hunger pangs.

NUTRITION EDUCATORS. All consumers, regardless of age, lifestyle, and cultural and socioeconomic background, need nutrition education. To reach everyone will require the enormous team approach of the following:

1. Government agencies.
2. Educational institutions.
3. Professional organizations.
4. Voluntary agencies.
5. Food industries.

Each of the above agencies or groups has its own special resources, facilities, programs, and opportunities to contribute to a national nutrition education program. Pertinent information about each of them follows:

Government Agencies.
The golden rule is that "those who have the gold make the rules." It follows that government programs have the greatest impact of any agency or group on foods and nutrition.

The two government agencies having most to do with foods and nutrition issues are the U.S. Department of Agriculture (USDA) and the U.S. Department of Health and Human Services—notably two of its agencies, the Food and Drug Administration (FDA) and The National Institutes of Health (NIH).

The USDA is involved in massive food and nutrition programs, such as the School Lunch Program, the School Breakfast Program, and the Special Milk Program—all for children; the special supplemental food program for Women, Infants, and Children (WIC); and the Food Stamp Programs. These food programs cost more than $21.7 billion in 1990. Additionally, through its Cooperative Extension Programs, the USDA has a vast network of home economists who reach a large segment of the consuming public and have a tradition of success in consumer education programs.

The FDA is charged with protecting the safety of the food supply. Through its dietary guidelines for a variety of food products, and through its regulation of "standards of identity," the FDA can also influence the food supply in very significant ways. More recently, the FDA has become involved in nutrition education efforts through its nutrition labeling program, as a consequence of which more and more foods appearing on supermarket shelves are nutritionally labeled. This program, which represents the first effort in this country to provide nutrition information for consumers, should be augmented with adequate education efforts.

The National Institutes of Health supports nutrition and health research.

A host of other government agencies have an impact, in one way or another, upon the nation's food and nutrition. The Federal Trade Commission regulates food advertising and the application of anti-trust laws. The Office of Education supports educational research.

Also, Congress has become interested and involved in food and nutrition issues. Thus, there are more and more hearings dealing with nutrition topics. The dietary goals prepared by the Senate Select Committee on Nutrition and Human Needs indicate the tremendous impact which Congress can have on nutrition programs of this country.

All of the above comments indicate that the government recognizes that foods and nutrition are clearly linked to health and, as an important factor in the maintenance of health, demand government attention.

(Also see FOOD STAMP PROGRAM; GOVERNMENT FOOD PROGRAMS; HOME ECONOMISTS; SCHOOL LUNCH PROGRAM; U.S. DEPARTMENT OF AGRICULTURE; and U.S. DEPARTMENT OF HEALTH AND HUMAN SERVICES.)

Educational Institutions.
Teachers (including physical education teachers and coaches, science teachers, and home economics teachers), doctors, nurses, dietitians, and social workers must receive adequate training in nutrition if they are to be competent educators in the area of health protection. Only through adequate training of these key personnel can sound nutrition education be imparted to students, patients, clients, and parents.

There is need for expanded nutrition instruction in higher education. For example, the professional preparation of most elementary school teachers does not include a single course in nutrition. It is noteworthy, however, that several American universities do have excellent postgraduate courses in nutrition.

In medical schools, some nutrition is taught as part of the undergraduate courses in biochemistry, physiology, pharmacology, pathology, internal medicine, pediatrics, obstetrics, surgery, and dentistry. However, there is a scientific core to nutrition which is unlikely to be covered except in one or more basic courses in nutrition. This includes such topics as the assessment of nutritional status, interactions of nutrients with one another and with diseases, recommended intakes of nutrients, food composition, food technology, world food problems, and psychological and sociological aspects of food habits.

Nurses should also have a sound elementary knowledge of nutrition and dietetics because of their responsibility in feeding patients.

Professional Organizations.
Professional organizations engaged in human nutrition work need to be active in nutrition education. A number of these organizations are listed in the last section of this article, under the heading "Nutrition Education Sources."

Voluntary Agencies.
Nutrition programs at the community level are greatly augmented by voluntary agencies; among them, parent-teachers associations, church organizations, the Salvation Army, children's camps, and day nurseries. Voluntary agencies are supported by private funds, such as the United Fund, foundations, and other means.

Food Industries.
Government alone cannot meet the need for nutrition education, nor can a single agency, organization, institution or group accomplish what is needed. The demands are too great.

Some early leaders of our basic food industries perceived that their success depended both on quality of production and on consumers who were well informed about the nutritive value of their products and how to use them. Thus, in the 1920s, organizations came into being which were concerned with production, processing, distribution and household use of specific foodstuffs—organizations that were variously named councils, boards, institutes, or foundations. They were primarily concerned with kinds of food, rather than brand names. Earliest among these organizations were the National Live Stock and Meat Board and the National Dairy Council. Other food industries followed in the ensuing years. The Nutrition Foundation, an organization of many industries, was formed in 1941 and "dedicated to the advancement of nutrition knowledge and its effective application in improving the health and welfare of mankind."

NUTRITION EDUCATION CHANNELS. The solution lies in instituting and/or enlarging nutrition education programs through the following channels:
1. Elementary and secondary schools.
2. Doctors, nurses, dietitians, teachers, and social workers.
3. News media and community programs.
4. Food Industries.
5. Regulatory controls.
6. Food Assistance.
7. Books and news media.

Nutrition Education Sources. Sound nutrition information should be disseminated through good books and the news media.

All consumers, and all those who counsel with them—doctors, dentists, nutritionists, health experts, and others in allied fields—should have a book shelf on which good nutrition books are readily available.

Today's superstar M.D. will be tomorrow's mediocre M.D. without a continuing effort to keep abreast of medical progress. There are several ways to do this, but reading good books is one of the best. Look at the books in your doctor's office. Are they new editions, or are they dusty with disuse and tattered with age?

Sound nutrition information should also involve the news media—newspapers, radio, and television. Food advertising, particularly on the television, has a powerful influence on food choices. Many foods are promoted for their convenience and ease of preparation or for their taste, rather than for nutritional value. Although convenience and good taste are important considerations, a balanced presentation should also consider nutritional value.

In addition to good books and the news media, recommended sources of information about foods and nutrition follow:

• **County Agricultural Extension (Farm Advisor)**—Information can be secured from this office in each county.

• **State Universities**—A list of available bulletins and circulars, including information regarding foods and nutrition, can be obtained by writing to (1) the Foods and Nutrition Department, or (2) the Agricultural and Home Economics Extension Service at each State University.

• **U.S. Department of Agriculture, Washington, D.C. 20250**—The following USDA services can be contacted for information:
Agricultural Research Service
Consumer and Marketing Service
Cooperative State Research Service
Economic Research Service
Federal Extension Service
Foreign Agricultural Service
International Agricultural Development Service
Office of Communication

• **U.S. Department of Health and Human Services, F St. between 18th & 19th Streets N.W., Washington D.C. 20006**—The following USDHS services (agencies) can be contacted for information:
Administration on Aging
Children's Bureau
Food and Drug Administration, 5600 Fishers Lane, Rockville, MD 20857
Maternal and Child Health Services

Office of Education
Public Health Service
National Institutes of Health, Bethesda, MD 20205

• **U.S. Department of State**—
Agency for International Development, U.S.A.I.D. Administration, 2401 E. St. N.W., Washington, DC 20523

• **Professional Societies and Voluntary Health Associations**—
American Academy of Pediatrics, P.O. Box 1034, Evanston, IL, 60204
American Council on Science and Health, 1995 Broadway, New York, NY 10023
American Dental Association, 211 E. Chicago Ave., Chicago, IL 60611
American Diabetes Association, 2 Park Ave., New York, NY 10016
American Dietetic Association, 430 North Michigan Ave., Chicago, IL 60611
American Heart Association, 7320 Greenville Ave., Dallas, TX 75231
American Home Economics Association, 2010 Massachusetts Ave. N.W., Washington, DC 20036
American Institute of Nutrition, 9650 Rockville Pike, Bethesda, MD 20814
American Medical Association, Council on Foods and Nutrition, 535 N. Dearborn St., Chicago, IL 60610
Council for Agricultural Science and Technology (CAST), 137 Lynn Ave., Ames, IA 50010-7197
Institute of Food Technologists, Suite 2120, 221 North LaSalle St., Chicago, IL 60601
National Academy of Sciences, National Research Council, 2101 Constitution Ave., Washington, DC 20418
Nutrition Today Society, 703 Giddings Ave., P.O. Box 1829, Annapolis, MD 21404
Society for Nutrition Education, 1736 Franklin St., Oakland, CA 94612

• **Foundations**—
Ford Foundation, 320 East 43rd St., New York, NY 10017-4801
Kellogg Foundation, 400 North Ave., Battle Creek, MI 49017-3398
Medical Education and Research Foundation, 1100 Waterway Blvd., Indianapolis, IN 46202
Vitamin Information Bureau, 612 N. Michigan Ave., Chicago, IL 60611
The Nutrition Foundation, 489 Fifth Ave., New York, NY 10017
Rockefeller Foundation, 1133 Avenue of the Americas, New York, NY 10036

• **Industry-Sponsored Groups**—
American Meat Institute (AMI), 1700 N. Moore St., Suite 1600, Arlington, VA 22209-1995
Cereal Institute, 1111 Plaza Drive, Schaumburg, IL 60195
National Dairy Council, 6300 North River Road, Rosemont, IL 60018
National Live Stock and Meat Board, 444 North Michigan Ave., Chicago, IL 60611

• **International Organizations**—
Food and Agriculture Organization (FAO) of the United Nations, Via delle Terme di Caracalla, 00100 Rome, Italy
League for International Food Education, 1126 Sixteenth St., N.W., Washington, DC 20036

United Nations Children's Fund (UNICEF), 331 East 38th St., New York, NY 10016

United Nations Scientific and Cultural Organization (UNESCO), 7, Place de Fontenoy, 75700 Paris, France

World Health Organization (WHO, Geneva, Switzerland), WHO Publication Centre, 49 Sheridan Ave., Albany, NY 12210

SUMMARY. Food choices are determined primarily by availability, personal and family likes and dislikes, marketing and advertising, and disposable income. There is need that the person who buys and prepares the food should give greater consideration to nutritional need.

Without sacrificing soundness and accuracy in the least, we need to emulate the purveyors of myths and misinformation when it comes to communicating effectively. Through whatever channel—books, news media, lecturers, schools, seminars, counselors—the chosen words and terms should always convey the intended meaning and concept. But a big word should not be used if it can be avoided. Choose words that are clear, specific, and simple, and that come alive. As an illustration, consider the words used by one of the masters of the English language, Sir Winston Churchill. During World War II, Churchill could never have rallied the British people to defend their country if he had called on them for "hemorrhage, perspiration, and lachrymation." Instead, he used the immortal words, "blood, sweat, and tears"—words that continue to live in history.

NUTRITION REVIEWS

This journal is published by the International Life Sciences Institute (formerly known as the Nutrition Foundation, Inc., which was formed in 1941 by the food and allied industries). The Institute's goal is to make essential contributions to the advancement of nutrition knowledge and its effective application, and thus serve the health and welfare of the public. *Nutrition Reviews* contains abstracts of current scientific literature in nutrition. Also, throughout the years, the Institute has published semipopular brochures on current topics in nutrition.

NUTRITIONIST

An individual who is trained in and able to apply a knowledge of foods and their relationship to growth, maintenance, and health. A nutritionist may apply his or her training and knowledge in clinics, consultation, medicine, public health, research, and teaching.

(Also see DIETITIAN and DIETETICS.)

NUTRITIVE RATIO (NR)

The ratio of digestible protein to other digestible nutrients in a food. It is the sum of the digestible protein, fat, and carbohydrate divided by the digestible protein.

NUTS

Contents

	Page
Origin and History	774
World and U.S. Production	775
Processing	775
Nutritive Values	775
Fatty Acid Content	776
Protein Quantity and Quality	776
Minerals and Vitamins	776
Nuts of the World	776

Fig. N-13. Assorted nuts—a good source of copper, zinc, magnesium, and the B-complex vitamins. (Courtesy, USDA)

This name is commonly given to the shell-encased seeds of nonleguminous trees, although various other seeds commonly called nuts may not grow on trees, and may even be legumes. (The peanut is the seed from a leguminous plant.)

ORIGIN AND HISTORY Archaeological evidence indicates that between 20,000 B.C. and 10,000 B.C. various nomadic peoples around the world began to form permanent settlements in certain areas where wild animals and plants were abundant. However, the numbers of large wild game species dwindled as the human populations grew. As a result, the settlers in the temperate regions came to depend upon small game and wild grains, legumes, and nuts to meet the greater part of their food requirements. Nuts could be handled and stored almost as well as cereal grains and legume seeds.

Soon after 10,000 B.C., some of the settlements began to cultivate the plants they favored. By Biblical times, almonds, pistachio nuts, and walnuts were grown in southwestern Asia; and cashew nuts and peanuts were grown in South America. The early agriculturalists experimented with various ways of processing nuts, which eventually resulted in the production of such items as *milks*, oils, and nut powders. The North American Indians made milks from hickory nuts and pecans; the southeastern Asians made milk from coconuts; and the Middle Eastern peoples made milks from almonds and walnuts. Similarly, oils were obtained from almonds, coconuts, and walnuts. Finally, the use of almond and pistachio nut powders in a variety of dishes appears to have been passed along by the Persians to the Arabs, who, in turn, spread the practice westward throughout the Mediterranean, and eastward to India. Today, food technologists are striving to perfect modern day counterparts of the nut products prepared by these ancient peoples.

The first Spanish explorers to reach the Americas found peanuts growing on some of the Carribean islands, to which they had apparently been brought by earlier migrations of South American Indians. These nuts were taken to Africa, where they soon became an important crop.

Through the years, several species of nuts have been improved by selection and breeding, and subjected to complex cultural systems. Today, most marketed nuts come from cultivated nut trees. But Brazil nuts are still produced only by wild trees, and wild trees contribute to the production of many other species, such as the pecan and filbert.

WORLD AND U.S. PRODUCTION.
World nut production is small compared to grain, oilseed, and legume production.

Fig. N-14 shows the leading nut crops of the world, and the annual production of each. Fig. N-15 shows the leading nut crops of the United States, and the annual production of each.

WORLD NUT PRODUCTION

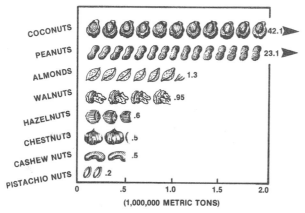

Fig. N-14. The world production of the leading nut crops. (Source: *FAO Production Yearbook 1990*, FAO/UN, Rome, Italy, Vol. 44)

Fig. N-14 shows that coconuts and peanuts are by far the most important nut crops of the world. Together, they account for about 94% of the world nut production. This situation prevails because coconut and peanut oils are among the leading ingredients of margarines and shortenings.

The production of nuts in the United States represents only a small fraction of the world production of these crops, except in the case of almonds, where the United States contributes almost one-fourth of the world's production.

U.S. NUT PRODUCTION

Fig. N-15. The U.S. production of nut crops. (Source: *FAO Production Yearbook 1990*, FAO/UN, Rome, Italy, Vol. 44, and *Agricultural Statistics 1991*, USDA, p. 221)

Fig. N-15 shows that the peanut crop accounts for most (about 68%) of the nut production in the United States. It is noteworthy that the per capita consumption of peanuts in the United States is the highest in the world.

PROCESSING.
Following harvesting, nuts must be shelled. Also, green nuts contain 50% or more of water when harvested; hence, they must be cured or semidried for proper storage. Certain types of nut meats are often

Fig. N-16. Harvesting almonds. (Courtesy, Anderson/Miller & Hubbard Consumer Services, San Francisco, Calif.)

roasted and salted, while others are packed without processing. The most common types of packaging are plastic bags, glass jars, and metal cans. However, some nuts are still sold unpacked or loose from bins or cannisters.

NUTRITIVE VALUES.
The Food Composition Table F-21 gives the nutrient composition of some of the commonly used nuts and nut products.

A few pertinent points concerning the nutritive value of nuts follow:

1. Most nuts contain about 500 or more Calories (kcal) per 100 g portion (140 or more kcal per oz) due to their high fat content.

2. The protein content of nuts is about the same as that of legumes (close to 20%, on the average), except that the protein-to-calorie ratio for most nuts (about 4 g of protein per 100 kcal) is only two-thirds of that for legumes.

3. Defatted products, such as almond meal and peanut flour, are much better sources of protein (they contain 40% or more) and substantially lower in calories than nondefatted products.

4. Nuts contain only one-quarter to one-third as much carbohydrate as grains and legumes. Hence, nuts may be useful for low-carbohydrate diets.

5. Almonds, Brazil nuts, and filberts are moderately good sources of calcium (100 g furnish from ⅕ to ¼ of the Recommended Dietary Allowance [RDA] for adults). Almost all nuts furnish at least ⅓ of the RDA for phosphorus, and are fair to good sources of iron.

6. Nuts are also good sources of the B-complex vitamins, as indicated by the values for thiamin, riboflavin, and niacin. However, roasting destroys much of the thiamin. Nuts also furnish biotin, pantothenic acid, vitamin B-6, and vitamin E.

7. Nuts are good sources of (a) magnesium, a macromineral that activates enzymes and aids functioning of nerves and muscles, (b) zinc, a component of various enzymes, and (c) copper, the mineral that functions in enzyme actions, hair growth and pigmentation, connective tissue development, and red blood cell formation.

Details regarding some of the nutritive values not indicated in Food Composition Table F-21 are discussed in the sections that follow.

Fatty Acid Content. The fats in all of the common edible nuts, with the exception of coconuts, contain 77% or more of unsaturated fatty acids. Coconut fat contains over 90% saturated fatty acids. Table N-4 lists the unsaturated and polyunsaturated fatty acid content, as well as the content of the essential fatty acid linoleic acid, in some common nuts.

TABLE N-4
FATTY ACID CONTENT OF SOME COMMON NUTS[1]

Nut	Saturated Fatty Acids	Polyunsaturated Fatty Acids	Linoleic Acid
	(g/100 g)	(g/100 g)	(g/100 g)
Almond	4.3	10.5	9.9
Brazil nut	17.4	25.7	25.4
Cashew	9.2	7.7	7.3
Coconut	31.2	.8	.7
Filbert (hazelnut)	4.6	6.9	6.6
Hickory nut	6.0	12.7	12.0
Peanut	8.6	13.3	13.3
Pecan	6.1	18.3	17.0
Pistachio	7.4	7.3	6.8
Walnut, black	5.1	40.9	36.8
Walnut, English	6.9	42.0	34.9

[1]Data from Table F-4, Fats and Fatty Acids in Selected Foods.

All of the nuts in Table N-4, except almonds, cashews, coconuts, filberts, and pistachios contain liberal amounts of linoleic acid—the essential polyunsaturated fatty acid. One ounce (*28 g*) of peanuts, Brazil nuts, or English walnuts provides about 4, 7, and 10 g of linoleic acid, respectively. Therefore, it may be better to rely on nuts rather than vegetable oils to supply the essential fatty acid linoleic acid because the nuts also furnish protein, minerals, and certain vitamins that are not supplied by the oils. Shelled nuts, like vegetable oils, should be kept refrigerated to prevent them from becoming rancid.

(Also see FATS AND OTHER LIPIDS.)

Protein Quantity and Quality. Like legumes, nuts contain about twice as much protein as the common cereal grains, and have amino acid patterns that generally complement those of the cereals. However, nuts contain less of the amino acid lysine than legumes. Hence, the age-old practice of combining nut milks with cooked cereals made a lot of sense when animal protein was scarce. The American version of these complementary protein mixtures is peanut butter sandwiches, which children should eat with milk to offset the mild deficiency of lysine in the sandwich.

(Also see PROTEIN[S], section headed "Sources.")

Minerals and Vitamins. In addition to the minerals listed in Food Composition Table F-21, nuts are also good sources of (1) chromium, a micromineral which renders insulin more effective in promoting the passage of nutrients into cells; (2) manganese, an element required for growth of bones and other connective tissues, insulin action, blood clotting, and enzyme actions in many aspects of metabolism; and (3) selenium, which is present in an enzyme that detoxifies dangerous peroxides formed during metabolism. In general, nuts contain from 1 to 7 times the amounts of these minerals present in the other types of fresh, minimally processed natural foods.

Nuts also supply choline, a vitaminlike factor required for the transport of fats from the liver. Normally, the body synthesizes choline, but sometimes the amounts may be insufficient to meet its needs.

In general, the nutritive values of nuts are sufficient to enable people to live healthfully on them (provided that 12 to 16 oz [*340 to 454 g*] are consumed daily), plus some fruits and/or vegetables to supply the missing vitamin A and vitamin C. However, the prices of nuts have recently become so high that it is now more economical to eat dietary combinations of animal foods, cereals, and legumes instead of nuts. Also, the other foods are lower in calories. Therefore, it seems that with the exception of a few undeveloped areas blessed with an abundance of wild nut trees, most people of modest means will have to settle for using nuts as occasional snacks and garnishes for dishes based mainly upon other foods.

NUTS OF THE WORLD. Each type of nut tree (or vine in the case of peanuts) originated in an area with a particular climate and certain soil condition. Also, the growing conditions, harvesting, and processing for the market may vary for each species, as may the most important food uses.

Fig. N-17. Pistachio nuts. (Courtesy, Field Museum of Natural History, Chicago, Ill.)

Finally, nutritive values vary over a spectrum ranging from fair to excellent with regard to the major nutrients. Information on these aspects is given in summary form in Table N-5. Additionally, the most important U.S. nuts—almonds, coconut, peanuts, pecans, and walnuts—are accorded narrative status in this book.

NOTE WELL: Where production figures are given in Table N-5, column 2 under "Importance," unless otherwise stated, they are based on 1990 data. Most world figures are from *FAO Production Yearbook 1990,* FAO/UN, Rome, Italy, Vol. 44. Most U.S. figures are from *Agricultural Statistics 1991,* USDA. Annual production fluctuates as a result of weather and profitability of the crop.

TABLE N-5
NUTS OF THE WORLD

Popular Name; Source; Scientific Name; Origin and History	Importance; Principal Areas; Growing Conditions	Processing; Preparation; Uses	Nutritive Value[1]
Almond (See ALMONDS.)			
Brazil Nut **Source:** The seed of the Brazil nut tree. **Scientific Name of the Brazil Nut Tree:** Family: *Lecythidaceae* Genus: *Bertholletia* Species: *B. excelsa* **Origin and History:** Originated in the Amazon valley in northern Brazil, the only region where it now grows.	**Importance:** An important native food in the areas where it is grown. Also, from 30,000 to 40,000 metric tons are exported to the U.S. and Europe. **Principal Areas:** The Amazon valley in Brazil. **Growing Conditions:** Grows wild in the forests of the Amazon valley. Little attempt is made to cultivate Brazil nut trees in their native area, and outside the Amazon valley they have survived poorly.	**Processing:** The nut-bearing fruits are allowed to fall to the ground when ripe. Then, the tough fruit capsules are cracked open and 15 to 30 nuts are removed. The hard shells may either be removed or left on the nuts prior to shipping and marketing. **Preparation:** No treatment is required for the shelled nuts. Sometimes, the kernels are crushed and oil is extracted. **Uses:** Snacks; ingredient of baked goods, confections, and nut mixtures. The oil may be used in salads.	Brazil nuts are similar in nutritive value to almonds, except that they contain more calories, essential fatty acids, phosphorus, and thiamin. Also, they contain less calcium, iron, magnesium, riboflavin, and niacin. The shelled nuts should be refrigerated or at least kept under cool conditions because their high oil content makes them very susceptible to rancidity. A 3½ oz (*100 g*) portion of Brazil nuts furnishes 715 kcal (204 kcal per oz).
Cashew Nut **Source:** The seed adhering to the cashew fruit. **Scientific Name of the Cashew Tree:** Family: *Anacardiaceae* Genus: *Anacardium* Species: *A. occidentale* **Origin and History:** Native to the West Indies and to Central and South America. Introduced into India and Africa by the Portuguese.	**Importance:** The fifth leading nut crop in the world, with an annual production of about 478,832 metric tons. **Principal Areas:** India, Brazil, Nigeria, Mozambique, and Tanzania. **Growing Conditions:** Dry, tropical climate. Well adapted to poor soils and sandy soils.	**Processing:** Nuts are allowed to fall to the ground. The shell around the nuts contains a caustic oil that blisters the hands. Hence, the shell oil is often extracted with solvents or by roasting or steaming after the nuts have been dried in the sun, cleaned, moistened, and allowed to stand in heaps. Then, the nuts are cracked and the kernels removed, after which the adhering skin is removed from the kernel. **Preparation:** Cashew nut kernels are usually consumed raw or after roasting and salting. **Uses:** Nut kernels are used as snacks, and as ingredients of baked goods and confections. Shell oil is used as a food and as a flavoring.	Cashew nuts are similar in nutritive value to almonds, except that they contain more thiamin. Also, they contain a little less energy (calories), protein, phosphorus, and iron; and much less essential fatty acids, calcium, riboflavin, and niacin. Oleic acid accounts for about 90% of the unsaturated fatty acids, and linoleic only 10%. Hence, cashew nuts are less likely to grow rancid than some of the other nuts which are much richer in linoleic acid. A 3½ oz (*100 g*) portion of cashew nuts contains 596 kcal (170 kcal per oz).
Chestnut **Source:** The seed of the chestnut tree. **Scientific Name of the Chestnut Tree:** Family: *Fagaceae* Genus: *Castanea* Species: *C. crenata* (Japanese), *C. dentata* (N. American), *C. mollissima* (Chinese), *C. sativa* (European) **Origin and History:** Different species native to each of 4 areas: Europe, Africa, Asia, and U.S.	**Importance:** The sixth leading nut crop in the world, with an annual production of about 488,161 metric tons. **Principal Areas:** China, South Korea, Italy, Japan, and Spain. **Growing Conditions:** Mild, temperate climate with few frosty nights. Does best on well drained soils. The N. American chestnut is the hardiest of the varieties, but it was practically wiped out by chestnut blight. Hence, hybrids of the American and Oriental chestnuts are now being grown.	**Processing:** The fallen burrs are gathered and the chestnuts removed. After roasting, the nuts may be left whole or ground into flour. **Preparation:** Roasting, boiling, making into a puree, grinding into a flour, preserving with sugar or syrup. **Uses:** Snacks; ingredients of confections, desserts, fritters, porridges, soups, stews, and stuffings.	Dried chestnuts differ greatly from all the other nuts in that they contain considerably more carbohydrate (3 to 4 times as much), and only fractions of the protein (about ⅓) and fat (about 1/16) content. Also, they furnish little more than ½ the calories of other nuts. The overall nutrient composition of chestnuts is quite close to that of dry cereal products made from corn or rice. A 3½ oz (*100 g*) portion of dried chestnuts contains 377 kcal (107 kcal per oz).
Coconut (coconut palm) (See COCONUT.)			
Hazelnuts (Filberts) **Source:** The seed of the hazel or filbert tree. **Scientific Names of the Filbert and Hazel Trees:** Family: *Betulaceae* Genus: *Corylus* Species: *C. maxima* (Filbert), *C. avellana* (Hazel) **Origin and History:** Various species are native to Europe, Asia, and N. America.	**Importance:** The seventh leading nut crop in the world, with an annual production of over 561,000 metric tons. **Principal Areas:** Turkey, Italy, and Spain. In the U.S., most commercial growing of filberts is in Oregon, Washington, and California. **Growing Conditions:** Moist, temperate climate. The filbert tree is usually larger and more hardy than the hazel tree.	**Processing:** Fallen nuts are gathered by hand or by mechanical sweepers. After shelling, the nuts may be dried and/or the oil extracted. **Preparation:** No treatment is required for shelled nuts. **Uses:** Snacks; ingredients of chocolates and other confections, and desserts. Filbert oil is used in the manufacture of cosmetics and perfumes.	Filberts are similar in nutritive value to almonds, except that they furnish more calories, fat, and thiamin; but less protein (only ⅔ as much), calcium, phosphorus, iron, and niacin. A 3½ oz (*100 g*) portion of filberts supplies 700 kcal (200 kcal per oz).

Footnote at end of table

(Continued)

TABLE N-5 *(Continued)*

Popular Name; Source; Scientific Name; Origin and History	Importance; Principal Areas; Growing Conditions	Processing; Preparation; Uses	Nutritive Value[1]
Macadamia Nut **Source:** The seed of the macadamia tree. **Scientific Name of the Macadamia Tree:** Family: *Proteaceae* Genus: *Macadamia* Species: *M. ternifolia* **Origin and History:** Native to Australia. Brought to Hawaii, where it has prospered.	**Importance:** One of the leading orchard crops of the Hawaiian Islands, where about 18,210 short tons are produced annually. **Principal Areas:** Australia and Hawaii. **Growing Conditions:** Moist semitropical climate. Does best on well-drained soils.	**Processing:** Fallen nuts are gathered and the outer husk is removed by a special machine. The unshelled nuts can be stored safely in a dry environment. Shelling is done by a cracking machine after the nuts have been air dried to a low moisture content. The shelled nuts are fried in deep fat, then salted and packaged. **Preparation:** No treatment is needed for packaged nuts. **Uses:** Snacks; ingredients of cakes, candy, and ice cream.	Macadamia nuts are very high in fat (75%), but low in protein (8%), calcium, phosphorus, iron, and the B vitamin complex. A 3½ oz (*100 g*) portion furnishes 776 kcal (259 kcal per oz).
Peanut (See LEGUMES, Table L-2, Legumes of the World; and PEANUTS.)			
Pecan (See PECAN.)			
Pine Nut (Pinon Nut) **Source:** The seeds of various species of pine trees. **Scientific Names of the Nut-bearing Pine Trees:** Family: *Pinaceae* Genus: *Pinus* Species: *P. cembra* (Swiss stone pine), *P. cembroides* (Mexican nut pine), *P. edulis* (pinon pine of southwestern U.S), and *P. pinea* (stone pine of the Mediterranean) **Origin and History:** Trees are native to the areas noted under Species.	**Importance:** Little of the crop enters the nut trade. The annual production in northern Mexico and southwestern U.S. averages 3,000 or more metric tons. **Principal Areas:** Northern Mexico, southwestern U.S., Europe, and Asia. **Growing Conditions:** Varies according to species. Generally, the nut-bearing species grow wild in the temperate zones of the Northern Hemisphere. The nut pines of N. America are adapted to an arid climate.	**Processing:** Most harvesting consists of picking the cones from the trees before the cone scales separate. Then, the cones are dried to free the nuts. Machines are used to shell the nuts. (American Indians gathered fallen nuts and cracked them with hand tools.) Some species require roasting to remove a turpentine flavor. **Preparation:** No further treatment is required for commercially processed nuts, but unprocessed items may require roasting. **Uses:** Snacks; ingredient of baked goods, confections, desserts, and vegetarian dishes.	The European types (generally called pignolias) are richer in protein (31%) but lower in carbohydrates (12%) than the American types (pinon nuts) which contain 13% protein and 20% carbohydrates. Compared to almonds, pinon nuts are richer in phosphorus, iron, and the vitamin B complex, but lower in calcium. All types of pine nuts contain fats that are highly susceptible to rancidity. Hence, they should be stored in a refrigerator. A 3½ oz (*100 g*) portion of pignolia nuts supplies 630 kcal (180 kcal per oz), whereas the same amount of pinon nuts furnishes 593 kcal (169 kcal per oz).
Pistachio Nut **Source:** The seed of the pistachio tree. **Scientific Name of the Pistachio Tree:** Family: *Anacardiaceae* Genus: *Pistacia* Species: *P. vera* **Origin and History:** Native to western Asia. Recently introduced to the central valley of California.	**Importance:** The eighth leading nut crop in the world, with an annual production of about 220,490 metric tons. **Principal Areas:** Iran, U.S., Turkey, Syria. **Growing Conditions:** Temperate to subtropical climate. Requires well-drained soils, and thrives on sandy loams. Tolerates drought well, and often grows wild on dry wastelands.	**Processing:** The nuts are harvested by picking the clusters or by beating them from the trees onto cloths. They may be husked immediately or after drying in the sun followed by soaking in water. Some of the nuts are roasted and salted in brine while in the shell, while others are cracked and the shells removed. **Preparation:** None required. **Uses:** Snacks; ingredient of confections, desserts, and ice cream.	Pistachio nuts are very similar in nutritional value to almonds, except that they are richer in iron and thiamin, and lower in calcium and niacin. A 3½ oz (*100 g*) portion contains 635 kcal (181 kcal per oz).
Walnut, English (See WALNUT.)			

[1]Additional details on the nutritive values are given in the Food Composition Table F-21, Table P-16 Proteins and Amino Acids in Selected Foods, and Table F-4, Fats and Fatty Acids in Selected Foods.

NYCTALOPIA

The medical term for night blindness.
(Also see NIGHT BLINDNESS.)

OATS *Avena sativa*

Contents

	Page
Origin and History	779
World and U.S. Production	779
Processing Oats	779
Nutritional Value of Oats	780
Oat Products and Uses	780

Fig. O-1. Oat panicles with many branches and spikelets. Note the loose-fitting hulls surrounding the grain, which are inedible by humans and must be removed when oats are processed for food. (Photo by J.C. Allen & Sons, West Lafayette, Ind.)

ORIGIN AND HISTORY. Oats, one of the most important food crops in the world, belong to the grass family, *Gramineae*, the same family as wheat, corn, rice, barley, and rye. Most cultivated oats have been developed from the genus *Avena*, species *A. sativa*.

Oats are grown in most temperate countries of the world for food for humans and livestock. They are believed to have developed from wild stocks that first grew in Asia. As a cultivated crop, oats were substantially later in origin than wheat. Not until about 2,000 years ago, at the beginning of the Christian era, are references to oats as a cultivated crop found in the literature. Early use of oats appears to have been medicinal, rather than as a food crop.

WORLD AND U.S. PRODUCTION. In 1990, 44 million metric tons of oats were produced in the world, almost exclusively in temperate North America and Europe. Oats rank sixth among the cereal grains, but account for only 2.2% of the total cereal grain produced globally. The ten leading oat-producing countries of the world by rank are: The Soviet Union, United States, Canada, Germany, Poland, Finland, Sweden, Australia, France, and Argentina.[1]

The United States produced 357,000,000 bushels of oats in 1990; and the ten leading oat-producing states, by rank, were: South Dakota, Minnesota, Wisconsin, Iowa, North Dakota, Ohio, Pennsylvania, Nebraska, Michigan, and Illinois.[2]

In the United States, oats rank fifth as a cereal crop in tonnage produced, being exceeded only by corn, wheat, grain sorghum, and barley.

PROCESSING OATS.

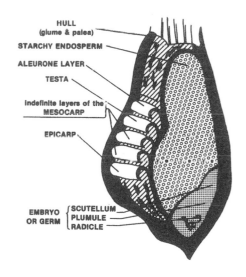

Fig. O-2. Covered kernel (caropsis) of oats.

---

[1]Based on data from *FAO Production Yearbook 1990*, FAO/UN, Rome, Italy, Vol. 44, p. 22. **Note well:** Annual production fluctuates as a result of weather and profitability of the crop.

[2]Based on data from *Agricultural Statistics 1991*, USDA, p. 41, Table, 51. **Note well:** Annual production fluctuates as a result of weather and profitability of the crop.

Only about 15% of the U.S. oat crop is processed for food. The oat kernel has a fibrous hull, which is inedible by humans. So, the goal in milling oats is to obtain the maximum yield of clean, uniform, sound, whole oat kernels which are free from hulls, floury material, extraneous matter, and undesirable flavors. The hulls range from 21 to 43%, with an average of 25%.

The milling of oats involves the following steps:

1. **Cleaning.** Foreign material and oats not suitable for milling are first removed.

2. **Drying and roasting.** The grain is subjected to heat and the moisture is reduced to somewhere between 7 and 8½%. This process develops flavor, improves keeping quality, and facilitates the subsequent separation of the hull from the groats.

3. **Cooling.** After drying, the oats are cooled by air circulation.

4. **Hulling.** The hulling machine separates the groat kernel from the surrounding hull. The following products are obtained from the hullers: hulls, groats, broken groats and meal, flour, and unhulled oats. These materials are separated by air aspiration and screening. The choicest, plumpest groats are used to make package-grade rolled oats, while the less choice groats make either bulk or feed rolled oats. The broken kernels are used as livestock feed.

5. **Cutting.** The purpose of cutting is to convert groats to uniform granules.

6. **Flaking.** The granules are flaked between rolls to produce a quick-cooking breakfast cereal.

NUTRITIONAL VALUE OF OATS.
The nutrient composition of oats and oat products is given in Food Composition Table F-21.

A few salient points regarding the nutritional value of oats follow:

1. Oat cereals are palatable, rich in carbohydrates (starch), a good source of thiamin and vitamin E, but deficient in minerals, especially calcium.

2. Oat groats contain more protein than wheat. The protein is not of the glutenin type; hence, oat flour is not suitable as the sole flour for breads. However, it may be used for cakes, biscuits, and breakfast foods.

3. Oats lack protein quantity and quality. While dry oatmeal is about 14% protein, cooked oatmeal is only about 2% protein. Moreover, it is deficient in the essential amino acid lysine.

4. The oat grain very closely resembles a kernel of wheat in structure. Unlike wheat, however, the nutritious bran and germ are not removed in the normal processing because oats are not refined. The most important nutritional advantage of oats is the soluble fiber which helps to lower the cholesterol in the blood. Oat bran is a major source of this fiber.

When oats are processed, the hull is removed, leaving the whole grain. After further processing, the whole grain oat will become rolled, steelcut, or instant oats. The outer covering of the whole grain oat is bran. In the production of oat bran, however, the bran plus 3 to 4 layers of cells called aleurone layers are included in the final product.

(Also see CEREAL GRAINS and PROTEIN[S].)

OAT PRODUCTS AND USES.
The products from the milling of oats are: oat hulls, oat groats, steel-cut oats, oatmeal, rolled oats, oat flakes, oat flour, and the feed by-products.

Fig. O-3. Many a school child has been raised on a breakfast of hot oatmeal porridge. (Courtesy, USDA)

Table O-1 presents in summary form the story of oat products and uses for human food.

(Also see BREAKFAST CEREAL[S]; CEREAL GRAINS; and FLOURS, Table F-13, Special Flours.)

TABLE O-1
OAT PRODUCTS AND USES FOR HUMAN FOOD

Product	Description	Uses	Comments
Oat flakes	The product made from the whole groats through (1) by-passing the cutting system, and (2) subjecting the whole kernel to steaming and flaking to a slightly greater thickness than regular rolled oats.	Porridge	This product is sometimes referred to as *old fashioned* or *5-minute* oat flakes.
Oat flour	The product produced by grinding and screening the oat flakes from the rolls.	As an antioxidant, as a constituent of baby foods, and in soaps and cosmetic preparations.	The antioxidant property of oat flour is due to the presence of caffeic acid derivatives which act as strong antioxidants. This explains why properly processed oatmeal is extremely stable and can be maintained in good quality in closed containers for many years.
Oatmeal	Uniform granules of a mealy texture, with a minimum of fine granules and flour.	Porridge	The decline in the consumption of oatmeal porridge is attributable to the greater convenience and speed of ready-cooked breakfast cereal. Oatmeal is made by subjecting groats to cutters, which convert them to uniform granules.
Rolled oats	The product resulting from rolling the groats after treating them with live steam, then passing the flakes through separators to remove all the fine material.	Breakfast foods, cookies, and breads. Granola (homemade or commercially prepared), eaten either— .1. As a ready-to-eat breakfast cereal; or 2. As a snack	Rolled oats and oatmeal contain about 1,850 calories (kcal) per lb (4,079 calories [kcal] per kg); hence, they are excellent food for the winter diet. Granola is a mixture of rolled oats, honey, vegetable oil, chopped nuts, shredded coconut, sunflower seeds, raisins, chopped dates, cinnamon, vanilla, and/or other ingredients. The raw granola is usually heated in an oven until crisp and slightly browned.

OATCAKES

These are thin, flat cakes made by mixing oatmeal with either water, milk, or sour milk, then cooked on a griddle.

OBESE

Overweight due to a surplus of body fat.
(Also see OBESITY.)

OBESITY

Contents   Page
Types of Obesity 781
Causes of Obesity 782
    Hereditary and Acquired Traits 782
        Body Builds (Somatotypes) 782
        Differences in the Utilization of Food Energy 782
        Hormonal Influences on Metabolism 783
        Hormonal Abnormalities in the Obese 783
    External Factors Which May Spur
        the Development of Obesity 783
Disorders Which Are Often Aggravated by Obesity 784
Ways of Detecting Borderline Obesity 785
    Formulas for Estimating Desirable Body Weights 785
    Height-Weight Tables 785
    An Index of Slenderness (Ponderal Index) 785
    Pinch Tests for Hidden Fat 786
    Weighing People Under Water 786
    Detection of Early Signs of Obesity in Children 786
Treatment of Obesity 786
    Principles of Weight Reduction 786
    Planning Diets and Selecting Foods 786
        Modified Diets 787
        Suggestions for Preparing Meals at Home 789
        Eating Away from Home 789
    Changing Eating Behavior 790
        Eating Behavior at Home 790
        Eating Behavior Away from Home 790
    Participation in Weight Control Groups 791
    Drugs Which Affect Appetite and Metabolism 791
    Exercise 791
    Surgery and Other Drastic Measures 793
Summary 793

This term means an excess of body fat beyond that needed for optimal maintenance of body functions. Obesity is *not* synonymous with overweight because (1) a person may have an optimal body weight, yet have an excess of fat and less than the normal amount of lean (non-fat) tissue; or (2) some overweight people may have above average amounts of bone and muscle, but only normal amounts of body fat.

Some of our leading medical people consider obesity to be the most widespread nutritional disorder in the United States.[3] An estimated 100 million Americans are obese. In 1990, Americans spent an estimated $33 billion on diets and diet-related products. Statistics gathered by life insurance companies show that obese people are more likely to develop degenerative diseases and die early. However, these statistics have limited value when applied to the U.S. population at large because (1) they are based upon the characteristics of people who purchased life insurance policies; (2) the distinction between obesity and overweight is not made, since no measurements of fatness, other than body weight, are made; (3) the high sickness and death rates

Fig. O-4. Down through the ages, people have been concerned with their weight. (Courtesy, The Bettmann Archive, Inc.)

of overweight people who have diseases that are aggravated by the gaining of weight, such as diabetes, heart disease, high blood fats, and high blood pressure, tend to distort the overall rates for overweight people who have none of these disorders, because unhealthy people are not considered separately from healthy people; and (4) the ranges of desirable weights for small, medium, and large frames were established arbitrarily, with no attempts to define criteria for assigning people to the appropriate categories.

In spite of the great uncertainty as to what constitutes the *ideal* weight for any particular person, the attainment of such a weight has long been promoted as a goal for those who wish to remain in good health. Some doctors devote their entire medical practice to the treatment of obesity. Also, great amounts of money are now being made by manufacturers of such items as dietetic foods, exercise devices, jogging attire, liquid and powdered formulas for losing weight, *rainbow* pills (mixtures of drugs for losing weight), and special suits which cause loss of weight by inducing a heavy loss of water as sweat.

TYPES OF OBESITY. Although all fat people may look alike to the untrained observer, there seems to be considerable variation in the physiological characteristics of obese people, judging from the many recent reports on this subject. Studies of the development of obesity in both animals and man have shown that there are two major types of obesity: early-onset and late-onset.[4] Descriptions follow:

[3]Sebrel, W. H., Jr., "Changing Concept of Malnutrition," *The American Journal of Clinical Nutrition*, Vol. 20, p. 653.

[4]Winick, M., "Childhood Obesity," *Nutrition Today*, Vol. 9, May/June 1974, p. 6.

• **Early-onset obesity**—This type is characterized by an abnormally large number of fat cells because it originates in childhood while the cells in the fatty tissues are still dividing. Early-onset obesity is often accompanied by extra nonfat tissue because the dietary, hereditary, and hormonal factors which spur the proliferation of fat cells also stimulate extra growth of bone and muscle.

• **Late-onset obesity**—This type, in contrast to early-onset type, is characterized by enlargement of normal numbers of fat cells because it usually begins after the division of cells in the tissues has ceased. Sometimes, adult obesity may be distinguished from that which began in childhood because the former is characterized by proportionately greater deposits of excess fat on the trunk of the body than on the arms and legs. Those who first became obese in adulthood are usually more successful in achieving long-term control of their body weight.

CAUSES OF OBESITY.   There are many different opinions as to the cause(s) of obesity. The simplest concept is that it is almost always due to a combination of overeating and underactivity. Hence, there tends to be an attitude of moral superiority in some people who flaunt their leanness at those who are obese. However, there may be many other reasons for obesity and leanness. Some of the current ideas on this subject follow:

Hereditary and Acquired Traits.   Each newborn baby has unique features which are determined mainly by heredity. However, the eventual development of the baby into a fat or lean adult depends upon a combination of both hereditary and environmental factors. Some well-known anthropologists, physicians, and psychologists believe that tendencies toward fatness or leanness may depend somewhat upon body build and other traits. Explanations follow.

BODY BUILDS (Somatotypes).   Dr. Sheldon, an American physical anthropologist, and his co-workers have described the temperaments they observed to be associated with each of three types of physique: ectomorph, mesomorph, and endomorph.[5] These types are shown in Fig. O-5.

ECTOMORPH   MESOMORPH   ENDOMORPH

Fig. O-5. Sheldon's three extreme types of body build.

Details follow:

• **Ectomorph**—This type of physique is tall and slender with long arms and legs, but the trunk is short compared to the rest of the body. The personality of the person who has such a physique is generally of the introvert type. Most of the introvert's drive originates from within, so he or she works well alone, and with minimal supervision. Ectomorphs are often characterized by other people as ambitious and persevering, but also as somewhat inhibited and tense. Statistics which have been compiled by life insurance companies show that people with this physique live the longest. Hence, the tables of ideal weights for various heights tend to be closer to those of ectomorphs than to those of endomorphs or mesomorphs.

• **Mesomorph**—This body build has traditionally been viewed as the epitome of masculinity, although some women may also have this physique yet be truly feminine in every respect. It is characterized by an athletic build with good proportions between the arms, legs, and trunk. Usually, the chest is more prominent than the abdomen. Mesomorphs are rarely extreme extroverts or extreme introverts, but they are energetic and perform well in such team sports as football and hockey. Although life insurance statistics suggest that mesomorphs do not usually live as long as ectomorphs in sedentary, urban societies, it might be possible for people with this body build to live longer than expected by regularly engaging in strenuous physical activities. For example, a study of San Francisco longshoremen showed that those who consistently performed the heaviest work in loading cargo had only about ½ the death rate and ⅓ the sudden death rate as those who did lighter loading work.[6]

• **Endomorph**—People with this body build usually have large abdomens, long trunks, but short arms and legs. They are closest in personality to the extrovert type. In contrast to the introvert, the extrovert's drive depends mainly upon stimulation by such external factors as good food, interaction with other people, and a comfortable environment. Others often see endomorphs as good natured, pleasure loving, talkative, and on the lazy side. Extroverts sometimes have difficulty working alone. Many endomorphs may appear to be obese, but this designation may often be made without taking into consideration their marked differences in body build from the ectomorphic physique.

DIFFERENCES IN THE UTILIZATION OF FOOD ENERGY.
It has long been known that energy production by the body is stepped up after a meal has been consumed. The extra energy is usually lost from the body in the form of heat. *Hence, this effect of eating food is called thermo-genesis (heat production) or specific dynamic action (SDA).* Thermogenesis is taken into consideration in caloric allowances since they are usually increased by 10% after the energy costs of (1) basal metabolism, and (2) the daily activities have been estimated. However, the amount of extra heat production after meals varies considerably for different people, depending upon circumstances such as those which follow:

[5]Sheldon, W. H., and S. S. Stevens, *The Varieties of Temperament,* Harper and Row, New York, N.Y.

[6]Paffenbarger, R. S., Jr., and W. E. Hale, "Work Activity and Coronary Heart Mortality," *New England Journal of Medicine,* Vol. 292, 1975, p. 545.

• **Degree of fatness or leanness**—Certain obese patients had lower metabolic rates after eating, whereas thin patients had a rise in heat production in response to the test meal.[7] These results suggest there may be great differences in the efficiency with which various people utilize food energy, and that the obese may be able to maintain their body weights on below average caloric intakes.

• **Level of physical activity**—Subjects who exercised before and after a high calorie breakfast lost twice as much heat after eating as those who did not exercise.[8]

Hence, exercising before and after meals may result in a greater expenditure of calories than performance of the same amount of exercise at other times.

• **Prior level of dietary calories**—People who have long tried to lose weight by dieting sometimes find it difficult, if not impossible, to maintain a steady pattern of weight loss after having had some initial success in this regard. A study showed that about one-third of a group of women who had participated in weight control groups for 6 months or more could not lose weight on a diet containing 1,350 Calories (kcal).[9] Generally, the women who failed to lose weight had subnormal basal metabolic rates and low total energy expenditures. It was suspected that the low basal metabolism was an adaptation to the long use of low calorie diets.

• **Cellular metabolism**—Not all fat people are guilty of gluttony and sloth. On this point increasing physiological evidence agrees with the anguished testimony of overweight men and women. By and large, obese people do not eat more than thin people. As for sloth, it is noteworthy that physical activity accounts for only about 20% of most adults' energy expenditure. Why, then, do some people gain weight on a diet that allows others to remain slim, even at the same level of physical activity? It is not that the metabolic engine runs steadily at a lower energy cost in fat people than in thin people; it does not. Rather, it is because people with a tendency to obesity apparently are deficient in a particular cellular mechanism that in other people switches on automatically to burn off excess calories.

On being overfed, most people do not gain weight in proportion to the amount of overfeeding; they dispose of the extra energy by generating heat—through nonshivering thermogenesis (the production of heat in the body by oxidation). However, diet-induced thermogenesis is significantly reduced in obese and formerly obese individuals. This diminished thermogenesis may be due to a metabolic defect in brown fat, a particular tissue long known to be associated with heat production in young animals, and found mainly in the chest, upper back, and near the kidneys of adult animals. Brown fat is brown because it has more iron containing cytochrome molecules than white fat. The cytochromes are components of the membrane of the mitochondria; the cells of brown fat are packed with these organelles which carry out cellular respiration by oxidizing nutrient molecules. In the mitochondria of brown-fat cells, a proton leak bypasses the coupling of oxidation with adenosine triphosphate (ATP) pro-

duction. As a result, the mitochondria, oxidizing local stores of fat, generate heat instead of ATP.

Although more experimental work is needed, it does seem clear that there is diminished capacity in many obese people for thermogenesis induced by dietary fat. The defect seems to be associated with brown fat. It could be that the amount of tissue is inadequate, or that there is a difference in the proton-leakage pathway, or that there is a failure at some stage of the switching-on process.

Of course, not all fat people suffer from a metabolic defect; many people do eat too much. However, because of the possibility of a metabolic defect, a physician should not too readily condemn his fat patients for lack of willpower while he, unwittingly, enjoys a physiological system that allows him to stabilize his own weight with effortless ease.

HORMONAL INFLUENCES ON METABOLISM. The knowledge of the effects of the various hormones is at best a science in its infancy, since most of the major discoveries in this field have been made in the 20th century. When the early breakthroughs were first publicized, many doctors had high hopes of being able to treat obesity and other disorders with specific hormones which might produce immediate results like those seen in diabetics given insulin. The failure of these expectations to materialize led some doctors to abandon their interest in endocrinology. However, other doctors persist in the belief that hormones do exert a very considerable influence on metabolism. Some of those subscribing to the latter school of thought believe that a careful study of a patient's features (body build, distribution and type of hair, facial characteristics, head shape and size, skin, and trunk) will usually reveal which of three endocrine glands is most active: the adrenals, the pituitary, or the thyroid.

HORMONAL ABNORMALITIES IN THE OBESE. It is well known that patterns of endocrine secretion may be altered by such factors as the climate, diet, emotions, and the amount of body fat. For example, the development of obesity often results in increases in the secretion of insulin from the pancreas and of steroid hormones from the adrenal cortex; and by a decreased rate of secretion of growth hormones from the pituitary. However, this does not mean that such abnormalities are the causes of obesity, since they often disappear when the body weight is reduced to normal. Nevertheless, once the abnormalities are established, they may make it difficult to lose weight.

External Factors Which May Spur the Development of Obesity. Many of the lean people who believe that they have stronger character traits than the obese may not owe their success in staying lean to their own efforts; rather, they may owe them to complex physiological processes which regulate appetite. There is evidence for both the short-term and the long-term regulation of food intake in species of animals which are similar to man in their physiology.[10] Details of appetite regulation are given elsewhere in this book.

(Also see APPETITE.)

[7]Robson, J. R. K., et al., "Metabolic Response to Food," *The Lancet,* December 24 & 31, 1977, p. 1367.

[8]Miller, D. S., P. Mumford, and M. J. Stock, "Gluttony: 2. Thermogenesis in Overeating Man," *The American Journal of Clinical Nutrition,* Vol. 20, 1967, pp. 1228-1229.

[9]Miller, D. S., and S. Parsonage, "Resistance to Slimming: Adaptation or Illusion?," *The Lancet,* April 5, 1975, p. 773.

[10]Van Itallie, T. B., N. S. Smith, and D. Quartermain, "Short-Term and Long-Term Components in the Regulation of Food Intake: Evidence for a Modulatory Role of Carbohydrate Status," *American Journal of Clinical Nutrition,* Vol. 30, 1977, p. 742.

Animal experiments have also shown that under various circumstances, appetite control may be interfered with, or bypassed, so that the animals will eat sufficient food for fattening to occur. For example, it is well-known that confining livestock in pens speeds up their gains in weight. Likewise, other procedures, such as changing the composition of rations and/or administering hormones, helps the farmer to fatten his animals. Therefore, it is reasonable to expect that similar factors may stimulate the development of human obesity.

The following external factors may spur the development of obesity:

1. **Birth-control pills (oral contraceptives).**—These agents may provoke elevated blood levels of insulin,[11] a condition which is known to favor the development of obesity.

2. **Climate.**   Natives of tropical regions have about 10% lower basal metabolic rates than people living in cool regions, whereas people in cold environments expend extra energy to maintain their body temperature. Food consumption should be governed accordingly.

3. **Cultural factors.**—Obesity was, and still is, considered to be very desirable by various peoples; for example, in certain parts of Africa women are still being fattened up prior to marriage, and a man's worth may be partly based upon the total weight of his wives.

4. **Eating in response to external cues.**—This refers to such cues as the sight and smell of food, social gatherings, or while viewing movies or television. Hence, these people are likely to overeat and become obese if they are regularly exposed to these cues.

5. **Emotional stresses.**—With some people, overeating and obesity may stem from chronic emotional disturbances.[12]

6. **Excessive dietary carbohydrates.**—The overconsumption of such highly refined and easily utilized forms of carbohydrates as table sugar and white flour may be a major cause of obesity.

7. **Excessive dietary fats.**—The low bulk and high calories of fat may cause some people to consume too much fat before feeling satisfied.

8. **Excessive dietary proteins.**—Test feedings consisting only of lean meat have been found to be potent stimulators of the copious secretion of the growth hormone, an agent favoring the development of obesity in growing children.

9. **Fatigue.**—Eating when excessively tired may lead to a habit of eating when tired even when the tiredness is not due to lack of food and/or low blood sugar.

10. **Lack of exercise.**—Low levels of physical activity are usually accompanied by consumption of excess energy and a gaining of weight.

11. **Lack of fiber.**—There appears to be a minimum requirement for bulk in the diet so that distention of the small intestine may make for a feeling of satiety.

12. **Meal patterns.**—Eating two or three large meals per day is more likely to result in excessive fattening than five or six smaller meals, even when the total consumption remains the same.[13]

13. **Nutritional deficiencies.**—Metabolic and nervous abnormalities may result from deficiencies of minerals and vitamins, so it seems reasonable to suspect that these nutritional deficiencies may lead to derangements in appetite control.

14. **Overfeeding of infants and children.**—Many bottle-fed babies are literally force fed, particularly when mothers enlarge the holes in the rubber nipples, then prop up the bottle so that the milk runs down the throats of their babies. Another practice is holding the infant and spooning in as much baby food as will be tolerated.

15. **Overseasoning of food.**—Heavy use of such seasoning agents as catsup, garlic, onions, mustard, pepper, and sugar may make food highly palatable, increase the flow of digestive juices, and generally promote overeating.

16. **Pregnancy.**—Many women gain weight after each pregnancy.

17. **Prenatal and early postnatal influences.**—Many recent research reports suggest that, if not genetic, the tendency towards lifelong obesity begins in a child at some time between the latter part of pregnancy and the end of the first year after birth.[14]

18. **Rapid eating.**—Rapid eating appears to lead to obesity, perhaps due to a delay in the operation of the various satiety mechanisms which allows rapid eaters to consume more than is required.

19. **Sexual and other frustrations.**—From time to time there are reports of studies which show that more obese than lean people complain of sexual frustrations.

20. **Social customs.**—Eating is often an important part of baptisms, bar mitzvahs or confirmations, business luncheons, funerals, professional meetings, and weddings. While these infrequent binges may seem to be unimportant, it may be difficult for some people to regain control of their eating.

21. **Types of food.**—Overeating is more likely to occur when the forms and types of food are high in calories, low in fiber, and do not require much chewing.

DISORDERS WHICH ARE OFTEN AGGRAVATED BY OBESITY.

Statistics from the Metropolitan Life Insurance Company show that obese people (those who are more than 20% overweight) have significantly higher death rates than the nonobese from such conditions as appendicitis, cancers of the gallbladder and liver, cirrhosis of the liver, diabetes, gallstones, heart disease, hernia and intestinal obstruction, kidney disease, stroke, and toxemia of pregnancy.

The statistical associations between obesity and various medical problems do not necessarily mean that the former is always the cause of the latter. For example, a tendency towards diabetes may make some people more susceptible to obesity and other problems, whereas similar problems may be absent in others who are obese. Furthermore, it is important to distinguish between those problems which are directly aggravated by obesity and those which are only indirectly affected, because the choice of the most effective

[11]Spellacy, W. N., K. L. Carlson, and S. A. Birk, "Carbohydrate Metabolic Studies after Six Cycles of Combined Type Oral Contraceptive Tablets. Measurement of Plasma Insulin and Blood Glucose Levels," *Diabetes*, Vol. 16, 1967, p. 590.

[12]Bruch, H., "Psychological Implications of Obesity," *Nutrition News*, Vol. 35, 1972, p. 9.

[13]Leveille, G. A., and D. R. Romsos, "Meal Eating and Obesity," *Nutrition Today*, Vol. 9, November/December 1974, p. 4.

[14]"Influence of Intrauterine Nutritional Status on the Development of Obesity in Later Life," *Nutrition Reviews*, Vol. 35, 1977, p. 100.

treatment for a disorder depends upon identification of the primary causative factor.

There is evidence that the following disorders are often aggravated by obesity:

1. Angina pectoris.
2. Breathing difficulties.
3. Coronary heart disease.
4. Diabetes.
5. Easy fatigability.
6. Enlargement of the heart.
7. Excessive numbers of red blood cells and abnormal clotting of the blood.
8. Gallstones.
9. Gout.
10. Hernia.
11. High blood pressure.
12. High levels of fats in the blood (hyperlipoproteinemias).
13. Increased sweating and susceptibility to heat exhaustion.
14. Infertility in men.
15. Kidney disease.
16. Low blood sugar (hypoglycemia).
17. Osteoarthritis.
18. Skin problems.
19. Sleepiness.
20. Stroke.
21. Sudden-unexpected death.
22. Surgical dangers.
23. Toxemia.
24. Troubles with the feet and legs.

WAYS OF DETECTING BORDERLINE OBESITY.

The standard practice of doctors and life insurance companies has been to consider their clients to be obese when their body weight is greater than their ideal weight (for their height and body build) by some arbitrary figure—commonly 20%. This arbitrary figure may be an unsatisfactory boundary between normal weight and obesity when (1) it is too low, as in the cases of people who have heavy bones and muscles; or (2) it is too high, as in the case of people with abnormally slender bones and muscles. Nevertheless, it is not wise to wait until someone is grossly obese before taking corrective measures, since the difficulty in achieving weight reduction increases with the degree of obesity. Therefore, it is worth noting the diagnostic standards which follow.

Formulas for Estimating Desirable Body Weights.

Doctors and dietitians commonly estimate desirable weights for (1) adult males by allowing 106 lb for the first 5 ft of height and 6 lb for each additional inch of height over 5 ft, and (2) adult females by allowing 100 lb for the first 5 ft of height and 5 lb for each additional inch. According to the formula for males, a man who is 6 ft 2 in. tall has a desirable weight of 106 lb + 14 × 6 lb (for each inch of height over 5 ft) = 190 lb.

Height-Weight Tables.

Different height-weight tables have been compiled, published, and used through the years. The first of these tables, one for men and the other for women, was compiled from data on people who purchased life insurance policies between 1885 and 1908. The Metropolitan Life Insurance Company introduced new tables in 1943. In 1959, data was published based on 26 life insurance companies in the United States and Canada pooling and analyzing their data on body weight and mortality which

had been compiled from records of several million policyholders over a period of 20 years.[15] Shortly thereafter, the Metropolitan Life Insurance Company issued new, revised tables of desirable weights, which were those found to be associated with the greatest longevity. These tables, like the ones which preceded them, give a range of weights for each of the three categories of build. It should be noted that the values for heights and weights which are presented in these tables represent those obtained from people wearing shoes and their usual indoor clothing. This means that the heights have been obtained from men wearing shoes with 1-in. heels, and from women wearing 2-in. heels; and that the total extra weight for clothing in the men was about 8 lb, while in women it was about 5 lb. Here again, it is necessary for the examiner to decide on a person's body build, since no directions are given for making such determinations. Even if body build were to be determined more precisely, there would still be considerable variation of the relative proportions of fat and lean tissues within each body build.

Therefore, desirable weights obtained from the life insurance tables, or from the formula given in the preceding section, give at best only rough indications of the most desirable body characteristics.

(Also see UNDERWEIGHT, section on "Body Weight for Height, Age, and Sex.")

An Index of Slenderness (Ponderal Index).

This derived quantity represents an attempt to express relative leanness as a function of height and weight. It is obtained through the use of the following formula:

$$\text{Ponderal Index (PI)} = \frac{\text{Height (In.)}}{\sqrt[3]{\text{Weight (lb)}}}$$

This quantity may also be calculated using metric units. However, the values so obtained cannot be compared with those obtained using English units because the ratios of the height to weight units differ between the two systems of measurement. The ponderal index has the applications that follow:

• **Prediction of the effect of body weight on health**—The higher the ponderal index, the greater the leanness and the lower the risk of cardiovascular disease since values under 12.0 are associated with higher rates of this type of disorder.[16] People who have ponderal indexes of 13 or more have very low risk for cardiovascular disease because they are more slender than average.

• **Estimation of trends towards fatness or leanness in growing children**—In this application, a ponderal index which increases with growth in height supposedly indicates a thinning out; while an index which decreases with growth indicates fattening. Although there has been little use of the ponderal index for this purpose, it has been suggested that

[15]*Build and Blood Pressure Study*, 1959, Society of Actuaries, Chicago, Ill., 1959.
[16]Seltzer, C. C., "Some Re-evaluations of the Build and Blood Pressure Study, 1959, as Related to Ponderal Index, Somatotype and Mortality," *New England Journal of Medicine*, Vol. 274, 1966, p. 254.

a value of less than 12.4 might indicate obesity in adolescent girls.[17]

Pinch Tests for Hidden Fat. Skinfold measurements are more precise versions of the long used *pinch* tests for obesity. In a pinch test, one selects an area of the body where fat is likely to be deposited and grasps a double fold of skin between his thumb and forefinger. Excessive fat is present when the thickness of the double fat is an inch (*2.5 cm*) or more. Special calipers are available for accurately measuring skinfold thicknesses. The use of these calipers is shown in Fig. O-6.

Fig. O-6. The use of skinfold calipers.

Weighing People Under Water. Anyone observing obese people swimming can hardly fail to notice that their bodies float higher in the water than those of nonobese people. The reason for the greater buoyancy in the obese is that a given volume of fatty tissue weighs only about 90% as much as an equal volume of lean tissue. It follows that differences in the body fat content of people may be detected by weighing them while they're immersed in water.

Although measurement of body density is generally limited to research studies, it is a useful tool for evaluating other more convenient measurements such as skinfold thicknesses.

Detection of Early Signs of Obesity in Children. Much might be done to prevent hard-to-manage obesity in adults if more efforts were made to detect and treat the early stages of fattening in children. However, children tend to grow at variable rates so many doctors are reluctant to advise dietary restriction for heavier than average children for fear of slowing their rates of growth.

One means of detecting fattening in children is to pay close attention to the long-term pattern of growth in each child, rather than attempting to make a judgment on the basis of one-time measurements of height and weight. The pattern of growth is determined by plotting the monthly values of a child's height and weight on a standard growth chart. The standard charts usually have curved lines corresponding to growth rates for the 95th, 90th, 75th, 50th, 25th, 10th, and 5th percentiles of the population of children. There are separate charts for boys and girls.

For example, an infant, who at the age of 6 months had a height and weight corresponding to the 75th percentile of the population, would usually be judged to be growing at a normal rate, because the statistical range of normal values extends from the 5th to the 95th percentiles. However, if the same infant had from birth up to age 5 months followed the curves for the 25th percentile of heights and weights, then there would be reason to suspect that the child's growth might have been excessively rapid between 5 and 6 months of age, because of the progression from the 25th to the 75th percentiles in a month. This finding should lead the doctor or other examiner to question the parent as to the feeding of the infant, since overfeeding may force rapid growth and ultimately produce an obesity that is difficult to correct. Likewise, a body weight which is disproportionately high (for example, in the 75th percentile) for height (which might be in the 25th percentile) should lead an examiner to suspect excessive fattening.

(Also see CHILDHOOD AND ADOLESCENT NUTRITION, section on "Childhood Growth and Its Measurement"; and INFANT DIET AND NUTRITION, section on "Measurement of Growth.")

TREATMENT OF OBESITY. While the prospects of developing any of the deadly disorders associated with obesity are enough to make anyone shudder, there are rarely good reasons for crash dieting, since extremely rapid rates of weight loss are metabolically equivalent to the eating of extraordinary quantities of dietary fat. Hence, the cure may be worse than the disease. Furthermore, many other metabolic abnormalities may accompany drastic changes in body weight. Therefore, the principles of weight control that follow are noteworthy.

Principles of Weight Reduction. It is often said that all a person has to do to lose weight is to eat fewer calories than needed to meet his or her requirements. While this statement may be true over a long period of time, short-term changes in body weight often appear to defy these scientific principles. For example, a very lean person who skips supper one night might lose as much as 2 lb (*0.9 kg*) while a very obese person may go without food for two whole days and yet lose no weight. Furthermore, even very obese people might lose as much as 2 lb (*0.9 kg*) per day, when their daily diet is restricted to 400 Calories (kcal) from fat or protein alone, or from mixtures of these nutrients. The secret behind such seemingly contradictory observations is that weight losses may be due not only to loss of fat, but also to losses of variable amounts of water and protein from the lean tissues of the body.

Planning Diets and Selecting Foods. Anyone with a tendency to become obese needs to plan a balanced diet

---

[17]Canning, H., and J. Mayer, "Obesity—Its Possible Effect on College Acceptance," *The New England Journal of Medicine*, Vol. 275, 1966, p. 1172.

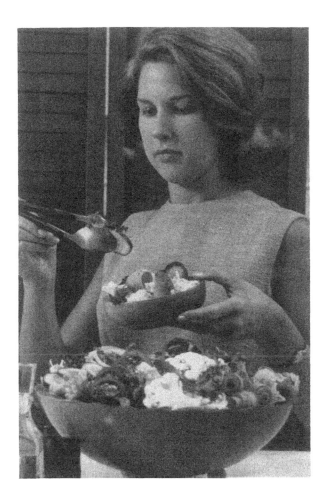

Fig. O-7. A low calorie salad. (Courtesy, USDA)

which provides the required nutrients, but which does not contain excess calories. It is easy to do such planning when one follows the procedures used by dietitians and others with expertise in planning such meals. Simplified versions of these procedures follow.

MODIFIED DIETS. This term is now used instead of the older designation of *special diets* because nutritionists want to advance the idea that one should start with a basic diet which is suitable for most people, then modify it so as to achieve whatever composition may be desired. The modifications which are usually made for low-calorie diets involve reduction in the number of items high in carbohydrates and fats, but retention of those which are good sources of proteins, minerals, and vitamins. For example, it might be desirable to avoid candy, cookies, cakes, ice cream, butter or margarine, and table sugar, and to substitute in their place, whole grain breads, fruits, vegetables, and salad oils. Also, the leaner cuts of meat such as round steak might be used in place of short ribs, porterhouse steak, and luncheon meats (which tend to have higher fat contents).

However, such dietary modifications may not in themselves be sufficient to produce loss of weight, because excessive quantities of the low-calorie foods may be eaten. Therefore, it is often necessary to calculate caloric requirements so that the dietary calorie content may be low enough to

produce the appropriate caloric deficit. Descriptions of such calculations follow:

• **Estimating requirements for calories**—The equation for this calculation is: Calories (kcal) in modified diet equal caloric requirement minus caloric deficit for weight loss.

Most people require 15 Calories (kcal) or less per pound of desirable weight. (Desirable weights may be estimated by the means described in the section headed "Ways of Detecting Borderline Obesity.")

The caloric deficit is 500 Calories (kcal) per day to lose 1 lb (0.45 kg) per week, or 1,000 Calories (kcal) to lose 2 lb. People who wish to lose weight more rapidly should only do so under a doctor's supervision.

• **Selecting foods to make up a diet**—It is not wise to select foods solely on the basis of their calorie contents (a practice that is sometimes called *counting calories*) because it may lead to dietary deficiencies. Instead, one should select items

Fig. O-8. Slim cuisine calls for lots of vegetables. (Courtesy, National Film Board of Canada)

from each of the various food groups so as to obtain a nutritionally balanced diet. Even then, the diet may not furnish all of the required nutrients unless special effort is made to select foods which are good sources of these nutrients, such as dairy products, eggs, fish, fresh or frozen fruits and vegetables, meats, and whole grain cereal products. However, the amounts of these highly nutritious items must be limited in order to keep the energy content of the diet low enough for weight loss to occur. Hence, it is best to use the exchange system for planning diets. Table O-2 gives some typical diets planned by this system.

Table O-2 gives typical menus for diets which are (1) low in carbohydrates (thought to be useful for abnormally high levels of blood triglycerides), (2) moderately low in fat content (the basis of the most commonly used standard diets), and (3) high in carbohydrates and very low in fats (this type of diet may be helpful in preventing heart disease). However, it was noted earlier that when the dietary energy deficit is great enough to bring about the metabolism of large amounts of body fat, the mixture of nutrients metabolized by the body will have a much higher fat content than the diet. Hence, even low-fat diets may make it necessary for the body to burn considerable amounts of fat, which is mostly saturated because it is derived mainly from the fatty tissues. Therefore, one should avoid crash diets which result in the loss of more than 2 lb per week because the metabolic consequences of such programs are uncertain.

TABLE O-2
TYPICAL MENUS FOR VARIOUS TYPES OF LOW-ENERGY DIETS

| Types of Food Exchanges[1] (used for each meal or snack) | Number of Exchanges Per Meal (based upon daily energy requirements) | | | | | |
| | Low Carbohydrate Diet | | Low-Fat Diet | | High Carbohydrate Very Low-Fat Diet | |
	1,200 kcal per day	1,500 kcal per day	1,200 kcal per day	1,500 kcal per day	1,200 kcal per day	1,500 kcal per day
Breakfast:						
Bread (or biscuits, cereals, crackers, muffins, pancakes, waffles, etc.)	1	1	1	1	1	2
Fat (bacon, butter, coffee whitener, margarine, mayonnaise, etc.)[2]	1	1	1	1	0	1
Fruit (or its equivalent in juice)	1	1	1	1	1	1
Lean meat (or low-fat cheese, fish, fowl, etc.)	1	1	1	1	0	0
Milk, nonfat[2]	0	0	0	0	1	1
Optional beverage: bouillon (fat free), clear broth, club soda, coffee, herb tea, tea, water	◄———— Any amount, but limit sweetener[3] ————►					
Mid-morning snack:						
Bread	0	0	1	1	1	1
Fat	0	1	0	0	0	0
Milk, nonfat	1	1	1	1	0	0
Optional beverage (see Breakfast)	◄———— Any amount, but limit sweetener[3] ————►					
Lunch:						
Bread	1	1	1	2	2	3
Fat	2	2	0	2	0	1
Fruit (or juice)	0	0	0	1	1	1
Meat, lean (see Breakfast)	3	3	3	3	2	2
Optional beverage (see Breakfast)	◄———— Any amount, but limit sweetener[3] ————►					
Raw vegetable, or salad without dressing	1	1	1	1	1	1
Mid-afternoon snack:						
Bread	0	0	1	1	0	1
Fat	0	1	0	0	0	0
Milk, nonfat	1	1	1	1	1	1
Optional beverage (see Breakfast)	◄———— Any amount, but limit sweetener[3] ————►					
Supper:						
Bread	1	1	1	2	2	2
Fat	3	3	1	2	0	1
Fruit	0	1	0	0	1	1
Meat, lean (see Breakfast)	3	4	3	3	3	3
Optional beverage (see Breakfast)	◄———— Any amount, but limit sweetener[3] ————►					
Nonstarchy vegetable, cooked	1	1	1	1	1	1
Raw vegetable, or salad without dressing	1	1	1	1	1	1
Late evening snack:						
Bread	0	1	0	0	1	1
Fat	0	1	0	0	1	0
Fruit (or juice)	0	0	1	1	0	0
Milk, nonfat	1	1	1	1	1	1

[1]Exchanges are portions of food which have been grouped together because they contain similar proportions of carbohydrates, fats, proteins, and calories. Portion sizes have been calculated so that the exchanges within each group may be substituted for one another because their nutritive values are approximately equal.
(Also see MODIFIED DIETS, section headed "Use of Exchange Lists in Planning Diets.")
[2]The distribution of fat at meals may have to be adjusted if the foods chosen from among the other groups of exchanges contain more or less fat than the amount which is typical for each of the groups. For example, two fat exchanges should be deducted from the amounts allowed each time that whole milk is used instead of the nonfat type, or one fat exchange deducted to compensate for the use of 2% milk or plain yogurt in lieu of nonfat milk. Similarly, one fat exchange should be deducted for each exchange of high-fat meat used instead of lean meat, or ½ fat exchange deducted for each exchange of medium-fat meat. Also, many doctors recommend that, wherever possible, the items used as fat exchanges be chosen from such sources of polyunsaturated fats as salad oils and soft margarines because ample amount of saturated fats are provided by the meat exchanges.
[3]These fluids contain few calories, so they may be taken as desired. However, the use of sugar and other sweeteners should be limited because 4 tsp (20 ml) of sugar furnish about the same amount of carbohydrates as one bread exchange, 1¾ fruit exchanges, one nonfat milk exchange, or three exchanges from the cooked vegetables.

The procedure for using the table follows:

1. Determine your desirable body weight by either referring to a standard height-weight table, or by using the appropriate formula given in the section headed "Formulas for Estimating Desirable Body Weights."

2. Calculate your caloric requirement for maintenance of weight by multiplying your ideal body weight by 15.

3. Choose a dietary calorie level by deducting a caloric deficit of either 500 Calories (kcal) (to lose 1 lb [0.45 kg] per week) or 1,000 Calories (kcal) (to lose 2 lb per week) from your caloric requirement calculated in step 2. Do not select a diet which furnishes less than 1,000 Calories (kcal) per day without consulting a physician.

4. Set up a daily menu by selecting exchanges from the food groups listed in Table O-2.

(Also see MODIFIED DIETS, section headed "Use of Exchange Lists in Planning Diets.")

EXAMPLE: An adult female who is 5 ft 4 in. (*163 cm*) tall and has a medium body build, weighs 140 lb (*63 kg*). She would like to get her weight down to normal as soon as possible. Her ideal body weight is 120 lb (*54 kg*) and her energy requirement for maintenance of this weight is 1,800 Calories (kcal) per day. It is not advisable for her to try to lose more than 1 lb (*0.45 kg*) per week because diets containing less than 1,200 Calories (kcal) per day are neither nutritious nor satisfying. Hence, it might take her about 17 weeks to reach her ideal weight if she sticks to her diet. (Her weight loss might be accelerated by extra activity.) Some typical meals and snacks for a 1,200 Calorie (kcal) low-fat diet are given in the column on the right. Generally, such low-fat, high-carbohydrate diets are more filling than low-carbohydrate, high-fat diets because the former contain more high-fiber foods like bread (if whole grain varieties are used) and fruits, while the latter are based upon foods where the calories are concentrated in a small amount of bulk.

(Also see FIBER.)

Fig. O-9. Weight watchers should snack on apples instead of candy bars. (Courtesy, Univ. of Tennessee, Knoxville, Tenn.)

Meal	Menu		Calorie (kcal)
Breakfast ..	Ready-to-eat cereal	¾ c ...	70
	with nonfat milk	¾ c ...	60
	and sliced fresh peaches .	1 whole peach	40
	Coffee	1 c	0
	Evaporated skim milk	2 Tbsp .	20
	Sugar	½ tsp ..	9
Mid-morning Snack ...	Graham crackers	two 2½ in. square	70
	Tea with lemon	1 c.....	0
Lunch	Sandwich made with whole-rye bread	2 slices	140
	boiled ham	2 slices	110
	lettuce, tomato, and mustard	no limit	0
	Clear, fat-free hot broth ...	1 c	0
	Apple	one 2 in. diameter ...	40
Mid-afternoon snack ...	Nonfat milk	1 c	80
Supper ...	Tossed salad made with chicory, chopped green peppers, mushroom slices, oregano (dried), and vinegar	no limit	0
	Flank steak	three 1 oz slices	165
	Horseradish sauce	2 tsp ..	0
	Mashed potato without butter or margarine	½ c ...	70
	Carrots, cooked	½ c ...	25
	Cantaloupe	¼ of 6 in. diameter fruit	40
	Whole-wheat bread	1 slice .	70
	Mint tea	1 c	0
Late-evening snack ...	Tortilla	one 6 in. diameter ..	70
	Low-fat yogurt, plain	1 c	125

Total food energy 1,204

SUGGESTIONS FOR PREPARING MEALS AT HOME. A good deal of study and planning is necessary in order for weight reduction to be achieved through dieting. One must learn (1) which foods are low in energy, yet are nutritious; (2) how to estimate correct portion sizes for various types of foods; and (3) ways of combining separate items in such dishes as casseroles, desserts, soups, and stews. Then, the necessary food ingredients must be purchased and kept available for meal preparation. In short, safe and successful weight reduction may require the expenditure of extra effort, money, and time.

EATING AWAY FROM HOME. Too many people who are frequently away from home may put on fat because they eat whatever is served to them. However, dieters should have a preplanned strategy for coping with the pitfalls of eating away from home. Suggestions follow:

• **Lunches Brought from Home**—It seems obvious that with careful planning it is possible both to stick to a diet and save money by bringing food from home for lunches, picnics, and short automobile trips.

• **Eating in Restaurants**—Although the staff in the better restaurants will usually make every effort to accommodate diners who are on diets, dieters must also be prepared to cope with much less flexibility in such places as fast food establishments where the foods and the cooking systems are designed for rapid turnover.

Changing Eating Behavior. This method of combating obesity, which is sometimes called *behavior mod* (mod is short for modification), differs from *crash* dieting in that it seeks permanent modification of the behavior that led to the obesity. Many people regain weight they lost by crash dieting because they failed to change their eating habits.

Fig. O-10. Changing eating behavior—say no to seconds. (Courtesy, Univ. of Tennessee, Knoxville, Tenn.)

However, before a person starts on such a program, it is a good idea to try and identify each of the factors which might be cues for his or her overeating. For example, some people habitually may eat, whether hungry or not, at such times and places as: (1) while engaged in such activities as conversation, reading, or watching television; or (2) in response to such emotions as anger, anxiety, boredom, excitement, frustration, happiness, or loneliness.[18] Hence, it might be necessary for a compulsive eater to keep a detailed diary for a week or so, noting both the circumstances (time, place, emotions, etc.) that prompt eating, and the amounts and types of foods which are eaten.

EATING BEHAVIOR AT HOME. Some do's and don'ts in changing eating behavior at home are:

1. Establish specific times and places for eating, then eat *only* under those circumstances. For example, this might mean that when one is home, eating is done only in the dining room at 7:30 a.m., 6:30 p.m., and 10:00 p.m.; and *never* while working in the kitchen or when people unexpectedly drop in between meals. Visitors may be served food,

but a dieter should learn to avoid eating just because others are doing so if the time is not in accordance with the new behavior pattern which is being established.

2. When eating, concentrate only on that activity. It might be necessary to ask other members of the family to cooperate by refraining from starting intense discussions at meals. One way to emphasize the importance of eating without distractions is to establish a regular etiquette for each meal and snack by using place settings and saying grace.

3. Do *not* put platters and bowls of food on the dining table so that diners may readily refill their plates; rather, put out only the servings of food needed for those present. It is rarely necessary to prepare more than one serving of foods per person, other than salads, since underweight members of the family would probably be better off eating extra food between meals; instead of eating more food at meals. However, the athletes or laborers in the family might go into the kitchen and serve themselves extra food, if such is absolutely necessary.

4. Put individual servings of all of the courses on the dining table before sitting down to eat. That way, everyone may see the total meal and thereby find it easier to avoid overeating the items which are normally served early in the meal. Using small dishes helps to emphasize the size of food portions.

5. Slow down your rate of eating by (a) breaking or cutting off only one small bite of food at a time, (b) chewing each bite of food thoroughly, and (c) swallowing each chewed bite before taking another. Development of a habit of chewing rather than bolting food may require the eating of such foods as dry breads, tough meats, and fibrous fruits and vegetables—all items which cannot be comfortably swallowed without thorough chewing.

6. The pace of eating may also be slowed by serving foods that require more time to eat like (a) fish or meat containing bones, (b) nuts with shells, and (c) fruits with rinds and seeds. Also, according to the rules of etiquette, one should *not* bring to the mouth a large piece of food, then chew off pieces; rather, he should use fingers (in the case of breads) or utensils to remove one bite size piece at a time.

7. When everyone has eaten, the leftovers should immediately be removed from the dining table and stored. Do *not* provide opportunities for nibbling after meals have been completed.

8. Allow sufficient time for meals and snacks each day so that food never has to be eaten on the run.

EATING BEHAVIOR AWAY FROM HOME. Some do's and don'ts in changing eating behavior away from home are:

1. Try to avoid conducting business or talking about work while eating, since emotionally disturbing discussions may (a) stimulate the flow of extra stomach acid and intensify movements of the digestive tract, two conditions which may increase the urge to eat more; (b) distract from the enjoyment of food so that the meal is completed without the usual satisfaction; and (c) provoke compulsive eating by susceptible people. There are very few people who cannot comfortably eat a moderate-sized meal in 30 minutes or less. Expenditure of this amount of time just for eating and small talk might in the long run prove to be a good investment, since a friendly relationship might be established at the meal so that unnecessary argumentation might be avoided when discussing business *after* the meal.

2. Planning in advance for uncertain eating conditions while traveling. For example, find out whether or not food will be served on an airline flight, so that one does not eat

---

[18]Leon, G. R., "A Behavioral Approach to Obesity," *The American Journal of Clinical Nutrition*, Vol. 30, 1977, p. 785.

both before and during the flight. Not all airline terminals have shops where nutritious, low-calorie snacks may be purchased. Some of those near large cities have various types of restaurants which serve a variety of salads, main dishes, and other items suitable for dieters. However, there are often long waits for service at the better restaurants, so at least an extra hour should be allowed for eating at airline terminals.

3. Learn to be firm when coaxed by dining companions to have more food or drink. One may be gracious to hosts by praising them for their hospitality, rather than by trying to eat every morsel which is served. The old notion that guests should be filled almost to bursting is now considered passe.

Participation In Weight Control Groups. It may be difficult for highly extroverted obese people to persist in dieting when the required sacrifices depend entirely upon their own initiative. Therefore, they might consider joining a dieting group which meets weekly to share experiences in the battle against the bulges. These groups differ in that (1) a few are affiliated with teaching hospitals whose doctors and dietitians provide guidance to members, (2) others operate as private, profit-making organizations which have paid nutrition consultants, and (3) still others are private, nonprofit clubs where there may be some voluntary participation by medical and/or nutrition professionals. It is desirable for weight control groups to have persons trained in food and nutrition as consultants because many members would benefit from professional evaluations of the new dietary concepts and the various dietetic foods which are promoted through the mass media.

Before joining a weight control group, a prospective member should obtain a thorough medical examination so as to rule out the possibility that a metabolic problem might be accompanying or contributing to his or her obesity. Then, the obese person might make certain that overeating is not a response to one or more emotional problems which are not related to foods. If the latter is the case, it might be necessary for the dieter to find some other means for coping with this situation. Finally, the prospective member should investigate the club to determine (1) whether the diet(s) recommended might be considered both safe and suitable by his or her doctor, and (2) if the techniques used by the group to modify behavior are compatible with the personality of the dieter. For example, some people might find it annoying to be chided, or even ridiculed, for dietary indiscretions, while others might find such tactics to be just what is needed to give a boost to their will power. However, certain members may so value the esteem of the group that a day or so prior to meetings they may resort to such drastic measures as depriving themselves of fluids and food, or taking enemas or strong laxatives. Hence, these groups might do well to advise against such practices.

People may find out about weight control groups in their local community by asking their doctor.

Drugs Which Affect Appetite and Metabolism. Certain types of drugs, taken alone or in combination with others, may help to speed loss of weight when they (1) reduce appetite (these are called anorexic agents), (2) promote loss of water through urination (diuretics), and (3) increase the proportion of food energy lost as heat (thyroid hormones). In addition to these types of substances, other drugs might be used to counteract some of the undesirable side effects of those which promote the loss of weight.

NOTE: Each of the drugs which follow may have harmful side effects. Hence, their use in the treatment of obesity should be done only under the close supervision of a physician. It would be much better to attempt to reduce the body weight by dieting and exercise, and to use drugs only as a last resort. Some reducing drugs follow:

• **Anorexic agents**—These drugs are appetite suppressants which are used as *part* of the treatment for obesity, augmenting the other part of the treatment consisting of a nutritious, calorie-restricted diet.

• **Atropine**—This antispasmodic drug slows excessive activity of the digestive tract, a condition sometimes resulting from the use of amphetamines. Hence, the two types of drugs may be combined in an antiobesity formulation.

• **Barbiturates**—These drugs, also known as *downers*, depress the activity of the central nervous system. Hence they may be combined with amphetamines.

• **Digitalis**—This substance slows the heart rate, so it may be used in combination with thyroid hormones, which speed up the heart rate. *The combination of these two types of drugs is considered to be both dangerous and unwarranted in the treatment of obesity.*[19]

• **Diuretics**—These agents increase the amounts of water lost from the body through urination. Some of the diuretic drugs also cause marked losses of potassium in the urine, a situation which might result in increased susceptibility of the heart muscle to the toxic effects of digitalis. Patients treated for congestive heart failure with combinations of these drugs are usually warned to obtain sufficient dietary potassium to compensate for urinary losses.

• **Hormones**—Thyroid hormones are often used to treat obesity because they increase the amount of food energy lost as heat so that the total energy metabolism of the body must be increased in order to meet the needs of the body tissues. However, the hormones may also cause palpitations of the heart. Therefore, many doctors believe that the use of these agents should be limited to people who have thyroid deficiencies.

• **Laxatives**—Drugs such as atropine and diuretics may make some people constipated, so laxatives may sometimes be added to such mixed medications. Some misinformed people also use excessive amounts of laxatives to cause weight loss which results in an increased loss of nutrients and water in the stool.

• **Rainbow pills**—These are multicolored pills which often contain mixtures of such drugs as amphetamines, barbiturates, digitalis, diuretics, laxatives, and thyroid hormones.

Exercise. Although most people would not be able to exercise vigorously enough to lose weight by this method alone, more and more programs for weight control and physical fitness emphasize the need for frequent and brief periods of strenuous activity, e.g., bicycling, jogging, tennis, or swimming, all of which result in deep breathing and a speeding of the heart rate. These activities should be engaged in for at least 4 hours a week, and if possible, at least 1 hour every day. However, some people may be in such poor physical condition that they would be risking excessive strain on their hearts, joints, and muscles if they were suddenly to begin such exercises. Therefore anyone who

[19]"No End to the Rainbow," *Nutrition Today,* Vol. 3, June 1968, p. 24.

wishes to start an exercise program should first obtain a thorough physical examination, then ask his or her physician how much should be done. People unaccustomed to exercise might begin with walking at a moderate pace for an extra 30 minutes to 1 hour each day.

It is noteworthy that regular participation in vigorous activity might slow weight loss because the building up of muscles involves both protein and water. However, one might determine whether or not fat is being lost by periodically taking measurements of the skinfold thickness at various sites on the body.

Many of the dangerous tendencies which may accompany obesity may be reduced or eliminated by regular exercise, even when the rate of weight loss may be very slow. Therefore, some health benefits of exercise follow:

• **Appetite reduction**—It was mentioned earlier in this article that appetite decreases when sedentary people take up moderate physical activity. Hence, a good time for jogging or similar exercise would be after work, and just before supper.

• **Clearing of fats from the blood**—Many studies have shown that blood triglycerides often drop sharply when a person undertakes a daily program of exercise. This improvement in the levels of blood fats may occur even when the energy expended is compensated for by eating extra food.[20] There may also be gradual, but significant, declines in blood cholesterol.

• **Extra energy expenditure and loss of weight**—About the most weight which may be lost through extra exercise is 1 lb per week, when an additional 500 Calories (kcal) are expended each day, and no extra food is taken. However, it is not difficult at all to expend an extra 250 Calories (kcal) per day by jogging for half an hour, which will result in the loss of 13 lb (5.9 kg) over 6 months. Extra exercise might also help to rid the body of excess weight due to the buildup of water.

• **Firming up of muscles and improvement in appearance**—Men and women who lose weight without exercise may retain a protruding pot-belly or drooping breasts because flabby muscles allow these tissues to sag. Although most experts state that one cannot preferentially reduce certain fatty spots in the body without equally reducing all such deposits, the firming up of adjacent muscles helps to prevent unsightly sagging of flesh.

• **Improved tolerance for heat stress**—It was recently found that the tolerance of men for heat stress was improved when they underwent a 6-week training period where they engaged in such vigorous activities as basketball, handball, jogging, and running for 1 hour per day, 5 times a week.[21]

• **Increased flow of blood to vital organs and tissues**—Exercise speeds the heart rate and the flow of blood around the body. However, if the flow of blood in the muscles is greatly increased, as in strenuous exercise, there is likely to be a decreased flow to the visceral organs. Hence, it is *not* wise to do such exercises right after meals, since it may result in digestive problems.

• **Lowering of blood pressure**—Two conditions which may contribute to a drop in blood pressure during exercise are (1) the opening up of small blood vessels in the muscles, and (2) a similar dilation of blood vessels in the skin as part of the body's means of dissipating heat.

• **Lowering of elevated blood sugar**—It is well known that exercise lowers blood sugar, even in diabetics, since there appears to be an insulinlike effect of strenuous activity. Hence, diabetics who are required to take injections of insulin may often lower the dosage of this hormone when they take extra exercise.

• **Opening up of "emergency" blood vessels in the heart muscle**—Regular exercise may be one of the best forms of insurance against dying of a heart attack, since moderate working of the heart muscle causes enlargement of the branches of the coronary artery. Hence, if there is an obstruction to the flow of blood in one part of this system, the chances for survival are greatly enhanced when one has a highly functional system of alternative blood vessels for supplying the heart muscle.

• **Prevention of varicose veins**—Frequent exercising of the legs during the day helps to force the blood in the veins back up to the heart. It is believed that the enlargement of the leg veins is at least partly due to the pooling of blood in them when insufficient exercise is taken.

• **Reduction in the loss of calcium from bones**—Among the other benefits which accrue to older people from exercise is a dramatic decrease in the loss of calcium from their bones, which is considerable when people take little exercises.[22] (Also see OSTEOPOROSIS.)

• **Relaxation of tensions**—Getting out and exercising may help compulsive eaters since it helps to dissipate some of the tension which may build up from emotional stresses. Remember, that man's physiology is oriented towards "fight or flight" when confronted by various threats.

• **Slowing of the resting heart rate**—One of the signs that a person has achieved a high level of physical fitness is the slowing of his or her heart rate to as low as 60 beats per minute. A slower resting heart rate means that the organ has more time to rest between each beat.

• **Thinning of the blood**—Some obese, middle-aged men have excessive numbers of red blood cells (polycythemia) which increases the thickness (viscosity) of their blood along with its tendency to clot. It is believed that one cause of this condition may be a lack of oxygen in certain tissues, a condition which stimulates the production of extra red cells by the bone marrow. However, it has been found that vigorous exercise may cause a drop in the red cell count.[23] The lowering of the blood sugar by exercise might also help to prevent thickening of the blood since excess sugar produces excess glycoproteins, substances which contribute resistance to the flow of blood in the small blood vessels. (Also see DIABETES MELLITUS.)

[20]Gyntelberg, F., et al., "Plasma Triglyceride Lowering by Exercise Despite Increased Food Intake in Patients with Type IV Hyperlipoproteinemia," *The American Journal of Clinical Nutrition*, Vol. 30, 1977, p. 716.

[21]Gisolfi, C., and S. Robinson, "Relations between Physical Training, Acclimatization, and Heat Tolerance," *Journal of Applied Physiology*, Vol. 26, 1969, p. 530.

[22]Sidney, K. H., R. J. Shephard, and J. E. Harrison, "Endurance Training and Body Composition of the Elderly," *The American Journal of Clinical Nutrition*, Vol. 30, 1977, p. 326.

[23]Yoshimura, H., "Anemia during Physical Training (Sports Anemia)," *Nutrition Reviews*, Vol. 28, 1970, p. 251.

Surgery and Other Drastic Measures. When all else fails to produce the desired loss in weight, drastic surgical measures may be taken. For example, one type of operation removes deposits of fat and some of the excess skin from the abdomen. Then, the wound is sewn together at the edges.

Another type of surgery is the removal of part of the small intestine so that the absorption of nutrients is greatly reduced. However, this operation may be followed by liver disorders, kidney stones, and other problems, and even death.[24] Hence, many doctors do not recommend it unless the risks of obesity are greater than those of the surgery.

Finally, a few people have had their jaws wired shut so that they might drink fluids, but could not take any solid foods. Dramatic losses of weight have occurred, but there has been no news as to whether these losses were sustained after the wires were removed.

SUMMARY. No matter what the reason may be for obesity, it is not to be desired and should be avoided or treated promptly if it develops. The two most important points are:

1. Total caloric consumption must be reduced, and a new permanent pattern of eating established—otherwise, the weight will be regained eventually.
2. Exercise should be increased on a daily basis.

OBESITY DRUGS

Generally, this refers to anorexic agents—those drugs which suppress the appetite. Originally, amphetamines were used for this purpose but now a number of other drugs are available; among them, diuretics, hormones, barbiturates, and atropine, used alone or in combination. Drugs may be used as part of the treatment for obesity, with the other part of the treatment consisting of a nutritious calorie-restricted diet. But they should only be used as a last resort, and then only under the close supervision of a doctor.

(Also see OBESITY, section on "Drugs Which Affect Appetite and Metabolism.")

OCTACOSANOL ($C_{28}H_{58}O$)

This substance, which exists in nature as a constituent of many plant oils and waxes, including the oil of raw wheat germ, is reputed to be important in physical fitness. Based on a long series of human experiments on physical fitness conducted by Dr. Thomas Kirk Cureton of the University of Illinois Physical Fitness Institute, it appears that octacosanol is one of the factors responsible for the beneficial effects of unrefined, unheated wheat germ oil. Vitamin E alone was not effective.

Dr. Cureton's subjects were tested for pulse rates, breath-holding ability, basal metabolism, heart action, reaction time, agility, bicycling, strength, and many other physical activities. Improvements were noted in stamina, speed, reflexes, agility, and heart action.

Octacosanol, which has been concentrated from wheat germ oil, will be listed on any product that contains it.

NOTE WELL: More experimental work is needed relative to the importance of octacosanol as part of a nutritional program for persons engaged in strenuous activities where agility, speed, and endurance are important. Also, persons eating plenty of whole grain cereals, unheated seeds and nuts, and other foods rich in this substance, do not need supplemental sources of octacosanol.

OFFAL

Formerly, the term offal, which literally means *off fall*, included all parts of the animal that fell off during slaughtering. As our knowledge of nutrition improved, the edible portions—liver, tongue, sweetbread, heart, kidney, brain, tripe, and oxtail—were upgraded and classed as *variety meats*. Hand in hand with this transition, the hide, wool, and tallow became known as by-products. Thus, in a modern packing plant, only the entrails and inedible trimmings are referred to as offal.

OIL

Although fats and oils have the same general structure and chemical properties, they have different physical characteristics. The melting points of oils are such that they are liquid at ordinary room temperatures.

(Also see FATS AND OTHER LIPIDS; and OILS, VEGETABLE.)

OIL CROPS

The main oil crops are soybeans, cottonseed, peanuts, and flaxseed. Sunflower, safflower, castor bean, and corn are also used for making oil.

OIL PALM *Flaeis guineensis*

Fig. O-11. Oil palm of West Africa, showing climber ascending tree to harvest the fruit.

The oil palm is one of the world's most important sources of edible and soapmaking oil. It yields more per acre than can be obtained from any other vegetable oil; hence, its importance is not likely to decline.

[24]Bray, G. A., and J. R. Benfield, "Intestinal Bypass for Obesity: A Summary and Perspective," *The American Journal of Clinical Nutrition*, Vol. 30, 1977, p. 121.

ORIGIN AND HISTORY. The oil palm is indigenous to tropical Asia and/or Africa, where it grows wild in great numbers. A large proportion of the commercial palm oil comes from the fruits collected from wild palm groves. But plantations have also been established in the Congo, in Tanzania, along the Ivory Coast, and in Malaysia and Indonesia.

WORLD PRODUCTION. The world production of palm oil totals about 11,044,000 metric tons; and the world production of palm kernel oil totals about 3,468,186 metric tons.[25] The leading palm oil producing countries, by rank, are: Malaysia, Indonesia, Nigeria, Colombia, and Thailand.[26]

PROCESSING. The primitive method of processing, which is still used by many small farmers, consists of leaving the bunches of fruit to ferment for a few days (usually in holes in the ground); stripping off the fruits, then boiling and pounding them; and skimming off the oil from the surface of the water. The shells of the nuts are broken between two stones or in a hand-operated cracking machine. The kernels are usually sold for export, with their oil extracted by the importing country.

But modern oil processing factories are gradually replacing the primitive method, and recovering a larger percentage of oil from the fruits than can be secured by hand processing. The trend is toward steam distillation, followed by centrifuging and even hydraulic expression. In industrialized nations, the kernels are either expressed or solvent-treated.

PREPARING. Although liquid when first extracted, palm oil sets about as thick as butter at normal temperatures. It has a yellowish color (due to the carotene pigment), and a pleasant aroma.

NUTRITIONAL VALUE. Palm oils vary; in some cases resembling kernel oil, and in others olive oil. But they are always a rich source of vitamin A; they contain from 37,300 to 128,700 mcg of beta-carotene equivalent per 100 g. All kernel oils are very similar—they're high in lauric acid and high in saturation. Food Composition Table F-21 lists the nutrient composition of palm oil and palm kernel oil.

USES. The leading producing countries use palm oil locally as a food in all kinds of cookery, and export the surplus.

When exported, palm oil is used chiefly for soapmaking and industrial purposes; but with suitable treatment it may also be used in margarine. Palm kernel oil, which is white or pale yellow, is largely used for margarine. Following extraction, the oilcake is used as a protein supplement for livestock.

As with many other palms, wine may be prepared from the oil palm by tapping and fermenting the sugary sap, which contains a useful content of B vitamins.

OILSEED

Seeds from which oil is extracted for commercial use are termed oilseeds, including castor bean, cottonseed, flaxseed, peanut, safflower, sesame, soybean, and sunflower.

(Also see OILS, VEGETABLE, Table O-5.)

---

[25]Based on data from *Agricultural Statistics 1991*, USDA, p. 135, Table 189. **Note well:** Annual production fluctuates as a result of weather and profitability of the crop.

[26]Based on data from *FAO Production Yearbook 1990*, FAO/UN, Rome, Italy, Vol. 44, p. 124, Table 48.

OILS, VEGETABLE

Contents	Page
History	794
Production and Processing	794
Extraction of Oils	795
Refining	796
Hardening (Hydrogenation)	796
Chemical Composition and Other Characteristics	797
Fatty Acid Composition	797
Chemical and Physical Properties	797
Nutritive Values	798
Essential Fatty Acids	798
Polyunsaturated vs. Saturated Fats	798
Medium-Chain Triglycerides (MCT Oil)	798
Vitamin E and Other Beneficial Substances	798
Antinutritional and/or Toxic Factors	799
Vegetable Oils	799

Vegetable oils are obtained from the seeds and fruits of various plants. Many of their characteristics are similar to those of animal fats, which is fortunate because the per capita supply of the latter is ever dwindling as the human population grows larger. In fact, the past century has been marked by a continual effort to modify vegetable oils so that they may serve as replacements for butter, cream, and lard. Recently, there has been considerable controversy regarding the possible roles of the various fatty foods in the prevention or provocation of such disorders as atherosclerosis, cancer, heart disease, and obesity. Apparently, there are no simple answers to these questions, since dietitians, doctors, and nutritionists disagree on the relative merits and demerits of the different products. Therefore, some noteworthy background material on the vegetable oils is presented in the sections that follow.

HISTORY. The first fats produced by primitive peoples were those rendered from animal carcasses. Later, butter was obtained by churning the milk from domesticated animals. However, certain areas of the world became too densely populated to support a large population of wild or domesticated animals, and so the people in these places came to depend upon plants as sources of fats and oils.

It appears that olives may have been cultivated in various places around the eastern end of the Mediterranean Sea for at least 6,000 years. Similarly, the sesame plant has been grown for thousands of years in the tropical belt extending from Africa to India. Other sources of oil, such as coconut palm and the oil palm, have long grown wild in the tropics, where people gathered the nuts and extracted the oil. These oils (olive, sesame, coconut, and palm) keep better without refrigeration than most of the other vegetable oils. Hence, the stable oils have long been shipped to various parts of the world. Other, more perishable oils, such as safflower, were not articles of worldwide commerce until recently, when certain technological developments made it possible to reduce the potential for spoilage, or breakdown during frying.

PRODUCTION AND PROCESSING. At the present time, vegetable oils contribute the major share of the world production of fats and oils. Fig. O-12 shows the contribution of each of the important fats and oils.

The per capita utilization of the major fat and oil products in the United States is shown in Fig. O-13.

WORLD PRODUCTION OF FATS & OILS

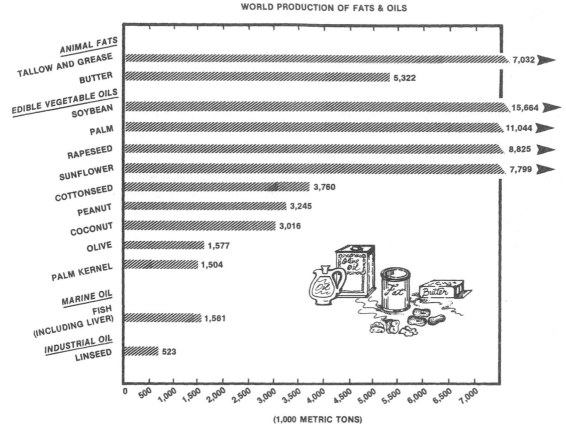

ANIMAL FATS	
TALLOW AND GREASE	7,032
BUTTER	5,322
EDIBLE VEGETABLE OILS	
SOYBEAN	15,664
PALM	11,044
RAPESEED	8,825
SUNFLOWER	7,799
COTTONSEED	3,760
PEANUT	3,245
COCONUT	3,016
OLIVE	1,577
PALM KERNEL	1,504
MARINE OIL	
FISH (INCLUDING LIVER)	1,561
INDUSTRIAL OIL	
LINSEED	523

(1,000 METRIC TONS)

Fig. O-12. World production of fats and oils. (Source: *Agricultural Statistics 1991*, USDA, p. 135, Table 190)

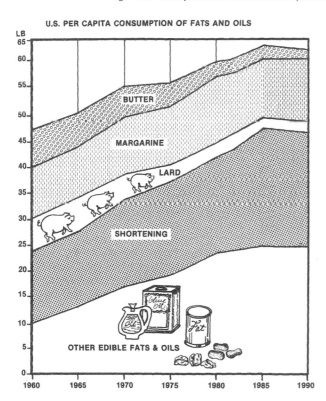

Fig. O-13. Per capita consumption of fats and oils in the United States between the years 1960 and 1990. (Source: *Statistical Abstract of the United States 1980*, p. 131, No. 211; and *Agricultural Statistics 1991*, USDA, p. 135, Table 190)

It may be seen from Fig. O-13 that the per capita consumption of both butter and lard has declined significantly in recent years, whereas the consumption of vegetable oil products has risen sharply. The categories of margarine, shortening, and other edible fats and oils represent mainly items made from vegetable oils.

Descriptions of the processes used in the production of vegetable oils and related products follow.

Extraction of Oils. Vegetable oils are enclosed within the cells of seeds or fruits. Hence, the plant tissues must be broken so that the oil may be extracted. Various processes have been developed to insure the maximum yield of oil because this product has generally brought considerably more profit than the extracted plant residue. Details of the more commonly used processes are noteworthy:

• **Cracking, crushing, flaking, or grinding of the oil-bearing tissue**—Depending upon the nature of the plant material, one or more of these processes are used to prepare the oil source for extraction; sometimes, the broken pieces are also cooked prior to extraction.

• **Pressing**—Various devices, ranging from small hand-operated presses to large mechanically driven expellers, are used to squeeze the oil from the pretreated pulverized plant tissue.

• **Solvent extraction**—Often, the residues from pressing are further treated with solvents such as ethylene dichloride or hexane to remove the remaining oil. Then, the oil and solvent mixture is heated to drive off the solvent.

(Also see SOYBEAN, Section headed "Extracting the Oil.")

Fig. O-14. Soybeans, world's leading source of vegetable oil. (Courtesy, American Soybean Assn., St. Louis, Mo.)

Fig. O-16. Cotton, the seed of which is the source of the fourth most consumed vegetable oil in the U.S. (Courtesy, USDA)

Fig. O-15. Coconuts, source of coconut oil, in demand by bakers and confectioners for use in coatings and fillings for baked goods and candy which may stand for a relatively long time between manufacture and consumption. (Courtesy, Field Museum of Natural History, Chicago, Ill.)

Fig. O-17. Soybean oil bottling line. (Courtesy, American Soybean Assn., St. Louis, Mo.)

Refining. Different processes may be used to refine each of the various oils, because the undesired constituents vary in each case:

• **Agitation with caustic soda**—The freshly extracted oil is stirred with alkali and heated in order to convert any free fatty acids to soaps. Then, centrifugation is used to remove the soaps, gums, sludge, and other undesirable materials. All traces of alkali are then removed by one or more washings with water, followed by centrifugation.

• **Bleaching (Decolorization)**—After alkali treatment, colored materials are removed from the oil by (1) the addition of fine particles of filtering clay and activated charcoal, and (2) passage of the mixture through a filter press to remove the colored materials absorbed into the clay and charcoal.

• **Deodorization**—Odoriferous materials are removed by the passage of steam through the heated oil in a vacuum chamber.

• **Winterization**—Certain oils contain fats which solidify (crystalize) at refrigerator temperatures and cause cloudiness. These harmless, but esthetically undesirable, components are removed by chilling the oil and filtering.

Hardening (Hydrogenation). This process is used to reduce the polyunsaturation of certain oils in order to (1) make them more solid at room temperature (for use in margarines and shortenings), and (2) increase their stability against deterioration when stored for long periods or when heated. It consists of (1) adding a trace amount of a nickel catalyst to the oil; (2) heating, and passing hydrogen gas through the oil in a converter kept under a vacuum; and (3) rebleaching, and removing the catalyst from the hydrogenated oil by filtration.

CHEMICAL COMPOSITION AND OTHER CHAR-
ACTERISTICS.   Both vegetable oils and animal fats are comprised mainly of mixtures of triglycerides, which are chemical compounds containing three molecules of fatty acids combined with one molecule of glycerol. Triglycerides differ from one another because their fatty acid content varies, while the glycerol in each triglyceride remains the same. Hence, the chemical and physical characteristics of vegetable oils depend ultimately on the types and amounts of chemically-combined fatty acids that they contain.

Fatty Acid Composition.   Fatty acids are made up of hydrocarbon chains that contain an organic acid (COOH) grouping at one end. The backbone of the chain consists of carbon atoms connected by either single (saturated) or double (unsaturated) linkages (valence bonds). Saturated linkages are more stable than unsaturated linkages because the latter have the ability to combine readily with oxygen or other substances. Furthermore, fatty acids which contain two or more unsaturated bonds per molecule (polyunsaturated fatty acids or PUFA) are prone to rancidity (oxidative deterioration).

Another variable factor in the makeup of fatty acid molecules is the number of carbon atoms. Fatty acids are designated as having (1) short chains when the number of carbon atoms is 6 or less, (2) medium chains when there are 8 or 10 carbon atoms, and (3) long chains when the number of carbon atoms is 12 or more.

The properties of the various fatty acids depend mainly upon the combination of (1) the number of unsaturated bonds present, and (2) the length of the carbon chain. Table O-3 shows how these chemical characteristics affect melting points. Melting points of fats are very important in nutrition, because fats that remain solid in the digestive tract are utilized poorly.

TABLE O-3
CHEMISTRY OF FATTY ACIDS
COMMONLY PRESENT IN VEGETABLE OILS

Name	Structure	Chain Length (no. C atoms)	Melting Point (°F)	Melting Point (°C)
Saturated:				
Caproic	$CH_3(CH_2)_4COOH$	6	18	*-8*
Caprylic	$CH_3(CH_2)_6COOH$	8	62	*16*
Capric	$CH_3(CH_2)_8COOH$	10	88	*31*
Lauric	$CH_3(CH_2)_{10}COOH$	12	112	*44*
Myristic	$CH_3(CH_2)_{12}COOH$	14	129	*54*
Palmitic	$CH_3(CH_2)_{14}COOH$	16	146	*63*
Stearic	$CH_3(CH_2)_{16}COOH$	18	157	*70*
Arachidic	$CH_3(CH_2)_{18}COOH$	20	170	*76*
Lignoceric	$CH_3(CH_2)_{22}COOH$	24	187	*86*
Mono-unsaturated:				
Palmitoleic	$CH_3(CH_2)_5CH = {}^1CH(CH_2)_7COOH$	16	31	*-1*
Oleic	$CH_3(CH_2)_7CH = CH(CH_2)_7COOH$	18	56	*13*
Poly-unsaturated:				
Linoleic	$CH_3(CH_2)_4CH = CHCH_2CH = CH(CH_2)_7COOH$	18	23	*-5*
Linolenic	$CH_3CH_2CH = CHCH_2CH = CHCH_2CH = CH(CH_2)_7COOH$	18	12	*-11*
Arachidonic	$CH_3(CH_2)_4CH = CHCH_2CH = CHCH_2CH = CHCH_2CH = CH(CH_2)_3COOH$	20	-57	*-50*

¹The symbol = indicates a double valence (unsaturated) bond between adjacent carbon atoms.

It may be noted from Table O-3 that there is a steady rise in melting points as the chain lengths of fatty acids increase. However, for a given chain length, the melting points decrease as the degree of unsaturation increases. The latter principle is exemplified in the fatty acids with 18 carbon atoms by the high melting point for stearic acid (157°F, or 69°C), which is saturated; compared to those for monounsaturated oleic acid (56°F, or *13°C*) and polyunsaturated linolenic acid (12°F, or *-11°C*).

Although it is not shown in Table O-4, it is noteworthy that melting points of fats containing both low- and high-melting fatty acids are intermediate between the two extremes.

The relative amounts of saturated, monounsaturated, and polyunsaturated fats present in each of the commonly used vegetable oils are given in Table O-4.

TABLE O-4
FATTY ACID CONTENT OF 1-TBSP PORTIONS
OF SELECTED VEGETABLE OILS¹

Oil	Total Fat (g)	Saturated (g)	Monounsaturated (g)	Polyunsaturated (g)	Other Minor Fats (g)
Coconut	13.6	11.8	0.8	0.2	0.8
Corn	13.6	1.7	3.3	8.0	0.6
Cottonseed	13.6	3.5	2.4	7.1	0.6
Olive	13.5	1.8	9.9	1.1	0.7
Peanut	13.5	2.3	6.2	4.3	0.7
Safflower	13.6	1.2	1.6	10.1	0.7
Sesame	13.6	1.9	5.4	5.7	0.6
Soybean	13.6	2.0	3.2	7.9	0.5
Sunflower, commercial	13.6	1.4	2.7	8.9	0.6
Sunflower, southern	13.6	1.4	6.2	5.5	0.5

¹*Composition of Foods*, Ag. Hdbk. No. 8–4, USDA, 1979.

Chemical and Physical Properties.   The wide variety of ways in which vegetable oils are utilized makes it necessary to have expeditious means for determining certain characteristics of these products. It is too expensive and time consuming for many commercial users of vegetable oils to assess the chemical and physical properties by the more comprehensive methods used in research studies. Some of the commonly conducted tests follow:

• **Iodine Number**—This test measures the degree of unsaturation of a fat. The higher the number, the greater the unsaturation. For example, the highly saturated coconut oil has an iodine value of only 9, whereas the highly unsaturated soybean oil has a value of 134. Susceptibility to oxidative rancidity increases, and the melting point decreases with the iodine value. It is noteworthy that this test is used to assess the amount of hydrogenation applied to vegetable oils. The value is usually given in the purchasing specifications for these products.

• **Saponification Value**—The name of this test is derived from the reaction (saponification) in which fats react with alkalis to form soaps, because the higher the value, the better the soap making potential. What is actually measured is the average length of the carbon chains in the fatty acids which constitute the fat. A high saponification value signifies a low melting point and a short chain length, and vice versa. For example, the value for coconut oil ranges between 250 and

264, because this product is rich in the fatty acids which contain 12 or fewer carbon atoms per molecule. On the other hand, the value for rapeseed oil ranges between 170 and 180, due to the high content of long-chain fatty acids.

• **Melting Point**—This is an important value for food processors because (1) consumers expect certain items to be solid at ordinary room temperatures, and others to be liquid; and (2) most shortenings should be at least semisolid at room temperature because flakiness of pastry depends upon the production of layers of solid fat. In common language, an item is called a fat if solid under ordinary conditions, or an oil if liquid. Ordinary conditions usually imply a temperature between 64°F (18°C) and 75°F (24°C). Products with a suitable melting point may be obtained by mixing high-melting items with low-melting ones.

NUTRITIVE VALUES.

The nutrient compositions of the common vegetable oils are given in Food Composition Table F-21, and their fatty acid contents are given in Table F-4, Fats and Fatty Acids In Selected Foods.

Vegetable oils, like other highly edible fats, furnish about 9.3 Calories (kcal) per gram. Hence, 1 Tbsp (13.6 g) furnishes about 127 Calories (kcal). However, the most valuable nutritional assets of these products are the content of polyunsaturated fats and vitamin E. Nevertheless, the merits of the former may have been exaggerated somewhat over the past 3 decades, as exemplified by the growing doubts as to whether polyunsaturated fatty acids are efficacious in the prevention of heart disease. Also, the consumption of large amounts of vegetable oils might under certain conditions be hazardous to health. Therefore, some of the basic principles that bear on these and other controversial issues are noteworthy.

Essential Fatty Acids.

The polyunsaturated fatty acid, linoleic acid, is required in the diet because it cannot be synthesized in the body. Linoleic appears to be needed in amounts of about 7.5 g per day.

Corn, cottonseed, safflower, soybean, and sunflower oils are excellent sources of linoleic acid, which comprises about one-half of the fatty acid content of these oils. Hence, about 1 Tbsp per day of any of them will provide the requirement for adults.

NOTE: The polyunsaturated fatty acid content of each of the oils listed in Table O-4 is mainly linoleic acid.

The essential fatty acids serve important functions in cell membranes, and as raw materials for the synthesis of hormonelike agents called prostaglandins. Deficiencies of these fatty acids result in skin disorders, reproductive difficulties, loss of hair, and perhaps even an increased susceptibility to high blood pressure. It is suspected that certain prostaglandins, that are synthesized from the essential fatty acids, help to control blood pressure.

(Also see FATS AND OTHER LIPIDS, section headed "Essential Fatty Acids.")

Polyunsaturated vs. Saturated Fats.

Most vegetable oils, with the exception of cocoa butter, coconut oil, and the palm oils, are more unsaturated (they have higher iodine numbers) than most fats from animals. Marine animals and fish may also be rich in unsaturated fats. Also, vegetable oils are free of cholesterol, which is suspected of contributing to the development of atherosclerosis and heart disease, when excessive amounts of the sterol are consumed by susceptible people over long periods of time. Furthermore, the substitution of substantial amounts of vegetable oils for

saturated animal fats such as butter, cream, lard, and meat drippings has sometimes led to (1) a modest (up to 20%) drop in blood cholesterol, and (2) a reduction in the tendency of the blood to form clots. However, arachidonic acid, a polyunsaturated fatty acid found mainly in peanut oil, promotes clotting.[27]

The effects of polyunsaturated fatty acids on the heart muscle are uncertain, since it has been found that the lifetime feeding of corn, cottonseed, or soybean oils produced more heart lesions in rats than beef fat, butter, chicken fat, or lard.[28]

Some of the other findings which call into question the wisdom of consuming large amounts of polyunsaturated vegetable oils have been summarized in a recent review.[29] They are as follows:

1. The drop in blood cholesterol may be due mainly to a shifting of the sterol from the blood to the tissues.
2. Polyunsaturated fats cause an increase in the amount of cholesterol secreted in the bile, a condition that sometimes leads to the formation of cholesterol gallstones.
3. Extra vitamin E is required to prevent the formation of toxic peroxides when the dietary level of polyunsaturated fats is raised. However, diets containing ample amounts of the essential trace mineral selenium (which is a part of the enzyme that breaks down the peroxides) may also help to offset this danger.

Medium-Chain Triglycerides (MCT Oil).

This special dietary product is made from coconut oil by (1) steam and pressure hydrolysis of the oil into free fatty acids and glycerol, (2) fractionation of the resulting hydrolysate into medium-chain and long-chain fatty acids, and (3) recombination of the medium-chain fatty acids with glycerol to form MCT oil, which is made up of about ¾ caprylic acid (a saturated fatty acid containing 8 carbon atoms) and about ¼ capric acid (a saturated fatty acid containing 10 carbon atoms). The special nutritive values of this product are as follows:

1. It is much more readily digested, absorbed, and metabolized than either animal fats or vegetable oils which contain mainly long-chain triglycerides. Hence, MCT oil is valuable in the dietary treatment of fatty diarrhea and other digestive disorders in which the absorption of fat is impaired.
2. Medium-chain triglycerides, unlike other saturated fats, do *not* contribute to a rise in blood cholesterol.

At the present time, MCT oil is used almost exclusively for special dietary formulations in the United States, but it is also available in various nonmedical consumer products in Europe.

Vitamin E and Other Beneficial Substances.

Vegetable oils and the seeds from which they are derived are the best food sources of vitamin E. For example, the amounts supplied by some typical oils (mg of vitamin E per 100 g [about 3½ oz] of oil) are as follows: wheat germ, 115; saf-

[27]"Dietary Essential Fatty Acids, Prostaglandin Formation and Platelet Aggregation," *Nutrition Reviews*, Vol. 34, 1976, p. 243.

[28]Kaunitz, H., and R. E. Johnson, "Influence of Dietary Fats on Disease and Longevity," *Proceedings of the Ninth International Congress of Nutrition (Mexico, 1972)*, Vol. 1, edited by A. Chavez, et al., published by S. Karger, Basel, Switzerland, 1975, p. 369, Table III.

[29]"The Biological Effects of Polyunsaturated Fatty Acids," *Dairy Council Digest*, Vol. 46, 1975, p. 31.

flower, 34; cottonseed, 32; peanut, 19; corn, 12; and soybean, 7.[30] It is noteworthy that these figures apply only to the amounts of alpha-tocopherol, which is the official standard for vitamin E activity. However, most vegetable oils also contain substantial quantities of related compounds (nonalpha-tocopherols) that have a reduced vitamin activity, but which, nevertheless, contribute to the total vitamin E value. Hence, some nutritionists have suggested that allowances also be made for the contributions of the other tocopherols.

Vegetable oils also contain plant sterols (phytosterols) that resemble cholesterol in structure, but which help to lower blood cholesterol by (1) interfering with its absorption, and (2) possibly accelerating its removal from the body.[31] It is noteworthy that vegetable oils which are very similar in fatty acid composition lower blood cholesterol by markedly different amounts. However, the anticholesterolemic effects of the oils are often in direct proportion to their content of plant sterols. For example, wheat germ oil, which contains approximately 5% phytosterols, has been shown to produce a much greater lowering of blood cholesterol than the other vegetable oils, which contain from 0.4% to about 1.0% phytosterols.[32] The higher plant sterol content of corn oil (about 1.0%) might also be the reason why this product is much more anticholesterolemic than soybean oil, which contains only about 0.4% phytosterols.

Antinutritional and/or Toxic Factors.
Under certain conditions, small amounts of potentially harmful substances may be formed from polyunsaturated fatty acids. These substances present hazards only when large amounts of the affected oil products are consumed. This is *not* to suggest that these products be avoided, but rather that they should be used judiciously, while keeping in mind the possible consequences of misuse. Consider:

• **Peroxides**—The highly polyunsaturated fatty acids that are abundantly present in corn, cottonseed, safflower, soybean, and sunflower oils may be oxidized to toxic peroxides when exposed to air, heat, light, and metals such as copper and iron. Also, the long reuse of frying fats that have been overheated repeatedly is likely to result in the production of peroxides and other toxic substances.

Peroxides destroy vitamin E and help to speed the breakdown of polyunsaturated fatty acids in cell membranes. It is even suspected that these chemical reactions are responsible for some of the degenerative changes in the aging process because, upon autopsy, the brains of some senile people have been found to contain oxidized fats from cell membranes.

Some of the means by which the formation of peroxides in oils may be minimized are: (1) the addition of trace amounts of antioxidants; (2) storage of the oils in tightly capped, dark brown bottles in a refrigerator; (3) careful heating during frying so that smoking does not occur (a thermometer may be helpful to inexperienced cooks); and (4) straining of the oil

after frying to remove food particles, because they increase the tendency of the oil to smoke.

Peroxide formation and the breakdown of polyunsaturated fatty acids in the body may be prevented by vitamin E (synthetic antioxidants such as BHA and BHT do *not* act in the cells of the body). Also, the essential trace mineral selenium is part of the enzyme which breaks down peroxides once they have been formed. Therefore, adequate dietary intakes of both vitamin E and selenium are the best forms of insurance against toxicity from peroxides, when the handling of vegetable oils leaves much to be desired.

• **Trans Fatty Acids**—The hardening of vegetable oils by hydrogenation converts the naturally-occurring *cis* fatty acids to *trans* fatty acids which do *not* function as essential fatty acids. Also, some researchers have found that (1) the *trans* fatty acids were not as effective as their *cis* analogs in lowering blood cholesterol, and (2) fats rich in *trans* fatty acids appeared to promote atherosclerosis.[33]

The content of *trans* fatty acids generally increases with the extent to which a vegetable oil has been hardened by hydrogenation. For example, hard sticks of vegetable oil margarines may contain from 25 to 35% of *trans* fats, whereas lightly hydrogenated liquid oils usually contain 5% or less of these fats.

(Also see TRANS FATTY ACIDS.)

VEGETABLE OILS. Noteworthy information on each of the commonly used vegetable oils is given in Table O-5.

NOTE WELL: Where production figures are given in Table O-5, column 1 under "Importance," unless otherwise stated, they are based on 1990 data. Most world figures are from *FAO Production 1990*, FAO/UN, Rome, Italy, Vol. 44. Most U.S. figures are from *Agricultural Statistics 1991*, USDA. Annual production fluctuates as a result of weather and profitability of the crop.

---

[33]"Newer Concepts of Coronary Heart Disease," *Dairy Council Digest*, Vol. 45, 1974, p. 33.

Fig. O-18. Several popular brands of vegetable oils. (Courtesy, American Soybean Assn., St. Louis, Mo.)

---

[30]Slover, H. T., *et al.*, "Vitamin E in Foods: Determination of Tocols and Tocotrienols," *Journal of the American Oil Chemists' Society*, Vol. 46, 1969, p. 420, Table VII.

[31]Konlande, J. E., and H. Fisher, "Evidence for a Nonabsorptive Antihypercholestrolemic Action of Phytosterols in the Chicken," *The Journal of Nutrition*, Vol. 98, 1969, p. 435.

[32]Alfin-Slater, R. B., 1961, "Factors Affecting Essential Fatty Acid Utilization," *Drugs Affecting Lipid Metabolism*, edited by S. Garattini, and R. Paoletti, Elsevier Publishing Company, Amsterdam, Netherlands, 1961.

TABLE O-5
VEGETABLE OILS

Oil; Source; Scientific Name; Importance	Processing; Preparation; Uses	Nutritional Value[1]
COCONUT OIL **Source:** The nut (fruit) of the coconut palm tree, one of the world's most important crop trees. **Scientific name of coconut palm:** Family: *Palmae* Genus: *Cocos* Species: *C. nucifera* **Importance:** The world production of coconut oil totals about 9.6 million metric tons annually. (Also see COCONUT.)	**Processing:** Most coconut oil is obtained from copra by pressing in mechanical screw presses, sometimes followed by solvent extraction. In some areas, hydraulic cage and box presses are still used. **Preparation:** Crude coconut oil is refined and deodorized to remove free fatty acids and flavors. **Uses:** Because of the high percentage of lauric acid (45%), a saturated fatty acid, confectioners and bakers use coconut oil in coatings and fillings for baked goods and candy which may stand for a relatively long time between manufacture and consumption. Major uses of coconut oil are in shortening and oleomargarine, deep fat frying, detergents, laundry and toilet soaps. Minor uses are in filled milk, imitation milk, prepared flours, and cake mixes; lotions, rubbing creams, and in pressurized toppings. Recently, coconut oil has served as the raw material for producing medium chain triglycerides, which are very useful in the treatment of certain digestive disorders.	The fatty acids of coconut oil are: % Lauric .... 44 Myristic .... 18 Palmitic .... 11 Oleic .... 7 Caprylic .... 6 Capric .... 6 Stearic .... 6 Linoleic .... 2 Coconut oil has the highest percentage of saturated fatty acids of all common food oils; 86% vs 11% for corn oil. Hence, it is very stable, either alone or in products. Iodine value: 9 (This indicates the degree of unsaturation; It is the number of grams of iodine absorbed by 100 cm of fat. The higher the iodine value the greater the degree of unsaturation.) The high content of lauric, caprylic, and capric acids is a nutritional asset because these fatty acids are very useful in the dietary treatment of certain digestive disorders. However, the high degree of saturation of coconut oil may raise the blood cholesterol in some people, even though the oil itself contains no cholesterol.
CORN OIL **Source:** Corn (Maize) grain **Scientific name of corn:** Family: *Gramineae*, the grass family Genus: *Zea* Species: *Z. mays* **Importance:** Corn has always been important in North America. It accounts for 48% of the total world production of 475 million metric tons of corn grain; the United States alone produced 42% of the world total. (Also see CORN.)	**Processing:** Each year, about 6% of the U.S. corn crop is milled. Corn is wet milled for the production of starch, sweeteners, and oil; and corn is dry milled for the production of grits, flakes, meal, oil, and feeds. **Preparation:** Corn oil is further refined before using for food. **Uses:** Corn oil is used for salad dressings and cooking oil; and in such products as margarine and shortening. Also it is used in paints, varnishes, soaps, glycerine, and linoleum.	Corn oil is rich in essential fatty acid and contains moderate amounts of vitamin E. The chief fatty acids of corn oil are: % Linoleic .... 54 Oleic .... 29 Palmitic .... 13 Stearic .... 4 Iodine value: 127
COTTONSEED OIL **Source:** Derived from the seed of the cotton plant. **Scientific name of cotton:** Family: *Malvaceae* Genus: *Gossypium* Species: There are about 20 species, of which the following four are cultivated: *G. hirsutum*, American upland cotton; *G. barbadense*, Egyptian and Sea Island cottons; *G. herbaceum* and *G. arboreum*, the Asiatic cottons. **Importance:** Today, cottonseed oil is one of the most important food oils. It is the fourth most widely consumed vegetable oil in the United States, exceeded only by soybean oil, rapeseed oil, and sunflower oil. A ton of seed yields an average of 336 lb (152 kg) of oil. (Also see COTTONSEED.)	**Processing:** The oil is removed by three methods. The methods along with the proportion of the U.S. cottonseed production processed by each, follows: screwpressing, 40%; prepress solvent extraction, 28%; and direct solvent extraction, 32%. **Preparation:** The crude oil extracted from the seed is usually subjected to the following processes: (1) refined, (2) bleached, (3) deodorized, and (4) winterized. **Uses:** The principal use of cottonseed oil in the United States is in salad and cooking oils. It is also used in shortening, margarine, and mellorine—a frozen dessert that is comparable to ice cream in appearance and nutritive value. Cottonseed oil is also used in emulsifiers, pharmaceuticals, insecticides and fungicides, cosmetics, rubber, plastics, and finishes for leather, paper, and textiles.	Two major technological breakthroughs have largely eliminated gossypol as a barrier to cottonseed oil in food: (1) the separation of the pigment glands in processing; and (2) glandless cottonseed. Cottonseed oil is classified as a polyunsaturated vegetable oil. Linoleic acid, its principal fatty acid, comprises about 47 to 50% of the total fatty acids. Its unique crystalline properties result from the presence of about 26% palmitic acid. Its good flavor is generally attributed to the absence of linoleic acid. Cottonseed oil is preferred human food oil because of its flavor stability. Iodine value: 109

Footnote at end of table

(Continued)

TABLE O-5 *(Continued)*

Oil; Source; Scientific Name; Importance	Processing; Preparation; Uses	Nutritional Value[1]

OLIVE OIL
Source: Derived from the olive, a fruit tree
Scientific name of olive tree:
  Family: *Oleaceae*
  Genus: *Olea*
  Species: *O. europae*
 	
Importance: About 9.2 million metric tons of olives and 1.7 million metric tons of olive oil are produced annually in the world.
In the United States, olive oil is a sideline. The major portion of the fruit is processed for eating.
(Also see OLIVE.)
	Processing: Olives are processed by crushing between stone or steel rollers. Then the crushed pulp is pressed, after which the oil is separated from the liquor.
	Preparation: Olive oil of good quality is ready to use following extracting, without refining.
	Uses: Olive oil is used chiefly in salad dressings and as a cooking oil. It is also used in soaps, perfumes, and medicines.
		Olive oil is one of the most digestible of the edible oils. The fatty acid content of olive oil is:
		
		                                                      %
		Oleic 75
		Palmitic 13
		Linoleic 9
		Stearic 2
		Palmitoleic 1
		Iodine value: 84
		A medium size green olive averages about 5.8 Calories (kcal). Olives are a good source of calcium, iron, and vitamin A.

PALM OIL
Source: Derived from the oil palm tree.
Scientific name of oil palm:
  Family: *Palmae*
  Genus: *Elaeis*
  Species: *E. guineensis*
 	
Importance: Palm oil is one of the world's most important edible and soap-making oils. Moreover, the yield per acre is higher than can be obtained by any other vegetable oil. Annually, the world production of palm oil totals over 11 million metric tons, plus 3 million metric tons of palm kernel oil.
(Also see OIL PALM.)
	Processing: The traditional method of processing by farmers consists of (1) fermenting the fruits, followed by boiling, pounding, and skimming off the oil; and (2) cracking the nuts and exporting the kernels for processing.
	Modern factories process by steam distillation, followed by centrifuging or hydraulic expression. In industrialized nations, the kernels are either expressed or are solvent extracted.
	Correctly speaking, the fruits yield palm oil and the kernels yield palm kernel oil. But usually both are referred to as palm oil.
	Preparation: Palm oil of good quality is ready for certain uses following extracting, without refining. Although it is liquid when first extracted, it sets about as thick as butter at normal temperature.
	Uses: The leading producing countries use palm oil locally as food in all kinds of cookery, and export the surplus.
	When exported, palm oil is used chiefly for soap-making and industrial purposes. But, with suitable treatment, it may also be used in margarine.
	Palm kernel oil, which is white or pale yellow, is largely used for margarine.
		Palm oils vary; in some cases, they resemble kernel oil, and in others olive oil.
		The fatty acid profile of palm oil is:
		
		                                         %
		Palmitic 48
		Oleic 38
		Linoleic 9
		Stearic 4
		Myristic 1
		The fatty acid profile of palm kernel oil is:
		
		                                         %
		Lauric 51
		Myristic 17
		Oleic 13
		Palmitic 8
		Capric 4
		Caprylic 3
		Stearic 2
		Linoleic 2
		Iodine value of palm oil: 51
		Iodine value of palm kernel oil: 16

Footnote at end of table

(Continued)

TABLE O-5 *(Continued)*

Oil; Source; Scientific Name; Importance	Processing; Preparation; Uses	Nutritional Value[1]
PEANUT OIL **Source:** Derived from the peanut; a member of the pea family. **Scientific name of peanut:** Family: *Leguminosae* Genus: *Arachis* Species: *A. hypogaea* **Importance:** Peanuts, which contain 47 to 50% oil, are an important crop, especially in the warm regions of the world. Annual world production totals about: (1) peanuts in the shell, over 23 million metric tons; (2) peanut oil, 3.76 million metric tons. About ⅔ of the world's peanut crop is crushed for oil. Peanuts supply about 5% of the world's fats and oils. (Also see PEANUTS.)	**Processing:** The oil from peanuts is extracted by one of three methods: (1) hydraulic extraction, (2) expeller extraction, or (3) solvent extraction. **Preparation:** Following extraction, peanut oil is refined. The major portion of the characteristic peanut aroma and flavor is retained in the oil. **Uses:** In the United States, only a small proportion of the crop is processed for oil and protein concentrate. But in the rest of the world, peanuts are primarily processed for their separate constituents—oil and protein. Peanut oil is used primarily for food purposes; for frying foods, in salad oils, dressings, and margarine, and along with other vegetable shortenings. Also, peanut oil is used in soaps, face powders, shaving creams, shampoos, paints, as machinery oil, and in making nitroglycerin.	Peanut oil contains the following fatty acids: 　　　　　　　　　　　　　% Oleic . . . . 61 Linoleic . . . . 22 Palmitic . . . . 6 Stearic . . . . 5 Behenic . . . . 3 Arachidic . . . . 2 Lignoceric . . . . 1 Iodine value: 101 The major portion of the characteristic peanut aroma and flavor is imparted by the oil.
RAPESEED OIL **Source:** Derived from the rape plant, a member of the mustard family. **Scientific name of Rape:** Family: *Cruciferae* Genus: *Brassica* Species: *B. napus* and *B. camtestris* **Importance:** Rape oil is the only oil derived from an oilseed crop grown successfully in all parts of the world. Annually, world production of rapeseed totals about 24.5 million metric tons, and world production of rapeseed oil totals over 8.8 million metric tons. (Also see RAPE.)	**Processing:** The processing of rapeseed to obtain oil is similar to that of other oilseeds. There are three methods: (1) mechanical pressing, (2) straight solvent, or (3) prepress solvent-extraction. The latter is the method of choice. **Preparation:** The crude oil must still be refined. **Uses:** The primary use of rapeseed oil is for human food. It's relatively low linolenic acid content permits it to compete with other vegetable oils in shortenings, margarine, salad oils, and frying oils. Oils which are almost free of erucic acid will open up new uses of rapeseed oil.	In the past, rapeseed oil was high (it contained 40 to 45%) in the long-chain fatty acid, erucic acid. Since the nutritional studies indicate that this substance, along with glucosinolates which the meal formerly contained, can be detrimental to human and animal health if consumed in substantial quantities, plant breeders were stimulated to make genetic changes in the composition of rapeseed—to lower or eliminate erucic acid and glucosinolate. This change has proceeded at different rates in different countries, but is now nearly complete in Canada—one of the leading rapeseed producers of the world. Oil made from the new varieties of rape is almost free of the long-chain fatty acid, erucic acid, while the oleic acid content is significantly higher. Except for the presence of linolenic acid, the composition of the new oil greatly resembles olive oil. Canada renamed the changed plant; it is called *canola*.
SAFFLOWER OIL **Source:** Safflower, a relative of the thistle. **Scientific name of safflower:** Family: *Compositae* Genus: *Carthamus* Species: *C. tinctorius* **Importance:** The importance of safflower oil stems from its unique fatty acid composition, which places it highest in polyunsaturates and lowest in saturates of all commercial fats and oils. Annually, world safflower seed production totals about 922,000 tons. (Also see SAFFLOWER.)	**Processing:** In the United States, extraction of the oil is largely by the continuous screw-press solvent extraction method. **Preparation:** The oil must be refined. Due to the susceptibility of safflower oil to oxidation, care is taken to exclude air during storage, transport, and packaging. **Uses:** Safflower oil is used principally in the production of margarine, salad oils, mayonnaise, shortening, and other food products. The oil is also used as a drying agent in paints and varnishes.	Safflower oil is (1) extolled as a preventative of cholesterol build-up in the blood, and (2) recommended in the diets of persons suffering from heart disease and hypertension. The value of polyunsaturated fatty acids for such treatments, however, remains a matter of controversy. Safflower oil averages about 6.6% saturated acids and 93.4% unsaturated acids, with the latter consisting of 77.0% linoleic acid and 16.4% oleic acid. Linolenic acid is absent.

Footnote at end of table

(Continued)

TABLE O-5 *(Continued)*

Oil; Source; Scientific Name; Importance	Processing; Preparation; Uses	Nutritional Value[1]
SESAME OIL **Source:** Derived from the sesame plant, an annual herb. **Scientific name of sesame:** Family: *Pedaliaceae* Genus: *Sesamum* Species: *S. indicum* **Importance:** Annually, the world production of sesame seed now exceeds 2 million metric tons. (Also see SESAME.)	**Processing:** In those areas where sesame is primarily processed for its oil content, the seed is not dehulled; rather, the entire seed is crushed. **Preparation:** Sesame oil is natural salad oil requiring little or no winterization and is one of the few vegetable oils that can be used without refining. These are factors of increasing importance as energy costs escalate. **Uses:** Sesame oil is popular because of its pleasant, mild taste and remarkable stability—it is the most stable naturally-occurring liquid vegetable oil. Sesame oil is much sought because of the presence of a natural antioxidant, sesamol, and because of its high content of unsaturated fatty acids, 40% oleic and 44% linoleic. The oil is used as a substitute for olive oil—primarily as a salad and cooking oil and for margarine and soap. Also, a significant quantity of oil is used by the cosmetic industry for softening and soothing purposes and by the pharmaceutical industry as a carrier for medicines.	Due to the presence of natural antioxidants in the crude oil, sesame oil is the most stable naturally occurring liquid vegetable oil. It will keep for several years without turning rancid. The fatty acid composition of sesame oil is: % Linoleic .... 44 Oleic .... 40 Palmitic .... 9 Stearic .... 5 Linolenic .... 2 Sesame oil is classified as polyunsaturated. Iodine value: 114
SORGHUM OIL **Source:** Derived from the grain of the sorghum plant. **Scientific name of sorghum:** Family: *Gramineae* Genus: *Sorghum* Species: *S. bicolor* **Importance:** Sorghum is the staple food in parts of Asia and Africa. Sorghum is the fifth ranking cereal food crop of the world, being exceeded by wheat, corn, rice, and barley. Annually, the world sorghum crop totals over 58 million metric tons. (Also see SORGHUM.)	**Processing:** As with corn, sorghum may be either dry milled or wet milled. However, most sorghum is dry milled, simply because it is easier to accomplish. Sorghum oil is extracted from the germ of the grain. **Preparation:** For food use, sorghum oil must be refined, much like corn oil. **Uses:** Cooking oil; and in such products as margarine and shortenings.	The oil of sorghum is important nutritionally, and it influences flavor and acceptability of sorghum products. The fatty acid composition of sorghum germ oil is similar to corn oil (see CORN OIL). The ether extract of sorghum bran is composed mainly of long-chain alcohols and esters with some long-chain hydrocarbons.
SOYBEAN OIL **Source:** Soybean oil is derived from seeds produced by soybean plants. **Scientific name of soybean:** Family: *Leguminosae* Genus: *Glycine* Species: *G. max* **Importance:** Soybean oil is the leading vegetable oil for human food in the United States. In the Far East, soybean oil is consumed extensively as a food. Annually, world production of soybeans totals about 108 million metric tons, and world production of soybean oil is about 15.7 million metric tons. (Also see SOYBEANS.)	**Processing:** There are three basic processing methods of obtaining oil from soybeans: solvent extraction, hydraulic extraction, or expeller extraction. Today, almost all of the oil is extracted by the solvent process. **Preparation:** About 93% of soybean oil is used for food. To be suitable for human consumption, the extracted crude oil must undergo further processing, which is generally referred to as refining. **Uses:** Soybean oil is widely used throughout the world for human consumption as margarine, salad and cooking oils, and shortening. Also, soybean oil is used in paints, varnishes, enamels, soaps, linoleum, pharmaceuticals, cosmetics, core oil, synthetic rubber, and printing ink.	The fatty acid profile of soybean oil follows: % Linoleic .... 54 Oleic .... 24 Palmitic .... 12 Linolenic .... 8 Stearic .... 2 Soybean oil is low in saturated fat and free of cholesterol. Iodine value: 134

Footnote at end of table

(Continued)

TABLE O-5 *(Continued)*

Oil; Source; Scientific Name; Importance	Processing; Preparation; Uses	Nutritional Value[1]
SUNFLOWER OIL **Source:** Sunflowers **Scientific name of sunflower:** Family: *Compositeae* Genus: *Helianthus* Species: *H. annuus* **Origin and history:** The sunflower was first domesticated in the United States, although there is evidence that this honor should be shared with Peru and Mexico. **Importance:** Today, sunflower oil holds undisputed claim to third place among edible vegetable oils. Annually, world sunflower seed production totals about 22 million metric tons, and world sunflower oil production is about 7.8 million metric tons. Twenty-nine percent of the world's sunflower seed is produced in the U.S.S.R. Eighty percent of the vegetable oil produced in the Soviet Union comes from sunflowers. (Also see SUNFLOWERS.)	**Processing:** The separation of the oil from the seed may be acheived (1) by direct solvent-extraction, (2) by prepress solvent-extraction, or (3) by mechanical means (screwpressing). In the United States, sunflower oil is processed primarily by the prepress solvent-extraction method; and only small quantities of seed are dehulled prior to extraction. **Preparation:** Crude sunflower oil is stored in tanks following extraction until further processing. After refining, it has an attractive color and a pleasant, faintly nutty flavor. **Uses:** Cooking oil, salad oil, shortening, margarine, or for frying foods, making potato chips, producing modified butter with improved low temperature spreadability, and blending with other vegetable oils.	The fatty acid composition of sunflower oil makes it desirable as an edible oil. It follows: % Linoleic .... 66 Oleic .... 21 Palmitic .... 8 Stearic .... 5 As noted, sunflower oil is relatively low in the saturated fatty acids, palmitic and stearic. The high linoleic acid content and the high ratio of polyunsaturated to saturated fatty acid prompts some nutritionists to believe that sunflower oil might be useful in the prevention and treatment of high blood cholesterol and heart disease. Iodine value: 134

[1]Food Composition Table F-21 and Table F-4, Fats and Fatty Acids In Selected Foods contain additional information.

OLD PROCESS

Pertains to the extraction of oil from seeds. Same as hydraulic process.
(Also see SOYBEAN, section on "Hydraulic Extraction.")

OLEIC ACID

An 18-carbon unsaturated fatty acid (one double bond) which reacts with glycerol to form olein.
(Also see FATS AND OTHER LIPIDS.)

OLFACTORY

Pertaining to the sense of smell.

OLIGOSACCHARIDE

A complex carbohydrate which contains 2 to 10 molecules of monosaccharides combined through glycoside bonds.

OLIGURIA

Scanty urination.

OLIVE *Olea europaea*

Olives, which are classed as a fruit but used as a vegetable, belong to the olive family, *Oleaceae*.

Fig. O-19. Olive tree showing olives being picked by hand—the way it is done. (Courtesy, California Olive Industry, Fresno, Calif.)

ORIGIN AND HISTORY. It is not known when the wild olive was first brought under cultivation. But all indications are that the limestone hills of Attica, the Greek peninsula, was the seat of its first cultivation. The Spanish brought the olive to California in 1769.

Fig. O-20. Primitive olive press. (Courtesy, Field Museum of Natural History, Chicago, Ill.)

WORLD AND U.S. PRODUCTION. About 9 million metric tons of olives and 1.6 million metric tons of olive oil are produced in the world, annually. Most of the production is in the countries bordering the Mediterranean Sea. The leading olive-producers, by rank, are: Spain, Italy, Greece, and Turkey.[34] Spain and Italy produce 50% of the world's olives and 54% of the olive oil.

Olives rank fourteenth among the vegetable crops of the United States. About 131,000 short tons of olives are produced in the United States, annually.[35] California accounts for 99% of the olive production of the United States.

PROCESSING. Olives were one of the first crops that man learned to adapt to his needs by developing special techniques for making the fruit edible and for extraction of the oil.

The methods of processing olives for the table vary widely. However, the three major commercial methods, known after the countries of their origin, are: (1) the Spanish method, in which unripe yellowish-green olives are fermented; (2) the American method, in which half-ripe, reddish fruit is used; and (3) the Greek method, in which the fully ripe, dark purple fruit is preserved. Additionally, there are numerous

local methods of processing, including pitting, stuffing, chopping, and spicing. In most processing methods, a weak solution of lye is applied for the purpose of neutralizing the bitter principle; the lye penetrates the olives and hydrolyzes the bitter phenolic glycoside, oleuropeen. The lye-treated olives are immediately rinsed and soaked in water with frequent changes in order to remove the lye. The washed olives are then placed in fermentation tanks and barrels and covered with brine.

Olive oil is produced by mechanical extraction of the ground fruit. Traditionally, this involves three steps: (1) crushing the fruit, (2) pressing (usually 2 to 4 times) the paste, and (3) separating the oil from the liquor. After the second pressing, the cake is usually solvent-extracted. Unlike most other vegetable oils, olive oil prepared from properly matured, harvested, and stored fruit is consumed without any refining treatment, thereby alleviating the need for added energy.

SELECTION, PREPARATION, AND USES. Most table olives are prepared commercially in ready-to-serve form; hence, selection by the consumer is largely alleviated. Olives are commonly prepared commercially by one of the following methods:

1. **Canned or bottled green fermented olives (Spanish method).** This refers to green olives that are fermented, then put in jars or barrels. Increasingly, green olives are being pitted and the pit cavity stuffed with either pimientos, onions, almonds, anchovies, or other edibles.

2. **Canned ripe olives (American method).** The production of canned ripe olives is centered in California. The olives are picked when straw yellow to cherry red in color. In the curing process, the olives darken progressively due to the oxidation of the phenolic substances in a basic envi-

Fig. O-21. Ripe olives are canned in brine and heat sterilized. (Courtesy, California Olive Industry, Fresno, Calif.)

[34]Data from *FAO Production Yearbook 1990*, FAO/UN, Rome, Italy, Vol. 44, p. 120, Table 46. **Note well:** Annual production fluctuates as a result of weather and profitability of the crop.

[35]Data from *Agricultural Statistics 1991*, USDA, p. 371, Table 548. **Note well:** Annual production fluctuates as a result of weather and profitability of the crop.

ronment. The cured olives are packed in enamel-lined cans, covered with salt solution, sealed, and sterilized. More and more canned ripe olives are being pitted and stuffed.

3. **Black, naturally ripe olives (Greek method).** Olives to be prepared by this method are kept on the trees until they are fully matured and completely dark. The fruit is brined, with frequent changes of water in order to hasten the destruction of the bitter principle. Pretreatment with lye solution before brining, such as is used in the Spanish method, will hasten debittering and fermentation.

In addition to the above three major commercial procedures, a large number of different techniques for preparing olives for table use exist. Olives are often pitted, stuffed, chopped, and spiced, and put in fancy packs.

Fig. O-22. Olives enhance a vegetable dish. (Courtesy, Olive Administrative Committee, Fresno, Calif.)

Olives are used as a relish, and as an ingredient of salads and other mixed dishes such as pizza. Increasingly, table olives are becoming a luxury product in countries with high standards of living.

Olive oil of good quality needs no refining. The oil is highly esteemed by the gourmet. It is used chiefly in salad dressings, as a cooking oil, and for canning sardines.

NUTRITIONAL VALUE. The nutrient compositions of olives and olive oil are given in Food Composition Table F-21.

A few salient points regarding the nutritional value of olives follow:

1. At full maturity, the fruit meat (mesocarp) contains 15 to 35% oil. The pit, which accounts for 15 to 30% of the weight of the fruit, contains about 5% oil. The characteristic bitter glycoside, oleuropein, present in fresh olives, is concentrated close to the peel (exocarp).

2. Olives vary in calories according to size and oil content: On a per olive basis, they average about as follows:

Olive/Size and Kind	Calories/Olive
Medium, green	5.8
Medium, ripe	9.0

3. Olives contain fair amounts of iron and calcium and some vitamin A.

(Also see OILS, VEGETABLE, TABLE O-5.)

–OLOGY

Suffix meaning science of, or study of.

OMOPHAGIA

The practice of eating foods, particularly flesh, raw. The primitive diet of Eskimos consisted of raw meat and fish.

ONION *Allium cepa,* **variety** *cepa*

The onion, like the other related vegetables (*Allium* genus), has long been classified as a member of the Lily family (*Liliaceae*), but now some botanists place the onion and its close relatives in a new family called the *Alliaceae.* The common onion is classified as the variety *cepa* to distinguish it from other varieties of onions such as the shallot, which is the *aggregatum* variety. Fig. O-23 shows typical onion bulbs.

Fig. O-23. Onions, the sixth leading vegetable crop of the world. (Courtesy, University of Minnesota, St. Paul, Minn.)

The underground bulb of the onion plant is the sixth leading vegetable crop of the world. It is distinguished by its pungency when raw and the appetizing flavor it imparts when cooked with various other foods. The odor is due to an oil, which readily forms a vapor and escapes into the air when onions are peeled or cut. It affects nerves in the nose connected with the eyes, and makes tears flow.

ORIGIN AND HISTORY. Onions originated in the central part of Asia. Prehistoric man gathered and cooked wild onions. Some time after the dawn of agriculture in the Middle East, onions were grown in ancient Chaldea and Egypt. They were used as early as 5,000 years ago in the First Egyptian Dynasty.

WORLD AND U.S. PRODUCTION. Onions rank sixth among the leading vegetable crops of the world, and fourth among the leading vegetable crops of the United States. About 27.9 million metric tons of onions are produced worldwide, annually. The leading onion-producing nations, by rank, are: China, India, United States, U.S.S.R., Turkey, Japan, and Spain.[36]

About 2.6 million short tons of onions are produced in the United States, annually. In descending order, the leading onion-producing states are: Texas, California, New Mexico, Georgia, Arizona, and Washington.[37]

PROCESSING. The most common means of processing are:

• **Canning**—White-skinned onions, which are the most desirable ones for canning, are processed by removal of the papery skin, blanching, packing into cans, acidifying, and processing in a boiling water bath. Onions are also canned in prepared soups that require only heating.

• **Dehydration**—This procedure involves (1) burning off the onion skin and pieces of adhering roots, (2) washing with a high pressure stream of water to remove the charred pieces of skin, (3) slicing the deskinned onions thinly, (4) drying the slices with hot air, and (5) packaging the dehydrated pieces. Sometimes, dehydrated onion pieces are ground finely to make onion powder and onion salt.

• **Freezing**—Onions are commonly frozen in forms such as chopped raw onions, French fried onion rings, prepared soups, and TV dinners.

• **Pickling**—Small, immature onion bulbs are pickled by soaking in several changes of a salt solution (brine), boiling in fresh brine, and packing in a well-seasoned mixture of vinegar and sugar. Pickled onions are quite popular as an appetizer, and in certain cocktails.

SELECTION, PREPARATION, AND USES. Professional chefs select types of onions according to how they will be utilized, since there are notable differences in flavor, pungency, and quality among the varieties commonly available in the United States. Hence, some guidelines for selecting and utilizing the various kinds of onions follow:

Dry Onions. Pertinent information relative to the types, quality, and utilization of dry onions follow:

• **Types**—Two general classes of dry onions grown in the United States are found in retail markets: the mild flavored types, either large and elongated, or flat; and the usually stronger flavored types, generally globe shaped and medium in size. The former are *Spanish* or *Bermuda* types, and the latter are known as *globe* or *late crop* types.

• **Judging quality**—Bright, clean, hard, well-shaped, mature onions with dry skins are usually of good quality.

• **Utilization**—Onions are used as follows:
Chopped or sliced raw, which go well with roasted meats and poultry, and with other raw vegetables in relishes and salads.
Stewed, which may be served with butter, cheese sauce, or white sauce.
Broiled, which make a good accompaniment to broiled fish, fried liver, hamburger, and steak.
Stuffed with breadcrumbs, cooked kasha (buckwheat groats), fish, meat, poultry, and/or rice and baked like other stuffed vegetables. Stuffing requires large onions.
Dehydrated, and used in place of either raw or cooked fresh onions, provided that they are rehydrated sufficiently.
Powdered, or prepared as onion salt, which makes an excellent seasoning, but some of the flavor components of fresh onion are missing.
Finally, the various forms of onions may be added to a wide variety of casseroles, soups, and stews.

Green Onions. These onions are usually early white or bulbless varieties that are harvested when the partially developed bulbs reach the desired size. Good quality green onions should have green, fresh tops, medium-sized necks well blanched for 2 or 3 in. (5 to 8 cm) from the root; and should be young, crisp, and tender.

Green onions may be served raw as an appetizer, or they may be chopped and added to cottage cheese, cream cheese, salads, and salad dressings. They are also good when pickled, stewed and served with a sauce, or when added to casseroles, soups, and stews.

NUTRITIONAL VALUE. The nutrient compositions of various forms of onions are given in Food Composition Table F-21.

Some noteworthy observations regarding the nutrient composition of onions follow:

1. Dry, mature onions are high in water content (89%) and low in calories (38 kcal/100 g), protein (1.5%), and most other nutrients. They are a fair source of potassium, but they contain barely enough vitamin C to prevent scurvy (about 10 mg of vitamin C per 100 g).

2. Dehydrated onion flakes and onion powder contain almost ten times the solids content of fresh onions. Hence, the flakes and powder are much better sources of calories, protein, and other nutrients than the fresh bulbs.

3. The leafy tops of green onions are nutritionally superior to the green bulbs and to mature bulbs in that the tops are (a) a better source of potassium, (b) an excellent source of vitamin A, and (c) a good source of vitamin C. Hence, green onions should be eaten with the tops whenever feasible.

4. Europeans have long believed that onions have medicinal properties such as (a) prevention of colds, (b) loosening of phlegm, (c) correction of indigestion, (d) inducement of sleep, (e) stimulation of the appetite, (f) disinfectant in wounds, and (g) driving parasites from the digestive tract. Furthermore, they have often been added to food preparations to retard spoilage. So far, there has been only limited scientific documentation for these beliefs, consisting mainly of findings that onions (1) have a mild antibacterial effect, and (2) contain a substance (adenosine) which stimulates the breakdown of fibrinogen to fibrin, a protein in the blood which is involved in blood clotting.

---

[36]Data from *FAO Production Yearbook 1990*, FAO/UN, Rome, Italy, Vol. 44, p. 141, Table 58. **Note well:** Annual production fluctuates as a result of weather and profitability of the crop.

[37]Data from *Agricultural Statistics 1991*, USDA, p. 157, Table 222. **Note well:** Annual production fluctuates as a result of weather and profitability of the crop.

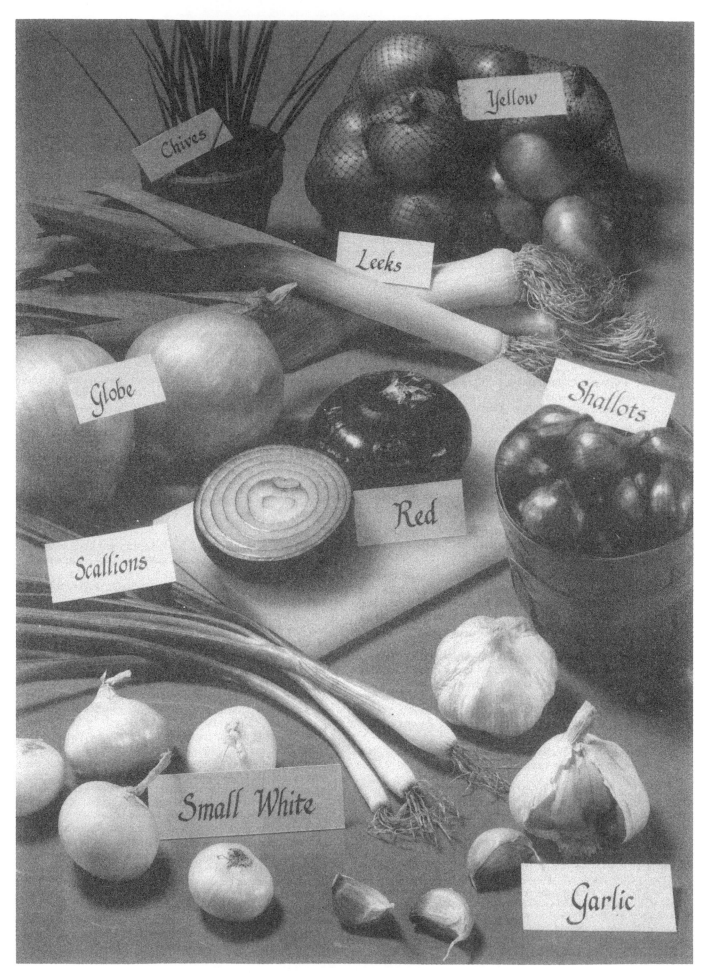

Fig. O-24. The onion and its close relatives. (Courtesy, United Fresh Fruit & Vegetable Assn., Alexandria, Va.)

ON-THE-HOOF

A term applied to a live animal; for example, the on-the-foot weight would be liveweight.

ON-THE-RAIL

A term applied to carcasses on the rail.

OPHTHALMIA

Severe inflammation of the conjunctiva (the membrane that lines the inner surface of the eyelids) or of the eyeball.

OPOSSUM

A marsupial, reaching the size of a rabbit, which abounds in certain regions of North America. The flesh is esteemed as a food in some sections and resembles that of rabbit.

OPSIN

A protein compound which combines retinal (vitamin A) to form rhodopsin (visual purple).
(Also see VITAMIN A, section headed "Functions".)

OPSOMANIA

Indicates a craving for special or certain foods—the pickles and ice cream appetite of pregnancy. In some cases, it may signal a deficiency of some essential element, for instance the periodic craving of chocolate may mean an insufficiency of chromium in the diet.
(Also see CISSA; and PICA.)

ORAL

Pertaining to the mouth.

ORANGE, BLOOD *Citrus sinensis*

Varieties of the sweet orange (a fruit of the *Rutaceae* family) in which the flesh of the fruit has a reddish tint due to the presence of anthocyanin pigments. It appears that the blood coloration develops only when the fruit is grown under hot, dry conditions such as those of the Mediterranean region. Hence, the major areas of production are Italy, Spain, Algeria, Morocco, and Tunisia. Blood oranges are usually consumed fresh, since the variability of the pigmentation presents difficulties in juice production and canning.
(Also see ORANGE, SWEET.)

ORANGE BUTTER

A mixture of butter or margarine, orange juice, grated orange peel, and/or other seasonings that is used as a sauce for dressing, cooked fish, meats, and poultry.

ORANGE JUICE

The fluid obtained from ripened oranges. Usually, it has been strained to remove excess pulp and seeds. Commercially canned orange juices may also contain juices from mandarin oranges (not more than 10%) and sour oranges (not more than 5%).

CAUTION: Some products may contain allergenic substances from the peel, but those made for infants contain little or none of these substances.

Fig. O-25. Orange juice, refreshing and high in vitamin C. (Courtesy, USDA)

The nutrient compositions of various forms of orange juice are given in Food Composition Table F-21.

Some noteworthy observations regarding the nutrient composition of orange juice follow:

1. A 1-cup (*240 ml*) serving of fresh orange juice provides 112 Calories (kcal) and 26 g of carbohydrates. It is an excellent source of potassium and vitamin C, and a fair to good source of vitamin A.

2. Canned, unsweetened orange juice is similar in composition to the fresh juice, except that it is about 20% lower in vitamin C.

3. Frozen orange juice concentrates contain over 3 times the nutrient level of the unconcentrated juice. Hence, it may make a significant nutritional contribution when added undiluted to sauces and toppings.
(Also see CITRUS FRUITS; and ORANGE, SWEET.)

ORANGE PEKOE

This refers to the grade of tea leaves from which a black tea is made. Orange pekoe leaves are the smallest grade of leaves, and they are generally from the tips of the branches of the tea plant.
(Also see TEA.)

ORANGE, SWEET *Citrus sinensis*

This citrus species is by far the leading fruit crop of the United States, and it is the second leading one of the world. (Grapes are first, and bananas, watermelons, and plantains follow in that order.)

There are two kinds of closely related oranges—the sweet orange, and the sour (Seville) orange. Both are members of the rue family, *Rutaceae*.

This article pertains to the sweet orange, which is the one commonly grown and eaten in the United States (for sour orange, see SEVILLE ORANGE).

Fig. O-26. Orange trees, showing ripe oranges being hand-picked. (Courtesy, USDA)

Sweet oranges grow best in subtropical climates. Hence, commercial orange production in the United States is limited to Florida, Texas, Arizona, and California.

ORIGIN AND HISTORY. The sweet orange originated in southern China, but it was domesticated so long ago that it is no longer known in the wild. Oranges were not introduced to the Mediterranean region until about 1500 A.D. They were brought to America by the Spanish and Portugese in the 1500s.

WORLD AND U.S. PRODUCTION. Worldwide, over 52 million metric tons of oranges are produced annually; and the leading orange-producing countries, by rank, are: Brazil, United States, China, Spain, Mexico, India, Italy, Egypt, Iran, and Pakistan.[38]

Fourteen percent of the world's orange crop comes from the United States, where oranges are the number one fruit crop. About 7.8 million short tons of oranges are produced annually in the United States. Florida is the leading state by a wide margin; it accounts for more than 84% of the nation's orange crop. In descending order, the ranking of the other states are: California and Arizona.[39]

PROCESSING. About 84% of the U.S. orange crop is processed. The leading item is frozen orange juice concentrate, which utilizes almost 80% of the oranges that are grown for processing.

---

[38]Data from *FAO Production Yearbook 1990*, FAO/UN, Rome, Italy, Vol. 44, p. 163, Table 71. **Note well:** Annual production fluctuates as a result of weather and profitability of the crop.

[39]Data from *Agricultural Statistics 1991*, USDA, p. 189, Table 278. **Note well:** Annual production fluctuates as a result of weather and profitability of the crop.

An additional 15% of the orange crop is used to make chilled fresh orange juice that is packed in glass bottles, plastic bottles, and paperboard cartons; while the remainder is converted into canned orange juice and orange juice blends, canned orange sections, and chilled fresh orange sections packed in glass or plastic containers.

SELECTION, PREPARATION, AND USES. Fresh oranges of the best quality are firm, heavy, have a fine-textured skin for the variety, and are well-colored.

Preparation of fresh oranges is relatively simple. It consists of peeling, then serving fresh orange halves or sections alone or with other fruit.

Fresh orange sections and freshly squeezed orange juice have tangy flavors that may not be present in processed orange products because heating during processing drives off some of the aromatic constituents of the fruit. Furthermore, certain types of processing also reduce the content of vitamin C and bioflavonoids (vitaminlike substances). Therefore, some suggestions for using fresh fruit and juice follow:

1. Orange sections go well with other, more bland-flavored fruits such as apples, apricots, avocados, bananas, blueberries, cherries, figs, melons, peaches, and pears. It might also be a good idea to add some orange juice to the fruit mixtures to prevent cut pieces of certain fruits from darkening.

2. The flavor of oranges or orange juice also enhances such vegetables as beets, carrots, squash, and sweet potatoes. A little honey with butter or margarine helps to bring about the blending of the citrus and vegetable flavors.

3. Weight watchers may make a low calorie snack or a light meal by combining orange sections with lowfat cottage cheese or plain (unflavored) yogurt. The fruit also goes well with custard, ice cream, ices, puddings, and sherbets.

4. More nutritious gelatin desserts and salads may be prepared from fresh oranges, orange juice, and unflavored gelatin than from flavored gelatin dessert mixes.

5. The fruit sections may be added to sauces for fish, meats, poultry, and seafood, or to salads which contain these ingredients.

6. Orange juice may be mixed into thick jams and marmalades in order to (a) make them more spreadable and (b) raise their nutrient levels.

NUTRITIONAL VALUE. The nutrient compositions of various forms of oranges are given in Food Composition Table F-21.

Some noteworthy observations regarding the nutrient composition of oranges and orange products follow:

1. Fresh oranges are low in calories (49 kcal per 100 g), a good source of fiber, pectin, and potassium; an excellent source of vitamin C, inositol, and bioflavonoids (vitaminlike substances); and a fair source of folic acid.

2. Fresh orange juice contains about the same levels of most of the nutrients present in fresh oranges, except for fiber, pectin, and the bioflavonoids, which are present mainly in the peel and the membranes which surround the segments of fruit.

3. Higher quantities of vitamin C and folic acid are likely to be retained in chilled fresh orange juice packed in bottles or cartons than in canned orange juice which is heated strongly during canning, then stored at room temperature for an extended period of time.

4. Frozen orange juice concentrate contains from 3 to 4 times the levels of calories and nutrients of an equal amount of fresh orange juice. Hence, the concentrate may

make a significant nutritional contribution when it is added undiluted to various beverages, salad dressings, sauces, and toppings.

5. Dehydrated pure orange juice crystals contain over 8 times the levels of calories and nutrients of equal amounts of fresh orange juice. At the present time the crystals are sold mainly by institutional food service suppliers. (The dry mixes sold in retail groceries usually contain considerable amounts of added sugar.)

6. Orange and apricot drink (contains 40% fruit juices) supplies about 3 times the vitamin A of orange juice, but less than half as much potassium and vitamin C.

(Also see BIOFLAVONOIDS; CITRUS FRUITS; FIBER; and VITAMIN C.)

ORGANIC

Substances derived from living organisms. Carbon-containing compounds.

ORGANIC ACID

Any organic compound that contains a carboxyl group (COOH).

ORGANICALLY GROWN FOOD

This phrase and the phrases *organic food, organic gardening, natural foods,* and *health foods* are all related, carrying similar implications. Natural and health foods are often the products of organic farming or gardening methods.

At present, there is no national definition of *organic,* and no nationwide program to monitor produce labeled *organically grown.* However, 25 states have established their own definitions. Also, the Farm Bill of 1990 requires that by October 1993, plans for national certification be in effect.

Although the term organic traditionally refers to compounds of carbon, J. I. Rodale, the father of the organic farming and food movement in the United States, defined organic as meaning production of crops without using pesticides and chemical fertilizers. Recently, the USDA defined organic farming as follows:

"Organic farming is a production system which avoids or largely excludes the use of synthetically compounded fertilizers, pesticides, growth regulators, and livestock feed additives. To the maximum extent feasible, organic farming systems rely upon crop rotations, crop residues, animal manures, legumes, green manures, off-farm organic wastes, mechanical cultivation, mineral-bearing rocks, and aspects of biological pest control to maintain soil productivity and tilth, to supply plant nutrients, and to control insects, weeds, and other pests."[40]

Often, the meanings for *organic, natural,* and *health* which people imply or conjure up in their minds, are misleading, harmful, and tend to polarize people. The growing interest of consumers in the safety and nutritional quality of the American diet is a welcome development. Regrettably, however, much of this interest has been colored by those who state or imply that the American food supply is unsafe or somehow inadequate to meet our nutritional needs.

[40]*Report and Recommendations on Organic Farming,* USDA, 1980.

Fig. O-27. How it used to be done—organically grown. (Courtesy, Field Museum of Natural History, Chicago, Ill.)

The Food and Drug Administration (FDA) has taken no position on use of the terms *organic, natural,* and *health* in food labeling, since the terms are often used loosely and interchangeably. The Federal Trade Commission in its proposed Food Advertising Rule would prohibit use of the words *organic* and *natural* in food advertising because of concern about the ability of consumers to understand the terms in the conflicting and confusing ways they are used. FTC also proposes to prohibit the term *health food* in advertising because it is undefined and may fool consumers into thinking one particular food will provide good health.

Food Quality. Much of the promotion of organically grown food, natural and health foods included, comes from the idea that these foods are nutritionally superior to foods grown by conventional agricultural methods. Moreover, organically grown foods are promoted as containing no pesticides or additives, while conventionally grown foods are pictured as being poisoned with pesticides and additives. The truth is, that once organically grown and conventionally grown foods are removed from the field they cannot be identified as to their origin. Plants convert inorganic compounds to organic compounds. Therefore, it is relatively immaterial whether conventional methods or organic methods are followed. Inorganic ions—nitrates, potassium, iron, phosphate—are taken up by the plant roots and manufactured into new organic materials. The nutrient content of a crop—the amount of protein, carbohydrate, fat, vitamins, and most minerals—is largely determined by plant genetics, weather, and time of harvest. Scientific experiments at the Michigan Experiment Station for 10 years; at the United States Plant, Soil and Nutrition Laboratory in Ithaca, New York for 25 years; and as well as a 34-year investigation of organic and chemical agriculture on a British experimental farm failed to show a nutritional superiority of organically grown foods compared to conventionally grown foods.

Organically grown foods contain pesticide residues just as often as do conventional foods, although they may be present in smaller amounts. However, all residues are within Federal tolerance levels, which are set low enough to protect consumers. As for poisons in the food supply, many common foods contain naturally-occurring toxicants, but these are usually present in low levels and pose no health hazards. Poison is a matter of dosage. Further, a poisonous substance made by a plant is no different than the same substance made in a laboratory.

(Also see ADDITIVES; section on "Government Controls of Additives"; and POISONS.)

Food Quantity. The use of chemical fertilizers has been partly responsible for the abundance of food available. If all farmers were to adopt organic methods, there would be a decline in productivity as shown in Table O-6.

TABLE O-6
ESTIMATED NATIONAL AVERAGE CROP YIELDS UNDER CONVENTIONAL AND ORGANIC FARMING[1]

Crop	Bushels per Acre	
	Conventional	Organic
Corn	98	49
Wheat	43	20
Soybeans	40	20
Other grains	57	17

[1]*Organic and Conventional Farming Compared*, Council for Agricultural Science and Technology (CAST), Report No. 84, October 1980, p. 24, Table 6.

Cost. One major difference between organically grown food and conventionally grown food is that it is often sold at relatively high prices to persons who are willing to tolerate imperfections in return for the suggestion that commercial fertilizers, pesticides, and additives have not been used. Numerous surveys have indicated that people will pay 30 to 100% more for organically produced foods than for their regular counterparts.

Individuals who desire organic foods—foods grown without the use of modern chemical technology—seem to be left with two choices: (1) purchase organic foods from a reputable dealer and pay the price; or (2) engage in organic gardening and grow their own.

Summary. Good farmers have always incorporated many of the so-called organic methods, and they have also incorporated new methods such as chemical fertilizer, pesticides, and the other chemical methods which ensure more food for all and at a reasonable price. Undoubtedly, world food production of the future will make use of a combination of both organic and inorganic fertilizers, with the nature and proportions of the combination for different farms and for different countries dependent on their access to fossil fuels, the availability and price of fertilizers, their soils, their food production requirements, their environmental control problems, and many other factors. Regardless of the combination of organic and inorganic fertilizers used, feed and food plants of adequate nutritional quality can be produced.

Furthermore, to meet world food needs, pesticides and other chemical methods cannot be altogether abandoned. Even with the use of pesticides, about one-third of the food produced is lost to pests.

All foods are organic, and all edible foods, when properly selected for a balanced diet, are conducive to physiological and psychological health regardless of whether they are organically grown (as defined) or conventionally grown.

(Also see ADDITIVES, section on "Government Controls of Additives"; DELANEY CLAUSE; HEALTH FOODS; PESTICIDES; PESTICIDES AS INCIDENTAL FOOD ADDITIVES; and POISONS.)

ORGAN MEAT

Any edible part of a slaughter animal that consists of, or forms a part of, an internal organ, such as the liver, kidney, heart, or brain; distinguished from carcass meat.

ORGANOLEPTIC

It describes the employment of one or more of the special senses. Subjecting food to an organoleptic test employs the sensations of smell, taste, touch, and vision. We accept or reject food based on its organoleptic properties. The food industry sells food based on its organoleptic properties. First the consumer assesses the appearance (vision) and aroma (smell), but most important are the in-the-mouth properties—flavor, texture, and consistency.

(Also see TASTE.)

ORGEAT

• A nonalcoholic beverage prepared from the sweetened juice of almonds and other flavorings (such as orange blossom essence, or rose water), usually served cold.

• A sweet almond-flavored nonalcoholic syrup used as a cocktail ingredient or food flavoring.

ORNITHINE-ARGININE CYCLE (UREA CYCLE)

A cyclic sequence of biochemical reactions in protein metabolism in which citrulline is converted to arginine, thence urea splits off from arginine, producing ornithine.

(Also see UREA CYCLE.)

ORTANIQUE *Citrus reticulata X C. sinensis*

A citrus fruit (of the family *Rutaceae*) that was discovered growing in Jamaica, West Indies in 1920. It is believed to be the result of a natural crossbreeding (hybridization) between the mandarin orange (*C. reticulata*) and the sweet orange (*C. sinensis*). It is about the size of an average orange and it is flattened at both ends like certain tangerines. The fruit is quite juicy and its flavor is distinctive. Also, the rind adheres tightly, but is readily peelable.

OSMOPHILIC YEAST

Food preservation by high concentration of sugars or salts, such as in jams and pickles, takes advantage of the inability of most types of microorganisms to grow under this condition—high osmotic pressure. Some yeasts—osmophilic yeasts—are adapted. They thrive under the condition of high osmotic pressure.

(Also see PRESERVATION OF FOOD.)

OSMOSIS

The passage of a solvent such as water through a semipermeable membrane from the side of the membrane where the solution is dilute, to the side where the solution is more concentrated.

OSMOTIC PRESSURE

The force acting upon a semipermeable membrane placed between two solutions of differing concentrations.

OSSEIN

When the mineral salts of bone are dissolved by a dilute acid, the remaining organic material is called ossein. (Also see PROTEIN[S].)

OSSIFICATION

The process of bone formation.

OSTEITIS

Inflammation of a bone.

OSTEO–

Prefix meaning bone.

OSTEOBLAST

Bone-forming cells.

OSTEODYSTROPHY (OSTEODYSTROPHIA)

Defective ossification of bone, usually associated with disturbed calcium and phosphorus metabolism and renal insufficiency.

OSTEOMALACIA

Osteomalacia is most prevalent in areas where either climate or clothing practices limit exposure of skin to sunlight, or where the diet does not supply correct proportions of calcium, phosphorus, and vitamin D.

Osteomalacia is an adult form of rickets caused by lack of vitamin D, inadequate intake of calcium or phosphorus, or an incorrect dietary ratio of calcium and phosphorus. It is characterized by a softening and deformity of bones, bone tenderness and pain, muscular weakness, and tetany.

The recommended treatment: administration of vitamin D (1,000 IU daily) and calcium (1-2 g per day). Care should be taken that vitamin D toxicity does not develop during treatment.

Prevention consists in exposure of skin to sunlight or sunlamps, fortification of milk with vitamin D, and provision of extra calcium when diets are high in oxalates or low in calcium (dietary phosphorus is usually adequate).

(Also see BONE, section on "Bone Disorders"; CALCIUM, section on "Calcium Related Diseases"; PHOSPHORUS, section on "Phosphorus Related Diseases"; VITAMIN D, section on "Deficiency Symptons"; and RICKETS.)

OSTEOPOROSIS

Contents   Page
Causes of Osteoporosis.............................813
    Dietary Deficiencies of Calcium......................813
    Dietary Imbalances of Calcium and Phosphorus.......813
    High Protein Intakes.............................813
    Deficiencies of Vitamin D.........................814
    Stresses...814
    Lack of Sufficient Exercise.......................814
    Postmenopausal Loss of Calcium....................814
    Bone Loss—A Natural Accompaniment of Aging.......814
Signs of Osteoporosis..............................814
Treatment and Prevention of Osteoporosis............814
    Health Hints for Preventing Bone Fractures..........814

A loss of bone mass which generally occurs with aging is called osteoporosis. When such loss is signficant (such as one-third of the original adult bone mass), bones will fracture easily and heal poorly. Recent research has shown that osteoporosis often occurs in the skeleton in the following order:[41]

1. The alveolar bone in the jaw.
2. The spinal column.
3. The long bones of the body.

CAUSES OF OSTEOPOROSIS. Many investigations conducted in man and animals have led to tentative explanations of why osteoporosis is a problem in the United States where, on the average, dietary calcium intakes are higher than elsewhere in the world. Some of the theories concerning this problem follow.

Dietary Deficiencies of Calcium. A dietary survey of the United States, made by the Department of Agriculture, showed that adult females had calcium intakes around two-thirds of their RDA (800 mg of calcium per day).[42]

Dietary Imbalances of Calcium and Phosphorus. Diets in the United States are believed to contain disproportionately high ratios of phosphorus to calcium (about 5 to 1), which have been shown in animal studies to result in significant loss of bone minerals.[43]

(Also see CALCIUM-PHOSPHORUS RATIO; and VITAMIN D.)

High Protein Intakes. Investigators at the University of Wisconsin found that the amounts of calcium needed to balance urinary and fecal losses in young adult males increased as their dietary protein levels were raised.[44] On the average, these young men required about 10 mg of calcium for every gram of protein.

Similar findings were obtained by researchers at the University of California at Berkeley, who also noted that the elevations in urinary calcium caused by high levels of dietary

[41]Lutwak, L., "Continuing Need for Dietary Calcium Throughout Life," *Geriatrics*, Vol. 29, 1974, p. 171.

[42]*Dietary Levels of Households in the United States, Spring, 1965*, Ag. Res. Serv. Bull. No. 62, 1968.

[43]Krook, L., et al., "Reversibility of Nutritional Osteoporosis: Physiochemical Data on Bones from an Experimental Study in Dogs," *Journal of Nutrition*, Vol. 101, 1971, p. 233.

[44]Linkswiler, H. M., C. L. Joyce, and C. R. Anand, "Calcium Retention of Young Adult Males as Affected by Level of Protein and of Calcium Intake," *Transactions of the New York Academy of Sciences*, Vol. 36, 1974, p. 333.

protein appeared to be independent of the levels of dietary calcium.[45] However, high protein intakes imply high phosphate intakes, because most of the nondairy sources of proteins contain liberal amounts of phosphate, but only small amounts of calcium. Thus, there is likely to be an imbalance in the ratios of phosphate to calcium in these diets.

Deficiencies of Vitamin D.

People who have moderate intakes of calcium, but who live in northern areas, seem to have more osteoporosis than people who eat diets low in calcium, but live in sunny, tropical areas. Furthermore, considerable effort is made in most countries to provide infants and children with adequate levels of the vitamin by either diet or exposure to sunlight, but there is often a lack of attention to the needs of adults, many of whom are kept indoors by either their occupations or the circumstances under which they live.

Stresses.

Many stresses—such as trauma, chilling, starvation, dehydration, surgery, and fear—are accompanied by great increases in the secretions of adrenal cortical hormones. The net effect of the hypersecretion of these hormones on bone is dissolution of both the mineral and protein components.

Lack of Sufficient Exercise.

There is some evidence that sedentary persons lose more calcium from their bones than active persons eating essentially the same diet. Also, it has been found that mechanical forces on bone, such as those which occur during exercise, produce a surface charge (a piezoelectric effect) which may stimulate the building up of bone.[46] Therefore, exercise may be an important factor in the development and maintenance of strong bones.

Postmenopausal Loss of Calcium.

Most cases of osteoporosis are diagnosed in women who are more than 50 years of age. Until recently, it has been assumed that the cessation of estrogen secretion after the menopause resulted in a loss of stimulation of bone remineralization. Now, it is not certain whether estrogen has more than a temporary effect in this regard.

The more rapid development of osteoporosis in women than in men might also be explained by the observation that women generally eat less calcium than men; therefore, they are more likely to have a chronic deficiency of the mineral, since the allowance is the same for both sexes. Both men and women consume, on the average, 30 mg of calcium per 100 Calories (kcal), but women consume far fewer calories.

Bone Loss—A Natural Accompaniment of Aging.

Some researchers have concluded that, irrespective of diet, bone loss is a general phenomenon which accompanies aging in man and may start as early as age 30.[47]

SIGNS OF OSTEOPOROSIS.

It has long been thought that the initial signs of this disorder are the occurrence of bone fractures when there has been little or no trauma. Since the long bones of the body do not readily lose bone mass until there has been loss in the jaw and the spinal column, it is difficult to detect early stages of osteoporosis by x rays of the long bones. However, research conducted at Cornell University showed that the early stages of this disorder might readily be observed in x rays of the jaw.[48] Thus, it would seem that dentists might be able to detect early stages of this disorder, particularly when patients seek treatment of periodontal disease caused by loss of bone in the jaw.

TREATMENT AND PREVENTION OF OSTEOPOROSIS.

A variety of therapeutic and prophylactic measures have been suggested, but each of them has its merits and demerits. Therefore, separate discussions of these measures follow:

• **Supplementation of dietary calcium**—It has been suggested that 1,000 mg of calcium per day might be an optimal intake for both the treatment of the early stages of the disease (when bone loss is detectable only in the jaw) and the prevention of the more severe stages.[49]

• **Combined therapy with fluoride, vitamin D, and calcium**—The supplementation of fluoride therapy with vitamin D and calcium apparently produced normal bone in patients with osteoporosis.[50]

• **Estrogen replacement therapy**—At one time it was thought that this therapy offered promise for the treatment of osteoporosis, but recent studies show that the beneficial effects of these hormones on bone may only be temporary.[51] Furthermore, physicians are cautious about the administration of estrogens for fear that long-term treatment with high doses of these hormones may lead to cancer in aging, estrogen-sensitive tissues.

Health Hints for Preventing Bone Fractures.

Every safe measure for the prevention of osteoporosis should be considered, because bone fractures are painful, incapacitating, and costly; and immobilization of the patient during recovery may result in other serious problems, such as further weakening of bones, formation of blood clots, and atrophy of muscles. Some practical recommendations for normal, healthy people follow:

1. Exercise regularly. This practice results in the improved utilization of calcium for bone remineralization.

2. Avoid exercises where sharp impacts are transmitted to bones and joints. For example, some joggers run in hard-soled shoes on hard pavement. It has been shown that this practice puts considerable force on the hip joints. Learn to be light on your feet and wear appropriate footwear.

3. Eat sufficient dairy foods and dark-green vegetables to obtain at least 800 mg of calcium per day (the amount in 3 cups [720 ml] of milk).

[45]Margen, S., et al., "Studies in Calcium Metabolism. 1. The Calciuretic Effect of Dietary Protein," The American Journal of Clinical Nutrition, Vol. 27, 1974, p. 584.

[46]Marino, A. A., and R. O. Becker, "Piezoelectric Effect and Growth Control in Bone," Nature, Vol. 228, 1970, p. 473.

[47]Garn, S. M., C. G. Rothman, and B. Wagner, "Bone Loss as a General Phenomenon in Man," Federation Proceedings, Vol. 26, 1967, p. 1729.

[48]Lutwak, L., et al., "Calcium Deficiency and Human Periodontal Disease," Israel Journal of Medical Sciences, Vol. 7, 1971, p. 504.

[49]Lutwak, L., "Dietary Calcium and the Reversal of Bone Demineralization," Nutrition News, Vol. 37, 1974, p. 1.

[50]Jowsey, J., et al., "Effect of Combined Therapy with Sodium Fluoride, Vitamin D and Calcium in Osteoporosis," American Journal of Medicine, Vol. 53, 1972, p. 43.

[51]Riggs, B. L., et al., "Short- and Long-Term Effects of Estrogen and Synthetic Anabolic Hormone in Postmenopausal Osteoporosis," Journal of Clinical Investigation, Vol. 51, 1972, p. 1659.

Fig. O-28. Dairy foods, important for prevention of osteoporosis, can be incorporated into many dishes. (Courtesy, United Dairy Industry Assn., Rosemont, Ill.)

4. Keep your protein intake down to around the RDA (63 g per day for males, and 50 g per day for females).

5. Add calcium supplements, such as dolomite, when either the protein intake is considerably higher than the RDA (like 90 g or more per day), or when circumstances prevent the consumption of adequate calcium in foods.

(Also see BONE MEAL; CALCIUM; and DOLOMITE.)

6. Obtain sufficient vitamin D by either diet or regular exposure to sunlight.

(Also see CALCIUM; MINERAL[S], Table M-25, Mineral Table, Calcium; and VITAMIN D.)

OTAHEITE APPLE (AMBARELLA)
Spondias cytherea.

The fruit of a tree (of the family *Anacardiaceae*) that is native to Polynesia. Otaheite is an old name for Tahiti. Otaheite apples are yellow, round fruits that range from 2 to 3 in. (*5 to 7 cm*) in diameter and contain a large stone. The fruit is usually eaten fresh, but is sometimes made into jam and jelly.

Otaheite apples are fairly low in calories (46 kcal per 100 g) and carbohydrates (12%). They are a good source of fiber and vitamin C.

OTAHEITE GOOSEBERRY (GOOSEBERRY TREE) *Phyllanthus acidus*

The fruit of a small tropical tree (of the family *Euphorbiaceae*) that is native to India and Madagascar.

Otaheite gooseberries are small, greenish, sour-tasting fruit about ¾ in. (*2 cm*) in diameter. They are usually eaten cooked with sugar and made into jam, jelly, pies, and tarts.

The fruit has a high water content (91%) and a low content of calories (37 kcal per 100 g) and carbohydrates (5%). It is an excellent source of iron, but only a fair source of vitamin C.

OVALBUMIN

This is egg albumin, the major protein of egg white representing 75% of the total egg white protein.
(Also see EGG[S].)

OVERWEIGHT

Persons who are 10 to 20% above the ideal weight for their height, age, build, and sex are said to be overweight. A person, especially an athlete, may be overweight without being obese since overweight does not imply fatness. Nevertheless, some use obesity and overweight interchangeably.
(Also see OBESITY.)

OVOMUCOID

A minor protein component of egg white containing large amounts of carbohydrate in its structure.
(Also see EGG[S].)

OXALIC ACID (OXALATE; $C_2H_2O_4$)

A naturally occurring toxicant present in such plants as beet leaves, cabbage, peas, potatoes, rhubarb, and spinach. There is some concern that oxalic acid may render calcium as well as some trace minerals less available for absorption from the gut. However, it is doubtful that oxalates pose a problem to man unless the intake of a mineral is marginal, and unusually large amounts of the oxalate-containing food are eaten. Poisoning could occur from the accidental ingestion of some cleaning compounds containing oxalic acid. Interestingly, oxalic acid is a normal constituent of urine derived from the metabolism of ascorbic acid (vitamin C) or glycine (an amino acid).
(Also see POISONOUS PLANTS.)

OXALOACETIC ACID

A 3-carbon ketodicarboxylic acid; an intermediate in the Krebs cycle.

OXALURIA

The presence of an excess of oxalic acid or of oxalates in the urine.

OXIDATION

Chemically, the increase of positive charges on an atom or the loss of negative charges or electrons. It also refers to the combining of oxygen with another element to form one or more new substances. Burning is one kind of oxidation.
(Also see METABOLISM.)

OXIDATIVE PHOSPHORYLATION

The principal function of the oxidation of carbohydrates and fatty acids is to make available to the cells the free energy released in the oxidation process, in a form physiologically usable for cellular energy processes, viz, ATP. This is accomplished by the process known as oxidative phosphorylation, whereby adenosine triphosphate (ATP) with three phosphate groups, two of which are held by high energy bonds, is formed from adenosine diphosphate (ADP) by the addition of phosphate.

(Also see METABOLISM, section headed "Carbohydrates, Catabolism, • ATP—energy currency.")

PALATABILITY

If they don't eat it, it won't do them any good. And that applies to all people.

The palatability of a food is the result of the following factors: taste, smell, texture, temperature, sound, and appearance. These factors are affected by the physical and chemical nature of foods.

(Also see FLAVORINGS AND SEASONINGS.)

PALM family *Palmae*

Palms are an ancient group of plants. Fossil (buried remains) palm leaves have been found that date from the age of dinosaurs.

The family *Palmae* embraces more than 2,600 kinds of palms, varying greatly in size and kind of flowers, leaves, and fruits they produce. They are of great economic importance; furnishing food, shelter, clothing, timber, fuel, building materials, sticks, fiber, paper, starch, sugar, oil, wax, wine, tannin, dyeing material, resin, and a host of other products, all of which render them most valuable to the natives and to tropical agriculture. Palms thrive in warm climates, especially in the tropics. They are most common in Southeast Asia, the Pacific Islands, and in tropical America. They grow wild along the coast of North Carolina, in Arizona, and in the deserts of southern California.

Most palms grow straight and tall, but a few do not. The trunks of some palms lie on the ground. Others have most of the trunk buried in the soil. Still others have slender, vinelike stems that are from 10 to 250 ft (3 to 76 m) long.

Because of their importance in food production, the following three palms are treated separately in this book:

Coconut (Coconut Palm) (*Cocos nucifera*)
Date(s) (*Phoenix dactylifera*)
Oil Palm (*Elaeis guineensis*)

PALMITIC ACID

A 16-carbon saturated fatty acid.
(Also see FATS AND OTHER LIPIDS.)

PANCREATIN

A commercially available preparation containing the enzymes of the pancreas of cattle or pigs. It is capable of breaking down starch, lipids, and proteins. Pancreatin may be used as an aid to digestion. However, the stomach environment destroys it; hence, coated pills are used to allow it to reach the small intestine.

PANCREATITIS

An acute or chronic inflammation of the pancreas whose cause is poorly understood. Cases may be mild to severe, and some may even cause death. Often its occurrence is associated with other diseases such as alcoholism, gallbladder disease, peptic ulcer, an infectious disease, or an acci-

dent. Symptoms of pancreatitis consist of a severe upper abdominal pain radiating into the back and stimulated by eating, tenderness above the stomach, distention, constipation, nausea, and vomiting. The basis of treatment is to limit pancreatic secretion to a minimum. Hence, during the first few days of treatment, intravenous feeding is employed, followed by a soft, bland, or low residue diet. When medical treatment is ineffective, surgery may be needed.

(Also see MODIFIED DIETS; DIGESTION AND ABSORPTION; and INTRAVENOUS FEEDING.)

PANDANUS FRUIT (SCREW-PINE FRUIT)
Pandanus adoratissimus

The fruit of a tree (of the family *Pandanaceae*) that grows throughout southeast Asia and the Pacific Islands. Pandanus fruits are usually yellow to red in color, round or pear-shaped, and range from 6 to 10 in. (15 to 25 cm) in diameter. The fleshy base of the fruit is usually the only part that is eaten. Some of the Polynesian peoples mix the fruit and coconut milk or grated coconut and bake the pasty mixture into flat cakes. The pandanus-coconut paste may also be made into a refreshing drink by diluting it with water.

Pandanus fruit is rich in calories (150 kcal per 100 g), carbohydrates (18%), fat (8%), protein (5%), and fiber (7%), but it is low in vitamin C.

PANGAMIC ACID (VITAMIN B-15)

Contents	Page
History	818
Chemistry, Metabolism, Properties	818
Measurement/Assay	818
Functions	818
Disorders Treated with Pangamic Acid	818
Deficiency Symptoms	819
Dosage	819
Toxicity	819
Sources of Pangamic Acid	819
Conclusion	820

The authors make this presentation for two primary purposes: (1) informational, and (2) stimulation of research.

The nutritional status and biological role(s) of pangamic acid await clarification. Its stimulation of such fundamental processes of transmethylation, cellular respiration, etc. may have physiological application. Certainly, there is sufficient indication of the value of pangamic acid as a therapeutic agent to warrant further study. In the meantime, pangamic acid should not be taken unless prescribed by a physician.

Pangamic acid (also known as vitamin B-15, calcium pangamate, and dimethylglycene) is a chemical substance of an organic nature. It is sometimes classed as a vitamin or vitaminlike substance. Obviously, however, pangamic acid does not meet the classical definition of a vitamin, which is: *A substance, organic in nature, necessary in the diet in small amounts to sustain life, in the absence of which* a *specific deficiency disease develops.* The proponents counter with the argument that: "It's not essential, strictly speaking, but it's helpful biologically under many circumstances."

The FDA classifies pangamic acid as a food additive, and, therefore, subject to the regulations requiring proof that it is nontoxic.

Pangamic acid has been widely studied and accepted in the U.S.S.R. as a necessary food factor with important physiological actions.

HISTORY.
In 1951, Krebs *et al.* reported the presence of a water-soluble factor in apricot kernels, which they subsequently isolated in crystalline form from rice bran and polish. Later, it was extracted from brewers' yeast, cattle blood, and horse liver. The name *pangamic acid* (*pan*, meaning universal; *gamic*, meaning seed) was applied to the substance to connote its seeming universal presence in seeds; and it was assigned the fifteenth position in the vitamin B series by its discoverers.

The claim that pangamic acid is a B vitamin is based primarily on its presence in B-vitamin rich foods and on the broad spectrum of physiological functions attributed to it. However, it is not known whether man or other animals have the capacity to synthesize pangamic acid, and no specific disease can be attributed exclusively to a deficiency of the substance. Thus, the designation of pangamic acid as a vitamin is not accepted by most U.S. scientists.

CHEMISTRY, METABOLISM, PROPERTIES.

- **Chemistry**—The chemical structure of pangamic acid is given in Fig. P-1.

Fig. P-1. Pangamic acid ($C_{10}H_{19}O_8N$).

But the chemical composition of pangamic acid varies from brand product to brand product, and even within the same brand. This, within itself, makes the substance controversial. Some products are a mixture of sodium (or sometimes calcium) gluconate, glycine, and diisopropylamine dichloroacetate. However, the Russians confine their product to pangamic acid, i.e., the form of vitamin B-15 in which it occurs in nature.

- **Metabolism**—Little is known about the metabolism of pangamic acid. Within 10 to 15 minutes after subcutaneous injection of the substance, it is found in the blood, brain, liver, heart, and kidney; the site of its highest concentration and longest duration in the body is the kidney, where it persists for at least 4 days.

Excessive amounts of pangamic acid are excreted in urine, feces, and perspiration.

- **Properties**—Pangamic acid is a white crystalline compound, very soluble in water.

MEASUREMENT/ASSAY.
The content of pangamic acid is usually expressed in milligrams (mg).

Chromatographic and spectrophotometric techniques have been developed for the quantitative determination of pangamic acid in biological material.

FUNCTIONS.
Numerous functions have been attributed to pangamic acid; among them, the following:

1. **Stimulation of transmethylation reactions.** Pangamic acid possesses methyl groups ($-CH_3$) which are capable of being transferred from one compound to another within the body, resulting in the stimulation of creatine synthesis in muscle and heart tissue. In turn, when ATP (adenosine triphosphate), a transitory form of energy, is in excess of immediate requirements, creatine forms phosphocreatine, a more permanent form of energy for storage in muscle. Then when the supply of ATP is insufficient to meet the demands for energy, more ATP is produced from phosphocreatine by the reverse reaction.

2. **Stimulation of oxygen uptake.** Pangamic acid stimulates tissue oxygen uptake, thereby helping to prevent the condition called *hypoxia*—an insufficient supply of oxygen in living tissue, especially in heart and other muscles. It does not add to the overall oxygen supply of the body; rather, it increases the efficiency with which the oxygen is delivered from the bloodstream to the cells.

3. **Inhibition of fatty liver formation.** Oral or injection of pangamic acid to rats and rabbits exerts a protective effect against fatty infiltration of the liver induced by starvation, protein-free diets, anesthetics, carbon tetrachloride, or cholesterol.

4. **Adaptation to increased physical activity.** Pangamic acid enables animals to adapt to increased exercise. After periods of enforced swimming, animals previously treated with pangamic acid demonstrate a better maintenance of oxidative metabolism and energy levels than untreated controls; moreover, these effects persist for several days.

Some horse trainers have been known to administer pangamic acid to racehorses because of its reputed, but undocumented, quality of enabling animals to run faster and tire less. Also, athletes have been taking it for years.

5. **Control of blood cholesterol levels.** In most cases, the administration of pangamic acid causes a fall in both cholesterol biosynthesis and blood levels.

DISORDERS TREATED WITH PANGAMIC ACID.
Most treatments with pangamic acid have been conducted in the U.S.S.R., where the principal field of application of the drug has been in the treatment of cardiovascular disorders associated with insufficiency of oxidative metabolism. The Russians also report good results in treatment of liver disorders, a number of skin diseases, and some toxicoses. In the U.S.S.R. and/or elsewhere pangamic acid has been reported to be successful in treating the following disorders:

1. **Cardiovascular disease.** Pangamic acid is considered to be safe and effective in the treatment of—

a. **Arteriosclerosis (hardening of the arteries),** characterized by headaches, chest pains, shortness of breath, tension, and insomnia. In patients with arteriosclerosis, pangamic acid strengthens the action of the heart muscles.

b. **Cardiopulmonary (heart-lung) insufficiency,** characterized by shortness of breath.

c. **Angina pectoris,** characterized by chest pain, due to the arteries being incapable of supplying sufficient oxygen-rich blood to the heart muscle.

d. **Congestive heart failure,** characterized by shortness of breath on slight exertion and dropsy (swelling around the ankles with pitting edema). This is a disease complex and group of symptoms caused by a failing heart, with congestion either in the lungs or the systemic circulation, or both. The chief causes are weakness of the heart muscles, high blood pressure, hardening of the arteries, and rheumatic or syphilitic disease of the heart valves. Treatment of patients suffering from congestive heart failure with pangamic acid increases diuresis (urination) and reduces high blood pressure and edema.

e. **Patients recovering from serious heart attacks,** as an aid in speeding their recovery.

It is noteworthy, too, that pangamic acid increases the effectiveness of strophanthin (a heart tonic) on heart function of patients with congestive failure and reduces the incidence of side effects associated with digitalis therapy.

2. **Fatigue.** Pangamic acid is used to reduce fatigue and increase energy—to alleviate that tired feeling. According to the proponents, it accomplishes this by extracting more oxygen from the bloodstream for better metabolism in the cells. Some athletes, both professionals and amateurs, claim that it increases their physical endurance.

3. **Hypoxia.** Pangamic acid is effective in treating hypoxia, characterized by an insufficient supply of oxygen in tissues. It increases the efficiency with which oxygen is delivered from the bloodstream to the cells.

4. **High cholesterol.** Several workers have observed significant reductions in serum cholesterol levels as a result of daily treatment with pangamic acid for 10 to 30 days.

5. **Liver function.** As supportive treatment for infectious hepatitis in children, pangamic acid given orally for 10 to 20 days leads to a more marked and rapid decline in fever, liver size, jaundice, and serum transaminase levels and a 5 to 10 day shorter hospitalization period. Similar findings have been reported in adults with acute or chronic hepatitis treated with pangamic acid.

6. **Skin disorders.** It is claimed that pangamic acid is effective in the treatment of various skin diseases. It is reported that patients with scleroderma (*hidebound skin*), a chronic disease in which the skin and subcutaneous tissues become fibrous, rigid, and thickened, respond to daily treatment with pangamic acid for 45 days—relief is associated with softening of the affected skin and often with new hair growth. Also, a reduction of inflammation, edema, and itching has been reported in cases of eczema, psoriasis (a chronic skin disease that disfigures the face and body with recurrent red scaly patches), hives, and other skin disorders treated daily for 15 to 30 days with 100 to 150 mg of pangamic acid.

7. **Tumors.** Pangamic acid offers promise in the treatment of certain kinds of tumors. When given to rats with experimentally induced solid tumors, it did not interfere with the effectiveness of standard chemotherapeutic agents, but it significantly reduced the incidence and severity of the toxicity produced by certain drugs. Also, in preliminary work with rats, it has been reported that pangamic acid reduced both the incidence and latency of drug-induced mammary cancer. These limited studies await confirmation.

DEFICIENCY SYMPTOMS. There is indication that a deficiency of pangamic acid may cause fatigue (a tired feeling), hypoxia (an insufficient supply of oxygen in blood cells), heart disease, and glandular and nervous disorders.

DOSAGE. The usual dosage of pangamic acid is from 150 to 300 mg daily. But pangamic acid should not be taken unless prescribed by a physician.

TOXICITY. Despite initial caution regarding the use of pangamic acid in conditions of hypertension and glaucoma, "it is now regarded that the substance is without demonstrable toxicity in these or other disease states."[1] It is noteworthy, too, that, beginning in 1965, the U.S.S.R. Ministry of Health approved widespread manufacture and distribution of pangamic acid. One of the largest U.S. purveyors of pangamic acid—Da Vinci Laboratories, a subsidiary of Food Science Laboratories—cites several studies in which laboratory rats were fed quantities of B-15 equivalent to more than 100,000 times the recommended dosage for human consumption, without any ill effects.[2]

Perhaps the toxicity of pangamic acid depends on the particular formula. On the basis of available evidence, the so-called *Russian formula* appears to be relatively free of toxicity. Nevertheless, the U.S. Food and Drug Administration does not consider pangamic acid safe for human consumption.

Until the product is standardized (potency, quality, and chemical composition) and the question of toxicity is resolved, the authors recommend that pangamic acid be taken only under the direction of a physician.

SOURCES OF PANGAMIC ACID. The best natural sources of pangamic acid are sunflower seed, pumpkin seed, yeast, liver, rice, whole grain cereals, apricot kernels and other seeds. Evidence suggests its occurrence wherever B-complex vitamins are found in natural foods.

Fig. P-2. Rice is a rich source of pangamic acid. (Courtesy, Department of Food Science and Human Nutrition, University of Hawaii)

[1]Stacpoole, Peter W., "Pangamic Acid (Vitamin B-15)," *World Review of Nutrition and Dietetics—Some Aspects of Human Nutrition,* S. Karger, Basel, Munchen, 1977, p. 153.

[2]Grosswirth, Marvin, "B-15: Is it Superpill?," *Science Digest,* September 1978, p. 12.

CONCLUSION. Until more conclusive evidence on the role of pangamic acid is available, the authors subscribe to the view so well expressed by biochemist Richard A. Passwater, Ph.D., as follows:

"While the processes through which vitamin B-15 works are being elucidated, natural wisdom suggests that we take care to see that our diets are rich in vitamin B-15."[3] (Also see VITAMINS[S], Table V-5.)

PANTOTHENIC ACID (VITAMIN B-3)

Contents	Page
History	820
Chemistry, Metabolism, Properties	820
Measurement/Assay	820
Functions	821
Deficiency Symptoms	822
Recommended Daily Allowance	822
Toxicity	822
Pantothenic Acid Losses During Processing, Cooking, and Storage	822
Sources of Pantothenic Acid	822

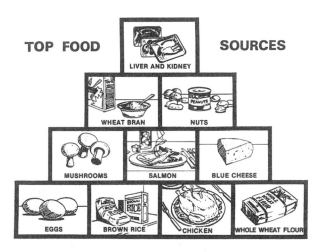

Fig. P-3. Top food sources of pantothenic acid.

Pantothenic acid, a member of the vitamin B complex, is a dietary essential for man and animals; and, as an important constituent of coenzyme A (CoA), it plays a key role in body metabolism.

In recognition of its wide distribution in foods, the name *pantothenic acid,* derived from the Greek word *pantothen*— meaning *everywhere,* was first given to it in 1933, by R. J. Williams, then of Oregon State, later at the University of Texas.

Before its structure was known, pantothenic acid was called by several other names, now obsolete, including: filtrate factor, chick antidermatitis factor, bios factor, the antigray hair factor, pantothen, factor H, factor 2, vitamin B-x, vitamin B-2, vitamin B-3.

HISTORY. The existence of pantothenic acid and its nutritive significance for the proliferation and fermentative activity of yeast cells first emerged from the studies of R. J.

Williams, beginning in 1919. In 1933, Williams fractionated this compound from yeast and called it pantothenic acid; and in 1939, he isolated pantothenic acid from liver. In 1939, Jukes concluded that the antidermatitis factor isolated from liver was one and the same thing as the factor found in yeast. In 1940, pantothenic acid was synthesized by Williams and two other laboratories, all working independently. Also, in 1940, pantothenic acid received widespread attention as a possible preventive for gray hair, since it had been observed that the black hair of a rat would turn gray when the animal was deprived of the vitamin; but subsequent studies did not reveal any such benefits to accrue to humans. In 1946, Lipmann and co-workers showed that coenzyme A is essential for acetylation reactions in the body; and in 1950 reports from the same laboratory showed pantothenic acid to be a constituent of coenzyme A.

CHEMISTRY, METABOLISM, PROPERTIES.

• **Chemistry**—Pantothenic acid is composed of pantoic acid and the amino acid beta-alanine, as shown in Fig. P-4.

Fig. P-4. Structure of pantothenic acid.

• **Metabolism**—Pantothenic acid, like the other B vitamins, is readily absorbed through the mucosa of the small intestine and enters the portal circulation. Within the tissues, most of the panothenic acid is used in the synthesis of coenzyme A (CoA); but a significant amount found in the cells is bound to a protein in a compound known as acyl carrier protein (ACP).

Pantothenic acid is present in all living tissue, with high concentrations in the liver and kidney. (Large amounts of CoA are found in the liver, with lesser amounts in the adrenal glands.)

The vitamin is excreted from the body by way of the kidneys.

• **Properties**—In pure form, pantothenic acid is a viscous yellow oil, soluble in water, quite stable in neutral solutions, but destroyed by acid, alkali, and prolonged exposure to dry heat (2 to 6 days—far longer than the usual cooking or baking procedures). Calcium pantothenate, the form in which it is commercially available, is a white, odorless, bitter, crystalline substance, which is water-soluble and quite stable. (Pantothenic acid is also available as the sodium salt.)

MEASUREMENT/ASSAY. The activity of pantothenic acid is expressed in grams and milligrams of the chemically pure substance.

[3]Passwater, Richard A., Ph.D., "B-15—New Ally In The Fight Against Major Diseases," *The Health Quarterly,* Vol. 3, No. 5, 1979, p. 16.

Pantothenic acid content may be determined by chemical methods (including gas chromatography), microbiologic procedures, and the chick and rat bioassay.

FUNCTIONS.   Pantothenic acid functions in the body as part of two enzymes—coenzyme A (CoA) and acyl carrier protein (ACP).

ACP is composed of a panthetheine linked through a phosphate group to protein. It, along with CoA, is required by the cells in the biosynthesis (the manufacture, or building up) of fatty acids (CoA, without ACP, is involved in their breakdown.)

Coenzyme A, meaning a coenzyme for acetylation, is one of the most important substances in body metabolism. It is a complex molecule containing the vitamin combined with adenosine 3-phosphate, pyrophosphate, and beta-mercaptoethylamine (a compound containing an -SH group), as shown in Fig. P-5.

Fig. P-5. Structure of CoA.

The way in which pantothenic acid is incorporated in the molecule of coenzyme A is shown in Fig. P-5. The sulphydryl (-SH) group of the molecule is extremely active, but the other groups are also involved in some of the functions of coenzyme A.

CoA functions in any reaction in which an acetyl group (-CH$_3$CO) is formed or is transferred from one substance to another. In addition to acetyl, other acyl radicals require coenzyme A; it is required whenever succinyl, benzyl, or fatty acid radicals are formed or transferred.

Coenzyme A participates in several fundamental metabolic functions; these include the following:

1. **The synthesis (the building up) of fatty acids.**   The most important function of coenzyme A (and, hence, of pantothenic acid) in metabolism is the transfer of radicals of acetic acid (or C$_2$) in the synthesis of fatty acids. During the process of digestion, the triglycerides which are present in the food are split by lipases of the pancreas and of the intestinal wall into glycerol, monoglycerides, and fatty acids which are simultaneously emulsified by the bile acids. The degree of absorption of the fatty acids through the wall of the small intestine depends on the length of the chain; the shorter the chain, the more complete the absorption. After passing through the intestinal wall, the short chain fatty acids are converted into longer chain fatty acids, the major step in the process being the combination of coenzyme A with acetic acid to form "activated acetic acid" or acetyl-coenzyme A.

Acetyl-coenzyme A is next converted to malonyl-coenzyme A (which contains one more carbon atom), a reaction which is catalyzed by an enzyme which contains biotin. Malonyl-coenzyme A then reacts with another activated fatty acid (the chain of which has also become longer by one carbon atom) yielding a product which is a fatty acid with a chain which has been lengthened by two carbon atoms. In this manner, for example, stearic acid (C$_{18}$) is formed from palmitic acid (C$_{16}$) in the body.

2. **The degradation (the breaking down) of fatty acids.**   In metabolism, both the synthesis and the degradation of fatty acids are required. Pantothenic acid, as a constituent of coenzyme A, participates in this breaking down process. The energy which is released in this process of degradation is gathered and transferred elsewhere by another system which is based on the (reversible) formation of adenosine triphosphate (ATP), which is high in energy, from two compounds having lower levels of energy, adenosine diphosphate (ADP) and monophosphate (AMP).

3. **The citric acid cycle.**   A very large part of the energy required in metabolism is supplied by the citric acid cycle (or the *Krebs' cycle*), in which compounds high in energy (carbohydrates, fats, and proteins) are continually being converted into compounds which are lower in energy; the energy released is again contained in the form of ATP.

Pantothenic acid, as a constituent of coenzyme A, is involved in several of the steps of the citric acid cycle; these include the synthesis of citric acid from oxalacetic acid and its salts, and the oxidation by decarboxylation of a-keto-acids.

4. **The acetylation of choline.**   Pantothenic acid is necessary for the formation of acetylcholine, the transmitter of nerve impulses.

5. **The synthesis of antibodies.**   Pantothenic acid stimulates the synthesis of those antibodies which increase resistance to pathogens.

6. **The utilization of nutrients.**   Coenzyme A is essential for the metabolism of fats, carbohydrates, and proteins; hence, a deficiency of pantothenic acid will necessarily impair the utilization of digestible nutrients. The function of coenzyme A is in the endogenous metabolism, not in the digestive tract. Hence, it cannot affect the digestible or metabolizable energy; only its utilization—the net productive energy which is most simply measured in terms of the retention of nutrients. This is true not only for those nutrients which supply energy, but also for those which supply protein.

7. **Other functions.**   Pantothenic acid also affects the endocrine glands, and the hormones they produce. Thus, a deficiency of pantothenic acid in rats reduces not only the rate of gain in weight but also the rate of basal metabolism. Also, it has been postulated that the influence which pantothenic acid exerts on the fertility of various animals may be due to some relationship between pantothenic acid and the synthesis of steroid hormones.

Other functions attributed to pantothenic acid (or coenzyme A) are:

a. It is necessary in the synthesis of porphyrin, a precursor of heme, of importance in hemoglobin synthesis.

b. It is necessary for the maintenance of normal blood sugar levels.

c. It may facilitate the excretion of sulfonamide drugs.

d. It influences the metabolism of some of the minerals and trace elements.

e. It can be used for detoxification of drugs, including the sulphonamides.

DEFICIENCY SYMPTOMS. A deficiency of pantothenic acid has been associated with the "burning foot syndrome" that occurred in Japan and the Philippines among prisoners during World War II.

Fig. P-6. Pantothenic acid deficiency in chick. Note the lesions at the corners of the mouth and on the eyelids and feet. (Courtesy, Department of Poultry Science, Cornell University)

Pantothenic deficiency has been produced in human volunteers (1) by feeding them semisynthetic diets low in pantothenic acid for a period of 10 to 12 weeks; and (2) by feeding them diets to which the pantothenic acid antagonist omega-methylpantothenic acid was added. (A vitamin antagonist is a substance so similar in structure to the vitamin that the body accepts it in place of the vitamin, but the antagonist is unable to perform the functions of the true vitamin.) The subjects developed a wide variety of symptoms including: irritableness and restlessness; loss of appetite, indigestion, abdominal pain, nausea; headache; sullenness, mental depression; fatigue, weakness; numbness and tingling of hands and feet, muscle cramps in the arms and legs; burning sensations in the feet; insomnia; respiratory infections; rapid pulse; and a staggering gait. Also, in these subjects there was increased reaction to stress; increased sensitivity to insulin, resulting in low blood sugar levels; an increased sedimentation rate for erythrocytes; decreased gastric secretions; and marked decrease in antibody production. All symptoms were cured by the administration of pantothenic acid.

A lack of pantothenic acid results in premature graying of the hair in piebald rats, foxes, and dogs. But, neither pantothenic acid nor any other nutritional factor has been shown to be involved in the graying of hair in humans.

RECOMMENDED DAILY ALLOWANCE. The amount of pantothenic acid required by human beings has not been determined; so, a recommended daily allowance for pantothenic acid has not been made by the Food and Nutrition Board of the National Research Council. Nevertheless, they do give "estimated safe and adequate intakes" based on proportional energy needs (see section on VITAMIN(S), Table V-5, Vitamin Table). Further, they suggest that a higher intake may be needed during pregnancy and lactation.

NOTE WELL:

1. Processing of food can, in some instances, result in appreciable losses of pantothenic acid. Thus, it is possible that unrecognized marginal deficiencies may exist, along with deficiencies of other B-complex vitamins.

2. Deficiencies of pantothenic acid occur in farm animals (especially chickens and swine), fed *natural rations*. As a result, pantothenic acid is commonly added to commercial poultry and swine rations.

These two facts point up the need for more studies on the pantothenic requirement of humans, and perhaps the need for dietary supplementation for buoyant good health.

• **Pantothenic acid intake in average U.S. diet**—The average American diet provides 5 to 10 mg of pantothenic acid per day, with an average of 6 mg.

TOXICITY. Pantothenic acid is a relatively nontoxic substance. As much as 10 g of calcium pantothenate per day was given to young men for 6 weeks with no toxic symptoms; other studies indicate that daily doses of 10 to 20 g may result in occasional diarrhea and water retention.[4]

PANTOTHENIC ACID LOSSES DURING PROCESSING, COOKING, AND STORAGE. Losses up to 50%, and even more, of the pantothenic acid content of foods may occur from production to consumption. Here are some of them:

1. About 50% of the pantothenic content of grains is lost in milling.

2. Up to 50% of the pantothenic content of fruits and vegetables is lost in canning or freezing, and storage, of fruits and vegetables.

3. From 15 to 30% of the pantothenic acid content of meat is lost in cooking or canning.

4. Losses in pantothenic acid from dry processing of foods may exceed 50%.

Pantothenic acid is reasonably stable in natural foods during storage, provided that oxidation and high temperature are avoided. Cereal grains may be stored for periods up to a year without appreciable loss.

Henry A. Schroeder, M.D., of the Dartmouth Medical School, conducted an extensive survey of the pantothenic acid content in hundreds of common foods, and found that modern processing was causing massive losses of the vitamin. His studies showed pantothenic acid losses as follows: From milling wheat and making all-purpose flour, 57.7%; from canning vegetables, 56 to 79%; from freezing vegetables, 48 to 57%; from canning meat and poultry, 26.2%; and from canning seafood, 19.9%.[5]

SOURCES OF PANTOTHENIC ACID. Pantothenic acid is widely distributed in foods, with considerable variation in content according to food and processing. A grouping and ranking of foods according to pantothenic content is given in the section on VITAMIN(S), Table V-5, Vitamin Table.

[4]Committee on Dietary Allowances, Food and Nutrition Board, *Recommended Dietary Allowances*, 10th ed., 1989, National Research Council, National Academy of Sciences, Washington, D.C., p. 172.

[5]Schroeder, Henry A., M.D., "Losses of Vitamins and Trace Minerals Resulting From Processing and Preservation of Foods," *The American Journal of Chemical Nutrition*, Vol. 24, May 1971, pp. 562–573.

Fig. P-7. Mushrooms, a good source of pantothenic acid. (Courtesy, USDA)

Intestinal bacteria also synthesize pantothenic acid, but the amount produced and the availability of the vitamin from this source are unknown.

For additional sources and more precise values of pantothenic acid, see Food Composition Table F-21 of this book. (Also see VITAMIN[S], Table V-5.)

PAPAIN (VEGETABLE PEPSIN)

This is an enzyme obtained from the green fruit and leaves of the papaya. It is proteolytic; it breaks down proteins such as the digestive enzymes pepsin and trypsin. Papain can be employed to tenderize meats, clear beverages, prevent adhesions during wound healing, and aid digestion.

(Also see PAPAYA)

PAPAYA (MAMAO) Carica papaya

Fig. P-8. The papaya—fruit of a tropical tree that looks like a palm tree with a tuft of large leaves at the top. (Courtesy, USDA)

The papaya tree, a member of the family Caricaceae, is a fast-growing, short-lived tree native to tropical America. It is a soft wooded palmlike evergreen with palmate leaves clustered at the top. In tropical countries around the world, the papaya is cultivated for its edible melonlike fruit, the papaya, which is round to oblong and may weigh as much as 20 lb (9.1 kg).

ORIGIN AND HISTORY. The papaya is native to Central America. Since it grows readily from seed, the papaya was distributed quite early by man all over the tropical and subtropical areas of the world.

WORLD AND U.S. PRODUCTION. Papayas rank about nineteenth in quantity produced among the fruit crops of the world. Worldwide, about 4.4 million metric tons are produced annually; and the leading producing countries of the world, by rank, are: Brazil, Mexico, Thailand, Indonesia, India, Zaire, China, Philippines, Colombia, and Vietnam.[6]

The United States produces about 34,000 short tons annually. Most of the U.S. crop is grown in Hawaii.[7]

PROCESSING. Papayas are usually consumed fresh like a cantaloupe or in salads, pies, and sherbets. There is little processing involved. However, they can be squeezed for juice, pickled, candied, or made into jellies.

SELECTION, PREPARATION, AND USES. Papaya fruit ranges in shape from globose to long-ovid, in weight from 4 oz (113.4 g) to 20 lb (9.1 kg) or more, and in color from light yellow through deep yellow, orange, and pink to red. The fruit should be firm. Papaya is usually prepared by washing, peeling, and slicing ripe fruit, then serving with or without a little sugar. Unripe fruit may be cooked and used like a squash.

Papaya may be used as a fruit or vegetable dish, and in fruit salads, jams, jellies, juices, pickles, or sherbets. Also, papaya contain the enzyme papain which is used to tenderize meat.

NUTRITIONAL VALUE. Papayas are listed in Food Composition Table F-21. They contain about 89% water and only 39 Calories (kcal) per 100 g (about 3½ oz) or about 120 Calories (kcal) per medium-sized fruit. Papayas are rich in vitamin A and contain some vitamin C.

PARA-AMINOBENZOIC ACID (PABA)

Para-aminobenzoic acid (PABA) is a constituent of foods, which is sometimes listed with the B vitamins.

In addition to having activity as a growth factor for certain bacteria, PABA has considerable folacin activity when fed to deficient animals in which intestinal synthesis of folacin takes place. For example, for rats and mice, it can completely replace the need for a dietary source of folacin. This explains why para-aminobenzoic acid was once considered to be a vitamin in its own right.

For man and other higher animals, PABA is an essential part of the folacin molecule. But it has no vitamin activity in animals receiving ample folacin, and it is not required

[6]Data from *FAO Production Yearbook 1990*, FAO/UN, Rome, Italy, Vol. 44, p. 169, Table 74. **Note well:** Annual production fluctuates as a result of weather and profitability of the crop.

[7]Data from *Agricultural Statistics 1991*, USDA, p. 177, Table 254. **Note well:** Annual production fluctuates as a result of weather and profitability of the crop.

in the diet; hence, it can no longer be considered a vitamin, contrary to its listing in many vitamin preparations on the market.

HISTORY. PABA was first identified as an essential nutrient for certain microorganisms. Later, it was shown to act as an antigray hair factor in rats and mice (but not people) and as a growth-promoting factor in chicks.

CHEMISTRY, METABOLISM, PROPERTIES.

- **Chemistry**—The chemical structure of PABA is given in Fig. P-9.

Fig. P-9. Structure of *para*-aminobenzoic acid.

- **Metabolism**—The body manufactures its own PABA if conditions in the intestines are favorable.

- **Properties**—PABA is a yellow, crystalline, slightly water-soluble substance.

MEASUREMENT. The activity of PABA, and of its sodium and potassium salts, is ordinarily expressed in grams of chemically pure substances.

FUNCTIONS. For man and other higher animals, PABA functions as an essential part of the folacin molecule.

As a coenzyme, PABA functions in the breakdown and utilization of proteins and in the formation of blood cells, especially red blood cells.

- **Human pharmaceutical uses**—PABA is sometimes used as a human pharmaceutical, not as a vitamin, in the following: as an antirickettsial; to counteract the bacteriostatic action of sulfonamides; and as a protective agent against sunburn.

1. **Antirickettsial.** PABA is sometimes used in the treatment of certain rickettsial diseases—diseases in man and animals caused by microscopically small parasites of the genus *Rickettsia,* notably typhus and Rocky Mountain spotted fever.

The therapeutic use of PABA in the treatment of rickettsial diseases is based on the concept of metabolic antagonism. PABA acts as an antagonist to a material essential to these organisms, para-oxybenzoic acid; hence, the rickettsial organisms are killed because PABA blocks their essential metabolite.

2. **Sulfonamide antagonist.** PABA has the ability to reverse the bacteriostatic effects of sulfonamides, thereby counteracting their action. This is an antimetabolite action, explainable on the basis of similarity of structures (see Fig. P-10). According to this theory, sulfonamides suppress bacterial

Fig. P-10. The similarity of the structures of PABA and sulfanilamide is obvious.

growth by replacing their chemical analog PABA in bacterial enzyme systems; PABA in excess reverses the effect.

3. **Sunscreen agent.** PABA is used for protection against sunburn; usually about 5% PABA is incorporated in an ointment, to be applied over exposed parts of the body. Currently, some dermatologists are recommending other sunscreen lotions; so, for protection against sunburn, see your dermatologist.

DEFICIENCY SYMPTOMS. Sulfa drugs may induce a deficiency of not only PABA, but of folic acid as well. The symptoms: fatigue, irritability, depression, nervousness, headache, constipation, and other digestive disorders.

(Also see DISEASES, Table D-10, Food-Related Infectious and Parasitic Diseases; and MALARIA.)

RECOMMENDED DAILY ALLOWANCE OF PABA. The need for PABA has not been established; hence, it follows that there is no recommended daily allowance. The ordinary dose of PABA is 2 to 6 g, but even larger doses have been administered, in a variety of disorders.

TOXICITY. It has generally been considered that PABA is essentially nontoxic in man. But, continued high doses may be toxic, causing nausea and vomiting.

SOURCES OF PABA. Food composition tables do not list PABA, but the following foods are generally recognized as the richest sources: brewers' yeast, fish, soybeans, peanuts, beef liver, eggs, wheat germ, lecithin, and molasses.

(Also see FOLACIN; and VITAMIN[S], Table V-5.)

Fig. P-11. Peanuts, a rich source of para-aminobenzoic acid. (Courtesy, USDA)

PARABENS

This is a general term referring to the methyl, ethyl, propyl, or butyl esters (acid alcohol combination) of p hydroxybenzoic acid (parabens)—a group of antimicrobial agents used in foods, cosmetics, and drugs. Specifically, these compounds may be called methyl, ethyl, propyl, or butyl hydroxybenzoic acid, or methyl, ethyl, propyl, or butyl paraben.

PARAPLEGIA

Commonly found in spinal cord diseases, or accidents in which the spinal cord is severed in the lumbar region, it is the paralysis or weakness of both legs. Nutritionally, paraplegia is known to occur in humans who ingest Lathyrus seeds over a long period of time. Some dietary modification may be necessary due to reduced activity.

(Also see LATHYRISM; and MODIFIED DIETS.)

PARASITE INFECTIONS

People may harbor a wide variety of internal and external parasites (organisms that live in or on people). They include fungi, protozoa (unicellular animals), arthropods (insects, lice, ticks, and related forms), and helminths (worms). Some parasites require only one host—an animal or person serving as residence—while others need more in order to complete their life cycle.

While in residence, parasites usually seriously affect the host, but there are notable exceptions. Among the ways in which parasites may do harm are (1) absorbing food, (2) sucking blood or lymph, (3) feeding on the tissue of the host, (4) obstructing passages, (5) causing nodules or growths, (6) causing irritation, and (7) transmitting diseases. Two common infestations of man by parasites from undercooked meats are trichinosis and beef tapeworm.

(Also see DISEASES, Table D-10, Food-Related Infectious and Parasitic Diseases; and MALARIA.)

PARASITES

Broadly speaking, parasites are organisms living in, on, or at the expense of another living organism.

(Also see PARASITE INFECTIONS.)

PARATHYROID GLAND

As its name implies, the parathyroid is in the neck in close association with the thyroid gland. Both are endocrine glands. Normally, there are four very small (each weighs only 0.02 g) parathyroid glands lying behind the thyroid gland. They secrete the hormone parathormone, which controls blood calcium levels.

(Also see ENDOCRINE GLANDS.)

PARBOIL

Partial cooking which is sometimes done before finishing the cooking in the oven. Bagels are parboiled before baking in the oven. Some of the cereals that are partially cooked are parboiled.

PARENTERAL

Introduction of nutrients other than by mouth; it may be accomplished (1) subcutaneously (beneath the skin), (2) intramuscularly (into the muscle), or (3) intravenously (into the vein.)

PARESTHESIA

A sensation of burning, numbness, pricking, or tingling, usually associated with an injury or irritation of a sensory nerve or nerve root.

PARIETAL CELLS

The cells of the gastric glands of the stomach which secrete hydrochloric acid (HCl).

PARKIN

This is a popular cake in Scotland. It is made with oatmeal, butter, molasses, baking powder, and spiced with ginger.

PARTS PER BILLION (PPB)

It is an expression of extremely minute quantities. One part per billion equals 0.0000001%, 0.907 mg/ton, 1 mcg/kg, and 1 mcg/liter.

PARTS PER MILLION (PPM)

It is an expression of minute quantities. One part per million is equivalent to the following: 0.454 mg/lb; 0.907 g/ton; 0.0001%; 0.013 oz/gal; 1 mg/liter; and 1 mg/kg.

PARTURITION

Giving birth.

PASTA

Pasta consists of pieces of dough that have been formed in various shapes, then dried.

(Also see MACARONI AND NOODLE PRODUCTS.)

PASTEURIZATION

This is a method of processing food by destroying the microorganisms that cause disease and spoilage. The process was developed by and named for Louis Pasteur, the French scientist who in the early 1860s demonstrated that wine and beer could be preserved by heating above 135°F (57.2°C). In the United States, milk, cheese, egg products, wine, beer, and fruit juice are pasteurized.

(Also see MILK AND MILK PRODUCTS, "Processing Milk"; PRESERVATION OF FOOD, Table P-10, Methods of Food Preservation; and ULTRAHIGH TEMPERATURE STERILIZATION.)

PÂTÉS

• The most common meaning for pâté is a meat or liver dish which has been finely minced and mixed with other ingredients to form a spread for crackers.

• A pastry filled with meat or fish.

PATH-, PATHO-, -PATHY

Each meaning disease; for example, pathological, apathy.

PATHOGENS

Those microorganisms responsible for producing disease. Many microorganisms are harmless.

(Also see DISEASES.)

PATHOLOGY

The science dealing with diseases; their essential nature, causes, and development, and the structural and functional changes produced by them.

PAVLOV POUCH

This is an experimental, surgical technique introduced by the famous Russian physiologist-psychologist, Ivan Petrovich Pavlov. A portion of the stomach is brought to the body wall and formed into a pouch from which samples may be taken. Secretions in the pouch are the same as those in the stomach.

PBI TEST

This test is a measure of thyroid activity by determining the amount of iodine that is bound to thyroxin and in transit in the plasma. The test is based upon the fact that the level of protein-bound iodine circulating in the blood is proportional to the degree of thyroid activity.

Since the hormone thyroxin constitutes the most important single factor influencing the rate of tissue oxidation, or basal metabolism, determination of the level of thyroid hormone in the bloodstream is used to detect the relative rate of basal metabolism. Good correlation with basal metabolism tests has been obtained by this method.

Thyroxin (tetraiodothyronine) is circulated in the blood in temporary union with blood proteins; hence, what the chemical test actually measures is the protein-bound iodine content of the blood, which, in turn, serves as an index of relative activity of the thyroid in releasing its hormone into the blood—hence, the approximate basal metabolic rate.

The PBI test has largely replaced the measurement of oxygen consumption by respiration apparatus (1) because it is much simpler for both patient and technician, and (2) because many doctors feel that it is better suited to clinical purposes. However, the basal metabolism test is still the method of choice for nutritional studies.

Fig. P-12. A cluster of peaches the juicy, but fuzzy fruit. (Courtesy, USDA)

PEACH AND NECTARINE *Prunus persica*

The peach is the fruit of a tree bearing the same name. Botanically, a peach is classified as a drupe—a fruit whose seed is contained in a hard pit or stone surrounded by soft, pulpy flesh with a thin skin. The peach tree belongs to the rose family, *Rosaceae*. As the genus name, *Prunus*, suggests, peaches are close relatives of the apricot, almond, cherry, and plum.

Peaches are round with yellow skin and edible flesh, though the skin may have areas of red. The edible flesh is either soft or quite firm. Peaches are classified as freestone or clingstone, according to how difficult it is to remove the pit from the fruit.

• **Nectarine**—The peach and the nectarine are essentially alike. Only the fuzz-free skin, the usually smaller size, greater aroma, and distinct flavor separate the two. The scientific name for nectarine is *Prunus persica*—the same as the peach. Nectarines have been called smooth-skin peaches. Herein, comments made regarding peaches also apply to nectarines, unless stated otherwise.

ORIGIN AND HISTORY. The peach is native to China, where it probably grew over 4,000 years ago. From China, it was taken to Persia (Iran), where it was known as the Persian apple. Thence, it spread to Europe, and from Europe to America.

WORLD AND U.S. PRODUCTION. Worldwide, about 8.7 million metric tons of peaches and nectarines are produced annually; and the leading producing countries, by rank, are: Italy, U.S.A., China, Greece, Spain, France, U.S.S.R., Turkey, Mexico, and Argentina. Italy and the United States are far in the lead; combined, they account for 34% of the world production.[8]

Peaches are the sixth leading fruit crop in the United States.

The U.S. produces about 1.3 million short tons of peaches annually; and the leading producing states, by rank, are: California, Georgia, South Carolina, Pennsylvania, Washington, New Jersey, Michigan, Texas, Arkansas, and Colorado. California accounts for almost 50% of the U.S. peach crop. In addition, the United States produces 211,000 short tons of nectarines.[9]

PROCESSING. Most of the U.S. peach crop is either marketed as fresh or canned produce. Sixty-seven percent is marketed as fresh fruit, while 28% is canned. The remaining 5% of the crop is distributed among the following processes: dried, 1%; frozen, 3%; and jams, preserves, brandy, and some other miscellaneous products, 1%. Nearly all nectarines are marketed as fresh fruit.

Before freezing or canning, peaches are inspected, graded, pitted, and peeled. Automatic pitters can remove pits from either freestone or clingstone varieties. Peaches are canned in water pack, juice pack, or in syrup from light to extra heavy. Frozen peaches are generally frozen in a syrup with a small amount of ascorbic acid to prevent browning.

Dried peaches are halved, pitted, peeled, sulfured, and then dried in the sun or in dehydrators. It takes 6 to 7 lb

---

[8]Data from *FAO Production Yearbook 1990*, FAO/UN, Rome, Italy, Vol. 44, p. 162, Table 70. **Note well:** Annual production fluctuates as a result of weather and profitability of the crop.

[9]Data from *Agricultural Statistics 1991*, USDA, p. 201, Table 303. **Note well:** Annual production fluctuates as a result of weather and profitability of the crop.

(2.7 to 3.2 kg) of fresh peaches to make 1 lb (0.45 kg) of dried fruit.

SELECTION, PREPARATION, AND USES. Fresh peaches should be fairly firm or becoming slightly soft. Peaches and nectarines can be enjoyed when eaten fresh out of the hand or they can be cooked in a variety of baked goods. Sliced fresh or canned peaches can be eaten alone as a dessert dish or in combination with other fruits in salads, or in fruit plates. Also, peaches are a component fruit of canned fruit cocktail. Dried fruit can be eaten as a snack or cooked. Additional uses of peaches include: baked goods, gelatin desserts, ice cream, jams, jellies, juices, pies, sherbets; and in the production of alcoholic beverages.

NUTRITIONAL VALUE. Fresh peaches contain 89% water. Additionally, each 100 g (about 3½ oz or one medium-sized peach) contain 38 Calories (kcal) of energy, 202 mg potassium, 1,330 IU vitamin A, and only 1 mg of sodium. The calories in fresh peaches are derived primarily from the natural sugars (carbohydrate) which give them their sweet taste. Canned peaches are slightly higher in calories due to the addition of syrup, and their vitamin A content is lower. Dried peaches contain only 33% water; hence, most of the nutrients are more concentrated, including calories. Each 100 g of dried peaches contain 237 Calories (kcal) of energy, 983 mg potassium, 3.9 mg of iron and 2,142 IU of vitamin A. Fresh nectarines are 82% water, and they contain 64 Calories (kcal) of energy, 294 mg potassium, 1,650 IU vitamin A, and 13 mg vitamin C in each 100 g (about two medium-size nectarines). More complete information regarding the nutritional value of fresh, canned, dried, and frozen peaches and fresh nectarines is presented in Food Composition Table F-21.

PEACH PALM (PEJIBAYE) *Guilielma gasipaes*

The fruit of a palm tree (of the family *Palmae*) that is native to Central America and northern South America, but is presently underutilized, considering that its production might be as profitable as that of the cereal grains in the temperate climates.

It is noteworthy that the egglike peach palm fruit develops a nutlike flavor when boiled in saltwater. The oily seeds of the fruit are also edible. They may also be made into flour or fermented into chicha (an alcoholic beverage made by allowing the previously chewed nuts to ferment).

The fruit is rich in calories (196 kcal per 100 g) and carbohydrates (42%). It also contains 2.6% protein (much more than most fruits), and is a good source of iron, vitamin A, and vitamin C.

PEACOCK (PEAFOWL)

This magnificent bird is prized for both its ornamental qualities and its gastronomical value.

In the Middle Ages, the peacock held sway at great state banquets. Its cooking called for all the skill of an experienced master-cook; and only the noblest at the feast had the right to carve it.

Peacocks are still regarded as a delicacy for special occasions. Culinary preparation is similar to pheasant.

PEA, GARDEN *Pisum sativum*

This legume is closely related to the field pea (*Pisum arvense*).

Fig. P-13 shows freshly harvested pods and seeds of the garden pea.

Fig. P-13. Close up of peas in a pod. (Photo by J. C. Allen & Son, Inc., West Lafayette, Ind.)

ORIGIN AND HISTORY. The garden pea appears to have been derived from the field pea by centuries of cultivation and selection for certain desired characteristics. Credence to this theory is lent by the fact that the garden pea is not found in the wild state, whereas the field pea grows wild in the Georgian Republic of the U.S.S.R.

The garden pea, like the field pea, is thought to have originated in Central Asia and Europe, with possibly secondary developments in the Near East and North Africa. Through the centuries, it spread westward and northward throughout Europe, southward into Africa, and eastward to India and China.

Peas were known to and used by the Chinese in 2000 B.C.; and the Bible mentions peas. Apparently, the Chinese were the first to use the green pods and seeds as food. (They also used other legumes similarly.) However, green peas did not appear on European menus until the 16th century, when they were popularized by French royalty. About this time, the agricultural writings began to distinguish between field peas and garden peas.

Peas appear to be the first crop that was scientifically bred to produce new varieties with more desirable characteristics. By the end of the 19th century, many cross breeding trials had been made, the most notable of which were those conducted by the Austrian monk Mendel.

The observations made by Mendel in his monastery garden at Brunn (now Brno, in Czechoslovakia) provided the foundation for the science of genetics.

Peas were brought to America about 1800.

WORLD AND U.S. PRODUCTION. Dry peas and green peas rank fourth and sixth, respectively, among the legumes of the world.

Worldwide, about 17.5 million metric tons of dry peas are produced annually; and the leading countries, by rank, are: U.S.S.R., France, China, Denmark, India, Hungary, Australia, U.K., Canada, and Czechoslovakia. The U.S.S.R. produces 49% of the world dry pea crop.[10]

---

[10]Data from *FAO Production Yearbook 1990*, FAO/UN, Rome, Italy, Vol. 44, p. 103, Table 34. **Note well:** Annual production fluctuates as a result of weather and profitability of the crop.

Worldwide, about 4.8 million metric tons of green peas are produced annually; and the leading countries, by rank, are: U.S.A., U.K., China, Hungary, India, U.S.S.R., Belgium/Luxembourg, France, Italy, and Romania.[11]

Green peas and dry peas rank fifth and sixth, respectively, among the legumes of the United States.

The United States produces about 528,150 short tons of green peas annually; and the leading states, by rank, are: Wisconsin, Minnesota, Washington, Oregon, New York, and Delaware.[12]

The United States produces about 108,000 metric tons of dry peas annually, virtually all of which is produced in the states of Washington and Idaho. It is noteworthy that dry peas represent only 20% of the U.S. pea crop.[13]

PROCESSING. Almost all of the U.S. crop of green peas is processed, since fresh peas spoil rapidly and there is little demand for them. Over half the crop is canned; most of the rest is frozen.

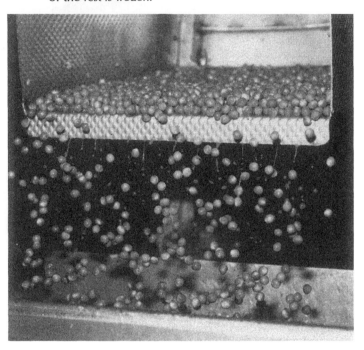

Fig. P-14. Processing peas. (Courtesy, University of Minnesota, St. Paul)

Dried peas may be utilized in the production of canned or dehydrated pea soups, or to a limited extent to make the following products: instant peas soup mixes, quick-cooking dried peas, pea flour, and pea protein concentrate.

SELECTION, PREPARATION, AND USES. Fresh peas of the best quality are young, tender, and sweet. The pods should be fresh, uniformly light-green in color, slightly velvety to the touch, and filled with well-developed peas.

Most homemakers use canned and/or frozen green peas (the leading frozen vegetable in the United States) in preparing meals for their families, since (1) the cooking of dried

[11]*Ibid,* p. 146, Table 61.

[12]Data from *Agricultural Statistics 1991,* USDA, p. 158, Table 225. **Note well:** Annual production fluctuates as a result of weather and profitability of the crop.

[13]Data from *FAO Production Yearbook 1990,* FAO/UN, Rome, Italy, Vol. 44, p. 103, Table 34.

Fig. P-15. Split pea soup, made from dried peas. (Courtesy, USDA)

peas (the most economical form of this vegetable) at home has declined steadily over the past 3 decades, and (2) fresh green peas are sold in pods and require shelling. Some suggestions for preparing the different products follow:

1. Canned peas need no further cooking since they are already tender. Therefore, they might be added to cold salads, or to piping hot dishes that are ready to be served.

2. Frozen peas are usually sufficiently tender after thawing to be used without cooking, but most people enjoy them more after a very brief cooking, followed by the addition of butter or margarine. Only the minimal amount of water to prevent burning or sticking to the pan should be used in cooking.

3. Fresh green peas should be cooked from 10 to 15 minutes after the water comes to a boil. (About 1 lb [0.45 kg] of peas in pods yields a cup [240 ml] of shelled peas.)

4. Dried peas may be cooked much more quickly than dried beans, so it is necessary to use care not to convert the peas to a mushy consistency by overcooking them.

5. Each pound (about 2 cups, or 0.45 kg) of dried peas yields about 3 lb (6 cups, or 1.4 kg) of cooked peas. Usually, this form of peas is the most economical for use in preparing soups.

Peas go well with most other food items such as cereal products, cheeses and other dairy products, fish and seafood, meats, poultry, and vegetables. They may be used in casseroles, salads, soups, stews, and vegetable side dishes. The number and types of pea dishes need be limited only by the imagination of the cook and the ingredients on hand.

NUTRITIONAL VALUE. The nutrient composition of peas is given in Food Composition Table F-21.

Some noteworthy observations regarding the nutrient composition of garden peas follow:

1. Cooked dry (mature) peas provide about twice as much solids, calories, carbohydrates, and proteins as cooked green (immature) peas. However, the green peas are a good to excellent source of vitamin A, and a good source of vitamin C, whereas dried peas are a poor source of both vitamins.

2. A 3½ oz (100 g) portion (about ½ cup) of cooked dry peas or about 1 cup of cooked green peas provide an amount of protein (8 g) equivalent to 1 oz (28 g) of cooked lean meat. However, the peas provide about twice as many calories per gram of protein.

3. Peas are much lower in calcium and phosphorus than beans.

4. Both dry and green peas are good sources of iron and potassium.

Protein Quantity and Quality. The following facts regarding the quantity and quality of pea protein are noteworthy:

1. The grams of protein per 100 Calories (kcal) provided by green and dry peas compared to other selected foods follow:

Food	Grams of Protein per 100 Calories (kcal) (g)
Green peas	7.5
Dry peas	7.0
Cottage cheese	13.2
Lean meat	12.7
Eggs	8.6
Navy beans	6.6
Sweet corn	3.9
Brown rice	2.1

As noted, peas have almost as much protein as eggs. The above figures also show that the amount of protein per calorie in pea dishes may be increased by the addition of a high protein animal food. Similarly, peas may be used to raise the protein content of cereal dishes.

2. Pea protein is moderately deficient in the amino acids methionine and cystine, which are supplied in ample amounts by cereal proteins. However, the legume protein contains sufficient lysine to cover the deficiency of this amino acid in grain proteins. Hence, combinations of peas and cereal products supply higher quality protein than either food alone.

PEANUT (GROUNDNUT; EARTH NUT; GOOBER; GOOBER PEA; GROUNDPEA; PINDA; MONKEY NUT; CHINESE NUT) *Arachis hypogaea*

Contents	Page
Origin and History	829
World and U.S. Production	829
Processing Peanuts	829
Nutritional Value of Peanuts	830
Peanut Products and Uses	830

Fig. P-16. How peanuts grow. (Courtesy, National Peanut Council, Washington, D.C.)

The peanut is the strange fruit of the peanut plant. It begins as a fertilized flower above ground, but the pod and the seed mature in the ground; it's the plant with an aerial flower and subterranean fruit.

Unlike the soybean, peanuts are highly esteemed for human food in the United States; and they are standard fare at American sporting events, circuses, and cocktail parties; and, in the form of peanut butter, they have nourished millions of American youths.

ORIGIN AND HISTORY. Peanuts are native to South America. They were introduced to Africa by European explorers, and they reached North America with the slave trade. Early in the 19th century, the value of peanuts as a food crop was recognized. Today, the plant is cultivated on 6 continents in more than 40 countries.

WORLD AND U.S. PRODUCTION. Peanuts rank second among the legume crops of the world. World production in the shell totals about 23.1 million metric tons, annually; and the leading peanut-producing countries, by rank, are: India, China, United States, Nigeria, Indonesia, Senegal, Myanmar, Zaire, Argentina, and Vietnam. More than half of the world crop is produced by India and China.[14]

Peanuts also rank second as a legume crop of the United States; yet, the U.S. produces only 7% of the world's peanuts. About 1.8 million short tons of peanuts are produced annually in the United States; and the leading states, by rank, are: Georgia, Texas, North Carolina, Alabama, Virginia, Oklahoma, Florida, New Mexico, and South Carolina.[15]

PROCESSING PEANUTS. Following picking, peanuts are delivered to warehouses, where they remain until processed. Processing varies somewhat according to market use, with the following operations involved:

• **Cleaning**—The cleaning operation consists of removing sticks, stems, small rocks, and faulty nuts by a series of screens and blowers.

• **Storing**—Cleaned peanuts are either (1) stored unshelled in silos or warehouses for subsequent shelling and delivery to users; or (2) shelled, then stored in refrigerated warehouses at 32° to 36°F (0° to 2° C) and 65% relative humidity, so as to protect against insects and rancidity.

• **In the shell**—Large, unshelled peanuts may be cleaned, polished, whitened, then marketed in the shell.

Some peanuts are salted and roasted in the shell. This involves soaking in salt water under pressure, then drying and roasting.

• **Shelling**—This consists of breaking the shells by passing the nuts between a series of rollers. Then, the shells (along with small, immature pegs) are separated by screens and blowers; and the discolored kernels are removed by hand and by electric eye. Although varying according to variety, shelling reduces the weight of peanuts by 30 to 60%, the space occupied by 60 to 70%, and the shelf life by 60 to 75%.

---

[14]Data from *FAO Production Yearbook 1990*, FAO/UN, Rome, Italy, Vol. 44, p. 109, Table 38. **Note well:** Annual production fluctuates as a result of weather and profitability of the crop.

[15]Data from *Agricultural Statistics 1991*, USDA, p. 118, Table 160. **Note well:** Annual production fluctuates as a result of weather and profitability of the crop.

• **Blanching**—This consists of removing the skins or seed coats, (and usually the hearts) prior to use in peanut butter, bakery products, confections, and salted nuts.

Blanching may be done with heat or with water:

1. Blanching with heat consists of embrittling the skins by exposure to 259 to 293°F (126 to 145°C) heat for 5 to 20 minutes, followed by rubbing the kernels between soft surfaces and removing the skins by blowers and the hearts by screens.

2. Blanching with water is accomplished in different ways. The newest and most rapid method consists of wetting the nuts with 140°F (60°C) water and removing the skins by rapidly revolving spindles. The kernels are dried to 7% moisture prior to storage or conversion into peanut products.

• **Dry roasting**—Peanuts for use in peanut butter, confections, or bakery products are dry-roasted to develop desirable color, texture, or flavor. Dry roasting is accomplished by heating unblanched peanuts to 399°F (204°C) for 20 to 30 minutes, followed by cooling and blanching.

• **Oil roasting**—Peanuts for salting are first roasted in coconut oil or partly hydrogenated vegetable oil at 300°F (148.9°C) for 15 to 18 minutes.

• **Salting**—Either blanched or unblanched peanuts are roasted in oil and salted. Finely ground salt and an oil-base binder are mixed with freshly cooked nuts, which are then placed in flexible bags and canned under vacuum.

• **Extraction of oil**—The oil from peanuts is extracted by one of three methods: hydraulic extraction, expeller extraction, or solvent extraction.

One hundred pounds of peanuts in the shell will yield about 32 lb of crude oil, 45 lb of meal, and 25 lb of hulls.

NUTRITIONAL VALUE OF PEANUTS. The nutritive value of the different forms of peanuts and peanut products is given in Fig. P-17, and in Food Composition Table F-21.

Peanuts are a healthful food. The energy value in 1 lb (0.45 kg) of peanuts (containing about 2,558 Calories [kcal]) is about equal to any of the following: 14 oz (397 g) of cooked round steak, 15 oz (425 g) of natural Cheddar cheese, 3 qt (2.8 liter) of milk, or eight eggs.

Peanuts are also a good source of protein, containing about 26% and a fair balance of essential amino acids. They make a good complementary protein. (See Table P-16, Protein and Amino Acids in Selected Foods.)

Raw or processed peanuts are excellent sources of riboflavin and niacin, and peanut skins are high in the B vitamins.

PEANUT PRODUCTS AND USES. Peanuts have much to contribute to the world's food supply, whether the need be for calories, fats, proteins, or certain vitamins.

In the United States, peanuts are almost entirely a human food and not an oilseed crop as such, since so small a proportion of the crop is processed for oil and protein concentrate.

In the rest of the world, peanuts are primarily processed for their separate constituents—oil and protein. Fortunately, some of the highest peanut-producing countries are in greatest need of their food value. The biggest problem in these countries is to protect the peanuts from molds, rodents, insects, and rancidity. In such countries where there coexists protein deficiency and availability of peanuts, there is much interest in incorporating peanut protein concentrate in the diet of children (see Table P-1, section on "Fortified Foods, Based on Peanut Products, For Developing Countries"). The main deterrent to this program to date has been the cost of peanut protein.

Fig. P-18. Here are three examples of peanut foods: (1) Presidential Peanut Pie; (2) peanut butter, lettuce and tomato sandwich; and (3) peanut butter parfait. (Courtesy, National Peanut Council, Washington, D.C.)

Table P-1 presents in summary form the story of peanut products and uses for human food. In addition to the products listed therein, the following human food products are among those that have not yet reached wide distribution: peanut protein, peanut milk, peanut cheese, boiled fresh peanuts, canned boiled peanuts—shelled and unshelled, frozen boiled peanuts, peanut bread, peanut cereals, and numerous products that may be developed from these.

(Also see FLOURS, Table F-13, Special Flours; and OILS, VEGETABLE, Table O-5.)

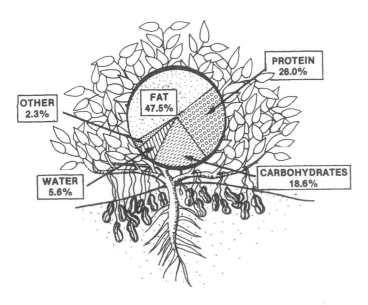

OTHER 2.3%

PROTEIN 26.0%

FAT 47.5%

WATER 5.6%

CARBOHYDRATES 18.6%

Fig. P-17. Food value of the peanut—the nut with skin.

TABLE P-1
PEANUT PRODUCTS AND USES FOR HUMAN FOOD

Product	Description	Uses	Comments
Whole, roasted-in-shell peanuts	Peanuts roasted inside the shells	Snacks	About ¼ of the peanuts marketed in the U.S. are in the form of roasted peanuts, primarily following shelling.
Defatted peanuts	This process removes 60 to 80% of the oil and ¾ of the calories, leaving intact a high-protein peanut with a flavor that is generally acceptable. Defatted peanuts are prepared as follows: shelled peanuts are reduced to a specific moisture content, then placed in a hydraulic press where most of the oil and calories are removed. Although the kernels are distorted due to the hydraulic pressure, they return to their original shape and size when soaked in water. Salt, sugar, spices and other flavors may be added during the reconstruction period. These low calorie nuts are then dried.	Defatted peanuts are consumed raw, roasted, or used in candy and confectionary products.	
Flour mix (In Congo)	A mixture of fish flour, millet, and peanut flour (50% protein). Made into biscuits.	For children ages 2 to 6 years	These biscuits, which have a good nutritive value and can be recommended for the prevention of malnutrition, are readily accepted by children.
Flour mix (In India)	A mixture of either (1) 25% roasted Bengal gram, 74% peanut flour (low fat), and 1% alfalfa (lucerne) powder; or (2) 25% roasted Bengal gram, 49% peanut flour, 25% low-fat sesame, and 1% alfalfa (lucerne) powder.	These mixtures are eaten with bread and jaggery (an unrefined brown sugar made especially from palm sap).	These vegetable protein diets are nearly as effective in controlling clinical manifestations of malnutrition as diets based on skim milk, but they are somewhat inferior to milk in restoring the level of blood protein (albumen) to normal.
Flour mix (In Senegal)	A mixture of 60% millet, 30% peanut flour, and 10% fish flour. Baked products.	Food for infants	This is a satisfactory food for infants.
Fresh peanuts	Fresh peanuts, not dried or roasted	As a vegetable	In tropical countries, peanuts are usually consumed in this form.
Granulated (chopped), artificially flavored peanuts	These are granulated, artificially flavored peanuts that have the identical flavor and taste of other nuts, especially black walnuts, almonds, and pecans.	Widely used for ice cream toppings, pies, cakes, and other confectionary products.	These artificially flavored nuts are now commercially available in many food stores.
High protein food (In India)	A mixture of equal parts of peanut protein isolate and skim milk powder	A high protein food suitable for treatment of protein malnutrition	
Peanut butter	A cohesive, finely ground product prepared from dry roasted, clean, sound, mature peanuts from which the seed coats and "hearts" are removed, and to which salt, hydrogenated fat, sugars (optional), antioxidants, and flavors have been added. About 90% of the peanut butter is stabilized with hydrogenated oil and antioxidants; 10% is unstabilized (and more flavorsome). Peanut butter contains about 50 to 52% fat, 28 to 29% protein, 2 to 5% carbohydrate, and 1 to 2% moisture.	Sandwiches, salads, desserts, ice cream, custards, confections, and in many baked goods	Peanut butter, consisting of shelled, ground, and parched peanuts, was first prepared as a kitchen product about 1890, as food for infants and invalids. Today, about half the peanuts consumed in the U.S. are made into peanut butter. By federal regulation, 90% of peanut butter must be peanuts in order to be labeled peanut butter. One lb (0.45 kg) of peanut butter has as much total food energy as any of the following: 1⅜ lb (0.6 kg) cheese, 2¼ lb (1 kg) of steak, 4 qt (3.8 liter) of milk, or 32 eggs. Peanut butter is an excellent source of protein (7.8 g per oz), contains no cholesterol and the fat (14.4 g per oz) is mostly unsaturated. An ounce of peanut butter contains 6.4 g of carbohydrate, making it acceptable for most diabetic diets. In a diabetic diet, 2 Tbsp (30 ml) of peanut butter is equivalent to one meat exchange, two fat exchanges, or one B vegetable exchange. However, as with any food, it should not be used indiscriminately.
Peanut flakes	After peanuts are blanched (skin removed) and cooked, they are ground to a slurry, then dehydrated. In the process of dehydration, most peanut flavor constituents volatize, leaving a "tasteless-faceless" peanut product.	Peanut flakes, which contain essentially the same nutritive value (protein and calories) as the peanut, offer virtually unlimited potential as a bland high protein extender for meat and bakery products without adversely affecting the flavor of the finished product.	The processing equipment for making peanut flakes is quite expensive. The worldwide demand for this new product is expected to increase substantially.
Peanut flour	In processing peanut flour, a high proportion of the oil is removed, leaving a product that contains about 60% protein, 22% carbohydrates, 5% minerals, and less than 1% fat.	When commecially available, it is expected that peanut flour will be widely used as a highly digestible protein extender in bakery and confectionary products.	Research has also demonstrated the possibility of using peanut protein fibers for manufacturing meat product substitutes.
Peanut oil	The oil obtained by crushing the nuts in hydraulic presses, or by using chemicals to dissolve the oil out of the nuts. The cold press method gives the lowest yield but the highest quality of edible oil.	Nearly 90% of the peanut oil used in the U.S is for food uses. For frying food. It smokes only	About ⅔ of the world's peanut crop is crushed for oil. Peanuts supply about 6% of the world's edible oil production. India produces 31% of the world's peanuts.

(Continued)

TABLE P-1 *(Continued)*

Product	Description	Uses	Comments
Peanut oil *(Continued)*		at high temperatures and does not absorb odors easily. In salad oils and dressings, margarine, and other vegetable shortenings	The large-podded, large-seeded varieties grown in the U.S. contain about 45% oil, and the small-podded, small-seeded kinds up to 50%. From the two types, a ton of nuts in the shell will yield about 480 lb and 580 lb*(218 and 263 kg)* of oil, respectively. Peanut oil contains at least 8 nutritionally essential fatty acids and 76 to 82% unsaturated fatty acids, of which 40 to 45% is unsaturated oleic acid and 30 to 35% polyunsaturated linoleic acid. The major portion of the characteristic peanut aroma and flavor is imparted by the oil.
Roasted peanuts *(following shelling)*	Whole nuts, following roasting and shelling. They are usually salted to improve their flavor, except where they go into peanut butter.	As snacks, eaten alone. In candies, salads, desserts, cookies, pies, and in other ways. In mixed, salted nuts. Peanut bread is made from ground peanuts.	There is a preference for the larger nuts for salted peanuts. About ¼ of all edible peanuts in the U.S. find their way into candies. Some of the most flavorsome, nutritious, and popular candies contain peanuts. Peanut candies are mostly bars and brittle.
FORTIFIED FOODS, BASED ON PEANUT PRODUCTS, FOR DEVELOPING COUNTRIES			
Biscuit *(In Uganda)*	A mixture of 41% peanuts, 26% corn meal, 12% sucrose, 6% cottonseed oil, and 15% dried skim milk. Baked into a biscuit containing 20% protein.	It can be mixed with water (1) into a gruel, or (2) into a drink (by adding more water).	When fed to infants and preschool children, it gives the same weight gain as a milk biscuit.

PEAR *Pyrus* spp

Fig. P-19. Pears—fruit of the ages and relative of the apple. Color may be yellow, russet, or red. (Photo by J. C. Allen & Son, Inc., West Lafayette, Ind.)

The pear which is a member of the rose family, *Rosaceae*, is classified as a pome fruit and is closely related to the apple and the quince. Generally pears are large and round at the bottom, tapering inward toward the stem. However, some are almost completely round and others are as small as a cherry. Pears are covered with a smooth thin skin that may be yellow, russet, or red when ripe. Its edible flesh is juicy, sweet, and mellow. The flesh of some pears has a sandy texture due to the presence of grit cells. Enclosed within the center of the fleshy portion is a core which contains as many as 10 seeds. There are hundreds of pear varieties. Depending upon variety, pears vary in shape, size, color, texture, flavor, aroma, time of ripening, and keeping qualities.

ORIGIN AND HISTORY. The pear is indigenous to western Asia; and it has long been cultivated there and in Europe. Pears were taken to America with some of the first colonists.

WORLD AND U.S. PRODUCTION. Worldwide, 9.8 million metric tons of pears are produced annually; and pears rank eighth among the fruit crops of the world. The leading pear-producing countries, by rank, are: China, Italy, United States, U.S.S.R., Japan, Spain, Turkey, Germany, France, and Argentina.[16]

About 963,800 tons of pears are produced annually in the United States; and the following states lead in production, by rank: Washington, California, and Oregon. These three states account for 97% of the U.S. pear crop.[17]

PROCESSING. About 48% of the pear crop is sold as fresh pears for consumption as fresh fruit and for home use. The remaining 52% is processed as canned or dried fruit. There are no recent estimates available as to the amounts canned or dried, though in past years, the amount dried has been less than 1% of the crop. Therefore, around 51% of the pear crop is probably canned.

---

[16]Data from *FAO Production Yearbook 1990*, FAO/UN, Rome, Italy, Vol. 44, p. 162, Table 70. **Note well:** Annual production fluctuates as a result of weather and profitability of the crop.

[17]Data from *Agricultural Statistics 1991*, USDA, p. 206, Table 313. **Note well:** Annual production fluctuates as a result of weather and profitability of the crop.

Preservation of pears by canning includes pears canned in water pack, juice pack, light syrup, heavy syrup or extra heavy syrup.

SELECTION, PREPARATION, AND USES.
As pears ripen their skin color changes from green to the color characteristic for the variety. Bartlett pears turn yellow. Fresh pears are best when they yield to gentle palm pressure. If still green skinned when purchased, pears can be ripened at home by storing them at room temperature, preferably three or more together in a closed paper bag. When pears begin to turn colors and pass the palm pressure test they should be refrigerated. Bartlett pears can be held in the refrigerator for several days either before or after ripening.

Pears can be enjoyed by eating fresh out of the hand like an apple, or they may be used in salads, main dishes or desserts, limited only by the imagination. Canned pears are used alone as a dessert dish or in combination with other fruits and/or in gelatin dishes. Some pears are canned with other fruits in fruit cocktails. Dried pears are eaten as a snack or they may be cooked. Also, pears may be used in the production of alcoholic beverages.

Fig. P-20. Pears are often used in fruit salad. (Courtesy, USDA)

NUTRITIONAL VALUE.
Fresh pears contain 83% water. Additionally, each 100 g (about 3½ oz) provides 61 Calories (kcal) of energy, 130 mg of potassium, and only 2 mg of sodium. The calories in pears are derived primarily from the sugars (carbohydrate) which give pears a sweet taste. Pears canned in a syrup contain more calories due to the addition of sugar. Also, canning decreases the concentration of some minerals and vitamins. Dried pears contain only 26% water; hence, many of the nutrients, including calories, are more concentrated. More complete information regarding the nutritional value of fresh, canned, and dried pears is presented in Food Composition Table F-21.

PEARLED

Dehulled grains reduced to smaller smooth particles by machine brushing.

PECAN *Carya illinoensis*

Fig. P-21. Pecans, an American crop requiring patience. Newly planted trees do not bear enough nuts to be profitable until they are about 11 years old.

The nut-bearing pecan tree is classified botanically as a hickory. The trees produce a popular U.S. nut—the delicately flavored, fat-rich pecan. The nut is the basis for pralines, pecan pie, and other foods originating in the South, where the tree grows and yields most abundantly.

ORIGIN AND HISTORY.
Pecans are native to North America—to the middle southern United States, and to Mexico. Most pecans are still grown in the South, either as wild trees or as cultivars bred for thinness of shell.

WORLD AND U.S. PRODUCTION.
Some wild pecan trees grow in Mexico; and some plantings of pecans have been made in Australia and South Africa. But no production figures for these countries are available.

Pecans are the fourth leading nut crop of the United States; outranked by peanuts, almonds, and walnuts. Most U.S. pecans are grown in the southern states. About 205,000 short tons of pecans are produced annually in the United States; and the leading states, by rank, are: Georgia, Texas, New Mexico, Louisiana, Alabama, Oklahoma, Florida, California, Mississippi, and Arkansas. Georgia alone produces 32% of the U.S. pecan crop. Wild, or native, trees produce 22% of the pecan crop, while improved varieties (cultivars), bred for thinness of shell, produce 78% of the crop.[18] The thin-shelled varieties are called *papershell*, because they can be cracked between the fingers.

PROCESSING.
Following harvest, pecans are often dried or *cured* for a few weeks. At processing centers the nuts are cleaned, graded, and packaged. More than 90% of the pecans are marketed as shelled nuts. Pecans in the shell may be stored at 25°F (-4°C) or lower for 2 years or more.

Pecan processing equipment for shelled pecans can size, crack, and separate meats and shells. After separation from the shells, meats are dried to 3 to 4% moisture for better storage. Meats are graded by electric eye and by hand

---

[18]Data from *Agricultural Statistics 1991*, USDA, p. 221, Table 345. **Note well:** Annual production fluctuates as a result of weather and profitability of the crop.

whereupon they are packaged for distribution to bakeries, retail outlets, confectioners, and ice cream manufacturers. Some pecans are salted.

SELECTION, PREPARATION, AND USES. Pecans in shells should be free from splits, cracks, stains, or holes. Moldy nuts may not be safe to eat. Nutmeat should be plump and fairly uniform in color and size. No further treatment following shelling is necessary. But pecans keep better if stored in tightly closed containers in the refrigerator or freezer.

Delicately flavored, sweet tasting pecans are eaten alone or in an array of dishes and products. It is estimated that there are 1,200 uses for pecans in prepared dishes.

Pecans are used to impart their qualities to such foods as baked goods, dairy products, confections, salads, desserts, fowl stuffings, puddings, souffles, meat combinations, cereals, and vegetable dishes. Pralines and pecan pie

Fig. P-22. Pecans enhance many dishes, like the salad pictured here. (Courtesy, USDA)

originated in the South where most pecans grow. The flavor of pecans is compatible with that of most foods, so that they may be used natural, sweetened, salted, or spiced. The texture is such that they may be used as halves or pieces of any desired size. They may be eaten raw or toasted.

NUTRITIONAL VALUE. Pecans are extremely nutritious. They contain only 3 to 4% water and they are loaded with energy. Each 100 g (about 3½ oz) of pecans contains 739 Calories (kcal), primarily due to their fat (oil) content of 71%. Additionally, each 100 g provides 9 g protein, 15 g carbohydrate, 2.4 mg of iron, and 4.1 mg of zinc. More complete information regarding the nutritional value of pecans is provided in Food Composition Table F-21.

PECTIN

Most plant tissue, fruits in particular, contain pectin—a polysaccharide which functions as a cementing material. Lemon or orange rind is one of the richest sources of pectin,

containing about 30% of this substance. Pectin, hemicellulose, and cellulose comprise a part of the diet which is often referred to as fiber or roughage. It is not digestible.

Chemically, pectin is a mixture of methyl esterified galacturonan—long chains of the uronic acid of galactose joined by methyl alcohol—plus some galactan and araban, depending upon the plant source. Hence, the complete breakdown of pectin into component parts yields galacturonic acid, methyl alcohol, and small amounts of the sugars arabinose and galactose.

In a pure form extracted from plant material, pectin is a powder, yellowish-white in color, almost odorless, and with a mucilaginous taste. Due to its ability to form gels in a water solution, pectin is an important commercial product used as a food additive by commercial and household food processors.

Many plant materials could be employed for the production of pectin, but apple pomace and citrus peel (lemon, lime, orange, and grapefruit) are the sources for the commercial extraction of pectin.

The type of pectin, hence, its uses in food, is governed by the amount of methyl alcohol joined to the galacturonic acid within the pectin molecule. Therefore, pectin is generally classified as high methoxyl pectin or low methoxyl pectin. Each type possesses different properties.

• **High methoxyl pectin**—When the degree of methylation is more than 50% for the high methoxyl pectin, half or more of the galacturonic molecules are associated with methyl alcohol. High methoxyl pectin requires 55 to 85% sugar and a pH of 2.5 to 3.8 (acid) to gel. This limits the use of high methoxyl pectin as a gelling agent in sweetened fruit products such as jams and jellies at a concentration of 0.1 to 1.5%.

• **Low methoxyl pectin**—Processing techniques can deesterify pectin (remove methyl alcohol) and create low-methoxyl pectin. This type of pectin requires no sugar or acid to gel. Therefore, low methoxyl pectin is used as a gelling agent in fruit products with a low sugar content or products containing no added sugar. The low methoxyl pectin has more and varied uses than high methoxyl pectin.

Pectin is a food additive, and since it is found naturally in plant material, and has always been part of man's diet, the Food and Drug Administration (FDA) accepts it as a generally recognized as safe (GRAS) additive. Moreover, pectin may be a beneficial constituent of the diet, even though it is not digested, since it has been shown to lower blood cholesterol levels.

(Also see FIBER, section on "Kinds of Fiber.")

PECTINASES

Like all living things, plants contain enzymes—protein molecules which catalyze chemical reactions. Pectinases are enzymes in plants or plant products which cause the chemical breakdown of the polysaccharide pectin, resulting in softening and viscosity changes in fruit products. Depending upon the food product, this may be desirable or undesirable. For example, pectins are desirable in tomato, orange, and apricot juices as they make the juice viscous and hold dispersed solids in solutions; hence, pectinase must be destroyed by heat during production. On the other hand, the fruit juices, such as apple and grape juice, a clear juice is desired so commercial pectinase is added. This ensures the production of a clear juice. Vinegar and jelly produced from depectinized juices are a brighter color. Pectinases are also used in making wines. Most commercial pectinases are

derived from *Aspergillus niger* fungi.
(Also see PECTIN.)

PEDIATRICS

Relating to the study and treatment of disease in children. A specialist in pediatrics is called a pediatrician. He is trained in both the medical and surgical diagnosis of the illnesses and disorders of infants and children.

PEKOE

This is a grade of tea leaves, from which a black tea is derived. Pekoe leaves are the middle-sized tea leaves, while orange pekoe leaves are the smallest.
(Also see TEA.)

PELLAGRA

Contents   Page
History of Pellagra.................................835
Causes of Pellagra.................................835
Signs of Pellagra...................................835
Treatment and Prevention of Pellagra.................835

This dietary deficiency disease is due to lack of the vitamin niacin or of the amino acid tryptophan (which is converted in the body to nicotinic acid). It usually afflicts people whose dietary protein comes mainly from maize (corn) (which is deficient in both nicotinic acid and tryptophan). Also, it occurs in areas of India where the diet consists mainly of a millet called *jowar* (*Sorghum vulgare*) without any animal foods. Some alcoholics and persons with disorders of absorption have also been found to have the disease.

HISTORY OF PELLAGRA. There is no mention of pellagra in early medical writing in Europe. It appears that the first cases in Europe were observed after the cultivation of maize, taken there from the Americas. Casal first described the disorder in Spain in 1730; and by the end of the 18th century, the disease had spread through Europe, North Africa, and Egypt. Casal found that a milk diet cured the disease. Cerri demonstrated the dietary cure of Italian pellagra victims in 1795; and in the early 1800s Buniva showed, by inoculation of humans with material from diseased persons, that pellagra was not infectious. Most of the European investigators recognized the importance of diet in the prevention and treatment of the disease. In general, the dietary prescription was to eliminate corn from the diet and to replace it with foods of animal origin such as meat, milk, and eggs.

Pellagra, the age-old deficiency disease, killed thousands of people until Dr. Conrad Elvehjem, and co-workers, at the University of Wisconsin, discovered that the vitamin niacin cured blacktongue in dogs, a condition recognized as similar to pellagra in man. Shortly thereafter, several investigators found that niacin was effective in the prevention and treatment of pellagra in humans. Following closely in period of time, it was found that the amino acid tryptophan was a precursor of niacin; and that it would also prevent and cure pellagra.
(Also see NIACIN, section on "History.")

CAUSES OF PELLAGRA. Although pellagra was long associated with reliance on corn as a dietary staple, it took investigators over 200 years to identify the specific nutritional conditions which cause this disease. Today, it is known that pellagra may be caused by any one, or a combination, of the following nutritional conditions:

1. Deficiencies of niacin (nicotinic acid).
2. Deficiencies of tryptophan.
3. Amino acid imbalances which reduce the utilization of tryptophan.
4. Chronic alcoholism.

SIGNS OF PELLAGRA. Clinical signs of pellagra are the characteristic three D's—dermatitis, diarrhea, and dementia. The skin disorder is usually one of the first clear-cut signs of the development of the disease. There is reddening of the areas exposed to the sun or of those areas subjected to friction (such as the elbows). A characteristic dermatitis around the collarbone is called Casal's necklace after the man who first described the condition in Spain. Another very early sign in development of the disease is soreness of the tongue which is swollen, raw, and extra sensitive to hot or seasoned foods (apparently part of an irritation of the entire gastrointestinal tract). The diarrhea, which is not always present, may contain blood and mucus. The mental symptoms range from irritability, anxiety, and depression to psychosis. (In the days of endemic pellagra in the southeastern United States, there were many admissions of pellagra victims to mental hospitals.) A victim of pellagra is shown in Fig. P-23.

Fig. P-23. An East African victim of pellagra. Note the inflammation and darkening of the skin around his neck. (Courtesy, FAO, Rome, Italy)

TREATMENT AND PREVENTION OF PELLAGRA. Persons suffering from severe forms of the disease may be given either nicotinamide (a derivative of niacin) or nicotinic acid in oral doses as high as 100 mg every 4 hours. Such high doses of nicotinic acid often cause flushing, which does not result when similar doses of nicotinamide are given.

Niacin is well absorbed, even when gastrointestinal disorders are present. Often, a pellagra victim suffers from multiple vitamin and nutrient deficiencies, and, therefore, may require supplemental sources of the vitamin B complex and protein. Brewers' yeast has often been used as such a supplement.

Diets used for the treatment and prevention of pellagra should include abundant sources of animal protein such as meat, fish, eggs, and milk since these foods are good sources of both niacin and tryptophan. Vegetable protein sources high in tryptophan are peanuts, sesame seeds, sunflower seeds, and whole wheat. The refining of grains results in the loss of much of the niacin content, so it is wise to use either whole grain products or those which have been enriched. It should be noted that gelatin, and meats which are high in connective tissue (such as chitterlings, ham hocks, hog jowls, and maws), are very low in tryptophan. It is noteworthy, too, that these were the types of meats commonly used in the South in the years when pellagra was endemic.

(Also see NIACIN.)

PEPPERONI (PEPERONI)

A highly seasoned dry sausage made with beef, or with pork and beef. Also, poultry and turkey may be added to pepperoni provided it is properly labeled.

PEPPERS Capsicum spp

The species of peppers which are native to the Americas have been enthusiastically adopted as commercial crops around the world. These shrubby plants, which are called capsicums to distinguish from the viny pepper plants (*Piper* spp) of the Old World, are members of the nightshade family (*Solanaceae*). Hence, the capsicum peppers are close relatives of the eggplant, Irish potato, and tomato. Although many botanists believe that the different types of capsicums are simply different varieties of a single species, it is customary to classify most types as either (1) chilies (*C. frutescens*), which are usually very pungent; or (2) sweet peppers (*C. annuum*), which have milder flavored, larger fruits.

Fig. P-24. Harvesting hot peppers. (Courtesy, Louisiana Cooperative Extension Service, Louisiana State University, Baton Rouge, La.)

ORIGIN AND HISTORY. It seems likely that the prehistoric peoples of Latin America gathered wild peppers in much the same way as certain varieties are gathered today. Pepper seeds dating from more than 5000 B.C. were found by archaeologists at Techuacan, Mexico. It is believed that these seeds came from wild varieties of sweet peppers.

Chili peppers may also have been domesticated independently between 2000 and 1000 B.C. by the Indians living on the Peruvian coast.

The rapid spread of sweet and chili peppers around the world was due in large part to the Spanish and Portuguese explorers who introduced them wherever they went on their ocean voyages.

WORLD AND U.S. PRODUCTION. Peppers rank tenth among the leading vegetable crops of the world, and thirteenth among the leading vegetable crops of the United States.

About 9.1 million metric tons of chilies and sweet (green) peppers are produced worldwide, annually. The leading pepper-producing countries of the world, by rank, are: China, Turkey, Nigeria, Spain, Mexico, Romania, Indonesia, Yugoslavia, Italy, and Egypt.[19]

About 230,000 metric tons of chilies and sweet peppers are produced annually in the United States, with Florida and California ranking first and second, respectively. Most of the peppers produced in the United States are sweet peppers.[20]

PROCESSING. Much of the U.S. pepper crop is utilized as a fresh vegetable because refrigerated storage and transportation facilities make it possible for the vegetable to be marketed around the country with a minimum of spoilage. Most of the remainder is utilized in such items as (1) canned products—corn and peppers, pimientos, sauces, stuffed peppers, and tomatoes and peppers; (2) dehydrated or dried preparations for reconstitution with water, or for making condiments; (3) frozen items—chopped peppers, pizza, and stuffed peppers; and (4) pickles which utilize chilies or sweet peppers. Some are also used in the preparation of prepackaged salads and in making the stuffed peppers sold in the meat sections of certain supermarkets.

Good cooks often enliven their dishes with various condiments made from chilies and/or sweet peppers. Some of the more popular peppers are: Cayenne pepper, chili powder, chili sauce, dried chilies, paprika, pimiento, red pepper, and Tabasco sauce.

SELECTION, PREPARATION, AND USES. Good quality sweet peppers are fresh, firm, bright in appearance, thick fleshed, and have a fresh, green calyx.

The pungent or hot varieties are sold in either the green or red stage of maturity, and vary in size from the small chili peppers to nearly as large as the bell type. Red-colored pods are generally the most desired. The same quality characteristics noted for sweet peppers also apply to fresh hot peppers.

Many people remove the seeds from peppers, but some leave them in and eat them right along with the fleshy part of the vegetable.

Fresh, frozen, or pickled peppers may be (1) served without cooking in relishes, salads, and sandwiches; or (2) cooked in casseroles; Chinese stir-fried dishes; fried onion garnishes for fish, meats, and poultry; omelets; pizzas; soups;

[19]Data from *FAO Production Yearbook 1990*, FAO/UN, Rome, Italy, Vol. 44, p. 139, Table 57. **Note well:** Annual production fluctuates as a result of weather and profitability of the crop.

[20]*Ibid.*

Fig. P-25. Several varieties of peppers. (Courtesy, Shell Chemical Company, Houston, Tex.)

and stews. Sweet peppers are also good when stuffed with bread crumbs, fish, meat, poultry, or rice, then baked. Chili peppers are used as seasoning, as an appetizer, or as a vegetable dish; and as an ingredient of casseroles, pizza, salads, sandwiches, soups, and stews. Chilies are suitable for making *rellenos* by stuffing them with cheese and either baking them in a sauce, or dipping them into an egg batter and deep-frying.

NUTRITIONAL VALUE. The nutrient compositions of various forms of peppers and pepper products are given in Food Composition Table F-21.

Some noteworthy observations regarding the nutrient composition of peppers and pepper products follow:

1. All types of raw peppers have a high water content (74 to 94%) which decreases with maturity, and a low caloric content which increases with maturity. These fresh vegetables are excellent sources of vitamin C and fair to excellent sources of vitamin A. The levels of both vitamins increase with the degree of ripeness. Red peppers have a significantly higher vitamin content than green peppers. Furthermore, chili peppers are richer in these vitamins than sweet peppers at comparable stages of ripeness. Canned, cooked, and dried peppers contain significantly less vitamin C, but approximately the same amount of vitamin A as raw peppers (on a moisture-free basis).

2. An average stuffed green pepper that weighs about 6½ oz (*185 g*) and is filled with a mixture of meat and bread crumbs is literally a nutritious meal in itself—supplying 314 kcal, 24 g protein, 78 mg calcium, 224 mg phosphorus, 477 mg potassium, 3.9 mg iron, 518 IU vitamin A, and 74 mg of vitamin C.

3. A 3½ oz (*100 g*) portion of canned pimientos contains 27 kcal, 2,300 IU of vitamin A, and 95 mg of vitamin C. Hence, this item makes an excellent addition to preparations based upon cereal products, cheeses, eggs, fish, meats, legumes, and poultry because it supplies the vitamins not furnished by the other foods.

4. The chili sauces made from chili peppers (the ones designated as "Hot chili sauces") usually contain only about one-fifth as many calories as the mild sauces made from tomato puree. Furthermore, the vitamin A content of the former preparations is usually several times that of the latter ones (about 1,630 IU vitamin A per Tbsp vs 238 IU vitamin A per Tbsp).

5. A teaspoon of paprika furnishes 1,212 IU of vitamin A, compared to the 800 to 900 IU provided by the same amount of chili powder or red pepper.

6. Although it is not shown in Food Composition Table F-21, it is noteworthy that peppers are excellent sources of bioflavonoids.

Certain types of hot chilies may contain potentially harmful amounts of a highly irritating substance called *capsaicin*, which can raise blisters on sensitive tissues or even cause ulceration. Also, peppers contain small amounts of solanine. (Also see SOLANINE.)

PEPSIN

The proteolytic enzyme present in the gastric juice. It acts on protein to form proteoses, peptones, and peptides.
(Also see DIGESTION AND ABSORPTION.)

PEPSINOGEN

The inactive precursor of pepsin, produced in the mucosa of the stomach wall and converted to pepsin by hydrochloric acid.
(Also see DIGESTION AND ABSORPTION.)

PEPTONE

An intermediate product of protein digestion.
(Also see DIGESTION AND ABSORPTION; and PROTEIN[S].)

PERINATAL

The period from just before to about 100 hours after birth.

PERISTALSIS

An advancing wave of circular constriction preceded by an area of relaxation in the wall of the gastrointestinal tract. This advancing wave propels the food from the esophagus to the anus. It is similar to the action obtained by wrapping one's fingers tightly around a plastic tube of shampoo and then sliding the fingers forward to expel the shampoo that is forced forward in front of the fingers.
(Also see DIGESTION AND ABSORPTION.)

PERNICIOUS

Denoting a severe disease that is usually fatal.

PERNICIOUS ANEMIA

(See ANEMIA, PERNICIOUS.)

PER ORAL

Administration through the mouth.

PER OS

Oral administration (by the mouth).

PEROXIDATION

Oxidation to the point of forming a peroxide—a compound (oxide) which contains an –0–0– group.

PEROXIDE VALUE

As fats decompose peroxides are formed. Chemically, peroxides are capable of causing the release of iodine from potassium iodide (KI). Therefore, the amount of iodine released from potassium iodide added to a fat is a rancidity test. The more peroxide present, the more iodine released; hence, the higher the peroxide value.

PERSIAN BERRY

The name given to the yellow to green dyes obtained from the native American fruits of the buckthorn family *Rhamnaceae*. Some of these berries are poisonous and have an extremely strong cathartic effect.

PESTICIDES

Some pesticides produce unplanned and undesirable side effects, particularly when they are not used properly. Among such effects are: reduction of beneficial species; drift; wildlife losses; honeybee and other pollinating insect losses; and pollution of air, soil, water, and vegetation. Farmers know that unless they follow federal and state regulations they risk having their products seized and condemned, or refused by food processors. It is, of course, essential that pesticide residues remaining on food following harvesting and processing be at levels that do not constitute a health hazard. Each pesticide is issued a tolerance for residues that may result from its use on food or feed crops. To this end, the Environmental Protection Agency (EPA) sets pesticide residue tolerances and the Food and Drug Administration (FDA) routinely monitors foods in an effort to enforce safe levels of pesticides on domestic food.

In recent years, several programs have evolved in both the public and private sectors to reduce the dependency on synthetic chemicals. These include: (1) breeding by both traditional and bio-engineered methods for crop resistance to pests; (2) integrated pest management, which stresses minimal use of pesticides; (3) organic production, which eliminates all synthetic chemical pesticides; (4) biological controls, which use either living organisms or toxic extracts from them; (5) low input sustainable agriculture, which applies integrated pest management principles to all areas of crop production.

No pest control system is perfect; and new pests keep evolving. So, research and development on a wide variety of fronts should be continued. We need to develop safer and more effective pesticides, both chemical and nonchemical. In the meantime, there is need for prudence and patience.

(Also see POISONS, Table P-2 Some Potentially Poisonous [Toxic] Substances.)

PETITGRAIN OIL

An essential oil extracted from the leaves and twigs of the bitter orange tree, the lemon tree, or the tangerine tree. Petitgrain oils are employed for flavoring purposes and are on the GRAS (generally recognized as safe) list of approved substances.

pH

A measure of the acidity or alkalinity of a solution. Values range from 0 (most acid) to 14 (most alkaline), with neutrality at pH 7.

(Also see ACID-BASE BALANCE.)

PHAGOCYTE

A cell capable of ingesting bacteria or other foreign material.

(Also see BACTERIA IN FOOD, section on"Bacterial Control Methods"; and BACTERIOPHAGE.)

PHAGOMANIA

An unusual or unreasonable fondness for eating. (Also see APPETITE.)

PHEASANT

The pheasant, which originated in the Orient, is classed as both a game bird and an ornamental bird. Connoisseurs of this savory bird prefer hen pheasants about 1 year of age. Pheasants are similar to chickens; so, they may be processed and cooked in a similar manner.

PHENOL OXIDASES (PHENOLASES)

A group of copper-containing enzymes responsible for the browning of the cut surfaces of certain fruits and vegetables. Browning reduces the acceptability of a food due to the off-color and off-flavor. Depending upon the intended use of a food, five different methods may be employed to prevent enzymatic browning by the action of phenol oxidases: (1) heat inactivation, blanching; (2) chemical inhibition with sulfur dioxide; (3) reducing agents such as ascorbic acid; (4) exclusion of oxygen during packaging; and (5) alteration of the phenol oxidase substrates by adding another enzyme.

(Also see PRESERVATION OF FOOD.)

PHENYLALANINE

One of the essential amino acids. (Also see AMINO ACID[S].)

PHENYLPYRUVIC ACID

An intermediate product in phenylalanine metabolism.

PHEOPHYTIN

A waxy, olive-brown pigment which results from the treatment of chlorophyll with an acid such as oxalic acid. The magnesium atom in the chlorophyll molecule is replaced by two hydrogen atoms. This accounts for the color change when green vegetables are cooked.

PHOSPHATE BOND, ENERGY RICH

Energy is required for practically all life processes—for the action of the heart, maintenance of blood pressure, muscle tone, transmission of nerve impulses, kidney function, protein synthesis, fat synthesis, secretion of milk, and muscular contraction. In the body, fats, proteins and carbohydrates represent potential energy sources. However, releasing all of the energy contained in these sources in one step would result in lost and wasted heat energy. Instead, energy is released in steps by a series of metabolic reactions. As the energy is released, it is captured and stored for use in the form of adenosine triphosphate (ATP). When this ATP is formed from ADP (adenosine diphosphate), 8 kcal/mole (molecular weight) must be put into the system. The ATP then acts as an energy rich storage compound. When energy is required for body functions, ATP is broken down to ADP and 8 kcal/mole of ATP are liberated.

(Also see METABOLISM.)

PHOSPHOLIPIDS

These are similar to triglycerides except one of the fatty acids attached to the glycerol molecule is replaced by phosphate and a nitrogen-containing compound. Lecithin, cephalin, phosphatidyl inositol, and phosphatidyl serine are all classified as phospholipids. Functionally, phospholipids are (1) structural components of membranes; (2) potential energy sources; (3) involved in the blood clotting mechanism; and (4) components of certain enzymes.

(Also see FATS AND OTHER LIPIDS.)

PHOSPHOPROTEIN

A conjugated protein that contains phosphorus; for example, casein found in milk.

PHOSPHORUS (P)

Contents   Page
History...839
Absorption of Phosphorus.........................839
Metabolism of Phosphorus.........................840
Excretion of Phosphorus..........................840
Functions of Phosphorus..........................840
Deficiency Symptoms..............................840
Interrelationships...............................840
Recommended Daily Allowance of Phosphorus........840
Toxicity...840
Phosphorus Related Diseases......................840
Calcium-Phosphorus Ratio and Vitamin D...........841
Sources of Phosphorus............................842

The era of phosphate formation

Fig. P-26. Primeval giants, such as rhinoceros (left) and mastodons (right) roamed Florida dulring the phosphate forming era.

Phosphorus is the Dr. Jekyll and Mr. Hyde of mineral elements—it can deal life or death. As white phosphorus, it's the flame of incendiary bombs; as red phosphorus, it's the heart of the common match; as organic phosphate, it's nerve gas and insecticides; and as organic combinations, it's a constituent of every cell and fluid of the body. Without phosphorus, no cell divides, no heart beats, and no baby grows.

Phosphorus is closely associated with calcium in human nutrition—to the extent that a deficiency or an overabundance of one may very likely interfere with the proper utilization of the other. Also, both calcium and phosphorus occur in the same major food source—milk; both function in the major task of building bones and teeth; both are related to vitamin D in the absorption process; both are regulated metabolically by the parathyroid hormone and calcitonin; both exist in the

blood serum in a definite ratio to each other; and both, as the chief components of bone ash, were used in many ancient medieval remedies.

Phosphorus comprises about 1% or 1.4 lb (650 g) of the adult body weight. That's about ¼ the total mineral matter in the body. Eighty percent of the phosphorus is in the skeleton (including the teeth) in inorganic combination with calcium, where the proportion of calcium to phosphorus is about 2:1. The remaining 20% is distributed in the soft tissues, in organic combination, where the amount of phosphorus is much higher than calcium. In the soft tissues, it is found in every living cell, where it is involved as an essential component in interrelationships with proteins, lipids, and carbohydrates to produce energy, to build and repair tissues, and to act as a buffer. Whole blood contains 35 to 45 mg of phosphorus per 100 ml; of this, about ½ is in the red cells, and, in adults, 2.5 to 4.5 mg per 100 ml is in the serum. In children, the serum phosphorus level is somewhat higher, ranging from 4 to 7 mg per 100 ml; the higher level during the growth years reflecting its role in cell metabolism. Four to nine mg of the whole blood phosphorus is inorganic phosphorus, which is readily affected by dietary intake and is in constant exchange with the organic phosphate of the blood.

HISTORY. Phosphorus, a nonmetallic element, was first identified in urine by Hennig Brand, a German alchemist, in 1669. It created much interest because, in the unnatural free form, it glowed in the dark and took fire spontaneously upon exposure to the air. Eventually, the name phosphorus (from the Greek for *light-bringing*) was appropriated to this element. Fortunately, phosphorus exists in nature only in combined forms, usually with calcium, in such sources as bone and rock phosphates.

ABSORPTION OF PHOSPHORUS. Phosphorus is more efficiently absorbed than calcium; 70% of the ingested phosphorus is absorbed and 30% is excreted in the feces, whereas only 20 to 30% of the ingested calcium is absorbed and 70 to 80% is excreted in the feces.

Since much of the phosphorus in foods occurs as a phosphate compound, the first step (prior to absorption) is the splitting off of phosphorus for absorption as the free mineral. The phosphorus is then absorbed as inorganic salts.

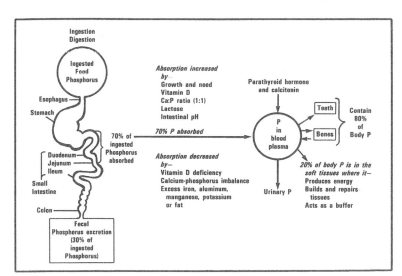

Fig. P-27. Phosphorus utilization. Note that healthy adults absorb 70% of the phosphorus in their food, and that 30% is excreted in the feces. Note, too, the factors that increase and decrease absorption.

Phosphorus is absorbed chiefly in the upper small intestine, the duodenum. The amount absorbed is dependent on several factors, such as source, calcium: phosphorus ratio, intestinal pH, lactose intake, and dietary levels of calcium, phosphorus, vitamin D, iron, aluminum, manganese, potassium, and fat. As is the case for most nutrients, the greater need, the more efficient the absorption. Absorption increases, although not proportionally, with increased intake.

METABOLISM OF PHOSPHORUS. Phosphorus absorbed from the intestine is circulated through the body and is readily withdrawn from the blood for use by the bones and teeth during periods of growth. Some incorporation into the bones occurs at all ages. It may be withdrawn from bones to maintain normal blood plasma levels during periods of dietary deprivation.

The plasma phosphorus level (which generally falls within the range of 2.5 to 4.5 mg per 100 ml in adults—higher in children), along with calcium, is regulated by the parathyroid hormone and thyrocalcitonin (calcitonin) and is inversely related to the blood calcium level.

Phosphorus metabolism can be disturbed in many types of disease, notably those involving the kidneys and bone.

EXCRETION OF PHOSPHORUS. The kidneys provide the main excretory mechanization for regulation of the serum phosphorus level. All of the plasma inorganic phosphate is filtered through the renal glomeruli. If the serum phosphorus level falls, the renal tubules return more phosphorus to the blood; if the serum phosphorus level rises, the renal tubules excrete more. Also, when the diet lacks sufficient phosphorus, the renal tubules conserve phosphorus by returning it to the blood. On an average diet, an adult excretes from 0.6 to 1.0 g of phosphorus every 24 hours.

FUNCTIONS OF PHOSPHORUS. Among the mineral elements essential to life none plays a more central role than phosphorus. It is found throughout the body in every cell; and it is involved in almost all, if not all, metabolic reactions. The diverse and most important functions of phosphorus are given in the section on MINERAL(S), Table M-25, Mineral Table.

DEFICIENCY SYMPTOMS. Dietary deficiencies of phosphorus in man are unlikely, since this mineral is present in nearly all foods. However, deficiencies are seen in people in certain clinical conditions; in persons receiving excessive antacids over long periods, and in certain stress conditions such as bone fractures.

Vegetarian diets, especially those low in milk products, may be deficient in phosphorus as well as in other elements.

Severe and prolonged deficiencies of phosphorus may be manifested in rickets and phosphorus related diseases (see subsequent section headed "Phosphorus Related Diseases").

(Also see section on MINERAL[S], Table M-25, Mineral Table.)

INTERRELATIONSHIPS. Phosphorus is involved in certain relationships, the most important of which follow:

• **Calcium-phosphorus ratio and vitamin D**—Adequate phosphorus and calcium nutrition is dependent upon three factors: a sufficient supply of each element, a suitable ratio between them, and the presence of vitamin D. Generally, human nutritionists recommend a calcium-phosphorus ratio of 1.5:1 in infancy, decreasing to 1:1 at 1 year of age, and remaining at 1:1 throughout the rest of life; although they consider ratios between 2:1 and 1:2 as satisfactory. With plenty of vitamin D, the ratio becomes of less importance, and more efficient utilization is made of the amounts of the element present.

• **Excess of calcium over phosphorus; excess of phosphorus over calcium**—If the diet contains an excess of calcium over phosphorus (that is, more calcium than can be absorbed in the first part of the small intestine) free calcium will be present at the points where phosphorus is absorbed. This excess calcium will combine with the phosphorus to form insoluble tricalcium phosphate, and thus interfere with the absorption of phosphorus.

An excess of dietary phosphorus over calcium will in the same way decrease the absorption of both calcium and phosphorus.

• **Excess intakes of iron, aluminum, and magnesium**—Large intakes of iron, aluminum, and magnesium may bind phosphorus in insoluble salts and inhibit its absorption.

• **Phosphorus present as phytin or phytic acid**—There is evidence that man utilizes phytin phosphorus poorly—that phytin is incompletely broken down in the digestive tract. As a result, the undigested phytate may depress the absorption of calcium and iron.

RECOMMENDED DAILY ALLOWANCE OF PHOSPHORUS. The National Research Council recommended daily dietary allowances, with provision for individual variation, of phosphorus are given in the section on MINERAL(S), Table M-25, Mineral Table.

Note that the recommended daily allowances for phosphorus range from 300 mg to 1,200 mg. Note, too, that the allowances vary according to age, and that provision is made for added allowances for pregnant and lactating females.

The recommended allowance of phosphorus in milligrams per day is the same as that for calcium, except for the young infant.

• **Phosphorus intake in average U.S. diet**—The intake of phosphorus in ordinary diets is usually higher than that of calcium and is thought to be adequate.

In recent years, the widespread use of phosphate food additives (mainly sodium salts of orthophosphates, pyrophosphates, and polyphosphates) as acidifiers, emulsifiers, chelators, leavening agents, and water-binders has significantly increased the phosphorus content of the food supply. Such additives currently appear to contribute an additional 0.5 to 1.0 g phosphorus per day to the diet of U.S. adults.

The average available phosphorus for consumption per capita per day in the United States is approximately 1,540 mg, the primary sources of which are: dairy products, 34.3%; meat, poultry, and fish, 29.2%; and flour and cereal products, 14.1%. Thus, except for unusual circumstances, a deficiency in man is unlikely.

TOXICITY. There is no known phosphorus toxicity *per se*. However, excess phosphate consumption may cause hypocalcemia (a deficiency of calcium in the blood) and result in enhanced neuroexcitability, tetany, and convulsions.

PHOSPHORUS RELATED DISEASES. Phosphorus is present in nearly all foods, with the result that a dietary deficiency is extremely unlikely to occur in man. Intake of this mineral is almost always, if not invariably, higher than that of calcium and is entirely adequate under most conditions. However, phosphorus deficiencies may occur (1) when there is excess intake of nonabsorbable antacids, (2) when infants are on cow's milk, and (3) when vegetarians are on high fiber diets produced on phosphorus-deficient soils. Phosphorus related diseases include rickets and osteomalacia.

(Also see RICKETS; and OSTEOMALACIA.)

• **Phosphorus depletion due to prolonged and excessive intake of nonabsorbable antacids**—The frequency of this syndrome is unknown. It is characterized by weakness, anorexia (reduced appetite), and pain in the bones. Treatment consists in discontinuing antacids and providing adequate dietary phosphorus.

• **Hypocalcemic tetany of infants**—Where infants are raised on cow's milk, the Ca:P ratio in cow's milk may contribute to the occurrence of hypocalcemic tetany during the first week of life. This postulation is based on the fact that the Ca:P ratio in cow's milk is approximately 1.2:1, compared with 2:1 in human milk. Because of this situation, the current recommendation is that in infancy the Ca:P ratio in the diet be 1.3:1, decreasing to 1:1 at 1 year of age.

• **Phosphorus deficiency of vegetarians on a high fiber diet produced on phosphorus-deficient soils**—Presently, this is without proof in humans, but there is abundant evidence that this condition occurs in animals.

At Washington State University, in a study with rabbits, the effect of soil phosphorus on plants, and, in turn, the effects of these plants on animals, was established.[21] Generation after generation, rabbits were fed on alfalfa, with one group receiving hay produced on low-phosphorus soils and the other group eating alfalfa grown on high-phosphorus soils. The rabbits in the low-phosphorus soil–alfalfa group (1) were retarded in growth—with 9.8% lower weaning weights, (2) required 12% more matings per conception, and (3) had a 47% lower breaking strength of bones than the rabbits on the high-phosphorus soil–alfalfa group. There is reason to believe that a phosphorus deficiency can affect people similarly, and that some people are consuming suboptimum levels of phosphorus—especially vegetarians on a high fiber diet produced on phosphorus-deficient soils.

In addition to the conditions noted above, a phosphorus deficiency may also interfere with the utilization of vitamin A and, in growing children, teeth as well as bones will be less dense and less completely mineralized.

• **Excess phosphorus syndrome**—Excess phosphorus may also create problems. For example, an inhibitory effect of excess dietary phosphorus on bone development in domestic animals has long been recognized. Chronic ingestion of high phosphate diets causes secondary hyperparathyroidism, bone resorption, and, in some cases, calcification of kidney and heart tissue. The level of dietary phosphate required to produce these effects is dependent upon the concentrations of calcium, magnesium, and other ingredients. Excess phosphate also causes the effects of magnesium deficiency to be more severe and increases the intake of calcium required to maintain normocalcemia.

The demonstration that adult animals fed high phospho-

Fig. P-28. Rabbit with bowed legs and enlarged joints resulting from eating alfalfa produced on low-phosphorus soils. There is reason to believe that vegetarians on high fiber diets may suffer similar consequences. (Courtesy, Washington State University)

rus diets, even in the presence of normally adequate concentrations of calcium, undergo an enhanced rate of bone resorption and net loss of bone mass has raised the possibility that diets high in phosphorus, particularly those that are concomitantly low in calcium, may contribute to *aging bone loss* or of *osteoporosis* in man.

CALCIUM-PHOSPHORUS RATIO AND VITA-MIN D. The proportional relation that exists between calcium and phosphorus is known as the calcium-phosphorus ratio. When considering the calcium and phosphorus requirements, it is important to realize that the proper utilization of these minerals by the body is dependent upon (1) a suitable calcium-phosphorus ratio (Ca:P), (2) an adequate supply of calcium and phosphorus in an available form, and (3) sufficient vitamin D to make possible the assimilation and utilization of the calcium and phosphorus. Generally speaking, a rather wide variation in Ca:P ratio in the diet is tolerated. However, nutritionists generally recommend a calcium-phosphorus ratio of 1.3:1 in infancy, decreasing to 1:1 at 1 year of age, and remaining at 1:1 throughout the rest of life; although they consider ratios between 2:1 and 1:2 as satisfactory. If plenty of vitamin D is present (provided either in the diet or by sunlight), the ratio of calcium to phosphorus becomes less critical. Likewise, less vitamin D is needed when there is a desirable calcium-phosphorus ratio.

It is noteworthy that there is much evidence indicating that calcium-phosphorus ratios of 1:1 to 2:1 for nonruminants (hogs and horses) and 1:1 to 7:1 for ruminants are satisfactory; but that ratios below 1:1 are often disastrous.

The dietary Ca:P ratio is particularly important during the critical periods of life—for children, and for women during the latter half of pregnancy and during lactation.

[21]Heinemann, W. W., M. E. Ensminger, W. E. Ham, and J. E. Oldfield, "Phosphate Fertilization of Alfalfa and Some Effects on the Animal Body," *Wash. Ag. Exp. Sta., Tech. Bull. 24,* June 1957.

SOURCES OF PHOSPHORUS. Good sources are meat, poultry, fish, eggs, cheese, nuts, legumes, and whole-grain foods. It is noteworthy, however, that much of the phosphorus in cereal grains occurs in phytic acid, which combines with calcium to form an insoluble salt that is not absorbed. Vegetables and fruits are generally low in phosphorus.

Fig. P-29. T-bone steak. Meat is a good source of phosphorus. (Courtesy National Live Stock & Meat Board, Chicago, Ill.)

Groupings, by rank, of common sources of phosphorus are given in the section on MINERAL(S), Table M-25, Mineral Table.

For additional and more precise values of phosphorus, see Food Composition Table F-21.

(Also see MINERAL[S], Table M-25, Mineral Table; and CALCIUM.)

PHOSPHORYLATE

To introduce a phosphate grouping into an organic compound; for example, glucose monophosphate produced by the action of phosphorylase.

PHOSPHORYLATION

The chemical reaction in which a phosphate group is introduced into an organic compound.

PHOTOSYNTHESIS

Life on earth is dependent upon photosynthesis. Without it, there would be no oxygen, no plants, no food, and no people.

As fossil fuels (coal, oil, shale, and petroleum)—the stored photosynthates of previous millennia—become exhausted, the biblical statement, "all flesh is grass" (Isaiah 40:6), comes alive again. The focus is on photosynthesis. Plants, using solar energy, are by far the most important, and the only renewable, energy-producing method; the only basic food-manufacturing process in the world; and the only major

Fig. P-30. Photosynthesis. With their tops in the sun and their roots in the soil, plants are green factories producidng energy. (Courtesy, University of Maryland, College Park, Md.)

source of oxygen in the earth's atmosphere. Even the chemical and electrical energy used in the brain cells of man are the products of sunlight and the chlorophyll of green plants. Thus, in an era of world food shortages, it is inevitable that the entrapment of solar energy through photosynthesis will, in the long run, prove more valuable than all the underground fossil fuels—for when the latter are gone, they are gone forever. So, all food comes directly or indirectly from plants which have their tops in the sun and their roots in the soil. Hence, we have the nutrition cycle as a whole—from the sun and soil, through the plant, thence to animals and people, and back to the soil again. In summary form, the story is—

$$\text{Green factories} + \text{4-stomached animals} \xrightarrow[\text{(store)}]{\text{(convert)}} \text{energy}$$
$$\text{(Plants)}$$

Technically, green factories are known as photosynthetic plants, and 4-stomached animals are called ruminants; together, they convert and store energy in a form available to man. Not only that! They are relatively free from political control; they are universally available on a renewable basis; and, properly managed, they enhance the quality of the environment. Besides, they're free of cancer-laden carcinogens.

The energy required by every living creature, man included, is derived from the photosynthetic process occurring in green plants in which light energy from the sun is converted to chemical energy, then trapped in newly made sugar molecules.[22] Ultimately, all the energy in our food comes from this source.

---

[22]Certain types of microorganisms, termed chemoautotrophs, get their energy from inorganic compounds, but aside from this minor exception, the energy that runs the life support system of the biosphere comes from photosynthesis.

The word photosynthesis means putting together with light. More precisely, photosynthesis may be defined as *the process by which the chlorophyll-containing cells in green plants capture the energy of the sun and convert it into chemical energy; it's the action through which plants synthesize and store organic compounds, especially carbohydrates, from inorganic compounds—carbon dioxide, water, and minerals, with the simultaneous release of oxygen.*

Photosynthesis is dependent upon the presence of chlorophyll, a green pigment which develops in plants soon after they emerge from the soil. Chlorophyll is a chemical catalyst —it stimulates and makes possible certain chemical reactions without becoming involved in the reaction itself. By drawing upon the energy of the sun, it can convert inorganic molecules, carbon dioxide (CO_2) and water (H_2O), into an energy-rich organic molecule such as glucose ($C_6H_{12}O_6$), and at the same time release free oxygen (O_2). It transforms solar energy into a form that can be used by plants, animals, and man. Because of this capability, chlorophyll has been referred to as the link between nonliving and living matter, or the pathway through which nonliving elements may become part of living matter.

Through the photosynthetic process, it is estimated that more than a billion tons of carbon per day are converted from inorganic carbon dioxide (CO_2) to organic sugars ($C_6H_{12}O_6$—glucose), which can then be converted into carbohydrates, fats, and proteins—the three main groups of organic materials of living matter.

Photosynthesis is a series of many complex chemical reactions, involving the following two stages (see Fig. P-31):

Stage 1—The water molecule (H_2O) is split into hydrogen (H) and oxygen (O); and oxygen is released into the atmosphere. Hydrogen is combined with certain organic compounds to keep it available for use in the second step of photosynthesis. Chlorophyll and light are involved in this stage.

Stage 2—Carbon dioxide (CO_2) combines with the released hydrogen to form the simple sugars and water. This reaction is energized (powered) by ATP (adenosine triphosphate), a stored source of energy. Neither chlorophyll nor light is involved in this stage.

The chemical reactions through which chlorophyll converts the energy of solar light to energy in organic compounds is one of nature's best-kept secrets. Man has not been able to unlock it, as he has so many of life's other processes. Moreover, photosynthesis is limited to plants; animals store energy in their products—meat, milk, and eggs—but they must depend upon plants to manufacture it.

Although photosynthesis is vital to life itself, it is very inefficient in capturing the potentially available energy. Of the energy that leaves the sun in a path toward the earth, only about half ever reaches the ground. The other half is absorbed or reflected in the atmosphere. Most of that which reaches the ground is dissipated immediately as heat or is used to evaporate water in another important process for making life possible. Only about 2% of the earthbound energy from the sun actually reaches green plants, and only half of this amount (1%) is transformed by photosynthesis to energy storage in organic compounds. Moreover, only 5% of this plant-captured energy is fixed in a form suitable as food for man.

With such a small portion of the potentially useful solar energy actually being used to form plant tissue, it would appear that some better understanding of the action of chlorophyll should make it possible to increase the effectiveness of the process. Three approaches are suggested: (1)

PHOTOSYNTHESIS

LIGHT REACTIONS:

WHERE: IN GREEN MESOPHYLL OF LEAVES.
REQUIRES: CHLOROPHYLL ENERGY FROM SUNLIGHT.
PRODUCES: ENERGY-RICH ATP.

DARK REACTIONS:

WHERE: IN STROMA OF CHLOROPLAST OF LEAVES.
REQUIRES: DOES NOT REQUIRE LIGHT OR CHLOROPHYLL, BUT REQUIRES ENERGY-RICH ATP.
PRODUCES: GLUCOSE-ENERGY-RICH CARBON COMPOUNDS.

Fig. P-31. Photosynthesis fixes energy. Diagrammatic summary of (1) photosynthesis, and (2) the metabolic formation of organic compounds from the simple sugars. This diagram shows the following:

1. Carbon dioxide gas from the air enters the green mesophyll cells of plant leaves.
2. Plants take up oxygen from the air for some of their metabolic processes and release oxygen back to the air from other metabolic processes.
3. Plants take up water and essential elements from the soil.
4. The energy essential to photosynthesis is absorbed by chlorophyll and supplied by sunlight.
5. For a net input of 6 molecules of carbon dioxide and 6 molecules of water, there is a net output of 1 molecule of sugar and 6 molecules of oxygen.
6. The process is divided into light and dark reactions, with the light reactions building up the energy-rich ATP required for the dark reactions.
7. In the process, 673 Calories (kcal) of energy are used.
8. The sugar (glucose) manufactured in photosynthesis may be converted into fats and oils, sugars and other carbohydrates, and amino acids and proteins.

increasing the amount of photosynthesis on earth, (2) manipulating plants for increased efficiency of solar energy conversion, and (3) converting a greater percentage of total energy fixed as chemical energy in plants (the other 95%) into a form available to man. Ruminant (4-stomached) animals are the solution to the latter approach: they can convert energy from such humanly inedible plant materials as grains and other high-energy feeds and protein supplements, crop residues, pasture and range forages, and harvested forages into food for humans (see Fig. P-32). Also, it is noteworthy that animals do not require fuel to graze the land and recover the energy that is stored in the grass. Moreover, they are completely recyclable; they produce a new crop each year and perpetuate themselves through their offspring. It would appear, therefore, that there is more potential for solving the future food problems of the world by manipulating plants for increased solar energy conversion (*genetic engineering*) and by using ruminants to make more plant energy available to man than from all the genetic and cultural methods combined.

(Also see ENERGY REQUIRED FOR FOOD PRODUCTION.)

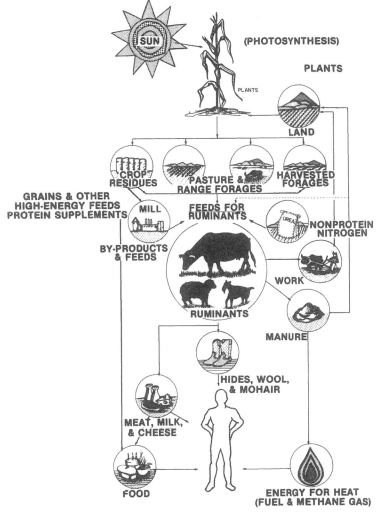

Fig. P-32. Ruminants step up energy. Their feed comes from plants which have their tops in the sun and their roots in the soil. Hence, we have the nutrition cycle as a whole—from the sun and soil, through the plant, thence to the ruminant (and man) and back to the soil again.

PHYLLOQUINONE

A form of vitamin K—K_1, which occurs in nature and is fat soluble.

(Also see VITAMIN K.)

PHYSICAL ACTIVITIES

The need for continuous, quick and strenuous movement of the body—physical activity—has declined over the years in our society. From youth to old age, we now spend most of our time engaged in moderate to very light activities. However, there has been a heightened interest in the need for physical activity coupled with sound nutrition as the two chief components of health.

The following major points indicate the need for physical activities:

1. Physical activity more than any other body process raises the body's energy requirements, and thus provides a means of weight maintenance and weight reduction.

2. Muscular tone and strength are increased by physical activity. This includes the heart and blood vessels.

3. A psychological benefit is derived from physical activities; it provides an outlet for tension and restlessness, and gives the individual a sense of well-being.

(Also see CALORIC [ENERGY] EXPENDITURE; and PHYSICAL FITNESS AND NUTRITION.)

PHYSICAL FITNESS AND NUTRITION

Contents

Physical Fitness Around the World . . . . Page 845
Important Aspects of Physical Fitness 845
  Ability to Perform Physical Work 845
  Athletic Prowess 845
  Bearing of Healthy Children 845
  Emotional Well-Being 845
  Mental Acuity 845
  Resistance to Infectious Diseases 846
  Slowing of the Physical Decline in Adulthood 846
Dietary Guidelines 846
  Modified Diets and Nutritional Supplements 846
Physical Activity as an Adjunct to Dietary Practices . . . 847
Summary 847

Many people equate fitness with the ability to perform either athletically or in feats of endurance and strength. However, health professionals consider it to be much broader in scope and to include functions such as (1) performance of work and other physical activities, (2) resistance to degenerative and infectious diseases, and (3) adaptation in beneficial ways to emotional and physical stresses.

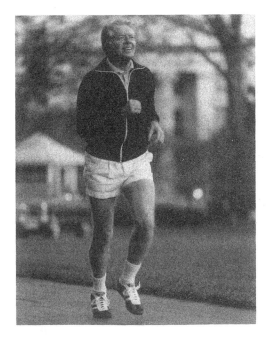

Fig. P-33. When he was president, Jimmy Carter kept fit by jogging. (Courtesy, Karl Schumacher of the Carter White House)

The treatment of nutrition as it relates to athletics is presented in the section on Athletics and Nutrition; hence, the reader is referred thereto.

PHYSICAL FITNESS AROUND THE WORLD.
Some well-known examples of exceptionally fit people, based on longevity and buoyant good health, in various parts of the world are:

1. The Ecuadorians of Vilcabamba.
2. The Georgians of the Soviet Union.
3. The Hunzas of Pakistan.
4. The Bulgarians.
5. The longshoremen in San Francisco.
6. The Masai tribesmen in Africa.
7. The Mormons.
8. The Seventh-Day Adventists.

Additional fit people, along with their cultural characteristics, follow:

1. **The Canadian Indians,** hunting and fishing tribes, in northern Canada near the Arctic Circle, whose physical fitness is attributed to a nutritious diet of meat and fish.
2. **The Eskimos of the Arctic,** who still live by hunting and fishing, are noted for a high aerobic capacity.
3. **The Polish farmers,** who live in the low mountains of southern Poland, enjoy a favorable climate, work hard, and are well fed, retain their fitness much longer than other Poles.
4. **The Swiss dairy farmers,** who consume considerable milk and cheese, are strong and healthy; many of the Pope's Swiss Guards come from this region.
5. **The Tarahumara Indians,** who live in the Madre mountains of northwestern Mexico, and whose diet is mainly vegetarian, are exceptionally good runners and have unusually high aerobic capacities.

Descriptions and life styles of each of the above groups are given in this book in the section on Life Expectancy and Nutrition; hence, the reader is referred thereto.

It is noteworthy that the health practices of some of the fittest groups of people around the world are characterized by (1) diets comprised of fresh, minimally processed foods, such as meats and fish, dairy products, fruits and vegetables, and whole grain breads and cereals; (2) limited consumptions of potentially harmful items such as strong alcoholic drinks, coffee and tea, and refined fats, starches, and sugars; and (3) engagement in moderate to vigorous physical activities.

IMPORTANT ASPECTS OF PHYSICAL FITNESS.
Although good overall fitness is to be desired by everyone, the particular circumstances that are unique to each person's life make it necessary to emphasize certain aspects, such as those which follow:

Ability to Perform Physical Work.
Certain types of jobs, such as construction work, firefighting, the harvesting of crops, military combat, and police work, require considerable physical exertion. Therefore, it is noteworthy that both the maximum rates of performing work and the times during which maximum performance may be sustained tend to decline with increasing age.

Athletic Prowess.
Many sports require agility, speed, and sufficient strength to make the required movements without excessive strain on the various organs and tissues. Therefore, the athletes who perform best owe their success to factors such as (1) development of the requisite neuro-muscular coordination, (2) optimal cardiovascular functioning, (3) a high aerobic capacity, (4) well developed muscles, (5) a minimal accumulation of excess body fat, and (6) willpower to persevere.

It is generally believed that optimal nutrition somehow enhances these performance factors although hard data is often difficult to obtain since the effects of mediocre nutrition may not be evident in early adulthood.

(Also see ATHLETICS AND NUTRITION.)

Bearing of Healthy Children.
Even primitive tribes have long recognized the importance of the maternal diet to the successful outcome of pregnancy, as evidenced by the practices which follow:

1. Eskimos, Peruvian Indians, Polynesians, and other peoples living near the sea or inland lakes gave expectant mothers fish eggs and other aquatic foods.
2. Dairying tribes in Africa did not allow young women to marry until the season when the cattle were fed on lush, green pastures.
3. Papuans of New Guinea reserved special foods such as dried large snakes for their expectant mothers.

Fortunately, mothers-to-be may now skip the fish eggs and snakes since much is known about meeting the nutritional requirements with other types of foods.

(Also see PREGNANCY AND LACTATION NUTRITION.)

Emotional Well-Being.
For many people, peace of mind depends upon their feeling well, since symptoms of an illness may provoke anxiety, depression, and other distressful emotions. Recent research on the effects of diet on the nervous system has shown that mental depression may also arise from a subnormal brain content of the nerve transmitter, serotonin, which in turn may result from (1) lack of sufficient amounts of the amino acid tryptophan in the diet, or (2) excessively high levels of dietary fat and/or protein that interfere with the entry of tryptophan into the brain where it is converted to serotonin. Another factor which affects the emotional state and sense of well being is the level of blood sugar, since the brain requires a continuous supply of glucose for optimal functioning, and subnormal blood sugar makes many people feel poorly. Hence, erratic and imbalanced dietary patterns may be responsible for poor mental health.

In general, people who are physically fit appear to be more optimistic and more able to accomplish both mental and physical tasks. Hence, they may have a greater sense of self-sufficiency than those who have let themselves become weak, flabby, and overly dependent on others.

Sometimes, the emotional well-being and ability to work are also impaired by chronic disorders such as adrenal cortical insufficiency, allergies, anemia, depression, high or low blood pressure, hypochondriasis (an exaggerated sensitivity to minor discomfort that may be accompanied by unfounded fear of developing a deadly disease), insufficient flow of blood to the brain, nutritional deficiencies, and thyroid diseases. Therefore, a person who feels poorly much of the time should see a doctor to rule out the possibility of a chronic disease before resorting to the excessive use of stimulants such as caffeine and various mood elevating drugs.

Mental Acuity.
This type of fitness is closely related to overall physical fitness because the brain and nerves are very sensitive to changes in the blood chemistry. For example, the performance of tasks which require mental and sensory acuity is enhanced by meal patterns which maintain

the blood sugar level at normal throughout the day. Skipping breakfast or having only coffee in the morning usually lowers the quality of work. Other dietary factors, particularly those relating to depression and emotional well being are also important.

Notable changes in a person's mental acuity may result from conditions or factors such as alcoholism, antihypertensive drugs, atherosclerosis, chronic stress, dehydration, depression, diabetes, fatigue, head injuries, insuffient blood flow to the brain, low blood pressure, low blood sugar (hypoglycemia), nutritional deficiencies, sedative or tranquilizer drugs, small strokes, and starvation. Sometimes these possibilities have been overlooked and the patient has been diagnosed as neurotic, psychotic, or senile and confined to an institution. However, some amazing improvements have resulted from the proper types of corrective measures.

Resistance to Infectious Diseases.

Both undernutrition and overnutrition impair the functioning of the body defenses against infection which are collectively called the immune system. Similarly, immunity may be reduced in disorders such as cancer and diabetes; and during severe emotional stress. Finally, some people fail to obtain the immunizations which are now available.

Scientists are now exploring the interactions between these and other factors that affect immunity. For example, there is evidence that the production of optimal immunity requires ample amounts of dietary selenium, zinc, vitamin A, and vitamin C; but only moderate amounts of dietary calories and proteins. Too little or too much of the latter nutrients may impair the functioning of the body's defense system. Hence, it should be possible in the near future to promote optimal functioning of the immune system by a balanced program of dietary measures, immunizations, medical treatment, and psychological counseling.

Slowing of the Physical Decline in Adulthood.

Mental and physical faculties must be used fully, but wisely, if they are to be kept operative at high levels of performance throughout adulthood. Furthermore, a vigorous life style helps to prevent obesity, which may hasten physical decline if not corrected soon enough. Finally, the progress of the various degenerative diseases that impair the faculties can be retarded by the proper diet and the use of corrective therapies when needed. This means that regular physical examinations and laboratory tests should be obtained to detect any early warning symptoms of degenerative diseases. It might also be a good idea to have tests of physical performance such as aerobic capacity and endurance.

Some of the symptoms that warrant dietary modification and/or other types of therapies are: diabetic glucose tolerance; high blood levels of cholesterol, triglycerides, and/or uric acid; low blood levels of hemoglobin, iron, red cells, and/or certain vitamins; and subnormal capacity for physical exertion, as evidenced by an electrocardiogram or stress testing on an exercise bicycle or a treadmill. Means of correcting these problems are described in the sections that follow.

DIETARY GUIDELINES.

In order to be certain of obtaining sufficient amounts of each of the required nutrients, one should consult a table showing the best food sources of these essential substances, such as the one given in Table N-3 of this encyclopedia.

Furthermore, meals can be planned with the aid of a food guide that groups commonly used items according to the nutrients which are furnished (see Food Groups).

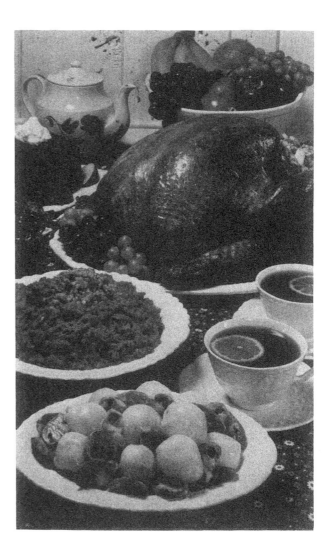

Fig. P-34. A wholesome diet, along with sufficient exercise, makes for physical fitness. (Courtesy, United Fresh Fruit & Vegetable Assn., Alexandria, Va.)

Finally, it appears that it would be beneficial to the fitness of many Americans to eat more food for breakfast and less at the evening meal. For example, breakfast should contain one or more sources of animal protein such as cheese, eggs, milk products, and meats, plus ample amounts of fresh fruit and whole grain breads or cereals. Lunch should also contain a good source of proteins, plus a milk product, one or more vegetables and whole grain bread. When two good meals have already been consumed, supper may consist of only small servings of protein foods with a bread or a cereal product, plus a large serving of vegetables in a salad or a freshly prepared soup. (Most canned soups have rather high salt contents, which may be harmful to people with tendencies to retain fluids and develop congestive heart failure or high blood pressure.)

(Also see FOOD GROUPS; NUTRIENTS: REQUIREMENTS, ALLOWANCES, FUNCTIONS, SOURCES; and RDA.)

Modified Diets and Nutritional Supplements.

The decision to make a radical change in one's dietary pattern should not be made without first consulting a doctor and/or a dietitian because many factors have to be considered before the appropriate changes can be made. For example, it may be dangerous for people with a family history of gout to lose weight rapidly.

PHYSICAL ACTIVITY AS AN ADJUNCT TO DIETARY PRACTICES. The need for following a carefully controlled diet is lessened somewhat by exercising enough to burn off at least 300 to 500 extra Calories (kcal) daily. Fig. P-35 shows that women who engaged in recreational physical activities were able to keep trim while consuming more calories than those whose activity levels were lower.

* AVERAGES OF REPORTED RECREATIONAL ACTIVITY
  (LITTLE OR NONE = 1, MODERATE = 2, MUCH = 3)

** ASSESSMENT OF BODY FATNESS FROM AVERAGE
   MEASUREMENTS OF TRICEPS SKINFOLD
   (13.5 MM = SLENDER, 20.8 MM = MEDIUM,
   30.8 MM = HEAVY)

Fig. P-35. The effects of caloric intake and recreational exercise on a measure of body fatness in women. (Based on data from Habicht, J-P, testimony before the U.S. House of Representatives Committee on Science and Technology, July 28, 1977, p. 376, Fig. 2)

Physical activity is also valuable as a means of rehabilitation from various degenerative conditions. For example, the cardiovascular fitness of seven patients with heart disease was improved greatly by a 6-week program of intensive physical training that was supervised by doctors.[23]

(Also see CALORIC [ENERGY] EXPENDITURE; and ATHLETICS AND NUTRITION.)

SUMMARY. Studies around the world have shown that most people can remain fit well into middle age and beyond, providing that they consume a wholesome diet and engage in sufficient exercise. In fact, it is even possible to have a dramatic recovery from disabling heart disease when suitable measures are taken.

PHYSIOLOGICAL SALINE

A salt solution (0.9% NaCl) having the same osmotic pressure as the blood plasma.

PHYTIC ACID

It is a hexaphosphoric acid ester of inositol present mainly in cereal grains, nuts, and legumes.

(Also see MINERAL[S], section headed "Naturally Occurring Mineral Antagonists in Foods.")

---

[23]Redwood, D. R., et al., "Circulatory and Symptomatic Effects of Physical Training in Patients with Coronary Artery Disease and Angina Pectoris," The New England Journal of Medicine, Vol. 286, 1972, pp. 959-965.

PHYTOCHEMICALS

Phytochemical ("phyto" is derived from the Greek word for plant) is a new term embracing a host of plant chemicals which offer hope and help for the prevention of cancer. They are abundant in fruits and vegetables. In the mid-1990s, the race was on to find, isolate, and study them; hopefully, to find a magic pill that would go beyond vitamins.

Among the multitude of phytochemicals, their sources, and their claims, are the following:

• **Allylic sulfide**—Plentiful in onions and garlic, which detoxifies carcinogens.

• **Capsaicin**—Abundant in hot peppers, which keeps toxic molecules from attaching to DNA.

• **Ellagic acid**—Present in strawberries, grapes, and raspberries, which neutralizes carcinogens before they can invade DNA.

• **Flavonoids**—Present in berries, citrus, yams, and cucumbers, which keeps cancer-causing hormones from attaching to the surface of cells.

• **Genistein**—Abundant in soybeans, which prevents breast, prostate, and other lumps from growing and spreading.

• **Indole-3-carbinol**—Present in cauliflower and cabbage, seems to lessen breast cancer by acting on a precursor to the female hormone estrogen and producing a harmless form of estrogen rather than the form linked to breast cancer.

• **P-coumaric acid and chlorogenic acid**—Found in many fruits and vegetables, including tomatoes, green peppers, pineapples, strawberries, and carrots, which prevents the formation of cancer-causing nitrosamines from nitric acids and amines.

• **Phenethyl isothiocyanate (PEITC)**—Found in cabbage and turnips, which inhibits lung cancer.

• **Sulforaphane**—Found in broccoli, cauliflower, Brussels sprouts, turnips, and kale, which lessens breast cancer in animals.

But phytochemicals are not omnipotent! Remember, too, that lifelong vegetarians get cancer. Much more research is needed before the role of phytochemicals can be established. In the meantime, a balanced diet, including fruits and vegetables, is recommended for buoyant good health.

PICA

An abnormal craving of appetite for substances not usually considered as food—starch, clay, hair, dirt, chalk, and wood.
(Also see ALLOPRIOPHAGY; CISSA; and GEOPHAGIA.)

PIGMENTATION

This refers to melanin, the pigment which imparts color (1) to the skin, hair, and eyes, and (2) to bile and blood.

In sunburn, the skin darkens due to the ultraviolet rays of the sun. In albinos, melanin is absent. Certain skin diseases manifest excessive or unevenly distributed pigment, including freckles, liver spots, melanosis, vitiligo, and leukoderma.

Bile and blood color are due to melanin formed in the body. The two main bile pigments are bilirubin (red) and biliverdin (green). The red blood pigment is hemoglobin, from which the pigments hemosiderin, methemoglobin, bilirubin, and biliverdin are derived.

PIG SOUSE (PIG'S-FOOT JELL)

Made from pig's feet (free from hair and with toes removed), along with hearts, tongues, and shoulder hocks, by boiling in water until the meat separates from the bones;

removing the bones; cutting the meat into small chunks; placing the meat back in the broth in which it was cooked; seasoning to taste with salt, pepper, and vinegar; pouring into shallow pans and setting away to chill and jell.

PINEAPPLE *Ananas comosus*

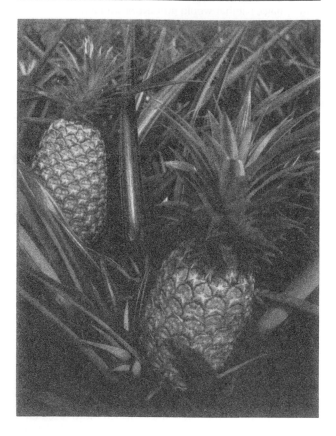

Fig. P-36. Pineapples ripening in the field. (Courtesy, Castle & Cooke, Inc.)

Pineapple is a tropical fruit plant, which grows from 2 to 3 ft (*61 to 91 cm*) tall, and produces a fruit weighing 4 to 8 lb (*2 to 4 kg*). It probably received its name because the fruit looks like a large pinecone. The pineapple is a member of the bromelaid family, *Bromeliaceae*.

The edible flesh of the pineapple is firm, pale yellow or white, and the flavor is a mixture of sweetness and tartness. Its flavor has been described as apple, strawberry, and peach mingled all at once.

ORIGIN AND HISTORY.

The pineapple is native to South America, where it was domesticated by the Indians long before the coming of the white man to the New World.

WORLD AND U.S. PRODUCTION.

Worldwide, about 9.7 million tons of pineapples are produced annually; and the leading pineapple-producing countries, by rank, are: Thailand, Philippines, China, Brazil, India, United States, Vietnam, Mexico, Indonesia, and South Africa.[24]

The United States ranks sixth among the leading pineapple-producing nations, with an annual production of 575,000 short tons. But pineapples are produced only in Hawaii.[25]

PROCESSING.

A considerable amount of pineapple is consumed fresh. However, most exports are canned. Every part of a pineapple is used in the canning process. The flesh is canned as sliced, chunked or crushed pineapple. The juice is also canned.

At the cannery, pineapples are first washed and graded. Once graded, a machine called a *Ginaca* cuts the core from the fruit, removes the shell, and cuts off both ends. The flesh left on the shell is then scraped by machine so that it may be used for crushed pineapple and juice.

Fig. P-37. Processing pineapple, showing the fruit being sliced mechanically and placed in cans. (Courtesy, Castle & Cooke, Inc.)

The trimmed fruit is carefully inspected, sliced by machine, and placed in cans. Broken pieces are sent to the shredder to be crushed. The shells and other trimmings are shredded, pressed, and dried to make pineapple bran to feed livestock. Other by-products of the canning process are sugar, alcohol, and vinegar. Sugar is recovered by ion exchange purification of the juice pressed from the cannery waste. However, alcohol and vinegar, instead of sugar, may be produced from the by-product juice through fermentation.

---

[24]Data from *FAO Production Yearbook 1990*, FAO/UN, Rome, Italy, Vol. 44, p. 167, Table 73. **Note well:** Annual production fluctuates as a result of weather and profitability of the crop.

[25]Data from *Agricultural Statistics 1991*, USDA, p. 177, Table 254. **Note well:** Annual production fluctuates as a result of weather and profitability of the crop.

• **Bromelin**—Another by-product in some operations is the enzyme, bromelin, which is used to tenderize meat, to chill-proof beer, and to produce protein hydrolyzates. Bromelin is found in the stem and fruit of the pineapple.

(Also see MEAT, section headed "Meat Tenderizing."

SELECTION, PREPARATION AND USES. Pineapple varieties vary in color, but most are yellowish. Ripe pineapples have a fruity, fragrant aroma. Usually the heavier the fruit is for its size, the better the quality.

Pineapple may be served fresh or canned, by itself, or it may be used in pies, ice cream, puddings, baked goods, and salads, or with meats. The flavor blends well with other flavors. Cookbooks contain long lists of ideas for the uses

Fig. P-38. A fruit salad attractively served in a pineapple shell. (Courtesy, Anderson/Miller & Hubbard Consumer Services, San Francisco, Calif.)

of pineapple. The fresh fruit cannot be used in molded gelatin, because it contains the enzyme bromelin that digests and softens the gelatin. This enzyme is destroyed by heating; hence, canned pineapple can be used in gelatin dishes.

NUTRITIONAL VALUE. Fresh pineapple is 85% water and each 100 g (about 3½ oz) contains 58 Calories (kcal) as well as 70 IU of vitamin A and 17 mg of vitamin C. The calories in pineapples are derived primarily from the sugar (carbohydrate) present. They contain approximately 15% sugar. Canned and frozen pineapples are similar to fresh pineapples, but those canned in a heavy syrup contain more calories, and some vitamin A and C is lost during the processing. The nutrient composition of pineapple juice is very similar to the fresh fruit.

Further information regarding the nutritional value of fresh, canned, frozen, and candied pineapple, and pineapple juice is presented in Food Composition Table F-21.

PINEAPPLE GUAVA (FEIJOA) *Feijoa sellowiana*

The fruit of a small South American tree (of the family *Myrtaceae*) that is related to the common guava (*Psidium guajava*).

Feijoa is green outside and white inside, and has a tart sweet flavor somewhat like unsweetened pineapple. As a fresh fruit, it is good in fruit salads; or when made into jam.

PINEAPPLE JUICE

The fluid extracted from fresh pineapple by pressing the pulverized fruit pulp, filtering the juice, adding back some of the pulp, heat processing at 140-145°F (60-63°C), pressure homogenizing to keep the pulp suspended, filling into cans, and pasteurizing at 190°F (88°C).

The nutrient compositions of various forms of pineapple juice are given in Food Composition Table F-21.

Some noteworthy observations regarding the nutrient composition of pineapple juice follow:

1. A 1-cup (*180 g*) serving of canned pineapple juice supplies 138 Calories (kcal) and is a good source of potassium and a fair to good source of vitamin C. However, some pineapple drinks and juice products contain sufficient added vitamin C to make them much better sources of this vitamin.

2. Frozen pineapple juice concentrate contains over 3 times the caloric and nutrient contents of the single strength canned juice. Hence, it may make a significant nutritional contribution when added undiluted to dishes such as ice cream and sauces for ham and other meats.

(Also see PINEAPPLE.)

PINOCYTOSIS

The process by which liquid droplets are taken into the cell through invagination and subsequent dissolving of part of the cell membrane. This process enables cells to absorb certain lipids and dissolve proteins intact. As can be seen in Fig. P-39, the material to be engulfed comes into contact with the cell membrane. The membrane then invaginates to surround the material. Once the material has been completely surrounded, the membrane fuses, and the invaginated section of the membrane is dissolved by lysosomal enzymes. Absorptive cells in the small intestine are capable of using this mechanism during digestion.

(Also see DIGESTION AND ABSORPTION.)

CELL
MEMBRANE

Fig. P-39. Pinocytosis involves the invagination of the cell membrane around the liquid to be engulfed. Once the liquid droplet has been completely engulfed, the membrane surrounding the material disintegrates and the material is incorporated into the cell.

PINTAIL (PINTAILED DUCK)

A river duck—a game bird.

PITH

In botany, this word has the following meanings:

• Soft white spongy material found in the center of certain plant stems. It is not usually consumed as a food.

• Another name for the albedo of citrus fruits, which is the white fibrous matter that lies between the fruit segments and the outer pigmented portion of the peel (flavedo). Citrus pith is rich in (1) bioflavonoids, and (2) pectin, a gelling agent and an ingredient in diarrhea remedies.

PITS

Stones or seeds of fruits such as cherries, plums, peaches, apricots, dates, or similar fruits.

PITUITARY GLAND

An endocrine gland, located at the base of the brain, which produces a number of hormones and regulates a large portion of the endocrine activity.

(Also see ENDOCRINE GLANDS.)

PIZZA

This is an Italian word meaning *pie*, which a pizza resembles. Pizza, the creation of which is credited to Naples, is flat leavened bread topped with a variety of such good things as tomatoes, mozzarella and parmesan cheeses,

Fig. P-40. Pizza, a popular fast food in America. (Courtesy, USDA)

sausages, mushrooms, anchovies, olives, capers, and a profusion of herbs and spices. A pizza is actually an open sandwich.

Although pizza varies nutritionally according to the ingredients, typical pizza contains about 15% protein, 27% fat, and 58% carbohydrate; hence, it's a fairly well balanced food.

(Also see CONVENIENCE AND FAST FOODS; and FAST FOODS.)

PLACEBO

An inert preparation given for its psychological effect; especially (1) to satisfy the patient, or (2) to serve as a control in an experiment.

PLACENTA

The organ of communication between the fetus and the mother. It is the means by which the fetal blood receives oxygen and nourishment and gives up waste material; hence, the placenta performs the functions of respiration, nutrition, and excretion.

PLANTAIN *Musa paradisiaca*

Plantains belong to the banana family, *Musaceae*, which also includes bananas. The plantain is *M. paradisiaca,* while the common banana is *M. paradisiaca,* variety *sapientum.* Although plantains look like bananas, they have a higher starch content and a lower sugar content than bananas.

ORIGIN AND HISTORY.   Plantains originated on the Malay Peninsula.

WORLD AND U.S. PRODUCTION. Statistically, plantains are never reliably distinguished from bananas. Nevertheless, they are an important fruit crop. World production totals 25,433 million metric tons; and the leading plantain-producing countries are: Uganda, Colombia, Rwanda, Zaire, Tanzania, Nigeria, Cameroon, and Cote Divoire.[26]

Plantain is not produced commercially in the United States.

PROCESSING. Plantains are processed much like bananas. However, they are not likely to be ripened after picking.

SELECTION, PREPARATION, AND USES. When purchased, plantains should be firm and free from bruises or other injury.

Because of their high starch content, it is necessary that plantains be cooked before eating. They are usually baked, boiled, fried, roasted, steamed, or made into a flour.

Plantains are generally used as a vegetable dish.

NUTRITIONAL VALUE.   The nutrient composition of plantains is given in Food Composition Table F-21.

Plantains are high in carbohydrates (over 30%); and they contain more energy than bananas (more than 115 Calories [kcal] per 100 g versus 85 to 100 Calories [kcal] per 100 g for bananas). Also, they are an excellent source of potassium (385 mg/100 g), and a fair source of vitamin C.

(Also see BANANA.)

PLANT PROTEINS

Those proteins derived from plant sources. This consists of proteins from the seeds, nuts, fruits, vegetables, leaves, stalks, and roots. Common sources of plant proteins are the cereal grains and their products, and legumes, oil seeds, and green leafy vegetables. Plant proteins are less concentrated than animal proteins, and often lack some essential amino acids giving them a lower biological value.

(Also see PROTEIN[S].)

PLAQUE

Any patch; for example, atherosclerotic plaque refers to a deposit of lipid material in the blood vessel.

PLASMA

The colorless fluid portion of the blood in which corpuscles are suspended. It is often used as a basis for measurement of bloodborne nutrients and their metabolites.

PLATELETS

Small, colorless, round or rod-shaped bodies in the circulating blood of mammals.

PLUCK

• The organs that lie in the thoracic cavity, consisting of the heart, lungs, gullet, and windpipe.

• To remove the feathers of poultry before cleaning.

---

[26]Data from *FAO Production Yearbook 1990,* FAO/UN, Rome, Italy, Vol. 44, p. 169, Table 74. **Note well:** Annual production fluctuates as a result of weather and profitability of the crop.

PLUM AND PRUNE *Prunus* spp

The plum is the fruit of a tree bearing the same name. Botanically, a plum is classified as a drupe—a fruit whose seed is contained in a hard pit or stone surrounded by soft, pulpy flesh with a thin skin. Plums belong to the rose family, *Rosaceae*. As the genus name, *Prunus*, suggests, plums are close relatives of the apricot, almond, cherry, and peach.

Plums vary in size from those as small as a cherry to those as large as a small peach. Their shape may be round or oval, and their smooth skin may be colored green, yellow, red, blue, or purple. The edible flesh is thick, juicy, and sweet.

Plum trees range from low, shrubby type to trees 30 ft (9 m) high. Its white flowers blossom in the spring before the leaves appear. Since there are many species and hundreds of varieties, plums enjoy widespread distribution in different climates and soil conditions.

Fig. P-41. The fruiting branch of a plum tree. (Courtesy, USDA)

• **Prune**—A prune is a dried plum, but more than this a prune is a variety of plum that is satisfactory for drying without removing the pit. Growers use the term prune to refer to both the fruit in its fresh state and to the dried product. Also the following saying helps make the distinction between plum and prune: "The prune is always a plum—but a plum is not always a prune." Prune plums generally have firmer flesh and a higher sugar content. Popular varieties of the European plum, *Prunus domestica*, used for

Fig. P-42. A prune (right) is a dried plum (left), but a plum is not always a prune.

prunes include: French or Agen (most popular), Imperial, Sugar, Robe de Sergeant, Burton, and Contes.

TYPES OF PLUMS. There are five main types of plums from which the important varieties are derived. These five types include the following: European, Japanese, American, damson, and ornamental.

• **European plum**—This plum is also called the garden or common plum. Its scientific name is *Prunus domestica*. Its fruits are blue or red and medium to large in size and may be eaten fresh or canned, though most are dried. European plums grown in the United States include: French or Agen, President, Tragedy, and Grand Duke.

• **Japanese plum**—This plum is also called the salicina plum. The fruits are yellow, crimson and purple, ranging in size from small to large. They are all very juicy and sweet, and they are eaten fresh, cooked, or canned—but never as prunes. The scientific name for the Japanese plum is *Prunus salicina,* and some common varieties include: Beauty, Burmosa, Santa Rosa, Wickson, Redheart, Duarte, Red Rosa, Kelsey, and Burbank.

• **American plum**—This group includes several species such as (1) the cold-resistant *Prunus americana* with amber skin and flesh; (2) the bushy and thorny Hortulan plum, *Prunus hortulana;* and (3) the sand cherry plum, *Prunus besseyi,* which grows in the Middle West and Canada. Some common names include De Soto, Pottawattomi, and Golden Beauty.

• **Damson plum**—These trees resemble the European type, but they are smaller and more cold resistant. Their fruits are tart and blue favorites for jellies and jams. The scientific name for the damson plum is *Prunus insititia.* Several common varieties include Bullaces, St. Juliens, and Mirabelles.

• **Ornamental plum**—These trees produce red foliage and red fruit which is suitable for jams and jellies. The scientific name for the ornamental plum is *Prunus cerasifera* which is also the myrobalan plum. The major value of the myrobalan plum is its use as a rootstock for other stone fruits.

ORIGIN AND HISTORY. Over 2,000 varieties of plums are known. The most important species for commercial production in the United States is *Prunus domestica,* the common, or European, plum, which originated in Europe and was introduced to the United States by the Pilgrims. It produces most of the dessert plums and all of the prune plums in the United States. The Japanese plum, *Prunus salincina,* believed to be native to China, was introduced to the United States by way of Japan in 1870.

WORLD AND U.S. PRODUCTION. Worldwide, about 5.7 million metric tons of plums and prunes are produced annually; and the leading producing countries, by rank, are: U.S.S.R., China, Romania, U.S.A., Yugoslavia, Germany, Turkey, Hungary, France, and Italy.[27]

The United States produces 488,800 short tons of plums and prunes annually, on a fresh basis, with the following states leading in production, by rank: California, Oregon,

[27]Data from *FAO Production Yearbook 1990,* FAO/UN, Rome, Italy, Vol. 44, p. 162, Table 70. **Note well:** Annual production fluctuates as a result of weather and profitability of the crop.

Washington, Idaho, and Michigan. California produces 90% of the nation's plums and prunes.[28]

PROCESSING. Plums are marketed as fresh, dried (prunes), and canned or frozen fruit. Approximately 56.3% is marketed fresh, 20.6% dried, 22.2% is canned, and 0.9% frozen. In California, about 66% of the plum crop is dried for prunes, and the majority of the remaining crop is utilized fresh.

Fig. P-43. Plums. (Courtesy, USDA)

A few prunes are still sun dried, but most are dehydrated. Sun drying requires more than 1 week for completion, and may be hampered by bad weather. Prior to dehydration, plums are immersed in hot water for a few seconds to remove the natural wax, and then dipped in lye to prevent fermentation. Next, the plums (prunes) are placed in single layers on trays, and placed in the dehydrator where temperature, air movement, and humidity are controlled. In efficient operations drying in a dehydrator is completed in 3 to 4 days, when the fruit reaches 18 to 19% moisture. Some larger dehydrators may only require 12 to 24 hours. The drying ratio is approximately 2½ lb (1.13 kg) of fresh fruit to 1 lb (0.45 kg) of dried fruit.

Canned plums and prunes are canned in water pack, light syrup, heavy syrup, or extra heavy syrup. Prune juice is a by-product of prune dehydration. It is made by leaching prunes with hot water or by using pectic enzymes on a slurry of whole ground prunes. Prune juice is canned, and yearly production is about 7,000,000 cases of No. 2 cans. Plums are frozen primarily in large containers for remanufacture into jams and preserves at a later date.

SELECTION, PREPARATION, AND USES. With so many varieties which differ widely in appearance and flavor, it is best to buy a few of the fresh fruits and be certain that the taste and appearance is appealing. Fresh plums should be selected on the basis of their color characteristic for the type. Additionally, ripe plums are firm to slightly soft, and taste juicy and sweet. Slight softening at the tip is also a good sign of maturity.

If not quite ripe when purchased, keep plums at room temperature a day or so, but watch closely. Plums can turn overripe very quickly. Ripe plums should be refrigerated.

Plums can be eaten fresh out of the hand or prepared in a variety of fruit dishes. Plums make good jelly, preserves, plum butter, and jam. Canned prunes can be served alone

---

[28]Data from *Agricultural Statistics 1991*, USDA, pp. 207, 208, Tables 318, 319. **Note well:** Annual production fluctuates as a result of weather and profitability of the crop.

as a side dish or a dessert or in combination with other fruits, or in baked goods.

Prunes are used in baked goods, confections, desserts, salad, and meat dishes. Many prunes are served as a breakfast or dessert fruit, stewed and served with or without cream. Prunes are an excellent ready-to-eat snack.

NUTRITIONAL VALUE. Fresh plums contain 78 to 87% water, depending upon the variety. Additionally, each 100 g (about 3½ oz) provides 48 to 75 Calories (kcal) of energy, 170 to 299 mg potassium, 250 to 300 IU of vitamin A, and only 1 to 2 mg of sodium, depending upon the type of plum. The calories in plums are derived primarily from natural sugars (carbohydrates) which give them their sweet taste. Some canned plums contain more calories due to the addition of sugar, but some canned varieties are also excellent sources of vitamin A, containing 1,180 to 1,250 IU per 100 g. Dried plums (prunes) contain only 32% water; hence, many of the nutrients, including calories, are more concentrated. Prunes contain 239 Calories (kcal), 754 mg potassium, 2.5 mg iron, and 1,994 IU vitamin A in each 100 g. More complete information regarding the nutritional value of fresh plums, canned plums, prunes and cooked prunes is presented in Food Composition Table F-21.

POACH

A method of cooking eggs in a shallow amount of near-boiling water. A favorite method for cooking breakfast eggs.

POI

This is a Hawaiian food made from the root of the taro which is baked, pounded, moistened, and fermented. Poi tastes slightly sour; hence, some individuals trying poi for the first time find it unpleasant.

(Also see VEGETABLE[S], Table V-2, Vegetables of the World; and TARO.)

POISONING

The act of administering a substance which, in sufficient quantities and/or over a period of time, kills or harms living things. Many substances are poisonous in massive doses while other substances require only minute amounts to be lethal.

(Also see POISONS.)

POISONOUS PLANTS

Contents	Page
History	853
Precautions and Cautions	853
Aid for Poisoned Persons	853
Poison Control Centers	854
Some Important Poisonous Plants of North America	854

Herbal medicine enthusiasts and stalkers of the wild edible plants beware! Plants can be dangerous. But one does not have to be an herbal medicine enthusiast or stalker of the wild to encounter poisonous plants. Many poisonous plants grow in homes, gardens, and recreational areas. Some of the most prized cultivated ornamentals are extremely poisonous. However, this does not mean that poisonous plants should be destroyed, or that laws should be passed to ban their use. If this were done, many valuable plants, some of them edible plants, would require destruction or legislation. Rather, people should familiarize themselves with the hazards in their immediate environment, and develop a respectful attitude toward the potential hazards of unfamiliar plants.

Poisonous plants are those plants that contain substance(s) which in sufficient quantities and/or over a period of time kill or harm man or animals. Primarily, the effects of poisonous plants are noted when the poisonous plant parts are eaten. However, there are some plants whose poisons may be inhaled or contacted by the skin.

HISTORY. Man's experience with plants over the ages has helped him to identify, use, and avoid poisonous plants. Primitive people obtained toxins from plants to use for hunting. African hunters tipped their arrows with ouabain from *Strophanthus gratus*—a poison powerful enough to stop an elephant. American Indians tipped blow gun darts with the powerful muscle relaxant curare from *Strychinos toxifera*.

Men used their knowledge of poisonous plants on other men. Plotters poisoned their enemies with deadly nightshade, *Atropa belladonna.* Some primitive tribes provided drinks from toxic plants to accused individuals as a trial—a survivor was held as innocent. Other cultures used poisonous plants to carry out death sentences. Socrates, about 400 B.C., was condemned to die by drinking a cup of poison hemlock.

The effects of some poisonous plants includes hallucinations. Through the ages men have used these plants for religious rites and to escape from everyday ills. The Aztecs revered a mushroom, *Psilocybe mexicana,* for its hallucinogenic properties, while the priests of India deified a toadstool, *Amanita muscaria,* for its intoxicating juices. Here in the United States, the Navajo Indians and other tribes employed the peyote cactus, *Lophophora williamsii,* to send users into a euphoric state.

No doubt, men watched their livestock get sick or die after eating certain plants. Stories of the Old West have made locoweeds famous because of their effects on animals. When eaten by dairy cows, White snakeroot transfers its poisonous alcohol, *tremetol,* to their milk. During the 19th century, this created the outbreaks of *milk sickness* in the Appalachians and the Midwest, and resulted in the loss of numerous lives.

Gradually, men learned to be cautious, to avoid and treat plant poisonings, and to identify poisonous plants.

PRECAUTIONS AND CAUTIONS. Even in a modern society, that has to a large extent removed itself from the natural vegetation, poisonous plants may still·be a hazard. Therefore, the following precautions and cautions will help avert the hazards of poisonous plants:

1. Become familiar with the poisonous plants in your home, garden, yard, and recreational areas; know them by sight and name.

2. Keep bright seeds, berries, and flowers out of the reach of small children, as they attract their curiosity.

3. Teach older children to keep unknown plants and plant parts out of their mouths; and instruct them as to the potential dangers of poisonous plants.

4. Know the plants that children use, or are apt to use, as play things.

5. Do not eat wild plants without *positive* identification.

6. Avoid all mushrooms and toadstools growing in the wild; their identification is too risky.

7. Never suppose a plant to be edible because it resembles a well-known edible plant.

8. Don't assume that a plant is safe for humans just because it is eaten by animals.

9. Be aware of the fact that some plants concentrate or confine the toxic substance to one part of the plant.

10. Be cognizant that in some cases cooking destroys or removes the poisonous substance, but that this is not the general rule.

11. Approach with caution the preparation of homemade herbal remedies from native or cultivated plants.

12. Remember that there are *no* absolute "tests" or "rules of thumb" which can be applied to distinguish edible from poisonous plants.

Fortunately, many poisonous plants must be eaten in relatively large amounts, and many are quite unpalatable and therefore will not be eaten in sufficient amounts to cause poisoning. Still, there are some plants that are very toxic and require only a bite or two, or a seed or two, to cause death. Also, some poisonous plants are not at all distasteful and may be eaten easily in large enough quantities to cause serious disturbances and even death.

AID FOR POISONED PERSONS. Despite everything, people will be poisoned by plants. When this occurs, time is the most important factor. The toxic substance must be removed from the body before it is absorbed. When possible the first step should be to call a physician or the Poison Control Center immediately. The person calling should be able to (1) identify or describe the plant that was eaten, (2) tell when and how much was eaten, and (3) describe any observable symptoms. When the victim is transported, a sample of the plant should be taken for identification or verification. Also, it is important to keep parts of a plant that may be present in the stool or vomit, if there is a question about which plant part was eaten. Samples—fruits, flowers, and leaves—of the plants in the area should be quickly collected and taken with the victim.

When a physician is not available, first aid should be administered and the victim transported to a hospital. The first-aid treatment will be determined by the poison. If no other advice has been given by a doctor or a Poison Control Center, take the following emergency steps:

1. If the victim is unconscious, administer artificial respiration if necessary, and head for the hospital immediately.

2. If the swallowed poison is a noncaustic substance (not a strong alkali or strong acid), proceed as follows:

   a. Dilute the poison by getting the victim to drink lukewarm soapsuds, soda water made with common baking soda, or salt water (2 Tbsp of salt in a glass of warm water); but *beware of the danger of choking.* It may be necessary to give six or more glasses of the liquid. A poison diluted with a large amount of liquid is absorbed less quickly than a concentrate, and vomiting can be induced more easily when the stomach is filled.

   b. Induce vomiting by (1) using the fingers to stimulate gagging, (2) giving 1 tsp of syrup of ipecac (which should be in every medicine cabinet) every 5 minutes for 3 or 4 doses, or (3) giving either soda water made with baking soda, or salt water. When the victim vomits, lay him on his stomach with his head hanging over the edge of the bed or over your knees. If possible, catch any material thrown up in basin or bowl and save for laboratory analysis.

   c. After the stomach has been thoroughly emptied by vomiting, give an antidote (a substance to counteract the poison), such as an all-purpose product containing activated charcoal, magnesium hydroxide, or tannic acid.

3. If the poison is a strong alkali, a strong acid, or a petroleum distillate rather than a poisonous plant, *do not induce vomiting* because of the danger of perforating the stomach or esophagus and/or aspirating the corrosive fluid

into the lungs. Instead, neutralize and dilute with a weak acid such as dilute lemon juice or vinegar, or with an alkali, such as baking soda, lime water, milk of magnesia, or chalk, then give milk, olive oil, egg white, or an all-purpose antidote.

4. As soon as possible, transport the victim via private car, ambulance, police, or paramedics to the nearest hospital and/or physician. Be sure to take samples of the suspected poisonous plant.

(Also see POISONS.)

POISON CONTROL CENTERS. Centers have been established in various parts of the country where doctors can obtain prompt and up-to-date information on treatment of poison cases, if desired. Poison Control Centers have information relative to thousands of poisonous substances, cross-indexed by brand name as well as generic or chemical name, and can give prompt, responsible advice about antidotes.

Local medical doctors have information relative to the Poison Contol Centers of their area, along with their telephone numbers. When the phone number is not known, simply ask the operator for the Poison Control Center. If this information cannot be obtained locally, call the U.S. Public Health Service at either Atlanta, Georgia, or Wenatchee, Washington.

SOME IMPORTANT POISONOUS PLANTS OF NORTH AMERICA. A description of all plants that are potentially poisonous is not within the scope of this book. Almost all plants are poisonous if the wrong part is consumed or if too much is consumed. For example, the stalks of rhubarb are a wholesome and delicious food while the leaves contain poisonous levels of oxalic acid; or large quantities of apple seeds, peach pits, or apricot pits may cause cyanide poisoning.

The listing given herein includes some of the most important, and often the most common, poisonous plants in North America. It is presented for informational purposes and with the hope of stimulating reader interest in learning more about the poisonous plants in his particular environment. More can be learned through the advice of experts and by reading publications dealing with poisonous plants common to a specific area.

Listed alphabetically by common name, followed by the scientific name, some of the most important poisonous plants in North America follow:

Baneberry, *Actaea* spp
Buckeye; Horsechestnut, *Aesculus* spp
Buttercup, *Ranunculus* spp
Castor bean, *Ricinus communis*
Chinaberry, *Melia azedarach*
Death camas, *Zigadenus paniculatus*
Dogbane (Indian hemp), *Apocynum cannabinum*
Foxglove, *Digitalis purpurea*
Henbane, *Hyoscyamus niger*
Iris (Rocky Mountain Iris), *Iris missouriensis*
Jasmine, *Gelsemium sempervirens*
Jimmyweed (Rayless goldenrod), *Haplopappus heterophyllus*
Jimsonweed (Thornapple), *Datura stramonium*
Lantana (Red Sage), *Lantana camara*
Larkspur, *Delphinium* spp
Laurel (Mountain laurel), *Kalmia latifolia*
Locoweed (Crazyweed), *Oxytropis* spp
Lupine (Bluebonnet), *Lupinus* spp
Marijuana (hashish, Mary Jane, pot, grass), *Cannabis sativa*

Mescal bean (Frijolito), *Sophora secundiflora*
Mistletoes, *Phoradendron serotinum*
Monkshood (Wolfsbane), *Aconitum columbianum*
Mushroom (toadstools), *Amanita muscaria, Amantia verna, Chlorophyllum molybdites.*
Nightshade, *Solanum nigrum, Solanum elaeagnifolium*
Oleander, *Nerium oleander*
Peyote (Mescal buttons), *Lophophora williamsii*
Poison hemlock (poison parsley), *Conium maculatum*
Poison ivy, *Rhus radicans*
Poison oak, *Rhus toxicondendron*
Pokeweed (Pokeberry), *Phytolacca americana*
Poppy (common poppy), *Papaver somniferum*
Rhododendron; Azaleas, *Rhododendron* spp
Rosary pea (precatory pea), *Abrus precatorius*
Snow-on-the-mountain, *Euphorbia marginata*
Skunkcabbage, *Veratrum californicum*
Tansy, *Tancacetum vulgare*
Waterhemlock, *Cicuta* spp
White snakeroot, *Eupatorium rogosum*

(Also see MEDICINAL PLANTS; and WILD EDIBLE PLANTS. **NOTE WELL:** For a more complete discussion of poisonous plants, including a detailed summary relative to each of the poisonous plants, see the two-volume *Foods & Nutrition Encyclopedia*, a work by the same authors as *Food For Health*.)

POISONS

Contents   Page
Poison Prevention and Treatment 855
Emergency Measures in Acute Poisoning 855
Poison Control Centers 856
Environmental Contaminants in Food 856
Some Potentially Poisonous (Toxic) Substances 856

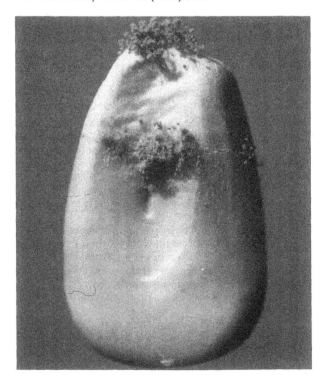

Fig. P-44. Sporulation of *Aspergillus flavus* on a kernel of corn. This fungus produces aflatoxin. Aflatoxin (1) is associated with a high indicence of liver cancer, and (2) may be involved in some types of acute poisoning of children. (Courtesy, C. W. Hesseltine, USDA, Agricultural Research, Peoria, Ill.)

Sola dosis facit venenum, means "only the dose makes the poison." Paracelsus, the noted German-born physician, said it in the 16th century. But it's just as true today as it was four centuries ago. For most food-related poisons, there is both a safe level and a poisonous level; and the severity of the effect depends upon (1) the amount taken, (2) the period of time over which the substance is taken (certain poisons are cumulative), and (3) the age and physical condition of the person.

Take potatoes, for example. They contain solanine, which can be toxic. In consuming an average of 119 lb (*54 kg*) of potatoes per year, the average American ingests 9,700 mg of solanine—enough to kill a horse. Yet it doesn't kill us because we don't eat it all at one time.

Or take lima beans! The average person eats about 1.85 lb (*0.84 kg*) of these tasty little morsels each year, not realizing that they contain hydrogen cyanide, a deadly poison. But they don't kill us simply because we can't eat enough at any one sitting to be deadly. Besides, most U.S. limas contain little hydrogen cyanide.

And so it goes, on and on through a long list of foods and drinks.

By definition, *a poison is a substance which in sufficient quantities and/or over a period of time kills or harms living things.* Many poisons are called toxins. The study of poisons is called *toxicology.* The discussion that follows and Table P-2 pertain primarily to food-related poisons that are eaten.

There are more than four million chemical compounds, of which more than 60,000 are commercially produced; and about 1,000 new ones are introduced each year. Some of these make their way into food and water. With the growth and use of chemicals, food supplies are subject to contamination from or treatment with chemicals in the course of growing, fertilizing, harvesting, processing, and storing.

Today, there are substantially fewer dangers than formerly from contamination of food and water by bacteria or from ingesting lead (now that paints are lead-free). Nevertheless, some two million ingestions of potentially dangerous substances occur in the United States each year; and there were 6,226 fatal poisonings in 1988.[29] Despite popular misconceptions, children are no longer the main victims of poisons. Today, only 1% of the deaths from poison involve children under 5 years of age; and poisoning now accounts for only 23% of the accidental deaths among children under age 5.[30] This decline is attributed primarily to the use of childproof safety caps in packaging poisonous agents; the storage of toxic substances separate and apart from other products and out of reach of children; and the banning of lead-containing paints, along with public health screening of children living in old housing areas in which lead-based paints may have been used years ago.

The general symptoms of poisoning are: nausea, vomiting, cramps, and stomach pains. If a corrosive poison (a strong alkali or a strong acid) has been taken, burns and stains may show on and around the mouth and tongue.

POISON PREVENTION AND TREATMENT. In recent years, physicians, pharmacists, and Poison Control Center personnel have joined forces in advocating the following simple measures to avert poisoning and provide lifesaving aid when poisoning occurs:

---

[29]*Statistical Abstracts of the United States 1991*, U.S. Department of Commerce, Bureau of the Census, p. 85, Table 124.

[30]*Healthy People,* the Surgeon General's report on health promotion and disease prevention, U.S. Department of Health, Education, and Welfare, 1979, p. 70.

1. **Packaging and storing potential poisons with care.** The use of child-resistant containers (opened only with difficulty and know-how), strip packages (with pills and capsules sealed individually in a plastic strip), and safe storage have reduced the incidence of accidental poisoning, especially among children.

2. **Keeping a supply of syrup of ipecac on hand at all times.** As a reliable (and usually rapid) means of inducing vomiting, it is the best.

3. **Having an all-purpose antidote on hand.** Antidotes are usually listed on the labels of potentially poisonous products, but finding the ingredients and mixing them takes up valuable time. So, it is best to have an all-purpose antidote readily available. The main ingredients in all-purpose antidotes are powdered activated charcoal, which has a strong capacity to absorb many chemicals; a weak buffering alkali such as magnesium hydroxide to help neutralize an acid poison; and a weak buffering acid such as tannic acid to help neutralize any alkali present. This type of antidote is particularly useful in cases of poisoning in which it is best not to induce vomiting.

EMERGENCY MEASURES IN ACUTE POISONING.
In case of *acute* poisoning, it is important to act swiftly, calmly and knowledgeably. The following emergency measures, taken in the order listed, may save a life:

NOTE WELL: Although these measures were prepared especially for counteracting the poisons listed in Table P-2, they are applicable to all poisons, including those from drugs and household products.

1. **Call a doctor, the nearest Poison Control Center, or a hospital.** Describe over the phone exactly what the poison was (read off the label), provided it is known. If the victim is unable to give information, perhaps a nearby container or the symptoms will identify the poison. Based on this information, the doctor or Poison Control Center may be able to recommend emergency treatment, thereby saving precious time.

2. **Apply first-aid treatment.** The first-aid treatment will be determined by the poison. If no other advice has been given by a doctor or a Poison Control Center, take the following emergency steps:

   a. If the victim is unconscious, administer artificial respiration if necessary, and head for the hospital immediately.

   b. If the swallowed poison is a noncaustic substance (not a stong alkali or strong acid), proceed as follows:

   (1) Dilute the poison by getting the victim to drink lukewarm soapsuds, soda water made with common baking soda, or salt water (2 Tbsp of salt in a glass of warm water); but *beware of the danger of choking.* It may be necessary to give six or more glasses of the liquid. A poison diluted with a large amount of liquid is absorbed less quickly than a concentrate, and vomiting can be induced more easily when the stomach is filled.

   (2) Induce vomiting by (a) using the fingers to stimulate gagging, (b) giving 1 teaspoonful of syrup of ipecac (which should be in every medicine cabinet) every 5 minutes for 3 or 4 doses, or (c) giving either soda water made with baking soda, or salt water. When the victim vomits, lay him on his stomach with his head hanging over the edge of the bed or over your knees. If possible, catch any material thrown up in a basin or bowl and save for laboratory analysis.

(3) After the stomach has been thoroughly emptied by vomiting, give an antidote (a substance to counteract the poison), such as an all-purpose product containing activated charcoal, magnesium hydroxide, or tannic acid.

c. If the poison is a strong alkali or a strong acid, *do not induce vomiting* because of the danger of perforating the stomach or esophagus and/or aspirating the corrosive fluid into the lungs. Instead, proceed as follows:

(1) For alkali poisoning: (a) neutralize and dilute the alkali with a weak acid such as dilute lemon juice or vinegar, then give milk; or (b) give an all-purpose antidote.

(2) For acid poisoning: (a) neutralize and dilute the poison with an alkali, such as baking soda, lime water, milk of magnesia, or chalk, then (b) give milk, olive oil, egg white, or an all-purpose antidote.

3. **Control shock.** Keep the victim warm.
4. **See the doctor.** A physician should always see a victim of poisoning as soon as possible; with transportation provided by private car, an ambulance, the police, or paramedics (fire department). Also, take the poison container and label along.

POISON CONTROL CENTERS.
Centers have been established in various parts of the country where doctors can obtain prompt and up-to-date information on treatment of poison cases, if desired. Poison Control Centers have information relative to thousands of poisonous substances, cross-indexed by brand name as well as generic or chemical name, and can give prompt, responsible advice about antidotes.

Local medical doctors have information relative to the Poison Control Centers of their area, along with their telephone numbers. When the phone number is not known, simply ask the operator for the Poison Control Center. If this information cannot be obtained locally, call the U.S. Public Health Service at either Atlanta, Georgia, or Wenatchee, Washington.

ENVIRONMENTAL CONTAMINANTS IN FOOD.
Environmental contaminants include organic chemicals, metals and their derivatives, and radioactive substances that inadvertently enter the human food supply through agriculture, mining, industrial operations, or energy production. To regulate them under the law, the Food and Drug Administration (FDA) defines environmental contaminants as "added, poisonous, or deleterious substances that cannot be avoided by good manufacturing practices, and that may make food injurious to health."

Unlike food additives, environmental contaminants inadvertently find their way into the human food supply (including sports fish and game).

Four factors determine whether, and how seriously, the environmental contamination of food will affect human health: (1) toxicity of the contaminant, (2) the amount of the substance in the food, (3) the amount of the contaminated food eaten, and (4) the physiological vulnerability of the individual consuming the food.

Fig. P-45. Pesticide being applied by airplane. When properly used, consumers benefit greatly from pesticides. It is important, however, to gauge the effect of each pesticide on the ecological food chain and human health.

Although the United States has escaped mass poisonings such as have occurred in most other industrialized nations, between 1968 and 1978, according to an Office of Technology Assessment (OTA) survey, 243 food-contamination incidents were reported in this country. During this same period (1968-1978), at least $282 million in food was lost to contamination—a conservative estimate that included only 30% of the known incidents and ignored such hidden costs as medical expenses and lost workdays.

The FDA sets permissible levels for all known contaminants. Then federal and state regulatory agencies monitor food to ensure that environmental contaminants do not exceed prescribed levels. Consequently, contamination involving an unregulated substance is rarely identified before it becomes a major problem. None of the major environmental contamination incidents in the United States—animal feeds in Michigan contaminated by polybrominated biphenyls (PBBs); the Hudson River contaminated by polychlorinated biphenyls (PCBs); Virginia's James River contaminated by kepone; and fat used to produce meat and bone meal feeds in a meat-packing plant in Billings, Montana, contaminated by PCBs from a damaged transformer—were initially discovered by ongoing monitoring programs. In each case, actual human or animal poisonings alerted authorities to the danger.

SOME POTENTIALLY POISONOUS (TOXIC) SUBSTANCES.
Table P-2 is a summary of some potentially poisonous (toxic) substances to people and animals.

(Also see POISONOUS PLANTS; and BACTERIA IN FOOD.)

NOTE WELL: For a more complete discussion of poisons, see the two-volume *Foods & Nutrition Encyclopedia,* a work by the same authors as *Food For Health.*

TABLE P-2
SOME POTENTIALLY POISONOUS (TOXIC) SUBSTANCES

Poison (Toxin)	Source	Symptoms and Signs	Distribution; Magnitude	Prevention; Treatment
Aflatoxins (See MYCOTOXINS in this table.)				
Aluminum (Al)	Food additives, mainly present in such items as baking powder, pickles, and processed cheeses. Aluminum-containing antacids. Aluminum cooking utensils.	Abnormally large intakes of aluminum irritate the digestive tract. Also, unusual conditions have sometimes resulted in the absorption of sufficient aluminum from antacids to cause brain damage. Aluminum may form nonabsorbable complexes with essential trace elements, thereby creating deficiencies of the trace elements.	**Distribution:** Aluminum is widely used throughout the world. **Magnitude:** The U.S. uses more aluminum than any other product except iron and steel. However, known cases of toxicity are rare.	**Prevention:** Based on the evidence presented herein, no preventative measures are recommended. **Treatment:** This should be left to the doctor.
Arsenic (As)	Consuming foods and beverages contaminated with excessive amounts of arsenic-containing sprays used as insecticides and weed killers. Arsenical insecticides used in vineyards exposing the workers (1) when spraying or (2) by inhaling contaminated dusts and plant debris. Arsenic in the air from three major sources: smelting of metals, burning of coal, and use of arsenical pesticides.	Burning pains in the throat or stomach, cardiac abnormalities, and the odor of garlic on the breath. Other symptoms may be diarrhea and extreme thirst along with a choking sensation. Small doses of arsenic taken into the body over a long period of time may produce hyperkeratosis (irregularities in pigmentation, especially on the trunk); arterial insufficiency; and cancer. There is strong evidence that inorganic arsenic is a skin and lung carcinogen in man.	**Distribution:** Arsenic is widely distributed, but the amount of the element consumed by man in food and water or breathed, is generally small and not harmful. **Magnitude:** Cases of arsenic toxicity in man are rather infrequent. Two noteworthy episodes occurred in Japan in 1955. One involved tainted powdered milk; the other, contaminated soy sauce. The toxic milk caused 12,131 cases of infant poisoning, with 130 deaths. The soy sauce poisoned 220 people. There are many scattered case reports of subacute to chronic arsenic poisoning.	**Prevention:** Avoid consuming (1) arsenic products, or (2) foods and beverages contaminated with excessive amounts of arsenic. **Treatment:** Induce vomiting and more vomiting, followed by an antidote of egg whites in water or milk. Afterward, give strong coffee or tea, followed by epsom salts in water or castor oil.
Chromium (Cr) (See CHROMIUM; and MINERAL[S].)				
Copper (Cu) (See COPPER; and MINERAL[S].)				
Ergot Although ergot is a small mycotoxin it is treated separately herein because it differs from the other mycotoxins in that it results from the consumption of a considerable amount of fungus tissue in which the toxin is found, whereas in the other mycotoxicoses the toxins are secreted into the substrate on or in which the fungus is growing, and very little fungus tissue as such is consumed. **Cause:** *Claviceps purpurea.* Ergotism is caused by overuse of ergot-containing foods or drugs.	Rye, wheat, barley, oats, and triticale. Ergot replaces the seed in the heads of cereal grains, in which it appears as a purplish-black, hard, banana-shaped, dense mass from ¼ to ¾ in. (6 to 19 mm) long.	When a large amount of ergot is consumed in a short period, convulsive ergotism is observed. The symptoms include itching, numbness, severe muscle cramps, sustained spasms and convulsions, and extreme pain. When smaller amounts of ergot are consumed over an extended period, ergotism is characterized by gangrene of the fingertips and toes, caused by blood vessel and muscle contraction stopping blood circulation in the extremities. These symptoms include cramps, swelling, inflammation, alternating burning and freezing sensations ("St. Anthony's fire") and numbness; eventually the hands and feet may turn black, shrink, and fall off. The amount of ergot in the food consumed determines how rapidly toxic effects will show. Ergotism is a cumulative poison, depending on the amount of ergot eaten and the length of time over which it is eaten.	**Distribution:** Ergot is found throughout the world, wherever rye, wheat, barley, oats, or triticale are grown. **Magnitude:** There is considerable ergot, especially in rye. But, normally, screening grains before processing alleviates ergotism in people.	**Prevention:** Consists of an ergot-free diet. Ergot in plants can not be controlled by treatment; and resistant varieties of cereal grains are not available. Control of ergot in cereal grains consists of using ergot-free seed, following a crop rotation, and using cultural practices which will reduce the incidence of ergot in plants and its spread to other crops. Ergot in food and feed grains may be removed by screening the grains before processing. In the U.S., wheat and rye containing more than 0.3% ergot are classed as *ergoty*. In Canada, government regulations prohibit more than 0.1% ergot in feeds. **Treatment:** An ergot-free diet; good nursing; treatment by a doctor.
Fluorine (F) (fluorosis) (See FLUORINE OR FLUORIDE; and MINERAL[S].)				

TABLE P-2 *(Continued)*

Poison (Toxin)	Source	Symptoms and Signs	Distribution; Magnitude	Prevention; Treatment
Lead (Pb)	Consuming food or medicinal products (including health food products) contaminated with lead. Inhaling the poison as a dust by workers in such industries as painting, lead mining, and refining. Inhaling airborne lead discharged into the air from auto exhaust fumes. Consuming food crops contaminated by lead being deposited on the leaves and other edible portions of the plant by direct fallout. Consuming food or water contaminated by contact with lead pipes or utensils. Old houses in which the interiors were painted with leaded paints prior to 1945, with the chipped wall paint eaten by children. Such miscellaneous sources as illicitly distilled whiskey, improperly lead-glazed earthenware, old battery casings used as fuel, and toys containing lead.	Symptoms develop rapidly in young children, but slowly in mature people. **Symptoms of acute lead poisoning:** Colic, cramps, diarrhea or constipation, leg cramps, and drowsiness. The most severe form of lead poisoning, encountered in infants and in heavy drinkers of illicitly distilled liquor, is characterized by profound disturbances of the central nervous system, and permanent damage to the brain; damage to the kidneys; and shortened life span of the erythrocytes. **Symptoms of chronic lead poisoning:** Colic, constipation, lead palsy especially in the forearm and fingers, the symptoms of chronic nephritis, and sometimes mental depression, convulsions, and a blue line at the edge of the gums.	**Distribution:** Predominantly among children in poverty-stricken neighborhoods, living in dilapidated housing built before 1945, where they may eat chips of lead-containing paints, peeled off from painted wood. Leaded paints are no longer used on the interior surface of houses. **Magnitude:** The Center for Disease Control, Atlanta, Georgia, estimates that (1) lead poisoning claims the lives of 200 children each year, and (2) 400,000 to 600,000 children have elevated lead levels in the blood. Lead poisoning has been reduced significantly with the use of lead-free paint.	**Prevention:** Avoid inhaling or consuming lead. Workers in industries where lead poisoning is an occupational hazard should follow the safety regulations that have been established. **Treatment:** **Acute lead poisoning:** An emetic (induce vomiting), followed by drinking plenty of milk and ½ oz (*14 g*) of epsom salts in a glass of water. **Chronic lead poisoning:** Remove the source of lead. Sometimes treated by administration of magnesium or lead sulphate solution as a laxative and antidote on the lead in the digestive system, followed by potassium iodide which cleanses the tracts. Currently, treatment of lead poisoning makes use of chemicals that bind the metal in the body and help in its removal.
Mercury (Hg)	Mercury is discharged into the air and water from industrial operations and is used in herbicide and fungicide treatments. Mercury poisoning has occurred where mercury from industrial plants has been discharged into water and has then accumulated as methylmercury in fish and shellfish. Accidental consumption of seed grains treated with fungicides that contain mercury, used for the control of fungus diseases of oats, wheat, barley, and flax.	The toxic effects of organic and inorganic compounds of mercury are dissimilar. The organic compounds of mercury, such as the various fungicides, (1) affect the central nervous system, and (2) are not corrosive. The inorganic compounds of mercury include mainly mercuric chloride, a disinfectant; mercurous chloride (calomel), a cathartic; and elemental mercury. Commonly the toxic symptoms are: corrosive gastrointestinal effects, such as vomiting, bloody diarrhea, and necrosis of the alimentary mucosa.	**Distribution:** Wherever mercury is produced in industrial operations or used in herbicide or fungicide treatments. **Magnitude:** Limited. But about 1,200 cases of mercury poisoning identified in Japan in the 1950s were traced to the consumption of fish and shellfish from Japan's Minamata Bay contaminated with methylmercury. Some of the offspring of exposed mothers were born with birth defects, and many victims suffered central nervous system damage. Still another outbreak of mercury toxicity occurred in Iraq, where more than 6,000 people were hospitalized after eating bread made from wheat that had been treated with methylmercury.	**Prevention:** Do not consume seed grains treated with a mercury-containing fungicide. Surplus of treated grain should be burned and the ash buried deep in the ground. Control mercury pollution from industrial operations. **Treatment:** This should be left to a medical doctor.
Mycotoxins ("myco", prefix from the Greek word *mykes*, meaning fungus), which may produce the toxicity syndromes referred to as mycotoxicoses. There are an estimated 800 species of fungi (yeasts and molds), but fortunately not all fungi produce metabolites that are toxic. Some mycotoxins are carcinogens—capable of causing malignant tumors (cancers). **Cause:** Toxin-producing molds: e.g. *Aspergillus flavus, Aspergillus parasiticus, Penicillium cyclopium, P. islandicum, P. palitans, Fusarium roseum, F. tricinctum.* The aflatoxins constitute a family of 14 naturally-occurring toxins. (Also see ERGOT which is presented alphabetically in this table.)	Contaminated foods. Mycotoxins can produce toxic compounds on virtually any food (even synthetic) that will support growth. It is important to know that mycotoxins remain in food long after the organism that produced them has died; thus, they can be present in food that is not visibly moldy. Moreover, many kinds of mycotoxins are relatively stable substances that survive the usual methods of processing and cooking. The possibility of mycotoxicoses resulting from the ingestion of previously contaminated meat and eggs remains uncertain, unless the intake level is extremely high. But there is abundant evidence that lactating cows and lactating mothers consuming diets containing more than 20 ppb of aflatoxin on a dry matter basis will produce milk which contains aflatoxin residues. High levels of phytic acid in soybeans prevents aflatoxin synthesis.	In animals, liver damage is the most charcteristic symptom. It is reasonable to assume that man would respond similarly and that liver disease, particularly liver cancer, would be associated with aflatoxin exposure. Studies conducted in Uganda, Thailand, and Kenya have shown that high incidence of liver cancer is associated with high aflatoxin intake. Practically nothing is known about the acute effects of aflatoxins in humans. However, evidence is accumulating that suggests that aflatoxins and mycotoxins may be involved in some types of acute poisoning in children. In the Thailand study mentioned above, it was found that the liver, brain, and other tissues of children dying of a form of acute encephalitis, characterized by degeneration of the liver and other visceral organs (known as Reye's syndrome), contained far higher amounts and frequencies of aflatoxin B$_1$ than the tissues from children from the same region dying of other causes.	**Distribution:** Mycotoxins are widely distributed throughout the world. It has been known for a long time that liver cancer, which is relatively uncommon in the U.S. and Europe, occurs at much higher frequency in central and southern Africa and Asia. This prompted several field studies which confirmed that elevated liver cancer incidence is associated with aflatoxin exposure. **Magnitude:** Mass poisoning of human population by mycotoxins occurred in the U.S.S.R. during World War II. (See column headed "History.") Also, there has long been a high incidence of aflatoxin toxicity in Uganda, Thailand, and Kenya. In contrast to the limited numbers of documented mycotoxicoses in humans, there have been hundreds of reports of toxicity symptoms in livestock.	**Prevention:** The best way to control aflatoxins is to control their production. This calls for the maintenance of food and feed quality during growth, harvest, transportation, processing, and storage. The most important means of controlling mold growth and subsequent aflatoxin production is preventing damage to crops during harvest and reducing post harvest moisture levels below those required for fungal growth. Moisture levels below 18.5% in cereal grains, below 9% in such oilseeds as peanuts, sunflower, and safflower, and below 6% in copra. Organic acids (acetic acidpropionic acid mixture or propionic acid alone) may be applied to inhibit the growth of molds. The high aflatoxin risk crops are: peanuts, cottonseed, and copra. Aflatoxin in cottonseed may be inactivated by treating with anhydrous ammonia. Ultraviolet irradiation will also reduce the toxicity of aflatoxin. Certain spices, i.e. pepper.

(Continued)

TABLE P-2 *(Continued)*

Poison (Toxin)	Source	Symptoms and Signs	Distribution; Magnitude	Prevention; Treatment
Pesticides are chemicals used to destroy, prevent, or control pests. Pesticides also include (1) chemicals used to attract or repel pests, and (2) chemicals used to regulate plant growth or remove or coat leaves. Pests can be classified into six main groups: 1. Insects (plus mites, ticks, and spiders) 2. Snails and slugs 3. Vertebrates, including rats, mice, and certain birds (starling, linnets, English sparrows, crows, and blackbirds) 4. Weeds 5. Plant disease 6. Nematodes (Also see PESTICIDES AS INCIDENTAL FOOD ADDITIVES.)	Pesticides are chemicals. When properly used, they are beneficial; when improperly used, they may be hazards. Pesticide poisoning may be caused by either (1) sudden exposure to lethal quantities, or (2) as a result of repeated exposure to nonlethal quantities (chronic poisoning) during a protracted period of time. Farmers know that unless they follow state and federal regulations they risk having their products seized and condemned, or refused by processors. Nevertheless, economics dictate that new products be used as soon as they prove useful. On the other hand, food faddists may feel that they are being poisoned; wildlife conservationists may be concerned over possible damage to songbirds and other animals; beekeepers may become unhappy if insectants kill honeybees; and public health officials may be concerned about contamination of soil and food supplies.	Knowing something of the toxicity and symptoms or signs of poisoning of each type of pesticide may result in getting medical advice quickly and in saving a life. The leading chemical groups, along with the toxicity symptoms of each follow: **Chemical Group:** Organophosphates—most toxic; they injure the nervous system. Carbamates—safer than the organophosphates. Fumigants—make a person seem drunk. Plant-derived pesticides—some are very toxic.	**Distribution:** In every pursuit of modern agriculture, more and more pesticides are being used. They are the first line of defense against pests that affect human health and well-being and attack crops, livestock, pets, and structures. **Magnitude:** Pesticides are used to control many of the estimated 10,000 species of harmful insects; more than 160 bacteria, 250 viruses, and 8,000 fungi known to cause plant diseases; 2,000 species of weeds and brush; and 150 million rats. **Symptoms or signs of toxicity:** *Mild poisoning:* Fatigue, dizziness, sweating, vomiting. *Moderate poisoning:* Inability to walk, constriction of pupil of eye. *Severe poisoning:* Unconsciousness, tremors, convulsions. They produce the same symptoms and signs as the organophosphates, but they respond more easily to treatment. Poor coordination, slurring words, confusion, sleepiness. Technical pyrethrum may cause allergic reaction. Some rotenone dusts irritate the respiratory tract. Nicotine is fast-acting nerve poison.	**Prevention:** The first and most important precaution to observe when using any pesticide is to read and heed the directions on the label. In the event of an accident, the label becomes extremely important in remedial measures. For public protection, all chemicals are rigidly controlled by federal laws. Each one is required to be registered by the Environmental Protection Agency before it can be sold in the United States, and each one is issued a tolerance for residues that may result from its use on food or feed crops. **Treatment:** Each label contains a "Statement of Practical Treatment." Read it before handling a pesticide. First aid procedures follow: **Happenstance: What to do:** Pesticide skin contact — Wash with soap or detergent and plenty of water. Pesticide inhalation — Get to fresh air right away. Pesticide in eyes — Flush eyes and face thoroughly with water. Pesticide in mouth or stomach — Rinse mouth with water. Do not induce vomiting. Go to physician immediately. *CAUTION:* Take the pesticide label to the physician; it will help him treat the problem.
Polybrominated biphenyls (PBBs), a fire retardant which may cause cancer when taken into the food supply.	All of the 1973 Michigan toxicity problem was traced to livestock feed which became contaminated with PBB when the fire retardant was shipped by mistake to a feed manufacturer.	People exposed to PBB in Michigan reported suffering from neurological symptoms such as loss of memory, muscular weakness, coordination problems, headaches, painful swollen joints, acne, abdominal pain, and diarrhea.	In 1973, the accidental contamination of animal feeds exposed many people in Michigan to PBB in dairy products and other foods. The Michigan incident eventually led to the slaughter and burial of nearly 25,000 cattle, 3,500 hogs, and 1.5 million chickens; and the disposal of about 5 million eggs and tons of milk, butter, cheese, and feed.	**Prevention:** Avoid human error such as substituting PBB for a livestock feed supplement. Avoid harmfully contaminated food. **Treatment:** Follow the prescribed treatment of a medical doctor.
Polychlorinated biphenyls (PCBs), industrial chemicals; they're chlorinated hydrocarbons which may cause cancer when taken into the food supply.	Sources of contamination to man include: 1. Contaminated foods. 2. Mammals or birds that have fed on contaminated food or fish. 3. Residues on foods that have been wrapped in papers and plastics containing PCBs. 4. Milk from cows that have been fed silage from silos coated with PCB-containing paint; and eggs from layers fed feeds contaminated with PCBs. 5. Absorption by humans beings of PCBs through the lungs, the gastrointestinal tract, and the skin.	The clinical effects on people are: an eruption of the skin resembling acne, visual disturbances, jaundice, numbness, and spasms. Newborn infants from mothers who have been poisoned show discoloration of the skin which regresses after 2 to 5 months.	**Distribution:** PCBs are widespread. They have, directly and indirectly, found their way into animal feeds and animal food products through water, paints, sealants used on the interior of silos, heat transfer fluids, and plastic and cardboard food-packaging materials.	**Prevention:** Comply with the law; do not use PCBs. Avoid harmfully contaminated food. **Treatment:** People afflicted with PCB should follow the prescribed treatment of their doctor.

Magnitude: PCB constitutes the first example of a widespread environmental hazard from an industrial chemical not classed as a pesticide. The wide usage and persistence indicate that PCB will be of considerable concern for many years.

In 1968, the accidental contamination of edible rice-bran oil by PCBs led to a poisoning epidemic among the Japanese families who consumed the oil. The disease, which afflicted 1,291 people, became known as Yusho or rice-oil disease—a condition marked by chloracne (a severe form of acne), eye discharges, skin discoloration, headaches, fatigue, abdominal pains, and liver and menstrual disturbances. In 1972, Japan banned the manufacture of PCB.

(Continued)

TABLE P-2 *(Continued)*

Poison (Toxin)	Source	Symptoms and Signs	Distribution; Magnitude	Prevention; Treatment
Polychlorinated biphenyls (PCBs) *(Continued)*		The following U.S. incidents dramatically illustrate the potential health hazard and economic harm that can be caused by PCBs: (1) the discharging of PCBs into the Hudson River with waste water from capacitor plants from 1946 to 1976, and the contamination of Hudson River fish; (2) in 1977, a fire in an animal-feed warehouse in Puerto Rico contaminated fish meal, which was subsequently fed to poultry and resulted in the destruction of 400,000 chickens and millions of eggs in the U.S.; and (3) in 1979, the fat used to produce meat and bone meal feeds in meat-packing plant in Billings, Montana, contaminated by PCBs from a damaged transformer, resulted (to October 1979) in the destruction of 600,000 to 700,000 chickens, several hundred thousand eggs, and 16,000 lb of fresh pork. In 1980, the EPA estimated that 91% of Americans carry a measurable quantity of PCB in their fatty tissues.		
Salt (NaCl—sodium chloride) poisoning (Also see MINERAL[S]; and SALT.)	Table salt (sodium chloride) and foods containing salt.	Excess dietary salt over a prolonged period of time is believed to lead to high blood pressure in susceptible people. Salt may be toxic (1) when it is fed to infants or others whose kidneys cannot excrete the excess in the urine, or (2) when the body is adapted to a chronic low-salt diet.	**Distribution:** Salt is used all over the world. Hence, the potential for salt poisoning exists everywhere. **Magnitude:** Salt poisoning is relatively rare.	**Prevention:** Do not deplete the body of salt such as can occur (1) when on a strict vegetarian diet without salt, or (2) when there has been heavy and prolonged sweating; followed by gorging on salt. The best preventative against salt poisoning is to provide plenty of water. **Treatment:** Drink large quantities of water.
Selenium poisoning (See MINERAL[S]; and SELENIUM.)				
Tin (Sn)	From acid fruits and vegetables canned in tin cans. The acids in such foods as citrus fruits and tomato products can leach tin from the inside of the can. Then the tin is ingested with the canned food. In the digestive tract tin goes through a methylation process in which nontoxic tin is converted to methylated tin, which is toxic.	Methylated tin is a neurotoxin—a toxin that attacks the central nervous system, the symptoms of which are numbness of the fingers and lips followed by a loss of speech and hearing. Eventually, the afflicted person becomes spastic, then coma and death follow.	**Distribution:** Tin cans are widely used throughout the world. **Magnitude:** The use of tin in advanced industrial societies has increased 14-fold over the last 10 years.	**Prevention:** Many tin cans are coated on the inside with enamel or other materials. Perhaps such coating should be a requisite when canning acidic fruits and vegetables. **Treatment:** This should be left to the medical doctor.

POISONS, CHEMICAL

The issue of chemical poisons is complicated, and sometimes emotional. We should remember that no chemical is totally safe, all the time, everywhere. Many chemicals, however, help make life more livable; hence, rather than banishing all chemicals, or discouraging the production of new ones, the obvious hazards must be eliminated. Then, our challenge for those chemicals with unknown hazards is their prudent and proper use and disposal coupled with monitoring their levels in food, water, and the environment. If we are able to do this, then we can reap the benefits derived from chemicals without suffering from their misuse.
(Also see PESTICIDES; and POISONS.)

POLLUTION CONTROL

Today, there is worldwide awakening to the problem of pollution of the environment (air, water, and soil) and its effects on human health and on other forms of life. Much of this concern stems from the sudden increase of animals in confinement.

We must be mindful that life, beauty, wealth, and progress depend upon how wisely man uses nature's gifts—the soil, the water, the air, the minerals, and the plant and animal life.

Certainly, there have been abuses of the environment (and it hasn't been limited to agriculture). There is no argument that such neglect should be rectified in a sound, orderly manner. But it should be done with a minimum disruption of the economy and lowering of the standard of living.
(Also see PESTICIDES; and POISONS.)

POLY-

Prefix meaning much or many.

POLYARTERITIS NODOSA (KASSMAULMAIER DISEASE)

A usually fatal disease affecting the connective tissue in the arteries of the kidneys, heart, lungs, liver, gastrointestinal tract, eyes, and joints. Diseased arteries allow blood to clot, become inflamed, and deteriorate. This causes functional impairment of the organs and tissues supplied by the affected arteries. Outward symptoms of the sufferer include fever, weight loss, abdominal and muscular pains, skin disturbances, and high blood pressure. The disease usually occurs in middle age, and affects three times as many men as women. The cause of polyarteritis is unknown, but it belongs to a class of diseases called collagen diseases. If detected early enough, steroids may provide temporary relief and slow the progess of the disease. During steroid therapy, dietary sodium may have to be restricted since many steroid preparations promote sodium retention.

POLYCYTHEMIA

Excess blood cells that contain a high concentration of hemoglobin. One type of polycythemia may be caused by an excess of cobalt in the diet of humans, resulting in stimulation of bone marrow accompanied by excessive production of red corpuscles and higher than normal hemoglobin.

POLYDIPSIA

An abnormal or excessive thirst due to the loss of body fluids, as observed in diabetes mellitus.

POLYNEURITIS

Neuritis of several peripheral nerves at the same time,

caused by metallic and other poisons, infectious disease, or vitamin deficiency. In man, alcoholism is also a major cause of polyneuritis.

POLYNEUROPATHY

A disease which causes the noninflammatory degeneration of many of the peripheral nerves. It may result from: (1) certain deficiency diseases, (2) metabolic diseases, (3) chemical poisoning, (4) some infective diseases, (5) a carcinoma, or (6) some rare genetic diseases. Symptoms consist of cramps, numbness, weakness, loss of reflexes, loss of positional senses, and partial paralysis. Usually the legs are affected first, and then the arms. Those deficiency diseases exhibiting polyneuropathy are beriberi, pellagra, chronic alcoholism, burning feet syndrome, and pyridoxine or pantothenic acid deficiency.

POLYPEPTIDES

Ten to one hundred amino acids chemically joined by a peptide bond are termed polypeptides. In the body, polypeptides occur as a result of protein digestion or protein synthesis.
(Also see PROTEIN[S].)

POLYPHAGIA

Continuous eating or excessive appetite.
(Also see APPETITE.)

POLYPHOSPHATES

The ability of the phosphates (PO_4) to chemically link up with each other and form long chains gives rise to a variety of phosphate compounds called polyphosphates. They have numerous uses in the food industry as nutrient and/or dietary supplements, sequestrants (chelating agents), flavor improvers in processed meats, and chemical aids to curing meats. The FDA considers most polyphosphates and their sodium, calcium, potassium and ammonium salts as GRAS (generally recognized as safe) additives.

POLYURIA

A medical term indicating excessive formation and discharge of urine. It is a sign of disease. For example, polyuria is noted in diabetics.

POMACE

The residue of a fruit pulp that remains after the fruit has been pressed to extract the juice. Large amounts of apple, citrus, and grape pomace are produced in the United States during the manufacturing of the respective juices. The apple and citrus pomaces are the raw materials used in the production of pectin, a gelling agent and an ingredient in diarrhea remedies. All three types of pomace are used as animal feed ingredients.

POME

A type of fruit that contains many seeds enclosed within a central cartilagenous core which, in turn, is surrounded by the fleshy portion and skin of the fruit. The best known pome fruits are apples, pears, and quinces.
(Also see FRUIT[S].)

POPULATION, WORLD

Population growth is the major determinant of demand for food.

It took from the dawn of man until 1,600 years after the birth of Christ for the number of people in the world to reach one-half billion. The population of the world first topped 1 billion in 1830. And it took another 100 years—until 1930—for the population to reach 2 billion. In 1960, it was 3 billion; in 1975, it was 4 billion; and in 1982, it was 4.5 billion. Even more frightening in the people-numbers game is the estimation of population experts for the years ahead. By the year 2000, it's predicted that the world will have 6.2 billion mouths to feed; and that by the year 2045, world population will be doubled. (See Fig. P-46.)

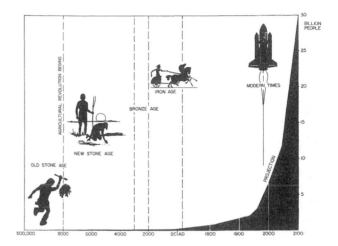

Fig. P-46. Human population growth over the past 1.5 million years, with projections to 2100 A.D.
NOTE: If the Old Stone Age were in scale, its base line would extend 8 feet to the left.

Assuming 6.2 billion (the estimated population in the year 2000) people standing shoulder to shoulder and occupying an average width of 2 ft, 6.2 billion people would circle the world 97 times, or reach to the moon and back (477,750 mi) 5 times. Today's one-year-old baby will likely see the world's population reach 12 billion—three times present numbers—within his or her lifetime.

Do these population figures foretell the fulfillment of the doomsday prophecy of Thomas Malthus, made in 1798, that world population grows faster than man's ability to increase food production? For almost 200 years, we proved Malthus wrong. Disease had a major role in holding population down; and as population increased, new land was brought under cultivation; and machinery, chemicals, new crops and varieties, and irrigation were added to step up the yields. Now science has given us the miracle of better health and longer life. Malthus was right!

The ominous challenge of our children is that the demand for food will triple in their lifetime. And that's not all! Seventy-five percent of the people of the world live in the developing countries where only 40% of the world's food is produced. Also, this is where most of the world's increase in population is occurring; population in the developing countries is now growing more than 2.1% annually compared to just under 0.6% in the developed countries. The developing countries now account for 89% of the world's annual population increase. More disturbing yet, rate of increased food production per capita in the developing countries is only two-thirds of that in the developed countries. So, people in some of the developing countries will starve to death because of population growth outrunning food production. Increased agricultural production in the developing countries requires more scientific knowledge than the older methods.

(Also see HUNGER, WORLD; MALNUTRITION, PROTEIN-ENERGY; MALTHUS; PROTEIN, WORLD PER CAPITA; and WORLD FOOD.)

PORK

Contents | Page
Qualities in Pork Desired by Consumers..............862
Federal Grades of Pork.............................862
Pork Cuts and How to Cook Them...................862
U.S. Pork Production..............................864
Per Capita Pork Consumption......................864
Pork as a Food...................................864

Fig. P-47. The crest of Lord Bacon (1561–1626)—noted English Viscount, lawyer, statesman, politician. According to some authorities, the word "bacon" was derived from Lord Bacon. (Courtesy, Picture Post Library, London, England)

Pork is the flesh of hogs, which are usually slaughtered at about 240 lb liveweight. About 70% of the pork carcass becomes salable retail cuts; the remaining 30% consists of fat trim, sausage trimmings, and feet, tail, and neckbones.

Only about ⅓ of the pork processed in the United States is sold as fresh product, the other ⅔ is cured and smoked. Originally, pork was cured to preserve it, but, today, pork is cured primarily to enhance flavor and provide variety and convenience.

QUALITIES IN PORK DESIRED BY CONSUMERS.
Consumers desire the following qualities in pork:

1. **Quality.** The quality of the lean is based on firmness, texture, marbling, and color.

2. **Firmness.** Pork muscle should be firm so as to display attractively. Firmness is affected by the kind and amount of fat. For example, pigs that are fed liberally on peanuts produce soft pork. Also, pork with small quantities of fat will contain more moisture and tend to be soft.

3. **Texture.** Pork lean that has a fine-grained texture and porous pinkish bones is preferred. Coarse-textured lean is generally indicative of greater animal maturity and less tender meat.

4. **Marbling.** This characteristic contributes to buyer appeal. Feathering (flecks of fat) between the ribs and within the muscles is indicative of marbling.

5. **Color.** Most consumers prefer pork with a white fat on the exterior and a greyish pink lean marbled with flecks of fat.

6. **Maximum muscling; moderate fat.** Maximum thickness of muscling influences materially the acceptability by the consumer. Also, consumers prefer a uniform cover of not to exceed ¼ in. (6 mm) of firm, white fat on the exterior.

7. **Repeatability.** The housewife wants to be able to secure a standardized product; pork of the same eating qualities as her previous purchase.

If these seven qualities are not met by pork, other products will meet them. Recognition of this fact is important, for competition is keen for space on the shelves of a modern retail food outlet.

FEDERAL GRADES OF PORK.
Grades of barrow and gilt carcasses are based on quality-indicating characteristics of the lean and expected yields of the four lean cuts (ham, loin, picnic shoulder, and Boston butt). Quality and yield are combined in one set of grades and are not kept separate, in contrast to the system used with beef and lamb.

From the standpoint of quality, two general levels are recognized—*acceptable* and *unacceptable*. Acceptability is determined by direct observation of the cut surface and is based on considerations of firmness, marbling, and color, along with the use of such indirect indicators as firmness of fat and lean, feathering between the ribs, and color. The degree of external fatness is not considered in evaluating the quality of the lean. Suitability of the belly for bacon (in terms of thickness) is also considered in quality evaluation, as is the softness and oiliness of the carcass. Carcasses which have unacceptable quality lean, and/or bellies that are too thin, and/or carcasses which are soft and oily are graded U.S. Utility.

If a carcass qualifies as acceptable in quality of lean and in belly thickness, and is not soft and oily, it is graded U.S. No. 1, 2, 3, or 4, based entirely on projected carcass yields of the four lean cuts; carcasses not qualifying for these four grades are graded U.S. Utility. The expected yields of each of the grades in the four lean cuts, based on using the U.S. Department of Agriculture standard cutting and trimming methods, are as given in Table P-3.

TABLE P-3
EXPECTED YIELDS OF THE FOUR LEAN CUTS
BASED ON CHILLED CARCASS WEIGHT, BY GRADE

Grade	Yield
U.S. No. 1	60.4% and over
U.S. No. 2	57.4% to 60.3%
U.S. No. 3	54.4% to 57.3%
U.S. No. 4	Less than 54.4%

Carcasses vary in their yields of the four lean cuts because of variations in their degree of fatness and in their degree of muscling (thickness of muscling in relation to skeletal size).

PORK CUTS AND HOW TO COOK THEM.
The method of cutting pork is practically the same in all sections of the United States. Fig. P-48 illustrates the common retail cuts of pork, and gives the recommended method or methods for cooking each. This informative figure may be

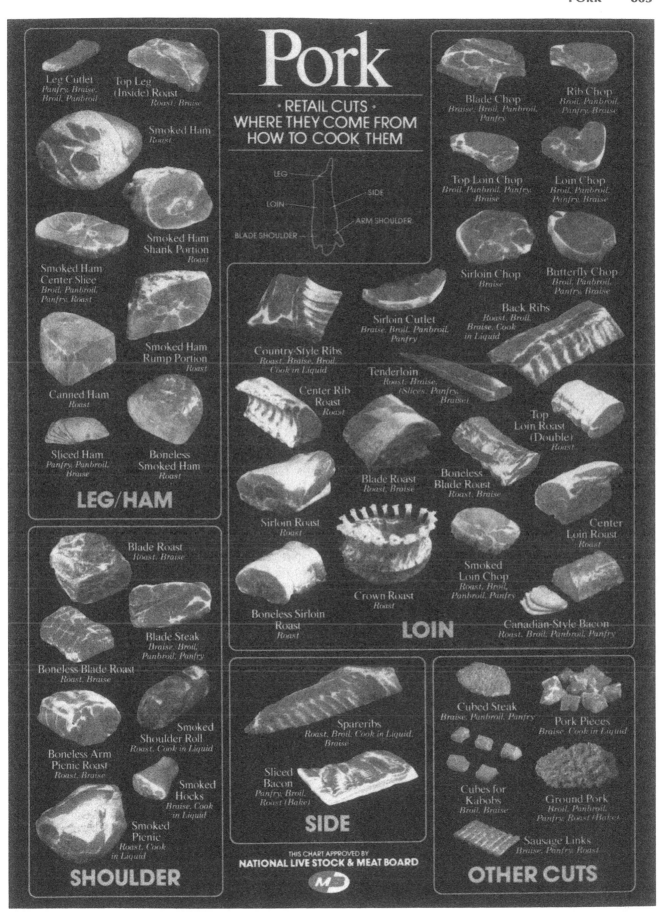

Fig. P-48. Pork Retail Cuts and how to cook them. (Courtesy, National Live Stock and Meat Board, Chicago, Ill.)

used as a guide to wise buying, in dividing the pork carcass into the greatest number of desirable cuts, in becoming familiar with the types of cuts, and in preparing each type of cut by the proper method of cookery.

U.S. PORK PRODUCTION. On December 1, 1990, there were 54,562,000 hogs in the United States; and 15.3 billion pounds *(7 bil kg)* of pork were produced in the calendar year 1990. In 1990, pork production accounted for 40% of all red meats produced in the United States, while beef accounted for 59%.

PER CAPITA PORK CONSUMPTION. In 1990, U.S. per capita consumption of pork, carcass basis, was 46.3 lb *(21 kg)*. That same year, U.S. per capita beef consumption was 112.3 lb *(51 kg)*.

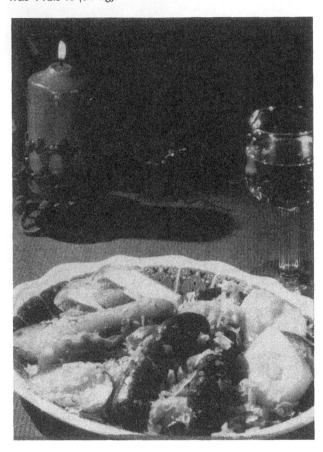

Fig. P-49. Pork sausage, rich in essential nutrients. (Courtesy, National Live Stock & Meat Board, Chicago, Ill)

Although comprising less than 5% of the world's human population and having only 6.3% of its hogs, the people of the United States eat 10% of the total world production of pork.

In general, pork consumption (and production) is highest in the temperate zones of the world, and in those areas where the human population is relatively dense. In many countries, such as China, pigs are primarily scavengers; in others, hog numbers are closely related to corn, barley, potato, and dairy production. As would be expected, the per capita consumption of pork in different countries of the world varies directly with its production and availability. Food habits and religious restrictions also affect the amount of pork consumed.

PORK AS A FOOD. Pork is an important food and a rich source of many essential nutrients. An average 3.5 oz *(99.4 g)* serving of cooked pork ham provides 37 g of protein (that's ⅔ of the recommended daily allowance of protein) and 8.8 g of fat, along with being an excellent source of minerals and vitamins. Its high-quality protein contains all the essential amino acids needed to build, maintain, and repair body tissues. Pork is rich in iron, and the iron is readily used in the formation and maintenance of red blood cells. Also, pork is a major dietary source of the B vitamins, especially thiamin, riboflavin, and niacin. Also pork is about 98% digestible.

Because of its high concentration of several key nutrients, many physicians and nutritionists regard pork as a desirable part of many special diets, such as those for peptic ulcer, diabetes, diseases of the liver, and in geriatric and pediatric conditions where it is important to maintain a good supply of high-quality protein and other body-building and restorative nutrients which patients need. The lean cuts of fresh and cured pork are excellent foods, which afford appetizing variety for the patient.

(Also see MEAT[S].)

PORPHYRIN

A derivative of porphin. Porphyrin is present in hemoglobin; heme is ferroprotoporphyrin, a porphyrin combined with iron (ferro- for Fe^{++}).

PORTAL

An entrance. When applied to blood, it may refer to—

• **Portal blood**—The blood in the portal vein passing from the gastrointestinal tract to the liver.

• **Portal circulation**—The circulation within the liver; blood is taken into the liver by the portal vein and carried out by the hepatic vein.

• **Portal system**—The portal vein and its branches through which the portal circulation takes place; the portal system begins and ends in capillaries.

• **Portal vein**—A vein carrying blood from the digestive organs and spleen to the liver where the nutrients are altered by liver cells before passing into the systemic circulation.

POSTOPERATIVE NUTRITION

Major surgery results in a generalized loss of tissue—a catabolic phase—placing the patient in a negative nitrogen balance state. This is in part due to the changes brought about by the neuroendocrine system stimulating the secretion of adrenal cortical hormones. However, the nutritional concerns for most patients immediately following surgery are fluids, electrolytes, and energy. Depending upon the type of surgery these can be given (1) orally, (2) by a nasogastric tube, (3) by a tube placed directly in the small intestine or stomach, or (4) intravenously. As soon as possible the patient should be returned to an adequate oral intake of energy and protein to support a positive nitrogen balance during the time new tissues are being synthesized—the anabolic phase. As normal convalescence proceeds, and if there are no complications, the patient's appetite provides a reliable guide to requirements.

Specific information regarding postoperative situations requiring specialized methods is explained in detail elsewhere.

(Also see MODIFIED DIETS; INTRAVENOUS [Parenteral] NUTRITION, SUPPLEMENTARY; LIQUID DIETS; and TOTAL PARENTERAL [Intravenous] NUTRITION.)

POTASSIUM (K)

Contents	Page
History	865
Absorption, Metabolism, Excretion	865
Functions	865
Deficiency Symptoms	865
Interrelationships	865
Recommended Daily Allowance of Potassium	866
Toxicity	866
Sources of Potassium	866

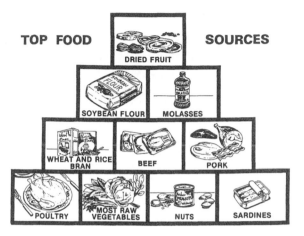

Fig. P-50. Top food sources of potassium.

Potassium is a silver white metallic element. It is never found as a pure metal—it is always combined with other substances.

Potassium is the third most abundant element in the body, after calcium and phosphorus; and it is present in twice the concentration of sodium.

Potassium constitutes about 5% of the mineral content of the body. It is the primary cation of intracellular (within the cells) fluids. Approximately 98% of the total body potassium is located intracellularly, where its concentration is 30 or more times that of the extracellular (between cells) fluid. The concentration of sodium in blood plasma is much higher than potassium. On the other hand, the potassium concentration in muscle tissue and milk is many times higher than sodium.

HISTORY. The history of potassium is closely linked to that of sodium. Materials containing their compounds, particularly carbonates and nitrates, were known to some of the earliest civilizations. Records show that they were used in Mesopotamia in the 17th century B.C. and in Egypt in the 16th century B.C. However, the ancient technicians and artisans who used these materials, having no knowledge of the chemical and physical methods of analysis and identification, did not distinguish between them. This problem was finally solved in 1807 by the brilliant young English chemist Sir Humphry Davy, who isolated the metal which he named potassium, and gave it the chemical symbol K, from *Kalium*, the Latinized version of the Arabic word for *alkali*. However, more than 100 years elapsed following discovery before McCollum, in 1938—using the rat, obtained positive proof that potassium is an essential nutrient, although this had been suggested earlier.

ABSORPTION, METABOLISM, EXCRETION. Absorption of dietary potassium is very efficient; more than 90% of ingested potassium is absorbed. Most of the absorption occurs in the small intestine.

Although the digestive juices contain relatively large amounts of potassium, most of it is reabsorbed and the loss in the feces is small.

The kidneys provide the major regulatory mechanism for maintaining potassium balance; and relatively wide variations in intake are not reflected in fluctuations in plasma concentration. Aldosterone, an adrenal hormone, stimulates potassium excretion. Also, alcohol, coffee, and excess sugar increase the urinary excretion of potassium.

Excessive potassium buildup may result from kidney failure or from severe lack of fluid.

FUNCTIONS. Potassium and sodium are closely interrelated in the maintenance of proper osmotic pressure within cells. Both minerals are involved in the maintenance of proper acid-base balance and the transfer of nutrients in and out of individual cells. The potassium ion relaxes muscle; hence, a high concentration in the heart relaxes the heart—action opposite to that of calcium, which is stimulatory.

Potassium is also required for the secretion of insulin by the pancreas, in enzyme reactions involving the phosphorylation of creatine, in carbohydrate metabolism, and in protein synthesis.

DEFICIENCY SYMPTOMS. A potassium deficiency may cause rapid and irregular heartbeats, and abnormal electrocardiograms; muscle weakness, irritability, and occasionally paralysis; and nausea, vomiting, diarrhea, and swollen abdomen.

Hypokalemia (decreased serum potassium) of dangerous degree may be caused by a prolonged wasting disease with tissue destruction and malnutrition; by prolonged gastrointestinal loss of potassium as in diarrhea, vomiting, or gastric suction; or by continuous use of diuretic drugs. Finally, the heart muscle may stop.

It should be noted that deficiencies of potassium rarely result from dietary lack of the mineral; rather, they result from crash diets, diarrhea, diabetic acidosis, vomiting, intense and prolonged sweating, body burns, and heavy urine losses induced by diuretic drugs (also known as "water pills").

INTERRELATIONSHIPS. A magnesium deficiency results in failure to retain potassium; hence, it may lead to a potassium deficiency.

Excessive levels of potassium interfere with magnesium absorption. Also, large excesses of potassium may slow the heart to a standstill when the kidneys are unable to excrete the surplus in the urine.

Because sodium and potassium must be in balance, excessive use of salt depletes the body's potassium supply. Although sodium intake may be the most important dietary determinant of blood pressure, variations in the sodium:potassium ratio in the diet affect blood pressure under certain circumstances.

Potassium is almost a constant component of lean body tissue, so much so that one method of estimating the amount of lean tissue in a person is by measuring the amount of potassium present. (This is accomplished by determining the amount of radioactive potassium that is naturally present in a constant ratio to ordinary potassium.) The need for potassium is increased when there is growth or deposition of lean tissue; and potassium is lost whenever muscle is broken down owing to starvation, protein deficiency, or injury.

Potassium is often prescribed for people with high blood pressure who are required to take diuretics to reduce excess water in the body because of the belief that diuretics deplete the body of potassium. However, newer knowledge indicates that diuretics do not seriously deplete natural potassium levels of the great majority of patients taking these drugs for high blood pressure. Since high blood potassium concentration can cause cardiac arrest (a heart attack) and death, potassium supplements should be taken with caution and only on the advice of a physician.

RECOMMENDED DAILY ALLOWANCE OF PO-TASSIUM. The exact potassium requirements are unknown. But it is known that the daily intake on typical diets far exceeds the requirements.

Growth (increase in lean body mass) and fecal losses are the major determinants of potassium needs during the early years of life. The Food and Nutrition Board of the National Research Council (FNB-NRC) estimated minimal potassium requirements for the infant and young child are 65 mg per day (Recommended Dietary Allowances, 10th ed., 1989, p. 257). Human milk contains 500 mg of potassium per liter, whereas commonly used commercial formulas contain slightly more and cow's milk contains about 1,364 mg per liter. Average intakes of potassium during the first year of life range from about 780 mg per day at age 2 months to about 1,600 mg per day at the end of the first year; hence, normal diets of infants and young children provide a surplus of potassium.

Under ordinary circumstances, the healthy adult can maintain potassium balance with an intake of 1,600 to 2,000 mg per day.

Estimated safe and adequate intakes of potassium are given in the section on MINERAL(S), Table M-25, Mineral Table. The levels of potassium suggested in this table were calculated from the sodium intakes in order to achieve equivalent amounts of potassium on a molar basis. Older individuals need relatively less potassium than the rapidly growing infant, but an equivalent intake of potassium appears to be somewhat protective against the blood pressure-elevating effects of a given level of sodium.

• **Potassium intake in average U.S. diet**—The normal adult dietary intake of potassium is about 2,500 mg per day, which is more than adequate.

TOXICITY. Acute toxicity from potassium (known as hyperpotassemia or hyperkalemia) will result from sudden increases of potassium to levels of about 18 g per day for an adult, provided the kidneys are not functioning properly and there is not an immediate and sharply increased loss of potassium from the body. Although no significant increase in intracellular potassium content or in total body potassium occurs in this condition, hyperkalemia may prove fatal due to cardiac arrest.

SOURCES OF POTASSIUM. Potassium is widely distributed in foods. Meat, poultry, fish, many fruits, whole grains, and vegetables are good sources. (Although meat, poultry, and fish are good sources of potassium, these items may be restricted when dietary sodium is restricted.)

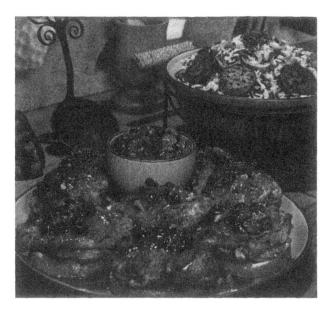

Fig. P-51. Chicken Valencia, with raisins. Chicken and raisins are good sources of potassium. (Courtesy, P. Sigel/Ketchum, San Francisco, Calif.)

Groupings by rank of common food sources of potassium are given in the section on MINERAL(S), Table M-25, Mineral Table.

NOTE WELL: Sources of potassium other than foods, such as potassium chloride, should be taken only on the advice of a physician or nutritionist.

For additional sources and more precise values of potassium, see Food Composition Table F-21 of this book.

(Also see MINERAL[S], Table M-25.)

POTATO (IRISH POTATO) Solanum tuberosum

This vegetable, commonly known as Irish or white potato, is grown around the world and ranks after the major grains (wheat, rice, and corn) in importance as a food. It is a member of the nightshade family of plants (Solanaceae), which includes the tomato, red pepper, tobacco plant, and eggplant. However, it is not related to the sweet potato or the yam. The scientific name Solanum is derived from the Latin word solamen, which means soothing.

Fig. P-52. Potatoes, a vegetable which feeds millions of people around the world. (Courtesy, Shell Chemical Co., Houston, Tex.)

The edible part of the potato plant is the tuber, which grows in the ground. Tubers are not part of the roots of the plant, but are formed from underground stems. The tubers of the modern types of potatoes range in size from as small as a pea to so large that they can hardly be lifted by a strong man.

ORIGIN AND HISTORY.
Native to the Andes mountains of Peru and Bolivia, the potato was first cultivated between 4,000 and 7,000 years ago. It, along with corn (maize), was a staple of the Inca Indian diet. Spanish explorers took potatoes to Europe in the 16th century.

WORLD AND U.S. PRODUCTION.
The potato is the leading vegetable crop of the world, by a wide margin. About 270 million metric tons are produced worldwide, annually. The leading potato-producers, by rank, are: U.S.S.R., Poland, China, United States, India, Germany, U.K., Netherlands, France, and Spain.[31]

The potato is also the leading vegetable crop of the United States. The United States produces about 20 million short tons annually. By rank, the leading states are: Idaho, Washington, Colorado, Wisconsin, Oregon, Maine, California, North Dakota, Minnesota, and Michigan. Idaho alone accounts for 27% of the nation's potatoes.[32]

PROCESSING.
The consumption of fresh potatoes has declined steadily since World War II, while the use of various forms of processed potatoes has risen to the point where over half the crop is processed. Over 97% of the processed products is represented by chips and shoestring potatoes, dehydrated items (mainly instant mashed potatoes), and frozen french fries and other frozen products.

Fig. P-53. Newly harvested potatoes. (Courtesy, USDA)

[31]Data from *FAO Production Yearbook 1990*, FAO/UN, Rome, Italy, Vol. 44, p. 90, Table 26. **Note well:** Annual production fluctuates as a result of weather and profitability of the crop.

[32]Data from *Agricultural Statistics 1991*, USDA, p. 159, Table 228. **Note well:** Annual production fluctuates as a result of weather and profitability of the crop.

SELECTION, PREPARATION, AND USES.
Good quality potatoes are sound, firm, relatively smooth, and well shaped.

Several hundred varieties of potatoes are grown in the world. They differ in time of maturity, appearance, skin colors, resistance to insects and diseases, yield, and cooking and marketing qualities. A variety that thrives in one area may do poorly in another area.

Most supermarkets carry many types of canned, dehydrated, or frozen products that enable homemakers to prepare potato dishes with a minimum expenditure of time. Fig. P-54 shows a dish made from fresh potatoes.

Fig. P-54. Baked and stuffed potatoes being browned in an oven. (Photo by J. C. Allen & Son, Inc., West Lafayette, Ind.)

Some suggestions for preparing various potato dishes follow:

1. There is no need to peel potatoes unless dirt is ground into the skins. The outer layer of the tuber is richer in nutrients than the inner layers. However, the skins should be cleaned thoroughly with a vegetable brush.

2. Microwave heating of baked potatoes is much faster than baking them in an ordinary oven. Also, the use of a microwave oven is a more efficient use of energy because it does not give off heat to the room in which it is located.

3. Cooking potatoes in a pressure cooker requires only a fraction of the time needed for boiling them in an open saucepan.

4. The oil or fat used to fry potatoes should *not* be heated to the smoking point because it may break down into bitter tasting and harmful products that are likely to be absorbed by the potatoes.

5. It is better to cook frozen french fries without any additional fat by placing them in an oven or under a broiler than it is to refry them because the second frying often gives the pieces a tough, poorly digested "skin" that contains more fat than most people need.

6. The fat content of homemade french fries may be kept low by tossing the raw pieces of potato in a little heated fat or oil (about 1 Tbsp per cup of potatoes), then baking them in an oven.

7. Yogurt or cottage cheese that has been blended or whipped until smooth may be substituted for butter, cheese, mayonnaise, or sour cream as a dressing for potatoes. The substitution works best when liberal amounts of seasonings such as chives, dried horseradish, mustard powder, oregano, pepper, and/or thyme are used.

NUTRITIONAL VALUE.
The nutrient composition of various forms of potatoes is given in Food Composition Table F-21.

Fig. P-55. Versatile, everyway potatoes. (Courtesy, United Fresh Fruit & Vegetable Assn., Alexandria, Va.)

Some noteworthy observations regarding the nutrient composition of potatoes follow:

1. An average size raw potato (7 oz or *200 g*) contains about 115 Calories (kcal), 3.2 g of protein, 80 mg of phosphorus, 1 mg of iron, and 30 mg of vitamin C (about the amount in one-half of an average size orange), but only small amounts of calcium and vitamin A. Boiled potatoes have a similar nutrient composition, except that some of the vitamin C is destroyed by cooking.

2. Baked potatoes contain about 25% more solids and proportionately higher levels of all nutrients than raw or boiled potatoes. Also, an average size baked potato is equivalent in nutritional value to a cup of mashed potatoes, and contains about 4 g of protein, or the amount present in about a cup of cooked cereal such as cornmeal or rice.

3. French frying reduces the water content of potatoes to the extent that the cooked product contains over twice the solids, 3 to 4 times the calories, and double the protein content of baked or boiled potatoes. Furthermore, each ounce (*28 g*) of french fries contains almost 1 tsp (*5 ml*) of fat.

4. Potato chips and shoestring potatoes are high in calories because of their very low water content (2%) and high fat content (40%). Therefore, they are concentrated sources of calories (2,400 to 2,600 kcal per lb, or *0.45 kg*), protein (24 to 29 g per lb), and salt (up to 5,000 mg of sodium per lb). However, salt-free potato chips are sold in many health food stores.

5. Dried potato products, such as instant mashed potato granules and potato flour, are similar in composition to cereal flours. Therefore, they may be used in breads, pancakes, waffles, and similar baked products. Moist forms of potatoes may be used to make these items if the liquid called for in the recipes is reduced somewhat. The Irish have long made potato breads with mashed potatoes.

6. A 1-cup (*240 ml*) serving of au gratin potatoes made with cheese contains slightly more calcium and protein than a cup of milk, and about twice as many calories.

7. Potato salads are usually high in calories due to the liberal use of fat-rich mayonnaise or salad dressing in their preparation. Nevertheless, 1 cup of potato salad made with chopped egg usually provides about twice as much protein as a cup of cooked cereal.

Potatoes that are green on the surface (due to exposure to light), spoiled, or sprouted may contain harmful amounts of solanine. (See SOLANINE.)

• **The use of potatoes as a staple food**—People living in certain regions where grains do not grow well have long depended upon potatoes as a staple food. Nevertheless, there is a widely held belief that the potato-dominated diets of the Peruvian Indians and the Irish were likely to have led to multiple nutritional deficiencies. Therefore, Fig. P-56 is presented to help dispel such a belief.

It may be seen from Fig. P-56 that the yield from 1 acre (*0.4 ha*) of potatoes meets both the energy and protein requirements for over ten people, whereas, the other major life-sustaining crops are not as balanced in these nutrients. The requirements for most of the other nutrients may be met if sufficient quantities of potatoes are consumed together with other foods that supply calcium and vitamin A. The Peruvian Indians obtained these supplemental nutrients from the leaves of amaranth (pigweed) and chenopodium (lambsquarters), but, it is likely that the Irish received them from Irish moss (a seaweed) and other plants.

• **Protein Quantity And Quality**—Potatoes, which contain only 2% protein on a fresh basis, are being tested in

Fig. P-56. Number of people whose requirements are met by the yield from 1 acre of the specified crop.

Peru and in other countries for use as a protein supplement. Some of the reasons for this interest in the tubers are: (1) dried potato contains 8% protein; (2) the protein-to-calorie ratio of potatoes is 2.3 g per 100 Calories (kcal), which is about the same as that for corn and rice; and (3) potato protein supplies ample lysine, the amino acid that is lacking in cereal grains. Hence, mixtures of potato and cereal proteins contain higher quality protein than either type of food alone.

(Also see FLOURS, Table F-13, Special Flours.)

POTATO SYRUP

A clear, viscous, syrup produced by the hydrolysis—splitting—of potato starch with heat and acids, or enzymes. It contains dextrins, maltotetrose, maltotriose, maltose, and dextrose (glucose). The amount of each depends upon the conditions and duration of the hydrolysis. Corn syrup, derived in the same manner from corn starch, is more common in the United States.

POULTRY

Contents
Kinds of Poultry 869
World Poultry Numbers, Production, and Consumption . . .870
Marketing Poultry 871
  Processing, Inspecting, Classifying, and Grading 871
    Federal Inspection 871
    Market Classes of Poultry 871
    Market Grades of Poultry 871
  Retailing Poultry 871
How to Buy Poultry 872
How to Store Poultry 872
How to Handle and Cook Poultry .873
U.S. Per Capita Consumption of Poultry Meat 874
Nutritive Qualities of Poultry Meat 874

The term *poultry* applies to a rather wide variety of birds of several species, and it refers to them whether they are alive or dressed. Poultry appears more frequently in the diet of people throughout the world than any other type of meat.

KINDS OF POULTRY. The most common kinds of poultry used for meat follow (the respective names of each kind of poultry also refer to the flesh of each of them used for food):

• **Chicken**—The common domestic fowl (*Gallus gallus*).

• **Turkey**—The large American bird (*Meleagris gallopavo*) derived primarily from a Mexican variety of the wild turkey and raised chiefly for their flesh.

Fig. P-57. Chickens, the most common type of poultry. (Courtesy, USDA)

• **Ducks**—Swimming birds of the family *Anatidae*, characterized by short neck and legs, a broad, flat bill, the sexes differing from each other in plumage, and smaller in size than swans and geese.

• **Geese**—Swimming birds of a distinct subfamily of *Anatidae*, characterized by a high somewhat compressed bill, legs of moderate length, and longer necks and larger size than ducks.

• **Guinea fowl**—A West African bird of the subfamily *Phasianidae* (now usually classed as a separate family, *Numididae*), characterized by a bare neck and head.

• **Pigeon**—A bird of the widely distributed family *Columbidae*, characterized by a stout body with rather short legs, a bill that is horny at the tip but soft at the base, and smooth, compact plumage.

WORLD POULTRY NUMBERS, PRODUCTION, AND CONSUMPTION.

Through the ages, poultry meat and eggs have been basic foods throughout the world. In the future, they will become increasingly important for feeding the hungry for the following reasons:

1. Poultry convert feed to food efficiently.
2. The poultry industry can adjust rapidly to a variety of economic factors, e.g., feed availability, demand for animal products, cost, etc.
3. Poultry feeds are not commonly used for human consumption.
4. Layers provide a continuous source of food.
5. Most vegetarians eat eggs.
6. Poultry products are inexpensive.
7. Poultry manure can be used as a fertilizer or as a feed.

• **World poultry numbers**—The production of poultry is worldwide. Figs. P-58, P-59, and P-60 show the leading poultry-producing countries of the world.

China holds a commanding lead in chicken numbers, followed by the United States and the U.S.S.R., respectively.

The United States ranks first in turkey production with the U.S.S.R. second, and France third.

China ranks first in duck numbers, followed by Bangladesh, Indonesia, Vietnam, Thailand, and France, respectively. The duck numbers given in Fig. P-60 are, as shown, from the *FAO Production Yearbook 1991*, United Nations.

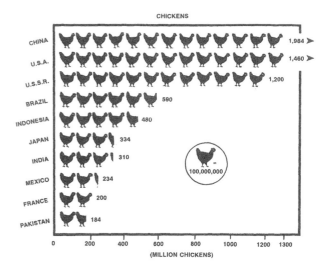

Fig. P-58. Leading chicken producers of the world. (Data from *FAO Production Yearbook 1990*, FAO/UN, Rome, Italy, Vol. 44, p. 197)

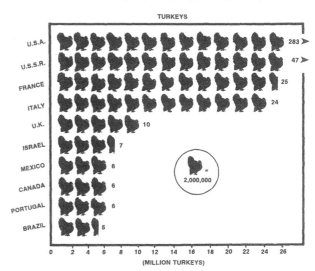

Fig. P-59. Leading turkey producers of the world. (Except for U.S. turkeys raised, data from *FAO Production Yearbook 1990*, FAO/UN, Rome, Italy, Vol. 44, p. 197. U.S. turkey data, showing turkeys raised, from *Agricultural Statistics 1991*, USDA, p. 347, Table 516)

Fig. P-60. Leading duck producers of the world. (Data from *FAO Production Yearbook 1990*, FAO/UN, Rome, Italy, Vol. 44, p. 198, Table 91)

Although not shown in the FAO report, it is noteworthy that the annual production of ducks in the United States is approximately 20 million.

• **World poultry meat production**—Table P-4 shows the poultry meat production of the ten leading countries of the world in 1990. With a production of 10,851 metric tons, the United States ranks first in total poultry meat production; that is 27% of the world total.

TABLE P-4
POULTRY MEAT PRODUCTION IN
TEN LEADING COUNTRIES OF THE WORLD[1]

Country (leading countries, by rank, of all poultry meats)	Total Poultry Meat Production 1990
	(1,000 MT)
U.S.A.	10,851
China	3,295
U.S.S.R.	3,280
Brazil	2,417
Japan	1,487
France	1,384
Italy	1,104
United Kingdom	981
Spain	825
Canada	699
World Total	39,862

[1]*FAO Production Yearbook*, Vol. 44, 1990, p. 212.

• **World poultry meat consumption**—United States, with a per capita consumption of 91.9 lb *(41.8 kg)*, is the leading poultry meat consumer, followed by Singapore with 81.8 lb *(37.2 kg)*.

Table P-5 shows the leading poultry eaters of the world.

TABLE P-5
LEADING POULTRY MEAT-EATING
COUNTRIES OF THE WORLD 1990[1]

Country (leading countries by rank)	Per Capita Consumption, Dressed Weight Basis	
	(lb)	(kg)
U.S.A.	91.9	41.8
Singapore	81.8	37.2
Israel	77.6	35.3
Hong Kong	70.8	32.2
Canada	61.5	28.0
Saudi Arabia	56.0	25.5
Australia	54.0	24.5
Taiwan	53.8	24.5
Hungary	51.2	23.3
Spain	50.3	22.9

[1]*Statistical Abstract of the United States 1991*, U.S. Dept. of Commerce, p. 843, Table 1452 (with bone in). Per capita consumption is on ready-to-cook basis.

MARKETING POULTRY. It is estimated that 100% of the broilers and 90% of the turkeys are produced under some kind of integrated or contract arrangement. This means that a limited number of big firms are organized in such manner as to control every level of production—all the way from producing broilers and turkeys, through the processing, and finally, the ultimate promotion and marketing of the finished product.

Processing, Inspecting, Classifying, and Grading. The slaughter and processing of broilers and turkeys is an assembly line operation conducted under sanitary conditions. Inspecting, classifying, and grading are a part of the processing operation.

FEDERAL INSPECTION. The Poultry Products Inspection Act was enacted on August 28, 1957, and became fully effective January 1, 1959. It requires inspection of poultry and poultry products by the U.S. Department of Agriculture. Birds are inspected twice: (1) live (antemortem); and (2) after slaughter (postmortem), when the carcasses and entrails are examined. Inspection assures U.S. consumers that the retail meat supply has been inspected to assure wholesomeness; sanitary preparation and handling; and freedom from disease, adulteration, and misbranding.

MARKET CLASSES OF POULTRY. The U.S. Department of Agriculture has established specifications for different kinds, classes, and grades of poultry. *"Kind" refers to the different species of poultry, such as chickens, turkey, ducks, geese, guineas, and pigeons. "Class" refers to kinds of poultry by groups which are essentially of the same physical characteristics, such as fryers or hens. These physical characteristics are associated with age and sex.* The kinds and classes of live, dressed, and ready-to-cook poultry listed in the U.S. classes, standards, and grades are in general use in all segments of the poultry industry.

MARKET GRADES OF POULTRY. Dressed and ready-to-cook poultry are graded for class, condition, and quality. These are most important since they are the grades used at the retail level. These grades are:

U.S. Grade A
U.S. Grade B
U.S. Grade C

These grades apply to dressed and ready-to-cook chickens, turkeys, ducks, geese, guineas, and pigeons.

Additionally, there are U.S. Procurement Grades, which are designed primarily for institutional use. These grades are: U.S. Procurement Grade 1 and U.S. Procurement Grade 2. In procurement grades, more emphasis is placed on meat yield than on appearance.

The factors determining the grade of carcass, or ready-to-cook poultry parts therefrom are: conformation, fleshing, fat covering, pinfeathers, exposed flesh, discoloration, disjointed bones, broken bones, missing parts, and freezing defects.

Retailing Poultry. Most of the poultry meat marketed consists of broilers and turkeys, although ducks, geese, guineas, and pigeons are also marketed.

• **Broilers**—In 1989, processors sold 18.3% of their broilers as whole carcasses, 50.4% as cut-up parts, 6.3% as further processed, and 25% in other forms. In 1992, approximately 40% of the broilers were further processed.

Fig. P-61. Fresh dressed turkey in the supermarket fresh meat case. (Courtesy, Cryovac Division, W. R. Grace & Co., Duncan, S.C.)

● **Turkeys**—About 80% of the turkey meat produced today is cut-up or used as further processed product. Whole bird processed turkeys, such as self-basting turkeys, comprise about 20% of the total.

Most turkeys marketed whole are Grade A. To the extent possible, Grades B and C and lower-priced parts are utilized in further processing.

More and more further processed and whole bird turkeys are being marketed directly to retail outlets and institutional outlets.

HOW TO BUY POULTRY. When buying any kind of poultry—chicken, turkey, duck, goose, guinea, or squab—in addition to price, the buyer should look for and select on the bases of the following:

1. **Look for the inspection mark.** Poultry must be officially inspected for wholesomeness before it can be graded for quality. Often the inspection mark and the grade shield are displayed together, as shown in Fig. P-62.

Fig. P-62. *Left:* the mark of wholesomeness. *Right:* the mark of quality.

2. **Select by grade (quality).** Look for the USDA shield, which certifies that the poultry has been graded for quality by a technically trained government grader and is the buyer's assurance of quality (see Fig. P-62, *right*).

The highest quality is Grade A. Grade A birds are—

● fully fleshed and meaty.

● well finished.

● attractive in appearance.

Grade B birds may be less attractive in finish and appearance and slightly lacking in meatiness.

3. **Select by class.** The grade of poultry does not indicate tenderness—the age (class) of the bird determines tenderness. Young birds are more tender than older one.

4. **Select to suit the occasion.** Poultry can be selected to fill every need.

● Chilled or frozen ready-to-cook poultry may be purchased in various sizes and forms to suit every occasion.

● Most kinds of ready-to-cook poultry are available as parts and in whole, halved, and quartered form. Some kinds are also available as boneless roasts and rolls.

HOW TO STORE POULTRY. All poultry is perishable. So, it is important that it be properly stored. To this end, the following pointers will be helpful.

For Refrigeration:

● Store poultry in the refrigerator only if you are going to use it within a few days. Temperatures between 35° and 40°F *(2° and 4°C)* should be maintained for refrigerator storage. Keep it in the coldest part of the refrigerator, which is usually near the ice cube compartment or in a special meat keeper.

● Wrap poultry properly for refrigeration. The special wrap on prepackaged poultry is designed to control moisture loss in the refrigerator. Raw poultry wrapped in paper should be unwrapped, placed on a platter, and then covered for refrigeration. Wrap and store giblets separately.

● Use fresh-chilled poultry within 1 to 2 days.

● Follow Table P-6, Poultry Storage Time Chart, relative to refrigerating poultry.

For Freezing:

● Keep frozen poultry at 0°F *(−18°C).*

● Keep frozen poultry hard-frozen until time to thaw, then cook promptly after thawing.

● Refreeze poultry only if it still contains ice crystals or if it is still cold and has not been held at refrigerator temperature for more than 1 to 2 days. Remember that refreezing may reduce the quality of a product.

● Follow Table P-6, Poultry Storage Time Chart, relative to freezing poultry.

TABLE P-6
POULTRY STORAGE TIME CHART

Product	Refrigerator 35°−0°F *(2°−4°C)*	Freezer 0°F *(−18°C)*
	(days)	(months)
Fresh Poultry:		
Chicken and turkey (whole)	1 to 2	12
Chicken (pieces)	1 to 2	6
Turkey (pieces)	1 to 2	6
Duck and goose (whole)	1 to 2	6
Giblets	1 to 2	3
Cooked Poultry:		
Pieces (covered with broth)	1 to 2	6
Pieces (not covered)	1 to 2	1
Cooked poultry dishes	1 to 2	6
Fried chicken	1 to 2	4

HOW TO HANDLE AND COOK POULTRY. Care and cleanliness should be used in the preparation, cooking, and serving of poultry products. The following pointers will be helpful:

- Wash poultry before preparing it for cooking.

- Wash your hands often when preparing poultry.

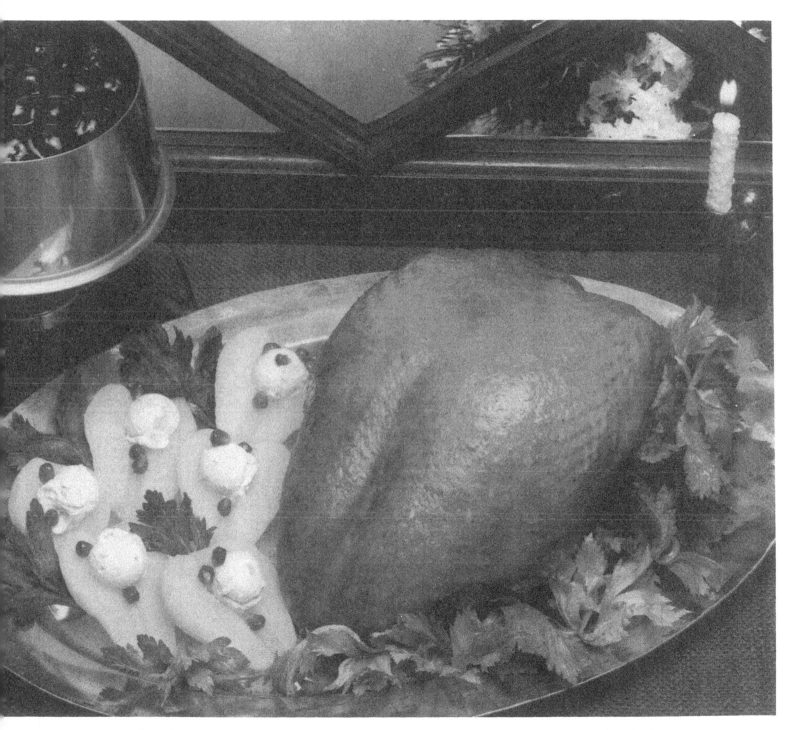

Fig. P-63. Roast turkey. Turkeys produce a higher proportion of edible meat to liveweight than any other species, and compare favorably with other meats as a source of amino acids. (Courtesy, National Turkey Federation, Reston, Va.)

• Completely cook poultry at one time. Never partially cook, then store, and finish cooking at a later date.

• It is safest to cook dressing outside the bird, but if you want to stuff it, do so right before roasting; don't stuff raw poultry and then refrigerate or freeze it. Commercially stuffed poultry should always be cooked without thawing.

• Left-over cooked poultry, broth, stuffing, and gravy should be separated, covered, and refrigerated. Use within 1 to 2 days. Freeze for longer storage.

U.S. PER CAPITA CONSUMPTION OF POULTRY MEAT.
Poultry meat is supplied chiefly by chickens and turkeys, although ducks, geese, guinea fowl, squabs (pigeon), and other fowl contribute thereto. Poultry meat is economical, and quick and easy to prepare and serve.

Fig. P-64 shows the U.S. per capita consumption of ready-to-cook poultry since 1965. Note that most of the increase occurred with broilers. In 1990, poultry meat consumption on boneless trimmed weight, or edible weight, totaled 63.8 lb (29 kg).

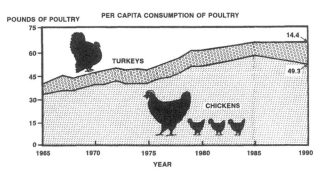

Fig. P-64. Per capita consumption of poultry—boneless, trimmed weight, or edible weight. (From: USDA)

Fig. P-65 shows how the consumption of poultry meat is fast catching up to beef and pork. From 1960 to 1990, the per capita consumption of poultry almost doubled. During this same period of time, beef and pork consumption both dropped 39%. If the recent trends shown in Fig. P-65 con-

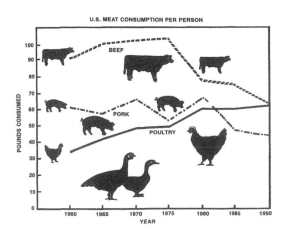

Fig. P-65. U.S. consumption of poultry, pork, and beef.

tinue, soon Americans will be eating more pounds of poultry meat than of beef or pork.

NUTRITIVE QUALITIES OF POULTRY MEAT.
Poultry meat has a number of desirable nutritional properties. Chicken and turkey meat is higher in quality protein and lower in fat than beef and pork. Additionally, the protein is a rich source of all the essential amino acids. The close resemblance of the amino acid content of poultry meat to the amino acid profiles of milk and eggs serves to emphasize the latter point.

Poultry meat is also a good source of phosphorus, iron, copper, and zinc. Additionally, it is a rich dietary source of vitamin B-12 and vitamin B-6, and it supplies appreciable amounts of vitamin A, biotin, niacin, pantothenic acid, riboflavin, and thiamin.

Poultry is a good, low-fat source of protein. However, poultry skin contains 32.35% saturated fat. Removing the skin reduces the amount of fat by approximately 5 grams per 2.5 oz breast. Because consumers are turning to foods that are lower in cholesterol and saturated fats, beginning in the early 1990s almost all U.S. restaurants included low fat, skinless chicken on their menus.

Although the RDAs are believed to be more than adequate to meet the requirements of most consumers, it is recognized that individuals vary greatly in their requirements—by as much as twofold. So, Table P-7 presents data showing the contribution of meat-poultry-fish combined, toward the total dietary intake of each of the nutrients.

TABLE P-7
CONTRIBUTIONS OF MEAT-POULTRY-FISH COMBINED, TO THE DIETARY INTAKE IN THE UNITED STATES[1,2]

Nutrition	Average Per Capita Consumption Per Day	Total Consumption Contributed by Meat-Poultry-Fish Combined
		(%)
Energy	3,600 kcal	18.8
Carbohydrate	425 g	0.1
Fat	168 g	31.7
Protein	105 g	43.3
Calcium	890 mg	3.9
Phosphorus	1,540 mg	29.2
Magnesium	339 mg	15.4
Copper	1.7 mg	17.5
Iron	17.1 mg	22.1
Zinc	12.7 mg	47.2
Vitamin A (R.E.)	1,630 mcg	20.6
Folate	284 mcg	9.8
Niacin	26 mg	44.8
Riboflavin	2.4 mg	22.2
Thiamin	2.2 mg	25.0
Vitamin B-6	2.2 mg	40.5
Vitamin B-12	9.1 mcg	76.1
Vitamin C	118 mg	2.2

[1]Agricultural Statistics 1991, USDA, pp. 474–479.

[2]The average yearly per capita consumption (carcass basis) amounted to 112.3 lb of red meat, 63.8 lb of poultry, and 15.4 lb of fish that was consumed in 1990.

PRECURSOR

A compound that can be used by the body to form another compound; for example, carotene is a precursor of vitamin A.

(Also see VITAMIN A.)

PREGNANCY AND LACTATION NUTRITION

Contents   Page
History...875
Changes in the Female Body During Pregnancy
    and Lactation...................................876
Nutrient Allowances 877
Dietary Guidelines.................................880
Common Nutrition-Related Problems of Pregnancy 881
Anemia...881
Constipation 881
Cravings for Unusual Foods and Nonfood
    Items (Pica)...................................881
Diabetes...881
Excessive Gain in Weight...........................881
Fluid Retention....................................882
Heartburn..882
High Blood Pressure (Hypertension).................882
Insufficient Gain in Weight........................882
Morning Sickness...................................882
The Decline of Toxemia as a Problem During
    Pregnancy......................................882
Breast-Feeding (Lactation).........................883
Maternal Diet During Lactation.....................883
Nutritional Supplementation........................883
Use of Drugs by the Lactating Mother...............883
Factors that May Warrant Special Counseling
    of Pregnant Women..............................883
Adolescence..883
Ethnic and/or Language Differences.................884
Habituation to Alcohol, Drugs, and/or Smoking......884
History of Obstetrical Complication.................884
Low Income...884
Psychological Condition............................884
Rh Factor Incompatibility..........................884
Several Previous Pregnancies.......................885
Use of an Oral Contraceptive.......................885
Vegetarianism 885

Women who consume a good diet before conception and during pregnancy and lactation greatly improve their chances of successful reproduction, including (1) absence of complications during pregnancy, (2) birth of a healthy full-term baby, and (3) having little or no trouble in meeting the nutritional needs of their infants, whether they are breast-fed or formula-fed. Although even poorly fed mothers may have healthy babies, this result is sometimes achieved at the expense of the mother's tissues and health. Therefore, proper prenatal care is a means of promoting optimal health in both the baby and the mother.

HISTORY. Many ancient cultures appear to have been concerned about the successful result of pregnancy, judging by the many depictions of this condition on the artifacts found at various archaeological sites. Furthermore, students of contemporary primitives, whose practices have changed little over the centuries, have often discovered that the diets which are culturally prescribed for pregnant women have higher nutritional values than some of the food consumption patterns of people in the developed countries.

It appears that primitive diets have often deteriorated in both quantity and quality as a result of factors such as (1) rapid growth of the population as a result of which the per capita supply of food decreased greatly; (2) people changing their diets from a wide variety of native foods to a few highly refined imported foods such as flour, lard, and sugar; and (3) migration of the people to new areas where the foods are unfamiliar and likely to have been rejected because of cultural biases. Nevertheless, some groups apparently made a gradual and healthful transition from nomads to settled agriculturalists, as demonstrated by the outstanding physiques of the Swiss dairy farmers in isolated high Alpine valleys.

A new era in nutrition was ushered in during the early 1900s when the first discoveries of the vitamins were made. Before that time it was assumed that diets were adequate if sufficient amounts of carbohydrates, fats, and proteins were provided. Nevertheless, the application of sound nutritional principles to pregnancy and lactation diets lagged behind the many advances in nutritional science which were made during the first half of the 20th century.

In the 1920s, many doctors believed that a calorie-restricted diet protected pregnant women against toxemia because there was a reduction in the incidence of this condition in Austria, Hungary, and Germany during World War I when scarcities of fats and meats led to smaller weight gains during pregnancy. However, few systematically planned observations were made of (1) the weights of the women before pregnancy, and (2) the effects of dietary restrictions on the health and survival of infants. Evidence concerning the dangers of inadequate diets during pregnancy was provided by studies conducted in the 1930s which showed that toxemia might result from low protein intakes, because subnormal levels of blood proteins sometimes allowed fluid from the blood to accumulate in the tissues.

After World War II, there was considerable interest in (1) the wartime experiences of pregnant women, and (2) studies which had been conducted in the United States and England during the 1940s. Most of the studies provided evidence that fetal and neonatal deaths, low-birth-weight infants, and congenital malformations were linked to dietary deficiencies during pregnancy. However, interest in the subject waned during the 1950s after other studies failed to find significant relationships between diet and the outcome of pregnancy. One of the main reasons for the contradictory findings in the various studies may have been the failure to obtain information on the eating patterns of the subjects prior to conception. Previously, well-fed women would have had more tissue reserves to carry them through their pregnancies.

Throughout the 1950s and 1960s many American doctors told pregnant women not to gain more than 15 to 20 lb (7 to 9 kg) because they believed that excessive weight gain rendered the women more susceptible to complications such as toxemia. Furthermore, salt was often restricted and diuretics were administered to prevent the buildup of fluids in the tissues. However, this period was also characterized by a growing trend towards prescribing mineral and vitamin supplements for pregnant women. Hence, the latter practice may have offset some of the consequences of dietary restriction.

In the decade following the National Academy of Science (NAS) report, infant mortality in the United States decreased by an average of about 5% annually, thanks to the efforts of an ever-growing number of dedicated doctors, educators, medical researchers, nurses, nutritionists, and public health workers who have applied nutritional principles for the benefit of pregnant and lactating women.

*CHANGES IN THE FEMALE BODY DURING PREG-
NANCY AND LACTATION.* Few people are aware of
the amount of tissue growth that occurs in women during
their reproductive years. For example, a woman who bears
no children has to produce blood and repair uterine lining
tissue to cover the loss of menstruation. This task alone is
equivalent to the production of about 110 lb (*50 kg*) of body
tissue between ages 15 and 45. However, a woman who
bears 6 children and nurses each one for 9 months builds
the equivalent of about 220 lb (*100 kg*) of body tissue during
her reproductive years.[33]

A depiction of the tissue growth during pregnancy is given
in Fig. P-66.

Details of the changes during pregnancy and lactation are
given in Table P-8.

[33]Toverud, K. U., et al., *Maternal Nutrition and Child Health, An Interpretive Review*, National Academy of Sciences, 1950, p. 116.

3 MONTHS   6 MONTHS   9 MONTHS

Fig. P-66. The growing fetus depends on its mother for nutrients and
removal of wastes. The mother's heart pumps blood through the uterine
arteries into the placenta; nutrients and oxygen diffuse through the placental
membranes into the fetal bloodstream by way of the umbilical cord; and
wastes pass in the reverse direction to the maternal bloodstream through
the uterine veins.

TABLE P-8
MAJOR CHANGES IN THE FEMALE BODY DURING PREGNANCY AND LACTATION

Part of the Body	Changes[1]	Biological Significance	Comments
Body Composition Protein	Gain of over 2 lb (*0.9 kg*), which is equal to about 8 lb (*3.6 kg*) of lean body tissue.	Used for enlargement of maternal tissues, production of additional blood plasma and blood cells, and growth of the placenta and fetus.	About ⅓ of the additional protein is converted to fetal tissue. The rest is used for the various "support systems."
Fat	Gain of almost 10 lb (*4.5 kg*), which is equivalent to 47,500 kcal.	Provides energy for both the mother and the fetus, and subsequent lactation.	A woman may build up considerable fat in several pregnancies, unless it is lost between pregnancies.
Water	Gain of about 15 lb (*7 kg*), or about 60% of the weight gained in pregnancy.	All newly synthesized body tissues in pregnancy contain considerable fluid. A reserve of fluid against blood loss in childbirth.	Over 40% of all pregnant women have some degree of excess fluid retention in their tissues, but it is usually harmless.
Circulatory System Heart	Volume of blood pumped per minute increases by ⅓. Heart rate increases by 40% (from 70 to 85 beats per minute).	The heart has to work harder to pump the extra blood formed during pregnancy to the enlarged body mass of the mother.	Sometimes, there is a temporary enlargement of the heart during pregnancy.
Total blood volume	Increases by about 30% (from 4,000 to 5,250 ml).	Extra blood is needed to serve the tissues added during pregnancy.	The increased flow of blood to the skin causes it to feel warm and moist.
Plasma	Plasma volume increases by about 40%.	Plasma is the vehicle for transporting nutrients and other substances within the body.	This increase ocurrs early in pregnancy. Hence, the blood is temporarily diluted.
Red blood cells	Increased by about 18%.	These cells carry oxygen to the tissues.	The ratio of red cells to total blood volume is usually about 10% subnormal in pregnancy.
Digestive System Stomach	Sphincter between the esophagus and stomach is relaxed and emptying of the stomach is slowed.	Food is more thoroughly mixed with gastric juice.	Sometimes, the stomach contents pass up into the esophagus, causing heartburn.[2]
Intestines	Motion is slowed. Towards the end of pregnancy, the fetus presses against the lower bowel.	The slower passage of food results in better digestion and absorption.	Pregnant women tend to be constipated unless the diet is rich in fiber and water.
Endocrine Glands[3] Adrenals	Increased secretion of the adrenal cortical hormone aldosterone.	Promotes sodium and water retention by the body, thereby meeting the special needs of pregnancy.	Progesterone from the placenta prevents the development of excessive fluid retention and high blood pressure.
Kidneys	Increased production of renin, which acts to increase the secretion of aldosterone.	Initiates the chain of events which leads to sodium and water retention by the body.	The renin-initiated reactions may induce high blood pressure in some pregnant women.
Ovary	At the beginning of pregnancy, hormones from the placenta induce the corpus luteum in the ovary to (1) survive much longer than it does in nonpregnant women, and (2) secrete hormones such as the estrogens and progesterone.	The estrogens and progesterone from the ovary help to maintain pregnancy and prevent spontaneous abortion during the first 8 weeks.	After the first 8 weeks of pregnancy, the placenta has usually developed sufficiently to provide the hormones that were secreted by the corpus luteum.
Pancreas	Insulin secretion is usually higher than normal in the second half of pregnancy because the placenta hormone HCS[4] antagonizes the effects of insulin and causes an elevation in the blood sugar.	Resistance to the effects of insulin in normal pregnant women results in (1) higher than normal maternal blood levels of nutrients, and (2) a continuously available supply of nutrients for the fetus.	The insulin-secreting capacity of some diabetic women may be overtaxed to the extent that both the mother and fetus are harmed.

Footnotes at end of table

(Continued)

TABLE P-8 *(Continued)*

Part of the Body	Changes[1]	Biological Significance	Comments
Pituitary	Secretion of prolactin increases throughout pregnancy. Increased secretion of MSH.[4] Increased secretion of TSH.[4]	Brings about milk production by the mammary glands. Causes pigmentation in various areas of the body. Stimulates the growth of the thyroid so that it becomes more efficient in the use of iodine.	There is competition between the maternal and fetal thyroids for the supply of iodine. The placenta efficiently takes up iodine and transfers it to the fetus.
Placenta	Secretion of HCG[4] at the onset of pregnancy. Secretion of HCS[4], estrogens, and progesterone throughout pregnancy. (The amounts of the latter hormones increase as the placenta grows larger along with the fetus.)	HCG maintains the functions of the corpus luteum so that pregnancy is maintained during the first 8 weeks. HCS acts as a growth hormone and stimulates development of functioning mammary glands. Estrogens and progesterone promote many of the maternal adaptations to pregnancy. (After the first 8 weeks, the major source of estrogens and progesterone is the placenta. Hence, poor development and growth of the placenta may lead to a poor outcome of pregnancy.)	
Thyroid	May become a little enlarged.	Enlargement allows the thyroid to use iodine more efficiently.	The formation of a goiter is a sign that the dietary supply of iodine is inadequate.
Fetus	Growth proceeds slowly until about 3 months after conception. The weight of the fetus is tripled during the third trimester (final 13 to 14 weeks).	Optimal growth of the fetus and accompanying placenta are best ensured by a good diet and a maternal weight gain of about 1 lb (0.5 kg) per week in the latter half of pregnancy.	Restriction of the maternal diet may jeopardize the growth of the fetus and placenta.
Kidneys	Secrete greater amounts of renin, usually without causing an increase in maternal blood pressure. The flow of blood to the kidneys is increased by about 40%, and the glomerular filtration rate by about 60%.	Renin initiates actions that bring about greater retention of sodium and water in the mother. Waste products are removed with high efficiency so that the fetus is protected against toxic accumulations of substances.	Extra sodium and water is required in the maternal body during pregnancy. The high rate of urination in early pregnancy decreases somewhat in the latter half of this period.
Lungs	Although the rate of breathing remains at the nonpregnancy level, the amount of air inspired and expired with each breath increases by more than 40%.	Abundant oxygen is provided for the increased metabolism of pregnancy. However, the blood levels of carbon dioxide become subnormal.	Resting pregnant women may feel lightheaded and have other symptoms that disappear when they increase their activity. (The symptoms are due to the subnormal blood levels of carbon dioxide.)
Mammary Glands	Preparation for lactation is stimulated during pregnancy by hormones from ovaries, adrenal glands, pituitary, and the placenta. After childbirth, the hormone prolactin from the pituitary stimulates milk production. Then, sucking by the infant brings about the letdown and release of milk through the action of the hormone oxytocin from the pituitary.	Little is wasted in human lactation because (1) milk production is delayed until after childbirth, (2) sucking stimulates continuation of production, and (3) lactation ceases when mothers either fail to nurse or stop nursing after having done it for a while.	

[1]Adapted by the authors from *Maternal Nutrition and the Course of Pregnancy*, National Academy of Sciences, 1970; and *Laboratory Indices of Nutritional Status in Pregnancy*, National Academy of Sciences, 1978.
[2]Also see HEARTBURN; and HIATUS HERNIA.
[3]Only the well established endocrine changes are presented. It seems likely that other glandular changes occur in pregnancy, but firm evidence of them is lacking.
[4]HCS is human chorionic somatomammotropin; MSH is melanocyte stimulating hormone; TSH is thyroid stimulating hormone; and HCG is human chorionic gonadotropin.

NUTRIENT ALLOWANCES. Although the nutrition of the fetus is aided by the metabolic alterations that take place in the mother's body during pregnancy, adequate levels of nutrients must be provided in the mother's diet to ensure that (1) the fetus is nourished adequately without depletion of maternal tissues, and (2) maternal reserves for childbirth and lactation are accumulated. Some examples of how certain metabolic changes affect dietary needs follow:

• **Calories**—The requirement for food energy is high during the latter half of pregnancy because the fetus is accumulating a reserve supply of fat for the period after childbirth and the mother is putting on fat in order to be able to supply the necessary food energy in her breast milk. Although some women have given birth to apparently healthy babies after having gained little or no weight during pregnancy, it is likely that there was some withdrawal of fat, protein, and other nutrients from the mother's tissues. Hormones secreted during pregnancy maintain high levels of sugar (glucose), amino acids, and fatty acids in the maternal blood by (1) slowing the rate of nutrient utilization by the maternal tissues, and (2) bringing about the withdrawal of these substances from the maternal tissues when the blood levels decline. However, overweight women should *not* try to lose weight by restricting their diets during pregnancy because harmful conditions such as ketoacidosis may result.

• **Protein**—Most of the growth of the fetus takes place in the latter half of pregnancy, when the rate of protein synthesis is high, as indicated by Fig. P-67.

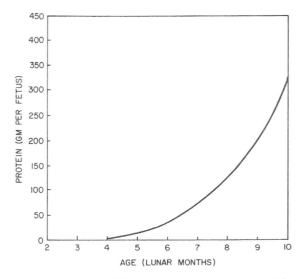

Fig. P-67. Protein accumulation in the fetus during pregnancy. (Adapted by the authors from Toverud, K. U., et al., *Maternal Nutrition and Child Health, An Interpretive Review*, National Academy of Sciences, 1950, p. 36, Fig. 3)

It is noteworthy that the placenta can extract amino acids from the maternal blood so efficiently that the fetal blood levels of these nutrients are usually considerably greater than those of the mother. However, the mother herself has higher than normal needs for protein which is used to enlarge her uterus, mammary glands, and the other tissues that grow to support pregnancy and lactation. Hence, the maternal blood levels of proteins such as plasma albumin may drop to subnormal levels. This condition may be dangerous to both the mother and the fetus because lack of sufficient blood protein may allow fluid from the blood to accumulate in the maternal tissues, causing waterlogging and/or an elevated blood pressure. There appears to be an association between subnormal levels of blood protein and the tendency to develop toxemia of pregnancy.

• **Calcium and phosphorus**—Fetal bone growth proceeds on the same schedule as the growth of the lean body soft tissues, except that the rate of calcium accumulation in the bones literally skyrockets during the last 2 months of pregnancy, as is shown in Fig. P-68.

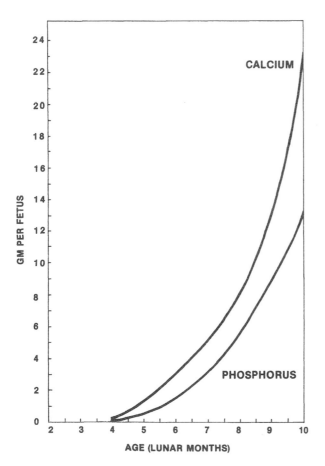

Fig. P-68. Calcium and phosphorus accumulation in the fetus during pregnancy. (Adapted by the authors from Toverud, K. U., et al., *Maternal Nutrition and Child Health, An Interpretive Review,* National Academy of Sciences, 1950, p. 55, Fig. 4)

It is noteworthy that the mother's diet should contain extra calcium from the beginning of pregnancy until the end of lactation, because the early accumulation of the mineral in the maternal tissues provides a reserve for later use, when it becomes almost impossible to consume enough to meet the needs. There is evidence to indicate that (1) dietary

calcium is about twice as well absorbed by pregnant women compared to nonpregnant women, and (2) adequate vitamin D during pregnancy enhances calcium utilization. The extra phosphorus that is required is usually provided in diets that contain sufficient protein. Infants that have received adequate calcium while in their mother's womb are less likely to develop rickets.

(Also see MINERAL[S]; CALCIUM; and PHOSPHORUS.)

• **Iodine**—The placenta extracts this essential mineral from the mother's blood so efficiently that fetal blood levels of iodine are usually several times those of the mother. However, the mother may develop a goiter when the supply of the mineral is inadequate to meet her needs. Enlargement of the maternal thyroid gland makes it more efficient in using iodine for the synthesis of thyroid hormones. Goitrous mothers have an increased risk of giving birth to infants that are cretins. Cretinism is a severe form of thyroid hormone deficiency in infants that is characterized by (1) failure of the fetal thyroid to develop, (2) mental retardation, (3) poor mineralization of the bones, and (4) dwarfism. Infants that are born with little or no functioning of the thyroid gland cannot be cured of this condition. Rather, they require the administration of thyroid hormone preparation for the rest of their lives.

(Also see MINERAL[S]; and IODINE.).

• **Vitamin E**—This vitamin does *not* readily pass across the placenta into the fetus. Hence, infants are born with low levels of vitamin E in their tissues. However, the consumption of adequate amounts of this vitamin by the pregnant mother is likely to be reflected in the amounts provided in her breast milk.

(Also see VITAMIN[S]; and VITAMIN E.)

• **Vitamin B-6**—Some pregnant women have developed a diabeticlike condition while consuming diets moderately high in protein, but low in vitamin B-6. Most of them were brought back to normal by giving them a supplement containing the vitamin.[34] It is noteworthy that the placenta concentrates the vitamin in the fetus, which may be a factor in the production of subnormal maternal blood levels of vitamin B-6 during pregnancy. Furthermore, women who have long used oral contraceptives prior to becoming pregnant may need extra amounts of this vitamin.

(Also see VITAMIN[S]; and VITAMIN B-6.)

Many other nutrients are required in greater amounts by pregnant and lactating women than by nonpregnant women. So, the Food and Nutrition Board of the National Academy of Sciences has established the allowances that are given in the section on Nutrient: Requirements, Allowances, Functions, Sources, Table N-3; hence, the reader is referred thereto.

Many other nutrients are required in greater amounts by pregnant and lactating women than by nonpregnant women. Hence, the Food and Nutrition Board of the National Academy of Sciences has established the allowances that are given in Table P-9.

---

[34]Coelingh Bennenk, H. T. J., and W. H. P. Schreurs, "Improvement of Oral Glucose Tolerance in Gestational Diabetes by Pyridoxine," *British Medical Journal,* July 5, 1975, pp. 13-15.

TABLE P-9
AVERAGE HEIGHTS, WEIGHTS, AND RECOMMENDED DAILY NUTRIENT INTAKES FOR PREGNANT OR LACTATING WOMEN OF VARIOUS AGES[1]

Category	Pregnant				Lactating			
	Age				Age			
	11–14 years	15–18 years	19–24 years	25–50 years	11–14 years	15–18 years	19–24 years	25–50 years
Weight[2] ....lb (kg)	101 (46)	120 (55)	128 (58)	138 (63)	101 (46)	120 (55)	128 (58)	138 (63)
Height ....in. (cm)	62 (157)	64 (163)	65 (164)	64 (163)	62 (157)	64 (163)	65 (164)	64 (163)
Energy ....kcal	2,500	2,500	2,500	2,500	2,700	2,700	2,700	2,700
Protein.......................g	76	76	74	74	66	66	64	64
MINERALS								
Calcium ....mg	1,200	1,200	1,200	1,200	1,200	1,200	1,200	1,200
Phosphorus ....mg	1,200	1,200	1,200	1,200	1,200	1,200	1,200	1,200
Sodium[3,4] ....mg	500	500	500	500	500	500	500	500
Chloride[4] ....mg	750	750	750	750	750	750	750	750
Magnesium ....mg	320	320	320	320	355	355	355	355
Potassium[4] ....mg	2,000	2,000	2,000	2,000	2,000	2,000	2,000	2,000
Chromium[4,5] ....mcg	50–200	50–200	50–200	50–200	50–200	50–200	50–200	50–200
Copper[4,5] ....mg	1.5–3.0	1.5–3.0	1.5–3.0	1.5–3.0	1.5–3.0	1.5–3.0	1.5–3.0	1.5–3.0
Fluoride[4] ....mg	1.5–2.5	1.5–2.5	1.5–4.0	1.5–4.0	1.2–2.5	1.5–2.5	1.5–4.0	1.5–4.0
Iodine ....mcg	175	175	175	175	200	200	200	200
Iron[6] ....mg	30	30	30	30	15	15	15	15
Manganese[4,5] ....mg	2.0–5.0	2.0–5.0	2.0–5.0	2.0–5.0	2.0–5.0	2.0–5.0	2.0–5.0	2.0–5.0
Molybdenum[4,5] ....mcg	75–250	75–250	75–250	75–250	75–250	75–250	75–250	75–250
Selenium[4,5] ....mcg	65	65	65	65	75	75	75	75
Zinc[8] ....mg	15	15	15	15	19–16	19–16	19–16	19–16
VITAMINS, FAT-SOLUBLE								
Vitamin A ....mcg RE	800	800	800	800	1,300	1,300	1,300	1,300
Vitamin D ....mcg	10	10	10	10	10	10	10	10
Vitamin E ....mgαTE	10	10	10	10	12	12	12	12
Vitamin K[4] ....mcg	65	65	65	65	65	65	65	65
VITAMINS, WATER SOLUBLE								
Biotin[4] ....mcg	30–100	30–100	30–100	30–100	30–100	30–100	30–100	30–100
Folate[7] ....mcg	400	400	400	400	280	280	280	280
Niacin ....mg	17	17	17	17	20	20	20	20
Pantothenic acid[4] ....mg	4–7	4–7	4–7	4–7	4–7	4–7	4–7	4–7
Riboflavin ....mg	1.6	1.6	1.6	1.6	1.8	1.8	1.8	1.8
Thiamin ....mg	1.5	1.5	1.5	1.5	1.6	1.6	1.6	1.6
Vitamin B-6 ....mg	2.2	2.2	2.2	2.2	2.1	2.1	2.1	2.1
Vitamin B-12 ....mcg	2.2	2.2	2.2	2.2	2.6	2.6	2.6	2.6
Vitamin C ....mg	70	70	70	70	95	95	95	95

[1]Adapted by the authors from *Recommended Dietary Allowances*, 10th ed., National Academy of Sciences, 1989.

[2]Average weight of nonpregnant, nonlactating women in each age group.

[3]There is no justification for the restriction of sodium in the diets of healthy women during pregnancy. Hence, the recommended intakes are those for nonpregnant women.

[4]These figures are given in the form of ranges of recommended intakes, because there is insufficient information to determine allowances.

[5]Since the toxic levels for many trace elements may be only several times average intakes, the upper values of the intakes should not be habitually exceeded.

[6]The increased requirement during pregnancy cannot be met by the iron content of habitual American diets nor by the existing iron stores of many women. Therefore, the use of 30 to 60 mg of supplemental iron is recommended. Iron needs during lactation are not substantially different from those of nonpregnant women, but continued supplementation of the mother for 2 to 3 months after childbirth is advisable in order to replenish stores depleted by pregnancy.

[7]The use of a folacin supplement appears to be desirable during pregnancy in order to maintain maternal stores and keep pace with the increased folacin turnover.

[8]For the first 6 mo. of lactation, the recommended allowance is 19 mg; and for the second 6 mo., it is 16 mg.

DIETARY GUIDELINES. If the mother does not have good nutrition during pregnancy, there is a greater chance of having a premature infant or one of low birth weight. There is also a greater chance of problems during pregnancy and delivery, infant death, infant defects, and failure of the infant to grow and develop normally. The benefits of eating well during pregnancy are illustrated in Fig. P-69.

* DIVIDE BY 2.2 TO CONVERT TO KG   * * MULTIPLY BY 2.54 TO CONVERT TO CM

Fig. P-69. The effect of dietary protein during pregnancy on the weights and heights of newborn infants. (Based upon data from Toverud, K. U., *et al., Maternal Nutrition and Child Health, An Interpretive Review,* National Academy of Sciences)

The rate at which the pregnant woman gains weight is also important. Weight should be gained slowly over the entire pregnancy. A good rate of weight gain is 1½ to 3 lb *(0.7 to 1.4 kg)* the first trimester (first 3 months), and then about ¾ lb *(0.3 kg)* a week until the end of pregnancy. Gaining too much weight too quickly may be harmful to mother and baby. Excess weight gain is often due to extra fluid which causes swelling (edema) of the mother's feet, hands, and face.

Excessive weight gain during pregnancy may cause complications or problems for the mother and baby. It also places a strain on the back and leg muscles that may cause pain and fatigue. Too much weight gain will make it difficult to get back to normal size and figure after the baby is born.

Weight reduction is not generally recommended during pregnancy. And crash diets or fad diets should never be used during pregnancy. They may be harmful to mother and baby because essential nutrients may be provided in limited amounts. Low calorie diets should usually be restricted to before pregnancy or following pregnancy.

The mother is eating for two—but that's not two adults! A slight increase of 300 Calories (kcal) per day the last half of pregnancy is recommended for the woman whose activity is light. This is for the growth of the fetus, the placenta, and other maternal tissue. The pregnant woman who is working at a job requiring physical activity or at home with small children may need more than 300 additional calories per day.

Fig. P-70. Mother and baby. (Courtesy, Gerber Products Co.)

COMMON NUTRITION-RELATED PROBLEMS OF PREGNANCY. The physiological changes that occur in the maternal body as a result of conception may bring on or aggravate certain conditions that tend to make pregnancy burdensome to the mother. However, there are nutritional ways of coping with these problems so that the well being of the mother and fetus is promoted.

Anemia. This condition is characterized by a subnormal blood level of hemoglobin and/or red blood cells. It is usually the result of a chronic dietary deficiency of iron, although it may also result from (1) deficiencies of other nutrients such as protein, copper, folacin, vitamin B-6, vitamin B-12, and vitamin C; or (2) losses of blood that have not been replaced fully.

The need for iron during pregnancy is quite high because it is used to (1) maintain the hemoglobin level of the mother, (2) maintain the mother's store of iron, (3) promote fetal development, and (4) provide a reserve in the infant for blood formation during early infancy before iron-rich foods are added to the diet. Furthermore, the iron stores in women of childbearing age are apt to be inadequate for pregnancy because of (1) loss of blood during menstruation, and (2) failure to consume sufficient amounts of iron-rich foods. Therefore, the Committee on Maternal Nutrition of the National Academy of Sciences has recommended that pregnant women take a daily supplement that contains from 30 to 60 mg of iron.[35]

(Also see IRON.)

One of the benefits of iron supplementation during pregnancy is shown in Fig. P-71.

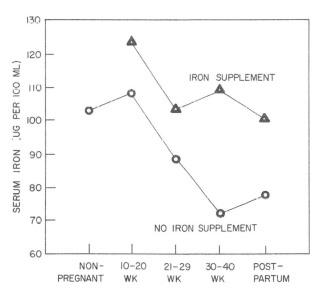

Fig. P-71. The effects of iron supplementation on the serum iron of the mother during and after pregnancy. (The points on the graphs represent averages of the different values given in *Laboratory Indices of Nutritional Status in Pregnancy,* National Academy of Sciences, 1978, p. 161, Table 7-1)

It may be noted from Fig. P-71 that the women who received the iron supplement maintained significantly higher levels of serum iron than the women who did not receive the supplement.

(Also see ANEMIA.)

Constipation. This condition may occur in late pregnancy. It may be due to pressure of the growing fetus on the digestive tract and to a decrease in the firmness (tone) of the muscles in the abdomen which slows down digestion. Thus, nutrients remain longer in contact with the walls of the intestines and nutrient absorption is increased.

Constipation may be prevented by drinking plenty of water and eating foods that add bulk to the diet such as fruits and vegetables (preferably raw) and whole grain cereals such as oatmeal, cracked wheat, all bran, etc.

(Also see CONSTIPATION.)

Cravings for Unusual Foods and Nonfood Items (Pica). Pregnant women sometimes have cravings for unusual items such as starch, clay, plaster, or chalk. The appetite for nonfoods is called pica. Eating of these substances should be discouraged because the practice may result in (1) filling of the digestive tract and reduction of the appetite for nutritious foods, (2) and poisoning by contaminants in the nonfoods. (Some types of plaster contain lead.)

Recently, evidence has been found that pica may be associated with iron deficiency anemia.[36] The abnormal cravings have often been reduced by treatment of the iron deficiency of pregnancy.

It is noteworthy that the craving for pastrami, pickles, or pizza in the wee hours of the morning is not pica and may be satisfied without harm, providing that pregnant women do not overindulge in these foods.

(Also see PICA.)

Diabetes. Diabetes poses a great risk to the developing fetus. The hormones of pregnancy (estrogen, human chorionic gonadotropin, and human placental lactogen) act to antagonize (counteract) the action of insulin in the body. Because of this antiinsulin activity, women are more prone to display diabetes when they are pregnant. To counteract this situation, the physician may prescribe increased insulin doses and put the mother on a special diet.

Problems with a fetus in a diabetic pregnancy include unexplained stillbirth, placental malfunction, and a large infant (macrosomia).

(Also see DIABETES MELLITUS.)

Excessive Gain in Weight. Weight, *per se,* is no indication of nutritional status. Food selections of some pregnant women who have gained weight too rapidly are likely to emphasize foods high in fats or carbohydrates and calories, yet be low in protein, minerals, or vitamins. Diets of other heavy women may be low in many nutrients, including calories. For the expectant mother with poor food practices, the prenatal period provides an opportunity to learn better food practices under medical supervision. There is little room in prenatal menus for foods that contribute mainly calories.

The *minimal caloric level recommended for the obese expectant mother is 1,500 Calories (kcal) per day.* It is very difficult to provide the needed nutrients below this caloric level. If insufficient calories are taken, protein may be used for the mother's energy needs instead of for building new tissue. Even though vitamin and mineral supplements may be given, these supplements do not furnish protein. Major efforts toward weight loss should not be attempted. Weight

[35]*Maternal Nutrition and the Course of Pregnancy,* National Academy of Sciences, 1970, p. 189.

[36]Gibbs, C. E., and J. Seitchik, "Nutrition in Pregnancy," *Modern Nutrition in Health and Disease,* 6th ed., edited by R. S. Goodhart and M. E. Shils, Lea & Febiger, Philadelphia, Pa., p. 748.

reduction in pregnancy has been associated with neuro-psychological abnormalities in the infant.

Guidelines released by the Institute of Medicine in 1990 indicate that weight gain of 25 to 35 lb during pregnancy is satisfactory.

Fluid Retention (Preeclampsia, Eclampsia). Many pregnant women accumulate a little fluid in their legs, which in itself is not usually harmful unless it is accompanied by (1) notable fluid retention in other parts of the body, (2) high blood pressure, and (3) passage of protein in the urine. These three conditions are signs of toxemia (preeclampsia) which may pose a threat to mother and fetus. However, it is noteworthy that salt restriction and the administration of diuretics may be hazardous in pregnancy and not advisable unless there is a clear cut threat of toxemia. Some doctors advise pregnant women who have some fluid retention and a moderate elevation in blood pressure to lie down for 30 to 45 minutes several times a day so that an increased amount of fluid is passed in the urine. Furthermore, they may allow the continued use of small amounts of salty foods, but it would be wise to avoid highly salted items such as pizza, potato chips, and other similar snacks.

The incidence of toxemia in American women has declined steadily in recent years, thanks to better diets and earlier prenatal care. Also, the condition is more severe when it develops in women who are markedly underweight at conception and who fail to gain weight normally during the early part of pregnancy.

Heartburn. This condition sometimes occurs when stomach contents are forced back into the esophagus. It usually happens during the latter part of pregnancy when the growing fetus puts pressure on the stomach. As with morning sickness, small, frequent meals may help. It is best to avoid greasy, fried foods and other foods that have previously caused digestive problems.

(Also see HEARTBURN; and HIATUS HERNIA.)

High Blood Pressure (Hypertension). Two forms of high blood pressure may be recognized during pregnancy: (1) chronic hypertension, which may cause fetal death and placental malfunction; or (2) acute hypertension, particularly in the last 12 weeks of pregnancy, which is indicative of toxemia. Toxemia can cause stroke, as well as blood-clotting difficulties in the mother, and lead to placental malfunction, separation of the placenta from the uterine wall, and stillbirth.

High blood pressure always warrants the attention of a physician. The blood pressure of the mother before pregnancy is an important factor to be considered in the diagnosis because many teenage girls have low blood pressure when not pregnant so that the rise which occurs may be much greater than suspected. Hence, high blood pressure is declared to be present when (1) the systolic blood pressure is 140 or greater, or the diastolic blood pressure is 90 or greater; or (2) the systolic reading has risen by 30 points or the diastolic reading has risen by 15 points.

Some moderately hypertensive pregnant women benefit by (1) staying off their feet as much as possible and/or (2) lying down for 30 to 45 minutes several times during the day. Many doctors refrain from restricting salt or giving diuretics to pregnant women for fear of creating an even greater problem. However, it would be wise for even a moderately hypertensive woman to avoid eating very salty foods.

(Also see HIGH BLOOD PRESSURE.)

Insufficient Gain in Weight. Recent studies of pregnant women show that markedly underweight expectant mothers have a greater than average incidence of toxemia and a strikingly increased incidence of prematurity of the infant. In subsequent pregnancies, these same expectant mothers on an adequate diet had a great reduction in the incidence of prematurity over an untreated control group.

This does not mean that the women benefited from an overabundance of calories, but rather that mothers who had various nutritional deficiencies profited from a correction of these shortcomings during pregnancy.

Morning Sickness. Nausea and vomiting, symptoms often called morning sickness, may occur in early pregnancy. This is usually due to changes taking place in the body. It may also be due to tension or anxiety. Morning sickness usually ends about the third or fourth month.

Morning sickness may be relieved by eating small, frequent meals rather than three large meals. Foods that are fairly dry and easily digested such as toast or crackers may be eaten by the mother before getting out of bed in the morning. Liquids should be taken between meals instead of with food.

THE DECLINE OF TOXEMIA (PREECLAMPSIA, ECLAMPSIA) AS A PROBLEM DURING PREGNANCY. Fortunately, the rate of maternal deaths from toxemia dropped by 88% between 1940 and 1965 (from a rate of 52.2 to 6.2 per 100,000 live births).[37] Nevertheless, many pregnant women may have a fear of becoming toxemic as a result of horror stories told by older female relatives. Therefore, some background on this increasingly uncommon condition follows.

First of all, the term *toxemia* is a misnomer in that there does not appear to be any toxic agent involved in the conditions it designates. Therefore, the Working Group on Nutrition and the Toxemias of Pregnancy (of the National Academy of Sciences Committee on Maternal Nutrition) has suggested that this term be replaced by preeclampsia and eclampsia. They also suggest (1) that the term preeclampsia be used only when the three abnormalities of fluid retention are present—(a) fluid retention throughout the body, not just in the legs, (b) high blood pressure, and (c) passage of protein in the urine; and (2) that the term eclampsia be used to designate the convulsive disorder that sometimes accompanies or follows the development of preeclampsia. These highly specific criteria for diagnosis are designed to prevent the overreporting of toxemia as a complication of pregnancy. (In the past, women were sometimes diagnosed as toxemic on the basis of a single abnormality such as fluid retention or high blood pressure.)

At one time, toxemia was thought to have resulted from the gain of excessive weight during pregnancy, but recent epidemiological studies have shown that it is much more likely to be associated with the conditions that follow:

1. Underweight at the time of conception, followed by failure to gain sufficient weight by the latter part of pregnancy.

2. Dietary deficiencies of calories and/or protein that produce a marked decrease in the blood level of protein.

3. High blood pressure that was present prior to conception.

4. Little or no prenatal care or counseling prior to the latter part of pregnancy.

5. First pregnancy, or a pregnancy following a multiple birth.

---

[37]*Maternal Nutrition and the Course of Pregnancy*, National Academy of Sciences, 1970, p. 177.

6. Inadequate consumption of calcium during pregnancy.

BREAST-FEEDING (Lactation). Lactation is the period after childbirth when a woman breast-feeds her baby. This is an important time for both the mother and her infant. Maternal nutrition plays a significant role during this time. A good diet is necessary for maternal tissue maintenance and replenishment of nutrient stores. In addition, a high-quality diet helps produce breast milk that is an excellent and natural source of nutrients for growth and development of the infant.

Inadequacies in the maternal diet during lactation may have severe consequences. The woman's health status may deteriorate as a result of large withdrawals of nutrients from her tissues. A poor maternal diet will also result in a decreased volume of breast milk. Also, there will be a reduction in vitamin levels. As a result, the infant may be short-changed with regard to calories and essential nutrients.

Since human milk has unique properties not found in other milks, it is the best initial food for infants. The use of breast milk will reduce the possibility of allergic reactions to protein or other components of cow's milk formulas. In addition, breast milk contains a smaller amount of casein, and therefore forms smaller and softer curds in the stomach. This makes it easier to digest by the infant whose digestive processes are not fully developed.

Another advantage of breast milk is that it contains antibodies against infectious microorganisms. This helps the infant resist infections during the first few months of life. In addition, breast milk has the advantage of being sterile. Hence, there is no problem with the type of contamination that can occur during formula preparation.

Breast-feeding is an excellent way to prevent overeating during infancy because the breast-fed baby usually consumes fewer calories than the infant who is bottle fed. The lactating woman simply allows her baby to feed at her breast until satisfied. On the other hand, a mother who bottle feeds her infant tends to see that the baby consumes the entire content of the bottle. Excessive caloric intake during infancy can lead to a permanently increased number of fat cells in the body. This can increase the chance of obesity in later life.

Besides its physiological advantages, breast-feeding has a psychological benefit for both mother and infant. Both experience close attachment. This is a natural situation for the mother to provide a soothing and loving atmosphere; it helps the infant develop desirable emotional characteristics.
(Also see BREAST FEEDING.)

Maternal Diet During Lactation. In order to ensure adequate infant nutrition, the maternal diet must meet the nutrient requirements for lactation. (See Table P-9.) Ethnic, social, and economic factors may influence what the mother eats. The health professional should consider these in counseling.

Contrary to a prevalent belief, the diet of a lactating woman must contain an even higher number of calories than during pregnancy. These additional calories are necessary for milk production. Furthermore, the mother should consume the equivalent of 2 to 3 qt (*1.9 to 2.8 liter*) of liquids daily. This fluid is essential to provide the liquid volume for the breast milk and to meet the other needs of the mother.
(Also see BREAST FEEDING.)

Nutritional Supplementation. During lactation, there is a need to supplement the maternal diet with certain nutrients. As in pregnancy, 30 to 60 mg of elemental iron should be taken daily by all women. This supplement should be continued for at least 2 or 3 months after childbirth. Iron supplementation during this period is necessary to replenish maternal iron stores depleted by pregnancy. Doctors may recommend other mineral and vitamin supplements for some patients.

Since breast milk does not contain an adequate amount of vitamin D, it is advisable to give the infant an oral supplement of 7.5 mcg daily.

The fluoride content of breast milk varies with the fluoride concentration of the community's water supply. If the fluoride content of the water exceeds 1.1 ppm, the infant will receive sufficient fluoride through breast milk. However, an oral supplement of 0.5 mg fluoride daily should be given to the infant if the water supply has less than 1.1 ppm fluoride.
(Also see BREAST FEEDING.)

Use of Drugs by the Lactating Mother. If oral contraceptives are taken sooner than 6 weeks after childbirth, the amount of breast milk a woman produces may be diminished. (Hormones such as estrogen and progesterone in these pills reduce milk production.) Clinical experience shows that the majority of women will not have difficulty producing an adequate amount of milk if they do not use oral contraceptives for 6 weeks after delivery. Hence, they should substitute another form of birth control.

In addition to oral contraceptives, other drugs such as barbiturates (sleeping pills), laxatives, and salicylates (aspirin and similar substances) can be transmitted through breast milk. The taking of any drug during lactation should be done only under a doctor's supervision.
(Also see BREAST FEEDING; and INFANT DIET AND NUTRITION.)

FACTORS THAT MAY WARRANT SPECIAL COUNSELING OF PREGNANT WOMEN. Experience has shown that even an affluent and educated woman can have nutritional problems throughout her pregnancy. Sufficient money to buy food is not a guarantee that the mother's diet will be nutritionally adequate. In some cases, her diet may consist of a limited variety of foods due to misinformation or adherence to unsound weight reduction diets. For example, the pregnant woman may not drink milk for fear that it will cause leg cramps, or she may eat only low-calorie foods to lose weight. Meals may be infrequent or eliminated with the result that the mother may be vulnerable to nutritional deficiencies. This type of deprivation may have harmful effects on the outcome of her pregnancy. Furthermore, certain factors related to the mother's status may classify her as a higher-than-average risk. These factors may be physiological, socioeconomic, and/or psychological. A high risk pregnant woman needs continuous nutritional assessment and in depth nutritional counseling. Some of the more common risk factors follow.

Adolescence. Girls younger than 17 years, who are pregnant before completion of their own growth, have greater nutritional requirements than adult women. For these patients, pregnancy creates a dual growth demand—that of the fetus, and of the patient herself. Compounding this, many adolescent girls may restrict their caloric intake in order to lose weight or may have poor dietary habits. Adolescent pregnancies have been associated with low birth weight, premature births, and a high death rate of newborn infants. Fig. P-72 shows how the mother's age is associated with infant death rates.

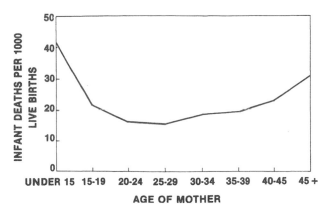

Fig. P-72. Infant mortality rates by age of mother. (Based upon data from *Maternal Nutrition and the Course of Pregnancy*, National Academy of Sciences, p. 144, Table 3)

Ethnic and/or Language Differences. The U.S. population is composed of many national, cultural, and regional groups. These groups have a variety of dietary practices, many of which are excellent, that are part of their cultural traditions. However, some families may be unfamiliar with compositions and nutritional values of many foods available in local markets. Unfortunately they frequently adopt poor food practices and replace nutritious foods with ones that contain large amounts of fats, starch, and sugars, but few other nutrients. For example, substituting soft drinks for fruit juice or potato chips for potatoes increases the calories consumed and reduces the intake of essential minerals and vitamins.

For certain individuals, English is not a familiar language. This may present a problem, if the woman is not able to read or understand information about the foods commonly eaten in this country. In many cases, ethnic and/or language differences have been associated with anemia and inadequate or sporadic weight gain during pregnancy.

Habituation to Alcohol, Drugs, and/or Smoking. Some of the current ideas on each of these harmful habits follow:

• **Alcohol**—Dr. David Smith of the University of Washington in Seattle, medical researcher who discovered that alcoholic mothers gave birth to infants with various defects, states that there is no *safe dose* of alcohol for a pregnant woman.[38] It is suspected that the consumption of alcohol during pregnancy may be the greatest single cause of birth defects and the most common cause of mental retardation in infants. Dr. Smith has named the complex of abnormalities the *fetal alcohol syndrome.*

Therefore, anyone who has a role in the counseling of pregnant women should strongly advise against the consumption of alcoholic beverages and help alcoholic mothers find the appropriate therapy means for breaking the habit.

• **Drugs**—The use of addictive drugs such as heroin and morphine by pregnant women has been reported to have neurological effects on their newborn infants. Sometimes, the infants are addicted at birth and have characteristic symptoms of withdrawal such as irritability, sleeplessness, and lack of interest in eating. The symptoms may occur as late as 7 days after delivery.

It is noteworthy that the identification and treatment of drug-addicted mothers early in pregnancy has resulted in great reduction in the harmful effects of the addiction on infants.

• **Smoking**—Women who smoke during pregnancy may give birth to a low-birth-weight infant. The reduction of weight in the fetus is due to the interference of carbon monoxide (from the smoke) with its oxygen supply. Therefore, women of childbearing age or younger should be discouraged from starting to smoke or given assistance in breaking the habit once it is established.

History of Obstetrical Complications. Previous obstetrical complications reflect potential problems which may recur in subsequent pregnancies. Nutrition related factors in the obstetrical history include inadequate weight gain, preeclampsia and/or toxemia, anemia, diabetes, bleeding prior to delivery, multiple pregnancy, premature or small-birth-weight infant, and fetal or neonatal death.

Low Income. Women from low-income families may find it difficult to obtain sufficient food and to consume adequate diets during pregnancy. Furthermore, the likelihood of prematurity rises sharply with a decrease in nutritional status. The lowest birth weights and the largest number of deaths in the neonatal period occur among the infants born to the most poorly nourished mothers.

In order to help improve the nutritional quality of the diet, the low-income person should be encouraged to participate in food programs such as Food Stamps and WIC (Special Supplemental Feeding Program for Women, Infants, and Children). These programs increase the buying power or provide specific high-quality foods.

Nutrition education and counseling are important to maximize the benefits from the limited resources available to low-income individuals. Information on food buying, storage, and preparation can be extremely useful. In many cases, existing community resources may help provide this support. It is important and valuable to coordinate these efforts.

Psychological Conditions. Depression, anorexia nervosa, and other mental problems may affect the course and outcome of a patient's pregnancy. These conditions may result in reduced caloric and nutrient intake, which is associated with poor weight gain of the fetus, low-birth-weight infants, and a high death rate of newborn babies.

RH Factor Incompatibility. This problem, which is also known as erythroblastosis fetalis, has little to do with nutrition, but it is presented here so that it will not be overlooked by prospective parents. (All obstetricians are well aware of this problem and the appropriate preventive measures, but some American women may have their babies with little or no prenatal advice from a doctor.)

The medical name of this condition is actually a description of the effects of the Rh incompatibility on the fetus. During pregnancy, a mother whose blood is Rh negative (from 12 to 15% of Americans have this type of blood) may develop antibodies against a fetus that is Rh positive. Some of the maternal antibodies may cross the placenta into the fetus, where they may destroy some of the red blood cells (this effect is called erythroblastosis). Pigment released from the blood cells may poison the infant's brain before or after birth. It is noteworthy that this condition was the cause of death in 1 of every 400 births at the Chicago Lying-In

---

[38]Iber, F. L., "Fetal Alcohol Syndrome," *Nutrition Today*, September/October 1980, pp. 4-11.

Hospital in the early 1940s (most of the deaths occurred within the first few days of life). Since then, the cause of the problem was studied intensively and appropriate preventative measures were developed. Fig. P-73 shows the most common combinations of parental Rh blood types that occur.

NO Rh TROUBLE   NO Rh TROUBLE   POSSIBLE FORMATION OF MATERNAL ANTIBODIES

Fig. P-73. Combinations of parental Rh blood types and the possibility of Rh factor incompatibility.

It may be seen from Fig. P-73 that only the combination of an Rh-negative mother and an Rh-positive father may lead to incompatibility. Even in these cases, the problem does not occur very often. Nevertheless, all couples should have their blood types checked to rule out the possibility of trouble. (A vaccine may be administered to prevent the formation of maternal antibodies if the potential for incompatibility is discovered early enough.)

Several Previous Pregnancies. Expectant mothers who have had many pregnancies in rapid succession are often in a depleted nutritional state. For example, if there is not enough time between pregnancies, the losses of iron and other nutrients will not be restored. Careful attention should be given to these dietary needs.

Use of an Oral Contraceptive. An estimated 10 to 11 million American women use oral contraceptives. Some have used them for many years prior to discontinuing their use in order to have a child. Researchers have found that these agents change the nutritional needs of some women. The levels of several vitamins and minerals in the blood are known to be, or suspected to be, influenced by *the pill.* As a result, some oral contraceptive users need increased amounts of certain nutrients in their diets to maintain normal blood levels (a sign of adequate nutrition). Folacin, a B vitamin, is of particular concern since many oral contraceptive users may need a supplement of folacin. In addition, adequate amounts of vitamin C (ascorbic acid), vitamin B-2 (riboflavin), vitamin B-6 (pyridoxine), vitamin B-12, and zinc are needed.

Therefore, pregnant women who previously used an oral contraceptive should ask their doctor whether they should take a mineral and vitamin supplement during pregnancy.

Vegetarianism. The types of diets used by people who abstain from eating meat vary widely in terms of the other foods that are used or not used.

Lacto-vegetarians do not consume meat, poultry, fish, or eggs. However, they do use milk, cheese, and other dairy products. The pure vegetarian (vegan) abstains from all animal protein foods and milk and milk products.

The fruitarian, unlike the majority of vegetarians, eats a more restricted diet of raw or dried fruits, nuts, honey, and oil. Even more restrictive is the macrobiotic diet. This regimen consists of ten diets, ranging from the lowest level which includes cereals, fruits, vegetables, and some animal products, to the highest level made up entirely of brown rice. At all levels, fluids are avoided as much as possible. Individuals who continue to follow the higher levels of this diet are in serious danger of developing nutritional deficiencies.

A vegetarian who eats a wide variety of grains, legumes, nuts, fruits, vegetables, and milk and milk products can have a nutritionally sound diet. However, the vegan and fruitarian diets do not contain any significant amount of vitamin B-12. Lacto-ovo vegetarians will obtain the Recommended Dietary Allowance for this vitamin by consuming the equivalent of four glasses of cow's milk daily. Soybean and goat's milk contain less vitamin B-12 than cow's milk. Therefore, both vegans and vegetarians who do not drink cow's milk will need to take a vitamin B-12 supplement. A deficiency of vitamin B-12 causes an anemia and eventually results in spinal cord degeneration. A high intake of folacin will mask this anemia and thus vitamin B-12 deficiency may go undiagnosed. The irreversible neurological changes will then be the first indication of a deficiency.

The omission of cow's milk or goat's milk from a strict vegetarian diet is also likely to lead to deficiencies of calcium and vitamin D since few other foods contain adequate quantities of these nutrients in forms that are utilized readily. Hence, supplements containing them should be taken.

Fig. P-74. Pregnancy and lactation can be great opportunities for the mother to give her child a good start in life, providing a wide variety of nutritious foods are consumed before and during these periods. (Courtesy, United Fresh Fruit & Vegetable Assn., Alexandria, Va.)

(Also see NUTRIENTS: REQUIREMENTS, ALLOWANCES, FUNCTIONS, SOURCES.)

PREGNANCY ANEMIAS

The total iron requirements of pregnancy are considerable; hence, it is not uncommon to detect an iron deficiency anemia during pregnancy because the iron content of habitual U.S. diets is not sufficient. Such an iron deficiency anemia has three contributing factors: (1) low iron storage which may result from iron loss during a previous pregnancy; (2) low dietary iron intake; and (3) greatly increased demand for iron for maternal and fetal hemoglobin synthesis. Iron deficiency anemia during pregnancy can be corrected, and even more important it can be prevented, by appropriate iron supplementation. A daily supplement of 15 mg of iron, averaged over the entire pregnancy, should satisfy the needs of most women. Also, pregnant women should eat foods rich in iron. Liver contains far more iron than any other food. Other meats, dried beans, dried fruit, green vegetables, eggs, and enriched cereals are good sources of iron.

Although not very common in the United States, megaloblastic anemia due to a folate deficiency does occur during pregnancy. In most cases it appears to be the result of long term use of an oral contraceptive, a marginal diet, vomiting, and increased demands of the developing fetus. Pregnant women have double the requirement of other persons for folic acid. The administration of folic acid causes prompt reversal of the symptoms. If not treated during pregnancy, the baby may show symptoms of megaloblastic anemia and require folic acid supplementation.

(Also see ANEMIA, section headed "Groups of Persons Susceptible to Nutritional Anemias"; ANEMIA, MEGALOBLASTIC; FOLIC ACID; INFANT ANEMIA; IRON; and PREGNANCY AND LACTATION NUTRITION.)

PRENATAL

Before birth.

PRESERVATION OF FOOD

Since the beginning of time, man has searched for ways to protect himself from hunger—for ways to save food in times of plenty for times of scarcity. No doubt, man learned that he could often collect more food than he could eat, but that it soon spoiled and became unfit for consumption—or even poisonous.

There are no records to indicate when some of the first methods of preservation were discovered. Like other discoveries, some of the preservation methods were based on a series of enlightening observations which led to the development of a process. The methods of preservation became refined and more effective as men like Anton van Leeuwenhoek, Lazaro Spallanzani, and Louis Pasteur contributed to the understanding of spoilage of food. Until men learned some of the modern methods of preservation, large cities could not develop. Nearby farms were required to feed the city people, since food spoilage was a major problem.

Spoilage of food is caused by growth of microorganisms, enzyme action, oxidation (chemical reactions), extremes in physical surroundings, and/or pests. Insofar as possible, the aim of preservation is to reduce, eliminate, and/or control the causes of spoilage. With the modern methods of preservation, seasonal harvests can be enjoyed year round, and food specifically grown in one area can be shipped across the country or even worldwide. Still, scientists continue to search for better methods of preservation that yield a nutritious, flavorful, and desirable product, with minimal long-term requirements.

Fig. P-75. Home-canned foods. (Courtesy, University of Minnesota, St. Paul)

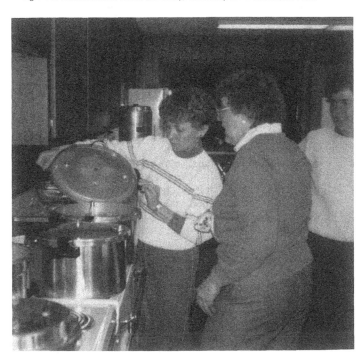

Fig. P-76. Pressure processing low-acid home canned foods is essential for food safety. (Courtesy, North Dakota State Cooperative Extension Service)

Table P-10, which follows, summarizes in a general way the methods of preservation which are applied to food. Often more details concerning the preservation of a food item may be found in the articles dealing specifically with that food. Foods are often preserved by a combination of the methods listed in Table P-10.

(Also see BACTERIA IN FOOD; FRUIT[S]; MEAT[S]; MILK AND MILK PRODUCTS; SPOILAGE OF FOOD; and VEGETABLE[S].)

TABLE P-10
METHODS OF FOOD PRESERVATION

Method(s); History	Description and Principle; Food Application	Length of Storage; Changes in Quality	Remarks
CANNING **History:** A Frenchman, Nicholas Appert, invented canning in response to a prize offered in 1795 by the French government for a food preservation method that would not seriously impair the natural flavor of fresh food. After 14 years of experimenting, Appert won the prize in 1809. His method involved heating foods in glass flasks sealed with corks. Pasteur later explained why the metod worked. An Englishman, Peter Durand, conceived the idea of using cans, and by 1839 tin-coated steel containers were widely used.	**Description and Principle:** Properly prepared foods are placed in the glass or can containers and sealed. Then depending upon the type of food (primarily the acidity [pH]), the cans or bottles are heated to a known temperature and then held at that temperature for a specified time. Low acid foods such as meats and vegetables are heated to 240° to 265°F (116° to 129° C) while acidic foods such as fruits are heated to about 212°F (100°C). The principle of canning is twofold: (1) to destroy spoilage, and disease-causing microorganisms, and (2) to keep air away from the food. **Food Application:** Meat, fish, poultry, vegetables, fruits, soups, orange juice, tomato paste, jams, jellies, and other manufactured foods.	**Length of Storage:** Most canned foods keep well for more than 1 year when properly canned. Commercially canned items are dated. **Changes in Quality:** This varies with the food and the process, but some general statements can be made: (1) color and texture often differ from fresh products; (2) no practical effect on protein, carbohydrates, and fats; (3) no effect on vitamins A, D, and riboflavin if protected from oxygen; (4) vitamin C destroyed by heat and exposure to oxygen; and (5) thiamin destruction dependent upon heat treatment and acidity of food.	The chief canning processes are: (1) the conventional retort method for most vegetables, fruits, fish and meat products; (2) preheating and hot filling for very acid foods like orange juice, tomato paste, jams and jellies; and (3) fast canning for foods such as baby foods, sauces, potted meats, and cream-style corn. Fast canning uses high temperatures of 250° to 280°F (121° to 138°C) for short periods of time. The major concern is proper processing to ensure the destruction of the spores of *Clostridium botulinum*—a toxin-producing bacterium. Commercially canned foods have been free of the botulism hazard for over 30 years.
CHEMICAL ADDITIVES **History:** Man's first uses of salt to preserve foods are lost in antiquity. Sugar has also enjoyed a long-time usage as a preservative. More recently, other chemicals have been tested and used in minute amounts to prevent spoilage and the growth of microorganisms. The uses of chemicals (food additives) in foods for preservative purposes are controlled by the government under the Food Additive Amendment of 1958. (Also see ADDITIVES; and NITRATES AND NITRITES.)	**Description and Principle:** A variety of chemicals will destroy microorganisms. Other than salt and sugar, chemicals used for this purpose must act at low levels. Some common chemical preservatives include ethyl formate, sodium benzoate, sodium and calcium propionate, sodium nitrate or nitrite, sorbic acid, and sulfur dioxide. Some chemicals act by preventing the growth of microorganisms, while others prevent oxidation or enzymatic deterioration of quality during manufacture and distribution. **Food Application:** Depends on the chemical and processing method; ranges from fruits and vegetables to meats to soft drinks; common examples include: BHA (butylated hydroxyanisole) and BHT (butylated hydroxytoluene) in beverages, cereals, chewing gum, and oil; calcium propionate in baked goods; and sodium nitrite in meats.	**Length of Storage:** This varies widely depending upon the chemical(s) used and the food product. Some chemical additives merely increase the shelf life a few days or weeks while others may increase storage time to an indefinite period. **Changes in Quality:** The prime reason for the addition of many chemicals is to maintain the quality of the food. Some chemicals actually increase the value of a food; for example, vitamin C (ascorbic acid) may be added to fruits to prevent browning.	Chemicals may be added to foods for reasons other than preservation. They are also added to enrich, or to improve a food. New chemicals added to food, regardless of the purpose, must undergo rigorous testing for safety.
DRYING AND DEHYDRATION **History:** Drying is one of the oldest methods of preserving food and is still a common practice in the world today. Prehistoric man probably dried foods in the sun or near fires. There is evidence that drying foods by the fire was an ancient practice in both the New and Old Worlds. Artificial drying (dehydration) via hot air was developed in France in 1795, where it was used to dry thin slices of vegetables. Technology gained during World War II resulted in the dehydration of more foods.	**Description and Principle:** Drying is accomplished via the sun in hot, dry areas of the world or via dehydrators which produce heat at certain temperatures, maintain humidity, and produce a flow of air. Whatever the method, the heated air carries away the moisture. Depending on the food, drying or dehydration may reduce foods to 2% moisture. Drying or dehydrating is effective because microorganisms require water to grow and many enzymatic and nonenzymatic reactions require water to proceed. **Food Application:** Eggs, milk and milk products, fruits, vegetables, juices, meats, spices, soups, coffee, tea, gelatin dessert mixes, and macaroni.	**Length of Storage:** This varies with the product, storage conditions, and the final moisture content. Some very dry items can be stored indefinitely if protected from moisture. Molds can grow on foods with as little as 5 to 16% moisture. Bacteria and yeast usually require a moisture content greater than 30%. **Changes in Quality:** Drying concentrates the nutrients. Protein, fat, and carbohydrate are present in larger amounts per unit weight. Often the reconstituted or rehydrated food is comparable to the fresh foods. However, the vitamin content of dried meat is much less than that of fresh meat.	Many foods dry as part of the natural process in their production on the plant; for example, legumes, nuts, and grains. Vacuum driers may be used for foods that are especially sensitive to heat. Besides the preservative action of drying, it also reduces the size and weight of foods making them easier to transport and store. There are many methods of drying (dehydrating) foods that are especially adapted to certain food products. Occasionally some toxin-producing bacteria may survive the process and cause poisoning when water is added.
FREEZE DRYING (Lyophilization) **History:** First application of freeze drying to a biological sample was reported in 1890 by the German histologist, Altmann. During the 1930s, the technology was applied to blood plasma. World War II stimulated the development of the technology to a commercial basis.	**Description and Principle:** The process removes water from food while the food is still frozen. After cooling food to about −20°F (−29°C), it is placed on trays in a vacuum chamber and heat is carefully applied. Water is evaporated without melting since it goes from the frozen-ice state to the gaseous state without passing through the liquid state—sublimation. The food is not exposed to high temperature until most of the water is removed. Without water, microorganisms cannot grow, and most enzymatic and nonenzymatic reactions cannot proceed. **Food Application:** Any food, raw or cooked, which can withstand freezing.	**Length of Storage:** Freeze-dried foods are often packaged in an inert gas such as nitrogen and they must be packaged in moisture-proof containers. When properly packaged, most products are acceptable after 1 year of storage at 100°F (38°C) —a temperature much higher than normal storage. Longer storage times can be expected at lower temperatures. **Changes in Quality:** Upon rehydration (adding water), freeze-dried foods are equal in acceptability to their frozen counterparts. There is a substantial retention of nutrients, color, flavor, and texture.	Freeze drying usually produces higher quality foods than the other drying methods, but it costs more. The first significant commercial application of freeze drying was in the production of an improved instant coffee.

(Continued)

TABLE P-10 *(Continued)*

Method(s); History	Description and Principle; Food Application	Length of Storage; Changes in Quality	Remarks
FREEZING **History:** Since early times, farmers, fishermen, and trappers, living in regions with cold winters, have preserved their meat and fish by freezing and storage in unheated buildings. About 1880, ammonia refrigeration machines were introduced for use in freezing fish. Clarence Birdseye stimulated the development of the business of packaging frozen foods by forming a company that began quick freezing foods in 1924. The frozen food business boomed when home freezers became available after World War II.	**Description and Principle:** Foods are subjected to temperatures below 32° to 25°F *(0° to −3°C)* where the water in the food turns to ice. Foods are best if rapidly frozen. Freezing methods may employ direct immersion in a cooling medium, contact with refrigerated plates in a freezing chamber, freezing with liquid air or nitrogen, or dry ice. Freezing does not kill microorganisms or destroy enzymes, it merely arrests or slows spoilage changes. **Food Application:** Most kinds of fruits and vegetables, meat, fish, poultry, and some dairy products; a variety of precooked foods (convenience foods) from french fries to complete dinners. Some foods such as tomatoes, cabbages, bananas, avocados, pears, and some shellfish are not well suited for freezing.	**Length of Storage:** Each food has its own recommended storage period, after which time undesirable off-flavors slowly develop. **Changes in Quality:** Many foods are pre-treated further to ensure quality during frozen storage. In general, freezing promotes the retention of nutrients and does not destroy nutrients. However, some destruction of vitamins may occur during the processing that proceeds freezing.	Frozen foods must be kept frozen since partial thawing and refreezing lowers the quality. Freezing and canning are the two most widely used methods of food preservation. Frozen foods must be protected from dehydration and freezer burn during storage.
FERMENTATION **History:** Some of the earliest writings describe the fermentation of grapes and other fruits. Most cultures devised some type of fermented beverage. Fermentation of milk has been known for ages; it is mentioned in the Old Testament. Nomadic tribes fermented milk. Sauerkraut is fermented cabbage—a product of ancient China. (Also see BEERS AND BREWING; MEAT[S], section headed "Meat Processing"; MILK AND MILK PRODUCTS; and WINE.)	**Description and Principle:** In fermentation, the growth of certain bacteria and yeasts which synthesize chemicals that aid in preservation of the food is encouraged. These products of fermentation are primarily lactic acid and ethyl alcohol. Lactic acid is important in the fermentation of cucumbers, olives, cabbage, meats, and milk. Ethyl alcohol is produced by the fermentation of fruits and cereals. **Food Application:** Fruits, vegetables, cereals, meats, and milk.	**Length of Storage:** Fermented products are often subjected to other preservation methods to increase storage time. Milk products are refrigerated while wines are pasteurized. **Changes in Quality:** Fermentation changes the quality of both plant and animal products to a marked degree. Some foods are made more edible, others are made more nutritious, and still others are made more flavorful.	Some fermented products require that individuals acquire a taste for their distinctive flavor. Fermentations create wine, beer, rum, hard liquor, breads, cheeses, pickles, salami sausage, olives, Sajur asin (vegetables and rice), poi (taro), yogurt, buttermilk, and acidophilus. Fermentations producing lactic acid are encouraged by the level of salt in the product.
PASTEURIZATION **History:** The process is named for Louis Pasteur, who, in the 1860s, found that heating liquids, especially wines, to a temperature of 140°F *(60°C)* improved their keeping qualities.	**Description and Principle:** This process is the application of mild heat for a specified time to a liquid, food, or beverage. Pasteurization destroys molds, yeasts, and the non-spore forming bacteria. The lower the temperature, the longer the application time, and vice versa. For example, milk can be pasteurized at 143°F *(62°C)* for 30 min. or 162°F *(72°C)* for 15 seconds. Time and temperature vary with the type of food. Foods are rapidly cooled following the treatment. **Food Application:** Milk, cheese, wine, beer, and other beverages such as fruit juices.	**Length of Storage:** This depends upon the type of storage following the pasteurization and upon the type of food. **Changes in Quality:** Since heat is less than boiling temperature and for a short time, the quality of the food is maintained. Foods which are pasteurized include those whose flavor and appearance may be adversely affected by high temperatures.	For milk, the times and temperatures employed are based upon the heat tolerance of *Mycobacterium tuberculosis*, one of the most heat resistant of the nonspore-forming pathogens (disease causing).
PICKLING **History:** Pickling of plant and animal foods is an ancient practice. Sauerkraut was manufactured in ancient China. In the 18th century, scientists recognized that vinegar (acetic acid) was good for preserving biological specimens. The pickling industry has grown rapidly since the 1930s.	**Description and Principle:** Preserving foods in vinegar and salt is called pickling. Vinegar is an acid, and microorganisms are sensitive to acids. Moreover, acids increase the lethality of heat on microorganisms. There are two methods of pickling: (1) fresh pack pickling, and (2) fermentation pickling. Both rely on brine and vinegar as the primary preservatives. The vinegar needs an acetic acid content of 4 to 6%, and either cider vinegar or white vinegar is acceptable. The addition of spices is a common practice. **Food Application:** Vegetables, fruits, meats, eggs, and nuts; most common include cucumbers, pears, peaches, plums, cured meats, cabbage, mushrooms, cauliflower, onions, tomatoes, beets, peppers, and watermelon.	**Length of Storage:** Pickling alone is not sufficient for long-term storage. Pickled products should be heat treated (canned) unless they are consumed soon after pickling. Even vinegar is often pasteurized for a longer shelf life. **Changes in Quality:** Pickling not only increases the keeping quality of a food, it adds an appealing flavor to the food. The nutritional quality depends on other preservative processes employed on the food.	Pickled foods are not likely to be suspect in a food poisoning incident—at least where sufficient vinegar is used. Vinegar is a chemical preservative used in other foods such as mayonnaise and salad dressing. Cucumbers are the most commonly pickled vegetable. Near perfect produce should be used for pickling.

(Continued)

TABLE P-10 *(Continued)*

Method(s); History	Description and Principle; Food Application	Length of Storage; Changes in Quality	Remarks
RADIATION **History:** The first patent related to the radiation treatment of food was taken out in France in 1930. Serious investigations in the U.S. were conducted during the 1950s. Radiation preservation of foods was considered to be in the experimental stages until 1981, when the ban on treating food with gamma rays was eased by the FDA. (Also see RADIATION PRESERVATION OF FOOD.)	**Description and Principle:** Ionizing radiation—moving subatomic particle or energetic electromagnetic waves— is used to treat foods. The radiation prevents the proliferation of microorganisms, insects, and parasites, and inactivates enzymes. The source of radiation may be cobalt-60, or cerium-137 or X rays. This process does *not* make the food radioactive. **Food Application:** Very adaptable; considered for use on fruits, vegetables, meats, fish, poultry, eggs, cereals, flours, and dried fruits; many potential foods tested.	**Length of Storage:** Radiation significantly increases the storage of any food in comparison to its preservation by conventional methods. **Changes in Quality:** Only minimal changes in texture, flavor, color, and odor are experienced; and nutrient destruction is no greater than that which occurs when food is preserved by more conventional methods.	At low levels of radiation, the process is comparable to pasteurization; at high levels, it is equivalent to sterilization. Some doses inhibit sprouting and delay ripening in foods. The process has great potential since irradiated foods can be stored for long periods without refrigeration. Also, it is very flexible, adapting to various sizes and shapes of food.
REFRIGERATION (Cold Storage) **History:** In ancient Rome, snow from the mountains was used to pack prawns and other perishables. Root cellars were constructed by many settlers to keep foods cool. The use of ice for refrigeration was introduced toward the beginning of the 19th century. It was not until 1890 that mechanical refrigeration came into use on a large scale.	**Description and Principle:** Refrigeration temperatures may range from 29° to 60°F $(-1.6°$ to $15.5°C)$. Commercial and household refrigerators run at about 40° to 45°F $(4.5°$ to $7.2°C)$. Refrigeration retards deterioration by slowing the growth of microorganisms and decreasing the rate of chemical reactions that deteriorate foods. **Food Application:** Meats, poultry, fish, eggs, fruits, vegetables, and all in different stages of processing—cooked, fresh, etc.; baked goods.	**Length of Storage:** The amount of time foods can be stored depends on the type of food and its age. At 40°F $(4.5°C)$, meats, fish, and poultry keep for 2 to 10 days depending upon their stage of processing, while some fruits and vegetables keep for weeks or months. **Changes in Quality:** Gradual changes in quality and nutrient composition continue during storage.	Large cold storage warehouses keep huge quantities of apples, apricots, pears, butter, cheese, and eggs for periods of 6 to 10 months. Some fruits and vegetables are harvested at full size, but unripe. These are shipped in refrigerated transports, and allowed to ripen or held in cold storage to ripen.
SALT CURING (Brining) **History:** Salting is an ancient method of food preservation. Dry salting and the pickling of meats and fish in brine were practiced before 2000 B.C. (Also see CABBAGE; CUCUMBER AND GHERKIN; MEAT[S], section headed "Meat Processing"; and OLIVE.)	**Description and Principle:** There are 4 basic methods of salt curing: dry salting, brining, low-salt fermentation, and pickling. Dry salting and brining require concentrated amounts of salt. When a food is impregnated with salt, water is drawn out of the cells of the microorganisms and their growth is inhibited. Microorganisms vary in their sensitivity to salt, but usually 15 to 20% dissolved in the water phase of a food is highly preservative. **Food Application:** Meat, fish, poultry, and almost any fruit or vegetable by one or more of the salt curing methods.	**Length of Storage:** Most produce will remain fit for consumption 3 weeks to several months when stored at 38°F $(3.3°C)$, though the actual salt-curing process may take several weeks. For longer periods of storage the food needs to be subjected to some other type of preservative process such as canning, drying, or smoking. **Changes in Quality:** Dry salt and brining require the greatest concentrations of salt, and generally the more salt used the better the food is preserved, but the greater the loss in food value. Heavily salted food must be soaked and rinsed to make it palatable—a process which further depletes vitamins and minerals.	Nowadays most people use a salt curing method for the distinctive flavor it imparts to foods such as cucumber pickles, sauerkraut, green olives, and sausage. Low-salt fermentation and pickling ensure that the proper microorganisms develop and suppress the development of undesirable microbial activity.
SUGAR **History:** Early man used some sugar for preserving. However, sugar (sucrose) availability in the world began increasing after the mid-1750s; hence, more sugar continued to be used in the preservation of foods.	**Description and Principle:** Concentrated sugar solutions draw water out of the microorganisms and bind water, thereby preventing their growth. Microorganisms differ in their sensitivity to sugar, but around 70% sucrose (sugar) is highly preservative. **Food Application:** Preservation of fruits in jellies, jams, conserves, marmalades, preserves, fruit butters, and candied and glacéd fruits.	**Length of Storage:** After the addition of sugar, many fruits require further heat processing. When properly bottled or stored, these fruits may be stored a year or more. Some may be stored without hermetic sealing, though such protection is useful to control mold growth, moisture loss, and oxidation. **Changes in Quality:** Due to the addition of sugar, these foods contain more calories than fresh fruit. Also, some of the processing may leach minerals and destroy vitamins. However, some unpalatable fruits may be given a pleasing taste with this method.	Flavoring, coloring agents, and additional pectin may be added to overcome any deficiencies in the fruit. Most fruits contain significant amounts of sugars, and dried fruits are preserved in part by the increased concentration of natural sugars. Sugar is added to meat in some curing processes.
SMOKING **History:** Early settlers relied on curing and smoking of meats for their preservation. The old smokehouse was a common sight on many early farms. This old method led to the development of sausages, hams, and bacon, which are still popular. (Also see MEAT[S], section headed "Meat Processing.")	**Description and Principle:** Chemicals contained in smoke help prevent the growth of microorganisms, but, smoking is combined with other preservation methods which cure the product. Curing may be accomplished with salt alone, but generally herbs, spices, and sugar are also added. Cured meats are stored in rooms which can be filled with smoke, derived from burning hard woods. The amount of smoke, the temperature, and time of smoking depend upon the desired product. **Food Application:** Meat and fish, and some meat and fish products.	**Length of Storage:** Smoking by itself, as a cool smoke, contributes little to the preservation of meats and fish. In times past, the addition of smoke was associated with heat and drying of the product. Thoroughly cured and smoked meats such as Smithfield hams are reported to keep for years. **Changes in Quality:** The primary use of smoking today is to give the food a desirable smoked taste—an improvement in quality.	Modern refrigeration has opened the way for mild, sweet cures in which the smoked flavor is more important than the long-term preservation by smoke. Many people cure and smoke their game catches.

PRESSED

Compacted or molded by pressure; having fat, oil, or juice extracted under pressure.
(Also see OILS, VEGETABLE.)

PRESSURE COOKING

A pressure cooker is a heavy pan with a tight lid which can be clamped down, a safety valve, and an instrument for measuring the pressure. As the pressure increases, so does the boiling point of water, which means that the food cooks at a higher temperature, thus shortening the length of time required. The pressure cooker is also used for processing canned meats and vegetables.

PRETZELS

The name came from the German word, *brezel*, which originally came from the Latin word, *brachiatus*, meaning *having branches like arms*. Pretzels are brittle, glazed and salted crackers made of ropes of dough typically twisted into forms resembling the letter "B." The dough consists of a mixture of flour, salt, yeast, vegetable shortening, water, and malt. In the final step of their production, they are dipped into a tank of coconut oil maintained at a temperature of about 225°F (*107°C*), which seals the pretzel and extends its shelf life by 300%. Pretzels are a commercial product, difficult to duplicate in the home kitchen. They are used mainly as a snack food.

PRICKLY PEAR (INDIAN FIG; TUNA)
Opuntia ficus-indica

The fruit of a cactus (of the family *Cactaceae*) which is native to Mexico and now grows throughout the dry, subtropical areas in the northern and southern hemispheres.

Prickly pears are a little larger than large plums, but they have a prickly outer skin and a sweet inner pulp. They may be eaten fresh, stewed, or made into jams.

The nutrient composition of the prickly pear (Indian fig) is given in Food Composition Table F-21.

Some noteworthy observations regarding the nutrient composition of the prickly pear follow:

1. The fresh fruit is moderately high in calories (67 kcal per 100 g) and carbohydrates (17%).

2. Prickly pears are a good source of fiber and iron, and a fair source of calcium and vitamin C.

PROCESSED FOODS, HIGHLY

Contents	Page
Current Trends	890
Cereal Products	890
Dairy Products and Eggs	890
Fish, Meats, Poultry, and Seafood	890
Fruits and Vegetables	891
Meal Replacements	891
Limitations of Enrichment and Fortification	891

Many species of animals collect and store food, but only *Homo sapiens* processes it. *Processing refers to all the physical and/or chemical operations that are applied to foodstuffs.* Food processing influences the nutritional value of foodstuffs—enhancing some, lowering others.

The term "highly processed foods" designates foods that have been altered significantly in appearance, culinary characteristics, nutritional value, structure, and/or texture as a result of a considerable amount of processing. Some of the reasons why foods are highly processed are: (1) to prevent spoilage; (2) to render them more storable and transportable, by reducing their bulk and water content; (3) to meet special dietary requirements; and (4) to produce concentrated sources of certain nutrients. Often, processing increases the levels of certain nutrients while reducing the levels of others. Hence, selected nutritional and economic aspects of highly processed foods are discussed in the sections which follow.

CURRENT TRENDS. Some of the most commonly employed types of processing tend to concentrate solids, carbohydrates, fats, and/or proteins at the expense of fiber, certain essential minerals, and certain vitamins. Examples of some highly processed items follow.

Cereal Products. The processing of grains has traditionally yielded products that are higher in starch than the original grains, but lower in most other nutrients. Enrichment of the products results in the restoration of only four of the more than 2 dozen essential nutrients that are depleted significantly by processing. Furthermore, many cereal products are made with substantial amounts of added fats and sugars. However, there has been a recent trend to restore some of the bran and wheat germ that was removed during processing.

Dairy Products And Eggs. These foods were once considered to be ideal for nourishing growing children and hard-working adolescents and adults. Now, many of the traditional items are being modified by (1) replacement of the animal fats with vegetable oil derivatives, (2) removal of much of the fat content without replacement, and (3) extension of certain expensive dairy and egg products with air, emulsifying agents, fillers, gummy binders, water, and less expensive animal or plant products that may possess analogous characteristics. Some noteworthy examples of these trends are cream substitutes (often called *nondairy blends*), filled milks, imitation cheeses, imitation eggs, imitation ice cream, imitation milk (this type of product may contain a soybean derivative plus vegetable oil), margarines, process cheeses, and sour cream substitutes. Producers often capitalize on the current concern over the alleged dangers of animal fats and cholesterol in promoting the modified items as being more beneficial to health than the traditional products.

Fish, Meats, Poultry, And Seafood. These items are steadily becoming more expensive, which makes it increasingly profitable to extend or even replace them with various plant products such as soybean derivatives or wheat gluten.

Fig. P-77. A hamburgerlike vegetetable protein product made from soybeans. (Courtesy, Worthington Foods, Inc.)

Fruits And Vegetables. Here again, binders and fillers come to the rescue of food processors beset with rising production costs. *Simulated* pieces of fruits or vegetables may be made from mixtures of fruit or vegetable pulp, vegetable gums, mineral salts, starch, sugar, artificial or natural coloring and flavorings, and water.

Meal Replacements. These products, which are often in the form of beverage powders or sweetened bars, are designed to be consumed with milk, water, or other beverage in lieu of the various mixtures of foods which usually constitute a meal. Federal regulations require that a typical serving of an item designated as a "meal replacement" provide at least 25% of the Recommended Daily Allowance (RDA) for certain specific nutrients. The macronutrients in these products are provided by various combinations of sugar, nonfat dry milk, soybean derivatives, and peanut butter and/or flour, whereas the required levels of micronutrients are usually met by fortification with selected minerals and vitamins. However, these products usually contain little or no fiber.

Limitations Of Enrichment and Fortification. In theory, it should be possible to add pure forms of the nutrients present in minimally processed foods to highly processed foods in order to obtain nutritional equivalency. However, the U.S. Food and Drug Administration specifies enrichment and fortification levels for only some of the nutrients known to be essential. There are no specified levels for other essential nutrients such as chromium, selenium, and vitamin K. It might even be illegal to add the purified forms of these nutrients to foods.

Secondly, foodborne nutrients may in certain cases be present in combinations which work together. For example, the metabolic activities of the members of the vitamin B complex are interrelated. Hence, the failure to provide sufficient dietary quantities of one or more of these vitamins may impair the functions of other B vitamins in the diet.

(Also see NUTRITIONAL SUPPLEMENTS; and PROCESSING OF FOOD.)

PROCESSING OF FOOD

Many foods are altered before they reach the table. Grains are ground into flour and other cereal foods. Milk is pasteurized, homogenized, or used in making cottage cheese, butter, or yogurt. Meat animals such as cattle and pigs are slaughtered, graded and cut into steaks, ground to hamburger, salted, dried, corned, or cured and smoked. Fruits and vegetables are sold fresh, canned, or frozen. All of these represent some stage or some portion of processing —treatment, preparation, or handling by some special method.

Processing starts on the whole raw material of whatever the food may be. Then, for some foods the whole raw material may be broken into parts which are futher processed, or the parts may be mixed with other components and used to form new products. The final stages of processing includes preservation so that the foods may be stored without spoiling. Finally processing includes some form of packing for shipment, storage, and consumer appeal.

(Also see NUTRITIONAL SUPPLEMENTS; and PROCESSED FOODS, HIGHLY.)

PROENZYME (ZYMOGEN)

Inactive form of an enzyme; e.g., pepsinogen.

PROGNOSIS

Prediction of the outcome of an illness based on the person's condition and on scientific knowledge concerning the usual course and results of such illnesses, together with the presence of certain symptoms and signs that indicate the expected outcome.

PROPHYLAXIS

Preventive treatment against disease.

PROPYL GALLATE (PG)

An FDA approved food additive which acts as an antioxidant in foods, fats, oils, and waxes. It is often used in combination with butylated hydroxyanisole (BHA) and butylated hydroxytoluene (BHT).

PROSTAGLANDINS

These are a class of hormonelike compounds, made in various tissues of the body from arachidonic acid (and other derivatives of linoleic acid), which are important in the regulation of such diverse reactions as gastric secretion, pancreatic functions, release of pituitary hormones, smooth muscle metabolism, and control of blood pressure. Some 16 prostaglandins have been identified.

PROSTHETIC GROUP

A protein conjugated with a nonamino acid substance; for example, hemoglobin is a conjugated (joined together) protein—each chain in the protein moiety, globin, is combined with a heme group to form the biologically functional molecule. Proteins are often described in terms of their prosthetic group; thus, hemoglobin is a *hemeprotein*, and proteins containing lipid, carbohydrate, or metals are called *lipoproteins*, *glycoproteins*, and *metalloproteins*, respectively.

PROTEASE

An enzyme that digests protein.

PROTEIN(S)

Contents	Page
History	892
Synthesis	893
Classification	894
Digestion, Absorption, and Metabolism	894
Functions	894
Required Protein Intake	895
Protein Allowance	895
Amino Acid Requirement	895
Sources	896
Quantity and Quality	896
Complementary Proteins	898
Factors Affecting Amino Acid Utilization	899
Protein Malnutrition	899
Protein for the World	901
Protein and Amino Acids in Selected Foods	901

Chemically, proteins are complex organic compounds made up chiefly of amino acids. For each different protein there are specific amino acids and a specific number of amino acids which are joined in a specific order. Since amino acids always contain carbon, hydrogen, oxygen, and nitrogen, so do proteins. Moreover, the presence of nitrogen provides a tool for chemically estimating the amount of protein in a tissue, food, or some other substance. Crude protein is routinely determined by finding the nitrogen content and

PROTEIN is needed for growth

THREE RATS FROM THE SAME LITTER, 11 WEEKS OLD

THIS RAT ATE FOODS THAT FURNISHED GOOD QUALITY PROTEIN, BUT NOT ENOUGH. IT WEIGHED ONLY 70 GRAMS.

THIS RAT ATE FOODS THAT FURNISHED PLENTY OF PROTEIN, BUT NOT THE RIGHT COMBINATION TO GIVE GOOD QUALITY. IT WEIGHED ONLY 65 GRAMS.

THIS RAT HAD PLENTY OF GOOD QUALITY PROTEIN FROM A VARIETY OF FOODS. IT HAD GOOD FUR, A WELL-SHAPED BODY, AND WEIGHED 193 GRAMS.

TOP FOOD SOURCES

BEEF & LAMB | PORK | POULTRY | FISH | EGGS

CHEESE | DRY BEANS PEAS, NUTS | CEREAL PRODUCTS | SOYBEAN PRODUCTS | MILK

Fig. P-78. Protein quantity and quality made the difference! The top rat ate too little of a good quality protein. The middle rat ate plenty of poor quality protein, while the bottom rat ate plenty of a good quality protein. (Adapted from USDA sources)

multiplying the result by 6.25 since the nitrogen content of all protein averages about 16% (100 ÷ 16 = 6.25). In addition, proteins usually contain sulfur and frequently phosphorus. Proteins are essential in all plant and animal life as components of the active protoplasm of each living cell.

In plants, the protein is largely concentrated in the actively growing portions, especially the leaves and seeds. Plants also have the ability to synthesize their own proteins from such relatively simple soil and air compounds as carbon dioxide, water, nitrates, and sulfates, using energy from the sun. Thus, plants, together with some bacteria which are able to synthesize these products, are the original sources of all proteins.

In animals, proteins are much more widely distributed than in plants. Thus, the proteins of the body are primary constituents of many structural and protective tissues—such as bones, ligaments, hair, fingernails, skin, and the soft tissues which include the organs and muscles. The total protein content of animal bodies ranges from about 10% in very fat, mature animals to 20% in thin, young animals. By way of further contrast, it is also interesting to note that, except for the bacterial action in the rumen of ruminants—cows, sheep, and goats—animals, humans included, lack the ability of the plant to synthesize proteins from simple materials. They must depend upon plants or other animals as a source of dietary protein. Hence, humans must have certain amino acids or more complete protein compounds in the diet.

Animals of all ages and kinds require adequate amounts of protein of suitable quality. The protein requirements for growth, reproduction, and lactation are the greatest and most critical.

The need for protein in the diet is actually a need for amino acids. Proteins in the food are broken down into amino acids by digestion. They are then absorbed and distributed by the bloodstream to the body cells, which rebuild these amino acids into body proteins.

The various amino acids are then systematically joined together to form peptides and proteins. The amino portion of one amino acid will combine with the carboxyl (acid) portion of another amino acid to form a peptide linkage. When several of these junctions occur, the resulting molecule is called a polypeptide. Generally polypeptide chains contain 50 to 1,000 amino acids.

HISTORY. In 1838, the Dutch chemist Geradus Johannes Mulder applied the name *proteins* to a group of complex organic compounds found in both plant and animal materials. It was because of their recognized importance to the structure and function of living matter that proteins were so named from the Greek word *proteios,* meaning *of first quality.*

In 1841, Justus von Liebig, a German scientist, published a paper on the analysis of proteins. Liebig thought that the protein values of different foods could be assessed on the basis of nitrogen content. Then, in 1881, while studying the changes in protein content in grain during germination and fermentation, Johan Kjeldahl found his progress blocked by lack of an accurate method for determining nitrogen. Kjeldahl concentrated all his efforts on developing an analytical method for determining nitrogen; hence, protein content of substances. The Kjeldahl method is still routinely used today.

Gradually, the constituents of proteins—the amino acids—were discovered. In 1902, Emil Fischer, determined the chemical structure of amino acids, the building blocks of protein; and he determined the nature of the chemical bond—the peptide bond—holding the amino acids together.

One of the first recommendations as to protein intake was made by Carl von Voit, a student of Liebig. He suggested in 1881 that 118 g of protein daily was desirable. In 1902, a student of Voit and an American, W. O. Atwater, recommended 125 g per day. However, in 1904, R. H. Chittenden of Yale University recommended only 44 to 53 g of protein per day. Then, between 1909 and 1921, T. B. Osborne and L. B. Mendel conducted numerous experiments on rats which involved protein quantity and quality. Between 1935 and 1955, W. C. Rose conducted experiments in humans to determine how much of each amino acid was needed.

Further studies of actual proteins increased scientists' knowledge of exactly how proteins function in the body. In about 1927, James B. Sumner demonstrated that an enzyme was a protein. Then, during the 1950s, Frederick Sanger made a major discovery when he described the actual amino acid sequence of the protein hormone insulin.

Other studies pointed to the relationships between DNA (deoxyribonucleic acid) and RNA (ribonucleic acid) and protein. In 1953, Francis Crick and James Watson described the structure of the DNA molecule, and gradually scientists have unraveled how cells manufacture specific proteins with specific amino acid sequences.

(Also see NUCLEIC ACIDS.)

SYNTHESIS. The basic structural components of protein are amino acids. Many of the amino acids can be synthesized within the body. These are called nonessential amino acids or dispensable amino acids. If the body cannot synthesize sufficient amounts of certain amino acids to carry out physiological functions, they must be provided in the diet; hence, they are referred to as essential or indispensable amino acids. Actually, it is not entirely correct to say that all indispensable amino acids need to be provided in the diet; rather, the requirement is for the preformed carbon skeleton of the indispensable amino acids, except in the case of lysine and threonine.

According to our present knowledge, the following division of amino acids as essential and nonessential seems proper for humans:

Essential (indispensable)	Nonessential (dispensable)
Histidine	Alanine
Isoleucine	Arginine
Leucine	Asparagine
Lysine	Aspartic acid
Methionine (some	Cysteine
used for the synthe-	Cystine
sis of cysteine)	Glutamic acid
Phenylalanine (some	Glutamine
used for the synthe-	Glucine
sis of tyrosine)	Hydroxyproline
Threonine	Proline
Tryptophan	Serine
Valine	Tyrosine

NOTE WELL: Arginine is not regarded as essential for humans, whereas it is for animals; in contrast to human infants, most young mammals cannot synthesize it in sufficient amounts to meet their needs for growth.

(Also see AMINO ACID[S].)

In order for a protein to be synthesized, all of its constituent amino acids must be available. If one amino acid is missing, the synthesis procedure is halted. When a particular amino acid is deficient, it is referred to as a limiting amino acid because it limits the synthesis of protein. This is why protein quality is so important in human nutrition. Upon digestion, high-quality proteins provide balanced supplies of the various amino acids which can subsequently be absorbed as precursors for protein synthesis. Plant foods often contain insufficient quantities of lysine, methionine and cystine, tryptophan, and/or threonine. Lysine is the limiting amino acid of many cereals, while methionine is the limiting amino acid of beans (legumes). In general, the proteins of animal origin—eggs, dairy products, and meats—provide mixtures of amino acids that are well suited for human requirements of maintenance and growth.

Protein synthesis is not a random process whereby a number of amino acids are joined together; rather, it is a detailed predetermined procedure. When proteins are formed three requirements must be met: (1) the proper amino acids, (2) the proper number of amino acids, and (3) the proper order of the amino acid in the chain forming the protein. Meeting these requirements allows the formation of proteins which are very specific and which give tissues

their unique form, function, and character.

Protein synthesis involves a series of reactions which are specific for *each* protein. The directions for manufacturing protein are received in coded form from the genetic material of the cell—the DNA (deoxyribonucleic acid).

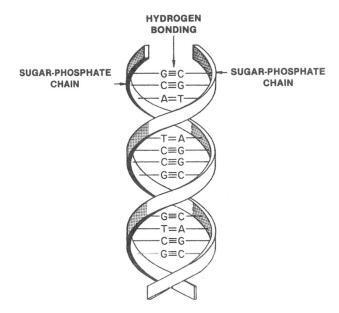

Fig. P-79. Deoxyribonucleic acid or DNA—the code of life. It's a double helix (a double spiral structure), with the sugar (deoxyribose)-phosphate *backbone* represented by the two spiral ribbons. Connecting the *backbone* are four nitrogenous bases (a base is the nonacid part of a salt): adenine (A) paired with thymine (T), and guanine (G) paired with cytosine (C); with the parallel spiral ribbons held together by bonding between these base pairs. The sequence of the bases codes for the synthesis of every protein in the body.

The code of the DNA is put into action by ribonucleic acid (RNA). There are three forms of RNA, each one with a special function. Messenger RNA (mRNA) transcribes the information from the DNA and transports it to the cytoplasm of the cell as a single strand. Transfer RNA (tRNA) are small molecules which act as carriers for the specific amino acids. The third type of RNA, ribosomal RNA (rRNA), is a major component of the ribosomes within the cytoplasm of the cell. The site of protein synthesis is ribosomes bound to mRNA. Ribosomes bound to the mRNA translate (decode) the genetic code. As the message from DNA is translated, tRNA supplies the proper amino acid. These amino acids are then linked by peptide bonds. The amino portion (NH_2) of one amino acid will combine with the carboxyl (COOH) portion of another amino acid releasing water (H_2O) and forming a peptide linkage—like railroad cars joining to form a train. When numerous of these junctions occur, the resulting molecule is called a protein.

Each sequence of amino acids is a different protein. Hence, different proteins are able to accomplish different functions in the body. With 22 amino acids, the different arrangements possible are endless, yielding a variety of proteins.

Since the DNA of each cell carries the master plan for forming proteins, genetic mutations often are manifested by disorders in protein metabolism. Many metabolic defects—inborn errors of metabolism—are the result of a missing or modified enzyme or other protein. The misplacement or omission of just one amino acid in a protein molecule can result in a nonfunctional protein. The disease called sickle-cell anemia results because the blood cells of afflicted individuals contains a faulty protein, with the result that the blood cells often collapse, causing anemia.

(Also see INBORN ERRORS OF METABOLISM; METABOLISM, section headed "Proteins"; and NUCLEIC ACIDS.)

CLASSIFICATION. Proteins occur in nature in a number of forms, each one possessing unique chemical properties. Based on chemical compostion, proteins may be divided into two main categories: (1) simple and (2) conjugated. Simple proteins consist of only amino acids or their derivatives, while conjugated proteins are joined to various nonprotein substances. Then, these two main categories are further subdivided. Table P-11 lists the various categories of proteins along with their distinguishing characteristics.

Possibly a third category—*derived proteins* may be added

to the two above. Essentially derived proteins are the product of digestion. They are fragments of various sizes. From largest to smallest, in terms of the number of amino acids, derived proteins are proteoses, peptones, polypeptides, and peptides.

Proteins may also be classified according to their structure which is very important to their function in the body. Some proteins are round to ellipsoidal and are called *globular proteins*. These include *enzymes, protein hormones, hemoglobin*, and *globulins*. Other proteins form long chains bound together in a parallel fashion, and are called *fibrous proteins*. These include *collagen, elastin*, and *keratin*—the proteins of connective tissue, elastic tissue, and hair. Many of the fibrous proteins are not very digestible, and if they are, they are poor-quality proteins.

TABLE P-11
CLASSIFICATION OF SOME COMMON PROTEINS

Type	Chemical Properties	General Comments
SIMPLE PROTEINS		
Albuminoids (scleroproteins)	Insoluble in water; highly resistant to enzymatic digestion; some become gelatinous upon boiling in water or dilute acids and bases.	Includes collagen, elastin, and keratin; common in supporting tissues; sometimes referred to as fibrous protein.
Albumins	Readily soluble in water; coagulate upon heating.	Present in egg, milk, and serum.
Globulins	Low solubility in water; solubility increases with the addition of neutral salts; coagulates upon heating.	Abundant in nature; examples are serum globulins, muscle globulins, and numerous plant globulins.
Glutelins	Insoluble in water; soluble in dilute acids or bases.	Abundant in cereal grains; an example is wheat gluten.
Prolamins	Insoluble in water, absolute alcohol or neutral solvents; soluble in 80% ethanol.	Zein in corn and gliadin in wheat are prolamins.
CONJUGATED PROTEINS		
Chromoproteins	Combination of a protein and a pigmented (colored) substance.	Common example is hemoglobin—hematin and protein.
Lecithoproteins	Combination of protein and lecithin.	Found in fiber of clotted blood and vitellin of egg.
Lipoproteins	Water soluble combination of fat and protein.	A vehicle for the transport of fat in the blood; all contain triglycerides, cholesterol, and phospholipids in varying proportions.
Metalloproteins	Proteins that are complexed with metals.	One example is transferrin, a metalloprotein that can bind copper, iron, and zinc. Various enzymes contain minerals.
Mucoproteins or glycoproteins	Contain carbohydrates such as mannose or galactose.	Examples are mucin from the mucus secretion, and some hormones such as that in human pregnancy urine.
Nucleoproteins	Combination of proteins and nucleic acids.	Present in germs of seeds and glandular tissue.
Phosphoproteins	Compounds containing protein and phosphorus in a form other than phospholipid or nucleic acid.	Casein in milk and ovovitellin in eggs, are examples.

DIGESTION, ABSORPTION, AND METABOLISM. Protein digestion—the breakdown into smaller units—begins in the stomach and continues to completion in the small intestine. Enzymes from the pancreas and the small intestine eventually split dietary protein into individual amino acids. These amino acids are absorbed from the intestine by the process of active transport whereupon they are distributed to the cells of the body via the blood. In the body, amino acids may be used for primarily protein synthesis or energy production, depending upon the (1) protein quality, (2) caloric level of the diet, (3) stage of development, including growth, pregnancy, and lactation, (4) prior nutritional status, and (5) stress factors such as fever, injury, and immobilization.

Occasionally, individuals are born with very specific disorders related to the metabolism of protein or amino acids. These are often referred to as inborn errors of metabolism. Albinism, maple syrup urine disease, and phenylketonuria are examples of some of the more familiar types.

(Also see DIGESTION AND ABSORPTION; INBORN

ERRORS OF METABOLISM, Table I-1, Inborn Errors of Amino Acid Metabolism; and METABOLISM.)

FUNCTIONS. Each specific protein performs a specific function in the body. One protein cannot and will not substitute for another. Rather broadly, proteins may be classified as performing the following five functions:

1. The formation of new tissues from infancy to adulthood—*anabolism*.

2. Maintenance of body tissues—a process called *protein turnover*. Protein maintenance involves replacing blood cells every 120 days, replacing the cells that line the intestine every 1½ days, and replacing the protein lost from the body in the perspiration, hair, fingernails, skin, urine, and feces.

3. *Regulatory*. The protein in the cells and body fluids provide regulatory functions. Thus, hemoglobin, an iron-bearing protein, carries oxygen to the tissues; The protein of the blood plasma regulates water balance and osmotic pressure; the proteins act as buffers controlling the acid-base balance of the body; many of the hormones which regulate body processes are proteins; enzymes, the catalysts for most

all chemical reactions of the body, are proteins; antibodies, which protect the body from infectious diseases are proteins; the blood clotting mechanism of the body is dependent upon proteins; and certain amino acids derived from dietary protein or synthesized in the body also perform important regulatory functions.

4. *Milk production.* Each liter of human milk contains about 12 g of protein.

5. *The production of energy*—the catabolism (breakdown) of amino acids yields energy.

REQUIRED PROTEIN INTAKE. The minimum protein requirement of humans (other animals, too) may be determined by nitrogen balance studies. Nitrogen is obtained from the consumption of protein. As indicated previously, each 100 g of protein contains approximately 16 g of nitrogen. Nitrogen is lost from the body each day in the feces, urine, skin, hair, nails, perspiration and other secretions. Dietary protein (nitrogen) must be sufficient to cover the daily losses. Normal, healthy adults should be in a state of nitrogen equilibrium—intake approximately equals output. When nitrogen intake exceeds nitrogen output, a positive nitrogen balance exists. Such a condition is present in the following physiological states: pregnancy, lactation, recovery from a severe illness, muscular buildup, growth, and following the administration of an anabolic steroid such as testosterone. A positive nitrogen balance indicates that new tissue is being formed. A negative nitrogen balance occurs when nitrogen intake is less than output. Starvation, diabetes mellitus, fever, surgery, burns or shock can all result in a negative nitrogen balance. This is an undesirable state since body protein is being broken down faster than it is being built up.

Protein Allowance. Using nitrogen balance, the recommended daily dietary allowance for protein may be derived as follows:

1. Obligatory urinary nitrogen losses of young adults amount to about 37 mg/kg of body weight.

2. Fecal nitrogen losses average 12 mg/kg of body weight.

3. Amounts of nitrogen lost in the perspiration, hair, fingernails, and sloughed skin are estimated at 3 mg/kg of body weight.

4. Minor routes of nitrogen loss such as saliva, menstruation, and seminal ejaculation are estimated at 2 mg/kg of body weight.

5. The total obligatory nitrogen lost—that which must be replaced daily—amounts to 54 mg/kg, or in terms of protein lost this is 0.34 g/kg.

6. To account for individual variation the daily loss is increased by 30%, or 70 mg/kg. In terms of protein, this is 0.45 g/kg of body weight.

7. This protein loss is further increased by 30%, to 0.6 g/kg of body weight, to account for the loss of efficiency when consuming even a high quality protein such as egg.

8. The final adjustment is to correct for the 75% efficiency of utilization of protein in the mixed diet of Americans. Thus, the daily recommended allowances for protein becomes 0.8 g/kg of body weight for normal healthy adult males and females, or 63 g of protein per day for a 174-lb *(79 kg)* man and 50 g per day for a 138-lb *(63 kg)* woman.

Growth, pregnancy, and lactation, and possibly work and stress, require an additional intake of protein. Aging may present some special considerations.

The Food and Nutrition Board (FNB) of the National Re-

search Council (NRC) recommended daily allowance of protein are given in Table P-12.

TABLE P-12
RECOMMENDED DAILY PROTEIN ALLOWANCES[1]

Group	Age	Weight		Height		RDA
	(years)	(lb)	(kg)	(in.)	(cm)	(g)
Infants	0.0–0.5	13	6	24	60	13
	0.5–1.0	20	9	28	71	14
Children	1–3	29	13	35	90	16
	4–6	44	20	44	112	24
	7–10	62	28	52	132	28
Males	11–14	99	45	62	157	45
	15–18	145	66	69	176	59
	19–24	160	72	70	177	58
	25–50	174	79	70	178	63
	51+	170	77	68	173	63
Females	11–14	101	46	62	157	46
	15–18	120	55	64	163	44
	19–24	128	58	65	164	46
	25–50	138	63	64	163	50
	51+	143	65	63	160	50
Pregnant						60
Lactating						65

[1]*Recommended Dietary Allowances*, 10th ed., 1989, NRC-National Academy of Sciences, p. 284.

Table P-13 presents the recommended protein allowance for the elderly on the basis of their energy intake.

TABLE P-13
RECOMMENDED DAILY PROTEIN ALLOWANCES FOR THE ELDERLY[1]

Group	Age	Weight		Height		Energy Needs	Protein
	(years)	(lb)	(kg)	(in.)	(cm)	(kcal)	(g)
Males	51+	170	77	68	173	2,300	63
Females	51+	143	65	63	160	1,900	50

[1]Based on 12% of the energy intake in the form of protein. Age, weight, height, and energy needs are from *Recommended Dietary Allowances*, 10th ed., 1989, NRC-National Academy of Sciences, p. 284.

Both dietary protein and energy supplies are adequate in the United States. Individuals in the United States derive about 14 to 18% of their calories from protein. Average consumption levels of proteins are quite generous: about 50 g/day in preschool children; 70 to 85 g in older children; 90 to 110 g in male, and 65 to 70 g in female adolescents and adults; and 75 to 80 g in men and 55 to 65 g in women over age 65.

Amino Acid Requirement. The requirement for protein carries a restriction. Protein must supply individuals with amounts of the essential amino acids at levels sufficient to meet the protein synthesizing needs of the body. Based on a number of studies on infants, children, and adults, some estimated daily requirements for the essential amino acids have been made; and these are listed in Table P-14.

Amino Acid	Requirement, mg/kg Body Weight/Day			
	Infants	Children		Adults
	(3–4 mo.)	(2 yr)	(10–12 yr)	
Histidine	16	19	19	11
Isoleucine	40	28	28	13
Leucine	93	66	44	19
Lysine	60	58	44	16
Total S-containing amino acids (methionine and cystine)	33	25	22	17
Total aromatic amino acids (phenylalanine and tyrosine)	72	63	22	19
Threonine	50	34	28	9
Tryptophan	10	11	(9)	5
Valine	54	35	25	13

[1]*Recommended Dietary Allowances*, 10th ed., 1989, NRC-National Academy of Sciences, p. 67.

These requirements are adequate only when the diet provides enough nitrogen for the synthesis of the nonessential amino acids so that the essential amino acids will not be used to supply amino groups for the nonessential amino acids via the process of transamination. Even in infants the essential amino acids make up only about 35% of the total need for protein. In adults, essential amino acids account for less than 20% of the total protein requirement. Most proteins contain plenty of dispensable amino acids; usually the concern is to meet the essential amino acid needs, particularly of infants and children.

SOURCES. The protein content of numerous foods is given in Food Composition Table F-21.

Unfortunately, meeting the protein requirement is more involved than just choosing foods which provide the daily amount of protein. As previously indicated, the requirement for protein is primarily a requirement for essential amino acids. In terms of amino acid composition, all proteins are not created equal. Some proteins of certain foods are low or completely devoid of some essential amino acids. Hence, meeting the protein requirement demands quality as well as quantity.

Fig. P-80. All proteins are not created equal. Animal proteins generally have a higher biological value than plant proteins. Shown: (left) T-bone steak, (Courtesy, National Live Stock & Meat Board, Chicago, Ill.); and (right) fried chicken. (Courtesy, USDA)

Quantity and Quality. Protein values such as those in Food Composition Table F-21 are crude protein values derived by the Kjeldahl method, which determines the nitrogen content of a substance and then assumes that each gram of nitrogen represents 6.25 g of protein. This procedure estimates protein quantity; and quantity is important. For example, boiled potatoes contain 1.9 g of protein per 3½ oz (*100 g*). To obtain just the extra 30 g of protein daily during pregnancy, a woman would have to consume about 3½ lb (*1.6 kg*) of potatoes, while only 3½ to 4 oz (*100 to 113 g*) of beef, fish, or poultry would supply the extra 30 g of protein. Still, crude protein values may be misleading. For example, gelatin ranks second among the top protein sources; yet, gelatin contains virtually no tryptophan and very low levels of other essential amino acids. Used as a protein source, gelatin will not support growth or life. Hence, some evaluation of protein quality is necessary.

In 1911, T. B. Osborne, of the Connecticut Agricultural Experiment Station, and L. B. Mendel, of Yale University, formed a brilliant partnership and pioneered in studies of protein quality. Their early work on the quality of protein resulted in proteins being classified as complete, partially incomplete, and incomplete. Complete proteins contain all of the essential amino acids in sufficient amounts to maintain life and support growth. Partially incomplete proteins can maintain life, but cannot support growth. Incomplete proteins cannot maintain life or support growth. Following the work of Osborne

Fig. P-81. Thomas B. Osborne (left) and Lafayette B. Mendel (right), who prioneered in studies of protein quality. (Courtesy, The Connecticut Agricultural Experiment Station, New Haven, Conn.)

and Mendel, there was a quest for some method of evaluating proteins in terms of their ability to meet the needs of the body. The following are descriptions of some of these methods:

• **Biological value**—This method assumes that the retained nitrogen—ingested nitrogen not recovered in the urine or feces—represents a perfect assortment of amino acids utilized by the body. Biological value is a measure of the efficiency of utilization of absorbed nitrogen, and it depends primarily on the amino acid composition of the dietary protein. Needless to say, determining the quality of protein in the diet by this means is time-consuming, and hence, not often used. Moreover, few animals other than rats will consume protein-free diets long enough to complete the trial. However, biological value is often used to describe protein quality.

(Also see BIOLOGICAL VALUE [BV] OF PROTEINS.)

Fig. P-82. High protein vs. low protein diets! The rat on the left had a diet with 18% casein. The other rat's diet consisted of 4% casein. This experiment was conducted by Thomas B. Osborne, of the Connecticut Agricultural Experiment Station, and Lafayette B. Mendel, of Yale University, who in 1911, formed a brilliant partnership and pioneered in studies of protein quality. (Courtesy, The Connecticut Agricultural Experiment Station, New Haven, Conn.)

• **Net protein utilization (NPU)**—Digestibility is estimated from the quantity ingested that is subsequently not recovered in the feces, expressed as a percentage of the intake. The multiplication of the biological value and the digestibility gives the net protein utilization, a measure of the efficiency of overall utilization of the ingested nitrogen. For proteins that are completely digested, net protein utilization and biological value are the same. For less digestible proteins or for foods containing large amounts of fiber, the biological value does not provide a measure of efficiency of utilization of the protein consumed—only the efficiency of utilization of that absorbed.

• **Protein efficiency ratio (PER)**—This is the simplest test of protein quality, and the one used widely in comparing the nutritive value of proteins in individual foods. The PER is defined as the weight gain of a growing animal divided by its protein intake over the period studied, often 10 days. Although simple in application, it does not provide protein evaluations that are directly proportional to quality, because no account is taken of loss of weight when animals are given no protein in the diet. It assumes that all protein is used for growth and no allowance for maintenance is made.

• **Net protein ratio (NPR)**—This method attempts to correct the fault of the PER by making some allowance for maintenance. The NPR accomplishes this by the inclusion of a control group of animals receiving no dietary protein. Thus, the NPR is the weight gain of a group of animals (rats) fed the test diet plus the weight loss of a similar group fed a protein-free diet, and the total divided by the weight of the protein consumed by the animals on the test diet.

• **Nitrogen balance index**—This method is similar to the biological value but there is no period of feeding a protein-free diet. It is simply the nitrogen in the food minus that in the urine and feces. Nitrogen balance has been used as a means of evaluating protein foods in the human diet and determining the amino acid requirements of humans. When proteins (amino acids) are adequate in the diet, the nitrogen balance is positive. W. C. Rose, using a diet consisting of sugar, starch, fat, and a mixed solution of purified amino acids in known amounts, was the first to determine the amount of each essential amino acid needed for young men. One at a time, the amino acids were left out of the solution and nitrogen balance was noted. When the omitted amino acid was essential, a negative nitrogen balance resulted. The method used for determining the nutritive value of proteins for humans is an extension of nitrogen balance and is termed a nitrogen balance index. The food to be evaluated is fed at several intakes below and slightly above nitrogen equilibrium. Then, values for nitrogen balance versus nitrogen absorbed or nitrogen intake are plotted on a graph, and a straight line connects these points and the slope of the line is determined. When nitrogen intake is used in place of nitrogen absorbed, the values obtained are equivalent to net protein utilization (NPU).

• **Slope ratio method**—This is a modification of the nitrogen balance index wherein various levels of protein are fed *ad libitum* to growing animals. Then growth rate and protein intake are plotted on a graph, and the slope of this line is taken as the measure of protein quality.

• **Chemical score (CS)**—By chemically examining the essential amino acid content of a food, some idea of its nutritive value may be gained. To determine the chemical score one must first have a reference pattern of amino acid concentrations with which to compare the dietary protein. Table P-15 indicates what type of amino acid pattern a high-quality protein should possess in order to meet the body's needs. This amino acid pattern is based on the essential amino acid requirements of children 10 to 12 years of age. Also, Table P-15 shows how the essential amino acid patterns of human milk and eggs compare to that of the requirement of children 10 to 12 years of age. Human milk and eggs are frequently used

TABLE P-15
AMINO ACID REQUIREMENT OF CHILDREN 10–12 YEARS COMPARED TO THE AMINO ACID PROFILES OF HUMAN MILK AND EGGS

Amino Acid	Amino Acid Requirement for Children 10–12 Years[1]	Human Milk[2]	Whole Egg[2]
	(mg/g protein)		
Histidine	19	26	22
Isoleucine	28	46	54
Leucine	44	93	86
Lysine	44	66	70
Methionine and cystine	22	42	57
Phenylalanine and tyrosine	22	72	93
Threonine	28	43	47
Tryptophan	9	17	17
Valine	25	55	66

[1]Recommended Dietary Allowances, 10th ed., 1989, National Academy of Sciences, p. 67, Table 6–5.
[2]Values for the protein and amino acid content are from Proteins and Amino Acids in Selected Foods Table p. 37.

as standards due to their amino acid patterns. The reference or standard pattern is assigned a value of 100, and the percentage by which each essential amino acid in the food or dietary protein differs from the value of the standard is calculated. The essential amino acid showing the greatest deficit is considered to be the amino acid limiting utilization of the protein. The amount of this amino acid present, expressed as a percentage of that in the standard, provides the chemical score. This score usually shows good agreement with biological evaluation of protein quality but tends to overestimate quality for proteins that are not well digested.

The following equation may be used to calculate the chemical score of each essential amino acid:

$$CS = \frac{\text{mg of amino acid per g test protein x 100}}{\text{mg of amino acid per g reference protein}}$$

Often, only the chemical scores for lysine, methionine and cystine, and tryptophan are calculated as one of these is commonly the limiting amino acid. The following is an example for potatoes:
1. Consider the reference pattern of whole eggs:
   Methionine and cystine .... 56 mg/g
   Lysine .... 68 mg/g
   Tryptophan .... 16 mg/g
2. List the amount of the same amino acids in the protein of potatoes:
   Methionine and cystine .... 27 mg/g
   Lysine .... 127 mg/g
   Tryptophan .... 35 mg/g
3. Calculate the chemical score for these amino acids as indicated by the above formula:
   Methionine and cystine 48
   Lysine 187
   Tryptophan 219

Obviously, the amino acids methionine and cystine—the sulfur-containing amino acids—are limiting in potatoes.

• **Net dietary protein calories percent (ND$_p$Cal%)**—This method of evaluating protein considers the percentage of the calories from protein in the food which is adjusted according to the net protein utilization (NPU) or quality. The following formula describes ND$_p$Cal%:

$$ND_pCal\% = \frac{\text{g of protein x 4 kcal x 100}}{\text{total kcal in food or diet}} \times NPU$$

One must bear in mind that each method for evaluating protein quality is only a tool to aid in identifying foods capable of meeting the protein—amino acid—needs of the body. These protein quality measurements are useful for comparing the nutritive value of different lots of a single protein, such as an infant formula, a processed food, or a uniform diet for man or animals. However, measurements made on individual foodstuffs do not give useful information about the protein quality of complex human diets. The only meaningful measure of the protein quality of a diet is one made on the total diet as consumed. To this end, there are some steps which can be taken to ensure that the total diet provides the needed amino acids.

Complementary Proteins. Fortunately, the amino acid deficiencies in a protein can usually be improved by combining it with another, and the mixture of the two proteins will often have a higher food value than either one alone. In other words proteins having opposite strengths and weaknesses complement each other. For example, many cereals are low in lysine, but high in methionine and cystine. On the other hand, soybeans, lima beans, and kidney beans are high in lysine but low in methionine and cystine. When

Fig. P-83. Complementary proteins of bun (low in lysine, but high in methionine and cystine) and beans (high in lysine, but low in methionine and cystine). (Courtesy, University of Minnesota, St. Paul)

eaten together, the deficiencies are corrected. Complementary protein combinations are found in almost all cultures. In the Middle East, bread and cheese are eaten together. Mexicans eat beans and corn (tortillas). Indians eat wheat and pulses (legumes). Americans eat breakfast cereals with milk. This kind of supplementation works only when the deficient and complementary proteins are ingested together or within a few hours of each other.

Other examples of complementary food proteins may include: (1) rice and black-eyed peas; (2) whole wheat or bulgur, soybeans, and sesame seeds; (3) cornmeal and kidney beans; and (4) soybeans, peanuts, brown rice, and bulgur wheat.

Factors Affecting Amino Acid Utilization. The presence of amino acids in a protein does not assure their utilization. So, in addition to the amino acid content of a food(s), the following related factors must be considered:

• **Digestibility**—The amino acids of most animal proteins are efficiently absorbed, but this is not necessarily so for many proteins of plant origin. Animal proteins are about 90 to 95% digestible, but the digestibility of some plant proteins may be as low as 73%.

• **Energy**—Protein is used inefficiently when the energy intake is grossly inadequate. When the energy intake is below a certain level, the nutritive value of the protein to the consumer diminishes. For example, an increase in the protein supply to undernourished subjects will not be fully effective if the energy content of the diet is restricted. This principle also applies to replenishment of body protein following malnutrition and during convalescence after injury or disease; full utilization of dietary protein for replenishment is best assured by an adequate calorie intake.

The influence of calorie intake on protein metabolism is shown by the rapid development of a negative nitrogen balance when the energy content of the diet is reduced below requirements—an unfavorable effect which persists if the sub-maintenance diet continues to be fed. From this, it can be concluded that an inadequate intake of energy will, by itself, cause a loss of protein from the body and will consequently aggravate protein deficiency in the diet. Hence, carbohydrates are said to have a protein-sparing effect. A daily intake of 50 to 100 g of digestible carbohydrates is required for protein-sparing.

• **Vitamins and minerals**—Any essential mineral or vitamin whose presence is needed for normal growth and metabolism can be presumed to affect utilization of dietary protein, in so far as deficiency of the vitamin or mineral leads to loss of body substance. For this reason, tests of protein quality are normally performed with diets containing adequate amounts of vitamins and minerals.

In treating persons with protein deficiency, the supply of some vitamins and minerals appears to be more critical than that of others. In particular enough niacin, potassium, and phosphorus should be given to ensure that they are not limiting factors in protein replenishment. Thus, in a niacin deficiency, not only is tryptophan converted to niacin when present in excess but there is evidence that some conversion may occur even when there is insufficient tryptophan for protein synthesis.

(Also see MINERAL[S]; and VITAMIN[S].)

• **Amino acid imbalances**—One type of amino acid imbalance arises when the addition of a single amino acid or mixture of amino acids to a diet reduces the utilization of the dietary protein. Even a small increase in the concentrations of certain amino acids can increase the needs for others when the total protein intake is low.

Utilization of one dietary amino acid may also be depressed by addition to the diet of another structurally related to it. The two best-known examples are the interference of an excess of leucine with the utilization of isoleucine and valine, and the interference of lysine with the utilization of arginine. Large amounts of single amino acids added to experimental diets may induce various toxic reactions, including depression of growth. The most toxic amino acids are methionine, tyrosine, and histidine, and their effect is most serious when the diet is low in protein.

Not enough is presently known about the practical bearing these observations may have in relation to human diets but they must be taken into account in studies of the biological effectiveness of essential amino acid patterns. Circumstances in which amino acid imbalances would occur are unlikely in individuals at normal levels of dietary protein.

• **Nonessential amino acid nitrogen**—The proportion of nonessential amino acid nitrogen has an influence on the essential amino acid requirements. If the ratio of essential amino acids to the total nitrogen in a food is too high, essential amino acids will be used as a source of nitrogen for the nonessential amino acids which, in spite of their being so designated, are necessary parts of the protein molecule and needed for protein synthesis. When the nonessential amino acids are in short supply, some of the essential amino acids furnish nitrogen (NH_2) more readily than others for the synthesis of the nonessential amino acids.

• **Food processing**—Heat and chemicals used during processing can affect amino acid availability. For example, loss of available lysine can occur from mild heat treatment in the presence of reducing sugars such as glucose or lactose.

Heating can also have favorable effects. Heating soybean flour improves the utilization of protein by making the amino acid methionine more available, and heating raw soybeans destroys the inhibitor of the protein digestive enzyme trypsin. Cooking eggs destroys the trypsin inhibitor ovomucoid in the white.

PROTEIN MALNUTRITION. Protein deficiency is common worldwide. This is primarily because in many underdeveloped areas of the world protein intake is marginal, but energy intake is so low that the protein eaten is not spared for its essential functions. The name, protein-energy (calorie) malnutrition is applied to a whole spectrum of protein and energy deficiencies. At one end of the spectrum is kwashiorkor—a severe clinical syndrome caused by a deficiency of protein. Pertinent protein malnutrition information follows:

• **Kwashiorkor**—Infants fed low-protein, starchy foods such as bananas, yams, and cassava after weaning may develop kwashiorkor. This imbalanced diet, which has a subnormal protein-to-calorie ratio for infants, prevents some of the adaptive mechanisms of the body from operating the way that they do in the case of starvation or marasmus—a wasting condition due to lack of food.

Fig. P-84. A 4½-year-old child from South India suffering from kwashiorkor. *Top:* Lukaiah before rehabilitation weighed 21½ lb (*9.8 kg*) and was in a state of severe kwashiorkor showing edema, hair and skin changes, and misery. *Bottom:* After 1 month of rehabilitation, Lukaiah weighed almost 24 lb (*10.9 kg*). The edema disappeared; the skin was normal, and the hair was recovering. Furthermore, Lukaiah was happy and full of mischief—like any child. (Courtesy, WHO, Geneva, Switzerland)

In kwashiorkor there is less breakdown of protein and release of amino acids from muscle than in marasmus since provision of adequate energy in the diet reduces the adrenal gland response and, consequently, the flow of amino acids from muscle is not sufficient to meet the synthesizing needs of the internal organs. Particularly critical is the loss of tissue cells from the gastrointestinal lining which leads to various types of malabsorption. Also, the severe shortage of protein results in a reduction in synthesis of digestive enzymes, normally secreted from the pancreas or present in the intestinal wall. Thus, there is likely to be diarrhea which may cause excessive loss of water or dehydration, loss of minerals salts, and an electrolyte imbalance.

Liver function is greatly reduced due to lack of protein for the synthesis of enzymes. Also, there is likely to be fat accumulation in the liver due to its inability to package fat with protein for transport in the blood. Low blood levels of albumin, along with anemia, result from the scarcity of amino acids for protein synthesis. Edema—swelling—is a consequence of the depletion of plasma proteins and electrolytes.

The production of antibodies is likely to be greatly reduced and the victim, therefore, will be prone to infectious diseases. In chronic cases, there is likely to be delayed eruption of teeth, poor enamel with many caries, and pale gums and mucous membranes due to anemia.

Besides the physical changes in kwashiorkor, there are also some mental changes such as general indifference, irritability, and apathy.

(Also see KWASHIORKOR; and MALNUTRITION, Protein-Energy.)

• **Ill and elderly**—Once protein is ingested, an individual's condition may preclude proper digestion or absorption of the intact protein source. Problems may include various types of inflammatory bowel disease, such as pancreatitis or enteritis, or a gastrointestinal fistula or bowel resection, or just the stress of an infection or surgery. Elderly individuals with low incomes, poor health, and poor appetites may not secure food with high-quality protein. In both the ill and the aged, the protein deficiency may be manifested by delayed convalescence or poor wound healing.

(Also see GERONTOLOGY AND GERIATRIC NUTRITION.)

• **Infants**—The diets of infants after weaning should be carefully planned to include a high-quality protein source since infants and children require suitable patterns of amino acids in their diets. This means that if grains are a major dietary staple, there should be supplementation with animal proteins—such as milk, eggs, fish, or if these are unavailable, with such legumes as soybean or peanuts (which may be in the form of a powder). None of the commonly used grains—corn, rice, wheat, and barley—are adequate as the major source of protein for growing children. Mental function disturbances may follow a severe protein deficiency occurring in infancy.

(Also see INFANT DIET AND NUTRITION.)

• **Pregnancy**—Many uncertainties exist concerning the storage and efficiency of protein utilization during pregnancy, but inadequate protein may have adverse effects on the mother and fetus. Pregnant adolescents and women from low-income groups ignorant of the essentials of a good diet may be particularly susceptible.

(Also see PREGNANCY AND LACTATION NUTRITION.)

• **Vegetarians**—Strict vegetarian diets which offer little variety such as the Zen macrobiotic diet can be dangerous. Some individuals on the Zen diet plan eat only brown rice, which fails to provide many essential nutrients. Fruitarians, who eat only raw or dried fruits, nuts, honey, and olive oil may also suffer from an inadequacy of essential nutrients. Although many adults may get along through well-planned vegetarian diets, the protein needs of infants and growing children may not be met. Indeed, children of some vegetarian cults have demonstrated retarded growth and poor nutrition.

(Also see VEGETARIAN DIETS; and ZEN MACROBIOTIC DIET.)

PROTEIN FOR THE WORLD. While there is adequate protein and energy available in the United Sates and other developed countries, many countries are low in calories and low in protein supplies. Moreover, countries low in total protein are also low in animal protein, which generally has a higher biological value than plant protein. Several approaches have been and are being taken to develop and supply good-quality proteins to undeveloped countries. The aim is to provide acceptable, nutritious, and inexpensive products from the foods available. Some of the approaches to feeding the world include the following:

1. **Food combinations.** Using foods common to an area, various nutritious food combinations have been developed and tried by agencies and governments throughout the world. Some examples of these include: (a) Incaparina, a mixture of whole grain corn, whole grain sorghum, cottonseed flour, torula yeast, calcium carbonate, and vitamin A developed for use in central America; (b) L' aubina, developed in Lebanon and containing chick peas, parboiled wheat, some dried skim milk, and bone ash; (c) corn-soy-milk (CSM) fortified with minerals and vitamins; and (d) whey soy drink mix (WSDM)—a mixture of sweet whey, full-fat soy flour, soybean oil, corn syrup solids, and a vitamin and mineral premix.

2. **Genetic improvement of grains.** Scientists at Purdue University developed a special corn called Opaque-2 in which the incomplete protein, zein, is reduced to one-half the normal amount, while the complete protein glutelin is doubled. Moreover, lysine and tryptophan concentrations are increased about 50%, and leucine and isoleucine are in a better balance. Feeding trials on both animals and humans have demonstrated the increased value of Opaque-2. Breeding programs have also developed (a) high lysine sorghum and barley, (b) high protein rice and wheat, and (c) a cross between wheat and rye, called triticale. These genetic improvements bring plant protein quality closer to that of the animal proteins, and are very promising.

(Also see CEREAL GRAINS; and CORN, section headed "High-Lysine Corn.")

3. **Protein concentrates.** Protein concentrates from oilseeds, nuts, and leaves provide another method of increasing the supply of quality protein.

Isolating protein from plant leaves may provide a protein source which could be used as a complementary protein in countries where it is too rainy to dry seed crops. Green leaves are among the best sources of protein. By pressing the leaves, a protein-containing juice can be obtained which may be coagulated and dried forming a product that is 50% protein. This product is called leaf protein concentrate (LPC).

(Also see COTTONSEED; PEANUTS; and SOYBEANS, Table S 12, Soybean Products and Uses, Soybean Protein Isolates, and Textured Soybean Proteins.)

4. **Single-cell protein.** This refers to protein obtained from single-cell organisms or simple multicellular organisms such as yeast, bacteria, algae, and fungi. The most popular and most familiar of these are brewers' yeast and torula yeast, both of which are marketed. The potential for producing single-cell protein (SCP) is tremendous. Furthermore, many of the industrial by-products with little or no economic value may be used as a growing media. Some of these by-products include petroleum products, methane, alcohols, sulfite waste liquor, starch, molasses, cellulose, and animal waste. Still with all its potential, problems such as (a) safety, (b) acceptability, (c) palatability, (d) digestibility, (e) nutrient content, and (f) economics of production need to be solved.

(Also see ALGAE; and SINGLE-CELL PROTEIN.)

5. **Amino acid supplementation.** Since lysine is the limiting amino acid in wheat and other grains, the addition of limited amounts of lysine to cereal diets improves their protein quality. Indeed, studies in Peru and Guatemala demonstrated that growing children benefited by this addition. However, in most countries there is more to be gained by focusing on increasing the food supply in general.

(Also see LYSINE.)

PROTEINS AND AMINO ACIDS IN SELECTED FOODS. Table P-16, Proteins and Amino Acids in Selected Foods, provides information on a number of foods. This information can be used to select foods and plan diets which yield a high-quality protein.

(Also see MALNUTRITION; MALNUTRITION PROTEIN-ENERGY; and VEGETARIAN DIETS.)

THERE IS PROTEIN AND PROTEIN!
AVERAGE GRAMS CONSUMPTION PER PERSON PER DAY

	ANIMAL PROTEIN	VEGETABLE PROTEIN	TOTAL
NORTH AMERICA	70.7	27.5	98.2
AUSTRALIA & NEW ZEALAND	63.4	31.0	94.4
ARGENTINA, PARAGUAY, & URUGUAY	57.4	36.8	94.0
WESTERN EUROPE	48.5	39.7	88.2
EASTERN EUROPE	35.8	55.1	90.9
USSR	35.6	56.6	92.2
JAPAN	31.8	45.1	76.9
LATIN AMERICA & CARIBBEAN	22.8	35.2	58.0
NEAR EAST	12.2	53.7	65.9
AFRICA	12.1	48.9	61.0
CHINA	9.8	47.8	56.6
SOUTH ASIA	6.3	42.5	48.8

SOURCE: FAO

Fig. P-85. Average grams protein consumption per person per day, with a breakdown into animal and vegetable protein, by geographic areas and countries. (From: *Ceres*, FAO/UN, Vol. 8, No. 3)

TABLE P-16
PROTEINS AND AMINO ACIDS IN SELECTED FOODS[1]

FOOD NAME— 100 Gram (3.5 oz) Portion	Water	Food Energy Calories	Total Protein	Animal Protein	Plant Protein	Amino Acids										
						Cystine	Histidine	Iso-leucine	Leucine	Lysine	Methi-onine	Phenyl-alanine	Threo-nine	Trypto-phan	Tyrosine	Valine
	(g)	(kcal)	(g)	(g)	(g)	(mg)	(mg)	(mg)	(mg)	(mg)	(mg)	(mg)	(mg)	(mg)	(mg)	(mg)
BAKERY PRODUCTS																
BREADS																
CRACKED WHEAT BREAD	34.9	263	8.7	0	8.7	—	—	373	582	234	130	426	252	104	—	400
FRENCH BREAD	30.6	290	9.1	0	9.1	—	—	255	695	205	115	495	260	110	—	385
RAISIN BREAD	35.3	262	6.6	Trace	6.6	—	—	278	434	173	95	321	130	78	—	300
RYE BREAD, DARK (pumpernickel)	34.0	246	9.1	0	9.1	—	—	390	606	371	143	425	334	100	—	471
RYE BREAD, LIGHT	35.5	243	9.1	0	9.1	—	—	391	613	291	139	439	291	100	—	473
WHITE BREAD, enriched	35.6	270	8.7	0.2	8.5	—	—	417	691	260	126	469	269	104	—	400
WHOLE WHEAT BREAD	36.4	243	10.5	—	—	—	—	460	721	308	160	508	313	126	—	491
CAKES																
ANGEL FOOD CAKE	34.0	259	5.7	4.4	1.3	—	—	262	439	131	74	314	165	68	—	245
DOUGHNUTS																
DOUGHNUT, cake type, plain	23.7	391	4.6	1.5	3.1	—	—	240	381	165	81	256	159	59	—	237
DOUGHNUT, raised type, plain	28.3	414	6.3	2.1	4.2	—	—	323	513	223	106	346	216	80	—	320
LEAVENINGS																
YEAST, DRY (active), baker's	5.0	282	36.9	0	36.9	—	—	2280	3970	3160	844	2090	2270	440	—	2900
MUFFINS																
BLUEBERRY	39.0	281	7.3	3.1	4.2	—	—	388	608	305	133	393	265	95	—	393
PLAIN, enriched	38.0	294	7.8	3.3	4.5	—	—	415	650	328	143	420	283	103	—	420
QUICK BREADS																
BAKING POWDER BISCUIT	27.4	369	7.4	0.2	7.3	—	—	342	571	171	97	408	214	88	—	320
BOSTON BROWN BREAD, canned	45.0	211	5.5	1.6	3.9	—	—	257	480	225	97	265	200	57	—	302
CORNBREAD OR JOHNNYCAKE	37.9	267	8.7	—	—	—	—	400	1130	260	160	400	350	52	—	440
POPOVER, home recipe	54.9	224	8.8	4.8	4.0	—	—	514	766	468	200	478	366	124	—	542
ROLLS																
CINNAMON BUN, w/raisins	32.0	275	6.9	1.6	5.3	—	—	315	527	157	88	377	198	82	—	293
DANISH PASTRY, plain w/o fruit or nuts	22.0	422	7.4	—	—	—	—	366	586	220	120	411	240	94	—	357
DINNER TYPE (pan), enriched	31.4	289	8.2	0.2	8.0	—	—	403	655	268	124	437	268	100	—	392
HARD ROLL, enriched	25.4	312	9.8	0.3	9.5	—	—	480	780	320	149	520	320	120	—	469
SWEET ROLL	31.5	316	8.5	2.0	6.5	—	—	436	698	309	142	464	291	109	—	431
CEREALS & FLOURS																
CEREALS																
BARLEY, pearled light, uncooked	11.1	349	8.2	0	8.2	—	—	343	568	279	114	429	279	100	—	411
BULGUR (parboiled wheat) canned, unseasoned	56.0	168	6.2	0	6.2	—	—	—	—	217	148	—	—	347	—	—
CORN GRITS, degermed, enriched, cooked	87.1	51	1.2	0	1.2	—	—	55	155	34	22	54	47	7	—	60
CORN MEAL, white or yellow, degermed, enriched, cooked	87.7	50	1.1	0	1.1	—	—	46	131	29	19	45	40	5	—	51
FARINA, quick cooking, enriched, cooked	89.0	43	1.3	0	1.3	—	—	489	689	200	121	521	279	—	—	439
MACARONI, any shape, enriched, cooked	72.0	111	3.4	0	3.4	—	—	171	226	110	51	179	134	41	—	196
NOODLES, EGG, cooked	70.4	125	4.1	2.6	1.5	—	—	202	273	137	70	198	173	46	—	243
OATMEAL, dry	8.3	390	14.2	0	14.2	309	261	733	1065	521	209	758	470	183	524	845
RICE, BROWN, long grain, cooked	70.3	119	2.5	0	2.5	—	—	119	218	99	45	127	99	28	—	177
RICE, WHITE, enriched, long grain, cooked	72.6	109	2.0	0	2.0	—	—	96	172	78	36	100	78	22	—	140
SPAGHETTI, enriched, cooked	72.0	111	3.4	0	3.4	—	—	170	225	109	51	177	133	41	—	194
WHEAT GERM, dry	11.5	363	26.6	0	26.6	—	—	1270	1840	1650	430	970	1430	270	—	1460
BREAKFAST CEREALS																
SHREDDED WHEAT	6.6	354	9.9	0	9.9	204	236	449	684	331	139	481	405	85	236	577
FLOURS																
FLOUR, WHITE, all purpose, enriched	12.0	364	10.5	0	10.5	210	210	483	809	239	138	577	302	129	359	453
DESSERTS & SWEETS																
GELATIN, unsweetened, dry powder	13.0	335	85.6	85.6	0	77	771	1357	2930	4226	787	2036	1912	6	401	2421
GELATIN, sweetened, dry powder	1.6	371	9.4	9.4	0	—	—	94	207	301	56	150	141	—	—	169
TOPPINGS, DESSERT																
non-dairy powdered	1.5	577	4.9	4.9	0	21	145	301	484	394	149	263	207	68	280	351
non-dairy pressurized	60.4	264	1.0	1.0	0	4	29	60	97	79	30	53	41	14	56	70
non-dairy frozen, semisolid	50.2	318	1.3	1.3	0	5	37	77	123	101	38	67	53	17	72	89
EGGS AND SUBSTITUTES																
EGG, WHOLE, fresh & frozen, raw	74.6	158	12.1	12.1	0	289	293	759	1066	820	392	686	596	194	505	874
EGG WHITE, fresh & frozen, raw	88.1	49	10.1	10.1	0	251	230	618	883	625	394	638	451	156	407	759
EGG YOLK, fresh, raw	48.8	369	16.4	16.4	0	291	394	939	1396	1110	417	714	890	241	706	1000
EGG OMELET, made w/butter and milk	76.3	148	9.3	9.3	0	208	227	581	883	641	294	519	453	147	394	666
EGG, WHOLE, POACHED	74.3	157	12.1	12.1	0	288	292	756	1062	816	391	683	594	193	503	870
EGG, WHOLE, DRIED	4.1	594	45.8	45.8	0	1093	1107	2876	4026	3094	1481	2588	2251	733	1907	3300
EGGNOG	74.4	135	3.8	3.8	0	38	95	230	369	298	87	182	175	54	182	253
DUCK EGG, whole fresh, raw	70.8	185	12.8	12.8	0	285	320	598	1097	951	576	840	736	260	613	885

Footnote at end of table

(Continued)

TABLE P-16 *(Continued)*

FOOD NAME—100 Gram (3.5 oz) Portion	Water	Food Energy Calories	Total Protein	Animal Protein	Plant Protein	Amino Acids										
						Cystine	Histidine	Iso-leucine	Leucine	Lysine	Methi-onine	Phenyl-alanine	Threo-nine	Trypto-phan	Tyrosine	Valine
	(g)	(kcal)	(g)	(g)	(g)	(mg)	(mg)	(mg)	(mg)	(mg)	(mg)	(mg)	(mg)	(mg)	(mg)	(mg)
FATS & OILS																
MARGARINE, regular, soft, whipped, or low sodium	15.5	720	0.6	0	0.6	–	–	50	70	60	20	40	40	10	–	50
FISH & SEAFOODS																
ANCHOVY, pickled, canned	58.6	176	19.2	19.2	0	–	–	1067	1590	1840	607	772	900	209	–	1109
CAVIAR, sturgeon, granular	46.0	262	26.9	26.9	0	–	–	1510	2210	1920	700	1190	1620	240	–	1650
COD, cooked	64.6	129	28.5	28.5	0	–	–	1400	2084	2421	800	1010	1210	273	–	1463
FLOUNDER, baked	58.1	140	30.0	30.0	0	–	–	1530	2250	2640	870	1110	1290	300	–	1590
HADDOCK, cooked	66.3	90	19.6	19.6	0	–	–	1000	1491	1725	568	725	843	196	–	1039
HALIBUT, cooked	66.6	130	25.2	25.2	0	–	–	1285	1915	2218	731	933	1083	252	–	1336
HERRING, pickled	59.4	223	20.4	20.4	0	–	–	1040	1530	1775	592	755	877	204	–	1081
HERRING, w/tomato sauce, canned	66.7	176	15.8	14.8	1.0	–	–	806	1185	1375	458	585	679	158	–	837
LOBSTER, whole, cooked	76.8	95	18.7	18.7	0	–	–	742	1558	1721	580	852	798	164	–	815
SALMON, broiled or baked	63.4	182	27.0	27.0	0	–	–	1350	2025	2349	783	999	1161	270	–	1431
SALMON, canned, Sockeye (red)	67.2	171	20.2	20.3	0	271	–	1025	1526	1771	588	750	876	200	546	1076
SARDINE, Pacific canned. w/mustard	64.1	196	18.8	18.8	0	238	–	898	1337	1552	515	657	767	176	479	943
SHRIMP, canned	70.4	116	24.2	24.2	0	–	–	1234	1839	2130	702	895	1041	242	–	1283
SWORDFISH, broiled	64.6	174	28.0	28.0	0	–	–	1428	2100	2464	812	1036	1204	280	–	1484
TUNA, canned w/oil or water	60.6	197	28.8	28.8	0	–	–	1469	2160	2534	835	1066	1238	288	–	1526
WHITEFISH, lake, cooked	63.2	125	13.7	13.7	0	–	–	775	1140	1338	441	562	654	152	–	806
WHITEFISH, smoked	68.2	155	20.9	20.9	0	–	–	1066	1568	1839	606	773	899	209	–	1108
FLAVORINGS & SEASONINGS																
BASIL, ground	6.4	251	14.4	0	14.4	159	287	588	1078	618	202	733	588	221	432	717
DILL SEED	7.7	305	16.0	0	16.0	–	320	767	925	1038	143	670	575	–	–	1120
FENNEL SEED	8.8	345	15.8	0	15.8	222	331	695	996	758	301	647	602	253	410	915
FENUGREEK SEED	8.8	323	23.0	0	23.0	369	668	1241	1757	1684	338	1089	898	391	764	1102
GARLIC POWDER	6.4	332	18.6	0	16.8	172	309	648	1027	578	336	484	468	215	215	712
GINGER, ground	9.4	347	9.1	0	9.1	42	158	266	387	299	67	236	187	63	102	382
MUSTARD SEED, YELLOW	6.9	469	24.9	0	24.9	582	762	1081	1783	1519	480	1067	1095	526	744	1325
ONION POWDER	5.0	347	10.1	0	10.1	181	136	293	327	467	86	249	199	120	232	238
POPPY SEED	6.8	533	18.0	0	18.0	453	528	905	1484	1099	470	882	905	255	681	1287
SESAME SEEDS, decorticated	4.0	500	26.4	0	26.4	523	677	1280	2150	831	806	1628	1180	473	1125	1478
THYME, ground	7.8	276	9.1	0	9.1	274	–	468	430	207	274	482	252	186	482	502
FRUITS																
APPLE, w/skin, raw	84.4	58	0.2	0	0.2	–	–	–	–	–	–	6	–	–	3	–
APRICOTS, raw	85.3	51	1.0	0	1.0	–	–	–	–	–	–	24	–	–	–	–
AVOCADO, raw	74.0	167	2.1	0	2.1	–	–	–	–	74	12	–	–	14	–	–
BANANA, raw	75.7	85	1.1	0	1.1	–	–	–	–	55	11	34	–	18	33	–
CANTALOUPE, cubed, raw	91.2	30	0.7	0	0.7	–	–	–	–	15	2	21	–	1	–	–
DATES, moisturized or hydrated	22.5	274	2.2	0	2.2	–	49	74	77	65	27	61	61	61	17	94
FIGS, dried, uncooked	23.0	274	4.3	0	4.3	–	–	–	–	–	–	107	–	–	179	–
GRAPEFRUIT, sections, all varieties, raw	88.4	41	0.5	0	0.5	–	–	–	–	6	0	–	–	1	–	–
GRAPES, GREEN, seedless raw, Thompson	81.4	67	0.6	0	0.6	11	25	6	14	15	23	18	19	3	11	19
HONEYDEW MELON	90.6	33	0.8	0	0.8	–	–	–	–	15	2	–	–	1	–	–
LIME, acid type	89.3	28	0.7	0	0.7	–	–	–	–	15	2	–	–	3	–	–
MANGO, raw	81.7	66	0.7	0	0.7	–	–	–	–	93	8	–	–	14	–	–
NECTARINES, raw	81.8	64	0.6	0	0.6	–	–	–	–	–	–	19	–	–	–	–
ORANGE, raw	86.0	49	1.0	0	1.0	–	–	–	–	24	3	12	–	3	21	–
PEACH, raw	89.1	38	0.6	0	0.6	9	17	13	29	30	31	12	27	4	15	40
PEAR, canned, water pack	91.1	32	0.2	0	0.2	–	–	–	–	–	–	6	–	–	10	–
PEAR, canned, heavy syrup	79.8	76	0.2	0	0.2	–	–	–	–	–	–	6	–	–	10	–
PINEAPPLE, raw	85.3	52	0.4	0	0.4	–	–	–	–	9	1	8	–	5	8	–
PINEAPPLE, canned, water pack	89.1	39	0.3	0	0.3	–	–	–	–	7	1	8	–	4	8	–
PRUNES, dried, uncooked	28.0	255	2.1	0	2.1	–	–	–	–	–	–	88	–	–	–	–
RAISINS, uncooked	18.0	289	2.5	0	2.5	–	–	–	–	–	–	36	–	–	–	–
STRAWBERRIES, fresh	89.9	37	0.7	0	0.7	7	16	18	42	32	1	17	25	9	26	23
TANGERINE, raw	87.0	46	0.8	0	0.8	–	–	–	–	28	4	–	–	5	–	–
WATERMELON, raw	92.6	26	0.5	0	0.5	–	–	–	–	–	–	12	–	–	12	–
JUICES																
GRAPEFRUIT, canned, unsweetened	90.0	39	0.5	0	0.5	–	–	–	–	6	0	11	–	1	6	–
ORANGE, canned, unsweetened	87.4	48	0.8	0	0.8	–	–	–	–	21	2	–	–	3	–	–
PINEAPPLE, canned, unsweetened	85.6	55	0.4	0	0.4	–	–	–	–	9	1	–	–	5	–	–
TOMATO, canned or bottled	93.6	19	0.9	0	0.9	–	–	26	37	38	6	25	30	8	–	25
MEAT																
BEEF																
dried, chipped, uncooked	47.7	203	34.3	34.3	0	–	–	1795	2809	2996	850	1410	1515	401	–	1903
flank steak, braised	61.4	196	30.5	30.5	0	–	–	1793	2808	2999	851	1409	1513	401	–	1904
hamburger (ground beef), cooked	54.2	286	24.2	24.2	0	–	–	1342	2101	2243	636	1055	1132	300	–	1424
kidney, braised	53.0	252	33.0	33.0	0	–	–	942	1678	1402	396	911	858	285	–	1130
liver, fried	–	229	26.4	26.3	0.1	–	–	1246	2197	1781	560	1200	1130	357	–	1495
stew w/vegetables, canned	82.5	79	5.8	–	–	–	–	282	427	459	126	235	241	67	–	308
tongue, braised	60.8	244	21.5	21.5	0	–	–	792	1286	1364	356	661	708	196	–	840

Footnote at end of table

(Continued)

TABLE P-16 (Continued)

FOOD NAME— 100 Gram (3.5 oz) Portion	Water	Food Energy Calories	Total Protein	Animal Protein	Plant Protein	Amino Acids										
						Cystine	Histidine	Iso-leucine	Leucine	Lysine	Methi-onine	Phenyl-alanine	Threo-nine	Trypto-phan	Tyrosine	Valine
	(g)	(kcal)	(g)	(g)	(g)	(mg)	(mg)	(mg)	(mg)	(mg)	(mg)	(mg)	(mg)	(mg)	(mg)	(mg)

MEAT (Continued)

CHICKEN

| chicken, canned | 65.2 | 198 | 21.7 | 21.7 | 0 | — | — | 1100 | 1507 | 1830 | 543 | 820 | 887 | 253 | — | 1023 |
| liver, simmered | 65.0 | 165 | 26.5 | 26.5 | 0 | — | — | 1533 | 2707 | 2197 | 690 | 1480 | 1393 | 440 | — | 1847 |

LAMB

| ground, cooked | 54.0 | 279 | 25.3 | 25.3 | 0 | — | — | 1332 | 1990 | 2081 | 617 | 1045 | 1176 | 332 | — | 1267 |
| leg roast, roasted | 54.0 | 279 | 25.3 | 25.3 | 0 | — | — | 1332 | 1990 | 2081 | 617 | 1045 | 1176 | 332 | — | 1267 |

LUNCHEON MEATS

BRAUNSCHWEIGER (smoked liver sausage)	52.6	319	15.4	15.4	0	187	458	754	1291	1200	320	700	668	172	471	956
CORNED BEEF, canned	59.3	216	25.3	25.3	0	—	—	1185	1857	1982	560	932	1000	264	—	1260
HEAD CHEESE	58.8	268	15.5	15.5	0	209	278	509	946	907	250	569	418	79	569	617
LIVERWURST	53.9	307	16.2	16.2	0	203	497	818	1400	1301	347	759	724	187	510	1037
POLISH SAUSAGE (kolbassi), cooked	53.7	304	15.7	15.7	0	—	—	1420	1580	1717	507	800	890	147	—	1080
SALAMI, beef	29.8	450	23.8	23.8	0	298	642	1159	1713	1923	505	872	979	203	776	1201
SPICED LUNCHEON MEATS, pork/ham type	54.9	294	15.0	15.0	0	241	479	741	1151	1252	362	570	610	143	575	775
VIENNA SAUSAGE	63.0	240	14.0	14.0	0	197	425	766	1133	1272	334	576	647	134	513	794

PORK

| Canadian bacon, cooked | 49.9 | 277 | 27.6 | 27.6 | 0 | — | — | 1471 | 2281 | 2481 | 719 | 1129 | 1210 | 286 | — | 1538 |
| Loin roast or chops, cooked | 45.8 | 362 | 24.5 | 24.5 | 0 | — | — | 1279 | 1833 | 2044 | 621 | 980 | 1155 | 323 | — | 1295 |

VEAL

| cutlet, braised or broiled | 60.4 | 216 | 27.1 | 27.1 | 0 | — | — | 1751 | 2429 | 2769 | 758 | 1346 | 1438 | 435 | — | 1714 |
| loin roast or chop, cooked | 58.9 | 234 | 26.4 | 26.4 | 0 | — | — | 1189 | 1657 | 1889 | 517 | 919 | 981 | 297 | — | 1169 |

| TURKEY, canned | 64.9 | 202 | 20.9 | 20.9 | 0 | — | — | 1109 | 1611 | 1904 | 586 | 837 | 900 | — | — | 1046 |

MILK & PRODUCTS

| BUTTER, regular or unsalted | 15.9 | 717 | 0.9 | 0.9 | 0 | 8 | 23 | 51 | 83 | 67 | 21 | 41 | 38 | 12 | 41 | 57 |

CHEESE

AMERICAN, pasteurized process	39.2	375	22.1	22.1	0	142	903	1024	1958	2198	573	1125	719	323	1212	1326
AMERICAN, spread, pasteurized process	47.7	290	16.4	16.4	0	—	509	833	1780	1507	538	931	628	—	890	1366
BLUE	42.4	353	21.4	21.4	0	108	759	1126	1922	1855	585	1089	786	313	1297	1559
BRICK	41.1	371	23.2	23.2	0	131	823	1137	2244	2124	565	1231	882	324	1115	1472
BRIE	48.4	334	28.8	20.8	0	114	716	1015	1929	1851	592	1158	751	322	1200	1340
CAMEMBERT	51.8	300	19.8	19.8	0	109	683	968	1840	1766	565	1105	717	307	1145	1279
CHEDDAR, shredded	36.8	403	24.9	24.9	0	125	874	1546	2385	2072	652	1311	886	320	1202	1663
COLBY	38.2	394	23.8	23.8	0	119	834	1475	2275	1978	622	1251	845	305	1147	1586
COTTAGE, creamed	79.0	103	12.5	12.5	0	116	415	734	1284	1010	376	673	554	139	666	773
COTTAGE, lowfat, 2% fat	79.3	90	13.7	13.7	0	127	457	808	1413	1111	413	741	609	153	732	851
CREAM	53.8	349	7.6	7.6	0	66	271	399	731	676	181	419	321	67	360	443
EDAM	41.6	357	25.0	25.0	0	—	1034	1308	2570	2660	721	1434	932	—	1457	1810
GOUDA	41.5	356	24.9	24.9	0	—	1032	1306	2564	2654	719	1431	930	—	1454	1806
GRUYERE	33.2	413	29.8	29.8	0	304	117	1612	3102	2710	822	1743	1089	421	1776	2243
LIMBURGER	48.4	327	20.0	20.0	0	—	578	1219	2093	1675	619	1116	739	289	1197	1439
MONTEREY	41.0	373	24.5	24.5	0	123	859	1519	2344	2037	641	1289	871	315	1182	1635
MOZZARELLA	54.1	281	19.4	19.4	0	116	731	931	1893	1972	542	1014	740	—	1123	1215
MUENSTER	41.8	368	23.4	23.4	0	132	829	1145	2260	2139	569	1240	888	327	1123	1482
PARMESAN, grated	17.7	456	41.6	41.6	0	274	1609	2202	4013	3843	1114	2234	1531	560	2319	2853
PIMIENTO, pasteurized process	39.1	375	22.1	22.1	0	142	902	1023	1956	2196	572	1124	718	323	1211	1325
PORT DU SALUT	45.5	352	23.8	23.8	0	—	686	1446	2482	1987	734	1323	876	343	1420	1707
PROVOLONE	41.0	351	25.6	25.6	0	116	1115	1091	2297	2646	686	1287	982	—	1520	1640
RICOTTA, made w/whole milk	71.7	174	11.3	11.3	0	99	459	589	1221	1338	281	556	517	—	589	692
RICOTTA, made w/part skim milk	74.4	138	11.4	11.4	0	100	464	596	1235	1353	284	562	523	—	596	700
SWISS	37.2	376	28.4	28.4	0	290	1065	1537	2959	2585	784	1662	1038	401	1693	2139

CREAM

| half & half (milk & cream, fluid) | 80.6 | 130 | 3.0 | 3.0 | 0 | 27 | 80 | 179 | 290 | 235 | 74 | 143 | 134 | 42 | 143 | 198 |
| heavy whipping | 57.7 | 345 | 2.0 | 2.0 | 0 | 19 | 56 | 124 | 201 | 163 | 51 | 99 | 93 | 29 | 99 | 137 |

CREAM SUBSTITUTES

| non-dairy liquid, w/hydrogenated vegetable oil & soy protein | 77.3 | 136 | 1.0 | 0 | 1.0 | 18 | 28 | 56 | 85 | 68 | 15 | 55 | 43 | 15 | 37 | 56 |
| non-dairy, powdered | 2.2 | 546 | 4.8 | 4.8 | 0 | 21 | 142 | 294 | 473 | 385 | 145 | 257 | 203 | .66 | 274 | 343 |

ICE CREAM

French vanilla ice cream, soft serve	59.8	218	4.1	4.1	0	42	109	244	393	317	102	195	188	58	194	269
vanilla ice cream, regular (10% fat) hardened	60.8	202	3.6	3.6	0	33	98	218	354	286	91	174	163	51	174	242
vanilla ice milk, hardened	68.6	140	3.9	3.9	0	36	107	238	386	312	99	190	178	56	190	264
orange sherbet	66.1	140	1.1	1.1	0	10	30	68	110	89	28	54	51	16	54	75

MILK, COW'S

| whole milk, 3.7% fat | 87.7 | 64 | 3.3 | 3.3 | 0 | 30 | 89 | 198 | 321 | 260 | 82 | 158 | 148 | 46 | 158 | 220 |
| lowfat milk, 2% fat | 89.2 | 50 | 3.3 | 3.3 | 0 | 31 | 90 | 201 | 326 | 264 | 84 | 161 | 150 | 47 | 161 | 223 |

Footnote at end of table

(Continued)

TABLE P-16 *(Continued)*

FOOD NAME— 100 Gram (3.5 oz) Portion	Water	Food Energy Calories	Total Protein	Animal Protein	Plant Protein	Amino Acids											
						Cystine	Histidine	Iso-leucine	Leucine	Lysine	Methi-onine	Phenyl-alanine	Threo-nine	Trypto-phan	Tyrosine	Valine
	(g)	(kcal)	(g)	(g)	(g)	(mg)	(mg)	(mg)	(mg)	(mg)	(mg)	(mg)	(mg)	(mg)	(mg)	(mg)
MILK & PRODUCTS *(Continued)*																
MILK, COW'S *(Continued)*																
skim milk	90.8	35	3.4	3.4	0	32	92	206	334	270	86	165	154	48	165	228
chocolate milk, whole, 3.3% fat	82.3	83	3.2	3.2	0	29	86	192	311	251	79	153	143	45	153	212
chocolate milk, 2% fat	83.6	72	3.2	3.2	0	30	87	194	314	255	81	155	145	45	155	215
buttermilk, cultured	90.1	40	3.3	3.3	0	31	95	204	329	277	81	174	158	36	139	243
whole milk, evaporated, canned	74.0	134	6.8	6.8	0	63	185	412	667	540	171	329	307	96	329	456
sweetened condensed milk, canned	27.2	321	7.9	7.9	0	73	214	479	775	627	198	382	357	112	382	529
nonfat dry milk powder, instantized	4.0	358	35.1	35.1	0	325	952	2124	3438	2784	880	1694	1584	495	1694	2349
hot cocoa, homemade w/whole milk	81.6	87	3.6	3.6	0	34	99	220	357	289	91	176	164	51	176	244
malted milk powder, natural flavor	2.6	411	13.1	—	—	261	261	391	749	356	180	439	335	140	362	445
malted milk beverage, chocolate flavor	81.2	88	3.5	—	—	11	28	61	100	79	26	51	47	14	50	71
milkshake, vanilla flavor, thick type	74.4	112	3.9	—	—	36	105	234	378	306	97	186	174	54	186	258
milkshake, chocolate flavor, thick type	72.2	119	3.0	—	—	28	83	185	299	242	76	147	138	43	147	204
MILK, GOAT'S, whole	87.0	69	3.6	3.6	0	46	89	207	314	290	80	155	163	44	179	240
MILK, HUMAN, whole	87.5	70	1.0	1.0	0	19	23	56	95	68	21	46	46	17	53	63
YOGURT																
plain	87.9	61	3.5	3.5	0	—	86	189	350	311	102	189	142	20	175	287
plain, lowfat	85.1	63	5.3	5.3	0	—	130	286	529	471	155	286	216	30	265	434
coffee & vanilla varieties, lowfat	79.0	85	4.9	4.9	0	—	122	269	497	442	145	269	202	28	249	408
fruit varieties, lowfat (10 g protein/8 oz)	74.5	102	4.4	—	—	—	108	238	440	392	129	238	179	25	221	362
NUTS & SEEDS																
ALMONDS	4.7	598	18.6	0	18.6	377	517	873	1454	562	259	1146	610	176	618	1124
BRAZIL NUTS	4.6	654	14.3	0	14.3	—	—	593	1129	443	941	617	422	187	—	832
CASHEW NUTS	5.2	561	17.2	0	17.2	—	—	1135	1410	740	327	877	688	430	—	1479
COCONUT, fresh	50.9	346	3.4	0	3.4	62	69	180	269	152	71	174	129	33	101	212
COCONUT, dried, shredded, sweetened	3.3	548	3.6	0	3.6	—	—	190	284	162	75	183	136	35	—	223
FILBERTS OR HAZELNUTS	5.8	634	12.7	0	12.7	165	288	853	939	417	139	537	415	211	434	934
PEANUTS, roasted, Spanish or Virginia	1.6	585	26.0	0	26.0	—	—	1355	2003	1176	290	1666	886	364	—	1639
PEANUT BUTTER	1.8	581	26.1	0	26.1	449	727	1228	1816	1066	263	1510	803	330	1071	1487
PECANS	3.4	687	9.4	0	9.4	216	273	553	773	435	153	564	389	138	316	525
PISTACHIO NUTS	5.3	594	18.9	0	18.9	385	471	881	1523	1080	367	1088	613	—	667	1344
PUMPKIN OR SQUASH SEEDS	4.4	553	29.0	0	29.0	—	—	1624	2291	1334	551	1624	870	522	—	1566
SUNFLOWER SEED KERNELS, hulled	4.8	560	23.0	0	23.0	464	586	1276	1736	868	443	1220	911	343	647	1354
WALNUTS, ENGLISH	3.5	651	15.0	0	15.0	320	405	767	1228	441	306	767	589	175	583	974
PICKLES & RELISHES																
BREAD & BUTTER PICKLES	78.7	73	0.9	0	0.9	—	—	28	39	40	9	21	24	6	—	31
DILL PICKLE, whole	93.3	11	0.7	0	0.7	—	—	22	30	31	7	16	19	5	—	24
OLIVE, ripe	73.0	184	1.2	0	1.2	—	24	53	79	15	17	45	40	—	38	60
PEPPER, HOT CHILI, green, canned	92.5	25	0.9	0	0.9	—	—	34	34	38	12	42	38	6	—	25
PICKLE, SWEET	60.7	146	0.7	0	0.7	—	—	22	30	31	7	16	19	5	—	24
PIMIENTOS, canned	92.4	27	0.9	0	0.9	—	22	37	60	48	6	45	43	—	22	49
SALADS																
COLESLAW, w/salad dressing	82.9	99	1.2	0	1.2	—	—	47	48	56	11	25	34	10	—	36
SNACK FOODS																
POPCORN, plain	4.0	386	12.7	0	12.7	—	—	593	1671	371	243	579	514	79	—	657
POPCORN, popped in coconut oil w/salt	3.1	456	9.8	0	9.8	—	—	461	1300	289	189	450	400	61	—	511
POTATO CHIPS	1.8	568	5.3	0	5.3	—	—	233	265	281	64	233	217	53	—	281
SOUPS & CHOWDERS																
BEEF BROTH, bouillon, consomme, canned, diluted w/water	5.8	13	2.1	2.1	0	—	—	27	70	76	14	47	79	1	—	58
BEEF NOODLE, canned, diluted w/water	92.8	29	1.6	—	—	—	—	80	65	156	56	65	75	14	—	77
CHICKEN NOODLE, canned, diluted w/water	93.1	27	1.5	—	—	—	—	80	75	180	43	75	65	13	—	55
CHICKEN W/RICE, canned, diluted w/water	94.7	22	1.3	—	—	—	—	60	65	176	47	52	50	11	—	40
CHICKEN VEGETABLE, canned, diluted w/water	92.3	30	1.7	—	—	—	—	62	92	87	21	47	53	18	—	56
CREAM OF ASPARAGUS, canned, diluted w/milk	85.2	69	2.7	—	—	—	—	159	254	181	57	119	155	28	—	173
CREAM OF CHICKEN, canned, diluted w/water	92.2	38	1.4	—	—	—	—	60	60	115	42	53	50	11	—	47
CREAM OF MUSHROOM, canned, diluted w/milk	83.4	92	2.6	—	—	—	—	151	213	227	85	124	109	31	—	152
CREAM OF SHRIMP, frozen, made w/milk	81.8	100	3.8	—	—	—	—	194	316	265	84	165	242	38	—	209
MINESTRONE, canned, diluted w/water	91.2	36	1.6	0	1.6	—	—	68	98	78	16	56	54	17	—	71
ONION, canned, diluted w/water	94.5	24	1.5	—	—	—	—	39	85	86	15	47	99	6	—	58
PEA, green, canned, diluted w/water	85.1	58	3.5	1.1	2.3	—	—	135	167	359	43	141	104	22	—	42
TURKEY NOODLE, canned, diluted w/water	92.7	32	1.6	—	—	—	—	92	108	81	23	53	58	16	—	70
TOMATO, canned, diluted w/water	90.5	36	0.8	0	0.8	—	—	33	28	75	32	37	27	6	—	25
VEGETABLE BEEF, canned, diluted w/water	91.9	33	2.1	—	—	—	—	71	170	234	36	92	94	21	—	124

Footnote at end of table

(Continued)

TABLE P-16 *(Continued)*

FOOD NAME— 100 Gram (3.5 oz) Portion	Water (g)	Food Energy Calories (kcal)	Total Protein (g)	Animal Protein (g)	Plant Protein (g)	Amino Acids										
						Cystine (mg)	Histidine (mg)	Iso-leucine (mg)	Leucine (mg)	Lysine (mg)	Methionine (mg)	Phenyl-alanine (mg)	Threo-nine (mg)	Trypto-phan (mg)	Tyrosine (mg)	Valine (mg)
VEGETABLES																
ASPARAGUS, green, cooked	93.6	20	2.2	0	2.2	—	—	79	97	103	33	68	66	26	—	106
white, canned	92.3	22	2.1	0	2.1	—	—	56	68	73	14	48	46	19	—	75
BEANS, common, red (kidney), unsalted, cooked	69.0	118	7.8	0	7.8	—	—	437	671	577	78	429	335	70	—	468
common, white, canned w/pork & sweet sauce	66.4	150	6.2	Trace	6.2	—	—	355	536	462	62	343	269	56	—	381
green, cut or french style, boiled	92.4	25	1.6	0	1.6	—	—	72	92	83	24	38	60	21	—	76
lima, frozen, cooked	73.5	99	6.0	0	6.0	—	—	429	614	496	118	437	348	67	—	466
sprouts (mung beans), cooked	91.0	28	3.2	0	3.2	—	—	179	291	218	35	154	99	22	—	189
BEET GREENS, cooked	93.6	18	1.7	0	1.7	—	—	71	109	92	29	99	65	20	—	85
BEETS, red, whole, cooked	90.0	32	1.1	0	1.1	—	—	34	37	59	4	18	22	9	—	33
BROCCOLI, raw	89.1	32	3.6	0	3.6	—	—	137	176	158	54	131	133	40	124	184
BRUSSELS SPROUTS, cooked	88.2	36	4.2	0	4.2	—	—	176	185	189	42	151	147	42	86	185
CABBAGE, raw	92.4	24	1.3	0	1.3	28	25	40	57	66	13	32	39	11	27	43
CABBAGE, RED, raw	90.2	31	2.0	0	2.0	-	—	78	80	94	18	50	56	16	41	60
CARROT, raw	88.2	42	1.1	0	1.1	29	17	46	65	52	10	36	43	10	17	56
CAULIFLOWER, raw, whole flowers	91.0	27	2.7	0	2.7	—	—	116	181	151	54	84	113	35	—	162
cooked	92.8	22	2.3	0	2.3	—	—	99	154	129	46	71	97	30	—	138
CELERY, raw	94.1	17	0.9	0	0.9	—	1	19	22	25	2	15	16	—	7	26
CHICORY GREENS, raw	92.8	20	1.8	0	1.8	6	24	—	—	52	16	—	—	24	40	—
COLLARDS, leaves & stems, cooked	90.8	29	2.7	0	2.7	—	—	84	151	140	32	86	78	38	—	135
CORN, sweet, cooked, cut off cob before cooking	76.5	83	3.2	0	3.2	—	—	—	—	—	—	—	—	—	—	—
sweet, cooked on cob, white/yellow	74.1	91	3.3	0	3.3	—	—	122	363	122	63	185	135	20	88	208
CUCUMBER, raw, pared	95.7	14	0.6	0	0.6	—	—	18	26	26	6	14	16	4	—	20
DANDELION GREENS, cooked	86.8	33	2.0	0	2.0	—	—	—	—	—	—	—	—	—	—	—
ENDIVE, curly, & escarole, raw	93.1	20	1.7	0	1.7	13	31	72	123	78	22	78	71	—	54	81
KOHLRABI, cooked	92.2	24	1.7	0	1.7	—	—	—	—	—	—	43	—	—	—	—
LENTILS, whole seeds, cooked	72.0	106	7.8	0	7.8	—	—	413	554	476	55	359	273	70	—	421
LETTUCE, Bibb or Boston, raw	95.1	14	1.2	0	1.2	—	—	—	—	70	4	—	—	12	—	—
MUSHROOMS, raw, *Agaricus campestris*	90.4	28	2.7	0	2.7	—	—	597	316	—	188	—	—	8	—	424
MUSTARD GREENS, leaves w/o stems & midribs, boiled	92.6	23	2.2	0	2.2	—	—	73	59	108	22	70	57	35	—	103
OKRA PODS, cooked	91.1	29	2.0	0	2.0	—	—	78	112	84	24	72	74	20	—	100
ONION, raw	89.1	38	1.5	0	1.5	—	14	21	37	64	13	38	22	21	42	31
PARSLEY, raw	85.1	44	3.6	0	3.6	—	—	—	—	230	18	—	—	72	—	—
PEAS, green, immature, raw	78.0	84	6.3	0	6.3	—	—	290	397	296	44	223	101	50	161	258
PEPPERS, green, sweet, raw	93.4	22	1.2	0	1.2	—	14	46	46	51	16	55	50	9	—	33
POTATO, baked in skin	75.1	93	2.6	0	2.6	—	101	463	486	330	70	295	225	90	84	420
pared, boiled	82.8	65	1.9	0	1.9	—	—	84	95	101	23	84	78	19	—	101
french fried	44.7	274	4.3	0	4.3	—	—	184	210	222	50	184	172	42	—	222
RADISHES, raw	94.5	17	1.0	0	1.0	—	—	—	—	34	2	—	59	5	—	30
SOYBEANS, mature seeds, cooked	71.0	130	11.0	0	11.0	—	—	649	935	759	165	594	423	165	—	638
SPINACH, raw	90.7	26	3.2	0	3.2	—	—	150	246	198	54	145	141	51	116	176
SQUASH, summer, cooked	95.5	14	0.9	0	0.9	—	—	29	41	34	12	24	21	7	—	33
winter, baked	81.4	63	1.8	0	1.8	—	—	58	81	68	23	49	41	14	—	66
SWEET POTATO, baked in skin	63.7	141	2.1	0	2.1	—	—	101	120	99	38	104	99	36	61	157
TOMATO, raw	93.5	22	1.1	0	1.1	—	15	29	41	42	7	28	33	9	24	28
TOMATO, GREEN, raw	93.0	24	1.2	0	1.2	—	15	29	41	42	7	28	33	9	14	28
TOMATO PASTE, canned, w/o salt	75.0	82	3.4	0	3.4	—	58	70	95	98	18	72	78	—	45	78
TOMATO PUREE, canned	87.0	39	1.7	0	1.7	—	—	49	70	71	12	48	56	15	—	48
TURNIP GREENS, frozen, chopped, cooked	92.7	23	2.5	0	2.5	—	—	81	156	99	40	100	95	35	71	112
TURNIP, cooked	93.6	23	0.8	0	0.8	—	—	14	—	42	9	17	—	—	21	—
WATER CRESS or **GARDEN CRESS,** raw	89.4	32	2.6	0	2.6	—	—	140	230	160	20	110	150	50	—	150

[1]The authors gratefully acknowledge that these food compositions were obtained from the HVH-CWRU Data Base developed by the Division of Nutrition, Highland View Hospital, and the Departments of Biometry and Nutrition, School of Medicine, Case Western Reserve University, Cleveland, Ohio.

PROTEIN, ANIMAL

Generally high quality protein derived from meat, milk, poultry, fish, and eggs, and their products.
(Also see PROTEIN[S].)

PROTEIN AS ENERGY SOURCE

In the body, protein functions in the (1) building of new tissues, (2) upkeep of tissues, (3) regulation of water and acid base balance, (4) production of enzymes, antibodies, hormones, and vitamins, (5) formation of milk, and (6) provision of energy. When more protein is eaten than is needed for the first five functions listed above, the excess protein is metabolized for energy. In addition, protein of the diet, along with tissue proteins, is burned for energy when the diet contains insufficient carbohydrates and fats. When this occurs, the building or repair processes of the body suffer. However, energy needs of the body have a higher priority. As a source of energy, protein yields about 4 kcal/g.

(Also see METABOLISM; PROTEIN[S]; and BIOLOGICAL VALUE [BV] OF PROTEINS.)

PROTEIN-BOUND IODINE (PBI)

The iodine that is bound to thyroxin and in transit in the plasma is known as protein-bound iodine. In the normal individual, PBI values range from 4 to 8 micrograms/100 milliliters of plasma.

PROTEIN, COMPLETE

Casein and egg albumin are examples of complete proteins. They contain all of the essential amino acids in sufficient amounts to maintain life and support growth.

(Also see PROTEIN[S], Table P-16.)

PROTEIN, CRUDE

This refers to all the nitrogenous compounds in a food. It is determined by finding the nitrogen content, as determined by the Kjeldahl process, and multiplying the result by 6.25. The nitrogen content of protein averages about 16% (100 ÷ 16 = 6.25).

(Also see PROTEIN[S].)

PROTEIN EFFICIENCY RATIO (PER)

By definition, *it is the weight gain of a young growing animal expressed in grams divided by the grams of protein eaten over a 4-week period, or some other predetermined time.* It provides a biological means for evaluating protein quantity. Casein and egg albumin yield the maximum values, while proteins like gliadin will not even support growth.

(Also see PROTEIN[S] section on Quantity and Quality; and BIOLOGICAL VALUE [BV] OF PROTEINS.)

PROTEIN FACTOR

A number used to convert the Kjeldahl nitrogen content of food to protein, since most of the nitrogen measured by the Kjeldahl method is derived from the protein contained in the food. On the average, protein contains about 16% nitrogen (16 g of nitrogen for every 100 g of protein); however, this can range from 15 to 18%. For most foods, to determine the crude protein content, the Kjeldahl nitrogen value is multiplied by 6.25 (100 ÷ 16). Where greater accuracy is desired, the specific factor for converting nitrogen to protein may be used. For example, the factor for most nuts is 5.30 while for milk and cheese it is 6.38.

(Also see KJELDAHL; and PROTEIN[S].)

PROTEIN HYDROLYSATE

A solution containing the amino acids derived from an artificially digested protein, usually milk or beef protein. Used extensively in medicine and surgery. Usually administered by a stomach tube or intravenous injection.

PROTEIN, INCOMPLETE

A protein that cannot maintain life or support growth is classified as incomplete; for example, zein (corn protein) and gelatin.

(Also see PROTEIN[S].)

PROTEIN MALNUTRITION

When the diet fails to meet the body's needs for protein quality and quantity, impaired function, growth and/or development result. The extreme case of protein malnutrition is kwashiorkor, which develops in infants consuming a deficiency of protein and marginal to adequate amounts of energy. Often there is an inadequate or imbalance of both protein and energy; hence, it is convenient to class the resulting disorders as various types of protein-energy malnutrition (PEM).

(Also see DEFICIENCY DISEASES, Table D-1, Major Dietary Deficiency Diseases—"Protein-energy malnutrition"; MALNUTRITION; and MALNUTRITION, PROTEIN-ENERGY.)

PROTEIN, MILK

When the term milk refers to cow or human milk, the proteins involved are casein and lactalbumin, both high quality proteins. In cow milk, casein is the predominate protein. In human milk, casein and lactalbumin are present in about equal amounts. However, human milk contains only about one third as much protein as cow milk.

(Also see MILK AND MILK PRODUCTS; and PROTEIN[S].)

PROTEIN, PARTIALLY COMPLETE

Gliadin, a protein found in wheat, is an example of a partially complete protein. It can maintain life, but it cannot support growth.

(Also see PROTEIN[S].)

PROTEIN SCORE (CHEMICAL SCORE)

A chemical means of evaluating a protein on the basis of its amino acid content. More often this is called the chemical score or amino acid score.

(Also see PROTEIN[S], section headed "Quantity and Quality.")

PROTEIN, SINGLE-CELL

Protein obtained from single-cell organisms, such as yeast, bacteria, and algae, and grown on specially prepared growth media. Dried brewers' yeast is one of the most familiar examples.

(Also see PROTEIN[S], section headed "Protein of the World.")

PROTEIN, VEGETABLE

Protein derived from plants. Most vegetable protein supplements, although not all, are the by-products that remain after extracting oil from soybeans, cottonseed, linseed, peanuts, safflower, sunflower, rapeseed, and coconut. Soybean is by far the most widely used vegetable protein.

(Also see PROTEIN[S].)

PROTEINURIA

Presence of protein in the urine.

PROTEIN UTILIZATION (NET PROTEIN UTILIZATION; NPU)

A procedure employed to determine protein quality. It is the proportion of the nitrogen in the food that is retained by the tissues of the body. In other words, it is the amount of nitrogen contained in the food consumed, minus the amount lost in the urine and feces, divided by the amount of nitrogen in the food.

(Also see PROTEIN[S], section headed "Quantity and Quality.")

PROTEIN, WORLD PER CAPITA

Worldwide, there are about 69 g of protein available each day for every individual—an amount which should be more than adequate. However, each individual does not receive this amount and the following must be considered: (1) distribution of protein, (2) the source of protein, and (3) the calories available.

(Also see HUNGER, WORLD; INCOME, PROPORTION SPENT FOR FOOD; MALNUTRITION; POPULATION, WORLD; PROTEIN[S]; and WORLD FOOD.)

PROTEOLYSIS

The breakdown of proteins or peptides into smaller units—polypeptides tripeptides, dipeptides and amino acids. Digestion of proteins entails proteolysis.

(Also see DIGESTION AND ABSORPTION; and PROTEIN[S].)

PROTEOSE

A derivative of protein formed during digestion.

PROTHROMBIN

One of four blood clotting proteins synthesized by the liver, the manufacture of which is regulated by vitamin K, present in the blood plasma, essential to clotting of blood.

(Also see VITAMIN K.)

PROTHROMBIN ACTIVATOR

A complex substance that splits prothrombin to form thrombin, an essential step in the blood clotting process.

PROTON

A particle of the nucleus of an atom that has a charge of plus one. A proton is a positive hydrogen ion (H^+).

PROTOPLASM

The living matter in all cells.

PROTOPORPHYRIN

The precursor of heme.

PROVITAMIN A

Carotene.
(Also see VITAMIN A.)

PROXIMAL

Next to or nearest the point of attachment or origin.

PRUNE *Prunus* spp

A type of plum that may be dried satisfactorily without removal of the pith. Only a few of the many cultivated varieties of plums can be so dried. Usually, the prune types of plums have a higher than average sugar content.

(Also see PLUM AND PRUNE; and PRUNE JUICE.)

PRUNE JUICE

A water extract of dried prunes that is produced by (1) pulverizing the prunes (this may be done with the aid of enzymes that break down the pectin which binds the fruit tissue), and (2) leaching the prune pulp with hot water. The resulting prune juice may be either (1) sweetened, or (2) left unsweetened, because the natural sugar content of prunes is generally high.

The nutrient composition of prune juice is given in Food Composition Table F-21.

Some noteworthy observations regarding the nutrient composition of prune juice follow:

1. A ½-cup serving (*125 g*) of unsweetened prune juice is moderately high in calories (96 kcal) and carbohydrates (24 g).

2. Prune juice is an excellent source of iron and potassium, but it is a poor source of vitamins A and C.

3. Many people find that the consumption of prune juice helps to promote regularity of bowel movement. However, the amounts required to produce this effect vary considerably among different people, from as little as a small glass to as much as a pint or more.

(Also see PLUM AND PRUNE; and PRUNE.)

PSORIASIS

A condition of the skin wherein red, itchy patches frequently appear on the scalp, knees, elbows, chest, abdomen, palms, and soles of the feet. These red patches tend to grow and join together creating extensive, unsightly and umcomfortable areas. However, psoriasis seldom causes long-term physical harm. Its cause is obscure but there is evidence to suggest it runs in families. No single treatment exists which will completely clear up the disease. Some treatments providing relief include ointments containing coal tar derivatives or mercury compounds, cortisone pills or ointment, x rays, and sunbathing. In the past, low taurine—a product of cysteine metabolism and normal constituent of animal protein—diets were designed, and recommended for sufferers of psoriasis.

PSYCHROPHILIC BACTERIA

These bacteria are "cold lovers," but despite their name they actually tolerate cold rather than prefer it. Most bacteria, mesophiles, have optimal growth at 97-111°F (*37-44°C*). Psychrophiles grow and reproduce at the usual refrigeration temperatures. In fact, some types can substantially multiply at 32°F (*0°C*). These bacteria are important as they are responsible for most of the fresh food spoilage that occurs. Low temperatures should not be relied on to destroy bacteria. Even in frozen foods some bacteria survive; hence, after thawing, frozen foods should not be allowed to stand at room temperature.

(Also see BACTERIA IN FOOD.)

PSYLLIUM *Plantago psyllium*

An annual herb grown in southern Europe and India. It bears a seed that has laxative qualities and is used in medicines. When the seed is moistened, it looks like gelatin.

PTOMAINES

The word ptomaine comes from the Greek word *ptoma*, meaning *dead body*. It refers to a group of extremely poisonous organic compounds formed during the microbial or enzymatic decomposition of animal proteins. Ptomaines are easily detected by the deteriorated appearance of the material (almost to a liquid state) and the putrid odor. Such food is hardly human fare!

Food poisoning by bacterial toxins, such as salmonellosis or staphylococcal intoxication, is sometimes called *ptomaine poisoning*, which is incorrect.

(Also see BACTERIA IN FOOD.)

PUBERTY

The age at which the sex organs begin to function and at which sexual features begin to appear. The average age of puberty in boys is from 13 to 16, in girls from 11 to 14.

PUERPERIUM

The period from delivery of the infant to the time when the uterus regains its normal size, usually about 6 weeks.

PULP

The solid residue remaining after extraction of juices from fruits, roots, or stems.

PULSES

Pulses are the seeds of leguminous plants. The following pulses are commonly used for human food: beans, chickpeas, cowpeas, field peas, peanuts, pigeonpeas, and soybeans.

PUMMELO (POMELO; SHADDOCK)
Citrus grandis

This fruit is the largest of the citrus fruits (which belong to the family *Rutaceae*) ranging in size between 4 and 12 in. *(10 and 30 cm)* in diameter.

The pummelo was brought from its native home of Malaysia to Barbados in the 17th century by Captain Shaddock of the British East India Company. The grapefruit (*C. paradisi*), which was discovered in Barbados in 1750, is believed to have arisen as either a mutation or a natural hybrid of this fruit.

Pummelos are low in calories (34 kcal per 100 g) and carbohydrates (8.5%). However, they are also a good source of vitamin C.

(Also see GRAPEFRUIT.)

PUMPERNICKEL BREAD

This is a sourdough bread made from unbolted (unsifted) rye flour.

PUMPKINS *Cucurbita* spp

The fruits of these plants provide both edible flesh and edible seeds. Also, the flowers are edible. Pumpkins belong to the gourd or melon family (*Cucurbitaceae*), which also includes squashes and cucumbers. The different species of pumpkins are *C. maxima, C. mixta, C. moschata,* and *C. pepo.* Certain varieties of fruits within each of these species are squashes rather than pumpkins. Although there is some confusion between the terms *pumpkin* and *squash,* pumpkins are generally considered to be the large, orange fruits that have a coarse, strongly flavored flesh. They are not usually served as table vegetables, but are used mainly for (1) pies, and (2) decorations during the holidays in the fall.

ORIGIN AND HISTORY. The pumpkin descended from wild ancestors in Mexico and Guatemala, where it is reasonable to conclude that it has been under cultivation for at least 9,000 years. Most likely, the modern varieties with abundant sweet flesh arose when the Indians selected mutant varieties for cultivation.

WORLD AND U.S. PRODUCTION. Statistics on the world production of pumpkins are not available. *The FAO Production Yearbook* groups pumpkins, squash, and gourds together. Since squashes are much more widely grown and utilized than pumpkins, the combined data reflects mainly the production of squashes (See SQUASHES). Likewise, U.S. production figures on pumpkins are not available.

PROCESSING. Pumpkins are processed by removal of the rind and seeds, followed by cooking of the pulp prior to canning, freezing, or baking into cakes, custards, or pies.

Other noteworthy, but less common, pumpkin products are (1) dehydrated pumpkin flakes that are made by drum-drying cooked pumpkin puree which has been mixed with starch and sugar; (2) dehydrated pumpkin pie mix which contains dehydrated pumpkin flakes, dried milk powder, dried egg, sugar, corn syrup solids, starch, flavorings, and a vegetable gum; (3) pumpkin pickles made from pumpkin cubes that have been cooked briefly, then mixed with sugar, vinegar, water, and spices, and canned in glass jars; and (4) pumpkin seeds that are sold raw or roasted and salted.

SELECTION, PREPARATION, AND USES. The rind of a pumpkin should be hard. However, slight variations in the color of the skin do not affect the quality.

For the most part, pumpkins are either cooked or baked. Some suggestions for utilizing the various parts of the pumpkin plant follow:

• **Pumpkin flesh**—The fruits should be cut into halves or small sections, the rind, fibrous matter, and seeds removed, and the remaining flesh cut into smaller pieces. Usually, a half hour to an hour of boiling is required to tenderize the flesh, but only 15 minutes of cooking in a pressure cooker is sufficient. Then, the cooked pumpkin may be mashed or pureed in a blender. The puree may require straining to remove residual fibers. Cooked pumpkin is usually made into custards, pies, and puddings. Canned, cooked pumpkin serves these purposes very well. However, Latin Americans utilize pumpkin mainly in soups and stews.

• **Pumpkin flowers**—These should be picked when open so that no bees will be trapped within the blossoms. The flowers are good when dipped in a batter and fried in deep fat.

• **Pumpkin seeds**—The seeds of pumpkins may be dried for a few days, after which any adhering tissue should be removed. Then, they may be roasted or fried in oil. Some

Fig. P-86. Pumpkins, a sign of autumn in North America. (Courtesy, New Jersey Department of Agriculture)

people boil the seeds briefly in salted water before roasting or frying.

NUTRITIONAL VALUE. The nutrient compositions of various forms of pumpkin are given in Food Composition Table F-21.

Some noteworthy observations regarding the nutrient composition of pumpkin follow:

1. Fresh and canned pumpkin contains over 90% water and is low in calories (about 33 kcal per 100 g). However, it is an excellent source of vitamin A.

2. A 4 oz (*114 g*) serving of pumpkin pie plus an 8 oz (*244 g*) glass of milk constitute a nutritious meal for a growing child, except that insufficient iron is provided unless dark molasses is used as the sweetener in the pie. The pie supplies ample calories and vitamin A, a moderate amount of protein, and fair amounts of calcium and phosphorus; whereas the milk is rich in protein, calcium, phosphorus, and the vitamin B complex.

3. Pumpkin flowers contain 95% water and are low in calories (16 kcal per 100 g) and most other nutrients, except that they provide fair amounts of phosphorus, iron, vitamin A, and vitamin C. It is suggested that calorie-conscious people eat pumpkin flowers in a salad, soup, or stew.

4. The seeds from pumpkins are very rich in calories (553 kcal per 100 g), protein (29%), iron (11.2 mg per 100 g), and phosphorus (1,144 mg per 100 g). Hence, the consumption of as little as 1 oz (*28 g*) of the seeds per day will make a significant nutritional contribution to the diet.

PURGATIVE

A strong laxative.
(Also see LAXATIVE.)

PURINES

These are nitrogen-containing substances with a ring structure. They are widely distributed in nature, and are components of the nucleic acids. Important purines are adenine, guanine, xanthine, and uric acid, which is the excretory form of the purines in humans. Another common purine is the stimulant caffeine. Persons suffering from gout or uric acid kidney stones should limit their dietary sources of purines. Some of the richest food sources of purines are anchovies, asparagus, brains, kidney, liver, mincemeats, mushrooms, sardines, and sweetbreads. Foods with a low purine content include breads, cereals, fats, cheese, eggs, fruits, milk, nuts, sweets, and most vegetables.
(Also see ARTHRITIS, section headed "Gout [gouty arthritis]"; CAFFEINE; and NUCLEIC ACIDS.)

PURIS

A puri is a very light, fried wheat cake which originated in India. Butter is the preferred fat in which to fry it. Because it is an expensive dish, it is usually reserved for festive occasions.

PUROTHIONIN

A small, easily digested protein of wheat. It has antibacterial properties, and in excess it may interfere with the rising of dough.
(Also see WHEAT.)

PUTREFACTION

The decomposition of proteins by microorganisms under anaerobic conditions.

PUTRESCINE

A foul-smelling chemical which arises in the bacterial fermentation of animal protein. It can be formed from ornithine or the amino acid arginine.

PYRIMIDINE

The parent substance of several nitrogenous compounds found in nucleic acids. The principal pyrimidines found in RNA are uracil and cytosine; in DNA they are thymine and cytosine.
(Also see NUCLEIC ACIDS.)

PYRUVIC ACID (PYRUVATE; $CH_3COCOOH$)

An organic 3-carbon acid that is a key intermediate in carbohydrate, fat, and protein metabolism. It can participate in several metabolic pathways in the body. These pathways include (1) complete oxidation to water and carbon dioxide in the tricarboxylic acid (Krebs) cycle, (2) formation of fatty acid, (3) reversible conversion to lactic acid, (4) conversion to the amino acid alanine, and (5) formation of glucose by reversal of the enzymatic sequence which normally breaks down glucose.
(Also see METABOLISM.)

QUADRIPLEGIA

Paralysis of both arms and both legs. It can be caused by disease—poliomyelitis, Landry's acute ascending paralysis (due to an acute infection of the spinal cord), diptheria, leprosy—or injury which severs the spinal cord at the fifth or sixth cervical vertebra. Quadriplegics require some special dietary considerations due to reduced activity and difficulties encountered in preparing and eating food.

(Also see MODIFIED DIETS; and DISEASES.)

QUAIL (CALIFORNIA QUAIL, GAMBEL'S QUAIL, MEARN'S QUAIL, MOUNTAIN QUAIL, SCALED QUAIL)

Quail is the name given to several different kinds of birds. In Europe, it refers to several kinds of game birds of the pheasant family. Americans use the name for the bobwhite.

The bobwhite quail and the Japanese quail are listed in separate sections of this book. Other kinds of American quail are the California quail, Gambel's quail, Mearn's quail, mountain quail, and scaled quail. These birds live in western and southwestern United States.

All of the different quails are highly prized game birds and furnish good eating.

QUICK BREADS

Contents   Page

History 911
Major Types of Quick Breads 912
Nutritive Values 912
    Improving the Nutritive Values of Homemade
    Quick Breads 914
Recent Trends in Quick Bread Technology 914

This term is usually applied to batters or doughs that are baked (in some cases, boiled, fried, grilled, or steamed) right after mixing, rather than being allowed to rise before baking. Hence, quick breads are not made with yeast. Instead, they are unleavened, or leavened with air, baking powder, a sourdough starter, or steam.

The elimination of the need for yeast leavening allows the use of a wide variety of flours that may contain little or no gluten—the elastic protein in wheat flour that helps to trap the leavening gas. Weaker flours may be used for quick breads because rising takes place in the oven, as the proteins in the dough are gradually made firm by baking. Furthermore, egg white has a greater strengthening effect on doughs that are baked right after mixing.

HISTORY. There is archaeological evidence that the first breads baked by primitive peoples were crude, unleavened forms of quick breads similar to those used today. These breads were made by mixing flours or meals with water and baking the dough on hot stones. Fig. Q-1 depicts the making of an early type of quick bread.

Thin, unleavened pancake-type breads became popular

Fig. Q-1. History's first pancake-maker: prehistoric man pouring dough on heated stones. At the right side, a boy is preparing the dough. On the left side, another boy is heating the baking stones. Stones were piled, one upon the other, until sufficient pancakes were made for the whole family. (Courtesy, The Bettmann Archive, Inc.)

in many places around the world, because they could be prepared more rapidly, and they were more palatable after baking. The thicker types of unleavened breads were likely to remain soggy inside even though the outside had been baked until brown, or burned. Some of the thin breads which have remained popular to the present day are: (1) Chinese unleavened pancakes (egg roll wrappers); (2) chapaties made from whole wheat flour in India; (3) tortillas made from corn in Mexico; (4) crepes, usually made from wheat, in France; (5) the crackerlike matzo—the only form of bread eaten by orthodox Jews during Passover; (6) Swedish crispbreads made by mixing rye meal with snow or powdered ice; and (7) bannocks made from oats and/or barley in Scotland. Even the American pioneers made an unleavened corn bread called a *hoe cake*, which was baked on the blade of a hoe placed over hot coals. Wheat was scarce in North America until the Great Plains were settled and hardy Northern European wheats were planted.

The first type of baking powder appears to have been pearl ash, a crude form of potassium carbonate derived from wood ashes, which was invented in America in the 1790s. (Carbonates react with acids and give off carbon dioxide gas, the most commonly used leavener. Sometimes, heating is required to bring about the leavening reaction.) However, pearl ash left an unpleasant taste that had to be masked with sugars and other ingredients. Later, in 1835, a mixture of cream of tartar (obtained from the residue in wine vats) and baking soda was developed. Shortly thereafter, the mixture was utilized in commercial baking powders. Also, a self-rising flour (one that contains baking powder) was developed in England in 1849. The marketing of these quick acting leavening agents soon led to the development of many new types of baked goods, such as baking powder biscuits, layer cakes, quick muffins, and soda bread. By the early 1900s, many home-makers were using self-rising flour and preleavened pancake mixes. Other preleavened baking mixes, which contained all required ingredients except a liquid, did not become popular until the 1940s, when new products were developed for the use of the military services.

At the present time, most people use preleavened baking mixes for preparing quick breads, cakes, and similar items.

(Also see BAKING POWDER AND BAKING SODA.)

MAJOR TYPES OF QUICK BREADS. Fig. Q-2 shows some of the more popular items.

Fig. Q-2. Cakes, cookies, muffins, and other highly desirable quick breads. (Courtesy, J. C. Allen & Son, West Lafayette, Ind.)

The basic ingredients used in all quick breads are flour, liquid, and a leavening agent. Many also contain shortening, eggs, sugar, salt, flavorings, and other ingredients. However, the characteristics of each baked product are determined mainly by the proportions of the ingredients, and the ways in which they are mixed. Some of the more common types of quick breads are: biscuits, cakes (angel food, chiffon, plain, pound, sponge), coating batter, cookies, cornbread, crackers (soda, hard tack, pilot bread), doughnuts, dumplings, gingerbread, muffins, pancakes, pastry (pie crust, puff), popovers, scones, soda bread, and waffles.

(Also see BREADS AND BAKING, sections headed "Quick Breads Made With Baking Powder or Baking Soda," and "Products Leavened With Air or Steam.")

NUTRITIVE VALUES. Most quick breads are higher in calories than most yeast-leavened breads because they are usually made with more sugar and fat. Since many people have to control their caloric intake in order to avoid becoming obese, careful selection of dietary breads is necessary if one is to obtain the optimal nutritional benefits from these products without putting on excess weight. It is noteworthy that both the American Diabetes Association and the American Dietetic Association have advised dieters not to include cakes, cookies, and pies in their meal plans without the permission of a diet counselor.[1] However, certain dessert items such as custard pie may equal or exceed the nutritive values of many plain breads (in terms of nutrients per calorie). What is most important is that portion sizes of high caloric items be chosen within appropriate limits.

The Food Composition Table F-21 presents the nutritive values for many of the most popular baked goods—biscuits, cakes, cupcakes, cookies, crackers, doughnuts, muffins, pancakes, pies, waffles, and wafers.

In general, the information in Food Composition Table F-21 shows that there are wide variations in the values for calories and other nutrients supplied by the various quick breads. However, a dieter should not judge the merits of an item solely by its caloric content because sometimes a rich but nutritious item might be substituted for one or more of the nonbread foods in the diet. For example, two popovers have the protein content of 1 oz (28.4 g) of meat. This type of substitution should not be carried too far, because popovers supply from 2 to 3 times as many calories as the lean types of meat.

A convenient way of assessing the protein contribution of quick breads is to consider items containing less than 2.0 g of protein per 100 Calories (kcal) as poor sources of protein, those supplying from 2.0 g to 2.9 g as fair sources, and those containing 3.0 g or more of protein per 100 Calories (kcal) as good sources. Therefore, some poor choices for dieters are most cakes with icing and/or filling, cookies, doughnuts, gingerbread, and pies. Fair choices are biscuits, uniced angelfood cake, peanut cookies, cheese crackers, muffins, coconut custard and custard pies, and waffles. Some good choices are the items made with plenty of eggs and milk, but with only small amounts of fat and sugar. The best examples of the latter types of products are pancakes and popovers. Dieters might also select whole rye wafers, which contain more than 3 g of protein per 100 Calories (kcal). However, the protein quality of items containing only cereal protein (rye wafers) is lower than those containing a mixture of cereal with eggs and milk (pancakes).

Most quick breads are low in minerals and vitamins because they are made from white flour. Even the use of enriched flour does little to correct these deficiencies because only iron, thiamin, riboflavin, and niacin are restored; whereas more than 2 dozen essential minerals and vitamins are removed during the production of white flour.

Fig. Q-3. Quick bread, made with baking powder and baked right after mixing. (Courtesy, USDA)

---

[1]*Exchange Lists for Meal Planning*, American Diabetes Association, Inc., and The American Dietetic Association, 1976, p. 4.

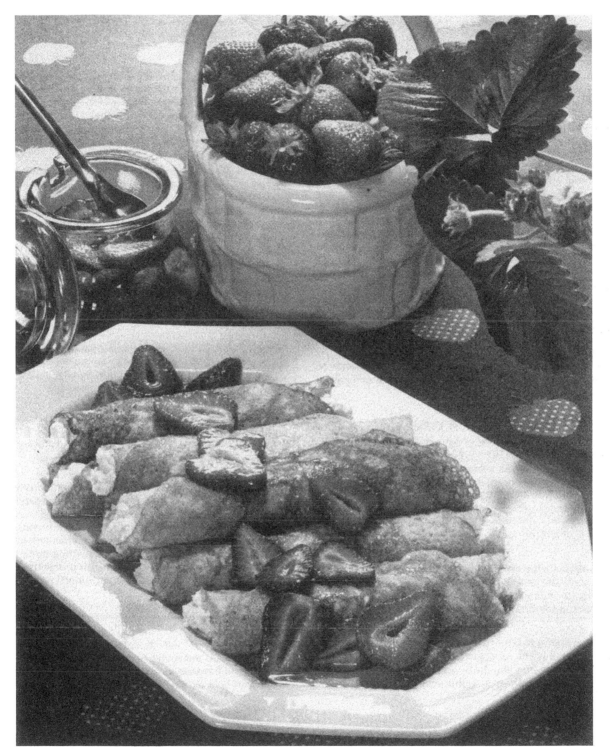

Fig. Q-4. A delectable *quick bread* dessert—Strawberry crepes—which is easier to prepare than it looks. (Courtesy, California Strawberry Advisory Board)

Improving The Nutritive Values Of Homemade Quick Breads. Most baked goods supply too many calories and too few other nutrients. Hence, it is necessary for modern homemakers to revise recipes for quick breads by (1) reducing fat and sugar content, and (2) adding other ingredients that are low in calories and rich in more essential nutrients. Some suggestions for modifying recipes follow:

1. Muffins and pancakes lend themselves readily to experimentation because (a) wide ranges of consistencies and textures are acceptable in these products, and (b) small amounts may be prepared so that wastage of food is minimized if an experiment is unsuccessful.

2. Low-calorie, high-protein items may be prepared from mixtures of whole wheat flour, nonfat dry milk, beaten egg white, and water.

3. The use of a nonstick frying pan or griddle for making pancakes eliminates the need for oiling the cooking surface.

4. About ⅓ cup of grated cheese may be added for each cup (*240 ml*) of flour. This addition raises the content of protein, calcium, and phosphorus.

5. Sprouted seeds (usually those of beans or grains) add protein, fiber, minerals, and vitamins (about ⅓ cup per cup of flour).

6. Extra eggs may be added to most baked products, providing that a soft flour is used. Too many eggs toughen doughs.

7. About ⅛ cup of soy flour or nonfat dry milk may be substituted for a cup of wheat flour.

8. The caloric content of dessert items may be reduced by mixing one part of cake or cookie mix with one part of biscuit mix.

RECENT TRENDS IN QUICK BREAD TECHNOLOGY.

The continuing rise in the employment of mothers has been accompanied by a reduction in the amount of time available for baking at home. Furthermore, the growing trends of eating out and of buying baked goods from bakeries have brought about a shortage of experienced bakers in certain areas. Hence, food technologists have recently developed some almost foolproof, time-saving products, such as those which follow:

• **Emulsified shortenings for packaged cake mixes**—The need to cream fats has been eliminated by the specially emulsfied shortenings used in prepared cake mixes. Air bubbles are trapped by the special shortenings during mixing.

• **Frozen batters**—After thawing, these products may be poured right from the package (which is usually a short, squat cardboard container like those used for cream or milk) onto a pancake griddle or a waffle iron.

• **Frozen quick breads**—These items are available in the forms of unbaked doughs and frozen baked products. The state of the art is such that few people can tell the difference between frozen and unfrozen products once they have been heated.

• **Fruit pie fillings**—Thickened fillings containing pieces of fruit have long been available in grocery stores. Now, a new line of pie fillings is made from gums, starches, sugar, and fruit purees. The new types of fillings make more efficient use of available fruit, so, hopefully, the cost of these products may be lower than those of the former types of fillings.

• **Gums**—These ingredients are often added to commercial baked goods and to packaged mixes because they strengthen weak flours without causing excessive toughening. Some other functions of these additives are: (1) thickening pie fillings and jellied ingredients, and (2) reducing the fat absorption of fried doughnuts.

Prebaked items—Most prebaked products require only a little warming and/or browning in an oven prior to serving. Hence, they are being utilized increasingly by both homemakers and commercial and institutional food service establishments.

Prepackaged items for microwave baking—Certain items packaged in cardboard trays may be heated in microwave ovens, but *not* those that are sold in metal foil pans. Metal foil may cause damage to microwave ovens.

• **Refrigerated doughs**—Most of these products require only (1) their placement in a suitable pan, and (2) baking. Cookie doughs also require slicing. The most popular items in this line are ready-to-bake biscuits, pastry, and rolls.

• **Separation of flours into high-protein and low-protein fractions**—In the past, high-protein hard wheats were used to make flours suitable for yeast-leavened doughs and low-protein soft wheats yielded flours for quick breads. Now, either type of wheat may be milled and separated into high-protein and low-protein fractions by swirling streams of air in a process called air classification. This procedure enables millers to produce flours to meet customers' requirements.

QUILLAJA (SOAP BARK)

This is the inner dried bark of *Quillaja saponaria*, which grows in Peru and Chile. It contains quillaic acid, quillajasaponin, sucrose, and tannin. Soap bark is used to manufacture saponin, and to produce foams in various products. It is also approved as a natural flavor which can be used in foods.

RABBIT

A small grayish brown mammal, native to southern Europe and northern Africa, which has been introduced into various other regions where it is often a pest because of its rapid reproduction. Under domestication, many breeds have been developed, with special adaptation for meat, fur, or show.

Ready-to-cook domestic rabbit is classed as either (1) fryer or young rabbit, or (2) roaster or mature rabbit. Fryers weigh from 1 ½ to 3 ½ lb (0.7 to 1.6 kg) and are usually less than 12 weeks of age; and the flesh is fine grained, tender, and a bright pearly pink color. Roasters usually weigh more than 4 lb (1.8 kg) and are 8 months of age or older; and the flesh is more coarse grained, slightly darker, less tender, and the fat more creamy in color than fryers.

RACCOON (RACOON)

A small wild animal, with longish grey fur, living in the forests of America. It is edible and cooked like wild rabbit.

RAD

In relation to radiation, it is a unit for measuring the radiation energy absorbed by a substance. One rad equals the absorption of 100 ergs of energy per gram.

(Also see RADIATION PRESERVATION OF FOOD.)

RADIATION (IRRADIATION)

The emission and propagation of energy in the form of waves or particles through space or matter.

• **In food**—It refers to ionizing radiation which kills off various microorganisms—so-called cold sterilization.

• **In health and disease**—It refers to the process in which any one or a combination of rays—sunshine, radioactive particles, x rays, for example—are used for diagnostic or therapeutic purposes.

RADIATION PRESERVATION OF FOOD (COLD STERILIZATION; IRRADIATION; RAD APPERTIZATION; RADICIDATION; RADURIZATION)

Ionizing radiation—fast moving subatomic particles or energetic electromagnetic waves which are strong enough to strip electrons from atoms or molecules—provides the basis for a new method of food preservation. By subjecting foods to ionizing radiation a number of desirable storage characteristics —sprouting inhibition, slowed ripening, pasteurization, and complete sterilization—may be induced, depending upon the dose of radiation. Since the radiaton must not make the food radioactive, the sources of radiation are limited to the isotopes cobalt-60 and cesium-137, and to x rays. The doses of ionizing radiation to which these sources subject foods are expressed by a unit called the rad—the energy absorption of 100 ergs per gram. Since one million rads only increases the temperature of food by about 2 °C, radiation processing of foods is sometimes called cold sterilization. Table R-1 indicates some of the potential uses of this process.

Despite the fact that some people are somewhat squeamish about accepting irradiated products, irradiation under approved processes does not make food radioactive; hence, it poses no danger to consumers.

The possible use of radiation preservation is especially important in tropical and subtropical countries, where the temperature is high and the humidity is often excessive. In India, for example, 10 to 50% of the food production is often lost because of spoilage.

(Also see PRESERVATION OF FOOD, Table P-10, Methods of Food Preservation; RADIATION.)

TABLE R-1
POTENTIAL APPLICATIONS OF RADIATION TO FOOD PROCESSING

Use	Foods	Action of Radiation	Rads of Exposure
Extension of storage life	Potatoes, onions, and other tubers and bulbs	Inhibits sprouting	Less than 20,000
	Fruits and some vegetables	Delays ripening and reduces the yeast and mold population	100,000 to 500,000
	Meat, fish, poultry, and other highly perishable foods	Reduces the population of microorganisms capable of growth below 37°F (3° C)	50,000 to 1 million
	Cereals, flours, dried fruits, and any other food prone to insect infestations	Kills or sexually sterilizes the insects preventing loss during storage or spread of the pests	10,000 to 50,000
	Meat, fish, poultry, and other highly perishable foods	Complete sterilization, destruction of any organism capable of causing spoilage thus preparing food for long-term storage when properly packaged	4 to 6 million
Prevention of *Salmonella* food poisoning	Frozen meat, eggs, poultry, and other foods liable to contamination	Destroys *Salmonella* bacteria	300,000 to 1 million
Prevention of parasitic diseases	Meat or any other parasite-carrying food	Destroys parasites such as *Trichinella spiralis* and *Taenia sagnata*	10,000 to 30,000

RADIATION THERAPY

The therapeutic use of x rays or radioactive substances such as radium and radioactive cobalt-60 isotope in the treatment of disease, primarily of a malignant nature.

RADIOACTIVE FALLOUT (CONTAMINATION)

As a safeguard to our health, the Food and Drug Administration (FDA), the U.S. Department of Agriculture (USDA), the U.S. Public Health Service (USPHS) and the Federal Radiation Council (FRC), are among the groups that periodically monitor the level of radioactivity in the food supply. The FRC has stipulated the lower levels of radioactive contamination of foods. Should the level exceed these limits, the Agency will recommend actions which should be taken.

RADIOACTIVE ISOTOPES

An isotope is an element which has the same atomic number as another element but a different atomic weight. A radioactive isotope is one that decomposes spontaneously, emitting alpha or beta particles or gamma rays through the disintegration of its atomic nuclei. Usually, when fed or injected, its course and concentration can be traced and tagged by use of a special instrument, a Geiger counter. More than 150 radioactive isotopes are used (some experimentally) in the treatment of various diseases; among them, radioactive iodine (sodium iodide I-131)—used in the treatment of hyperthyroidism and other ailments, radioactive gold, radioactive phosphorus, and radioactive iron.

RADIOACTIVE I¹³¹ TEST

This is a test of thyroid function by using a radioactive isotope of iodine, I^{131}. After administering the test dose, the uptake and utilization of iodine by the thyroid gland is measured by tracing the I^{131}.

RAFFINOSE

A trisaccharide containing the three hexoses, glucose, fructose, and galactose. Raffinose is found in molasses, cottonseed meal, and Australian manna.
(Also see CARBOHYDRATE[S].)

RAGOÛT

This dish is essentially the same as a stove-top stew. The meat is browned, then the onion, garlic, parsley, stock, and vinegar are added along with seasonings, and the mixture is simmered until the meat is tender. Duck, chicken, rabbit, hare, lamb, or young partridge maybe prepared in this way. The gravy is usually thickened at the end. The name comes from the French word *ragoûter*, which means *to awaken the senses*.

RAISIN(S)

These are the dried fruit produced from certain varieties of grapes. The word raisin comes from the Latin *racemus*, which means a *cluster of grapes or berries*. All of the U.S. production, and nearly half of the world's supply of raisins comes from the San Joaquin Valley in California. Other raisin-producing countries include Turkey, Australia, Greece, Iran, South Africa, and Spain.

Many varieties of grapes are produced in California, but only four principal varieties have been found suitable for drying into raisins. These four varieties are the Thompson Seedless, Muscat, Sultana, and Zante Currant, and they are either sun-dried or mechanically dehydrated.

Fig. R-1. Raisin pie. (Courtesy, Phil Seagle/Ketchum, San Francisco, Calif.)

Raisins contain only 17% water whereas grapes contain about 80% water. Therefore, many of the nutrients in grapes are more concentrated in raisins. Each 100 g (about 3 ½ oz) of raisins provides 289 Calories (kcal) of energy, 2.8 mg of iron, and 678 mg of potassium. The calories in raisins are derived primarily from the naturally occurring sugars (carbohydrates). More complete information regarding the nutritional value of raisins is presented in Food Composition Table F-21.
(Also see GRAPE [RAISINS].)

RAISINE

A preserve made with either pears or grapes plus quinces, and cooked slowly in sweet wine or cider.

RAISIN OIL

The oil that is extracted from the seeds of muscat grapes before they are dried to make raisins. It gets its name from its use as a coating on raisins to prevent them from (1) sticking together, (2) becoming excessively dried out, and (3) being attacked by insects.

RAMP (WILD LEEK) *Allium tricoccum*

The ramp is a wild leek that is native to North America. An onionlike odor is emitted by the leaves, which are often eaten with the bulbs.

Data on the nutrient composition of ramp is not readily available. However, it is believed that they are a fair to good source of vitamin C, and that the green leaves may supply valuable amounts of vitamin A.

RAMPION *L. rapa*

This now-neglected vegetable is of the same family as the turnip. It has an edible root which can be eaten raw in salads and tastes like a walnut. The leaves are used raw in salads, or cooked like spinach.

RANCID

A term used to describe fats that have undergone partial decomposition.
(Also see FATS AND OTHER LIPIDS.)

RANGPUR (RANGPUR LIME; MANDARIN LIME)
Citrus limonia

A highly-acid citrus fruit (of the family *Rutaceae*) that is native to India and which (1) resembles the mandarin orange (*C. reticulata*) and (2) is highly acid. It is *not* related to the lime (*C. aurantifolia*). Rather, it is thought to be a sour mutant of the mandarin orange. The rangpur is used mainly as an ornamental plant and a rootstock for bearing the grafts of other citrus fruit trees.

RAPE (CANOLA) *Brassica napus; B. campestris*

Rape is a general term applied to members of the *Cruciferae* family. It ranks number two among the edible vegetable oils globally; and it is the only oilseed grown successfully in all parts of the world.

Development of new varieties of rape which contain little or no erucic acid and glucosinolate, two antinutritional factors, has made for expanded production and utilization of rapeseed, a cool climate oilseed. This has had a significant impact on producer countries, lessening their dependence on soybeans, sunflowers, peanuts, or cottonseed.

• **Canola**—The name "canola" is registered by the Western Canadian Oilseed Crushers Association; it stands for CANadian Oil Low Acid. To qualify as canola, rapeseed must yield oil with less than 2% erucic acid and meal with less than 30 micromoles of glucosinolate per gram.

ORIGIN AND HISTORY. It is probable that the primary types of rape came from Asia, the Mediterranean, and western Europe. It has a long history, with earliest references found in Indian Sanskrit writings of 2000 to 1500 B.C.

WORLD AND U.S. PRODUCTION. World production of rapeseed totals about 24.5 million metric tons annually; and the leading producing countries, by rank, are: China, India, Canada, Germany, and France.[1]

U.S. production of rapeseed in 1990 totaled 67,000 metric tons. The Food and Agricultural Organization of the United Nations reports that a mere 1,000 metric tons of seed are produced in the United States, annually.

PROCESSING RAPESEED. The processing of rapeseed to obtain oil and meal is similar to that of other oilseeds. There are three methods: (1) mechanical pressing, (2) straight solvent, or (3) prepress solvent extraction. The latter is the method of choice. In the prepress solvent extraction method, a large portion of the oils is removed from the seeds by mechanical presses (expellers), then the remaining oil is extracted by using the organic solvent n-hexane. After discharging from the extractor, the meal, which contains about 1% oil, is dried to about 10% moisture, cooled, and placed in storage. The crude oil must still be refined.

NUTRITIONAL VALUE OF RAPE. The seeds contain 40 to 45% oil and 25% protein. When processed, rapeseed yields about 40% oil and 50% oil meal which ranges from 32 to 40% protein.

[1]Based on data from *FAO Production Yearbook 1990*, FAO/UN, Rome, Italy, Vol. 44, p. 113, Table 41. **Note well:** Annual production fluctuates as a result of weather and profitability of the crop.

High glucosinolate rapeseed meal can cause goiter—an enlargement of the thyroid gland. Here again the best solution to the problem is to lower or eliminate the causative substance from the seed. This has been accomplished.

Plant breeders in Canada and northern Europe have been highly successful in changing rapeseed species to meet consumer needs. Today, these new products are being promoted as canola products, rather than rapeseed products.

It is noteworthy that the reduction or elimination of glucosinolates from rapeseed meal also results in more effective use of the favorable amino acid balance of rapeseed protein.

Oil made from the new varieties of rape is almost free of the long chain fatty acid, erucic acid, while the oleic acid content is significantly higher. Except for the presence of linolenic acid, the composition of the new oil greatly resembles olive oil.

Rapeseed meal, the residue remaining after removal of oil from the seed, contains 35 to 40% protein, 20 to 25% carbohydrate, 12 to 16% crude fiber, and 5 to 7% ash. Due to its high content of lysine, methionine, and cysteine, rapeseed protein has a higher nutritive value than any other known vegetable protein. Based on growth studies with rats and on a nitrogen balance study on student volunteers, the nutritive value of rape protein is as high as that of good animal protein. With the exception of a negative effect on the zinc balance in rats, which can be compensated, no negative findings have been recorded.

RAPESEED PRODUCTS AND USES. The primary use of rapeseed oil is for human food. Its relatively low linolenic acid content permits it to compete with other vegetable oils in shortenings, margarines, salad oils, and frying oils. Oils which are almost free of erucic acid will open up new uses of rapeseed oil.

Until now, rapeseed meal has been used almost solely as a protein supplement for animals. With the development of glucosinolate-free rapeseed meal, it is expected that rapeseed protein will join soybean protein as an ingredient in meat analogs, meat extenders, dairy products, bakery goods, and other processed foods.

(Also see CANOLA [RAPE]; and OIL, VEGETABLE, Table O-5.)

RASHER

A portion, usually 2 or 3 slices, of bacon.

RATAFIA

A liqueur made by infusing and usually not distilled, flavored with plum, peach, and apricot kernels and bitter almonds, and supplied with a base of brandy and fruit juices. There are many different ratafias in different countries; obviously, they are of great potential value to cooks.

RAVIOLI

This is a favorite pasta dish of Italy, and Italians around the world. It is made by rolling out the pasta dough and dropping a spoonful of special cheese filling on each 2-inch square (*13 cm²*) of pasta. This is covered with another sheet of pasta, sealed, and cut with a ravioli cutter. These little squares are cooked in rapidly boiling water and served with or without a tomato sauce, grated Parmesan cheese, or butter.

(Also see MACARONI AND NOODLE PRODUCTS.)

RAW SUGAR

This is the stage in cane sugar manufacturing at which

sugar is transferred from the sugar mills to the refineries. Raw sugar is about 96 to 98% sucrose covered with molasses, but containing extraneous materials which prevent its use without further refining. FDA regulations prohibit the sale of raw sugar unless impurities such as dirt and insect fragments are removed.

(Also see SUGAR, section on "Processing.")

RDA (RECOMMENDED DIETARY ALLOWANCES)

The first edition of *Recommended Dietary Allowances* (RDA) was published in 1943. The current RDA was published in 1989.

Ideally, these allowances should be determined through experimentation on humans. However, for a number of reasons—some quite obvious—this is not always possible. Therefore, estimates of nutrient requirements are determined by the following six methods, some rather indirect: (1) determination of nutrient intake in apparently normal, healthy people; (2) biochemical studies; (3) studies of clinical cases of nutrient deficiencies; (4) balance studies that determine nutrition status in relation to nutrient intake; (5) experiments that maintain individuals on diets low or deficient in a nutrient followed by correction of the deficiency; and (6) animal studies.

The RDAs are estimates, not standards, and should be employed as a goal—not ideal nor optimal, but acceptable—for meeting the nutritional needs of a group of people. If nutrients are consumed at the level of the RDA, the needs of nearly all healthy members of the group will be satisfied, since the RDAs are not averages (except energy), but calculated for individuals with the highest requirements. However, the RDAs do not allow for the special needs of inherited metabolic disorders, infections, and chronic diseases. Furthermore, the RDAs are not intended for use to evaluate individual nutritional status.

In the 1989 edition of *Recommended Dietary Allowances,* compiled by the Committee on Dietary Allowances and the Food and Nutrition Board, nutrient allowances are given by age, weight, height, sex, pregnancy, and lactation.

• **Dietary recommendations for Americans**—The National Academy of Sciences has also called on the American public to make the following changes in its eating habits:

1. Reduce total fat consumption to 30% or less of total daily calorie intake.

2. Reduce cholesterol intake to less than 300 mg a day.

3. Increase the amount of starches and other complex carbohydrates in the daily diet.

4. Increase the amount of fiber in the diet.

5. Increase the amount of fruits and vegetables eaten daily.

6. Avoid drinking alcohol, especially during pregnancy. If abstaining is impossible, drink in moderation.

7. Reduce daily intake of sodium to 6 g or less.

8. Maintain desirable body weight through prudent diet and regular exercise.

NOTE WELL: Table N-3 in section headed "Nutrients: Requirements, Allowances, Functions, Sources" of this encyclopedia gives the United States Recommended Daily Allowances (RDA) of each nutrient, *plus* its major function(s) in the body and best food sources; hence, the reader is referred thereto. A comparison of United States RDAs with the recommended daily dietary guidelines of selected countries and FAO follows.

TABLE R–2
COMPARATIVE DAILY DIETARY GUIDELINES FOR ADULTS IN SELECTED COUNTRIES AND FAO[1]

Country	Sex	Age	Weight		Activity[2]	Calories	Protein	Calcium	Iron	Vitamin A (retinol eq.)	Thiamin	Ribo-flavin	Niacin eq.	Vitamin C
		(yr)	(lb)	(kg)			(g)	(g)	(mg)	(mcg)	(mg)	(mg)	(mg)	(mg)
United States	M	25–50	174	79	MA	2,900	63	0.8	10	1,000	1.5	1.7	19	60
(1989)	F	25–50	138	63	MA	2,200	50	0.8	15	800	1.1	1.3	15	60
Canada	M	19–35	154	70	MA	3,000	56	0.8	10	1,000	1.5	1.8	20	30
	F	19–35	123	56	MA	2,100	41	0.7	14	800	1.1	1.3	14	30
United Kingdom	M	18–34	143	65	MA	2,900	63	0.5	10	750	1.0	1.6	18	30
	F	18–34	120	55	MA	2,150	54	0.5	12	750	0.9	1.3	15	30
FAO	M	25	143	65	MA	3,000	37[3]	0.4–0.5	5–9	750	1.2	1.8	19.8	30
	F	25	120	55	MA	2,200	29[3]	0.4–0.5	14–28	750	0.9	1.3	14.5	30

[1]Daily dietary guidelines for Canada, United Kingdom, and FAO provided by Z. I. Sabry, Director, Food Policy and Nutrition Division, Food and Agriculture Organization of the United Nations, Rome, Italy.
[2]Moderate activity.
[3]Expressed in reference protein.

RECOMMENDED DAILY DIETARY GUIDELINES OF SELECTED COUNTRIES AND FAO. Table R-2 gives a comparison of the daily dietary guidelines for adults in selected countries and in all countries as made by the Food and Agriculture Organization (FAO) of the United Nations. The FAO dietary guidelines are intended to apply to people in all countries when translated in terms of local foods.

1. They are set for population groups that live under different environmental conditions (climate, occupation, and activity) and that have different dietary practices.

2. Those who prepare the standards give varied interpretations to scientific data; differences which have narrowed with increased research and knowledge of the nutrient needs of human beings.

All the standards include recommended intakes of only those nutrients for which the requirements for human beings have been established. In order to assure that the nutrients that are known to be essential, but for which the requirements are as yet unknown, will be included in the diet, the selected diet should include as much variety of foods as practical, acceptable, and palatable.

(Also see NUTRIENTS: REQUIREMENTS, ALLOWANCES, FUNCTIONS, SOURCES.)

RECTUM

The last 6 in. *(15 cm)* of the large intestine (colon) before ending at the anal opening.

(Also see DIGESTION AND ABSORPTION.)

RED MEAT

Meat that is red when raw. Red meats include beef, veal, pork, mutton, and lamb.

(Also see MEAT[S].)

REDUCING DIETS

The principle is to decrease the calorie intake. However, a good reducing diet must still provide the required nutrients. At the same time it should be acceptable and palatable, offer a variety, be economically feasible, and promote a sense of well-being.

Basically, the modifications which are usually made in low-calorie diets involve reduction in the number of items high in carbohydrates and fats, but retention of those which are good sources of minerals and vitamins. However, it is often necesary to calculate caloric requirements so that the dietary calorie content may be low enough to produce the appropriate caloric deficit. Generally a weight loss of 1 to 2 lb (0.45 to 0.91 kg) per week is ideal. This requires a caloric deficit of 500 to 1,000 Calories (kcal) per day.

(Also see FOOD GROUPS; and OBESITY, section headed "Treatment of Obesity.")

REFERENCE MAN

A healthy man of specified age, weight, environmental temperature, and physical activity permitting the formulation of standard calorie allowances. In the 1989 *Recommended Dietary Allowances*, the average mature male used for the basis of energy recommendation is engaged in light activity, is presumed to live in an area where the mean environmental temperature is 68°F (20°C), and has the desirable weight of 174 lb (79 kg). The energy requirements for the average mature male are given for two age groups: (1) 25 to 50 years of age; and (2) 51 years and over.

REFERENCE PROTEIN (CHEMICAL SCORE)

Proteins differ in nutritive value mainly due to their amino acid composition. A reference protein provides all of the essential amino acids in sufficient quantities and balances so as to meet the body's requirements without an excess.

A hypothetical reference pattern of amino acids expressed as milligrams per gram of protein, along with the amino acid patterns of human milk and eggs, is given under Protein, section on "Quantity and Quality," Table P-15; hence, the reader is referred thereto.

(Also see AMINO ACID[S]; and PROTEIN[S].)

REFERENCE WOMAN

A healthy woman of specified age, weight, environmental temperature, and physical activity permitting the formulation of standard calorie allowances. In the 1989 *Recommended Dietary Allowances*, the average mature female used for the basis of energy recommendation is engaged in light activity, is presumed to live in a climate where the mean environmental temperature is 68°F (20°C), and has a desirable weight of 138 lb (63 kg). Her energy requirements are given for two age groups: (1) 25 to 50 years of age, and (2) 51 years and over.

REGURGITATION

The return of food from the stomach, gullet, or duodenum (first portion of the small intestine) after eating, without vomiting. Patients with this affliction can usually lessen regurgitation (1) by sleeping with the head and chest higher than the rest of the body, and (2) by drinking lots of water.

REINDEER

The flesh of the reindeer provides venison, which, however, is inferior to that of the roebuck or deer. The venison from reindeer is prepared in the same way as venison from roebuck or deer. The meat of young reindeer is delicate; the meat from old reindeer is less tender and needs marinating.

RELAPSE

The return of the symptoms of a disease after convalescence has begun.

RELIGIONS AND DIETS

Religion and food have always played an important role in the lives of men. Often, food was the most precious and scarce of man's possessions. Not surprisingly, religious rituals or customs became associated with food. Religions have (1) decreed what foods man could or could not eat, (2) prescribed the preparation of foods, (3) indicated what food may or may not be eaten on certain days or certain occasions, and (4) symbolized foods and drink, and their ingestion. Furthermore, the religious idea of defiling or polluting objects influenced food choices. Since food is taken into the body, many religions held beliefs that certain foods and drinks were polluting or defiling.

Fig. R-2. Symbolic representation of man's religions.

Through the ages, man has had many different religions, 11 of which have survived. While there are countless religious groups in the world, the broad groupings of these 11 living religions represents the wellspring of religious thought.

These religions developed from various areas of the world and then spread. Hinduism, Buddhism, and Jainism originated in India. Confucianism and Taoism originated in China. Shintoism started in Japan. Christianity, Islam, Judaism, and Zoroastrianism arose from the Middle East. Sikhism is a hybrid of Islam and Hinduism from India. Some of these religions prescribe very specific dietary practices while others do not appear to do so. Nevertheless, the philosophy of a religion may have subtle influences over food selection. For this reason, Table R-3 presents the basic tenets of the 11 living religions as well as foods or drinks which are specifically taboo or encouraged by the religion.

TABLE R-3
LIVING RELIGIONS OF THE WORLD, AND THE DIET PRESCIBED BY EACH

Religion; Origin; Adherents	Chief Scripture(s); Basic Tenets	Food Taboos	Foods Encouraged by Beliefs
BUDDHISM **Origin:** More than 2,500 years ago in India with the teachings of Prince Siddhartha Guatama, the Buddha. **Adherents:** About 245 million, primarily in Asia. About 60,000 in the United States.	**Chief Scriptures:** The *Pali Canon* **Basic Tenets:** The way beyond sorrow and suffering is a middle path between striving and spiritual contemplation. Buddhists believe in the rebirth of persons through transmigration of the soul. Also, Buddhists abstain from killing or injuring any living being.	Strict Buddhists are vegetarians. They will not eat the flesh of any animal and many will not eat milk and eggs. Part of rite action includes a vow to abstain from intoxicating beverages.	Buddhist monks do not eat after midday, but they are allowed to drink tea and coconut milk. Every morning Buddhist monks make the rounds in the village with their beggar's bowls and receive the very best of the rice, curry, or fruit available. Abstaining from taking life limits the use of livestock, but encourages the growing of rice and the introduction of new food crops. Eating fish is not viewed as killing but simply removing the fish from water.
CHRISTIANITY **Origin:** Began with the birth of Jesus Christ about 2,000 years ago. **Adherents:** About 1 billion throughout the world, of which about 135 million are in the United States.	**Chief Scriptures:** The *Bible* (Old and New Testament) **Basic Tenets:** Jesus Christ—the Savior—was the Son of God. He suffered for man's sins. He was crucified and then resurrected. Jesus Christ will return. Teachings and life of Christ are contained in the New Testament of the Bible. Christ taught love of God and love of fellow men.	Overall, Christianity does not teach that any specific foods are taboo, but several of the major Christian religions prohibit some foods: • **Greek Orthodox**—On Fridays and Wednesdays, the use of meat, fish, poultry, eggs, and dairy products is restricted. These foods are also restricted during the first and last weeks of the Greek Orthodox Lent. Devout members may abstain from olive oil. • **Mormons (Latter-day Saints)**—The Mormons prohibit the use of alcoholic beverages, tea, coffee, and tobacco, and encourage the sparing use of meat. • **Roman Catholic**—Historically, the most notable dietary law in Christianity was the Roman Catholic ban on eating meat on Friday. During the reign of Pope John XXIII, this ban was lifted. • **Seventh-day Adventists**—Adventists are taught to avoid meat, fish, poultry, alcohol, tea, coffee, ripened cheese, excess sugar, and excess refined grains, as well as irritating spices and too much salt.	Several of the major Christian religions encourage certain types of foods or diets: • **Greek Orthodox**—Dietary laws have been interpreted more liberally in recent years. On fast days—days meat and dairy products are restricted—shell fish may be eaten. Koliva, a boiled whole wheat dish, is important in memorial services for the dead. • **Mormons (Latter-day Saints)**—The Mormon "Word of Wisdom" encourages the use of grains, fruits, and vegetables. • **Roman Catholic**—Local customs vary with regard to foods allowed on fast days and days of abstinence. • **Seventh-day Adventists**—In general, Adventists are ovolactovegetarians, and allow the use of eggs and milk. Furthermore, they encourage the use of nuts, legumes, and whole grains. The Adventists have encouraged the development of meat analogs such as the soy protein products.
CONFUCIANISM **Origin:** Developed around the philosophy of Confucius (K'ung Fu-tzu) who lived in China from 551 to 479 B.C. **Adherents:** About 276 million, primarily in China and Taiwan.	**Chief Scriptures:** The *Analects* (Conversations or sayings) **Basic Tenets:** Confucianism is primarily a group of ethics or proper conduct: kindness, righteousness, decorous behavior, wisdom, and uprightness. The chief ethic is benevolence. Confucianism encourages ancestor worship.	Confucianism does not seem to have any specific food taboos.	Confucianism does not seem to encourage the use of any certain foods, other than those that may be part of ancestor worship.
HINDUISM **Origin:** No founder, but developed gradually over 5,000 years. **Adherents:** Over 500 million in India and in Indian communities throughout the world.	**Chief Scriptures:** The *Vedas* which includes the *Rig Veda*, the *Upanishads*, the *Bhagavad-Gita*, and many other writings **Basic Tenets:** Hinduism does not have a common creed or one doctrine to bind Hindus together. Membership in an Indian caste is encouraged by Hinduism. Brahman is the unifying spirit, and Hindus may worship many gods. Hindus believe in reincarnation or rebirth. They will not injure or kill an animal.	Some Hindus reject pork because of the value they place on life and because of the filth of pigs. Hindus view the cow as sacred; hence, beef is not a food, though there are numerous cows in India. Both eggs and poultry are rejected by the Hindus. Strict Hindus will not eat the flesh of any animal. Strict Hindus also avoid onions, garlic, turnips, and mushrooms, and for some, the association of blood with the color of some lentils and tomatoes has made these vegetables unacceptable.	Strict Hindus are vegetarians. Ghee (clarified butter) and milk from the cow are sacred foods. The coconut is also considered sacred. To insure success, new enterprises may be started by breaking a coconut. Following a fast, which may be either a complete fast or at least abstaining from cooked foods, a Brahman—high priest of Hinduism—can be fed ghee, coconuts, fruits, or some uncooked food.
ISLAM **Origin:** Founded by Mohammed, who was born in Mecca (now Saudi Arabia) and who lived from about 570 to 632 A.D. **Adherents:** About 700 million worldwide, with about 1 million in the United States.	**Chief Scriptures:** The *Koran*, A compilation of revelations Mohammed received from Allah. **Basic Tenets:** The simple article of faith of Islam is that they worship one God (Allah) and Mohammed was his messenger. Moslems believe that good deeds will be rewarded in paradise and evil deeds will be punished in hell. Duties of Islam include: 5 daily prayers, observance of Ramadan, giving of alms, and a pilgrimage to Mecca at least once in a lifetime.	Eating pork and drinking intoxicating beverages is forbidden. The observance of the month of Ramadan involves fasting during the daylight hours. With the exception of fish and locusts, no animal food is considered lawful unless it has been slaughtered according to the proper ritual. At the instant of slaughter, the person killing the animal must repeat, "In the name of God, God is great."	During the month of Ramadan, nights are often spent in feasting, possibly beginning with an appetizer like dates and a sherbetlike drink. Then following Ramadan there may be a feast lasting up to 3 days. This is called Eid-al-Fitr, where chicken or veal sauteed with eggplant and onions and simmered in pomegranate juice and spices may be served. But the highlight is a whole lamb stuffed with a rich dressing of dried fruits, cracked wheat, pine nuts, almonds, onions, and spices. The feast is concluded with rich pastries and candies, some of which are taken home and used over a period of time as reminders of the festival.

(Continued)

TABLE R-3 (Continued)

Religion; Origin; Adherents	Chief Scripture(s); Basic Tenets	Food Taboos	Foods Encouraged by Beliefs
JAINISM **Origin:** Founded in the 6th century B.C. in India by Prince Vardhamana who later became known as Mahavira or "the Great Hero." **Adherents:** About 2¼ million, mostly in the western part of India.	**Chief Scriptures:** None **Basic Tenets:** Jainism stresses nonviolence to all living creatures, a doctrine called *ahismsa*. Jains also believe in reincarnation or transmigration of the soul. Jainism does not recognize a supreme diety.	Since Jainism stresses nonviolence to all living creatures, any food that requires the taking of life is forbidden. Also, Jains refrain from eating certain fruits, honey, wine, and root vegetables which carry organisms on their skin.	Jains are vegetarians. A Jain monk carries a broom representing the sacredness of a life and a bowl for begging. The broom is used to sweep insects from his path so he will not step on them, and the bowl for begging represents the asceticism or self denial.
JUDAISM **Origin:** From the time of Abraham in about 1800 B.C. **Adherents:** Worldwide, about 15 million; in the United States about 6 million.	**Chief Scriptures:** The *Torah,* containing the first 5 books of the *Old Testament,* and the *Talmud.* **Basic Tenets:** The Jews believe in one God. Their life is guided by the commandments contained in the *Torah.* Judaism teaches that man is responsible for his actions and that he has a choice between right and wrong. The Jewish Sabbath is from sunset Friday to sunset Saturday. Judaism is a way of life, which assumes that life is significant, has value, and that the individual is important. Jews look forward to the coming of the Messiah—Savior.	Eating unclean animals is forbidden. The *Torah* provides the distinction between animals considered as clean and those which are unclean (see Leviticus Chap. 11 and Deuteronomy Chap. 14). Unclean foods include pork, camel, horse, most winged insects, reptiles, creeping animals such as the mouse, and birds of prey. All shellfish and eels are eliminated from the diet. Blood is taboo for human consumption, as is the internal fat from an animal. Meat and dairy foods can't be eaten in the same meal. Trefah is the term used to indicate an unclean food or a food not correctly prepared.	Clean foods which are permitted to be eaten include cow, sheep and goat, and fish (see Leviticus Chapter 11 and Deuteronomy Chapter 14). Dietary laws are further expanded in the *Talmud* which contains the Laws of Kashrut. Kosher foods are those which are permitted according to the *Torah (Bible)* and foods which have been processed in a prescribed manner. A rabbi must supervise the killing of all animals which is performed by a trained person called a shohet. A swift deep slash at the throat renders the animal unconscious and allows the blood to drain from the animal. Then truly kosher meat is soaked in cold water, drained, sprinkled with salt, washed in cold water, and cooked.

The *Talmud* forbids Jews to eat certain sinews of the hindquarters which presents some problems since the sinews are difficult and costly to remove. Many of the processed foods contain the symbols U or K on the label which refers to the Union of Orthodox Jewish Congregations and O.K. Laboratories, respectively. These symbols indicate that the food is kosher.

Food plays a major role in the symbolism of the Jewish holidays and festivals. Familiar foods during these times include honey, honey cake, carrot tzimmes, holishkes, strudel, potato latkes, potato krugel, St. John's Bread, fruit, nuts, raisins, pastry, wine, cheese blintzes, cheese kreplach, egg and unleavened wheat bread (matzo).

(Also see MEAT, section headed, "Kosher Meats.")

Religion; Origin; Adherents	Chief Scripture(s); Basic Tenets	Food Taboos	Foods Encouraged by Beliefs
SHINTO **Origin:** Originated in ancient times with the beginnings of the Japanese culture. **Adherents:** About 63 million Japanese.	**Chief Scriptures:** None **Basic Tenets:** The word *shin-to* or in Japanese Kami-no-michi means "the way of the gods." The gods and goddesses that are worshipped are the forces of nature which may reside in rivers, trees, rocks, mountains, certain animals, and particularly in the sun and moon. Today, Shinto sects stress world peace and brotherhood but no universal prophetic message.	No foods seem to be forbidden by the Shinto religion.	The only foods that may be considered as encouraged are those associated with ceremonies dealing with abundant harvest or good health. Gifts of cakes are offered at some public shrines. In traditional villages, the diet tends to be largely vegetarian with an occasional fish or chicken. These food habits may be the influence of Shinto and/or other religions which entered Japan over the years.
SIKHISM **Origin:** Founded by Guru Nanak about 1500 A.D. **Adherents:** About a million worldwide, primarily in India. Sikhs are easily identified by their turbans.	**Chief Scriptures:** *Granth Sahib* **Basic Tenets:** Sikhism combines the beliefs of Islam and Hinduism. Nanak taught that there is one God, and opposed the caste system, uniting his followers into one class.	Sikhs vow not to smoke or to drink alcoholic beverages.	No foods seem to be encouraged by the Sikh beliefs. There are certainly some traditional foods which may in part reflect the Hindu and Moslem influence.
TAOISM **Origin:** Founded about 2,600 years ago by Lao-tzu. **Adherents:** About 30 million worldwide. In China a person may be a Confucian, a Taoist, and a Buddhist all at the same time.	**Chief Scriptures:** *Tao te ching* **Basic Tenets:** In Chinese, the word "tao" means way or path. Taoism stresses quiet contemplation, and the elimination of all striving and strong passions which allows a man to live in harmony with the principles that underlie and govern the universe. Taoism divided all reality into male and female principles of *yang* and *yin,* respectively. Taoism developed beliefs of an afterlife with a heaven and hell.	No specific food taboos seem to be taught.	Yang and yin philosophy may have some influence in the food choices, as may the philosophy of living close to nature. Some Taoist groups have included special diets as part of their search for immortality. Early Taoists practiced alchemy—mixing elixirs designed to ensure immortality.
ZOROASTRIANISM **Origin:** Founded in the 6th century B.C. by Zoroaster, a religious teacher and prophet of Persia (now Iran). **Adherents:** Only about 200,000, primarily in limited areas of Asia.	**Chief Scriptures:** The *Avesta* or *Zend-Avesta* **Basic Tenets:** Zoroaster taught monotheism—the belief in one god. Also, this religion teaches that the world is ruled by two forces—good and evil. A lie is a great evil in Zoroastrianism.	No specific foods are limited by the teachings of Zoroastrianism.	Zoroastrians have a ritual drink called hoama. There are traditional foods which are associated with religious festivals. Zoroaster encouraged people to till the soil, raise grain, grow fruits, irrigate and weed crops, reclaim wasteland, and to treat animals kindly.

RELISH, CHUTNEY, PICCALILLI

All three of these words mean essentially the same thing: a mixture of chopped vegetables, with a blend of spices and vinegar.

REMISSION

An interval in the course of a disease in which the symptoms subside or abate.

RENAL GLUCOSURIA

A term which implies the excretion of glucose (sugar) in the urine while blood glucose remains within the limits of normal concentrations. It is caused by a disorder in kidney (renal) function and is sometimes called renal diabetes. Glucosuria or the excretion of sugar in the urine is often, but not always, associated with diabetes mellitus. However, in diabetes mellitus the blood sugar is elevated far beyond normal concentrations.

(Also see INBORN ERRORS OF CARBOHYDRATE METABOLISM, Table I-2.)

RENAL THRESHOLD

The level of concentration of a substance in the blood beyond which it is excreted in the urine. For example, the renal threshold of glucose is about 180 mg per 100 ml; diabetics excrete glucose because this level is exceeded.

RENDERING

• The process of liberating the fat from the fat cells, as in the production of lard and tallow.

• The processing of inedible animals and meats into livestock feeds, industrial fats and oils, and fertilizers.

RENNET

An extract of the stomach of certain mammals, which contains the enzyme rennin. It is used in making most cheeses and junkets.

RENNET, VEGETABLE

The name given to proteolytic enzymes derived from plants, such as bromelin (from the pineapple) and ficin (from the fig).

RENNIN (CHYMOSIN)

The milk-coagulating enzyme found in the gastric juice of the fourth stomach compartment of certain mammals.

REPLETION

To fill up; the act of overeating or the state of being overfed.

RESECTION

Removal of part of an organ.

RESERPINE

An active alkaloid that is extracted from the root of shrubs of the genus *Rauwolfia*, and used as a tranquilizer or sedative. It is also used as an antihypertensive drug, in the treatment of high blood pressure, various mental diseases, and tension.

RESIDUE

Left over, remaining; dietary residue refers to the content remaining in the intestinal tract after digestion of food—including fiber and other unabsorbed products.

RESORPTION

A loss of substance; for example, the withdrawal of calcium from bone.

RESPIRATION (BREATHING)

Commonly used to mean the taking in of oxygen and the throwing off of carbon dioxide and water vapor through the lungs, but it also includes the metabolism of the cells.

RESPIRATORY CHAIN (ELECTRON TRANSPORT SYSTEM)

The series of chemical reactions in the oxidation system of the cell that transfer hydrogen ions or electrons to produce ATP (high-energy phosphate compounds.)

(Also see METABOLISM, section on "Catabolism, • ATP—energy currency.")

RESPIRATORY SYSTEM

The body's apparatus for inhaling oxygen and exhaling carbon dioxide and water vapor; it consists mainly of the lungs and bronchial tubes, and its movements are activated by the midriff and the muscles between the ribs.

RESTAURANT

The word *restaurant* is a French word meaning *restorative*. It was originally applied to establishments that served soups. The soups, with their variety of ingredients, were thought to be restorative of one's health. In 16th-century France, one enterprising person advertised his soup establishment as a *restaurant*. Soon others copied him, and the name eventually became popular around the world.

In America, changing life-styles have contributed to the popularity of eating out, whether it be at a restaurant serving haute cuisine, or a fast-food chain. According to the National Restaurant Association, the average American family spent 39.5% of its food dollar in restaurants in 1983.

RETICULOCYTE

A young red blood cell.

RETICULOENDOTHELIAL SYSTEM

Groups of cells, except leukocytes, with phagocytic properties, functioning to rid the body of debris.

RETINAL

Retinal (or retinene) is the aldehyde form of retinol (formerly called vitamin A); it is one of the three biologically active forms of retinol (retinol is also active as an alcohol and as an acid).

(Also see VITAMIN[S], Table V-5; and VITAMIN A.)

RETINOIC ACID

Retinol (formerly called vitamin A) is biologically active as an alcohol, as an aldehyde, and as an acid. The acid form is usually referred to as retinoic acid. Retinoic acid can partially replace retinol in the rat diet; it promotes growth of bone and soft tissues and sperm production, but it cannot be used in the visual process and will not permit maturation of embryos. Retinoic acid is converted by the rat to an unknown form that is several times as active as the parent compound in the usual vitamin A nutritional assays.

(Also see VITAMIN[S], Table V-5; and VITAMIN A.)

RETINOL

One of the two forms of vitamin A as such; the other form is dehydroretinol. Retinol (formerly called vitamin A), which is found as an ester (retinyl palmitate) in ocean fish oils and fats, and in liver, butterfat, and egg yolk, is biologically active as an alcohol, as an aldehyde, and as an acid. The alcohol, the most common form, is usually referred to as retinol, the aldehyde as retinal or retinene, and the acid as retinoic acid.

(Also see VITAMIN[S], Table V-5; and VITAMIN A.)

RETINOPATHY

Degenerative disease of the retina.

RETROGRADATION

Starch is insoluble in cold water, but in a heated suspension the granules swell and form a paste or gel. However, upon aging or freezing, the amylose portion of starch aggregates, forming an insoluble substance and reducing the water-holding ability of the starch. This process, which is termed retrogradation, is undesirable in foods. In the food industry, monoglycerides are added to starch to reduce the tendency of retrogradation.

RHAMNOSE

A hexose monosaccharide found combined in the form of plant glycosides. It occurs in free form in poison sumac.

RHEUMATIC FEVER

A childhood disease that is always preceded by an infection of *Streptococcus* bacteria or scarlet fever. It occurs in acute and chronic forms, and it is one of the leading causes of chronic illness in children. Rheumatic fever belongs to a class of diseases termed collagen diseases.

(Also see MODIFIED DIETS.)

RHODOPSIN (VISUAL PURPLE)

The pigment in the rods of the retina that contains vitamin A. When light strikes the normal retina, this pigment is bleached to another pigment known as retinaldehyde (visual yellow). As a result of this change, images are transmitted to the brain through the optic nerve. Vitamin A is required for regeneration of rhodopsin.

(Also see VITAMIN A, section headed "Functions • **Vision.**")

RHUBARB (PIEPLANT) *Rheum rhaponticum*

Rhubarb is believed to be a native of Asia minor. It is a perennial plant which forms large fleshy rhizomes and large leaves with long, thick petioles. It is one of the few vegetables in which the petiole is the part consumed. The leaf portion contains a toxic substance. Rhubarb's popularity in preserves, pies, and desserts causes many people to think of it as a fruit; however, it is a vegetable.

RIBOFLAVIN (B-2)

Contents	Page
History	924
Chemistry, Metabolism, Properties	924
Measurement/Assay	924
Functions	924
Deficiency Symptoms	925
Recommended Daily Allowance of Riboflavin	925
Toxicity	926
Riboflavin Losses During Processing, Cooking, and Storage	926
Sources of Riboflavin	926

RIBOFLAVIN
PROMOTES HEALTH BY HELPING BODY CELLS USE OXYGEN TO RELEASE ENERGY FROM AMINO ACIDS, FATTY ACIDS, AND CARBOHYDRATES.

RIBOFLAVIN MADE THE DIFFERENCE! SAME RAT BEFORE (LEFT) AND AFTER (RIGHT). LEFT: THIS RAT 28 WEEKS OLD, HAD NO RIBOFLAVIN. IT SOON BECAME SICK, AND LOST HAIR, ESPECIALLY ABOUT THE HEAD. IT WEIGHED ONLY 63 g. RIGHT: THE SAME RAT 6 WEEKS LATER. AFTER RECEIVING FOOD RICH IN RIBOFLAVIN. IT RECOVERED ITS FINE FUR AND WEIGHED 169 g. (COURTESY, USDA.)

TOP FOOD SOURCES

Fig. R-3. The riboflavin story.

Riboflavin is present in virtually all living cells; and, like niacin, it has an essential role in the oxidative mechanisms in the cells.

As early as 1879, the existence of a yellow-green fluorescent pigment in milk whey was recognized. Subsequently, other workers found this pigment in such widely varying sources as liver, heart, and egg white. This pigment, which possessed fluorescent properties, was called *flavin*. But, at the time, the biological significance of the pigment was not understood.

By 1928, it became evident that what had been called vitamin B was not a single vitamin. Numerous investigators found that a growth-promoting substance remained after heat had destroyed the beriberi-preventive factor (thiamin, or vitamin B-1) in yeast. This unknown substance was called *vitamin G* by U.S. research workers and *vitamin B-2* by British scientists. At the time, it was thought to be only one vitamin; later, it was found that the heat-stable fraction was composed of several vitamins.

HISTORY. The first serious attempts to isolate the long-known and widely-distributed fluorescent pigment was undertaken by workers in Germany and Switzerland in the early 1930s. In 1932, two German scientists, Warburg and Christian, isolated the *yellow enzyme*, part of which was later identified as flavin mononucleotide (FMN)—riboflavin phosphate. In the following year (1933), Kuhn, working at the University of Heidelberg, isolated pure riboflavin from milk; and, in 1935, he elucidated the structure and synthesized the vitamin. Independently, Swiss researchers, Karrer and co-workers, accomplished the same feat that year (in 1935). Karrer[2] named it riboflavin, because it was found to have a pentose side chain—ribitol (similar to the sugar, ribose)—attached to a flavinlike compound; and, in 1952, the name was adopted by the Commission on Biochemical Nomenclature.

Riboflavin is widely distributed in both plant and animal tissues. It is formed by all higher plants, chiefly in the green leaves. Also, the bacteria in the intestinal tract may be a considerable, but variable, source of it for man and other animals, just as with thiamin. So, higher animals must rely on food for their riboflavin.

CHEMISTRY, METABOLISM, PROPERTIES.

• **Chemistry**—Chemically, riboflavin is composed of an

Fig. R-4. Structure of riboflavin.

---

[2]Karrer, who died in 1971, was also the first to synthesize carotene. In 1937, he was awarded the Nobel Prize for his important discoveries in nutrition.

alloxine ring linked to an alcohol derived from the pentose sugar ribose (see Fig. R-4)

• **Metabolism**—Riboflavin is absorbed in the upper part of the small intestine by passive diffusion, which controls the amount of the vitamin taken up by the cells of the intestinal mucosa. It is phosphorylated in the intestinal wall and carried by the blood to the tissue where it may occur as the phosphate or as a flavoprotein.

The body has limited capacity for storing riboflavin, although higher concentrations are found in the liver and kidneys than in other tissues. So, day-to-day tissue needs must be supplied by the diet. Excretion is primarily via the urine, with the amount excreted related to uptake. When the intake is high, urinary excretion is high; when the intake is low, excretion is low. Some riboflavin is excreted in the feces. All mammals secrete riboflavin in their milk.

• **Properties**—In pure form, riboflavin exists as fine orange-yellow crystals, which are bitter tasting and practically odorless. In water solutions, it imparts a greenish-yellow fluorescence. It is sparingly soluble in water. (It is much less soluble in water than thiamin.) It is heat-stable in neutral or acid solutions, but it may be destroyed by heating in alkaline solutions. It is easily destroyed by light, especially ultraviolet light; for example, it may be destroyed by sunlight striking milk kept in glass bottles. (Milk in cartons is protected against such losses.) Synthetic riboflavin should always be kept in dark bottles. Because of its heat stability and limited water solubility, very little riboflavin is lost in processing and cooking foods.

MEASUREMENT/ASSAY. Riboflavin is measured in terms of the metric weight of pure riboflavin; human requirements are expressed in milligrams, and food content in milligrams or micrograms.

Although the growth of rats and chicks may occasionally be used to assay riboflavin in mixed diets, the biologic method of assay has been generally superseded by microbiological and chemical methods.

The assessment of riboflavin nutriture in man is determined by urinary excretion or by blood analysis.

FUNCTIONS. Riboflavin functions as part of a group of enzymes called flavoproteins. *Flavin mononucleotide (FMN)* and *flavin adenine dinucleotide (FAD)* operate at vital reaction points in the respiratory chains of cellular metabolism. The structure of these two compounds is shown in Fig. R-5.

FMN and FAD function as coenzymes in a number of different flavoprotein systems. They play a major role with thiamin- and niacin-containing enzymes in a long chain of oxidation-reduction reactions by which energy is released. In the process, hydrogen is transferred from one compound to another until it finally combines with oxygen to form water. Thus, riboflavin functions in the metabolism of amino acids, fatty acids, and carbohydrates. During this process, energy is released gradually and made available to the cell. In addition, riboflavin, through its role in activating pyridoxine (vitamin B-6), is necessary for the formation of niacin from the amino acid tryptophan. Also, riboflavin is thought to be (1) a component of the retinal pigment of the eye; (2) involved in the functioning of the adrenal gland; and (3) required for the production of corticosteroids in the adrenal cortex.

FLAVIN MONONUCLEOTIDE (FMN)

FLAVIN ADENINE DINUCLEOTIDE (FAD)

Fig. R-5. The structure of FMN and FAD.

DEFICIENCY SYMPTOMS. Unlike some of the other vitamins, riboflavin deficiency does not cause any serious disease in human beings; nevertheless, clinical signs associated with riboflavin deficiencies are found among persons of all ages in developing countries.

Manifestations of riboflavin deficiency center around the following symptoms:

1. **Fatigue.** The most important of the nonspecific symptoms is fatigue and inability to work.

2. **Wound healing.** Even minor wounds become aggravated and do not heal easily.

3. **Mouth.** Cheilosis (sores around the mouth) develops. The lips become swollen, reddened, and chapped, and characteristic cracks develop at the corners of the mouth (angular stoma).

4. **Nose.** Cracks and irritation develop at nasal angles.

5. **Tongue.** Glossitis (inflammation of the tongue) develops. The tongue becomes swollen, reddened, fissured, and painful.

6. **Eyes.** Extra blood vessels develop in the cornea (the eyes become bloodshot—a condition known as corneal vascularization), and the eyes become sensitive to light and easily fatigued. There is also blurring of the vision, and the eyes burn, itch, and tear. Cataracts have been observed in rats, mice, chickens, pigs, and monkeys after prolonged deficiency of riboflavin.

7. **Skin.** Seborrheic dermatitis (oily crusts and scales) may develop. The skin may become scaly, and greasy eruptions may develop—especially in the skin folds, around the nose, and on the scrotum in males.

Fig. R-6. Riboflavin deficiency in a young chick. Note the curled toes and the tendency to squat on the hocks; this condition known as *curled toe paralysis*, is caused by degenerated nerves. (Courtesy, Department of Poultry Science, Cornell University)

8. **Anemia.** Anemia has been produced experimentally in human subjects given galactoflavin, a riboflavin antagonist. The anemia, along with other symptoms, responded rapidly to riboflavin therapy.

9. **Pregnancy.** Deficiency of riboflavin during pregnancy leads to skeletal abnormalities of the fetus, including shortened bones; deformed growth between ribs, toes, and fingers; and short fingers with fewer joints.

All of the above conditions are common, especially in children in the developing countries where the supply of meat, milk, and eggs is poor. The same conditions are sometimes seen among the elderly in well-fed countries.

Also, it is noteworthy that riboflavin deficiencies seldom occur alone. Instead, they are apt to be found along with deficiency diseases of the other members of the B complex.

RECOMMENDED DAILY ALLOWANCE OF RIBOFLAVIN. Prior to 1980, the Recommended Dietary Allowances of the Food and Nutrition Board (FNB), National Research Council–National Academy of Sciences, related the allowances for riboflavin to (1) protein allowances (1958), (2) energy intake (1964), and (3) metabolic body size (1968). In the ninth edition, 1980, the FNB concluded that the information available does not support strongly any one of these over another. In the tenth edition, 1989, the FNB listed the following factors as being known to affect the riboflavin requirements: nitrogen balance, energy (work) expenditure, and pregnancy–lactation.

The current FNB Recommended Daily Allowances are given in the section on VITAMIN(S), Table V-5, Vitamin Table.

• **Adults and children**—The riboflavin allowances in Table V-5 have been computed on the basis of 0.6 mg/1,000 Calories (kcal) for people of all ages. However, for elderly people and others whose calorie intake may be less than 2,000 Calories (kcal), a minimum intake of 1.2 mg/day is recommended.

• **Pregnancy**—The recommended allowances call for an additional 0.3 mg/day of riboflavin during pregnancy.

• **Lactation**—An additional daily intake of approximately 0.5 mg is recommended during lactation.

• **Riboflavin intake in average U.S. diet**—According to the U.S. Department of Agriculture, the riboflavin available for civilian consumption in the United States was 2.4 mg per person per day. Further, 79.5% of the riboflavin in the U.S. diet is supplied by three food groups: (1) meat, fish, and poultry; (2) milk and dairy products; and (3) flour and cereal products.

A U.S. enrichment program for white flour (with standards for riboflavin, thiamin, niacin, and iron) was initiated in 1941. Subsequently, several other products have been designated for enrichment, and the levels of added nutrients have been changed.

Currently, the riboflavin supplementation of flour and cereal products from synthetic and/or fermentation sources contributes an average of 0.33 mg of riboflavin per person per day, which is about one-fifth of the requirement of an adult male—a very significant amount. However, not all states require an enrichment program; hence, where products are not involved in interstate commerce, they need not be enriched. Besides, average consumption figures do nothing for those persons whose consumptions are below the recommended allowances, or who have higher than normal requirements. Chronic or borderline riboflavin deficiencies are most likely to occur to a variable extent in persons with inadequate intake of animal proteins (especially among those not consuming adequate milk) and enriched foods, in alcoholics, in the aged, in women taking oral contraceptives or who are pregnant or lactating. Also, it is recognized that riboflavin deficiencies, usually along with deficiencies of vitamin A, folacin, calcium, iron, and sometimes vitamin D, are commonplace in many other countries.

TOXICITY. There is no known toxicity of riboflavin.

RIBOFLAVIN LOSSES DURING PROCESSING, COOKING, AND STORAGE. The two properties of riboflavin that may account for major losses are (1) that it is destroyed by light, and (2) that it is destroyed by heat in alkaline solution. The following facts relative to riboflavin losses during processing, cooking, and storage are noteworthy:

• **Milk**—As much as 20% of the riboflavin content of fluid milk is destroyed in pasteurization, evaporation, or drying. Also, one-half or more of the riboflavin in milk may be lost in 2 hours if it is exposed to light; as happened prior to 1960 when clear glass bottles of milk were delivered on the doorsteps of U.S. homes—a practice still common in many countries. The use of opaque cartons or dark bottles for the distribution of milk markedly reduces the loss of riboflavin.

• **Blanching**—In the blanching processes used prior to canning or freezing certain foods, losses of 5 to 20% of riboflavin occur.

• **Drying**—Dehydration and freeze-drying processes have little effect on the vitamin content of foods. However, the practice of sun-drying of foods, such as fish and vegetables in tropical countries, results in considerable destruction of riboflavin.

• **Cooking**—Because riboflavin is heat stable and only slightly soluble in water, little of it is lost in home cooking or commercial canning. Average losses of riboflavin in cooking are 15 to 20% in meats, 10 to 20% in vegetables, and 10% in baking bread.

• **Alkali**—Riboflavin may be destroyed by heating in alkaline solutions; hence, sodium bicarbonate should not be used in the cooking of vegetables.

SOURCES OF RIBOFLAVIN. Riboflavin is widely distributed in most foods. However, there is wide variation in levels, due primarily to source, harvesting, processing, enrichment, and storage. For example, whole cereal grains contain useful amounts, but much of this is removed in milling; hence, highly milled cereals and breads contain very little riboflavin unless they are enriched. Riboflavin is the only vitamin present in significant amounts in beer; beer drinkers may find consolation in knowing that 1 liter (1.06 qt) of beer daily almost meets the recommended intake. Green leafy vegetables vary greatly, but some are fair sources of riboflavin. Fruits, roots and tubers, with few exceptions, contain relatively small amounts of riboflavin. Pure sugars and fats are entirely lacking in riboflavin.

Fig. R-7. Cheese fondue, cheese is a good source of riboflavin. (Courtesy, USDA)

A grouping and ranking of foods according to normal riboflavin content is given in the section on VITAMIN(S), Table V-5, Vitamin Table.

The average person is not apt to get an optimum amount of riboflavin unless he/she consumes a generous amount of milk. Each quart (950 ml) of milk contains 1.66 mg of riboflavin, which just meets the recommended daily allowance of adult males. Of course, the addition of riboflavin in the enrichment of flour, bread, and certain other products has helped to raise the average intake.

For additional sources and more precise values of riboflavin (B-2), see Food Composition Table F-21 of this book.

(Also see VITAMINS, Table V-5.)

RIBONUCLEIC ACID (RNA)

Molecules in cell cytoplasm which serve for the transfer of the amino acid code from nuclei and the synthesis of protein.

(Also see METABOLISM; and NUCLEIC ACIDS.)

RIBOSE

A 5-carbon sugar—a pentose—synthesized by the body in all animals, including man. Hence, it is not essential in the diet, but in the body ribose plays an important role. When it is joined with pyrimidines—cytosine, thymine, and uracil; and purines—adenine and guanine—nucleosides are formed. When phosphoric acid is esterified with the nucleosides, nucleotides are formed. These compounds are then used in the formation of ribonucleic acid (RNA) and deoxyribonucleic acid (DNA). The nucleotides of adenosine monophosphate (AMP), adenosine diphosphate (ADP), and adenosine triphosphate (ATP) are compounds that are essential to cellular metabolism. Ribose is also a constituent of the vitamin riboflavin.

(Also see CARBOHYDRATE[S]; and METABOLISM.)

RIBOSOMES

The site of protein synthesis in the cell.
(Also see METABOLISM.)

RICE *Oryza sativa*

Contents   Page
Origin and History................................927
World and U.S. Production.........................927
Processing Rice...................................928
Nutritional Value................................929
Nutritional Losses...............................929
Enrichment.......................................930
Protein Quality..................................930
Better Rice for Millions.........................930
Rice Products and Uses...........................931

Fig. R-8. Rice. Yields of flooded rice are higher than those of upland rice. (Courtesy, USDA)

Rice is an annual grass of the family *Gramineae*, genus *Oryza*. Species *O. sativa* is the kind most commonly cultivated for food.

Rice is one of the most important cereal grains of the world. It provides most of the food for over half the human population of the earth, most of them living in the Orient. In some countries of the Orient, the per capita consumption of rice is 200 to 400 lb (*90 to 180 kg*) per year; by contrast, the yearly per capita consumption of rice in the United States is less than 8 lb (*3.6 kg*). In several national languages and local dialects, the word *eat* means *to eat rice*. American children associate rice with chopsticks.

Fig. R-9. Japanese children eating rice, precious grain of the Orient. (Courtesy, *Sankei Kawashima*—newspaper, Tokyo, Japan.)

Rice is the only major cereal that is largely consumed by man directly as harvested (after hulling, and usually polishing). Wheat is usually milled to flour, then baked; and corn (maize) is largely fed to animals for the production of meat, milk, and eggs.

(Also see CEREAL GRAINS.)

ORIGIN AND HISTORY. Rice originated from wild species indigenous to southeastern Asia, where wild types still persist, although there is some evidence of an African center of domestication. Primitive man collected wild rice. Eventually, he cultivated it and selected mutant types with larger nonshattering grains. Hybridization with various wild species followed, contributing to great variability in the plant.

WORLD AND U.S. PRODUCTION. Rice is the third ranking cereal grain of the world. It is a cheap and abundant crop, easy to grow on little land, and well adapted to the climatic conditions of the East Indies and the Asian Coast from India to Japan. Asia produces 92.3% of the world's rice crop.

World production of rice totals about 518.5 million metric tons annually; and the ten leading rice-producing countries of the world are: China, India, Indonesia, Bangladesh, Thailand, Vietnam, Myanmar, Japan, Philippines, and Korea.[3]

The United States, which produces less than 1% of the world crop, usually accounts for a third of the world rice trade. Over half of the U.S. crop of about 7.7 million short tons annually is exported, largely to the Asian countries.

---

[3]Data from *FAO Production Yearbook 1990*, FAO/UN, Rome, Italy, Vol. 44, p. 72, Table 17. **Note well:** Annual production fluctuates as a result of weather and profitability of the crop.

The leading rice-producing states of the United States, by rank, are: Arkansas, California, Louisiana, Texas, Mississippi, and Missouri.[4]

PROCESSING RICE.   The main parts of the rice grain are the hull (husk, lemma), the seed coat (pericarp), the embryo (germ), and the endosperm (starchy endosperm).

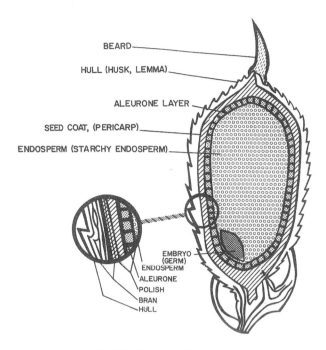

BEARD
HULL (HUSK, LEMMA)
ALEURONE LAYER
SEED COAT, (PERICARP)
ENDOSPERM (STARCHY ENDOSPERM)

EMBRYO (GERM)
ENDOSPERM
ALEURONE
POLISH
BRAN
HULL

Fig. R-10. Structure of the rice grain.

The seed coat consists of six layers of differentiated types of cells, the innermost one of which is the aleurone layer.

Table R-4 shows that, together, the pericarp, aleurone, and scutellum (which is part of the germ) contain 79% of the

---

[4]Data from *Agricultural Statistics 1991*, USDA, p. 22, Table 28. **Note well:** Annual production fluctuates as a result of weather and profitability of the crop.

TABLE R-4
THIAMIN CONTENT OF FRACTIONS OF THE RICE GRAIN[1]

Part of Grain	Proportion of Grain	Thiamin Content	Proportion of the Total Thiamin of the Grain
	(%)	(mcg)	(%)
Pericarp and aleurone ...	5.95	31	} 35.2
Covering to germ	.20	12	
Epiblast	.27	78	3.9
Coleorhiza	.20	94	3.5
Plumule	.31	46	2.7
Radicle	.17	62	2.0
Scutellum	1.25	189	43.9
Outer endosperm	18.80	1.3	} 8.8
Inner endosperm	73.10	.3	

[1]Davidson, Sir Stanley, *et al., Human Nutrition and Dietetics*, 7th ed., The Williams and Wilkins Company, Baltimore, Md., 1979, p. 173, Table 17.5.

---

total thiamin present in the grain, although constituting only 6.2% of the weight. By contrast, the endosperm, which represents 92% of the grain by weight, contains only 8.8% of the thiamin.

• **Rice milling**—The purpose of milling rice is to separate the outer portions from the inner endosperm with a minimum of breakage. In modern mills, the rough rice passes through several processes in the mill: cleaning (usually parboiling), hulling, pearling, polishing, and grading.

Dramatic changes of immense health importance occur when rice is milled, the most important of which is the loss of much of the thiamin. Fig. R-11 shows the thiamin content of rice at different stages of milling. As noted, highly milled rice is almost devoid of thiamin. This loss has been responsible for much beriberi among people whose diet consists almost entirely of white rice.

HUSKED ONLY:	4.0 MCG/G
ONCE POLISHED:	1.8 MCG/G
TWICE POLISHED:	1.0 MCG/G
THRICE POLISHED & READY FOR MARKET:	0.7 MCG/G

Fig. R-11. Thiamin content of rice at different stages of milling.

In the United States, most rice undergoes a polishing process, during which the outer, colored, protein-rich parts (12%) become livestock feed, and the starchy white endosperm alone is left for human consumption. Polished rice is approximately 92% carbohydrates and has only 2% additional materials of nutritional value.

Dried rice that still has its hull is referred to as *rough* or *paddy* rice. Following the removal of foreign material by means of vibrating sieves and air currents, the rough rice passes into an awning machine where the awns are removed from the grain. The grain then passes through a series of shellers, and the hulled grain is separated from the unhulled in the paddy machine—a complex sifter. At this stage, the hulled rice, which represents approximately 79% of the weight of rough rice, is referred to as *brown rice*.

In the next milling step, the brown rice is conveyed to the hullers, which scour off the outer bran coats and germ from the rice kernels. The term *hullers* is a misnomer because these machines remove the bran, not hulls, from the rice kernel. Loosened bran and smaller pieces of kernel pass through the huller screen and are later separated by aspiration and screening. The rice bran is sold as a by-product for livestock feed.

Thence, the undermilled rice is passed through the brushing machine (polisher) which removes most of the inner bran coat or polish, producing a by-product known as rice polishings; which is sold for animal feeds, and sometimes as an ingredient of baby foods. The rice resulting from this operation is termed *polished rice*, which consists of the white, starchy endosperm, together with fragments of the aleurone layer.

Rice may be sold either as (1) polished or uncoated rice, or (2) coated with talc and glucose, an inert, harmless coating used to give it a high gloss or sheen, which is desired by the Puerto Rican market.

Throughout the entire milling process, a number of kernels are unavoidably broken. So, a series of machines classify the different size kernels as head rice (whole and ¾ kernels), second heads (¾ to ⅓ size grains), screenings (⅓ to ¼ length grains), or brewers' rice (under ¼ length fragments).

In the United States, there are the following grades of rice: U.S. No. 1, U.S. No. 2, U.S. No. 3, U.S. No. 4, U.S. No. 5, U.S. No. 6, and U.S. Sample grade. The grades are determined by size of kernels, number of heat-damaged kernels and objectionable seeds, percent of red rice and damaged kernels, percent of chalky kernels, percent of broken kernels, and color.

• **Parboiled rice**—In southern Asia, rice is parboiled before milling. Approximately 20% of the world's rice is treated in this manner. Essentially, the process consists of soaking dry rough or paddy rice (in the hull) in warm water, steaming under pressure, and drying before milling.

In the United States, there are several patented methods of parboiling. When parboiled, the rice retains more minerals and vitamins (some of the minerals and vitamins present in the bran and embryo move into the grain proper), breaks less in subsequent processing, and stores better.

After parboiling, rice is milled in the conventional manner. It requires a longer cooking period (25 min) than regular rice (20 min). Because parboiling eliminates the surface starch common to regular rice, it ensures a separateness of grain that's especially desirable for kitchens preparing rice in large quantities.

NUTRITIONAL VALUE.
Rice without hulls contains about 80% starch, 8% protein, and 12% water.

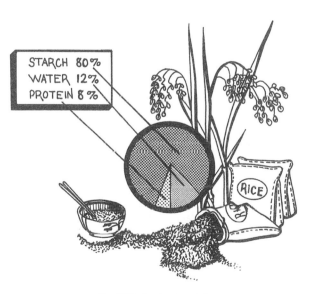

Fig. R-12. Food value of rice.

Because rice is such a predominant food in the diets of Oriental people, its nutrient composition is relatively more important than the nutritional composition of wheat in Western countries. In most wheat-eating countries, the cereals contribute only 30 to 40% of the average food-caloric consumption, whereas in the rice-eating nations some 60 to 80% of the calories come from rice. Thus, at best, the Oriental diet includes only 20 to 40% vegetables, meats, fish, fruits, and all other foods that could furnish the missing vitamins and minerals. Under such dietary conditions, it is easy to understand how the nutrient composition of rice determines the health of those who subsist largely on it.

In its natural state, rice has good nutritional values, comparing favorably with those of the other major cereals used as food staples around the world. It is better than corn and approximately as good as wheat. Brown rice—rice freed only of its chafflike hulls—has about the same caloric content, vitamins, and minerals as whole wheat; somewhat less proteins, but better quality proteins; and more fats and carbohydrates. Compared with corn, it has the advantage of carrying liberal amounts of the antipellagra vitamin, niacin. White rice—brown rice that has been milled and polished to remove the bran and germ—loses a portion of its best protein and most of its fat, vitamins, and minerals, especially if it is cooked in an excess of water which is discarded. Thus, where white rice is the main food, growth retardation, kwashiorkor, marasmus, vitamin A deficiency, and beriberi (thiamin deficiency) are commonplace.

Nutritional Losses. In most of the rice-eating areas of the world, there is chronic undernourishment, malnutrition, low vitality, impairment of general health and physical development, and a high incidence of diseases resulting from insufficient and improper diet. Ranking high among these diseases as a killer of mankind is beriberi.

(Also see BERIBERI.)

Fig. R-13. Probable beriberi regions of the world. The field for action in the fight against beriberi is most of the rice-growing and rice-eating area of the Orient. Although health statistics are largely lacking, there are probably high death rates from beriberi in all the regions where ordinary white rice is the chief food.

High beriberi death rates and high consumption of rice, occurring in the same areas, are not the result of coincidence. The kind of rice most of the people prefer is white—the kind that deprives them of the natural food substances that would prevent or cure beriberi.

The rice grown by these people has many of the vitamins and minerals they need, but by the time it is milled and polished, then washed and cooked, it has lost most of them. The taste of cooked white rice appeals to their palates; the bulk appeases their hunger; the protein, carbohydrate, and calorie content give them the strength to live and work; but the grain, robbed of many of its health-giving elements and reinforced with little other food, brings disease and death.

As early as 1897, Eijkman, working in Java, showed that beriberi was due to the continued consumption of polished rice. Thus, it might seem that the rice-eating nations of the world have been either negligent or indifferent. This is not the case, however, for many of the nations have made intensive studies of the problem. Some have attempted to reform milling practices or to bring about wider consumption of other foods, only to find that these solutions could not be effected because of very real economic or human factors. Important among these is the cost of other and more varied foods. Even more important is the extraordinary resistance of the people to changes in their established dietary habits and food practices.

Washing and cooking practices in many Asian countries are also obstacles to better rice. In general, the more water used in cooking, and the more water thrown away, the more vitamins and minerals lost—whether from brown, parboiled, white, or enriched rice. However, somewhat more thiamin, riboflavin, niacin, and iron are retained in all improved forms of rice, even with the worst cooking practices. Fortunately, the Filipinos, Cantonese, and Malays customarily cook rice with relatively small amounts of water which is fully absorbed by the swelling of the grains. Elsewhere, it is to be hoped that education may in time improve cooking methods so that the vitamin-laden water will not be thrown away.

Enrichment. Enrichment is another possible solution for the rice-eating peoples who insist on the pure white product. Enrichment of white rice is a process in which selected vitamins and minerals are sprayed on rice grains. The treated rice is then coated with a film of edible substances, which protects the added elements against deterioration and reduces losses of vitamins during washing prior to cooking. The *premix* thus produced is as white as the untreated milled grain. When it is mixed with white rice in the proportion of one part of the premix to 200 parts of ordinarily milled rice, the resulting enriched rice contains approximately the same amounts of thiamin, niacin, and iron as brown rice.

While the addition of riboflavin, another vitamin lost in milling, is perfectly feasible, it is presently being omitted because it colors the premix grains. Confirmed white-rice eaters would undoubtedly be suspicious of these yellow grains, and sort them out and throw them away. However, in those rice-eating countries where it is desirable, riboflavin or other nutrients can be included. With proper public education, the occasional yellow grain might become the hallmark of enrichment.

Table R-5 shows the enrichment levels for rice in the United States.

TABLE R-5
ENRICHMENT LEVELS FOR MILLED RICE

Nutrient	Level	
	(mg/lb)	(mg/100 g)
Thiamin	2.0–4.0	.44–.88
Riboflavin[1]	1.2-2.4	.26-.53
Niacin	16–32	3.53–7.05
Iron	13–26	2.87–5.73
Calcium[2]	500-1,000	110-220

[1]The addition of riboflavin is feasible but the practice was stopped many years ago since its addition colors the rice yellow; hence, rice may or may not be enriched with riboflavin.

[2]The addition of calcium is optional.

Protein Quality. Table R-6 presents a comparison between the essential amino acid content of brown rice, white rice, and milk—a high quality protein. Like the other cereal

TABLE R-6
PROFILE OF THE ESSENTIAL AMINO ACIDS OF BROWN RICE AND WHITE RICE COMPARED TO MILK—A HIGH-QUALITY PROTEIN[1]

Amino Acid	Brown Rice	White Rice	Cow's Milk
	◄——————— (mg/g protein) ———————►		
Histidine	26	25	27
Isoleucine	40	46	47
Leucine	86	89	95
Lysine	40	39	78
Methionine and cystine	36	40	33
Phenylalanine and tyrosine ...	91	87	102
Threonine	41	36	44
Tryptophan	13	13	14
Valine	57	63	64

[1]*Recommended Dietary Allowances*, 10th ed., 1989, NRC-National Academy of Sciences, p. 67, Table 6–5. The essential amino acid requirement of infants can be met by supplying 0.79 g of high-quality protein per pound of body weight (1.7 g/kg of body weight per day).

grains, rice—brown and white—is deficient in the amino acid lysine; hence, lysine is the limiting amino acid. Infants and small children in particular need their essential amino acid requirement fulfilled. Therefore, due to the amino acid deficiency, and other vitamin and mineral deficiencies, brown or white rice alone is not a diet on which children will thrive or survive.

(Also see PROTEIN[S], section headed "Sources.")

The nutritive value of 1 cup *(240 ml)* of each of the different forms of rice is given in Food Composition Table F-21. Of the whole rice forms, brown rice ranks highest and unenriched white rice lowest. Of course, enriched rice can be fortified to any desired level. As shown, rice polish is a rich source of thiamin, which explains why it will prevent and cure beriberi.

BETTER RICE FOR MILLIONS. While the dramatic results of the Bataan experiment (See ENRICHMENT, section headed "The Bataan Experiment")—a full-scale enrichment program which reduced beriberi among a rice-eating population—gave world health authorities another means for bringing better rice and better health to millions of people, the difficulties of actually adopting this measure should not be underestimated. Most of these are the obstacles which have stood in the way of adopting undermilling and parboiling, the earlier known measures. Principal among them have been the lack of popular understanding of the need for better rice (as portrayed in Fig. R-14), inadequately financed governments, authorities too busy with other problems to enforce the necessary inspection systems, and an inadequate supply of technically trained and administratively effective personnel to perform required services for the protection of the masses.

Enrichment avoids one additional barrier that has hindered adoption of undermilling and parboiling—the distaste of many people for any but white forms of rice. This is an enormous gain, but the money shortage may be a serious deterrent to importing vitamins for fortification.

Perhaps the greatest significance of the Bataan project was to point up the possibility of doing something about beriberi immediately. The ready and rapid introduction of enrichment in the Philippines produced far more striking evidence than could be cited for any measure which by its nature requires slow introduction, bit by bit, over a period of years. This success furnishes a basis for reconsidering the merits of all the possible measures.

Where the difficulties of undermilling or parboiling appear insurmountable, rice enrichment may be the answer, as was demonstrated in the Philippines.

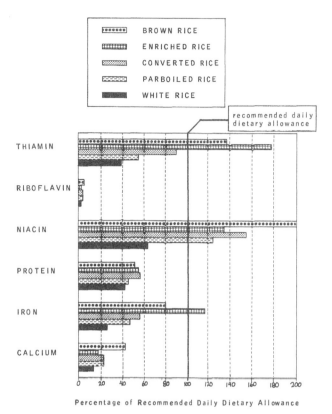

Fig. R-14. Nutritional values of rices. One pound per day of any of the rices shown here will yield these proportions of recommended daily dietary allowances. Of the "natural" forms, brown rice ranks highest, milled white rice lowest, with undermilled rice (not shown) ranging between the direct proportion to the degree of undermilling. The enriched rice is of the strength used in the Philippines, but the formula is adjustable to any level. "Converted" rice is a patented method of parboiling in the United States; it is highly milled, polished rice of greatly improved nutritive value.

RICE PRODUCTS AND USES. The whole grain of rice, with the hulls on, is known as rough rice or paddy rice. Milled rice (raw or parboiled), consisting of the kernels, is the major rice product. But rice milling produces a number of other products (see Table R-7).

Fig. R-15. Rice is used in many dishes, from soup to dessert. (Courtesy, USDA)

Milled rice is used mainly for human food; it may be sold as raw, quick-cooking, or canned rice. Rice is also used extensively as breakfast food—as puffed rice, rice flakes, or rice crispies. Broken rice is used as food or in the manufacture of alcoholic beverages. Rice flour is used in various mixes. Hulls are used as fuel, insulation, and certain manufactured products. The bran is used mainly as livestock feed. The products and uses of rice for human food are summarized in Table R-7.

TABLE R-7
RICE PRODUCTS AND USES FOR HUMAN FOOD

Product	Description	Uses	Comments
Brown rice	Rice with the hulls removed, but with the bran left on.	To provide more protein and more of the antipellagra vitamin, thiamin, than can be obtained from white rice.	Because of its color and relatively long cooking period (45–60 min.), this highly nutritious rice is a specialty food in the U.S. It is available in many health food stores and in some supermarkets.
Milled rice	The kernels, after the milling is completed and the hulls, bran layers, and germ are removed.	Cooking, mostly by boiling.	The U.S. per capita consumption of milled rice averages about 7.5 lb *(3.4 kg)*. Milled rice with the hulls removed, either polished or parboiled, is the major rice product.
Polished rice	Kernels from which the outer bran coats and germ have been removed, followed by polishing. Polished rice consists of the white, starchy endosperm, with fragments of the aleurone layer.	Cooking, mostly by boiling.	For the Puerto Rican market, polished rice is coated with talc and glucose, an inert, harmless coating to give it a high gloss or sheen.
		Quick-cooking rice (instant rice), a convenience food, some types of which can be cooked in 5 minutes.	There are many approaches to making raw milled rice into a quick-cooking product. Essentially, the process consists of selecting premium grade rice, and, under carefully controlled conditions, soaking, cooking, cooling, and drying it.
		Canned rice.	The canning process consists of (1) soaking the rice to a moisture of 30 to 35%; (2) cooking in excess water for 4 to 5 min.; (3) draining and packing the rice in cans; and (4) vacuum sealing and retorting. "Clumping" of the kernels and difficulty in emptying the contents of the can have been overcome by using surface-active agents to produce a free-flowing product, or freeze-processing, with the processed rice frozen in the can, then thawed prior to labeling.

(Continued)

TABLE R-7 *(Continued)*

Product	Description	Uses	Comments
Parboiled rice	Rough or paddy rice (in the hull) soaked in warm water, steamed under pressure and dried before milling.	Parboiled rice is used in cooking much the same as polished rice. Parboiling retains higher portions of the vitamins and lessens beriberi among people eating it.	In the United States, there are several patented methods of parboiling.
Enriched rice	Polished or parboiled rice enriched by spraying with a watery solution of vitamins (usually thiamin and niacin, and sometimes riboflavin) and sometimes iron and other minerals. .	In rice-eating areas to prevent beriberi and other nutritional diseases.	See Table R-5 for the federal standard for enriched rice.
Ang-khak	A product prepared in China by culturing a red pigment-producing fungus *(Monascus purpureus)* on whole rice kernels until they are thoroughly permeated with mycelia which produce the color. They are then dried and powdered.	Used as a food coloring agent in China, Taiwan, and the Philippines, for coloring cheese, fish, red wine, and other foods.	The Chinese are very secretive about the method of preparation of ang-khak Apparently the amount of water used in producing ang-khak is quite critical; the secret of success is to have the rice grains just moist enough to permit the fungus to grow.
Bran	Rice bran is the layer beneath the hull, containing outer bran layers and parts of the germ.	As a water-soluble fiber, similar to oat bran. Cholesterol responds to water-soluble fibers, which combine with bile acids and decrease the body's absorption of fat. Rice bran is also used as a livestock feed.	In addition to being high in water-soluble fiber, rice bran is rich in protein and B-vitamins. Rice oil is extracted from rice bran.
Breakfast foods: Puffed rice Rice crispies Rice flakes	Made from quality milled rice, usually of the short-grain type, which is generally precooked, dried, flaked, foamed and/or puffed or expanded, then toasted.	Ready-to-eat breakfast cereals.	Preparations of different products vary in cooking time, steam pressure, temperature, and the addition of malt and nutrients.
Brewers' rice (Chipped rice, broken rice)	Small, broken rice segments, separated out in the grading process.	Much of it is mixed back with whole grain rice and sold as low-grade rice. As the raw material for rice flour.	Brewers' rice has the same chemical composition as polished rice.
Flour, of which there are two types: 1. Flour made from regular rice, raw or parboiled	Fine, powdery, white particles.	Primarily as a thickening agent in gravies and sauces.	Because rice protein does not contain gluten, its use for confectionery and baking purposes is limited.
2. Waxy rice flour. (Also see FLOURS, Table F-12, Major Flours.)	Waxy rice flour is made from waxy glutinous rice. It has little starch or amylose; it is essentially amylopectin.	As a thickening agent for white sauces, gravies, and puddings—preventing liquid separation (syneresis) when these products are frozen, stored, and subsequently thawed.	Both types of rice flour are available commercially in the United States.
Rice baby foods	Flaked rice.	Baby foods.	Most rice baby foods are prepared by cooking ground, broken rice kernels and rice polishings; adding nutrients; and drum drying the rice slurry.
Rice oil	A high-grade oil extracted from rice bran, followed by refining and bleaching.	Rice oil is used in margarine, salad oil, and cooking oil.	Rice bran contains 14 to 18% oil. When refined, bleached, and deodorized, rice oil has greater stability than other vegetable oils.
Rice polishings	Inner white bran, protein-rich aleurone layers, and starchy endosperm; obtained in the milling operation by brushing the grain to polish the kernel.	In many baby foods. To prevent and cure beriberi.	Rice polishings are easily digested, low in fiber, rich in thiamin, and high in niacin.
Wild rice (Also see WILD RICE.)	The kernels are slender, round, purplish-black, and starchy. It's high in protein and low in fat.	Highly esteemed for its unique taste.	Normally, wild rice sells at 2 to 3 times the price of regular white rice.
FORTIFIED FOOD, BASED ON RICE PRODUCTS, FOR DEVELOPING COUNTRIES			
Protex	Defatted mixture of rice bran, germ, and polish. Contains between 17% and 21% protein, plus minerals and vitamins present in the outer layers of the rice grain.	Acceptable at levels of 5%, 10%, and 15% in yeast-leavened and quick breads. Base for breakfast cereals, pasta, and milklike beverages.	The process for producing Protex commercially was developed by Riviana Foods, Inc. at their Abbeville, Louisiana plant.
BEVERAGES (Also see BEERS AND BREWING; DISTILLED LIQUORS)			
Brewers' rice (Chipped rice, broken rice)	Small, broken rice segments, separated out in the grading process.	Used by the brewing industry, where it is mixed with barley. Used for the production of arrak in the Orient, a liquor of high alcoholic content distilled from a fermented mash of rice and molasses.	
Polished rice	Cooked, whole grains of white rice.	Japanese rice beer (sake), which has an alcohol content of 14 to 16%. (It is sometimes called *rice wine.*)	Sake is prepared by (1) conversion of the starch in cooked rice to sugar by the fungus *Aspergillus oryzae;* (2) a multistage conversion of the rice sugar to alcohol; (3) filtration, settling, heating, and aging; and (4) bottling for distribution. It is usually warmed before serving.

RICE PAPER

A thin delicate material resembling paper made by cutting the pith of the rice-paper tree into one roll or sheet and flattening under pressure. It is edible; so, macaroons and similar biscuits are baked on it and the paper can be eaten with the biscuit.

RICE, WHITE (POLISHED RICE)

Brown rice from which the outer bran layers up to the endosperm and germ have been removed.
(Also see RICE.)

RICKETS; OSTEOMALACIA

Contents	Page
History	933
Causes of Rickets and Osteomalacia	933
Rickets	933
Osteomalacia	933
Signs and Symptoms of Rickets and Osteomalacia	933
Rickets	934
Osteomalacia	934
Treatment and Prevention of Rickets and Osteomalacia	934

The most common disorder of bone formation in growing children is rickets. It is characterized by softening and deformation of the bones. The adult counterpart of rickets is called osteomalacia. Both of these conditions result from subnormal mineralization of the protein structure of bone, which is also called the bone matrix.
(Also see OSTEOMALACIA.)

A third bone disorder, osteoporosis, differs from rickets and osteomalacia in that it is characterized by a normal amount of mineralization of the bone matrix, but there is a reduction in the total bone mass and a tendency for fractures rather than deformation of bone.
(Also see OSTEOPOROSIS.)

HISTORY. In 1924, Steenbock and Black, of the Wisconsin Experiment Station, discovered that irradiation of liver and a rickets-producing diet with ultraviolet light transformed these materials into cures for rickets. In 1936, the German chemist, Windaus, demonstrated that provitamin D (7-dehydrocholesterol) is found in the skin, and that it is converted into vitamin D (calciferol) by the action of sunlight. The demonstration of Windaus led to an argument as to whether vitamin D might better be considered a hormone, rather than a vitamin, since its effect on bone is like that of a hormone. There are present in the body some related hormones, such as parathormone and calcitonin, which act on bones; and these hormones, together with vitamin D, regulate the net flow of calcium in and out of bones. It might also be argued that vitamin D is not actually a vitamin since it can be made in the body when the skin is exposed to ultraviolet light. In the absence of sufficient sunlight, however, it is necessary to provide an adequate source of vitamin D in the diet.

CAUSES OF RICKETS AND OSTEOMALACIA. These disorders may result from dietary deficiencies and/or poor utilization of calcium, phosphorus, and/or vitamin D; or from an incorrect ratio of the two minerals. However, different circumstances may be responsible for the development of each disorder. Also, rickets is a disease of children, while osteomalacia occurs in adults. Therefore, separate discussions of these disorders follow.

Rickets. This disease is most commonly caused by lack of vitamin D since it is usually found in children whose skin has had only a limited exposure to sunlight. The ultraviolet rays in sunlight convert provitamin D on the surface of the skin to vitamin D. Even children living in tropical areas may not have sufficient exposure to sunlight if clothing covers most of their skin, or they are kept indoors by their parents. In some areas, it is desirable to have a light-colored skin, so parents try to protect their children from the skin-darkening effects of sunlight.

Recently, it has been found that even when sufficient vitamin D may be obtained from the diet and/or its synthesis in the body, some persons may be unable to convert efficiently the vitamin to its physiologically active forms.[5] Vitamin D is converted to its 25-hydroxy derivative in the liver; then, the derivative is carried through the blood to the kidneys, where it is converted to 1, 25-dihydroxy-vitamin D. Sometimes rickets occurs in conditions such as uremia where the kidney fails to synthesize the 1, 25-dihydroxy form of the vitamin.

Exposure to sunlight or oral administration of the vitamin may fail to be effective in preventing or curing the disease when there are great reductions in the amount of calcium absorption from the gastrointestinal tract (either as a result of severe dietary deficiencies of the mineral, or due to malabsorption). For example, diets consisting mainly of whole grain cereals which contain phytates (phosphorus compounds found in the bran) may result in reduced absorption of dietary calcium, due to the binding of calcium by the phytates.

Osteomalacia. This disease is more likely to be found in women than men because (1) women usually wear more clothing than men, with the result that they have less synthesis of vitamin D in their skin; (2) pregnancy and lactation greatly increase the requirements of women for calcium (it is mobilized from their bones when it is not provided in adequate amounts in the diet); and (3) the average bone mass of an adult woman is about one-third less than that of an adult man, so women have a lower margin of safety against deformation of bone due to physical stresses.

Like rickets, osteomalacia may also be caused by interference with calcium absorption by phytates in cereals, disorders of absorption, excessive excretion of calcium and/or phosphate in the urine, and subnormal synthesis of the physiologically active form of vitamin D.

SIGNS AND SYMPTOMS OF RICKETS AND OSTEOMALACIA. Although these two disorders have essentially the same causes, the features of the two diseases differ somewhat because rickets occurs while the bones are still growing, while osteomalacia develops after growth of the

---

[5]"Recent Developments in Vitamin D," *Dairy Council Digest*, Vol. 47, No. 3, May-June 1976, p. 13.

long bones has ceased. Therefore, the signs and symptoms of these conditions will be considered separately.

Rickets.   The characteristic signs of rickets are protruding abdomen, beaded ribs or *rachitic rosary,* bowed legs, knock knees, cranial bossing (thickening of parts of the skull), pigeon chest (the breastbone or sternum is pushed backwards as it descends, forming a depression between the ribs), and enlargement of the epiphyses.

The epiphyses are the regions at the ends of the long bones which are separated from the shaft of bone or diaphysis by a layer of cartilage called the epiphyseal plate. These regions are mineralized during growth and eventually become part of the shaft of bone.

Teeth may erupt in an abnormal manner and the enamel may be defective. Malformed pelvic bones in women cause serious difficulties in childbirth. Permanent crippling results if severe cases are not treated. Other abnormalities associated with rickets are weak and flabby muscles, and muscle spasms (in severe cases). The deformities of rickets (curving and twisting of the bones from their normal shapes) are likely to be found in the bones that bear the most weight or stress, such as the legs.

Fig. R-16. An Indian boy suffering from rickets. Note the protruding abdomen, knock knees, enlarged skull, and enlarged wrists. (Courtesy, FAO, Rome, Italy.)

Osteomalacia.   One of the worst features of this disorder in a woman is the deformation of the pelvic bones so that she has great difficulty in childbirth. Another common feature of this disorder is a curvature of the spine (kyphosis) which gives the person a bent or hump-backed shape. Other features are pains in the bones, muscle weakness, a waddling gait, and muscle spasms (a mild form of tetany due to a subnormal level of serum calcium).

TREATMENT AND PREVENTION OF RICKETS AND OSTEOMALACIA.   Treatment of these disorders is based upon the administration of therapeutic doses of vitamin D (1,000 to 5,000 IU per day) together with the provision of adequate levels of dietary calcium (ranging from 800 mg [milligrams] per day for infants and children, to 2,400 mg per day for adults suffering from severe osteomalacia). Periodic x-rays may be used to determine whether healing is taking place as expected. There may be cases of either rickets or osteomalacia that are resistant to therapeutic doses of vitamin D. It has sometimes been necessary to administer massive doses of the vitamin (50,000 to 100,000 IU per day). However, prolonged administration of such large doses of the vitamin may result in toxic effects.

Prevention of rickets and osteomalacia is usually achieved by the provision of adequate dietary calcium (a daily amount equivalent to that in 1½ pints [720 ml] of milk) and making certain that vitamin D is obtained from the diet or from the effect of sunlight on the skin. There is sufficient phosphorus present in the diets of people that a deficiency occurs only when there is an abnormally high level of urinary excretion of this element. Therefore, there have been worldwide efforts to promote the fortification of all milk with 400 IU of vitamin D per quart (950 ml). Should this prophylactic measure be unfeasible in some instances, calcium and vitamin D might be administered in the form of a mineral and vitamin supplement. It is noteworthy that in some areas of the Middle East, infants are fitted with miniature goggles (to protect their eyes) and placed out in the sun in bassinets.

(Also see BONE, section on "Bone Disorders"; CALCIUM, section on "Calcium Related Diseases"; OSTEOMALACIA; PHOSPHORUS, section on "Phosphorus Related Disease"; and VITAMIN D, section on "Deficiency Symptoms.")

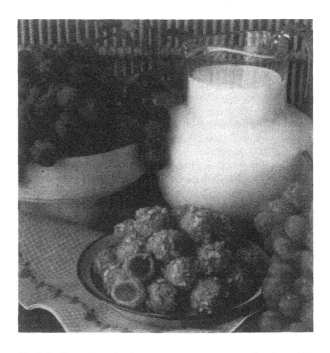

Fig. R-17. Prevention of rickets and osteomalacia is usually achieved by using fortified milk, a good source of both calcium and vitamin D. (Courtesy, California Grape Commission, Fresno, Calif.)

RIGOR MORTIS

Within a few hours after death the muscles become stiff and rigid. Since the blood no longer delivers oxygen to the tissues, anaerobic metabolism produces lactic acid which lowers the pH. As the ATP (adenosine triphosphate) is depleted, the muscles harden—contract.

RISSOLES

• These are like little turnovers, filled with a highly seasoned mixture of chopped chicken, ham, or other tender meat moistened with a little sauce, dropped onto pastry, which is sealed, then fried in deep fat.

• A patty of ground meat or fish, rolled in crumbs and dipped in egg, and more crumbs, then fried. The English name for the American croquette.

ROASTED GRAIN DRINKS (CEREAL COFFEE, CEREAL BEVERAGE, COFFEE SUBSTITUTE)

By roasting certain grains and grain products in combination with other flavor sources, a product may be obtained which can be mixed with hot water yielding an aromatic beverage similar to coffee. These cereal beverages are commercially available or they may be homemade. One popular product, developed over 75 years ago by C.W. Post, helped form the foundation for the General Foods Corporation.

Grains and grain products in the commercially prepared products are generally barley, wheat, rye, malt, and bran, while additional flavor is contributed by molasses, chicory, carob, cassia bark, allspice, and star anise, depending upon the manufacturer. A homemade roasted grain drink can be prepared by baking a combination of wheat bran, eggs, cornmeal, and molasses. There are even those cereal beverages which can be percolated like coffee.

Roasted grain drinks are caffeine-free. With the recent publicity involving the health effects of coffee and caffeine, these cereal grain beverages have gained in popularity.

ROASTERS (ROASTING PIGS)

Roasters are fat, plump, suckling pigs, weighing 30 to 60 lb (14 to 27 kg) on foot. They are dressed shipper style (with the head on), and are not split at the breast or between the hams. When properly roasted and attractively served with the traditional apple in the mouth, roast pig is considered a great delicacy for the holiday season.

ROCAMBOLE *Allium scorodoprasum*

European leek, usually cultivated like the shallot and used in the same way.

ROLLS

These can be defined as yeast-leavened miniature loaves of bread. There are as many kinds of rolls as the cook can devise. Some are the same basic recipe, but formed into different shapes. Others have slight variations in the proportion of fat or sugar to flour; and eggs may or may not be added. And the choices of added ingredients such as raisins, other fruits, spices, nuts, and poppy, sesame, or caraway seeds give added dimensions. A good basic roll recipe is like a good basic suit in the wardrobe—it can be used for all occasions. Any bread recipe can be used for rolls. Some of the more familiar rolls are: bagels, bread sticks, brioche, butterflake, cinnamon, clover-leaf, crescents (butterhorns),

Fig. R-18. Christmas bread roll. (Courtesy, New Mexico State Univ., Las Cruces, N.M.)

croissants, English muffins (crumpets), hamburger buns, hot cross buns, Parker House (pocket rolls), pecan, submarine, twists, and wiener buns.

Rolls are as nutritious as the ingredients in them. See Table F-21, Food Composition Table, under "Bakery Products," for the nutritive qualities of rolls.

(Also see BREADS AND BAKING.)

ROPE

A condition in breads caused by *Bacillus mesentericus* or *Bacillus subtilis*. Spores of these bacteria can survive baking and then germinate under proper moisture and temperature surroundings. The bread develops sticky yellow patches that can be pulled into ropelike threads; hence, the name rope. Obviously, the bread is inedible. Food additives such as calcium acetate, calcium propionate, and sodium diacetate retard this bacterial growth.

(Also see ADDITIVES.)

ROSE APPLE (POMARROSA) *Eugenia jambos*

The fruit of a small tree (of the family *Myrtaceae*) that is native to the Indo-Malaysian region, and which was introduced into the New World tropics where it now grows wild. The rose-centered fruits are egg-shaped, about 1½ in. (4 cm) in diameter, yellowish-white or pink with a yellow flesh and one or two seeds. The fruit is eaten fresh or made into jam or pies.

The nutrient composition of rose apples is given in Food Composition Table F-21.

Some noteworthy observations regarding the nutrient composition of rose apples follow:

1.   The raw fruit is low to moderately high in calories (56 kcal per 100 g) and carbohydrates (14%).

2. Rose apples are a good source of iron and a fair to good source of vitamin C.

ROSE HIPS *Rosa* spp

The fleshy fruit at the base of the rose bloom of plants of the family *Rosaceae*.

Rose hips are not eaten as such; rather, they are concentrated as a powder, made into jams, jellies, and syrups, or brewed as a tea. Rose hips are used as an ingredient of many vitamin C products, often in combination with ascorbic acid.

(Also see VITAMIN C, section on "Sources of Vitamin C.")

ROSELLE

An East Indian herb, cultivated for its fleshy calyxes, which are used to make tarts and jellies, and a very acid drink.

ROUX

A mixture of fat and flour which is used as a base for thickened sauces. It is prepared by melting the butter or other fat, adding the flour, and stirring until it is browned. This produces some dextrin which gives the sauce an enjoyable flavor, that gourmet cooks feel is superior to other methods of making white or other sauces.

ROWE ELIMINATION DIETS

Widely used elimination diets formulated by Dr. Albert H. Rowe. These diets are used to determine food allergies by removing certain types of foods from the diet. In general, elimination diets are tedious and time-consuming and should be employed only when dietary history, skin tests, and provocative food tests fail to detect the causative food. Rowe diets are organized on the basis of food items that have proven the least likely to produce an allergic response.

(Also see ALLERGIES, section on "Elimination Diets; Rowe Elimination Diets.")

ROYAL JELLY

This is the milky white food that is prepared exclusively for the queen bee in a beehive. Royal jelly, which is available as a supplement in capsule form, is a rich source of the B-complex vitamins, along with 20 amino acids. But about 3% of this substance made by bees still defies analysis.

Queen bees look different from worker bees: they are about twice the size; they live up to 8 years—fully 40 times longer than the normal lifespan of worker bees; and they lay about 2,000 eggs per day, whereas female worker bees are infertile. What makes queen bees so different? All the eggs start out the same, fed on a rich protein diet secreted from the glands of the worker bees. After 2 days, the eggs that are to become worker bees are changed to a diet of honey. Meanwhile, the egg that is to become the queen is fed royal bee jelly throughout her growth stage. Royal jelly makes the difference!

The implication of the promotors of royal jelly is that it will do as much for humans as for queen bees; that it will increase size, longevity, and fertility.

No effect has been demonstrated in humans, except a tendency to produce wakefulness. Royal jelly is judged to be of no practical value in human nutrition because of the very large amounts required for any definite effect.

RUMEN (PAUNCH)

The large first compartment of the stomach of a ruminant from which the food is regurgitated for cud-chewing and in which cellulose is broken down by bacterial and protozoan symbionts.

RUSKS (ZWIEBACK)

Zwieback is the German word meaning *twice-baked*. Originally, rusks were developed to take on ships for long voyages. They are made from either plain or sweet bread which is baked, then sliced and baked again to produce a hard crisp product. Rusks are a favorite food on which the infant can chew. They are also good for taking on camping or fishing trips, instead of trying to keep fresh bread on hand.

RYE *Secale cereale*

Contents	Page
Origin and History	937
World and U.S. Production	937
Processing	937
Nutritional Value	937
Products and Uses	937

Fig. R-19. Rye, ripe and ready to harvest. (Courtesy, North Dakota State University, Fargo, N.D.)

The cereal grain, rye, *Secale cereale* of the grass family, *Gramineae*, is closely related to wheat and is grown as a bread grain, for distillation into grain alcohol spirits, and as a livestock feed. The most winter-hardy of the cereals, rye is the main bread grain of the Scandinavian and eastern European countries. Although nutritious, and palatable to some people, rye bread is not comparable to wheat bread in crumb quality and bold appearance of the loaf. As living standards rise, the consumption of rye bread falls, and the consumption of wheat bread rises.

(Also see CEREAL GRAINS.)

ORIGIN AND HISTORY. Rye was domesticated relatively recently, about the 4th century B.C., in Germany, and later in southern Europe. According to N. I. Vavilov, the Russian plant scientist, cultivated rye originated from wild species that occurred as weeds in wheat and barley crops, and rye was introduced into cultivation simultaneously and independently at many localities in central Asia or Asia Minor.

During the Middle Ages, the poorer people of England ate bread made from rye, or from a mixture of rye and wheat, known as *maslin*.

WORLD AND U.S. PRODUCTION. World production of rye is about 37 million metric tons, only about 6% of the world production of wheat, and only 2% of the world production of cereals.[6] Rye is more important in Europe and Asia than in the western hemisphere. The Soviet Union is the leading world producer, followed by Poland, Germany, China, Canada, Czechoslovakia, Denmark, Austria, Sweden, and Spain.[7]

Rye production in the United States has steadily declined since World War I. Annually, about 245 thousand metric tons are harvested from about 373,000 acres, with an average yield of 27.1 bu per acre. The U.S. wheat crop averages 36.7 bu per acre (2,069 *kg/ha*). The leading rye-producing states, by rank, are: South Dakota, Georgia, Minnesota, North Dakota, Nebraska, South Carolina, Michigan, Wisconsin, and North Carolina.[8]

---

[6]Data from *FAO Production Yearbook 1990*, FAO/UN, Rome, Italy, Vol. 44, p. 81, Table 21. **Note well:** Annual production fluctuates as a result of weather and profitability of the crop.

[7]*Ibid.*

[8]Data from *Agricultural Statistics 1991*, USDA, p. 16, Table 19. **Note well:** Annual production fluctuates as a result of weather and profitability of the crop.

PROCESSING. The processing of rye involves cleaning, conditioning, and milling.

The cleaning principles and machinery are similar to those used in wheat cleaning, complicated by two primary differences: (1) The grain is more variable in size; and (2) the grain is more subject to ergot (a poisonous fungus), which should be separated out before it is broken up and rendered difficult to remove (see ERGOT).

Conditioning, or tempering, consists of bringing the grain to a moisture content of about 14.5%.

The milling process resembles wheat flour milling, but it deviates from the latter because of two important differences between the grains; (1) the endosperm of rye breaks up into flour fineness more readily than the endosperm of wheat, and (2) the separation of the endosperm from the bran of rye is more difficult.

NUTRITIONAL VALUE. Food Composition Table F-21 gives the nutrient compositions of rye grain and rye products.

The food value of rye is similar to wheat. It is a good source of starch (carbohydrates) and a fair source of protein. However, the protein, in terms of essential amino acids, falls short in lysine compared to high-quality proteins. Therefore, like wheat and the other cereal grains, lysine is the limiting amino acid. Of the vitamins, rye is devoid of vitamins A, C, and B-12.

PRODUCTS AND USES. Rye is used for human food, fermented products, livestock feed, and for certain other purposes.

The uses of rye and rye products are summarized in Table R-8.

(Also see CEREAL GRAINS, and BREADS AND BAKING.)

TABLE R-8
RYE PRODUCTS AND USES FOR HUMAN FOOD

Product	Description	Uses	Comments
Rye	Whole grain, unground	Hot breakfast cereal. Served with main dish.	The whole grains may be cooked like rice by (1) soaking overnight in water (2-3 c of water per c [*240 ml*] of rye) then boiling until tender; or (2) cooking in a pressure cooker, using a standard recipe for rice.
Rye	Ground, dry whole rye grain. Rye meal, consisting of any extraction rate less than 100%.	Rye crisp Rye crisp	Rye crisp is a popular bread served with most meals in Sweden. Traditionally, rye crisp bread in Sweden is made by mixing rye meal with snow or powdered ice, then the expansion of the small air bubbles in the ice cold foam raises the dough when it is placed in the oven.
Rye flakes	Rolled whole rye	Hot breakfast cereal	
Rye flour (Also see FLOURS, Table F-12, Major Flours.)	Produced by milling rye, much like milling wheat. It does not contain as much gluten as wheat flour.	Bread. In the U.S., rye bread is usually made from a combination of rye flour and wheat flour. Biscuits and crackers (usually made from mixtures of about 10% rye flour and 90% wheat flour). Rye pancakes. Filler for sauces, soups, and custard powders.	In most countries, rye is used as a bread grain—for human food. Yeast does not raise rye dough as easily as wheat dough, because of lack of gluten in rye flour. When bread is made from straight rye flour, it is black, soggy, and rather bitter. For centuries, a large portion of the population of Europe lived mainly on schwarzbrot, made from rye flour, which is still rather common in rural Germany, Poland, and the U.S.S.R. Whole rye and mixed rye-wheat breads have a longer shelf life than wheat bread. Rye contains more pentosans than other cereals; hence, it is used in reducing diets because: (1) the pentosans gelantinize and swell in the stomach, imparting a feeling of satisfaction; and (2) digestion of the polysaccharides is slow, with the result that the blood sugar level rises slowly but is maintained for 5 to 6 hours, thereby controlling appetite.
FERMENTED BEVERAGES			
Rye grain	Cleaned whole grain	Rye whiskey. Industrial alcohol.	Rye whiskey is made from a fermented mash containing a minimum of 51% rye.
Rye bread	Brown rye bread	Russian rye bread beer (kvass or quass)	A mash made by pressure cooking the bread is treated with rye malt, then fermented by a mixture of yeast and *Bacillus lactis*. The beer, which contains only about 0.7% alcohol, is usually dispensed cold from tank trucks.

Fig. R-20. Besides using rye flour for rye bread, and pumpernickel, rye flour can be substituted for some of the wheat flour in such quickbreads as steamed brown bread, muffins, and gingerbread (as above). (Courtesy, California Foods Research Institute for Cling Peach Advisory Board, San Francisco, Calif.)

SACCHARIMETER

An apparatus for determining the concentration of sugar in a solution, based on the optical activity of the sugar; especially a polarimeter adapted for distinguishing different kinds of sugar in a solution.

SACCHARIN

A nonnutritive, noncaloric synthetic sweetener which is 300 to 500 times sweeter than table sugar. It was on the first GRAS (generally recognized as safe) list published in 1959. However, in 1972, following the ban on cyclamates, saccharin was removed from the GRAS list and given a provisional food additive status.

Today, saccharin is still being used on a limited basis.

(Also see ADDITIVES, Table A-3 Common Food Additives; ARTIFICIAL SWEETENERS; CANCER; and DELANEY CLAUSE.)

SACCHAROMETER

A floating apparatus, a special hydrometer, used to determine the specific gravity of sugar solutions. It can be graduated to read directly the percentage of sugar in solution. It is distinct from a saccharimeter.

SAFETY OF FOOD

Food safety involves a number of concerns, the most important of which are: (1) pesticides, and (2) harmful bacteria.

1. Pesticide residues in and on foods are the public's number one food safety fear. Because pesticides cannot be seen or smelled, consumers feel that they have little control over their level of exposure. So, we need to develop safer and more effective pesticides, and to use less pesticides on fruits and vegetables.

(Also see PESTICIDES AND INCIDENTAL FOOD ADDITIVES; and POISONS.)

2. Without bacteria in food there would be no yogurt, no blue cheese, and no sourdough bread. However, other bacteria left to grow unchecked can cause spoilage and even food-related illness. Making sure that this does not happen is the job of producers, workers, supervisors, and quality control specialists from farm to food counter, and finally to the consumer. Despite some concerns and scares from time to time, America's cornucopia of foods remains the envy of the world, both in quantity and quality.

(Also see DISEASES, sections on "Sanitary Measures Which Help to Prevent the Spread of Diseases" and "Health and Nutrition Functions of Government Agencies"; FOOD—BUYING, PREPARING, COOKING, AND SERVING; BACTERIA IN FOOD; and FOOD POISONING.)

SAFFLOWER *Carthamus tinctorius*

The safflower is a relative of the thistle. Its scientific identity: family, *Compositae*; genus, *Carthamus*, species, *C. tinctorius*.

The seed, which resembles a small sunflower seed, is com-

Fig. S-1. Field of safflowers in bloom in the Sacramento Valley of California. The safflower is a relative of the thistle family. (Courtesy, Pacific Oilseeds Incorporated, Woodland, Calif.)

posed of a thick fibrous white hull encasing a yellow kernel.

In the 1950s, interest in safflower oil as a food ingredient skyrocketed as a result of studies relating saturated fats in the diet to atherosclerotic heart disease. Safflower's unique fatty acid composition places it highest in polyunsaturates and lowest in saturates of all commercial fats and oils.

ORIGIN AND HISTORY. Safflower is native to Southeastern Asia, but it has long been cultivated in India, Egypt, China, and Northern Africa.

WORLD AND U.S. PRODUCTION. The world production of safflower seed totals 922,000 metric tons annually. India is far in the lead globally with a production of about 491,000 metric tons—over half of the world production. The United States with 170,000 metric tons ranks second, and Mexico with a production of about 159,000 metric tons ranks third. Lesser quantities are produced in Spain.[1]

Total U.S. seed production reached a high of 350,000 metric tons in 1967, but has since declined to around 170,000 metric tons.

PROCESSING SAFFLOWER SEED. Normally, the steps in processing consist of: (1) cleaning the seed by screening and aspiration; (2) grinding the seed; (3) cooking under steam pressure; (4) extracting the oil by the continuous screw press-solvent extraction method; (5) grinding and screening the cake to produce meal; and (6) refining the oil.

NUTRITIONAL VALUE OF SAFFLOWER SEED. The commercial varieties of safflower grown in the United States average 35 to 40% hull and yield 39 to 40% oil and 15% protein. However, varieties with up to 50% oil content have been developed.

The oil is extremely variable, perhaps reflecting its mixed heredity. California studies on safflower procured from many parts of the world showed great variation in the content of

---

[1]Data from *FAO Production Yearbook 1990*, FAO/UN, Rome, Italy, Vol. 44, p. 117, Table 44. **Note well:** Annual production fluctuates as a result of weather and profitability of the crop.

polyunsaturated fats, with iodine numbers ranging from 87 to 149. It averages about 6.6% saturated acids and 93.4% unsaturated acids, with the latter consisting of 77.0% linoleic acid and 16.4% oleic acid (The high linoleic acid content makes the oil susceptible to rancidity). Linolenic acid is absent. Genetically modified varieties of safflower with greater resistance to rancidity have been introduced that contain 80% oleic, 15% linoleic, and 5% saturated fatty acids. Safflower oil is listed in Food Composition Table F-21 and Table F-4, Fats and Fatty Acids in Selected Foods.

The extracted cake is ground to yield a 20 to 42% protein meal, with the higher protein content obtained from decorticated (dehulled) seed. Meal produced by the solvent method contains about 1% fat vs 5% when the expeller process is used. Safflower meal is of good quality, although somewhat deficient in lysine and methionine.

SAFFLOWER PRODUCTS AND USES. Safflower oil is a bland, almost colorless product. Due to its susceptibility to oxidation, care is taken to exclude air during storage, transport, and packaging. Interest in the oil as a food ingredient stems largely from its high percentage of unsaturated (polyunsaturated) fatty acids.

Safflower oil is used principally in the production of margarine, salad oils, mayonnaise, shortening, and other food products. The oil is also used as a drying agent in varnishes and paints.

High protein content safflower meal, made from dehulled seed and containing about 42% protein, is now considered a protein source for human consumption. However, most safflower meal is used as a protein supplement for livestock. The meal is not very palatable when used alone.

(Also see OILS, VEGETABLE, Table O-5.)

SAINT ANTHONY'S FIRE

A thousand years ago, this name was given to the condition known now as ergotism—a poisoning due to the consumption of foods containing the fungus ergot. Symptoms of ergotism include alternating burning and freezing sensations; hence, the name Saint Anthony's Fire.

(Also see POISONS, Table P-2, Some Potentially Poisonous [Toxic] Substances, "Ergot.")

SALAMI

A highly seasoned sausage made of pork and beef in various proportions; either (1) air-dried, hard, and of good keeping qualities, or (2) fresh, soft, and requiring refrigeration until consumed.

SALEP

A starchy foodstuff from the Middle East which consists of the dried tubers of various orchids. It is easily digested and highly esteemed in the East.

SALINE

It refers to anything containing or consisting of salt. Physiological saline is a solution containing 0.9% NaCl (salt). Since it is compatible with the blood, it is sometimes administered intravenously.

SALIVA

A clear, somewhat viscid solution secreted into the mouth by three pairs of salivary glands—the parotid, the sublingual, and the submaxillary.

The functions of saliva are manyfold including the following: lubrication, enzymatic activity, buffering capacity, taste, and moistening the membranes of the mouth and providing some antibacterial action.

Secretion of saliva is stimulated by the sight, the smell, and even the mere thought of food.

(Also see DIGESTION AND ABSORPTION, section on "Mouth.")

SALLY LUNN

This bread came to America from Bath, England, the home of Sally Lunn. It is baked in a tube-center pan, and it should have a porous, cakelike texture.

SALMONELLA

A genus of bacteria responsible for one of the most common foodborne infections in the United States—salmonellosis. *Salmonella* bacteria grow rapidly in cooked foods such as meats, eggs, custards, and salads which have been left unrefrigerated for several hours. It may also be transmitted by sewage-polluted water.

(Also see BACTERIA IN FOOD, Table B-2, Food Infections [Bacterial], "Salmonellosis.")

SALT (SODIUM CHLORIDE; NaCl)

Contents | Page
Historical Importance of Salt 940
Functions of Salt 941
  Physiological 941
    Salt Deficiency 941
    Salt Toxicity 941
  Commercial Uses 941
    Food Industry 941
    Chemical Industry 941
    Other Industries and Businesses 941
  Excess Salt; Labeling 941
  Sources of Salt 942

Salt is created by the combination of the soft silvery-white metal sodium (Na), and the yellow poisonous gas chlorine (Cl). Sodium and chlorine are vital elements found in the fluids and soft tissues of the body. Salt also improves the appetite, promotes growth, helps regulate the body pH, and is essential for hydrochloric acid formation in the stomach.

HISTORICAL IMPORTANCE OF SALT. Throughout history, salt has occupied a unique position. Wars have been fought over it, empires have been founded on it and have collapsed without it, and civilizations have grown up around it.

Mosaic law prescribed the use of salt with offerings made to Jehovah; and there are frequent Biblical references to the purifying and flavoring effects of salt. The Greeks used salt as the medium of exchange in buying and selling slaves; a good slave was said to be *worth his weight in salt.* The word salary is derived from the Latin *salarium,* referring to the salt which was part of a Roman soldier's pay. In medieval England, royal banquet halls had imposing salt cellars; and the seating arrangement in relation thereto served as a status symbol. Important persons were invited to *sit above the salt,* where they could use the salt on their food freely; those of lesser importance were seated *below the salt.* In some parts of Africa, gold was once evaluated in terms of how much salt it would buy—rather than the reverse. Salt caravans, part of the ancient lore of Africa, still ply certain desert areas to this day. By common use, the expression *salt of the earth* refers to a really good person. These and other salt lore give ample evidence that in olden times salt was a valuable and relatively scarce commodity.

FUNCTIONS OF SALT. Most people think of salt as a seasoning, but it is probably used in greater quantities and for more applications than any other chemical. In fact, it is estimated that over 14,000 uses are made of salt.

Physiological. Sodium (Na) and chlorine (Cl) are essential parts of the human diet. (Salt as such isn't essential, since other sources of Na and Cl are satisfactory; for example, Na_2CO_3 and KCl.)

In body solution, salt dissociates into two ions—sodium and chloride, both of which are normal and necessary constituents.

(Also see CHLORINE OR CHLORIDE [Cl]; and SODIUM [Na].)

SALT DEFICIENCY. Salt deficiency is rarely a problem since most Americans consume many times more salt than is recommended. Fortunately, the body adjusts.

If there is a low intake, the excretion rate is low. Deficiencies may occur from strict vegetarian diets without salt, or when there has been heavy, prolonged sweating, diarrhea, vomiting, or adrenal cortical insufficiency—Addison's disease.

A deficiency of salt (sodium) may cause reduced growth, loss of appetite, loss of body weight due to loss of body water, reduced milk production of lactating mothers, muscle cramps, nausea, diarrhea, and headache. In cases of prolonged or severe vomiting, or indiscreet use of diuretics, chloride losses may exceed that of sodium—resulting in a state of metabolic alkalosis. To restore the acid-base balance of the body, adequate chloride must be provided.

SALT TOXICITY. Excess dietary salt, over a long period of time, is believed to contribute to the development of high blood pressure in susceptible individuals.

In addition, salt may be toxic (1) when ingested in large quantities, especially with a low water intake; (2) when the body has been depleted by a salt-free vegetarian diet or excess sweating, followed by gorging on salt; (3) when large amounts are fed to infants or others afflicted with kidney diseases, whose kidneys cannot excrete the excess in the urine; or (4) when the body is adapted to a chronic low-salt diet, followed by ingesting large amounts.

(Also see HIGH BLOOD PRESSURE; MINERAL[S]; POISONS, Table P-2, Some Potentially Poisonous, [Toxic] Substances; and SODIUM.)

Commercial Uses. The majority of the millions of tons of salt produced each year is used by a variety of industries and businesses; it's used commercially.

FOOD INDUSTRY. Its first uses in foods are lost in antiquity. Thus, not surprisingly, salt is on the FDA list of GRAS (generally recognized as safe) food additives. In foods, salt performs three primary functions: (1) as a flavor enhancer or seasoner; (2) as a preservative; and (3) as a solubilizing agent.

CHEMICAL INDUSTRY. Much of the salt is utilized in the manufacture of other chemicals. By passing an electric current through salt—electrolysis—it can be broken up into sodium metal and chlorine gas. The sodium can be used as a catalyst, or it can combine with other elements to form new chemicals such as sodium carbonate, sodium bicarbonate, and lye (sodium hydroxide). The chlorine formed by electrolysis can also be used to make other chemicals, or it can be employed in bleaching paper and textiles, or in disinfecting water supplies. Many of the chemicals derived from salt find their way into numerous other industries.

OTHER INDUSTRIES AND BUSINESSES. Salt is important to a number of other industries and businesses. The leather industry uses salt for the preservation of hides. Salt can be employed to soften water. Railroads use salt to keep their switches from icing up. Highway departments use salt on icy roads, and on secondary roads to stabilize the soil and control the dust. Farmers feed salt to livestock. It is also employed in heat-treating, smelting, and refining metals. The list of uses seems endless for this basic chemical.

EXCESS SALT; LABELING. For years, the subject of salt and its relation to hypertension has been debated by consumers, federal officials, and food industry representatives. Finally, in 1981, for the first time, the food industry was asked by the Food and Drug Administration to lower the salt content in foods, voluntarily and not by regulation. In support of the request, Richard S. Schweiker, Secretary of Health and Human Service, had the following to say:

"Sixty million Americans suffer from hypertension, which can lead to strokes and heart attacks. Hypertension can be significantly controlled."

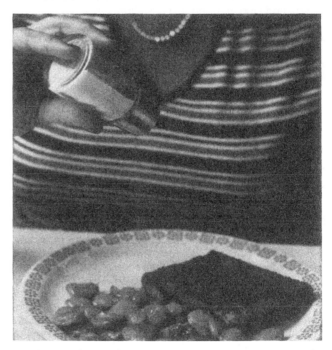

Fig. S-2. Use salt sparingly. (Courtesy, University of Tennessee, Knoxville, Tenn.)

At about the same time, a bill was introduced in the U.S. Congress that would *require* food processors and manufacturers to disclose, through a label, the salt content of their products if the salt is above a certain level. Also, it was revealed that the FDA had drafted a regulation that would *require* sodium labeling, but it would have to be approved by the Office of Management and Budget before it could be instituted.

The FDA, which has jurisdiction over about 80% of all processed foods, indicated that labeling would apply to both natural and added salt in a product.

Simultaneously, the U.S. Department of Agriculture announced that the department plans to encourage sodium disclosure for meat products under its jurisdiction through similar action to that of the FDA.

(Also see SODIUM, section headed "Recommended Daily Allowances of Sodium.")

SOURCES OF SALT. The United States is by far the leading salt-producing country. The states of Louisiana, Texas, New York, and Ohio produce 88% of U.S. salt. In 1985, 39.5 million short tons of salt, worth $741.8 million, were taken from the seas, mines, and wells in the United States.

(Also see ADDITIVES; CHLORINE OR CHLORIDE; MINERAL[S]; and SODIUM.)

SALT DIET, LOW

Low salt diet is a rather misleading term. Since table salt is by far the most important source of sodium in the diet, most sodium-restricted diets begin with the elimination or at least restricted use of table salt on food. Hence, sodium-restricted diets are sometimes called low salt diets. A mild sodium restriction which is used as a maintenance diet in cardiac and renal diseases limits daily sodium intake to 2,000 to 3,000 mg. This means no salty foods and no salt used at the table. Other sodium-restricted diets include moderate sodium restriction (1,000 to 1,500 mg daily), strict sodium restriction (500 mg), and severe sodium restriction (250 mg). All of these diets require limited or no use of salt at all stages of food preparation. The strict and severe sodium restriction necessitates careful selection of foods for all sources of sodium, not just salt.

(Also see MODIFIED DIETS, Table 28, Modified Diets, "Sodium restricted diet"; and SODIUM.).)

SALT FOODS, LOW

Table salt—sodium chloride—is the most important source of sodium in the diet. Salt may be added to foods during processing, during cooking, and at the table by the consumer. When an individual requires the restriction of daily sodium intake, the first step involves limiting the intake of salt. Depending upon the level of sodium restriction, foods may need to be carefully selected in regards to the amount of salt used in processing. However, salt is not the only source of sodium in foods, so if a strict or severe restriction of sodium is needed, all sources of sodium in foods must be considered. Other sodium sources include baking soda, baking powder, sodium benzoate, sodium citrate, monosodium glutamate (MSG), sodium propionate, and sodium alginate. When selecting foods for sodium restricted diets, labels should be noted for the words *salt* and *sodium*. Furthermore, some foods possess a naturally high level of sodium and must be used in measured amounts in sodium restricted diets.

(Also see MODIFIED DIETS; SALT, section on "Excess Salt; Labeling"; and SODIUM.)

SALT-FREE DIETS

A more accurate description of these diets is sodium restricted or low sodium. Since much of the sodium added to foods comes from table salt—sodium chloride—the first step in a sodium-restricted diet is the elimination of table salt. Hence, these diets are commonly called salt-free.

Sodium-restricted diets, varying in the amount of sodium restriction, are prescribed for the elimination, control, and prevention of edema—water accumulation in the tissues—which accompanies such things as congestive heart failure, cirrhosis, kidney disorders, and corticosteroid therapy.

Sodium restriction is also helpful in control of some cases of high blood pressure.

For persons on sodium restricted diets, a 1:1 mixture of table salt (NaCl) and potassium chloride (KCl) for salting foods will lessen the sodium intake without a change in taste. However, excessive use of dietary potassium should be avoided due to toxic effects at high levels.

(Also see MODIFIED DIETS; HIGH BLOOD PRESSURE; and SODIUM.)

SALTINE

This is a thin crisp cracker covered with salt. All saltines are soda crackers, but not all soda crackers are saltines.

SALTPETER

A name sometimes applied to potassium nitrate; sodium nitrate is called Chile saltpeter.

(Also see NITRATES.)

SAMBAL

A condiment common to Indonesia and Malaya, eaten with curry and rice. Typically, it contains peppers, pickles, grated coconut, salt fish, or fish roe.

SAMP

Coarse hominy or a boiled cereal made from it.

SAPONIFICATION

The formation of soap and glycerol from the reaction of fat with alkali.

SAPONINS

A group of heart-stimulating glycosides derived from plants such as foxglove, squill, and legumes. They can be broken down into a sugar and a steroid. All saponins foam when shaken with water, and all are surface-active agents. When injected into the bloodstream they are capable of bursting (hemolyzing) red blood cells. Saponins are used as an emulsification agent for fats and oils, as a soap or detergent, and as a subject of research on steroid sex hormones.

SARCOMA

A tumor of fleshy consistency—often highly malignant.

SATIETY

Full satisfaction of desire; may refer to satisfaction of appetite.

(Also see APPETITE.)

SATURATED

A state in which a substance holds the most of another substance that it can hold.

SATURATED FAT

A completely hydrogenated fat—each carbon atom is associated with the maximum number of hydrogens; there are no double bonds.

(Also see FATS AND OTHER LIPIDS.)

SAUCES

The word *sauce* comes from the Latin word *salsa* meaning *salted*. A sauce can be defined as a *liquid food, which*

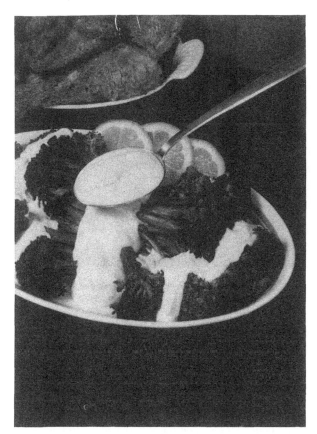

Fig. S-3. Sauce—the crowning touch. (Courtesy, Carnation Co., Los Angeles, Calif.)

is poured over a solid food. It can appear on the menu from appetizer to dessert. It can range from liquid to very thick; from no seasoning to highly seasoned; from cold to piping hot; from sour to extremely sweet; from simple to complicated and rich; and from the same every time to never the same way twice.

Sauces do not take any great amount of time to make. Nevertheless, they should not be a neglected art, for, in the final analysis, the sauce often lifts up the dish: it glamorizes the simplest food; it gives the recipe distinction; it increases the appetite appeal; it is the crowning touch.

There are so many kinds of sauces that it is difficult to organize them into any kind of order; for instance, a white sauce can be used either for fish or for dessert, depending on the added flavorings and seasonings.

Some of the basic sauces are:

• **For hors d'oeuvres and salads**—French dressing, mayonnaise, and salad dressings.

• **For main dishes**—Barbecue sauce, butter sauce, cheese sauce, fruit sauce, gravies, hollandaise sauce, tartar sauce, tomato sauce and ketchup, and white sauce.

• **For desserts**—Chocolate sauce, custard sauce, fruit sauces, hard sauce, and white sauce.

SAUERKRAUT

This is salted sliced cabbage that has undergone fermentation by lactic acid bacteria. In the fermentation process, the sugars of the cabbage are converted primarily to lactic and acetic acids, ethyl alcohol, and carbon dioxide.

The first description of sauerkraut manufacture, comparable to the methods used commercially today, was given by James Lind in 1772, in his treatise on scurvy. Barrels, kegs, and stone crocks were the common containers. Wooden vats were introduced in the United States about 1885. Recently, reinforced concrete with plastic coatings and glazed tile vats have been introduced.

In most modern kraut factories, salt is weighed and applied at a 2.25% level to the shredded cabbage. Ordinarily, sauerkraut is left in the vats until completely fermented.

Very little kraut is retailed in bulk today. Most of it is canned, although the use of plastic and glass containers is increasing.

Sauerkraut is often eaten as an adjunct to other foods, thereby making them more appetizing and digestible. It is generally recognized that when sauerkraut is cooked with other foods, particularly meats, it enhances their palatability.

(Also see CABBAGE.)

Fig. S-4. Sauerkraut—fermented cabbage—is often eaten as an appetizing adjunct to other foods. (Courtesy, University of Minnesota, St. Paul, Minn.)

SAUSAGE

Highly seasoned finely ground meat (usually pork or beef, but other meats may be used), most of which is stuffed in casings, used either fresh or cured.

SAUSAGE CASINGS

Natural casings are made from the middle wall of the small and large intestines of cattle, hogs, sheep, and goats. Also, the lung, bladder, and (in the case of hogs) the stomach are used as containers for special sausages.

Additionally, there are: cellulose casings, made from cotton linters; collagen casings, made from a collagen source, such as the corium layer of beef hide; and plastic netting made of polyethylene threads.

SCHOOL LUNCH PROGRAM

Contents   Page
History 944
Provisions of the National School Lunch Act 944
Benefits of the School Lunch 944
School Lunch Guidelines 945
Goal of the School Lunch Program 945

Fig. S-5. School Lunch Program. (Courtesy, University of Minnesota, St. Paul, Minn.)

The National School Lunch Program, which is administered by the U.S. Department of Agriculture (USDA), was established for the purposes of (1) safeguarding the health and well-being of American children, and (2) encouraging the consumption of nutritious agricultural commodities and other foods. It operates by assisting the states, through grants-in-aid and other means, in the establishment, maintenance, conduct, and expansion of nonprofit school lunch programs. Hence, the costs of providing lunches are shared by the USDA, state and local government, and the recipients of the meals.

A good understanding of the school lunch program is important to parents and teachers who influence the child's eating habits.

HISTORY. United States surplus agricultural commodities were distributed on a limited basis for free lunches as early as 1932. In 1935, the U.S. Department of Agriculture was authorized by law to purchase and distribute the commodities on a much larger scale. In 1942, surplus foods were used to feed 6.2 million children.

By 1948, however, the supplies of surplus agricultural commodities were greatly reduced as a result of wartime demands. Hence, the USDA started a cash reimbursement program to pay schools for a part of the food purchased locally. But, the participating schools were required to (1) meet certain nutritional standards in the lunches served, (2) provide free meals for the children unable to pay, and (3) operate their lunch programs on a nonprofit basis. These provisions formed much of the basis for the National School Lunch Act of 1946, which placed the basic responsibility for the administration of school lunch programs in the hands of state education agencies. Since then, the provisions of the original act have been modified several times by new laws that were enacted to bring about the serving of greater numbers of free and reduced-price lunches to needy children. Participating schools are provided (1) cash assistance, (2) donation of surplus food commodities, and (3) technical assistance in the purchase and use of foods and in the equipping and managing of the school lunchroom.

During the 1970s, authorizations and regulations were established for a Nutrition Education and Training Program (NET) that had the objectives of (1) teaching students about the relationships between food, nutrition, and health, (2) training food service personnel in the principles and skills of food service management, and (3) instructing teachers in sound principles of nutrition education.

On a typical day in 1990, lunches were served to 24.6 million children and adolescents. That year, the federal cost of school lunches was $3.2 billion.

PROVISIONS OF THE NATIONAL SCHOOL LUNCH ACT. The enactment of the National School Lunch Act of 1946 placed the school lunch program of the USDA on a firm foundation.

Eligibility for participation in the National School Lunch Program is contingent upon (1) operating the program on a nonprofit basis, (2) providing free or reduced-price lunches to needy children, (3) making lunches available without regard to race, color, or national origin, (4) providing kitchen and dining facilities, (5) avoiding public identification or segregation of needy children, and (6) serving nutritious lunches that conform to USDA guidelines.

BENEFITS OF THE SCHOOL LUNCH. Some of the main benefits of the National School Lunch Program follow:

1. Since the 1940s most of the children attending public and private schools in all 50 states, the District of Columbia, Puerto Rico, the Virgin Islands, and Guam have had the opportunity to obtain a school lunch.

2. Many needy children have been served nutritious lunches which they would not likely have received through other means. Fig. S-6 shows that the numbers of children receiving free or reduced-price lunches increased steadily throughout the 1980s.

3. About 50% of the families with children receiving lunches free or at reduced prices had incomes below $10,000.

Children in the National School Lunch Program

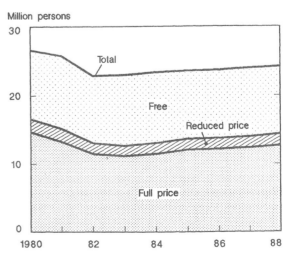

Fig. S-6. The number of children in the National School Lunch Program during the 1980s. (Source: *1989 Agricultural Chartbook*, USDA Ag. Hdbk. No. 684, p. 70)

4. School lunches have provided about one-third of the Recommended Dietary Allowances for school children on a daily basis.

5. Recent modifications in the food patterns have provided pupils with experiences in (1) eating foods used mainly by minority ethnic groups (items such as bulgur wheat and corn grits), and (2) new types of food preparations (such as meat dishes extended with soy protein).

6. Teachers have been motivated by students' comments on their lunches to spend more time discussing foods, health, and nutrition subjects.

7. Children, parents, teachers, and community groups have had opportunities to contribute viewpoints regarding the types of foods served. Some of these groups have been successful in making significant improvements in their local programs. For example, whole wheat pizza, sprouts, yogurt, and other highly nutritious foods are now served in certain local schools.

8. Poorly equipped and staffed schools have received both financial and technical assistance in developing better cooking and serving facilities and in upgrading the education and training of their personnel.

9. Surplus commodities have been utilized profitably and have helped to keep the prices of the lunches within the means of the students.

10. Local food suppliers have benefited from the purchases made for the program.

11. States have been induced to contribute some of their revenues to the program. At first, the state contributions came mainly from the payments of school children for the lunches.

SCHOOL LUNCH GUIDELINES.
Published U.S. Department of Agriculture school lunch guidelines are given in Table S-1.

GOAL OF THE SCHOOL LUNCH PROGRAM.
The goal of the school lunch program should be twofold:

1. The lunches offered to children should be nutritionally adequate and designed to develop and encourage the child's liking for a wide variety of the protective foods which frequently are not included in home diets.

2. The meals should be closely linked to the school's education program for better health.

These objectives merit further implementation. They need to be carried out by informed food service managers, cooks, and helpers in thousands of participating schools in every section of the country.

In addition to school lunches, it is noteworthy that, today, more and more children eat at least one meal outside the home—in day nurseries, child-care centers, day-care centers, or nursery schools. Usually, the noon meal is provided.

TABLE S-1
SCHOOL LUNCH GUIDELINES[1]
(Minimum amounts of foods listed by food components to serve students of various age/grade groups)

Food Component and Description	Preschool Children		Elementary School Students		Secondary School Students
	Group I (1 and 2 years of age)	Group II (3 and 4 years of age)	Group III (5-8 years of age; or Grades 1 to 3)	Group IV (9-11 years of age; or Grades 4 to 6)	Group V (12 years of age and over; or Grades 7 to 12)
	(amount per day)	(amount per day)	(amount per day)	(amount per day)	(amount per day)
Meat and meat alternates[2] Meat—a serving (edible portion) of cooked lean meat, poultry, or fish. Alternates: The following meat alternates may be used alone or in combination to meet the meat/meat alternate requirement:[3] Cheese (1 oz = 1 oz of meat), eggs (1 large egg = 1 oz of meat), cooked dry beans or peas (½ cup [120 ml] = 1 oz of meat),[4] peanut butter (2 Tbsp [30 ml] = 1 oz of meat)	1 oz (28 g)	1½ oz (43 g)	1½ oz (43 g)	2 oz (57 g)	3 oz (85 g)
Vegetables and fruits 2 or more servings consisting of vegetables or fruits or both. (A serving of full strength vegetable or fruit juice can be counted to meet not more than ½ of the total requirement.)	½ cup (120 ml)	½ cup (120 ml)	½ cup (120 ml)	¾ cup (180 ml)	¾ cup (180 ml)
Bread and bread alternates[5] 1 serving (1 slice) of enriched or whole-grain bread, or a serving of biscuits, rolls, muffins, etc. made with whole-grain or enriched meal or flour, or a serving (½ cup [120 ml]) of cooked enriched or whole-grain rice, macaroni, noodles, and other pasta products.[6]	5 servings per week	8 servings per week	8 servings per week	8 servings per week	10 servings per week
Milk, fluid[7] Unflavored fluid, lowfat milk, skim milk, or buttermilk must be offered as a beverage. If a school serves whole or flavored milk, it must offer unflavored fluid lowfat milk, skim milk, or buttermilk as a beverage choice.	¾ cup (180 ml)	¾ cup (180 ml)	½ pint (240 ml)	½ pint (240 ml)	½ pint (240 ml)

[1]Adapted by the authors from *Code of Federal Regulations*, Title 7, Revised January 1, 1980, Part 210, Sec. 210.19b.

[2]It is recommended that in schools not offering a choice of meat/meat alternate each day, no one form of meat (ground, sliced, pieces, etc.) or meat alternate be served more than three times per week. Meat and meat alternates must be served in a main dish, or in a main dish and one other menu item.

[3]When it is determined that the serving size of a meat alternate is excessive, the particular meat alternate shall be reduced and supplemented with an additional meat/meat alternate to meet the full requirement.

[4]Cooked dry beans or peas may be used as the meat alternate or as part of the vegetable/fruit component, but not as both food components in the same meal.

[5]One-half or more slices of bread or an equivalent amount of bread alternate must be served with each lunch with the total requirement being served during a 5-day period. Schools serving lunch 6 or 7 days per week should increase the quantity specified for a 5-day week by approximately 20% for each additional day.

[6]Enriched macaroni products with fortified protein may be used as part of a meat alternate or a bread alternate, but not as both food components in the same meal.

[7]One-half pint of milk may be used for all age/grade groups if the lesser specified amounts are determined by the school authority to be impractical.

A sound school lunch program can help to improve the diets and food habits of children, and, thus, eventually lead to better diets and food habits for the population as a whole. This goal merits the support of federal, state, and local governments, and of parents and school officials.

(Also see CHILDHOOD AND ADOLESCENT NUTRITION, section on "School Lunch and Breakfast Programs"; and GOVERNMENT FOOD PROGRAMS.)

SCLERODERMA

One of the first signs of scleroderma is tight, firm skin, and eventually a diffuse thickening and rigidity of the skin and subcutaneous tissue. These changes in the skin are reflected in the name of this disease, which means *hard skin*. Sometimes the disease is called *hidebound skin*. As the disease progresses, it involves the internal organs, specifically the intestinal tract, the heart, and the kidneys.

The cause of scleroderma is unknown. It belongs to a larger group of diseases sometimes classed as collagen diseases. In scleroderma, the changes in the skin and internal organs seem to be due to the overproduction of collagen—connective tissue.

There is no specific treatment for scleroderma. Corticosteroid therapy may slow the disease, or produce improvement in the early stages. This may require a sodium-restricted diet to prevent sodium retention during corticosteroid therapy. As the disease progresses, swallowing difficulties and a malabsorption syndrome may be encountered, which require some dietary adjustments.

SCRAPPLE

Made from head meat, feet, hearts, tongues, shoulder, spare ribs, fresh picnic shoulders, or any pork trimmings that contain some fat. Liver may be used if desired. Twenty percent of the meat may consist of beef or veal, but all pork is preferable.

Cook the meat in sufficient water to keep it covered; drain off the liquid when the meat separates readily from the bones. Remove the bones, then grind the meat. Place the ground meat and the liquor in which it was cooked together in a kettle and bring it to a boil.

Mix meal or flour (cornmeal, oatmeal, buckwheat flour, or soybean flour) with some water or some of the meat juice, add slowly and work the cereal to avoid lumps. Pour the diluted meal or flour into the cooked meat, season with condiments and herbs, and cook for another 30 minutes. Pour into a mold to cool, and serve sliced and fried.

SCREENED

A ground material that has been separated into various-sized particles by passing over or through screens.

SCREENINGS

By-products obtained from screening grains and other seeds.

SCURVY

Contents   Page
History of Scurvy...................................946
Causes of Scurvy...................................947
Signs and Symptoms of Scurvy......................947
    Special Features of Scurvy in Infants.................947
    Treatment and Prevention of Scurvy.................947

A severe deficiency of ascorbic acid (vitamin C) results in a specific disease called scurvy. The major lesions of scurvy are believed to be due to an impairment in the formation of connective tissue which requires ascorbic acid for its synthesis and repair. This impairment leads to pathologies of bone, teeth, skin, muscle, joints, adrenal glands, and blood vessels. Also, scurvy may be accompanied by hemorrhages of the adrenal glands and a gross impairment of adrenal function. If scurvy is not promptly treated, internal hemorrhages of increasing severity occur and death may follow.

Fig. S-7. This shows British Sailors who ate little except salt meat and biscuits when on long voyages 200 years ago, collapsing and dying of scurvy. (Reproduction with permission of *Nutrition Today*, P.O. Box 1829, Annapolis Maryland, 21404, 1979)

HISTORY OF SCURVY. Although disorders resembling scurvy were described in ancient medical writings (such as *Papyrus Ebers* in Egypt about 1550 B.C., and in Greek and Roman writings), the causative agent was frequently attributed to a plague or other infectious diseases. Perhaps this happened because scurvy may be accompanied by lesions that resemble those associated with certain contagious diseases.

Most long sea voyages suffered large losses of men from scurvy. The greatest number of outbreaks occurred, however, after the change from oar-driven vessels which traveled only short distances from land, to sail-powered ships which traversed the high seas. There was not enough time during short voyages for depletion of the vitamin C stored in the tissues. Fresh fruits were not generally carried aboard such ships, since it was feared that they might spoil, and thereby cause diarrhea.

The first production of scurvy in an experimental animal was accomplished in 1907 by Holst and Frolich. These Norwegian investigators showed that it was necessary to supplement a ration of oats, rye, and rice, with vegetables and fruits, in order to protect guinea pigs against scurvy. This was a landmark experiment because only guinea pigs, monkeys, and man require vitamin C in their diets; other species are able to synthesize this substance in their livers.

These investigators also found that substances which did not normally protect against scurvy (dried grains and legumes—such as oats, barley, peas, beans, and lentils) developed the ability to protect against scurvy when they were germinated into sprouts.

The concept of vitamins was born in 1912 when Hopkins, in England, showed that animals apparently required in their diet small amounts of what he called *accessory food factors*. At the same time, Funk, a Polish scientist working in London, developed the concept of special organic compounds called *vitamines* (later changed to vitamins). In 1928, Szent-Gyorgyi, a Hungarian biochemist, working in Hopkin's laboratory, isolated "hexuronic acid" from orange and cabbage juices and adrenal glands of animals. That same year, King, at the University of Pittsburgh, showed that hexuronic acid was vitamin C. In 1933, the British chemist, Haworth, determined the structure of ascorbic acid and Reichstein, of Switzerland, synthesized ascorbic acid. These investigations made possible the development of a commercial synthesis of vitamin C from glucose. As a result, vitamin C can now be made in very large quantities and at a low cost.

(Also see VITAMIN C, section on "History.")

CAUSES OF SCURVY.
Scurvy is caused by a severe deficiency of vitamin C.

In recent times, very few persons are known to have died from scurvy, except, perhaps, prisoners of war and malnourished infants. However, mild cases of scurvy may occur in infants, especially when they're raised on cow's milk; in adults who eat few vegetables or fruits; and in older persons who live on reduced incomes and under stress.

Fig. S-8. Scurvy can be prevented by a daily serving of citrus fruit or juice-grapefruit (shown above) oranges, lemons or limes. (Courtesy, USDA)

SIGNS AND SYMPTOMS OF SCURVY.
Prior to the appearance of specific signs of the disease, the victim may feel feeble and listless, be short of breath, and have slow healing of wounds. Next, there develops the characteristic feature of swollen, bleeding gums (gingivitis), which may become at least slightly infected and impart a foul odor to the breath. Sometimes, there is loosening or loss of teeth. Another early sign of scurvy is the minute hemorrhages (petechiae) which appear around the hair follicles on the abdomen, buttocks, arms, and legs. As the disease progresses, these petechiae merge into larger hemorrhagic areas or bruises, which are unusual in that they occur in the absence of trauma. At this stage of the disease, the victim usually experiences severe pain in his or her bones, joints, and muscles.

Fig. S-9 depicts the afflictions which often accompany scurvy.

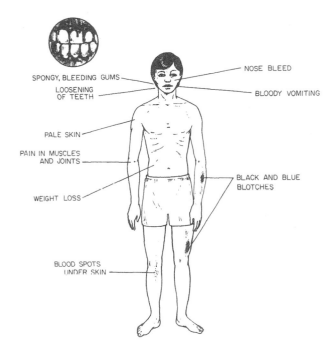

Fig. S-9. Signs and symptoms of scurvy.

Special Features of Scurvy In Infants. The first signs and symptoms of scurvy in infants differ from those in adults because the infant has special requirements for ascorbic acid to sustain growth and development. Thus, early features of infant scurvy may be retarded growth, irritability, lack of appetite, and shrinking from the touch of others in anticipation of pain. Soon, the infant may assume the *frog-legs* position (lying on its back, with the thighs spread widely apart and the legs flexed at the knees). There may also be bone malformations such as *bayonet chest* (a condition in which the breastbone or sternum gradually slopes out from the collarbone, then is indented sharply at its tip due to the sinking of the chest wall just above the belly) and beading of the ribs like that which occurs in rickets. Sometimes, there may be bloody stools or urine and discoloration of the eyelids.

TREATMENT AND PREVENTION OF SCURVY.
Sudden death sometimes occurs in severe scurvy, so prompt and aggressive treatment is necessary. The usual treatment consists of ascorbic acid in doses as high as 250 mg each, four times a day, for a week. The purpose of such high levels is to achieve rapid saturation of the body fluids with ascorbic acid so as to hasten the healing of diseased tissues. Excesses of the vitamin are excreted in the urine, so there is little

danger of large doses over a *short* period of time. The long-term effects of such dosage may be hazardous. When the crisis has passed, sufficient amounts of the vitamin may be provided by diets which contain one or more servings per day of fresh fruits or vegetables. Vitamin tablets may need to be provided for persons who do not eat such foods.

Although it has been estimated that as little as 10 mg per day of ascorbic acid will prevent scurvy, such a low intake is not likely to provide a reserve against such unexpected stresses as disease, trauma, and chilling. The Recommended Dietary Allowances (RDAs)[2] are 60 mg per day of ascorbic acid for adult males and for females who are neither pregnant nor lactating.

Infants should receive 30 mg of vitamin C from birth to 6 months of age, and 35 mg from 6 to 12 months of age. Therefore, at least one rich source of ascorbic acid (such as orange juice) should be given to a baby each day. In the event that citrus juices are not tolerated well, there are vitamin C drops which may be put in the drinking water.

(Also see DEFICIENCY DISEASES, Table D-1 Major Dietary Deficiency Diseases; and VITAMIN C.)

SEAFOOD

This refers to marine fish and shellfish used as food. A great variety of food is prepared from many species of fish and shellfish. Seafood products are preserved by refrigerating, freezing, canning, salting, smoking, pickling, dehydrating, or by combinations of these processes. In many countries, seafoods serve as a principle source of protein and an important source of fat, minerals, and vitamins in the diet.

(Also see FISH AND SEAFOOD[S].)

SEA KALE *Crambe maritiima*

This is a European perennial herb that has a fleshy branching rootstalk and is sometimes cultivated for its large ovate long-stalked leaves which are used as a potherb.

SEAWEED (KELP)

Any plant that grows in the sea is a seaweed. Botanically, seaweeds are algae.

For centuries, seaweed has been used as a food for humans and animals; prized for its minerals and vitamins.

• **Many names and varieties of seaweed**—One of the most confusing things about seaweed is the various names by which it is known; among them, seaweed, agar-agar, algae, carrageenan, dulse, Iceland moss, Irish moss, kelp, laver, rockweed, and sea lettuce. It has been estimated that there are some 2,500 varieties of marine plants; so, seaweed embraces a great variety of plants of many hues, shapes, and sizes. It is noteworthy, too, that seaweed is one of the few members of plant life that has remained unchanged for centuries, since its cultivation and growth is still controlled by the elements—not man.

Seaweeds are algae, of which there are four principal groups:

> Brown algae
> Red algae
> Green algae
> Blue-green algae

When the botanist speaks of seaweed, he usually means one of the larger brown or red algae. The seaweeds of cold waters are chiefly brown algae; those of the tropics are mainly red algae.

• **Composition of seaweed**—The Norwegian Seaweed Institute reports the following proximate analysis of seaweed:

Component	Percent
Protein	5.7
Fat	2.6
Fiber	7.0
Nitrogen-free extract	58.6
Ash	15.4
Moisture	10.7
	100.0

Contrary to claims that are sometimes made, seaweeds are low in protein, and the protein is of very low biological value.

The Norwegian Seaweed Institute has found an assortment of 60 different mineral elements in seaweed, all harnessed from the sea. Additionally, seaweed contains carotene, vitamin D, vitamin K, and most of the water-soluble vitamins, including vitamin B-12.

CAUTION: Dried kelp (seaweed) is so rich in iodine that consumption of a large amount for a prolonged period may be harmful; hence, it should be taken according to directions.

(Also see IODINE, section headed "Toxicity.")

• **Uses**—Seaweed has many important uses. Mention has already been made of seaweed as a staple food for humans and animals. For human food, kelp can be used in a variety of ways. In powdered form, it can be added to soups, salads, cottage cheese, tomato juice, fruit juices, or sprinkled on baked potatoes. Some persons use it as a salt substitute. Also, it is made into tablets and sold as a mineral-vitamin supplement in health food stores. About 25% of all food consumed in Japan consists of one form or another of seaweed (sea-vegetable), prepared and served in many forms.

Seaweed has many important uses, in addition to food. It has long been harvested and made into fertilizers. During World War I, giant kelp was harvested and made into explosives. Chemists extract large amounts of iodine and algin from seaweeds. Algin has many commercial uses because it can hold several different liquids together. In ice cream, it keeps the water in the milk from forming crystals. In the food and bakery industry, it is used for thickening, gelling, or binding products. It is also used in salad dressings, chocolate milk, aspirin, and in other foods and drugs.

Seaweed is sometimes promoted for its therapeutic properties in alleviating constipation, gastric catarrh, mucous colitis, and other disorders. But these claims need to be substantiated by more properly conducted and controlled experiments.

Some years ago, scientists became interested in the longevity of the population of Hizato situated in the Nagano province of Japan. Nearly 10% of the villagers were over 70 years of age; and a survey showed that there were 250% more 70-year-olds than in any other Japanese village. After extensive study, the conclusion was reached that the diet of the inhabitants of Hizato was a contributing factor to their longevity. The food consumed by these villagers consisted of all kinds of vegetables, including dried seaweed.

• **Sea farming and world food shortage**—There exist almost unlimited opportunities for increased sea farming. The water area of the world is many times greater than the land space; and phenomenal yields of seaweed are obtained

---

[2]*Recommended Dietary Allowances*, 10th rev. ed., Food and Nutrition Board, NRC-National Academy of Sciences, 1989.

—as much as 60 tons per acre. Not only that, seaweed farming does not suffer from drought or loss of crop through pests and disease; and seaweed requires no planting, weeding, or fertilizing. Some scientists predict that by the turn of the century the world's sea crops will have to be farmed to ensure the survival of our teeming population.

With one of the world's densest populations and a long coastline, the Japanese already have great expertise in the art of sea farming. Teams of girls, all expert swimmers and skin divers, play a part in this form of cropping, which is also known as aqua-culture. These underwater laborers are specially trained and equipped to carry out the cutting of seaweed from cultivated beds off the seashores.

(Also see ALGAE; and CARRAGEENAN.)

SEEDS

Lots of seeds are eaten for human nutrition. A seed is the primary reason for which whole plants exist. It is the next generation of plants, since it contains a new plant in the form of an embryo plus stored food to nourish the embryo as it develops. Since the plant embryo requires proper nutrition from the stored food, seeds are good sources of protein, carbohydrate, fat (oil), minerals, and vitamins.

Seed types are as diverse as the plant kingdom. Seeds come in all shapes and all sizes, from the dustlike seeds of the orchid to the large seed contained in the coconut.

Man has learned to use a variety of seeds for the food stored in them.

Specific information about seeds consumed by humans can be located under individual articles or in Food Composition Table F-21.

SELECTIVE PERMEABILITY

When only certain substances are permitted to pass through the membrane, and others are rejected.

SELENIFEROUS

Areas in which the soil is high in the mineral selenium. (Also see SELENIUM.)

SELENIUM (Se)

Contents   Page
History..949
Selenium in Soils...................................949
　Regional Differences in the Selenium Content of
　　Foods...950
Absorption, Metabolism, Excretion...................950
　Factors Which May Affect Dietary Selenium.........950
Functions of Selenium..............................951
　Interrelationship of Selenium and Vitamin E.......951
Deficiency Symptoms................................951
　Evidence of Selenium Deficiencies.................951
Associations Between Certain Diseases and
　Selenium...952
Treatment and Prevention of Selenium Deficiency....954
Interrelationships.................................954
Recommended Daily Allowance of Selenium............954
Toxicity...954
Sources of Selenium................................955
Summary..955

This comparatively rare, nonmetallic element, which makes up less than 0.0001% of the earth's crust, has recently been found to be essential. Fortunately, it is needed only in minute amounts, because (1) only traces are present in most foods, and (2) poisoning may result when the dietary level of this element is 0.0003% or greater. However, it seems that adequate dietary selenium is one of the keys to the maintenance of health under stressful conditions such as (1) prematurity at birth, (2) protein-energy malnutrition, and (3) the tissue disorders which accompany aging.

HISTORY. This element also occurs as a trace contaminant of substances which contain sulfur, because the chemical properties of the two elements are similar. Therefore, it is noteworthy that in 1817 the Swedish chemist Berzelius discovered the element selenium while testing the residue which remained after sulfur had been burned to make sulfuric acid.

In the 1950s, indirect evidence of the beneficial effects of dietary selenium accumulated slowly, but steadily. During the 1940s, German scientists had tested the European type of brewers' yeast for use as a protein supplement and found that it sometimes produced a liver disease in rats which was prevented by feeding wheat germ or other sources of vitamin E. Then, in 1951, the German medical researcher Schwarz, who at the time was a visiting scientist at the National Institutes of Health (NIH), discovered that the American type of brewers' yeast contained an unidentified *Factor 3* which apparently acted along with vitamin E and sulfur-containing amino acids in protecting the liver against damage due to certain types of diets. Schwarz stayed in the United States to continue his research, and, in 1957, he and a co-worker, Foltz, reported that Factor 3 contained selenium.

Other studies by the group at NIH, and by Scott and his co-workers in the Department of Poultry Husbandry at Cornell, showed that the addition of selenium salts to the diets of chicks prevented certain disorders which resulted from vitamin E deficiencies. The pace of discovery quickened as other investigators showed that selenium protected calves and lambs against white muscle disease (nutritional muscular dystrophy). Finally, in 1973, Rotruck and his co-workers at the University of Wisconsin reported that selenium acted as a co-factor for a recently-discovered enzyme (glutathione peroxidase) which breaks down toxic peroxides, most of which are formed from the oxidation of polyunsaturated fats. Hence, the link between selenium and vitamin E was shown; with selenium participating in the breaking down of these highly toxic compounds, and with vitamin E preventing their formation. Nevertheless, it may be that new chapters will be added to the story on selenium, as recent research suggests that there are other roles for this essential element.

SELENIUM IN SOILS. In large areas of the United States, the soils contain little available selenium. Plants produced in these areas are low in this element, and selenium deficiency in livestock (and perhaps in people) is a serious problem. It would appear, therefore, that if foodstuffs grown in these areas are to exert their maximum benefit, some means of combatting their selenium deficiency must be achieved, and this must be accomplished without hazard.

In the Great Plains and Rocky Mountain States (especially in parts of the Dakotas and Wyoming), however, some of the soils are so rich in available selenium that the plants produced thereon are so high in selenium that they are poisonous to animals that eat them.

Fig. S-10 is a map showing areas of the United States where forages and grains have been found (1) low, (2) variable, and (3) adequate or high in selenium. As noted, deficient and high regions are scattered around the country, and some states contain two or more levels of concentration.

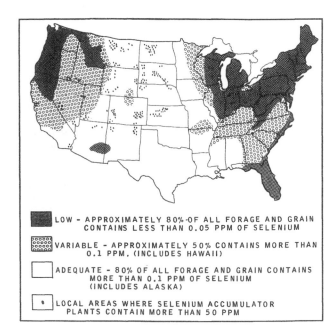

LOW - APPROXIMATELY 80%-OF ALL FORAGE AND GRAIN
CONTAINS LESS THAN 0.05 PPM OF SELENIUM

VARIABLE - APPROXIMATELY 50% CONTAINS MORE THAN
0.1 PPM. (INCLUDES HAWAII)

ADEQUATE - 80% OF ALL FORAGE AND GRAIN CONTAINS
MORE THAN 0.1 PPM OF SELENIUM
(INCLUDES ALASKA)

LOCAL AREAS WHERE SELENIUM ACCUMULATOR
PLANTS CONTAIN MORE THAN 50 PPM

Fig. S-10. Geographic distribution of low, variable and adequate or high selenium areas in the United States. (Source: Kubota, J. and W. H. Allaway, "Geographic Distribution of Trace Element Problems," *Micronutrients in Agriculture*, Soil Science Society of America, 1072.)

Regional Differences in the Selenium Content of Foods.

The average selenium content of the crops used for human foods and livestock feeds is highest in the region of the United States which lies west of the Mississippi River and east of the Rocky Mountains. On the other hand, the areas which yield the crops that are very low in selenium are (1) the northern states between the Atlantic Ocean and the Mississippi River; (2) the southeastern coastal plain, including most of the state of Florida; and (3) the western sections of Washington, Oregon, and northern California. Hence, it is noteworthy how these regional differences affect the selenium content of grains and meats, which are the main sources of the mineral in the American diet.

ABSORPTION, METABOLISM, EXCRETION.

Ingested selenium is absorbed in the intestine, mainly in the duodenum. Thence, it is bound to a protein and transported in the blood to the tissues, where it is incorporated into tissue protein as selenocysteine and selenomethionine; in the latter process, selenium replaces the sulfur in the amino acids cysteine and methionine. Excretion of selenium is largely by way of the kidneys, although small amounts are excreted in the feces and in sweat.

Many factors may affect dietary selenium; some enhancing it, others reducing it. These factors are elucidated in the section that follows:

Factors Which May Affect Dietary Selenium.

It is not surprising that deficiencies of selenium are sometimes hard to detect, because vitamin E and the sulfur-containing amino acids (cysteine and methionine) may act as partial substitutes for the mineral in some of its functions. Other nutrients, such as fat and protein, may also affect the body's need for selenium. Therefore, some of the major factors which influence selenium nutrition follow:

• **Biological availability of selenium from various foods** —All of the selenium in wheat appears to be biologically available whereas only about ⅓ of that in herring and tuna is available.[3] It might be that the selenium in certain species of ocean fish—which contain more of this element than most other foods—has a reduced availability because some of it is chemically bound to mercury. This idea is supported by the following evidence: (1) Many ocean fish contain small amounts of methylmercury; (2) selenium compounds have been shown to bind tightly with methylmercury; and (3) the toxicity of the methylmercury in tuna appears to be reduced greatly by the presence of selenium.[4]

• **Dietary protein**—Increasing the dietary protein intake reduces the chances of selenium poisoning when foods containing potentially hazardous amounts of the mineral are consumed. (High-selenium foods are produced only in the areas of the United States around the Rocky Mountains and in the Upper Missouri River Valley of the Dakotas.) However, most people are more likely to consume too little selenium rather than too much; so, protein foods are also valuable in this regard, because they are generally the best sources of selenium.

• **Form and source of the mineral**—Although seafood has a high selenium content, its value as a dietary source is reduced because of the poor availability of its selenium. For example, the selenium in tuna is only 33 to 50% available, possibly because it tends to bind to heavy metals such as mercury.

The selenium content of plant foods reflects the content of the soil in which they are grown. The soil content also influences the amount of selenium in muscle, eggs, or milk of animals raised on crops grown in the soil. Organ meats, muscle meats, cereals, and dairy products rank in descending order as good sources of selenium. The amount of selenium present in food tends to parallel the protein content of the food. Selenium is reduced in the milling process and is lost as vegetables are boiled.

Inorganic salts of selenium, called selenites, should be taken only under the direction of a medical doctor.

• **Intestinal microorganisms**—*E. coli, Streptococcus faecalis* and certain species of *Clostridia* and *Salmonella* metabolize selenium by (1) incorporating it into their enzymes, (2) utilizing selenium amino acids instead of sulfur amino acids, and/or (3) converting soluble forms of selenium into insoluble forms. The insoluble selenium is not likely to be absorbed, if it is produced in the intestines of people who harbor these microorganisms. Therefore, the dietary allowances for selenium should take into account the possibility that intestinal bacteria might somehow interfere with the utilization of selenium by animals and man.

• **Iron-deficiency anemia**—This disorder may be accompanied by a reduction in the activity of a selenium enzyme (glutathione peroxidase) in human red blood cells. Supplementation of the diet with iron sometimes restores the enzyme activity to normal, providing that the dietary supply of selenium is adequate. Hence, the optimal utilization of selenium may depend upon adequate iron nutrition.

• **Malnutrition**—Certain malnourished children who were given therapeutic formulas based upon skim milk powder did not respond well and had (1) subnormal blood levels of selenium, (2) a persistent anemia, and (3) very poor

[3]*Selenium and Interrelationships in Animal Nutrition*, Hoffman-La Roche, Inc., Nutley, N.J., 1974, p. 4.

[4]*An Assessment of Mercury in the Environment*, National Academy of Sciences, 1978, pp. 84-86.

growth; all of which were corrected by giving them a selenium supplement.[5] However, one of the children required extra vitamin E to bring about the restoration of red blood cells. These findings show the need for (1) administering selenium supplements when rehabilitation diets contain only minimal amounts of this mineral; and (2) providing sufficient vitamin E, which enhances some of the effects of selenium.

- **Polyunsaturated fats from vegetable oils**—A diet rich in polyunsaturated fatty acids (PUFA) but poor in vitamin E might raise the requirement for selenium because (1) PUFA may be readily converted to toxic peroxides by various metabolic processes unless there is sufficient vitamin E to prevent this conversion, and (2) selenium is needed to activate the enzyme (gluthathione peroxidase) which destroys the peroxides.

- **Processing and preparation of foods**—The losses of selenium during the milling of grains into flours are smaller than those for the other essential minerals, because selenium is distributed throughout the kernel, whereas the other minerals are concentrated in the outer layers of the grain which are removed during milling.

Most cooking procedures have been found to cause little loss of selenium from most foods, except that the boiling of asparagus and mushrooms resulted in significant losses of the mineral (29% and 44%, respectively).[6]

- **Stresses**—Pigs which had been subjected to social and environmental stresses such as crowding, chilling, and overheating developed the severe disorders associated with selenium deficiency much sooner than unstressed animals.[7] Hence, it seems likely that highly stressed people might have above average requirements for selenium, since pigs and people are very similar in the ways that their bodies respond to various stresses.

- **Sulfur-containing amino acids**—Certain disorders which result from selenium-deficient diets are prevented in part by the sulfur-containing amino acids methionine and cysteine because they are converted in the body to glutathione, a substance that has a limited ability to carry out some of the functions of the selenium enzyme glutathione peroxidase. Hence, the need of the human body for selenium is greater when the dietary levels of the sulfur-containing amino acids are low. It is noteworthy that vegetarian diets may be low in the amino acids and low in selenium, unless large amounts of high-selenium wheat are consumed.

- **Vitamin E**—The metabolic roles of selenium and vitamin E overlap, so that each nutrient may replace the other to a limited extent in preventing certain types of disorders. However, there are also unique functions for each nutrient, so each must be supplied in the diet to ensure good nutrition. Furthermore, the experimental supplementation of animal feed with vitamin E resulted in an improvement in selenium absorption, and a doubling of the amount of the mineral which accumulated in the liver.[8]

[5]Hopkins, L. L., Jr., and A. S. Majaj, "Selenium in Human Nutrition," *Selenium in Biomedicine*, edited by O. H. Muth, The Avi Publishing Company, Inc., Westport, Conn., 1967, pp. 205-206.

[6]Higgs, D. J., V. C. Morris, and O. A. Levander, "Effect of Cooking on Selenium Content of Foods," *Journal of Agricultural and Food Chemistry*, Vol. 20, 1972, p. 678.

[7]Ullrey, D. E., "The Selenium-Deficiency Problem in Animal Agriculture," *Trace Element Metabolism in Animals-2*, edited by W. G. Hoekstra, et al., University Park Press, Baltimore, Md., 1974, pp. 276-277.

[8]Combs, G. F., Jr., "Influences of Vitamin A and Other Reducing Compounds on the Selenium-Vitamin E Nutrition of the Chicken," *Proceedings of the 31st Distillers' Feed Conference*, 1976, p. 44, Table VIII.

FUNCTIONS OF SELENIUM. The essential functions of selenium have been proven conclusively in many species.

The best known biochemical function of the element is its role as part of the enzyme glutathione peroxidase, which protects vital components of the cell against oxidative damage. Selenium also functions as a protective agent against toxic substances. It appears that inorganic forms of selenium somehow bind with toxic minerals such as arsenic, cadmium, and mercury—and render them less harmful. However, the detoxification of these poisonous elements might tie up some of the selenium which would ordinarily be utilized by the body. This effect may be desirable when dietary selenium is excessive, but undesirable when it is marginal or deficient. Additionally, selenium functions in an interrelationship with vitamin E. Details relative to the selenium-vitamin E relationship follow.

Interrelationship of Selenium and Vitamin E. During the 1950s, an interrelationship between selenium and vitamin E was established. It was found that selenium prevented exudative diathesis in vitamin E-deficient chicks and liver necrosis in vitamin E-deficient rats. Subsequent research demonstrated that both selenium and vitamin E protect the cell from the detrimental effects of peroxidation, but each takes a distinctly different approach to the problem. Selenium functions throughout the cytoplasm to destroy peroxides, while vitamin E is present in the membrane components of the cell and prevents peroxide formation. This explains why the biological need for each nutrient can be offset, at least partially although not totally, by the other.

DEFICIENCY SYMPTOMS. There are no clear-cut signs or symptoms which indicate when people are mildly to moderately deficient in selenium. However, many tests on diverse species of animals and man have shown that blood and tissue levels of selenium are in most cases accurate reflections of the selenium nutritional status. But the interpretation of laboratory findings and the diagnosis of selenium deficiency are strictly jobs for health professionals, even though certain laboratories may analyze specimens for lay persons. Hence, the information which follows is presented so that the patient and his or her doctor might consider whether such tests are advisable.

- **Blood tests**—The selenium content of red blood cells is higher than the levels in either whole blood or plasma, because more of the selenium enzyme glutathione peroxidase is present in the red cells than in the other fractions of blood. The selenium concentration in the red blood cells of normal vs sick children is about as follows: healthy children, 0.4 mcg/ml; malnourished children, 0.2 mcg/ml.

- **Analysis of hair and fingernails**—The selenium contents of the hair and fingernails appear to be rough indicators of the selenium content of the rest of the body, so clippings of these materials might be a painless way of testing people for their nutritional status. However, the tests may be invalid if the patient has recently used a shampoo or a medication which contains selenium. Selenium is the active ingredient in preparations used to treat dandruff and fungus infections of the skin.

Evidence of Selenium Deficiencies. There is no hard evidence of human selenium deficiency in the United States. Nevertheless, a growing number of nutrition researchers believe that the well-documented and severe disorders which occur in all species of selenium-deficient farm animals show the need for closer examination of the circumstantial evidence for human deficiencies. Also, it is noteworthy that several studies have shown that there are low levels of selenium in the diet and/or in the blood of severely malnourished children in developing countries.

Associations Between Certain Diseases And Selenium.

Studies of the geographical distribution of certain diseases have shown that the disease rates are lower in areas where the selenium content of crops is higher than average. Such findings do not necessarily prove that selenium prevents the diseases, or that high rates of the diseases indicate selenium deficiencies; although they give reason to probe further into the association between certain diseases and selenium.

According to the National Research Council, National Academy of Sciences, "One could reasonably suppose that selenium is involved in such human medical problems as cancer, cataracts, diseases of the liver, cardiovascular or muscular diseases, and the aging process."[9] Additional health problems which appear to result from a selenium deficiency are discussed in this section, also.

Selenium participates in some of the important metabolic processes which take place in many of the body's tissues. Therefore, the disruption of these processes by deficiencies of selenium might be expected to result in various disorders if the deficiencies are not corrected promptly.

NOTE: This is *not* to say that selenium deficiency is the one and only cause of the disorders which follow, but rather that the possibility of such a deficiency ought to be taken into account in the planning of measures to prevent or to treat these conditions.

• **Cancer**—Certain forms of this disease might result from both the effects of potent carcinogens and the weakening of the body's defenses by deficiencies of nutrients such as selenium.

Some human population data and laboratory experiments indicate that selenium may have anticancer properties. An increased incidence of cancer in human beings has been correlated with decreased levels of selenium in the blood. Also, selenium has been found to have some inhibitory effect on the development of tumors in rodents given certain carcinogens.

The cancer rates are generally higher in the states which have low levels of selenium in their crops.[10] In many cases, however, the states which have high cancer rates have other potentially health-threatening conditions such as (1) heavy industrialization, (2) mining and ore processing, and/or (3) large populations.

Shamberger and Frost reported that human cancer mortality might bear an inverse relationship to selenium distribution.[11] Also, controlled animal studies conducted by Shamberger at the same time showed an inhibitory effect of selenium on cancer.[12]

Other studies showed that (1) certain male cancer patients had subnormal blood levels of selenium; and (2) administration of supplemental selenium slowed the growth of tumors which were induced in animals by various carcinogens.[13]

One of the current theories regarding the prevention of cancer by selenium postulates that this element acts by (1) counteracting poisonous substances which may cause cells to mutate; and (2) stimulating the defenses of the body that act against abnormal cells.

Professor Gerhard N. Schrauzer of the University of California, San Diego, considers selenium to be a promising candidate for human cancer protection. He has been studying the effects of selenium in animals for 15 years. Using mice infected with viral particles which cause breast cancer, he showed that only 10% of the animals given nontoxic amounts of selenium in the drinking water or the diet developed breast cancer as compared to 80-90% in the controls.[14] His work was confirmed by Dr. Daniel Medina,[15] Baylor College of Medicine, in Houston.

Currently research on the cancer-protecting effects of selenium is being conducted in laboratories all over the world.

• **Heart disease and selenium**—A review of the effects of selenium deficiency reveals a repetitious undercurrent of vascular-type lesions suggesting a heart involvement in selenium disorders. Sudden death associated with selenium deficiency in newborn or rapidly growing lambs and calves apparently results from weakening of the heart muscle, commonly called white muscle disease. Selenium-deficient monkeys also have heart lesions. This disorder is characterized by white streaks in muscles, and in some cases, by abnormal electrocardiograms. Often, the animals affected with this disease die suddenly when they are subjected to moderate stresses.

Fig. S-11. Heart of a 6-week-old calf afflicted with white muscle disease. Note abnormal white areas in the heart muscles. (Courtesy, Oregon Agricultural Experiment Station, Corvallis, Ore.)

Some researchers have suggested that combined deficiencies of selenium and vitamin E might contribute to heart disease in man because these nutrients help maintain adequate levels of coenzyme Q in the heart muscle. (*Coenzyme Q is a vital cofactor in energy metabolism.*) If this cofactor is lacking, the production of energy in the heart and in other muscles may fall off to the extent that these tissues can no longer carry their workloads. This hypothesis is supported by the finding that biopsy samples from the hearts of people with heart disease had subnormal levels of coenzyme Q. Hence, it is noteworthy that the administration of selenium to farm animals raised the levels of coenzyme Q in their hearts.

[9]*Selenium*, Subcommittee on Selenium, National Research Council, National Academy of Sciences, Washington, D.C., 1976, p. 152.

[10]Schrauzer, G. N., "Trace Elements, Nutrition and Cancer: Perspectives of Prevention," *Inorganic and Nutritional Aspects of Cancer*, edited by G. N. Schrauzer, Plenum Press, New York, N.Y., 1978, p. 326, Fig. 1.

[11]Shamberger, R. J., and D. V. Frost, "Possible Protective Effect of Selenium Against Human Cancer," *Canadian Medical Association Journal*, Vol. 100, 1969, p. 682.

[12]Shamberger, R. J., "Relation of Selenium to Cancer. 1. Inhibitory Effect of Selenium on Carcinogenesis," *Journal of National Cancer Institute*, Vol. 44, 1970, pp. 931-936.

[13]"Selenium and Cancer," *Nutrition Reviews*, Vol. 28, 1970, pp. 75-80.

[14]Schrauzer, G. N., "Trace Elements, Nutrition and Cancer: Perspectives of Prevention," *Inorganic and Nutritional Aspects of Cancer*, edited by G. N. Schrauzer, Plenum Press, New York, N.Y., 1978, p. 330, Scheme 1.

[15]Medina, D. and F. Shepherd, *Cancer Lett.* 8, 241-245, 1980.

A sure sign of a dietary deficiency of selenium and/or of vitamin E in poultry and pigs is damaged blood vessels that leak into the surrounding tissues.[16] It is believed that this disorder is prevented by (1) selenium which circulates in the blood as part of the enzyme glutathione peroxidase, and (2) vitamin E which protects the membranes in the cells lining the blood vessels. While it does not appear that people develop the acute disorder which occurs in farm animals, it may be that the development of atherosclerosis is hastened by selenium deficiency, because the selenium-containing enzyme has been found to break down the cholesterol complexes that build up in lesions within the walls of arteries.

The following studies indicate that selenium is likely involved in heart disease:

1. The death rates from heart disease have consistently been highest in the low-selenium states.[17] However, these states are also the most densely populated and heavily industrialized states—so this evidence is open to other interpretations.

2. A comparison of the maps of early heart mortality and cardiovascular-related deaths of different areas of the United States revealed an inverse relationship between selenium levels and the mortality pattern.[18]

3. It has been hypothesized that selenium deficiency (rather than manganese deficiency as was first thought), which is prevalent all over Finland, may contribute to the unusually high death rates from heart disease and cancer in Finland.[19]

4. Lesions of selenium deficiency in rats have been associated with vascular abnormalities.[20]

5. Selenium is required by the enzyme that breaks down toxic peroxides which may damage the heart muscle.[21]

6. In May 1980, in a report before the Second International Symposium in Biology and Medicine, held at Texas Tech University, Lubbock, Texas, Dr. G. R. Yang, Chinese Academy of Medical Sciences, Peking, reported that selenium was effective in preventing heart muscle disease in certain parts of China. The malady, known as keshan, is common to certain areas of China where the soil is low in selenium. In a study involving 45,000 Chinese—the most massive study ever done on selenium deficiency in people—supplementation with selenium wiped out a heart disorder affecting 40 out of every 1,000 children in certain areas of China. Although there is no question that selenium was a pivotal factor in the Chinese study, Dr. Yang hastened to add that it is unlikely that it is the only factor.

• **Cloudiness in the lens of the eye (cataracts)**—Normally, the selenium content of the lens of the eye increases steadily from birth to death, but it has been found that lenses with cataracts may contain less than one-sixth of the normal amounts of this element.[22] It is suspected that lack of selenium to activate the enzyme glutathione peroxidase impedes the destruction of peroxides in the lens of the eye, and that the peroxides may then accumulate in amounts sufficient to damage the lens.

It is noteworthy, too, that cataracts are one of the selenium deficiency symptoms noted in rats.[23] Many other injurious agents and nutrient deficiencies might also be associated with the formation of cataracts, so it is uncertain whether selenium deficiency is a frequent cause of this disorder.

• **Growth failure during the rehabilitation of malnourished children**—So far, no evidence has been found to indicate that any normal, apparently healthy children suffer from a deficiency of selenium. However, several groups of children who were suffering from protein-energy malnutrition were found to have very low blood levels of selenium.[24] Selenium and protein are found together in both the diet and the body, so it is not surprising that children who lacked protein were also deficient in selenium. The standard rehabilitation procedures —which are based upon providing a good source of protein such as skim milk, extra fat and/or sugar, and certain vitamins and minerals—produced only limited improvement until selenium supplements were given.

• **Hemolytic anemia of newborn infants**—Newborn infants are more prone than other people to have abnormally short-lived red blood cells because blood levels of vitamin E—which together with selenium protects the membranes of red cells from premature disintegration—are often subnormal at birth. Hence, it is suspected that selenium deficiency might tend to aggravate this condition, since the selenium enzyme glutathione peroxidase protects against this disorder. It is noteworthy, therefore, that cow's milk contains only half as much selenium as human milk, and that both types of milk are relatively poor sources of the mineral.

• **Increased susceptibility to infections**—The germ-eating (phagocytic) white blood cells constitute one of the body's major defenses against infectious disease, because their formidable cells ingest living microorganisms, enclose them in killing chambers (phagocytic vacuoles) and kill them with blasts of highly potent forms of oxygen. The selenium enzyme glutathione peroxidase might extend the lifespan of these white blood cells by breaking down the toxic peroxides which may escape from the "killing chambers" into the more vulnerable parts of the cells. It is noteworthy that phagocytic white cells from selenium-deficient rats had a reduced ability to kill yeast cells, even though they had no difficulty ingesting the yeast.

So far, it has been difficult to confirm that selenium deficiency might be responsible for chronic susceptibility to infections in people, because many other factors affect this condition.

• **Infertility**—Studies on various species of farm animals and laboratory animals have shown that selenium deficiency causes infertility which is characterized in the male by fluid accumulation in the testicles, and feeble and broken sperm cells; and in the female by death of the fetus during early embryonic development. Although no similar investigations have been made in man, it is well known that people are subject to most of the disorders of fertility which have been observed in selenium-deficient animals.

[16]*Selenium*, National Academy of Sciences, 1976, pp. 84-88.

[17]Sauer, H. I., and F. R. Brand, "Geographic Patterns in the Risk of Dying," *Environmental Geochemistry in Health and Disease*, edited by H. L. Cannon and H. C. Hopps, The Geological Society of America, Inc., Boulder, Colo., Memoir 123, 1971, p. 137.

[18]Frost, D. V., "Selenium Has Great Nutritional Significance for Man; Should Be Cleared for Feed," *Feedstuffs*, Vol. 44 (8), 1972, pp. 58-59.

[19]Marjanen, H., and S. Soini, "Possible Causal Relationship Between Nutritional Imbalances, Especially Manganese Deficiency and Susceptibility to Cancer, in Finland," *Ann. Agric. Fenn*, Vol. 11, 1972, pp. 391-406.

[20]Sprinker, L. H., et al., "Selenium Deficiency Lesions in Rats Fed Vitamin E Supplemented Rations," *Nutr. Rep. Int.*, Vol. 4, 1971, pp. 335-340.

[21]Hoekstra, W. G., "Biochemical Role of Selenium," *Trace Element Metabolism in Animals-2*, edited by W. G. Hoekstra, et al., University Park Press, Baltimore, Md., 1974, p. 61.

[22]Ganther, H. E., et al., "Selenium and Glutathione Peroxidase in Health and Disease—A Review," *Trace Elements in Human Health and Disease*, edited by A. S. Prasad and D. Oberleas, Academic Press, Inc., New York, N.Y., 1976, p. 205.

[23]*Selenium*, Subcommittee on Selenium, National Research Council, National Academy of Sciences, 1976, p. 143.

[24]Hopkins, L. L., Jr., and A. S. Majaj, "Selenium in Human Nutrition," *Selenium in Biomedicine*, edited by O. H. Muth, The Avi Publishing Company, Inc., Westport, Conn., 1967, pp. 203-214.

• **Liver disease**—The liver is vulnerable to damage by the toxic peroxides generated during fat metabolism, unless it is supplied with sufficient amounts of selenium, vitamin E, and/or sulfur-containing amino acids (methionine and cysteine) to prevent the buildup of peroxides.

• **Pancreatic disease**—The reason for the degeneration of the pancreas in selenium-deficient animals is not clear. It does *not* appear to be related to the level of activity of the selenium enzyme glutathione peroxidase. Furthermore, an abundant dietary supply of vitamin E does not protect the pancreas when selenium is lacking. Eventually, deficiencies of vitamin E and other fatborne nutrients may develop, when the damage to the pancreas is great enough to interfere with the secretion of enzymes which digest the fatty components of foods.

• **Poisoning by normally harmless amounts of toxic substances**—In addition to the counteraction of the toxic effects of arsenic, cadmium, and methylmercury by inorganic selenium, selenium in the enzyme glutathione peroxidase may protect against the effects of poisonous organic compounds such as carbon tetrachloride (a common dry-cleaning agent) and certain drugs.

• **Sudden infant death syndrome (crib or cot death)**—Infants that are fed cows milk receive only ½ as much selenium and 1/10 as much vitamin E as infants that are breast fed. It follows that several researchers have advanced the hypothesis that combined deficiencies of selenium and vitamin E in cow's milk may be contributory to *crib death*, where an infant that had appeared to be healthy is found to have died suddenly in its bed.[25]

However, the tentative explanation of crib deaths which is entertained by many doctors is that (1) an infant may develop antibodies (become sensitized) to cow's milk after bringing up part of a feeding and inhaling it into its bronchial tubes (the air passages which lead from the windpipe into the lungs), and (2) repetitions of these episodes lead to such a severe allergy to cow's milk that eventually one such event leads to a fatal allergic shock. Other doctors believe that the severe allergic shock may be triggered by a variety of allergens, or perhaps even by a virus.

The explanations are all compatible because severe allergic reactions are accompanied by (1) a marked enlargement (dilation) of the blood vessels, and a great drop in the blood pressure; and (2) a speeding up of the heart rate as the heart attempts to maintain the normal circulation of blood. Deficiencies of selenium and vitamin E may weaken the heart muscle so that it cannot cope with the increased workload brought about by a state of allergic shock.

TREATMENT AND PREVENTION OF SELENIUM DEFICIENCY.
As a result of extensive animal experiments, in 1974 the Food and Drug Administration (FDA) approved the use of selenium for livestock, with certain restrictions.[25]

In 1989, for the first time ever, the Food and Nutrition Board (FNB), National Research Council (NRC), included selenium in its *Recommended Dietary Allowances*, 10th Edition.

INTERRELATIONSHIPS.
The major selenium relationships are:

1. The functions of selenium are closely related to those of vitamin E and the sulfur-containing amino acids.

[25]*Selenium*, National Academy of Sciences, 1976, pp. 93–94.

2. Selenium protects against the toxic effects of arsenic, cadmium, copper, mercury, and silver. Likewise, these elements counteract the toxic effects of selenium.

3. A diet high in protein or high in sulfate provides some protection against selenium poisoning.

4. The optimal utilization of selenium may depend upon adequate iron nutrition.

5. Diets rich in polyunsaturated fatty acids but poor in vitamin E may raise the requirements for selenium.

RECOMMENDED DAILY ALLOWANCE OF SELENIUM.
Tables S-1a and M-25 give selenium intakes that the FNB–NRC deems to be appropriate for various age groups.

TABLE S–1A
RECOMMENDED SAFE AND ADEQUATE
DAILY DIETARY INTAKES OF SELENIUM[1]

Group	Age	Selenium Intake
	(years)	(mcg)
Infants	0.0–0.5	10
	0.5–1.0	15
Children	1–3	20
	4–6	20
Adolescents	7–10	30
	11+	40
Adults		55–70

[1]Adapted by the authors from *Recommended Dietary Allowances*, 10th ed., 1989, NRC-National Academy of Sciences, p. 285.

Certain researchers, such as Dr. Schrauzer and his co-workers at the University of California at San Diego, believe that many cases of cancer might be prevented by selenium intakes that are approximately double the level recommended by the Food and Nutrition Board.[26]

• **Selenium intake in average U.S. diet**—Analyses of national food composites in the United States indicate that the overall adult mean dietary selenium intake was 108 mcg/day between 1974 and 1982. The daily mean for each year ranged from 83 to 129 mcg.

TOXICITY.
Confirmed cases of selenium poisoning in man are not often found because (1) the foods and beverages which are consumed by people are not likely to contain toxic excesses of the element, and (2) well-nourished people are protected by metabolic processes that convert selenium into harmless substances which are excreted in the urine, or in the breath. Nevertheless, a few cases of poisoning sometimes occur under unusual circumstances such as (1) very high levels of the element in drinking water, or (2) the presence of malnutrition, parasitic infestation, or other factors which may make people highly susceptible to selenium toxicity. Therefore, descriptions of some poisonous effects of selenium follow:

• **Abnormalities of the hair, nails, and skin**—It appears that some of the toxic effects of selenium might be due to its interference with the normal structures and functions of proteins rich in sulfur-containing amino acids. Hence,

[26]Schrauzer, G. N., "Trace Elements, Nutrition and Cancer: Perspectives of Prevention," *Inorganic and Nutritional Aspects of Cancer*, edited by G. N. Schrauzer, Plenum Press, New York, N.Y., 1978, pp. 330-331.

poisoning by this element is characterized by abnormalities in the hair, nails, and skin; tissues which are rich in sulfurous proteins. For example, a group of children living in a high-selenium area of Venezuela had loss of hair, discolored skin, and chronic digestive disturbances.[27] However, most of the children were infested with intestinal parasites, which might have weakened them so that they had increased susceptibility to the toxic effects of selenium.

• **Garlic odor on the breath**—Normally, people who have consumed large excesses of selenium excrete it as trimethyl selenide in the urine, and/or as dimethyl selenide on the breath. The latter substance has an odor resembling garlic.

• **Intensification of selenium toxicity by arsenic or mercury**—Usually, the toxicity of selenium is reduced by the presence of either arsenic or mercury. However, the methyl derivatives of selenium—which are formed when animals or people are chronically exposed to excessive amounts of the mineral—may become highly toxic when they react in the body with arsenic or mercury. So far, these reactions have been observed only in animals in laboratory experiments. Hence, it is not certain whether there are circumstances under which they might occur in man.

• **Promotion of tooth decay**—Various studies have shown the rates of tooth decay to be slightly higher in high-selenium areas than in low-selenium areas. Research on laboratory animals has produced evidence that excesses of selenium interfere with the formation of the protective tooth enamel. However, this research has usually been conducted with much higher levels of the mineral than are likely to be encountered by man.

SOURCES OF SELENIUM. The selenium content of plant products varies according to the amounts and availability of the mineral in the soil. Likewise, the selenium content of animal products depends upon the content of the element in livestock feeds; for example, the Ohio Station reported that the addition of 0.1 ppm selenium to the animal's diet increased the selenium content of beef liver by 72%.[28] Also, there are cooking losses in some cases, although these are minor for most foods. Hence, the data regarding the amounts of the mineral supplied by various foods *cannot* be taken to indicate a constant content of selenium for any particular food. Nevertheless, such data are useful in that they show which foods are likely to be good sources of the mineral.

Fortunately, food eaten in the United States is generally varied in nature and origin. Nevertheless, it appears likely that most diets barely meet minimal requirements; hence, many diets are bound to be deficient.

Groupings by rank of some common food sources of selenium are given in the section on MINERAL(S), Table M-25, Mineral Table.

For additional sources and more precise values of selenium, see Table S-2, Selenium Content of Some Foods, which follows.

SUMMARY. It might be desirable for people who live in low-selenium areas of the United States—which are generally located east of the Mississippi River and west of

TABLE S-2
SELENIUM CONTENT OF SOME FOODS

Food	Selenium
	(mcg/100 g)
Fish flour	193
Butter	146
Eulachon (smelt)	123
Torula yeast	123
Wheat germ	111
Lobster	104
Brazil nuts	103
Brewers' yeast	91
Cider vinegar	89
Pork kidneys	64
Wheat, whole grain	63
Wheat bran	63
Clams	55
Whole wheat flour	53
Crab	51
Oysters	49
Pork	42
Rye, whole grain	37
Kidney beans	36
Lamb	30
Soybean flour	30
Turnips	27
Swiss chard	26
Blackstrap molasses	26
Garlic	25
Oats, whole grain	21
Beer	19
Barley, whole grain	18
Eggs	16
Skim milk (dehy)	13
Mushrooms	13
Soybeans	11
Cheese	8
Corn	7
Orange juice	6
Grape juice	4
Cow's milk, whole 3.7% fat	3
Pecans	3
Filberts (hazelnuts)	2
Almonds	2
Carrots	2
Cabbage	2

the Rocky Mountains—to make certain that they get enough selenium by eating ample amounts of foods rich or good in selenium. However, undesirable effects or toxicities might result from consuming too much selenium, so one should consult with a doctor or a dietitian before taking special, selenium-rich supplements.

(Also see MINERAL[S], Table M-25; and POISONS, Table P-2, Some Potentially Poisonous [Toxic] Substances.)

SEMIDISPENSABLE AMINO ACID

An amino acid which is essential only under certain circumstances or which may replace part of one of the essential amino acids. Arginine, cystine, and tyrosine fall into this group.

(Also see AMINO ACID[S].)

SEMIPERMEABLE MEMBRANE

A membrane that is permeable to some small molecules (like water and inorganic salts) but bars the passage of larger particles (like protein molecules).

SEPTICEMIA

A diseased condition resulting from the presence of pathogenic bacteria and their associated poisons in the blood.

[27]Burk, R. F., "Selenium in Man," *Trace Elements in Health and Disease,* edited by A. S. Prasad and D. Oberleas, Academic Press, Inc., New York, N.Y., 1976, p. 121.

[28]Moxon, A. L. and D. L. Palmquist, "Selenium Content of Foods Grown or Sold in Ohio," *Ohio Report,* Ohio State University, Columbus, Ohio, January-February 1980, pp. 13-14, Table 1.

SERENDIPITY BERRY
Dioscoreophyllum cumminsii

A very sweet tasting berry that is native to West Africa. The substance which is responsible for the sweet taste has been isolated and named monellin.

SEROTONIN

A derivative of the essential amino acid tryptophan which plays a role in brain and nerve function.

SERUM

The colorless fluid portion of blood remaining after clotting and removal of corpuscles. It differs from plasma in that the fibrinogen has been removed.

SERUM CHOLESTEROL

The level of the sterol cholesterol in the blood.

SERUM TRIGLYCERIDE

The level of fat in the blood.

SESAME *Sesamum indicum*

The sesame plant belongs to the family *Pedaliaceae;* genus *Sesamum;* species, *S. indicum.*

Sesame has been grown in tropical countries since time immemorial, primarily to obtain seed for use on bakery goods and food delicacies and for edible oil production. It appears to be the earliest condiment used and the oldest crop grown for edible oil. The magic words *open sesame* found in the Arabian Nights are thought to have been inspired by the characteristic bursting open of the sesame pods when the grain is ripe, a nettelsome trait that has necessitated hand harvesting of the crop.

1)cross section of flower

2)white seed   3)black seed

4)ripe fruit

Fig. S-12. Sesame plant.

Production of sesame is limited to countries where labor is plentiful and inexpensive, and will so remain until and unless nonshattering varieties with satisfactory yields and/or improved mechanical harvesting techniques are developed. Intense breeding and engineering research programs to alleviate these problems are in progress.

ORIGIN AND HISTORY. Although sesame has not been identified in the wild state, it is believed to have originated in the Indian archipelago. Archeological evidence indicates that sesame was cultivated in Palestine and Syria around 3000 B.C., and by the civilization of Babylonia in 1750 B.C. An Egyptian tomb bears a 4,000-year-old drawing of a baker adding sesame to bread dough. In 1298, Marco Polo observed the Persians using sesame oil for cooking, body massage, medicinal purposes, illumination, cosmetics, and lubricating primitive machinery.

WORLD AND U.S. PRODUCTION. World production of sesame seed totals about 2,000,000 metric tons annually; and the six leading nations, by rank, are: India, China, Myanmar, Mexico, Nigeria, and Sudan.[29]

The United States produces only about 1,000 metric tons annually. However, this country is a large importer of sesame seed; the U.S. imports close to 43,000 metric tons each year.

NUTRITIONAL VALUE OF SESAME. Sesame seed averages about 50% oil, which is highly resistant to oxidation, and 25% protein, which has a unique balance of amino acids. Dehulled, defatted meal contains 50 to 60% protein and is bland. Food Composition Table F-21 gives the nutrient composition of sesame seeds and sesame oil.

The unique quality of sesame protein is the presence of a high level of the sulfur-containing amino acids, methionine and cystine. The limiting amino acid of sesame protein is lysine (Table S-3). It is also noteworthy that sesame does not contain some of the objectionable characteristics found in soy protein, particularly with regard to the trypsin inhibiting factor.

---

[29]Data from *FAO Production Yearbook 1990*, FAO/UN, Rome, Italy, Vol. 44, p. 114, Table 42. **Note well:** Annual production fluctuates as a result of weather and profitability of the crop.

TABLE S-3
PROFILE OF ESSENTIAL AMINO ACIDS IN DEHULLED SESAME PRODUCTS COMPARED TO MILK—A HIGH-QUALITY PROTEIN[1]

Amino Acid	Meal	Isolate	Cow's Milk[2]
	(mg/g protein)		
Histidine	24	21	27
Isoleucine	47	36	47
Leucine	74	66	95
Lysine	35	21	78
Methionine and cystine	56	37	33
Phenylalanine and tyrosine	106	79	102
Threonine	39	33	44
Tryptophan	19	18	14
Valine	46	46	64

[1]Adapted from Johnson, L.A., T. M. Suleiman, and E. W. Lusas, "Sesame Protein: A Review and Prospectus," *Journal of the American Oil Chemists' Society*, Vol. 56, No. 3, March 1979, p. 465, Table 3.

[2]*Recommended Dietary Allowances*, 10th ed., 1989, National Academy of Sciences, p. 67, Table 6–5. The essential amino acid requirements of infants can be met by supplying 0.79 g of a high-quality protein per pound of body weight (*1.7 g/kg body weight*) per day.

PROCESSING SESAME SEED. In those areas where sesame is primarily processed for its oil content, the seed is not dehulled; rather, the entire seed is crushed. However, in areas such as India, where the meal is an important food, dehulling is necessary.

Table S-4 shows the average analysis of crude sesame oil. Due to the presence of natural antioxidants in the crude oil, sesame oil is the most stable naturally-occurring liquid vegetable oil. It will keep for several years without turning rancid. Sesame oil is classified as polyunsaturated; it contains approximately 44% of the essential fatty acid, linoleic acid, and 40% oleic acid.

(Also see FATS AND OTHER LIPIDS; and OILS, VEGE-TABLE, Table O-5.)

TABLE S-4
ANALYSIS OF CRUDE SESAME OIL (AVERAGE)[1]

Free fatty acid (as oleic)	1.3%
Color	35y/1.2r
Iodine Value	110
Peroxide Value	13
AOM	24 hr
Percent unsaponifiable	2.3%
Smoke point	330 °F
Specific gravity @ 25°C	.918
Saponification Value	185.8
Fatty acids (natural oil)	**(%)**
Oleic	40
Linoleic	44
Other unsaturates	1
Palmitic	9
Stearic	5

[1]Source: USDA Southern Regional Laboratories.

SESAME PRODUCTS AND USES. Sesame in many forms is now offered to the food industry for human consumption; among them, the following:

Dry, cleaned, unhulled sesame seed
White, hulled sesame seed
Toasted sesame seed
Toasted, flaked sesame seed
Partially defatted sesame flour
Solvent extracted sesame flour
Sesame butter made from toasted sesame
    seed

In the United States, sesame is used primarily in the bakery industry, where whole, hulled sesame seed is used as a decoration and flavoring agent on specialty breads, rolls, buns, candies, and other delicacies. Sesame is also used in high-protein snack foods and granola.

The protein (meal) of sesame has become increasingly important for human food due to the following unique properties: (1) the presence of a high level of the sulfur-containing amino acids, methionine and cystine; (2) its freedom from the trypsin inhibiting factor, an objectionable characteristic of soy protein; and (3) its pleasant flavor. The meal is very palatable to humans. In India, it is used extensively in human foods; and in India and Java, it is sometimes fermented for food.

Sesame oil (also known as gingili, benne, or til) is straw-colored, with a pleasant, mild taste and remarkable stability. It is a natural salad oil requiring little or no winterization and is one of the few vegetable oils that can be used without refining. These are factors of increasing importance as energy costs escalate. It is popular because of the presence of a natural antioxidant, sesamol, and because of its high content of polyunsaturated fatty acids, 43% oleic and 43% linoleic. The oil is used as a substitute for olive oil—primarily as a salad and cooking oil and for margarine and soap. Also, a significant quantity of oil is used by the cosmetic industry for softening and soothing purposes, and by the pharmaceutical industry as a carrier for medicines.

In addition to the uses listed above, sesame is being used to enhance the flavor of fried or baked snacks, high-protein beverages, cereals, seasoning blends; and as a garnish for vegetables, high-protein breads, and pies.

Blends of peanut/chickpea, wheat/chickpea, rice/chick-pea, peanut/soybean, sunflower/maize, and cowpea/rice have all shown improved nutritional qualities with supplementation of sesame meal. Even more significant, however, is the finding that a simple blend of one part sesame and one part soy protein has about the same protein nutritive value as casein, the main protein of milk. The high-lysine and low-methionine content of soy protein is complementary to sesame protein.

(Also see OILS, VEGETABLE, Table O-5, Vegetable Oils.)

SEVEN FOODS PLAN (BASIC 7 FOODS PLAN)

The different types of foods are sometimes divided into the following seven groups: *Group 1*—Green and yellow vegetables; *Group 2*—Oranges, tomatoes, grapefruit, or raw cabbage or salad greens; *Group 3*—Potatoes and other vegetables, and fruits; *Group 4*—Milk and milk products; *Group 5*—Meat, poultry, fish, or eggs, or dried beans, peas, nuts, peanut butter; *Group 6*— Bread, flour, and cereals; and *Group 7*—Butter and fortified margarine. A well-balanced diet should include food from each group every day. A more common plan is the selection of foods from the basic four food groups—(1) the meat, poultry, fish, and beans group; (2) the milk and cheese group; (3) the vegetable and fruit group; and (4) the bread and cereal group.

(Also see FOOD GROUPS.)

SEVILLE ORANGE (SOUR OR BITTER ORANGE) *Citrus aurantium*

This citrus fruit is rarely seen in the fresh form in the United States because it is grown mainly in Seville, Spain, for making marmalade. The Seville orange, like the other citrus fruits, belongs to the rue family (*Rutaceae*). It is distinguished from the Sweet orange (*C. sinensis*) by its much rougher skin and very sour taste.

Fig. S-13. The Seville orange, a sour-flavored fruit which is used mainly for making marmalade and flavorings.

ORIGIN AND HISTORY. The sour orange originated in Indonesia and was brought to India in ancient times. The first orchards were planted on the Arabian peninsula in the beginning of the 10th century. Later, the trees were planted in Mesopotamia and Syria. The spread of Islam and Arabic culture across North Africa and into Spain led to the introduction of the sour orange into Seville, Spain, in the 11th century.

WORLD AND U.S. PRODUCTION. Most of the world's crop of sour oranges is produced in the area around Seville, Spain, although production figures are not available.

PROCESSING. Most of the sour orange crop is processed because the fruit is too sour to eat raw. Large quantities of the oranges are exported to Great Britain, where they are used to make a marmalade that has a strong distinctive flavor that is not found in similar products made from sweet oranges.

In Spain, the peel is scraped from the oranges and pressed to yield an oil that is much used as a flavoring agent. The clarified juice is used to impart a strong orangy flavor to highly diluted beverages such as orangeades and other fruit drinks. The U.S. Standards of Identity permit the use of up to 5% sour orange juice in the production of frozen orange juice concentrate.

SELECTION, PREPARATION, AND USES. One is not likely to see fresh sour oranges in the United States except in a few orchards in Florida. However, sour oranges are grown rather extensively in the Mediterranean region, especially Spain. Bitter orange marmalade and flavorings are made from it. Also, bitter oranges are grown for two additional purposes: (1) they are used as rootstock seedlings on which to bud-graft sweet oranges and other citrus fruits; and (2) in France, the flowers of the bitter orange are distilled for their perfume—eau de cologne is made from it.

NUTRITIONAL VALUE. The fruit and juice of the sour orange are rarely consumed alone, but are usually mixed with other citrus products. Hence, data on the nutrient composition of this fruit are not readily available. However, it might be expected to be similar to the sweet orange in its content of potassium, vitamin C, and bioflavonoids.

Citrus peel contains citral, an aldehyde which antagonizes the effects of vitamin A. Hence, people should not consume large amounts of peel-rich products such as candied peel and marmalade without making certain that their dietary supply of this vitamin is adequate.

(Also see CITRUS FRUITS.)

SHEA BUTTER

This is a fat obtained from the kernel of the fruit of the African tree, *Butyrospermum parkii*, also called butter tree or shea tree. Natives use the fat as a substitute for butter.

SHORTENING

Fat substances given this name are used to *shorten foods*. Their function is accomplished by forming films or clumps of fat throughout the food thereby preventing the protein and carbohydrate from cooking into one continuous hard mass. Thus, shortenings tenderize foods. Also, cake batters and icings often contain tiny air bubbles trapped in shortening that help to develop a fine delicate structure. It is the plastic consistency of shortenings that helps hold air in the

Fig. S-14. Frying doughnuts in shortening. (Courtesy, American Soybean Assn., St. Louis, Mo.)

food. The amount of shortening varies according to the product. Breads and rolls contain 1 to 2%, cakes 10 to 20%, and piecrusts contain over 30%. When shortenings are used in frying they serve for heat transfer and as antisticking agents, and provide some tenderness and richness of flavor.

Rendered animal fats and butter are shortening agents, but the term generally refers to fats which have been especially processed. For a number of years, lard or beef tallow were the main shortenings. Then with the development of hydrogenated vegetable oils came the increased use and acceptance of vegetable shortenings. Hydrogenation—adding hydrogen to the unsaturated fatty acids—can convert naturally liquid fats to semisolids. These shortenings may be prepared in all degrees of stiffness, from liquid to solid, depending upon their melting point. Melting point is determined by their degree of saturation. Then, in order to achieve specific physical, chemical, and biological properties, it is common to blend three or more different fats. Vegetable oils used include corn oil, cottonseed oil, soybean oil, olive oil, palm oil, peanut oil, safflower oil, and sesame oil. Thus, there are several general types of shortening meeting different needs.

• **General purpose**—These shortenings meet a variety of uses and are the household shortenings sold in grocery stores in metal containers. General purpose shortenings are semisolids when temperatures range from 60 to 90°F (*16 to 32°C*). They may be employed by the home cook in biscuits, cookies, pie crusts, breads and rolls, pancakes, and pan frying. Also, these shortenings usually contain monoglycerides (glycerol molecules with only one fatty acid) so they may also be employed in cakes, icings, frostings, and other recipes wherein relatively large amounts of fat and water must be mixed without danger of separation. Monoglycerides act as emulsifiers.

• **Frying**—Shortenings used in frying must resist oxidation at high temperatures. Oxidation causes fat to darken, become thicker, foam during frying, and develop unpleasant flavors and odors. To resist oxidation at high temperatures, frying fats are composed of saturated fatty acids—low iodine value, and cannot contain mono- or diglycerides as these produce smoke at high frying temperatures. Moreover, fats whose fatty acids are short chain such as coconut oil are not suitable for deep fat frying because they break down and foam. Frying fats generally contain some antioxidants and antifoaming agents.

• **Special emulsifying**—These shortenings contain at least four emulsifying agents which complement each other and provide some particular characteristic. They provide the special properties necessary in prepared baking mixes, whipped toppings, icings, fillings, and the like.

• **Liquid oils**—Vegetable oils are required in mayonnaise and salad dressings, and may be used as a shortening agent in rolls, breads, and other baked goods. They are a notable convenience in many kitchens. Also, the vegetable oils are the basis of margarine which may also be a shortening agent.

(Also see FATS AND OTHER LIPIDS; LARD; MARGARINE; and OILS, VEGETABLE.)

SHREDDED

Cut into long, narrow pieces.

SIFTED

Materials that have been passed through wire sieves to separate particles of different sizes. The common connotation is the separation of finer material than would be done by screening.

(Also see FLOURS.)

SILICON (Si)

Silicon is one of the most abundant elements on earth—present in large amounts in soils and plants. The highest concentrations in animal tissue are found in the skin and its appendages; for example, the ash of feathers is more than 70% silicon.

HISTORY. Although the early chemists considered silica (SiO_2) an elementary substance, Antoine Lavoisier, founder of quantitative chemistry, suspected in 1787 that it was an oxide of an undiscovered element. In 1823, Jons J. Berzelius, the Swedish chemist, discovered the element. The name *silicon* is derived from the Latin *silex* or *silicis*, meaning flint—appropriately indicating its hardness.

In 1972, Dr. E. Carlisle, nutritionist at the University of California, Los Angeles, reported that the trace element silicon (Si) is needed in microgram amounts for normal growth and skeletal development of chicks and rats.

The free element silicon is not found in nature, but it occurs either as the oxide silica (SiO_2) in such forms as sand and quartz, or as silicates in such materials as granite. Silicon is important in plant and animal life; it is present in the ashes of plants and in human skeletons.

ABSORPTION, METABOLISM, EXCRETION. Silicon is absorbed readily. Even over a wide range of intake, the concentration in the blood remains relatively constant—not more than 1 mg per 100 ml. It is excreted easily—via both the feces and urine.

FUNCTIONS OF SILICON. Silicon appears to be essential for chicks and rats; its absence from the diet impedes normal growth and skeletal development. These findings suggest that it may also have essential functions in man.

DEFICIENCY SYMPTOMS. The production of deficiencies in chicks and rats, characterized by growth retardation and skeletal alterations and deformities, especially of the skull, depends on strict control of dietary and environmental contamination; this suggests a low requirement that is easily met by the amount in the environment.

Deficiencies of silicon have not been produced in man. However, it has been found that the silicon content of the aorta, skin, and thymus decrease significantly with age; this is probably related to the fact that the mucopolysaccharide (the substance that binds with water to form the thick gelatinous material which cements cells together and lubricates joints and bursas) content of body tissues also declines with aging.

INTERRELATIONSHIPS. Silicon appears to take part in the synthesis of mucopolysaccharides and is a component of the mucopolysaccharide-protein complexes of connective tissue.

Normally, urinary silicon is eliminated efficiently. However, high levels of silicon in the diet of farm animals may be detrimental. In a manner not yet fully understood, part of it is sometimes deposited in the kidney, bladder, and urethra to form calculi (stones).

A common nondietary form of silicon toxicity is a lung condition known as silicosis, due to inhalation of airborne silicon oxide dust. The amount of silicon in the blood and urine increases in silicosis.

RECOMMENDED DAILY ALLOWANCE OF SILICON. A human requirement of silicon cannot be established on the basis of available knowledge.

• **Silicon intake in average U.S. diet**—Human intake of silicon has been estimated at approximately 1 g per day.

TOXICITY. Silicon does not appear to be toxic in the levels usually found in foods.

SOURCES OF SILICON. Because of the abundance of silicon in all but the most highly purified foods, it is of little concern in human diets.

The best sources of silicon are the fibrous parts of whole grains, followed by various organ meats (liver, lungs, kidneys, and brain) and connective tissues. Much of the silicon in whole grains is lost when they are milled into highly refined products.

(Also see MINERAL[S], Table M-25.)

SIMPLE SUGARS

The *monosaccharides*. The most important simple sugars are *fructose, galactose, glucose,* and *mannose.*

(Also see CARBOHYDRATES, section on "Monosaccharides.")

SINGLE-CELL PROTEIN (SCP)

This refers to protein obtained from single-cell organisms, such as yeast, bacteria, and algae, that have been grown on specially prepared media. Production of this type of protein can be attained through the fermentation of petroleum derivatives or organic waste, or through the culturing of photosynthetic organisms in special illuminated ponds.

CURRENT PROBLEMS ASSOCIATED WITH SINGLE-CELL PROTEIN. Although single-cell protein appears to be an excellent alternative or supplemental source of protein, several problems must be overcome before it becomes a widely used food; among them, palatability, digestibility, nucleic acid content, toxins, protein quality, and economics.

As the world's human population increases, there will be an increasing demand for cheap protein. This demand could dry up the sources of traditional protein, thereby opening the way for intensive development of alternative sources of proteins.

(Also see PROTEIN, section on "Protein For The World.")

SIPPY DIET

A type of diet offered to bleeding peptic ulcer patients, originated by an American physician, Bertram Sippy, in 1915. It broke the previous practice of initial starvation treatment and established the beginning principles of continuous control of gastric acidity through diet and alkaline medication. To start with, the Sippy Diet consists of hourly feedings of 3 oz (*150 ml*) of whole milk, skim milk, or equal parts of cream and milk. After 2 or 3 days, up to 3 oz (*85 g*) of soft eggs, or strained cereals are added to the milk. Bland, easily digested foods may be occasionally added to break the monotony. Never does the total bulk of one feeding exceed 6 oz (*300 ml*).

(Also see LENHARTZ DIET; MODIFIED DIETS; and ULCERS, PEPTIC.)

SKIMMED

Material from which floating solid material has been removed. It is also applied to milk from which fat has been removed by centrifuging.

SKIN

• Outer coverings of fruits or seeds, as the rinds, husks, or peels.

• Dermal tissue of man and animals.

SLOE (BLACKTHORN) *Prunus spinosa*

A small, wild European plum (of the family *Rosaceae*) that is blackish purple with a very sour, green-colored flesh.

The sloe plum is used mainly in the production of sloe gin, jam, and a highly esteemed liqueur. Few people care to eat it raw.

SNACK FOODS

Snacks are those small portions of food eaten between meals; hence, no certain food is really a snack food. Individuality and need determine snack foods.

Snacking can be good or bad, depending upon the snack food and the frequency of snacking. Unfortunately, many people choose foods loaded with sugar and fat—calories—but low in protein, vitamins, and minerals. In keeping with the objective of good nutrition for good health, it just makes sense to eat nutritious snacks; for example, an orange, an apple, peanut butter and crackers, milk, nuts, raw vegetables, juice, etc. However, the type of snack depends on age and activity. Small children often fail to consume the amount of food in regular meals that will add up to their nutritional needs. A slice of cheese, a wheat cracker, or a banana at various times helps supply the added energy they need. The active growing bodies of teenagers need nutritious snacks which provide extra protein, vitamins, minerals, and energy; for example, nuts, yogurt, milk, fruit salads, and leftovers. Senior citizens often have chewing or digestion problems which interfere with regular eating habits. Here again nutritious snacks are helpful to meet daily needs.

Nutritious snacks are, however, no excuse for overindulgence. A calorie is still a calorie, and lots of little snacks add up. If more energy is consumed than used, the excess is stored as fat. Dieters should choose nutritious, but filling, snacks such as fresh fruits and raw vegetables.

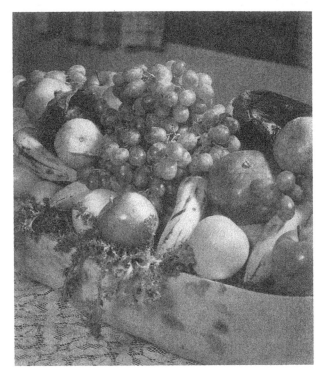

Fig. S-15. America is blessed with a distribution system that provides a wide variety of fresh fruits and vegetables the year round. (Photo by J. C. Allen and Son, West Lafayette, Ind.)

The best way to ensure nutritious snacking is to keep the nutritious snacks on hand and available. Ideas for nutritious snacks are endless, and may be chosen by evaluating foods in terms of energy, protein, mineral and vitamin content as given in Food Composition Table F-21 of this book.

(Also see CONVENIENCE AND FAST FOODS; and FAST FOODS.)

SNIPE

Any of several game birds (genus *Capella*) that are widely distributed in the New and Old Worlds, especially in marshy areas. The snipe is best prepared by roasting.

SOAP

Chemically, soap is the alkali salt of fatty acids. It is formed by a process called saponification. Specifically, soap is a mixture of the sodium salts of the following fatty acids: stearic acid ($C_{17}H_{35}COOH$); palmitic acid ($C_{15}H_{31}COOH$); and oleic acid ($C_{17}H_{33}COOH$). Soft soaps are the potassium salts of these fatty acids. Hence, saponification is accomplished with either sodium hydroxide or potassium hydroxide. Glycerin or glycerol is a by-product of saponification, since the fatty acids are triglycerides—three fatty acids joined to glycerol. Saponification releases the fatty acids from glycerol. During World War II, housewives saved used cooking fats to be used in making soap, and more importantly glycerin for explosives. In a broad sense, soap is any salt of long-chain fatty acids. In the body, fatty acids which leave via the feces are almost entirely soaps.

(Also see FATS AND OTHER LIPIDS.)

SODA WATER

Water charged with carbon dioxide is termed soda water. It bubbles and fizzes as the gas is released. Soda water is added to syrups in soft drinks and to mixed alcoholic drinks. It does not possess any medicinal properties.

(Also see SOFT DRINKS [Carbonated Beverages].)

SODIUM (Na)

Contents	Page
History	961
Absorption, Metabolism, Excretion	961
Functions of Sodium	961
Deficiency Symptoms	961
Interrelationships	961
Recommended Daily Allowances of Sodium	961
Toxicity	962
Sources of Sodium	962
High Sodium Sources	963

Most of the sodium in the diet is in the form of sodium chloride (NaCl), a white granular substance used to season and preserve food. It is created by combining the elements sodium (Na), a soft silvery-white metal, and chlorine (Cl), a poisonous, yellow gas; neither of which is desirable for use alone.

The need for sodium in the diet has been recognized since the classic experiment of Osborne and Mendel, reported in 1918.[30]

The body contains about 0.2% sodium, or about 100 g in a 154-lb (70 kg) person. About 50% of the sodium is in the extracellular (between cell) fluids, 40% is in the skeleton, and only about 10% is within the cells. In the blood, most of the sodium is in the plasma, which contains 320 mg per 100 ml.

HISTORY. Compounds of sodium were known and used extensively during ancient times, but the element was not isolated until 1807, when Sir Humphrey Davy, the English chemist, used electricity (in a process called electrolysis) to extract the pure metal from sodium hydroxide. Although sodium had long been believed to be a dietary essential, final proof was not obtained until 1918 when Osborne and Mendel conducted laboratory animal experiments.

ABSORPTION, METABOLISM, EXCRETION. Virtually all sodium ingested in the diet is readily absorbed from the gut, following which it is carried by the blood to the kidneys, where it is filtered out and returned to the blood in amounts needed to maintain the levels required by the body.

Excess sodium, which usually amounts to 90 to 95% of ingested sodium, is, for the most part, excreted by the kidneys as chloride and phosphate and controlled by aldosterone, a hormone of the adrenal cortex. A deficiency of aldosterone, known as Addison's disease, leads to excessive losses of sodium in the urine.

The levels of sodium in the urine reflect the dietary intake; if there is high intake of sodium the rate of excretion is high, if there is low intake the excretion rate is low.

In hot weather, sodium excretion through the skin is important because each liter of sweat contains 0.5 to 3.0 g of sodium.

FUNCTIONS OF SODIUM. Sodium is the major positively charged ion in the fluid outside the cells, where it helps to maintain the balance of water, acids, and bases. Also, it is a constituent of pancreatic juice, bile, sweat, and tears.

Sodium is associated with muscle contraction and nerve functions. Sodium also plays a specific role in the absorption of carbohydrates.

(Also see section on MINERAL[S], Table M-25, Mineral Table.)

DEFICIENCY SYMPTOMS. Deficiencies of sodium are rare because nearly all foods contain some sodium. However, deficiencies may occur from strict vegetarian diets without salt, or when there has been heavy prolonged sweating, diarrhea, vomiting, or adrenal cortical insufficiency.

Excess perspiration and salt depletion may be accompanied by heat exhaustion. Salt tablets taken with a liberal amount of water may be advised under these conditions.

(Also see section on MINERAL[S], Table M-25, Mineral Table.)

INTERRELATIONSHIPS. Sodium, potassium, and chlorine are closely related metabolically. They serve a vital function in controlling osmotic pressures and acid-base equilibrium; and they play important roles in water metabolism. Both sodium and potassium ions occur in the body chiefly in close association with the chloride ion; therefore, a sodium or potassium deficiency is rarely found in the absence of a chlorine deficiency.

Excess dietary salt, over a long period of time, is believed to lead to high blood pressure in susceptible people. Also, excess sodium may cause edema, evidenced by swelling of such areas as the legs and face. The simplest way to reduce sodium intake is to eliminate the use of table salt.

RECOMMENDED DAILY ALLOWANCES OF SODIUM. The salt requirements of the infant and young child are estimated to be about 120–300 mg per day. Human milk contains 161 mg of sodium per liter, whereas commonly used bottle formulas contains between 161 and 391 mg per liter, and cow's milk contains about 483 mg per liter. Average intake of sodium during the first year of life ranges from about 300 mg per day at 2 months of age to about 1,400 mg per day at 12 months—far in excess of needs.

The healthy adult can maintain sodium balance with an intake of little more than the minimum requirements of the infant. When doing hard physical labor in high ambient temperatures, an adult may lose 8.0 g of sodium per day in sweat. Whenever more than 3 liter of water per day are required to replace sweat losses, extra salt (NaCl) should be provided; somewhere between 2 and 7 g of sodium chloride per liter of extra water loss, depending on the severity of losses and the degree of acclimatization. For those with a family history of high blood pressure, the food intake of salt should be limited to 500 mcg per day.

Pregnancy increases the sodium requirement, by an estimated 69 mg of sodium per day.

Estimated safe and adequate intakes of sodium are given in the section on MINERAL[S], Table M-25, Mineral Table.

• **Sodium intake in average U.S. diet**—The average daily American intake of salt is 10 to 15 g, or about 5 g of sodium (salt is approximately 40% sodium and 60% chlorine by weight).

[30]Osborne, T., and L. Mendel, "Inorganic Elements in Nutrition," *Journal of Biological Chemistry*, Vol. 34, 1918, p. 131.

• **Salt intake and high blood pressure**—There is strong evidence linking too much salt to high blood pressure, which affects 20 million Americans. Also, there is evidence that high salt intake increases the incidence of hypertension among various societies and cultures of the world (Altschal, A. M. and J. K. Grommet, "Sodium Intake and Sodium Sensitivity," *Nutrition Reviews*, Vol. 38, December 1980, No. 12, pp. 393-402).

Under ordinary circumstances, children and adults require no more than 0.2 g—about 1/10 tsp—of salt per day. With strenuous exercise or labor accompanied by profuse sweating, only 2 g—just under 1 tsp— is necessary. Normally, therefore, an allowance of 1 g of salt per day is sufficient. Yet, studies reveal that the average American consumes about 10 g of salt per day—ten times more than the recommended allowance.

People with high blood pressure should drastically curtail their salt intake. Fortunately, salt appetite can be readily revised downward, for there are no withdrawal pains from giving up salt. This can be accomplished by using the salt shaker sparingly, both in the kitchen and in the dining room; and by avoiding salty foods, such as olives, popcorn, potato chips, french fries, and sauerkraut.

It is noteworthy that cultural factors appear to determine the salt consumption. For example, the Japanese commonly consume 25 to 50 g of salt per day, whereas the Eskimos, Lapps, Australian aborigines, numerous tribes of New Guinea, South America and Africa, and many midwestern American Indian tribes consume from a few hundred milligrams to no more than 5 g of salt per day.

(Also see SALT.)

TOXICITY. Salt may be toxic when (1) a high intake is accompanied by a restriction of water, (2) when the body is adapted to a chronic low salt diet, or (3) when it is fed to infants or others whose kidneys cannot excrete the excess in the urine.

SOURCES OF SODIUM. In addition to salt (sodium chloride) used in cooking, processing, and seasoning, sodium is present in all foods in varying amounts. Generally more sodium is present in protein foods than in vegetables and grains. Fruits contain little or no sodium. The salt added to these foods in preparation may be many times that found naturally in foods. Also, the sodium content of the water

Fig. S-16. Grapes and other fruits contain little or no sodium. (Courtesy, USDA)

supply varies considerably; in some areas of the country, the amount of sodium in the water is of sufficient quantity to be of significance in the total daily intake.

Generally speaking, the need is for low sodium diets, rather than high sodium diets; especially by people with high blood pressure. Table S-5 has been prepared to meet this need.

TABLE S-5
GROUPINGS OF SOME LOW SODIUM FOODS

SODIUM LEVEL						
0-5 mg/100 g		5-10 mg/100 g	10-20 mg/100 g	20-40 mg/100 g	40-60 mg/100 g	
Almonds	Oranges	Apples, dried	Avocados	Artichokes	Arrowroot	
Apples	Peaches	Apricots, dried	Broccoli	Brown sugar	Black pepper	
Asparagus	Pears	Chestnuts	Brussels sprouts	Chinese cabbage	Beef	
Apricots	Pineapples	Cucumbers	Cabbage	Coconut	Carrots	
Bananas	Potatoes	Dock	Cantaloupe	Cream	Chicken giblets	
Blackberries	Raspberries	Green onions	Cauliflower	Hot chili peppers	Dandelion greens, cooked	
Cherries	Rhubarb	Honey	Collards	Passion fruit	Egg yolks	
Currants	Rye flour	Kohlrabi	Dried figs	Potato flour	Kale	
Dates	Shortening	Lettuce	Endive	Red cabbage	Milk	
Eggplants	Soybean flour	Nectarine	Garden cress	Sunflower seeds	Parsley	
Figs	Squash	Parsnips	Garlic	Turnips	Pork pancreas	
Fresh fruit juices	Strawberries	Peaches, dried	Mangoes		Rabbit meat	
Grapefruit	Sugar	Peanut flour	Mature onions		Red beets, boiled	
Grapes	Sweet corn	Peanuts	Molasses		Sablefish	
Green beans	Tomatoes	Pears, dried	Mushrooms		Sole	
Green peas	Walnuts	Persimmons	Mustard greens		Sunflower seed flour	
Guavas	Watermelon	Pigeonpeas	Pecans		Tamarinds	
Lard	Wax beans	Waxgourd	Prunes		Tuna	
Lima beans	Wheat flour	Wheat bran	Radishes		Turkey	
Oils	Wheat germ		Raisins		Walleye pike	
Okra	Whole wheat flour		Split peas		Watercress	
			Sweet potatoes		Yogurt, whole milk	
			Turnip greens			
			Yeast			

NOTE WELL: Many processed foods contain high levels of sodium due to the addition of salt and other sodium-containing additives; among them, sodium alginate, sodium aluminum sulfate, sodium benzoate, sodium citrate, sodium diacetate, sodium erythorbate, sodium nitrate, sodium nitrite, sodium propionate, and sodium sorbate.

For additional sources and more precise values of sodium, see Food Composition Table F-21.

HIGH SODIUM SOURCES. In order that consumers may intelligently cut back on sodium intake, it is important that they know which foods are high in sodium. Table S-6 lists some of the high sodium foods and gives the precise sodium content of each.

(Also see MINERAL[S], Table M-25; and SALT.)

TABLE S-6
SOME HIGH SODIUM FOODS[1]

Food	Sodium
	(mg/100 g)
Canadian bacon	2,555
Olives, green	2,400
Shrimp, canned	2,300
Caviar	2,200
Corned beef	1,740
Pretzels	1,680
Parmesan cheese	1,602
Pasteurized process cheese	1,430
Cucumber pickles	1,428
Bologna	1,300
Oat cereal	1,267
Luncheon meat	1,200
Soda crackers	1,100
Cured ham	1,100
Frankfurters	1,004
Wheat flakes cereal	1,032
Bacon	1,021
Bran cereal	1,012
Corn flakes cereal	1,005
Potato chips	1,000
Rice flakes cereal	987
Sausage	958

[1]Data from the Food Composition Table F-21.

SODIUM BICARBONATE (NaHCO₃)

A chemical compound which is known to function as a buffer and pH agent, maintaining sufficient alkaline reserves (buffering capacity) in the body fluids to ensure normal physiological and metabolic functions.

(Also see ACID-BASE BALANCE; and ANTACIDS.)

SODIUM DIACETATE

An antimicrobial food additive which controls both rope (bacterial action) and mold in breads. The FDA allows flour used in bakery goods to contain 0.4% sodium diacetate.

SODIUM PHYTATE

A chelating agent that is formed from phytic acid. In the past, it has been employed as part of a treatment for calcium phosphate kidney stones.

(Also see PHYTIC ACID.)

SODIUM-POTASSIUM RATIO

Sodium and potassium are closely interrelated. Although sodium intake may be the most important dietary determinant of blood pressure, variations in the sodium: potassium ratio in the diet affect blood pressure under certain circumstances. In rats with high blood pressure due to a high intake of sodium, blood pressure may be lowered to a more normal level by increasing the potassium intake and lowering the sodium intake. It seems that a 1:1 ratio may be somewhat protective against the blood pressure-elevating effects of a given level of sodium. This ratio can be achieved by increasing the intake of potassium, or lowering the intake of sodium, or both. Good dietary sources of potassium include meats, milk, dried dates, bananas, cantaloupe, apricots, tomato juice, and the dark green leafy vegetables.

(Also see HIGH BLOOD PRESSURE; POTASSIUM; and SODIUM.)

SOFT DRINKS (CARBONATED BEVERAGES)

This name is given to the familiar American beverages, often called *soda pop* or *pop*. The name *soft* is used to distinguish these beverages from *hard* or alcoholic beverages. Soft drinks are generally, but not always, carbonated, and contain a sweetening agent, edible acids, and natural or artifical flavors. Today, everywhere people work or play, soft drinks are available and their yearly consumption reflects their availability—and the advertising budget of the major producers.

Fig. S-17. In moderation, the occasional soft drink at a party can be part of the fun.

HISTORY. In 1772, the Englishman, Joseph Priestley, described a pleasant-tasting, sparkling water which he produced by introducing carbon dioxide (CO₂) into the water. Thus, carbonated beverages were born; however, Priestley is better known for his discovery of oxygen.

In 1806, the real business of selling soft drinks was started by Benjamin Silliman, a chemistry professor at Yale College in Connecticut. In New Haven, he bottled and sold soda water—water charged with carbon dioxide. Sometime after 1830, flavored soda water became popular, and by 1860, the census reported 123 plants producing carbonated drinks. In 1886, a druggist and former Confederate soldier named John Styth Pemberton created a drink which is still favored. He added an extract from the African kola nut to an extract from cocoa. Another enterprising pharmacist named Hires, introduced bottled root beer in 1893. Currently, the popular flavors are cola, lemon-lime, orange, ginger ale, root beer, and grape, but every conceivable type of flavor has been produced. Cola drinks are the most popular.

CONSUMPTION. In 1850, the United States per capita consumption of soft drinks was only about 1 pt (*470* year. In 1990, the per capita consumption was about 47.5 gal—more than 3.5 times the consumption in 1960 (see Fig. S-18). But that's not all! Ponder these figures:

• In 1990, the average man, woman, and child in the United States drank 760 8-oz soft drinks (in comparison to only 25.4 gal of milk)—that's an average of 1½ 12-oz cans per day.

• Soda pop now provides about 8% of the calories consumed daily by the average person, with virtually all these calories coming from sugar.

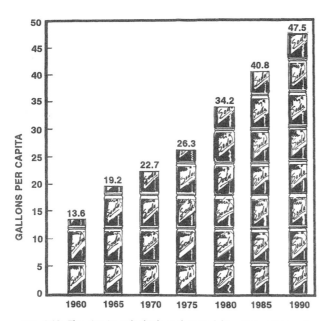

Fig. S-18. The rise in soft drink—carbonated beverage—consumption. (Sources: USDA; J. C. Maxwell, *Beverage Industry,* February 1991)

Yearly, consumption of soft drinks surpasses milk, beer, or coffee. This popularity is probably due to their availability and the tremendous advertising efforts of the major companies. The ad expenditures of the soft drink companies total over $200 million each year—all for convincing people to drink "the real thing," "the uncola," or to join a certain "generation" or to become a "pepper."

INGREDIENTS. The precise formulaton of most soft drinks is not well known. However, certain aspects of soft drinks are governed by standards of identity enforced by the Food and Drug Administration (FDA).

• **Soda water**—According to the standards of identity, soda water is the class of beverages made by absorbing carbon dioxide in potable water. The amount of carbon dioxide used is not less than that which will be absorbed by the beverage at a pressure of one atmosphere and at a temperature of 60°F *(15.5°C).* It either contains no alcohol or only such alcohol as is contributed by the flavoring ingredient used, but not in excess of 0.5% by weight of the finished beverage.

• *Cola or pepper*—Soda water designated by any name which includes the word *cola* or *pepper* shall contain caffeine from kola nut extract and/or other natural caffeine-containing extracts. Also, caffeine may be added to any soda water. The total caffeine content in the finished food shall not exceed 0.02% by weight.

Soda water may contain any safe and suitable optional ingredient, except that vitamins, minerals, and proteins added for nutritional purposes and artificial sweeteners are

not suitable for food encompassed by this standard.

The optional ingredients include: sweetening agents, acids, flavors, colors, preservatives, emulsifying, stabilizing or viscosity-producing agents, and foaming agents. Sweetening agents may consist of the following nutritive sweeteners (calorie-containing): dry or liquid sugar (sucrose), invert sugar, dextrose, fructose, corn syrup, glucose syrups, or sorbitol, singly or in combination.

With the sudden popularity of diet soft drinks in the early 1960s, low-calorie sweeteners, once used primarily by diabetics and other individuals with special dietary restrictions, came into vogue. Since that time, consumer demand for reduced-calorie beverages has increased greatly, with a corresponding increase in the use of low-calorie sweetening agents, chiefly aspartame. The present status of each of three well-known low-calorie sweeteners follows:

• **Saccharin**—Used as a sweetener by diabetics. Also used in some cosmetics, toothpastes, and cough syrups.

• **Cyclamate**—Cyclamate was taken off the U.S. market completely in 1970.

• **Aspartame**—In the United States, both aspartame alone and aspartame-saccharine mixtures are used as sweeteners in diet sodas. In Canada, aspartame is the only sweetener used in soft drinks. In 1991, aspartame had 75% of the U.S. sweetener market.

(Also see SWEETENING AGENTS.)

Flavoring agents in a carrier of ethyl alcohol, glycerin, or propylene glycol may include fruit juices, including concentrates and natural flavoring derived from fruits, vegetables, bark, roots, leaves, and similar plant materials. Artificial flavoring may also be used, as well as natural or artificial color. The edible acids which may be used singly or in combination are acetic, adipic, citric, fumaric, gluconic, lactic, malic, tartaric, and phosphoric acids. Table S-7 shows how some of these ingredients make up the familiar soft drinks.

TABLE S-7
SOFT DRINK INGREDIENTS

Soft Drink	Flavors	Color	Sugar (%)	Edible Acid	Carbon Dioxide (CO_2) (volume of gas)[1]
Cola	Extract of kola nut, lime oil, spice oils, caffeine.	Caramel	11–13	Phosphoric	3.5
Orange	Oil of orange and orange juice.	Sunset yellow FCF with some Tartrazine.	12–14	Citric	1.5–2.5
Ginger Ale	Ginger root, oil of ginger, and lime oil.	Caramel	7–11	Citric	4.0–4.5
Root beer	Oil of wintergreen, vanilla, nutmeg, cloves or anise.	Caramel	11–13	Citric	3
Grape	Methyl anthranilate and oil of cognac, sometimes grape juice.	Amarinth and Brilliant Blue FCF.	11–13	Tartaric	1.0–2.5

[1] A volume of gas is equivalent to 15 lb per square inch at sea level and 60°F *(15.5°C).* Correct carbonation results in pungent taste.

Soft drink plants are subject to state and federal laws and regulations which are designed to protect the consumer and to serve as an operational guide for the manufacturer.

NUTRITIONAL VALUE. Probably the best thing that can be said about soft drinks is that their flavor encourages people to drink water. Soft drinks containing nutritive sweeteners contribute empty calories to the diet. Each 12-oz (360 ml) can supplies about 147 Calories (kcal); hence, excessive usage may contribute to tooth decay and overweight. Diet soft drinks, which became popular during the 1960s, are essentially flavored water. Also, sparkling water (club soda) is carbonated water without sugar (calories), and sometimes with added sodium bicarbonate or sodium sulfate in an effort to simulate natural mineral water. Despite the nutritional shortcomings of carbonated beverages, many people indulge primarily because they enjoy the taste, the tingling sensation of the carbonation, and the mildly stimulating effects. Food Composition Table F-21 compares some of the common soft drinks.

Considering the alternatives, water costs less and quenches the thirst, and milk provides a few more calories but better nutrition. Moreover, 12 oz of milk costs less than 12 oz of a soft drink. Fruit juices are also viable alternatives providing substantially more nutrition at a cost comparable to soft drinks.

SOLANINE

A naturally occurring toxicant found in the nightshade family, which includes such useful plants as the potato, tomato, ground cherry, red pepper, and eggplant. In large enough doses it acts on the nervous system since it is an inhibitor of acetylcholinesterase, an enzyme involved in the transmission of nervous impulses. Also, the feeling persists (without research proof) that consumption of vegetables in the nightshade family may aggravate arthritis in susceptible people. Potatoes contain the highest levels of solanine. However, a person would have to eat, at one sitting, 4.5 lb (2.0 kg) of potatoes containing 200 mg (100 ppm) of solanine to produce the initial effects of poisoning—drowsiness and itchiness behind the neck.

CAUTION: When potatoes are exposed to light, or when they sprout, their solanine content increases; it's concentrated chiefly in the sprouts and to a lesser extent in the skin. Thus, potatoes that are green, spoiled, or sprouted may contain harmful amounts of solanine.

(Also see POISONOUS PLANTS; and POTATO, section headed "Selection.")

SOLIDS, MILK

The solids, or total solids, of milk includes all the constituents of milk except water. Milk contains an average of 12% total solids.

SOLIDS-NOT-FAT, MILK

This includes the carbohydrate, protein, minerals, and water-soluble vitamins of milk—it excludes the fat. Milk contains an average of 8.6% solids-not-fat.

SOLUTES

Particles in solution in body water. Three types of solutes influence internal shifts and balances of body water: (1) electrolytes, (2) plasma proteins, and (3) organic compounds of small molecular size.

SOLVENT-EXTRACTED

Fat or oil removed from materials (such as oilseeds) by organic solvents. Also called *new process.*

(Also see OILS, VEGETABLE, section on "Extraction of Oils"; and SOYBEAN, section headed "Extracting the Oil.")

SORBIC ACID (SORBATE; $C_6H_8O_2$)

A food additive possessing antimicrobial activity especially toward yeast and molds. It is commonly used in the form of *sodium sorbate* and *potassium sorbate.* In the body, all evidence indicates that sorbate is metabolized like other fatty acids. It is on the GRAS (generally recognized as safe) list of approved food additives. Food applications of sorbic acid are numerous, among them, baked goods, beverages, cheese and cheese products, dried fruits, fish, fruit juices, jams, jellies, meats, pickled products, salads, and wines.

SORGHUM (GUINEA CORN; KAFFIR; MILO)
Sorghum bicolor

Contents	Page
Origin and History	965
World and U.S. Production	965
Kinds	966
Processing	966
Nutritional Value	966
Products and Uses	966

Fig. S-19. Hybrid grain sorghum, headed out and nearing harvest. (Courtesy, National Grain Sorghum Producers Assn., Abernathy, Tex.)

Sorghum, a member of the grass family, *Gramineae,* is of great importance in dry and arid lands. The genus *Sorgum* embraces hundreds of varieties and many hybrids which are commonly lumped as the many-faceted species, *Sorghum bicolor.* Different varieties are grown for different uses—grain, syrup, forage, and broom fiber.

ORIGIN AND HISTORY. Sorghum has been cultivated in Africa and Asia for 4,000 years. It is said to have been a principal item in the diet of the ancient city of Ninevah on the Tigris.

WORLD AND U.S. PRODUCTION. Sorghum is the fifth ranking cereal crop of the world, being exceeded by wheat, corn, rice, and barley. The world sorghum crop averages 58 million metric tons. Ten leading producing countries, by rank, are: United States, India, Mexico, China, Nigeria, Argentina, Sudan, Ethiopia, Australia, and Burkino Faso.[31]

---

[31]Data from *FAO Production Yearbook 1990,* FAO/UN, Rome, Italy, Vol. 44, p. 85, Table 24. **Note well:** Annual production fluctuates as a result of weather and profitability of the crop.

The United States produces about 25% of the world's sorghum grain. The U.S. *sorghum belt* is the Southern Great Plains. The ten leading states, by rank, are: Kansas, Texas, Nebraska, Missouri, Arkansas, Oklahoma, Illinois, South Dakota, Colorado, and Louisiana. Kansas and Texas account for 56% of the nation's production.

KINDS. The cultivated sorghums can be classified into the following four broad classifications according to use:

1. **Grain sorghums.** These sorghums are bred specially for grain production. Most types in the United States are less than 5 ft tall. Common grain varieties include kaffir, milo, feterita, Durra, shallu, koaliang, and hegari.

2. **Sweet sorghums (sorgos).** These have tall, sweet, juicy stalks that are used for forage, syrup, and sugar production.

3. **Grass sorghums.** These have thin stems, narrow leaves, and numerous tillers—characteristics which make them useful for hay or grazing by animals. Sudangrass and Johnsongrass belong to this group.

4. **Broomcorn sorghums.** These have panicle branches which are suitable for making brooms.

Today, most of the sorghums grown in the United States are hybrids, developed by crossing varieties. For grain production, crosses involving kaffir and milo are most common.

PROCESSING. As with corn, sorghum may be either dry milled or wet milled. However, most sorghum is dry milled, simply because it is easier to accomplish.

• **Dry milling**—The objective in dry milling of sorghum grain is to separate the endosperm, germ, and bran from each other.

The endosperm is processed into grits and flour. The other products of dry milling are bran, germ, and hominy feed. Sorghum grits from dry milling are used for brewing and industrial purposes.

(Also see CORN, section headed "Dry Milling.")

• **Wet milling**—The wet milling of sorghum closely resembles the wet milling of corn. However, sorghum is more difficult to wet mill than corn because (1) it is difficult to separate the starch and protein, (2) it is difficult to extract the sorghum germ from the kernel, (3) pigments discolor the starch and must be removed by bleaching, and (4) recovery of starch is lower than with corn. It is noteworthy, however, that new hybrids with improved wet milling qualities have been developed.

(Also see CORN, section headed "Wet Milling.")

NUTRITIONAL VALUE. Generally speaking, sorghum and corn are interchangeable in the diet; with corn preferred and grown where adapted, and sorghum grown in hot, dry regions where corn cannot be grown successfully. So, both the nutritionist and the consumer need to com-

pare the two foods. Table S-8 gives a comparison of selected nutrients of sorghum grain and corn. Also, Food Composition Table F-21 gives the nutrient compositions of corn and sorghum, and their products.

TABLE S-8
COMPARISON OF SELECTED NUTRIENTS IN SORGHUM GRAIN AND CORN

Nutrient		Sorghum Grain	Corn
		◄--- (per 100 g) ---►	
Calories	kcal	366.0	348.0
Carbohydrates	g	73.0	71.1
Fat	g	3.3	3.9
Protein	g	11.0	9.6
Calcium	mg	28.0	30.0
Phosphorus	mg	287.0	270.0
Iron	mg	4.4	2.0
Niacin	mg	3.9	3.0
Riboflavin	mg	.15	.13
Thiamin	mg	.38	.21

It is recognized that compositional data provide only an approximate comparison because the influence of environment is very striking, affecting both the physical and chemical properties of both sorghum and corn. Nevertheless, the following comparisons of the two grains appear to be justified: sorghum grain is higher in protein but lower in fat than corn.

The amino acid composition of sorghum grain and corn is strikingly similar. The most limiting amino acid is lysine—a feature common to all grains.

Although not shown in Table S-8, cognizance is taken of the following additional differences between sorghum and corn: yellow corn contains more carotene (vitamin A) than sorghum; sorghum has slightly more starch than corn; sorghum contains 0.1 to 0.3% waxes, which is 50 times the quantity of waxes in corn; tannins are present in the pericarp and testa of sorghums, especially in bird-resistant varieties, but absent in corn; and high yields of sorghum are generally inversely related to the protein content of the grain.

Other than the differences noted above, the nutrient composition of sorghum grain and corn are similar.

PRODUCTS AND USES. Sorghum is used for food, fermentation products, feed, and industrial purposes.

The sweet stalks of sorghum are chewed by the natives in various countries; and sorghum is the staple food and basis of beverages in much of Asia and Africa. It is usually consumed as a porridge or stiff paste prepared by adding pounded flour to hot water. Sometimes a flat cake is prepared. Also, the grain may be parched, popped, or boiled whole.

In the United States, the vast majority of grain sorghum is used as livestock feed; hence, in this country sorghum enters the human diet primarily as meat, milk, and eggs. Some sorghum is also used for industrial products.

Table S-9 presents in summary form the use of sorghum products for human food.

(Also see BEERS AND BREWING; BREADS AND BAKING; CEREAL GRAINS; and OILS, VEGETABLE, Table O-5.)

---

[32]Data from *Agricultural Statistics 1991*, USDA, p. 52, Table 67. **Note well:** Annual production fluctuates as a result of weather and profitability of the crop.

TABLE S-9
SORGHUM PRODUCTS AND USES FOR HUMAN FOOD

Country/Product	Description	Uses	Comments
In the United States: Sorghum flour—wheat flour, blended	An off-white flour	Muffins, bread, and griddle cakes	Only about 1.3% of U.S. sorghum grain is used for food, alcohol, and seed; The rest is used for livestock feed. Nevertheless, sorghum has many potential uses in food products.
Sorghum oil	Oil extracted from the germ of the grain	Cooking oil; and in such products as margarines and shortening	
Sorghum starch	An off-white powdery material (varieties with dark colored outer layers may stain the starch)	Thicken puddings	
Sorghum syrup	A distinctly flavored, mild, sweet, light amber-colored syrup. Sorghum syrup is a good source of carbohydrates, calcium, potassium, and iron.	A speciality product, especially in southeastern U.S. As a carbohydrate source, it is about equal to maple syrup and molasses.	The technology for satisfactorily converting sweet sorghum beyond the syrup stage—into sugar— is known. Thus, when the price is right, sweet sorghum can serve as a supplementary crop for the sugar cane and sugar beet industries.
Africa and Asia: East Africa	Pounded sorghum flour	Porridge, made by adding pounded sorghum flour to hot water	
Ethiopia and Sudan	Sorghum flour. Sorghum grain, whole.	Flat cakes. Parched, popped, or boiled white.	
India	Sorghum grain, ground or cracked Sorghum grain	Prepared into a dough and baked as flat, unleavened bread (called *rotti*, or cooked like rice. Whole grain is parched or "fried" in a hot pan, then ground and mixed with salt, buttermilk, or molasses. Special sorghums for popping or roasting are grown in small quantities.	White, pearly grains are preferred for bread.
Nigeria	Sorghum flour Immature sorghum heads Sorghum (special varieties)	A food called *tuwo*, prepared by stirring flour in hot water, then allowing the thick paste to cool and gel. Roasted, much like sweet-corn in the U.S. Popped, similar to popcorn.	Pieces of the cooled gel are eaten with soup.
Uganda	Sorghum grain, whole	The grain is malted and sprouted, the radicle (root portion) removed, and the grain dried.	Some of the pigment and the bitter principle are removed. The sugars produced make the porridge sweet.
West Africa	Sorghum flour	Porridge or stiff paste made from stirring flour in hot water. These may be allowed to cool and gel, or the paste may be fried to make pancakes.	

FERMENTATION PRODUCTS

Country and beverage:	Description	Uses	Comments
Country and beverage: United States Alcohol	The whole grain is ground, cooked, hydrolyzed by enzymes or acid, fermented by yeast, and the alcohol distilled off. Beverage alcohol production is similar, except malt enzymes are preferred for hydrolysis.	1. Ethyl alcohol for beverages. 2. Denatured alcohol for industrial purposes. 3. Gasohol (10% anhydrous ethanol/90% gasoline)—a motor fuel.	In the U.S., the use of sorghum depends on its price relative to other grains.
Sorghum grits Sorghum starch	The starchy portion of the endosperm. An off-white powdery material.	Brewing and distilling, making industrial alcohol Brewing and distilling, making industrial alcohol	
Africa (Bantu tribe or family)	Sorghum beer	Beer, made by malting and brewing sorghum (preferably red grains)	It differs from beer made from barley malt in that it is an opaque reddish-colored liquor, contains 5 to 6% solids, has a yeasty sour taste, has a low alcohol content, and is high in nutritional value.
Central Africa; tribes in Zambia, Zaire, and the Central African Republic	Homemade sorghum beer; made from sorghum alone, or from a mixture of sorghum with maize (corn) and/or millet.	As a nutritious beer	The grain is soaked in water, sprouted, dried, ground, mixed with water, and allowed to ferment in clay pots. The crude beer may or may not be filtered prior to consumption. Sprouting and fermentation of sorghum results in a more nutritious product than the original grain because additional amounts of the vitamin B complex are produced during the processes.
China Mao-tai	About a 53% alcoholic beverage made from grain sorghum.	At Chinese banquets, Mao-tai is used for toasts.	The Chinese claim that "it's easy on the throat, and it doesn't go to the head." But *first timers* are admonished to approach it with caution. It's potent! (As a result of experiencing the traditional Chinese toasts on 5 different visits to China, two of the authors of this book, the Ensmingers, concluded that the only reason Mao-tai is not used for airplane fuel is that it is readily combustible.)

SOUPS

Probably the first dish ever cooked in a kettle over the open fire was a soup, and because this method of cooking was the only one for countless years, soups were standard fare.

In France, the evening meal is called *la soupe*; and in English, our word *supper* came from the word *soup*.

Each nation has its own special soups, rich in meat, chicken, or fish; health-giving vegetables; and nourishing barley, rice, or macaroni. At one time, soups were thought to be restorative of one's health, and in 16th-century France, one enterprising person advertised his soup establishment as a *restaurant*—alluding to its reputed health-restoring properties. Soon others copied him. Eventually, the name *restaurant* came to be used for all establishments that serve any kind of food.

The purpose of soup in the meal is twofold: (1) to improve digestion and stimulate appetite; and (2) to increase the variety of foods served at the meal. The clear, stock soups serve the first purpose, and the cream soups serve the latter. A heavy meal such as steak or roast, should be preceded by a clear soup, but a light meal of fish, eggs, or salad, is best with a heartier soup.

The number of good soup recipes would, literally speaking, stretch around the world. Usually the soup is named after the ingredients, but there are some world-famous soups with which everyone should become acquainted, unless they are already old friends. Some of these, the recipes for which can be found in the local cookbook library, follow:

• **Bird's nest soup (Chinese)**—The Asian swiftlets coat their nests with a translucent gelatinous material (their saliva), which is high in protein (11.9 g/100 g). It is packaged and available in Chinese specialty shops. First, the nest is soaked, then simmered in chicken stock. "Chicken velvet" is added and it is served garnished with ham. The chicken velvet is made by finely mincing the chicken and mixing it with cornstarch and beaten egg whites. When added to soup, this becomes light and frothy. Bird's nest soup is considered a delicacy.

• **Bisque**—A cream soup made from fish.

• **Borsch (Russian)**—This soup, of which there are several different recipes, is popular in both the U.S.S.R. and in Poland. Basically, it is a beet soup which is thickened with cornstarch. It may or may not have added vegetables and meats.

• **Bouillabaisse (French)**—This is a fish soup with a hot pepper sauce.

• **Bouillon**—Meat and/or meat scraps are simmered slowly to produce the stock or bouillon. It is usually served clear, but it can also be served with added vegetables and pieces of meat.

• **Chowder**—This word comes from the French word, *chaudiere*, meaning *caldron*. The chowder of today has not changed much from the chowder of yesterday. It consists of pieces of different vegetables, or of fish and potatoes, with seasonings, all cooked in milk, and with crackers added just before serving.

• **Cock-a-leekie**—This soup is made by boiling chicken and leeks together.

• **Consomme**—From the French word meaning *to boil down*. Basically, there is little difference between consomme and bouillon. But bouillon is usually thought of as strictly beef stock, whereas consomme may be beef, veal, and/or chicken stock.

• **Cream soups**—A thin white sauce is combined with any desired vegetables or meats. When vegetables or meats are pureed to make a soup, this can be classified as a cream soup, also.

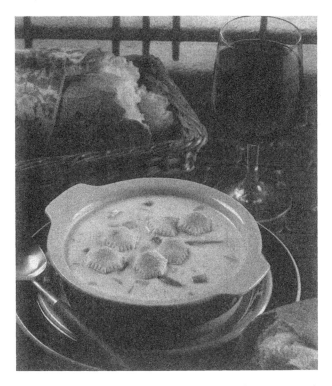

Fig. S-20. Plain and fancy—creamy vegetable soup is welcome at any meal. (Courtesy, Pet Incorporated, St. Louis, Mo.)

• **Egg drop soup (Chinese)**—A clear stock soup, into which is dropped a slightly beaten egg. The egg is cooked in the soup the same way that a poached egg is cooked.

• **Gazpacho (Spain)**—This soup, which originated in Spain, and is popular in Mexico and the southwestern United States, is a highly-seasoned tomato soup with added vegetables, which is served chilled. It may also be jellied.

• **Minestrone (Italy)**—A mixed vegetable soup with beans and macaroni.

• **Mulligatawny (East Indies)**—Chicken, carrots, green pepper and apple are cooked together. The vegetables and apple are pureed and returned to the pot, seasonings, including curry, cloves, and mace, are added.

• **Printanier soup**—This name comes from the French word *printemps*, meaning *spring*. The soup is made from fresh spring vegetables.

• **Vichyssoise (French)**—Potatoes and leeks are cooked together in a chicken broth, then blended, or put through a sieve. Thick cream is added and the mixture is served chilled with a garnish of chopped chives.

Soups can be served with inumerable garnishes, plus a raft of soup accessories. All good cookbooks will furnish the recipes for these.

The nutritive value of soups depends on the ingredients. See Table F-21, Food Composition Table, under the section, "Soups and Chowders" for the analysis of some of the more popular soups.

SOYBEAN *Glycine max*

Contents	Page
Origin and History	969
World and U.S. Production	969
Processing Soybeans	970
Extracting the Oil	970
Solvent Extraction	970
Hydraulic Extraction	970
Expeller Extraction	970
Refining Soybean Oil	970
Nutritional Value	971
Soybean Products and Uses for Human Food	972

Fig. S-21. Cultivating soybeans. (Courtesy, Deere & Co., Moline, Ill.)

Soybeans belong to the pea family, *Leguminosae*. It is classed as genus *Glycine*, species *G. max*.

Soybeans were cultivated in China long before written history. The ancient Chinese used the soybean as food and made medicines from it.

Today, U.S. soybeans supply food for human beings, feed for animals, and many raw materials for industry.

ORIGIN AND HISTORY.
The soybean plant is native to China, where it emerged as a cultivated crop in the 11th century B.C. Records of methods of culture, varieties for different purposes, and numerous usages indicate that the soybean was among the first crops grown by man. The ancient Chinese considered the soybean their most important crop, and one of the five sacred grains necessary for living. Soybeans were taken from China to Japan in the 7th century.

Soybeans were taken to Europe in the 17th century. They were introduced into the United States in 1804. However, scant attention was given to soybeans in North America until the 1930s. At that time, the U.S. Department of Agriculture and the agricultural experiment stations of the states and the Canadian provinces cooperated in developing improved soybean varieties through hybridization and selection.

WORLD AND U.S. PRODUCTION.
Soybeans are the leading legume crop of the world, and the leading legume crop of the United States.

Fig. S-22. A soybean plant. Soybeans are the leading legume crop in the world. (Courtesy, Field Museum of Natural History, Chicago, Ill.)

World production of soybeans totals about 107,767,000 metric tons annually; and the ten leading soybean-producing countries of the world, by rank, are: United States, Brazil, China, Argentina, India, Italy, Paraguay, Indonesia, Canada, and U.S.S.R.[33] The United States is, by far, the leading soybean-producing country, producing more than 49% of the world's soybeans.

The United States produces about 1.9 billion bu of soybeans annually; and the ten leading states, by rank, are: Illinois, Iowa, Minnesota, Indiana, Ohio, Missouri, Arkansas, Nebraska, South Dakota, and Kansas.[34]

---

[33]Data from *FAO Production Yearbook 1990*, FAO/UN, Rome, Italy, Vol. 44, p. 107, Table 37. **Note well:** Annual production fluctuates as a result of weather and profitability of the crop.

[34]Data from *Agricultural Statistics 1991*, USDA, p. 123, Table 169. **Note well:** Annual production fluctuates as a result of weather and profitability of the crop.

PROCESSING SOYBEANS. About 59% of the soybeans produced in the United States are crushed and processed domestically, 35% are exported as whole beans, 5% are used for seed, and 1% have miscellaneous uses.

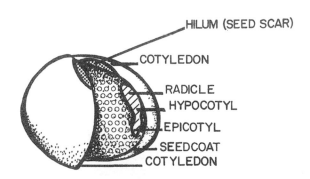

Fig. S-23. The soybean seed. There are three basic parts: (1) the seedcoat, which protects the embryo from fungi and bacteria before and after planting; (2) the embryo, consisting of (a) the radicle (which becomes the primary root), (b) the hypocotyl (which is the main stem and growing point); and (3) the cotyledons, which account for most of the bulk and weight of the seed and contain nearly all the oil and protein.

The processing of soybeans into oil and meal began in the Orient. The oil was used primarily for human food, while the meal was used for animal feed.

In the United States, soybeans are processed primarily to obtain (1) oil for use in shortenings, margarines, and salad dressings; (2) soybean protein products for direct human consumption; and (3) soybean meal for use as a protein supplement for livestock.

Extracting the Oil. Oil, primarily for human food, is extracted from several rich oil-bearing seeds; among them, soybeans, coconuts, cottonseed, linseed, peanuts, rapeseed, safflower seed, sesame seed, and sunflower seed. The oil is extracted from these seeds by one of the following basic processes or modifications thereof: solvent extraction, hydraulic extraction, or expeller extraction. For purposes of illustration, all three processes are discussed relative to soybeans, since soybeans are the most widely used oilseed in the United States, and since the principles involved in processing all oilseeds are essentially the same. However, it should be noted that little of the U. S. soybean crop is processed via hydraulic or expeller extraction.

SOLVENT EXTRACTION. In this method, the soybeans are first cracked, then heated to 140°F (60°C) for about 10 minutes. After the cracked seeds are heated, they proceed through a series of grinding rollers where they are flaked. The flakes are allowed to cool to about 113°F (45°C) and are then moved to the extraction equipment where the oil is removed by a petroleum solvent—usually hexane.

The extracted flakes then proceed to driers where the solvent is completely volatilized. From the drier, the flakes are conveyed to a toaster, thence they are cooled and ground.

The normal processing yields are: crude oil, 17 to 19.3%; and meal and feed products, 80.7 to 83%. One bushel (60 lb) of soybeans gives approximately 48 lb (22 kg) of meal and 11 lb (5 kg) of oil. The maximum yield of crude oil is always sought. Generally, less than 2% oil remains in the meal after solvent extraction. The hulls, accounting for about 5% of the beans, are ground and may be added back to the meal for the production of meal of a specific protein content.

HYDRAULIC EXTRACTION. In the hydraulic extraction

procedure, raw soybeans are cracked, ground, and flaked. The flakes (called meats) are then transported to cookers where they are exposed to both dry and steam heat. The cooking stage takes about 90 minutes.

After cooking, the meats are formed into cakes and wrapped in heavy cloth whereupon they are placed in hydraulic presses for the mechanical extraction of the oil. This procedure takes about 1 hour. Following extraction, the cakes are ground. Hydraulic press cake may have 5 to 8% residual oil.

Since this form of extraction is labor intensive and inefficient in the removal of oil, very few soybeans are processed by this method today.

EXPELLER EXTRACTION. In this type of extraction, raw soybeans are initially cracked and dried to about 2% moisture. The dried soybeans are then transported hot to a steam-jacketed tempering apparatus which is directly above the expeller apparatus. The tempering apparatus stirs the cracked soybeans for about 10 to 15 minutes so that the seeds are heated uniformly.

From the tempering bin, the soybeans are fed into the expeller barrel (screw presses). A central revolving worm shaft creates pressure within the expeller barrel thereupon extracting the oil from the ground soybeans. The extracted soybeans leave the expeller in the form of flakes, which are subsequently ground.

The expeller process tends to extract less oil than the solvent process; consequently, it is used less frequently. Generally, expeller-extracted soybean meal contains 4 to 5% oil while solvent-extracted soybean meal contains less than 1%.

Refining Soybean Oil. About 93% of soybean oil is used for food. To be suitable for human consumption, the extracted crude oil must undergo further processing, which is generally referred to as refining.

The steps in processing are:

1. **Degumming.** This consists in the removal of the soybean phospholipids, which are dried and sold as soybean lecithin. The crude lecithin, which is dark colored, may be bleached with hydrogen peroxide to produce a light, straw-colored product. Fatty acids may be added to make a fluid product, or the lecithin may be treated with ammonia to form a firm, plastic consistency. More refined grades (pharmaceutical lecithins) are produced by acetone extraction of the entrained soybean oil.

2. **Alkali refining.** The degummed soybean oil is next alkali-refined (treated with caustic soda, soda ash, or a combination of the two) to neutralize the free fatty acids. The neutralized free fatty acids, known as soap stock, may be removed from the oil and sold to soap manufacturers, fatty acid industries, or feed manufacturers in either (a) the alkaline form, or (b) the acidulated form after treatment with acid.

3. **Bleaching (decolorizing).** Bleaching agents (activated earth or carbon) are used to remove the pigments present in the oil.

4. **Hydrogenating.** Hydrogenation is achieved by treating the oil with hydrogen gas at an elevated temperature and pressure in the presence of a catalyst.

5. **Winterizing.** Hydrogenated soybean oil that is to be used as a salad oil must remain clear at refrigerator temperatures. This is accomplished by winterization—cooling the oil and filtering off the cloudy haze that forms.

6. **Deodorizing.** Undesirable flavors and odors are removed under high temperature and vacuum, with steam injected to assist volatization of the undesirable components.

Fig. S-24. Soybean oil, widely used throughout the world for human consumption, as margarine, salad and cooking oils, and shortening. (Courtesy, USDA)

TABLE S-10
PROFILE OF ESSENTIAL AMINO ACIDS IN SOYBEANS COMPARED TO MILK—A HIGH-QUALITY PROTEIN

Amino Acid	Soybeans	Cow's Milk[1]
	← (mg/g protein) →	
Histidine	28	27
Isoleucine	50	47
Leucine	85	95
Lysine	70	78
Methionine and cystine	28	33
Phenylalanine and tyrosine	88	102
Threonine	42	44
Tryptophan	14	14
Valine	53	64

[1]*Recommended Dietary Allowances*, 10th ed., 1989, NRC–National Academy of Sciences, p. 67, Table 6–5. The essential amino acid requirements of infants can be met by supplying 0.79 g of a high-quality protein per pound of body weight (1.7 g/kg body weight) per day.

Fig. S-25 compares the levels of essential amino acids of soybeans and certain other foods.

V. R. Young and N. S. Scrimshaw, Clinical Research Center and Department of Nutrition and Food Science, Massachusetts Institute of Technology, make the following authoritative statement relative to the nutritional value of processed soy protein in human nutrition:

> When well-processed soy products serve as the major or sole source of protein intake, their protein value approaches or equals that of foods of animal origin, and they are fully capable of meeting the long term essential amino acid and protein needs of children and adults...For feeding of the newborn, the limited data available suggest that supplementation of soy-based formulas with methionine may be beneficial.[35]

NUTRITIONAL VALUE. The nutritive value of different forms of soybeans and soybean products is given in Food Composition Table F-21.

The soybean seed is 13 to 25% oil, 30 to 50% protein, and 14 to 24% carbohydrate. The fatty acid composition of the oil is: linoleic acid, 55%; oleic acid, 21%; palmitic acid, 9%; stearic acid, 6%; and other fatty acids, 9%. The ratio of polyunsaturated to saturated fatty acids (P/S ratio) is 82:18, which is conducive to lowering blood cholesterol. The soybean contains more protein than beef, more calcium than milk, and more lecithin than eggs. Also, it is rich in minerals, vitamins, and amino acids.

In terms of essential amino acids—those not produced by the body, and which must be eaten—soybeans are an excellent source as shown in Table S-10. Indeed soybeans and soybean products are good complementary protein sources when the diet consists largely of cereal grains.

---

[35]Young, V. R., and N.S. Scrimshaw, "Soybean Protein in Human Nutrition: An Overview," *Journal of the American Oil Chemists' Society*, Vol. 56, No. 3, March 1979, p. 110.

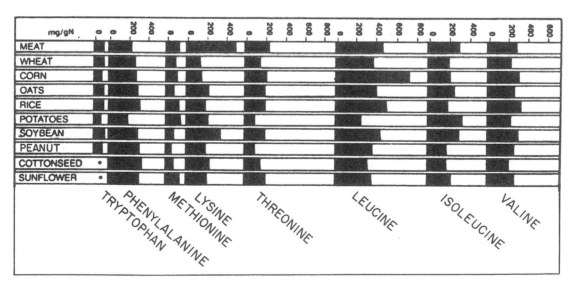

Fig. S-25. Comparative levels of essential amino acids (in mg/g of nitrogen) of soybeans and other foods. (Source: *Vegetable Protein: Products and the Future,* Food Protein Council, Washington, D.C.)

Perhaps more important than the individual nutritional strength of soy protein is its ability to complement the biological quality of other protein sources. For example, corn meal, which is the basis of many human diets around the world, is of low protein quality when used alone because of certain amino acid deficiencies. But, when blended with soy flour, the resulting nutritive value is comparable to milk protein (casein). This is accomplished because the low lysine content of corn meal is offset by the naturally high lysine content of soy protein. Conversely, the limited methionine content in soy products is offset by the excess of methionine in corn products. The results, when the two are combined at levels that do not inhibit palatability, is a combined meal nutritionally superior to corn meal alone. This same complementary relationship occurs when soy products are used with wheat and other foods (see Fig. S-25).

The cereals and meals based on corn and wheat are a vital basis of diets in the less developed countries. But these grains do not provide adequate nutrition. If combined with soybeans, however, these grains can boost the general nutritional value of the foods eaten in the less developed countries.

Different nutritional needs prevail in industrialized economies. For the affluent, meat analogs made from textured soy protein are largely free of cholesterol and high in polyunsaturates. For example, one kind of sausage made from textured soy protein contains 40% more polyunsaturated fats than saturated fats, while traditional pork sausage contains 400% more saturated than polyunsaturated fats.[36]

Evidence relative to the effectiveness of a soy protein diet in reducing serum cholesterol concentrations was dramatically corroborated by an Italian research team, whose work was reported in *The Lancet*, a respected British journal of medicine. A group of patients found to have high blood serum cholesterol levels was put consecutively on two kinds of diets: (1) a low-fat animal protein diet, and (2) a predominantly soy protein diet. The study showed that the soy protein reduced serum cholesterol concentrations in the patients by about 20%.[37]

It is noteworthy, too, that among the leading users of soy protein products are mass feeding institutions, including schools, hospitals, and military units, where nutritional standards must be maintained in the face of cost restrictions. The Federal Government's National School Lunch Program is the largest consumer of soy protein for food served in institutional settings. Hydrated soy protein may constitute up to 30% by weight of meat products served in that program. Over 50 million lb of soy protein are used each year in the school lunch program.

• **Savings in protein costs**—Savings realized by food service operators and feeding institutions explain why the applications of soy protein continue to expand and diversify. Also, the individual consumer is becoming increasingly aware of these savings. (See Table S-11.)

To determine relative protein costs from Table S-11, multiply the cost of a product by its cost multiplying factor. *Example:* Soy flour at 20¢ per pound has a protein cost of 40¢ per pound (20¢ x 2); whereas beef at 80¢ per pound has a protein cost of $4.80 per pound (80¢ x 6).

---

[36]*Vegetable Protein: Products For the Future*, Food Protein Council, Washington, D.C.
[37]*Ibid.*

TABLE S-11
RELATIVE COST INDICATORS FOR LEADING PROTEIN FOODS

	Approximate Percent Protein	Cost Multiplying Factor
Soy isolates	90.	1
Sodium caseinate	90	1
Soy concentrates	65	1.5
Soy flour	50	2
Dried skimmed milk	36	3·
Cheese (Cheddar)	25	4
Dried beans	22	5.5
Fish/chicken	20	5.5
Beef (chuck—35% fat)	16	6
Dried whey	13	8
Wheat flour	12	8.5
Corn meal (whole)	9	11
Rice	7	15
Whole milk	3.5	28.5

• **Improved nutritive value of infant products**—Soy protein-based infant formulae are a universal staple for babies who are (1) allergic to milk protein, or (2) unable to utilize lactose. Soy protein infant products provide the same order of growth and weight gain as cow's milk. Also, soy protein is used in baby cereal blends and canned baby foods to improve nutritive value.

• **Improved nutritive value of breakfast cereals and pastas**—New emphasis on nutrition in breakfast cereals has meant more use of soy protein to increase their protein quantity and value. Today, soy proteins are used extensively as additives to hot cereal mixes, and as components of granola bars and compound breakfast bars. Also, U.S. standards of identity for pasta products permit fortification with soy protein.

SOYBEAN PRODUCTS AND USES FOR HUMAN FOOD.
In the People's Republic of China, Taiwan, Japan, Korea, and Indonesia, soybeans are a regular part of the diet and a major source of protein and oil. Soybean milk, soybean curd, soy sauce, fermented soybean paste, soybean sprouts, immature green seed, and deep-fried mature seed are common foods in the Orient.

Soybean oil is widely used throughout the world for human consumption as margarine, salad and cooking oils, and shortening.

In the United States, about 30% of the soybean production is used for food and industrial products, and soybean products are the leading protein supplements for livestock. In addition to the use of soybean oil, soy protein is revolutionizing the modern food system; its use as an ingredient, extender, and analog has spread to products in every category of food available to the public—in home and away from home.

Soy protein is available to the food industry in a multitude of forms, most of which may be classified according to protein content as: (1) soy flour and grits, containing 40 to 50% protein; (2) soy protein concentrates, with about 70% protein; (3) soy protein isolates, with 90 to 95% protein; and (4) textured soy protein, which is made from one or more of the other three types. Fig. S-26 shows an attractive meat-substitute dish.

Fig. S-26. Textured soybean hamburger made from spun soy fiber. (Courtesy, Worthington Foods, Worthington, Ohio)

• **Soy proteins in foreign-feeding programs**—Soy flour is included in many commodities used in foreign-feeding programs, such as soy-fortified bulgur, soy-fortified bread flour, corn-soy milk, corn-soy blend, and wheat-soy blend. In particular, soy protein fortification improves the amino acid composition of the foods to which it is added.

Low cost, good nutrition, and functional versatility account for the ever-increasing quantities of soy proteins in food programs in the less-developed countries. Soy flour is the major ingredient in blended foods with flour and grits added to corn, sorghum, and oat products distributed overseas. Whey-soy drink mix is a new product recently added to the list of commodities available for distribution in feeding programs. It was developed as a replacement for dry milk. The finished product—41.4% sweet whey, 36.5% full fat soybean flour, 12.1% soybean oil, and 9.5% corn sugar—when mixed with water at 15% solids, provides a drink with nearly the same energy and protein content as whole milk.

Table S-12 presents in summary form the story of soybean products and uses for human food.

(Also see FLOURS, Table F-13, Special Flours; LEGUMES, Table L-2, Legumes of the World; OILS, VEGETABLE, Table O-5; and VEGETABLE[S], Table V-2, Vegetables of the World.)

TABLE S-12
SOYBEAN PRODUCTS AND USES FOR HUMAN FOOD

Product	Description	Uses	Comments
Soybeans	Whole soybeans	As mature, dry soybeans are used (1) in recipes like other dry beans, or (2) as roasted soybeans—a snack food. As immature, green seed served hot as a vegetable, like peas or corn.	In the Orient, soybeans (including whole soybeans) are the principal source of protein in the diet. Soybeans are also rich in minerals, vitamins, and amino acids. Direct food usage of soybeans is expected to increase dramatically around the world due to population pressure on the food supply.
Chee-fan	A type of sufu made by salting small cubes of soybean curd, innoculating with a species of *Mucor*, and allowing them to ferment for about 7 days. After fermentation, allowed to age in yellow wine for a year.	Consumed much like cheese is eaten in the western world.	
Fermented soybeans	Prepared from black soybeans in China, using a species of *Mucor* as the fermentation organism. After fermentation, they are aged for 6 months or longer in sealed earthenware jars, then seasoned with salt, spices, and wine or whiskey.	As an appetizer	
Hamanatto (A similar product called Meit anza or tu su is prepared in China. In the Philippines, it is known as taosi)	A liquid, fermented soybean product prepared from whole soybeans by using a starter (koji) made from roasted wheat or barley treated with spores of the mold *Aspergillus oryzae*.	For flavoring many foods	Hamanatto is expensive and very dark in color. Nagoya University, in Japan, reported the following analyses of hamanatto: water, 38%; protein, 25%; carbohydrate, 25%; salt and calcium, 12%.
Ketjap	A type of soy sauce made from black soybeans in Indonesia	As a flavoring agent for many foods	
Metlauza	A product prepared from the insoluble protein and other materials which are separated out from the soybean milk in the preparation of tofu. These residual solids are pressed into cakes; fermented with *Actinomucor elegans* for 10 to 15 days; then sun dried.		
Miso	A fermented food paste; made by innoculating trays of rice with *Aspergillus oryzae*, and leaving it to mold abundantly. Then a ground preparation of cooked soybeans and salt is mixed in, and the mass is allowed to ferment and age for several days before being ground into the paste of the consistency of peanut butter.	As a food in Japan. It is used primarily as flavoring material, added to soups and vegetables. Miso is also spread thinly on cucumbers or substituted for meat sauce on spaghetti.	Miso originated in China. A Japanese missionary who was sent to China in the 7th century learned how it was made and modified it into a product suited to the Japanese taste. The entire process of making miso requires from 10 to 40 days, varying according to the temperature at which the yeast fermentation is conducted. The qualities and proportions of amino acids in miso compare quite favorably with those of casein and soybean meal.

(Continued)

TABLE S-12 *(Continued)*

Product	Description	Uses	Comments
Natto	A fermented whole protein food, fermented by *Bacillus natto*. In this product, the shape of cooked whole soybeans is retained.	Natto is served with soy sauce (shoyu) and mustard.	Natto originated in Japan. In 1087, a ruler in the northern part of Japan discovered natto to be part of the local farmers' diets.
Soybean flour	An extremely fine powdered product, containing 40 to 50% protein, the particles of which are 100 or finer mesh size (U.S. standard screen)	Bakery products: breads, rolls, and buns; doughnuts; sweet goods; cakes and cakes mixes; pancake and waffle mixes; specialty crackers and cookies. Meat products: sausages, luncheon loaves, patties, canned meats in sauces. Breakfast cereals. Infant and junior foods. Confectionery items. Dietary foods.	Soybean flour is produced like soybean meal, except that the hulls are carefully removed before extraction. Flours are differentiated from grits on the basis of particle size. In Israel, 10% soybean flour is included with wheat flour in baking bread.
Soybean grits	A granular product containing 40 to 50% protein, the particles of which range from 10 to 80 mesh size (U.S. standard screen)	Same as for soybean flour	Soybean grits are produced like soybean meal, except that the hulls are carefully removed before extraction.
Soybean lecithin	The phospholipids which are removed from the crude oil during the degumming process	Baked goods, cake mixes, instant foods, candies, and drugs. As an antioxidant, emulsifier, and softener in food manufacturing.	Crude soybean oil contains 1 to 3% lecithin.
Soybean milk	The product obtained by soaking beans in water, grinding the soaked beans, and filtering the insoluble pieces out.	Consumed primarily by the Chinese. As the starting material for the preparation of a soybean curd, probably the most important and popular soybean food in the Orient. In soybean-milk-based infant formulas.	The soft cheeselike curd, known as "toufu," "tofu," "tubu," and other local names, contains about 53% protein, 26% fat, 17% carbohydrate, and 4% fiber, minerals, and vitamins. Curd has a bland taste and can be flavored with seasoning or blended with other foods. It is made into a variety of products by frying, dehydrating, or freezing and is consumed daily in the same manner as high-protein foods in the U.S.
Soybean oil	The crude oil which is extracted from soybeans, then refined.	Margarine, shortening (cooking oil), and salad oil. The hydrogenated oil finds much use in restaurant chains and institutions for deep fat frying of foods. In the Far East, oil is consumed extensively as a food.	About ⅕ of the bean is oil. More than 90% of the oil is refined and used for food—for margarine, shortening, and salad oil. All edible oils are deodorized. Hydrogenation varies, depending upon the final use of the oil.
Soybean protein concentrates	Soybean protein products containing 70% protein. They are prepared from defatted flakes or flours by removing water soluble sugars, ash, and other minor constituents.	Bakery products: bread, biscuits, and buns; cakes and cake mixes. Meat products: sausages, luncheon loaves, poultry rolls, patties, meat loaves, canned meats in sauces. Breakfast cereals. Infant foods. Dietary foods.	
Soybean protein hydroslysates	Soybean proteins that are partially hydrolyzed by a number of agents such as enzymes, acids, alkalis, or steam, or by yeasts, molds, or bacteria. Hydrolysates are soluble in water over the entire pH scale.	Soy sauce and other flavorings, foaming agents, and whipping agents	
Soybean protein isolates (Also see "textured soybean proteins")	Soybean protein isolates, which contain 90 to 95% protein, are the purest form of soybean protein marketed. They are produced by (1) washing out the proteins from dehulled, defatted soybean flakes, (2) precipitating the protein out of the liquid solution with a mild alkali, and (3) drying by a spray process.	As the basic ingredient of spun fiber types of meat analogs. Meat products: sausages, luncheon loaves, poultry rolls. Dairy-type foods: whipped toppings, coffee whiteners, frozen desserts, beverage powders. Infant foods. Dietary foods.	Soybean protein isolates are required for spinning.
Soybean sprouts	Sprouts of soybeans about 4 in. (*10 cm*) long. Prepared by germinating whole soybeans under optimum moisture conditions at a temperature of 72-86°F (*22-30°C*) for 4 to 7 days.	As a fresh vegetable, or in soup	Soybean sprouts are a rich source of vitamin C. On a moisture-free basis, soybean sprouts run about 50% protein content.
Soy sauce (shoyu in Japan, chiang-yu in China)	A salty brown sauce made by fermenting soybeans—a liquid food. The *starter* (koji) for making soy sauce is prepared from rice inoculated with the spores of the molds of either *Aspergillus oryzae* or *Aspergillus soyae*. In Japan, about equal quantities of soybeans and wheat are used in making soy sauce. In China, more soybeans and less wheat is used.	As a flavoring agent for many foods—especially in Chinese and Japanese cooking	Soy sauce originated in China some 2,500 years ago and was introduced in Japan in the 7th century by Buddhist priests. The average annual per capita consumption in Japan is about 3 gal. (*11.4 liters*). Good soy sauce contains large amounts of amino acids, especially glutamic acid.
Tao-cho	Prepared from light colored soybeans. They are soaked, dehulled, and boiled, then mixed with rice flour and roasted. Next, they are inoculated with *Aspergillus*, then fermented for 3 days and sun-dried. The cakes are dipped in brine, arenga sugar and a paste of glutinous rice, following which they are exposed to the sun for a month or more.	As an appetizer	

TABLE S-12 *(Continued)*

Product	Description	Uses	Comments
Tao-si	A soybean product, made from unhulled soybeans. Soybeans are soaked and boiled; wheat flour is added; the mix is inoculated with *Aspergillus oryzae* and incubated for 2 to 3 days; thence placed in earthenware jars and salt brine is added; and the product is finished in about 2 months.	As an appetizer	
Taotjo	A soybean product of the East Indies. Boiled soybeans are mixed with roasted wheat meal or glutinous rice; inoculated with *Aspergillus oryzae* for 2 to 3 days; then placed in brine for several weeks, with palm sugar added at intervals.	As a condiment	
Tempeh	An Indonesian food made entirely from soybeans; prepared by using a phycomycete, *Rhizopus oligosporus*	Tempeh is used as a main dish, rather than as a flavoring agent for other foods. Raw tempeh is sliced and then fried or baked.	A similar product called *tempeh bongkrek* is sometimes made from coconut meat. Tempeh made from coconut meat may become poisonous, but tempeh made from soybeans never does. The *Rhizopus* mold synthesizes vitamin B-12; hence, tempeh is one of the few vegetable products that contain significant amounts of this vitamin.
Textured soybean proteins (analogs) (Also see "soybean protein concentrates" and "soybean protein isolates.")	Further processing of the basic forms—flours and grits, concentrates, and isolates— is now practiced to give soybean proteins a texture that resembles specific types of meat; these items range from extenders to be used with ground meats to complete meat analogs. Two basic types of textured soybean protein products are now available: 1. *Extruded*, made by cooking soybean flour and other ingredients and forcing the mixture through small holes into a chamber of lower temperature and pressure. Granules of these expanded materials are available in both colored and uncolored and flavored and unflavored forms.	Ground meat extenders. Meat analogs (baconlike bits, etc.)	Textured items, mainly of the extruded type, are the fastest growing segment of the edible soy protein business. Production of meat analogs either by extruding soybean flour or by spinning protein isolate into fibers results in products that look and taste like beef, pork, and other kinds of meat. Soy protein can augment whole fish, improve the texture and eating characteristics of finished consumer fish products, and replace the functional fish protein. It can, for example, build the visual content of processed shellfish without altering the distinctive flavor and mouthfeel of such products. Soy protein seems assured of increased use in dairy product analogs made for people who have special health or religious diet requirements. The extruded soy process is patented and licensing is required.
	2. *Spun*, made by spinning the fibers and then flavoring, coloring, and forming them in shapes which resemble pieces of meat, poultry, and fish.	Meat analogs: baconlike bits, simulated sausages, simulated ham chunks, simulated chicken chunks, simulated bacon slices. Meat extenders.	The basic patent for spun protein isolates has expired, but companies now engaged in this business hold patents on improved methods for converting the fiber into meat analogs.
Tofu and Sufu (soybean cheese)	In producing sufu, a product called tofu, the Oriental equivalent of curdled milk, is produced. It is made by soaking, grinding, and straining soybeans to produce soybean milk; then boiling to coagulate the proteins and pressing to remove water. At this stage, which is prior to inoculation with fungi, it is called tofu; hence, tofu is not fermented. Sufu, a fermented soybean product, is often referred to as Chinese cheese, because it closely resembles cheese of the western world. At least 5 species of fungi have been isolated from sufu, but *Actinomucor elegans* appears to be the organism of choice.	Both tofu and sufu are sliced and eaten much like cheese is consumed in the western world.	Tofu has been one of the most important foods in the Orient for centuries. Sufu, which originated in China in the 5th century, is widely used in China today.
BEVERAGES			
Soy protein shake	A beverage containing soy protein	As shakes at fast food outlets	Shakes with soy protein have been widely marketed.
Whey-soy drink	A beverage containing whey and soy protein. The base product (41.4% sweet whey, 36.5% full fat soybean flour, 12.1% soybean oil, and 9.5% corn sugar) is mixed with water at 15% solids.	As a drink with nearly the same energy and protein content as whole milk	This is a new product available for distribution in the developing countries.

SPANISH GOAT

The primary meat-type goat, represents a cross of many dairy goat breeds. The meat from this type of goat is known as *cabrito*, meaning little goat in Spanish.

SPANISH LIME (MANONICILLO)

Melicocca bijuga

The fruit of a tropical American tree (of the family *Sapindaceae*) that is related to the litchi of Asia. Spanish limes, which are not related to the citrus fruits, are greenish plumlike fruits that have a tart sweet flesh and a large pith. They may be eaten fresh or when made into preserves.

The fruit is moderately high in calories (59 kcal per 100 g) and carbohydrates (20%). It is a poor source of vitamin C.

SPECIFIC DYNAMIC ACTION (SDA)

The increased production of heat by the body as a result of a stimulus to metabolic activity caused by ingesting food.

SPECIFIC GRAVITY

The ratio of the weight of a body to the weight of an equal volume of water.

SPECIFIC HEAT

• The heat-absorbing capacity of a substance in relation to that of water.
• The heat expressed in calories required to raise the temperature of 1 g of a substance 1 °C.

SPHINCTER

A muscle surrounding a bodily opening or channel, which is able to contract and close it.

SPHINGOMYELIN

A lipid composed of a fatty acid such as stearic or palmitic, phosphate, choline, and the amino alcohol, sphingosine. It is found primarily in nervous tissue.

SPICES

The word *spice* is of French origin, meaning *fruits of the earth. Spice is the name given to food seasonings made from plants.* They have a sharp taste and odor—they're more pungent than herbs. Some spices are valued for their taste, others for their smell. Common spices include pepper, nutmeg, cloves, ginger, allspice, mace, mustard, and cinnamon.

Fig. S-27. Fruiting nutmeg branch. Nutmeg is a common spice. (Courtesy, Field Museum of Natural History, Chicago, Ill.)

Spices have little in common except for their use. They usually come from tropical plants; and different parts of various spice plants may be used. For example, cloves come from the bud, cinnamon comes from the bark, pepper comes from the fruit, ginger and horseradish come from the root, and mustard comes from the seed.

Some people grow spice plants such as sage, marjoram, thyme, and others in their gardens; others grow them in pots in sunny windows. They then dry the plants for later use.

Spices have little food value, but they do increase the appetite and stimulate the organs of digestion. Before foods were canned or refrigerated, spices were used to make tainted foods taste better.

(Also see FLAVORINGS AND SEASONINGS.)

SPOILAGE OF FOOD

Although food spoilage is often thought of as something caused by bacteria, molds, or yeast, it is actually any change which renders a food unfit for human consumption. Here a distinction should be drawn between unfit and undesirable. Food may be undesirable due to the dislikes of an individual, but this does not mean it is unfit for everyone. Furthermore, food spoilage cannot be equated to danger to the consumer. Spoilage results in abnormal colors, flavors, odors, texture, or other unacceptable changes. The causes of food spoilage can be grouped into five broad categories: (1) microbial spoilage, (2) enzymatic spoilage, (3) chemical spoilage, (4) physical spoilage, and (5) pest spoilage.

The aim of modern food production, processing, and preservation is to, so far as is possible, reduce and/or eliminate all sources of food spoilage. By the time a food reaches the consumer, proper heat and/or chill treatments, canning, drying, freezing, sanitation, and food additives may all play a role in eliminating spoilage. Then the control of spoilage is passed on to the consumer who must exercise proper storage and preparation methods.

(Also see ADDITIVES; BACTERIA IN FOODS; POISONS; and PRESERVATION OF FOOD.)

SPORE

A resting reproductive form of certain microorganisms.

SPRAY-DEHYDRATED

Material which has been dried by spraying onto the surface of a heated drum. It is recovered by scraping it from the drum.

SPRUCE BEER

This is a fermented drink made from an extract of spruce needles and twigs boiled with molasses or sugar. At one time it was used as a diuretic and antiscorbutic.

SPRUE

A collective term for a group of nutritional deficiency diseases characterized by impaired absorption of nutrients from the small intestine, especially fats, glucose, and vitamins. Although all sprues exhibit the same general clinical manifestations of intestinal malabsorption and steatorrhea (fatty diarrhea), the following three etiologic classifications are presented in this book:

1. **Celiac disease,** a rare metabolic disorder of children, which results from a sensitivity to gluten.

2. **Nontropical sprue,** the term commonly applied to adults exhibiting a sensitivity to gluten.

3. **Tropical sprue,** caused by a deficiency of folic acid and vitamin B-12.

(Also see CELIAC DISEASE; SPRUE, NONTROPICAL; and SPRUE, TROPICAL.)

SPRUE, NONTROPICAL (CELIAC-SPRUE)

The term nontropical sprue is commonly applied to adults exhibiting a sensitivity to gluten. It is due to an intolerance to wheat gluten, a main constituent of wheat flour, which is also present to a small extent in rye, barley, and oats, but not in rice. Symptoms are provoked by the ingestion of any of the numerous foods containing wheat gluten. Characteristic symptoms of sprue include steatorrhea (fatty diarrhea), weight loss, and lesions of the small intestine. As the disease progresses, a multitude of other symptoms may appear due to the loss of fat and other nutrients in the stool.

Treatment consists of a gluten-free diet. This means the exclusion of all cereal grains except corn and rice and the use of potato and soy flours.

(Also see ALLERGIES, section headed "Wheat Allergy"; MODIFIED DIETS—GLUTEN-FREE DIET; and SPRUE.)

SPRUE, TROPICAL

A malabsorption syndrome affecting mainly individuals living in the West Indies, Central America, and the Far East. Even individuals from temperate climates visiting these countries may develop tropical sprue during or after their visit. Although somewhat similar to nontropical sprue, tropical sprue responds to different therapy. Both types of sprue are characterized by the typical flat mucosa of the small intestine, secondary enzyme deficiencies in the intestinal mucosa, and steatorrhea. However, in tropical sprue, a macrocytic anemia occurs as a manifestation of folic acid and vitamin B-12 deficiencies. Indeed, marked improvement of the sufferer is noted following the administration of folic acid and vitamin B-12. Diet therapy consists of a diet high in calories and protein. Substitution of fat with medium-chain triglycerides has alleviated the steatorrhea (fatty diarrhea), and weight gain. Unlike nontropical sprue, response to a gluten-free diet is minimal. Although no infective agent has been demonstrated, broad-spectrum antibiotics have proven beneficial.

(Also see MODIFIED DIETS; FOLACIN; and SPRUE.)

SQUAB

A young pigeon that is about 4 weeks old and weighs 1 lb. Squabs should be slaughtered just before leaving the nest. They are processed and cooked much like any other poultry.

SQUASHES *Cucurbita* spp

The fruits of these plants provide both edible flesh and edible seeds. Also, the flowers are edible. Squashes belong to the gourd or melon family (*Cucurbitaceae*), which also includes pumpkins and cucumbers. Most summer squashes are classified as *C. pepo* and bear relatively small fruits that are eaten while immature. Winter squashes are *C. maxima* and *C. moschata*. Winter squashes are eaten when mature and tend to be more nutrituous than summer squashes.

ORIGIN AND HISTORY. The wild ancestors of the squashes appear to have originated in the vicinity of the border between Mexico and Guatemala. The first use of these vegetables as food appears to have occurred around 8000 B.C. At that time the Indians gathered the wild plants mainly for the seeds because the fruits contained only small amounts of bitter-tasting flesh. From Mexico and Central America, squashes spread throughout North and South America.

WORLD AND U.S. PRODUCTION. Production figures on squashes alone are not available. However, the worldwide production of pumpkins, squashes, and gourds totals about 6.8 million metric tons annually; and the leading producing countries, by rank, are: China, Romania, Egypt, Argentina, Turkey, Japan, Italy, Spain, Indonesia, and Thailand. Pumpkins, squashes, and gourds rank eleventh among the leading vegetables of the world.[38]

The United States produces a mere 1,000 tons of pumpkins, squashes, and gourds annually.[39] Without doubt, the actual production is higher than this figure would indicate, but this does show that they are produced primarily in home gardens rather than for commercial purposes.

PROCESSING. Summer squashes are often sliced and cooked briefly in preparation for canning or freezing, whereas winter squashes are usually cooked much longer before being similarly processed.

SELECTION, PREPARATION, AND USES. The culinary characteristics of summer squash and of winter squash, and of the varieties within each group, vary considerably. Hence, the cook should know the differences between them.

Types of Squashes. Descriptions follow:

• **Summer squash**—A bush-type cucurbit that produces many fruit which are usually harvested in the summer, while they are immature (the rind is still quite soft) and very watery.

The most common white or creamy-white variety of this type is the White Bush Scallop which is disk-shaped and smooth with scalloped edges. It is also called Cymling and Patty Pan.

---

[38]Data from *FAO Production Yearbook 1990*, FAO/UN, Rome, Italy, Vol. 44, p. 134, Table 54. **Note well:** Annual production fluctuates as a result of weather and profitability of the crop.

[39]*Ibid.*

Fig. S-28. Squashes for all seasons, including the pumpkin, a favorite at Thanksgiving and Christmas. (Courtesy, Shell Chemical Co., Houston, Tex.)

Fig. S-29. Summer squash featured in a vegetable dish. (Courtesy, National Film Board of Canada)

Most yellow summer varieties are elongated-bulbous in shape with a rough warty rind and are designated as either Straight or Crookneck.

Green, green-black or green-striped varieties, such as Zucchini and Italian Marrow, are elongated-cylindrical in shape.

• **Winter squash**—A bush-type, semivining, or vining cucurbit that is harvested in the fall when fully mature. The flesh of these squashes is usually less coarse and less fibrous, darker in color, milder flavored, and sweeter than that of pumpkins. They are often used in baked vegetable dishes and are sometimes used instead of pumpkin in making custards and pies.

Fig. S-30. Winter squash stuffed with ground meat. (Courtesy, Agriculture Canada, Ottawa, Canada)

Common fall and early winter varieties are the green-colored corrugated Des Moines, Acorn, or Danish, the buff-colored Butternut, and the green or golden Delicious.

Vegetable spaghetti is an unusual variety of winter squash that has a large cylindrical fruit which contains a spaghetti-like pulp.

Most of the late winter squash varieties are relatively large and may have either a light-green, dark-green, bluish-green, or orange-colored rind.

Indicators of Quality. Summer squash should be fresh, fairly heavy in relation to size, free from cuts or noticeable bruises, crisp, and tender.   The rind of winter squash should be hard.

Some suggestions for utilizing the various forms of squashes follow:

• **Squash flowers**—These should be picked when open so that no bees will be trapped within the blossoms. The flowers are good when dipped in a batter and fried in deep fat.

• **Squash seeds**—The seeds of winter squashes may be dried for a few days, after which any adhering tissue should be removed. Then, they may be roasted or fried in oil. Some

people boil the seeds briefly in salted water before roasting or frying.

• **Summer squash**—Usually these fruits are left unpeeled and cooked whole, sliced, cubed, or grated. The cooked item may be served with butter, margarine, or a special sauce. Some other ways in which summer squash may be prepared are (1) stuffed with ingredients such as bread crumbs, cheese, chopped eggs, ground meat, nuts, seasonings, and/or sour cream, and then baked; (2) peeled, grated, mixed with eggs, flour, baking powder, salt, and seasonings, shaped into patties, and fried on a griddle or baked in a bread; (3) sliced, dipped in batter, and fried in deep fat; (4) cooked, pureed, and used as a soup stock; or (5) sliced and baked in casseroles with cheese and other ingredients.

• **Vegetable spaghetti**—This special type of winter squash is usually baked or boiled whole until the rind softens, then it is cut open lengthwise, and the spaghettilike strands removed for serving alone with butter or margarine as a substitute for spaghetti, or it may be added to sauces or soups.

• **Winter squash**—These fruits should be cut in half and the fibrous matter and seeds removed. Then, they may be baked, broiled, or steamed and served with butter, margarine, and/or seasonings. Most varieties are equal or superior to pumpkin in recipes calling for the latter. The mashed, cooked vegetable is also good when it is (1) used to make breads, cakes, muffins, and pancakes, or (2) added to soups and stews near the end of the cooking periods to give them more body and to enhance their nutritional values.

NUTRITIONAL VALUE. The nutrient compositions of various forms of squashes are given in Food Composition Table F-21.

Some noteworthy observations regarding the nutrient composition of squashes follow:

1. Squash flowers contain 95% water and are low in calories (16 kcal per 100 g) and most other nutrients, except that they provide fair amounts of phosphorus, iron, vitamin A, and vitamin C. It is suggested that calorie-conscious people eat squash flowers in a salad, soup, or stew.

2. Boiled summer squashes contain more than 95% water and are very low in calories (14 kcal per 100 g) and most other nutrients. However, they are fair sources of potassium, vitamin A, and vitamin C. (The yellow and dark green varieties are generally richer in vitamin A than the white and pale green varieties.)

3. Baked winter squashes have a lower water content (81%) than most other cooked vegetables. They contain more than 4 times the calories, twice the potassium, and 10 times the vitamin A content of summer squashes. Although all winter squashes are excellent sources of vitamin A, the content is greatest in the varieties with deeply-colored yellow-orange flesh.

4. The seeds from winter squashes are very rich in calories (553 kcal per 100 g), protein (29%), iron (11.2 mg per 100 g), and phosphorus (1,144 mg per 100 g). Hence, the consumption of as little as 1 oz (28 g) of the seeds per day will make a significant nutritional contribution to the diet.

SQUIRREL

A rodent of the family *Sciuridae*, characterized by a bushy tail and long strong hind limbs. In some countries, squirrel is highly esteemed as game. It is cooked in the same way as rabbit.

STABILIZED

Made more resistant to chemical change by the addition of a particular substance.

STABILIZER

Any substance which tends to make a compound, mixture, or solution resistant to changes in form or chemical nature. In the food industry stabilizers are used to thicken, to keep pigments and other components in emulsion form, and to prevent the particles in colloidal suspensions from precipitating. Algin, methyl cellulose, carrageenan, and propylene glycol alginate are but a few of the stabilizers utilized.

(Also see ADDITIVES; and EMULSIFYING AGENTS.)

STACHYOSE

A tetrasaccharide composed of two molecules of galactose and one molecule of fructose and glucose. It occurs in many beans, beets, peas, soybeans, and tubers. Since the intestinal tract does not produce enzymes able to split stachyose, bacteria in the intestine act on it—forming gas. Hence, it is believed to be responsible for the gas-producing properties of beans.

(Also see BEAN[S], COMMON, section headed "Dried Beans.")

STALING

This term generally relates to bakery products where it is characterized in breads by (1) a firming of the crumb due to loss of water, (2) loss of flavor, and (3) loss of crispness in the crust resulting in leatheriness. The staling of breads is a complex combination of changes. Methods have been developed to detect these changes. However, the method most commonly employed to study staling is the detection of changes in the compressibility of a loaf of bread, since consumers associate softness with freshness. Softness of a bread may be increased by the inclusion of monoglycerides (emulsifiers), but this does not necessarily retard staling. Storing bread at refrigerator temperature increases the rate of staling while freezing bread arrests staling. Furthermore, staling can be reversed by heat; hence, stale bread makes good toast. No bakery products stale as fast as bread, but even rich cakes stale if kept long enough. Also, dry bakery goods like crackers stale, but they stale because they pick up moisture from the air.

STAR-APPLE (CAINITO) *Chrysophyllum cainito*

The fruit of a tree (of the family *Sapotaceae*) that is native to the West Indies and Central America. The fruit is apple-like, purple, round, up to 4 in. (*10 cm*) in diameter, with a white, sweet, edible pulp. The pulp is usually eaten fresh after removal of the skin.

The ripe fruit pulp is moderately high in calories (68 kcal per 100 g) and carbohydrates (14.5%). It is a fair source of vitamin C.

STARCH

Contents | Page
In the Diet..980
Natural Starch...980
Commercial Starch....................................980
  Production..980
  Characteristics......................................980
    Gelatinization....................................980
    Modifications......................................980
  Uses...981
Nutritional Value.......................................981

Fig. S-31. Corn, a major source of starch. (Courtesy, USDA)

Starch is a polysaccharide—long chains of the simple sugar, glucose—formed by the process of photosynthesis. It is the storage carbohydrate, stored as granules in seeds, roots, tubers, or stems of the higher plants. The chemical formula of starch is $(C_6H_{10}O_5)_n$. The "n" outside of the parentheses indicates an indefinite number of repeating glucose molecules. Furthermore, these glucose units may be linked in either of two patterns: (1) a straight-chained pattern called amylose; or (2) a highly branched pattern called amylopectin. The amount of amylose and amylopectin in starch depends on the source, but on the average starch is about 27% amylose and 73% amylopectin. Fig. S-32 illustrates the structures of amylose and amylopectin.

Unlike cellulose, which is also a long chain of glucose molecules, starch can be split by the digestive enzymes of the body. When either amylose or amylopectin are split, dextrins are formed. Dextrins may vary in length from five or more glucose molecules, and the final product of enzymatic splitting of starch is glucose.

G—G—G—G—G—G—G—G—G—G
AMYLOSE

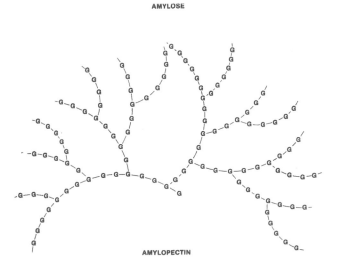

AMYLOPECTIN

Fig. S-32. The structure of amylose and amylopectin of starch. Each "G" represents a glucose molecule.

IN THE DIET. Since starch is a product of photosynthesis, many foods contain starch. It is the primary digestible carbohydrate in foods. Furthermore, starch and starch products are employed in many foods to facilitate processing or to give foods specific properties. Hence, starch and starch products are food additives.

Natural Starch. Very little of the available carbohydrate occurring naturally in foods occurs as the disaccharides sucrose, lactose, or maltose or as the monosaccharides glucose or fructose. Cereal grains, roots, tubers, legumes, nuts, and some fruits and vegetables provide carbohydrate to the diet mainly as starch. The major sources of starch are the seeds of corn, rice, rye, sorghum, and wheat. Potatoes,

Fig. S-33. Potatoes, a good source of starch. (Courtesy, USDA)

which are tubers, are also a good source of starch. Dry mature beans and peas also contain starch, while fresh beans and peas contain starch but at much lower levels. Carbohydrate values given in Food Composition Table F-21 for the cereal grains, roots, tubers, nuts, dried beans, and peas can be considered as approximating the starch content of these foods.

Worldwide, natural starch is an important energy food. The staple food of a country provides the primary source of starch; for example, corn in Mexico; roots, tubers, and sorghum in parts of Africa; and rice in India, China and Indonesia. Starch is an economical food; hence, it contributes a large portion to the diets where income is low. As income increases, the amount of starch in the diet decreases while consumption of fat, animal protein, and sugar increase.

Aside from the starch which is natural in many foods, starch is an important commercial product, and it is extracted from several plant sources which produce it in sufficient quantities.

Commercial Starch. Commercially extracted starch, due to some unique chemical and physical properties, low cost, and availability, may be added to foods during processing—as a food additive. While many plants produce starch, only a few produce it in quantities to make extraction feasible.

PRODUCTION. Corn is the major source of commercial starch in the United States, followed by sorghum, wheat, and rice. Other sources include oats, barley, rye, potato, sago, arrowroot, and tapioca (cassava).

Starch production from corn is achieved by the wet milling process. Corn kernels are soaked in warm water containing sulfur dioxide until they become soft. Then, the kernels are cracked and the germ is removed. Next, the kernel fragments are ground and screened down to starch granules and gluten. The gluten is removed, and the starch is filtered, washed, dried, and packaged.

A newer process for obtaining wheat starch involves the milling of wheat into flour, followed by preparing a flour and water dough. Starch granules are then washed from the dough.

Starch is obtained from rice by soaking the grain in an alkaline chemical which dissolves the gluten, leaving the starch.

Starch from roots and tubers such as potato, arrowroot, and cassava is prepared by grinding the root or tuber to free the starch granule, screening to remove the fiber and skins, soaking, and separating the starch from other water-soluble material by settling or centrifugation. Starch granules are insoluble in cold water.

(Also see CEREAL GRAINS; CORN; POTATO; SORGHUM; and WHEAT.)

CHARACTERISTICS. Although starch is a white powder to the unaided eye, it appears as granules or grains under the microscope, which vary in size and shape depending upon the plant source. Furthermore, the content of amylose and amylopectin vary depending upon the source. These characteristics of the starches affect their gelatinization, modification, and uses.

Gelatinization. Starch gelatinizes, when mixed with water and heated beyond 133° to 167°F (56° to 75°C)—the critical temperature. This temperature differs depending upon the type of starch. As starch is heated, the chemical bonds (hydrogen bonds) holding the granules together start to weaken. This permits the water to penetrate the granules causing them to swell to many times their original size—a change referred to as gelatinization. As gelatinization occurs the clarity and the viscosity (thickness) of the solution increases, and the starch granules lose their unique microscopic shape by rupturing and releasing amylose and amylopectin.

The characteristics of these cooked viscous solutions vary from starch to starch. After cooling to room temperature, the starch from roots are clearer and more fluid, while starch from the cereal grains yield a cloudy less fluid paste that tends to be jellylike. These characteristics are dependent upon the amylose and amylopectin content of the starch and upon the size of the amylose and amylopectin molecule. Some hybrids—waxy hybrids—of corn and sorghum have been developed which yield starch that is almost entirely amylopectin, while other hybrids have a high amylose content. Overall, the tendency to thicken or gel upon cooling, and to become opaque is caused by the presence of amylose.

Modifications. Besides the property of gelatinization, starch may be modified to yield some unique and useful properties which are used advantageously by the food industry and others; syrups and sugars are well-known products of modification.

• **Syrups and sugars**—Starch from any starch source—rice, corn, or other grains, or potatoes—may be broken down into its component sugar either enzymatically or with dilute acids. The most familiar example in the United States is corn syrup (Karo). Complete hydrolysis of starch yields glucose—dextrose.

Corn syrup is the product of the incomplete hydrolysis of starch. It is a viscous liquid containing dextrose, maltose, and dextrins. Unlike sugar, it has a distinct flavor other than sweetness. The degree of conversion is expressed by the *dextrose equivalent* (D.E.) which is, in effect, the measure of sweetness in the syrup, or in other words the amount of starch converted to glucose. Even high conversion syrups are substantially less sweet than sugar. A special corn syrup on the market is a *high fructose corn syrup* (HFCS) which is made by treating high conversion corn syrup with enzymes. The enzymes convert some of the glucose to fructose, which is much sweeter than glucose.

Corn sugar is glucose (dextrose) recovered by crystallization from hydrolyzed starch. Two types of refined dextrose are commercially available: dextrose hydrate, containing 9% by weight of water of crystallization, and anhydrous dextrose, containing less than 0.5% of water. Dextrose hydrate is most often used by food processors.

USES. Starch and modified starches are considered food additives, and they are employed in a wide variety of foods. As additives they contribute calories to foods, but starch and modified starches are primarily used in foods to facilitate processing or impart specific properties to foods. The following are some examples of the uses of starch in its various forms:

• **Dry granular starch**—One pound of cornstarch granules has a surface area of about 3,500 sq ft (*315 m²*). Dried starch or modified starch serve as moisture absorbing agents in many products. Starch is added to baking powder to absorb moisture and keep the ingredients from losing their activity. Confectioners sugar is protected from lumping during storage by the addition of cornstarch. Aside from food, the absorbing capacity of starch is sometimes employed on the human body as a substitute for powder, including baby diapers.

Candy and gum molds are formed from powdered cornstarch or modified starch.

Dry starch derivatives are used in foods to maintain their flow properties.

Starch powders are used as bulking or diluting agents in enzyme preparations or flavorings so that a larger unit of measurement may be used.

• **Starch pastes**—Most of the uses of starch as a food additive are for thickening and gelling. The proper starch or modified starch is chosen which yields the desirable properties. In the canning industry starches are used in baby foods, soups, sauces, pie fillings, gravies, vegetables in sauces, chow mein, chili, spaghetti, stews, and cream style corn. In the baking industry, starch, cross-linked starch and pregelatinized starches are employed in flour, fruit pie fillings, imitation jellies, whipped topping stabilizers, icing stabilizers, cream fillings, custards, salad dressings, candies, and gums.

• **Starch films**—The ability of starch to form films is utilized in covering foods with decorative and protective coatings, binding foods and providing a matrix to carry food sub-

stances. Starches are used as binders in meat products, and pet food, while modified starches are used as a matrix to carry flavor oils. Futhermore, starch coatings are oil resistant and can be used on nut and chocolate confections to prevent the migration of oil.

Starch coatings are found on chewing gums or other confections with hard sugar coatings.

• **Starch syrups and sugar**—Since corn syrups can (1) control crystallization, (2) retain moisture, (3) ferment, (4) produce a high osmotic pressure solution for preservation, and (5) aid browning; they are employed in many food products including the following: baby foods, bakery products, canned fruits, carbonated beverages, confections, dry bakery mixes, fountain syrups and toppings, frozen fruits, fruit juice drinks, ice cream and frozen desserts, jams, jellies and preserves, meat products, pickles and condiments, and table syrups (Karo syrup).

Due to its browning, fermentability, flavor enhancement, osmotic pressure, sweetness, humectancy (prevention of drying), hygroscopicity (moisture absorption), viscosity, and reactivity properties, starch sugar—dextrose—is utilized in many food products. The major uses of dextose are the confection, wine, and canning industries.

In medicine, various concentrations of glucose (dextrose) are utilized for intravenous administration.

NUTRITIONAL VALUE. Starch contributes energy—calories—to the diet. On the average, each gram of starch supplies 4 Calories (kcal) of energy. Upon ingestion, enzymes in the saliva begin breaking starch down to its component sugar, glucose. This process continues in the small intestine where the glucose is absorbed into the bloodstream and used by the cells of the body for energy production, stored as glycogen, or converted to chemicals used in the synthesis of fatty acids. Raw starch, as it occurs naturally in food, is contained in granules, and cooking of starch softens and ruptures these granules making the starch-containing food more palatable and more digestible.

Over the past 70 years, the value of starch in American nutrition has been decreasing. As shown in Fig. S-34, starch now supplies about 21% of the calories in the American diet. The decreased intake of starch has been countered by an increase in fat intake, particularly unsaturated fats.

(Also see ADDITIVES; CARBOHYDRATE[S]; CEREAL GRAINS; DIGESTION AND ABSORPTION; and METABOLISM.)

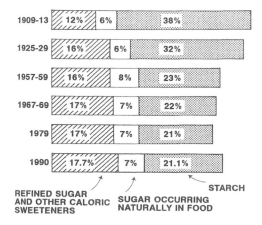

1909-13	12%	6%	38%
1925-29	16%	6%	32%
1957-59	16%	8%	23%
1967-69	17%	7%	22%
1979	17%	7%	21%
1990	17.7%	7%	21.1%

REFINED SUGAR AND OTHER CALORIC SWEETENERS   SUGAR OCCURRING NATURALLY IN FOOD   STARCH

Fig. S-34. Changes in the percentage of calories in the American food supply derived from sugars and starch. (Courtesy, USDA)

STARCHES, WAXY

Properties of starch are influenced by the ratio of amylose (linear chains of glucose molecules) to amylopectin (branched chains of glucose molecules). Common starches contain 17 to 27% amylose, and 73 to 83% amylopectin. Waxy starches contain nearly 100% amylopectin. These do not form gels except at high concentrations; they remain as relatively clear solutions. Waxy starches can be derived from corn, grain sorghum, and rice.

(Also see STARCH.)

STARCH, PREGELATINIZED

None of the raw starches will form a paste with cold water unless they are precooked and then rolled or spray-dried. Therefore, pregelatinized starch is starch which has been cooked and dried. It can then be used in instant puddings, fillings, imitation jellies, whipped toppings, etc.

(Also see STARCH.)

STARVATION

Contents	Page
Causes of Starvation	982
Adaptation to Starvation	982
Effects of Starvation	983
Signs of Starvation	983
Treatment and Prevention of Starvation	983
Programs for the Prevention of Starvation	984

This condition usually results from the prolonged and drastic deprivation of food. It is characterized by wasting of tissues and other alterations of normal physiology. If continued sufficiently long, there may be disability, or even death. Situations where large numbers of people starve are called *famines.*

Failure of children to get sufficient food results in growth failure, marasmus, and other serious disorders developing much sooner than in adults because children have proportionately higher requirements for nutrients to support their rapid rate of growth.

(Also see MALNUTRITION, PROTEIN-ENERGY.)

CAUSES OF STARVATION. The ultimate effect of starvation is the deprivation of the cells of the body of their essential nutrients. Although lack of food is the main cause of this condition, other factors may produce starvation at the cellular level.

The main causes of starvation are:
1. Unavailability of food.
2. Restriction of food intake due to such things as surgery or deprivation of food imposed for coercion or punishment.
3. Self-denial of food for purposes of weight reduction, protest, or religious fasting.
4. Failure to eat due to lack of appetite because of physiological or psychological reasons, including anorexia (lack of appetite), acidosis or ketosis, uremic poisoning in kidney failure, or emotional aversions to food such as in *anorexia nervosa.*
5. Disorders of the digestive tract due to reduced nutrient absorption in conditions such as diarrhea, malabsorption, vomiting, or surgical removal of part of the digestive tract.
6. Metabolic abnormalities caused by such conditions as fever, diabetes, hyperthroidism, Cushings' syndrome, extreme stress, or wasting diseases such as cancer.

ADAPTATION TO STARVATION. Persons, who are otherwise healthy, can adapt to deprivation of nutrients for short periods of time by means of various physiological processes which have helped the human race to survive many famines.

Prolonged dietary deficiency of energy results in an eventual reduction in voluntary activities and in the metabolic rate, so that it is possible to survive on about one-half of the energy which had been the normal intake in the diet. It is thought that the reduction in metabolic rate may be due in part to a shortage of energy-yielding nutrients and in part to subnormal blood levels of thyroid hormones. Likewise, body temperatures of starved persons are often subnormal.

Energy is supplied first from stored fat, then from body protein. Stored fat, in the form of triglycerides, is split into fatty acids and glycerol by hormone-sensitive lipase enzymes in adipose tissue. The fatty acids are then released into the blood where they immediately form complexes with plasma albumin, while the glycerol travels free in the blood to the liver where it may be converted to glucose or recombined into triglycerides. Many tissues, such as muscles, are able to utilize fatty acids as their sole sources of energy, although the metabolism of carbohydrates when they are available, requires about 10% less oxygen to produce the same amount of energy.

Blood glucose (critical for the brain and nervous system) is maintained at close to a normal level by the breakdown of protein in muscle to amino acids which travel through the blood to the liver where they are converted to glucose. The breakdown of body protein is limited, however, and so it is fortunate that the brain can adapt to the utilization of ketone bodies which are formed from fat in the liver when glucose becomes scarce.

Catabolism of body tissue for energy results in rapid weight loss, particularly when fat stores have been exhausted, since each pound of protein loss results in a corresponding loss of 3 lb of water which are eliminated from the body.

Starved persons vary in their susceptibility to low blood sugar and ketosis. Some become weak and feel dizzy after only a day of starvation, while others feel few effects after weeks of total fasting; hence, the use of starvation to obtain weight reduction should be done only under the supervision of a physician.

If fluids are limited, the kidneys set into play a mechanism which conserves water. They secrete renin, an enzyme which initiates the conversion of a bloodborne protein (an alpha-globulin) to angiotension II, a hormonelike substance. The latter substance stimulates the adrenal cortex to secrete aldosterone—a hormone which causes the retention of water and sodium by the kidney and a reduction in the amount of urine volume.

The edema of starvation results from a flow of water from the blood to the tissues because the ability of blood to retain water is reduced when proteins in the blood are depleted.

Diarrhea results from the loss of absorptive cells in the intestine which occurs as part of the wholesale catabolism of body tissues in severe starvation.

Minerals—such as potassium, magnesium, and phosphate ions—are released into the blood from the catabolism of tissues, thereby delaying the appearance of clinical deficiencies of these essential nutrients. However, there are likely to be deficiencies of nutrients which are not stored in the body and need to be provided on a regular basis in the diet, such as water-soluble vitamins, chloride, and other ions. Also, the absorption of vitamins and minerals is increased and their urinary excretion is decreased. Thus, there is more efficient utilization of vitamins and minerals by the body in starvation.

(Also see FASTING.)

EFFECTS OF STARVATION.   The effects of food deprivation are dependent upon the severity and length of the deprivation, and the status of the victim prior to starvation. Thus, one cannot draw up an exact timetable of survival. There have been reports of survival after 8 months of total fasting, but the survivors were obese adult women at the beginning of starvation.

Death results from the loss of function of vital organs such as the digestive tract, which may become almost transparent and nonfunctional due to wasting, and the heart, where muscle wasting often results in circulatory failure.

Some crude generalizations about survival from starvation are as follows: Young adults may survive much longer than children (due to the proportionately higher nutrient requirements of the latter) or older persons; women survive longer than men because of their greater proportions of body fat to lean tissue which results from the action of female hormones; and the obese survive longer than lean persons. Thus, the now frowned-upon practice of trying to fatten children had a basis in the more precarious times of our ancestors.

Some of the specific effects of starvation follow:

1. **Physical defects.** Low birth-weight babies from food-deprived mothers are unlikely to show any long-term physical defects as a result of maternal starvation, *unless* the mother was subjected to both insufficient food and malnutrition.

2. **Impaired mental development.** Nutritional deprivation will most likely have a lasting effect on children if it occurs during the first year of life. Studies on the brains of infants who died from starvation at less than 1 year of age showed a significant reduction (15 to 20%) in the number of brain cells.[40] Smaller head circumferences of children who suffer from marasmus early in life are indicative of proportionate reductions in brain size. Itelligence tests have shown that such children do not seem to catch up in mental development by 6 years of age. However, when starvation occurs after 2 or 3 years of age, the effect on learning is temporary, since brain tissue is not affected once complete development of the brain has occurred.

3. **Stunted growth.** Children subjected to acute or chronic deprivation of water, calories, and protein in fetal life, or in the years between birth and school age, may show permanent retardation in growth and development or even disability and death.

4. **Increased susceptibility to disease.** Both children and adults who have been starved are more susceptible to, and suffer more severely from, diseases such as diarrhea, measles, plague, influenza, tuberculosis, and cholera. This is due to the exhaustion of nutrient reserves and concomitant reduced immunity.

5. **Abnormal behavior.** In the beginning stages of starvation, the victims become preoccupied with food and are increasingly restless, irritable, undisciplined, and nervous. Eventually there is increasing apathy, depression, and decreasing sociability until the victims become totally withdrawn and inactive. During rehabilitation from severe starvation, the patients go through a rebellious period that may last until recovery.

6. **Reduced work capacity.** Studies of worker produc-

tivity in the coal mines and steel mills of Germany showed that the average output per worker declined in proportion to the decrease in food energy which occurred as a result of decreasing food supplies during the course of war.[41] The work output returned to the original levels when it became possible for the workers to consume additional food.

SIGNS OF STARVATION.   Signs of this disorder are weight loss, wasting of tissue, loose skin, ketone breath, ketones in urine, dehydration, low blood sugar, weakness, diarrhea, low blood pressure, edema, irritability, and antisocial behavior. It is usually easier to detect starvation in children than in adults, since one can readily observe in the former the great reduction in activity, smaller body weight for height and age, and other signs of malnutrition.

Fig. S-35. African children suffering from starvation. (Courtesy, FAO, Rome, Italy)

TREATMENT AND PREVENTION OF STARVATION. Although many victims of starvation are ravenously hungry, their digestive functions may be severely impaired. Therefore, care must be taken not to feed foods known to cause or aggravate diarrhea (fats, fried foods, beans, and high-fiber foods), which is one of the main causes of death from starvation. It may be necessary to provide small, but frequent feedings of skim milk. Supplements of vitamins and minerals may need to be provided when there is likely to have been depletion of nutrients. Vitamins A and C are needed for optimal healing of tissues, while mineral elements such as potassium, sodium, and magnesium are critical for cardiac function and electrolyte balance. Persons who cannot eat normally by mouth may have to be fed intravenously or by tube feedings.

(Also see INTRAVENOUS FEEDING; and TUBE FEEDING.)

[40]Keller, W.D., and H.A. Kraut, "Work and Nutrition," *World Review of Nutrition and Dietetics*, G.H. Bourne, Editor, S. Karager Ag, Basel, Switzerland, Vol. 3, 1959, p. 69.

[41]Keller, W. D., and H. A. Kraut, "Work and Nutrition," *World Review of Nutrition and Dietetics*, G. H. Bourne, Editor, S. Karager Ag, Basel, Switzerland, Vol. 3, 1959, p. 69.

Programs For The Prevention Of Starvation follow:

• **Distribution of food**—Although this is the most direct approach, it may well be the least satisfactory means of long-term assistance for chronically underfed populations, since these people may come to depend upon free or low-cost food from outside sources, rather than developing their own food supply. However, increasing food production takes time; in the meantime, it may be necessary to provide food for such vulnerable groups as infants and children, pregnant mothers, handicapped persons, and the aged.

• **Provision of money or scrip to buy food**—When a local supply of food is available, its sale through commercial channels may help to encourage local enterprises such as farming and marketing. However, the money or scrip may be used to purchase items other than food, as evidenced by abuses of the food stamp program in the United States (for example, their use to purchase liquor followed by their illicit redistribution at discounts). Furthermore, such doles to the able-bodied unemployed may lead to their rejection of employment when it is available.

• **Price controls**—The government may attempt to control food prices in order to prevent those with more money from buying large quantities of food for resale at higher prices. When controls are stringent, this type of policy sometimes makes food production and distribution unprofitable in the private sector, thereby creating more problems than it solves. Also, just about every attempt at price control or rationing has been accompanied by diversion of foods into *black market* channels. Thus, there must be adequate provision for enforcement of price controls.

• **Planning for the relief of famines**—Famines are often presaged by such events as bad weather during the planting, growing, and harvesting of crops; outbreaks of plant or animal diseases; steady increases in unemployment and reduction of food production; and political unrest. Failure to heed such warnings has frequently resulted in considerable wasted effort and materials as public and private groups have rushed aid to famine-stricken areas.

(Also see DEFICIENCY DISEASES; HUNGER, WORLD; and WORLD FOOD.)

STASIS

A slowing or stoppage of the normal flow of fluid or semifluid material in an organ or vessel in the body, as (1) slowing the flow of blood in arteries and veins, or (2) reduced motility of the intestines accompanied by retention of feces.

STEARIC ACID

An 18-carbon saturated fatty acid which occurs in tallow and other animal fats, and in cocoa butter and other hard vegetable fats, and which reacts with glycerol to form stearin. (Also see FATS AND OTHER LIPIDS.)

STEATORRHEA

Describes a greasy or fatty bowel movement (stool), a characteristic of many malabsorption syndromes such as celiac disease and sprue.

(Also see CELIAC DISEASE; MALABSORPTION SYNDROMES; and SPRUE.)

STENOSIS

A constriction or narrowing of a channel.

STERILIZATION

A process used to destroy all living microorganisms. (Also see PRESERVATION OF FOOD.)

STEROIDS

Any of a group of fat-related organic compounds. They include cholesterol, numerous hormones, precursors of certain vitamins, bile acids, alcohols (sterols), and certain natural drugs and poisons (such as the digitalis derivatives).

STEROL

One of a class of complex, fatlike subtances widely distributed in nature.

STILLAGE

The mash from fermentation of grains after removal of alcohol by distillation.

STOMATITIS

Inflammation of the mucous membrane of the mouth due to a variety of causes, including a riboflavin deficiency.

STOOL

Fecal material; that which is evacuated from the digestive tract following the digestive process.

STRAWBERRY *Fragaria* spp

This fruit is borne by a small herbaceous plant (of the family *Rosaceae*) that grows close to the ground; embracing both *Fragaria virginiana*, the wild strawberry, and *Fragaria chiloensis*, from which most common cultivated strawberries descend.

Fig. S-36. Strawberries are often grown with a straw mulch or plastic sheeting between the rows, both of which keep the berries clean. (Courtesy, University of Minnesota, St. Paul, Minn.)

Strawberries are the world's leading berry-type fruit, although they are not true berries from a strict botanical point of view because the tiny seeds are carried on the outside of the fleshy part of the fruit, whereas, true berries have seeds enclosed within the flesh.

ORIGIN AND HISTORY. Strawberries are indigenous over a very wide area—to Chile; to North America from California to Alaska; and to the U.S.S.R. Cultivation of the strawberry as a garden plant began in the 13th century in France.

Most of the great advances in strawberry culture occurred after certain New World varieties of the fruit were introduced into Europe by the returning explorers. In the latter part of the 16th century, the explorer Cartier brought Canadian strawberries back to France, and Sir Francis Drake brought the Virginia strawberry to England. Both types are believed to have been varieties of *Fragaria virginiana.*

Strawberries remained a luxury during the first part of the 19th century and only the most affluent could enjoy them. The problem was that they spoiled readily and could not be shipped very far by the means of transport then available. By the middle of the century, this problem had been solved in the United States by the initiation of railway shipment, followed in short order by the development of a high-yielding variety of the hybrid strawberry. Today, fresh strawberries are carried thousands of miles in refrigerated railcars and trucks.

WORLD AND U.S. PRODUCTION. Worldwide, about 2.4 million metric tons of strawberries are produced annually; and the leading producing countries, by rank, are: United States, Poland, Japan, and Spain. These four countries produce more than 52% of the strawberries of the world.[42]

The United States produces about 570,300 metric tons of strawberries, annually; and the three leading strawberry states, by rank, are: California, Florida, and Oregon. California alone produces more than 78% of the U.S. strawberry crop.[43]

---

[42]Data from *FAO Production Yearbook 1990,* FAO/UN, Rome, Italy, Vol. 44, p. 171, Table 75. **Note well:** Annual production fluctuates as a result of weather and profitability of the crop.

[43]Data from *Agricultural Statistics 1991,* USDA, p. 210, Table 323. **Note well:** Annual production fluctuates as a result of weather and profitability of the crop.

Fig. S-37. Picking strawberries. (Courtesy, University of Minnesota, St. Paul, Minn.)

Strawberries that are to be sold fresh in distant markets are often picked before they are fully ripe in order to minimize spoilage during shipment, but the fruit intended for nearby markets or processing plants may be allowed to attain optimal ripeness.

Much of the crop is still picked by hand, although the labor requirement has been reduced somewhat in recent years by the development of mechanical pickers and other harvesting aids. The fruit keeps best during shipment if it is precooled within 2 hours of picking, followed by shipment or storage at a temperature below 40°F *(4°C).*

PROCESSING. Over ⅓ of the U.S. strawberry crop is processed. About 95% of processed fruit is frozen and most of the rest is made into preserves. However, it is noteworthy that sometimes freezing merely serves as a means of pre-serving the highly perishable berries until they can be utilized by the manufacturers of baked goods, ice cream, jams and jellies, and other items.

The most common ways in which frozen strawberries are packed for retail sales are (1) sliced and packed with sugar at a ratio of 80% by weight of fruit to 20% sugar, and (2) frozen whole berries without added sugar.

Frozen strawberries that are destined for manufacturing into various processed food products are usually packed with sugar at the ratio of 80% fruit to 20% sugar.

SELECTION, PREPARATION, AND USES. Fresh strawberries should be bright red, plump, medium size, free from dirt, and well rounded with attached stems and caps. Unripe berries will not ripen after picking. Excessively large berries may have an overly bland flavor, while small, misshapen berries may be bitter.

Some preparation tips follow:

1. Fresh strawberries may be kept in a refrigerator for several days if they are removed from their store containers and placed in a single layer in a shallow container.

2. The berries should not be washed until just before using. Then, they should be rinsed with caps and stems intact under a gentle stream of cold water.

3. After washing, the caps and stems may be removed and the berries patted dry with paper towels.

4. Fresh strawberries are good in dishes such as breakfast cereals, cheesecake, compotes, cottage cheese salads, crepes, custards, French toast, gelatin salads, ice cream, milk shakes, pancakes, pies, puddings, sherbets, sundaes, and waffles. Thawed frozen strawberries may also be used in many of these dishes.

5. The syrup from thawed, sweetened frozen strawberries may be mixed with unflavored carbonated water, or plain water flavored with a little lemon juice, to make a refreshing drink.

6. Some of the flavorings which enhance strawberry dishes are almond extract, brandy, cinnamon, citrus juices (particularly when diluted and sweetened a little), fruit-flavored liqueurs, and vanilla extract.

In summary, the main uses of strawberries are: desserts, fruit dishes, baked goods, gelatin desserts, ice cream, ices, jams, jellies, pies, sherbets, soft drinks, and syrups; and in the production of liqueurs and wines.

NUTRITIONAL VALUE. The nutrient compositions of various strawberry products are given in Food Composition Table F-21.

Some noteworthy observations regarding the nutrient

Fig. S-38. Strawberries and cream an all-time favorite. (Courtesy, USDA)

composition of the more common strawberry products follow:

1. The raw fruit and the frozen unsweetened berries are fairly low in calories (37 kcal per 100 g) and carbohydrates (8%). They are good sources of fiber, potassium, and iron; and an excellent source of vitamin C, and bioflavonoids.

2. Frozen sweetened strawberries contain about 3 times the caloric and carbohydrate levels of the raw fruit and the frozen unsweetened fruit, but the levels of the other nutrients are similar to those of the unsweetened products.

3. Strawberry ice cream is rich in calories (188 kcal per 100 g) and carbohydrates (24%). It is a good source of calcium, phosphorus, and potassium, and a fair to good source of vitamin A.

4. Pie made from strawberries is rich in calories (198 kcal per 100 g) and carbohydrates (31%). It is a fair to good source of potassium and vitamin C.

5. The red color of strawberries is due mainly to the anthocyanin pigment pelargonidin 3-monoglucoside (which belongs to the large family of substances called bioflavonoids).

STRAWBERRY GUAVA Psidium cattleianum.

A specie of guava, sometimes planted in the West Indies.

The nutrient composition of the strawberry guava is given in Food Composition Table F-21.

(Also see FRUIT[S], Table F-23, Fruits of the World— "Guava.")

STRAWBERRY PEAR Hylocereus undatus

The red egg-shaped fruit of a cactus that is native to the West Indies. It is fairly low in calories (54 kcal per 100 g) and carbohydrates (13%), and is a good source of iron, but only a fair source of vitamin C.

STRAWBERRY TOMATO (TOMATILLO)
Physalis pubescens

A member of the husk tomato fruits (Physalis species of the family Solanaceae) which is native to Mexico and is related closely to the Cape Gooseberry and the ground cherry.

It is noteworthy that the Indians of Mexico used this plant for food long before they used the related plant that was developed into the common tomato (Lycopersicum esculentum). Some botanists suspect that the latter was eventually domesticated because it resembled the strawberry tomato.

The strawberry tomato is low in calories (40 kcal per 100 g) and carbohydrates (9%). It contains much less vitamin A and vitamin C than the common tomato.

(Also see CAPE GOOSEBERRY; and TOMATO.)

STREPTOCOCCAL INFECTIONS, FOOD-BORNE

Streptococcus pyogenes may get into foods from infected handlers since they are carried on airborne droplets from the respiratory tract of infected people who may sneeze or cough on food. The disease caused by this bacteria is commonly called strep throat. Other Streptococcal bacteria can get into the food and cause scarlet fever. However, this is uncommon in the United States today. These diseases are characterized by fever, vomiting, and sore throat. To prevent their spread, food should be protected from contamination by infected handlers. Once contracted, the disease responds to penicillin and other antibiotics. Occasionally streptococcal infections produce complications such as rheumatic fever and glomerulonephritis.

(Also see DISEASES, Table D-10, Food-Related Infections and Parasitic Diseases—"Strep Throat" and "Scarlet Fever.")

STRESS

Contents   Page
The Stress Response................................987
Nutritional Demands of Stress......................987
  Physical Stress..................................987
    Injuries and Infections........................987
    Environmental Extremes.........................988
  Psychological or Emotional Stress................988
Coping with Stress or Distress.....................988
Diet-Stress Connection.............................988

Stress is one of those words often used and assumed to be understood—until one is required to become precise about its meaning. Then stress becomes different things to different people. Stated simply, stress is a strain, a force, or a great pressure acting on an individual resulting from physical factors—illness, injury, and environment—or psychological factors—prolonged fear, anger, and anxiety.

Stresses of many kinds affect people; among such external forces are previous nutrition, abrupt diet changes, a large number of people crowded together, changing houses or offices, irregularity, travel, excitement, presence of strangers, fatigue, previous illness, temperature, and abrupt weather changes.

Due to our individuality, what stresses one person may have no effect on another. Physiologically, stress is characterized by increased blood pressure, increased muscle tension, rapid heart rate, rapid breathing, and altered endocrine gland function. In the whole scheme, the nervous system and the endocrine system are intimately involved in the response to stress and the effects of stress. In turn, changes produced by these two systems—collectively the neuroendocrine system—can alter nutritional processes and increase the needs of tissues for nutrients. Moreover, stress and nutrition interact since (1) malnutrition itself may produce a

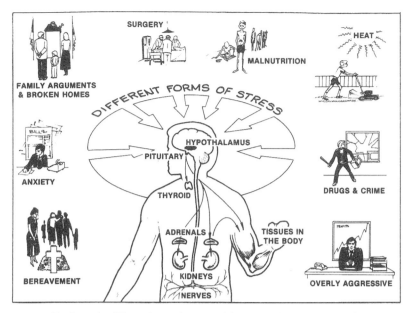

Fig. S-39. The different forms of stress and the action of stress on the body.

energy, and the body may fall victim to other disorders.

NUTRITIONAL DEMANDS OF STRESS.
Nutrient requirements of the stressed individual are affected by the previous nutritional status, and the type and duration of the stress. The 1989 edition of *Recommended Dietary Allowances* (RDAs), compiled by the Committee on Dietary Allowances and the Food and Nutrition Board, provides for the nutrient needs of healthy people living under usual environmental stresses. (See Table N-3.) The RDAs do not allow for the special needs of infections and chronic diseases—both forms of stress. Due to the individuality of the response to stress, persons under severe stress must be individually identified. Stress severe enough to increase the body's needs for essential nutrients may come from physical factors, or from psychological or emotional factors.

Physical Stress. Stress due to physical, tangible, and specific events is well understood, and results in some definite nutritional recommendations.

INJURIES AND INFECTIONS.
Injuries, surgery, burns, and infections result in increased energy expenditure via the neuroendocrine system. The magnitude of the changes elicited depends upon (1) the prestress nutritional status of the individual, (2) age of the individual, (3) the severity and duration of the stressing event.

Energy and protein requirements of physical stresses have been the prime concern. Little information exists concerning vitamins, minerals, and electrolytes.

• **Energy**—Stress associated with minor surgery increases the energy requirements of the body by less than 10%, while multiple bone fractures may increase energy expenditure by 10 to 30%. Some infections increase the basal metabolic rate (BMR); hence, energy expenditure, by 20 to 50%. The largest energy drain on the body is incurred by third degree burns, where the need for energy increases by 40%. When burns are complicated by infection, the requirement for energy is further increased. Some patients may require 4,000 to 8,000 kcal/day.

Often, individuals subjected to injuries, surgery, and/or infections have a depressed food intake which may complicate the nutritive requirements. Once carbohydrate stores of energy (muscle and liver glycogen) are depleted—usually within 24 hours—energy must be provided by fat and protein stores. Amino acids from the catabolism—breakdown—of skeletal muscle are an important energy source. If prolonged, this results in massive tissue wasting.

Energy replacement in some cases may be accomplished by oral feeding, tube feeding, or intravenous feeding, thus preventing or lessening tissue catabolism.

• **Protein**—As indicated, protein and energy metabolism are closely related. Injury, burns, surgery, and infection all induce protein breakdown via glucocorticoid secretion. Protein catabolism results in a negative nitrogen balance—the hallmark of stress which stimulates the catabolic mechanism. Adequate protein, with a high biological value, is necessary to prevent tissue wasting during a stressful illness, and to encourage convalescence following diminution of the stress.

stress response, (2) response to stress is determined by the nutritional state of an individual, and (3) stress can produce nutritional deficiencies or aggravate already existing deficiencies. Ultimately, through the neuroendocrine system, the nutrition and stress interactions are mediated via hormones released or inhibited by the body's response to stress.

THE STRESS RESPONSE.
Normally, the secretion of the hormones is not viewed as a single event, but as a concert. As one hormone comes into play, another may fade out; or one hormone may cause the secretion of another, or the action of one may complement the action of another. Furthermore, the brain or nervous system in many cases acts as the conductor by signaling the proper time for increased or decreased secretion of a hormone. In particular, the hypothalamus acts like a "switchboard" by "plugging in" the proper hormone in response to a variety of nervous stimuli received by the brain. Thus, such things as stress, nutritional status, emotions, nursing, time of day, season of the year, etc. may manifest themselves as disruptions in bodily function.

The hormones involved in the stress response alter the metabolic reactions of the body, and hence, increase the nutritional needs of the body, possibly for protein, energy, fat, carbohydrates, minerals, and vitamins.

The response to stress is characterized as occurring in three stages, even when the cause is nonspecific:

1. **Alarm.** The body recognizes a stress and prepares for flight or fight. This preparation occurs via the release of hormones from the endocrine glands, primarily the adrenal medulla. These hormones cause increased heart rate, respiration, blood sugar, perspiration, dilated pupils, and slowed digestion. During the alarm stage, the burst of energy or alertness may be used for flight or fight. However, resistance is down.

2. **Resistance.** During this stage the body adapts to, and the body reaches, a heightened level of resistance or preparedness. The body attempts to repair any damage caused by the stress. The resistance stage cannot be maintained indefinitely; if the stress is not removed, then the body plunges into the third stage.

3. **Exhaustion.** Continued stress depletes the body's

• **Vitamins, minerals, and electrolytes**—Studies have shown that during moderate to severe stresses, more zinc, copper, magnesium, and calcium are lost in the urine. Furthermore, stress results in altered blood levels of vitamins A and C, and of zinc and iron. Also, part of the response to stress includes water and sodium retention, via vasopressin and aldosterone secretion. As for the water-soluble vitamins—thiamin, riboflavin, niacin, pyridoxine (B-6), pantothenic acid, folic acid, and vitamin C—stress increases their requirement. However, no dietary recommendations are made for these nutrients for individuals under stressful situations. Still, it seems wise to supply some supplementation before deficiency symptoms appear.

ENVIRONMENTAL EXTREMES. Overall, man has learned to protect himself from extreme heat or cold. However, nutrient requirements are increased during adaptation to environmental temperatures above 97°F (37°C). Protein, energy, and water needs are elevated while losses of minerals and electrolytes (calcium, iron, sodium, and potassium) may be anticipated, all depending upon the amount of physical work performed. Also, vitamin C requirements may be increased by stress of a hot environment. Exposure to a cold environment increases the metabolic rate, and hence, the energy requirement of unprotected individuals. Warm clothing and dwellings have minimized the effects of cold on man.

Psychological or Emotional Stress. Day-to-day psychological or emotional stress is normal, and it is inevitable—a part of life. In our society, stressful situations may include family arguments, death of a loved one, drug addiction, crime, unemployment, broken homes, and executive positions requiring aggressive personalities. The degree each individual is stressed depends on how the individual perceives day-to-day situations. Some individuals react excessively, causing hormonal and metabolic responses which tax the body's protective mechanisms. Indeed, some studies indicate a relationship between emotional stress and diseases such as cardiovascular disease, high blood pressure, peptic ulcers, cancer, and streptococcal throat infections, which in turn may increase or alter nutritional requirements.

Nutritionally, emotional stress may have a variety of effects. Prolonged fear, anxiety, anger, and tension all may stimulate hormonal and metabolic responses thus depleting energy, and increasing the need for energy, protein, vitamins, and/or minerals. Overeating, extreme undereating (anorexia nervosa), and eating binges with self-induced vomiting—bulimarexia—may be the result of psychological stress.

Without question, more research is needed in this area—especially in a society such as ours—so that stressed individuals may be identified and specific recommendations made. It seems reasonable to assume that persons experiencing emotional stresses need nutrient adjustments similar to those of individuals experiencing physical stresses. Most of the hormones altering the metabolism are influenced by either emotional or physical stimuli. Ultimately the brain determines the body's response to any stimuli.

COPING WITH STRESS OR DISTRESS. Stress is unavoidable. Therefore, the best protection comes from recognizing it, and coping with it. Recognizing stress requires individual evaluation. Stress is different for every individual. Coping with stress involves relaxation, regular exercises, and periodic evaluation of priorities. Moreover, one should recognize that stress may be turned into a positive force since it can stimulate some form of favorable action on the part of an individual. Therefore, people should

differentiate between stress and distress in their lives. Thus, distress—not being able to adapt—is responsible for the harmful effects. Health professionals are beginning to recognize stress as a legitimate health concern, which may produce damaging or predisposing responses in the body, though clear cut connections between stress and disease are difficult to establish.

Diet-Stress Connection. It is noteworthy that most people have certain foods that they instinctively turn to when under stress—foods that create a feeling of security. The most common security food is milk, perhaps as a symbol of the more protected days of childhood and a desire to return to them. But the food or foods that comfort one person may trigger anxiety in another. So, individuals are best advised to study their own responses to stress, and to evaluate the effectiveness of their coping strategies, possibly by keeping a diary. If eating custard makes one feel better able to cope, then custard it should be. But because a certain approach works for one person it should not necessarily be urged upon the entire population. A moderate, well-balanced diet with emphasis on the stress-related nutrients and regular but not excessive meals is one approach that can be recommended for all.

(Also see ANOREXIA NERVOSA; BULIMAREXIA; DISEASES; ENDOCRINE GLANDS; HEART DISEASE; HIGH BLOOD PRESSURE; METABOLISM; and ULCERS, PEPTIC.)

STROKE (APOPLEXY)

An interruption in the normal circulation of the blood through the brain leading to a sudden loss of consciousness and some degree of paralysis, which may be temporary or permanent depending on the severity of the oxygen deprivation of the brain cells. Most strokes can be traced to previously existing conditions of atherosclerosis, hypertension, or arterial aneurysm.

The three immediate causes of stroke are:

1. **Cerebral thrombosis**, in which normal blood circulation through the brain is cut off from a part of it by a clot in an atherosclerotic artery.

2. **Cerebral embolism**, in which a traveling blood clot, fat, or an embolus of air (a bubble of air) settles in one of the cerebral arteries and chokes off the circulation.

3. **Cerebral hemorrhage**, in which there is a rupture of a blood vessel within the brain, usually an artery with a thin spot (aneurysm) in its wall. The latter type of stroke may be triggered by overexertion, overeating, great stress, or a violent coughing fit.

The most effective prevention of stroke consists in having a complete physical checkup once a year, following the doctor's instructions, and eliminating emotional stress and pressure.

STRONTIUM (Sr)

This is a soft, silvery metal with physical and chemical properties similar to those of calcium. There is no evidence as yet that strontium plays an essential role in man and other higher animals; so, it must be considered as a nonessential element at this time. It is noteworthy, however, that it has been reported that the omission of this element from the diet of rats and guinea pigs on a purified diet resulted in (1) growth depression, an impairment of the calcification of bones and teeth (but this report has been neither confirmed nor invalidated); and (2) a higher incidence of carious teeth.

The strontium content of bone has attracted particular interest because of the affinity of this tissue for strontium and, therefore, its relevance to the problem of strontium retention from radioactive fallout.

SUB–

Prefix denoting beneath, or less than normal.

SUBCUTANEOUS

Situated or occurring beneath the skin.

SUBSTRATE

The base on which an organism lives; a substance upon which an enzyme acts; a source of reactive material.

SUCCINIC ACID

An intermediate product in the Krebs cycle.

SUCCOTASH

A mixture of corn and beans with some milk and butter added. The beans may be either string, butter, or lima. It makes a good combination of cereal and beans together, and a more complete protein.

SUCRASE

An enzyme present in intestinal lining cells which acts on sucrose to produce glucose and fructose.
(Also see DIGESTION AND ABSORPTION, Table D-8.)

SUCROSE

A disaccharide having the formula $C_{12}H_{22}O_{11}$. It is hydrolyzed to glucose and fructose. Commonly known as cane, beet, or table sugar.
(Also see ADDITIVES; CARBOHYDRATE[S]; and SUGAR.)

SUCROSE ESTERS

Sucrose, table sugar, can form esters with fatty acids like lauric acid and stearic acid. These esters are surface active agents—surfactants, used as additives. An example is sucrose monostearate.

SUGAR

Contents   Page
Sources 989
  Sugarcane 989
    Origin and History 989
    World and U.S. Production 990
    Processing 990
    By-Products 990
  Sugar Beet 990
    Origin and History 991
    World and U.S. Production 991
    Processing 991
    By-Products 991
  Nutritional Value of Sugar 991
  Other Sources of Sugar 991
  Consumption and Use of Sugar 992
  Sugar as a Factor in Human Diseases 994

To most people, sugar is sucrose, the white granular sweetener, obtained from cane and beets, and sold in 1, 5, 10, and 25 lb bags at the supermarket. It contributes virtually pure energy to the diet—16 Calories (kcal) per teaspoonful. To the chemist, sugar includes many carbohydrates such as ribose, glucose, fructose, galactose, lactose, maltose, and sucrose—the chemical combination of the two sugars glucose and fructose. Since sugar is usually recognized as sucrose, the following discussion deals specifically with sucrose; hence, the terms *sugar* and *sucrose* are used inter-

changeably. Other sugars, when mentioned, are specified.

SOURCES. Although sucrose is widely distributed in nature in green plants, it is obtained commercially in large quantities from the sugarcane and sugar beet plants.

In 1990, world production of cane and beet sugar totaled over 109 million metric tons; and the 10 leading producing countries, by rank, were: India, U.S.S.R., Cuba, Brazil, China, United States, France, Germany, Thailand, and Australia.[44] That year, these 10 countries produced 60% of the world's cane and beet sugar. The United States produces more than 5.9 million metric tons of sugar annually, it imports 1.8 million metric tons and exports 463,000 metric tons.

In 1990, the United States per capita consumption of refined sugar totaled 64.2 lb *(29.2 kg).*[45]

Sugarcane. Sixty to sixty-two percent of the world's refined sugar is derived from sugarcane—a coarse grass of tropical and semitropical climates. Sugarcane produces stalks 7 to 20 feet *(2 to 7 m)* high and about 2 in. *(5 cm)* in diameter which contain a sugary juice. Most of the cultivated cane today is probably an ancestor of *Saccharum officinarum,* a native of New Guinea.

Fig. S-40. Sugar plantation in Brazil. (Courtesy, Field Museum of Natural History, Chicago, Ill.)

ORIGIN AND HISTORY. It is believed that sugarcane culture began in New Guinea. From there, it eventually spread throughout the islands of the South Pacific, and then to Indonesia, Asia, China, and the Philippines. On the second voyage of Christopher Columbus to the New World, he carried sugarcane. However, his transplants failed. But, other explorers who soon followed were able to introduce success-

---

[44]Data from *FAO Production Yearbook 1990,* FAO/UN, Rome, Italy, Vol. 44, p. 160, Table 69. **Note well:** Annual production fluctuates as a result of weather and profitability of the crop.
[45]*Agricultural Statistics 1991,* USDA, p. 479, Table 672.

fully sugarcane to the West Indies, Brazil, and Mexico. The first sugar mill in the New World was built in 1508 near Santo Domingo. Soon tropical America became the world's greatest sugar producing area. In 1751, Jesuit missionaries brought sugarcane from Haiti to New Orleans, Louisiana. By 1791, the commercial production of sugar had begun in a mill set up by Antonio Mendez. Then in 1795, Etienne de Bore produced the first granulated sugar in what was soon to become part of the United States. With the increased planting and processing of sugarcane in the New World and the development of the sugar beet industry, sugar became accessible to everyone.

WORLD AND U.S. PRODUCTION. In 1990, the worldwide production of sugarcane totaled over one billion metric tons. In the United States, all of the sugarcane is grown in four states—Florida, Hawaii, Louisiana, and Texas—where the yield per acre is about 34 tons on about 791,200 acres (320,324 ha).

PROCESSING. At the sugar mill the cane is washed, shredded, crushed, and passed through a series of heavy rollers (mills) under great pressure to extract the cane juice.

Fig. S-41. An early *sugar factory* where sugar was made from sugarcane and poured into molds. (Courtesy, The Bettmann Archive, Inc., New York, N.Y.)

Each set of rollers (*mills*) crushes the mat of cane stalks a little harder, pressing out more juice. At the end of the series of rollers, nothing remains but the fibrous part of the cane, termed bagasse. This is used to fuel the mill furnaces or to produce other by-products. Next, the juice is strained, and then pumped to large tanks for clarification by the addition of lime which precipitates impurities. After clarification, the juice is filtered and then transferred through a series of evaporators which remove water. The resulting syrup is then transferred to heated vacuum pans where it is reduced to a mixture of sugar crystals and molasses—massecuite. After the vacuum pans, the thick syrup is cooled and further crystallization occurs. Next this mass is moved to centrifugals which spin at high speeds and separate the sugar and molasses. The sugar is dried, cooled, weighed, and then readied for shipment as raw sugar, about 96% sucrose. The molasses is collected and reworked until no more sugar can be profitably extracted. At this point, usually the third extraction, the molasses—blackstrap molasses—becomes a by-product of the sugarcane industry.

Some raw sugar, in which part of the natural molasses remains in the crystals, is marketed and consumed as raw sugar. But most of it is further refined to produce granulated white sugar.

Raw sugar is an economical form in which to ship sugar in bulk to refineries where it is washed to remove the molasses, and dissolved in water. This syrup is then filtered, decolorized, crystallized in vacuum pans, and centrifuged to remove the sugar crystals which are then dried, screened, and packaged as the familiar granulated white sugar. The crystallization process is repeated as long as extractable sugar remains in the syrup. Brown sugar consists of the sugar crystals in the molasses syrup remaining after the granulated white sugar has been removed, although some commercial brown sugar is prepared by adding molasses back to the refined white sugar. Brown sugar is actually a mass of fine crystals covered with a film of brown, highly refined, molasses-flavored syrup. Brown sugar is not usually made from sugar beets due to the strong flavor of sugar beet molasses.

BY-PRODUCTS. The principal by-products of sugarcane processing are bagasse and molasses. Bagasse is used as a fuel for the sugar mill furnaces, and for the manufacture of paper products and building boards. In some areas, bagasse is ammoniated and used as cattle feed. Also, bagasse has been used effectively as a carrier of molasses for cattle feed. Molasses may contain up to 50% sugar. It is used either for livestock feed, for alcohol production, or for yeast production, which is subsequently manufactured into food for humans and livestock. Also, some molasses is sold for human consumption, and some is used to make rum. Cane wax and aconitic acid are also recovered during the processing of sugarcane. Cane wax is used in the manufacture of polishes, cosmetics, and paper coatings, while a derivative of aconitic acid is used in the manufacture of plastics.

(Also see MOLASSES; and SINGLE-CELL PROTEIN.)

Sugar Beet. Sugar beets rank second to sugarcane as a source of sugar. The sugar beet is a biennial plant. During the first year, it produces the fleshy root which contains 13 to 22% sugar. Hence, sugar beets are harvested and the sugar extracted following the first growing season. The scientific name for the sugar beet is *Beta vulgaris*, a newcomer as a world sugar source.

Fig. S-42. The sugar plant. (Courtesy, Field Museum of Natural History, Chicago, Ill.)

ORIGIN AND HISTORY. The sugar beet was cultivated in ancient times in southern Europe and North Africa. During the middle of the 18th century, a practical method for extracting sugar from beets was developed. Following this discovery, sugar beet culture spread into France, Austria, Hungary, and Russia. The world's first sugar beet factory was built in Germany in 1803.

During the middle of the 19th century, the sugar beet was brought to the United States, where the first successful beet sugar processing plant was built in 1870 at Alvarado, California, by E. H. Dyer. Today, sugar beets are grown in 17 states where the soil and climate are favorable.

WORLD AND U.S. PRODUCTION. In 1990, the worldwide production of sugar beets totaled 305.9 million metric tons, of which the United States accounted for only 25 million metric tons, or 8%. The leading U.S. sugar beet-producing states, by rank, are: Minnesota, Idaho, California, Michigan, and North Dakota.

PROCESSING. The major difference between processing sugarcane and sugar beets is that sugar beet processing is a single, continuous process. Upon arrival at a sugar factory, the beets are thoroughly washed and then shredded. The shreds are soaked in tanks of circulating hot water. This allows the diffusion of about 97% of the soluble sugar into the water. Next the shreds are pressed to squeeze out all of the sugary water. The squeezed shreds are called pulp, which is a by-product of the sugar beet industry. The sugar solution or *raw juice* contains about 10 to 15% sugar and many nonsugar substances. Clarification of the sugar solution is accomplished by the application of lime, carbon dioxide, and eventually sulfur dioxide. The sugar solution is filtered several times to remove sediments. Next the juice is concentrated by boiling, which promotes crystallization of the sucrose. Centrifuges separate the sugar crystals from the liquid portion or molasses. The molasses is subjected to additional crystallizations, but eventually it becomes a by-product. The sugar at this point is raw sugar and it is subjected to further refining which involves repeated washings and recrystallizations and decolorization until it is pure white and nearly 100% sucrose. At this stage, beet sugar and cane sugar are identical. However, for years beet sugar was considered inferior, and sold at a price disadvantage.

BY-PRODUCTS. All of the by-products of the sugar beet industry can be used as livestock feeds. The resulting beet pulp can be fed wet, or it can be ensiled or dried. Molasses is often added to beet pulp to increase the energy content. Furthermore, beet tops and crowns—very top portion of the beet—are relished by livestock. These can be fed fresh, dried, or ensiled.

NUTRITIONAL VALUE OF SUGAR. During digestion, sugar—sucrose—is split by the digestive enzyme, sucrase, into its two single sugars, glucose and fructose. Both of these sugars are easily used by the body for the production of energy since fructose is eventually converted to glucose. One ounce (28 g) of refined granulated sugar provides the body with 109 Calories (kcal) of energy and little else. Refined sugar is a concentrated form of energy. One ounce of brown sugar also provides nearly as much energy (106 kcal/oz) and small amounts of calcium, phosphorus, sodium, magnesium, potassium, and iron—hardly enough to contribute to our requirements.

Assuming our yearly consumption of sugar to be about 64.5 lb (29.3 kg), then each day we supply approximately 306 Calories (kcal) of our daily energy requirement from refined sugar—sucrose. For an average man, that is 10.5% of the recommended daily intake; for an average woman, it is 14%.

OTHER SOURCES OF SUGAR. Production and consumption of beet and cane sugar are so enormous that other sources are dwarfed, but several other important sugar sources or sweetening agents do exist. Other sugar sources are: maple sugar, palm sugar, sorghum, honey and plant nectar, and starch hydrolysis. A discussion of each of these follows:

• **Maple Sugar**—Making sugar from the sugar maple, *Acer saccharum*, was practiced by the American Indians long before there were any settlers. Early in the spring, the Northeast Indians tapped the hard maple trees by gashing them with their tomahawks, and then collecting the sap in a birch bark dish. By continually adding heated rocks, they evaporated the sap down to a thick, dark syrup. It was a rather crude, but effective, process. Upon arriving in the New World, the early settlers soon learned the skill from the Indians, and improved upon the Indian system by using iron drill bits to tap the tree and copper or iron kettles to evaporate the sap to syrup and sugar.

Even today, in the late winter or early spring when the sap flow begins, sugar maple trees—a common forest tree of the northeastern United States—are tapped. Large stands of maple trees are called *sugar bushes.* The sap of the sugar maple is about 95% water, and 15 to 20 gal (57 to 76 liter) of sap yields about 2 qt (1.9 liter) of maple syrup. Sugar in the sap is mainly sucrose. During the winter, some of the starch that the tree made the previous summer and stored in its roots is converted to sugar. When the sap begins to rise in the spring, the sugar is carried along. Sap flows for an average of 34 days when warm sunny days are followed by cool crisp nights.

Although collecting and processing sap have changed some since colonial days, gathering sugar from the maple is picturesque and truly American. Today, power drills are used to bore holes 2 to 3 in. (4.1 to 7.6 cm) deep in the trunk of the maple about 3 ft (1 m) above the ground. The number of holes varies with the size of the tree. Large trees, over 2 ft (0.6 m) in diameter, can be tapped in four or more locations. These holes usually heal by midsummer. Once the hole is drilled a spike is inserted and a pail hung below it. Sometimes the rate of flow is better than one drop per second. Collection of sap is a daily operation by hand using a sap-yoke with pails hung from its ends, by sleds, by wagons, or by gravity piping using plastic pipes connected to the hole in the tree and transporting to a single collection point.

The final stages of processing occur in the sugarhouse. Immediately after the sap is collected, it is strained. Then it is boiled in shallow aluminum evaporating pans which are 20 to 30 ft (6.2 to 9.2 m) long. The sap is boiled down to a specific thickness for maple syrup. Still more boiling produces maple sugar. A good tree yields up to 40 gal (152 liter) of sap in a season, but 30 to 50 gal (114 to 190 liter) of sap are required to make 1 gal (3.8 liter) of syrup, depending upon the quality of the sap. It is the boiling which gives maple syrup its characteristic flavor.

Nowadays, maple sugar and syrup are rather a treat and a luxury, but for the Indians and colonists they were important and abundant. Production of maple sugar increased in the colonies until about 1860, when cane sugar began to take its place. But, even today, about 1 million gal (3.8 million liters) of maple syrup, plus 1 million lb (0.45 million kg) of maple sugar are produced in the United States every year—some for home use and some for commercial use—almost entirely from wild trees.

The composition of maple sugar and maple syrup are given in Food Composition Table F-21. Maple sugar contains about 98 Calories (kcal) of energy per ounce and several macrominerals, while maple syrup contains about 71 Calories (kcal) of energy per ounce and several macrominerals. Both are slightly less concentrated sources of sugar than refined sugar.

(Also see MAPLE SYRUP.)

• **Palm sugar**—In tropical regions of the world, obtaining sugar from several species of palm is an important village industry. Collecting the sap from the sugar palm is an old practice in India and in the eastern tropics. The sap is collected from the stalk of the male flower rather than from a hole in the trunk as practiced with maple trees. At the time of collection, the sap is 10 to 16% sucrose. To process the sap, it is evaporated much the same way as maple sap until it becomes a thick syrup containing sucrose crystals. It may be crudely centrifuged or poured into molds to form small cakes, a product known as *jaggery*.

• **Sorghum**—Generally, sorghum is considered a grain, forage, or silage crop, but some sorghum is sweet. Sweet sorghum, or sorgo, has a tall, sweet, juicy stalk that can be used for syrup and sugar production. In some countries, the sweet stalks of sorghum are chewed by the natives. The juice contains about 12% sugar. In the United States, sorghum syrup is a distinctly flavored, mild, sweet, light amber-colored syrup. It is a specialty product of the southeastern United States. Although seldom carried beyond the syrup stage, the technology exists for converting sweet sorghum to sugar. Thus, sweet sorghum could serve as a supplementary crop for the sugarcane and sugar beet industries.

The composition of sorghum syrup is given in Food Composition Table F-21. It contains about 73 Calories (kcal) of energy per ounce, several of the macrominerals, and some iron.

(Also see SORGHUM.)

• **Honey and plant nectars**—Honey is essentially an invert sugar, a mixture of equal portions of the monosaccharides, glucose, and fructose. Bees collect plant nectar which contains sucrose, contribute an enzyme to invert the sucrose, and evaporate the water to produce the honey they store in combs. Man has learned to exploit this instinct of the honeybees.

Nutritionally, honey provides energy—about 86 Calories (kcal) per ounce—plus traces of minerals and vitamins, and 14 to 19% water. The composition of honey is given in Food Composition Table F-21.

(Also see HONEY.)

• **Starch hydrolysis**—Starch from any starch source—rice, corn, and other grains, and potatoes—may be broken down into its component sugar, either enzymatically or with dilute acids. The most familiar example in the United States is corn syrup (Karo). Complete hydrolysis of starch yields glucose or dextrose. Corn syrup and dextrose (corn sugar) are widely used in food products. Crystallized dextrose is pure energy. One ounce supplies 120 Calories (kcal) of energy. Each year, the foods we eat contain about 71.9 lb *(32.7 kg)* of corn sugar and 22 lb *(10 kg)* of corn starch.

(Also see CORN MILLING, WET MILLING AND DRY MILLING; STARCH; and SYRUPS.)

• **Naturally occurring**—Of course, all sugar occurs naturally in some form or another before being processed. Many raw fruits and some vegetables contain sugar—sucrose, fructose, and glucose.

Also, one other sugar—a nonplant sugar—is derived from an important food source. Milk sugar or lactose is present as 4.5% of whole cow's milk and 7.5% of human breast milk.

(Also see CARBOHYDRATE[S]; and MILK AND MILK PRODUCTS, section on "Carbohydrates in Milk.")

CONSUMPTION AND USE OF SUGAR. Sugar comes from a variety of sources, and the overall consumption of sugar from all sources is about 137.5 lb *(63 kg)* per person per year. Of this total, about 64.2 lb *(29 kg)* comes from sucrose or table sugar and 73 lb *(33 kg)* from corn syrup, honey, and other sweeteners. That translates into a per capita consumption of 658 empty Calories per day!

Most homes maintain a supply of sugar for use in baking, cooking, and sweetening, but a large share of the sugar consumed is derived from food products. Sugar is classified as a food additive, and on the basis of intake it is the number one food additive. Sugar—glucose, fructose, sucrose, corn syrup, and invert sugar—is employed as a food additive to make the aroma or taste of a food more agreeable or pleasurable. Sugar of one kind or another is commonly used in the following: cereals, baked goods, candies, processed meats, processed foods, and soft drinks. Table S-13 shows the level of sugar contained in some common foods.

Fig. S-43. Some uses of sugar. (Courtesy, Hershey Food Corp., Hershey, Pa.)

TABLE S-13
SUGAR CONTENT OF
SOME CANNED OR PACKAGED FOODS[1]

Food	Sugar
	(%)
Nondairy creamer	57-65
Ready-to-eat cereals[2]	1-56
Milk chocolate candy	44-51
Brownies	50
Chocolate cake	36
Salad dressings	7-30
Ketchup	29
Ice cream	21
Peaches, canned	7-18
Yogurt	14
Crackers	12
Fruit juice drink	12
Corn, canned	11
Cola-type beverages	9-10
Peanut butter	9

[1]Values from *CNI Weekly Report*, Vol. 4, May 2, 1974, and *Consumer Reports*, March 1978, pp. 136-141.

[2]For more information also see BREAKFAST CEREALS, section headed "The Sugared-Breakfast-Cereal Binge."

Today, the use of sugar in households is less than one-half of what it was at the beginning of this century, but the use of sugar by industry in processing foods and beverages is three times greater. Industry uses about 70% of the refined sugar.

Sugar is available in a variety of forms the more common of which follow:

• **Granulated sugar**—The white, refined sugar used in the home and in commerce comes from sugarcane and sugar beets. It is 99.9% pure, and keeps indefinitely. Granulated sugar is classified according to crystal size as fine, or ultrafine. Three other forms of granulated sugar are produced mainly for industrial use: very fine, medium-coarse, and coarse.

• **Powdered sugars**—Powdered sugars are a form of sucrose made by grinding the sugar crystals. They are used for icings, frostings, uncooked candies, and for dusting on finished products. Confectioners' sugars are usually packed with small amounts of cornstarch to prevent caking.

• **Raw sugar**—This is the tan to brown product obtained from the evaporation of sugarcane juice. FDA regulations prohibit the sale of raw sugar unless the impurities—dirt, insect fragments, etc.—are removed.

• **Turbinado sugars**—A partially refined sugar which is similar in appearance to raw sugar. Turbinado sugar is sold for consumption without further refining.

• **Brown sugar**—So called *soft* sugar or brown sugar is a mass of fine crystals covered with a film of highly-refined, colored, molasses-flavored syrup. It is valued primarily for flavor and color. Four grades are commonly available for food manufacturing—Numbers 6, 8, 10, and 13. The higher numbers are darker and more flavorful. Brown sugar contains 91 to 96% sucrose.

• **Invert sugar**—When a solution of sugar is heated in the presence of an acid or treated with enzymes, the sugar breaks up into the two sugars of simpler chemical structure that characterize sucrose as a disaccharide. One is glucose, commercially called dextrose. The other is fructose, com-mercially called levulose. This mixture of dextrose and levulose in equal weights, is known as invert sugar.

Invert sugar is sold only in liquid form and is sweeter than sucrose. It helps prolong the freshness of baked goods and confections and is useful in preventing food shrinkage.

• **Liquid sugars**—Sugar syrups or liquid sugars are clear solutions that contain a highly purified sugar, of which there are many grades.

• **Levulose**—Fructose, or levulose as it is known commercially, is one of the two components of invert sugar. It is intensely sweet and highly soluble, and is produced in small quantities mainly for pharmaceutical applications.

• **Dextrose (Glucose)**—Dextrose is the commercial name for glucose, the second component of invert sugar. Dextrose is also called corn sugar. It is made commercially from starch by the action of heat and acids, or enzymes. Two types of refined dextrose are available commercially: dextrose hydrate, containing 9% by weight of water of crystallization; and anhydrous dextrose, containing less than 0.5% of water. Dextrose hydrate is most often used by food processors. It is 74% as sweet as sugar.

• **Lactose**—Milk sugar, or lactose, is generally made from whey and skim milk. Compared to sugar, it is only slightly sweet and markedly less soluble in water. It is used primarily in pharmaceuticals.

• **Maltose**—Malt sugar is formed from starch by the action of yeast. Maltose is much less sweet than sugar. Preparations containing maltose, often in mixture with dextrose, are used in bread-baking and in infant foods.

• **Corn syrup**—This is a viscous liquid containing maltose, dextrin, dextrose, and other polysaccharides—glucose chains of various lengths. Unlike sugar, it has a distinct flavor other than sweetness. Corn syrup is the product of the incomplete hydrolysis of starch; it is usually obtained by heating corn-starch with a dilute acid or by enzymatic action. The degree of conversion is expressed by the *dextrose equivalent* (D.E.) which is, in effect, the measure of sweetness in the syrup. Even high conversion syrups are substantially less sweet than sugar. Corn syrup can control crystallization in candy-making, and has moisture-containing properties. A special corn syrup on the market is a *high fructose corn syrup* (HFCS) which is made by treating high conversion corn syrup with enzymes. The enzymes convert some of the glucose to fructose.

• **Molasses**—Molasses consists of concentrates extracted from sugar-bearing plants, such as the thick liquid produced in the refining of sugar. It contains other substances that occur naturally in the plants as well as sugar. The highest grade is edible molasses. It is most often seen as a table syrup or as an ingredient in a blend of table syrups. It is suitable for use in gingerbread, spice and fruit cakes, rye and whole wheat breads, cookies, baked beans, and certain candies. Edible molasses provides nutrition in the form of sugar and certain minerals—primarily iron. Blackstrap molasses is the final molasses in the sugar manufacturing process. (Also see MOLASSES.)

• **Honey**—Honey is essentially an invert sugar, but it contains a slight excess of fructose. It is used in food products where its distinctive flavor is desired.

• **Maple sugar and syrup**—Both are products of the condensation of the sap of the maple tree. However, their characteristic flavor is not manifested until the sap has been boiled. Maple sugar and syrup are also used in food products for their distinctive taste and aroma.

Whether sugar is utilized in the home or in the food industry, its properties have made it (1) the world's most widely used sweetener, and (2) one of the most versatile ingredients in food preparation. Six of sugar's qualities are particularly important; namely—
1. Sweetening.
2. Moistening—sugar absorbs and holds moisture.
3. Solubility—sugar is completely soluble in water.
4. Coloring (caramelization)—sugar turns dark brown when heated and thus adds to the browning of baked goods.
5. Preservation—a concentrated sugar solution acts as a preservative; it inactivates microorganisms.
6. Crystallization.
(Also see ADDITIVES; and CARBOHYDRATE[S].)

SUGAR AS A FACTOR IN HUMAN DISEASES.
Recently, charges have been leveled at sugar as a causative factor in some human diseases. Some studies have claimed that sugar intake is related to the development of coronary heart disease, diabetes mellitus, obesity, and dental caries. To date, however, any causative role of sugar in coronary heart disease is far from certain, and as for diabetes mellitus no proven direct links exist. As for obesity, any time the amount of energy flowing into our body exceeds the amount flowing out, the remainder is deposited as triglycerides in adipose tissue—fat. Excess energy derived from fats, carbohyrates, and proteins may all contribute to the deposition of fat. No reliable evidence exists that implicates any specific nutrient as contributing excess energy—rather, excessive intake creates the problem. Dental caries—tooth decay—is influenced by such factors as structural resistance of the teeth, oral hygiene, oral microflora, salivary flow, and composition of the diet. Nevertheless, indirect proof has been obtained implicating sugar. In China and Ethiopia, where the consumption of sugar is low, the incidence of tooth decay is also very low, while in Australia, Hawaii, and French Polynesia where sugar consumption is high, so is the incidence of dental caries. A number of studies have shown a firm link between sugar exposure and tooth decay. However, the manner of exposure and dental hygiene make a large contribution to the development and prevention of dental caries. Frequent exposure to a solid and/or sticky form of sugar between meals has resulted in a high incidence of dental caries, while taking sugar with meals, followed by quick brushing or rinsing, prevents the accumulation of bacteria and plaque, and removes the substrates for acid production by bacteria.
(Also see DENTAL HEALTH, NUTRITION, AND DIET; DIABETES MELLITUS; HEART DISEASE; and OBESITY.)

Until more definitive evidence can be gained, the real danger of sugar seems to be its attractiveness to the human taste buds—its palatability. Hence, excesses are often eaten, displacing more nutritious foods from the diet and possibly creating a diet low in other essential nutrients.
(Also see CARBOHYDRATE[S].)

SUGAR DOCTOR

Sucrose crystallizes. Since many foods, especially confectionery products like creams, fondants, and fudge, are dependent upon a balance between sugar crystals and sugar syrup for their texture characteristics, a sugar doctor helps prevent, or helps control, the degree of sucrose crystalliza-tion, thus extending the shelf-life by maintaining the desired consistency of the product. Often invert sugar is employed as a sugar doctor. Invert sugar may be added during the candy making process or developed from sucrose during the candy making process. Invert sugar "doctors" because it crystallizes more slowly and forms smaller crystals than sucrose. Glycerine, sorbitol, and the polyhydric alcohols also are used to control crystal formation.
(Also see INVERT SUGAR; and SUGAR, section on "Consumption and Use of Sugar.")

SUGAR, ICING

Icing for cakes is made from a grade of sugar termed powdered sugar or confectioners' sugar; hence, the name icing sugar.
(Also see DESSERTS; and SUGAR, section on "Consumption and Use of Sugar.")

SUGARING OF DRIED FRUITS

The whitish deposits of sugar that are sometimes seen on the surface of dried fruits are formed during drying when water containing dissolved sugars flows from the inside of the fruit to the surface. Hot air at the surface of the fruit causes the water to evaporate, leaving behind the sugars in the form of tiny crystals.

SUGARING OFF

The process of making maple sugar by boiling off the water from the maple sap. The early spring of the year, when warm days follow cool nights, is the sugaring off season in the northeastern and north central states and eastern Canada. During this time of year, the sap begins to flow and maple trees are tapped for collection. Traditionally, sugaring off gatherings where held at the sugarhouse—the shed where the sap is boiled—there neighbors helped in the making and sampling of the maple sugar. For the production of 1 gal (3.8 liter) of maple syrup, 30 to 50 gal (114 to 190 liter) of maple sap are required; and 1 gal of maple syrup yields about 8 lb (3.6 kg) of maple sugar.
(Also see MAPLE SYRUP.)

SULFA DRUGS (SULFONAMIDE DRUGS)

Any of a group of compounds characterized by the presence of both sulfur and nitrogen, with high specificity for certain bacteria. Among the best known are sulfanilamide, sulfadiazine, sulfapyridine, sulfamerazine, and sulfasoxazole.

SULFHYDRYL GROUP

The -SH radical that forms high-energy bonds in chemical compounds, which are similar to the high-energy bonds formed by phosphates in compounds such as ATP.

SULFUR

Sulfur is a nonmetallic element that occurs widely in nature; it is found in every cell of the body and is essential for life itself, mostly as a component of three important amino acids—cystine, cysteine, and methionine. Also, it is a part of two vitamins—thiamin and biotin, and it is present in saliva and bile, and in the hormone, insulin. Approximately 0.25% of the body weight (or 175 g in the adult male) and 10% of the mineral content of the body are sulfur. It is sometimes referred to as nature's *beauty mineral*, because it is reputed to keep the hair glossy and the complexion clear and youthful.

HISTORY.
The name is derived from the Latin word *sulphurum*. Sulfur has been used since ancient times. It was often called brimstone (burning stone); and ignited sulfur is mentioned in the earlier records of many countries as having been used in religious ceremonies and for purifying (fumigating) buildings. The early medical books of Dioscorides of Greece and Pliny the Elder mention sulfur; and the Romans used it in medicine and in warfare. Alchemists recognized sulfur as a mineral substance that could be melted and burned. It was first classified by Antoine Lavoisier in 1777.

ABSORPTION, METABOLISM, AND EXCRETION.
The small intestine is the major site of sulfur absorption.

During digestion, the sulfur-containing amino acids are split off from protein and taken into the portal circulation. Sulfur is stored in every cell of the body, with the highest concentration found in the hair, skin, and nails.

Excess sulfur is excreted in the urine and in the feces. About 85 to 90% of the sulfur excreted in the urine is in the organic form, derived almost entirely from the metabolism of the sulfur amino acids. Since inorganic sulfates are poorly absorbed, it follows that the fecal excretion of sulfur is about equal to the inorganic sulfur content of the diet.

FUNCTIONS OF SULFUR.
Sulfur has an important relationship with protein. It is a necessary component of the sulfur-containing amino acids methionine, cystine, and cysteine. Sulfur is present in keratin, the tough protein substance in the skin, nails, and hair; and it appears to be necessary for the synthesis of collagen.

As a component of biotin, sulfur is important in fat metabolism; as a component of thiamin and insulin, it is important in carbohydrate metabolism; as a component of coenzyme A, it is important in energy metabolism; as a component of certain complex carbohydrates, it is important in various connective tissues. Insulin and glutathione, regulators of energy metabolism, contain sulfur. Also, sulfur compounds combine with toxic substances such as phenols and cresols and convert them to a nontoxic form, following which they are excreted in the urine.

DEFICIENCY SYMPTOMS.
Sulfur deficiencies are manifested primarily in retarded growth because of sulfur's association with protein synthesis.

INTERRELATIONSHIPS.
Sulfur is related to the amino acids methionine, cystine, and cysteine, and to biotin, thiamin, insulin, coenzyme A, certain complex carbohydrates, insulin, and glutathione.

RECOMMENDED DAILY ALLOWANCE OF SULFUR.
Sulfur requirements are primarily those involving amino acid nutrition.

There are no recommended allowances for sulfur because it is assumed that the sulfur requirements are met when the methionine and cystine intakes are adequate.

TOXICITY.
Except in rare, inborn errors of metabolism where the utilization of the sulfur-containing amino acids is abnormal, excess organic sulfur intake is essentially nonexistent. However, inorganic sulfur can be dangerous if ingested in large amounts.

SOURCES OF SULFUR.
Inorganic sulfur is poorly utilized by man and other monogastrics. So, the sulfur needs of the body are largely met from organic complexes, notably the amino acids of proteins, rather than from inorganic sources. The sulfur content of protein foods varies from 0.4 to 1.6%, depending on the quality of the protein. The

Fig. S-44. Poultry—a good source of sulfur. (Courtesy, USDA)

average mixed diet contains about 1% sulfur. Good food sources of sulfur are: cheese, eggs, fish, grains and grain products, legumes, meat, nuts, and poultry.

Unfortunately, few foods have been analyzed for sulfur, with the result that sulfur content is not given in Food Composition Table F-21. So, a special table, Table S-14, was prepared for this book; this gives some good sources of sulfur.

(Also see MINERAL[S], Table M-25.)

TABLE S-14
SOME FOOD SOURCES OF SULFUR AND THEIR CALORIC CONTENT

Food	Sulfur	Energy
	(mg/100g)	(kcal/100 g)
Soybean flour	410	386
Brewers' yeast	380	283
Peanuts, roasted, salted	380	582
Molasses, blackstrap	350	230
Pork chops, lean, roasted	300	254
Brazil nuts	290	715
Turkey, light or dark meat, roasted	290	190
Sardines, canned in oil, drained solids	310	246
Beef, variety of lean cuts	270	220
Chicken, light or dark meat, fried	255	210
Lamb, shoulder roast, lean	240	205
Wheat germ	240	391
Navy beans, dry	230	340
Cheese, natural cheddar	230	402
Soybeans, dry	220	405
Wheat bran	220	353
Salmon, canned	220	124
Oats, whole grain	210	283
Beans, lima, dry	200	359
Wheat flour	190	345
Rice bran	180	276
Rice polish	170	265
Wheat, whole grain	160	360
Barley, whole grain	150	305
Almonds	150	598
Sorghum, whole grain	150	339
Eggs, chicken	140	157
Corn, whole grain	120	348
Cabbage, raw	110	24
Buttermilk, dehydrated	80	387
Alfalfa leaf meal, dehydrated	60	215
Peas, raw	50	84
Rice, whole grain	50	363
Sweet potato	40	82
Turnip greens, raw	40	28
Turnip, raw	40	28
Milk, skim	30	34
Rutabaga, raw	30	46
Beet, common, red, raw	20	43
Carrot, raw	20	42
Potato, raw	20	76
Apple, raw	10	58

SULFUR AMINO ACIDS

The two amino acids, cystine and methionine, contain sulfur in their chemical structure. Methionine is an essential amino acid; cystine is not.

(Also see AMINO ACID[S]; and PROTEIN[S].)

SULFUR DIOXIDE (SO₂)

A colorless, nonflammable gas derived from burning sulfur. Its odor is pungent and suffocating. The ancient Egyptians and Romans used the fumes of burning sulfur in their wine making. Thus, they were utilizing the antimicrobial properties of sulfur dioxide. It also prevents enzymatic and nonenzymatic decolorization of some foods. In the food industry, sulfur dioxide is applied to dehydrated fruits and vegetables; to increase storage life, preserve color and flavor, and aid in the retention of ascorbic acid and carotene. In wine making, it is employed first to sanitize the equipment, then at several stages as an antioxidant, and finally during bulk storage to prevent bacterial spoilage.

SULTANAS

This term is used differently in the United States and Europe. Hence, the two major meanings follow:

• In the United States, the term designates the raisins produced *without* sulfuring from sultana grapes. They are used mainly by commercial food manufacturers and food service establishments.

• In Europe, this term refers to golden seedless raisins which are produced from Thompson Seedless grapes or a related variety of grape by (1) washing the freshly harvested fruit, (2) dipping it in a solution to crack the skin, (3) sulfuring the grapes by exposure to the fumes of burning sulfur, and (4) drying the sulfured fruit in heating chambers. In the Mediterranean region, a delicate white wine is also made from sultana grapes.

SUN-CURED

Material dried by exposure in open air to the direct rays of the sun.

SUNFLOWER *Helianthus annuus*

Contents	Page
Origin and History	996
World and U.S. Production	996
Processing Sunflower Seed	996
Nutritional Value of Sunflower Seed	997
Sunflower Products and Uses for Human Food	998

Fig. S-45. Sunflowers at bloom stage. (Courtesy, Sunflower Association of America, Fargo, N.D.)

Sunflowers belong to the composite family *Compositae*. The common annual is genus *Helianthus*, species *H. annuus*.

United States sunflower production has spiraled in recent years largely because of two breakthroughs which vastly improved oil production: (1) the development in the 1960s of sunflower varieties with an oil content of more than 40%—a ⅓ increase over earlier varieties; and (2) the development of hybrid sunflowers in the 1970s, which boosted yields another 25%.

Sunflower oil is higher in polyunsaturates than corn oil and is much more stable than safflower oil. Thus, it has an edge over these two competitors for use in premium grade margarine and in cooking and salad oils.

ORIGIN AND HISTORY. The sunflower is one of the few annual cultivated plants which was first domesticated in the United States, although there is evidence that this honor should go to, or be shared with, Peru and Mexico. Its culture for food by the Indians was at an advanced state when the colonists came to America.

In the Soviet Union, superior varieties were selected, and in the 1830s a method was developed for obtaining oil from the seed. The present-day interest in sunflowers in Europe, and elsewhere, stems primarily from the work of the Russian scientist, Dr. V. S. Pustovoit, at the Sunflower Research Institute, near Krasnodar, U.S.S.R. Today, some of the newer varieties in the Soviet Union produce up to 60% oil, and 95% of the nation's sunflower seed averages 50% oil content. Eighty percent of the vegetable oil produced in the Soviet Union comes from sunflowers.

Although U.S. production of sunflowers for oil started in the 1940s, present-day interest in sunflowers began in 1962 when new high-oil varieties were introduced.

WORLD AND U.S. PRODUCTION. The global growth in the 1960s of sunflower seed production and of the sunflower seed and oil trade is the most impressive of any in the fat and oil industry. In 1990, sunflowers ranked third in the world among sources of edible vegetable oils, exceeded only by soybean oil and rape oil.

World production of sunflower seed totals about 22 million metric tons annually; and the 10 leading sunflower-producing countries, by rank, are: U.S.S.R., Argentina, France, China, Spain, United States, Turkey, Hungary, South Africa, and Romania. The U.S.S.R. accounts for ⅓ of the world's sunflower production.[46]

In addition to being the world's leading producer of sunflower seed, the U.S.S.R. is the world's largest exporter of sunflower oil. Sunflowers are grown throughout a wide geographical area of the country, and generally occupy about 11.4 million acres (*4.6 mil ha*).

The United States produces about one million metric tons of sunflower seed annually; and the four leading sunflower-producing states, by rank, are: North Dakota, South Dakota, Minnesota, and Kansas.[47]

PROCESSING SUNFLOWER SEED. The sunflower seed is a four-sided, flattened fruit or achene. It has a dry, brittle hull which ranges in color from white, white and grey

---

[46]Data from *FAO Production Yearbook 1990*, FAO/UN, Rome, Italy, Vol. 44, p. 112, Table 40. **Note well:** Annual production fluctuates as a result of weather and profitability of the crop.

[47]Data from *Agricultural Statistics 1991*, USDA, p. 129, Table 179. **Note well:** Annual production fluctuates as a result of weather and profitability of the crop.

striped, white and black striped, to black. The hull encloses a whitish kernel which has a thin, translucent skin covering or coat.

High-oil-type sunflower seeds are processed to obtain the oil and meal. The separation of the oil may be achieved (1) by direct solvent extraction, (2) by prepress solvent extraction, or (3) by mechanical means (screw pressing).

The steps in processing are: (1) cleaning, (2) dehulling, (3) rolling and cooking, and (4) extracting the oil.

NUTRITIONAL VALUE OF SUNFLOWER SEED.
Composition data on sunflower seed kernels, flour, and oil are given in the Food Composition Table F-21.

A few salient features of the nutritional value of sunflower products follow:

• **Composition of nonoil sunflower seeds for human food**—The seeds are a concentrated source of many nutrients. The oil is desirable in the human diet due to the high linoleic fatty acid content, the essential fatty acid. Their protein content is sufficient to recommend them as a meat substitute; and the nonoil varieties are substantially higher in lysine, an essential amino acid, than the high-oil varieties. Sunflower seeds contain 31% more iron than raisins—a popular source of iron; and they are a good source of the vitamins thiamin and niacin.

• **Composition of the hulls**—The fibrous hulls have only a small percentage of oil and crude protein and contain about 50% crude fiber. Chemically, the hulls are largely lignin, pentosans, and cellulosic constituents.

• **Composition of sunflower meal (sun meal)**—The composition of defatted sunflower meal varies considerably, depending primarily on the method of processing. A typical commercial sunflower meal contains approximately 9.0% water, 45% protein, 3.5% fat, 9.5% fiber, 7.0% ash (mineral), and 26% carbohydrate.

Sunflower meal compares favorably with other oilseed meals as a source of calcium and phosphorus; it runs about 0.46% calcium and 1.47% phosphorus.

The meal is superior to other oilseed meals in vitamin content. It is richer in B-complex vitamins than soybean meal; it is equal in nicotinic acid content to peanut meal, which is rated as an outstanding source; the pantothenic content is similar to soybean meal; and it is a rich source of vitamin A. Sunflower flour is similar to sunflower meal.

The toxic compounds present in several vegetable proteins have not been found in sunflower meal. Also, sunflower meal products cause considerably less flatus (gases) in the digestive tract than soybean meal products.

• **Protein quality**—Table S-15 shows the essential amino acid composition of the protein of sunflower seed kernels and of a typical sunflower meal. Kernels and meal are deficient in lysine, but adequate in the other essential amino acids. It is noteworthy, however, that the amino acids of sunflowers vary with varieties and planting locations.

The main disadvantage of sunflower meal as a source of high-protein human food is that it turns off-color (generally to green or brown) during processing. These off-colors are caused by the reaction of chlorogenic acid at high pH values.

The commercial use of sunflower flour for human food is dependent on the development of low chlorogenic acid varieties and hybrids and efficient procedures for dehulling them.

• **Composition of sunflower oil (sun oil)**—The fatty acid

TABLE S-15
PROFILE OF ESSENTIAL AMINO ACIDS IN SUNFLOWER SEEDS COMPARED TO MILK—A HIGH-QUALITY PROTEIN

Amino Acid	Sunflower Seeds		Cow's Milk[1]
	Kernels	Meal	
	(mg/g protein)		
Histidine	25	17	27
Isoleucine	55	52	47
Leucine	75	62	95
Lysine	38	38	78
Methionine and cystine	39	34	33
Phenylalanine and tyrosine	81	80	102
Threonine	40	40	44
Tryptophan	15	13	14
Valine	59	52	64

[1]*Recommended Dietary Allowances*, 10th ed., 1989, NRC–National Academy of Sciences, p. 67, Table 6–5. The essential amino acid requirements of infants can be met by supplying 0.79 g of a high-quality protein per pound of body weight (1.7 g/kg body weight) per day.

composition of sunflower oil makes it desirable as an edible oil. Of all oils produced in the United States, sunflower oil strikes the most ideal compromise between the amount of polyunsaturated fatty acids and stability. Its polyunsaturated fatty acid composition is superior to all oils except safflower oil; hence, it offers a popular approach to the prevention and/or cure of cardiovascular diseases. It contains only trace amounts of linolenic acids, which makes it a fairly stable oil. The stability imparts the capacity of the oil to maintain its flavor (not go rancid) and to resist change in viscosity (not congeal or leave deposits on cooking vessels) after prolonged periods of high temperature. The latter characteristic is particularly important to the rapidly growing fast food and snack industries where deep fat frying is used. Oils which can be reused without flavor deterioration and which allow maximum shelf life of products are desired.

TABLE S-16
COMPOSITION OF COMMERCIAL SUNFLOWER OIL

Nutrient		Amount	
		100 g	Tablespoon (13.6 g)
Calories	kcal	884	120
Vitamin E	mg	44.9	6.1
Palmitic acid[1]	g	5.9	.8
Stearic acid[1]	g	4.5	.6
Oleic acid[2]	g	19.5	2.7
Linoleic acid[3]	g	65.7	8.9

[1]Saturated fatty acid.
[2]Monounsaturated fatty acid.
[3]Polyunsaturated fatty acid and the essential fatty acid.

As shown in Table S-16 sunflower oil is relatively low in the saturated fatty acids, palmitic and stearic, but it is rich in the unsaturated fatty acids, oleic and linoleic. The lack of linolenic acid is primarily responsible for its good storage qualities. The oleic and linoleic acid contents of sunflower oil are quite variable, ranging from 13.9 to 60.0% for oleic and from 29.9 to 76.4% for linoleic acid. Seeds grown in cold climates (such as those in central U.S.S.R., Canada, and northern U.S.) contain less oleic acid and more linoleic acid than those grown in warmer areas (such as southern U.S.).

The high linoleic acid content and the high ratio of polyunsaturated to saturated fatty acids suggest that sunflower oil may be useful in the prevention of high blood cholesterol and heart disease. It is noteworthy, however, that the degree of unsaturation of sunflower oil is affected by plant location, climatic conditions during the growing season, and genetics.

SUNFLOWER PRODUCTS AND USES FOR HUMAN FOOD.

Table S-17 presents in summary form the story of sunflower products and uses.

(Also see FLOURS, Table F-13, Special Flours; and OILS, VEGETABLE, Table O-5.)

Fig. S-46. Sunflower kernels used in salad. (Courtesy, Sunflower Assn. of America, Fargo, N.D.)

SUPERGLYCINERATED FATS

These are mono- and diglycerides which are the products of interesterification. They are employed as emulsifiers by the food industry.

(Also see INTERESTERIFICATION.)

SUPEROXIDE DISMUTASE (SOD)

An enzyme contained in most cells of the body and in most organisms. It catalyzes the breakdown of superoxide free radicals—oxygen joined by an extra electron—to oxygen and hydrogen peroxide. This protects cells from the toxic effects of oxygen, since these superoxide radicals damage DNA, and age and destroy cells. Deactivation of these superoxide radicals is called dismutation; hence, the name superoxide dismutase. Some chemicals, poor nutrition, shock, and radiation therapy may increase the formation of superoxide radicals. Scientists feel that gaining an understanding of this enzyme may unlock secrets which (1) control the rate of human aging, (2) block tumor formation, and (3) maintain the body's immune system. As an injectable drug, SOD could be used to fight inflammatory diseases. Currently, the FDA has approved SOD for use by veterinarians. Human use is for investigational purposes only. Several European nations permit its use for rheumatoid arthritis, osteoarthritis, urological disorders, and the side effects of radiation treatment. In the United States, claims are sometimes made that SOD will reverse aging and degeneration. However, no firm evidence demonstrates that taking SOD will increase the life span.

TABLE S-17
SUNFLOWER PRODUCTS AND USES FOR HUMAN FOOD

Product	Description	Uses	Comments

Whole seed, with hulls	Whole seed may be:
	1. Dehulled with fingernails or teeth and eaten raw.
	2. Roasted by heating (a) in an oven for 15 min. at 350°F (177°C) or (b) in a frying pan while stirring.
	3. Salted by—
	  a. Soaking the whole seed overnight in salt brine (2 Tbsp salt to 1 cup water), boiled for a few minutes then dried with heat;
	  b. boiling in brine for at least 20 minutes; or
	  c. frying the salted whole seed in edible oil.
		Snack food	Consumers prefer a large seed and kernel with a loose hull to facilitate cracking. Also, they like uniformity, and bright black and white stripes.
			Roasted seeds have a relatively short shelf life because of high oil content and high proportion of polyunsaturated fatty acids.
			Many snack foods contribute little to the diet except calories. Raw or roasted sunflower kernels, with or without salt, are nutritious. They are a valuable source of unsaturated fat, quality protein, and some vitamins and minerals.
			Roasted seeds are very popular in the U.S.S.R.

Whole seed, hulled	Seed dehulled commercially in an impact dehuller, following which the mixed hulls and meats are separated by screening and aspiration. Then the dehulled seed is either— 1. Marketed raw as dehulled sunflower kernels, 2. roasted in sunflower or other vegetable oil, or 3. dry roasted, salted, and vacuum packed.	Considerable attention has been given to the use of hulled seed in candies, salads, cereals, bakery goods, and health foods. As a nut substitute in a number of confectionery and bakery formulas.	Dehulled kernels must be stored under refrigeration.
Sunflower meal and flour	Sunflower meal is the product obtained by grinding the residue that remains after extraction of most of the oil from the seed. Sunflower flour is similar to sunflower meal, except that it is finer and more of the hulls have been removed. Sunflower flour has high protein content, bland flavor, and contains no antinutritive factors.	Sunflower meals and flours have not been used commercially as a protein source for human nutrition in the U.S. Experimentally, sunflower flour has been added to wheat flour at 3 to 20% levels to make breads, griddle cakes, cupcakes, pies, dips, rolls, and candies; without loss in attractiveness, volume, or acceptance, and enhanced by a pleasant distinctive, nutty flavor.	Sunflower proteins are deficient in lysine and isoleucine, but they contain adequate levels of the other essential amino acids for humans. The vitamin B-complex content of defatted sunflower seed flour is superior to the products prepared from wheat germ, corn germ, and soya. A major problem in using sunflower meal in human foods is the presence of hulls and chlorogenic acid in the meal, both of which may cause undesirable discoloration. Hulls also cause excessive bulk and fiber in a food product.
Sunflower oil	Oil of an attractive color and pleasant, faintly nutty flavor	1. Cooking and salad oil. 2. In Europe, it is used extensively in shortening and margarine. 3. For frying foods, popping corn, and other culinary processes that require a liquid oil with a high smoke point. 4. For blending with other vegetable oils. 5. As a cooking fat for potato chip frying. 6. For producing a modified butter with improved low temperature spreadability.	Sunflower oil is a high-quality edible oil, high in the polyunsaturated acid linoleic acid. The oil is considered equal to olive oil and almond oil for table use. Oil from sunflowers produced in the South has a lower linoleic acid content than oil from sunflowers produced in the North; hence, it is more stable. Because of the high ratio of polyunsaturated to saturated fatty acids, sunflower oil possesses serum cholesterol-lowering abilities.

SUPPLEMENTS

Food supplements are foodstuffs used to improve the value of basal foods.

(Also see NUTRITIONAL SUPPLEMENTS.)

SURFACE AREA

The area covered by the exterior of cubes, spheres and cylinders is fairly easy to calculate. However, the area covered by the exterior of the body is not quite as easy to figure due to the variety of body shapes, heights, and weights. Charts and mathematical formulas are available for determining body surface area in relationship to height and weight. It is the body surface area which determines heat loss from the body; therefore, basal metabolism is correlated to surface area.

(Also see BASAL METABOLIC RATE; BODY SURFACE; and METABOLIC RATE.)

SURFACTANT (SURFACE ACTIVE AGENT)

Surface tension is that inward force acting on the surface of a liquid. It is an effect of the forces of attraction existing between the molecules below the surface. Water has a high surface tension compared to alcohol, and mercury has the highest surface tension of any known liquid. The surface tension of a liquid is related to (1) the tendency of a liquid to spread or wet a solid surface, and (2) the ability of a liquid to blend with other liquids. Hence, surface tension plays a role in a variety of processes; for example, emulsification, bubble formation, fabric cleaning, foams, and adhesives. Surface active agents or surfactants reduce the surface tension of a liquid, thereby increasing the wetting or blending ability. There are three categories of surfactants: (1) detergents, (2) wetting agents, and (3) emulsifiers. A variety of surfactants are employed as food additives by the food industry.

SWEETBREADS (THYMUS GLAND)

The sweetbreads, which are considered to be the most delicate of the variety meats, are located in the neck area of young bovines (veal); as the animal matures, they disappear.

SWEET CORN Zea mays saccharata

Varieties of corn grown chiefly for human food and harvested at an immature stage.

(Also see CORN.)

SWEETENING AGENTS

Sweet is a taste sensation—a pleasurable sensation for which man relies upon for his food selection. To satisfy his desire to stimulate the sweet taste sensation, man discovered and learned to keep honey bees. Later, he cultivated sugarcane and sugar beets. But the search for sweet tasting substances continues; primarily, for low-calorie or no-calorie substances which satisfy the taste sensation and are safe to eat.

There are over 200 chemicals known to taste sweet, but many sweet-tasting chemicals present problems such as bitter aftertaste, lack of stability, and toxicity.

Many people desire, or need, to limit their intake of sugar—and not just table sugar (sucrose) alone. Diabetics and individuals cutting down on calories also need to control their intake of monosaccharides such as glucose and fructose.

Any new sweetener should (1) possess a flavor that is clean and without aftertaste, (2) be made for a price competitive with sugar (sucrose) on a cost-per-sweetness basis, (3) be adequately soluble and stable, and (4) be subjected to and pass the lenthy and costly program of rigorous safety testing required by the U.S. Food and Drug Administration. Very few meet all of these requirements.

Table S-18 gives a comparison of the common sweet-tasting substances, ranked in descending order of sweetness. Note that sucrose is used as the standard; hence, it is given a sweetness rating of 1.0 and all other products are compared to it. Note, too, that Table S-18 gives the following information relative to each substance: (1) the caloric value, where available, and (2) the identity.

TABLE S-18
COMPARISONS OF SOME SWEET-TASTING SUBSTANCES

Substance	Sweetness[1]		Substance Identification
	(Calories)	(kcal/g)	
Neohesperidin dihydrochalone	1,500	0	Nonnutritive; artificial
Saccharin	500	0	Artificial sweetener
Aspartame (Nutrasweet)	180	4	Two amino acids
Cyclamate	100	100	Artificial sweetener
Tryptophan	30	0	Amino acid
Fructose (levulose)	1.7	4	Monosaccharide
Invert sugar[2]	1.3	0	Monosaccharide
Sucrose	1.0	4	Disaccharide
Maltitol	0.9	0	Sugar alcohol
Xylitol	0.9	4	Sugar alcohol
Glycine	0.8	4	Amino acid
Glucose (dextrose)	0.7	4	Monosaccharide
Mannitol	0.7	2	Sugar alcohol
Sorbitol	0.6	0	Sugar alcohol
Inositol	0.5	0	Sugar alcohol
Maltose	0.4	0	Disaccharide
Galactose	0.3	0	Monosaccharide
Lactose	0.2	0	Disaccharide
Raffinose	0.2	0	Trisaccharide
Starch	0	0	Polysaccharide

[1]The sweetness of sucrose taken as the standard.

[2]Invert sugar is formed by the splitting of sucrose. Thus, it is a mixture of half glucose and half fructose. Honey has the same sweetness as invert sugar.

Additional pertinent information relative to each of the sweeteners listed in Table S-18 follows, with the products listed in the same order as in Table S-18.

• **Neohesperidin dihydrochalone**—This is a new sweetener derived from citrus peel. It is slow to elicit the taste sensation, but it has a lingering licoricelike aftertaste. Potential uses include chewing gum, mouthwash, and toothpaste.

• **Saccharin**—Saccharin was discovered in 1879. Initially, FDA recognized it as safe (GRAS). But in 1972, saccharin was removed from the GRAS list. Presently, saccharin-containing foods and beverages are available under a Congressional moratorium requiring warning labels on packages.

No food energy is derived from saccharin, so its calorie count is zero. It is valuable in managing diabetes and obesity.

There is no evidence to indicate that, at conventional levels of use, it causes cancer in humans, or any other disease.

(Also see SACCHARIN.)

● **Aspartame (Nutrasweet)**—Discovered in 1965, Aspartame was introduced in the United States in 1981 as a low-calorie sweetener. Aspartame consists of three substances: two amino acids (phenylalanine and aspartic acid), and methanol (methyl alcohol). Aspartame's greatest virtue is its taste, which is very similar to sugar, and there is no unpleasant aftertaste.

Because aspartame contains phenylalanine, it can be consumed only in carefully controlled amounts by people with certain medical problems such as phenylketonuria; hence, labeling is required indicating that phenylalanine is present in aspartame-sweetened foods and beverages. Aspartame is unstable if subjected to prolonged heating, so it is not suitable for use in cooking or baking. Also, it gradually breaks down in beverages, but it can be used successfully in soft drinks if distribution and storage conditions are well controlled.

● **Cyclamate**—This noncalorie substance was discovered in 1937 and used as a dietetic sweetener in a variety of foods and drinks in the early 1950s. In 1969, in response to scientific evidence implicating cyclamate as a possible cancer-causing agent in rats, FDA removed the sweetener from the GRAS list; and in 1970, it was taken off the U.S. market completely. Presently cyclamate is being reevaluated by FDA, based on evidence that it may not pose a cancer hazard.

● **Tryptophan**—A sweet-tasting essential amino acid.
(Also see AMINO ACID[S].)

● **Fructose (levulose)**—This is a carbohydrate—a monosaccharide. It is found naturally in fruits, and it makes up about 50% of the sugar in honey. Commercially, it is found in high-fructose syrups and invert sugars. It is used in beverages, baking, and canned goods—wherever invert sugar or honey may be used.
(Also see FRUCTOSE.)

● **Invert sugar**—This is a mixture of glucose (dextrose) and fructose (levulose). Invert sugar resists crystallization and has moisture-retention properties. It prolongs the freshness of baked goods and confectionery.
(Also see INVERT SUGAR; and SUGAR, section on "Consumption and Use of Sugar.")

● **Sucrose (sugar)**—This is the standard with which other sweet substances are compared; hence, in Table S-18 it is given a sweetness rating of 1.0. Sucrose is widely used in many beverages and processed foods, and in homes. It is one of the oldest, most popular, and most available sweetening agents.
(Also see SUGAR.)

● **Maltitol**—A sugar alcohol, used to add sweetness to *sugarless* chewing gum and some dietetic foods.

● **Xylitol**—This is a sugar alcohol or polyhydric alcohol (polyol). It occurs naturally in some fruits and vegetables. Commercially, it is produced from plant parts containing xylans—oat hulls, corncobs, and birch wood chips. Xylitol is used in *sugarless* chewing gums and dietetic foods.
(Also see XYLITOL.)

● **Glycine**—This is a sweet-tasting nonessential amino acid. It is used to modify the taste of some foods.
(Also see AMINO ACID[S].)

● **Glucose (dextrose)**—It is commercially produced from the conversion of starch (corn starch) by the action of heat and acids, or enzymes. Primarily, glucose is used in the confection, wine, and canning industries; and in intravenous solutions.
(Also see GLUCOSE; and SUGAR, section on "Consumption and Use of Sugar.")

● **Mannitol**—This is a sugar alcohol or polyhydric alcohol (polyol). Mannitol occurs naturally in pineapples, asparagus, carrots, and olives. Commercially, it is prepared by the hydrogenation of mannose or glucose. Mannitol is used in candies, chewing gums, confections, and baked goods. Also, it is used in dietetic foods and in medicine—as a diuretic and to test kidney function.

● **Sorbitol**—This is a sugar alcohol or polyhydric alcohol (polyol). Sorbitol occurs naturally in many fruits. Commercially, it is prepared by the hydrogenation of glucose. It is used in chewing gum, dairy products, meat products, icings, toppings, and beverages.

● **Inositol**—Inositol is widely distributed in many foods and is closely related to glucose.

● **Maltose**—Malt sugar is formed from starch by the action of yeast. It is much less sweet than sugar. Preparations containing maltose, often in mixtures with dextrose, are used in bread-baking and in infant foods.
(Also see MALTOSE.)

● **Galactose**—This is a hexose sugar (a monosaccharide) obtained along with glucose from lactose, found in milk.
(Also see GALACTOSE.)

● **Lactose**—This disaccharide is composed of one molecule of galactose and one molecule of glucose. It is the only sugar not found in plants. It is found in milk; hence it is called milk sugar.
(Also see LACTOSE.)

● **Raffinose**—This is a trisaccharide containing the three hexoses—glucose, fructose, and galactose. Raffinose is found in molasses, cottonseed meal, and Australian manna.
(Also see RAFFINOSE.)

● **Starch**—Starch (as such) is tasteless (not sweet). But starch from any starch source—cereal grains or potatoes—may be broken down into its component sugar, either enzymatically or with dilute acids. The most familiar example in the United States is corn syrup (Karo). Complete hydrolysis of starch yields glucose (dextrose), which is sweet.
(Also see STARCH.)

(Also see SUGAR.)

SWEET POTATO *Ipomoea batatas*

This tropical plant, which forms nutritious tubers, is a member of the morningglory family (*Convolvulaceae*).

ORIGIN AND HISTORY. The sweet potato is native to the American tropics. It was once an important component of the Aztec diet. Remains of sweet potatoes estimated to be between 10,000 to 20,000 years old were found in a cave in Peru. However, it is not certain whether the

Fig. S-47. Loading sweet potatoes in the field. (Courtesy, Louisiana Cooperative Extension Services, Louisiana State University, Baton Rouge, La.)

Fig. S-48. Washing sweet potatoes. (Courtesy, Louisiana Cooperative Extension Services, Louisiana State University, Baton Rough, La.)

ancient tubers had grown wild or been cultivated. Europeans introduced the plant to their continent in the 16th century, and it later spread to Asia.

WORLD AND U.S. PRODUCTION.

Sweet potatoes rank second among the leading vegetables of the world, and tenth among the vegetables of the United States.

Worldwide, about 132 million metric tons of sweet potatoes are produced annually; and the leading sweet potato-producing nations in descending order are: China, Indonesia, Vietnam, Uganda, Japan, India, Rwanda, Brazil, Philippines, and Kenya.[48]

The United States produces about 651,000 short tons of sweet potatoes annually; and the leading sweet potato-producing states in descending order are: North Carolina, Louisiana, California, Georgia, Alabama, Mississippi, South Carolina, Texas, New Jersey, and Virginia.[49]

PROCESSING.

Tubers which are to be stored keep better if they are cured after harvesting. Curing consists of holding the sweet potatoes in a facility where the temperature is about 85°F (30°C) and the relative humidity is about 85%. These conditions promote the healing of small wounds in the tubers.

Various means of processing have been developed, the two most common of which follow:

• **Canning**—This process is used mainly for small whole tubers or cut pieces of larger tubers, which are first cooked, then packed in syrup. Only small amounts of canned sweet potato puree are produced, most of which is packed in glass jars by baby food manufacturers.

• **Dehydration**—The sun-drying of raw sweet potato chips or slices has long been one of the major means of preservation utilized in the developing countries of the tropics. However, dehydration in the United States usually involves the drum drying of a puree made from the cooked tubers. The retail market for this type of product is limited because most people use either fresh or canned sweet potatoes. Nevertheless, the dried powder is used by the armed forces, in school lunch programs, and by bakers who produce sweet potato cakes and pies.

SELECTION, PREPARATION, AND USES.

There are two general types of sweet potatoes. One type has soft, moist flesh when cooked and a high sugar content (sweet potatoes of this type are commonly, but incorrectly, called yams). The second type, when cooked, has a firm, dry, somewhat mealy flesh, which is usually light yellow or pale orange in color, as contrasted to the usually deeper yellow or distinctly orange-red colored flesh of the moist type. The skin of the dry type is usually light yellowish-tan or fawn colored, while the skin of the moist-fleshed varieties may vary in color from whitish-tan to brownish-red. Varieties of each type vary considerably in shape, but most moist-fleshed varieties are usually more plump in shape than most dry-fleshed varieties. A mixture of types or a mixture of varieties within types is undesirable because of lack of uniformity in cooking, as well as differences in flavor and color of flesh.

Best quality sweet potatoes are clean, smooth, well-shaped, firm, and bright in appearance.

Many people like sweet potatoes prepared with a sweetener such as honey, molasses, sugar, or a syrup. However, the yam type of tuber provides much of its own sweetening when it is cooked slowly, because it contains an enzyme (beta-amylase) that converts starch to maltose and dextrins. Heating the tuber activates the enzyme, which is eventually inactivated when the internal temperature becomes too high. Prior to cooking, the tubers should be scrubbed well with a vegetable brush so that they may be cooked with the skins intact. Removal of the skin before cooking may result in darkening of the flesh. Also, the skins loosen and are easier to remove after cooking.

---

[48]Data from *FAO Production Yearbook 1990*, FAO/UN, Rome, Italy, Vol. 44, p. 92, Table 27. **Note well:** Annual production fluctuates as a result of weather and profitability of the crop.

[49]Data from *Agricultural Statistics 1991*, USDA, p. 164, Table 237. **Note well:** Annual production fluctuates as a result of weather and profitability of the crop.

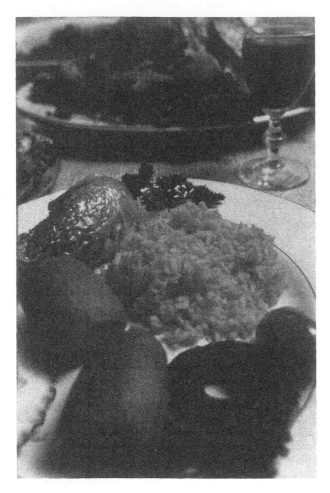

Fig. S-49. Sweet potatoes enhance the meal. (Courtesy, USDA)

Sweet potatoes may be cooked in much the same ways as Irish potatoes, but they require only about half as much cooking time. In addition, they are good in casseroles when accompanied by apple slices or pineapple rings and seasoned with spices such as allspice and cinnamon. Finally, mashed cooked sweet potato may be used to make biscuits, breads, cakes, cookies, custards, muffins, and pies.

NUTRITIONAL VALUE. The nutrient compositions of various forms of sweet potatoes are given in Food Composition Table F-21.

Some noteworthy observations regarding the nutrient composition of sweet potatoes follow:

1. Compared to cooked Irish potatoes, cooked sweet potatoes contain about an equal number of calories, less protein and vitamin C, and much more vitamin A—they are one of the leading plant sources of the latter vitamin.

2. A 1-cup serving (about 9 oz or 225 g) of mashed cooked sweet potatoes provides 275 Calories (kcal) and 5 g of protein, which is almost double the calories and about equal to the protein furnished by a cup of a cooked cereal such as corn or rice.

3. Candied sweet potatoes supply almost 50% more calories than the unsweetened cooked tubers.

4. The quantity and quality of protein is barely adequate to meet the needs of adults, and is inadequate for the needs of children. Fortunately, many tropical peoples supplement their diets with small amounts of protein-rich foods such as fish, legumes, and wild game.

SWEETSOP (SUGAR APPLE) Annona squamosa

The fruit of a small tree (of the family *Annonaceae*) that is native to the American tropics. The yellowish-green heart-shaped fruit are 3 to 4 in. (8 to 10 cm) in diameter, white, and have a sweet granular custardlike pulp. They are used as a dessert fruit, and are closely related to the other *Annona* species of fruit (the most common ones are the cherimoya, custard apple, and soursop). Their flavor is best when fully ripe, a stage at which the fruit is very perishable.

The nutrient composition of sweetsop (sugar apple) is given in Food Composition Table F-21.

Some noteworthy observations regarding the nutrient composition of sweetsop follow:

1. The raw fruit is moderately high in calories (94 kcal per 100 g) and carbohydrates (24%).

2. Sugar apples are a good source of fiber, potassium, and vitamin C, and a fair source of iron.

SYMPTOM

A complaint by patients which may lead them to seek medical advice.

SYN–

Prefix meaning with, or together.

SYNDROME

A combination of symptoms occurring together.

SYNERGISM

The joint action of agents in which their combined effect is greater than the sum of their separate actions; each agent enhances the action of the other.

SYNTHESIS

The process of building up a chemical compound by a reaction or a series of reactions.

SYNTHETIC

An artificially produced product that may be similar to the natural product.

SYRUPS

Although there is a large variety of syrups, they all contain sugar and water. There are those syrups which come from sources used to manufacture sugar and glucose, such as corn syrup, maple syrup, molasses, and sorghum syrup. Specialty syrups can be created from sugar, water, and a variety of flavors such as almond, cherry, chocolate, currant, orange, strawberry, and other fruit flavors. Plain sugar and water syrups are used for canning fruits. Also, syrup serves as a carrier of medicines as in cough syrup.

(Also see CARBOHYDRATE[S], section headed "Sources"; MAPLE SYRUP; MOLASSES; SORGHUM; and SUGAR.)

SYSTEMIC

Pertaining to the body as a whole, as distinguished from local.

TACHYCARDIA

An abnormally rapid heart beat, faster than 100 beats per minute. It occurs during exercise, in fevers, in certain heart conditions, and in a number of diseases.

TALLOW

The fat extracted from adipose tissue of cattle and sheep. (Also see FATS AND OTHER LIPIDS.)

TANGERINE (MANDARIN ORANGE)
Citrus reticulata

The names *tangerine* and *mandarin* are now used interchangeably in the United States. The tangerine, or mandarin, refers to citrus fruit which resembles the sweet orange (*C. sinesis*), but which are orange-red and have loose skins that peel easily. Tangerines, like the other citrus fruits, belong to the rue family (*Rutaceae*).

Fig.T-1. The tangerine, a species of citrus fruit that has a loose, easily peeled skin and a distinctive sweet flavor.

Of the many tangerine hybrids, the most widely produced in the United States are the tangelo, the tangor, and the temple orange. Details follow.

• **Tangelo**—This term designates the hybrids of the grapefruit and the tangerine. Usually, the fruits have the loose skin and sweetness of the latter parent, but they tend to have a distinctive flavor of their own.

• **Tangor**—Most lay people are unfamiliar with this name, which refers to hybrids of the tangerine or the sweet orange.

• **Temple orange**—This naturally occurring tangor (a hybrid of tangerine and grapefruit) was discovered in Jamaica. It is by far the most important hybrid of the mandarin.

ORIGIN AND HISTORY. Tangerines are native to China, where they were grown for 3,000 years. They did not reach Europe and the Americas until the 1800s. Today, tangerines and their hybrids are grown in most of subtropical and tropical areas of the world. As they spread throughout the world, many new varieties developed as a result of (1) mutations, and (2) accidental or intentional crossbreeding (hybridization) with the local varieties.

WORLD AND U.S. PRODUCTION. Worldwide, about 8.8 million metric tons of tangerines (and their hybrids) are produced annually; and the leading producing countries, by rank, are: Japan, Spain, Brazil, Korea, Italy, and Pakistan. Japan alone accounts for approximately 23% of the world tangerine production.[1]

The United States produces about 359,000 short tons of tangerines (and their hybrids) annually. Most of the U.S. crop (79% of it) is produced in Florida; California and Arizona rank second and third, respectively.[2]

PROCESSING. A little over half of the U.S. crop of tangerines is processed into single-strength juices and frozen concentrates. Much of the fruit juice is blended into orange juice and frozen orange concentrates because the tangerines contribute a deep orange color. However, U.S. regulations limit the amount of tangerine juice in orange juice blends to 10%. Greater amounts impart off-flavors to the orange juice products.

Some of the tangerine crop is made into single-strength tangerine juice and frozen tangerine juice concentrate. The latter product is by far the major one, because it is much more stable during storage.

Small amounts of tangerine segments are used in prepared citrus salads and gelatin desserts. It is noteworthy that almost all of the canned tangerine segments used in the United States are imported from the Orient.

SELECTION, PREPARATION, AND USES. Fresh tangerines are thin-skinned, and are usually oblate or decidedly flattened at the ends. The skin is easily removed; there is little coarse fibrous substance between the skin and the flesh, and the segments of the fruit separate readily. The flavor is distinctive; the aroma is pungent and pleasant. Because of the looseness of the skin these oranges are likely to feel puffy; therefore, judgment as to quality should be based mainly on weight for size and deep yellow or orange color of the skin.

Tangerines are at their best when served fresh in appetizers, desserts, or toppings for frozen desserts and puddings.

---

[1]Data from *FAO Production Yearbook 1990*, FAO/UN, Rome, Italy, Vol. 44, p. 163, Table 71. **Note well:** Annual production fluctuates as a result of weather and profitability of the crop.

[2]Data from *Agricultural Statistics 1991*, USDA, p. 189, Table 278. **Note well:** Annual production fluctuates as a result of weather and profitability of the crop.

However, their juice and/or their peel may be used to flavor ice cream, ices, sherbets, and sauces.

NUTRITIONAL VALUE. The nutrient compositions of various types of tangerine products are given in Food Composition Table F-21.

Some noteworthy observations regarding the nutrient composition of tangerines follow:

1. Fresh tangerines, fresh tangerine juice, and canned tangerine segments have similar nutritional values in that they are (a) moderately low in calories; (b) good sources of potassium and vitamin C; and (c) fair sources of vitamin A. Also, they all contain the vitaminlike bioflavonoids.

2. Frozen tangerine juice concentrate contains about 4 times the nutrient levels of the fresh, unconcentrated juice. Hence, the concentrate may make a significant nutritional contribution when it is added undiluted to drinks, sauces, and other preparations.

3. Canned tangerine juice has a nutritional value comparable to fresh juice, except that the canned product contains only two-thirds as much vitamin C.

4. Raw tangelos contain only half as much vitamin C as tangerines, but they supply about the same amount of potassium.

TANNIA (YAUTIA; MALANGA; COCOYAM)
Xanthosoma sagittifolium

This plant of the aroid family (*Araceae*) yields edible leaves and tubers. However, it is sometimes confused with taro (another aroid plant) because of certain similar characteristics.

Fig. T-2. Tannia (malanga) tubers in a Cuban market. (Photo by Audrey Ensminger)

The various names given to the plant are also responsible for cases of mistaken identity. Therefore, some definitions of the more common names follow:

• **Tannia and Yautia**—These common names designate all varieties of the species *X. sagittifolium*.

• **Malanga**—This Latin American name is often applied to tannia, but it may also refer to similar tuberous plants of different species.

• **Cocoyam**—Some people give this general name to various tuberous plants grown alongside cocoa (*Theobroma cacao*) to provide shade for the young plants. Africans distinguish between tannia and taro by calling the former the *new* cocoyam, and the latter the *old* cocoyam.

Only the name tannia will be used in the remainder of this article.

PREPARATION. In West Africa, tannia is used to prepare *fufu*, a pasty dough made by mashing boiled peeled tubers. People in other tropical areas of the world use the plant in much the same ways as they have long used cassava, taro, and yams. For example, the tubers may be baked, boiled, fried, or roasted; and the cooked young leaves may be used as a green vegetable.

CAUTION: Neither tannia roots nor tannia leaves should be consumed without cooking them thoroughly because some varieties contain high levels of calcium oxalate crystals, which are very irritating to the digestive system. The latter possibility may be alleviated by boiling the roots or leaves for about 15 minutes in water containing a small amount of baking soda (sodium bicarbonate), discarding the used water, rinsing the vegetable thoroughly, and boiling again in clear water.

NUTRITIONAL VALUE. The nutrient compositions of various forms of tannia are given in Food Composition Table F-21. Note that—

1. Compared to Irish potatoes, tannia tubers furnish about twice as many calories, the same amount of protein, and less than half as much vitamin C. The protein to calorie ratio (1.6 g per 100 kcal) of the tubers is barely high enough to meet the protein requirements of adults. So, children fed a tannia-rich diet are likely to develop protein-energy malnutrition unless they are given supplemental protein foods in their diets. Finally, tannia tubers are deficient in many vitamins and minerals.

2. Tannia leaves are nutritionally superior to the tubers with respect to protein, calcium, phosphorus, iron, vitamin A, and vitamin C. They are an excellent source of vitamin A. Hence, they compensate for most of the nutritional deficiencies of the tuber. However, the leaves may contain much higher levels of highly irritating calcium oxalate crystals than the tubers. Therefore, the leaves should not be consumed raw by people or livestock, unless it is certain that they contain little or no oxalates. The leaves may be rendered safe by boiling.

TANNINS

This refers to a complex group of chemicals found naturally in many plants. They are responsible for the astringent taste of coffee and tea. Tannic acid, a commercial tannin, is used widely in industry.

TAPIOCA

This is a starch derived from the roots of the tropical cassava plant. It is extracted by washing, peeling, and grinding the roots to fine pulp which is passed over a series of screens to remove root fibers. In some processes it may be further refined in settling basins or centrifuges. The moist starchy mass is dried; and, during the drying process, it forms small, uneven milky white balls known as pearl tapioca. High grade tapioca has a brilliant white color.

Tapioca is used like starch from other sources. It swells and thickens the liquid in which it is cooked. It is used in its natural state, as well as in modified and pregelatinized versions like other starches. The most familiar use of tapioca is the popular pudding which contains small lumps or granules. Special cooking, grinding, and screening methods produce this type of tapioca. Pearl tapioca requires prior soaking and yields larger lumps in puddings.

(Also see CASSAVA; and FLOURS, Table F-13, Special Flours)

TAPIOCA-MACARONI

A product developed in India. It may be a mixture (1) of 80 to 90 parts of tapioca flour and 10 to 20 parts peanut flour; or (2) of 60 parts of tapioca, 15 parts peanut, and 25 parts semolina. It is baked into shapes resembling rice grains or macaroni.

TARTAR

• A mineral deposit on the teeth, mainly calcium phosphate.

• A substance consisting basically of cream of tartar found in the juice of grapes, and often deposited in wine casks during the production of wine.

TARTAR EMETIC

This is another name for the chemical antimony potassium tartrate. Under hospital supervision, its slow intravenous administration is a dangerous but effective treatment used to eliminate certain parasites from the body.

TARTARIC ACID (HOOC[CHOH]$_2$–COOH)

A strong dicarboxylic acid. It is widely found in plants—especially fruits—both free and combined with salts.

TARTAZINE

This is an alternate name for the yellow color, FD & C Yellow no. 5, which is approved for use in food, drugs, and cosmetics. It may also be employed to dye wool and silk.

(Also see COLORING OF FOODS, Table C-19, Synthetic Color Additives Approved for Food Use.)

TASTE

This is a chemical sense detected by receptors—the taste buds—in the mouth, primarily on the tongue. Upon entering the mouth, certain dissolved chemicals create nervous impulses in the taste buds. These impulses travel to the brain where they are interpreted as (1) salty, (2) sweet, (3) bitter, or (4) sour. Humans have about 12,000 taste buds, and each taste bud possesses a greater degree of sensitivity to 1 or 2 of the taste sensations. In general, the tip of the tongue is the most sensitive to sweet, the sides to sour, the back to bitter, while salt sensitivity is distributed over most of the tongue.

Taste plays a major role in the acceptance or rejection of foods. Most of the time, salty foods are considered pleasant, but very salty foods will be rejected. Sweet is, of course, considered pleasant, and is associated with the sugar content of a food. However, much effort has been directed toward finding substances such as saccharin, which will *fool* the sweet taste buds, without contributing to calories as do the sugars. Foods with a bitter taste are generally rejected. This may have been a protective mechanism for early man, since many poisonous wild plants taste bitter due to alkaloid compounds. The sour taste may be considered as pleasing or objectionable. For example, foods containing dilute vinegar solutions are very palatable, while extremely sour foods are rejected. Sourness is a measure of the degree of acidity—

hydrogen ion concentration—of foods. Although, taste is important in food acceptance, most of the pleasurable experience associated with food is known as flavor. The flavor of foods results from the interaction of taste and smell—another chemical sense. Humans can detect as many as 10,000 different smells. Much of what is called taste is actually smell. This is emphasized during a cold when foods seem to lose their *taste*.

(Also see SUGAR; and SWEETENING AGENTS.)

TASTE BUDS

The sense organs for taste located in the epiglottis, palate, pharynx, and tongue. Taste buds are excited by chemical substances in the food we eat. We can detect four types of taste: (1) salty, (2) sweet, (3) bitter, and (4) sour. Taste sensations from these taste buds help regulate the diet. Much of what we usually call taste is actually smell.

(Also see TASTE.)

TAURINE (C$_2$H$_7$NO$_3$S)

This chemical is present in bile, combined with cholic acid.

TEA

The word *tea* refers to (1) a common drink made by pouring boiling water over the dried and prepared leaves of an oriental evergreen tree, (2) the dried and prepared leaves from which the drink is made, or (3) the tree upon which these leaves grow. Tea may also mean (1) a brew made from kinds of leaves other than tea, (2) an afternoon or early evening meal in Britain, or (3) a large reception at which tea is served. Herein, the word tea will refer to the dried leaves of the plant and the drink made from them. Tea is the most popular beverage in the world.

Fig. T-3. Tea for two, or three, or four—the pause that refreshes. (Painting by an unknown master, English. Courtesy, The Bettmann Archive, New York, N.Y.)

ORIGIN AND HISTORY. It is believed that the tea plant originated in a region encompassing Tibet, western China, and northern India. The exact origin of tea culture and brewing tea leaves into a drink is obscure. But, according to an ancient Chinese legend, tea was discovered by the Chinese emperor, Shen-Nung, in 2737 B.C. Leaves from a wild tea bush accidentally fell into the water that he was boiling; and he discovered that the drink was good. However, the earliest mention of tea in Chinese literature appeared about 350 A.D. Then, in 780 A.D. Lu Yu's *The Classic of Tea*, published in China, described the cultivation, processing, and use of tea. By the 9th century A.D., tea was growing in Japan after being introduced by Chinese Buddhist monks. Eventually, tea culture spread to Java, the Dutch East Indies, and other tropical and subtropical areas. Traders from Europe sailing to and from the Far East in the 16th century began to introduce Europeans to the unusual oriental beverage called tea, a word derived from the Chinese local Amoy dialect word *t'e,* pronounced *tay.* By the 18th century, tea had become the national beverage of England. Tea also was popular in the American colonies, but in 1767 the British government placed a tax on the tea used by the colonists. The rest of the story Americans know as the Boston Tea Party of 1773—a contributing factor to the Revolutionary War.

Fig. T-4. Teapot from Tibet. (Courtesy, Field Museum of Natural History, Chicago, Ill.)

WORLD AND U.S. PRODUCTION.

No longer is it "all the tea in China," since India ranks as the number one producer of tea. Sri Lanka (previously Ceylon) is, however, the major supplier of tea to the United States. Worldwide, about 2.5 million metric tons of tea are produced annually; and the leading tea-producing countries, ranked in descending order, are: India, China, Sri Lanka, Kenya, Indonesia, Turkey, U.S.S.R., Japan, Iran, and Bangladesh.[3]

Propagation and Growing. The tea plant is an evergreen tree with the botanical name of *Camellia sinensis.* In the wild, the tree may grow to a height of 30 ft (*10 m*), but on tea plantations, or so-called gardens or estates, the tea plants are constantly pruned to keep their height at about 3 ft (*1 m*) to encourage new growth and to keep at a convenient picking height.

Tea plants grow best in warm climates where the rainfall averages 90 to 200 in. (*220 to 500 cm*) yearly, and the plants

Fig. T-5. A pruned tea plant with abundant foliage, and a twig showing the camellialike flowers. It is the two leaves at the end of the twig which are plucked for tea.

grow at altitudes from sea level to 7,000 ft (*2,200 m*). The best teas are produced at the higher altitudes where plants grow more slowly in the cooler air and their leaves yield a better flavor.

Tea plants may be started from cuttings or seeds. The tea seedlings or small plants are placed in the soil about 3 to 5 ft (*1.0 to 1.5 m*) apart, and then, depending upon the altitude, 2½ to 5 years are required before the tea plants produce leaves for commercial production. Young plants are often shaded to protect them from the heat of the sun. Once in production a plant may produce for almost a century.

Harvesting. Tea plants produce abundant foliage, a camellialike flower, and berries containing one to three seeds. However, it is the smallest and youngest leaves that are plucked for tea. Two leaves and a bud at the top of each young shoot are picked. A growth of new shoots is termed a flush. At lower altitudes a new flush may grow every week, while at higher altitudes more time is required. Flushes are picked off the tea plant by hand by workers called *tea pluckers.* Good pluckers harvest about 40 lb (*18 kg*) per day, or enough to make 10 lb (*4.5 kg*) of manufactured tea.

Fig. T-6. Picking tea in Hangchow, China. (Photo by Audrey Ensminger)

---

[3]Data from *FAO Production Yearbook 1990,* FAO/UN, Rome, Italy, Vol. 44, p. 177, Table 80. **Note well:** Annual production fluctuates as a result of weather and profitability of the crop.

PROCESSING. Although the kind and quality of leaves vary somewhat according to the local conditions and particular strains, the real difference in color, aroma, and flavor is caused by the way in which the leaves are processed. There are three different types of tea: black, green, and oolong. Most green and oolong tea comes from China, Japan, and Taiwan, while all tea-producing countries manufacture black tea.

• **Black tea**—First the leaves are transported from the plantation to the factory as rapidly as possible. The leaves are spread on withering racks and air is blown over the leaves to remove excess moisture. This removes about one-third of the moisture, and the leaves become soft and pliable. After this they are rolled to break the cells and release the juices, which are essential for the fermenting process. Then the leaves are spread out and kept under high humidity to promote fermentation, which develops the rich flavor of black tea, and the leaves become a coppery color. Finally, the leaves are hot-air dried (fired) until the moisture is removed, leaving the leaves brownish black.

• **Green tea**—In making green tea, the plucked leaves are steamed as quickly as possible. Steaming prevents the leaves from changing color and prevents fermentation. After steaming, the leaves are rolled and dried in the same way as black tea until they are crisp.

• **Oolong tea**—Oolong tea is somewhat like both green tea and black tea. It is semifermented. Fermentation is stopped before the process is complete. This produces a greenish-brown leaf with a flavor richer than that of green tea but more delicate than that of black tea.

GRADES OF TEA AND BLENDING. Tea is graded according to the size of the leaf, which has nothing to do with the quality of the tea. Large and small, broken and unbroken leaves are sorted by passing them across screens with different size holes. The largest leaves, classified in order of size, are orange pekoe, pekoe, and pekoe souchong. These large leaves are generally packaged as loose tea. The smaller or broken leaves are classified as broken orange pekoe, broken orange pekoe fannings, and fannings. These smaller grades of tea are generally used in tea bags.

Since tea grown in different countries, or even different parts of the same country, varies in taste, flavor, and quality, tea companies employ a tea taster. The tea taster selects the tea to purchase so that a company can sell tea with the flavor for which it is known. To achieve a particular flavor, teas are often blended. Nevertheless, some unblended varieties are popular.

BREWING TEA. A cup of tea is brewed by pouring fresh, boiling water over 1 tsp (5 ml) of loose tea or one tea bag. The best flavor is obtained by allowing the tea to steep (soak) for 3 to 5 minutes before being served. If tea is steeped too long, it will have a bitter taste from the excess tannins that are released. Also, the longer steeping times yield tea with a higher caffeine content.

Iced tea is made with 50% more tea leaves, to allow for the melting ice. The tea is poured into tall glasses that are two-thirds full of ice. Sliced lemons are served with it, and each person sweetens the tea to his taste. Flavorings of cloves, grated orange peel, or mint are sometimes added.

• **Instant tea**—Instant tea is an innovation developed in the United States. A strong brew of tea is made in large vats, and then the water is removed by a drying process. Consumers can make hot or cold tea by merely adding water to the powder which remains after the drying process.

NUTRITIONAL VALUE. Without sugar, cream, or other additions, tea provides little nutrition other than a few calories, and trace amounts of some minerals as indicated in Food Composition Table F-21. Green tea is, however, an excellent source of vitamin K.

Caffeine in tea accounts for its mildly stimulating properties, while the tannins contribute pungency, slight astringency, and color to tea. An essential oil contributes the flavor. The amount of caffeine in a cup of tea depends upon the steeping time.

(Also see CAFFEINE.)

TEASEED OIL

This is a vegetable oil extracted from *Camellia sinensis*, the tea-oil plant, and cultivated in China. Teaseed oil is used for salad oil and frying. It is claimed to be similar to olive oil.

TEFF *Eragrostis abyssinica*

This is an important cereal food of Ethiopia and the eastern African highlands. It is a leafy, quick-maturing annual plant that grows up to 4 ft in height. Teff is adapted to dry areas with a short rainy season.

The teff of Africa is grown by primitive methods. Fields are cultivated by crude plows drawn by oxen, followed by hand sowing and another tilling to cover the seed. When ripe, the grain is harvested by small hand sickles and threshed by treading with animals.

Women grind the white or black grain (grain color depends on variety) on flat stones to make a flour, which is baked into large, flat pancakes called injara.

Nutritional data on teff grains and meal are listed in the Food Composition Table F-21.

Teff is used as an annual grass hay as well as a food grain. Following threshing, the straw may be mixed with mud for constructing huts.

Table T-1 tells about *faffa*, a special fortified food based on teff.

TABLE T-1
A FORTIFIED FOOD BASED ON TEFF

Product	Description	Uses	Comments
Faffa	A finely ground mixture of teff, chickpeas, skim milk powder, sugar, salt, and vitamins. Contains from 14% to 15% protein.	May be mixed with water to make a porridge for infants. Leavened or unleavened breads. Mixed with the false banana (*Ensete ventricosum*).	The Ethiopian and Swedish governments established the Children's Nutrition Unit in Addis Ababa to conduct the development, production, and test marketing of Faffa. Sufficient vitamins are added to the product to ensure that 3½ oz (*100 g*) supplies the daily allowances for all of the essential vitamins. A large scale feeding trial with children aged 6 months to 10 years produced dramatic decreases in the incidence and severity of goiter, rickets, and skin infections; and a noticeable increase in muscle mass. However, the cereal did not cure anemia as rapidly as it did the other deficiency diseases.

TEMPLATE

A mold, pattern, or guide.

TENDERLOIN

The most tender muscle of the carcass (beef, lamb, and pork), located inside the loin and running nearly the entire length of the loin.

TERRAMYCIN AS PRESERVATIVE

Terramycin is a brand name for the broad-spectrum antibiotic oxytetracycline. Although low levels of antibiotics such as Terramycin effectively extend the shelf life of food, the direct addition of antibiotics to foods is not permitted in the United States. After some initial studies and limited usage of antibiotics as food preservatives, the FDA withdrew their approval because of concern that repeated exposure of the consumer to low concentrations of antibiotics could lead to the development of antibiotic-resistant strains of bacteria. Furthermore, there was wariness lest antibiotics provide a substitute for good sanitary practices.

(Also see BACTERIA IN FOOD, section headed "Bacterial Control Methods"; and ANTIBIOTICS, section headed "Antibiotics in Foods.")

TETANY

Spontaneous muscular spasms in the wrists and ankles, resulting when the parathyroid secretion is deficient, causing the amount of calcium in the blood to drop and the phosphorus to increase.

TEXTURE

Food is accepted or rejected on the basis of its chemical and physical properties as perceived by the senses—sight, sound, touch, smell, and taste. Therefore, the texture of a food is important. Texture is a physical quality of food perceived by the sense of touch in the mouth—mouth feel. It is described by such adjectives as chewy, tacky, stringy, hard, soft, rough, smooth, gritty, firm, crisp, coarse, etc. Through experience, people learn the customary texture of foods in their diet, and when the texture is different than expected, the food is judged to be of lower quality. Often, unacceptable foods feel slimy, stringy, and greasy, or they contain unchewable parts, grit, or foreign particles.

THEINE

This is an alternative name for caffeine.

THERAPEUTIC

Pertaining to the medical treatment of disease.

THERAPEUTIC DIETS

Diets designed to treat a disease or metabolic disorder. They are diets adapted or modified to meet special needs of specific diseases. Some common examples of diseases requiring therapeutic or modified diets are diabetes mellitus, phenylketonuria, kidney diseases, and galactosemia. Depending upon the disorder, these diets may be modified in energy content, in consistency or bulk, or in the kinds and amounts of nutrients such as vitamins, minerals, fats, proteins, or fluids. Despite the modification, the therapeutic diet should remain nutritionally adequate.

(Also see MODIFIED DIETS.)

THERAPY

The medical treatment of disease.

THERMAL

Refers to heat.

THERMODURIC

A term used to describe microorganisms (bacteria) that are heat tolerant. Thermodurics survive for short periods of time at high temperatures, but they do not grow at high temperatures. For example, thermodurics survive pasteurization of milk for short periods of time.

THERMOGENESIS

The production of heat, especially within the body.

THERMOPHILES

Bacteria that are *heat lovers*. These bacteria grow at temperatures of 131°F (55°C) to as high as 167°F (75°C). They are somewhat of a curiosity, since many of the proteins of other organisms would coagulate at these high temperatures. Thermophiles are commonly found in hot springs and compost heaps, and do not pose any real health problems. However, thermophiles are responsible for the *flat sour* which occurs in some home-canned food.

(Also see FLAT SOURS.)

THI– (THIO–)

Prefix meaning sulfur-containing.

THIAMIN (VITAMIN B-1)

Contents	Page
History	1009
Chemistry, Metabolism, Properties	1009
Measurement/Assay	1010
Functions	1010
Deficiency Symptoms	1011
Clinical Effects of Thiamin Deficiency	1011
Recommended Daily Allowance of Thiamin	1011
Toxicity	1012
Thiamin Losses During Processing, Cooking, and Storage	1012
Antithiamin Factors in Food	1012
Sources of Thiamin	1012

Thiamin (vitamin B-1)—the antiberiberi, or antineuritic, or antipolyneuritis vitamin—was the first of the B complex vitamins to be obtained in pure form; hence, the name B-1, a name proposed by the British in 1927. Various other names were used for short periods along the way, including antineuritic factor, antiberiberi factor, water-soluble B, aneurin, and simply vitamin B.

The mystery of beriberi, an ancient disease among rice-eating peoples in the East, was eventually unraveled as a deficiency disease caused by a lack of thiamin.

Thiamin is required by all species of animals. They must have a dietary source, unless it is synthesized for them by microorganisms in the digestive tract, as in the case of ruminants.

A patient with wet beriberi—lying in bed breathless, waterlogged, and apparently dying—may recover in 1 or 2 hours after being given an injection of thiamin; this is perhaps the most dramatic cure in medicine.

THIAMIN (vitamin B-1) HELPS THE BODY CELLS CONVERT

CARBOHYDRATES (STARCHES AND SUGARS) OF FOOD TO ENERGY, AND KEEPS THE NERVOUS SYSTEM, APPETITE, MUSCLE TONE AND ATTITUDE BUOYANT.

ONE OF THE MOST DRAMATIC CURES IN MEDICINE! SAME RAT BEFORE (LEFT) AND AFTER (RIGHT). LEFT: THIS RAT, 24 WEEKS OLD, HAD PRACTICALLY NO THIAMIN. IT LOST THE ABILITY TO COORDINATE ITS MUSCLES. RIGHT: THE SAME RAT 24 HOURS LATER, AFTER RECEIVING FOOD RICH IN THIAMIN—FULLY RECOVERED. (COURTESY, USDA)

TOP FOOD SOURCES

LEAN PORK	CORN FLAKES (ENRICHED)	WHEAT BRAN	KIDNEYS	WHITE FLOUR (ENRICHED)
NUTS	WHOLE WHEAT FLOUR	CORNMEAL, SELF-RISING (ENRICHED)	RICE (ENRICHED)	WHITE BREAD (ENRICHED)

Fig. T-7. The thiamin (vitamin B-1) story.

HISTORY. The history of thiamin begins with a study of the age-old disease beriberi.

Beriberi, which affects the nervous system, was known to the Chinese as early as 2600 B.C. The word *beriberi* means *I cannot,* referring to the fact that persons with the disease cannot move easily. But the cause of beriberi in man, and of polyneuritis—the counterpart in poultry—remained elusive for centuries.

Recognition of the cause of beriberi, and its possible cure by better diet, is a landmark in the history of nutrition, a chronological record of which follows:

In 1873, Van Lent was the first to conclude that the type of diet had something to do with the origin of beriberi, based on reducing the ration of rice in the diet of sailors in the Dutch navy.

In 1882, Kanehiro Takaki, a Japanese medical officer, cured beriberi in sailors of the Japanese navy by giving them less rice and more meat and milk.

In 1897, Christiaan Eijkman, a Dutch physician, working in a military hospital in the East Indies, produced polyneuritis, a condition resembling beriberi, in chickens, pigeons, and ducks by feeding polished rice; which he wrongfully attributed to too much starch.

In 1901, G. Grijns, another Dutch physician, who continued the work of Eijkman at the same hospital, concluded that beriberi in birds and man resulted from lack in the diet of an essential nutrient.

In 1912, Casimir Funk, working at the Lister Institute in London, coined the term *vitamine* and applied it to the antiberiberi substance.

In 1916, Elmer V. McCollum of the University of Wisconsin, designated the concentrate that cured beriberi as *water-soluble B.*

In 1926, B. C. P. Jansen and W. P. Donath, in Holland, isolated the antiberiberi vitamin, which, at first, was called aneurin because of its specific action on the nervous system.

In 1936, Robert R. Williams, an American, determined the structure, synthesized it, and gave it the name *thiamine* because it contains sulfur (from *thio,* meaning sulfur-containing) and an amine group. Subsequently, the "e" was dropped and the spelling *thiamin* came to be preferred.

CHEMISTRY, METABOLISM, PROPERTIES.

• **Chemistry**—Thiamin is made up of carbon, hydrogen, oxygen, nitrogen, and sulfur (see Fig. T-8). It consists of a molecule of pyrimidine and a molecule of thiazole linked by a methylene bridge.

Fig. T-8. Structure of thiamin hydrochloride, the white, crystalline, stable form in which thiamin is usually marketed.

• **Metabolism**—The thiamin ingested in food is available (1) in the free form, or (2) bound as thiamin pyrophosphate (also called thiamin diphosphate), or (3) in a protein-phosphate complex. The bound forms are split in the digestive tract, following which absorption takes place principally in the upper part of the small intestine where the reaction is acid. It is noteworthy, however, that the absorption of thiamin is impaired in alcoholics with folate deficiency.

Following absorption, thiamin is transported to the liver where it is phosphorylated under the action of ATP to form the coenzyme thiamin diphosphate (formerly called thiamin pyrophosphate or cocarboxylase), (see Fig. T-9); although this phosphorylation occurs rapidly in the liver, it is noteworthy that all nucleated cells appear to be capable of bringing about this conversion.

Fig. T-9. Structure of thiamin diphosphate.

Thiamin is the least stored of all the vitamins. The adult human body contains approximately 30 mg. Of the thiamin stored in the body, about 80% is thiamin pyrophosphate, about 10% is thiamin triphosphate, and the remainder is thiamin monophosphate. The liver, kidneys, heart, brain, and skeletal muscles have somewhat higher concentrations than the blood. If the diet is deficient, tissues are depleted of their normal content of the vitamin in 1 to 2 weeks, so fresh supplies are needed regularly to provide for maintenance of tissue levels. Body tissues take up only as much thiamin as they need; with the need increased by metabolic demand (fever, increased muscular activity, pregnancy, and lactation) or by composition of the diet (carbohydrate increases the need for thiamin, while fat and protein spare thiamin). Because thiamin is water soluble, most of the vitamin not required for day-to-day use is excreted in the urine. This means that the body needs a regular supply, and that unneeded intakes are wasted. With a well-balanced diet, approximately 0.1 mg is normally excreted every 24 hours. However, the amount excreted in the urine decreases as the intake becomes inadequate and increases as the intake exceeds body needs; because of this, the most widely used biochemical method to assess thiamin status in individuals is the measurement of the vitamin in the urine.

• **Properties**—Synthetic thiamin is usually marketed as thiamin hydrochloride, which is more stable than the free vitamin. It is a crystalline white powder, with a faint yeastlike odor and a salty nutlike taste. It is stable when dry but readily soluble in water, slightly soluble in alcohol, and insoluble in fat solvents. Heating in solutions at 248°F (120°C) in an acid medium (pH 5.0 or less) has little destructive effect. But cooking foods in neutral or alkaline reaction is very destructive. Also, autoclaving and ultraviolet light destroy thiamin.

Thiamin mononitrate is more stable in heat than thiamin hydrochloride; for which reason it is often used for the thiamin fortification of cereal products that have to be cooked. Derivatives of thiamin, thiamin propyl disulfide and thiamin tetrahydrofurfural disulfide, have been synthesized. These products are recommended for oral administration when there is evidence of thiamin deficiency, because they are absorbed more rapidly than thiamin hydrochloride.

MEASUREMENT/ASSAY. The thiamin content of foods is expressed in milligrams or micrograms. It is usually determined by rapid chemical or microbiological methods, which have largely replaced the older bioassay methods in which pigeons, rats, and chicks were used.

FUNCTIONS. Thiamin is essential as a coenzyme (or cofactor) in energy metabolism, as a coenzyme in the conversion of glucose to fat, in the functioning of the peripheral nerves, and in such indirect functions as appetite, muscle tone, and a healthy mental attitude. A discussion of each of these functions follows:

• **As a coenzyme in energy metabolism**—Without thiamin, there could be no energy.

The major functioning form of thiamin is as thiamin diphosphate (formerly called thiamin pyrophosphate or cocarboxylase)—thiamin combined with two phosphate groups, in which it serves as a coenzyme in a number of enzyme systems.

In the metabolism of carbohydrates, thiamin diphosphate is needed in the conversion of pyruvic acid and the subsequent formation of acetyl coenzyme A, which in turn enters the Krebs cycle and produces vital energy. This is one of the most complex and important reactions in carbohydrate metabolism. In addition to thiamin diphosphate, it also requires the following cofactors: coenzyme A, which contains pantothenic acid, nicotinamide adenine dinucleotide (NAD), which contains niacin; magnesium ions; and lipoic acid.

Oxidative decarboxylation (removal of CO_2) in carbohydrate metabolism is also involved in the Krebs cycle in the conversion of alpha-ketoglutaric acid to succinic acid. Because fats and amino acids, as well as carbohydrates, can contribute to alpha-ketoglutaric acid, it follows that thiamin is involved in the metabolism of all three energy producing units.

In thiamin deficiency, pyruvic and alpha-ketoglutaric acids tend to accumulate in the body; sometimes they are measured as a means of determining thiamin status.

• **As a coenzyme in the conversion of glucose to fats—the process called transketolation (keto-carrying)**—Thiamin diphosphate is also a coenzyme with the enzyme transketolase in the important reaction which provides active glyceraldehyde through the pentose shunt. This is a key link providing activated glycerol for lipogenesis for the conversion of glucose to fat. Thiamin diphosphate is the key activator which provides the high energy phosphate bond. Ionized magnesium (Mg^{++}) is another cofactor present.

• **In the functioning of peripheral nerves**—Thiamin is involved in the functioning of the peripheral nerves. In this role, it has value in the treatment of alcoholic neuritis, the neuritis of pregnancy, and beriberi.

• **In indirect functions**—Because of its primary role in carbohydrate metabolism, thiamin appears to have several indirect functions in the body; among them, the maintenance of normal appetite, the tone of the muscles, and a healthy mental attitude.

DEFICIENCY SYMPTOMS. The numerous symptoms of thiamin deficiency vary with the severity and duration of the deprivation of the vitamin.

Fig. T-10. Polyneuritis (inflamation of the nerves) in the chick, the counterpart of beriberi in man, caused by thiamin (B-1) deficiency. Note the characteristic head retraction. (Courtesy, H. R. Bird, Department of Poultry Science, University of Wisconsin)

Moderate symptoms may be caused by poor diet, liver damage, or alcoholism. (In many cases, the latter is due to impaired absorption of thiamin caused in part by a deficiency of folacin.) Other persons at risk are kidney patients who are undergoing long-term dialysis treatment, patients fed intravenously for long periods of time, and patients with chronic febrile infections. Individuals consuming large amounts of tea, which contains a thiamin antagonist, or large amounts of raw fish, which contains thiaminase, an enzyme which inactivates the thiamin molecule by splitting it into two parts, may also run increased risk of developing a deficiency.

Symptoms of moderate thiamin deficiency include fatigue; apathy (lack of interest in affairs); loss of appetite; nausea; moodiness; irritability; depression; retarded growth; a sensation of numbness in the legs; and abnormalities of the electrocardiogram.

Clinical Effects of Thiamin Deficiency.

If a deficiency of thiamin is not corrected (if thiamin is not present in sufficient amounts to provide the key energizing coenzyme factor in the cells), the clinical effects will be reflected in the gastrointestinal system, the nervous system, and the cardiovascular system. Severe thiamin deficiency of long duration will culminate in beriberi, the symptoms of which are polyneuritis (inflammation of the nerves), emaciation and/or edema, and disturbances of heart function.

• **Gastrointestinal system symptoms**—The gastrointestinal symptoms of a thiamin deficiency are loss of appetite, indigestion, severe constipation, gastric atony (lack of tone), and decreased hydrochloric acid secretion. These manifestations likely result from insufficient energy from glucose being available for the smooth muscles and the glands of the intestinal tract to do their work of digestion.

• **Nervous system symptoms**—The central nervous system depends solely upon glucose for its energy. So, without thiamin to provide this need, neuronal activity is impaired, alertness and reflex responses are diminished, and fatigue and apathy (lack of interest) follow. If thiamin deficiency

persists, degeneration of myelin sheaths of nerve fibers in the central nervous system and the peripheral nerves occurs, resulting in nerve irritation, which produces pain and prickly or deadening sensations. Unchecked progressive degeneration of the nervous system may cause paralysis and muscle atrophy (wasting).

• **Cardiovascular system symptoms**—If the thiamin deficiency persists, the heart muscles weaken and heart failure may result. Also, the smooth muscles of the vascular system may be involved, causing peripheral vasodilation. As a result of cardiac failure, peripheral edema may be observed in the extremeties.

• **Beriberi**—The various forms and symptoms of beriberi follow:

1. **Dry beriberi (wasting of tissues).** In humans, this form is characterized by numbness or tingling in the feet and toes, stiffness of ankles, soreness in and wasting (atrophy) of the muscles of the legs, a decrease in the reflex of the knee, a drop in the muscles that support the toes and foot, and difficulty in walking. As the disorder advances, the arms and the other parts of the body are affected because there is nerve degeneration and lack of muscle coordination.

2. **Wet beriberi (collection of fluid in the tissues).** The presence of edema, especially of the legs, distinguishes this form from dry beriberi. Other symptoms are loss of appetite, breathlessness, and disorders of the heart.

3. **Infantile beriberi.** This acute disorder is common between the second and fifth months of life in children who are being suckled by mothers subsisting on beriberi-producing diets. Symptoms of the disease are weakness of voice during bawling (complete lack of sound in severe cases), lack of appetite, vomiting, diarrhea, rapid pulse, cyanosis (dark blue coloration of the skin and mucous membrane), and sudden death.

(Also see BERIBERI.)

• **Other clinical symptoms of thiamin deficiency**—Additional symptoms of thiamin deficiency include: low excretion of thiamin in the urine; electrocardiogram changes; reduced transketolase activity of the red blood cells; and an increase of pyruvic acid in the blood.

RECOMMENDED DAILY ALLOWANCE OF THIAMIN.

The Food and Nutrition Board (FNB) of the National Research Council (NRC) recommended daily allowances of thiamin are given in the section on VITAMIN(S), Table V-5, Vitamin Table.

In general, the recommended allowances are based (1) on assessments of the effects of varying levels of dietary thiamin on the occurrence of clinical signs of deficiency, (2) on the excretion of thiamin or its metabolites, and (3) on erythrocyte transketolase activity. Most studies have been conducted on subjects fed diets with ratios of carbohydrate and fat similar to those commonly consumed in the United States. There is evidence that dietary fat *spares* thiamin to some extent.

Because thiamin is essential for key reactions in energy metabolism, particularly carbohydrate metabolism, the requirement for thiamin has usually been related to energy intake.

• **Allowances for infants, children, and adolescents**—As shown in Table V-5, during the growth periods of infancy, childhood, and especially adolescence, thiamin needs are increased.

• **Allowances for adults**—The National Research Council recommended thiamin allowances for adults is 1.5 mg/day for males and 1.1 mg/day for females. Because there is some evidence that older persons use thiamin less efficiently, it is recommended that they maintain a minimum intake of 1 mg/day even if they consume fewer than 2,000 Calories (kcal) daily.

• **Allowance for pregnant and lactating women**—Increased thiamin needs accompany gestation and pregnancy. An allowance of 0.9 mg of thiamin per 1,000 Calories (kcal) is recommended throughout pregnancy, or about an additional 0.4 mg/day. To account for both the thiamin loss in milk and the increased energy consumption during lactation, an allowance of 1.0 mg/1,000 Calories (kcal) is recommended, amounting to an additional 0.5 mg/day.

Increased thiamin needs exist (1) during times of stress, fevers, infections, chronic illness, or surgery; (2) in old age; (3) with increased muscular work; and (4) in chronic alcoholism (due to high intake of energy in the form of alcohol and to reduced thiamin absorption).

• **Thiamin intake in average U.S. diet**—According to the U.S. Department of Agriculture, there is sufficient thiamin in foods available for consumption in the United States to provide an average intake of 2.2 mg per person per day. Of this total, the contribution of food groups is as follows: meats, poultry and fish, 25%; dairy products, 7.5%; and flour and cereal products, 43.3%.

TOXICITY. There are no known toxic effects from thiamin.

THIAMIN LOSSES DURING PROCESSING, COOKING, AND STORAGE. The thiamin content of foods is influenced by the practices employed in harvesting, handling, processing, cooking, and storing. Factors contributing to thiamin losses are pH, heat, oxidation, inorganic bases, enzymes, metal complexes, and radiation. The bond between the pyrimidine and thiazole rings is weak with the result that the vitamin is easily destroyed, particularly in an alkaline medium or by heat in the presence of moisture. Thiamin is sensitive to both oxidation and reduction. Sulfur dioxide, sulfites, etc., easily break the bond and vitamin activity is lost.

Pertinent facts about thiamin processing and cooking losses follow:

• **Parboiled (converted) rice**—Rice that is soaked and parboiled (converted) contains much more thiamin than rice that is milled from the raw state. The parboiling causes the thiamin and other water soluble nutrients to move from the outer layers to the inner layers of the rice kernel. As a result, fewer of them are removed in the milling process.

(Also see RICE.)

• **Processing losses**—Canning causes thiamin losses because of the solubility of the vitamin in the canning fluid; hence, the more canning fluid drained away, the greater the loss.

Drying (dehydrating) food, except fruit, results in only small losses. Thiamin is destroyed rapidly by sulfite, a fact which may explain the loss of thiamin in dried fruits, such as apricots and peaches, treated with sulfur.

Fresh frozen vegetables maintain in storage the thiamin content that is present following blanching to destroy the enzyme activity.

Irradiation destroys thiamin. Thus, the thiamin content of pork is virtually depleted by irradiation in food preservation.

• **Normal cooking losses**—Normal cooking of an ordinary mixed diet results in the loss of about 25% of the thiamin.

• **Moist heat**—Boiling for not more than an hour causes little destruction of this vitamin. But, because thiamin is soluble in water, as much as one-third of the original thiamin content may be lost if cooking water is liberal and is discarded.

• **Dry heating**—High dry-heat temperatures, as in toasting bread, can cause considerable loss.

• **Cooking breakfast cereals**—Little thiamin is lost in the cooking of breakfast cereals since they are cooked at moderate temperatures and the cooking water is absorbed.

• **Alkali medium**—Thiamin is readily destroyed in an alkali medium. Hence, the practice of adding baking soda (an alkali) to cooking water to preserve the color of green vegetables is not recommended.

• **Meat losses**—Generally, the thiamin losses in cooking meats are greater than in cooking other foods, ranging from 30 to 50% of the raw value; with the least loss occurring in frying.

• **Bread baking losses**—In baking bread, 15 to 20% of the original thiamin content is lost.

• **The same principles apply to both ascorbic acid and thiamin**—When the principles for the retention of ascorbic acid are observed in food processing and cooking, the maximum thiamin content will be preserved, also.

There is little destruction of thiamin from exposure to air at ordinary temperatures. Hence, storage losses of thiamin in foods are minimal.

ANTITHIAMIN FACTORS IN FOOD. Certain raw fish and seafood—particularly carp, herring, clams, and shrimp—contain the enzyme thiaminase, which inactivates the thiamin molecule by splitting it into two parts. This effect has been seen in mink and fox fed 10 to 25% levels of certain raw fish, giving rise to a thiamin deficiency disease known as Chastek paralysis. This action can be prevented by cooking the fish prior to feeding, thereby destroying the thiaminase. Of course, humans seldom eat sufficient thiaminase-containing raw fish or seafood to produce a thiamin deficiency.

Other agents known to affect thiamin levels in the body are: a large amount of live yeast in the diet of man, which reduces the amount of thiamin absorbed from the intestine; drinking large amounts of tea or chewing fermented tea leaves, a common practice in some parts of Asia; or imbibing alcohol in excess, which contains an antithiamin substance. Also, it is noteworthy that thiamin-splitting bacteria have been found in the intestinal tract of some Japanese with symptoms of beriberi, but the significance of this is not known at present.

SOURCES OF THIAMIN. Some thiamin is found in a large variety of animal and vegetable products but is abundant in few. Therefore, a deficiency of thiamin is a distinct possibility in the average diet, especially when calories are curtailed.

In addition to food sources, synthetic thiamin is also available; mostly as thiamin hydrochloride, although thiamin mononitrate is often used for the thiamin fortification of cereal products that have to be cooked.

The thiamin content of sources varies according to food and may be affected by harvesting, processing, enrichment, and storage. A grouping and ranking of foods according to normal thiamin content is given in the section on VITAMIN(S), Table V-5, Vitamin Table.

For additional sources and more precise values of thiamin, see Food Composition Table F-21.

Whole grains and enriched grain products are the best daily sources of thiamin. A quart of milk each day contributes 0.29 mg of the thiamin intake.

Foods lacking in thiamin are man-made—refined rice and cereal flours (from which almost all the natural store of the vitamin has been removed by the millers), refined sugar, separated animal and vegetable oils and fats, and alcoholic beverages. None of the thiamin in yeast used for fermentation is present in beers, wines, and spirits that enter normal commerce, although home-brewed beers and country wines may contain significant amounts. Indeed, there are communities in Africa and Latin America which derive the major part of their thiamin from native beers.

Enrichment of flour (bread) and cereals, with thiamin, riboflavin, niacin, and iron (with calcium enrichment optional), which was initiated in 1941, has been of special significance in improving the dietary level in the United States. On the basis of the average per capita consumption of flour and bread in the United States, slightly more than 40% of the daily thiamin requirement is now supplied by these foods.

Rice enrichment has been practiced for years in some of the rice-eating countries; for example, much of the rice used in Japan is enriched.

NOTE WELL: The thiamin requirements are usually met if the diet does not contain disproportionate amounts of the following:
1. Refined, unfortified cereal
2. White sugar
3. Alcohol

Since these three foods are usually matters of personal choice, nutrition education is important.

(Also see RICE; WHEAT, section on "Enriched Flour"; and VITAMIN[S], Table V-5, Vitamin Table.)

THIAMIN DIPHOSPHATE (TDP)

The activating coenzyme necessary for the transketolation reaction in the hexose monophosphate shunt (glucose oxidation) by which active glyceraldehyde is formed for synthesis of fats.

THIRST

Thirst (in humans) is the conscious desire for water. It is the primary means of regulating water intake, and generally, the thirst sensation ensures that water intake meets or exceeds the body's requirement for water. The sensation of thirst is caused by nerve centers in the hypothalamus of the brain which monitor the concentration of sodium (osmolarity) in the blood. When the sodium concentration of the blood increases above the normal 310 to 340 mg/100 ml (*136 to 145 mEq/1*), cells in the thirst center shrink. This shrinking causes more nervous impulses to be generated in the thirst center, thereby creating the sensation of thirst.

An increase of only 1% in the osmolarity of the blood is sufficient to evoke the sensation of thirst. Increased osmolarity of the blood is primarily associated with water loss from the extracellular fluid—blood—through low cardiac output, hemorrhage, or intracellular dehydration may stimulate the thirst sensation. As water is lost, the sodium concentration of the remaining fluid increases. When the individual drinks, the sensation disappears.

Increased osmolarity of the blood simultaneously stimulates other nerve centers in the hypothalamus causing the release of antidiuretic hormone (ADH) from the posterior pituitary. ADH release results in the formation of less urine by the kidneys; hence, conserving body water. Thus, drinking water and excreting water are controlled by centers in the hypothalamus which maintains the water balance of the body within narrow limits.

Constant thirst is a classical symptom of diabetes mellitus.

(Also see ENDOCRINE GLANDS; WATER; and WATER AND ELECTROLYTES.)

THREONINE

One of the essential amino acids.
(Also see AMINO ACID[S].)

THROMBIN

An enzyme that facilitates the clotting of blood by promoting the conversion of fibrinogen to fibrin.

THROMBOSIS

The obstruction of a blood vessel by the formation of a blood clot.
(Also see HEART DISEASE, section headed "Coronary Occlusion [Coronary Thrombosis]".)

THYMINE

One of the four nitrogenous bases in nucleic acid.
(Also see NUCLEIC ACID.)

THYROCALCITONIN

Another name for calcitonin, the hormone secreted by the thyroid gland which acts to lower blood calcium levels.
(Also see CALCITONIN; and ENDOCRINE GLANDS.)

THYROID GLAND

An endocrine gland which is located in the neck on top of the windpipe. It secretes the iodine-containing hormones thyroxine (T_4) and triiodothyronine (T_3), and the calcium regulating hormone calcitonin. The thyroid gland may grow to enormous proportions in the condition of goiter.
(Also see ENDOCRINE GLANDS; GOITER; and IODINE.)

THYROIDITIS

Inflammation of the thyroid gland.

THYROID-STIMULATING HORMONE (TSH)

A hormone secreted by the anterior pituitary gland that regulates uptake of iodine and synthesis of thyroxin by the thyroid gland.

THYROTOXICOSIS

Overactivity of the thyroid gland, causing exophthalmic goiter, which is characterized by nervousness, rapid pulse, bulging eyes, high basal metabolism, and loss of weight.

THYROXIN (T₄)

One of the two iodine-containing hormones secreted by the thyroid gland. It contains four iodine atoms per molecule. Hence, it is often called T₄. Thyroxin is less active than its counterpart triiodothyronine. Thyroxin's actions include increased metabolic rate, increased nervous system activity, stimulated protein synthesis, and increased motility and secretion of the gastrointestinal tract.

(Also see ENDOCRINE GLANDS.)

TIMBALE

• A creamy mixture of chicken, lobster, cheese, or fish cooked in individual molds or cups.

• A small pastry shell fried with a timbale iron and filled with a cooked timbale mixture, or served with fruit sauce, or dusted with powdered sugar.

TISSUE

A collection of cells, usually of a particular kind which form a definite structure, such as connective tissue, epithelium, muscle, and nerve.

(Also see BODY TISSUE.)

TISSUES OF THE BODY

Tissues are an aggregate of similar cells together with their intercellular substances. In turn tissues combine to form the organs of the body. The major tissues of the body are (1) nervous tissue, (2) muscles, (3) skin, (4) blood cells, (5) fat or adipose, and (6) bone.

TOASTED

Brown, dried, or parched by exposure to a fire, or to a gas or electric heat.

TOCOPHEROL

Any of four different forms of an alcohol. Also known as vitamin E.

(Also see VITAMIN E.)

TOCOTERIENOL

Compounds found in nature that have vitamin E activity, but which are less potent than alpha-tocopherol.

(Also see VITAMIN E.)

TOLBUTAMIDE

An oral drug, a derivative of sulfonylurea, capable of lowering blood glucose (sugar) when a functioning pancreas is still present. Since it acts by stimulating the release of insulin, it may be used in the treatment of diabetes when there is still some residual function of the pancreas.

(Also see DIABETES MELLITUS, section headed "Sulfa-type Drugs [Sulfonylureas].")

TOMATO *Lycopersicon esculentum*

The tomato is a member of the Nightshade family (*Solanaceae*), which also claims as members capsicum peppers, eggplant, and Irish potatoes.

Fig. T-11. Tomatoes, the second leading vegetable of the U.S. (Courtesy, National Film Board of Canada)

ORIGIN AND HISTORY. The modern tomato descended from a wild ancestor, the cherry tomato, native to Peru and Ecuador. It was first domesticated by the Indians in Central America. But it had spread to both North and South America long before Columbus arrived.

WORLD AND U.S. PRODUCTION. Tomatoes rank fourth among the leading vegetables of the world. Worldwide, about 69 million metric tons of tomatoes are produced annually; and the leading tomato-producing countries, ranked by descending order, are: United States, U.S.S.R., Turkey, China, Italy, Egypt, Spain, Romania, Brazil, and Greece.[4]

Tomatoes rank second among the leading vegetables of the United States, with a production of about 16.8 million short tons annually; and the leading tomato-producing states are: Florida, California, and South Carolina. Florida alone produces 43% of the U.S. tomato crop.[5]

PROCESSING. About 75% of the U.S. tomato crop is processed; still, the demand for tomato products can hardly be met, as evidenced by the various tomato extenders that are used in certain products. Among the popular tomato products are the following: canned tomatoes, catsup (ketchup), chili sauce, juice, juice crystals, paste, powder, puree, salad dressings, sauces, soups, and vegetable juice cocktails.

---

[4]Data from *FAO Production Yearbook 1990*, FAO/UN, Rome, Italy, Vol. 44, p. 131, Table 52. **Note well:** Annual production fluctuates as a result of weather and profitability of the crop.

[5]Data from *Agricultural Statistics 1991*, USDA, p. 165, Table 240. **Note well:** Annual production fluctuates as a result of weather and profitability of the crop.

Fig. T-12. Closeup of a tomato plant. (Courtesy, Louisiana State University, Baton Rouge)

SELECTION, PREPARATION, AND USES.

Good quality tomatoes are well formed and plump, of uniform red color, free from bruise marks, and not overripe or soft.

Fresh tomatoes and canned tomato products are very convenient foods for the busy homemaker because (1) the raw vegetable lends itself well to cold soups, relishes, and salads; (2) only a few minutes of simmering is needed to cook the raw vegetable; (3) canned tomato juice may be used to make a wide variety of molded gelatin salads, (4) canned tomato paste or puree may be used when concentrated sauces are desired, and (5) the flavor and color of the tomato blends well with those of many other ingredients. Some examples of easy-to-prepare tomato dishes and a few gourmet-type preparations follow:

• **Aspic**—This type of appetizer or side dish is usually made from tomato juice, flavored or unflavored gelatin, and seasonings such as lemon juice, sugar, salt, pepper, and Worcestershire sauce.

• **Chinese stir-fried dishes**—Tomatoes may be added near the end of the cooking period to stir-fried dishes

• **Eastern European tomato salad**—In Serbia (a region of Yugoslavia) this type of salad is made from tomatoes, raw onions, chili peppers, goat cheese, and baked sweet peppers.

• **Gazpacho soup**—This chilled tomato soup, which originated in Spain, contains small pieces of peeled and seeded tomatoes, other vegetables that have been finely chopped or pureed, olive oil, wine vinegar, and various seasonings.

• **Green tomato pie**—Although this dessert is very similar to apple pie with respect to ingredients and preparation, it differs in that deskinned, green tomato wedges are substituted for sliced apples.

• **Guacamole**—Deskinned, deseeded, chopped tomatoes are mixed with mashed avocado pulp, chopped onions, hot green chili peppers, coriander leaves or parsley, and a pinch of salt to make this popular Mexican sauce.

• **Italian pasta dishes**—Many of these items are served with a rich tomato sauce that is easily prepared if one starts with a canned tomato product such as tomato paste or tomato sauce.

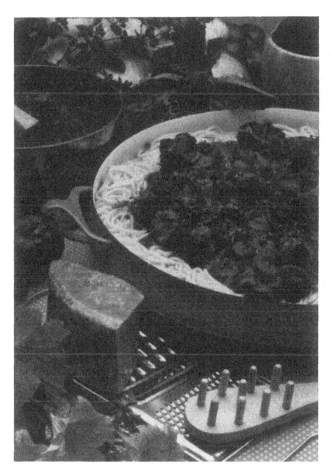

Fig. T-13. Tomatoes in a casserole. (Courtesy, National Film Board of Canada)

• **Lobster with tomato sauce**—Peeled and chopped tomatoes are simmered with white wine, fish stock, brandy, butter, chopped shallots, crushed garlic, parsley, Cayenne pepper, salt, and black pepper to make a delectable French sauce for lobster.

• **Piccalilli**—This popular relish is commonly made from green tomatoes, green peppers, red chili peppers, vinegar, sugar, salt, horseradish, ground mustard, cinnamon, and ginger.

• **Spanish rice**—Quick-cooking types of rice and canned tomato sauce make it possible to prepare this dish in less than one-half hour. Other ingredients are ground beef, chopped onion, chopped green pepper, chili sauce, butter or margarine, salt, Worchestershire sauce, and black pepper.

• **Stuffed tomato**—Fillings such as bread crumbs, cheese, chopped egg, crumbled bacon, fish or seafood, legumes, meats, nuts, poultry, and/or cooked rice may be used to stuff tomatoes. However, the fillings should be cooked first, so that the stuffed tomato shells will require only a few minutes of simmering in a sauce.

• **Tomato chutney**—Chutneys are condiments or relishes that appear to have spread to the rest of the world from India and Indonesia. A typical tomato chutney may contain chopped peeled tomatoes plus chili pepper, chopped dried fruit, vegetable oil, salt, and other seasonings.

Fig. T-14. Tomatoes in a salad. (Courtesy, Anderson/Miller & Hubbard Consumer Services, San Francisco, Calif.)

NUTRITIONAL VALUE. The nutrient compositions of various forms of tomatoes are given in Food Composition Table F-21.

Some noteworthy observations regarding the nutrient composition of tomatoes follow:

1. Fresh tomatoes and tomato juice are high in water content (about 94%) and low in calories. Both items are good sources of vitamins A and C, except that the unfortified juice has only about two-thirds the vitamin C content of raw, ripe tomatoes. However, tomato juice may be fortified to bring the level of the vitamin up to 10 milligrams per fluid ounce (30 ml).

2. Ripe (red) tomatoes contain from 3 to 4 times as much vitamin A as mature green tomatoes. Otherwise, green and red tomatoes are about equal in nutritional value.

3. Canned tomatoes contain only about three-fourths the vitamin C content of fresh ripe tomatoes.

4. Tomato puree and plain types of tomato sauce (those without added ingredients such as cheese, meat, mushrooms, etc.) have about twice the solids content and about double the nutritional value of fresh tomatoes and tomato juice.

5. Tomato paste, which has about four times the solids content of fresh tomatoes, is a concentrated source of nutrients. Table T-2 compares the composition of this product with canned corn.

TABLE T-2
NUTRIENT COMPOSITIONS OF TOMATO PASTE AND CANNED CORN[1]

Nutrient or Other Constituent		Tomato Paste	Canned Corn[2]
Water content	(%)	75	76
Calories	(kcal)	82	83
Protein	(g)	3.4	2.5
Phosphorus	(mg)	70	73
Potassium	(mg)	888	97
Iron	(mg)	3.5	0.5
Vitamin A	(IU)	3,300	350
Vitamin C	(mg)	49	5
Thiamin	(mg)	0.20	0.03
Riboflavin	(mg)	0.12	0.06
Niacin	(mg)	3.10	1.10

[1]Nutrients per 3½ oz (100 g) portion.
[2]Yellow sweet corn, vacuum packed, solids and liquid.

Table T-2 shows that tomato paste provides considerably greater amounts of nutrients per calorie than canned corn. A similar nutritional superiority is noted when the paste is compared to other grains and cereal products. Therefore, this product may make a valuable contribution when it is used in the preparation of pastas, pizzas, and other dishes.

6. Catsup and chili sauce are about equal in nutritional value, since each item is made with similar ingredients and contains about 32% solids (about 5 times the content of fresh tomatoes and tomato juice). However, the nutrients provided per calorie by these products are significantly less than those furnished by tomato paste, because the solid contents and caloric values of the two condiments are boosted by added salt and sugar.

Tomatoes contain small amounts of solanine. (See SOLANINE.)

(Also see VEGETABLE[S], Table V-2, Vegetables of the World.)

TOMATO JUICE

The juice extracted from tomatoes that is produced by chopping fresh fruit which may or may not have been preheated, pressing the fruit pulp through a series of screens, adding back some of the pulp to the screened juice, canning, and pasteurization.

The nutrient composition of various forms of tomato juice is given in Food Composition Table F-21.

Some noteworthy observations regarding the nutrient composition of tomato juice follow:

1. 1-cup (250 g) of bottled or canned tomato juice provides only 38 Calories (kcal), and is a good source of potassium, iron, vitamin A, and vitamin C. Most juice products often contain added sodium (about 400 mg per cup), but low sodium products are also available.

2. Frozen tomato juice concentrate contains about 4 times the caloric and nutrient contents of the single strength canned juice. Hence, it will make a significant nutritional contribution when added undiluted to dishes such as sauces and soups.

(Also see TOMATO.)

-TOMY

Suffix meaning to cut into; e.g., gastrectomy.

TONIC

• A drug or agent said to improve the health or vigor of an organ, or of the body.

• When referring to muscles, it can mean the state of muscular tension or continuous muscular contraction.

• A bottled quinine water used to make the drink "Gin & Tonic." This drink originated in the tropics where it was used to alleviate fevers, either real or imagined.

TOPHUS

Sodium urate deposits in the fibrous tissues near the joints or in the cartilage of the external ear; present in gout.

TORULARODIN

Carotenoid pigment in *Torula ruba* and *Rhodotorula mucilaginosa* yeasts, with vitamin A activity.

TORULA YEAST

Torula yeast (*Torulopsis utilis*) is a hardy type of yeast that can be propagated on a variety of substrates, such as press liquor obtained during the manufacture of dried citrus pulp, molasses, sulphite waste liquor from the paper industry, saccharified wood (both hexoses and pentoses can be used), and fruit wastes such as coffee beans, apples, etc. While brewers' yeast is truly a by-product food, torula yeast is cultured specifically as a foodstuff for man and animals.

Dried torula yeast is an excellent source of high-quality protein (50 to 62% crude protein), minerals, B vitamins (including vitamin B-12), unidentified factors, and of vitamin D if irradiated. Unlike the bitter-tasting brewers' yeast, torula yeast is tasteless, thereby facilitating its use as a palatable foodstuff.

Torula yeast is available in powder and tablets.

TOTAL CRUDE PROTEIN

The equivalent amounts of crude protein in the food represented by the total nitrogen present.
(Also see ANALYSIS OF FOODS.)

TOTAL MILK SOLIDS

Primarily milk fat, proteins, lactose, and minerals.
(Also see MILK AND MILK PRODUCTS.)

TOTAL PARENTERAL (INTRAVENOUS) NUTRITION (TPN) (CENTRAL INTRAVENOUS NUTRITION)

Contents	Page
Indications of TPN.	1017
Use of TPN at Home.	1018
Composition of Infusions	1018
Benefits	1018
Hazards and Precautions	1018
Contraindications	1019
Summary	1019

This nutritional procedure can provide all or most of a patient's nutrient requirements by intravenous means. Total parenteral nutrition is used when the patient cannot be fed through the gastrointestinal tract because of the danger of aggravating an abnormal condition, or because some part of the tract is functioning poorly. Usually, it involves infusion of nutrients by means of a catheter that is passed through a large vein into the vena cava, as is shown in Fig. T-15.

Fig. T-15. Total parenteral (intravenous) nutrition. The concentrated solution of nutrients flows through a sterile tubing attached to a catheter passed through the subclavian vein into the vena cava. This procedure results in rapid mixing of the nutrients with blood returning to the heart, which then pumps the nutrient-rich blood throughout the body.

It is noteworthy that the solutions used in total parenteral nutrition have much higher solute contents than those used in ordinary intravenous feedings (about 30% solute in TPN vs 5% solute in an IV) because (1) nutrient needs cannot be met completely by the latter, and (2) the means by which the former is administered (into a large central vein where rapid mixing with the blood occurs) is sufficiently safe, whereas concentrated (hypertonic) solutions cannot be infused safely into the peripheral veins. Phlebitis and clots are more likely to result from the injection of hypertonic solutions into peripheral veins because the flow of blood in them is slow. Therefore, TPN is sometimes referred to as *central intravenous nutrition* in order to distinguish it from peripheral IVs.

INDICATIONS FOR TPN. The foremost types of conditions requiring total parenteral nutrition are those in which the patient's nutritive needs are high, and cannot be met by more commonly used procedures such as oral, tube, or peripheral intravenous feeding. However, it may also be indicated even when some food can be taken orally or through a nasogastric tube because the nutritional requirements may be so high that they cannot be met via the digestive tract alone. Some of the most common indications for TPN follow:

1. Burns over large areas of the body.
2. Cancer.
3. Chronic diarrhea.
4. Fistulas of the digestive tract.
5. Inflammatory bowel disease.
6. Kidney failure.
7. Liver failure.
8. Refusal to eat.
9. Surgery on the digestive tract.
10. Undernourished and very-low-birth-weight infants.

Use of TPN at Home. The success of TPN in hospital settings led to the development of techniques and equipment that could be used at home so that patients requiring this therapy would not have to be kept in a hospital. Generally, the most likely candidates for home TPN are patients with reasonable chances of benefiting from it because their major nutritional problems are due to gastrointestinal disorders such as (1) a loss of considerable amounts of small intestine, (2) inflammatory bowel disease, and (3) digestive disorders resulting from chemotherapy and/or radiation treatments.

Home TPN is rather expensive because it requires a special pump or a gas pressure device, sterile tubing, special valves, and the nutrient solutions. It is noteworthy that the valves permit patients to be disconnected from the apparatus for 8 to 14 hours so that many normal acitivties may be carried out. Hundreds of patients have been placed on home TPN, and some have received this therapy for 5 years or more. It is expected that the cost will be reduced as the number of people on home TPN increases. At any rate, it is usually less expensive than receiving care in an institution.

Composition of Infusions. The TPN solutions that were first used on an experimental basis by Dr. Dudrick and his co-workers at the University of Pennsylvania School of Medicine contained approximately 20% dextrose (glucose), 5% protein, and 5% minerals and vitamins. Hence, the total solute concentration was about 6 times that of an ordinary 5% dextrose (glucose) intravenous feeding. Since then, many modifications of Dudrick's original TPN solutions have been made to accommodate the particular needs of individual patients. However, the commercial TPN preparations that are available in the United States do not always contain essential fatty acids. Therefore, it is noteworthy that deficiencies of these nutrients have been observed in patients receiving parenteral nutrition. This risk is reduced to zero if fat emulsions are infused on a regular basis (1 to 2 times per week). No doubt there are other essential nutrients that should be provided by supplemental means when patients are on TPN for long periods of time.

BENEFITS. Some of the recoveries that were brought about by TPN have been so remarkable that one eminent physician has ranked the development of this procedure on a par with other great medical innovations such as anesthesia, antiseptics, and antibiotics. What TPN has done is to demonstrate very dramatically the pivotal role played by nutrition in healing. Typical benefits of TPN follow:

1. The need to search all over the body for intact veins through which to make peripheral intravenous infusions is eliminated by TPN because an indwelling catheter may remain in place for long periods of time when the proper safeguards are observed.

2. Wasting of tissues after burns, injury, surgery, or other extremely stressful occurrences is minimized by TPN, which provides the high levels of nutrients needed for healing.

3. Bypassing of the digestive tract does away with the needs to cope with diarrhea, flatulence, nausea, vomiting, and other gastrointestinal reactions that interfere with nourishment. Hence, the tract may be rested and allowed to heal without disturbance.

4. Nutrient patterns may be tailormade to suit the needs of the patient because the infusions are made directly into a major blood vessel for delivery to the tissues of the body by the pumping of the heart.

HAZARDS AND PRECAUTIONS. TPN is not a routine nutritional procedure because special precautions must be taken in its administration to prevent undesirable consequences. The main hazards and precautions of TPN are:

• **Aggravation of disorders of metabolism**—When people obtain their nourishment via the digestive tract the absorbed nutrients are conveyed by the portal system of blood vessels to the liver, where excess sugars are converted into fats and glycogen (a complex carbohydrate that is stored in the liver and muscles), fats are combined with proteins for safe transport in the blood, and amino acids are converted into proteins and waste products such as urea.

However, in TPN the regulation of the blood levels of nutrients by the liver is bypassed to a great extent because the infused substances flow to the heart, where they are pumped through the arteries to the body tissues. The latter situation may be dangerous to diabetics because they utilize nutrients more slowly than normal people. Hence, they may develop excessively high blood levels of nutrients, which can lead to (1) loss of excessive amounts of nutrients and water through urination (osmotic diuresis), (2) dehydration (due to excessive loss of water), and/or (3) coma. On the other hand, people with some types of kidney disease may fail to excrete a normal amount of water and consequently become waterlogged (overhydrated) to the point of developing congestive heart failure.

Finally, some people who have been on TPN for 5 months or longer have developed bone demineralization and bone pain. Therefore, patients receiving TPN for the first time should be monitored carefully by blood chemistries, blood gas analyses, urine collection and analyses, and other laboratory tests for metabolism so that the appropriate therapies may be administered for any abnormal conditions that may develop.

• **Injuries resulting from the procedure**—The experts who have used TPN on a regular basis consider the placement of the catheter to be a serious surgical procedure and proceed accordingly. Failure to observe certain precautions may result in mishaps such as puncturing of the vein and passage of the catheter into neighboring blood vessels or organs such as the heart and lungs, and consequences such as bleeding, production of irregular heartbeats, the introduction of air or fluids into the lungs, or inflammation of the catheterized vein so that a spasm occurs or a clot is formed. Therefore, it is best to place the patient in an inclined position with the head downward (15% Trendelenburg position) so that the subclavian vein sticks out and is more easily entered. Also, an x ray should be taken after the catheter has been placed and before it is used to make certain that its location is correct for the patient. It is noteworthy that complications are rare when these precautions are observed strictly.

• **Microbiological contamination**—Infectious microorganisms may get into the body via (1) the point where the vein is punctured, (2) the surface of the catheter, (3) the inside of the tubing and/or the catheter. Also, inside the vena cava the tip of the catheter may serve as a place for bloodborne bacteria to lodge and multiply, then detach themselves and spread through the bloodstream to susceptible body tissues. Therefore, a sudden unexplainable rise in the body temperature of a patient should lead the attending doctors and nurses to suspect a catheter-induced infection, and/or any of the other sources of infection that are likely under the circumstances.

Some of the antiseptic precautions that are commonly used to prevent a catheter-induced infection of the patient consist of (1) shaving the hair from the chest and defatting of the skin with ether or acetone; (2) washing of the shaved and defatted area with sterile gauze sponges wetted with iodine solution; (3) handling of the catheterization assembly with sterile gloves while making sure that it does not come into contact with nonsterile items; (4) application of an antibiotic ointment and a sterile dressing on the area of catheterization, followed by the application of tincture of benzoin around the dressing; (5) periodic changing of the dressing and inspection of the catheter site for redness, infection, leakage, or other problems (ideally, the same nurse should perform this function so that she will readily recognize any changes); and (6) taking care that the solutions are changed under strictly aseptic conditions. These measures ensure a continued sterility of the catheter for months, or even years. The longevity of the preparation is important because there are only four sites (right and left subclavian veins, and right and left jugular veins) that are highly suited for this procedure as it is normally carried out. (Other means of central venous infusion are more complicated and dangerous.)

• **Nutritional deficiencies**—Recent reports in the medical literature have indicated that certain nutritional deficiencies may result from the long-term use of some proprietary brands of TPN infusions, unless the nutrients not present in the solutions are provided by other means. The deficiencies reported up to now have been those of essential fatty acids, the essential minerals chromium and selenium, and the vitamin biotin. Normally, biotin is synthesized by intestinal bacteria, but the administration of broad spectrum antibiotics in the case reported eliminated that source of the vitamin.

It is the responsibility of the doctor who orders the TPN procedure to make certain that all of the essential nutrients are provided to the patient. This may be done readily by adding the required substances to the TPN infusion.

• **Psychological impact**—Many patients may be quite apprehensive of being put on TPN because (1) they may not have known anyone else who received this type of therapy, (2) the providing of nutrients parenterally for more than a few weeks may cause concern over whether normal eating will ever be resumed, and (3) their lives may depend upon the apparatus functioning properly. Therefore, it may be necessary for the doctor, nurse, a knowledgeable member of the family, or a trusted friend to give the patient repeated assurances about the efficacy of the procedure and the possibility of eventually resuming many normal activities, even though it should be necessary to continue TPN for a long time at home.

The expression of anger, disbelief, grief, and/or depression is common for patients dependent upon complex treatments delivered by the medical care team, who may sometimes deliberately or inadvertently create a mystique about their procedures. Some type of psychiatric consultation may be required to help the patient cope with the situation and cooperate in the healing process.

CONTRAINDICATIONS. TPN carries some risks, and should not be used when (1) it is possible to meet all or most of a patient's nutritional needs by oral or tube feeding (some supplementation may be given by peripheral intravenous feeding); (2) the patient is moderately healthy and will be required to abstain from eating for only a few days; (3) there is little hope for improvement of the patient; and (4) certain disorders (cardiovascular, metabolic, or organ disfunctions) are present that are likely to be aggravated by TPN. In the latter cases, it may be possible to use TPN after the patient's condition has been stabilized.

SUMMARY. Total parenteral nutrition is a recently developed procedure that carries some risks, but may help to get seriously ill patients back on their feet, provided that the proper precautions are observed.

TOX-

Prefix meaning poison.

TOXEMIA

A condition produced by the presence of poisons (toxins) in the blood.

TOXEMIA OF PREGNANCY (PREECLAMPSIA AND ECLAMPSIA)

A cardinal indication of toxemia of pregnancy is a sudden weight gain sometime during the last trimester. It is a serious disorder of uncertain origin involving decreased kidney function. Numerous factors including endocrine, metabolic, and nutritional may be responsible for the disorder, and not a toxic substance as the name implies. Other features of the condition consist of high blood pressure, blurred vision, protein in the urine, and puffy neck, ankles, and face—edema. It is most often observed in (1) first pregnancies before the age of 30 years; (2) twin pregnancies; (3) economically underprivileged segments of the population—the malnourished; and (4) women with prior kidney or vascular disease. Preeclampsia and eclampsia are two stages of toxemia of pregnancy. Preeclampsia is characterized by those items mentioned above, while eclampsia indicates that the symptoms have intensified—circulatory failure, convulsions, and coma—possibly resulting in the death of the mother and the baby.

The most effective treatment is delivery of the baby. However, during the course of pregnancy, routine urine analysis and blood pressure checks by a doctor indicate the onset of toxemia. Once the onset of toxemia is noted, control measures such as sodium restriction, diuretics, sedatives, and bed rest may be initiated. Nevertheless, toxemia becomes increasingly difficult to manage the last month of pregnancy, and delivery may be induced slightly before term to prevent harmful effects to both the mother and the child.

Although the involvement of dietary factors in the development of toxemia is not clear, three points deserve mention. First, a vitamin B-6 deficiency during pregnancy may have a role in the development of toxemia. Second, while sodium restriction and diuretics are traditional approaches to the control of toxemia, there is also an increased demand for sodium during pregnancy; moreover, the Committee on Maternal Nutrition discourages the routine use of salt restriction and diuretics during pregnancy. Third, well-balanced diets provide the best protection against the development of complications during the course of pregnancy.

(Also see PREGNANCY AND LACTATION NUTRITION, section on "Fluid Retention.")

TOXINS

The poisons produced by certain microorganisms. They are products of cell metabolism. The symptoms of bacterial diseases, such as diptheria, tetanus, botulism, and staphylococcal food poisoning, are caused by toxins.

(Also see FOODBORNE DISEASE; and POISONS.)

TOXINS, BACTERIAL, FOODBORNE

Food poisoning is caused by the ingestion of bacterial toxins that have been produced in the food by the growth of specific kinds of bacteria before the food is eaten. The powerful toxins are ingested directly, and the symptoms of food poisoning develop rapidly, usually within 1 to 6 hours after the food is eaten. Two common bacteria that produce toxins in food are *Clostridium botulinum* (botulism) and *Staphylococcus aureus* (staphylococcus food poisoning). Each type is quite different in growth habits and in the symptoms of poisoning.

(Also see BACTERIA IN FOOD, Table B-3, Food Poisonings [Bacterial Toxins]; and POISONS.)

TRANSAMINATION

A metabolic process involving the transfer of an amino group (NH_2) from one compound to another. This is one process which makes possible the synthesis of a limited number of amino acids—nonessential amino acids. Carbon skeletons for this process are produced through various intermediates of carbohydrate metabolism. A new amino acid can be produced when an amino group (NH_2) is transferred from an amino acid to the carbon skeleton. The deaminated molecule—past amino acid—can be used as an energy source.

(Also see METABOLISM; section on " Proteins-Catabolism;" and PROTEIN[S].)

TRANS FATTY ACIDS

The hardening of vegetable oils by hydrogenation (the chemical addition of hydrogen) converts the naturally-occurring *cis* fatty acids to *trans* fatty acids. Generally speaking, it follows that the more an oil has been hardened, the higher the content of trans fatty acids.

(Also see FATS AND OTHER LIPIDS, section headed "Fatty Acids"; and OILS, VEGETABLE, section headed "Antinutritional and/or Toxic Factors.")

TRANSFERASE

Any of various enzymes that promote a transfer reaction; for example, transaminase.

TRANSFERRIN (SIDEROPHILIN)

Iron-binding protein for transport of iron in the blood.
(Also see IRON.)

TRANSKETOLASE

The enzyme is a transketolase and the process is called transketolation (keto-carrying). Transketolase uses thiamin diphosphate as a coenzyme to bring about the transfer of a 2-carbon unit from one sugar (a 2-keto sugar) to aldoses (monosaccharides with the characteristic aldehyde group [-CHO]).

TRAUMA

• A wound or injury.

• A psychological or emotional stress.

TREE TOMATO *Cyphomandra betacea*

The fruit of a small tree (of the family *Solanaceae*) that is native to Peru and is now grown throughout the tropics.

Tree tomatoes resemble the plum type of common tomato (*Lycopersicum esculentum*) in many respects in that they are usually yellow to red in color, 2 to 3 in. (5 to 7.5 cm) long, and contain many seeds. They may be eaten raw, stewed, or when made into jam, jelly, or juice.

The fruit ranges from low to moderately high in calories (50 kcal per 100 g) and carbohydrates (10%). It is a good source of fiber and vitamin C, and a fair source of iron.

TRIGLYCERIDES

Chemical compounds which contain three (tri-) molecules of fatty acid combined with one molecule of glycerol. Triglycerides differ from each other because the types of fatty acids linked to the glycerol vary. Fig. T-16 illustrates the formation of a triglyceride. The R in Fig. T-16 indicates a carbon chain which varies according to the fatty acid, while the COOH in Fig. T-16 is the acid portion of the fatty acid. When fatty acids combine with glycerol, water (H_2O) is formed.

(Also see FATS AND OTHER LIPIDS; and OILS, VEGETABLE.)

Fig. T-16. The formation of a triglyceride.

TRIPE

Most tripe is made from the first and second (rumen and reticulum) stomachs of cattle by: washing thoroughly; removing the stomach lining by scalding and scraping; and either (1) pickling in a 60°F (16°C) salt brine, or (2) cooking and pickling in a weak salt and vinegar brine.

TRIPEPTIDES

These are three amino acids chemically joined together by peptide bonds. They may be a step in the synthesis of new protein, or a step in the breakdown of dietary protein.

(Also see DIGESTION AND ABSORPTION; and PROTEIN[S].)

TRISTEARIN

A triglyceride of stearic acid.

TRITICALE *Triticum X Secale*

This is a hybrid cereal derived from a cross between wheat (*Triticum*) and rye (*Secale*), followed by doubling the chromosomes in the hybrid. The objective of the cross: to combine the grain quality, productivity, and disease resistance of wheat with the vigor and hardiness of rye. The first such crosses were made in 1875 when the Scottish botanist Stephen Wilson dusted pollen from a rye plant onto the stigma of a wheat plant. Only a few seeds developed and germinated; and the hybrids were found to be sterile. In 1937, the French botanist Pierre Givaudon produced some fertile wheat-rye hybrids with the ability to reproduce. Intensive experimental work to improve triticale was undertaken at the University of Manitoba, beginning in 1954.

Fig. T-17. Triticale grains. (Photo by International Development Research Center, Ottawa, Canada)

In comparison with wheat, triticale (1) has a larger grain, but there are fewer of them in each head (spike), (2) has a higher protein content, with a slightly better balanced amino acid composition and more lysine, and (3) is more winter hardy. Nutritional information on triticale flour is presented in the Food Composition Table F-21.

Earlier varieties of triticale did not offer much promise for milling and baking; they resembled their rye parent more than their wheat parent. However, bread of good quality has been made from more recent triticale selections. So far, the main use for triticale has been as a feed grain, pasture, green chop, and silage crop for animals.

(Also see CEREAL GRAINS.)

TRUFFLE *Tuber* spp

The truffle is a pungent wild fungus that grows underground, which is highly prized as a food. The fungus is believed to have a symbiotic relationship with the roots of oak and beech trees, near which it is usually found. Agronomists have not been successful in cultivating truffles. Depending on the species, they vary from white to brown or black in color. They range from ¼ to 4 in. (0.6 to 10 cm) in diameter; and they grow in clusters 3 to 12 in. (7.7 to 30 cm) below the ground. They resemble an acorn, a walnut, or a potato in shape. Most culinary truffles are found in western Europe. *Tuber melanosporum* is the famed black truffle of France, while *Tuber magnatum* is the more pungent and odoriferous white truffle that grows in the Italian Piedmont. Also, North Africa produces a truffle (*Terfezios*) in some quantity and they are occasionally found along the Pacific Coast in the United States. Trained dogs or pigs are usually used to sniff out the location of truffles (See Fig. T-18). From 300 to 500 tons are harvested annually.

(Also see MUSHROOMS.)

Fig. T-18. Hunting for truffles in France. (Courtesy, The Bettmann Archive)

TRYPSIN

A digestive enzyme formed in the small intestine when another enzyme, enterokinase, acts on trypsinogen, an inactive secretion of the pancreas. Trypsin cleaves polypeptides or proteins at peptide bonds adjacent to the amino acids arginine or lysine.

(Also see DIGESTION AND ABSORPTION.)

TRYPSINOGEN

Inactive form of trypsin.

TRYPTOPHAN

One of the essential amino acids, isolated and described by E. G. Hopkins of Cambridge University, England.

(Also see AMINO ACID[S].)

TUBE FEEDING

Among the circumstances which may necessitate tube feeding are surgery of the head and neck, esophageal obstruction, severe burns, gastrointestinal surgery, anorexia nervosa, and coma. Three types of tube feeding may be employed depending upon the situation: (1) a nasogastric tube (from the nose to the stomach) or a nasoduodenal tube (from the nose to the duodenum); (2) a surgically formed opening (stoma) in the stomach; or (3) a surgically formed opening (stoma) in the small intestine. Obviously, the nasogastric tube or the nasoduodenal tube is the method of choice but instances may arise when it is impossible to pass a tube from the nose to the stomach or the duodenum, or when there is gross disease of the stomach or the duodenum. All three types of tube placements are capable of supplying all types of adequate nutrients in a liquid form.

Since the nasogastric tube or the nasoduodenal tube is the most common, it will serve to illustrate the principles of tube feeding.

As the name implies, the tube is passed down the nose through the esophagus, and then to the stomach or duodenum. Extreme caution must be exercised in ensuring that the tube is in the stomach or duodenum, and not in the lungs. Also, in precoma or in comatose patients, the danger of aspiration pneumonia as a result of regurgitation of digestive juices must be kept in mind. Because there is less hazard of aspiration from nasoduodenal tube feeding than from nasogastric tube feeding, it is the method of choice of most physicians.

In order for tube feeding to be effective, it must meet several criteria. It must be: (1) nutritionally adequate; (2) easily digested without reactions such as diarrhea or constipation; (3) simply prepared; (4) well tolerated without inducing vomiting; and (5) low cost. There are three tube feedings which meet these criteria and are commonly used: (1) milk-base; (2) milk-base with suspended solids from strained or blenderized foods; and (3) synthetic low-residue formulas. Recipes are available for home preparation of these types of tube feedings. Commercial preparations are also available. All tube feedings generally contain ½ to 1½ kcal/ml.

Foods introduced into the tube are generally warmed to body temperature, and flow into the tube by gravity, or by the use of a food pump. The first tube feedings should be frequent and in small volumes, then if tolerated the volume of feedings may be gradually increased. Two liters per day is a common volume.

Several potential hazards of tube feedings are apparent: (1) tube feeding is dangerous when the patient is vomiting; (2) contaminated food leads to gastrointestinal infections; (3) concentrated feeding (high sodium or high protein) may lead to diarrhea, dehydration, and elevated blood urea; and (4) high concentrations of sugar, particularly lactose, are apt to produce diarrhea.

TUBER

A short, thickened, fleshy stem, or terminal portion of a stem or rhizome that is usually formed underground, bears minute leaves each with a bud capable under suitable conditions of developing into a new plant, and constitutes the resting stage of various plants such as the potato and Jerusalem artichoke.

(Also see VEGETABLE[S].)

TUBERIN

A globulin constituting the principal protein of the potato tuber.

TUN

Liquids, especially wine, ale, or beer, may be stored in a large cask or barrel called a *tun*. At one time, wine, liquor, and some other liquids were measured in terms of a tun, which equals 252 gallons (*958 liters*).

TURBINADO SUGAR

Sometimes this sugar is viewed erroneously as a raw sugar (sucrose), but turbinado sugar goes through a refining process to remove impurities and most of the molasses. It is produced by separating raw sugar crystals and washing them with steam. If produced under proper conditions, it is edible.

(Also see RAW SUGAR; and SUGAR.)

TURTLE

Reptiles with bodies encased in a bony shell. Both land turtles and water turtles are eaten, but it is the water turtle that is made into the famous turtle soup. The English hold turtle soup in high esteem; it is often served at their great diplomatic dinners and ceremonial repasts.

TWADDELL

This is a hydrometer used for measuring density (specific gravity) of industrial liquids which are greater than one. Measures of specific gravity between 1 and 2 are divided into 200 equal parts, and each division is 1 degree. The specific gravity of a solution equals the number of degrees on the Twaddell scale multiplied by 5 and divided by 1,000. The Twaddell is named for its inventor.

TYRAMINE

A pressor amine that has action similar to epinephrine; found in mistletoe, certain cheeses, some wines, and ergot, and also obtained from tyrosine by strong heating or by bacterial action.

TYROSINASE

A copper-containing enzyme responsible for the conversion of tyrosine to the dark pigment, melanin. Skin color depends, in part, upon the concentration of melanin. A genetic absence of tyrosinase results in albinism. Tyrosinase is also the enzyme responsible for the browning of cut surfaces of certain fruits and vegetables; as such, it is described by the generic term phenolase.

(Also see PHENOL OXIDASES.)

UDO

A Japanese plant grown for its tender young shoots which are eaten as a vegetable and in salads.

ULCERS, PEPTIC

Contents	Page
Occurrence	1023
Duodenal	1023
Gastric	1023
Causes	1023
Symptoms	1023
Diagnosis	1023
Treatment	1024
Physical	1024
Psychological	1024
Diet	1024
Prevention	1024

An ulcer is any open sore, other than a wound. Peptic ulcers are open sores or erosions of the surface lining of the digestive tract, usually in the stomach or duodenum. Their appearance is roughly comparable to that of canker sores of the mouth—for those unfortunate sufferers familiar with the appearance of canker sores. The term peptic ulcer is employed since it appears that these ulcers develop from the lessened ability of the lining of the digestive tract to withstand the digestive action of pepsin and hydrochloric acid (HCl). Ulcers can occur in any area of the digestive tract which is exposed to the action of these two substances. However, a vast majority of ulcers occur in two locations: (1) the first part of the small intestine, or duodenum, before the point of entry of the alkaline secretions of the pancreas; and (2) the stomach or gastric portion of the digestive tract, usually along the lesser curvature near the pylorus—opening from the stomach into the duodenum.

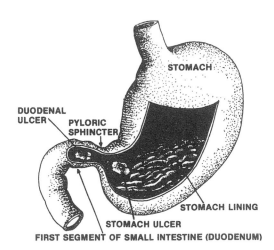

Fig. U-1. The major sites where peptic ulcers occur.

OCCURRENCE. Overall, peptic ulcers occur in about 10% of all Americans at some time of their lives. Between the two locations, duodenal ulcers are far more common than gastric ulcers. Unfortunately, both are characterized by high recurrence rates.

Duodenal. About 80% of all peptic ulcers are duodenal ulcers. These occur most often in men between the ages of 20 and 50. Thus, after puberty and before menopause the ratio of affected males to females varies from 3:1 to 10:1. For some unexplained reasons, duodenal ulcers are now less common than they were 20 years ago.

Gastric. Again, more males than females are affected by a ratio of about 3.5:1, and, as indicated, gastric ulcers occur less frequently than duodenal ulcers—the ratio being about 1:4. Interestingly, about 20% of the gastric ulcers occur in individuals who have, or who had, a duodenal ulcer.

CAUSES. The basic causes of peptic ulcers remain obscure. It is known that physiologically there is a disturbance in the acid-pepsin secretion and the tissue resistance of the digestive tract lining. Other factors, such as chemicals, heredity, and emotions also seem involved in some cases. Hence, peptic ulcers are likely to be caused by multiple factors.

SYMPTOMS. The hallmark of peptic ulcers is the sequence of pain-food-relief—the upper abdominal pain associated with ulcers when the stomach is empty, which is relieved by ingesting food or bland liquids. Thus, pain usually relates to eating patterns, and usually occurs several hours after eating and between midnight and 2:00 a.m. Interestingly, this pain is not present before breakfast. Occurrence of the pain relative to eating habits is an important point of differential diagnosis, as persons suffering from gallbladder diseases experience pain that is triggered by eating, rather than relieved. The pain experienced by ulcer sufferers is described as a steady, gnawing, burning, aching, or hungerlike discomfort. However, some victims have no pain even before serious complications develop, and, in some cases, gastric ulcer symptoms may be worsened by eating. A good rule is to have any persistent abdominal pain checked by a physician.

Ulcers that bleed slowly cause the stool to be black and may produce anemia. Other ulcers may bleed more profusely and lead to vomiting of dark brown blood.

DIAGNOSIS. Once a physician suspects an ulcer, x rays are taken after barium is swallowed. These x rays outline the stomach and small intestine and will clearly demonstrate a *crater* in most ulcer victims. This x ray procedure is commonly called an *upper GI series*. In most duodenal ulcers, an *upper GI series* is all that is required. Gastric ulcers, however, require some further tests since a small percentage of gastric ulcers are cancers that look like ulcers by x ray. Hence, gastric ulcer patients will probably be subjected to gastroscopy in an effort to rule out malignant disease. This involves passing a flexible lighted tube into the stomach for the direct visual inspection of the ulcer.

TREATMENT. Treatment, of course, varies with each individual; however, there are some general practices and recommendations which deserve mention. The facets of treatment involve tangible or physical approaches and intangible or psychological approaches.

Physical. Since psychological factors such as emotions and personality can have a large role in development of ulcers, treatment should stress individualized attention to the whole person and not just the ulcer. Nevertheless, some physical manipulations have proven helpful in the treatment of ulcers. These include diet, drugs, and surgery.

Psychological. Many ulcer patients improve within 24 to 48 hours with a change in environment; hence, the psychological side of treatment cannot be ignored. Both physical and psychological rest promote the healing of peptic ulcers. To this end sedatives may be prescribed. Often victims are urged to curtail their business and social responsibilities. Furthermore, those treating peptic ulcer patients need to demonstrate support and reassurance.

DIET. Table U-1 provides some general guidelines from which various diet plans may be derived for each individual during convalescence from a peptic ulcer.

PREVENTION. Peptic ulcers tend to recur. Those that recur and do not respond to treatment, or that recur frequently, may require surgery. Of course, at the first inkling of recurrent symptoms active therapy should be initiated.

A peptic ulcer provides one more reason for giving up smoking, since it appears to aggravate and contribute.

Being able to relax—although difficult to determine for each individual—would seem to offer some prevention due to the implication of emotions and personality type in the development of peptic ulcers. Each individual should determine the best form of relaxation based on needs and experiences.

(Also see ANTACIDS; MODIFIED DIETS; SIPPY DIET; and STRESS.)

TABLE U-1
GUIDELINES FOR SELECTING FOODS FOR PERSONS WITH PEPTIC ULCERS

Food Category[1]	Foods to Use	Foods to Avoid
Bakery products	Enriched white bread; toast; soda crackers; zwieback; melba toast; hard rolls; sponge cakes; other plain cakes; sugar cookies.	Whole wheat and other grain breads; hot bread; graham or coarse crackers; cakes and pastries containing dried fruit or nuts; spice cakes.
Beverages	Postum; decaffeinated coffee.	Alcohol; carbonated drinks; coffee; tea; chocolate.
Cereals and flours	Refined cereals; all ready-to-eat cereals; rice; oatmeal; macaroni; spaghetti; noodles.	Whole grain cereals; cereals containing bran; shredded wheat.
Desserts and sweets	Plain custard; rennet; plain puddings; gelatin desserts; clear plain jelly; honey; moderate amounts of sugar.	Excesses or concentrated sources like candy.
Eggs and substitutes	Boiled; poached; coddled; or scrambled; plain omelets.	Fried eggs.
Fats and oils	Margarine	All fried foods.
Fish and seafoods	Fresh or frozen; boiled; broiled; or baked; scalded canned tuna or salmon; oysters fresh or canned.	Smoked, preserved or pickled fish, crab, lobster, or sardines; other salted fish or fatty fish such as herring or mackerel.
Flavorings and seasonings	Salt	Most all, such as pepper, horseradish, catsup, mustard, vinegar, spices.
Fruits	Applesauce, baked apples without peel; ripe bananas; papayas; stewed or canned pears, peaches, apricots—all without peel.	Dried fruits, skins and peel of all fruits; raisins.
Gravies and sauces	Creamed sauces.	Gravies made from meat extracts.
Juices	Most all, but dilute half and half with water or drink after eating other foods.	
Meats	Tender cuts of beef, lamb, and veal; sliced chicken; liver; crisp bacon.	Fried, smoked, pickled, or cured meats; fat; sausages; ham; salami; wieners.
Milk and Milk products	Buttermilk; cream; milk; plain ice cream; plain milk shakes; cottage cheese; mild cheeses; butter; sour cream; yogurt.	Chocolate; strong cheeses.
Nuts and seeds		All nuts and seeds.
Pickles and relishes		All pickles and relishes.
Soups and chowders	Cream soups using such vegetables as asparagus, beets, carrots, peas, green beans, potatoes, tomatoes, and spinach.	All meat soups, soups from any other vegetables; all canned soups; dehydrated soups; broths; bouillon.
Vegetables	Raw cabbage, well-cooked or canned asparagus, carrots, peas, beets, peeled tomatoes, squash, spinach, green beans, mashed or baked potatoes without skins; sweet potatoes without skins; mushrooms.	All gas-forming vegetables which include cauliflower, Brussels sprouts, broccoli, cucumbers, onions, radishes, turnips; salad, coleslaw.

[1]Food categories are those listed in the Food Composition Table F-21 of this book.

ULTRAHIGH TEMPERATURE STERILIZATION (UHT STERILIZATION)

Rapid heating of a food to temperatures in the range of 205° to 307°F (96° to 153°C), then holding at these temperatures for three seconds or less. In combination with aseptic packaging, this process is designed for the manufacture of products with improved keeping qualities.

(Also see PRESERVATION OF FOOD.)

ULTRASONIC HOMOGENIZER

The use of high-frequency—over 20,000 vibrations per second—pressure waves of the same nature as sound waves to disintegrate fat globules or break up particles. This process is employed in a variety of food emulsifications, such as fruit and vegetable purees, milk homogenization, and peanut butter.

ULTRAVIOLET LIGHT (ULTRAVIOLET RADIATION)[1]

Light is a form of radiant energy. It travels in vibrations or waves. Each color of light vibrates with a different wavelength and carries a specific amount of energy.

The range of these wavelengths is measured in nanometers (nm), a unit of length equal to:

$$\frac{1}{100,000,000} \text{ mm}$$

The range of electromagnetic waves of some radiations follows:

Radiation	Approximate Wavelength (nanometer [nm])
X ray	50
Ultraviolet	300
Visible light	600
Infrared rays	1000

All radiations are important, but the discussion that follows will be limited to ultraviolet light because of its importance in human health and nutrition.

Ultraviolet light is generally divided into the near (400 to 300 nm), the far (300 to 200 nm), and the vacuum (200 to 4 nm) ultraviolet regions. The last wavelengths, which are particularly harmful to life, are strongly absorbed by the earth's atmosphere.

Ultraviolet light is created by the same processes that generate visible light—transitions in atoms in which an electron in high-energy state returns to a less energetic state.

USES OF ULTRAVIOLET LIGHT. Ultraviolet light serves mankind in numerous ways; among them, the following:

1. **Manufacture of vitamin D in the skin.** If the skin is exposed to the sun or to a sunlamp, the ultraviolet rays act on the provitamin 7-dehydrocholesterol in the skin and transform it into vitamin D_3.
(Also see VITAMIN D.)
2. **Formation of vitamin D in plant foods.** If ergosterol (the provitamin found in plants) is irradiated with ultraviolet light, vitamin D_2 is produced. This is the well-known Steenbock Irradiation Process, which was patented by Dr. Harry Steenbock of the University of Wisconsin.
3. **Sterilization.** Certain ranges of ultraviolet rays can kill bacteria. This property is used in sterilizing water, milk and other foodstuffs, and medicinal facilities and equipment. Also, modern food and drug plants often use germicidal lamps to disinfect their products and the containers.
4. **Industrial uses.** Ultraviolet rays have many important industrial uses, including use in fluorescent lamps and sun lamps, in testing materials, in identifying ores in mining, and

in lighting the instrument panels of aircraft. Scholars use the fluorescent effect of ultraviolet rays to examine old documents.

Ultraviolet rays, within moderation, are believed to have an invigorating effect upon the human body. But excessive exposure may be harmful, especially to the eyes and skin. Fair skinned people and those who sunburn easily and do not tan are especially susceptible to ultraviolet light-induced cancer of the skin.

NOTE WELL: Currently, there are many sunscreen lotions available. Some contain para-aminobenzoic acid (PABA). However, some dermatologists are recommending other sunscreen lotions; so, for protection against sunburn, see your dermatologist.
(Also see PARA-AMINOBENZOIC ACID.)

UMBLES

The entrails of an animal used as a food.

UNDERWEIGHT

Contents	Page
Effects On Health	1025
Beneficial Effects	1025
Potentially Harmful Effects	1026
Factors That May Be Responsible For Underweight	1026
Dietary Practices	1026
Other Factors	1026
Assessment Of Weight Status	1026
Body Weight For Height, Age, and Sex	1026
Skinfold Thicknesses	1028
Midarm Muscle Circumference	1028
Measures for Gaining Weight	1028
Summary	1028

This term means different things to different people. To some, it refers to mere slenderness, whereas to others it signifies undernourishment to the extent that a risk to health may be present. This article considers underweight to be a body weight that is notably low for a person's age, height, sex, and other personal characteristics. It is also considered to be an indication that the body contents of fat and lean tissue may be too low to ensure optimal responses to stresses such as chilling, diseases, fasting, injuries, prolonged and/or strenuous physical activities, and surgery.

EFFECTS ON HEALTH. The desirable or undesirable effects of underweight on the health of different people depend somewhat upon factors such as their life-styles (active or sedentary), their overall states of physical fitness, and the presence or absence of certain other health conditions. This means that some people are likely to be healthiest when they are moderately underweight, whereas others might be fittest for the important activities of their lives when they are average weight, or a little overweight. Therefore, the various effects of underweight are noteworthy.

Beneficial Effects. The people who may benefit the most from being moderately underweight are those who engage in predominantly intellectual work (artists, office workers, students, teachers, writers, etc.) and live otherwise sedentary lives because it is generally believed that slender people are less likely to become sleepy and tired when physical activity is limited. Runners and other athletes whose sports require the ability to maintain a rapid pace may also benefit from being a little underweight. Finally, people living in tropical climates or those troubled with one or more of the following conditions may be better off underweight than overweight: adult onset diabetes, angina pectoris, athero-

---

[1]This section was authoritatively reviewed by, and helpful suggestions were received from, Stephen A. Book, Ph.D., California Environmental Protection Agency, State of California, Sacramento.

sclerosis, breathing difficulties, chronic tiredness, congestive heart failure, coronary heart disease, enlargement of the heart, excessive numbers of red blood cells (polycythemia), gallstones, and troubles with the legs and feet.

Potentially Harmful Effects. In the early 1900s, underweight people were charged higher life insurance premiums than other policy holders because they were more susceptible to tuberculosis, which at that time was often fatal. Today, tuberculosis is rare, except in certain poverty-stricken areas, and leanness is extolled as a virtue. (Now, some obese policyholders may be charged higher premiums.) Nevertheless, there are certain other potentially harmful effects of a pronounced underweight, which may sometimes be the result of chronic undernutrition. Some of the more common undesirable effects are:

1. Cessation or irregularity of menstrual periods in females.
2. Growth retardation in growing infants and children.
3. Increased susceptibility to chilling and infections.
4. Lack of vigor, endurance, and sexual drive.
5. Low resistance to certain stresses.
6. Slow healing of injuries and surgical wounds.
7. Tendencies to be apathetic, irritable, listless, mentally depressed, nervous, restless, and sleepless.
8. Weak muscles.

FACTORS THAT MAY BE RESPONSIBLE FOR UNDERWEIGHT.
Lack of sufficient dietary calories is only one of the causes of underweight. Hence, attempts to bring about a gain in weight by merely increasing the caloric content of the diet may fail when there are other contributing factors that are unrecognized. Therefore, some of the major causes of underweight are noteworthy.

Dietary Practices. Many people would like to be able to eat more food without gaining weight, yet there are some who appear to have difficulty eating enough to maintain a normal body weight. Furthermore, the problem of consuming sufficient calories is more common in infancy, childhood, and adolescence than it is in adulthood because much more food energy per pound of body weight is required during growth.

Other reasons why some people may find it difficult to consume sufficient calories follow:

1. The diet may contain too much fibrous vegetable matter, which is very filling. Distention of the small intestine with bulky food curbs the appetite in many people. It is noteworthy that, on the average, vegetarians tend to be more slender than meat eaters.
2. Excessive amounts of fluids may be taken with meals. Hence, filling occurs before sufficient solid food is consumed. A similar effect occurs when the foods themselves are very high in water content, as in the cases of gelatin desserts or salads, thin soups and stews, and various vegetable dishes.
3. Oversalted foods may cause feelings of fullness in some people because the salt draws fluid from the intestinal wall into the intestinal contents, thereby producing a bulking effect.
4. Too much fat may be consumed. Fatty foods delay the emptying of the stomach and may make some people lose their appetites.

Finally, the diets of some underweight people may be lacking in the minerals and vitamins that promote hearty appetites. For example, a deficiency of the essential trace mineral zinc leads to loss of taste, and, an insufficient amount of the vitamin thiamin can lead to loss of appetite.

Other Factors. Even when adequate amounts of calories are consumed, various factors that alter the normal utilization of food by the body may prevent the gaining of weight; among them, the following:

1. Digestive disturbances.
2. Febrile illnesses.
3. Glandular disorders.
4. Parasitic infestations.
5. Excessive physical activity.

ASSESSMENT OF WEIGHT STATUS.
It may not always be easy to determine when a person is sufficiently underweight to require that remedial measures be taken because there is considerable variation in body weights among people of equal heights and the same sex.

Body Weight for Height, Age, and Sex. One procedure for assessing the weights of adults is to consider that the ranges of normal weights for males or females extend from the 10th to the 90th percentiles of the average weights for selected heights of persons aged 25 to 34. The young adult group is taken as a standard because it is widely believed that weight gained after the attainment of maturity is a detriment rather than a benefit to health. These data are given in Table U-2.

The problem with the data in Table U-2 is that there is no specification of body build. Hence, an adult with a large frame might be underweight with a weight falling between the 10th percentile and the average weight. Therefore, some other measures of the *fleshiness* of a person are needed.

NOTE: Weight norms for infants, children, and adolescents are given in the sections on CHILDHOOD AND ADOLESCENT NUTRITION; and INFANT DIET AND NUTRITION.

Fig. U-2. Slenderness is considered attractive, but being too thin may be hazardous to health.

TABLE U-2
AVERAGE WEIGHTS (without clothing) FOR PERSONS AGED 25 TO 35 YEARS[1,2]

Height (without shoes)			Men						Women						
			Light[3] (10th percentile)		Average		Heavy[4] (90th percentile)		Light[3] (10th percentile)		Average		Heavy[4] (90th percentile)	
(ft)	(in.)	(cm)	(lb)	(kg)	(lb)	(kg)	(lb)	(kg)	(lb)	(kg)	(lb)	(kg)	(lb)	(kg)
4	10	147	—	—	—	—	—	—	97	44	119	54	147	67
4	11	150	—	—	—	—	—	—	97	44	126	57	151	69
5	0	152	—	—	—	—	—	—	97	44	125	57	171	78
5	1	155	—	—	—	—	—	—	103	47	130	59	172	78
5	2	157	—	—	—	—	—	—	107	49	135	62	172	78
5	3	160	110	50	151	69	196	89	109	50	138	63	178	81
5	4	163	126	57	151	69	195	89	110	50	141	64	188	86
5	5	165	128	58	155	71	194	88	115	52	143	65	187	85
5	6	167	134	61	160	73	191	87	118	54	146	67	190	86
5	7	170	136	62	168	77	199	91	123	56	153	70	191	87
5	8	173	140	64	166	76	196	89	129	59	162	74	217	99
5	9	175	148	68	176	80	225	102	124	56	153	70	177	81
5	10	178	148	68	185	84	215	98	—	—	—	—	—	—
5	11	180	142	65	178	81	212	96	—	—	—	—	—	—
6	0	183	157	72	190	86	224	102	—	—	—	—	—	—
6	1	185	165	75	195	89	224	102	—	—	—	—	—	—
6	2	188	158	72	191	87	226	103	—	—	—	—	—	—

[1]Adapted by the authors from *Weight by Height and Age for Adults 18-74 Years: United States 1971-74*, U.S. Dept. of Health, Education, and Welfare, 1979, pp. 18, 21, Tables 2 and 3.
[2]**Note:** These weights are not what are commonly called "ideal" or "desirable" weights because recent studies have cast doubt upon the basis for the latter.
[3]Most people who weigh less than this amount are likely to be underweight.
[4]Most people who weigh more than this amount are likely to be obese.

Fig. U-3. Underweight people don't need to worry about eating too many calories. Sweet and Sour Broccoli Chicken. (Courtesy, Castle & Cooke, Inc.—Bud of California Fresh Vegetables.)

Skinfold Thicknesses. About one-half of the body fat content is distributed under the skin. Therefore, the thicknesses of various skinfolds are measures of body fatness. For example, it is generally considered that a triceps skinfold (which is usually measured at a point about halfway between the shoulder and the elbow) smaller than ½ in. (*12.5 mm*) for men or less than ⅝ in. (*16 mm*) for women is indicative of insufficient subcutaneous fat or underweight condition.

Midarm Muscle Circumference. This calculated value gives an approximate estimate of the muscle mass, which is useful in determining whether an underweight person is lacking fat alone (measured by skinfold thickness) or is short on both fat and muscle. The value for the midarm muscle circumference (MAMC) is derived from measurements of the triceps skinfold (TSF) and the midarm circumference (MAC) by the following formula:

$$MAMC = MAC - (3.14 \times TSF)$$

For example, a man who has a midarm circumference (measured by placing a tape measure around the upper arm where the triceps skinfold is measured) of 11 in. (*279 mm*), and a skinfold thickness of ½ in. (*12.5 mm*) would have a midarm muscle circumference of 9 %₁₆ in. (*240 mm*). These measurements would indicate that while he has just enough subcutaneous fat, he is lacking some muscle since the standard values for muscle circumferences are 10 in. (*254 mm*) for males and 9 in. (*232 mm*) for females.

MEASURES FOR GAINING WEIGHT. The selection of the measures to be used for gaining weight should take into account whether additional fat alone is needed, or a combination of extra muscle and fat is required. A diet high in calories will be suitable for the former objective, but the latter objective requires a diet that is high in both calories and protein.

Some suggestions for using high-calorie foods to gain weight are given in MODIFIED DIETS, Table M-28, Modified Diets/High-calorie diet.

Suggestions for using high-protein foods as sources of additional calories and protein are given in MODIFIED DIETS, Table M-28, Modified Diets—High protein, high calorie diet.

SUMMARY. Ideal body weights for different people vary according to age, sex, body build, life-style, and other factors. Hence, some feel better when slender, whereas others may benefit from a little extra weight. Therefore, a doctor should be consulted before one attempts to gain or lose a considerable amount of weight.

UNESCO

United Nations Educational, Scientific, and Cultural Organization, with headquarters in Paris, France.

UNIDENTIFIED FACTORS

In addition to the vitamins as such, certain unidentified or unknown factors are important in nutrition. They are referred to as *unidentified* or *unknown* because they have not yet been isolated or synthesized in the laboratory. Nevertheless, rich sources of these factors and their effects have been well established. A diet that supplies the specific levels of all the known nutrients but which does not supply the unidentified factors may be inadequate for best perform-ance. There is evidence that these factors exist in dried whey, marine and animal products, distillers' solubles, antibiotic fermentation residues, yeasts, alfalfa meal, and certain green leafy vegetables.

UNSATURATED FAT

A fat having one or more double bonds; not completely hydrogenated.

(Also see FATS AND OTHER LIPIDS.)

UNSATURATED FATTY ACID

Any one of several fatty acids containing one or more double bonds, such as oleic, linoleic, linolenic, and arachidonic acids.

(Also see FATS AND OTHER LIPIDS.)

UREA (NH₂CONH₂)

Before amino acids (protein) can enter energy metabolic pathways, the amino group (NH₂) must first be removed. This is accomplished by a process called deamination which produces a carbon skeleton that can be used for energy, and ammonia (NH₃). To facilitate elimination from the body, this ammonia is converted to urea in the urea cycle in the liver. Urea circulates in the blood, and is removed from the blood by the kidneys. It is the main nitrogenous constituent of urine. Normally 20 to 35 g of urea are excreted in the urine every 24 hours.

(Also see METABOLISM; PROTEIN[S]; and UREA CYCLE.)

UREA CYCLE (KREBS-HENSELEIT CYCLE)

Most of the ammonia (NH₃) produced through deamination—removal of amino groups (NH₂) from amino acids—is converted to urea in the liver for excretion by the kidneys. To facilitate elimination, 1 mole of ammonia (NH₃) combines with 1 mole of carbon dioxide (CO₂), another metabolic waste product. This compound is then phosphorylated to produce carbamyl phosphate. Carbamyl phosphate then combines with ornithine to form citrulline—an intermediate in the urea cycle. The amino acid, aspartic acid, contributes another amino group (NH₂), and citrulline is then converted to the amino acid arginine. Urea splits off from arginine forming ornithine, and the cycle is completed. Fig. U-4 illustrates the urea cycle.

(Also see METABOLISM; PROTEIN[S]; and UREA.)

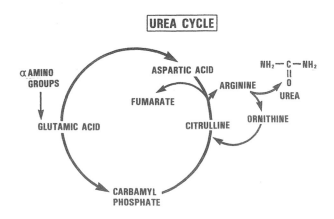

Fig. U-4. Urea cycle (nitrogen excretion).

UREASE

An enzyme which acts on urea to produce carbon dioxide and ammonia. It is found in the human digestive tract where it acts on the urea in the digestive secretions.

UREMIA

A toxic accumulation of urinary constituents in the blood.

URIC ACID

The main end product of purine—adenine, guanine and xanthine—metabolism in birds, reptiles, and man, which is excreted in the urine. Uric acid is formed from purines consumed in the diet, and from body purines derived from the breakdown of nucleic acids. Mammals other than man further metabolize uric acid to allantoin, which is excreted in the urine. However, man lacks the enzyme, urate oxidase, necessary for this conversion. Therefore, about 0.5 to 1.0 g of uric acid is lost in the urine each day. A high level of uric acid in the blood is associated with the development of gout. In addition, uric acid may form kidney stones, especially in individuals suffering from gout.

(Also see ARTHRITIS, section headed "Gout [gouty arthritis]"; PURINES; and URIC ACID KIDNEY STONES.)

URIC ACID KIDNEY STONES

Kidney stones formed from uric acid—the excretory product of purines in humans. Stones develop when (1) blood levels of uric acid are elevated, (2) urinary uric acid excretion is high, and (3) an acid urine is produced. Individuals suffering from gout are a thousand fold more likely to develop uric acid kidney stones than the general population. Formation of uric acid kidney stones may be minimized by keeping the urine alkaline, increasing the fluid intake, and reducing the purine intake. Most fruits and vegetables contribute to the formation of an alkaline urine, while meat, fish, and poultry contribute to the formation of an acid urine. A purine-restricted diet begins with the elimination of foods with high purine content such as anchovies, asparagus, brains, kidneys, liver, meat extracts, mincemeat, sardines, and sweetbreads.

(Also see ARTHRITIS, section headed "Gout [gouty arthritis]"; MODIFIED DIETS; and KIDNEY STONES.)

URICASE (URATE OXIDASE)

A copper-containing enzyme which all mammals possess with the exception of man and other primates. It is responsible for the conversion of uric acid to allantoin.
(Also see URIC ACID.)

URINARY CALCULI (WATER BELLY, KIDNEY STONES, UROLITHIASIS)

Mineral deposits which occur in the urinary tract. These deposits may block the flow of urine, followed by rupture of the urinary bladder and death. In severe cases of some duration, watery swellings (edema) of the lower abdomen may develop.

(Also see KIDNEY STONES; and URIC ACID KIDNEY STONES.)

URINARY SYSTEM

The organs engaged in the excretion of urine. They consist of the kidneys, the ureters, the bladder, and the urethra.

URINE

Liquid or semisolid matter produced in the kidneys and discharged through the urinary organs. Normally, it is a clear, transparent, amber-colored, slightly acid fluid.
(Also see KIDNEYS; UREA; and UREA CYCLE.)

USDA

The abbreviation commonly used for the United States Department of Agriculture.
(Also see U.S. DEPARTMENT OF AGRICULTURE.)

U.S. DEPARTMENT OF AGRICULTURE (USDA)

In 1796, President George Washington recommended the creation of a national board of agriculture, which eventually developed into the United States Department of Agriculture. The USDA and the FDA (Food and Drug Administration) share the responsibility for assuring the safety of the nation's food supply. Meat and meat products, and poultry and poultry products are the main food concerns of the USDA, operating under the authority of the Federal Meat Inspection Act (FMIA) and the Poultry Products Inspection Act (PPIA). Also, the USDA is involved in research, educational, nutritional, and health programs. The following eight divisions of the USDA have primary responsibilities in the areas of human health and nutrition: (1) the Agriculture Research Service; (2) the Animal and Plant Health Inspection Service; (3) the Consumer and Marketing Service; (4) the Cooperative State Research Service; (5) the Federal Extension Service; (6) the Food and Nutrition Service; (7) the Labeling and Registration Section; and (8) the Veterinary Services Division.

(Also see DISEASES, section headed "U.S. Department of Agriculture.")

U.S. DEPARTMENT OF HEALTH AND HUMAN SERVICES (FORMERLY, HEALTH, EDUCATION, AND WELFARE)

An agency of the federal government with responsibility for public health in its broadest aspects. This agency is under the jurisdiction of the U.S. Public Health Service (USPHS) and includes the following: (1) Alcohol, Drug Abuse and Mental Health Administration; (2) Center for Disease Control (CDC); (3) Food and Drug Administration (FDA); (4) Health Resources Administration; (5) Health Services Administration; and (6) National Institutes of Health (NIH).

(Also see DISEASES, section headed "U.S. Department of Health and Human Services.")

USP (UNITED STATES PHARMACOPIA)

A unit of measurement or potency of biologicals that usually coincides with an international unit. (Also see IU.)

U.S. PUBLIC HEALTH SERVICE (USPHS)

This service is in the U.S. Department of Health and Human Services and is concerned with the prevention and treatment of disease. It works in the areas of vector control, pollution control, and control of communicable diseases of man. In addition to its own research program, the USPHS provides grants for health-related research at many universities and research institutes in the United States.

A part of this important complex is the National Institutes of Health (NIH), which was formed in 1930, and is composed of the following nine sister institutes:

The National Cancer Institute

The National Heart Institute

The National Institute of Allergy and Infectious Diseases

The National Institute of Arthritis and Metabolic Diseases

The National Institute of Dental Research

The National Institute of Mental Health

The National Institute of Neurological Diseases and Blindness (including multiple sclerosis, epilepsy, cerebral palsy, and blindness)

The National Institute of Child Health and Human Development

The National Institute of General Medical Science

U.S. RECOMMENDED DAILY ALLOWANCES (U.S. RDA)

These allowances are guides to the amounts of vitamins and minerals an individual needs each day to stay healthy. They were set by the Food and Drug Administration (FDA) as nutritional standards for labeling purposes. The U.S. RDAs are *based* on the Recommended Dietary Allowances established by the Food and Nutrition Research Council, and they replace the Minimum Daily Requirements (MDR) which were used for years as guidelines for labeling products. For practical purposes, the many categories of dietary allowances for males and females of different ages were condensed to as few as nutritionally possible for labeling. Generally, the highest values for the ages combined in a U.S. RDA were used. For example, the U.S. RDAs for adults and children over 4 years are representative, generally, of the dietary allowances recommended for a teenage boy.

There are four groupings of the U.S. RDAs, shown in Table U-3. The best known, and the one that will be used on most nutrition information panels and most vitamin and mineral supplements, is for adults and children over 4 years of age. The second is for infants up to 1 year, and the third is for children under 4 years. The latter two are to be used on infant formulas, baby foods, and other foods appropriate for these ages, as well as vitamin-mineral supplements intended for their use. The fourth is for pregnant women or women who are nursing their babies.

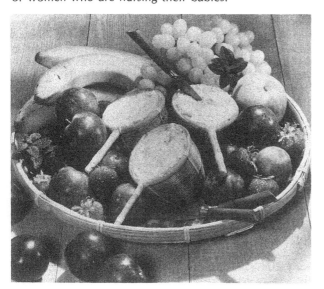

Fig. U-5. Fruit and cheese—common contributors to RDAs. (Courtesy, The California Plum Commodity Committee, Sacramento, Calif.)

TABLE U-3
U.S. RECOMMENDED DAILY ALLOWANCES (U.S. RDA)[1]

Nutrient	Adults and Children 4 Years or Older	Infants from Birth to 1 Year	Children Under 4 Years	Pregnant or Lactating Women
Required on labels:				
Protein ....g	50–63	13–14	16	60/65
Vitamin A ...mcg RE	500–1000	375	400	800/1,300
Vitamin C ....mg	60	30–35	40	70/95
Thiamin ....mg	1.5	0.3–0.4	0.7	1.5/1.6
Riboflavin ....mg	1.7	0.4	0.8	1.8
Niacin ....mg NE	20	5–6	9	17–20
Calcium ....mg	1,200	600	800	1,200
Iron ....mg	15	6–10	10	30/15
Optional on labels:				
Vitamin D ......mcg	10	7.5–10	10	10
Vitamin E ....IU	30	3–4	6	10–12
Vitamin B-6 .....mg	2	0.3–0.6	1	2.2
Folate ....mcg	200	25–35	50	400/280
Vitamin B-12 ..mcg	2	0.3–0.5	0.7	2.2/2.6
Phosphorus .....mg	1,200	400–600	800	1,200
Iodine ....mcg	150	40–50	70	175/200
Magnesium .....mg	350	40–60	80	320/355
Zinc ....mg	15	5	10	15/19
Copper ....mg	2	0.6	1	2
Biotin..........mcg	30–100	10–15	20	30/100
Pantothenic acid mg	4–7	2–3	3	4–7

[1]*Recommended Dietary Allowances*, NRC, 1989, pp. 284–285.

• **Labels and U.S. RDA**—Whenever a food product is labeled "enriched," or a product has added nutrients, or a nutritional claim is made for a product, the FDA requires that the nutritional content be listed on the label. In addition, many manufacturers put nutrition information on products when not required to do so. The lower part of the nutrition label must give the percentages of the U.S. Recommended Daily Allowances (U.S. RDA) of protein and of seven vitamins and minerals in a serving of the product, in the following order: protein, vitamin A, vitamin C, thiamin, riboflavin, niacin, calcium, and iron. As shown in Table U-3, listing is optional relative to the percentage of 12 other vitamins and minerals. Likewise, listing of cholesterol, fatty acid, and sodium content is optional—for now. Nutrients present at levels less than 2% of the U.S. RDA may be indicated by a zero or an asterisk which refers to the statement, "contains less than 2% of the U.S. RDA of these nutrients." Nutrition labels also list how many calories and how much protein, carbohydrate, and fat are in a serving of the product.

Unfortunately, some confusion has arisen since the National Research Council's *Recommended Dietary Allowances* have for years been called the RDAs; hence, U.S. RDA should always refer to the U.S. Recommended *Daily* Allowances used for labeling purposes.

(Also see ENRICHMENT [Fortification, Nutrification, Restoration]; LABELS, FOOD; NUTRIENTS: REQUIREMENTS, ALLOWANCES, FUNCTIONS, SOURCES; and RDA [Recommended Dietary Allowances].)

UTERUS (WOMB)

The organ that receives the fertilized egg from the fallopian tube and holds it during the 9 months of its growth until the infant is expelled at childbirth. It is situated deep in the pelvic cavity between the bladder and the rectum.

VACUUM COOLING

A means of rapidly cooling fruits and vegetables to prevent deterioration. Produce is sprayed with water, then while still wet it is subjected to a vacuum which causes the water to evaporate; thereby rapidly chilling the produce. Some vacuum coolers are railroad car size.
(Also see HYDROCOOLING.)

VACUUM-DEHYDRATED

Freed of moisture after removal of surrounding air while in an airtight enclosure.

VACUUMIZATION (VACREATION)

In milk processing, the subjection of heated milk to vacuum for the purpose of removing volatile off-flavors and odors and/or moisture.
(Also see DEAERATION.)

VALENCE

The power of an element or radical to combine with (or to replace) other elements or radicals. The valence number of an element is the number of atoms of hydrogen with which one atom of the element can combine.

VALINE

One of the essential amino acids.
(Also see AMINO ACID[S].)

VANADIUM (V)

Vanadium was first identified in 1831 by Sefstrom, of Sweden, who named it after the Norse goddess of beauty, Vanadis (because of the beautiful color of its compounds). However, it was not obtained in pure form until 1927. It has many industrial uses, especially in Vanadium steel.

In 1970, Dr. Klaus Schwarz demonstrated that the trace element Vanadium (V) is needed by higher animals. Vanadium-deficient diets resulted in retarded growth, impaired reproduction, increased packed blood cell volume and iron in the blood and bone of rats, and increased hematocrit in chicks.

VANASPATI

This is a term applied in India to hydrogenated fat made from vegetable oils. It is a butter substitute, like margarine in the United States.

VASCULAR

Pertaining to the blood vessels of the body.

VEGANS

A vegan is an individual who is a strict vegetarian. All foods of animal origin—meat, poultry, fish, eggs, and dairy products—are excluded from the diet of vegans. In addition, vegans share a philosophy and life-style.
(Also see VEGETARIAN DIETS.)

VEGETABLE(S)

Contents	Page
History	1032
World and U.S. Production	1033
Nutritive Values	1033
Effects of Processing and Preparation	1034
Vegetables of the World	1036

Fig. V-1. Common vegetables. (Courtesy, Shell Chemical Co., Houston, Texas)

In the phrase *the vegetable kingdom,* the word *vegetable* refers to the entire world of plants. In an 1893 decision, the U.S. Supreme Court held, in effect, that a plant or plant part generally eaten as part of the main course of the meal is a vegetable, while a plant part which is generally eaten as an appetizer, as a dessert, or out of hand is a fruit. As used herein, the term refers to the edible part of a plant that is consumed in raw or cooked form with the main course of a meal. Some typical parts of plants used as vegetables are: bulbs (garlic, onions), flowers (broccoli, cauliflower), fruits (pumpkins, squashes, tomatoes), leaves (lettuce, spinach), roots (beets, carrots), seeds (beans, corn, peas), stalks (celery), stems (asparagus), and tubers (Irish potatoes, yams). Although green land plants provide most of the vegetables for human diets, certain fungi and algae may also be used for vegetables. The different types of vegetables are shown in Fig. V-2.

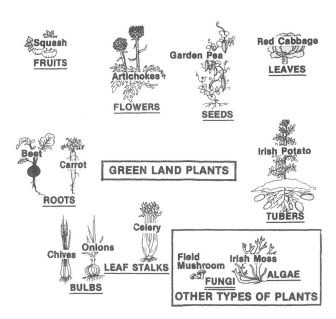

Fig. V-2. Common types of vegetables.

HISTORY. It has been difficult for archaeologists, botanists, geographers, and other researchers to trace the history of all of the vegetables people have used through the ages because (1) few remains of the more succulent parts of plants have been found at archaeological excavations since these parts decay very readily; (2) most of the identifiable plant fragments have been found in relatively dry areas, while evidence of vegetables used in the humid tropics is scarce; and (3) some of the vegetables that were once consumed by man may have become extinct, and others may have evolved into forms that bear little resemblance to the ancestral species. Nevertheless, enough evidence has been found to support some tentative conclusions about the first cultivation of the important vegetable crops.

Long before the development of the first agricultural societies some 10,000 to 12,000 years ago, most people lived in nomadic bands that obtained their food by fishing, hunting, and gathering wild plants. This period of human history lasted approximately 2 million years. It is not known why some of the nomads eventually settled in certain areas, but this change appears to have been accompanied by (1) the herding of animals and (2) a regular pattern of harvesting wild plants. The first cultivation of local plants appears to have occurred in the regions where systematic gathering had long been practiced. In many cases, the first long-term settlements were made in places where diverse geographical features such as mountains, foothills, valleys, and rivers were in close proximity. Hence, the varied habitats favored the growth of numerous plant species. A wide variety of plant foods would have been needed to form a nutritious diet when fish and game became scarce, or the early settlers would have succumbed to malnutrition. Fig. V-3 shows the locations of the regions where many of the important vegetables were first cultivated.

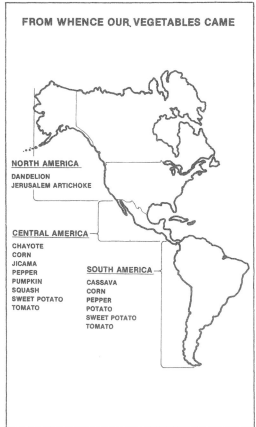

Fig. V-3. Places of origin and/or domestication of vegetables in the Old World (left) and in the New World (right).

Details on the history of individual vegetables are given in Table V-2 of this article, and in the separate articles covering each of the leading vegetable crops.

WORLD AND U.S. PRODUCTION.

The production of vegetables is an important part of the American agricultural scene. Market gardeners live near centers of population. Truck farmers live farther away; they grow vegetables and ship them in refrigerated trucks or cars to all parts of the nation. Home gardeners are everywhere; their vegetables provide (1) healthful exercise, and (2) cheap food and the maximum freshness and nourishment from crops that lose some of their goodness after harvesting.

Fig. V-4 shows the leading vegetable crops of the world, and the annual production of each. Fig. V-5 shows the leading vegetable crops of the United States, and the annual production of each.

Fig. V-4 shows that starchy roots and tubers (Irish potatoes, cassava, and sweet potatoes) are by far the leading vegetable crops of the world.

Fig. V-5 shows that the five leading U.S. vegetable crops, by rank, are: tomatoes, sweet corn, lettuce, onions, and Irish potatoes.

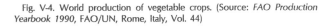

Fig. V-4. World production of vegetable crops. (Source: *FAO Production Yearbook 1990*, FAO/UN, Rome, Italy, Vol. 44)

Fig. V-5. U.S. production of vegetable crops. (Source: *Statistical Abstract of the United States*, U. S. Dept. of Commerce, 1991, p. 669, Table 1167, and *Agricultural Statistics 1991*)

NUTRITIVE VALUES.

"Eat your vegetables" is the admonition of mothers and doctors alike; and with reason. They know that vegetables are good—and good for you. They give bulk to the diet, which aids the digestive process; and they are rich sources of certain essential minerals and vitamins.

Vegetables are high in water content; hence, they are considerably lower in calories and proteins than most of the grains and legumes. However, most vegetables are fair to excellent sources of fiber (poorly digested carbohydrate which stimulates movements of the digestive tract), various essential macrominerals and microminerals, vitamins, and vitaminlike factors. The nutrient composition of the commonly used vegetables is given in Food Composition Table F-21. Table V-1 groups vegetables according to the amounts of vitamins A and C that they supply.

TABLE V-1
AMOUNTS OF VITAMIN A, VITAMIN C, AND CALORIES FURNISHED BY 100 G OF SELECTED VEGETABLES[1,2]

Nutritional Group	Vegetable	Vitamin Content A	C	Food Energy
		(IU)	(mg)	(kcal)
High in vitamins A and C	Parsley (raw)	8,500	172	44
	Spinach	8,100	28	23
	Collards	7,800	76	33
	Kale	7,400	62	28
	Turnip Greens	6,300	69	20
	Mustard Greens	5,800	48	23
	Cantaloupes	3,400	33	30
	Broccoli	2,500	90	26
High in vitamin A	Carrots (raw)	11,000	8	42
	Carrots (cooked)	10,500	6	31
	Sweet Potatoes	8,100	22	141
	Swiss Chard	5,400	16	18
	Winter Squash	4,200	13	63
	Green Onions	2,000	32	36
High in vitamin C	Peppers (immature green)	420	128	22
	Brussels Sprouts	520	87	36
	Cauliflower	60	55	22
	Kohlrabi	20	43	24
	Cabbage	130	33	20
	Chinese Cabbage	150	25	14
	Asparagus	900	26	20
	Rutabaga	550	26	35
	Radishes (raw)	322	26	17
	Tomatoes (ripe, raw)	900	23	22
	Tomatoes (ripe, cooked)	1,000	24	26
Other green vegetables	Beans, green	540	12	25
	Celery	540	9	17
	Lettuce (leaf)	1,900	18	18
	Lettuce	330	6	13
	Okra	490	20	29
	Peas (garden)	540	20	71
	Beans (lima)	280	17	111
Starchy vegetables	Corn, sweet (yellow)	400	9	91
	Onions (dry)	40	10	38
	Peas (field, southern)	350	17	108
	Potatoes (baked in skin)	Trace	20	93
Other vegetables	Beets	20	6	32
	Cucumbers	250	11	15
	Eggplants	10	3	19
	Pumpkins	1,600	9	26
	Rhubarb	80	6	141
	Summer Squash	440	11	15
	Turnips (roots)	Trace	22	23

[1]Data from Food Composition Table F-21.

[2]Figures are for amounts of vitamins and calories per 100 g sample of cooked vegetables (unless normally eaten raw). 100 g are equal to about ½ cup. Vitamin C values are generally higher if the vegetable is eaten raw; for example, 100 g of cabbage contains 33 mg of vitamin C when cooked, but 47 mg when raw.

Effects of Processing and Preparation. The maximum amounts of essential nutrients are supplied by raw, fresh vegetables that are harvested at just the right stage of maturity. However, many people find some vegetables unappealing when raw, and it is not practical to use only fresh vegetables, which are available only at certain times of the year. Furthermore, raw vegetables contain more undigestible carbohydrates (fiber) than cooked vegetables, and might have a strong laxative effect on people unaccustomed to so much fiber. Therefore, the effects of the various processing and preparation procedures should to be taken into account when planning menus that furnish all of the required nutrients.

NOTE: The information which follows is based on relatively few studies and may not be valid in all cases because (1) different vegetables are affected in different ways; (2) the effects depend partly upon the condition of the vegetables—whether immature or mature, fresh or taken from long storage, whole or in pieces, and untreated or treated; and (3) information is lacking on some of the effects of commonly used procedures. Nevertheless, pertinent information pertaining to nutrient losses in processing and preparation follow:

• **Artificial ripening**—Most commercially grown tomatoes are picked before full maturity in order to avoid excessive softness and oversusceptibility to damage and spoilage. However, one study showed that tomatoes ripened on the vine contained up to one-third more ascorbic acid than those allowed to ripen after picking.[1]

• **Baking**—Various studies of baked Irish potatoes and baked sweet potatoes have shown that baking results in vitamin losses as follows: carotene (provitamin A), 24%; thiamin, 24%; riboflavin, 11%; niacin, 18%; pantothenic acid, 23%; pyridoxine (vitamin B-6), 20%; and ascorbic acid (vitamin C), 20%.[2] Much lower losses of these vitamins were observed when the potatoes were boiled instead of baked.

• **Blanching**—This procedure is used to stop enzyme action in vegetables prior to canning, drying, or freezing. There are considerable losses (up to 80%) of water soluble minerals and vitamins when the vegetables are blanched in boiling water, then cooled by immersion in cold water.[3] Apparently, the nutrients are leached out into the water. These losses may be reduced greatly by (1) blanching with steam instead of boiling water, (2) cooking with air instead of water, or (3) blanching with microwaves. It is noteworthy that the vitamin losses from unblanched vegetables during frozen storage are from 2 to 3 times as great as those from blanched vegetables.

• **Boiling**—The losses of minerals and water-soluble vitamins increase as the amount of cooking water increases. Hence, greater mineral and vitamin retention results in steaming and waterless cooking.

[1]Pantos, C. E., and P. Markakis, "Ascorbic Acid Content of Artificially Ripened Tomatoes," *Journal of Food Science*, Vol. 38, 1973, p. 550.

[2]Lachance, P. A., "Effects of Food Preparation Procedures on Nutrient Retention with Emphasis on Food Service Practices," *Nutritional Evaluation of Food Processing*, 2nd ed., edited by R. S. Harris and E. Karmas, The Avi Publishing Company, Inc., Westport, Conn., 1975, pp. 508-509.

[3]Fennema, O., "Effects of Freeze-Preservation on Nutrients," *Nutritional Evaluation of Food Processing*, 2nd ed., edited by R. S. Harris and E. Karmas, The Avi Publishing Company, Inc., Westport, Conn., 1975, pp. 246-247.

Fig. V-6. Raw, fresh vegetables ready-to-eat. Shown here are cucumbers, green onions, tomatoes, watercress, cabbage, radishes, broccoli, cauliflower, carrots, celery, and green peppers—ideal finger foods when cut and served in interesting shapes like these. (Courtesy, U.S. Department of Agriculture)

• **Canning**—Blanching, plus the strong heat treatments applied to nonacid vegetables, appears to be responsible for the large vitamin losses in canning, because much less loss occurs in the more mildly processed tomato products. The nutrient losses may be reduced or offset by (1) utilization of newer procedures, such as microwave blanching, and (2) restoration of nutrients lost during processing (enrichment or fortification).

• **Chopping, dicing, grating, mashing, mincing, or slicing**—Any process which breaks many vegetable cells is likely to be responsible for significant losses of certain vitamins. The longer the broken pieces of vegetables are held, the greater the losses. Therefore, vitamins may be conserved by (1) cutting up vegetables just prior to serving, and (2) using a plastic knife for cutting, because metals speed up the destruction of vitamin C.

• **Drying**—Moderate amounts of carotene (provitamin A) and vitamin C may be destroyed in the process, unless the vegetables are sulfured before drying. However, sulfuring destroys most of the thiamin content. Vitamin losses are greater during slow processes, such as sun drying, than they are for more rapid processes like freeze drying.

• **Fermenting and/or pickling**—Pickles, olives, and sauerkraut are the only fermented and pickled vegetables used to any extent in the United States. Vitamin losses in these products may vary widely as a result of factors such as heat developed during fermentation, volume of pickling solution, and temperatures under which the pickles are canned and stored.

• **Freezing**—Most of the losses of vitamins in frozen vegetables occur during blanching, since freezing itself is responsible for only small losses. The vitamin losses during frozen storage may be considerably greater when blanching is omitted prior to freezing. The use of less drastic blanching procedures may help to prevent vitamin losses in frozen vegetables.

• **Frying**—Vitamin losses during the frying of vegetables vary greatly as a result of factors such as (1) time and temperature of frying, (2) type of frying (shallow fat, deep fat, or stir frying), and (3) fat or oil that is used (those which break down during frying may increase the rate of vitamin destruction).

• **Holding on a steam table**—Losses of vitamin C, thiamin, and riboflavin appear to increase with the length of time that vegetables are kept warm on a steam table.

• **Microwave cooking**—The vitamin losses during this procedure are small, and are comparable to those which occur when vegetables are cooked with little or no water.

• **Peeling, scraping, and/or trimming**—Removal of the outer leaves, peels, and/or skins of certain vegetables may result in disproportionately high losses of nutrients because the outer layers are often richer in minerals and vitamins than the inner layers. Therefore, these measures, which are often performed mainly for cosmetic reasons, should be used only to the extent that is absolutely necessary.

• **Pressure cooking**—This method of cooking vegetables results in a greater retention of minerals and vitamins than boiling in a saucepan. However, best results are obtained when the vegetables are placed on a rack above the water so that they are cooked only by steam.

• **Steaming**—Vegetables cooked with steam, but not in contact with water, have better nutrient retention than those cooked in boiling water.

• **Sulfuring**—This process, which consists of either exposing vegetables to fumes from burning sulfur, or dipping them in a sulfite solution, is used to prevent discoloration and spoilage. It is generally utilized before vegetables are to be dried. Sulfuring helps to prevent the losses of both carotene (provitamin A) and vitamin C, but it destroys thiamin. Fortunately, vegetables are utilized more as sources of the first two vitamins, than as a source of thiamin.

VEGETABLES OF THE WORLD. A relatively few species of vegetable plants account for most of the world's vegetable production.

Most of the better-known vegetables of the world are listed in Table V-2, Vegetables of the World. The *most important* (those produced in greatest quantity), worldwide and/or in the United States, are accorded detailed narrative coverage, alphabetically, whereas vegetables of lesser importance are presented in summary form only in Table V-2, Vegetables of the World.

TABLE V-2
VEGETABLES OF THE WORLD

Popular and Scientific Name(s); Origin and History	Importance; Principal Areas; Growing Conditions	Processing; Preparation; Uses	Calories; Nutritive Value[1]
Artichoke, Globe (See ARTICHOKE, GLOBE)			
Asparagus *Asparagus officinalis* **Origin and History:** Originated around the eastern Mediterranean. The ancient Egyptians cultivated asparagus. First used as a medicinal plant.	**Importance:** The U.S. produces much of the world's asparagus—about 110,000 tons annually. **Principal Areas:** U.S., Europe, Mexico, and Taiwan. **Growing Conditions:** Temperate climate and well-drained soil. Propagated by suckers from the root crown of the parent plant, which may have originally grown from seed.	**Processing:** The freshly cut spears are washed, bunched, and packed. About ⅔ of the U.S. asparagus crop is processed, with about twice as much canned as frozen. **Preparation:** Poached in water. Usually served with butter, sauce, or in an omelet. Slices may be stir-fried. **Uses:** Vegetable dish, entree (when served with a sauce and chopped hard boiled eggs on toast), soups, and salads.	High in water (92%), low in calories (26 kcal per 100 g), and a good source of vitamin A. Also, it is a fair source of thiamin, riboflavin, and niacin. Herbal lore has it that asparagus juice eases the discomforts of arthritis, neuritis, and rheumatism.[2]
Beets *Beta vulgaris* **Origin and History:** One of the vegetables derived in pre-Christian times from the wild beet that is native to Europe and Asia.	**Importance:** Beets are a minor commercial vegetable crop. **Prinicpal Areas:** U.S. and European countries. **Growing Conditions:** Temperate climate.	**Processing:** When ripe, they are pulled from the ground by hand or by machine, and the tops are removed. About 95% of the beet roots are canned or pickled, while the greens are sold fresh. **Preparation:** Beets are peeled by steaming, following which the skins are washed away with high velocity jets of water. The peeled roots may be canned whole, sliced, or cut in narrow strips; or the root may be pickled in vinegar. Tops may be eaten raw. **Uses:** Roots—vegetable dish, pickles, baby foods, beet soup (Borsch). Greens—ingredient of salads and soups.	Boiled beets are high in water content (91%) and low in calories (32 kcal per 100 g). Beet greens are much richer in calcium, iron, and vitamins A and C than beet roots. The heavy consumption of beets may lead to the passage of red-colored urine (beeturia), a condition generally regarded as harmless. Beets and beet juice have long been used as herbal remedies for a wide variety of ailments.[2]
Broccoli (See BROCCOLI)			
Brussels Sprouts *Brassica oleracea* It is closely related to cabbage, which it resembles; and is a member of the mustard family. **Origin and History:** Believed to have first been developed from wild cabbage sometime around the 15th century in the area that is now Belgium.	**Importance:** A minor U.S. vegetable crop. **Principal Areas:** U.S. and European countries. **Growing Conditions:** Mild, temperate climate. Benefits from nitrogenous fertilizers.	**Processing:** The lower sprouts are usually picked first. Some of the crop is sold fresh, while the rest is frozen. **Preparation:** Boiled briefly until tender, but crisp. May be served with butter or a special sauce. **Uses:** Vegetable dish.	High in water (88%), low in calories (36 kcal per 100 g), a good source of vitamin C and a fair source of vitamin A. Brussels sprouts contain small amounts of goitrogens. (See GOITROGENS)
Cabbage (See CABBAGE)			
Carrot (See CARROT)			
Cassava (manioc, yucca) (See CASSAVA)			
Cauliflower (See CAULIFLOWER)			
Celeriac (turnip rooted celery) *Apium graveolens* Celerica, which is closely related to celery, is a member of the parsley family, *Umbelliferae.* **Origin and History:** Native to the Mediterranean region.	**Importance:** A minor vegetable that is popular in certain European countries. Little used elsewhere. **Principal Areas:** European countries. **Growing conditions:** Moist, temperate climate. Plants are started from seed in greenhouses or hotbeds, then transplanted when the weather is suitable for growth in a field.	**Processing:** The plant is harvested by pulling it up when mature. The root is the edible portion. **Preparation:** Steamed, then served whole, diced, or pureed (Europeans often mix it with hot, mashed potatoes and serve it with butter or a cream sauce.) **Uses:** Vegetable dish; ingredients of salads, soups, and stews.	Low in calories (40 kcal per 100 g) due to its high water content (88%). Also contains about twice as much protein (1.8%) and fiber (1.3%) as ordinary celery. Herbalists have long claimed that celeriac is a cure for dropsy because it is thought to increase the flow of urine.[2]
Celery (See CELERY)			

Footnotes at end of table

(Continued)

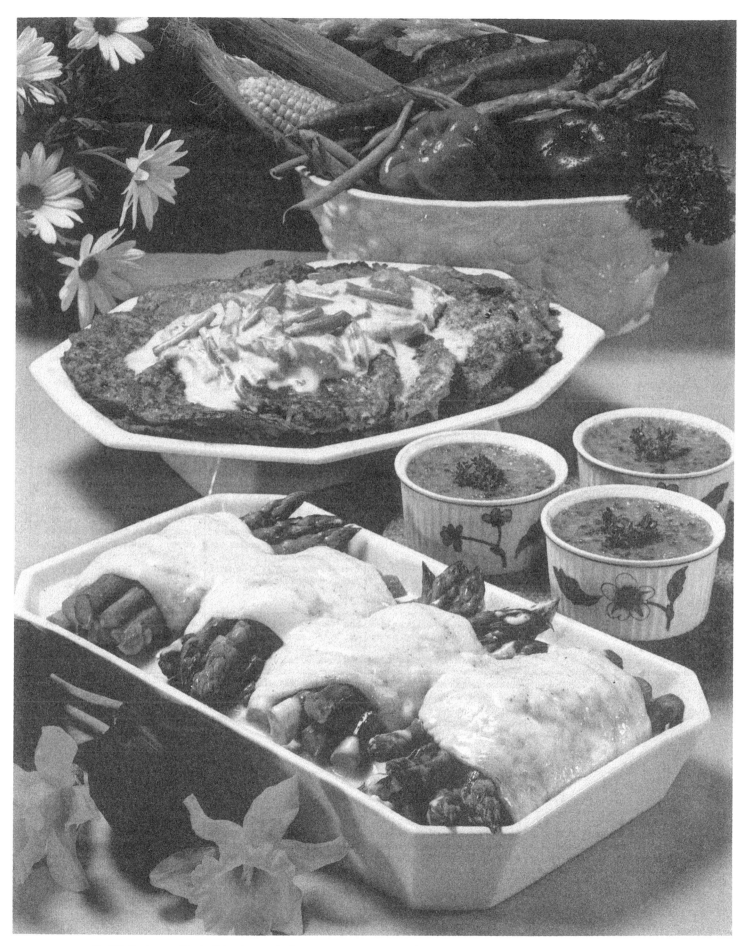

Fig. V-7. Fresh vegetables in dishes that speak of springtime. (Courtesy, United Fresh Fruit and Vegetable Assn.)

TABLE V-2 *(Continued)*

Popular and Scientific Name(s); Origin and History	Importance; Principal Areas; Growing Conditions	Processing; Preparation; Uses	Calories; Nutritive Value[1]
Chayote *Sechium edule* Chayote is a member of the gourd and melon family, *Cucurbitaceae*. **Origin and History:** Native to southern Mexico and Central America. Spread throughout Caribbean by local Indians.	**Importance:** A minor, but popular, vegetable in the tropical areas of the New World. **Principal Areas:** Mexico, Central America, and West Indies. **Growing Conditions:** Subtropical or tropical climate (freezing kills the plant); requires large amounts of water.	**Processing:** Usually, the fruits are picked when ripe and used immediately. Some of the fruit may be preserved by pickling. The large tuberous roots may also be harvested for food use. **Preparation:** Boiled, steamed, baked (after stuffing), or dipped in batter or fried. **Uses:** Vegetable dish (served with butter or cream sauce), ingredient in salads, and relish (when pickled).	Chayote fruit resembles squash in water content (92%) and caloric value (28 kcal per 100 g). However, the mineral and vitamin content of this vegetable is low. (It is a fair source of potassium and vitamin C.)
Chicory *Chicorium intybus* Chicory belongs to the composite family *Compositae*. **Origin and History:** Thought to have originated in the Mediterranean region. Certain varieties have been developed as green leafy vegetables, while others have been selected mainly for their roots.	**Importance:** A minor, but popular, vegetable in Italy, France, Belgium, and Holland. **Principal Areas:** It grows wild in Europe, Asia, and North America. Chicory is cultivated in the U.S. and southern Canada. **Growing Conditions:** Temperate climate. It is often grown in greenhouses or in heated beds in cellars. (Growing in the dark produces light colored leaves).	**Processing:** Blanching (the production of light colored leaves) may be done in the field by tying the outer leaves together around the head of each plant so that light is excluded. The darker outer leaves may be removed after harvesting. Roots may be chopped, roasted until brown, and ground into a finely ground coffee. **Preparation:** Greens—served raw or boiled briefly. Roots—roasted or boiled. **Uses:** Greens—vegetable dish, ingredient of salads, soups, and stews. Roots—vegetable dish; extender or substitute for coffee, when roasted.	The leaves are high in water (93%) and low in calories (20 kcal per 100 g). Unblanched (green) chicory is an excellent source of potassium and vitamin A, but the blanched forms contain only traces of vitamin A. Sometimes used as a tonic for the digestive system, and to treat disorders of the skin. In the skin treatment, the juice of the leaves is mixed with a little vinegar and applied externally.[2]
Chinese Cabbage *Brassica chinensis* and *Brassica pekinensis* Chinese cabbage belongs to the mustard family, *Cruciferae*. It decends from the wild turnip and rape. **Origin and History:** Thought to have originated in Asia Minor and to have been brought to China, where it was developed over thousands of years.	**Importance:** Major vegetable crop in eastern Asia, where about one-half of the world's crop of various cabbages is produced. **Principal Areas:** China and Japan. **Growing Conditions:** Cool temperate climate. It is usually grown in southern China during the winter, and elsewhere in China during the fall.	**Processing:** In China, the mature heads are harvested when winter approaches. After removal of loose outer leaves, the heads are stored in a cold cellar. Some of the crop is preserved by pickling. **Preparation:** Boiled briefly, or minced and stir fried with other vegetables, meats, poultry, and/or seafoods. **Uses:** Vegetable dish; ingredient of salads, soups, and various Chinese dishes.	Very high in water (95%) and low in calories (14 kcal per 100 g). Green types are good to excellent sources of calcium, potassium, and vitamin A, and fair sources of iron and vitamin C. Chinese cabbage contains small amounts of goitrogens. (See GOITROGENS.)
Chives *Allium schoenoprasum* A member of the amaryllis family, *Amaryllidaceae*, closely related to the onion. **Origin and History:** Native to the temperate zones of Asia and Europe.	**Importance:** Commercial production is limited to that utilized by a few processors. Very popular among home gardeners. **Principal Areas:** Temperate zones of Asia, Europe, and North America. **Growing Conditions:** Cool temperate climate. Propagated by bulbs or by seeds.	**Processing:** Common ways of processing are: (1) chopping and flash freezing; (2) chopping and blending into cottage cheese, cream cheese, or salad dressings; or (3) drying. **Preparation:** Chopped and added to various dishes. **Uses:** Decorative flavoring or garnish for cottage cheese, cream cheese, omelets, salads, sandwich fillings, and soups.	Low in calories (28 kcal per 100 g) due to high water content. A fair source of calcium; and a good source of iron and potassium. An excellent source of vitamin A and a fair source of vitamin C.
Collards *Brassica oleracea acephala* Collards belong to the mustard family, *Cruciferae*, and are closely related to cabbage. **Origin and History:** Developed from the primitive leafy, nonheading cabbage of Europe. (Collards and kale are very closely related.)	**Importance:** A minor vegetable crop. **Principal Areas:** Southeastern U.S. **Growing Conditions:** Temperate climate. One of the hardiest of the cabbage vegetables, in that it is quite resistant to cold and heat.	**Processing:** Small, young heads or large, mature heads may be harvested. Much of the crop is chopped and frozen; some is canned. **Preparation:** Boiled with salt pork; stir fried chopped greens with bits of bacon and chopped onions. **Uses:** Vegetable dish; ingredient of casseroles, salads, soups, and stews.	The cooked leaves with stems are like other greens—they're high in water content (91%) and low in caloric content (29 kcal per 100 g), but they contain about twice as much protein (about 3.4%). Also a good source of calcium and iron, and an excellent source of potassium; and excellent source of vitamin A, and a very good source of vitamin C. Collards contain small amounts of goitrogens. (See GOITROGENS.)

Footnotes at end of table

(Continued)

TABLE V-2 *(Continued)*

Popular and Scientific Name(s); Origin and History	Importance; Principal Areas; Growing Conditions	Processing; Preparation; Uses	Calories; Nutritive Value[1]
Corn, Sweet *Zea mays saccharata* **Origin and History:** Probably developed from older varieties of corn indigenous to Mexico and South America. (Also see CORN, section on "Kinds of Corn," "Sweet Corn.")	**Importance:** The second ranking vegetable crop in the U.S., which produces about 4.0 million short tons annually. The leading states, by rank, are: Wisconsin, Minnesota, Washington, and Oregon. **Principal Areas:** U.S., Mexico, and Canada. **Growing Conditions:** Mild temperate climate with a hot, humid summer. Usually requires the application of fertilizers containing nitrogen, phosphorus, and potassium.	**Processing:** Once sweet corn reaches its flavor peak, it must be harvested, husked, and processed rapidly to avoid excessive loss of its sweetness. Hence, the fresh ears cannot be stored for long. Much of the crop is canned and frozen. **Preparation:** Boiled, steamed with husk on but with silk removed, or roasted. **Uses:** Vegetable dish; ingredient of casseroles, fritters, salads, soups, stews, and succotash.	Fresh corn cooked on the cob has a high water content (74%) and a fairly high caloric content (91 kcal per 100 g) and protein content (3.3%). Yellow corn is a fair source of vitamin A.
Cress *Lepidium sativum* Cress belongs to the mustard family, *Cruciferae*. **Origin and History:** Appears to have originated in Western Asia and/or Ethiopia. Cultivated for many centuries in various European countries.	**Importance:** A minor vegetable crop that is grown mainly by home gardeners. **Principal Areas:** All over the world. **Growing Conditions:** Cool, temperate climate. Seeds germinate quickly when exposed to light.	**Processing:** May be harvested within a few weeks after sowing, and should be picked before the arrival of hot weather, when it quickly goes to seed. **Preparation:** No cooking is required, but it may be added to various hot dishes. **Uses:** Ingredient of salads, sandwiches, and soups.	High in water (89%) and low in calories (32 kcal per 100 g). The raw vegetable is an excellent source of vitamin A and a good source of vitamin C. Cress contains small amounts of Goitrogens. (See GOITROGENS.)
Cucumber (See CUCUMBER and GHERKINS)			
Dandelion *Taraxacum officinale* The dandelion is a member of the daisy family, *Compositae*. **Origin and History:** There are many related species native to various parts of the world, including Europe, North Africa, Central Asia, and North America.	**Importance:** A very common wild plant or weed that is cultivated on a small scale. **Principal Areas:** U.S. and European countries. **Growing Conditions:** Temperate climate.	**Processing:** The leaves should be picked while they are still young and tender (before flowers appear). Roots are harvested by digging up the entire plant. **Preparation:** The leaves need not be cooked, but the roots should be boiled or roasted. Roasted roots may be ground to make an extender or substitute for coffee. **Uses:** Greens—vegetable dish, ingredient of salads and soups. Roots—vegetable dish and coffee substitute.	Boiled dandelions are high in water (90%) and low in calories (33 kcal per 100 g). Dandelions are an excellent source of iron and potassium, and a good source of calcium; and an excellent source of vitamin A, and a good source of vitamin C. Through the years, dandelion greens have been used by some advocates as a laxative, as a tonic to aid digestion, or as a help to rid the body of excessive water.[2]
Eggplant (aubergine) (See EGGPLANT)			
Endive (escarole) *Cichorium endivia* This lettucelike vegetable is a member of the sunflower family, *Compositae*. **Origin and History:** Native to the eastern Mediterranean region.	**Importance:** Most of the winter crop is grown in Florida, and most of the summer crop is grown in New Jersey and Ohio. **Principal Areas:** Italy produces about half of the world's crop, while France contributes one-quarter. **Growing Conditions:** Mild temperate climate. May be started from seed in a hotbed or a greenhouse, then transplanted when there is no danger of frost.	**Processing:** The leaves are harvested by cutting them off near the surface of the soil. Endive is not processed—it is marketed exclusively as a fresh vegetable. **Preparation:** May be served raw, stuffed, baked, or tossed in hot bacon drippings or melted butter. **Uses:** Vegetable dish, ingredient of salads and sandwiches, and added to soups and stews.	Similar to other greens in high water content (93%), and low caloric content (20 kcal per 100 g). Endive is a good source of calcium, phosphorus, iron, and potassium; an excellent source of vitamin A, and a fair source of vitamin C. The bitter flavor is believed to stimulate the flow of digestive juices.[2]

Footnotes at end of table

(Continued)

Fig. V-8. Vegetables; sweet peppers, beets, tomatoes, summer squashes, and corn. Vegetables are rich sources of certain essential minerals and vitamins. (Courtesy, University of Minnesota, St. Paul, Minn.)

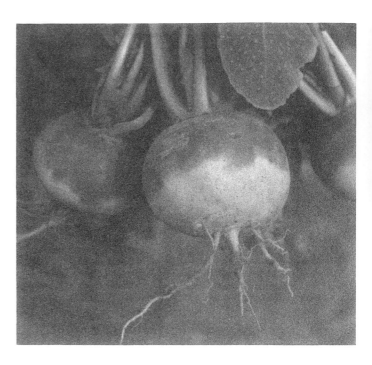

Fig. V-10. Turnips may be boiled and mashed or cooked with other items in mixed dishes. (Courtesy, USDA)

Fig. V-9. Brussels sprouts, used as a vegetable dish. A good source of vitamin C. (Courtesy, USDA)

Fig. V-11. Radishes, used as an appetizer, ingredient of mixed vegetable dishes, relishes, and salads. (Courtesy, Field Museum of Natural History, Chicago, Ill.)

TABLE V-2 *(Continued)*

Popular and Scientific Name(s); Origin and History	Importance; Principal Areas; Growing Conditions	Processing; Preparation; Uses	Calories; Nutritive Value[1]
Fennel (Florence Fennel) *Foeniculum vulgare* Fennel belongs to the parsley family, *Umbelliferae*. **Origin and History:** Originated in southern Europe. The variety with the fleshy leaf stalks (Florence fennel) was developed in Italy.	**Importance:** A minor vegetable used mainly as a flavoring agent and a medicinal plant. **Principal Areas:** European countries and U.S. **Growing Conditions:** Mild, temperate climate. It is often planted in the summer for harvesting in the fall.	**Processing:** The plant should be pulled from the ground when the leaf stalks are between 3 and 6 in. *(8 to 15 cm)* long. Both the thickened leafstem bases and the leaves are used. **Preparation:** Leaves and stalks may be served raw, boiled, or parboiled with butter, cheese, and seasonings. **Uses:** Leaves and stalks—vegetable dish; ingredient of casseroles, salads, soups, and stews. Oil is extracted from the seeds and used as a flavoring agent.	The stalks are high in water (90%) and low in calories (28 kcal per 100 g). The green stalks contain a fair amount of vitamin A. The leaves are a good source of calcium, and an excellent source of iron and potassium; and an excellent source of vitamin A, and a good source of vitamin C. Long used as a medicinal plant since it is believed to stimulate the digestive processes.[2] The ancient Greeks believed that it increased the fitness in their athletes.[2]
Garlic (See GARLIC)			
Horseradish *Amoracia rusticana* It is a member of the mustard family, *Cruciferae*. **Origin and History:** Horseradish is a native of Europe, where it is commonly cultivated. It was domesticated about the time of Christ.	**Importance:** A minor vegetable crop used mainly as a seasoning agent. **Principal Areas:** Temperate regions around the world. **Growing Conditions:** Temperate climate, and a rich, deep, moist, fertile soil. Propagated by root cuttings (usually those that are trimmed from the sides of the main root).	**Processing:** The highest quality horseradish is obtained by stripping the smaller side roots from the main root during the growing period. When the roots are harvested, the leaves and side roots are trimmed off. Fresh horseradish should be grated as soon as possible. **Preparation:** Grated horseradish is usually mixed with vinegar or beet juice, but it may also be mixed with sweet or sour cream. Also powdered horseradish is sold in grocery stores and served in restaurants. **Uses:** Condiment.(One of the 5 bitter herbs of the Jewish Passover).	The raw root has a moderately high water content and caloric content (87 kcal per 100 g). Also it is a good source of potassium and vitamin C. However, the pungent flavor limits the amount which may be eaten. It may make a valuable nutritional contribution by stimulating the appetite and enhancing the enjoyment of food. When eaten, the pungency is thought to help clear out congested breathing passages.[2] Used externally as a counter-irritant.
Jerusalem Artichoke *Helianthus tuberosis* This tuber-bearing plant is a member of the daisy family, *Compositae*, and is closely related to the sunflower. **Origin and History:** Native to North America; it was cultivated by the Indians.	**Importance:** It is a minor crop. **Principal Areas:** Europe, Asia, and the Americas. **Growing Conditions:** Adapted to a wide variety of conditions in temperate climates. Propagated by pieces of tubers.	**Processing:** The long, woody stems must be cut off before the tubers can be harvested. The most common type of processing consists of drying and grinding the tubers into flour. **Preparation:** The flour may be used to make dietetic baked goods and pastas. Also, it may be served unpeeled and raw, cooked with other vegetables, pureed and served like mashed potatoes, or fried, (after dipping the cooked vegetable in a batter). **Uses:** Baked products, and vegetable dishes and substitute for Irish potatoes. It can be used in the diet of diabetics.	Composed mainly of water (80%) and carbohydrates (17%). Most of the latter consists of inulin, a substance not utilized by the body. Hence, the caloric value of the fresh vegetable is negligible. However, the inulin is slowly converted to fructose (fruit sugar) when the tubers are left in the ground. This conversion tends to raise the caloric value. Jerusalem artichoke tubers are deficient in calories, minerals, and vitamins.
Jicama (Yam bean) *Exogonium bracteatum* It is a member of the morning glory family, *Convolvulaceae*, and is related to the sweet potato. **Origin and History:** Originated in Mexico.	**Importance:** A minor vegetable crop that is much appreciated where it is available. **Principal Areas:** Mexico and California. **Growing Conditions:** Subtropical climate. Requires rich, moist soil. Propagated by tubers.	**Processing:** After the vine has flowered, the tubers are harvested by digging around the base of the vine. There is little processing. However, they may be canned, dried, or frozen like potatoes. **Preparation:** Peeled and served raw, or pan fried like potatoes. **Uses:** Appetizer, ingredient of raw salads, or as a substitute for waterchestnuts in various Oriental dishes.	Raw jicama is high in water content (88%) and low in calories (55 kcal per 100 g) and protein (1.4%).

Footnotes at end of table

(Continued)

TABLE V-2 *(Continued)*

Popular and Scientific Name(s); Origin and History	Importance; Principal Areas; Growing Conditions	Processing; Preparation; Uses	Calories; Nutritive Value[1]
Kale *Brassica oleracea acephala* Kale, which belongs to the mustard family, is similar to collards, except that it has leaves with curly edges and is less tolerant to hot weather. **Origin and History:** Believed to have been developed from the wild cabbage that is native to the Mediterranean region.	**Importance:** A minor vegetable crop. **Principal Areas:** U.S. and European countries. **Growing Conditions:** Cool, temperate climate (Kale differs from collards, in that it does not tolerate hot weather.) Requires well-drained fertile soil.	**Processing:** In commercial production, the leaves are harvested by cutting the plant off at soil level with a large knife. Much of the crop is frozen, but some is canned. **Preparation:** Boiled briefly, or stir fried (in preparation of Chinese dishes). Often served with butter or a cheese sauce. **Uses:** Vegetable dish, ingredient of various mixed dishes, or in soups.	Kale is high in water (88%), low in calories (39 kcal per 100 g), high in protein (4.5%), high in calcium, iron, and potassium, but low in phosphorus; and very high in vitamin A, and moderately good in vitamin C. The eating of moderately large quantities of this vegetable may give some people gas pains. Kale contains small amounts of goitrogens. (See GOITROGENS.)
Kohlrabi *Brassica oleracea caulorapo* Kohlrabi is closely related to cabbage. It is a member of the mustard family, *Cruciferae*. **Origin and History:** Kohlrabi is native to Europe. Selections from a cabbage-like ancestral form were made near the end of the Middle Ages.	**Importance:** A minor vegetable crop which is more appreciated in Eurpe (especially Germany) than in the U.S. **Principal Areas:** European countries and U.S. **Growing Conditions:** Cool temperate climate. Requires a rich soil, or one to which fertilizer has been added.	**Processing:** Kohlrabi is usually harvested when the swollen stem is 2 to 3 in. (5 to 8 cm) in diameter. The root is cut off after harvesting. Kohlrabi is marketed fresh. **Preparation:** May be eaten raw, or after steaming. Eastern European cooks scoop out the center and stuff the hollow with various mixtures of bread crumbs, meat, seasonings, etc. **Uses:** Vegetable dish; ingredient of salads, soups, and stews.	Similar to the other vegetables of the cabbage species in water content (92%), caloric value (24 kcal per 100 g), protein content (l.7%), and potassium and vitamin C content (a good source). Kohlrabi contains small amounts of goitrogens. (See GOITROGENS.)
Leek *Allium porrum* This onion relative differs from onion in having flat leaves instead of tubular and relatively little bulb development. **Origin and History:** Native to southern Asia.	**Importance:** Grown to a much greater extent in Europe market gardens than in U.S. The leek is also the national flower of Wales. **Principal Areas:** Northwestern European countries. **Growing Conditions:** Temperate climate. May be transplanted from greenhouse or hotbed.	**Processing:** Often blanched (lightened in color) during growth by the banking of soil around the plants. (Some growers wrap the the stems with cardboard before piling up the soil.) There is little processing of leeks other than canning or drying small amounts of soup preparations. **Preparation:** Boiled, braised, cooked in mixed dishes similar to those which utilize asparagus, or used raw as relishes. **Uses:** Appetizer; ingredient of quiches, salads, soups, and stews.	Like most other vegetables in water content (85%), caloric value (52 kcal per 100 g), and protein content (2.2%). Leeks are a good source of vitamin C.
Lettuce (See LETTUCE)			
Mushroom, Cultivated *Basidiomycetes (club fungi)* *Ascomycetes (truffles)* Mushrooms are fungi. There are 38,000 known species of mushrooms. **Origin and History:** Wild mushrooms were consumed by primitive peoples long before the dawn of agriculture. However, it appears that the common mushroom was first grown commercially in France around 1700.	**Importance:** Commercial mushroom production is important in Asia and the United States. In 1981, the U.S. produced about 337,000 metric tons. **Principal Areas:** U.S., and European countries. **Growing Conditions:** Commercial production usually involves the planting of specially grown spores in a bed of compost inside a growing shed where the temperature and humidity may be controlled.	**Processing:** Young mushrooms are usually plucked from the bed shortly after they have emerged and while they are still in the button stage. Much of the crop is canned, but some is frozen, dried, or sold fresh. **Preparation:** May be sliced and served raw, stuffed and baked, fried or stewed alone or with other items, or cooked in sauces. **Uses:** Vegetable dish; ingredient of casseroles, quiches, salads, soups, and stews.	High in water content (over 90%) and low in caloric value (around 28 to 35 kcal per 100 g). Contains between 2 and 3% protein. Mushrooms are a good source of phosphorus and potassium, but low in calcium; and they're a good source of niacin, moderately low in thiamin and riboflavin, and very low in vitamins A and C. People who have gout and/or high blood pressure are usually advised to avoid eating mushrooms and other foods that are rich in nucleic acids.

Footnotes at end of table

TABLE V-2 *(Continued)*

Popular and Scientific Name(s); Origin and History	Importance; Principal Areas; Growing Conditions	Processing; Preparation; Uses	Calories; Nutritive Value[1]
Mustard Greens (Indian Mustard) *Brassica juncea* Mustard is a member of the family *Cruciferae*. It is grown for its pungent seed, and for its leaves, which are eaten as greens. **Origin and History:** Native to southern Europe and southwestern Asia. Cultivated since ancient times.	**Importance:** A minor vegetable crop that is much appreciated in the southeastern U.S. **Principal Areas:** Southeastern U.S., central and southern Europe, North Africa, and western Asia. **Growing Conditions:** Temperate climate. Must be planted at the beginning of spring, or in the fall, because high temperatures make the crop go to seed.	**Processing:** Leaves may be harvested anytime from the time they emerge to just before the seedstalks develop. Much of the crop is sold fresh, although mustard greens are frozen. **Preparation:** Steamed, boiled with salt pork, or stir-fried chopped greens with bits of bacon and chopped onions. **Uses:** Vegetable dish; ingredient of casseroles, salads, soups, and stews.	Boiled mustard has a nutritional value like that of most greens (92% water; 23 kcal per 100 g; and 2 to 3% protein). Mustard greens are high in calcium, iron, and potassium; and an excellent source of vitamins A and C. Mustard greens contain small amount of goitrogens. (See GOITROGENS.)
Okra *Hibiscus esculentus* Okra is a member of the mallow family, which includes cotton. **Origin and History:** Thought to be native to Ethiopia. Used in Egypt for many centuries. Brought to Louisiana from France.	**Importance:** A minor vegetable crop. **Principal Areas:** Southern U.S., Europe, Africa, and Asia. **Growing Conditions:** Warm temperate climate. The fruit develops during the summer.	**Processing:** Continuous harvesting of the young pods insures maximum yield. (Old pods are virtually inedible.) A large part of the crop is frozen, while much of the remainder is canned. **Preparation:** May be dipped in egg and cornmeal, then fried; boiled or stewed, alone or with other items. **Uses:** Vegetable dish; ingredient of soups and stews. The Creole cuisine of New Orleans includes various *gumbos*, or okra-containing stews.	Boiled okra contains about 90% water and around 30 kcal per 100 g. Okra is a fair to good source of calcium, potassium, vitamin A, and vitamin C. The mucilaginous material from cooked okra has been used to soothe irritations of the digestive tract.[2]
Olive (See OLIVE)			
Onion (See ONION)			
Parsley *Petroselinum crispum* It is a member of the parsley family, *Umbelliferae*. **Origin and History:** Parsley is native to southern Europe.	**Importance:** A minor vegetable crop. **Principal Areas:** European countries and U.S. **Growing Conditions:** Cool, temperate climate. However, the seeds need warmth to germinate. Hence, it is often transplanted from greenhouses and hotbeds.	**Processing:** Usually, the outer leaves of the plant are harvested when large enough. Some of the crop is dried or frozen, and some is sold fresh. **Preparation:** Served raw, stir-fried with other items, or boiled briefly in mixed dishes. **Uses:** For garnishing and flavoring. Ingredient of casseroles, quiches, relishes, salads, soups and stews. It is essential to French cooking.	It is lower in water (85%) and higher in calories (44 kcal per 100 g) than most other leafy vegetables. Also, it is very rich in protein, calcium, iron, potassium, and vitamins A and C.
Parsnip *Pastinaca sativa* Parsnips belong to the parsley family *Umbelliferae*, which also includes carrots, celery, and other vegetables. **Origin and History:** It is native to Eurasia and was cultivated by the Greeks and Romans.	**Importance:** A minor vegetable crop. **Principal Areas:** European countries and U.S. **Growing Conditions:** Requires a rich, moist soil, that does not pack down too readily. Usually planted in the early spring.	**Processing:** Harvested in the late fall by digging, and/or loosening the roots by plowing the ground around them. Eating quality is improved by storage at 34 °F (1 °C) for 2 weeks. Parsnip roots are generally processed as cooked food. However, they are also used for making parsnip wine. **Preparation:** Parboiled or steamed. Cooked parsnips may be mashed, formed into cakes, and fried. **Uses:** Vegetable dish; ingredient of soups and stews.	A typical root vegetable in water content (80%), and caloric value (76 kcal per 100 g). Parsnips are a good source of potassium, but only a fair source of vitamin C.
Pepper, Chili (See PEPPERS)			
Pepper, Sweet (See PEPPERS)			
Potato (Irish Potato) (See POTATO)			
Pumpkin (See PUMPKIN)			

Footnotes at end of table

(Continued)

TABLE V-2 *(Continued)*

Popular and Scientific Name(s); Origin and History	Importance; Principal Areas; Growing Conditions	Processing; Preparation; Uses	Calories; Nutritive Value[1]
Radish *Raphanus sativus* The radish is a member of the mustard family, *Cruciferae*, grown for its edible root. **Origin and History:** The radish was developed from a wild plant in the cooler areas of Asia. From here, it spread to the Mediterranean region before the Greek era, thence to the New World early in the 16th century.	**Importance:** A minor, but popular, vegetable. Often grown in home gardens and kept for family use. **Principal Areas:** Temperate and sub-tropical areas of the world. **Growing Conditions:** Requires warm, light, well-drained soils that are highly fertile.	**Processing:** The roots are harvested when they reach marketable size. There is little or no processing except in the Orient, where they may be made into fermented and pickled preparations. **Preparation:** Served raw in salads, or steamed with other ingredients and seasonings. **Uses:** Appetizer; ingredient of mixed vegetable dishes, relishes, and salads.	High in water (94%) and low in calories (17 to 19 kcal per 100 g) and protein (1%). Oriental radishes, which have a milder flavor and can be eaten in larger quantities than common radishes, are lower in potassium and higher in vitamin C than common radishes. Long used to stimulate digestive processes and to help rid the body of excessive water.[2]
Rhubarb (Pieplant) *Rheum rhabarbarum* The rhubarb is a member of the buckwheat family, *Polygonaceae*. **Origin and History:** Originated in Mongolia; used as a medicine in China in 2700 B.C.; introduced into U.S. in late 1700s.	**Importance:** A minor vegetable crop, grown chiefly in home gardens. **Principal Areas:** In Zurich, where it seems to be more appreciated than elsewhere, every small family vegetable garden displays at least one large rhubarb plant. **Growing Conditions:** Adapted to cool, moist climates and cold winters.	**Processing:** Rhubarb is frozen and canned, but many people prefer to cook the fresh stalks. **Preparation:** The thick stalks may be either stewed or baked. **Uses:** Rhubarb is classed as a vegetable, but used as a fruit in America. It is used as a side-dish or as a dessert—often in pie fillings and sauces. In France, it is used in jams and compotes.	The nutrient composition of rhubarb is given in Food Composition Table F-21. Rhubarb contains some vitamin C, and has laxative properties. The leafy portions and the roots contain toxic substances, including oxalic acid, and should not be eaten. (See OXALIC ACID [OXALATE]).
Rutabaga (Swede) *Brassica napobrassica* Rutabagas belong to the mustard family, *Cruciferae*. **Origin and History:** A hybrid of the turnip and cabbage, which originated in Europe in the Middle Ages.	**Importance:** A minor vegetable for human consumption, but an important feed and forage crop for livestock. **Principal Areas:** Europe and North America. **Growing Conditions:** Cool, temperate climate.	**Processing:** Usually, the roots are harvested in the fall. Then some of the crop is stored in a cold, damp environment. The roots keep better if they are covered with a thin layer of wax before storage. **Preparation:** Boiled and mashed, or cooked with other items in mixed dishes. **Uses:** Vegetable dish; ingredient of soups and stews; and feed for livestock.	Similar in nutritive value to other root vegetables (90% water, 35 kcal per 100 g, and 1% protein). It is noteworthy that a mashed mixture of rutabaga and potatoes is lower in calories than mashed potatoes alone. Rutabagas are a good source of vitamins A and C. Rutabagas contain small amounts of goitrogens. (See GOITROGENS.)
Salsify (Oysterplant) *Tragopogon porrifolius* Salsify is a member of the sunflower family, *Compositae*. **Origin and History:** Native to the Mediterranean countries.	**Importance:** A minor vegetable crop. **Principal Areas:** Soviet Union and other European countries. **Growing Conditions:** Temperate climate. Requires a well-drained crumbly soil.	**Processing:** There is little or no processing of salsify. Some of the crop is usually stored. **Preparation:** Boiled; sauteed (after breading), served raw; or roasted to make a coffee substitute. **Uses:** Vegetable dish; ingredient of salads, soups, and stews; and low-carbohydrate substitute for starchy vegetables. (Most of the carbohydrate in salsify is inulin, which is not metabolized by the body.)	Very similar to the Jerusalem artichoke in water content (78%) and carbohydrates (18%). Most of the latter consists of inulin, a substance not utilized by the body. Hence, the caloric value of the fresh vegetable is negligible. However, the caloric value increases when some of the inulin is converted into fructose after long storage of the roots in the ground. Salsify roots are a good source of iron and potassium, a fair source of calcium and phosphorus; a fair source of vitamin C, but a poor source of vitamin A.

Footnotes at end of table

(Continued)

TABLE V-2 *(Continued)*

Popular and Scientific Name(s); Origin and History	Importance; Principal Areas; Growing Conditions	Processing; Preparation; Uses	Calories; Nutritive Value[1]
Shallot *Allium cepa* Like the onion, which it resembles, the shallot belongs to the *Alliaceae* family. **Origin and History:** Believed to have originated in Central Asia as a modification of the ordinary type of onion. Onions produce larger and fewer bulbs.	**Importance:** Of little commercial importance. Grown mainly in home gardens. **Principal Areas:** European countries and the U.S. **Growing Conditions:** Temperate climate. Grown in the winter for green onions, and in the summer for dry, mature onions.	**Processing:** The *green onions* are trimmed and skinned, whereas, the dry bulbs are treated like ordinary onions. There is little or no processing of shallots because most of the crop is marketed as a fresh vegetable. **Preparation:** Chopped and served raw; sauteed; or boiled. **Uses:** Appetizer; ingredient of relishes, salads, soups, and stews.	Like other members of the onion family in water content (80%) and caloric value (72 kcal per 100 g). Shallot bulbs are similar in nutrient composition to the Irish potato. However, only small amounts of shallots are usually consumed.
Sorrel (Dock) *Rumex acetosa* Sorrel belongs to the buckwheat family, *Polygonaceae*. **Origin and History:** Native to Asia and Europe. Cultivated in France and Italy since the 14th century.	**Importance:** A minor vegetable crop. **Principal Areas:** Popular in Europe. **Growing Conditions:** Temperate climate. Requires rich, moist soil and abundant sunshine.	**Processing:** Almost all of the sorrel is consumed fresh. Sometimes the leaves are dried for storage. **Preparation:** Served raw, or boiled briefly. **Uses:** Seasoning; and an ingredient of casseroles, salads, soups, and other mixed dishes.	Similar in nutritional value to other green leafy vegetables (91% water, 28 kcal per 100 g, 2% protein); a good source of iron and potassium, and a fair source of calcium and phosphorus; and an excellent source of vitamins A and C. Sorrel is rich in oxalates. (See OXALIC ACID [OXALATE]).
Spinach *Spinacia oleracea* Spinach is a member of the goosefoot family, *Chenopodiaceae*. **Origin and History:** Native to Southwestern Asia (believed to have originated in Persia).	**Importance:** Spinach is a popular garden vegetable, which is widely cultivated throughout the world. **Principal Areas:** The U.S., the Netherlands, and the Scandinavian countries are the major producers. **Growing Conditions:** Cool, temperate climate (goes to seed readily in hot weather.)	**Processing:** Leaves may be harvested as soon as they are large enough. Some spinach is consumed fresh, but ¾ of the crop is processed—mostly by canning or freezing. **Preparation:** Cooked briefly in as little water as possible; served raw in salads; baked in casseroles and quiches. **Uses:** Vegetable dish; main dish (when served with chopped eggs and/or a cheese sauce on toast); and an ingredient of casseroles, quiches, salads, and soups. Small quantities are used to make baby foods, spinach noodles, and other special items.	Like other greens in water content (91%), caloric value (26 kcal per 100 g), and protein content (3%). Spinach is an excellent source of magnesium, iron, and potassium, and a good source of calcium; and an excellent source of vitamin A, and a good source of vitamin C. Spinach is rich in oxalates. (See OXALIC ACID [OXALATE]).
Squash, Summer (See SQUASHES)			
Squash, Winter (See SQUASHES)			
Sweet Potato (See SWEET POTATO)			
Swiss Chard (seakale beet) *Beta vulgaris* Swiss chard belongs to the goosefoot family, *Chenopodiaceae*. **Origin and History:** One of the vegetables derived in pre-Christian times from the wild beet that is native to the shores of the Mediterranean.	**Importance:** A minor, but popular vegetable in Europe and the U.S. **Principal Areas:** Swiss chard is grown more widely in Europe than in U.S. **Growing Conditions:** Temperate climate. May be started from seed in a greenhouse or a hotbed, then transplanted when the danger of hard frost is over.	**Processing:** Leaves may be picked from the plants until fall frost occurs. Little chard is processed. It is used mainly as a fresh vegetable. However, it may be canned or frozen like spinach. **Preparation:** Served raw as greens, or leaves and stems cooked (usually boiled or fried) separately, since the stems require longer cooking. **Uses:** Vegetable dish; entree (with cheese sauce and hard boiled eggs on toast); ingredient of salads, soups, and mixed dishes (such as chard fried with meat and onions or garlic).	Very high in water (91%) and low in calories (25 kcal per 100 g). A good source of iron and vitamin C, and an excellent source of vitamin A. The protein-to-calorie ratio of Swiss chard (9.6 g per 100 kcal) is about equal to that of skim milk. Hence, significant amounts of dietary protein may be supplied by this vegetable, if sufficiently large amounts are consumed. Has a high oxalic acid content which may reduce utilization of calcium, iron, and other essential minerals.

Footnotes at end of table

TABLE V-2 *(Continued)*

Popular and Scientific Name(s); Origin and History	Importance; Principal Areas; Growing Conditions	Processing; Preparation; Uses	Calories; Nutritive Value[1]
Taro (Cocyam, Dasheen, Eddoe) *Colocasia esculenta* Taro is a tropical plant of the Arum family, *Aracea*. **Origin and History:** Native to southeastern Asia, where it was first cultivated 4,000 to 7,000 years ago. Spread throughout the Pacific Basin by the ancestors of the Polynesians.	**Importance:** Among the leading tuberous plants grown and consumed in Africa. **Principal Areas:** Moist, tropical areas around the world. **Growing Conditions:** Warm, moist climate. Requires a highly fertile soil.	**Processing:** Small tubers may be stored more successfully than large ones. Only small amounts of taro are processed commercially; most of the crop is consumed where it is produced. **Preparation:** Baked, boiled, or fried (like potato chips). **Uses:** Starchy vegetable; used to make *poi*, a fermented starchy paste, in Hawaii.	Boiled taro tubers are similar in nutritional value to other root and tuber vegetables (67% water, 124 kcal per 100 g, and 1.9% protein). The nutritional value of taro tubers is similar to that of potatoes, except that the taro is much lower in vitamin C. The leaves are more nutritious than the tubers. Neither the roots nor the leaves of taro should be consumed raw, because some varieties contain potentially harmful amounts of calcium oxalate crystals.
Tomato (See TOMATO)			
Turnip *Brassica rapa* The turnip belongs to the mustard family, *Cruciferae*. **Origin and History:** One type is native to Europe, while another is native to central Asia. Turnips were first cultivated 4,000 years ago in the Near East.	**Importance:** A minor vegetable crop that is more important in Europe than in the U.S. **Principal Areas:** European countries. Most of the U.S. crop is produced in the southern states. **Growing Conditions:** Cool, temperate climate. Grows best during spring and early fall since it does not do well in hot weather.	**Processing:** Although fresh greens and roots are still marketed, much of the crop is now frozen. **Preparation:** Turnip roots are boiled and mashed, or cooked with other items in mixed dishes. Southern cooks prepare turnip greens as a hot vegetable, often mixed with bacon or pork. **Uses:** Vegetable dish; ingredient of salads, soups, and stews.	Boiled turnip roots have a high water content (94%) and are low in calories (23 kcal per 100 g) and protein (0.8%). They are a fair source of vitamin C. Compared to the roots, the cooked greens contain about the same amounts of water and calories, but twice as much protein, iron, and vitamin C, and many times as much vitamin A (of which they are an excellent source). Turnips contain small amounts of goitrogens. (See GOITROGENS.)
Waterchestnut, Chinese (Ma-tai, Pi-tsi) *Elocharis tuberosa*, and *Elocharis dulcis* The Chinese waterchestnut is not related to the waterchestnut *(Tropanatans)*. It is a member of the sedge family, *Cyperaceae*. **Origin and History:** Native to southern China.	**Importance:** Minor, but much appreciated vegetable. **Principal Areas:** China, Japan, and East Indies. **Growing Conditions:** Temperate climate. Grows wild in shallow water near the shores of lakes, and in marshes. Cultivated in China (grown in rice paddies).	**Processing:** The corms are dug from the mud by hand after the rice paddies have been drained. Much of the Chinese crop is canned for export. Sometimes starch is extracted by washing grated raw tubers on a fine screen. **Preparation:** Boiled briefly (after peeling). **Uses:** Ingredient of mixed Oriental dishes, salads, and soups.	Has a nutritional value like that of the other tuberous vegetables. Compared to Irish potatoes, Chinese waterchestnuts contain about the same calories, but only ⅔ the protein, and ⅕ the vitamin C.
Watercress *Nasturtium officinale* Watercress is a member of the mustard family, *Cruciferae*. **Origin and History:** Watercress is a native of Europe, but it has become naturalized wherever there is clear, cold, shallow water.	**Importance:** Minor crop, which also grows wild. **Principal Areas:** Germany, France, and the United Kingdom. **Growing Conditions:** Temperate climate. Requires an abundant supply of running water which is alkaline and contains nitrates.	**Processing:** None of the crop is processed; all of it is sold fresh. **Preparation:** Served raw in salads and sandwiches, or cooked with other vegetables to add flavor. **Uses:** Ingredient of salads, sandwiches, and soups.	Very similar to other greens in nutrient composition (93% water, 19 kcal per 100 g, and 2.2% protein). Watercress is high in calcium and iron, and an excellent source of vitamins A and C. Herbal remedy for minor disorders (lack of appetite, mild depression, and waterlogging).[2] Watercress contain small amounts of goitrogens. (See GOITROGENS.)

Footnotes at end of table

(Continued)

TABLE V-2 *(Continued)*

Popular and Scientific Name(s); Origin and History	Importance; Principal Areas; Growing Conditions	Processing; Preparation; Uses	Calories; Nutritive Value[1]
Yam (Asiatic Yam) *Dioscorea alata* Yams belong to the family, *Dioscoreaceae*. Although the starchy tubers resemble sweet potatoes, the two plants are not even distantly related. **Origin and History:** Originated in Africa, where they were cultivated 11,000 years ago; and in southeast Asia, where they were cultivated 10,000 years ago.	**Importance:** Among the leading tuberous plants grown and consumed in the tropics. Worldwide, production of yams is about 29.4 million metric tons per year. **Principal Areas:** Moist, tropical areas of Africa, Asia, and the Americas. About half of the yams of the world are grown in western Africa. **Growing Conditions:** Warm, moist climate. Do not grow well below 68°F *(20°C)*. Require fertile soils and at least 40 in. *(250 cm)* of rainfall during the 8 months growing season.	**Processing:** Toxic varieties must be soaked or boiled in water to remove the harmful substances. Some of the crop is made into flour, which stores better than the whole tubers. **Preparation:** Baked; boiled and mashed; or fried (after slicing). **Uses:** Staple food, starchy vegetable, and ingredient of soups and stews.	Somewhat like the potato in nutritional value, but yam tubers contain 50% more calories, and about the same amount of protein. Yams are a fair to good source of calcium, phosphorus, and iron; and a fair source of vitamin C, but they contain little carotene.

[1]See Food Composition Table F-21 for additional details.

[2]No medical claims are made for any of the vegetables listed in this table. The various beliefs regarding the beneficial effects of certain items are presented (1) for historical and informational purposes, and/or (2) to stimulate further investigation of the claimed merits.

Fig. V-12. Potatoes, the leading vegetable crop of the world and of the U.S. (Courtesy, USDA)

Fig. V-13. Swiss chard, used as a vegetable dish, entree, and ingredient of salads, soups, and mixed dishes. (Courtesy, New Jersey Department of Agriculture)

NOTE WELL: Where production figures are given in Table V-2, column 2 under "Importance," unless otherwise stated, they are based on 1990 data. Most world figures are from *FAO Production Yearbook 1990*, FAO/UN, Rome, Italy, Vol. 44. Most U.S. figures are from *Agricultural Statistics 1991*. Annual production fluctuates as a result of weather and profitability of the crop.

Although beans are often classed as vegetables, especially those that are used as immature green pods (snap beans or green beans), all beans are legumes. So, in order to avoid double listing, the species of beans commonly used in the United States are listed and summarized in Table L-2, Legumes of the World.

VEGETABLE BUTTERS

Due to their fatty acid composition, the lipids extracted from some plants are solids—fats—at room temperature and, hence, exhibit the consistency of butter and even the yellow color. Most lipids extracted from plants are liquids—oils—at room temperature. Two familiar vegetable butters are cocoa butter and coconut butter.

• **Cocoa butter**—The fat pressed from the roasted cacao bean is known as cocoa butter. It is yellow and possesses a slight chocolate smell and flavor. Cocoa butter is mainly used in drugs and cosmetics. As a food it is pleasant to the taste and easily digestible.

• **Coconut butter**—This is derived by pressing the dried coconut meats or copra, and then after bleaching and various other processes the oil is transformed into a product of firm consistency that looks like butter. Coconut butter is common in France.

Other less known vegetable butters include Shea butter from the African plant *Butyrospermum parkii*, Mowrah fat or illipe butter from the Indian plant *Bassia longifolia*, and nutmeg butter. By broadening the definition of butter to include foods that are spread like butter, peanut butter, hazelnut butter, and walnut butter are included as vegetable butters.

VEGETABLE OILS

These oils are pressed or extracted from a variety of plant seeds. Of primary importance as sources of edible oil on a world basis are soybeans, cottonseed, peanuts, corn germ, olives, coconut, rapeseed (canola), sesame, sunflower, safflower, cocoa beans, and various oil palms.

(Also see OILS, VEGETABLE; and FATS AND OTHER LIPIDS, section headed "Animal and Vegetable.")

VEGETARIAN DIETS

Contents   Page
Types of Vegetarians..............................1048
Prevalence of Vegetarianism........................1048
Problems of Vegetarianism..........................1049
Healthful Vegetarian Diets.........................1050
Meat Substitutes 1051
Possible Health Benefits of Vegetarianism 1052

In its purest form, a vegetarian diet consists of only foods of plant origin—no meat, fish, or other animal products are allowed. Most vegetarian diets allow vegetables, fruits, cereals and breads (often whole grain), yeast, dry beans, peas and lentils, nuts and peanut butter, seeds, vegetable oils, sugars and syrups, and possibly some rather unusual foods. Often a vegetarian diet is assumed to mean only no meat in the diet. For some types of vegetarianism this may be true.

TYPES OF VEGETARIANS. Persons following vegetarian patterns fall into one of the four basic types of vegetarianism.

1. *Ovolactovegetarians.* These individuals consume a diet of plant origin supplemented with milk, milk products, and eggs.

2. *Lactovegetarians.* These individuals eat a diet of plant origin supplemented with milk and milk products only.

3. *Pure vegetarians or vegans.* These individuals eat all foods of plant origin, and no animal-originating foods, dairy products, or eggs. In addition, the term vegan may embody a philosophy which discourages the use of anything that even indirectly requires the taking of animal life.

4. *Fruitarians.* These individuals consume diets consisting of raw or dried fruits, nuts, honey, and olive oil. Some may supplement with grains and legumes.

Each type of vegetarianism may run some risk of dietary inadequacy—some more than others.

PREVALENCE OF VEGETARIANISM. The practice of vegetarianism is not new. Throughout history there have been various individuals, groups or cults that have subscribed to a vegetarian diet. Of the five major religions of the world—Christianity, Judaism, Islam, Hinduism, and Buddhism—all place restrictions on some food of animal origin. Because of the belief in the transmigration of souls, the preservation of animal life is a basic tenet of Hinduism, Brahmanism, Buddhism, and Jainism.

Organized vegetarianism was brought to the United States by the Reverend William Metcalfe and 41 members of his church, who landed in Philadelphia in 1817. Their numbers grew, and their cause was furthered by the influence of several prominent men who became convinced of the advantages of the fleshless diet. Among these were Reuben D. Mussey (1780-1866), fourth president of the American Medical Association and Edward Hitchcock (1793-1864), professor of Science and then president of Amherst College, whose "Lectures on Diet, Regimen, and Employment" were published in 1831. Also, contributing to the vegetarian movement in 1829 was Sylvester Graham, a young Presbyterian who believed he had regained his health on a vegetarian diet, and who began writing and speaking on moderation in eating. Graham is best remembered for advocating the use of unbolted *graham* flour and the *graham* cracker. In 1858, Dr. James Caleb Jackson opened Our Home Hygienic Institute in Danville, New York, which emphasized, among other things, the Graham-type diet. Mother Ellen Harmon White, spiritual leader of the Seventh-day Adventist Church, founded the Western Health Reform Institute at Battle Creek, Michigan, in 1866. Later, Dr. John Harvey Kellogg was hired as superintendent, and the name was changed to the Battle Creek Sanitarium or "San" where, in an effort to provide a wholesome, palatable nonflesh diet for the sanitarium patrons, Dr. Kellogg and his brother, W. K. Kellogg, developed cereal foods. From this vegetarian movement sprang two breakfast cereal businesses—Kelloggs of Battle Creek and General Foods Corporation. C. W. Post, the founder of General Foods, had been a patient-guest at the "San."

While the history of vegetarian movements can be traced, it is important to remember that in much of the world, adherence to a vegetarian diet is not by choice but by economics and availability. For centuries large populations of the world have survived on vegetarian diets of one type or another out of necessity.

Vegetarianism has more or less always existed. However, in the United States, there is a renewed movement toward vegetarianism—some for religious convictions, and some for personal convictions as an attitude of the future. In recent years, various religious cults such as Zen Buddhism, Hare Krishnas, and some yogic groups, which promote vegetarianism, have gained popularity in the United States. As a movement of the future, vegetarianism seems to have roots in various other movements; for example, ecology, counterculture, environmental pollution, and population studies suggesting increased heart disease, cancer, and various other ills with increased meat consumption. Many animal products have been promoted as unhealthy due to additives, preservatives, cholesterol, fat, and lack of fiber. Hardest hit have been eggs, cured meats, and red meats. Moreover, proponents of animal rights make animal production methods seem inhumane; hence, eating animal products becomes an inhumane activity. Still other individuals say cereal grains could be better used to feed the hungry world than to feed cattle and pigs. So, much of the move toward vegetarianism may actually be a move toward what many feel is a more healthful diet—at least based upon popularized theories, or a move toward some feeling of a right action.

PROBLEMS OF VEGETARIANISM. Whatever the reason for choosing a vegetarian life-style, some nutritional problems are inherent unless the person practicing possesses a sound knowledge of nutrition. Furthermore, the level of vegetarianism is important. Those at the greatest risk are the vegans or pure vegetarians and fruitarians, and among these individuals their dietary practices are the most hazardous to pregnant or lactating women, infants, growing children, adolescents, and people who are ill or recovering from disease. Dietary concerns for fruitarians and vegans include energy, protein, calcium, iron, zinc, vitamin D, riboflavin, and vitamin B-12.

Fig. V-14. A wide variety of vegetables provides adventures in good eating and rich sources of certain essential minerals and vitamins. (Courtesy, United Fresh Fruit and Vegetable Assn.)

HEALTHFUL VEGETARIAN DIETS. Planning nutritious total vegetarian diets requires care and knowledge of the strengths and weaknesses of the various foods. By combining the information contained in Table V-3 and using the following guidelines, healthful vegetarian diets may be planned:

1. Reduce substantially all high calorie, low nutrient density foods (empty-calories). Instead, use unrefined foods as far as practical, which, on a caloric basis, supply their share of nutrients.

2. Replace meat by the protein-rich group of legumes, seeds, and nuts.

3. Increase intake of whole grain breads and cereals, legumes, nuts, and seeds to maintain energy intake.

4. Use a variety of legumes and whole grains, with some seeds and/or nuts, in meals each day to achieve good protein complementation; for example, beans with corn or rice, cereals with legumes and green, leafy vegetables, and peanuts with wheat.

5. Use a variety of fruits and vegetables, but be sure to include food high in ascorbic acid *in each meal* to enhance iron absorption.

6. Replace the nutrients lost by deleting milk and dairy products by (a) taking a modest amount of nutritional yeast (brewers' or torula), (b) eating dried fruits, (c) providing supplemental vitamin D by daily exposure to the sun, and (d) providing supplemental vitamin B-12.

7. Secure added food energy (calories) from such foods as sweeteners (sugar and syrups), margarine, oils, and shortenings; used either as ingredients in recipes, or at the table.

8. Eat sufficient amounts to maintain ideal weight for age and height. Larger servings and additional foods such as nuts or margarine may be necessary.

9. Consult the Food Composition Table F-21 when planning vegetarian diets to ensure the selection of nutritious foods.

TABLE V-3
THE VEGETARIAN FOUR FOOD GROUPS—THE BASIC FOUR[1]

Group	Foods Included	Amount Recommended	Contribution to the Diet	Comments
Protein rich	Dry beans, dry peas, soybeans, lentils, nuts, including peanuts and peanut butter; meat analogs.	Choose five servings every day. Count as a serving: ½ c. of cooked dry beans, dry peas, soybeans, or lentils; 2 Tbsp peanut butter; ¼ to ½ c. nuts, sesame, or sunflower seeds. Count one serving as equivalent to 1 oz of lean meat, poultry, or fish.	Foods in this group are valued for their protein. They also provide iron, thiamin, riboflavin, niacin, and phosphorus.	Commercially prepared plant protein products are not essential for a well-balanced diet. However, when changing from a nonvegetarian to a vegetarian diet, meat analogs are helpful because they replace the accustomed entree without further changes in the menu being necessary. They are convenience-type foods. Labels should be checked for nutritional data. Nuts are concentrated foods contributing flavor and unsaturated fat to the diet.
Milk and cheeses	Fortified soy milk and soy cheese (tofu).	Some milk every day for everyone. Recommended amounts are given below in terms of 8-oz cups: (8-oz cup) Children under 9 .... 2 to 3 Children 9 to 12 .... 3 or more Teenagers .... 4 or more Adults .... 2 or more Pregnant women .... 3 or more Nursing mothers .... 4 or more	Provides calcium, and vitamin B-12.	Children, in particular, need fortified soybean milk. Large servings of greens like collards, kale, mustard greens, turnip greens, and dandelion greens contribute to meeting the calcium requirement. Vitamin D normally obtained from milk should be supplemented by daily exposure to sunlight. Vitamin B-12 must be provided in fortified foods such as soy milk or yeast grown on B-12 enriched media; or provided as a supplement. Nursing mothers need supplemental vitamin D and B-12.
Vegetables and fruits	All vegetables and fruits including dried fruits; with emphasis on those that are valuable sources of vitamin C and vitamin A.	Choose 4 or more servings every day, including: 1 serving of a good source of vitamin C or 2 servings of a fair source; 1 serving, at least every other day, of a good source of vitamin A. If the food chosen for vitamin C is also a good source of vitamin A, the additional serving of a vitamin A food may be omitted. The remaining 1 to 3 or more servings may be of any vegetable or fruit, including those that are valuable for vitamin C and vitamin A. Count as one serving: ½ c. of vegetable or fruit; or a portion as ordinarily served, such as 1 medium apple, banana, orange, or potato, ½ a medium grapefruit or cantaloupe or the juice of 1 lemon.	Fruits and vegetables are valuable chiefly because of the vitamins and minerals they contain. In this plan, this group is counted on to supply nearly all the vitamin C needed and over ½ of the vitamin A. In addition, some members of this group supply fiber. Dark-green and deep-yellow vegetables are good sources of vitamin A. Most dark-green vegetables, if not overcooked, are also reliable sources of vitamin C, as are citrus fruits. Dark-green vegetables are valued for riboflavin, folacin, iron, and magnesium as well. Nearly all vegetables and fruits are low in fat, and none contain cholesterol. Unpeeled fruits and vegetables and those with edible seeds supply fiber. Dried fruits are good sources of iron.	A food high in vitamin C (ascorbic acid) at each meal enhances iron absorption. For sweetening cereals use raisins, dates, or sliced bananas. Avocados supply some fat. Dried fruits concentrate the mineral content.
Breads and cereals	All breads and cereals should be whole-grain or enriched products; whole grains include: wheat, oats, corn, rye, millet, barley, rice, wild rice.	Choose 5 or more servings daily; or if no cereals are chosen, have an extra serving of breads or baked goods, which will make at least 5 servings from this group daily. Count as 1 serving: 1 slice of bread; 1 oz ready-to-eat cereal; ½ to ¾ c. cooked cereal, cornmeal, grits, macaroni, noodles, rice, spaghetti, etc.	Foods in this group furnish worthwhile amounts of protein, iron, several of the B vitamins, and food energy. Whole-grain products also contribute magnesium, folacin, and fiber. Yeast fermentation of whole-grain flours, such as in breadmaking, lowers the phytates and significantly increases the availability of zinc.	Includes all products made with whole-grains or enriched flour, or meal. Example: bread, biscuits, muffins, waffles, pancakes, cooked or ready-to-eat cereals, cornmeal, flour, grits, macaroni and spaghetti, noodles, rice, rolled oats, barley, and bulgur. Most breakfast cereals are fortified at nutrient levels higher than those occurring in the natural grain. Use whole grains to complement the protein-rich group.

[1]In ovolactovegetarian or lactovegetarian diets, eggs and/or milk and milk products are used in place of some of the servings of food from the protein-rich group. Milk recommendations follow those set forth in the milk and cheese group. To convert to metric, see WEIGHTS AND MEASURES.

10. Make use of a number of good vegetarian cookbooks for preparing tasty and nutritious dishes.

NOTE WELL: Pure vegetarian diets are not recommended for infants and small children. Also, pure vegetarian diets for pregnant or lactating women should be planned with great care.

(Also see INFANT DIET AND NUTRITION; and PREGNANCY AND LACTATION NUTRITION.)

MEAT SUBSTITUTES. Vegetarians can have their meat and eat it too! There is a whole line of products developed primarily from plant proteins that look like meat and taste like meat. Similar to the vegetarian movement of the late 1800s which spawned the development of Kelloggs and General Foods, the vegetarian movement of the late 1900s has created a market for meat analogs or substitutes. These are manufactured by two companies, primarily—Loma Linda Foods and Worthington Foods. Although these products are not necessary for a well-balanced vegetarian diet, they do help when an individual is shifting from a nonvegetarian diet to a vegetarian diet. Furthermore, meat analogs add variety to a vegetarian diet. They can be produced to simulate the appearance, texture, and flavor of meat, poultry, and fish; and they can be made to meet the known nutrients in animal sources. Individuals purchasing these meat analogs should realize that they are convenience foods. Also, nutritional information on the labels must be checked to determine their adequacy.

(Also see MEAT[S], section headed "Meat Substitutes.")

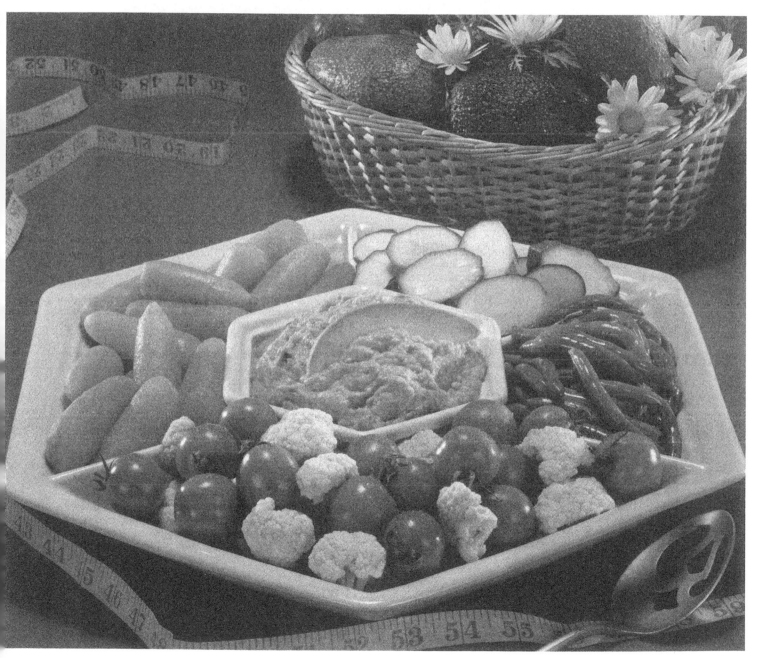

Fig. V-15. Vegetables measure up to *good nutrition.* (Courtesy, Harshe-Rotman & Druke, Inc., Public Relations, Los Angeles, Calif.)

Fig. V-16. A textured vegetable protein patty which can be fried, baked, or grilled to taste like a hamburger. (Courtesy, Worthington Foods, Worthington, Ohio)

POSSIBLE HEALTH BENEFITS OF VEGETARIAN-ISM. Vegetarianism has been credited with minimizing a variety of diseases including obesity, cancer, coronary heart disease, dental caries, diabetes, and diverticular disease of the colon. However, in many cases, lack of complete data prohibits sound conclusions.

(Also see CHOLESTEROL; FIBER; FOOD GROUPS; MEATS; WORLD FOOD; and ZEN MACROBIOTIC DIET.)

VENISON

The name venison comes from the Latin *venatio*, meaning *game* or *hunting*. Presently, the term connotes the flesh of any antlered animal.

Venison should be aged before eating or freezing. This may be accomplished by hanging the carcass in a walk-in cooler at 34° to 36°F (1° to 2°C); aging young deer for 1 week, and older deer for 2 to 3 weeks.

Venison lends itself well to corning (corned venison); to curing, drying, and smoking (dried venison); to sausage making, when mixed with 50% or more of fat pork trimmings, and prepared as summer or smoked sausage; and to freezing. The most widely used method of preservation is by freezing at 0°F (-18°C) or lower. Ground meat may be stored in the freezer for 2 to 3 months, and roasts and steaks for 8 to 12 months.

Venison may be cooked like beef—steaks and chops broiled or sauteed, legs roasted, and the less tender cuts potroasted. The best tenderizer for game is a tasty marinade. Any fat should be cut off and replaced with other fat, such as bacon or salt pork. Salt pork is excellent for larding large pieces of meat; smaller pieces may be wrapped in bacon. Venison, like beef, may be served rare, medium, or well done.

VERJUICE

The two different meanings of this term are as follows:

• The juice extracted from either unripe apples or unripe grapes, which was once used to make various drinks, but is now used mainly to impart a tart flavor to various sauces in European cuisine.

• Any tart juice, such as that expressed from cress, crabapples, gooseberries, or sorrel, which may be either fermented or unfermented.

VERMICELLI

This refers to a pasta that is much like spaghetti—only thinner.

(Also see MACARONI AND NOODLE PRODUCTS.)

VERMOUTH

A white wine flavored with wormwood or other herbs such as anise, cinnamon, bitter orange peel, cloves, and elderberries. There are two types: (1) a dark or reddish, richly flavored (sweet) Italian variety; and (2) a pale yellow or light, dry French variety. Vermouth can be used as a liqueur or in cocktails.

(Also see WINE.)

VETCH

The name applied to several different legumes (pulses) some of which are used for food. The starchy root of the tuber-vetch is roasted, while the chick-vetch is prepared in the same manner as chick peas.

VICILIN

A globulin (simple protein) associated with a legume, such as peas, lentils, and broad beans.

VIENNA BREAD

A bread that is eaten in France mainly as toast or croutons, or in sandwiches. It has a very thin crust; and it is baked in square pans.
(Also see BREADS AND BAKING.)

VILLI

Small threadlike projections attached to the interior side of the wall of the small intestine.
(Also see DIGESTION AND ABSORPTION.)

VINEGAR

The word vinegar is derived from the French word *vinaigre*, literally meaning sour wine. Indeed, sour wine was the original source of vinegar. Vinegar was a by-product of wine makers and brewers until about the 17th century when vinegar making became a separate industry in France.

By definition, vinegar is a sour liquid containing 4 to 12% acetic acid. Any product that will yield alcoholic fermentation—apples, grapes, pears, peaches, plums, figs, oranges, berries, honey, sugar, syrups, hydrolyzed starchy materials, beers and wines—is acceptable for the production of vinegar. However, wine and cider are the best raw materials for vinegar production. In addition to the acetic acid, other organic acids and esters are present in vinegar giving it a flavor and aroma characteristic of the material from which it was derived. Thus, there are a variety of vinegars: wine vinegar, apple-cider vinegar, beer vinegar, and malt vinegar.

In 1864, Louis Pasteur demonstrated that bacteria caused the conversion of alcohol to acetic acid. Hence, vinegar making is dependent upon the action of yeast and bacteria on the raw product. Fig. V-17 illustrates the chemical reactions involved in the production of vinegar. Cider and wine

$$C_6H_{12}O_6 \xrightarrow{\text{YEAST}} 2\,C_2H_5OH + 2\,CO_2$$

HEXOSE   ETHYL ALCOHOL   CARBON DIOXIDE
(SUGAR)

+

O_2
OXYGEN

↓ ACETOBACTER BACTERIA

$$2\,CH_3COOH + 2\,H_2O$$
ACETIC ACID   WATER
(VINEGAR)

Fig. V-17. Biochemical changes of vinegar making.

are the best raw materials for vinegar production. In the United States, cider is most frequently used. Basically, apples are mechanically crushed and pressed, and adjusted to contain not less than 8%, and not more than 20%, fermentable sugar. Following fermentation by yeast, the sugar content should preferably be below 0.3%. Next, this alcohol-containing substrate is subjected to the action of acetobacter organisms until the alcohol is converted to acetic acid, following which the vinegar is usually clarified, possibly pasteurized, then bottled.

USES OF VINEGAR. In the home, vinegar is used for baking, candies, pickling, production of sour milk, salad dressings, and sweet-and-sour sauces.

Commercially, vinegar is used for mayonnaise, mustard, pickled foods, salad dressings, and tomato products.
(Also see PRESERVATION OF FOODS, Table P-10, Pickling.)

Fig. V-18. Pickles (cucumbers) preserved in vinegar. (Courtesy, University of Minnesota, St. Paul, Minn.)

VIRAL INFECTIONS, FOODBORNE

Infectious hepatitis is the most common foodborne viral disease. The virus is found in excreted stools and blood of infected persons. It may be spread by poorly handled food or seafood taken from sewage contaminated water. In addition to hepatitis, there is some suggestion that the viruses responsible for epidemic diarrhea of the newborn, intestinal flu, and poliomyelitis may be transmitted in contaminated foods.
(Also see DISEASES, section on "Food-Related Infectious and Parasitic Diseases.")

VIRUS

A disease-producing agent that (1) is so small that it cannot be seen through an ordinary microscope (it can be seen by using an electron microscope), (2) is capable of passing through the pores of special filters which retain ordinary bacteria, and (3) propagates only in living tissues.

VISCOSITY

A term which indicates the condition or the property of being thick like syrup or glue. Viscous fluids resist flow or motion.

VISUAL PURPLE (RHODOPSIN)

Photosensitive vitamin A-containing pigment found in the rods of the retina.
(Also see VITAMIN A, section headed "Functions.")

VITAMERS

A substance structurally related to a certain vitamin, which possesses some biological activity—although usually less than the true vitamin, and which may relieve a particular vitamin deficiency.

VITAMIN(S)

Contents   *Page*
  History/Discovery of Vitamins **1054**
  Definition. **1055**
  Nomenclature and Classification **1055**
    Fat-Soluble vs Water-Soluble Vitamins. **1055**
    Vitamin B Complex **1056**
    Vitaminlike Substances **1056**
  Determination of the Vitamins in Food **1057**
    Biological (Animal) Assay 1057
    Microbiological Assay 1057
    Chemical Assay 1057
    Physical Assay 1057
    Human Assay 1057
  Factors Influencing the Utilization
    of Vitamins 1057
    Availability 1057
    Antivitamins (Vitamin Antagonists,
      or Pseudovitamins) 1057
    Provitamins 1057
    Synthesis in the Gut 1057
    Interaction of Nutrients 1058
  Vitamin Content of Foods 1058
  Vitamin Supplementation of the Diet 1058
  Vitamin Table 1059

Fig. V-19. This shows an artist's dramatic impression of how British sailors, who ate little except salt meat and biscuits when on long voyages 200 years ago, suddenly collapsed and died of scurvy. (Reproduced with permission of *Nutrition Today*, P.O. Box 1829, Annapolis, MD 21404, ©, 1979)

For proper physiological function, the human body requires some 40 to 50 dietary essentials, of which 14 are vitamins.

Throughout history, vitamin deficiencies have been a major cause of disease, morbidity, and death. Pellagra, scurvy, and beriberi decimated armies, ships' crews, and nations; they even reshaped the course of history. The importance of dietary factors in the genesis of diseases became recognized in the 18th century. But the significance of these observations was not fully understood until early in the 20th century, when scientists found it desirable in many types of investigations to use the biological approach—the use of laboratory animals (largely white, albino rats and mice; guinea pigs; and chicks) fed on purified diets using pure protein such as casein or albumen, pure fat such as lard, and pure carbohydrate such as dextrin, plus minerals to supplement chemical analyses in measuring the value of food. These diets were made up of relatively pure nutrients (proteins, carbohydrates, fats, and minerals) from which the unidentified factors were largely excluded. With these purified diets, all researchers shared a common experience—the animals not only failed to thrive, but they even failed to survive if the investigations were continued for any length of time. At first, many investigators explained such failures on the basis of unpalatability and monotony of diets. Finally, it was realized that these purified diets were lacking in certain factors, minute in amounts, the identities of which were unknown to science. These factors were essential for the efficient utilization of the main ingredients of the food and for the maintenance of health and life itself. The discovery, synthesis, and commercial production of vitamins followed. With these developments, the vitamin era of science was ushered in and the modern approach to nutrition was born.

Fig. V-20. The belief that certain factors in very minute amounts are needed by people and animals was difficult to conceive and hold in the early 1900s. Like many episodes of science, it was a mystery story. Yet, finally it was solved by the biological approach—the use of laboratory animals in controlled feeding experiments.

In 1909, Thomas B. Osborne and Lafayette B. Mendel of the Connecticut Agricultural Experiment Station (New Haven), began investigations with purified foodstuffs. (Although different mixtures were used, the basic ingredients of their purified diets were: casein, starch, sugar, and lard.)

The two white rats pictured above were the result of pioneering vitamin studies conducted by Osborne and Mendel during the years 1913 to 1917. *Top:* Rat on a diet deficient in B vitamins; note the rough, scraggly coat, and the emaciation. *Bottom:* The same rat after 12 days on a diet that included B vitamins; note the marked changes in the coat. (Courtesy, The Connecticut Agricultural Experiment Station, New Haven, Conn.)

HISTORY/DISCOVERY OF VITAMINS. Until the early 1900s, if a diet contained proteins, fats, carbohydrates, minerals, and water, it was considered to be a complete diet. True enough, the disease known as beriberi made its appearance in the rice-eating districts of the Orient when milling machinery was introduced from the West, having been known to the Chinese as early as 2600 B.C.; and scurvy was long known to occur among sailors fed on salt meat and biscuits. However, for centuries these diseases were thought to be due to toxic substances in the digestive tract caused by pathogenic organisms, rather than food deficiencies; and more time elapsed before the discovery of vitamins.

Largely through the trial-and-error method, it was discovered that specific foods were helpful in the treatment of certain of these maladies. Hippocrates (460-377 B.C.), the Greek, known as the father of medicine, advocated liver as a cure for night blindness 400 years before the birth of Christ. At a very early date, the Chinese also used a concoction rich in vitamin A as a remedy for night blindness; and cod-liver oil was used in treating or preventing rickets long before anything was known about the cause of the disease. In 1747, James Lind, a British naval surgeon, in a study involving 12 sailors with scurvy on board the "Salisbury," showed that the juice of citrus fruits was a cure for the disease. Nicholas Lunin, as early as 1881, while a student of von Bunge at the University of Dorpat, had come to the conclusion that certain foods, such as milk, contain, beside the principal ingredients, small quantities of unknown substances essential to life. In 1882, Kanehiro Takaki, Director-General of the Japanese Navy, greatly reduced the number of beriberi cases among naval crews by adding meat and evaporated milk to their diet of rice. In 1897, Christiaan Eijkman, a Dutch medical officer, working in Java, had satisfied himself that the disease beriberi was due to the continued consumption of a diet of polished rice.

In 1912, Dr. Casimir Funk, a 28-year-old Polish biochemist working in London, coined the word *vitamine*. Funk postulated, as others had before him, that beriberi, scurvy, pellagra, and possibly rickets, were caused by a lack in the diet of "special substances which are of the nature of organic bases, which we will call vitamines." Presumably, the name vitamines alluded to the fact that they were vital to life, and that they were chemically of the nature of amines (nitrogen-containing). The name caught the popular fancy and persisted, despite the fact that the chemical assumption was later proved incorrect, with the result that the "e" was dropped in 1920; hence, the word *vitamin*. In 1922, Funk's book entitled *The Vitamins* was published.

A chronological summary of the discovery/isolation and synthesis of the various vitamins is given in Table V-4.

The actual existence of vitamins, therefore, has been known only since 1912, and it was much later before it was possible to see or touch any of them in a pure form. Previously, they were merely mysterious invisible *little things* known only by their effects. In fact, most of the present fundamental knowledge relative to the vitamin content of both human foods and animal feeds was obtained through measuring their potency in promoting growth or in curing certain disease conditions in animals—a most difficult and tedious method. For the most part, small laboratory animals were used, especially rats, guinea pigs, pigeons, and chicks.

Today, there are 14 known vitamins. Additionally, there are at least nine other vitaminlike substances that have been proposed as a result of various experiments, but it is unlikely that all of them are distinct essentials. Yet, the probability that there are still undiscovered vitamins is recognized. Each of the vitamins and vitaminlike substances is alphabetically listed and discussed in a separate section of this book

DEFINITION. Vitamins are organic substances that are essential in small amounts for the health, growth, reproduction, and maintenance of one or more animal species, which must be included in the diet since they either cannot be synthesized at all or in sufficient quantities in the body.

Each vitamin performs a specific function; hence, one cannot replace, or act for, another. In general, the human body cannot synthesize them, at least in large enough amounts to meet its needs. However, vitamin D is an exception; when a person is exposed to ultraviolet rays, the vitamin is synthesized from its precursor, which is found in the skin.

TABLE V-4
CHRONOLOGY OF THE
DISCOVERY/ISOLATION AND SYNTHESIS OF VITAMINS

Year	Discovery/Isolation	Synthesis
1849[1]	Choline	
1866–67		Choline
1913	Vitamin A[2]	
1926	Thiamin (B-1)	
1929	Vitamin K[3]	
1931	Vitamin A[2]	
1932	Vitamin C (Ascorbic Acid)	
	Vitamin D$_2$	
1933	Riboflavin (B-2)	Vitamin C (Ascorbic Acid)
1935		Riboflavin (B-2)
1936	Biotin	Thiamin (B-1)
	Vitamin E	
1937	Niacin (Nicotinic acid)	
1938	Vitamin B-6 (pyridoxine)	Vitamin E
1939	Pantothenic acid (B-3)	Vitamin B-6 (pyridoxine)
	Vitamin K[3]	Vitamin K[3]
1940		Pantothenic acid (B-3)
1943		Biotin
1945	Folacin (Folic acid)	Folacin (Folic acid)
1947		Vitamin A[2]
1948	Vitamin B-12	
1952		Vitamin D$_3$
1955		Vitamin B-12

[1]In 1844 and 1846, Gobley isolated a substance from egg yolk, which he called lecithin. In 1849, Strecker isolated a compound from hog bile, to which he subsequently (in 1862) applied the name choline.

[2]A more detailed chronology of vitamin A is: in 1913, it was discovered, independently, by McCollum and Davis of the University of Wisconsin, and Osborne and Mendel of the Connecticut Experiment Station; in 1931, its chemical formula was determined by P. Karrer, a Swiss researcher; and, in 1947, it was synthesized by Isler, working in Switzerland.

[3]Vitamin K was discovered in 1929, isolated in 1939, and synthesized in 1939.

NOMENCLATURE AND CLASSIFICATION. There is no universal agreement on the nomenclature of the vitamins. But the modern tendency is to use the chemical name, particularly in describing members of the B complex. In this book, the most common designations are used.

Today, vitamins are generally classed as (1) fat-soluble vs water-soluble vitamins, (2) vitamin B complex, and (3) vitaminlike substances.

Fat-Soluble vs Water-Soluble Vitamins. Many phenomena of vitamin nutrition are related to solubility—vitamins are soluble in either fat or water. Consequently, it is important that both nutritionists and consumers be well informed about solubility differences in vitamins and make use of such differences in programs and practices. Based on solubility, vitamins may be grouped as follows:

The Fat-Soluble Vitamins	The Water-Soluble Vitamins
Vitamin **A**	Biotin
Vitamin **D**	Choline
Vitamin **E**	Folacin (folic acid)
Vitamin **K**	Niacin (nicotinic acid; nicotinamide)
	Pantothenic acid (vitamin B-3)
	Riboflavin (B-2)
	Thiamin (vitamin B-1)
	Vitamin B-6 (pyridoxine; pyridoxal; pyridoxamine)
	Vitamin B-12 (cobalamins)
	Vitamin C (ascorbic acid; dehydroascorbic acid)

It is noteworthy that vitamin C is the only member of the water-soluble group that is not a member of the B family.

The two groups of vitamins exhibit the following several differences that distinguish them both chemically and biologically:

• **Chemical composition**—The fat-soluble vitamins contain only carbon, hydrogen, and oxygen, whereas the water-soluble B vitamins contain these three elements plus nitrogen.

• **Occurrence**—Vitamins originate primarily in plant tissues; with the exceptions of vitamins C and D, they are present in animal tissues only if an animal consumes foods containing them or harbors microorganisms that synthesize them. Fat-soluble vitamins can occur in plant tissue in the form of a pro-vitamin (or precursor of a vitamin), which can be converted into a vitamin in the animal body. But no provitamins are known for any water-soluble vitamin. Also, the B vitamins are universally distributed in all living tissues, whereas the fat-soluble vitamins are completely absent from some.

• **Absorption**—Fat-soluble vitamins are absorbed from the intestinal tract in the presence of fat. It follows that any factor that increases the absorption of fat, such as small particle size or the presence of bile, will also increase the absorption of fat-soluble vitamins. Generally speaking, the absorption of the water-soluble vitamins is a simpler process because there is constant absorption of water from the intestine into the bloodstream.

• **Storage**—The fat-soluble vitamins are stored in appreciable quantities in the body, whereas the water-soluble vitamins are not. Any of the fat-soluble vitamins can be stored wherever fat is deposited; and the greater the intake, the greater the storage. By contrast, the water-soluble B vitamins are not stored in any appreciable amount. Moreover, the large amounts of water which pass through the body daily tend to carry out the water-soluble vitamins, thereby depleting the supply. Hence, they should be supplied in the diet on a daily basis. However, because all living cells contain all the B vitamins, and because the body conserves nutrients that are in short supply by using them only in vital reactions, deficiency symptoms do not appear immediately following their removal from the diet.

• **Excretion**—The fat-soluble vitamins are excreted exclusively in the feces via the bile. The water-soluble vitamins are excreted primarily in the urine, although limited amounts may be present in the feces. This difference in pathway of excretion reflects the difference in solubility.

• **Physiological action**—The fat-soluble vitamins are required for the regulation of the metabolism of structural units, and each vitamin appears to have one or more specific roles. Collectively, the water-soluble B vitamins are primarily concerned with the transfer of energy.

• **Deficiency symptoms**—The absence of one or more vitamins in the diet may lead to failure in growth or reproduction, or to characteristic disorders known as deficiency diseases. In severe cases, death may follow.

The sign of a fat-soluble vitamin deficiency can sometimes be related to the function of the vitamin. For example, vitamin D is required for calcium metabolism and a deficiency results in bone abnormalities. On the other hand, the signs of a B-vitamin deficiency are much less specific and are difficult to relate to function in most cases. Most B-vitamin deficiencies result in dermatitis, rough hair, and poor growth. Deficiencies of some produce loss of pigment in the hair, whereas deficiencies of others cause anemia. It is noteworthy, too, that not all animals suffer from the same deficiency diseases. Thus, man, the monkey, and the guinea pig get scurvy on diets that provide no vitamin C in the diet, whereas rats, fowl, and ruminants make this vitamin in their bodies and do not need it in their food.

The short supply of a vitamin may be more serious than a short supply of food. However, such vitamin deficiencies are less widespread throughout the world than hunger itself, for starvation has always stalked across much of the world, being referred to as famine only when the numbers dying approach the millions.

• **Toxicity**—Excesses of fat-soluble vitamins A and D can cause serious problems, whereas the water-soluble vitamins are relatively nontoxic.

Vitamin B Complex. With the exception of vitamin C, all of the water-soluble vitamins can be grouped together under the vitamin B complex, of which there are the following nine:

Biotin
Choline
Folacin (folic acid)
Niacin (nicotinic acid; nicotinamide)
Pantothenic acid (vitamin B-3)
Riboflavin (vitamin B-2)
Thiamin (vitamin B-1)
Vitamin B-6 (pyridoxine; pyridoxal; pyridoxamine)
Vitamin B-12 (cobalamins)
(Also see VITAMIN B COMPLEX.)

Vitaminlike Substances. Certain substances, although not considered true vitamins, closely resemble vitamins in their activity, and are sometimes classed with the B-complex vitamins. They're commonly refered to as *vitaminlike substances*. When listed in vitamin preparations, usually the names of these substances are followed by an asterisk, which is tied in with another asterisk at the bottom of the label reading: "Need in human nutrition has not been established."

The nutritional status and biological role(s) of each of the vitaminlike substances require further clarification; indeed, some of them are very controversial. In the meantime, the authors discuss each of them for two primary purposes: (1) historical and informational, and (2) stimulation of research. Additionally, cognizance is taken of the fact that there is usually a considerable time lag between the scientific validation and the acceptance of a vitamin, essential nutrient, and/or medical treatment. A list of the vitaminlike substances follows:

Bioflavonoids
Carnitine (vitamin B-T)
Coenzyme Q (ubiquinone)
Inositol
Laetrile (vitamin B-17), amygdalin, nitrilosides)
Lipoic acid
Pangamic acid (vitamin B-15)
Para-aminobenzoic acid (PABA)
Vitamin B-13 (orotic acid)

Each of these vitaminlike substances is alphabetically listed and discussed in this book.

DETERMINATION OF THE VITAMINS IN FOOD.

The potency, or vitamin content, of foods may be determined by biological, microbiological, chemical, physical methods, and/or human assay.

Biological (Animal) Assay. Before the chemical nature of vitamins was known, their potency could be measured by their ability to promote growth or cure a deficiency when test doses were fed to experimental animals—usually rats, mice, guinea pigs, pigeons, or chicks. Such measurement is known as *biological assay* or *bioassay* and is expressed in terms of units. Vitamins A, D, and E are still measured in international units (IU). (Also, vitamin A values are given in RE [mcg retinol equivalent]; vitamin D is given in micrograms of cholecalciferol; and vitamin E is given in milligrams of alpha-tocopherol equivalents.)

In most common methods of bioassay, laboratory animals are first depleted of the particular vitamin being studied by being fed a diet lacking in it. Then some of the depleted animals are divided into different groups and fed graded doses of the known vitamin, with each group receiving one of the doses in the series. The response of the animals to the intake, in growth and/or other appropriate criteria, is measured and recorded in a standard response curve. Simultaneously, a second set of depleted animals is divided into different groups, with each group fed increasing amounts of the food being assayed; and the responses of each of these groups is recorded. Then, the vitamin potency of the product being tested is estimated by comparing the responses of the second set of animals with the standard response curve of the first group.

Biological assays are laborious, time-consuming, and costly. Large numbers of samples are needed to produce statistically reliable results; the animals should be of approximately the same age, sex, and weight; it takes time to produce nutrient-deficient animals; and quite often the data obtained are highly variable. In addition, data from one species are not always relevant to another species.

The fundamental value of the biological assay is that it gives positive proof of biological activity. It is still the basis of comparison or standardization for newer microbiological or chemical methods.

Microbiological Assay. This assay is a measure of the ability of a vitamin to promote the growth of a microorganism. A microorganism is selected that is known to require the vitamin, or other nutrient, in question. Therefore, if the vitamin is not present, the selected microorganism will not grow. Actually, the procedure is the same as the biological assay; the only difference is that some suitable microorganism is used in place of rats or other animals. Growth media is prepared so that it is nutritionally complete except for the vitamin to be tested. Graded levels of the vitamin are then added to the media and a growth response curve is prepared. Then, the sample to be assayed can be tested and compared to the growth response curve to determine the concentration of the nutrient. The B vitamins may be assayed in this manner.

The microbiological assay requires less time than the biological assay. Its main disadvantage is that the vitamin being tested must first be extracted from the foodstuff before being added to the growth medium used for the microorganism.

Chemical Assay. Today, foods may be analyzed by highly sophisticated chemical procedures, with the vitamin content expressed in units of weight—milligrams (mg) or micrograms (mcg).

Chemical assays are much faster than biological assays, but, at intervals, they must be compared with bioassays in order to rule out the possibility of assaying as a vitamin some substance that does not function as a vitamin in the body.

Physical Assay. Each of the vitamins may now be determined by physical methods—measurement of absorption spectra, chromatography, fluorescence, turbidity, etc.

Human Assay. Actually, this is a special biological type of assay designed for humans. Experiments with people are very costly and difficult to control. Besides, it is necessary to protect the health and rights of the subjects, and to avoid psychic factors in interpreting the results. The use (1) of *placebos* as controls (pills that, unknown to the subject, contain an inactive substance such as sugar), and (2) of double blind tests (in which none of the participants—the subjects, the investigator, the person giving the diets or supplements, and the diagnostician—know the composition of the dietary variables) is standard procedure in human nutrition studies. In such studies, a code may be used for various test groups, the disclosure of which is not revealed until all tests are completed.

FACTORS INFLUENCING THE UTILIZATION OF VITAMINS.

Vitamin deficiency is usually caused by an insufficiency of the nutrient in the diet being consumed. However, the utilization of vitamins may be influenced by availability, antivitamins, provitamins, synthesis in the gut, and interactions of nutrients.

Availability. Not all the vitamins in foods are in absorbable form. For example, (1) niacin in many cereals is bound to a protein and cannot be absorbed through the intestinal wall, unless the food is treated with an alkali to release the vitamin from the inaccessible complex; (2) fat-soluble vitamins may fail to be absorbed if the digestion of fat is impaired; and (3) vitamin B-12 requires a factor produced in the stomach (the intrinsic factor) for its absorption.

Antivitamins (Vitamin Antagonists, or Pseudovitamins). Antivitamins, which are present in some natural foods, are compounds that do not function as vitamins even though they are chemically related to them. As a result, they may cause vitamin deficiencies if the body (1) is unable to distinguish between them and true vitamins, and (2) incorporates them into essential body compounds.

Provitamins. These are nonvitamin substances that occur in foods which can be converted into vitamins in the body. Well-known examples of provitamins are: (1) beta-carotene, which is converted to vitamin A in the intestinal wall; (2) 7-dehydrocholesterol in the skin, which is converted to vitamin D_3 by ultraviolet light (sunlight); (3) ergosterol of plants, which is converted to vitamin D_2 by ultraviolet light; and (4) the amino acid tryptophan, which can be converted to niacin. (Because of the poor efficiency—60 mg of tryptophan is required to produce 1 mg of niacin—tryptophan is not utilized for its provitamin value.)

Synthesis In The Gut. Although some species of bacteria can synthesize vitamins, other species compete with the host for the vitamins ingested in food and retain them until they are excreted in the feces.

In health, the small intestine of man is usually sterile. The large intestine carries a heavy load of bacteria, but usually absorption from the large intestine is limited to water and salts. It is unlikely, therefore, that bacterial synthesis in the gut affects the amounts of most vitamins available to the healthy human body.

When an intestinal disorder is present, especially if there is diarrhea, the small intestine may harbor large numbers of bacteria. These are likely to reduce, rather than increase, the amounts of available vitamins.

Interaction Of Nutrients. Several vitamins are closely linked to other nutrients. For example, (1) if the diet is rich in carbohydrates or alcohol, more thiamin is needed for metabolism; and (2) if the intake of polyunsaturated fats is high, more vitamin E is required. There are several other similar interactions. For this reason, the nutritive value of a diet in respect to a given vitamin may differ from the chemical analysis of its vitamin content. For this reason, too, the diet should supply a well-balanced mixture of all the nutrients, including the vitamins, such as is provided by eating a wide variety of traditional foods—meat, milk, eggs, fruits (including citrus fruits), vegetables (including green leafy vegetables), whole grain cereals and bread, and butter or fortified margarine.

Fig. V-21. Vitamin-rich salad. (Courtesy, California Iceberg Lettuce Commission, Monterey, Calif.)

VITAMIN CONTENT OF FOODS. The vitamin content of foods is given in Food Composition Table F-21. However, in selecting foods to furnish vitamins in the diet, it is well to keep in mind the following:

1. The vitamin losses of vegetables due to storage tend to parallel the degree of wilting. Such losses are progressive in the long storage of fresh vegetables and fruits.

2. The effects of processing and preparation of foods on vitamin retention must be clearly understood. For example, the fat-soluble vitamins (A, D, E, and K) are not easily lost by ordinary cooking methods, and they do not dissolve out in the cooking water. On the other hand, the water-soluble vitamins (B complex and C) are dissolved easily in cooking water, and a portion of the vitamins may be destroyed by heating; therefore, cooking food only until tender and in as little water as possible is, in general, the best procedure.

3. It is important to determine (a) how often any given food will be included in the diet, and (b) the size serving that will normally be eaten.

4. Economic factors such as availability and cost must be considered.

VITAMIN SUPPLEMENTATION OF THE DIET. Ideally, there should be enough of all the vitamins in the natural diet selected on the basis of the Six Food Groups to prevent disease and promote health. When we get our vitamins from such a diet, we usually also obtain the various other nutrients essential to health. Also, when vitamins are consumed in ordinary foods, they are never at a toxic level. The rule, therefore, is: Eat *enough* of *proper* foods *consistently* and chances are you'll get your full complement of needed vitamins. But for many Americans these three words—*enough, proper,* and *consistently*—are the catch. In the first place, lots of folks don't know what constitutes a good diet; and, worse yet, altogether too many people neglect or ignore the rules even if they know them. In either case, the net result is always the same: We shortchange ourselves on the right amounts of good, nutritious food, including vitamins. As a result, more and more doctors and nutritionists are recommending vitamin preparations as supplements. Additional arguments advanced in favor of judicious vitamin supplementation are:

1. **The vitamin content of animal and plant species and parts differ.** The vitamin content of foods varies widely; it's affected by species (for example, pork, is higher in thiamin than beef), and by parts (leaf, stalk, and seed of plants; and organ meats vs muscle meats of animals).

2. **Vitamin losses occur in processing, cooking, and storing.** Unfortunately, many of the vitamins naturally present in foods are destroyed by sunlight, oxidation, mold growth, and heat, to which they are subjected in processing, cooking, and storing. So, by the time they reach the table, much of their vitamin value has disappeared. By contrast, during olden times, for the most part man ate whole or natural foods which were subjected to little processing except for field preparation prior to consumption. Then came the food processing and refinement era, followed by the convenience, fabricated, and fast-food era.

Although it is apparent that raw foods supply adequate amounts of vitamins and other micronutrients, it's unlikely that today's processed, cooked, and stored foods provide adequate amounts of certain vitamins and other trace nutrients, especially for certain critical periods of life, unless they are enriched or supplemented.

One authoritative study revealed the following losses of vitamins from the processing and preservation of foods:[4]

a. The losses in seven vitamins from milling wheat and making flour ranged from 50 to 86.3%. Half of the pantothenic acid, 71.8% of the vitamin B-6, 86.3% of the alpha-tocopherol, and 66% of the folacin were removed.

b. The average losses by food groups in vitamin B-6 and pantothenic acid from canning and freezing were as follows:

Food Group	Frozen		Canned	
	Vitamin B-6	Pantothenic Acid	Vitamin B-6	Pantothenic Acid
	(% loss)	(% loss)	(% loss)	(% loss)
Fish and seafood	17.3	20.8	48.9	19.9
Meat and poultry		70.2	42.6	26.2
Vegetables, roots		36.8	63.1	46.1
Vegetables, legumes	55.6	57.1	77.4	77.8
Vegetables, green	36.7	48.2	57.1	56.4
Fruit and fruit juices	15.4	7.2	37.6	50.5

[4]Schroeder, Henry A., M.D., "Losses of Vitamins and Trace Minerals Resulting From Processing and Preservation of Foods", *The American Journal of Chemical Nutrition*, Vol. 24, May 1971, pp. 562-573.

The appalling processing and preservation losses revealed above underscore the dietary need for whole grains and unprocessed foods of many varieties. They also suggest that the vitamin enrichment and/or supplementation of certain refined and processed foods may be necessary in order to meet the recommended allowances of certain vitamins.

3. **Poor eating habits are commonplace.** Although there is an abundance of good food available, strange as it may seem many folks have fallen into poor eating habits.

But for the millions of Americans with poor eating habits supplemental vitamins are in the nature of *diet insurance* —they're a way of making sure that they are getting enough vitamins each day. That's the reason many of today's doctors and nutritionists recommend vitamin supplementation of the food we eat.

Fig. V-22. People who eat a variety of the proper foods consistently, generally get their full complement of vitamins. (Courtesy, Castle and Cooke, Inc.—Bud of California Fresh Vegetables)

4. **Several vitamins are not stored.** Several of the vitamins cannot be accumulated and stored in the body for any considerable period of time; they are used up quickly and therefore must be constantly replaced or renewed with a continuing supply from the outside.

5. **Fewer calories usually mean fewer vitamins.** With increased mechanization and decreased manual labor, fewer calories are consumed—and with it fewer vitamins.

6. **More vitamins required for growth, pregnancy, and lactation.** Vitamin intakes need to be increased during the critical periods of growth, pregnancy, and lactation. Babies are given supplemental vitamins from the time they are born until they are at least a year old and sometimes longer. Also, vitamin intakes are usually increased during pregnancy and lactation.

7. **Increased vitamins may be needed for therapeutic reasons.** Vitamin supplementation may be needed when a person is unable to consume an adequate diet due to illness, allergies, or emotional upsets. Likewise, vitamin supplementation may be prescribed when a serious dietary deficiency has occurred because of ignorance or poor eating habits; however, such supplementation should never replace the correction of factors leading to the dietary inadequacy. At such times, vitamin supplementation should be taken on the recommendation of physicians to help control particular conditions diagnosed by them.

8. **More vitamins needed during stress.** There are times when vitamin intakes may need to be increased, such as during infections; chronic disease; heavy drinking and/or smoking; regular use of birth control pills; restrictive dieting; environmental stress; and regular use of drugs which interfere with vitamin function. In these situations, professional advice should be sought relative to the given need, and proper adjustment should be made in the dietary pattern or the diet should be supplemented with a vitamin or a vitamin-mineral preparation.

9. **Lack of vitamins not easily recognizable.** *Too much* or *too little* of some food nutrients may soon show up on the bathroom scale as a pound or two gained or lost. But a lack of a vitamin(s) isn't so immediately recognizable.

Initially, it was thought that each vitamin was primarily a protective substance, which prevented a specific disease. While it is true that a specific disease can result from lack of a specific vitamin, specific deficiencies of a single vitamin are comparatively rare today. Instead, it is recognized that vitamins are interdependent of each other and on other nutrients for buoyant good health; hence, people are likely to have a deficiency of a group of vitamins, some consequences of which may be: a tired, run-down feeling; not quite up to par; insomnia; easily irritated; or lowered resistance to minor infection. Such signs and illnesses do not necessarily indicate vitamin deficiency, but, of course, the physician should be consulted. Also, everyone should understand the essential role of vitamins in maintaining health.

10. **Vitamin needs vary with individuals.** Vitamin needs vary according to age, kind of work, and general physical condition. Moreover, different individuals have different needs for different vitamins.

11. **We can learn from animals.** Much of the research on vitamins has been done with laboratory animals—rats, mice, guinea pigs, chicks, monkeys, dogs, etc. This makes sense when one realizes that many generations of white mice, for example, can be studied over a period of a few months or years, whereas if such clinical research were limited to humans, it would take 50 years or longer to make comparable tests with successive generations. Therefore, many conclusions concerning the role of vitamins in nutrition have been reached through basic research with animals.

So, today, more and more physicians and nutritionists rely on vitamin supplements, which in many cases are chemically pure sources that need to be taken only in very minute amounts, and which cost only a small fraction of the total food bill. Also, informed physicians and nutritionists no longer prescribe vitamins merely to prevent deficiency symptoms, or to meet minimum requirements, but, rather, to impart buoyant good health.

Undoubtedly, the use of vitamin supplements far exceeds the need for them in many cases. The potency of preparations on the market varies widely, and the number of nutrients present differs from one product to another. Moreover, the consumer may be unable to interpret the label information in terms of his/her own requirements.

Although water-soluble vitamins in excess of body needs will be excreted in the urine—with no harm done, except to the pocketbook, excessive intakes of vitamins A and D may be toxic. So, large amounts of vitamins should be taken under the direction of a physician or nutritionist.

VITAMIN TABLE. Table V-5 is a summary of each of the 14 known vitamins and of the 9 vitaminlike substances, totaling 23. Note that the 23 factors are divided into 3 appropriate groups—(1) fat-soluble vitamins, (2) water-soluble vitamins, and (3) vitaminlike substances—with the members within each group listed alphabetically. Note, too, that information relative to each of the 23 substances is given under the following headings: functions; deficiency and toxicity symptoms; recommended daily allowances; sources; and comments.

Additionally, a separate and more complete presentation relative to each vitamin and vitaminlike substance appears alphabetically in the book under its respective name.

TABLE V-5
VITAMIN TABLE

Functions	Deficiency and Toxicity Symptoms	Recommended Daily Allowance				Sources	Comments
		Age Group	Years	Vitamin			

FAT-SOLUBLE VITAMINS:
Vitamin A

Functions	Deficiency and Toxicity Symptoms	Age Group	Years	Vitamin A[1] (mcg RE)	(IU)	Sources	Comments
Helps maintain normal vision in dim light—prevents night blindness.	**Deficiency symptoms**—Night blindness (nyctalopia), xerosis, and xerophthalmia.	Infants	0.0–0.5	375	1,250	**Rich sources**—Liver and carrots.	The forms of vitamin A are: alcohol (retinol), ester (retinyl
Prevents xerophthalmia, an			0.5–1.0	375	1,250	**Good sources**—Dark-green leafy	palmitate), aldehyde (retinal or
eye condition which may lead to	Stunted growth of children.	Children	1–3	400	1,333	vegetables; beet greens, collards,	retinene), and acid (retinoic acid).
blindness in extreme vitamin A	Slowed bone growth, abnormal		4–6	500	1,667	dandelion greens, kale, mustard	Retinol, retinyl palmitate, and
deficiency.	bone shape, and paralysis.		7–10	700	2,333	greens, spinach, Swiss chard, turnip	retinal are readily converted from
Essential for body growth.	Unsound teeth, characterized by	Males	11–14	1,000	3,333	greens. Yellow vegetables: pumpkins,	one form to another, but retinoic
Necessary for normal bone	abnormal enamel, pits, and decay.		15–18	1,000	3,333	sweet potatoes, squash (winter). Yel-	acid cannot be converted to other
growth.	Rough, dry, scaly skin—a condition		19–24	1,000	3,333	low fruits: apricots, peaches. Some	forms. Retinoic acid fulfills some
Necessary for normal tooth	known as follicular hyperkeratosis (it		25–50	1,000	3,333	seafoods: crab, halibut, oysters, sal-	of the functions of vitamin A, but
development.	looks like "gooseflesh"); increased		51+	1,000	3,333	mon, swordfish, whale meat.	it does not function in the visual
Helps keep the epithelial tissues	sinus, sore throat, and abscesses in	Females	11–14	800	2,667	**Fair sources**—Butter (regular	cycle.
of the skin, and of the lining of	ears, mouth, or salivary glands;		15–18	800	2,667	salted), cantaloupe, cheese, cream,	
the nose, throat, respiratory and	increased diarrhea and kidney and		19–24	800	2,667	egg yolk, lettuce, margarine (forti-	
digestive systems, and genito-	bladder stones.		25–50	800	2,667	fied), tomatoes, whole milk.	
urinary tract, healthy and free of	Reproductive disorders, including		51+	800	2,667	**Negligible sources**—Breads and	
infection.	poor conception, abnormal embryonic	Pregnant		800	2,667	cereals (except yellow corn), chicken,	
May help prevent miscarriage	growth, placental injury, and death of	Lactating		1,200	4,000	cottage cheese (not creamed), dry	
of women during the first three	the fetus.					beans, muscle meats, potatoes, skim	

of women during the first three months of pregnancy.

New information suggests that vitamin A (1) acts in a coenzyme role, as for instance in the form of intermediates in glycoprotein synthesis; and (2) functions like steroid hormones, with a role in the cell nuclei, leading to tissue differentiation.

Other functions of vitamin A: necessary for (1) thyroxine formation and prevention of goiter; (2) protein synthesis; and (3) synthesis of corticosterone from cholesterol, and the normal synthesis of glycogen.
(Also see VITAMIN A.)

Toxicity—Toxicity of vitamin A is characterized by loss of appetite, headache, blurred vision, excessive irritability, loss of hair, dryness and flaking of the skin (with itching), swelling over the long bones, drowsiness, diarrhea, nausea, and enlargement of the liver and spleen.

[1]*Recommended Dietary Allowances*, 10th ed., NRC–National Academy of Sciences, 1989, p. 285.

The recommended daily allowances (RDA) are given in both International Units (IU) and Retinol Equivalents (RE). An International Unit (IU) of Vitamin A is defined on the basis of rat studies as equal to 0.344 mcg of crystalline retinylacetate (which is equivalent to 0.3 mcg of retinol, or to 0.6 mcg of beta-carotene). These standards are based on experiments which show that in rats only about 50% of the beta-carotene is converted to vitamin A.

milk.

Supplemental sources—Synthetic vitamin A, cod and other fish liver oils.

For additional sources and more precise values of vitamin A, see Food Composition Table F-21.

Vitamin D

Functions	Deficiency and Toxicity Symptoms	Age Group	Years	Vitamin D[1] (mcg)	(IU)	Sources	Comments
Increases calcium absorption from the small intestine.	**Deficiency symptoms**—Rickets in infants and children, characterized	Infants	0.0–0.5	7.5	300	Vitamin D is more sparse in foods than any other vitamin.	Vitamin D includes both—D$_2$ (ergocalciferol, calciferol, or vios-
Promotes growth and mineral-	by enlarged joints, bowed legs, knocked		0.5–1.0	10	400	**Rich sources**—Fatty fish (bloater,	terol) and D$_3$ (cholecalciferol).
ization of the bones.	knees, outward projection of the	Children	1–3	10	400	herring, kipper, mackerel, pilchard,	Vitamin D is unique among
Promotes sound teeth.	sternum (pigeon breast), a row of		4–6	10	400	salmon, sardines, tuna) and fish roe.	vitamins in three respects: (1) it
Increases absorption of phos-	beadlike projections on each side of		7–10	10	400	**Fair sources**—Liver, egg yolk,	occurs naturally in only a few
phorus through the intestinal wall,	the chest at the juncture of the rib	Males	11–14	10	400	cream, butter, cheese.	common foods (mainly in fish oils
and increases resorption of phos-	bones and joining (costal) cartilage		15–18	10	400	**Negligible sources**—Muscle	and a little in eggs and milk), (2)
phates from the kidney tubules.	(called rachitic rosary), bulging fore-		19–24	10	400	meats, milk (unfortified), fruits, nuts,	it can be formed in the body and
Maintains normal level of citrate	head, pot belly, and delayed eruption		25–50	5	200	vegetables, grains.	in certain foods by exposure to
in the blood.	of temporary teeth and unsound		51+	5	200	**D-fortified foods**—Milk	ultraviolet rays, and (3) the active
Protects against the loss of	permanent teeth.	Females	11–14	10	400	(400 IU/qt) and infant formulas. Other	compound of vitamin D (1, 25-
amino acids through the kidneys.	Osteomalacia in adults, in which		15–18	10	400	foods to which vitamin D is often	(OH)$_2$-D$_3$) functions as a hormone.
(Also see VITAMIN D.)	the bones soften, become distorted,		19–24	10	400	added include: breakfast and infant	
	and fracture easily.		25–50	5	200	cereals, breads, margarines, milk	
			51+	5	200	flavorings, fruit and chocolate	
Tetany, characterized by muscle twitching, convulsions, and low serum		Pregnant		10	400	beverages, and cocoa.	
calcium.		Lactating		10	400		

Toxicity—Excessive vitamin D may cause hypercalcemia (increased intestinal absorption, leading to elevated blood calcium levels), characterized by loss of appetite, excessive thirst, nausea, vomiting, irritability, weakness, constipation alternating with bouts of diarrhea, retarded growth in infants and children, and weight loss in adults.

[1]*Recommended Dietary Allowances*, 10th ed., NRC–National Academy of Sciences, 1989, p. 285.
400 IU of vitamin D = 10 mcg cholecalciferol.

Supplemental sources—Fish liver oils (from cod, halibut, or swordfish); irradiated ergosterol or 7-dehydrocholesterol such as viosterol.
Exposure to sunlight or sunlamp—Manufacture in the skin.
For additional sources and more precise values of vitamin D, see Food Composition Table F-21.

Vitamin E (Tocopherols)

Functions	Deficiency and Toxicity Symptoms	Age Group	Years	Vitamin E[1] (mg αTE)	(IU)	Sources	Comments
As an antioxidant which (a) re-	**Deficiency symptoms**—Innumer-	Infants	0.0–0.5	3	4.47	**Rich sources**—Salad and cooking	There are 8 tocopherols and
tards rancidification of fats in plant	able vitamin E deficiency symptoms		0.5–1.0	4	5.96	oils (except coconut oil), alfalfa seeds,	tocotrienols, of which alphato-
sources and in the digestive tracts	have been demonstrated in animals;	Children	1–3	6	8.94	margarine, nuts (almonds, Brazil nuts,	copherol has the greatest vitamin
of animals, and (b) protects body	and they are highly variable from spe-		4–6	7	10.43	filberts, peanuts, pecans), sunflower	E activity.
cells from toxic substances formed	cies to species. But vitamin E defi-		7–10	7	10.43	seed kernels.	
from the oxidation of unsaturated	ciency symptoms as such rarely					**Good sources**—Asparagus, avo-	

(Continued)

TABLE V-5 *(Continued)*

Functions	Deficiency and Toxicity Symptoms	Recommended Daily Allowance			Sources	Comments
		Age Group	Years	Vitamin		

Vitamin E *(Continued)*

Functions	Deficiency and Toxicity Symptoms	Age Group	Years	Vitamin	Sources	Comments
fatty acids. As a powerful antioxidant, vitamin E readily oxidizes itself (it combines with oxygen), thereby minimizing the destruction by oxidation of unsaturated fatty acids and vitamin A in the intestinal tract and in the tissues. **As an essential factor for the** integrity of red blood cells. **As an agent essential to cellular** respiration, primarily in heart and skeletal muscle tissues. **As a regulator in the synthesis of** DNA, vitamin C, and coenzyme Q. **As a protector of lung tissue** from air pollution (smog). **As a sometimes replacement** for selenium. (Also see VITAMIN E.)	occur in humans. Newborn infants (especially the premature), suffering from a deficiency of vitamin E (caused by shortened life span of red blood cells), characterized by edema, skin lesions, and blood abnormalities. **Patients unable to absorb fat like** those suffering from sprue or from fibrocystic disease of the pancreas) have low blood and tissue tocopherol levels, increased red blood fragility and shortened red blood cell life span, and increased urinary excretion of creatine. **Toxicity**—Vitamin E is relatively nontoxic. Some persons consuming daily doses of more than 300 IU of vitamin E have complained of nausea and intestinal distress. Excess intake of vitamin E appears to be excreted in the feces.	Males Females Pregnant Lactating	11–14 15–18 19–24 25–50 51+ 11–14 15–18 19–24 25–50 51+ 	10 14.90 10 14.90 10 14.90 10 14.90 10 14.90 8 11.92 8 11.92 8 11.92 8 11.92 8 11.92 10 14.90 12 16.39	cados, beef and organ meats, blackberries, butter, eggs, green-leafy vegetables, oatmeal, potato chips, rye, seafoods (lobster, salmon, shrimp, tuna), tomatoes. **Fair sources**—Apples, beans, carrots, celery, cheese, chicken, liver, peas. **Negligible sources**—Most fruits, sugar, white bread. **Supplemental sources**—Synthetic dl-alphatocopherol acetate, wheat germ, wheat germ oil. Most of the commercial vitamin E is synthetic dl-alphatocopherol acetate. It is the least expensive source of the vitamin, but, unlike natural food sources, it provides no other essential nutrient. For additional sources and more precise values of vitamin E (alphatocopherol), see Food Composition Table F-21.	

¹*Recommended Dietary Allowances*, 10th ed., NRC–National Academy of Sciences, 1989, p. 285.

The recommended daily allowances (RDA) are given in both International Units (IU) and alpha-tocopherol equivalents. 1 mg d-tocopherol = 1 alpha-TE. See section on Vitamin E (tocopherols) for variation in allowances and calculation of vitamin E activity of the diet as alpha-tocopherol equivalents.

Vitamin K

Functions	Deficiency and Toxicity Symptoms	Age Group	Years	Vitamin K¹ (mcg)	Sources	Comments
Vitamin K controls blood coagulation; recent research suggests that it acts in some way to convert precursor proteins to the active blood clotting factors. **Vitamin K is essential for the** synthesis in the liver of four blood clotting proteins: 1. Factor I, prothrombin. 2. Factor VII, proconvertin. 3. Factor IX, Christmas factor. 4. Factor X, Stuart-Power. (Also see VITAMIN K.)	**Deficiency symptoms**— 1. Delayed blood clotting. 2. Hemorrhagic disease of newborn. Vitamin K deficiency symptoms are likely in— 1 Newborn infants, especially if premature and breast-fed. 2. Infants born to mothers receiving anticoagulants. 3. Obstructive jaundice (lack of bile). 4. Fat absorption defects (celiac disease, sprue). 5. Anticoagulant therapy or toxicity. **Toxicity**—The natural forms of vitamin K₁ and K₂ have not produced toxicity even when given in large amounts. However, synthetic menadione and its various derivatives have produced some toxic symptoms in rats and jaundice in human infants when given in amounts of more than 5 mg daily.	Infants Children Males Females Pregnant Lactating	0.0–0.5 0.5–1.0 1–3 4–6 7–10 11–14 15–18 19–24 25–50 51+ 11–14 15–18 19–24 25–50 51+ 	5 10 15 20 30 45 65 70 80 80 45 55 60 65 65 65 65	Vitamin K is fairly widely distributed in foods and is available synthetically. **Rich sources**—Tea (green), turnip greens, broccoli, lettuce, cabbage, beef liver, spinach. **Good sources**—Asparagus, watercress, bacon, coffee, cheese, butter, pork, liver, oats. **Fair sources**—Peas (green), wheat (whole), beef fat, ham, beans (green), eggs, pork tenderloin, peaches, ground beef, chicken, liver, raisins. **Negligible sources**—Applesauce, bananas, bread, cola, corn, corn oil, milk (cow's), oranges, potatoes, pumpkin, tomatoes, wheat flour. **Supplemental sources** Chiefly synthetic menadione. The vitamin K values of some common foods is given in Table V-/.	There are 2 naturally occurring forms of vitamin K: K₁ (phylloquinone, or phytylmenaquinone), and K₂ (menaquinones), multiprenyl-menaquinones. Vitamin K₁ occurs only in green plants. Vitamin K₂ is synthesized by many microorganisms, including bacteria in the intestinal tracts of human beings and other species. There are several synthetic compounds, the best known of which is menadione, formerly known as K₃. The intake for young infants is based on 1 mcg/kg, assuming no intestinal synthesis. Thus, the amount provided by current formulas of 4 mcg/100 kcal should be ample for normal infants. The suggested intake of 10 mcg/day is also in the range supplied by breast milk (15 mg/liter).

¹Estimated Safe and Adequate Daily Intake from *Recommended Dietary Allowances* 10th ed., NRC-NAS, 1989, p. 285

WATER-SOLUBLE VITAMINS:

Biotin

Functions	Deficiency and Toxicity Symptoms	Age Group	Years	Biotin¹ (mcg)	Sources	Comments
Biotin is required for many reactions in the metabolism of carbohydrates, fats, and proteins. It functions as a coenzyme mainly in decarboxylation-carboxylation and in deamination. Biotin serves as a coenzyme for transferring CO₂ from one compound to another (for decarboxylation—the removal of carbon dioxide; and for carboxylation—the addition of carbon dioxide). Numerous decarboxylation and carboxylation reactions are involved in carbohydrate, fat, and protein metabolism; among them, the following:	**Deficiency symptoms**—The deficiency symptoms in man include: a dry scaly dermatitis, loss of appetite, nausea, vomiting, muscle pains, glositis (inflammation of the tongue), pallor of skin, mental depression, a decrease in hemoglobin and red blood cells, a high cholesterol level, and a low excretion of biotin; all of which respond to biotin administration. There is now substantial evidence that seborrheic dermatitis (an abnormally oily skin, which results in chronic scaly inflammation) of infants under 6 months of age is due to nutritional biotin deficiency. **Toxicity**—There are no known toxic effects.	Infants Children and adolescents Adults	0.0–0.5 0.5–1.0 1–3 4–6 7–10 11+ 	10 15 20 25 30 30–100 30–100	**Rich sources**—Cheese (processed), kidney, liver, soybean flour. **Good sources**—Cauliflower, chocolate, eggs, mushrooms, nuts, peanut butter, sardines and salmon, wheat bran. **Fair sources**—Cheese (natural), chicken, oysters, pork, spinach, sweet corn, whole wheat flour. **Negligible sources**—Refined cereal products; most fruits and root crops. **Supplemental sources**—Synthetic biotin, yeast (brewers', torula), alfalfa leaf meal (dehydrated). Considerable biotin is synthesized by the microorganisms in the intestinal tract; and much of it is absorbed, as evidenced by the fact that 3 to 6 times more biotin is excreted in the urine and feces than is ingested. For additional sources and more precise values of biotin, see Food Composition Table F-21.	Biotin is closely related metabolically to folacin, pantothenic acid, and vitamin B-12. It is noteworthy that the amount of avidin in raw egg white exceeds the amount of biotin in the whole egg. But, since avidin is destroyed by cooking, the usual diet includes little of the biotin-interfering substance.

NOTE WELL: the biotin values are "Estimated Safe and Adequate Daily Intakes," and *not* RDA; from *Recommended Dietary Allowances*, 10th ed., NRC–National Academy of Sciences, 1989, p. 284.

(Continued)

TABLE V-5 *(Continued)*

Functions	Deficiency and Toxicity Symptoms	Recommended Daily Allowance			Sources	Comments
		Age Group	Years	Vitamin		

Biotin *(Continued)*

1. Interconversion of pyruvate and oxaloacetate. The formation of oxaloacetate is important because it is the starting point of of the tricarboxylic acid cycle (TCA), known as the Krebs cycle, in which the potential energy of nutrients (ATP) is released for use by the body.
2. Interconversion of succinate and propionate.
3. Conversion of malate to pyruvate.
4. Conversion of acetyl CoA to malonyl CoA, the first step in the formation of long chain fatty acids.
5. Formation of purines, essential part of DNA and RNA, and for protein synthesis.
6. Conversion of ornithine to citrulline, an important reaction in the formation of urea.

Biotin also serves as a coenzyme for deamination (removal of-NH$_2$) reactions that are necessary for the production of energy from certain amino acids (at least aspartic acid, serine, threonine); for amino acids to be used as a source of energy, they must first be deaminated--the amino group must split off.

(Also see BIOTIN.)

Choline

Choline has several important functions; it is vital for the prevention of fatty livers, the transmitting of nerve impulses, and the metabolism of fat.

1. It prevents fatty livers through the transport and metabolism of fats. Without choline, fatty deposits build up inside the liver, blocking its function and throwing the whole body into a state of ill health.

2. It is needed for nerve transmission. Choline combines with acetate to form acetylcholine, a substance which is needed to jump the gap between nerve cells so that impulses can be transmitted.

3. By a phenomenon known as transmethylation, it serves as a source of labile methyl groups, which facilitate metabolism.

(Also see CHOLINE.)

Deficiency symptoms—Poor growth and fatty livers are the deficiency symptoms in most species except chickens and turkeys. Chickens and turkeys develop slipped tendons (perosis). In young rats, choline deficiency produces hemorrhagic lesions in the kidneys and other organs.

Toxicity—No toxic effects have been observed.

The Food and Nutrition Board of the National Research Council does not recommend any human allowance for choline because of lack of evidence of need. The Committee of the American Academy of Pediatrics recommends that choline be added to infant formulas in amounts equivalent to breast milk. Human milk contains about 145 mg of choline per liter, nearly 0.1% of total solids. It is estimated that the average mixed diet for adults in the U.S. contains 400 to 900 mg per day of choline and betaine, or about 0.1 to 0.18% of the diet.

Rich sources—Egg yolk, eggs, liver (beef, pork, lamb).

Good sources—Soybeans, potatoes (dehydrated), cabbage, wheat bran, navy beans, alfalfa leaf meal, dried buttermilk and dried skimmed milk, rice polish, rice bran, whole grains (barley, corn, oats, rice, sorghum, wheat), hominy, turnips, wheat flour, blackstrap molasses.

Negligible sources—Fruit, fruit juices, milk, vegetables.

Supplemental sources—Yeast (brewers', torula), wheat germ, soybean lecithin, egg yolk lecithin, and synthetic choline and choline derivatives.

Also, the body manufactures choline from methionine, with the aid of B-12. So, the needs for choline are supplied in two ways: (1) by dietary choline, and/or (2) by body synthesis through transmethylation.

It is noteworthy that choline has been known for a very long time. It was isolated in 1849, named in 1862, and synthesized in 1866-67. But the compound did not attract the attention of nutrition investigators at the time.

The classification of choline as a vitamin is debated because it does not meet all the criteria for vitamins, especially those of the B vitamins.

Folacin (Folic Acid)

In the body, folic acid is changed to at least five active enzyme forms, the parent form of which is tetrahydrofolic acid. Folacin coenzymes are responsible for the following important functions:

1. The formation of purines and pyrimidines which, in turn, are needed for the synthesis of the nucleic acids DNA and RNA, vital to all cell nuclei. This explains the important role of folacin in cell division and reproduction.

2. The formation of heme, the iron-containing protein in hemoglobin.

3. The interconversion of the three-carbon amino acid serine from the two-carbon amino acid glycine.

4. The formation of amino acids tyrosine from phenylalanine and glutamic acid from histidine.

5. The formation of amino acid methionine from homocysteine.

6. The synthesis of choline from ethanolamine.

7. The conversion of nicotinamide to N-methylnicotinamide, one of the metabolites of niacin that is excreted in the urine.

(Also see FOLACIN.)

Deficiency symptoms—Megaloblastic anemia (of infancy), also called macrocytic anemia (of pregnancy), in which the red blood cells are larger and fewer than normal, and also immature. The anemia is due to inadequate formation of nucleoproteins, causing failure of the megaloblasts (young red blood cells) in the bone marrow to mature. The hemoglobin level is low because of the reduced number of red blood cells. Also, the white blood cell, blood platelet, and serum folate levels are low.

Other symptoms include a sore, red, smooth tongue (glossitis), disturbances of the digestive tract (diarrhea), and poor growth.

Toxicity—Normally, no toxicity.

Age Group	Years	Folate[1] (mcg)
Infants	0.0–0.5	25
	0.5–1.0	35
Children	1–3	50
	4–6	75
	7–10	100
Males	11–14	150
	15–18	200
	19–24	200
	25–50	200
	51+	200
Females	11–14	150
	15–18	180
	19–24	180
	25–50	180
	51+	180
Pregnant		400
Lactating		280

[1]*Recommended Dietary Allowances*, 10th ed., NRC–National Academy of Sciences, 1989, p. 285.

The RDA are expressed in terms of "total" folacin; that is, the amount of folic acid activity available from all food folates.

Rich sources—Liver and kidney.

Good sources—Avocados, beans, beets, celery, chickpeas, eggs, fish, green leafy vegetables (such as asparagus, broccoli, Brussels sprouts, cabbage, cauliflower, endive, lettuce, parsley, spinach, turnip greens), nuts, oranges, orange juice, soybeans, and whole wheat products.

Fair sources—Bananas, brown rice, carrots, cheese (Cheddar), cod, halibut, rice, and sweet potatoes.

Poor sources—Chicken, dried milk, milk, most fruits, muscle meats (beef, pork, lamb), products made from highly refined cereals (including white flour), and most root vegetables (including Irish potatoes).

No folacin—Fats and oils, and sugar supply no folacin.

Supplemental sources—Yeast, wheat germ, and commercially synthesized folic acid (pteroylglutamic acid, or PGA).

Intestinal bacterial synthesis of folacin may be important in man, but the amount produced has not been determined.

For additional sources and more precise values of folacin (folic acid), see Food Composition Table F-21.

There is no single vitamin compound with the name *folacin*; rather, the term folacin is used to designate folic acid and a group of closely related substances which are essential for all vertebrates, including man.

Ascorbic acid, vitamin B-12, and vitamin B-6 are essential for the activity of the folacin coenzymes in many of their metabolic processes; again and again pointing up to interdependence of various vitamins.

Folacin deficiencies are thought to be a health problem in the U.S. and throughout the world. Infants, adolescents, and pregnant women are particularly vulnerable.

The folacin requirement is increased by tropical sprue, certain genetic disturbances, cancer, parasitic infection, alcoholism, and oral contraceptives.

Raw vegetables stored at room temperature for 2 to 3 days lose as much as 50 to 70% of their folate content.

Between 50 and 95% of food folate is destroyed in cooking.

(Continued)

TABLE V-5 *(Continued)*

Functions	Deficiency and Toxicity Symptoms	Recommended Daily Allowance — Age Group	Years	Calories (kcal)	Protein (g)	Niacin[1] (mg NE)	Sources	Comments
Niacin (Nicotinic acid; nicotinamide) The principal role of niacin is as a constituent of two important coenzymes in the body: nicotinamide adenine dinucleotide (NAD) and nicotinamide adenine dinucleotide phosphate (NADP). These coenzymes function in many important enzyme systems that are necessary for cell respiration. They are involved in the release of energy from carbohydrates, fats, and protein. NAD and NADP are also involved in the synthesis of fatty acids, protein, and DNA. Also, niacin is thought to (1) have a specific effect on growth, (2) reduce the levels of cholesterol, and (3) protect to some degree against nonfatal myocardial infarction. However, because of possible undesirable side effects, massive doses should be under the direction of a physician. (Also see NIACIN.)	**Deficiency symptoms**—A deficiency of niacin results in pellagra, the symptoms of which are: dermatitis, particularly of areas of the skin which are exposed to light or injury; inflammation of mucous membranes, including the entire gastrointestinal tract, which results in a red, swollen, sore tongue and mouth, diarrhea, and rectal irritation; and psychic changes, such as irritability, anxiety, depression, and in advanced cases, delirium, hallucinations, confusion, disorientation, and stupor. **Toxicity**—Only large doses of niacin, sometimes given to an individual with mental illness, are known to be toxic. However, ingestion of large amounts may result in vascular dilation, or *flushing* of the skin, itching, liver damage, elevated blood glucose, elevated blood enzymes, and/or peptic ulcer.	Infants	0–.5	650	13	5	**G**enerally speaking, niacin is found in animal tissues as nicotinamide and in plant tissues as nicotinic acid. Both forms are of equal niacin activity. **Rich sources**—The best food sources of niacin are liver, kidney, lean meats, poultry, fish, rabbit, corn flakes (enriched), nuts, peanut butter. **Good sources**—Milk, cheese, and eggs, although low in niacin content, are good antipellagra foods, because of their high tryptophan content, and because their niacin is in available form. Other good sources are: bran flakes, sesame seed, sunflower seed. **A**lso, enriched cereal flours and products are good sources of niacin. **Negligible sources**—Cereals (corn, wheat, oats, rice, rye) tend to be low in niacin.	**A**lthough nicotinic acid was prepared and named in 1867, it took another 70 years before it was known that it would cure blacktongue in dogs and pellagra in humans. **A**n average mixed diet in the U.S. provides about 1% protein as tryptophan. Thus, a diet supplying 60 g of protein contains about 600 mg of tryptophan, which will yield about 10 mg of niacin (on the average, 1 mg of niacin is derived from each 60 mg of dietary tryptophan). **N**iacin is the most stable of the B-complex vitamins. Cooking losses of a mixed diet usually do not amount to more than 15 to 25%. **B**orn of centuries of experience, (1) in Mexico, corn has long been treated with lime water before making tortillas; and (2) the Hopi Indians of Arizona have long roasted sweet corn in hot ashes. Both practices liberate the nicotinic acid in corn.
			.5–1	850	14	6		
		Children	1–3	1,300	16	9		
			4–6	1,800	24	12		
			7–10	2,000	28	13		
		Males	11–14	2,500	45	17		
			15–18	3,000	59	20		
			19–24	2,900	58	19		
			25–50	2,900	63	19		
			51+	2,300	63	15		
		Females	11–14	2,200	46	15		
			15–18	2,200	44	15		
			19–24	2,200	46	15		
			25–50	2,200	50	15		
			51+	1,900	50	13		
		Pregnant		+300	60	17		
		Lactating		+500	65	20		

[1]*Recommended Dietary Allowances*, 10th ed., NRC–National Academy of Sciences, 1989, p. 285. Calorie values from p. 33, Table 3–5 of the same report.
On the average, 1 mg of niacin is derived from each 60 mg of dietary tryptophan.

Moreover, 80 to 90% of the niacin is in the bran layer (outer seed coat) and removed in milling; and the niacin may be present in bound form (i.e. niacytin) and unavailable.

Fruits, roots, vegetables (other than mushrooms and legumes), butter, and sugar (white) are insignificant sources of niacin.

Supplemental sources—Both synthetic nicotinamide and nicotinic acid are commercially available. For pharmaceutical use, nicotinamide is usually used; for food nutrification, nicotinic acid is usually used. Also, yeast is a rich natural source of niacin.

For additional sources and more precise values of niacin, see Food Composition Table F-21.

Functions	Deficiency and Toxicity Symptoms	Recommended Daily Allowance — Age Group	Years	Pantothenic Acid[1] (mg)	Sources	Comments
Pantothenic Acid (Vitamin B-3) Pantothenic acid functions in the body as part of two enzymes—coenzyme A (CoA) and acyl carrier protein (ACP). **C**oA functions in the following important reactions: 1. The metabolic processes by which carbohydrates, fats, and proteins are broken down and energy is released. 2. The formation of acetylcholine, a substance of importance in transmitting nerve impulses. 3. The synthesis of porphyrin, a precursor of heme, of importance in hemoglobin synthesis. 4. The synthesis of cholesterol and other sterols. 5. The steroid hormones formed by the adrenal and sex glands. 6. The maintenance of normal blood sugar, and the formation of antibodies. 7. The excretion of sulfonamide drugs. **A**CP, along with CoA, required by the cells in the biosynthesis (the building up) of fatty acids. (CoA, without ACP, is involved in the breakdown of fatty acids.) (Also see PANTOTHENIC ACID.)	**Deficiency symptoms**—Pantothenic acid deficiency has been produced in human volunteers by either (1) feeding semi-synthetic diets low in pantothenic acid, or (2) adding a pantothenic antagonist to the diet. The symptoms: irritableness and restlessness; loss of appetite, indigestion, abdominal pains, nausea; headache; sullenness; mental depression; fatigue, weakness; numbness and tingling of hands and feet, muscle cramps in the arms and legs; burning sensation in the feet; insomnia; respiratory infections; rapid pulse; and a staggering gait. Also, in these subjects there was increased reaction to stress; increased sensitivity to insulin, resulting in low blood sugar levels; increased sedimentation rate for erythrocytes; decreased gastric secretions; and marked decrease in antibody production. **Toxicity**—Pantothenic acid is relatively nontoxic. However, doses of 10 to 20 g per day may result in occasional diarrhea and water retention.	Infants	.0–0.5	2	**Rich sources**—Organ meat (liver, kidney, and heart), cottonseed flour, wheat bran, rice bran, rice polish. **Good sources**—Nuts, mushrooms, soybean flour, salmon, blue cheese, eggs, buckwheat flour, brown rice, lobster, sunflower seeds. **Fair sources**—Chicken, broccoli, sweet peppers, whole wheat flour, avocados. **Negligible sources**—Butter, corn flakes, white flour, fats and oils, margarine, precooked rice, sugar. **Supplemental sources**—Synthetic calcium pantothenate is widely used as a vitamin supplement. Yeast is a rich natural supplement. **I**ntestinal bacteria synthesize pantothenic acid, but the amount and availability is unknown. **F**or additional sources and more precise values of pantothenic acid, see Food Composition Table F-21.	**C**oenzyme A, of which pantothenic acid is a part, is one of the most important substances in body metabolism. The following two facts indicate that there may be need for dietary supplementation of pantothenic acid for buoyant good health: 1. Losses of up to 50%, and even more, of the pantothenic content of foods may occur from production to consumption. 2. Deficiencies of pantothenic acid occur in farm animals fed natural rations (which are much less refined than human foods); hence, pantothenic acid is commonly added to commercial poultry and swine rations.
			.5–1.0	3		
		Children and Adolescents	1–3	3		
			4–6	3–4		
			7–10	4–5		
			11+	4–7		
		Adults		4–7		

[1]**NOTE WELL:** The pantothenic acid values are "Estimated Safe and Adequate Daily Intakes," and *not* RDA; from *Recommended Dietary Allowances*, 10th ed., NRC–National Academy of Sciences, 1989, p. 284.

(Continued)

TABLE V-5 *(Continued)*

Functions	Deficiency and Toxicity Symptoms	Age Group	Years	Vitamin	Sources	Comments
Riboflavin (Vitamin B-2) Riboflavin has an essential role in the oxidative reductions in all body cells by which energy is released. Thus, riboflavin functions in the metabolism of amino acids, fatty acids, and carbohydrates. Riboflavin, through its role in activating pyridoxine (vitamin B-6), is necessary for the formation of niacin from the amino acid tryptophan. Riboflavin is thought to be (1) a component of the retinal pigment of the eye; (2) involved in the functioning of the adrenal gland; and (3) required for the production of corticosteroids in the adrenal cortex. (Also see RIBOFLAVIN.)	**Deficiency symptoms**—Unlike all the other vitamins, riboflavin deficiency is not the cause of any severe or major disease of man. Rather, riboflavin often contributes to other disorders and disabilities such as beriberi, pellagra, scurvy, keratomalacia, and nutritional megaloblastic anemia. Riboflavin deficiency symptoms are: sores at the angle of the mouth (angular stomatitis); sore, swollen, and chapped lips (cheilosis); swollen, fissured, and painful tongue (glossitis); redness and congestion of the cornea of the eye; and oily, crusty, scaly skin (seborrheic dermatitis). **Toxicity**—There is no known toxicity of riboflavin.			**Riboflavin[1]** **(mg)**	**Rich sources**—Organ meats (liver, kidney, heart). **Good sources**—Corn flakes (enriched), almonds, cheese, eggs, lean meat (beef, pork, lamb), mushrooms (raw), wheat flour (enriched), turnip greens, wheat bran, soybean flour, bacon, cornmeal (enriched). **Fair sources**—Chicken (dark meat), white bread (enriched), rye flour, milk, mackerel and sardines, green leafy vegetables, beer. **Negligible sources**—Fruits (raw), roots and tubers, white sugar, fats and oils (butter, margarine, salad oil, shortening). **Supplemental sources**—Yeast (brewers', torula). Riboflavin is the only vitamin present in significant amounts in beer. For additional sources and more precise values of riboflavin (B-2), see Food Composition Table F-21.	Body storage of riboflavin is very limited; so, day-to-day needs must be supplied in the diet. Two properties of riboflavin may account for major losses: (1) it is destroyed by light; and (2) it is destroyed by heat in an alkaline solution. Because riboflavin is heat stable in neutral or acid solutions and only slightly soluble in water, little of it is lost in home cooking or commercial canning.
		Infants	0.0–0.5	0.4		
			0.5–1.0	0.5		
		Children	1–3	0.8		
			4–6	1.1		
			7–10	1.2		
		Males	11–14	1.5		
			15–18	1.8		
			19–24	1.7		
			25–50	1.7		
			51+	1.4		
		Females	11–14	1.3		
			15–18	1.3		
			19–24	1.3		
			25–50	1.3		
			51+	1.2		
		Pregnant		1.6		
		Lactating		1.8		

[1]*Recommended Dietary Allowances*, 10th ed., NRC–National Academy of Sciences, 1989, p. 285.

Functions	Deficiency and Toxicity Symptoms	Age Group	Years	Calories (kcal)	Thiamin[1] (mg)	Sources	Comments
Thiamin (Vitamin B-1) As a coenzyme in energy metabolism. Without thiamin, there could be no energy. As a coenzyme in the conversion of glucose to fats—the process called transketolation (Keto-carrying). In the functioning of the peripheral nerves. In this role, it has value in the treatment of alcoholic neuritis, the neuritis of pregnancy, and beriberi. In the direct functions in the body, including (1) maintenance of normal appetite, (2) the tone of the muscles, and (3) a healthy mental attitude. (Also see THIAMIN.)	**Deficiency symptoms**—Moderate thiamin deficiency symptoms include fatigue, apathy (lack of interest); loss of appetite; nausea; moodiness; irritability; depression; retarded growth; a sensation of numbness in the legs; and abnormalities of the electrocardiogram. More advanced thiamin deficiency is reflected in the gastrointestinal system, the nervous system, and the cardiovascular system. Severe thiamin deficiency of long duration culminates in beriberi, the symptoms of which are polyneuritis (inflammation of the nerves), emaciation and/or edema, and disturbances of heart function. Additional symptoms of thiamin deficiency include low excretion of thiamin in the urine; electrocardiogram measurement; reduced transketolase activity of the red blood cells; and increase of pyruvic acid in the blood. **Toxicity**—None, for there are no known toxic effects from thiamin.					Thiamin is found in a large variety of animal and vegetable products but is abundant in few. **Rich sources**—Lean pork (fresh and cured), sunflower seeds, corn flakes (enriched), peanuts, cottonseed flour, safflower flour, soybean flour. **Good sources**—Wheat bran, kidney, wheat flour (enriched), rye flour, nuts (except peanuts, which are a rich source), whole wheat flour, cornmeal (enriched), rice (enriched), white bread (enriched), soybean sprouts. **Fair sources**—Egg yolk, peas, turkey (hamloaf), beef liver, luncheon meat, crab mackerel (fried), salmon steak (broiled), roe (cod, herring), lima beans, refried beans, lentils. **Negligible sources**—Most fruits, most vegetables, polished rice, white sugar, animal and vegetable fats and oils, milk, butter, margarine, eggs, alcoholic beverages. **Supplemental sources**—Thiamin hydrochloride, thiamin mononitrate, yeast (brewers', torula), rice bran, wheat germ, and rice polish. Enriched flour (bread) and cereal, which was initiated in 1941, has been of special significance in improving the dietary level of thiamin in the U.S. For additional sources and more precise values of thiamin, see Food Composition Table F-21.	A patient with wet beriberi—lying in bed breathless, waterlogged, and apparently dying—may recover in one to two hours after being given an injection of thiamin. This is perhaps the most dramatic cure in medicine. Synthetic thiamin is usually marketed as thiamin hydrochloride or thiamin mononitrate. Antithiamin factors are present in certain raw fish and seafood, live yeast, tea and fermented tea leaves, and alcohol.
		Infants	0.0–0.5	650	0.3		
			0.5–1.0	850	0.4		
		Children	1–3	1,300	0.7		
			4–6	1,800	0.9		
			7–10	2,000	1.0		
		Males	11–14	2,500	1.3		
			15–18	3,000	1.5		
			19–24	2,900	1.5		
			25–50	2,900	1.5		
			51+	2,300	1.2		
		Females	11–14	2,200	1.1		
			15–18	2,200	1.1		
			19–24	2,200	1.1		
			25–50	2,200	1.1		
			51+	1,900	1.0		
		Pregnant		+300	1.5		
		Lactating		+500	1.6		

[1]*Recommended Dietary Allowances*, 10th ed., NRC–National Academy of Sciences, 1989, p. 285.
1 kcal = 4.184 kj (kilojoules)
1,000 kj = 1 mj (megajoules)
Because the principal functions of thiamin are concerned in energy metabolism, its requirement by the body bears a direct relation to the energy intake.

Functions	Deficiency and Toxicity Symptoms	Age Group	Years	Vitamin B-6[1] (mg)	Sources	Comments

Vitamin B-6 (Pyridoxine; pyridoxal; pyridoxamine)
Vitamin B-6, in its coenzyme forms, usually as pyridoxal phosphate but sometimes as pyridoxamine phosphate, is involved in a large number of physiologic functions, particularly:
1. In protein (nitrogen) metabolism, including—
  a. Transamination
  b. Decarboxylation
  c. Deamination
	Deficiency symptoms—In adults, the deficiency symptoms are: greasy scaliness (seborrheic dermatitis) in the skin around the eyes, nose, and mouth, which subsequently spread to other parts of the body; a smooth, red tongue; loss of weight; muscular weakness; irritability; mental depression.
	In infants, the deficiency symptoms are: irritability, muscular twitchings, and convulsions.
					Rich sources—Rice bran, wheat bran, sunflower seed.
					Good sources—Avocados, bananas, corn, fish, kidney, lean meat, liver, nuts, poultry, rice (brown), soybeans, whole grain.
					Fair sources—Eggs, fruits (except bananas and avocados, which are good sources), milk, vegetables.
					Negligible sources—Cheese (Cheddar, cottage), fat, milk,
						In rats, the three forms of vitamin B-6 have equal activity; and it is assumed that the same applies to man.
						Processing or cooking foods may destroy up to 50% of the B-6.
						Because vitamin B-6 is limited in many foods, supplemental B-6 with synthetic pyridoxine hydrochloride may be indicated, especially for infants and during preg-

		Infants	0.0–0.5	0.3		
			0.5–1.0	0.6		
		Children	1–3	1.0		
			4–6	1.1		
			7–10	1.4		
		Males	11–14	1.7		
			15–18	2.0		
			19–24	2.0		
			25–50	2.0		
			51+	2.0		

(Continued)

TABLE V-5 *(Continued)*

Functions	Deficiency and Toxicity Symptoms	Recommended Daily Allowance			Sources	Comments
		Age Group	Years	Vitamin		
Vitamin B-6 *(Continued)* d. Transsulfuration e. Tryptophan conversion to nicotinic acid. f. Hemoglobin formation. g. Absorption of amino acids. 2. In carbohydrate and fat metabolism, including— a. The conversion of glycogen to glucose-1-phosphate. b. The conversion of linoleic acid to arachidonic acid. 3. In clinical problems, including— a. Central nervous system disturbances. b. Autism, a mental and emotional affliction in children. c. Anemia that is iron-resistant. d. Kidney stones. e. Tuberculosis, in countering the antagonistic drug isonicotinic acid used in its treatment. f. Physiologic demands in pregnancy. g. Oral contraceptives. (Also see VITAMIN B-6.)	**Toxicity**—B-6 is relatively nontoxic. But large doses may result in sleepiness and be habit-forming when taken over an extended period.	Females Pregnant Lactating	11–14 15–18 19–24 25–50 51+	1.4 1.5 1.6 1.6 1.6 2.2 2.1 *[1]Recommended Dietary Allowances*, 10th ed., NRC–National Academy of Sciences, 1989, p. 285.	sugar, white bread. **Supplemental sources**—Pyridoxine hydrochloride is the most commonly available synthetic form; and yeast (brewers', torula), rice polish, and wheat germ are used as natural source supplements. For additional sources and more precise values of vitamin B-6 (pyridoxine), see Food Composition Table F-21.	nancy and lactation.
Vitamin B-12 (Cobalamins) In the body, vitamin B-12 functions in two coenzyme forms: coenzyme B-12, and methyl B-12. Vitamin B-12 coenzymes perform the following physiological roles at the cellular level, especially in the cells of the bone marrow, nervous tissue, and gastrointestinal tract: 1. Red blood cell formation and control of pernicious anemia. 2. Maintenance of nerve tissue. 3. Carbohydrate, fat, and protein metabolism. 4. Synthesis or transfer of single carbon units. 5. Biosynthesis of methyl groups (-CH₃), and in reduction reactions such as the conversion of disulfide (S-S) to the sulfhydryl group (-SH). In the following therapeutic uses: 1. The control of pernicious anemia. 2. In the treatment of sprue. (Also see VITAMIN B-12.)	**Deficiency symptoms**—Vitamin B-12 deficiency in man may occur as a result of (1) dietary lack, which sometimes occurs among vegetarians who consume no animal food; or (2) deficiency of intrinsic factor due to pernicious anemia, total removal of the stomach by surgery, or infestation with parasites such as the fish tapeworm. The common symptoms of a dietary deficiency of vitamin B-12 are: sore tongue, weakness, loss of weight, back pains, tingling of the extremities, apathy, and mental and other nervous abnormalities. Anemia is rarely seen in dietary deficiency of B-12. In pernicious anemia, the characteristic symptoms are: abnormally large red blood cells, lemon-yellow pallor, anorexia, dypsnea, prolonged bleeding time, abdominal discomfort, loss of weight, glossitis, and unsteady gait, and neurological disturbances, including stiffness of the limbs, irritability, and mental depression. Without treatment, death follows. **Toxicity**—No toxic effects of vitamin B-12 are known.	Infants Children Males Females Pregnant Lactating	**Vitamin B-12[1]** **(mcg)** 0.0–0.5 0.5–1.0 1–3 4–6 7–10 11–14 15–18 19–24 25–50 51+ 11–14 15–18 19–24 25–50 51+	0.3[2] 0.5 0.7 1.0 1.4 2.0 2.0 2.0 2.0 2.0 2.0 2.0 2.0 2.0 2.0 2.2 2.6 *[1]Recommended Dietary Allowances*, 10th ed., NRC–National Academy of Sciences, 1989, p. 285. [2]The recommended dietary allowance for vitamin B-12 in infants is based on average concentration of the vitamin in human milk. The allowances after weaning are based on energy intake (as recommended by the American Academy of Pediatrics) and consideration of other factors, such as intestinal absorption.	**Rich sources**—Liver and other organ meats—kidney, heart. **Good sources**—Muscle meats, fish, shellfish, eggs, and cheese. **Fair sources**—Milk, poultry, yogurt. **Negligible sources**—Bread (both whole wheat and white), cereal grains, fruits, legumes, vegetables. **Supplemental sources**—Cobalamin, of which there are at least three active forms, produced by microbial growth; available at the corner drugstore. Some B-12 is synthesized in the intestinal tract of human beings. However, little of it may be absorbed. For additional sources and more precise values of vitamin B-12, see Food Composition Table F-21.	Plants cannot manufacture vitamin B-12. Hence, little is found in vegetables, grains, legumes, and fruits. Vitamin B-12, like so many other members of the B-complex, is not a single substance; rather, it consists of several closely related compounds with similar activity. Vitamin B-12 is the largest and the most complex of all vitamin molecules. It is noteworthy (1) that vitamin B-12 is the only vitamin that requires a gastrointestinal tract secretion for its absorption (intrinsic factor); and (2) that the absorption of vitamin B-12 in the small intestine requires about 3 hours (compared to seconds for most of the other water-soluble vitamins).
Vitamin C (Ascorbic acid) Formation and maintenance of collagen, the substance that binds body cells together. So, vitamin C makes for more rapid and sound healing of wounds and burns. Metabolism of the amino acids tyrosine and tryptophan. Absorption and movement of iron. Metabolism of fats and lipids, and cholesterol control. Sound teeth and bones. Strong capillary walls and	**Deficiency symptoms**—Early symptoms, called latent scurvy: loss in weight, listlessness, fatigue, fleeting pains in the joints and muscles, irritability, shortness of breath, sore and bleeding gums, small hemorrhages under the skin, bones that fracture easily, and poor wound healing. **Scurvy**: swollen, bleeding, and ulcerated gums; loose teeth; malformed and weak bones, fragility of the capillaries with resulting hemor-	Infants Children Males Females 	**Vitamin C[1]** **(mg)** 0.0–0.5 0.5–1.0 1–3 4–6 7–10 11–14 15–18 19–24 25–50 51+ 11–14 15–18	30 35 40 45 45 50 60 60 60 60 50 60	Natural sources of vitamin C occur primarily in foods of plant origin—fruits (especially citrus fruits) and leafy vegetables. **Richest natural sources**—acerola cherry, *camu-camu*, and rose hips. **Excellent sources**—Raw, frozen, or canned citrus fruit or juice: oranges, grapefruit, lemons, and limes. Guavas, peppers (green, hot), black currants, parsley, turnip greens, poke greens, and mustard greens.	Actually, several compounds have vitamin C activity. So, the term vitamin C is a combined name for all of them. The terms ascorbic acid and dehydroascorbic acid should be used when specific reference is made to them. All animal species appear to require vitamin C, but dietary need is limited to humans, guinea pigs, monkeys, fruit-eating bats, redvented bulbul birds, certain fish,

TABLE V-5 *(Continued)*

Functions	Deficiency and Toxicity Symptoms	Recommended Daily Allowance			Sources	Comments
		Age Group	Years	Vitamin		

Vitamin C *(Continued)*

healthy blood vessels.

Metabolism of folic acid. (Also see VITAMIN C.)

Deficiency and Toxicity Symptoms: rhages throughout the body; large bruises; big joints, such as the knees and hips, due to bleeding into the joint cavity; anemia; degeneration of muscle fibers; including those of the heart; and tendency of old wounds to become red and break open. Sudden death from severe internal hemorrhage and heart failure is always a danger.

Toxicity—Adverse effects reported of intakes in excess of 8 g per day (more than 100 times the recommended allowance) include: nausea, abdominal cramps, and diarrhea; absorption of excessive amounts of iron; destruction of red blood cells; increased mobilization of bone minerals; interference with anticoagulant therapy; formation of kidney and bladder stones; inactivation of vitamin B-12; rise in plasma cholesterol; and possible dependence upon large doses of vitamin C.

	Years	Vitamin
	19–24	60
	25–50	60
	51+	60
Pregnant		70
Lactating		95

[1]*Recommended Dietary Allowances*, 10th ed., NRC–National Academy of Sciences, 1989, p. 285.

0.05 mg of ascorbic acid = 1 IU of vitamin C.

Sources:

Good sources—Green leafy vegetables: broccoli, Brussels sprouts, cabbage (red), cauliflower, collards, kale, lamb's-quarter, spinach, Swiss chard, and watercress. Also, cantaloupe, papaya, strawberries, and tomatoes and tomato juice (fresh or canned.)

Fair sources—Apples, asparagus, bananas, blackberries, blueberries, Irish potatoes, lima beans, liver, peaches, pears, sweet potatoes.

Negligible sources—Cereal grains and their by-products, cow's milk, eggs, fats, fish, meat, nuts, poultry, sugar.

Supplemental sources—Vitamin C (ascorbic acid) is available wherever vitamins are sold. For additional sources and more precise values of vitamin C, see Food Composition Table F-21.

Comments:

and perhaps certain reptiles.

Of all the vitamins, ascorbic acid is the most unstable. It is easily destroyed during storage, processing, and cooking; it is water-soluble, easily oxidized, and attacked by enzymes.

VITAMINLIKE SUBSTANCES:

Bioflavonids (Vitamin P)

Bioflavonoids function as follows:

1. They influence capillary fragility and permeability, perhaps together with vitamin C; increasing the strength of the capillaries and regulating their permeability. These actions help prevent hemorrhages and ruptures in the capillaries and connective tissues and build a protective barrier against infection.

2. They are active antioxidant compounds in food, ranking second only to fat-soluble tocopherols in this regard.

3. They possess metal-chelating capacity; and they affect the activity of the enzymes and membranes.

4. They have a synergistic effect on ascorbic acid, and they appear to stabilize ascorbic acid in human tissues.

5. They possess a bacteriostatic and/or antibiotic effect, which is sufficiently high to account for measurable anti-infectious properties of normal daily food.

6. They possess anticarcinogenic activity in two ways; a cytostatic effect against malignant cells and a biochemical protection of the cell from damage by carcinogenic substances.

(Also see BIOFLAVONOIDS.)

Deficiency symptoms—The symptoms of bioflavonoid deficiency are closely related to those of a vitamin C deficiency. The tendency to bleed (hemorrhage) or bruise easily are especially noted.

Toxicity—Bioflavonoids are non-toxic.

There are no NRC-National Academy of Sciences recommended daily allowances (RDA) of bioflavonoids.

The average daily intake of flavonoids in the American diet amounts to about 1 g, of which one-half (0.5 g) is absorbed from the gut.

The label on a rather typical vitamin C-bioflavonoid supplement reads as follows:

Each tablet contains:	**(mg)**
Rose hip powder	**500**
Vitamin C	**500**
Lemon bioflavonoid	**500**
Rutin (buckwheat)	**50**

NOTE: Need in human nutrition not established.

Rich sources—Citrus peel (especially orange and lemon peel), white pulp of citrus fruits, tangerine juice, rose hips, buckwheat leaves.

Good sources—Onions with colored skins, leafy vegetables, fruit, coffee, tea, cocoa, red wine, beer.

Poor sources—Frozen orange juice, most root vegetables.

NOTE: At this time, no evidence exists that bioflavonoids serve any useful role in human nutrition or in the prevention or treatment of disease in humans; hence, this presentation is for two purposes (1) informational, and (2) stimulation of research.

Bioflavonoids are a group of natural pigments in vegetables, fruits, flowers, and grains. They appear as companions of vitamin C, but they are not present in synthetic vitamin C.

Bioflavonoids do not fill the two prerequisites of a vitamin; they are not essential food constituents, and deficiency symptoms which can be cured by their administration are unknown.

Carnitine (Vitamin B_T)

Carnitine plays an important role in fat metabolism and energy production in mammals. It functions as follows:

1. *Transport and oxidation of fatty acids.* Carnitine plays an important role in the oxidation of fatty acids by facilitating their transport across the mitochondrial membrane. Carnitine is part of the shuttle mechanism whereby long-chain fatty acids are made into acyl carnitine derivatives and transported across the mitochondrial membrane, which is impermeable to long-chain fatty acids *per se* and to their coenzyme A esters. Once across the membrane, the acyl carnitines are reconverted to their fatty acid CoA form and undergo B-oxidation to liberate energy.

2. *Fat synthesis.* Although this role is controversial, carnitine appears to be involved in transporting acetyl groups back to the cytoplasm for fatty acid synthesis.

3. *Ketone body utilization.* Carnitine stimulates acetoacetate oxidation; thus, it may play a role in ketone body utilization. (Also see CARNITINE.)

Deficiency symptoms—If the body's supply of carnitine is low, the cells cannot get the fatty acids they need for energy production and growth. This is rare, but life-threatening. The symptoms of carnatine deficiency include muscle weakness, failure to thrive, poor growth, recurrent infections, cardiomyopathy, and hypoglycemia.

Toxicity—Not known.

Under normal conditions, there is no dietary requirement for carnitine. However, where a metabolic abnormality exists which inhibits synthesis, interferes with use, or increases catabolism of carnitine, illness may follow, which is sometimes relieved by a dietary supplement.

There is need for further research on the role of carnitine in human health and disease.

Generally speaking, carnitine is high in animal foods and low in plant foods. Few foods have been assayed for carnitine, but based on available data, the following evaluation of dietary sources of carnitine may be helpful.

Rich sources—Muscle meat, liver, heart, yeast (torula and brewers'), chicken, rabbit, milk, whey.

Good sources—Avocado, casein, wheat germ.

Poor sources—Cabbage, cauliflower, peanuts, wheat.

Negligible sources—Barley, corn, egg, orange juice, spinach.

Carnitine, a vital coenzyme in animal tissues and involved in fat metabolism, is another vitaminlike substance that has received much attention recently. It is similar to a vitamin with the exception that under normal conditions higher animals synthesize their total needs within their bodies; hence, no need appears to exist to supply this substance in food on a daily basis.

The lower level of carnitine in plant foods in comparison with animal foods is explainable on the basis that plant materials are most likely deficient in the essential amino acids, lysine and methionine, precursors of carnitine. Thus, a vegetarian diet will likely be low in carnitine, in both preformed carnitine and the amino acid precursors of carnitine.

Coenzyme Q (Ubiquinone)

Coenzyme Q functions in the respiratory chain in which energy is released from the energy-yielding nutrients as ATP.

There is evidence that specific ubiquinones function in the remission (prevention) of some of the symptoms of vitamin E deficiency. (Also see COENZYME Q.)

Deficiency symptoms—Non-specific.

Toxicity—Unknown.

Coenzyme Q is synthesized in the body. So, dietary allowance seems unimportant.

The entire series of ubiquinones has been prepared synthetically.

Quinones occur widely in aerobic organisms, from bacteria to higher plants and animals. Because they are synthesized in the body, they cannot be considered a true vitamin.

Coenzyme Q, or ubiquinone, is a collective name for a number of ubiquinones—lipidlike compounds that are chemically somewhat similar to vitamin E.

The importance of coenzyme Q as a catalyst for respiration im-

parts status as an essential metabolite. It may have other significant roles. For man and other higher animals a simple precursor substance with an aromatic ring may have vitaminlike status, but dietary ubiquinone seems, on the whole, to be unimportant unless it provides the aromatic nucleus for endogenous (body) synthesis.

(Continued)

TABLE V-5 *(Continued)*

Functions	Deficiency and Toxicity Symptoms	Recommended Daily Allowance	Sources	Comments
Inositol The functions of inositol are not completely understood, but the following roles have been suggested: 1. It has a lipotropic effect (an affinity for fat, like choline). In this role, inositol aids in the metabolism of fats and helps reduce blood cholesterol. 2. In combination with choline, inositol prevents the fatty hardening of arteries and protects the heart.	**Deficiency symptoms**—Myoinositol is a *growth factor* for certain yeasts and bacteria, and several species of fish. Earlier experiments indicated that a deficiency of inositol caused retarded growth and loss of hair in young mice, and loss of hair around the eyes (spectacled-eyes) in rats. But these symptoms are now being questioned because the experimental diets used were partially deficient in certain other vitamins. **Toxicity**—There is no known toxicity of inositol.	Need for inositol in human nutrition has not been established. So, no recommended daily allowance is given. Most authorities feel that people should consume about the same amount of inositol as choline. Therapeutic doses, which should be under the supervision of a physician, range from 500 to 1,000 mg daily.	Inositol is abundantly present in nature. **Rich sources**—Kidney, brain, liver, yeast, heart, wheat germ, citrus fruits, and blackstrap molasses. **Good sources**—Muscle meat, fruits, whole grains, bran of cereal grains, nuts, legumes, milk, and vegetables.	There is no evidence that human beings cannot synthesize all the inositol needed by the body. So, its classification as a vitamin is disputed. More properly perhaps, it should be classified as an essential nutrient, rather than a vitamin, for certain species of bacteria and animals.

In animal cells, inositol occurs as a component of phospholipids. In plant cells, it is found as phytic acid, an organic acid that binds calcium, iron, and zinc in an insoluble complex and interferes with their absorption.

3. It appears to be a precursor of the phosphoinosities, which are found in various body tissues, especially in the brain.
(Also see INOSITOL.)

Functions	Deficiency and Toxicity Symptoms	Recommended Daily Allowance	Sources	Comments
Laetrile (Vitamin B-17, amygdalin, nitrilosides) According to its advocates, Laetrile, or amygdalin, is a highly selective substance that attacks only cancerous cells. They explain this phenomenon as follows: Upon being absorbed by normal body cells, the enzyme rhodanese detoxifies the cyanide, which is	**Deficiency symptoms**—The advocates claim that prolonged deficiency of amygdalin may lead to lowered resistance to malignancies. **Toxicity**—None reported. But not more than 1.0 g should be taken at one time because of the hazard of hydrogen cyanide poisoning.	The usual dosage is 0.25 to 1.0 g taken at meals. Cumulative daily amounts of more than 3.0 g are sometimes taken, *but more than 1.0 g should never be taken at any one time.* According to the advocates, 5 to 30 apricot kernels eaten through the day may be effective as a preventive of cancer, but should not be eaten all at one time.	**A** concentration of 2 to 3% Laetrile is found in the whole kernels of most fruits, including apricots, apples, cherries, peaches, plums, and nectarines.	**NOTE WELL**; Laetrile has no known value for humans. The authors present this section relative to the controversial compound for informational purposes only. Further, it is listed with the vitaminlike substances because it is sometimes erroneously designated

then excreted through the urine. But cancer cells are completely deficient in rhodinese; instead, they are surrounded by another enzyme, beta-glucosidase, which releases the bound cyanide from the Laetrile at the site of the malignancy. So, Laetrile is believed to attack only the malignant cells.

Those who oppose Laetrile submit evidence that it is not effective as a treatment.

Those who support the use of Laetrile do so primarily on the basis that it is a preventive of cancer, rather than as a treatment of cancer.
(Also see LAETRILE.)

as vitamin B-17. Both the pros and cons relative to Laetrile are given, then the reader may make a judgement.

Regardless of the validity of the claims and counter claims pertaining to Laetrile, it is noteworthy that more and more medical authorities are accepting the nutrition approach in the prevention and treatment of cancer. Although the results of such nutrients as Laetrile, vitamin A (carotene), folic acid, and vitamin C, along with a general supplemental program of minerals and vitamins are significant for cancer treatment, perhaps even more important in the long run is the apparent success of nutrients of many kinds in strengthening the body's immune system.

Functions	Deficiency and Toxicity Symptoms	Recommended Daily Allowance	Sources	Comments
Lipoic acid Lipoic acid functions as a coenzyme. It is essential, together with the thiamin containing enzyme, pyrophosphatase (TPP), for reactions in carbohydrate metabolism which convert pyruvic acid to	**Deficiency symptoms**—No characteristc deficiency symptoms have been produced. **Toxicity**—Not known.	Because the body can synthesize the needed lipoic acid, no dietary requirement for humans and animals has been established.	Yeast and liver	Lipoic acid is not a true vitamin because it can be synthesized in the body and is not necessary in the diet of animals. However, it functions in the same manner as many of the B-complex vitamins.

aceytl-coenzyme A. Lipoic acid, which has two sulfur bonds of high-energy potential, combines with TPP to reduce pyruvate to active acetate, thereby sending it into the final energy cycle. It joins the intermediary products of protein and fat metabolism in the Krebs cycle in the reactions involved in producing energy from these nutrients. A metal ion (magnesium or calcium) is involved in this oxidative decarboxylation, along with lipoic acid and four vitamins: thiamin, pantothenic acid, niacin, and riboflavin.
(Also see LIPOIC ACID.)

Functions	Deficiency and Toxicity Symptoms	Recommended Daily Allowance	Sources	Comments
Pangamic acid (Vitamin B-15) Numerous functions have been attributed to pangamic acid; among them, the following: 1. Stimulation of transmethylation reactions. 2. Stimulation of oxygen intake. 3. Inhibition of fatty liver formation. 4. Adaptation to increased physical activity. 5. Control of blood cholesterol levels. (Also see PANGAMIC ACID.)	**Deficiency symptoms** There is indication that a deficiency of pangamic acid may cause fatigue (a tired feeling), hypoxia (an insufficient supply of oxygen in blood cells), heart disease, and glandular and nervous disorders. **Toxicity**—The so-called *Russian formula* appears to be relatively free of toxicity.	No NRC National Academy of Sciences recommended daily allowances (RDA) have been established. But the usual dosage of pangamic acid is from 150 to 300 mg daily.	**Rich sources**—All cereals and cereal products, with corn and rice particularly rich. Also, apricot kernels and brewers' yeast. **Good sources**—Wherever B-complex vitamins are found in natural foods.	**NOTE:** The authors make this presentation for two primary purposes: (1) informational, and (2) stimulation of research. The nutritional status and biological role(s) of pangamic acid await clarification. Its stimulation of such fundamental processes as transmethylation, cellular respiration, etc. may have physiological application. Certainly, there is sufficient indication of the value of pangamic acid as a therpeutic agent to war-

rant further study. In the meantime, pangamic acid should not be taken unless prescribed by a physician.

Pangamic acid is a vitaminlike substance, rather than a vitamin.

FDA classifies it as a food additive.

It has been widely studied and accepted in the U.S.S.R. as a necessary food factor (with important physiological actions). The Russians have used pangamic acid in treating cardiovascular disorders associated with insufficiency of oxidative metabolism, and in the treatment of liver disorders, a number of skin diseases, and some toxicoses.

(Continued)

TABLE V-5 *(Continued)*

Functions	Deficiency and Toxicity Symptoms	Recommended Daily Allowance	Sources	Comments
Para-aminobenzoic acid (PABA) For man and other higher animals, PABA functions as an essential part of the folacin molecule. **Human pharmaceutical uses**— PABA is sometimes used as a human pharmaceutical, not as a vitamin, in the following: as an antirickettsial; to counteract the bacteriostatic action of sulfonamides; and as a protective agent against sunburn.	**Deficiency symptoms**—Sulfa drugs may induce a deficiency of not only PABA, but of folic acid as well. The symptoms: fatigue, irritability, depression, nervousness, headache, constipation, and other digestive disorders. **Toxicity**—PABA is not known to be toxic to man. But continued high doses may result in nausea and vomiting.	The need for PABA has not been established; hence, it follows that there is no recommended daily allowance.	Food composition tables do not list PABA, but the following foods are generally recognized as the richest sources: brewers' yeast, fish, soybeans, peanuts, beef liver, eggs, wheat germ, lecithin, molasses.	In addition to having activity as a growth factor for certain bacteria, PABA has considerable folacin activity when fed to deficient animals in which intestinal synthesis of folacin takes place. For example, in rats and mice, it can completely replace the dietary source of folacin. This explains why para-aminobenzoic acid was once considered to be a vitamin in its own right. For man and other higher animals, PABA has no vitamin activity in animals receiving ample folacin, and it is not required in the diet; hence, it can no longer be considered a vitamin.

1. **Antirickettsial**. PABA is sometimes used in the treatment of certain rickettsial diseases—diseases in man and animals caused by microscopically small parasites of the genus *Rickettsia*, notably typhus and Rocky Mountain spotted fever.

2. **Sulfonamide antagonist**. PABA has the ability to reverse the bacteriostatic effects of sulfonamides, thereby counteracting their action. This is an antimetabolite action, explainable on the basis of similarity of structures. According to this theory, sulfonamides suppress bacterial growth by replacing their chemical analog PABA in bacterial enzyme systems; PABA in excess reverses the effect.

3. **Sunscreen agent**. PABA is used for protection against sunburn; usually about 5% PABA is incorporated in an ointment, to be applied over exposed parts of the body.

(Also see FOLACIN; and PARA-AMINOBENZOIC ACID.)

Functions	Deficiency and Toxicity Symptoms	Recommended Daily Allowance	Sources	Comments
Vitamin B-13 (Orotic acid) Vitamin B-13 has been found to stimulate the growth of rats, chicks, and pigs under certain conditions. Orotic acid is utilized by the body in the metabolism of folic acid and vitamin B-12. Also, it appears to aid the replacement or restoration of some cells. There is indication that vitamin B-13 may be helpful in the treatment of multiple sclerosis. (Also see VITAMIN B-13.)	**Deficiency symptoms**—Deficiency symptoms have not been proven. But it is believed that a deficiency may lead to liver disorders, cell degeneration, and premature aging; and to degenerative symptoms in multiple sclerosis victims. **Toxicity**—Not reported.	Dietary requirements are not known.	Vitamin B-13 is found in such natural sources as distillers' solubles, whey, sour or curdled milk, and root vegetables. Also, this nutrient is available in supplemental form as calcium orotate.	It is highly possible that the so-called vitamin B-13 is a growth promotant and a preventive of certain disorders. At this time, however, it is not known whether it plays an essential role in an otherwise adequate diet.

VITAMIN A (CAROTENE)

Contents	Page
History	1068
Chemistry, Metabolism, Properties	1069
Measurement/Assay	1070
Functions	1070
Deficiency Symptoms	1072
Recommended Daily Allowance of Vitamin A	1073
Toxicity (Hypervitaminosis A)	1074
Vitamin A Losses During Processing, Cooking, and Storage	1074
Sources of Vitamin A	1074

Vitamin A is probably the most important of all vitamins, if any vitamin can be singled out and ranked. More than any other vitamin, deficiencies of vitamin A are still widespread throughout most developing countries of the world and involve millions of people, especially children.

Vitamin A is required by man and all other animals. It is strictly a product of animal metabolism—present in all species of mammals, birds, and fish, no vitamin A being found in plants. The counterpart in plants is known as carotene, which is the precursor of vitamin A. Because the animal body can transform carotene into vitamin A, this compound is often spoken of as *provitamin A*.

Carotene, which derives its name from the carrot, from which it was first isolated over 100 years ago, is the yellow-colored, fat-soluble substance that gives the characteristic color to carrots and many other vegetables and to several fruits.

Thus, the ultimate source of all vitamin A is the carotenes which are synthesized by plants. Man and other animals convert a considerable proportion of the foods they eat into vitamin A.

HISTORY. At a very early date, the Chinese used a concoction rich in vitamin A as a remedy for night blindness; and, in ancient Greece, Hippocrates prescribed various forms of liver, a good source of preformed vitamin A, as a treatment for night blindness. But it remained for laboratory experiments conducted in the early 1900s to identify vitamin A as a distinct nutrient.

In 1913, vitamin A was discovered by Elmer V. McCollum and Marguerite Davis of the University of Wisconsin, and by Thomas B. Osborne and Lafayette B. Mendel of the Connecticut Experiment Station. Working independently, each research team demonstrated the presence of an essential dietary substance in fatty foods. McCollum and Davis found it in butterfat and egg yolks; Osborne and Mendel discovered it in cod-liver oil. These researchers believed that only one factor, which they called fat-soluble A, was needed to supplement purified diets. They described the condition as the "type of nutritive deficiency exemplified in the form of an infectious eye disease prevalent in animals inappropriately fed." In 1915, McCollum and Davis also noted that a deficiency of fat-soluble A caused night blindness. (It is noteworthy that Miss Marguerite Davis, a young biologist who had just obtained her bachelor's degree from the University of California, volunteered to do the rat work for McCollum without salary.)

VITAMIN A...

needed for growth, healthy eyes, skin and other tissues

TWO RATS FROM SAME LITTER, 11 WEEKS OLD

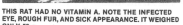

THIS RAT HAD NO VITAMIN A. NOTE THE INFECTED EYE, ROUGH FUR, AND SICK APPEARANCE. IT WEIGHED ONLY 53 g.

THIS RAT HAD PLENTY OF VITAMIN A. IT HAS BRIGHT EYES, SLEEK FUR, AND APPEARS ALERT AND VIGOROUS. IT WEIGHED 123 g.

TOP FOOD SOURCES

| LIVER | CARROTS | DARK-GREEN LEAFY VEGETABLES | SWEET POTATOES | SQUASH (WINTER) |
| PEACHES AND APRICOTS | BUTTER AND MARGARINE (FORTIFIED) | EGG YOLK | CHEESE | TOMATOES |

Fig. V-23. Vitamin A made the difference! *Left*: Rat on vitamin A-deficient diet. *Right*: Rat that received plenty of vitamin A. (Adapted from USDA sources.) Note, too, top ten sources of vitamin A.

In 1919, Steenbock and co-workers at the University of Wisconsin noted that an unknown substance, present in sweet potatoes, carrots, and corn, later to be identified as carotene, would support normal growth and reproduction.

In 1920, the English scientist Drummond proposed that this compound be called vitamin A.

In 1930, Moore, of England, demonstrated that this pro-vitamin A activity was beta-carotene.

In 1931, P. Karrer, a Swiss researcher, isolated the active substance in halibut-liver oil and determined the chemical formula for vitamin A—the first vitamin to have its chemical structure determined. For this, and for his work with riboflavin, he received the Nobel Prize. However, vitamin A was not produced (by Karrer) in crystalline form from fish liver oils until 1937, and it did not become available in synthetic form (by Isler working in Switzerland) until 1947. Today, pure, inexpensive synthetic forms of vitamin A are readily available.

CHEMISTRY, METABOLISM, PROPERTIES.

• **Chemistry**—*Vitamin A* may be a misleading term because it sounds as if only one chemical compound has vitamin A activity. Actually, several forms of vitamin A exist, with each possessing different degrees of activity.

There are two forms of vitamin A as such, retinol and dehydroretinol. Retinol (formerly called vitamin A), which is found as an ester (retinyl palmitate) in ocean fish oils and fats, and in liver, butterfat, and egg yolk, is biologically active as an alcohol, an aldehyde, and an acid. The alcohol, the most common form (see Fig. V-24 for formula), is usually referred to as retinol, the aldehyde as retinal or retinene, and the acid as retinoic acid.

Dehydroretinol, or vitamin A_2, differs from retinol in that (1) it has an extra double bond, and (2) it has about 40% the biological value (activity). It is found only in freshwater fish and in birds that eat these fish; hence, it is of limited interest.

Today, the general term *vitamin A* is used for both *retinol* and *dehydroretinol*.

In addition to the actual forms of vitamin A, related compounds, known as *carotenes*, are found in several fruits and vegetables. Carotene is also called (1) provitamin A, because it can be converted to vitamin A in the body; and (2) precursor of vitamin A, because it precedes vitamin A. At least 10 of the carotenoids found in plants can be converted with varying efficiencies into vitamin A. Four of these carotenoids alpha-carotene, beta-carotene, gamma-carotene, and cryptoxanthine (the main carotenoid of corn)—are of particular importance due to their provitamin A activity. Of the four, *beta-carotene* (see Fig. V-25) has the highest vitamin A activity and provides about two-thirds of the vitamin A necessary for human nutrition.

VITAMIN A

H_3C — C — H_2C — H_2C — CH_2 ... C—C=C—C=C—C=C—C=C—CH_2OH ... C—CH_3

Fig. V-24. The structure vitamin A (retinol).

β-CAROTENE

Fig. V-25. The structure of beta-carotene.

• **Metabolism**—Foods supply vitamin A in the form of vitamin A, vitamin A esters, and carotenes. Almost no absorption of vitamin A occurs in the stomach. In the small intestine, vitamin A and beta-carotene are emulsified with bile salts and products of fat digestion and absorbed in the intestinal mucosa. Here, much of the conversion of beta-carotene to vitamin A (retinol) takes place. There are wide differences in species and individuals as to how well they utilize the carotenoids. Their absorption is affected by several factors, including the presence in the small intestine of bile, dietary fat, and antioxidants. Bile aids emulsification; fat must be absorbed simultaneously; and antioxidants, such as alpha-tocopherol and lecithin, decrease the oxidation of carotene. Also, the presence of enough protein of good quality enhances the conversion of carotene to vitamin A—a matter of great importance in developing countries where protein is limited in both quantity and quality.

The absorption of vitamin A is adversely affected by the presence of mineral oil in the intestinal tract. Since mineral oil is not absorbed, and since it holds carotene and vitamin A in solution, carotene and vitamin A are lost through excretion. Therefore, mineral oil should never be used as a substitute for regular fats in food preparation (for example, as a salad dressing); neither should it be taken at mealtime when it is used as a laxative. It is noteworthy, too, that intestinal parasites adversely affect the absorption of vitamin A and carotene and the conversion of carotene to vitamin A—a factor of importance in tropical regions where intestinal parasite infections are common.

In the blood, vitamin A esters are transported in association with a retinol-binding protein, whereas the carotenoids are associated with the lipid-bearing protein. Storage of vitamin A is largely in the liver, but small amounts are also stored in the lungs, body fat, and kidneys. The amount of body storage of vitamin A tends to increase with age, but of course this depends on the quantity in the diet and the amount absorbed. It is estimated that a normal adult may store sufficient vitamin A in his/her liver to meet the needs for 4 to 12 months. Infants and children do not build up such reserves; hence, they are much more susceptible to deficiencies.

From the liver, vitamin A enters the bloodstream as a free alcohol, whereupon it travels to the tissues for use.

No vitamin A is excreted in the urine because it is not water-soluble, but considerable unabsorbed carotene is normally found in the feces.

• **Properties**—Vitamin A (retinol) is an almost colorless (pale yellow) fat-soluble substance. It is insoluble in water; hence, there is no loss by extraction from cooking. Although the esters of vitamin A are relatively stable compounds, the alcohol, aldehyde, and acid forms are rapidly destroyed by oxidation when they are exposed to air and light. Since vitamin A occurs in the stable form (the ester) in most foods, normal preparation procedures do not destroy much vitamin A activity. However, fats that undergo oxidative rancidity, can lose their vitamin A rapidly. We depend mainly on storage in a cool, dark place (refrigeration) and on added antioxidants, such as vitamin E, to protect fats and oils from vitamin A loss.

The carotenoid pigments (which are referred to as carotene in this discussion) are deep red color, but in solution they are bright yellow or orange yellow. They impart the color to many fruits and vegetables, including apricots, peaches, carrots, sweet potatoes, squash, pumpkins, and yellow corn. As a general rule of thumb, the more intense the pigmentation in these foods, the higher the provitamin A content. Green vegetables, particularly dark green and leafy ones, are also rich in provitamin A. They contain carotenoids, although their color is masked by that of the green pigment, chlorophyll. The properties of solubility and stability of carotene and carotenoid pigments are similar to those of vitamin A.

The yellow color of beta-carotene is so intense that it is widely used as a coloring agent by the food industry. For such purposes, beta-carotene is valued more for its aesthetic contribution than its nutritive value, although the potential vitamin A value of the food it colors is also increased.

MEASUREMENT/ASSAY. The assay of vitamin A is accomplished by two basic methods: biological, or chemical. The bioassay procedure is based on a biological response such as growth of rats or chicks deficient in vitamin A. It measures the total vitamin A, including provitamin A, present. But, because of the difficulties and time factor in bioassays, chemical assays are usually used.

Until recently, dietary allowances of vitamin A were stated in terms of either International Units (IU) or United States Pharmacopeia (USP) units, which are equal. An International Unit (IU) of vitamin A is defined on the basis of rat studies as equal to 0.344 mcg of crystalline retinylacetate (which is equivalent to 0.300 mcg of retinol, or to 0.60 mcg of beta-carotene). These standards were based on experiments that showed that in rats only about 50% of the beta-carotene is converted to vitamin A. In man, however, beta-carotene is not as available as in the rat, due to poorer absorption in the intestines and other factors, with the result that various factors have been used to compensate for this when vitamin A activity of foods and diets have been expressed in IU.

In order to quantify vitamin A values for humans within the metric system, therefore, international agencies have now introduced the biological equivalent of 1 microgram (1 mcg) of retinol as the standard. In 1967, the FAO/WHO proposed that vitamin A allowances be expressed as the equivalent weight of retinol, and that the use of IU be discontinued. The United Kingdom adopted this change and coined the term *Retinol Equivalent* (RE). The U.S. Food and Nutrition Board (National Academy of Science, National Research Council) and Canada followed suit. The RE system of measurement takes into account the amount of absorption of the carotenes as well as the degree of conversion to vitamin A; hence, it is more precise than the IU system.

In terms of International Units, beta-carotene (by weight) is ½ as active, and the other provitamin A carotenoids are ¼ as active, as retinol. Also, retinol is completely absorbed by the intestine, but only about ⅓ of the intake of provitamin A carotenoids is absorbed. Of the absorbed carotenoids, only ½ of the beta-carotene and ¼ of the other provitamin A carotenoids are converted to retinol. It follows that beta-carotene is only ⅙ as active, and the other carotenoids ¹⁄₁₂ as active, as retinol.

The vitamin A values in the food composition tables of this book are in International Units.[5]

But the recommended daily dietary allowances (RDA) are expressed in both IU and RE.

Functions. Vitamin A is essential for a number of physiological processes: among them, (1) vision, (2) growth, (3) bone development, (4) tooth development, (5), maintenance of body epithelial tissues, (6) protective effect

---

[5]Both the FAO/WHO and the FNB recommend that, in the future, foods be analyzed separately for (1) retinol, (2) beta-carotene, and (3) other provitamin A carotenoids, for inclusion in food composition tables, and that (4) total RE, as micrograms, also be listed in the tables. Until this information is available, we will primarily employ the still widely used IU values.

against cancer, (7) reproduction, and (8) coenzyme and hormone roles.

It is noteworthy that the basic physiological functions of vitamin A are, in general, common to man and all other animals. These follow:

• **Vision**—The best understood function of vitamin A is related to the maintenance of normal vision in dim light — the prevention of night blindness. The retina of the eye — the light-sensitive inside layer at the back of the eye which may be likened to the film of a camera—contains two kinds of light receptors: the rods for vision in dim light, and the cones for vision in bright light and color vision. (The terms *rods* and *cones* are derived from the shape of the cells.) *Rhodopsin* (*visual purple*) is the pigment in the rods that contains vitamin A; iodopsin is the main pigment in the cones that contains vitamin A. All of the pigments are composed of the same vitamin A fraction (retinol) but of different proteins. Opsin is the protein of rhodopsin.

The visual cycle proceeds as follows: When light strikes the normal retina, this pigment—called rhodopsin (visual purple)—is bleached to another pigment known as retinaldehyde (visual yellow). As a result of this change, images are transmitted to the brain through the optic nerve. Rhodopsin is rebuilt in the dark, but some vitamin A is lost in the reactions, and rod vision is impaired unless sufficient vitamin A is supplied by the blood from the diet or body stores to replace it. The ability of the eyes to adjust to changes from bright light to dim light is reduced. This condition is known as *night blindness*, or nyctalopia. Fig. V-26 shows, in simplified form, the role of vitamin A in vision.

The cones are not as sensitive to changes in the amount of vitamin A as the rods. Thus, vision in bright light and color vision are not affected to any extent in the early stages of vitamin A deficiency.

Night blindness is the earliest symptom of vitamin A deficiency in humans. It first manifests itself as a slow, dark adaptation, thence it progresses to total night blindness. In man, the *dark adaptation test*, which measures the eyes' ability to recover visual activity in dim light, is used as a means of determining vitamin A status. However, night blindness or dark adaptation tests are not always reliable in diagnosing a deficiency. Measurements of serum vitamin A and carotene values in comparison with known standards are most useful. Serum levels of 10 to 19 mcg of vitamin A, or of 20 to 39 mcg of carotene, per 100 ml are considered low.

Night blindness can usually be cured in a half hour or so by the injection of vitamin A. It is noteworthy that both retinol and retinaldehyde are effective in the visual process, but retinoic acid has no visual function; if retinoic acid is the only form of vitamin A fed to experimental animals, blindness results.

In 1967, the Nobel Prize in Medicine was awarded to Dr. George Wald of Harvard University, who clarified the role of vitamin A in vision, which, to that time, had been an extremely complicated picture.

Night blindness has a long medical history, associated with diet. It is common in Newfoundland among fishermen working in open boats in bright sunshine with their eyes exposed to the glare on the water. There, it is said that if a man cannot see at night his vision will be restored by the next night if he eats the liver of a codfish or of a sea gull.

Vision can also be impaired because of *xerophthalmia* (from the *Latin* words for *dry* and *eye*), another manifestation of a vitamin A deficiency. In this condition, the conjunctiva (the covering of the eye) dries out, the cornea becomes inflamed, and the eyes become ulcerated. It may progress to blindness in extreme deficiency of vitamin A. Xerophthalmia is a very common disease in infants and undernourished children in those parts of the world where the deficiency is prevalent; but it may be prevented by including in the diet a good source of vitamin A.

• **Growth**—Vitamin A aids in the building and growth of body cells; hence, it is essential for body growth. Because of this, prior to the time that vitamin A was isolated and its chemical formula determined, its potency in food was measured by growth rate. Since animals on a vitamin A-free diet stop growing when their bodies are depleted of its stores, foods to be assessed were fed to these animals in

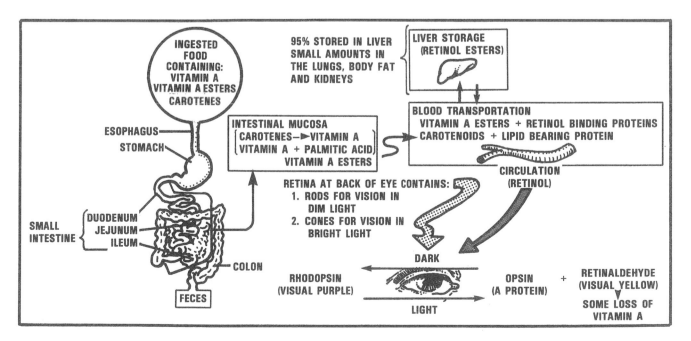

Fig. V-26. Metabolism of vitamin A and its role in vision in dim light.

known amounts for a prescribed period of time and their growth response noted. Better growth meant a higher vitamin A content in the food. Also, it has long been known that vitamin A is essential for the growth of children and for the normal development of babies before birth.

Although the relationship of vitamin A to growth was the first function studied, its role in the growth process is poorly understood. Experimental animals on vitamin A-deficient diets lose their appetites and fail to grow. The loss of appetite is attributed to the loss of the sense of taste because of keratinization (drying out) of the taste buds. When supplements of retinol or of retinoic acid are given in amounts to bring about growth, sense of taste returns to normal.

• **Bone development**—Vitamin A is especially needed for bone growth. If the intake is not sufficient, the bones will stop growth before the soft tissues are affected. Thus, if the brain and spinal cord grow too fast for the stunted skull and spinal column, injury of brain and nerves may occur and paralysis and various other neurological signs may result. In some cases, there is a constriction of the bone canal through which the optic nerve passes, thereby resulting in blindness. It is noteworthy, however, that a deficiency of vitamin A may cause degeneration of the nervous tissue without causing bone malformation.

• **Tooth development**—Vitamin A is essential for normal tooth development. Like other epithelia, the enamel-forming cells are affected by lack of vitamin A; instead of an even protective layer of enamel, fissures and pits will be present and the teeth will tend to decay. Also, when there is a shortage of vitamin A, the odontoblasts which form dentine become atrophied.

• **Maintenance of healthy epithelial tissues**—Epithelial tissues are of two kinds: (1) those that cover the outer surface of the body—the resistant, protective skin (epidermis); and (2) those that line all the tubular system—the secretory mucous membranes. Without vitamin A, the epithelial cells become dry and flat, and gradually harden to form scales that slough off. This process is known as keratinization. The skin may become rough, dry and scaly; and the membranes lining the nose, throat, and other air passages, and the gastrointestinal and genitourinary tracts, may become keratinized. Whenever these tissue changes occur, the natural mechanism for protection against bacterial invasion is impaired and the tissue may become infected easily. It follows that vitamin A deficiency commonly manifests itself in those parts that are particularly susceptible to infections. Vitamin A is also necessary for the health of the membranes lining the stomach and intestinal wall, and for the health of the sex glands, uterus, and the membranes that line the bladder and urinary passages. Renal calculi (kidney stones) may be related to the keratinization of the urinary tract.

Although adequate vitamin A is necessary to maintain healthy epithelial tissues, excessive intakes will not increase resistance to infections which enter through the epithelium.

• **Protective effects against cancer**—Vitamin A, either as retinol or carotene, appears to play an important nutritional role in keeping the body free of certain kinds of cancer, especially cancers of epithelial tissue (the skin and various membranes that line the mouth, internal passages, and hollow organs). This does not call for megadoses of vitamin A; rather, it indicates that getting sufficient vitamin A in the daily diet is likely a very important anticancer measure.

• **Reproduction**—In most animals, the absence of vitamin A in the diet will dramatically reduce reproductive ability. In the male rat, spermatogenesis stops; in the female, there may be an abnormal estrus cycle, fetal resorption, or congenital malformations. In rabbits, there is a reduction in fertility and an increase of abortion in pregnant does. Hatchability is significantly reduced when hens are fed a vitamin A-deficient ration. Lack of vitamin A in the diet of pregnant women during the first trimester (the first 3 months) may cause miscarriage.

Quite often, retinoic acid, the acid form of vitamin A, is used to study the effects of vitamin A deficiency on reproduction, because it can be substituted freely for retinol and retinaldehyde (the aldehyde form of vitamin A) for all physiological functions except reproduction and vision.

• **Coenzyme and hormone roles**—New information suggests that vitamin A (1) acts in a coenzyme role, as for instance in the form of intermediates in glycoprotein synthesis; and (2) functions like steroid hormones, with a role in the cell nuclei, leading to tissue differentiation.

• **Other functions of vitamin A**—Studies have indicated that vitamin A deficiency reduces the rate of thyroxine formation and increases the incidence of goiter in people. Also, protein synthesis is adversely affected by vitamin A deficiency. Animal studies indicate that vitamin A functions in the synthesis of corticosterone from cholesterol, resulting in a decrease in the body's capacity to synthesize glycogen.

DEFICIENCY SYMPTOMS. A deficiency of vitamin A may be due to a dietary lack of vitamin A and/or provitamin A, or to poor absorption. A diet that contains an insufficiency of vitamin A activity will, in due time, cause night blindness, xerosis, or xerophthalmia; stunted growth; slowed bone growth; unsound teeth; rough, dry, scaly skin; increased sinus trouble, sore throat, and abscesses in the ears, mouth, or salivary glands; and increased diarrhea and kidney and bladder stones.

It is noteworthy that vitamin A deficiency symptoms, and their prominence, vary from one species to another, as is true for other vitamins. Common deficiency symptoms follow:

• **Night blindness, xerosis, and xerophthalmia**—The earliest symptom of a vitamin A deficiency is the inability to see in dim light—the condition known as night blindness (nyctalopia). Night driving—facing the bright lights of oncoming vehicles—is difficult, and even dangerous, for persons whose eyes are slow to adjust.

The next symptom (following night blindness) to appear is usually xerosis (dryness) of the conjunctiva (the delicate membrane that lines the eyelids and covers the exposed surface of the eyeball), in which there may be (1) wrinkling, pigmentation, and accumulation of debris, and (2) loss of transparency.

When associated with generalized xerosis in children, Bitot's spots (named for the French physician who first discovered them)—which appear as small plaques of silvery gray, usually with a foamy surface, on the conjunctiva—are usually caused by a deficiency of vitamin A. However, in adults these spots may have another cause.

Extreme vitamin A deficiency over a long period of time may cause xerophthalmia, characterized by the following stages: (1) the cornea (the transparent membrane that coats the outer surface of the eye) becomes dry, then inflamed and edematous; (2) the eyes become cloudy and infected, which leads to ulceration; and (3) keratomalacia, a softening and keratinizing of the cornea, culminating in permanent blindness if the disease is not arrested. Xerophthalmia occurs most frequently in undernourished infants and children in India, the Middle East, Southeast Asia, and parts of Africa and South America. It has been estimated that throughout the world (exclusive of China) about 80,000 children become blind each year from vitamin A deficiency, and that about one-half of these die.

Fig. V-27. Keratomalacia, characterized by softening and ulceration of the cornea of the eye, due to a vitamin A deficiency. Unless vitamin A or carotene is provided, blindness may follow. (Courtesy, FAO, Rome, Italy)

• **Stunted growth**—A deficiency of vitamin A results in stunted growth of children.

• **Slowed bone growth**—When young animals are deprived of vitamin A, the bones fail to lengthen, and the remodeling processes necessary for the formation of compact bone cease to operate. Abnormalities in bone shape result. Bone growth in the cranium and spine slows, while nerve tissue continues to grow; sometimes this results in overcrowding of the skull and spine, and mechanical compression damage to nerve tissue with paralysis and degeneration.

• **Unsound teeth**—If a child gets too little vitamin A when his teeth are developing, the enamel forming cells become abnormal and pits are formed. Such pits may harbor food deposits, which may ferment and form acids that etch the enamel and lead to decay.

• **Rough, dry, scaly skin; increased sinus trouble, sore throat, and abscesses in ears, mouth, or salivary gland; increased diarrhea and kidney and bladder stones**—A deficiency of vitamin A injures the epithelial tissues throughout the body, and leads to a peculiar type of horny degeneration called keratinization. The epithelial cells form the outer layer of the skin and the mucous membranes that line the mouth and digestive, respiratory, and genitourinary tracts. Instead of being soft and moist, they become hard and dry. As a result, (1) the skin, especially in the areas of the arms, legs, shoulders, and lower abdomen, may become rough, dry, and scaly—a condition known as follicular hyperkeratosis or phrynoderma (it looks like *gooseflesh*); and (2) bacteria have easy access to the mucous membranes, with the result that there is increased susceptibility to infections such as sinus trouble, sore throat, and abscesses in the ears, mouth, or salivary glands. Also, certain other troubles, noninfective in character, increase as a result of the damaged epithelium, including diarrhea and the formation of kidney and bladder stones.

• **Reproductive disorders**—Deficient animals of all species studied—rat, fowl, pig, cow, sheep, dog, guinea pig, and others—show the following reproductive disorders: poor conception, abnormal embryonic growth, placental injury; and, in severe deficiency, death of the fetus. The same deficiency symptoms have been described in humans.

RECOMMENDED DAILY ALLOWANCE OF VITAMIN A.
The Food and Nutrition Board (FNB) of the National Research Council (NRC) recommended daily allowances of vitamin A are given in the section on VITAMIN(S), Table V-5, Vitamin Table.

• **Infants and children**—The daily allowance for infants from birth to 6 months is based on the average retinol content of human milk, which is about 49 mcg per 100 ml. Thus, an infant consuming 850 ml of breast milk would receive approximately 420 mcg of retinol (RE). The allowance for infants from 6 months to 1 year of age, who are fed solid foods in addition to milk, is reduced to 375 RE (280 as retinol, 95 as beta-carotene).

• **Adults**—The FNB-NRC recommended daily allowance of vitamin A for adult men is set at 1,000 mcg RE (3,333 IU). The allowance is based on the assumption that the average American diet provides half the total vitamin A activity as retinol and half as provitamin A carotenoids. In terms of Retinol Equivalents (RE), this is 750 mcg retinol (1 RE = 3.33 IU retinol) and 250 Retinol Equivalents as beta-carotene (1 RE = 10 IU beta-carotene) for a total of 1,000 RE.
Because the body size of women is usually smaller than that of men, the RDA of vitamin A for adult women is set at 80% of that of men, or 800 mcg RE (2,667 IU).

• **Pregnancy and lactation**—During pregnancy, the RDA remains the same as for adult women, but during lactation, the RDA is increased to 1,200 mcg RE (4,000 IU) to compensate for the vitamin A secreted in breast milk.

• **Therapeutic uses**—Retinol is valuable in the treatment of night blindness and xerophthalmia, malabsorption syndrome or obstructive jaundice, and malnourished people who show Bitot's spots or follicular keratosis.
For the prevention of blindness in vitamin A-deficient children, massive doses of vitamin A (200,000 IU) by mouth at 6-month intervals have been effective.

● **Vitamin A intake in average U.S. diet**—The vitamin A value of foods available in the United States amounts to about 1,630 RE per person per day: 44.3% of which is from fruits and vegetables; 31.7% from fats, oils, and dairy products; 24.6% from meat, fish, and eggs; and 4% from miscellaneous foods. This does not account for subsequent losses in processing, cooking, and storage.

Vegetarians will need to increase their intake of provitamin A carotenoids in order to meet the recommended allowances.

Vitamin A is efficiently stored in the liver, and well-nourished persons have several months' supply that the body can utilize.

TOXICITY (Hypervitaminosis A). Excessive intake of preformed vitamin A, called hypervitaminosis A, may cause serious injury to health. However, massive intakes of carotene are not harmful because they are not converted to vitamin A rapidly enough to cause toxicity. The excess carotene will merely produce a yellow coloration of the skin, which disappears when the intake is reduced.

Symptoms of vitamin A toxicity are: loss of appetite, headache, blurred vision, excessive irritability, loss of hair, drying and flaking of the skin (with itching), swelling over the long bones, drowsiness, diarrhea, nausea, and enlargement of the liver and spleen. The most direct and positive diagnosis of hypervitaminosis A is the determination of the vitamin A concentration in the plasma or serum from a fasting blood sample. Values higher than 100 mcg per 100 ml (normal 20 to 60 mcg/100 ml) can be considered suspect, whereas values greater than this indicate toxicity.

Poisoning of men and dogs from eating polar bear liver has been reported in the Arctic since 1596. Polar bear liver contains 13,000 to 18,000 IU of vitamin A per gram. It is estimated that a hungry Arctic explorer may eat about 500 g of liver per day, containing about 9,000,000 IU of vitamin A.

Chronic toxic symptoms (hypervitaminosis A) may occur in adults who receive doses of vitamin A in excess of 50,000 IU daily over a prolonged period. Lesser doses will produce symptoms in children; infants who receive 18,500 IU daily may show signs of toxicity within 12 weeks. Acute toxicity occurs in adults who are given massive doses of 2 to 5 million IU daily, and in infants from doses as low as 75,000 to 300,000 IU daily. Vitamin C can help prevent the harmful effects of vitamin A toxicity. When excess intake of vitamin A is discontinued, recovery is usually rapid and complete; in some cases, the toxicity symptoms disappear within 72 hours.

On the basis of the evidence presently available, prolonged daily doses of vitamin A to adults in excess of 50,000 IU, and to infants in excess of 18,500 IU, should be under the supervision of a qualified physician or nutritionist.

VITAMIN A LOSSES DURING PROCESSING, COOKING, AND STORAGE. The carotene losses of vegetables following harvest tend to parallel the degree of wilting. In order to conserve their maximum carotene value, they should be stored at low temperatures or be quick frozen.

Freezing and freeze-drying causes little loss. But drying of eggs, vegetables or fruits, with exposure to air, sunlight, or high temperatures, may cause serious loss of vitamin A value.

Because vitamin A and the carotenes are insoluble in water and stable to heat at ordinary cooking temperatures, it was once thought that little vitamin A activity was lost from foods in cooking and processing, *unless* they were exposed

to air. However, cooking or canning vegetables produces a rearrangement of atoms in the carotene molecule, resulting in carotenes of substantially lower vitamin A value. It is estimated that, on the average, the vitamin A value of cooked green vegetables is decreased by 15 to 20% and the value of yellow vegetables by 30 to 35%.

Both vitamin A and the carotenoids are easily oxidized and rapidly destroyed on exposure to ultraviolet light, with the rate of destruction influenced by the associated substances and the temperature and moisture conditions. Butter exposed in thin layers in air at 122°F (*50°C*) loses all of its vitamin A potency in 6 hours, but in the absence of air there is little destruction at 248°F (*120°C*) over the same period. Yellow corn has been reported to lose as much as 60% of its carotene in 7 months' storage.

Animal fats should be kept in a cold, dark place; and fish liver oils should be protected from light by being kept in dark bottles. Rancid fat destroys both vitamin A and carotene. Also, minerals such as iron oxide, charcoal, sulfur, ground limestone, bone meal, manganese, and iodine, contribute to the destruction of vitamin A in foods.

SOURCES OF VITAMIN A. Grouping by rank of common sources of vitamin A is given in the section on VITAMIN(S), Table V-5, Vitamin Table.

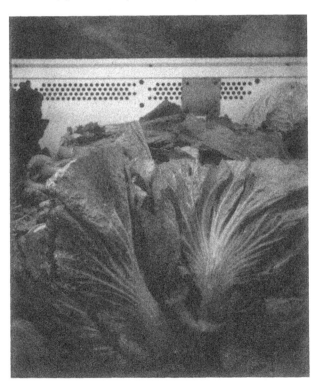

Fig. V-28. Green mustard cabbage. Dark green leafy vegetables are a good source of vitamin A. (Courtesy, Department of Food Science and Human Nutrition, University of Hawaii)

For additional sources and more precise values of vitamin A, see Food Composition Table F-21.

Only animal foods contain vitamin A as such. Although they are usually classed as food supplements rather than foods, the fish liver oils are the richest natural sources. Fish eat smaller fish or crustaceans, which, in turn, have fed on marine plants that contain provitamins A. Herbivorous animals eat green plants and convert these substances into the vitamin itself in their bodies. Carnivorous animals get the vitamin from feeding on plant-eating animals. The cow and the hen are efficient in converting the provitamins A in plant foods into the vitamin A in milk fat and eggs, respectively; and into the vitamin A in their body tissues. But some provitamin A in the diet escapes their conversion, with the result that milk fat, egg yolk, and other animal products contain a mixture of vitamin A and its plant precursors. The proportion of each depends partly on the animal species, or even the breed, and partly on the feed consumed. For example, the *golden* color of Guernsey milk is due to the high content of provitamin A, while whiter Holstein milk contains a higher proportion of vitamin A. But both milks have about the same amount of total vitamin A activity.

Also, the vitamin A value of animal foods varies widely according to the vitamin A value of the feed of the animals that produced them. For example, livers from older animals and from animals on green grass are higher in vitamin A than the livers of younger animals on dry, bleached feeds; and the butterfat in milk is usually yellower and of higher vitamin A value when the cows are grazing on green pastures than when they are confined to a corral.

The principal source of vitamin A in the diet is likely to be from the carotenes, which are widespread in those plant foods that have high green or yellow colorings. There is a direct correlation between the greenness of a leaf and its carotene content; dark-green leaves, such as beet greens, collards, dandelion greens, kale, mustard greens, spinach, Swiss chard, and turnip greens, are rich in carotene, but pale leaves, like cabbage and lettuce, are insignificant sources. The yellow vegetables and fruits, such as carrots, apricots, cantaloupe, peaches, pumpkins, squash (winter), sweet potatoes, and yellow corn, are rich in provitamin A carotenoids.

• **Synthetic vitamin A**—Today, synthetic forms of vitamin A are the most potent and inexpensive sources. They are just as effective and safe as the natural forms, but it must be remembered that they contain no other nutrients.

(Also see VITAMIN[S], Table V-5.)

VITAMIN A DEFICIENCY

This refers to an insufficiency of vitamin A, which may be due to either (1) a dietary lack of vitamin A and/or provitamin A, or (2) poor absorption.

(Also see VITAMIN A, sections headed "Deficiency Symptoms;" and "Blindness Due to Vitamin A Deficiency.")

VITAMIN ANTAGONIST

A substance so similar in structure to the vitamin that the body accepts it in place of the vitamin is known as a vitamin antagonist. But the antagonist is unable to perform the functions of the true vitamin.

VITAMIN B-6 (PYRIDOXINE; PYRIDOXAL; PYRIDOXAMINE)

Contents	Page
History	1076
Chemistry, Metabolism, Properties	1076
Measurement/Assay	1076
Functions	1076
Deficiency Symptoms	1077
Recommended Daily Allowance of Vitamin B-6	1077
Toxicity	1078
Vitamin Losses During Processing, Cooking, and Storage	1078
Sources of Vitamin B-6	1078

VITAMIN B-6

PROMOTES GROWTH AND FEATHERING IN CHICKS.
IT'S ESSENTIAL FOR HUMANS, TOO.

VITAMIN B-6 MADE THE DIFFERENCE! LEFT: CHICK SHOWS RETARDED GROWTH AND ABNORMAL FEATHERING DUE TO VITAMIN B-6 DEFICIENCY. RIGHT: NORMAL, CONTROL CHICK. (COURTESY, G.F. COMBS, UNIVERSITY OF GEORGIA)

TOP FOOD SOURCES

LIVER AND KIDNEY — FISH — SOYBEANS — POULTRY — BROWN RICE — NUTS — BANANAS — LEAN MEAT (BEEF, PORK, LAMB) — AVOCADOS — WHOLE GRAIN

Fig. V-29. The vitamin B-6 story.

By official action of the Society of Biological Chemists and the American Institute of Nutrition, vitamin B-6 is now the approved collective name for three closely related naturally occurring compounds with potential vitamin B-6 activity: pyridoxine, pyridoxal, and pyridoxamine. Pyridoxine is found largely in vegetable products, whereas the pyridoxal and pyridoxamine forms occur primarily in animal products. There is no information on the relative biological activity of the three compounds in man, but in rats they are equally active if given parenterally (injected intramuscularly or intravenously).

The need for vitamin B-6 was first demonstrated in rats, but it is now established that it is also a dietary essential for the human, pig, chick, dog, and other species, including microorganisms.

HISTORY.

In 1926, Goldberger and Lillie conducted an experiment designed to produce pellagra in rats. A severe dermatitis resulted, which was believed to be analogous to pellagra. In 1934, the Hungarian scientist, Gyorgy, produced a cure for this condition with an extract from yeast; but the curative compound of the extract was neither thiamin, niacin, nor riboflavin, but a substance which he called *vitamin B-6*. In 1938, five different laboratories, working independently, isolated the vitamin in crystalline form, but credit for obtaining the first crystals is generally given to Lepkovsky of the University of California. In 1939, Stiller *et al.*, of Merck and Co., established the chemical structure of the vitamin; and Harris and Folkers, of Merck, along with Kuhn, the German scientist, synthesized the compound.

When the vitamin was first isolated, the German biochemists gave it the name of adermin, while American researchers called it pyridoxine. Because the compound has a pyridine ring (five carbon and one nitrogen) and three hydroxy groups, Gyorgy, the Hungarian, expressed preference for the name pyridoxine, which became widely adopted.

In 1942, Snell discovered the vitamin B-6 activity of two closely related substances in natural products, which he named pyridoxal and pyridoxamine. Umbreit followed in 1945 with a report of the coenzyme functions of the vitamin in phosphate forms.

CHEMISTRY, METABOLISM, PROPERTIES.

• **Chemistry**—Vitamin B-6 is found in foods in three forms which are readily interconvertible—pyridoxine, pyridoxal, and pyridoxamine. Also vitamin B-6 is found in physiological systems in the forms of pyridoxal phosphate and pyridoxamine phosphate. The structural formulas of the three naturally occurring free forms of these compounds are given in Fig. V-30.

Fig. V-30. Structure of pyridoxine, pyridoxal, and pyridoxamine.

• **Metabolism**—In the free form, vitamin B-6 is absorbed rapidly from the upper part of the small intestine, thence it enters the body by the portal vein. It is present in most body tissues, with a high concentration in the liver. It is

secreted into milk and excreted primarily via the urine. Measurement of B-6 in the urine is the assay method used in nutrition surveys.

• **Properties**—Vitamin B-6 is readily soluble in water, quite stable to heat and acid, but easily destroyed by oxidation and exposure to alkali and ultraviolet light. All three forms are white crystalline substances.

Of the three forms, pyridoxine is more resistant to food processing and storage conditions and probably represents the principal form in food products.

MEASUREMENT/ASSAY.

No international standard or unit system of vitamin B-6 is in current usage. Analytical results are expressed in weight units of pyridoxine hydrochloride. One milligram of pyridoxine hydrochloride is equivalent to 0.82 mg pyridoxine or 0.81 mg pyridoxal, or 0.82 mg pyridoxamine.

The vitamin B-6 content in foods and tissues is determined by microbiological assay, chemical methods, and animal bioassays. The animal bioassays, using either the rat or chick, are time consuming, expensive, and variable; therefore, they have been generally replaced by microbiological and chemical methods.

FUNCTIONS.

Vitamin B-6, in its coenzyme forms, usually as pyridoxal phosphate but sometimes as pyridoxamine phosphate, is involved in a large number of physiologic functions, particularly in protein (nitrogen) metabolism, and to a lesser extent in carbohydrate and fat metabolism. Also, it appears to hold a key role in a number of chemical problems. Discussion follows:

• **Vitamin B-6 in protein metabolism**—Vitamin B-6 in its phosphorylated forms is active metabolically in the following types of reactions in amino acid metabolism:

1. **Transamination.** It is involved in shifting an amino group (NH_2) from a donor amino acid to an acceptor acid to form another amino acid. This reaction is important in the formation of nonessential amino acids. Transamination is illustrated in Fig. V-31.

Fig. V-31. Vitamin B-6 in transamination.

2. **Decarboxylation.** It is active in the removal of the carboxyl groups (COOH) from certain amino acids to form another compound. Decarboxylation is necessary for the synthesis of serotonin, norepinephrine, and histamine from tryptophan, tyrosine, and histidine, respectively.

3. **Deamination.** It aids in deamination. By removing amino groups (NH_2) from amino acids not needed for growth, it helps to render carbon residues available for energy.

4. **Transulfuration.** It aids in the transfer of sulfhydryl-group (HS) from the amino acid methionine to another amino acid (serine) to form the amino acid cysteine.

5. **Tryptophan conversion to niacin (nicotinic acid).** It assists in forming niacin (nicotinic acid) from tryptophan, thereby playing a role in the niacin supply.

6. **Hemoglobin formation.** It is necessary for the formation of a precursor of porphyrin compounds which is part of the hemoglobin molecule.

7. **Absorption of amino acids.** Vitamin B-6 is believed to play a role in the absorption of amino acids from the intestine.

• **Vitamin B-6 in carbohydrate and fat metabolism**—Vitamin B-6 in its phosphorylated forms also plays a role in carbohydrate and fat metabolism, although this role is less important than protein metabolism:

1. **Catabolism of glycogen.** It is an essential part of phosphorylase, the enzyme that brings about the conversion of glycogen to glucose-1-phosphate in muscle and liver.

2. **Fatty acid metabolism.** It takes part in fat metabolism, although the exact mode of action is unknown; for example, it is believed to be involved in the conversion of the essential unsaturated fatty acid, linoleic acid, to another fatty acid, arachidonic acid.

• **Clinical problems of pyridoxine deficiency**—It seems apparent that vitamin B-6 is also involved in a number of clinical problems; among them, the following:

1. **Central nervous system disturbances.** It assists in energy transformation in brain and nerve tissue; hence, the functioning of the central nervous system. When vitamin B-6 is lacking, convulsive seizures occur in both human infants and experimental animals.

2. **Autism.** This is a severe disturbance of mental and emotional development in young children, characterized chiefly by withdrawal from reality and lack of responsiveness or interest in other people in the normal activities of childhood. Although more experimental work is needed, nine studies reported in the world literature show that megadoses of vitamin B-6 are helpful in treating autism.[6]

3. **Anemia.** In studies on human subjects, vitamin B-6 has been effective in treating anemia that does not respond to iron—so-called iron-resistant anemia.

4. **Kidney stones.** It has been reported that a deficiency of vitamin B-6 causes increased urinary excretion of oxalates; thus, a lack of the vitamin may result in the formation of kidney stones.

5. **Tuberculosis.** The drug isoniazid (isonicotinic acid hydrazide, INH), which is used as a chemotherapeutic agent in treating tuberculosis, has been shown to be an antagonist to pyridoxine, causing a side effect of neuritis in some patients. Treatment with large doses (50 to 100 mg daily) of pyridoxine prevents this effect.

6. **Physiologic demands in pregnancy.** Pyridoxine deficiencies during pregnancy have been demonstrated and subsequently corrected by supplementation with vitamin B-6.

7. **Oral contraceptives.** Some women on estrogen-progesterone oral contraceptives appear to require additional vitamin B-6.

• **Other functions**—It appears from experimental studies that vitamin B-6 is involved in—

1. Antibody formation, for protection against infectious diseases, although this role is unproved.

2. Messenger RNA synthesis.

3. Nucleic acid metabolism.

4. Endocrine gland functions.

5. Biosynthesis of coenzyme A.

DEFICIENCY SYMPTOMS. It is difficult to produce a dietary deficiency of vitamin B-6 in human beings because it is so widely distributed in foods. The characteristic symptoms are: skin lesions—especially at the tip of the nose, anemia, convulsive seizures, and reduced antibody production. In adults, there may also be depression and confusion. If the vitamin B-6 deficiency is prolonged, the symptoms may include dizziness, nausea, vomiting, and kidney stones.

Experimentally, vitamin B-6 deficiency in adults has been produced by using a low vitamin B-6 diet along with one of the antagonists of the vitamin—desoxypyridoxine. After 2 to 3 weeks, the following symptoms were noted: greasy scaliness (seborrheic dermatitis) in the skin around the eyes, nose, and mouth, which subsequently spread to other parts of the body; a smooth, red tongue; loss of weight; muscular weakness; irritability; and mental depression. Administration of as little as 5 mg of pyridoxine, pyridoxal, or pyridoxamine daily corrected all the abnormalities in a few days.

The effects of vitamin B-6 deprivation appear to be more dramatic in infants than in adults.

RECOMMENDED DAILY ALLOWANCE OF VITAMIN B-6. The Food and Nutrition Board (FNB) of the National Research Council (NRC) recommended daily allowances of vitamin B-6 are given in the sections on VITAMIN(S), Table V-5, Vitamin Table.

The establishment of allowances for vitamin B-6 is complicated by the following facts: (1) the requirement varies with dietary protein intake—there is increased need for vitamin B-6 with increased intakes of protein; (2) the uncertainty of the availability of the vitamin in the diet; and (3) the uncertainty as to the extent of intestinal bacterial synthesis of the vitamin, and the degree to which it is utilized by the body. Also, there is evidence of increased need of the vitamin in pregnancy and lactation, in the elderly, and in various pathologic and genetic disturbances. Nevertheless, the NRC has set recommended allowances to assure a safety margin and to make a deficiency unlikely under most circumstances. Discussion follows:

• **Recommended allowance for infants**—Although based on limited information, a recommended dietary allowance of 0.3 mg of vitamin B-6 per day is considered adequate for the young infant. For older infants (0.5 to 1.0 year of age) consuming a mixed diet, a daily allowance of 0.6 mg of vitamin B-6 is recommended.

• **Recommended allowance for children and adolescents**—The recommended allowances for children range from 1.0 mg to 1.4 mg per day, depending on age.

• **Recommended allowance for adults**—For males 11–14 years = 1.7 mg; 15 years and over = 2.0 mg. Females 11–14 years = 1.4 mg; 15–18 years = 1.5 mg; 19 and over = 1.6 mg.

---

[6]Rimland, Bernard, Ph.D., Letters (to the editor), *Science News*, Vol. 119, No. 16, April 18, 1981, p. 243.

• **Recommended allowance for pregnancy**—Theoretically, several factors imply an increased need for vitamin B-6 in pregnancy: (1) because vitamin B-6 requirements increase with increasing protein in the diet, the extra protein allowance for the pregnant woman necessitates a modest increase in intake of the vitamin; (2) all forms of vitamin B-6 cross the placenta readily and are concentrated in the fetal blood; and (3) estrogens apparently increase tryptophan oxygenase activity, which will result in need for additional vitamin B-6.

An additional allowance of 0.6 mg of vitamin B-6 per day (for a total allowance of 2.2 mg/day) is recommended during gestation.

• **Recommended allowance for lactation**—The content of vitamin B-6 in milk appears to reflect the nutritional state of the mother with respect to the vitamin.

An additional allowance of 0.5 mg of vitamin B-6/day (for a total allowance of 2.1 mg/day) is recommended during lactation.

• **Vitamin B-6 and oral contraceptive agents**—Recent studies indicate that the vitamin B-6 requirement for most oral contraceptive users is approximately the same as that for nonusers; thus, the current evidence does not appear to justify the routine supplementation of the dietary vitamin B-6 with pyridoxine. However, some women report that depression occurs when they are taking oral contraceptives, probably as a result of the failure to convert tryptophan to serotonin, a neurotransmitter in the brain. When this problem occurs, the physician may suggest higher levels of vitamin B-6 (about 30 mg daily) in order to normalize tryptophan metabolism.

• **Vitamin B-6 intake in average U.S. diets**—According to the U.S. Department of Agriculture, the overall average per capita availability of vitamin B-6 per day in the United States is 2.2 mg, not including waste or cooking losses. Also, it is noteworthy that 40.5% of the available vitamin B-6 is provided by meat, poultry, and fish.

TOXICITY. Although vitamin B-6 is relatively nontoxic, side effects, such as sleepiness, may follow injection of large doses. Also, it may be habit-forming when taken in large doses over an extended period; a vitamin B-6 dependency has been induced in normal human adults given a supplement of 200 mg of pyridoxine daily for 33 days while on a normal diet.

VITAMIN LOSSES DURING PROCESSING, COOKING, AND STORAGE. More than 75% of the vitamin B-6 content of wheat is lost in milling white flour. Although vitamin B-6 is not added in white flour enrichment programs, perhaps it should be.

Canning and freezing result in considerable losses of vitamin B-6 with the losses being smaller in frozen foods. Freeze-dehydration and subsequent storage appear to have no adverse effect on the vitamin B-6 content of meat and poultry.

Considerable losses of vitamin B-6 occur during cooking. Beef loses 25 to 50% of its raw vitamin B-6 content in cooking, with the losses higher from oven braising than from oven roasting. Home cooking of fruits and vegetables results in a loss of about 50% of the vitamin B-6.

Storage losses appear to be minimal. Studies have shown that the storage of potatoes for as long as 6 months at 40°F (4.4°C) results in no loss of the vitamin.

Henry A. Schroeder, M.D., of the Dartmouth Medical School, conducted an extensive survey of the B-6 content in hundreds of common foods, and found that modern processing was causing massive losses of the vitamin. His studies showed B-6 losses as follows: From milling wheat and making all-purpose flour, 82.3%; from canning vegetables, 57 to 77%; from freezing vegetables, 37 to 56%; from canning meat and poultry, 42.6%; and from canning seafood, 48.9%.[7]

SOURCES OF VITAMIN B-6. In animal tissues and yeast, vitamin B-6 occurs mainly as pyridoxal and pyridoxamine. In plants, all three members of the vitamin are found, but pyridoxine predominates. The occurrence of vitamin B-6 in various forms has complicated the task of determining the content of the vitamin in foods.

Although vitamin B-6 is widely distributed in foods, many sources provide very small amounts. Thus, there is concern, as well as ever-increasing evidence, that in more than a few instances normal diets may be borderline or low in this vitamin.

Groupings by rank of common food sources of vitamin B-6 are given in the section on VITAMIN(S), Table V-5, Vitamin Table.

Fig. V-32. Bananas, a good source of vitamin B-6. (Courtesy, Department of Food Science and Human Nutrition, University of Hawaii)

For additional sources and more precise values of vitamin B-6 (pyridoxine), see Food Composition Table F-21.

Fat and sugar, which supply over ⅓ of the energy intake of the average American, are devoid of vitamin B-6. Generally speaking, processed or refined foods are much lower in vitamin B-6 than the original food; thus, white bread, rice, noodles, macaroni, and spaghetti are all quite low in vitamin B-6.

There is evidence that intestinal bacteria produce vitamin B-6. But the extent of this source, and the degree to which the bacterial-synthesized vitamin is utilized by the body, are undetermined.

Because vitamin B-6 is important for buoyant good health, and because it is limited in many foods, supplemental vitamin B-6 may be indicated, especially for infants and during pregnancy and lactation.

(Also see VITAMIN[S], Table V-5.)

[7]Schroeder, Henry A., M.D., "Losses of Vitamins and Trace Minerals Resulting From Processing and Preservation of Foods," *The American Journal of Chemical Nutrition*, Vol. 24, May 1971, pp. 562-573.

VITAMIN B-12 (COBALAMINS)

Contents	Page
History	1079
Chemistry, Metabolism, Properties	1080
Measurement/Assay	1080
Functions	1081
Deficiency Symptoms	1081
Recommended Daily Allowance of Vitamin B-12	1081
Toxicity	1082
Vitamin B-12 Losses During Processing, Cooking, and Storage	1082
Sources of Vitamin B-12	1082

Vitamin B-12, like so many other members of the B complex, is not a single substance; rather, it consists of several closely related compounds with similar activity. The term *cobalamins* is applied to this group of substances because all of them contain cobalt. Vitamin B-12, which is the most active member, is cyanocobalamin, named after the cyanide ion in the molecule. Other chemically related compounds known to have vitamin B-12 activity include hydroxocobalamin, nitritocobalamin, and thiocyanate cobalamin.

The most distinguishing characteristics of vitamin B-12 are: (1) unlike any other vitamin, the inability of higher plants to synthesize it (although it can be synthesized by animals); and (2) its most important deficiency state—Addisonian pernicious anemia, named after Thomas Addison, a physician working in London, who first described the malady in 1849. The anemia progressed slowly and ended with the death of the patient in 2 to 5 years. So feared and fatal was its course that it became known as pernicious anemia.

HISTORY. For 77 years (from 1849 to 1926) following the description of pernicious anemia by Thomas Addison of England, there was no hope for victims of the disease. Finally, step by step, scientists evolved with the treatment for Addisonian pernicious anemia and the discovery of vitamin B-12, a chronological record of which follows:

1. In 1925, George Hoyt Whipple, the Dean of the School of Medicine and Dentistry, University of Rochester, from 1921 to 1953, showed that liver was a great benefit in blood regeneration in dogs rendered anemic by bleeding.

2. In 1926, Minot and Murphy of the Harvard Medical School reported that feeding large amounts of raw liver (¼ to ½ lb per day) restored the normal level of red blood cells in cases of pernicious anemia. For this discovery, they shared a Nobel Prize with Whipple.

Following the report of Minot and Murphy, liver concentrates were developed, alleviating the necessity of eating large quantities of this food; and biochemists began a long series of studies to isolate the active component present in liver, which, at the time, was called the *antipernicious anemia factor*.

3. In 1929, W. B. Castle of Harvard showed that pernicious anemia could be controlled by feeding patients beef muscle incubated in normal gastric juice, although neither beef muscle nor gastric juice was effective alone. This finding

VITAMIN B-12
IT'S ESSENTIAL FOR HUMANS, TOO!

VITAMIN B-12 MADE THE DIFFERENCE! THE BIGGER CHICK AT THE RIGHT AND HIS SMALLER COMPANION ARE BOTH 3½ WEEKS OLD. <u>LEFT</u>: THE SMALL CHICK, FED A RATION DEFICIENT IN VITAMIN B-12, WEIGHED 157 g. <u>RIGHT</u>: THE LARGER CHICK, FED THE SAME RATION PLUS VITAMIN B-12, WEIGHED 280 g. (COURTESY, MERCK AND COMPANY, RAHWAY, N.J.)

TOP FOOD SOURCES

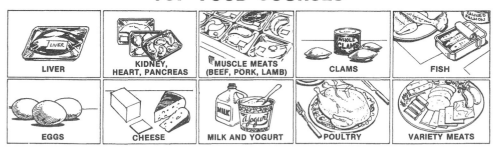

Fig. V-33. The vitamin B-12 (cobalamins) story.

led him to postulate that two factors were involved: one an *extrinsic factor* in food, and the other an *intrinsic factor* in normal gastric secretion; which, given together, caused red blood cell formation in pernicious anemia.

4. In 1948, two groups of researchers working independently, Rickes and co-workers of Merck and Co., Inc. of New Jersey, and Smith and Parker of England, isolated from a liver concentrate a crystalline, red pigment, which they called vitamin B-12.

5. In 1948, R. West, of Columbia University, New York, showed that injections of vitamin B-12 induced a dramatic beneficial response in patients with pernicious anemia.

6. In 1955, the structure of vitamin B-12 (cyanocobalamin) was determined by Dorothy Hodgkin and co-workers, at Oxford. Later (1964), Hodgkin was awarded the Nobel Prize.

7. In 1955, Woodward's group at Harvard synthesized vitamin B-12 using a very complicated and expensive procedure. Fortunately, soon thereafter, it was found that highly active vitamin B-12 concentrates can be produced from cultures of certain bacteria and fungi grown in large tanks containing special media; and this remains the main method of commercial production.

CHEMISTRY, METABOLISM, PROPERTIES.

• **Chemistry**—Vitamin B-12 is the largest and the most complex of all vitamin molecules. The main part of the molecule consists of a porphyrin ring containing cobalt as the central element. A cyanide (-CN) group may be attached to the cobalt, in which case the compound is called cyanocobalamin (or vitamin B-12); the commercially available form of the vitamin, little of which occurs naturally. The cyanide group attachment to the cobalt can be replaced by a hydroxy group (-OH), giving hydroxocobalamin, the common naturally occurring form of the vitamin; or it can be replaced by a nitrite group (-NO$_2$), giving nitritocobalamin, a form found in certain bacteria.

A coenzyme form of vitamin B-12 contains an adenosine (a nucleoside, which consists of a purine [adenine] combined with a pentose sugar, ribose) molecule in place of the cyanide and is thought to be the most common form in foods. Methylcobalamin is another form of the vitamin with a coenzyme role. All these forms have approximately equal vitamin B-12 activity in the diet.

The structure of vitamin B-12 ($C_{63} H_{90} O_{14} N_{14} PCo$) is shown in Fig. V-34.

Fig. V-34. Structure of vitamin B-12.

Vitamin B-12 occurs as a protein complex in animal proteins. The ultimate source, however, is the microorganisms in the gastrointestinal tract of herbivorous animals. Such microorganisms are found in large amounts in the rumen (the first stomach) of cows and sheep. Apparently, some synthesis occurs in the intestinal bacteria of man, also; but the amount supplied from this source is small and unknown.

• **Metabolism**—It is noteworthy (1) that vitamin B-12 is the only vitamin that requires a specific gastrointestinal tract secretion for its absorption (*intrinsic factor*); and (2) that the absorption of vitamin B-12 in the small intestine requires about 3 hours (compared to seconds for most other water-soluble vitamins). The absorption of vitamin B-12 involves the following five steps:

1. First, vitamin B-12 is released from the protein (the peptide bonds) to which it is linked in foods by the action of hydrochloric acid and intestinal enzymes.

2. Next, vitamin B-12 is bound to a highly specific glycoprotein, Castle's intrinsic factor, which is secreted in the stomach.

3. The vitamin B-12 intrinsic factor forms a complex with calcium and passes through the upper part of the small intestine to receptor sites in the ileum through which absorption of vitamin B-12 takes place.

4. In crossing the intestinal mucosa, vitamin B-12 is freed from the complex (the B-12-intrinsic-factor-calcium complex).

5. In the intestinal cells, vitamin B-12 is transferred to a plasma transport protein known as transcobalamin II, for transport in the blood circulation.

In normal persons, from 30 to 70% of the vitamin B-12 is absorbed as outlined above, in comparison with 1 to 3% absorbed by simple diffusion. Pernicious anemia results from a complete failure to absorb the vitamin, a condition caused by gastric abnormality (usually lack of intrinsic factor). Hence, therapeutic doses of B-12 given to pernicious anemia patients usually must be administered intramuscularly.

Intrinsic factor regulates the amount of absorption of vitamin B-12 to about 1.5 to 3.0 mcg daily. Absorption decreases with age (it drops to about 5% in the elderly), and with iron and vitamin B-6 deficiencies; and it increases with pregnancy. Infant levels are approximately twice that of the mother. Also, absorption is greater if the vitamin is provided in three meals than if all of it is provided in a single meal.

The liver is the principal site of storage of vitamin B-12; normally, it contains 2,000 to 5,000 mcg, sufficient to take care of the body needs for 3 to 5 years. Small amounts of vitamin B-12 are stored in the kidneys, muscle, lungs, and spleen. Storage in the bone marrow is limited, amounting to only 1 to 2% of that in the liver.

Vitamin B-12 is excreted by way of the kidneys and in the bile.

The most useful measurement for the detection of a vitamin B-12 deficiency is the serum vitamin B-12 level. Normal serum levels of vitamin B-12 range from 200 to 700 picograms (1 pg = 10^{-12} g per milliliter).

• **Properties**—The deep-red needlelike crystals are slightly soluble in water, stable to heat, but destroyed by light and by strong acid or alkaline solutions. There is little loss (only about 30%) of the vitamin by ordinary cooking procedures.

Vitamin B-12 is remarkably potent. It has a biologic activity 11,000 times that of the standard liver concentrate formerly used in the treatment of pernicious anemia.

MEASUREMENT/ASSAY. No International Units have been defined for the biological activity of vitamin B-12.

However, pure cobalamin can be used as a standard substance. Vitamin B-12 is measured in micrograms or picograms (pg, micromicrograms).

High potency preparations of vitamin B-12 are usually assayed by spectrophotometry. Also, vitamin B-12 may be assayed colorimetrically or fluorometrically. Some assays involve measurement of cobalt. However, food sources are usually assayed for vitamin B-12 by either (1) the microbiological method, or (2) the biological method, using chicks or rats.

FUNCTIONS. In the human body, vitamin B-12 is converted to a coenzyme form, if it is not already in such form. There are two active coenzyme forms: Coenzyme B-12 (adenosylcobalamin), and methyl B-12 (methylcobalamin). Coenzyme B-12 has an adenosine ribonucleoside attached to the cobalt atom in the vitamin B-12 molecule in place of the cyanide group, whereas methyl B-12 contains a methyl group in place of the cyanide group. The conversion of vitamin B-12 to coenzyme forms requires many nutrients, including riboflavin, niacin, and magnesium.

Vitamin B-12 coenzymes perform the following physiological roles at the cellular level, especially in the cells of the bone marrow, nervous tissue, and gastrointestinal tract:

1. **Red blood cell formation and control of pernicious anemia.** Vitamin B-12 is essential for the blood-forming organs of the bone marrow to function properly. Without sufficient B-12 coenzymes, the red blood cells do not mature normally, with the result that large, immature cells (megaloblasts) form and are released into the blood, causing megaloblastic anemia.

2. **Maintenance of nerve tissue.** Vitamin B-12 is essential to the health of the nervous system. Vitamin B-12 coenzymes are necessary for the synthesis of myelin, a lipoprotein, in the nervous tissue; but it is not known whether the vitamin is involved in the synthesis of the lipid of the protein part of myelin.

3. **Carbohydrate, fat, and protein metabolism.** Since coenzyme B-12 is necessary for the conversion of methylmalonate to succinate, it is required for normal carbohydrate and fat metabolism. It is also involved in protein metabolism, since the requirement for B-12 increases as protein intake increases.

4. **Synthesis or transfer of single carbon units.** Vitamin B-12 is thought to be required for the synthesis of single carbon units, whereas folacin participates in their transfer. It follows that B-12 takes part in most of the same reactions as folacin, including—
   a. The interconversion of serine and glycine
   b. The formation of methionine from homocysteine
   c. The formation of choline from ethanolamine

5. **Other functions.** Vitamin B-12 also serves as a coenzyme in the biosynthesis of methyl groups (-CH₃), and in reduction reactions such as the conversion of disulfide (S-S) to the sulfhydryl group (-SH).

• **Therapeutic uses of vitamin B-12**—Vitamin B-12 is being used in the treatment of the following maladies:

1. **Pernicious anemia.** The discovery that B-12 would control pernicious anemia was a great clinical breakthrough. Now a patient can be given intramuscular injections of 15 to 30 mcg of B-12 daily during a relapse, then maintained afterward by an injection of about 30 mcg every 30 days.

2. **Sprue.** Vitamin B-12 is effective in the treatment of sprue, especially if used in conjunction with folic acid. Actually, the role of B-12 may be indirect—to facilitate the action of folic acid.

DEFICIENCY SYMPTOMS. Vitamin B-12 deficiency in man may occur as a result of (1) dietary lack, which sometimes occurs among vegetarians who consume no animal foods; or (2) deficiency of intrinsic factor, due to pernicious anemia, total or partial removal of the stomach by surgery, or infestation with parasites such as the fish tapeworm.

The common symptoms of a dietary deficiency of vitamin B-12 are: sore tongue, weakness, loss of weight, back pains, tingling of the extremities, apathy, and mental and other nervous abnormalities. Anemia is rarely seen in dietary deficiency of B-12.

In pernicious anemia, the characteristic symptoms are: abnormally large red blood cells (macrocytes), lemon-yellow pallor, anorexia, dyspnea (short of breath), prolonged bleeding time, abdominal discomfort, loss of weight, glossitis (inflammation of the tongue), an unsteady gait, and neurological disturbances, including stiffness of the limbs, irritability, and mental depression. Without treatment, death follows. Only injections of vitamin B-12 can alleviate efficiently the symptoms of pernicious anemia.

• **Dietary deficiencies**—Dietary deficiencies of vitamin B-12 may occur under the following circumstances:

1. **Among vegans and ovolactovegans.** People who live exclusively on plant foods (vegans) may be seriously deficient in vitamin B-12. Vegetarianism is, for religious reasons, common among Hindus in India and elsewhere, but most of them are ovolactovegetarians. (They consume animal products other than flesh foods.) Yet, for large numbers of Hindus, the intake of animal foods, and therefore of vitamin B-12, falls far short of the recommended allowances.

2. **In some developing countries.** Vitamin B-12 deficiencies are not uncommon in developing countries where foods of plant origin predominate, especially among pregnant and lactating women. For example, very low intakes of the vitamin have been reported in Peru and in parts of Africa.

3. **Where consumption of animal products by mothers is low.** In areas of the world where intakes of animal products by mothers is low, vitamin B-12 deficiency in infants may occur.

When and where animal foods are in short supply, vitamin B-12 now makes it possible to rectify the above situations, and to use plant and cereal foods much more wisely in the human diet.

RECOMMENDED DAILY ALLOWANCE OF VITAMIN B-12. The Food and Nutrition Board (FNB) of the National Research Council (NRC) recommended daily allowances (RDA) of vitamin B-12 are given in the section on VITAMIN(S), Table V-5, Vitamin Table.

These recommended daily allowances provide for a margin of safety to cover variance in individual needs, absorption, and body stores. However, in using this table as a nutritional guide, the following facts should be noted: (1) exact daily human requirements of vitamin B-12 cannot be given because it is synthesized by intestinal flora; (2) in the absence of intrinsic factor (e.g., pernicious anemia), the vitamin is not absorbed; and (3) it is assumed that at least 50% of the vitamin B-12 in food is absorbed.

• **Infants and children**—The recommended allowance for infants up to 6 months of age is 0.3 mcg per day. This is based on the average concentration of the vitamin in human milk. For infants receiving commercial formulas, the Committee of Nutrition of the American Academy of Pediatrics recommends a daily vitamin B-12 intake of 0.15 mcg/100 kcal; thus, a 1-year-old child weighing 10 kg and receiving 1,000 kcal should receive 1.5 mcg of vitamin B-12 per day. The recommended daily allowances given in Table V-5 for older infants and preadolescent children have been calculated on the basis of average energy intakes by using the latter formula.

● **Adults**—The recommended daily allowance for both males and females over 10 years of age is 2.0 mcg of vitamin B-12. This value will maintain adequate vitamin B-12 nutrition and a substantial reserve body pool in most normal persons.

● **Pregnancy and lactation**—The recommended dietary allowance for pregnant women is 2.2 mcg/day; and for lactating women it is 2.6 mcg/day.

Only the effects of gross deficiency of vitamin B-12 are known; hence, it is possible that a minor degree of deficiency, especially if of long duration, may prevent buoyant good health, even in well-fed populations.

● **Vitamin B-12 intake in average U.S. diet**—According to the U.S. Department of Agriculture, the amount of vitamin B-12 in the U.S. food supply averages 9.1 mg per person per day. (But the intake may range from a low of 1 to a high of 100 mg per day.) Meat, fish, and poultry contribute 76.1% of the 9.1 mg of vitamin B-12 available in the typical U.S. daily diet; dairy products contribute 18.5%; eggs, 3.7%; and 1.7% comes from other sources.

TOXICITY.   No toxic effects of vitamin B-12 are known.

VITAMIN B-12 LOSSES DURING PROCESSING, COOKING, AND STORAGE.   About 30% of the vitamin B-12 activity of foods is lost during ordinary cooking.

Although only about 10% of the vitamin B-12 activity in milk is lost by pasteurization, from 40 to 90% of the B-12 is destroyed by evaporating milk.

It is noteworthy that in the presence of ascorbic acid vitamin B-12 withstands less heat.

Vitamin B-12 is destroyed by light.

SOURCES OF VITAMIN B-12.   The sole source of vitamin B-12 in nature is synthesis by microorganisms. It is synthesized by the many microorganisms in the rumen and intestine of herbivorous animals. The vitamin B-12 bound to a protein in animal foods results from such synthesis. This explains why vitamin B-12 is found in all foods of animal origin.

Plants cannot manufacture vitamin B-12; hence, except for trace amounts absorbed from the soil (because of the bacteria, soil is a good source of B-12) by the growing plant, very little is found in plant foods—in vegetables, grains, legumes, fruits, etc.

A classification of the B-12 content of food sources is given in the section on VITAMIN(S), Table V-5, Vitamin Table.

For additional sources and more precise values of vitamin B-12, see Food Composition Table F-21.

Some vitamin B-12 is formed by microorganisms in the intestinal tract of human beings. However, the synthesis is so far down in the colon that little of it may be absorbed.

The story of vitamin B-12 lends support to all developing country programs designed to improve animal production and to increase the supply and consumption of animal protein. Also, it is recognized that vitamin B-12 deficiency is not an uncommon consequence of many diseases and of

Fig. V-35. Slices of top round steak. Muscle meats are a good source of vitamin B-12. (Courtesy, National Live Stock and Meat Board, Chicago, Ill.)

surgical operations on the stomach and small intestine. (Also see VITAMIN[S], Table V-5.)

VITAMIN B-13 (OROTIC ACID)

It is highly possible that the so-called vitamin B-13 is a growth promotant and a preventative of certain disorders. At this time, however, it is not known whether it plays an essential role in an otherwise adequate diet; hence, this presentation is for two purposes; (1) informational, and (2) stimulation of research.

HISTORY.   This compound, called B-13, was first obtained from distillers' solubles. Subsequently, one of its constituents, orotic acid, has been synthesized in Europe and used to treat multiple sclerosis.

CHEMISTRY.   Vitamin B-13 is a compound of unknown structure which appears either to contain orotic acid or to yield it on decomposition (see Fig. V-36).

Fig. V-36. Orotic acid.

FUNCTIONS.   Vitamin B-13 has been found to stimulate the growth of rats, chicks, and pigs under certain conditions.

Orotic acid is utilized by the body in the metabolism of folic acid and vitamin B-12. Also, it appears to aid the replacement or restoration of some cells.

There is indication that vitamin B-13 may be helpful in the treatment of multiple sclerosis.

DEFICIENCY SYMPTOMS.   Deficiency symptoms have not been proved. But it is believed that a deficiency may lead to liver disorders, cell degeneration, and premature aging; and to degenerative symptoms in multiple sclerosis victims.

RECOMMENDED DAILY ALLOWANCE OF VITAMIN B-13.   Dietary requirements are not known.

SOURCES OF VITAMIN B-13.   Vitamin B-13 is found in such natural sources as distillers' solubles, whey, soured or curdled milk, and root vegetables. Also, this nutrient is available in supplemental form as calcium orotate.
(Also see VITAMIN[S], Table V-5.)

Fig. V-37. White turnips. Root vegetables are a good source of vitamin B-13. (Courtesy, Dept. of Food Service and Human Nutrition, University of Hawaii)

VITAMIN B COMPLEX

With the exception of vitamin C, all of the water-soluble vitamins can be grouped together under the vitamin B complex.

The story of the B vitamins begins with the study of the age-old disease beriberi.

In 1873, Van Lent was apparently the first to conclude that the type of diet had something to do with the origin of beriberi. By reducing the ration of rice in the diet of the sailors in the Dutch navy, he was able to eradicate beriberi almost entirely.

In 1882, Kanehiro Takaki, a Japanese Naval medical officer, reported that he had cured beriberi in sailors of the Japanese Navy by giving them less rice and more meat, milk, and vegetables. Takaki attributed the cure to the protein content of the diet.

Fifteen years later (1897), Christiaan Eijkman, a Dutch physician assigned to a prison hospital in the East Indies, observed beriberi among the inmates and sought the answer through experiments with chickens. To save money, he fed the birds scraps—mostly polished rice—from the patients' meals. The chickens unexpectedly developed a bad nerve ailment, which resulted in paralysis.

Later, the unsympathetic director of the hospital withheld permission to use scraps, and Dr. Eijkman had to buy natural (unmilled) rice for the chickens he used in his experiment. The ailing birds improved after they began eating the natural rice.

Dr. Eijkman then began a series of experiments that led to the first clear concept of disease due to nutritional deficiency. He fed polished white rice to chickens, pigeons, and ducks. They developed the paralysis that he had observed previously, then recovered when he fed them natural (unmilled) rice. Birds fed whole rice remained well.

Eijkman noted that the disease in birds which resulted from a polished rice diet resembled beriberi in man. He theorized that rice contained too much starch, which poisoned nerve cells, and that the outer layers, removed from the grain in milling, were an antidote.

Another Dutch physician, Dr. G. Grijns, continued the work of Eijkman. But he interpreted Eijkman's findings differently. In 1901, he concluded that beriberi in birds and man was due to a deficiency or absence of an essential nutrient from the diet.

From then on, chemists in many countries tried to concentrate the substance in rice that prevented beriberi in order to obtain it in pure form. Among them was Casimir Funk, of the Lister Institute, London, who, in 1912, coined the term *vitamine* and applied it to the antiberiberi substance.

In 1916, Dr. Elmer V. McCollum of the University of Wisconsin designated the concentrate that cured beriberi as *water-soluble B*, to distinguish it from the antinightblindness factor (called vitamin A) which had been found in carrots and butterfat of milk. At that time, McCollum thought that the antiberiberi substance was one factor only.

In 1926, B. C. P. Jansen and W. P. Donath, in Holland, isolated the antiberiberi vitamin, and in 1936 Robert R. Williams, an American, determined the structure and synthesized it.

As research continued (1919-1922), it was found that vitamin B was not a single substance, that it actually consisted of several factors. Collectively, they came to be known as the *vitamin B complex*, but each factor was given a separate designation. Some members of the group came to be referred to by subscript numbers as vitamins B_1, B_2, etc.; others became known by their chemical names; still others received both a number and a chemical designation. These vitamins differ in both chemical structure and specific functions. Yet, there are similarities. All of them are water-soluble; all of them are abundant in liver and yeast, and they often occur together in the same foodstuffs; each of them contains carbon, hydrogen, oxygen, and nitrogen; some of them contain mineral elements in their molecules (thiamin and biotin contain sulfur, and vitamin B-12 contains cobalt and phosphorus); most of them are part of a coenzyme molecule concerned with the breakdown of carbohydrate, protein, and fat in the body; the actions of many of them are interrelated; few of them are stored in large amounts in the body, so they must be provided daily; certain organs, particularly the liver, contain higher concentrations of them than others; and they are excreted from the body by way of the kidneys.

excreted from the body by way of the kidneys.

Another noteworthy characteristic of the B vitamins is that they are synthesized by microbial fermentation in the digestive tract, especially by ruminants (cattle and sheep) and herbivorous nonruminants (horse and rabbit). Some animals eat their own feces (coprophagy), thus recycling the vitamins synthesized in the microbial fermentation in the large intestine and cecum; rabbits, in particular, are known to do this on a routine basis. Unlike ruminants and herbivorous nonruminants, however, man, pigs, and poultry have only one stomach and no large cecum like the horse and rabbit. As a result, they do not synthesize enough of most B vitamins. Consequently, for man and other monogastrics, the B vitamins must be provided regularly in the diet in adequate amounts if deficiencies are to be averted.

At the present time, 9 fractions of the vitamin B complex are generally recognized, and others are postulated. Those discussed in this book (alphabetically under their name designations) are: biotin, choline, folacin (folic acid), niacin (nicotinic acid; nicotinamide), pantothenic acid (vitamin B-3), riboflavin (vitamin B-2), thiamin (vitamin B-1), vitamin B-6 (pyridoxine, pyridoxal; pyridoxamine), and vitamin B-12 (cobalamins).

A lack of B-complex vitamins is one of the forms of malnutrition that occur often throughout the world. Because the B vitamins are usually found in the same foodstuffs, a deficiency of several factors is usually observed rather than a deficiency of a single factor.

Many physiologic and pathologic stresses influence the need for the B vitamins. Larger amounts are needed during growth and in pregnancy and lactation than in maintenance of health in adult life. The requirement may be increased by diseases that elevate metabolism and by conditions associated with poor absorption, improper utilization, or increased excretion. Administration of antibiotics may lead to vitamin deficiency in some circumstances; in others, antibiotics spare vitamin requirements.

(Also see BERIBERI; and VITAMIN[S].)

VITAMIN C (ASCORBIC ACID; DEHYDROASCORBIC ACID)

Contents	Page
History	1084
Chemistry, Metabolism, Properties	1085
Measurement/Assay	1086
Functions	1086
Determining Vitamin C Status	1088
Deficiency Symptoms	1088
Recommended Daily Allowance of Vitamin C	1089
Toxicity	1090
Losses of Vitamin C During Processing, Cooking, and Storage	1090
How to Conserve Vitamin C in Foods in the Home	1090
Sources of Vitamin C	1091

Vitamin C—also called *ascorbic acid, dehydroascorbic acid, hexuronic acid,* and the *antiscorbutic vitamin*—is the very important substance, first found in citrus fruits, which prevents scurvy, one of the oldest scourges of mankind. All animal species appear to require vitamin C, but a *dietary need* is limited to humans, guinea pigs, monkeys, bats, certain fish, and perhaps certain reptiles. These species lack the enzyme L-gulonolactone oxidase which is necessary for vitamin C synthesis from 6-carbon sugars.

HISTORY. Scurvy, now known to be caused by a severe deficiency of vitamin C, has been a dread disease since ancient times. It was once common among sailors who ate little except bread and salt meat while on long voyages.

The historical incidence and conquest of scurvy constitute one of the most thrilling chapters in the development of nutrition as a science. A chronological summary of the saga of scurvy and vitamin C follows:

As early as 1550 B.C., scurvy was described by the Egyptians on medical papyrus rolls (man's first writing paper, made by the Egyptians from the papyrus plant as early as 2400 B.C.), discovered in Thebes by George Moritz Ebers,

VITAMIN C
helps to build healthy gums, teeth, and bones
TWO GUINEA PIGS OF SAME AGE

THIS GUINEA PIG HAD NO ASCORBIC ACID AND DEVELOPED SCURVY. NOTE CROUCHED POSITION DUE TO SORE JOINTS.

THIS GUINEA PIG HAD PLENTY OF ASCORBIC ACID. IT IS HEALTHY AND ALERT; ITS FUR IS SLEEK AND FINE.

TOP FOOD SOURCES

CITRUS FRUIT AND JUICE · GUAVAS · PEPPERS · BLACK CURRANTS · GREEN LEAFY VEGETABLES

CANTALOUPE · GOOSEBERRIES · PAPAYA · STRAWBERRIES · TOMATOES AND TOMATO JUICE

Fig. V-38. Vitamin C (ascorbic acid) made the difference! *Left:* Guinea pig on vitamin C-deficient diet. *Right:* Guinea pig that received plenty of vitamin C. Note, too, top sources of vitamin C. (Adapted from USDA sources.)

a German Egyptologist and novelist, who, in 1874, edited it in a romantic historical novel on medicine which he titled *Papyrus Ebers*).

In the Old Testament (which was written over a long period of time, thought to be from 1100 B.C. to 500 B.C.), reference is made to this disease.

About 450 B.C., Hippocrates, the Greek *father of medicine* described the symptoms of the malady—gangrene of the gums, loss of teeth, and painful legs in soldiers.

In 1248-54, Jean Sire de Joinville, the French chronicler, accompanied Louis IX of France to Cyprus and Egypt. In 1309, he completed in final form the *History of Saint Louis*, an account of the Crusade, in which he told of a disease (scurvy) "which attacked the mouth and the legs."

In 1497, when Vasco da Gama, Portuguese navigator, sailed around the Cape of Good Hope and established the first European trading colony on the coast of Malabar in India, 100 of his crew of 160 men perished from scurvy on the voyage.

During the winter of 1535 in Canada, Jacques Cartier, the daring explorer who laid claim to Canada for France, recorded in his log that the lives of many of his men dying of scurvy were saved "almost overnight," when they learned from the Indians that drinking a brew made from the growing tips of pine or spruce trees cured and prevented the malady. (It is now known that the *brew* contained vitamin C.)

Fig. V-39. Friendly Huron-Iroquois Indians shown in Quebec in 1535, (1) making a broth from pine branches, and (2) serving it to Jacques Cartier and his men to cure scurvy. (Reproduced with permission of *Nutrition Today*, P.O. Box 1829, Annapolis, MD, 21404, © 1979)

In the 15th and 16th centuries, scurvy was a scourge throughout Europe, so much so that medical men wondered if all diseases might stem from it. It was particularly prevalent and severe on long voyages of sailing ships, in cities, and in times of crop failures. During this period, there was also a tendency to associate scurvy and venereal disease; some authorities of the day believed that both diseases were brought from abroad by sailors. Mercury was sometimes used as a treatment, with disastrous results.

In 1600-1603, Captain James Lancaster, English navigator, recorded that on the long voyage to the East Indies he kept his crew hearty merely by the addition of a mandatory "three spoonfuls of lemon juice every morning."

In 1747, James Lind, an English naval surgeon, tested six remedies on 12 sailors who had scurvy and found that oranges and lemons were curative. His classical studies, the results of which were published in 1753, are generally credited as being the first experiments to show that an essential food element can prevent a deficiency disease. But another 50 years elapsed before the British Navy required rations of lemons or limes on sailing vessels.

On two historic voyages, each of three-years duration, from 1768 to 1771 and from 1772 to 1775, British Captain James Cook, avoided scurvy—hitherto the scourge of long sea voyages. He had his ship stocked with concentrated slabs of thick brown vegetable soup and barrels of sauerkraut. Of the sauerkraut he said: "It is not only a wholesome vegetable food, but, in my judgment, highly antiscorbutic, and spoils not by keeping." In addition, he sent seamen ashore at every port visited to gather all sorts of fresh fruits and green vegetables (including grasses), which the crew prepared, served, and ate. As a result, not one of the crew died from scurvy.

In 1795 (one year after Lind's death), by Admiralty Order, the British Royal Navy began providing 1 oz of lime juice daily in every sailor's food ration; from this date forward, British sailors were stuck with the nickname *limeys*.

In 1907, Holst and Frolich, of Norway, produced scurvy experimentally in guinea pigs by feeding them a diet deficient in foods containing ascorbic acid.

In 1928, Szent-Gyorgy, a Hungarian scientist, working in Hopkins' laboratory at Cambridge University, in England, isolated a substance from the ox adrenal glands, oranges, and cabbage leaves, which he called hexuronic acid; but he did not test it for antiscorbutic effect.

In 1932, Charles Glen King and W. A. Waugh, at the University of Pittsburgh, isolated from lemon juice a crystalline material that possessed antiscorbutic activity in guinea pigs; this marked the discovery of *vitamin C*, a deficiency of which caused the centuries old scourge of scurvy.

In 1933, vitamin C was synthesized by Reichstein, a Swiss scientist.

In 1938, *ascorbic acid* was officially accepted as the chemical name for vitamin C.

CHEMISTRY, METABOLISM, PROPERTIES.

• **Chemistry**—Ascorbic acid is a compound of relatively simple structure, closely related to the monosaccharide sugars. It is synthesized from glucose and other simple sugars by plants and by most animal species (see Fig. V-40).

Fig. V-40. Metabolic relation of glucose to ascorbic acid. In man, the absence of oxidase prevents this reaction, making the intake of preformed ascorbic acid in food necessary.

Man, monkeys, guinea pigs, fruit-eating bats, and red-vented bulbul birds (the latter two are native to India), cannot make the conversion from glucose to ascorbic acid, because these species lack the necessary enzyme (oxidase). Scurvy, then, in man can really be classed as a disease of distant genetic origin—an inherited metabolic error; a defect in carbohydrate metabolism due to the lack of an enzyme, which, in turn, results from the lack of a specific gene.

Two forms of vitamin C occur in nature: ascorbic acid (the reduced form), and dehydroascorbic acid (the oxidized form).[8] Their structural formulas are shown in Fig. V-41.

Fig. V-41. Structural formulas of vitamin C.

Although most of the vitamin C exists as ascorbic acid, both forms appear to be utilized similarly by the human. Also, the body efficiently utilizes either synthetic L-ascorbic acid or the vitamin in its natural form as in orange juice.

Ascorbic acid is easily oxidized to dehydroascorbic acid, which is just as easily reduced back to ascorbic acid. However, dehydroascorbic acid may be irreversibly oxidized, particularly in the presence of alkali, to diketogulonic acid, which has no antiscorbutic activity (see Fig. V-42.)

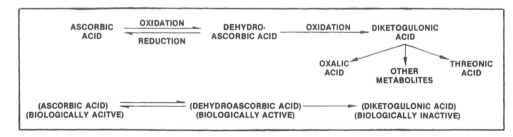

Fig. V-42. Relationship of ascorbic, dehydroascorbic, and diketogulonic acids.

Certain derivatives of vitamin C (for example, erythrobic acid and ascorbyl palmitate) are used as antioxidants in food products to prevent rancidity, to prevent browning of fruit, and to cure meat. Erythrobic acid (D-araboascorbic acid) is poorly absorbed and has little antiscorbutic activity.

• **Metabolism**—Pertinent facts about the absorption, storage, and excretion of vitamin C follow:

---

[8]Actually, several chemical compounds have vitamin C activity. So, it is now recommended that the term *vitamin C* be used as the combined name of all compounds having the biological activity of ascorbic acid, and that the terms *ascorbic acid* and *dehydroascorbic acid* be used only when specific reference to them is made.

1. **Absorption.** Vitamin C is readily and rapidly absorbed from the upper part of the small intestine into the circulatory system. Thence, it is taken up unevenly by the tissues; the adrenal gland and the retina of the eye contain an especially high concentration of vitamin C, but the liver, spleen, intestine, bone marrow, pancreas, thymus, pituitary, brain, and kidney also contain appreciable amounts. Blood cells contain more than blood serum.

2. **Storage.** Unlike the majority of the water-soluble vitamins, limited stores of vitamin C are held in the body. Thus, the signs of scurvy do not appear for some weeks in humans receiving no vitamin C.

3. **Excretion.** Vitamin C is largely excreted in the urine, with the amount excreted controlled by the kidney tubules. When the tissues are saturated, a large amount is excreted; but when the tissue reserves are depleted, only a small amount is excreted. Some vitamin C is always excreted by the kidneys even when the tissues are severely depleted.

• **Properties**—Ascorbic acid is a white, odorless crystalline powder, which is quite stable when dry.

Of all the vitamins, ascorbic acid is the most unstable when in solution. It is highly soluble in water, but not in fat. The oxidation (destruction) of ascorbic acid is accelerated by air, heat, light, alkalies, oxidative enzymes, and traces of copper and iron. It is markedly destroyed by cooking, particularly where the pH is alkaline. Cooking losses also result because of its solubility. The destruction of ascorbic acid is slowed down by foods that are acidic, by refrigeration, and by protection from exposure to air.

MEASUREMENT/ASSAY. The concentration of ascorbic acid in tissues and foods is expressed in milligrams. One IU is the activity of 0.05 mg of ascorbic acid.

Ascorbic acid is generally determined by chemical assay, accomplished by taking advantage of its reducing properties.

For bioassay work, guinea pigs are the preferred experimental animals because of their susceptibility to a deficiency of vitamin C. Thus, they are still used for demonstration of deficiency of the vitamin and to make comparative assays.

FUNCTIONS. The specific biochemical functions of vitamin C are not clearly understood. Nevertheless, it is established as a very important substance for body welfare because of being implicated in the following roles: (1) formation and maintenance of collagen, which makes for more rapid and sound healing of wounds and burns; (2) metabolism of the amino acids tyrosine and tryptophan; (3) absorption and movement of iron; (4) metabolism of fats and lipids, and cholesterol control; (5) as an antioxidant in the protection of vitamins A and E and the polyunsaturated fatty acids; (6) sound teeth and bones; (7) strong capillary walls and healthy blood vessels; (8) metabolism of folic acid; and perhaps in a number of other roles. Details follow:

• **Collagen formation**—The most clearly established functional role of vitamin C is the formation and maintenance of collagen, the substance that binds body cells together in much the same manner as mortar binds bricks.

Collagen is a fibrous protein that contains large amounts

of the amino acids proline and hydroxyproline. Postulation is that vitamin C is essential for making hydroxyproline in the body as follows: Vitamin C activates the enzyme prolyl hydroxylase, which, in turn, effects the conversion of proline (by the addition of an "OH" group) to hydroxyproline in the formation of collagen.

Vitamin C is also required for the conversion of lysine to hydroxylysine, another amino acid that is an essential part of collagen. The reaction is brought about by the enzyme lysine hydroxylase. The role of vitamin C in the formation of hydroxylysine is thought to be similar to its role in the formation of hydroxyproline.

Both of the above reactions require vitamin C, and both are essential for collagen formation. In turn, failure to synthesize collagen results in delayed healing of wounds and burns. So, the administration of vitamin C makes for more rapid and sound healing of wounds.

• **Metabolism of tyrosine and tryptophan**—Vitamin C is necessary in the metabolism of the amino acids tyrosine and tryptophan.

In the metabolism of tyrosine, a deficiency of vitamin C will result in the build-up and excretion of the intermediary product, P-hydroxyphenylpyruvate, as a result of inactivating the enzyme P-hydroxyphenylpyruvic acid oxidase. When large amounts of tyrosine are being metabolized, vitamin C protects the enzyme P-hydroxyphenylpyruvic acid oxidase from inactivation (rather than activates as was formerly thought), and enhances the synthesis of norepinephrine, a neurotransmitter, from tyrosine.

Also, vitamin C is required for the conversion of tryptophan to 5-hydroxytryptophan, the first step in the formation of the serotonin, a compound that raises blood pressure through vasoconstrictor action.

• **Iron utilization**—When the two nutrients are ingested simultaneously, dietary vitamin C increases the absorption of iron by converting ferric iron to the more readily absorbed ferrous form. Vitamin C is also necessary for the movement of transferritin (a combination of ferric iron and the protein transferrin) to the liver, and for the formation of the iron-protein compound ferritin for storage in the liver, spleen, and bone marrow.

(Also see IRON.)

• **Metabolism of fats and lipids**—There is some evidence that vitamin C affects fat and lipid metabolism as follows:

1. Vitamin C serves as a cofactor, along with ATP and magnesium ions, in the inactivation of the enzyme *adipose tissue lipase*, the enzyme that mobilizes the free fatty acids from adipose tissue to meet the energy demands of the body. When the body's energy needs have been met, vitamin C (along with the other two controlling agents—ATP and magnesium ions) inactivates the adipose tissue lipase.

2. Vitamin C may have a role in the metabolism of cholesterol. The levels of cholesterol in the liver and blood serum appear to rise during a deficiency of the vitamin, and to fall with the administration of the vitamin. The increased accumulation of cholesterol appears to be due to a decrease in the rate of conversion of cholesterol to bile acids when vitamin C intake is inadequate.

Vitamin C also appears to be involved in the metabolism of cholesterol in another way. Through its sulfated metabolite, *ascorbic acid sulfate*, it appears to bring about the formation of cholesterol sulfate, a water-soluble compound that is excreted in the urine. By this means, cholesterol may be mobilized from the body tissues, with the result that there is a lowering of blood cholesterol levels.

• **Antioxidant in the protection of vitamins A and E and the polyunsaturated fatty acids**—Ascorbic acid is an important antioxidant; thus, it has a role in the protection of vitamins A and E and the polyunsaturated fatty acids from excessive oxidation.

• **Sound teeth and bones**—Vitamin C is required for the normal development of odontoblasts, a layer of cells that forms dentin in teeth. It follows that a deficiency of vitamin C may cause defects in tooth dentin, especially during the critical period of tooth formation.

Also, vitamin C is necessary for the proper calcification and soundness of bone.

• **Strength of capillary walls and blood vessels**—Vitamin C is necessary for maintaining strength of capillary walls, especially of the small blood vessels. Shortages of the vitamin result in weakened and inelastic capillary walls, which may rupture and hemorrhage, evidenced by easy bruising, pinpoint peripheral hemorrhages, bone and joint hemorrhages, easy bone fracture, fragile bleeding gums with loosened teeth, and poor wound healing.

• **Metabolism of folic acid (folacin)**—Vitamin C is required for the conversion of the inactive form of the vitamin, folic acid (folacin), to its active form, folinic acid. When there is an insufficiency of vitamin C in the diet, the metabolism of folic acid (folacin) is impaired and the megaloblastic anemia that occurs in scurvy, and sometimes in infancy, may result.

In addition to the above functions of vitamin C, investigators have suggested a number of other roles for vitamin C. Although the evidence for most of these functions is not conclusive, a brief summary of each of them follows in the hope that it will stimulate further research:

• **Synthesis or release of the steroid hormones; stress**—It has been observed (1) that the normally high concentration of ascorbic acid in the adrenal glands is depleted with the synthesis of steroid hormones, and (2) that there is an increased requirement of vitamin C in all forms of stress—extremely high or low temperature, shock, fatigue, injury, burns, surgery, cigarette smoking, toxic levels of heavy metals (such as lead, mercury, and cadmium, etc.). Hence, it has been theorized that vitamin C is involved with either the synthesis or release of the steroid hormones by the adrenal glands; and that the greater the stress, the higher the vitamin C requirement.

Dr. James Cason, Professor of Chemistry, University of California, Berkeley, cites the following item in support of the role of ascorbic acid in resisting stress: A 150-lb goat makes about 13 g of ascorbic acid daily under ordinary circumstances, but if a goat is put under stress it will produce up to twice this amount.[9] It seems reasonable, therefore, that ascorbic acid would help man to resist stress, too.

• **Protection in infection and fever**—There are decreased tissue and blood levels of ascorbic acid during infections and fever, indicating either increased need for this vitamin or increased destruction of it. Thus, higher than normal intakes of vitamin C may be needed to provide maximum protection against infections and fevers.

• **Prevention and cure of colds and flu**—There is much controversy concerning the effectiveness of massive doses of vitamin C in the prevention and cure of the common cold and flu. Linus Pauling, a respected chemist with the rare

[9]Cason, James, "Ascorbic Acid, Amygdalin and Carcinoma," *The Vortex*, June 1978, pp. 9-23.

distinction of receiving two Nobel Prizes—one for science and the other for peace, is the most enthusiastic advocate of using vitamin C as a drug. His book *Vitamin C and the Common Cold and the Flu*, published in 1970, has had great influence. As a result of his advocacy, many people throughout the world began taking tablets of ascorbic acid. Nevertheless, as a result of studies on thousands of people, the only reasonable conclusion that can be drawn is that vitamin C has no effect on the number of colds people get, but in some people it lessens the severity of cold symptoms.

There should be an awareness that large doses of vitamin C are known to increase (1) the urinary output of oxalic acid and uric acid, and (2) the intestinal absorption of iron. Thus, massive doses of vitamin C may be hazardous to those with a liability to kidney stones or iron-storage disease.

• **Phagocyte activity and formation of antibodies**— Ascorbic acid may have a stimulating effect on the phagocytic activity, and on the formation of antibodies.

• **Reducing the requirements for certain vitamins**— Ascorbic acid (the reduced form of vitamin C) lessens animal requirements for thiamin, riboflavin, pantothenic acid, folacin, vitamin A, and vitamin E; likely, it has a similar role in man.

• **Smoking**—Recent research confirms the long-time claim that smoking lowers the blood level of vitamin C, although it is not known whether this is due to actual destruction or reduced availability of ascorbic acid. But there is no evidence that heavy smokers need more vitamin C than that supplied by the recommended dietary allowances.

• **Cancer**—Dr. Linus Pauling declares that vitamin C is beneficial to cancer victims, according to Dr. James Cason, Professor of Chemistry, University of California, Berkeley, who reported on the following cancer-vitamin C study in which Dr. Pauling was involved.[10]

In 1976, Cameron and Pauling published the results of an ongoing investigation of cancer therapy with vitamin C at a hospital in Scotland, where Dr. Cameron is physician in residence. The study involved 1,100 cancer victims who had been diagnosed as *terminal*. One hundred of these (matched for sex, age, and type of cancer for comparison with a control group) were given massive doses of 10 g/day of vitamin C in double blind placebo procedure, and the remaining 1,000 were the comparison group. The vitamin C-treated patients lived an average of four times longer than the comparison group. At the time of publication in 1976, all the comparison group were dead, while 16 of the 100 who received vitamin C were alive. Moreover, 1 year after publishing the results, Dr. Pauling reported, "13 of these 'hopeless' patients are still alive, some as long as 5 years after having been pronounced untreatable, and most of them are in such good apparent health as to suggest that they now have normal life expectancy."

But a controlled study by Dr. Edward T. Creagan, Mayo Clinic, showed no anticancer effect or improved survival as a result of megadoses of vitamin C. Likewise, Drs. J. Roberto Moran and Henry L. Greene, researchers at Vanderbilt University School of Medicine

concluded from a study of the benefits and risks of vitamin C megadoses that "at present no strong evidence can be found to support the routine prophylactic (preventive) use of ascorbic acid in well-nourished people."

• **Removal of ammonia in the deamination of proteins and peptides**—Vitamin C appears to accelerate the deamination of proteins and peptides and the conversion of ammonia (NH_3) to urea for excretion. Some have even conjectured that these processes (the oxidative deamination and the urea cycle) affect aging and longevity.

• **Antihistamine**—Ascorbic acid is an antihistamine; hence, it may be effective in treating respiratory infections due to a histamine.

• **Detoxifying drugs**—Vitamin C appears to be involved in a set of biochemical reactions responsible for detoxifying drugs, and for eliminating them from the body. Specifically, the vitamin may facilitate steps in which iron is introduced into the heme groups that subsequently become part of the proteins that carry out the detoxification reactions.

• **Longevity**—In 1992, the UCLA School of Public Health reported in the *Journal of Epidemiology* that (1) daily intake of 300 mg of vitamin C from food and supplements may increase life expectancy in men by 6 years, and (2) daily intake of 150 mg vitamin C from food, without supplements, may increase life expectancy in men by 2 years. The study focused on 11,000 adults, 25 to 74 years of age.

DETERMINING VITAMIN C STATUS. Vitamin C status in people may be determined by clinical signs and by blood levels of the vitamin. Evidence of capillary bleeding in the skin (perifolliculosis) and in the gums are clinical signs that may indicate a vitamin C deficiency.

Different investigators have employed a variety of tests for estimating the vitamin C nutrition of man and animals. The simplest of these is measurement of the L-ascorbic acid content of serum or plasma. Approximate values of L-ascorbic acid of man which may be used for guide purposes follow:

Nutritional Status	Serum or Plasma Concentration (mg/100 ml)
Well nourished	over 0.60
Adequate	0.40–0.59
Low	0.10–0.39
Deficient	under 0.10

In a study of the ascorbic acid level per 100 ml of blood plasma of 48 women and 41 men, the senior author of this book, Audrey H. Ensminger, found 14 subjects, or 15.7%, at the scurvy level.

DEFICIENCY SYMPTOMS. When deprived of a dietary source of vitamin C for a sufficient length of time, man, along with other primates and several other species, develops scurvy, a potentially fatal disease.

• **Early symptoms, called latent scurvy**—Early symptoms of vitamin C deficiency, called latent scurvy, include: loss of weight, listlessness, fatigue, fleeting pains in the joints and muscles, irritability, shortness of breath, sore and/or bleeding gums, small hemorrhages under the skin, bones that fracture easily, and poor wound healing.

---

[10]Cason, James, "Ascorbic Acid, Amygdalin and Carcinoma," *The Vortex*, June 1978, pp. 9-23.

Fig. V-43. Scorbutic gums. (Reproduced with permission of *Nutrition Today*, P.O. Box 1829, Annapolis, MD, 21404, ©, 1979)

• **Scurvy**—A severe deficiency of vitamin C results in acute scurvy, characterized by: swollen, bleeding, and ulcerated gums; loose teeth; malformed and weak bones; fragility of the capillaries with resulting hemorrhages throughout the body; large bruises; big joints, such as the knee and hip, due to bleeding into the joint cavity; anemia; degeneration of muscle fibers, including those of the heart; and tendency of old wounds to become red and break open. Sudden death from severe internal hemorrhage and heart failure is always a danger.

RECOMMENDED DAILY ALLOWANCE OF VITAMIN C. Many studies have been conducted to determine the human vitamin C requirements. Consideration has been given to the effect of age, environment, physical exertion, infections, and fevers. Also, various measurements for determining the adequacy of vitamin C have been devised, including: (1) the daily intake of vitamin C necessary to prevent the symptoms of scurvy; (2) the amount of vitamin C required to saturate whole blood, blood plasma, white blood cells, or body tissue; and (3) the intake of vitamin C necessary to maintain blood and urinary ascorbic acid levels within normal range. The results of these studies vary widely.

A daily intake of 10 mg of ascorbic acid will prevent scurvy. But this should be regarded as a minimum level. In order to provide for individual differences and margins of safety, the Food and Nutrition Board of the National Academy of Sciences recommends the following allowances: 30–35 mg for infants, 40–45 mg for children, 60 mg for males and females over 14 years, 70 mg for pregnancy, and 95 mg for lactation. (See the section on VITAMIN[S], Table V-5, Vitamin Table.)

It is noteworthy that the joint FAO/WHO Expert Committee makes somewhat lower recommendations than those given in Table V-5: 20 mg for infants and children up to 13 years of age, 30 mg for adults (males and females over 13 years), and 50 mg during pregnancy and lactation.

It is recognized, however, that these allowances are not necessarily adequate to meet the additional requirements of persons depleted by disease, traumatic stress, or prior dietary inadequacies. Also, the recommended levels given in Table V-5 may not be sufficient to assure vigorous good health.

• **Recommended allowances for infants, children, and adolescents**—A dietary allowance of 30 mg per day is recommended from birth to six months, and 35 mg per day from six months to twelve months. This is based on the fact that (1) human milk contains 30 to 55 mg/liter of vitamin C, although it varies with the mother's dietary intake of the vitamin; and (2) the breast-fed infant receives approximately 850 ml of milk per day. However, newborn infants, especially if they are premature, may have an increased requirement for the metabolism of tyrosine during the first week of life.

For children 1 to 3 years, an allowance of 40 mg/day is recommended; up to the age of 11 years, an allowance of 45 mg/day of vitamin C is recommended. For older children, an allowance of 50 mg/day is recommended as adequate to meet individual needs and to provide a margin of safety.

• **Recommended allowance for adults**—A dietary allowance of 60 mg of vitamin C per day is recommended for adults of both sexes. This will maintain an ascorbate body pool of 1,500 mg—a body pool of sufficient magnitude to protect against signs of scurvy in the adult male for a period of 30 to 45 days, and allow for an ascorbate catabolism rate of 3 to 4% and an average ascorbate absorption efficiency of approximately 85%.

• **Recommended allowances for pregnancy and lactation**—During pregnancy, plasma vitamin C levels fall. It is not known whether this is due to a physiological response to pregnancy and/or to increased demands of pregnancy. It is known that the placenta normally transmits sufficient ascorbic acid from mother to fetus to result in fetal levels 50% greater than maternal levels at birth. To provide for this fetal need, an additional allowance of 10 mg of ascorbic acid per day is recommended for pregnant women, particularly during the second and third trimester of pregnancy.

Human milk from well-nourished women is relatively high in ascorbic acid, but it varies with the mother's dietary intake of the nutrient. During lactation, a daily loss of 25 to 45 mg of vitamin C may occur in the secretion of 850 ml of milk. So, for lactating women, an additional allowance of 35 mg/day of vitamin C is recommended for the first six months in order to assure a satisfactory level of the vitamin in breast milk.

• **Massive doses of ascorbic acid**—Intakes of ascorbic acid in excess of 1,000 mg/day (1 g/day) or more have been reported to have some effect in reducing the frequency and severity of symptoms of colds and flu. To date, the results of research work have generally shown that the benefits of large doses of vitamin C are too small to justify recommending routine intake of large amounts. But further studies are needed.

Large doses of ascorbic acid have been reported to lower serum cholesterol in some hypercholesterolemic subjects, but not in others.

Ascorbic acid supplements can prevent the reduced platelet and plasma concentrations of ascorbic acid observed in aspirin-treated rheumatoid-arthritis patients.

The use of massive doses of vitamin C to improve the performance of athletes has long been a controversial issue. Present findings indicate that the vitamin is ineffective for this purpose, and that large doses may have a negative effect on athletic performance by disturbing the equilibrium between oxygen transport and oxygen utilization.

Large doses of ascorbic acid have generally been considered nontoxic, except for gastrointestinal symptoms experienced by some people. However, a number of adverse effects of excessive intakes of ascorbic acid have been reported, such as acid-induced uricosuria, absorption of excessive amounts of food iron, and impaired bactericidal activity of leucocytes.

Since many of the claims of significant beneficial effects of large intakes of ascorbic acid have not been sufficiently substantiated, and since excessive intakes may have some adverse effects, routine consumption of large intakes of ascorbic acid is not recommended without medical advice.

● **Vitamin C intake in average U.S. diet**—According to the U.S. Department of Agriculture, the foods available for civilian consumption in the United States provide 120 mg of vitamin C per person per day. Of this amount, fruits (especially citrus fruits) and vegetables provide 91% of the total.

TOXICITY. Doses of up to 2 g per day (which is more than 30 times the recommended daily allowance) of ascorbic acid are nontoxic to adults. However, in amounts of 2 to 8 g per day, caution should be exercised; and there is clear evidence that intakes in excess of 8 g per day (more than 100 times the recommended daily allowance) may be distinctly harmful.

A number of adverse effects of excessive intakes of vitamin C have been reported, such as: nausea; abdominal cramps and diarrhea; absorption of excessive amounts of food iron; destruction of red blood cells; increased mobilization of bone minerals; interference with anticoagulant therapy; formation of kidney and bladder stones; the inactivation of vitamin B-12; a rise in plasma cholesterol; and possible dependence upon large doses of vitamin C (small doses no longer meet nutritional needs). It is also noteworthy that undesirable side effects may be greater in certain physiological states (e.g. pregnancy).

Since excessive intakes of vitamin C may be hazardous, routine consumption of large amounts (above 2 g daily by adults) of the vitamin is not recommended without medical advice.

LOSSES OF VITAMIN C DURING PROCESSING, COOKING, AND STORAGE. Of all the vitamins, ascorbic acid is the most unstable. It is easily destroyed during harvesting, processing, cooking, and storage, because it is water-soluble, easily oxidized, and attacked by enzymes. Thus, a warm environment, exposure to air, solubility in water, heat, alkalinity, and dehydration are detrimental to the retention of ascorbic acid in foods. Also, cutting of vegetables releases an enzyme and increases the leaching by water. Hence, foods may lose much of their original vitamin C content from the time they are harvested until they are eaten. Details follow:

● **Processing losses**—The method of preparing fruits and vegetables affects the amount of the vitamin. Much vitamin C is lost when the products are washed slowly, cut up into small pieces, and soaked after peeling. In preparing foods for quick freezing, canning, or drying, a brief blanching with steam favors retention of vitamin C, because this process destroys the enzymes that hasten destruction of the vitamin in raw foods. The least loss of vitamin C occurs when foods are preserved by quick freezing; the most loss occurs when foods are preserved by drying, especially if they are exposed to sunlight. Losses of vitamin C from drying may be lessened by sulfuring before drying and by rapid dehydration (away from sunlight). Manufacturers of canned and frozen fruits

and vegetables should take special care to use products of high quality, then process them quickly. If this precaution is taken—if the products reach the cannery fresh from nearby fields and are heated quickly in vacuum-sealed cans—commercially canned fruits and vegetables may compare favorably in vitamin C content with home-cooked products. The vitamin C content of canned fruit juices varies considerably, unless they are specially protected in processing or fortified with vitamin C.

● **Cooking losses**—There is a great deal of variation in the amount of vitamin C lost in home cooking, depending on the nature of the food, the reaction (acid or alkaline), the length of time and the degree of heating, and the extent to which the food is exposed to water and air in the cooking process.

To retain a maximum of the ascorbic acid, frozen fruits should be used promptly, and frozen vegetables should be plunged directly into boiling water for immediate cooking. Losses may be minimized by cooking with peel left on or with the product in large pieces, by cooking with as much exclusion of air as possible (e.g., using a tightly covered vessel or pressure cooker); by boiling the cooking water for a minute before adding the food; by shortening the boiling time; by using a small amount of water; and by consuming the cooking water.

Increased cooking losses of vitamin C result from cooking in copper or iron utensils (the ions of which inactivate the vitamin); from adding baking soda to vegetables to retain the green color (as an alkaline medium facilitates oxidation); from mashing the food and leaving it in a hot place or exposed to air; or from holding cooked foods warm for prolonged periods of time, such as on hot plates or on steam tables in cafeterias.

● **Storage losses**—Losses of vitamin C occur during prolonged storage, whether at home or in a market, especially if the product is damaged or is in a warm place. New potatoes, which contain about 30 mg/100 g of ascorbic acid, may lose 75% of the vitamin during 9 months of storage. Leafy vegetables (with large surface areas) lose more vitamin C in storage than do roots and tubers. Refrigeration during storage reduces losses. In markets, more of the vitamins of vegetables are retained when they are kept in crushed ice than when they are kept in a refrigerator.

Citrus fruit juices stored in the refrigerator lose negligible amounts of the vitamin; the acid content of the juice helps preserve vitamin C.

HOW TO CONSERVE VITAMIN C IN FOODS IN THE HOME. Since vitamin C is essential for health, its maximum value should be conserved in foods. Losses may be minimized by keeping in mind that vitamin C is water-soluble and easily destroyed by oxidation, and that heat, alkalinity, and exposure to air hasten its destruction.

Practical suggestions for conserving vitamin C in foods in the home follow:

● Buy fresh fruits and vegetables in small quantities so that they will be used promptly. Store them in the refrigerator.

● Prepare foods immediately before they are to be served raw or cooked; do a minimum of chopping and cutting, and cook with the skins left on when possible; and do not allow foods to be exposed to air or stand in water before cooking.

● Use frozen foods promptly.

● Do not thaw frozen vegetables before cooking; keep them in the refrigerator until ready to cook, then, in their frozen state, plunge them directly into a limited amount of boiling water for immediate cooking.

● Cook in a small quantity of water, for as short a period of time as feasible, in a tightly covered cooking vessel; cook by steaming or broiling (instead of boiling).

● Never cook in copper or iron pans, and do not add soda in cooking; copper, iron, and soda hasten vitamin C destruction.

● Serve vegetables as soon after cooking as possible.

● Prepare fresh fruit juices immediately before serving. It is noteworthy, however, that acid juices (orange, grapefruit, tomato) may be left in a covered glass container in the refrigerator for several days with little loss in vitamin C.

SOURCES OF VITAMIN C.

Vitamin C occurs primarily in foods of plant origin—fruits (especially citrus fruits) and vegetables; those that may be eaten fresh, uncooked, or previously frozen are the best sources.

Contrary to common opinion, it takes 3 times as much tomato juice as citrus juice to supply the same amount of vitamin C.

If the mother's diet has contained sufficient vitamin C, human milk will contain 4 to 6 times as much ascorbic acid as cow's milk and will protect the infant from scurvy.

It is noteworthy that the vitamin content of plant foods varies greatly, depending on such factors as variety, climate, amount of sunshine, stage of maturity, part of plant (little is found in dry seeds), and length of storage. In general, the more sunshine to which a plant is exposed, the higher the vitamin C content; and the more mature the plant, the lower the vitamin C content.

A good rule to follow in order to assure sufficient vitamin C in the diet is to include a daily serving of citrus fruit or juice.

Fig. V-44. Rich, natural sources of vitamin C. (Courtesy, Louisiana Cooperative Extension Service, Louisiana State University, Baton Rouge, La.)

● **Richest natural sources**—The richest natural sources of vitamin C are the acerola cherry, *camu-camu*, and rose hips. Rose hips, which form the base of the rose bloom, are not eaten as such; rather, they are either made into a syrup (or extract) or brewed as a tea. During World War II food rationing in England, rose hip syrup was issued by the British Ministry of Food to help fortify the vitamin C intake of the English people. The acerola cherry, commonly called the Barbados cherry or West Indian cherry, which is grown in Florida, Hawaii, and Puerto Rico, has the highest ascorbic acid content of any known food. The *camu-camu* is native to the jungles of Peru. Also, pine needles, which are rich in vitamin C, have long been extracted (brewed) and used to prevent scurvy by the Indians of Canada and the northern Russians.

● **Synthetic ascorbic acid**—Pure ascorbic acid is available wherever vitamins are sold, at a cost of as little as ½¢ or less per 100 mg (which is more than the recommended daily allowance). It is less expensive than an equivalent amount in natural foods, but natural foods also supply a variety of minerals and other vitamins; hence, vitamin C supplements should be used to augment, rather than replace, natural food sources. Also, the satiety derived from eating fresh strawberries is not experienced when swallowing a tablet or a capsule. Nevertheless, the body utilizes synthetic vitamin C as effectively as it does the vitamin C in foods.

Groupings by rank of foods according to vitamin C values are given in the section on VITAMIN(S), Table V-5, Vitamin Table.

For additional sources and more precise values of vitamin C, see Food Composition Table F-21.

(Also see VITAMIN[S], Table V-5.)

VITAMIN D

Contents	Page
History	1092
Chemistry, Metabolism, Properties	1093
Measurement/Assay	1094
Functions	1094
Deficiency Symptoms	1095
Recommended Daily Allowance of Vitamin D	1096
Toxicity (Hypervitaminosis D)	1096
Vitamin D Losses During Processing, Cooking, and Storage	1097
Sources of Vitamin D	1097

The importance of vitamin D—the sunshine vitamin—in human nutrition lies in the role of regulating calcium and phosphorus metabolism. Vitamin D promotes intestinal absorption of calcium and phosphorus and influences the process of bone mineralization. In the absence of vitamin D, mineralization of bone matrix is impaired, resulting in rickets in children and osteomalacia in adults. Although rickets is rare in the United States, it is still prevalent in many countries.

A bone disorder, which we now call rickets, has been known since 500 B.C. But the disease was first properly described in London about 300 years ago. The word *rickets* is derived from the Old English word *wrikken*, meaning to bend or twist.

Vitamin D is unique among the vitamins in two respects: (1) it occurs naturally in only a few common foods (mainly in fish oils, and a little in liver, eggs, and milk), and (2) it can be formed in the body by exposure of the skin to ultraviolet rays of the sun—light of short wavelength and high frequency; hence, it is known as the *sunshine vitamin*.

Ten Top Sources

Fig. V-45. Factors that inhibit the sunshine vitamin—that screen out light and prevent the formation of vitamin D. Note, too, top food sources of vitamin D.

HISTORY. The history of rickets as a deficiency disease is much older than our knowledge of how to prevent it. During the Industrial Revolution in England in the 1600s, the disease became very prevalent in children in the crowded slums. Industrial smoke and high tenement buildings shut out the sunlight; so, as industrial cities grew, rickets spread. But no one blamed lack of sunshine (vitamin D) for the crippling disease. Instead, it was attributed to bad home environment and hygiene, and it became known as "a disease of poverty and darkness."

In 1824, cod-liver oil, long known as a folk medicine, was found to be important in the treatment of rickets. But the remedy lost favor with the medical profession because physicians could not explain its action.

As early as 1890, Palm, an English physician, observed that where sunshine was abundant, rickets was rare, but where the sun seldom shone, rickets was common.

In 1918, Sir Edward Mellanby of England demonstrated that rickets was a nutritional deficiency disease. He produced rickets in puppies, then cured it by giving them cod-liver oil. But Mellanby incorrectly attributed the cure to the newly discovered fat-soluble vitamin A.

In 1922, McCollum at Johns Hopkins University found that, after destruction of all the vitamin A in cod-liver oil, (oxidation, by passing heated air through cod-liver oil) it still retained its ricket-preventing potency. This proved the existence of a second fat-soluble vitamin, carried in liver oils and certain other fats, which he called *calcium-depositing vitamin*. It is of interest to note that, though McCollum discovered the existence of vitamin D, he did not call it by this name until after this designation was in common use by others.

In 1924, the mystery of how sunlight could prevent rickets was partially solved. Dr. Harry Steenbock of the University of Wisconsin and Dr. A. Hess of Columbia University, working independently, showed that the antirachitic activity could be produced in foods and in animals by ultraviolet light. The process, known as the Steenbock Irradiation Process was patented by Steenbock, with the royalties assigned to the Wisconsin Alumni Research Foundation of the University of Wisconsin. Subsequent research disclosed that it was certain sterols in foods and animal tissues that acquired antirachitic activity upon being irradiated. Before irradiation, the sterols were not protective against rickets.

By the late 1920s, it had been established that rickets could be prevented and cured by exposure to direct sunlight (ever since, vitamin D has been popularly called *the sunshine vitamin*), by irradiation with ultraviolet light, by feeding irradiated food, or by feeding cod-liver oil. Later, the natural vitamin D of fish liver oils was identified as the same substance that is produced in the skin by irradiation.

In 1932, crystals of pure vitamin D_2 (ergocalciferol) were isolated from irradiated ergosterol by Windaus of Germany and Askew of England; and in 1936, crystals of pure vitamin

D_3 (cholecalciferol) were isolated from tuna liver oil by Brockmann of Germany.

In 1952, the first total synthesis of a form of vitamin D (in this case vitamin D_3) was accomplished by R. B. Woodward of Harvard. He was awarded the Nobel Prize in Chemistry in 1965 for this and other similar achievements.

CHEMISTRY, METABOLISM, PROPERTIES.
Although about 10 sterol compounds with vitamin D activity have been identified, only two of these, known as provitamins D or precursors, are of practical importance today from the standpoint of their occurrence in foods—ergocalciferol (vitamin D_2, calciferol, or viosterol) and cholecalciferol (vitamin D_3); the name cholecalciferol of the latter is a reflection of its cholesterol precursor. Because these substances are closely related chemically, the term vitamin D is used collectively to indicate the group of substances that show this vitamin activity.

• **Chemistry**—Fig. V-46 shows the structures of vitamins D_2 and D_3.

IN VITAMIN D_2 (ERGOCALCIFEROL) R = CH_3–CH=CH–CH–CH=CH with CH_3 / CH_3 branch

IN VITAMIN D_3 (CHOLECALCIFEROL) R = CH_3–CH–CH_2–CH_2–CH_2–CH with CH_3 / CH_3 branch

Fig. V-46. Structure of vitamin D_2 and vitamin D_3.

Ultraviolet irradiation of the two provitamins—ergosterol and 7-dehydrocholesterol—will produce vitamins D_2 and D_3, respectively. Ergosterol is found in plants (in yeasts and fungi), whereas 7-dehydrocholesterol is found in fish liver oils and in the skin of man and other animals. Therefore, people or animals that are exposed to sunlight for extended periods of time do not need dietary supplementation of vitamin D. Both forms of vitamin D, D_2 and D_3, have equal activity for people and most other mammalian species. But chickens, turkeys, and other birds are exceptions—they utilize vitamin D_3 more efficiently than vitamin D_2.

• **Metabolism**—Vitamin D is unique in that man and animals normally obtain it from two sources; formation in the skin, and by mouth. The steps involved in the metabolism of vitamin D follow:

1. **Formation of vitamin D_3 in the skin and its movement into the circulation.** The unique mechanism for the synthesis, storage, and slow, steady release of vitamin D_3 from the skin into the circulation is shown in Fig. V-47.

When the skin is exposed to the ultraviolet radiation of sunlight, part of the store of 7-dehydrocholesterol undergoes a photochemical reaction in the epidermis and the dermis and forms previtamin D_3. Once previtamin D_3 is formed in the skin, it undergoes a slow temperature-dependent transformation to vitamin D_3, which takes at least 3 days to complete. Then, the vitamin D-binding protein transports D_3 from the skin into the circulation.

2. **Absorption.** Vitamin D taken by mouth is absorbed with fats from the small intestine (from the jejunum and ileum), with the aid of bile. Vitamin D formed in the skin by irradiation of the provitamin present there is absorbed directly into the circulatory system.

3. **Utilization.** Cholecalciferol—obtained either from the diet or from the irradiation of the skin—is transported by a specific vitamin D carrier protein (a globulin) to the liver where it is converted to 25-hydroxycholecalciferol (25-OH-

Fig. V-47. Diagram showing the sequence of steps in the formation of vitamin D_3 in the skin and its transport into the circulation.

D₃).[11] From the liver 25-OH-D₃ is transported to the kidneys where it is converted to 1,25-(OH)₂-D₃, the most active form of vitamin D in increasing calcium absorption, bone calcium mobilization, and increased intestinal phosphate absorption. The active compound 1,25-(OH)₂-D₃ functions as a hormone, since it is a vital substance made in the body tissues (the kidneys) and transported in the blood to cells within target tissues. This physiological active form of vitamin D₃ is then either transported to its various sites of action or converted to its metabolite forms of 24,25-dihydroxycholecalciferol or 1,24,25-trihydroxycholecalciferol (see Fig. V-48).

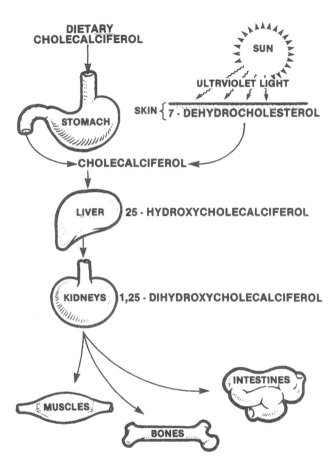

Fig. V-48. Fate of cholecalciferol in the body.

Although most of the research on vitamin D metabolism has been conducted on cholecalciferol, studies by DeLuca on ergocalciferol indicate that it is metabolized similarly to cholecalciferol; that it is changed to a similar active metabolite in the liver—25-hydroxyergocalciferol (25-OH-D₂).

4. **Storage.** The major storage sites of vitamin D are the fatty tissues and skeletal muscle. Some of it is also found in the liver, brain, lungs, spleen, bones, and skin. But body storage of vitamin D is much more limited than the storage of vitamin A.

5. **Excretion.** The main pathway of excretion of vitamin D is by way of the bile into the small intestine, thence the feces. Less than 4% of the intake is excreted in the urine.

[11]Identified by DeLuca and co-workers of the University of Wisconsin. First reported in Proc. Nat. Acad. Sci., 61:1503, 1968.

• **Properties**—Pure D vitamins are white, crystalline, odorless substances that are soluble in both fats and fat solvents (such as ether, chloroform, acetone, and alcohol). They are insoluble in water, and they are resistant to heat, oxidation, acid, and alkali.

Although the precursors are activated by ultraviolet light, excessive irradiation results in the formation of slightly toxic compounds that have no antirachitic activity.

MEASUREMENT/ASSAY. Vitamin D potency is expressed in International Units (IU) and United States Pharmacopeia Units (USP), which are equal. One IU, or one USP, of vitamin D is defined as the activity of 0.025 mcg of pure crystalline vitamin D₃ (cholecalciferol).

The ultraviolet light absorption property of vitamin D may be used for the assay of pure preparations free of irrelevant absorption. But it does not distinguish between vitamin D₂ and D₃.

Both vitamin D₂ and D₃ give a yellow-orange color with antimony trichloride. This color reaction forms the basis of the USP XVIII method for vitamin D. Since color reactions are subject to interferences from many sources, they should be limited to high potency pharmaceutical preparations or fortified foods. Combinations of column and thin-layer chromatographic purification steps with the antimony trichloride reaction have been successfully used.

Gas-liquid chromatography provides a means of combining qualitative and quantitative assays.

Food and feed samples, which contain vitamin D in very low concentrations, are not usually assayed by chemical methods. For these substances, bioassays are the only means available for the assessment of vitamin D activity. Rats and chicks are used as test animals; rats respond equally well to D₂ and D₃, whereas chicks respond only to D₃. The assays measure the alleviation (curative test) or the development (prophylactic test) of vitamin D deficiency in terms of the degree of rickets produced.

A bioassay method, known as the *line test*, uses stained longitudinal sections of the distal end of radius bones to evaluate calcification. Usually the test animal is the rat, although the chick must be used if the vitamin D activity of a sample intended for poultry nutrition is to be determined. Young rats from mothers having a deficient supply of vitamin D are kept on a rachitogenic diet so that no calcification occurs in the ends of the long bones. When a test material is fed to these vitamin D-deficient rats, its value as a source of vitamin D is measured by the amount that must be fed for 7 to 10 days to produce a good calcium line (line test) in the ends of the long bones. Standard cod-liver oil is fed to a similar group of animals and is used as a basis of comparison.

FUNCTIONS. Vitamin D is primarily associated with calcium and phosphorus. It influences the absorption of these minerals and their deposit in bone tissue. Research is continuing to unfold the relationship of vitamin D to calcium and phosphorus metabolism. Although there are many gaps in our knowledge relative to the exact mechanism by which vitamin D carries out its various physiological functions, the current thinking is as follows:

• **Calcium absorption**—It has been clearly established that vitamin D increases calcium absorption from the small intestine, and that a vitamin D deficiency produces large calcium losses in the feces.

• **Phosphate level**—Adequate vitamin D enhances the levels of phosphates in the body, because of (1) improved absorption of phosphorus through the intestinal wall, independent of calcium absorption; and (2) increased resorption of phosphates from the kidney tubules. When sufficient vitamin D is not available, urinary excretion of phosphate increases and the blood level drops.

Maintenance of a satisfactory phosphate level, and of the vital balance between calcium and phosphorus in the blood, is essential (1) to the process of bone calcification and (2) to the prevention of tetany.

• **Bone and teeth metabolism**—The growth and proper mineralization of the bones and teeth require vitamin D.

Lack of vitamin D or lack of exposure to sunlight of children results in weak bones and overgrowth of the softer tissues (cartilage) at the ends of the bones. The joints enlarge, and bowed legs, knock knees, beaded ribs, and skull deformities may result. Lack of vitamin D in adults may result in osteomalacia, in which changes occur in the shafts of bones and bone structure softens. Adequate vitamin D is also important during reproduction and lactation. Extreme deficiencies may result in congenital malformations of the newborn and injury to the skeleton of the mother.

With lack of vitamin D, the enamel and the dentin of the teeth, which are composed almost entirely of calcium and phosphorus, may not develop normally. Thus the teeth of rachitic children and animals have thin, poorly calcified enamel, with pits and fissures, and are especially prone to decay.

The withdrawal of calcium and phosphorus storage from bone (resorption) is stimulated by the action of vitamin D. In this way, vitamin D helps to maintain the blood levels of the two minerals.

• **Citrate metabolism**—The level of citrate in the blood decreases in a vitamin D deficiency. Citrate is an important organic acid involved in many metabolic functions, including mobilization of minerals from bone tissue and removal of calcium from the blood.

The effect on citrate metabolism is thought to be due to changes in mineral metabolism brought about by the absence or presence of vitamin D rather than to a direct action of the vitamin on the formation of citrate.

• **Amino acid levels in the blood**—Vitamin D is involved in the amino acid levels in the blood, by protecting against the loss of amino acids through the kidneys. When there is a deficiency of vitamin D, the excretion of amino acids in the urine is increased.

DEFICIENCY SYMPTOMS. A deficiency of vitamin D leads to inadequate absorption of calcium and phosphorus from the intestinal tract and to faulty mineralization of the bones and teeth, followed by skeletal malformations. The major deficiency symptoms follow:

• **Rickets**—Lack of vitamin D will cause rickets in infants and children, even though the diet is adequate in calcium and phosphorus. (Rickets may also be caused by lack of either calcium or phosphorus, or an incorrect ratio of the two minerals.) This disease is caused from failure of the bones to calcify normally (meaning that the deposition of calcium and phosphorus salts is not normal). As a result, they are soft and pliable and become deformed. The weight of the body causes the ends of the long bones of the legs

to flatten and mushroom outward, giving the appearance of enlarged knee and ankle joints. Other bone changes occurring in rickets include enlarged wrist joints; bowed legs; knocked knees; outward projection of the sternum (*pigeon breast*); a row of beadlike projections on each side of the chest at the juncture of the rib bones and joining (costal) cartilage (called *rachitic rosary*); delayed closure of the fontanel of the skull, causing a bulging of the forehead; narrowing of the pelvis so as to make childbirth difficult in subsequent years; and spinal curvature.

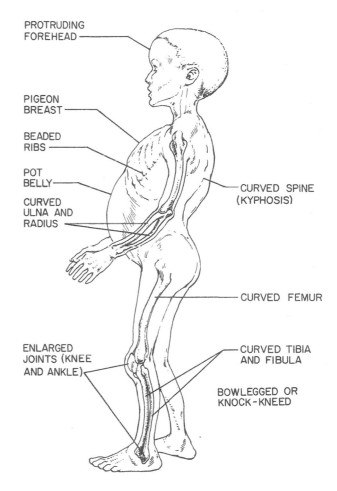

PROTRUDING FOREHEAD

PIGEON BREAST

BEADED RIBS

POT BELLY

CURVED ULNA AND RADIUS

CURVED SPINE (KYPHOSIS)

CURVED FEMUR

CURVED TIBIA AND FIBULA

BOWLEGGED OR KNOCK-KNEED

ENLARGED JOINTS (KNEE AND ANKLE)

Fig. V-49. Symptoms of severe rickets.

Sometimes muscles are poorly developed. Often rickets is associated with protrusion of the abdomen, or *pot belly*, due to weakness of the abdominal muscles.

Eruption of the temporary teeth is delayed; and the permanent teeth may be thin, pitted, and grooved, with the result that they decay easily.

Since the 1930s, the incidence of rickets in the United States has declined, largely because of the addition of vitamin D supplements to breast-fed infants and of the fortification of infant formulas and fluid and evaporated milk with vitamin D. Since 1968, nonfat dry milk for use in various food programs has also been fortified with vitamin D (and vitamin A).

The best protection against rickets and the most favorable bone growth are secured when calcium and phosphorus are supplied in approximately equal amounts (as in milk) and when liberal quantities of vitamin D are available.

When preventive measures are not taken, rickets occurs in northern regions, especially in dark, crowded cities during the winter months, where the ultraviolet rays of sunshine cannot penetrate through the fog, smoke, and soot. Dark-skinned children are more susceptible to rickets than those of the white race; and premature infants are more susceptible than full-term infants, because of the additional demands for vitamin D imposed by increased growth rate and calcification of the bones.

Rickets is still a serious problem in many tropical and semitropical countries and among the poor in overcrowded sections of large cities. In Africa, rickets may be caused by the custom of not exposing babies to sunlight, little vitamin D or calcium in the diet, intestinal parasites, and recurrent diarrhea.

• **Tetany**—A deficiency of vitamin D may cause tetany, though it is not the only cause. Tetany may also result from insufficient absorption of calcium or from a disturbance of the parathyroid gland. Tetany is characterized by muscle twitching, cramps, convulsions, and low serum calcium—less than 7 mg per 100 ml.

• **Osteomalacia**—This is the adult form of rickets, caused by depletion of the bone stores of calcium and phosphorus, as a result of a lack of vitamin D or exposure to sunlight, or lack of either calcium or phosphorus. It is most common during pregnancy, lactation, and old age. Since longitudinal bone growth has stopped in adults, only the shafts on long and flat bones, such as the pelvis, are affected. The bones soften, become distorted, and fracture easily. Also, osteomalacia generally causes rheumatic-type pain in the bones of the legs and lower back.

RECOMMENDED DAILY ALLOWANCE OF VITAMIN D.

The minimum vitamin D requirements are not known, mainly because information on the amount of cholecalciferol formed by the action of sunlight on the skin is not available. The amount formed varies with the length and intensity of exposure to sunlight; the color of skin (dark-pigmented skins can prevent as much as 95% of the sun's ultraviolet light from penetrating into the skin and forming cholecalciferol); the amount of window glass and clothing, which prevent ultraviolet rays from reaching the skin; the season (sunlight is usually scarce during the winter months, when there are cloudy and foggy days); and the amount of atmospheric pollution (smog, smoke, dust).

The Food and Nutrition Board (FNB) of the National Research Council (NRC) recommended dietary allowances of vitamin D are given in the section on VITAMIN(S), Table V-5, Vitamin Table.

• **Infants and children**—The vitamin D content of human milk has long been regarded as inadequate for infant needs. Thus, an allowance of 7.5 to 10 mcg should be provided to both breast-fed and bottle-fed infants, either as an oral supplement or included in the formula, particularly since exposure to sunlight of infants is often inadequate.

• **Adults**—Young adults, ages 19–24, may still be growing, therefore their requirements remain at 10 mcg. With the cessation of skeletal growth, calcium needs decrease, and with it, the need for vitamin D. So, the recommended allowance for vitamin D is reduced to 5 mcg after the age of 25 years. Actually, the requirements for the normal adult can usually be met by adequate exposure to sunlight. However, solar radiation may be inadequate under certain climatic conditions or because of chronic air pollution, under which circumstances a dietary source may be necessary.

• **Pregnancy and lactation**—Calcium needs increase during gestation. Moreover, vitamin D and its active metabolites cross the placenta readily. The recommended allowance for both pregnancy and lactation is 10 mcg.

Because vitamin D is a potentially toxic substance, and because there is a lack of evidence that amounts above the recommended allowance confer health benefits, intakes should closely approximate the recommended allowances. Only individuals with diseases affecting vitamin D absorption or metabolism should take vitamin D in excess of the recommended allowances; even then, such treatment should be on the recommendation, and under the supervision, of a physician or a nutritionist.

• **Vitamin D intake in the United States**—It is difficult to estimate the average daily intake of vitamin D per person in the United States because there is no way in which to determine the amount manufactured in the body by the action of sunlight. However, because of the widespread use of fortified milk and other foods, intakes of at least 10 mcg daily per person appear probable.

TOXICITY (HYPERVITAMINOSIS D).

There is wide individual difference in the tolerance to vitamin D; hence, it is impossible to give the level at which vitamin D intake becomes toxic. But, generally speaking, the intake of vitamin D should not exceed a total of 400 IU per day. Excessive amounts of vitamin D (above 2,000 IU/day) can lead to hypercalcemia (toxicity caused by elevated blood levels of calcium) as a result of increased intestinal absorption of calcium. Symptoms of mild toxicity include loss of appetite, excessive thirst, nausea, vomiting, irritability, weakness, constipation (which may alternate with bouts of diarrhea), retarded growth in infants and children, and weight loss in adults. Chronic hypercalcemia results in abnormal deposition of calcium in the soft tissues (including the heart, blood vessels, lungs, and tubules of the kidneys), with particular damage to the kidneys. If massive doses of vitamin D are continued, widespread calcification of the soft tissues may ultimately prove fatal.

Excessive intakes of vitamin D during pregnancy and early infancy may cause idiopathic hypercalcemia, or the hypercalcemic syndrome, a condition characterized by narrowing of the aortic valve of the heart, a peculiar facial appearance, and mental retardation.

The ingestion of excess vitamin D is due mainly to the use of dietary supplements that contain large amounts of the vitamin. Because of the vitamin D added to milk and infant formulas, as well as the vitamin D produced by exposure to sunlight, parents should consider all these sources before giving their children vitamin D supplements.

Adults should also be warned about taking massive doses of vitamin D (as well as about taking massive doses of vitamin A) over a long period of time unless it is under the supervision of a physician or nutritionist. Even moderate overdosage of vitamin D is not wise for the elderly.

The concurrent administration of large quantities of vitamin A with potentially toxic levels of vitamin D will reduce the toxicity of the latter.

VITAMIN D LOSSES DURING PROCESSING, COOKING, AND STORAGE.
Vitamin D in foods and food supplements is remarkably stable. Processing does not affect its activity; and because vitamin D is stable to heat and insoluble in water, there is little loss in cooking foods. Foods containing vitamin D can be stored for long periods of time with little deterioration.

The commercial forms of vitamin D_2 and D_3 supplements, which come in either oily solutions or powders, are destroyed relatively rapidly by light, oxygen, and acids. Therefore, they must be stored in opaque, hermetically-sealed containers from which the air is displaced by an inert gas, such as nitrogen. The crystalline compounds are relatively stable to heat.

SOURCES OF VITAMIN D.
Vitamin D is found in foods more sparsely than any other vitamin. But, fortunately, nature ordained that man could generate some of his supply of vitamin D by sunlight.

Groupings by rank of common sources of vitamin D are given in the section on VITAMIN(S), Table V-5, Vitamin Table.

For additional sources and more precise values of vitamin D, see Food Composition Table F-21. However, few vitamin D values are available, because foods do not lend themselves well to chemical analysis for this vitamin. For foods, costly and time-consuming bioassays with rats or chicks are the only means available for the assessment of vitamin D activity.

Among animal foods, fatty fish are rich sources of vitamin D. They obtain the vitamin by feeding on plankton which live near the surface of the sea exposed to sunlight. Liver, egg yolk, and butter contain useful amounts of vitamin D, but the potency of these foods varies widely according to the extent to which the animals have been exposed to sun-

light or ultraviolet light and to the feeds given them. Muscle meats contain only traces. Both human milk and cow milk are a poor source of vitamin D.

Vegetables, grains and their products, and fruits have little or no vitamin D activity. Although it is usually assumed that living plants do not contain vitamin D activity, an important exception is certain tropical and subtropical shrubs and plants of the *Solanum* family,[12] native to the West Indies, recently found to contain such large amounts of the vitamin D hormone as to be extremely toxic to animals and cause extensive economic losses.

Vitamin D is added to most commercial milk and infant formulas. These two foods are considered most suitable for vitamin D fortification because of their calcium and phosphorus content and the role of vitamin D in the absorption and utilization of the two minerals. Also, milk is consumed in large quantities by the young, who especially need vitamin D for skeletal development.

Practically all whole, low-fat, and nonfat fluid milks on the market today are fortified with 400 IU of vitamin D per quart. Either vitamin D_2 (ergosterol) or D_3 (irradiated cholecalciferol) is added, with the form of the added vitamin listed on the label. Evaporated milk is required by federal law to be fortified with sufficient vitamin D to provide 400 IU per quart when it is reconstituted. Although the fortification of nonfat dry milk is optional, most of it is fortified at a level that supplies 400 IU per reconstituted quart.

Although milk is the only food for which vitamin D fortification is recommended by the Food and Nutrition Board of the National Research Council and the Council of Foods and Nutrition of the American Medical Association, other foods to which vitamin D is added include breakfast cereals, infant cereals, breads, margarine, milk flavorings, fruit and chocolate beverages, and cocoa. Concern has been expressed about the possibility of overconsumption of vitamin D as a result of widespread fortification of foods with the vitamin. Obviously, when vitamin D-enriched milk is used in the amount of 1 qt daily, no other source of vitamin D is required.

All fish liver oils (cod liver, halibut, and swordfish liver oils) are rich sources of vitamin D, but they are not part of the usual American diet. Preparations containing them are available for use as supplementary sources of vitamin D, particularly for infants. Other vitamin D concentrates are made by irradiating pure ergosterol or 7-dehydrocholesterol; these are available in liquid and tablet form. Viosterol, which is a solution of irradiated ergosterol dissolved in neutral oil, is an example. Such preparations are labeled with the units per dose or tablet and are prescribed accordingly.

The alternative to oral sources of vitamin D is its manufacture in the skin. The vitamin can be produced in the skin if it is exposed to the short wavelengths of the sun's rays (or to a sunlamp), the ultraviolet rays, which act on the provitamin 7-dehydrocholesterol in the skin and transform it into vitamin D_3. But the amount of these rays that act on the skin is affected by atmospheric clouds, smoke, fog, dust, window glass, clothing, skin pigmentation, and season; hence, it is highly variable and cannot either be determined or relied upon. So, some other source of vitamin D is usually needed.

(Also see VITAMIN[S], Table V-5.)

Fig. V-50. Tuna fish, a rich source of vitamin D. (Courtesy, Department of Food Service and Human Nutrition, University of Hawaii)

---

[12]Potatoes and eggplant belong to the *Solanum* family, but they are not known to be a source of vitamin D activity. The *Solanum*-toxic plant known as day-blooming jessamine, wild jessamine, or King-of-the-Day has been introduced into the United States and cultivated for ornamental purposes.

VITAMIN E (TOCOPHEROLS)

Contents	Page
History	1098
Chemistry, Metabolism, Properties	1098
Measurement/Assay	1099
Functions	1099
Interrelationship of Vitamin E and Selenium	1100
Deficiency Symptoms	1100
Recommended Daily Allowance of Vitamin E	1100
Toxicity	1101
Vitamin E Losses During Processing, Cooking, and Storage	1101
Sources of Vitamin E	1101

For many years, vitamin E was known as the *antisterility* vitamin, a name taken from early work with rats, in which species a dramatic improvement in fertility was observed when vitamin E was added to a diet deficient in the substance.

Today, vitamin E is recognized as an essential nutrient for higher animals, including man. In humans, however, sexual potency has never been linked with vitamin E.

HISTORY. In 1922, Evans and Bishop, of the University of California, discovered that a fat-soluble dietary factor (then called *factor X*) in lettuce and wheat germ was essential for successful reproduction in rats. In 1924, Sure, of the University of Arkansas, named the factor vitamin E. In 1936, Evans and co-workers isolated crystalline vitamin E from wheat germ oil and named it tocopherol, from the Greek words *tokos* (offspring) and *pherein* (to bear), meaning *to bear offspring*. In 1938, the vitamin was first synthesized by the Swiss chemist, Karrer.

CHEMISTRY, METABOLISM, PROPERTIES.

• **Chemistry**—Eight tocopherols and tocotrienols with vita-

min E activity, collectively called vitamin E, have been identified. Differing from each other in the number and position of the methyl (CH_3) groups around the ring of the molecule, they are: alpha- (see Fig. V-51), beta-, gamma-, and delta-tocopherol; and alpha-, beta-, gamma-, and delta- tocotrienol. Alpha-tocopherol has by far the greatest vitamin E activity; the other tocopherols have biological activities ranging from 1 to 50% that of alpha-tocopherol. Nevertheless, the nonalpha-tocopherol compounds in foods normally consumed contribute vitamin E activity equivalent to about 20% of the indicated alpha-tocopherol content of a mixed diet.

Fig. V-51. Structure of alpha-tocopherol.

• **Metabolism**—A discussion of the absorption, transportation, storage, and excretion of vitamin E follows.

1. **Absorption.** As with other fat-soluble vitamins, the presence of both bile and fat are required for the proper absorption of vitamin E. Absorption takes place in the small intestine, where 20 to 30% of the intake passes through the intestinal wall into the lymph.

2. **Transportation.** Once absorbed, vitamin E is transported attached to the beta-lipoprotein fraction of the blood. In normal adults in the United States, the total tocopherol content of the plasma ranges from 0.5 to 1.2 mg/100 ml.

TOP FOOD SOURCES

Fig. V-52. *Left:* White muscle disease in a lamb, showing characteristic whitish areas or streaks in the heart. (Courtesy, Oregon State University) *Right:* A chick with nutritional encephalomalacia (crazy chick disease), due to a lack of vitamin E. Note head retraction and loss of control of legs. (Courtesy, Department of Poultry Science, Cornell University) Note, too, top food sources of vitamin E.

A level below 0.5 mg is considered undesirable. The ratio of total tocopherol to the total lipids in the plasma appears to be more important than the actual tocopherol level; a ratio of 0.8 mg of total tocopherol to 1 g of total lipids in the plasma is accepted as an indication of adequate vitamin E nutritional status. The predominant form of vitamin E in both the plasma and red cells is alpha-tocopherol, which accounts for 83% of the total tocopherol; gamma-tocopherol accounts for most of the remainder.

3. **Storage.** Adipose (fatty) tissue, liver, and muscle are the major storage sites, although small amounts of vitamin E are stored in most body tissues. Relatively high amounts are found in the adrenal and pituitary glands, heart, lungs, testes, and uterus. Vitamin E is deposited in the fat of tissues and is mobilized from them with fat.

There is little transfer of vitamin E across the placenta to the fetus; hence, newborn infants have low tissue stores.

4. **Excretion.** The major pathway of excretion of vitamin E is by way of the feces, although a small amount is excreted in the urine.

● **Properties**—The tocopherols and tocotrienols are light yellow, viscous oils, soluble in alcohol and fat solvents, but insoluble in water. They are stable to acids and heat, but they are destroyed upon exposure to oxygen, ultraviolet light, alkali, and iron and lead salts. Their ability to take up oxygen gives them important antioxidant properties. Vitamin E is not destroyed to any great extent by normal cooking temperatures, but there are appreciable losses of vitamin E activity in oils heated long periods of time at high temperature, such as in deep-fat frying, because rancidity develops.

Currently, the bulk of commercial vitamin E is synthetic di-alpha-tocopherol acetate, an ester of alpha-tocopherol, a form more stable to heat and oxidation than the free alcohol form, but with the same vitamin E activity. This stable form of alpha-tocopherol is added to vitamin preparations, various medicinal products, foods, and animals feeds.

MEASUREMENT/ASSAY. By agreement of international committees, it is now preferred to use "milligrams of alpha-tocopherol equivalents" as a summation term for all vitamin E activity. In this book, however, both International Units (IU) and alpha-tocopherol equivalents will be used because (1) food composition tables generally give values in IU, and (2) IU is still used for labeling most food products. Table V-6 shows the relationship between milligrams of alpha-tocopherol equivalents and IU.

For purposes of calculating the total vitamin E activity of mixed diets, the milligrams of beta-tocopherol should be multiplied by 0.5, those of gamma-tocopherol by 0.1, and those of alpha-tocotrienol by 0.3. (These are the only vitamins with significant activity that may be present in U. S. diets.) When these are added to the milligrams of alpha-tocopherol, the sum is the total milligrams of alphatocopherol equivalents. If only alpha-tocopoherol in a mixed diet is reported, the value in milligrams should be increased by 20% (multiply by 1.2) to account for the other tocopherols that are present, thus giving an approximation of the total vitamin E activity as milligrams of alpha-tocopherol equivalents.

A colorimetric method is the most commonly used assay method for estimating vitamin E activity. Additionally, spectrofluorometric and gas-liquid chromatographic methods are sometimes used. Animal tests are available, but are now rarely used.

TABLE V-6
RELATIONSHIP BETWEEN ALPHA-TOCOPHEROL EQUIVALENTS AND INTERNATIONAL UNITS (IU)

Compound (1 mg)	Alpha-tocopherol Equivalents in 1 mg	IU Activity in 1 mg
	(mg)	(IU)
d-alpha-tocopherol[1]	1.0	1.49
d-alpha-tocopheryl acetate	0.91	1.36
d-alpha-tocopheryl acid succinate	0.81	1.21
dl-alpha-tocopherol	0.74	1.10
dl-alpha-tocopheryl acetate[2]	0.67	1.00
d-gamma-tocopherol[3]	0.10	0.15

[1]The natural form in foods and the standard for *alpha-tocopherol equivalents,* usually just called alpha-tocopherol.

[2]The common commercial form of vitamin E—a synthetic and stable form. Naturally occurring di-alpha-tocopheryl acetate is considered to have the same potency as the synthetic product. This is the standard for International Units.

[3]The most abundant form in food oils.

FUNCTIONS. The primary function of vitamin E (tocopherols) is to help protect the integrity of cellular and intracellular structures and to prevent destruction of certain enzymes and intracellular components. But there is much controversy as to how this function is carried out.

The authors of this book are of the opinion that vitamin E is more important in human nutrition than is now known; based on current evidence, they subscribe to the following thinking relative to the functions of vitamin E:

1. **As an antioxidant which (a) retards the rancidification of fats in foods and in the digestive tract, and (b) protects cells from toxic substances formed from the oxidation of unsaturated fatty acids.** As a powerful antioxidant, vitamin E readily oxidizes itself (it combines with oxygen), thereby minimizing the destruction by oxidation of unsaturated fatty acids and vitamin A in the intestinal tract and in the tissues.

As an antioxidant, vitamin E prevents rancidity of fats in plant sources and in the animal's digestive tract. But the primary biological role of vitamin E is as an antioxidant inhibiting the oxidation of polyunsaturated fatty acids in tissue membranes, especially at the cellular level—in the membranes that surround the cells, the subcellular particles, and the erythrocytes. It stabilizes the lipid (fat) parts of cells and protects them from damage from toxic free radicals formed from the oxidation of polyunsaturated fatty acids; it reacts with the peroxides, converting them to forms that are not harmful to the cells. By functioning as a natural inhibitor of the destruction of cells, and by protecting tissue from breaking down, vitamin E may have a role in preventing a number of degenerative disorders—including aging.

In its antioxidant role, vitamin E also protects vitamin A (and carotene), vitamin C, sulfur-containing enzymes, and ATP from being oxidized, thereby enabling these essential nutrients to perform their specific functions in the body.

It is noteworthy that when fat is added to the diet it will destroy the vitamin E in both the diet and the digestive tract if rancidity occurs. For this reason, the quantitative relationship between vitamin E and the amount and kind of dietary fat is of practical importance; the higher the consumption of polyunsaturated fats, the higher the vitamin E requirement.

2. **As an essential factor for the integrity of red blood cells.** Evidence has been accumulating that vitamin E functions in the body in maintaining the integrity of the red blood cells. Thus, it is noteworthy that hemolytic anemia, an abnormality of the red blood cells in premature babies, may be corrected with vitamin E. Also, full-term babies with this same abnormality make a more rapid and complete recovery when fed with human milk (rather than cow's milk); and laboratory tests show that human milk contains from 2 to 4 times as much vitamin E as cow's milk.

3. **As an agent essential to cellular respiration.** Alpha-tocopherol appears to be necessary in cellular respiration, primarily in heart and skeletal muscle tissues. Thus, muscular dystrophy has been produced experimentally in various animals on E-deficient rations; however, vitamin E supplements have not been effective in treating people with muscular dystrophy.

4. **As a regulator in the synthesis of body compounds.** The tocopherols appear to be involved in the biosynthesis of DNA, probably by regulating the incorporation of pyrimidines into the nucleic acid structures.

Vitamin E also appears to act as a cofactor in the synthesis of vitamin C, and in the synthesis of coenzyme Q—a factor that is essential in the respiratory mechanism of cells that releases energy from carbohydrates and fats.

5. **As a protector of lung tissue from air pollution.** Recent studies on rats have shown that vitamin E may protect lung tissue from smog—from damage by such oxidant components of air pollution as nitrogen dioxide (NO_2) and ozone. At the present time, it is not known whether or not vitamin E protects human lungs from damage by air pollution.

6. **As a sometimes replacement for selenium.** Both vitamin E and selenium are antioxidants, although each takes a different approach. Nevertheless, vitamin E has a sparing or replacement effect on selenium.

The belief among some people that vitamin E is *the sex vitamin* and will reduce sterility and increase potency in man is not based on scientific fact; it is one of the falsities of nutrition information which is given wide publicity.

INTERRELATIONSHIP OF VITAMIN E AND SELENIUM.

During the 1950s, an interrelationship between vitamin E and the element selenium was established. It was found that selenium prevented exudative diathesis (a hemorrhagic disease) in vitamin E-deficient chicks and liver necrosis in vitamin E-deficient rats. Subsequent research demonstrated that both selenium and vitamin E protect the cell from the detrimental effects of peroxidation, but each takes a distinctly different approach to the problem. Vitamin E is present in the membrane components of the cell and prevents free-radical formation, while selenium functions throughout the cytoplasm to destroy peroxides. This explains why selenium will correct some deficiency symptoms of vitamin E, but not others.

There is also some indication that vitamin E and selenium work together to protect cell membranes, cell nuclei, and chromosomes from carcinogens—substances that can cause cancer. But much more experimental work is needed on this subject.

DEFICIENCY SYMPTOMS.

Many dietary factors seem to contribute to the development of vitamin E deficiency symptoms; among them, total fat, unsaturated fats, cod-liver oil (which is high in unsaturated acids), amount of protein, choline, cystine, inositol, cholesterol, vitamin A, and minerals have been implicated, at one time or another, as causing or aggravating deficiency symptoms.

Vitamin E deficiency symptoms as such rarely occur in humans because vitamin E (1) is widely distributed in foods, (2) is stored in almost all body tissues, and (3) is retained in the body for relatively long periods. However, clinical evidence of deficiency has been observed in infants, especially those born prematurely and formula-fed. Also, vitamin E deficiency occurs in individuals suffering from kwashiorkor (a protein deficiency), and in children and adults who have impaired fat absorption.

Newborn infants (especially the premature) have low plasma levels of vitamin E (the vitamin E concentration in full-term newborn infants is about one-third that of adults, and that of premature infants is even lower), because transfer of vitamin through the placenta to the fetus is limited. As a result, hemolytic anemia (caused by shortened life span of red cells) may occur in the early weeks of life. In this condition, the membranes of the red blood cells are weakened by the action on them of the products of peroxidation of polyunsaturated fats and the cells rupture easily, producing a condition characterized by edema, skin lesions, and blood abnormalities. Supplements of vitamin E bring about increases in blood levels of the vitamin, decreases in red blood cell hemolysis, and a return to normal hemoglobin levels.

Individuals suffering from kwashiorkor, due to a severe protein deficiency, have very little tocopherol in their serum; the fragility of their red blood cells is greater than normal, and they frequently suffer from an accompanying anemia.

The symptoms of vitamin E deficiency produced by artificial diets high in unsaturated fatty acids or observed in patients unable to absorb fat (sprue, fibrocystic disease of the pancreas) include low blood and tissue tocopherol levels, increased red blood cell fragility and shortened red blood cell life-span, and increased urinary excretion of creatine (the latter is indicative of muscle damage). Marked improvement in these conditions is noted following administration of alpha-tocopherol.

RECOMMENDED DAILY ALLOWANCE OF VITAMIN E.

The Food and Nutrition Board (FNB) of the National Research Council (NRC) recommended daily allowances of vitamin E are given in the section on VITAMIN(S), Table V-5, Vitamin Table.

The requirements of vitamin E for humans is known to vary with other ingredients in the diet; for example, the presence of large amounts of polyunsaturated fatty acids (PUFA), such as linoleic acid, markedly increases the requirement. This is important in today's diets, in which large amounts of vegetable oils are used. So, individuals consuming diets high in PUFA need more vitamin E than the amounts listed in Table V-5, Vitamin Table; those whose diets are low in PUFA need less. Also, the presence of rancid fats, oxidizing substances, and selenium modify the requirements for vitamin E.

• **Infants**—The vitamin E content of human milk, 2 to 5 IU (1.3 to 3.3 mg d-alpha-tocopherol equivalent) per liter (1.06 qt), is considered to be adequate for full-term infants. Table V-5 lists the daily allowances in mg d-alpha-TE of infants, as follows: birth to 6 months, 3 mg; from 6 months to 1 year of age, 4 mg.

Cow's milk differs from human milk in two ways: (1) it contains only $\frac{1}{10}$ to $\frac{1}{2}$ as much vitamin E, varying with the feed consumed by the cow; and (2) it is much lower in polyunsaturated fats, containing about $\frac{1}{2}$ as much.

Increased vitamin E is required for both premature and full-term infants fed commercial formulas made from vegetable oils that are high in polyunsaturated fats, like linoleic acid. According to the Committee on Nutrition of the Academy of Pediatrics, full-term infants require at least 0.7 IU (about 0.47 mg) of vitamin E per gram of linoleic acid in a formula preparation.

• **Children**—Table V-5 lists the daily allowances in mg d-alpha-TE of children as follows: from 1 to 3 years of age, 6 mg; from 4 to 10 years of age, 7 mg.

• **Adults**—Table V-5 lists the daily allowances in mg d-alpha-TE of adults as follows: males, 10 mg; females, 8 mg.

• **Pregnancy and lactation**—During pregnancy and lactation, the Food and Nutrition Board recommends that the daily allowance of vitamin E be increased to compensate for the amount deposited in the fetus and secreted in the milk; that the RDA be 10 mg α TE during pregnancy, and 12 mg α TE during lactation.

• **Vitamin E intake in average U.S. diet**—Limited reports on the vitamin E intake in the U.S. diet show a daily intake of 16.7 mg α tocopherol. To this must be added the additional forms of vitamin E other than alpha-tocopherol, particularly gamma-tocopherol from soybean oil, which provide significant amounts of vitamin E activity in usual diets in the United States. Generally, nonalpha-tocopherol forms of vitamin E in a mixed diet are considered to supply about 20% of the total vitamin E activity. For the latter reason, calculations based only on alpha-tocopherol underestimate the amount of dietary vitamin E activity.

TOXICITY. Compared with vitamins A and D, vitamin E is relatively nontoxic. Excess intakes of vitamin E are excreted in the feces.

VITAMIN E LOSSES DURING PROCESSING, COOKING, AND STORAGE.

Food processing, storage, and packaging cause considerable loss of vitamin E. The tocopherols are subject to destruction by oxygen; and oxidation is accelerated by exposure to light, heat, alkali, and the presence of certain trace minerals such as iron (Fe^{+++}) and copper (Cu^{++}).

The milling of grains removes about 80% of the vitamin E; for example, in converting whole wheat to white flour, and in processing corn, oats, and rice.

Various methods of processing cause considerable destruction of vitamin E. Dehydration causes 36 to 45% loss of alpha-tocopherol in chicken and beef, but little or none in pork. Canning causes losses of 41 to 65% of the alpha-tocopherol content of meats and vegetables. An 80% destruction occurs during the roasting of nuts.

Deep-fat frying of foods causes vitamin E losses of 32 to 75%. However, normal home baking or cooking in water do not involve large losses of tocopherol. Also, being insoluble in water, tocopherols are not drained off with water.

There is a great loss of tocopherol in the storage of potato chips and french fried potatoes. One study showed after-manufacture losses of tocopherol in potato chips stored at 73°F (23°C) of 71% in 1 month and 77% in 2 months. When the chips were frozen at 10°F (-12°C), the losses were 63% in 1 month and 68% in 2 months. The after-manufacture losses of french fries stored at 10°F (-12°C) were 68% in 1 month and 74% in 2 months.

SOURCES OF VITAMIN E. Vitamin E (the tocopherols) occurs mainly in a variety of plant materials, especially oil seed crops (vegetable oils), some grains, nuts, and green leafy vegetables. The amount in plant foods is affected by species, variety, stage of maturity, season, time and manner of harvesting, processing procedures, and storage time. Animal tissues and animal products are usually low to poor sources of vitamin E and are influenced by the dietary consumption of tocopherols by the animal.

Fig. V 53. Salad dressings, a rich source of vitamin E. (Courtesy, American Soybean Assn., St. Louis, Mo.)

Groupings by rank of common food sources of vitamin E are given in the section on VITAMIN(S), Table V-5, Vitamin Table.

For additional sources and more precise values of vitamin E (alpha-tocopherol), see Food Composition Table F-21.

Wheat germ and wheat germ oil, which are used as supplements, are the richest natural sources of vitamin E, followed by the vegetable oils (almond, corn, cottonseed, olive, palm, peanut, rapeseed, safflower, soybean, and sunflower—coconut oil is a poor source of vitamin E). It follows that margarine and cooking and salad oils are major sources in the diet; fortunately, the hydrogenation process used in their manufacture has little, if any, effect on the vitamin E content.

Refined grain products contain little vitamin E—most of the vitamin is removed in the milling process.

Human milk contains 2 to 4 times as much vitamin E as cow's milk.

(Also see VITAMIN[S], Table V-5.)

VITAMIN K

Contents

	Page
History	1102
Chemistry, Metabolism, Properties	1102
Measurement/Assay	1103
Functions	1103
Vitamin K Antagonists (Anticoagulants)	1103
Deficiency Symptoms	1104
Recommended Daily Allowance of Vitamin K	1104
Toxicity (Hypervitaminosis K)	1104
Vitamin K Losses During Processing, Cooking, and Storage	1104
Sources of Vitamin K	1104

Vitamin K, known as the *antihemorrhagic vitamin*, is necessary for the synthesis of prothrombin and other blood clotting factors in the liver. Presently, the term vitamin K is used to describe a chemical group of quinone compounds, rather than a single entity, which have characteristic antihemorrhagic effects.

Defective blood coagulation is the only well-established symptom of vitamin K deficiency. Deficiencies are rare in humans, but considerable use is made of synthetic substances that act as antagonists of vitamin K (such as dicoumarol, an anticoagulant) to prevent clotting of blood in patients with certain circulatory disorders.

HISTORY. In 1929, Professor Carl Peter Hendrik Dam, biochemist at the University of Copenhagen, Denmark, observed that certain experimental diets produced fatal hemorrhages in chicks. Bleeding could be prevented by giving a variety of foodstuffs, especially alfalfa (lucerne) and fishmeal. Further, it was found that the active principle in these materials could be extracted with ether; thus, a new fat-soluble factor was discovered. In 1935, Dam named it the Koagulation vitamin (the Danish word for coagulation), from which the shortened term vitamin K (from the first letter of Koagulation) was derived. In 1939, Dam and Karrer isolated vitamin K in pure form; and, that same year, Almquist and Klose synthesized vitamin K. In 1943, Dam received the Nobel Prize in physiology and medicine for his brilliant work.

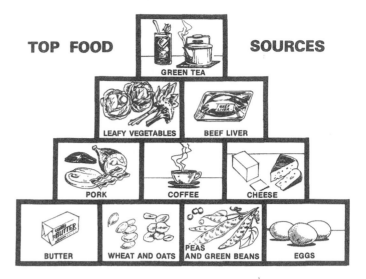

TOP FOOD   SOURCES

Fig. V-54. Top sources of vitamin K for common foods.

In 1941, Link and co-workers at the University of Wisconsin discovered dicoumarol—an antimetabolite of vitamin K.

CHEMISTRY, METABOLISM, PROPERTIES.

• **Chemistry**—A number of chemical compounds possessing vitamin K activity have been isolated or synthesized. There are two naturally occurring forms of vitamin K; vitamin K_1 (phylloquinone or phytylmenaquinone) which occurs only in green plants, and K_2 (menaquinones or multiprenylmenaquinones), which is synthesized by many microorganisms, including bacteria in the intestinal tracts of human beings and other species. Additionally, several synthetic compounds have been prepared that possess vitamin K activity, the best known of which is menadione (2-methyl, 1, 4-naphthoquinone), formerly known as K_3. Menadione,

CHEMISTRY. METABOLISM. PROPERTIES.

Fig. V-55. Vitamin K utilization.

which is converted to vitamin K_2 in the body, is 2 to 3 times as potent as K_1 and K_2. The structural formulas of K_1, K_2, and menadione, all of which are called vitamin K, are shown in Fig. V-56.

Fig. V-56. Structures of vitamin K, vitamin K_2, and menadione.

• **Metabolism**—Normally, vitamin K_1 is taken into the body in the diet, vitamin K_2 is synthesized in the intestine, and menadione is taken as a vitamin supplement.

1. **Intestinal synthesis**. Vitamin K_2 is synthesized by the normal bacteria in the small intestine and the colon; hence, an adequate supply is generally present. Since the intestine of a newborn infant is sterile at birth, however, the supply of vitamin K is inadequate until normal bacterial flora of the intestine develops on about the third or fourth day of life.

The intestinal synthesis of vitamin K reduces the food dietary requirements for the vitamin in man and other mammals (but not birds; birds have such a short intestinal tract and harbor so few microorganisms that they require a dietary source of vitamin K), although it appears that little of the vitamin K produced in the lower gut is absorbed. It is noteworthy, however, that animals that practice coprophagy, such as the rabbit, can utilize much of the vitamin K that is eliminated in the feces.

2. **Absorption**. Since the natural vitamins K (K_1 and K_2) are fat soluble, they require bile and pancreatic juice in the intestine for maximum absorption. Thus, anything that interferes with the normal absorption of fat, interferes with the absorption of natural vitamin Ks. By contrast, some of the synthetic vitamin K compounds are water soluble and more easily absorbed. Absorption takes place mainly in the upper part of the small intestine. Normally, 10 to 70% of the vitamin K in the intestines is absorbed. However, there is uncertainty as to how much of the vitamin K_2 that is synthesized in the colon is absorbed, based on the fact that absorption of nutrients in general from the large intestine appears to be limited because of the nature of the epithelial lining.

3. **Transportation**. Vitamin K passes unchanged from the small intestine into the lymph system. Thence, it is carried to the thoracic duct, where it enters the bloodstream. In the blood, it is attached to beta-lipoproteins and transported to the liver and other tissues.

4. **Metabolism**. Whether the vitamin Ks' functions are unchanged or are transformed to other metabolically active forms has not been determined. It is known that menadione must be converted to K_2 in the bodies of both animals and humans for it to be biologically active.

5. **Storage**. Vitamin K is stored only in small amounts. Modest amounts are stored in the liver, with the skin and muscle following in concentration. About 50% of the vitamin K found in the human liver is K_1 from the diet; the other 50% is K_2 from bacterial synthesis in the intestine.

6. **Excretion**. Excess vitamin K is excreted in the feces and urine.

• **Properties**—The naturally occurring forms of vitamin K are yellow oils; the synthetic forms are yellow crystalline powders. All K vitamins are resistant to heat and moisture, but they are destroyed on exposure to acid, alkali, oxidizing agents, and light—particularly ultraviolet light. Since natural vitamin K is stable to heat and not water soluble, little of it is lost in normal cooking processes. However, some of the synthetic forms of vitamin K are soluble in water.

MEASUREMENT/ASSAY. Vitamin K can be measured in micrograms of the pure synthetic compound (menadione), and the vitamin K activity of other substances can be expressed in similar terms.

The potency of low concentration samples, such as occurs in most foods, is commonly determined by bioassay, using young chicks, and is based on the minimum dose that will maintain the normal coagulation of the blood at the end of 1 month. In pure solutions, vitamin K may be assayed by U.V. spectrophotometry or colorimetric methods. Other accepted techniques are the oxidimetric assay after catalytic reduction to the hydroquinone and the polarographic determination.

FUNCTIONS. Vitamin K is essential for the synthesis by the liver of four blood clotting proteins—factor II, or prothrombin; factor VII, or proconvertin; factor IX, Christmas factor; and factor X, Stuart-Power factor. The exact way in which vitamin K functions in the synthesis of these proteins is unknown. Recent research suggests that vitamin K acts in some way to convert precursor proteins to the active blood clotting factors. Without vitamin K, or when an antagonist is given, the level of the blood clotting proteins in the blood is reduced and clotting time is prolonged.

At this time, there is no evidence that vitamin K has any functions in humans or animals other than in the blood coagulation process, although vitamin K-dependent proteins have been identified in the bone, kidney, and liver.

VITAMIN K ANTAGONISTS (ANTICOAGULANTS). The discovery of vitamin K antagonists stemmed primarily from investigations into a disorder in cattle known as *sweet clover disease*, in which there is a loss of clotting power of the blood. The causative agent of *sweet clover disease* was identified by Campbell, et al., (1941) of the University of Wisconsin, as dicoumarol, an oxidative product of coumarin, found in sweet clover hay that had undergone spoilage during harvesting or storage.

The discovery of the cause of *sweet clover disease* opened up an entirely new field. The synthesis of several dicoumarol derivatives followed. These antivitamins, or antagonists, decrease the ability of the blood to clot by inhibiting the synthesis of prothrombin and the other K-dependent blood clotting factors. They are used medically to reduce clotting of the blood in patients with certain circulatory disorders, especially some forms of atherosclerosis, phlebitis, and thrombosis.

DEFICIENCY SYMPTOMS. The symptoms of vitamin K deficiency are increased clotting times and hemorrhaging. Two laboratory tests are commonly used to assess vitamin K status: (1) the *prothrombin time*—the measurement in a plasma sample of the speed of conversion of prothrombin to thrombin; and (2) the *blood clotting time*—the placing of freshly drawn blood in a clean test tube, tilting it once each minute, and determining the time required for the clot to form (for normal blood the clotting time is about 10 minutes).

Fowl are very susceptible to vitamin K deficiency. When vitamin K is absent, they develop delayed blood clotting, causing hemorrhages under the skin and serious internal bleeding, followed by death if not corrected. Deficiencies have also been produced in cattle, pigs, rats, dogs, and all other species studied.

Although a hemorrhagic syndrome, due to a vitamin K deficiency, is rare in adult humans, it can be caused by the following:

1. Some defect in the absorption as in obstructive jaundice, or in malabsorption due to celiac disease or sprue.
2. Sterilization of the bowels by an antibiotic or a sulfa drug, thereby reducing the synthesis of vitamin K.
3. Anticoagulant therapy for certain circulatory disorders.

Hemorrhagic disease, due to a vitamin K deficiency, may occur in newborn babies—particularly those born prematurely and breast fed, those on unfortified soy formulas, or those whose mothers have been taking anticoagulants—between the second and fifth days of life; causing bleeding either in the skin, nervous system, peritoneal cavity, or alimentary tract. Because human milk provides less vitamin K than cow's milk, vitamin K deficiency is more common in breast-fed than in formula-fed babies. Hemorrhagic disease due to a vitamin K deficiency can be prevented by giving a dose of 1 mg of vitamin K_1 intramuscularly immediately after birth. It is noteworthy, however, that vitamin K deficiency is not always the cause of hemorrhagic disease; injury at birth is undoubtedly responsible in some cases.

Vitamin K has proved ineffective in treating hemophilia, an inherited condition causing abnormal hemorrhaging in man.

RECOMMENDED DAILY ALLOWANCE OF VITAMIN K. Because of the synthesis of vitamin K by intestinal bacteria in healthy individuals (except newborn babies), no specific recommended allowance is made for this vitamin by the Food and Nutrition Board of the NRC–National Academy of Sciences. However, because the adequacy of the intestinal synthesis over a long period is uncertain, NRC does give an estimated adequate dietary intake of vitamin K. (See section on VITAMIN[S], Table V-5, Vitamin Table.)

The Table V-5 estimated safe and adequate daily intakes of vitamin K, in mcg, are as follows: Infants birth to 6 months, 5 mcg; infants 6 months to 1 year, 10 mcg; children 1 to 3 years, 15 mcg; children 4 to 6 years, 20 mcg; children 7 to 10 years, 30 mcg; males 25 to 50 years, 80 mcg; females 25 to 50 years, 65 mcg, with no increase in pregnancy or lactation.

In 1989, the Committee on Nutrition of the American Academy of Pediatrics recommended that infant formulas include a minimum of 4 mcg per 100 kcal of vitamin K for milkbase formulas and 8 mcg per 100 kcal for soy isolate and other milk substitute formulas. Vitamin K hypoprothrombinemia (due to K deficiency) occurs in infants fed certain soy protein isolates, meat base formula, or formula containing hydrolyzed casein. Such diets may be low in vitamin K.

The intake suggested for young infants is based on 2 mcg/kg, assuming no intestinal synthesis. Therefore, the amount provided by current formulas of 4 mcg/100 kcal should be ample for normal infants. The suggested intake of 12 mcg/day is also in the range supplied by breast milk (15 mcg/liter).

• **Vitamin K intake in average U.S. diet**—The average mixed diet in the United States supplies 300 to 500 mcg of vitamin K per day; hence, there is little danger of insufficiency under normal conditions.

TOXICITY (HYPERVITAMINOSIS K). The natural forms of vitamins K_1 and K_2 have not produced toxicity even when given in large amounts. However, synthetic menadione and its various derivatives (formerly called K_3) have produced toxic symptoms in rats and jaundice in human infants when given in amounts of more than 5 mg daily. Consequently, the U.S. Food and Drug Administration does not allow any menadione in any food supplements, including prenatal vitamin capsules.

Vitamin K_1 preparations are preferred for the treatment of such medical conditions as liver disease, severe malabsorption syndromes, and anticoagulant overdosage.

VITAMIN K LOSSES DURING PROCESSING, COOKING, AND STORAGE. For maximum vitamin K, food should be fresh. Frozen foods tend to be deficient in vitamin K. Ordinary cooking processes destroy very little natural vitamin K, since the vitamin is stable to heat and not water soluble. However, some of the synthetic compounds are soluble in water and subject to greater loss in cooking. Sunlight destroys vitamin K_1; and all vitamin K compounds tend to be unstable to alkali.

SOURCES OF VITAMIN K. Vitamin K is fairly widely distributed in foods.

Also, synthesis in the intestine is an important source of vitamin K_2; and vitamin K supplements (chiefly the synthetic menadione) are available.

Fig. V-57. Lettuce, a good source of Vitamin K. (Courtesy, California Iceberg Lettuce Commission, Monterey, Calif.)

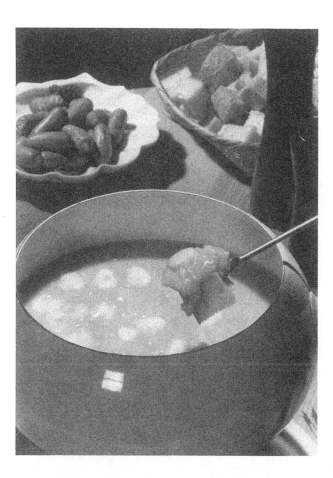

Fig. V-58. Cheese fondue, a good source of vitamin K. (Courtesy, California Dry Bean Advisory Board, Dinuba, Calif.)

Fig. V-59. Egg custard is another good source of vitamin K. (Courtesy, American Egg Board, Park Ridge, Ill.)

Groupings by rank of common sources of vitamin K are given in the section on VITAMIN(S), Table V-5, Vitamin Table.

No vitamin K values are listed in Food Composition Table F-21, for the reason that few foods have been assayed for this vitamin.

The vitamin K content of some common foods is given in Table V-7.

(Also see VITAMIN[S], Table V-5.)

TABLE V-7
VITAMIN K AND CALORIE CONTENT OF SOME COMMON FOODS[1]

Food	Vitamin K	Energy
	(mcg/100 g)	(kcal/100 g)
Tea, green	712	1
Turnip greens	650	28
Broccoli	200	26
Lettuce	129	15
Cabbage	125	24
Beef liver	92	222
Spinach	89	26
Asparagus	57	20
Watercress	57	19
Bacon	46	300
Coffee	38	3
Cheese	35	402
Butter	30	717
Pork liver	25	241
Oats	20	390
Peas, green	19	71
Wheat, whole	17	357
Beef fat	15	902
Ham	15	350
Beans, green	14	24
Eggs	11	157
Pork tenderloin	11	254
Peaches, raw	8	38
Ground beef	7	301
Chicken liver	7	165
Raisins	6	289

[1]Adapted from Goodhart, Robert S. and Maurice E. Shils, *Modern Nutrition in Health and Disease*, 6th ed., Lea & Febiger, 1980, p. 172, Table 6C-1; data taken from the studies of Dam and Glavind, Richardson, Doisy and Matschiner and Doisy.

VITAMIN L

Factors L_1 and L_2, which are found in yeast, are said to be necessary for lactation. However, they have not become established as vitamins; hence, there is no such thing as vitamin L at this time.

VITAMIN SUPPLEMENTS

Rich synthetic or natural food sources of one or more of the complex organic compounds, called vitamins, that are required in minute amounts by people and animals for normal growth, production, reproduction, and/or health.

(Also see VITAMIN[S].)

VOLATILE OILS

These are obtained from flowers, stems, leaves, and often the entire plant, and they are used for perfumery and flavorings. Often these are called essential oils. They vaporize quickly.

(Also see ESSENTIAL OILS.)

VOMITING

The forcible expulsion of the contents of the stomach through the mouth.

Fig. V-60. Most vitamins originate in plants; and they come in many different shapes, sizes, and colors. (Courtesy, California Dry Bean Advisory Board, Dinuba, Calif.)

WALNUT (BLACK WALNUT; ENGLISH WALNUT)
Juglans spp

Fig. W-1. English walnuts approaching maturity. (Courtesy, Sun-Diamond Growers of California, Stockton, Calif.)

The term walnut refers to several varieties of nuts. Two are of importance: (1) the English walnut (Persian walnut), *Juglans regia*; and (2) the black walnut, *Juglans nigra*. Of these two, the English walnut tree bears walnuts that are the most valuable commercially.

ORIGIN AND HISTORY.
English, or Persian walnuts, as they are also called, are a native of southeastern Europe and western Asia—not England as the name suggests. Perhaps, it gained the name English walnut because it was brought to America on English ships. English walnuts are commercially produced in orchards.

Black walnuts are native to the Central Mississippi Valley and the Appalachian region of North America. However, they have been planted beyond their natural range and they grow over a wide area on good agricultural soils. Black walnuts were an important source of food for the American Indians and early settlers.

Black walnut production is limited to trees in forests and farmsteads. In some areas, they are collected, shelled, and sold.

WORLD AND U.S. PRODUCTION.
Worldwide, about 946,626 metric tons of English walnuts are produced annually; and the leading producing countries, by rank, are: United States, Turkey, China, Romania, Iran, France, and Yugoslavia. The U.S. produces 21% of the world walnut crop.[1]

The entire U.S. commercial crop of English walnuts is produced in California, where about 210,000 metric tons are produced annually.[2]

No reliable production figures on the production of black walnuts are available.

PROCESSING.
At packing houses, those English walnuts to be marketed in the shells (about 40%) are sized, bleached, and bagged for shipment. Others are machine-shelled, and the kernels are graded for color and size, and packaged for the retail trade. Some of the poorer grades may be used to make walnut oil and shell flour.

It is difficult to remove the hulls of black walnuts, so they are usually allowed to rot away. The nutmeats are sold in broken pieces because it is difficult to remove whole nuts from the shell of black walnuts.

Fig. W-2. English walnuts with their husks removed are ready for selling or cracking. (Courtesy, Sun-Diamond Growers of California, Stockton, Calif.)

SELECTION, PREPARATION, AND USES.
Walnuts in shells should be free from splits, cracks, stains, or holes. Nutmeats should be plump and fairly uniform in color and size.

The dried nuts will store for at least a year under fairly variable conditions, up to 2 years if the storage temperature is kept below 40°F (4.8°C).

No preparation of walnut meats is required.

---

[1]Data from *FAO Production Yearbook 1990*, FAO/UN, Rome, Italy, Vol. 44, p. 173, Table 77. **Note well:** Annual production fluctuates as a result of weather and profitability of the crop.

[2]Data from *Agricultural Statistics 1991*, USDA, p. 219, Table 339. **Note well:** Annual production fluctuates as a result of weather and profitability of the crop.

Walnuts (kernels) are eaten alone as a snack, or they are used in endless ways in baking and confectionery.

The few black walnuts that are available can be used in baked goods, candy, and ice cream.

NUTRITIONAL VALUE. Walnuts—English and black—are extremely nutritious. They contain only 3 to 4% water and they are *supercharged* with energy. A 3½ oz (*100 g*) portion of black walnuts supplies 678 kcal (194 kcal per oz), whereas the same amount of English walnuts supplies 694 kcal (198 kcal per oz). Walnuts are high in calories, primarily because they are about 60% fat (oil). Additionally, each 100 g also provides 15 to 20 g of protein, 3 mg of iron and zinc, and only 2 mg of sodium.

Among the leading nuts, walnuts are the richest source of essential polyunsaturated fatty acids. Hence, they should be kept refrigerated to prevent rancidity.

More complete information regarding the nutritional value of walnuts is provided in Food Composition Table F-21.

WARBURG'S YELLOW ENZYME

A protein isolated in 1932 by two German scientists, Warburg and Christian. It played a role in the discovery of the vitamin riboflavin.

(Also see RIBOFLAVIN.)

WATER (H₂O)

Contents	Page
Importance of Water to Life	1108
Water Balance	1108
Requirement for Water	1109
Water Sources	1109
Drinking Water	1109
Foods	1109
Metabolic Water	1110

Fig. W-3. Plenty of cool, fresh, clean water, Lake Walcott canal, Minidoka County, Idaho. (Photo by Jayne Parker)

Chemically, water is the combination of two gases—hydrogen (H) and oxygen (O)—which are joined in the ratio of two hydrogen atoms to one oxygen. Thus, the chemical formula for water is H_2O. It is the most abundant chemical, and it performs endless functions in one of its three forms—liquid, solid, gas.

IMPORTANCE OF WATER TO LIFE. Water is a nutrient. It is one of the most vital of all nutrients. In fact, water is the only substance necessary to all life. Many organisms can live without air, but none can live without water. Humans and animals can survive for a longer period without food than without water. Only oxygen is more important to human life. Fortunately, water is usually provided in abundance and at little cost. In most U.S. cities it can be delivered to the kitchen for about 15¢ per ton.

Water is the most abundant body constituent. The younger the individual the more water the body contains. It accounts for about 98% of the embryo, for about 75% of infants, and for about 50 to 65% of the body weight of adults. In general, as the fat content of the body increases, the water content decreases.

In the body, water performs the following important functions:

1. It is necessary to the life and shape of every cell, and a constituent of every body fluid.

2. It acts as a carrier for various substances, serving as a medium in which nourishment is carried to the cells and waste products are removed therefrom.

3. It assists with temperature regulation in the body, cooling the individual by evaporation from the skin as perspiration.

4. It is necessary for many important chemical reactions of digestion and metabolism.

5. It lubricates the joints, as a constituent of the synovial fluid; it acts as a water cushion for the nervous system, in the cerebrospinal fluid; it transports sound, in the perilymph in the ear; and it is concerned with sight and provides a lubricant for the eye.

6. It acts as a solvent for a number of chemicals which can subsequently be detected by taste buds.

7. It aids in gas exchange in respiration by keeping the alveoli of the lungs moist.

The total body water involved in all of these functions is contained in two major compartments in the body: (1) the extracellular or water outside the cells (about 20% of the body weight), and (2) the intracellular or the water inside each cell (about 45% of the body weight).

Deficits or excesses of more than a few percent of the total body water are incompatible with health; large deficits, of about 20% of the body weight lead to death. Under normal circumstances, thirst ensures that water intake meets or exceeds the requirement for water. Surplus water is excreted from the body, principally in the urine, and to a lesser extent in the perspiration, feces, and water vapor from the lungs.

Water Balance. In healthy individuals, total body water remains reasonably constant. An increase or decrease in water intake brings about an appropriate increase or decrease in water output to maintain the balance. Water enters the body as a liquid, and as a component of the food—including metabolic water derived from the breakdown of food. Water is lost from the body (1) by the skin as perspiration, (2) by the lungs as water vapor in expired air, (3) by the kidneys as urine, and (4) by the intestines as feces. Therefore, under normal conditions the intake of water from the various sources is approximately equal to output of water by the various routes. For example, if a large volume of water is drunk, then the volume of urine excreted increases, or if water drinking is severely limited then the volume of urine formed is drastically reduced.

Water intake for a normal adult during a 24-hour period averages about 2.1 to 2.9 qt (*2 to 2.7 liter*) and water output via all four routes averages the same, assuming light activity and no visible perspiring, as Table W-1 shows.

TABLE W-1
DAILY WATER LEDGER

Water In		Water Out	
Source	Amount	Source	Amount
	(ml)		(ml)
Fluids	1,500	Skin	700
Water in foods	850	Lungs	350
Metabolic water	250	Kidneys	1,400
		Feces	150
Total	2,600	Total	2,600

The body is equipped with a number of mechanisms for regulating body water within narrow limits. Important among these mechanisms are nerve centers in the hypothalamus of the brain which control the sensation of thirst and water output by the kidneys. Stimulation of the thirst center in the hypothalamus results when water loss amounts to about 1% of the body weight, and creates the conscious desire for water. If water is not drunk, discomfort increases—heart rate increases, body temperature rises, and working and thinking abilities deteriorate. Heat exhaustion is certain if physical work is attempted when water loss is 10% of the body weight. Stimulation of other nerve centers in the hypothalamus causes the release of antidiuretic hormone (ADH) from the posterior pituitary. Release of ADH results in the formation of less urine, thereby conserving body water. The urine formed appears more concentrated, dark, and cloudy than when water intake is adequate. Despite the controls to maintain water balance, several factors can influence water balance and the requirement for water.

Requirement for Water. The requirement for water to maintain the water balance can be influenced by age, physical activity, heat, diet, illness, and injury. In general, the requirement for water can best be taken care of by allowing free access to plenty of clean, fresh water at all times. A discussion of each of the factors influencing the requirement for water follows:

● **Age**—Water needs for infants and children are proportionately higher than those for adults. The normal turnover rate of water per day is about 6% of the total body water in the adult and about 15% in the young infant.

Under ordinary circumstances, a reasonable allowance is 1 ml/kcal for adults and 1.5 ml/kcal for infants. Table W-2 presents daily water allowances in terms of the recommended dietary allowances for energy.

TABLE W-2
DAILY WATER ALLOWANCE FOR INFANTS AND ADULTS[1]

Category	Age	Weight		Energy Needs	Water Needs		
	(years)	(lb)	(kg)	(kcal)	(ml/kcal)	(ml)	(8 oz cups)
Infants	0.0–0.5	13	6	650	1.5	975	4.1
	0.5–1.0	20	9	850	1.5	1,275	5.4
Males	25–50	174	79	2,900	1.0	2,900	12.3
Females	25–50	138	63	2,200	1.0	2,200	9.4
Pregnancy				2,500	+30 ml	2,530	10.8
Lactating				2,700	+750 ml	3,450	14.7

[1]Based on *Recommended Dietary Allowances*, 10th ed., 1989, National Academy of Sciences–National Research Council, pp. 247–250.

● **Physical activity**—Even in a comfortable environment, physical activity increases the loss of water from the body through perspiration and water vapor from the lungs. Al-

though considerable amounts of sodium are lost in the perspiration, the loss of water significantly exceeds that of sodium. Hence, water depletion is the major problem. If water loss is not replenished then physical performance begins to deteriorate when the water deficit exceeds 3% of the body weight. Despite this fact, extremely ill-advised practices—withholding water, wearing rubberized apparel, and inducing vomiting—are sometimes used to meet weight ranges in competitive sports such as wrestling.

● **Heat**—The combination of high temperature and increased physical activity may increase water losses from the skin and lungs threefold to tenfold. These increased water losses must be covered by increased intake, and free access to water is essential. Water needs for individuals living in hot dry climates are increased as losses from the lungs and skin are elevated 50 to 100%. Failure to replace water losses arising from prolonged sweating may result in heat exhaustion. Furthermore, providing salt (sodium chloride) without free access to water may lead to the development of significant hypernatremia—elevated blood sodium. Symptoms of acute water lack during hard work in a hot environment appear rapidly, and recovery following water ingestion is even more rapid.

(Also see HEAT EXHAUSTION.)

● **Diet**—High-protein diets require extra water for excretion of urea. Special attention must be given to the water needs of infants on high-protein formulas, since the concentrating ability of the infant's kidneys is not well developed.

● **Illness**—In persistent vomiting large amounts of water are lost. Prolonged diarrhea also results in excessive loss of water. Fever accelerates the loss of body water though increased perspiration and the excretion of body waste. Illnesses in which a fever, vomiting, and/or diarrhea are present, can rapidly result in dehydration especially in infants, children, and older individuals.

In the opposite direction, some conditions are associated with excessive retention of water or edema. Often edema is noted in the following diseases: congestive heart failure, cirrhosis of the liver, nephritis, and nephrosis. In all of these diseases there is a reduction in sodium excretion which promotes the retention of water.

● **Injury**—During the postoperative period adequate fluid is essential since large fluid losses may occur due to vomiting, hemorrhage, exudates, diuresis, or fever. Abnormal loss of water also occurs from burns.

Water Sources. The water required to replace body losses is available from three sources: (1) drinking water and other beverages, (2) water contained in solid foods, and (3) metabolic water resulting from the breakdown (catabolism) of fats, carbohydrates, and proteins.

DRINKING WATER. Nearly three-fourths of the earth's surface is covered with water, but only 3% of this is *fresh* water—fit drinking water. Today, our drinking water comes from underground wells and reservoirs. Some is transported through miles of pipelines in huge quantities to places where the demand is great.

FOODS. Many solid foods contain a high water content, and contribute to meeting our water requirement. Even such *dry* foods as crackers, ready-to-eat cereals, and nuts provide some water, while some fruits and vegetables contain over 90% water. Table W-3 illustrates, over a wide range, the water content of foods.

TABLE W-3
WATER CONTENT OF FOODS[1]

Foods	Water
	(%)
Fruits and vegetables	70-95
Milk	87
Cooked cereals	80-88
Fish and shellfish	60-86
Cheese, cottage	79
Meat and poultry	40-75
Eggs	74
Cooked beans	69
Cheese, Cheddar	39
Bread	36
Butter	16
Crackers	2-6
Nuts	2-5

[1]Approximate values from Food Composition Table F-21.

METABOLIC WATER. Metabolic water is produced from the catabolism—breakdown—of nutrients. When 100 g of carbohydrates are oxidized, 60 g of water are produced. The oxidation of proteins yields 42 g of water for every 100 g of protein. Fats can be said to be "wetter than water." For every 100 g of fats that are oxidized, close to 110 g of water are produced. However, there are some losses of water in the oxidation of both proteins and fats. Water must be used to excrete nitrogen in the deamination process of protein—thus lowering the net availability of water. In fact, it requires more water to excrete nitrogen as urea than is formed in the deamination process. The oxidation of fats requires increased respiration. Water is lost from the lungs during this increased respiration, and the net yield of water produced from fat is less than that from the oxidation of carbohydrates.

On the average, 13 g of metabolic water are formed for every 100 Calories (kcal) of metabolizable energy in the typical human diet.

(Also see MINERAL[S], section on "Contamination of Drinking Water, Foods, or Air with Toxic Minerals"; THIRST; WATER ELECTROLYTES; and WATER BALANCE.)

WATER, AERATED

Municipal water works may aerate water by spraying into the air to remove disagreeable tastes and odors. Also, well water can be improved by aeration, particularly if it contains carbon dioxide, hydrogen sulfide, or iron. The term aerated water is synonymous to mineral water in British usage. However, mineral water in Britian is water that has been aerated or charged with carbon dioxide—soda water, or soft drink.

WATER AND ELECTROLYTES

Contents   Page
Body Fluid Compartments 1110
Regulation of Water and Electrolyte Balance 1110
Movement of Water and Electrolytes 1111
Factors Regulating Movement 1111
Imbalances of Water and Electrolytes 1111
Water Depletion 1111
Water Excess 1112
Sodium Depletion 1112
Sodium Excess 1112
Potassium Depletion 1112
Potassium Excess 1112
Chloride 1112
Acid-Base Balance 1113

Thoughts of constant, never ending motion are almost inseparable from thoughts of water—the water of the sea, of the rivers, and even household water. The water of our bodies is no different. It is constantly in motion—entering the body, leaving the body, and moving in and around the cells of the body. Life depends upon the movement of water and substances dissolved in the water which bathes the cells. Water is one of a few inorganic chemicals which exists as a liquid at the temperature of life processes. It is the chemical of life. Among the many important properties of water is its ability to act as a solvent for a variety of organic and inorganic chemicals. Of the inorganic chemicals, some of these—for example, acids, alkalis, and salts—will separate into ions when dissolved in water. Ions are atoms or molecules that carry an electrical charge—positive or negative. Such compounds which separate into positive and negative ions in solutions are called electrolytes. Those with a positive charge are cations, while those with a negative charge are anions.

Table W-4 lists ions which are found dissolved in the water of the body. Sodium, potassium, and chloride are the primary electrolytes which are most often discussed due to their important relationships in body fluids.

TABLE W-4
IMPORTANT IONS OF THE BODY

Cations			Anions		
Name	Formula	Charge	Name	Formula	Charge
Sodium	Na^+	+1	Chloride	Cl^-	-1
Potassium	K^+	+1	Bicarbonate	HCO_3^-	-1
Calcium	Ca^{++}	+2	Phosphate	$HPO_4^=$	-1
Magnesium	Mg^{++}	+2	Sulfate	$SO_4^=$	-2

Hence, electrolytes become chemical compounds such as sodium chloride (table salt), potassium chloride, calcium phosphate, magnesium sulfate, sodium bicarbonate, and so on.

Electrolytes may be dissolved in water at different concentrations. This introduces another concept. Electrolytes are measured according to the total number of particles in solution, not their total weight. Thus, the unit of measure for electrolytes is milliequivalents, which is abbreviated as mEq. It refers to the number of ions—cations and anions—in solution in a given volume, generally one liter. Hence, milliequivalents are expressed as mEq/liter. The concentration of electrolytes in the compartments of the body determines the flow of water within the body.

BODY FLUID COMPARTMENTS. Fluids—water, electrolytes, and other dissolved substances—are contained in two major compartments within the body. In order to gain this concept of body compartments, all the cells of the body must be thought of as a whole. Then, all fluid outside of the cells is termed extracellular fluid, while all fluid within the cells is termed intracellular fluid. Fluids in each compartment differ in composition.

(Also see WATER BALANCE, Fig. W-4.)

REGULATION OF WATER AND ELECTROLYTE BALANCE. Basically, water (1) enters the body via the digestive tract as liquid or food, (2) moves into the blood and tissue, and (3) leaves via the kidneys, skin, lungs, or feces. Water entering and leaving the body is under rigid control, and under normal conditions water leaving the body equals the water entering the body—a condition of water balance. Moreover, the shifts of water between various body compartments and water balance are controlled by the concentration and distribution of the electrolytes.

Movement of Water and Electrolytes. In the body, water and electrolytes are moved across cell membranes by one or more of five processes: (1) osmosis, (2) diffusion, (3) active transport, (4) filtration, and (5) pinocytosis.

Putting these five processes into perspective requires the realization that the body is a dynamic system. Adjustments are constantly taking place. Water and electrolyte balance is no different. Water enters the body as a liquid, and as a component of the food—including metabolic water derived from the breakdown of food. In the digestive tract water and the many substances dissolved in water—including electrolytes—are transported across the membrane lining the digestive tract into the blood. Then the blood plasma—a water solution—transports nutrients and other substances to the cells, and picks up the waste products from the cells. The kidneys act to regulate the composition of the body fluids by conserving some substances and excreting others into the urine. Furthermore, the kidneys regulate water loss from the body. Additional water is lost from the body by the skin, the lungs, and the feces. As water and the substances dissolved in it move through the body, some or all five of these processes—osmosis, diffusion, active transport, filtration, and pinocytosis—are at work maintaining balance within the body. In order to maintain the composition of the extracellular fluid and the intracellular fluid, the volume of water intake is very nearly equal to the volume of water lost. Whenever body fluids become too concentrated, water moves into this area and dilutes the body fluid. However, a number of factors act to regulate the movement of water.

(Also see DIGESTION AND ABSORPTION, section on "Water.")

Factors Regulating Movement. The body requires water. To ensure that this requirement is fulfilled, the sensation of thirst creates a conscious desire for water. The sensation of thirst is caused by nerve centers in the hypothalamus of the brain which monitors the concentration primarily of sodium in the blood. When the sodium concentration, and hence the osmolarity of the blood, increases above the normal 310 to 340 mg/100 ml (136 to 145 mEq/liter), cells in the thirst center shrink. They shrink because the increased osmotic pressure of the blood pulls water out of their cytoplasm. This shrinking causes more nervous impulses to be generated in the thirst center, thus creating the sensation of thirst. Increased osmolarity of the blood is primarily associated with water loss from the extracellular fluid. As water is lost the sodium concentration of the remaining fluid increases. When water is drunk, it moves across the membrane lining the gut into the blood thereby decreasing the sodium concentration—osmolarity—of the blood. In turn, the cells of the hypothalamus take on water and return to their normal size. This time water moves back into these cells via osmosis in the opposite direction.

Increased osmolarity of the blood simultaneously stimulates other nerve centers in the hypothalamus causing the release of antidiuretic hormone (ADH) from the posterior pituitary. This hormone makes a portion of the kidney permeable to water. Thus, water originally destined to be urine moves, by osmosis, back into circulation. ADH release results in the formation of less urine by the kidneys; hence, conserving body water and helping dilute the body fluids and decrease the sodium ion concentration. Thus, drinking water and excreting water are controlled by centers in the brain which help maintain body water within appropriate ranges, and in turn control the extracellular sodium ion concentration. While many changes in the body fluids are compensated for by shifting water, electrolytes are also regulated to some degree.

Sodium and potassium, the major electrolytes of the extracellular and intracellular fluid, respectively, are controlled by the hormone aldosterone which comes from the cortex (outside layer) of the adrenal gland. Of prime importance is the regulation of the potassium ion in the extracellular fluid since both nerve and muscle function are dependent on its close regulation. Aldosterone also acts to conserve sodium. Secretion of the hormone aldosterone is controlled by a chain of events called the renin-angiotensin system. Briefly, renin is released from the kidney in response to (1) changes in blood pressure, (2) blood levels of potassium, or (3) nervous stimulation. It then converts angiotensinogen, a protein in the blood, to angiotensin which stimulates the secretion of aldosterone. Aldosterone causes the reabsorption of sodium from the urine and the loss of potassium in the urine. Potassium is exchanged for sodium. This conservation also causes water to be reabsorbed via osmosis. Under some stressful situations such as surgery, the release of adrenocorticotropin hormone (ACTH) from the anterior pituitary may cause the secretion of aldosterone.

It is evident that a marvelous organ, the kidney, is largely responsible for regulating the water and electrolyte balance of the body. Each kidney contains about one million minute functional units called nephrons. As the blood passes through these nephrons, they select, reject, conserve, and eliminate water, electrolytes, and other substances in order to maintain the volume and the composition of the extracellular fluid. Each day they rejuvenate about 50 gal (190 liter) of blood. Moreover, they function to regulate red blood cell production, aldosterone secretion, blood pressure, and calcium metabolism. So important is their function that without it death results in 8 to 14 days.

IMBALANCES OF WATER AND ELECTROLYTES.
When the day-to-day and individual-to-individual variations in water and electrolyte intake are considered, maintaining the consistent water and electrolyte levels of the body is truly an amazing feat. However, there are times when the controlling mechanisms are disrupted and excesses or deficiencies result. Of prime concern are water and the major electrolytes, sodium, potassium, and chlorine. So intimate are the relationships between water and the major electrolytes that separation is often difficult.

Water Depletion. Life without water is short. When water is unavailable, or when water is lost faster than it can be replaced, events occur in the following order:
1. Sensation of thirst, when water loss amounts to about 1% of body weight.
2. Thirst accompanied by vague discomfort and loss of appetite.
3. Tingling and numbness in arms and hands.

4. Increase in pulse rate, respiratory rate, and body temperature.

5. Weakness, spastic muscles, and mental confusion.

6. Increase in concentration of the blood (hemoconcentration), decreased blood volume, and difficult circulation.

7. Cracked skin and cessation of urine formation.

8. Death when dehydration weight loss becomes greater than 20% of the initial weight.

Various stages of the above may occur when losses of water are incurred by (1) evaporation, (2) vomiting and diarrhea, (3) hemorrhage, or (4) burns.

Water Excess. Sometimes excessive water accumulates in the tissues, specifically the interstitial compartment. Outwardly, this condition is noted as swelling and it may occur in any area of the body. It is called *edema*, though some may still call it *dropsy*. In general, there are four causes of edema: (1) elevated fluid pressure in the capillaries as in heart failure; (2) low osmotic pressure in the blood due to decreased blood protein in such conditions as liver cirrhosis, kidney disease, severe burns, and starvation; (3) blockage of the lymph vessels as caused by the parasitic worm, filariae, in the disease elephantiasis; or (4) increased capillary permeability due to the release of histamine in allergic reactions. Edema may require the restriction of dietary sodium and/or diuretics.

• **Diuretics**—These are drugs which act to increase the output of sodium and water in the urine, and are often used to treat disorders of the heart, kidneys, or liver which cause edema. A majority of the diuretics act upon the kidneys by depressing the sodium reabsorbed. Thus, sodium remaining in the urine carries more water out of the body with it. Diuretics include thiazides, furosemide, and ethacrynic acid. Xanthine diuretics are mild diuretics but they are used by many people—perhaps unknowingly—since they are the caffeine, theophylline, and theobromine present in tea, coffee, cola and other soft drinks, cocoa, and many over-the-counter pain relievers. Moreover, water and ethyl alcohol (alcoholic beverages) can act as diuretics by inhibiting the release of ADH.

• **Water intoxication**—When water intake is more rapid than urine formation, the extracellular compartment fluid is diluted and water moves into the cells—cellular edema. Swelling of the cells of the brain causes drowsiness and weakness, convulsions, and coma. Water intoxication may be observed in (1) patients given excessive amounts of intravenous glucose and water, (2) individuals who absorb water from the colon during enemas or colon irrigations, (3) individuals who absorb water from wounds or burns treated with wet dressings, or (4) individuals who have impaired antidiuretic hormone.

Sodium Depletion. Sodium is the major positively charged ion—cation—in the extracellular fluid. Depletion is rare. Urinary output reflects the dietary intake. However, strict vegetarian diets without salt, heavy prolonged sweating, diarrhea, and vomiting or adrenal cortical insufficiency—lack of aldosterone—may result in sodium depletion. Continued depletion results in loss of appetite, muscle cramps, mental apathy, loss of body water, headache, and reduced milk production in lactating mothers. Sodium depletion/heat exhaustion occurs most frequently in persons unacclimatized to working in a hot environment who replace water losses but fail to replace electrolyte losses. It is characterized by fatigue, nausea, giddiness, vomiting, and exhaustion. The blood volume decreases, kidney blood flow is impaired, and cellular edema occurs.

Sodium Excess. Most of the sodium ingested is excessive, and for the most part it is excreted by the kidneys in combination with bicarbonate or phosphate. However, under some circumstances sodium accumulates in the extracellular fluid and causes edema since the retention of sodium is accompanied by water retention. Such conditions include (1) cardiac or renal failure, (2) adrenal tumors which secrete excessive cortical hormones, and (3) adrenocorticotropic hormone (ACTH) or steroid hormone therapy. In these conditions, individuals benefit from sodium-restricted diets.

Excessive sodium may be harmful (1) when ingested in large quantities, especially with a low water intake; (2) when the body has been depleted by a salt-free vegetarian diet or excess sweating, followed by gorging on salt; (3) when large amounts are fed to infants or individuals afflicted with kidney diseases, whose kidneys cannot excrete the excess in the urine; or (4) when the body is adapted to a chronic low-salt diet, followed by ingesting large amounts of salt. A dramatic example of excessive sodium (salt) is provided by imagining a sailor lost at sea who decides to drink sea water. The sea water is a much more concentrated salt solution than the extracellular fluids. Drinking sea water would only worsen a stranded sailor's condition.

(Also see SALT; and SODIUM.)

Potassium Depletion. Deficiencies of potassium rarely result from dietary lack of the mineral. Potassium is lost whenever muscle is broken down owing to starvation, malnutrition, or injury since it is tied to protein inside cells. Crash diets, diarrhea, vomiting, gastric suction, diabetic acidosis, and burns also induce potassium loss from the body. Also lean tissue growth increases the need for potassium. A potassium-depleted individual may display irregular heart function, muscle weakness, irritability, paralysis, nausea, vomiting, diarrhea, and swollen abdomen.

(Also see MALNUTRITION, PROTEIN-ENERGY; and STARVATION.)

Potassium Excess. The kidneys provide the major regulatory mechanism for maintaining potassium balance. Hence, a potassium buildup in the blood—hyperkalemia—is the frequent complication of kidney failure. Other causes of potassium excess include adrenal insufficiency, severe dehydration, or shock after injury wherein the potassium in the cells leaks into the blood. Symptoms of an excess are muscular weakness, mental apathy, and irregular heart action. Both an excess and depletion of potassium affect the heart muscle. Indeed, if either becomes severe enough the heart will stop. Although, it would be nearly impossible to ingest enough potassium-containing foods to create an excess, ill advised individuals who take a supplemental source such as potassium chloride run the risk of increasing potassium to dangerous levels.

(Also see POTASSIUM.)

Chloride. Excess chlorine in the diet is excreted via the urine accompanied by excess sodium or potassium and sometimes ammonia. Loss of chloride generally parallels that of sodium. However, deficiencies of chloride may develop from vomiting, diarrhea, stomach pumping, injudicious use of diuretic drugs, or strict vegetarian diets used without salt. A deficiency of chloride causes an increase in the pH of the body. This condition is called alkalosis and it is characterized by slow and shallow breathing, listlessness, muscle cramps, lack of appetite, and sometimes convulsions. (Also see CHLORINE OR CHLORIDE.)

ACID-BASE BALANCE. Aside from their involvement in body fluid distribution, water and electrolytes are involved in the acid-base balance of the body. This is important since the body tolerates only very minor shifts in this balance which refers to the hydrogen ion concentration of the body fluids. An acid is a chemical that can release hydrogen ions, whereas a base, or alkali, is a chemical that can accept hydrogen ions. The degree of acidity is expressed in terms of pH. A pH of 7 is the neutral point between an acid and alkaline (base). Substances with a lower pH than 7 are acid, while substances with a pH above 7 are alkaline. The normal pH of the extracellular fluids of the body is 7.4, with a range of 7.35 to 7.45. Maintenance of the pH within this narrow range is necessary to sustain the life of cells. The extremes between which life is possible are 7.0 to 7.8. To help maintain this narrow range the body has other chemicals called buffers. A buffer protects the acid-base balance of a solution by rapidly offsetting changes in its ionized hydrogen concentration—a chemical sponge. It protects against added acid or base. The three most important chemical buffers are the bicarbonate buffer, phosphate buffers, and hemoglobin and protein buffers.

The lungs and kidneys also have an important role in maintaining the acid-base balance. The extent to which they are involved depends upon the amount of adjustment necessary.

WATER BALANCE

Water is the major constituent of the body; it accounts for 50 to 75% of the body weight. Next to oxygen, it is the most important constituent for life itself. A person can live for several weeks without food but for only a few days without water. Dehydration (water loss) will kill far quicker than starvation.

Large deficits or excesses in the body water are reason for concern. In healthy individuals, the total amount of body water remains reasonably constant. Therefore, an increase or decrease in water intake brings about an appropriate increase or decrease in water output to maintain the balance. Fig. W-4 illustrates the intake of water, the routes of water output, and the movement of fluid between the compartments of the body. Water enters the body as a liquid, and as a component of the food—including metabolic water derived from the breakdown of food. Water is lost from the body by (1) the skin as perspiration, (2) the lungs as water vapor in expired air, (3) the kidneys as urine, and (4) the intestines in the feces. As Fig. W-4 shows, under normal conditions total water intake is approximately equal to total water output by the various routes.

The body is equipped with a number of mechanisms for regulating body water within narrow limits. Important among these mechanisms are nerve centers in the hypothalamus which control the sensation of thirst and water output by the kidneys. Stimulation of the thirst center in the hypothalamus creates the desire for water, while stimulation of other nerve centers in the hypothalamus causes the release of antidiuretic hormone (ADH) from the posterior pituitary. Release of ADH results in the formation of less urine; hence, conserving the body water. Water needs of the body are disrupted by fever, high protein diets, dry hot climates, high altitudes, vomiting, diarrhea, and injury.

(Also see BODY FLUIDS; ENDOCRINE GLANDS; METABOLISM; METABOLIC WATER; and WATER AND ELECTROLYTES.)

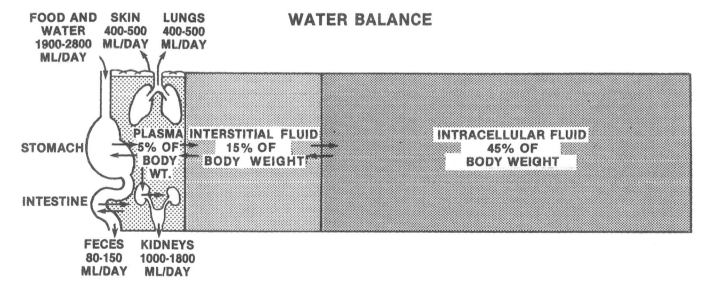

Fig. W-4. Water balance in humans showing the daily amount of water input and output by the various routes, and the compartments of the body which contain water. Arrows represent fluid movement.

WATERBORNE DISEASES

Many of the same diseases that are transmitted by contaminated food may also be transmitted by contaminated water. Important among these diseases are amebic dysentery, bacillary dysentery, cholera, intestinal flu, infectious hepatitis, poliomyelitis, salmonellosis, and typhoid fever. In addition, certain parasitic diseases, including beef tapeworms, fish tapeworms, liver flukes, and round worms, may be spread to man, either directly or indirectly, through contaminated water.

Because of the importance of water to the normal function of the body, it is imperative that our water supplies be safeguarded. Large cities sterilize water by the addition of chlorine, while questionable water in the home may be boiled vigorously for 1 to 3 minutes. However, chlorine or boiling does not eliminate toxic levels of industrial chemicals or metals that find their way into water supplies.

(Also see DISEASES, Table D-10, Infectious and Parasitic Diseases Which May Be Transmitted By Contaminated Foods and Water; and WATER.)

WATER, DEMINERALIZED (DEIONIZED WATER)

By passing water through two ion exchange resins, all mineral salts can be removed—both anions (negatively charged elements) and cations (positively charged elements). This is demineralized water. It is as pure as water can be. Frequently, demineralized water is utilized in scientific investigation to prevent the introduction of interfering substances during sensitive analyses of minute quantities.

(Also see WATER.)

WATER, EXTRACELLULAR

That water in the body which is found outside of the cells of the body. The blood plasma and interstitial fluid make up the extracellular water.

(Also see BODY FLUIDS; and WATER BALANCE.)

WATER GLASS

This is a common term for the chemical, *sodium silicate*, a solution of which may be used to preserve eggs by sealing the pores in the shell. Other uses include fireproofing of fabrics, waterproofing, a detergent in soaps, and adhesives.

WATER HARDNESS

Hard water contains the bicarbonate (HCO_3) and sulfate (SO_4) salts principally of calcium (Ca) and magnesium (Mg) leached from mineral deposits in the earth, though aluminum and iron salts are sometimes involved. Water hardness due to calcium and magnesium bicarbonate is known as temporary or bicarbonate hardness since boiling the water results in the decompositon of the bicarbonates and the precipitation of calcium and magnesium carbonate (CO_3) or hydroxide (OH). Boiling produces no change in water containing calcium and magnesium sulfate. Hence, the hardness caused by sulfates is known as permanent or noncarbonate hardness.

Following a chemical analysis, water hardness is expressed in parts per million (ppm) of an equivalent amount of calcium carbonate. This method is used for expressing the amount of magnesium as well as the amount of calcium, and for expressing the noncarbonate as well as the carbonate hardness. Total hardness of water varies with locality and source. A water with a total hardness of less than 100 ppm of calcium carbonate is generally considered soft, while a water with a total hardness about 300 ppm is considered very hard.

Hard water requires more soap because of the formation of insoluble salts of calcium and magnesium with soap. Furthermore, hard water leaves deposits in pipes and appliances. Some municipal water supplies have reported 2,000 ppm and some are known to be as high as 4,400 ppm.

(Also see WATER.)

WATER, INTRACELLULAR

That water in the body which is contained within the cells of the body. It represents about 45% of the body weight.

(Also see BODY FLUIDS; and WATER BALANCE.)

WATER LEMON (YELLOW GRANADILLA)
Passiflora laurifolio

This type of passion fruit (of the family *Passifloraceae*) grows wild in the hot, humid lowlands of the West Indies and northeastern South America. The fruits of this vine have an orange-yellow peel and a seedy, white pulp. They may be eaten fresh or made into confections, ice cream, ices, jams, jellies, juices, and sherbets. Its excellent flavor has made it popular throughout the tropical areas of the world.

(Also see FRUIT[S], Table F-23, Fruits of the World—"Passion Fruit.")

WATERMELON *Citrullus vulgaris*

Fig. W-5. Watermelons. (Courtesy, University of Maryland, College Park, Md.)

Watermelons are the fruit of an annual prostrate vine with multiple stems, branching out 12 to 15 ft (*4 to 5 m*). They may weigh 5 to 85 lb (*2.3 to 38.3 kg*), and vary in shape from round to oval to oblong-cylindrical. On the outside the hard rind of watermelons may be very light to very dark green with stripes or mottling, while the inside edible flesh (pulp) is red, pink, orange, yellow, or white. Red is the most familiar color in the United States. Most watermelons contain white, brown, or black seeds, but there are seedless watermelons.

The pumpkin, squash, muskmelon, and cucumber are relatives of the watermelon and all are members of the cucumber family, *Cucurbitaceae*.

ORIGIN AND HISTORY. Watermelons originated in Africa, where they still occur in the wild, and where they have been cultivated for 4,000 years. From Africa, they spread to Asia, thence throughout the world. They were introduced in America soon after its discovery.

WORLD AND U.S. PRODUCTION. Watermelons rank fifth among the leading fruit crops of the world.

Worldwide, about 28.9 million metric tons of watermelons are produced annually; and the leading producing countries, by rank, are: China, U.S.S.R., Turkey, Egypt, and United States.[3]

The United States produces about 1.3 million short tons of watermelons each year. Florida and Texas are the leading producers.[4]

PROCESSING. Almost all of the watermelon crop is consumed fresh after being chilled. Only small amounts are used to make processed fruit products.

SELECTION, PREPARATION, AND USES. Judging the quality of a watermelon is very difficult unless it is cut in half or quartered. When cut, indicators of a good watermelon include firm, juicy flesh with good red color, free from white streaks, and seeds which are dark brown or black.

When selecting an uncut watermelon, a few appearance factors are helpful, though not totally reliable. The watermelon surface should be relatively smooth; the rind should have a slight dullness (neither shiny nor dull); the ends of the melon should be filled out and rounded; and the underside, or *belly*, of the melon should have a creamy color.

Preparation is simple; after being chilled, watermelons are generally sliced and served with a fork or spoon. Some people prefer a little salt to enhance the sweetness. Also, the seeds can be removed from fresh watermelon, following which it can then be used in fruit dishes and desserts.

Sometimes, the rinds of watermelons are pickled or candied. In the U.S.S.R., a fermented drink is made from the juice. Sometimes, Orientals preserve chunks of watermelon in brine.

In some countries of the Middle East and in China, the seeds are also eaten. They are roasted and eaten like popcorn, or preserved in salt.

Fig. W-6. Sliced watermelon; simple preparation; significant amount of vitamin A. (Courtesy, USDA)

[3]Data from *FAO Production Yearbook 1990*, FAO/UN, Rome, Italy, Vol. 44, p. 149, Table 63. **Note well:** Annual production fluctuates as a result of weather and profitability of the crop.

[4]*Ibid.*

NUTRITIONAL VALUE. The nutritional values of fresh watermelon and watermelon seeds are listed in Food Composition Table F-21.

Fresh watermelon is true to its name. It contains about 93% water, and only 26 Calories (kcal) per 100 g (about 3 1/2 oz). The calories are primarily derived from the naturally-occurring sugar which gives watermelon its sweetness. Watermelons contain 6 to 12% sugar (carbohydrate) depending upon the variety and growing conditions. Vitamin A is present in significant amounts—about 590 IU/100 g.

The nutritional composition of watermelon seeds is typical of many other seeds and nuts—high in protein, fat, carbohydrates, and calories. Each 100 g of seeds contains about 23 g protein, 40 g of fat, and 27 g of carbohydrate for a whopping 536 Calories (kcal) per 100 g.

WATER-SOLUBLE VITAMINS

The large amounts of water which pass through most animals daily tend to carry out the water-soluble vitamins of the body, thereby depleting the supply. Thus, they must be supplied in the diet on a day-to-day basis. The water-soluble vitamins are: vitamin C (ascorbic acid), B-1 (thiamin), B-2 (riboflavin), B-6 (pyridoxine), niacin (nicotinic acid), pantothenic acid, biotin, folic acid (folacin), choline, B-12, and inositol.

(Also see VITAMIN[S].)

WAX, APPLE

The waxy material present in apple skin which is sometimes removed by dewaxing with hot isopropyl alcohol vapor in order to reduce the amount of peel lost during the processing of apples into canned slices or juice. Dewaxing of apples makes it possible to remove the peel with a lye solution.

(Also see APPLE.)

WAX GOURD (CHINESE PRESERVING MELON)
Benincasa hispida

The fruit of a vine (of the family *Cucurbitaceae*) that is native to Malaysia, but is now grown throughout the tropics of Asia.

Wax gourds are large, heavy, oblong fruits that range from 6 to 8 in. *(15 to 20 cm)* in diameter, and from 8 to 14 in. *(20 to 35 cm)* long. The fruits may be boiled or candied, and the seeds fried. Also, the young leaves and flower buds are eaten as vegetables.

The nutrient composition of the wax gourd is given in Food Composition Table F-21.

Some noteworthy observations regarding the nutrient composition of the wax gourd follow:

1. The raw fruit has a very high water content (96%) and is very low in calories (13 kcal per 100 g) and carbohydrates (3%).

2. Wax gourds are a fair to good source of potassium and vitamin C.

(Also see GOURDS.)

WEIGHT AND HEIGHT-AGE PERCENTILE STANDARDS

Charts or tables used to determine deviations in growth patterns—weight and height—from normal growth patterns. They are based on careful measurements of selected populations of children over a period of years, and they are expressed in percentiles—distribution of ranked values for weight and height divided into hundreths. Thus, the 50th percentile on the table or chart represents the median, or in other words the central tendency of all children. These weight and height-age percentile standards or growth charts may be used to spot undernutrition or overnutrition in children or adolescents.

(Also see CHILDHOOD AND ADOLESCENT NUTRI-TION, Section on "Childhood Growth and It's Measurement"; and INFANT DIET AND NUTRITION, section on "Measurement of Growth.")

WEIGHTS AND MEASURES

Weights and measures are the standard employed in arriving at weights, quantities, and volumes. Even among primitive people, such standards were necessary; and with the growing complexity of life, they became of greater and greater importance.

COMMON FOOD WEIGHTS AND MEASURES. In preparing foods, it is often more convenient for the cook to measure the ingredients, rather than weigh them. Table W-5, Common Food Weights and Measures, will serve as a useful guide when preparing foods by measure.

TABLE W-5
COMMON FOOD WEIGHTS AND MEASURES
(All measurements are level)

ABBREVIATIONS COMMONLY USED

tsp	=	teaspoon	oz	=	ounce or ounces
Tbsp	=	tablespoon	lb	=	pound or pounds
c	=	cup	sq	=	square
pt	=	pint	min	=	minute or minutes
qt	=	quart	hr	=	hour or hours
gal	=	gallon	mod.	=	moderate or moderately
pk	=	peck	doz	=	dozen
bu	=	bushel			

MEASUREMENTS

3 tsp	= 1 Tbsp		1 c	=	½ pt
4 Tbsp	= ¼ c		2 c	=	1 lb
5⅓ Tbsp	= ⅓ c		2 c	=	1 pt
8 Tbsp	= ½ c		2 pt (4 c)	=	1 qt
16 Tbsp	= 1 c		4 qt (liquid)	=	1 gal
1 oz	= 2 Tbsp		8 qt (solid)	=	1 pk
1 gill	= ½ c		4 pk	=	1 bu
8 oz	= 1 c				
16 oz	= 1 lb				

BUTTER OR MARGARINE

2 Tbsp	=	1 oz
½ c	=	¼ lb = 1 stick
2 c	=	1 lb

CEREALS

Rice	1 c	=	½ lb
Rice	1 c raw precooked	=	2 c cooked
Rice	1 c raw converted	=	3 to 4 c cooked
Rice	1 c raw long-grain	=	4 c cooked
Noodles	1 c	=	1¼ c cooked
Macaroni	1 c	=	2¼ c cooked

COFFEE, GROUND

1 lb	=	80 Tbsp or 5 c
	=	½ c makes 10 c beverage

EGGS

5 eggs	=	about 1 c
8 to 10 egg whites	=	1 c
12 to 15 egg yolks	=	1 c

FLOUR

All-purpose	4 c sifted	=	1 lb
Cake	1 c sifted	=	1 c all-purpose flour less 2 Tbsp
Cornmeal	3 c	=	1 lb
Potato flour (for thickening)	1 Tbsp	=	2 Tbsp flour
Cornstarch (for thickening)	1 Tbsp	=	2 Tbsp flour
Arrowroot (for thickening)	2 tsp	=	5 tsp flour

FRUITS

Apples	1 lb	=	3 med. or 3 c sliced
Candied fruit	1 lb	=	1½ c
Lemon (whole)	1	=	2 to 3 Tbsp juice
Lemon (grated rind)	1	=	about 1½ to 2 tsp
Orange (whole)	1	=	6 to 8 Tbsp or ⅓ to ½ c juice
Orange (grated rind)	1	=	about 1 Tbsp
Raisins	1 lb	=	3 c

MARSHMALLOWS

¼ lb	=	16

MILK AND CHEESE

Milk	1 c	=	½ c evaporated milk + ½ c water
Milk	1 c	=	4 Tbsp powdered whole milk + 1 c water
Cream	1 c	=	2 c whipped cream
Cheese (grated)	4 to 5 c	=	1 lb
Cottage cheese	1 c	=	½ lb
Cream cheese	3-oz package	=	7 Tbsp

NUTS

Unshelled	1 lb	=	2 c nut meats
Shelled	1 lb	=	3 to 4 c nut meats

SUGAR

Granulated	1 lb	=	2 c
Brown	1 lb	=	2¼ c
Confectioners	1 lb	=	3½ c
Powdered	1 lb	=	2⅓ c

CONTENTS OF CANS

Size (No.)	Weight	Measure (c)
¼	4 to 4.5 oz	½
½	7.5 to 8 oz	1
Picnic		1¼
No. 1 short or No. 300	10 to 13 oz	1¾
No. 1 tall or No. 303	1 lb	2
No. 2	1 lb 4 oz	2½
No. 2½	1lb 14 oz to 15 oz	3½
No. 3	2 lb to 2 lb 1 oz	4
No. 10	6 lb 8 oz to 8 lb 12 oz	12 to 13

TEMPERATURES

	Fahrenheit (°F)	Centigrade (°C)
Simmering point	180	82
Boiling point of water at sea level	212	100
Ovens		
Very slow	200 - 250	93 - 121
Slow	300	149
Moderately slow	325	163
Moderate	350	177
Moderately hot	375	191
Hot	400	204
Very hot	450 - 500	232 - 260
Extremely hot	over 500	260

Fig. W-7. Scales were developed by the ancient Egyptians to weigh grains.

ENGLISH SYSTEM. In about 1300, London merchants adopted a weight system called *avoirdupois* (from the old French term, *aveir-de-peis*), meaning *goods of weight*. This system, which was used to weigh bulky goods, is based on a pound of 7,000 grains or 16 oz. It is still used in many English-speaking countries.

U.S. CUSTOMARY SYSTEM. On May 29, 1830, the Senate passed a resolution directing the Secretary of the Treasury to make a comparison of the weights and measures in use in the principal customhouses. The study was made, followed by real progress toward the unification of weights and measures in the United States through the subsequent distribution of uniform standards to the customhouses based on the following: (1) the yard of 36 in.; (2) the avoirdupois pound of 7,000 grains; (3) the wine gallon of 231 cu. in.; and (4) the Winchester bushel of 2,150.42 cu.in. These units are still in use in the United States.

METRIC SYSTEM. The Metric System is a decimal system based on multiples of ten.

The basic metric units are the *meter* (length/distance), the *gram* (weight), and the *liter* (capacity). The units are then expanded in multiples of 10 or made smaller by ⅒. The

prefixes, which are used in the same way with all basic metric units, follow:

''milli-''	=	1/1,000
''centi-''	=	1/100
''deci-''	=	1/10
''deca-''	=	10
''hecto-''	=	100
''kilo-''	=	1,000

CONVERSIONS OF U.S. CUSTOMARY AND METRIC.
A comparison of U.S. Customary and Metric Systems is shown in Figs. W-9 and W-10.

Fig. W-9. Inches-centimeter scale for direct conversion and reading.

Fig. W-10. Farenheit-Celsius (Centigrade) scale for direct conversion and reading.

The following tables will facilitate conversion from U.S. Customary to Metric units, and vice versa:

TABLE W-6
WEIGHT EQUIVALENTS

1 lb	= 453.6 g = .4536 kg = 16 oz
1 oz	= 28.35 g
1 kg	= 1,000 g = 2.2046 lb
1 g	= 1,000 mg
1 mg	= 1,000 mcg = .001 g
1 mcg	= .001 mg = .000001 g
1 mcg per g or 1 mg per kg is the same as ppm

TABLE W-7
WEIGHT-UNIT CONVERSION FACTORS

Units Given	Units Wanted	For Conversion Multiply By	Units Given	Units Wanted	For Conversion Multiply By
lb	g	453.6	kcal/kg	kcal/lb	0.4536
lb	kg	0.4536	kcal/lb	kcal/kg	2.2046
oz	g	28.35			
kg	lb	2.2046	ppm	mcg/g	1.
kg	mg	1,000,000.	ppm	mg/kg	1.
kg	g	1,000.	ppm	mg/lb	0.4536
g	mg	1,000.	mg/kg	%	0.0001
g	mcg	1,000,000.	ppm	%	0.0001
mg	mcg	1,000.	mg/g	%	0.1
			g/kg	%	0.1
mg/g	mg/lb	453.6			
mg/kg	mg/lb	0.4536			
mcg/kg	mcg/lb	0.4536			
Mcal	kcal	1,000.			

Fig. W-8. Accurate measurements make for a successful cook. (Courtesy, American Egg Board, Park Ridge, Ill.)

TABLE W-8
CONVERSION FACTORS
U.S. CUSTOMARY TO METRIC

Symbol	When You Know	Multiply By	To Find	Symbol
U.S. Customary			Metric	
LENGTH				
in.	inches	2.5	centimeters	cm
ft	feet	30.	centimeters	cm
yd	yards	0.9	meters	m
mi	miles	1.6	kilometers	km
AREA				
in.²	square inches	6.5	square centimeters	cm²
ft²	square feet	0.09	square meters	m²
yd²	square yards	0.8	square meters	m²
mi²	square miles	2.6	square kilometers	km²
	acres	0.4	hectares	ha
MASS (weight)				
oz	ounces	28.	grams	g
lb	pounds	0.45	kilograms	kg
	short tons (2000 lb)	0.9	metric ton	t
VOLUME				
tsp	teaspoons	5.	milliliters	ml
Tbsp	tablespoons	15.	milliliters	ml
in.³	cubic inches	16.	milliliters	ml
fl oz	fluid ounces	30.	milliliters	ml
c	cups	0.24	liters	liter
pt	pints	0.47	liters	liter
qt	quarts	0.95	liters	liter
gal	gallons	3.8	liters	liter
ft³	cubic feet	0.03	cubic meters	m³
yd³	cubic yards	0.76	cubic meters	m³
TEMPERATURE (exact)				
°F	degrees Fahrenheit	5/9 (after subtracting 32)	degrees Celsius	°C (Centigrade)

TABLE W-9
CONVERSION FACTORS
METRIC TO U.S. CUSTOMARY

Symbol	When You know	Multiply By	To Find	Symbol
Metric			U.S. Customary	
LENGTH				
mm	millimeters	0.04	inches	in.
cm	centimeters	0.4	inches	in.
m	meters	3.3	feet	ft
m	meters	1.1	yards	yd
km	kilometers	0.6	miles	mi
AREA				
cm²	square centimeters	0.16	square inches	in.²
m²	square meters	1.2	square yard	yd²
km²	square kilometers	0.4	square miles	mi²
ha	hectares (10,000 m²)	2.5	acres	
MASS (weight)				
g	grams	0.035	ounces	oz
kg	kilograms	2.2	pounds	lb
t	metric ton (1000 kg)	1.1	short tons	

VOLUME

ml	milliliters	0.03	fluid ounces	fl oz
ml	milliliters	0.06	cubic inches	in.³
liter	liters	2.1	pints	pt
liter	liters	1.06	quarts	qt
liter	liters	0.26	gallons	gal
m³	cubic meters	35	cubic feet	ft³
m³	cubic meters	1.3	cubic yards	yd³

TEMPERATURE (exact)

°C	degrees Celsius (Centigrade)	9/5 (then add 32)	degrees Fahrenheit	°F

WHALE OIL

A yellowish-brown oil made by (1) boiling the blubber of whales, and (2) skimming off the oil. After hardening by hydrogenation some of it is used for making margarine and soap.

WHEAT, family *Gramineae*; genus *Triticum*

Contents	Page
Origin and History	1119
World and U.S. Production	1119
Kinds of Wheat	1119
Milling Wheat	1120
Flour	1121
Grades of Flour; Test Baking	1121
Kinds of Flours	1121
Enriched Flour	1122
Nutritional Value of Wheat	1123
Wheat Products and Uses	1124

Fig. W-11. Harvesting wheat. (Courtesy, Deere & Co., Moline, Ill.)

Wheat, the world's most important grain crop, is a member of the grass family, *Gramineae*, and of the genus *Triticum*.

Wheat provides more nourishment for more people throughout the world than any other food; indeed, it is the staff of life. While rice is the common food in the Orient, wheat is basic to the diet of Europe, Africa, North and South

America, Australia, and a large part of Asia. One-third of the world's population depends on wheat as its main staple. Total world production of wheat is about 250 g per day per person, enough to supply 800 Calories (kcal) and 30 g of protein daily if it were evenly distributed and unrefined. In many developing countries, wheat supplies 40 to 60% of the available energy and protein.

Wheaten foods provide generously of carbohydrates, protein, and certain minerals and vitamins. Of all the cereals, wheat alone could meet minimum protein requirements if used as the sole cereal product, with the exception of infants and young growing children (who need supplemental lysine when fed wheat-based diets) and possibly pregnant and lactating females. A diet with large amounts of wheat and only small amounts of protein from animal sources provides adequate quantity and quality protein.

(Also see CEREAL GRAINS.)

ORIGIN AND HISTORY. The development and progress of civilization can be linked to the history of wheat. Prior to recorded history, man cultivated wheat.

No one knows where the wheat plant originated, although it was cultivated where modern man is supposed to have first appeared—in southwestern Asia. The common ancestor of all wheats is believed to be a species called *wild einkorn* (meaning *one seed*), found in excavated ruins in the upper reaches of the Tigris-Euphrates basin in Southwestern Asia—called the "Fertile Crescent," the presumed birthplace of our civilization. The 14- and 21-chromosome species of wheats are believed to have developed as natural hybrids of the original einkorn.

Fig. W-12. The fertile crescent.

Man probably used wheat as food 10,000 to 15,000 years before Christ. In 1948, archaeologists from the University of Chicago found kernels of wheat, believed to be about 6,700 years old, preserved in the ruins of an ancient village in Iraq.

The first wheat food was probably the grain itself, stripped of its husk, or glumes, and chewed, a way in which it is still eaten in many parts of the world. But a form of bread was found in the remains of a Stone Age village of Swiss lake dwellers.

The reverence in which people held wheat and bread through the ages still lives in the Lord's Prayer: "Give us this day, our daily bread." In the Hebrew faith, the eating of unleavened matzoh during the Passover also marks the significance of bread in religion. In Europe, bread is commonly referred to as "the staff of life."

Fig. W-13. Throughout the ages, wheat and bread have lived in the Lord's Prayer.

WORLD AND U.S. PRODUCTION. Wheat is the leading cereal grain of the world. Farmers of the world grow approximately 595 million metric tons of wheat each year.[5] The Soviet Union, which produces about 18% of the world crop, is the leading wheat-producing nation of the world. But the U.S.S.R. does not grow enough wheat for its own needs.

The United States ranks third among the wheat-producing countries of the world. The ten leading wheat-producing countries of the globe are: U.S.S.R., China, United States, India, France, Canada, Turkey, Australia, Germany, and Pakistan.[6]

In the United States, wheat ranks second only to corn among the cereal grains in total acreage and production. Wheat is grown in every state of continental United States, although production in New England is minor. The greatest acreage is in the Central Plains and north central states.

The United States produces 74.5 million metric tons of wheat annually; and the ten leading wheat-producing states of the United States, by rank, are: Kansas, Oklahoma, Washington, Texas, North Dakota, Montana, Illinois, Nebraska, Colorado, and Ohio.[7]

KINDS OF WHEAT. Wheat is classified according to climatic adaptation, color of kernel, relative hardness of kernel, species, and varieties.

• **Winter wheat vs spring wheat**—There are two broad kinds of wheat, winter wheat and spring wheat.

Winter wheats, which are adapted to the Middle Great Plains, are planted in the fall and harvested in the following June and July.

[5]Data from *FAO Production Yearbook 1990*, FAO/UN, Rome, Italy, Vol. 44, p. 70, Table 16. **Note well:** Annual production fluctuates as a result of weather and profitability of the crop.

[6]*Ibid.*

[7]Data from *Agricultural Statistics 1991*, USDA, p. 6, Table 8. **Note well:** Annual production fluctuates as a result of weather and profitability of the crop.

Spring wheats are planted in the spring after the threat of frost is over and the ground is dry enough to work, and they ripen in the summer of the same year, usually a few weeks after winter wheat. Each group includes varieties of both hard and soft wheat.

• **Color of kernel**—Both winter and spring wheat produce grain that is red or white, with various shades of yellows or amber.

• **Texture of the ripened grain**—Wheat is classed as hard or soft. Hard wheats tend to be higher in protein content than soft wheats, and are primarily used in bread flour. Durham wheats are also hard wheats, used for macaroni production. Softer wheats are lower in protein and are chiefly milled into flour for cakes, cookies, pastries, and crackers.

• **Species of wheat**—All the wheats grown throughout the world belong to one of 14 species.

Of the 14 species, only three—common, club, and durum—account for 90% of all wheat grown in the world today.

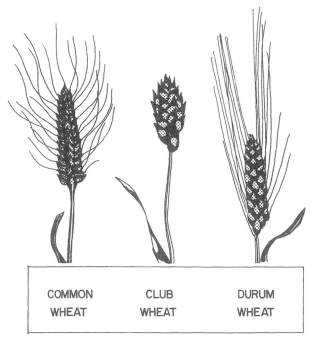

Fig. W-14. Heads of three most common types of wheat.

Common wheat is bread wheat. It probably originated in Turkey and southern U.S.S.R. Over 200 varieties of common wheat have been described, of which about 100 are now cultivated. They may be either red or white, hard or soft, and spring or winter type.

Club wheat is grown primarily for flour, in the Pacific Northwest. It may be either winter or spring type.

Durum wheat is grown for spaghetti, macaroni, and noodles. The kernels are white or red. The varieties of this species grown in North America are all spring wheats, grown in North Dakota, South Dakota, Montana, Minnesota, and parts of Canada.

MILLING WHEAT.   The main parts of the wheat grain are:

1. **The endosperm**, which constitutes about 83% of the kernel and is the source of white flour.

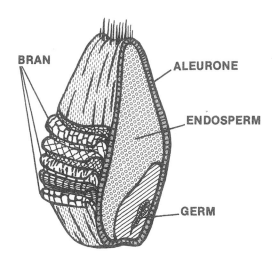

Fig. W-15. Structure of the wheat grain.

2. **The coat** (bran or coarse bran; and aleurone or fine bran), which constitutes about 14.5% of the kernel and is the source of bran, used primarily for animal feed.

3. **The germ**, which constitutes 2.5% of the kernel and is the source of wheat meal and wheat germ oil, used for human food or animal feed.

On threshing, the husk is removed and the grain is made naked. The wheat endosperm is covered with two kinds of fibrous coatings; the coarsest outer layer is called bran, and the less fibrous layer is the aleurone. At the base of the kernel is the germ. The objective of milling is to separate the starchy endosperm from the other parts of the grain. Whole wheat yields about 72% white flour, and 28% by-products, consisting of bran and germ. (By contrast, in China, by official ruling, the minimum extraction rate permissible for wheat flour is 81%; hence, the wheat flour of China is less refined than in the United States.) Since an average bushel of wheat weighs 60 lb (27 kg), this means that approximately 2.3 bu of wheat are required to produce 100 lb (45 kg) of flour and 38 lb (17 kg) of millfeed.

Whole wheat flour, as the name would indicate, contains the finely ground endosperm, bran, and germ. In comparison with products made from white flour, whole wheat products have a distinctive flavor and a coarser texture, and, because of the higher fat content of the germ, they are more difficult to keep and sometimes become rancid.

Since prehistoric times, the goal of milling has been the separation of the bran and germ from the endosperm, to make white flour.

• **Modern milling**—In modern milling, wheat passes through about 2 dozen processes before it is made into table flour. Fig. W-16 is a simplified diagram showing what happens in milling.

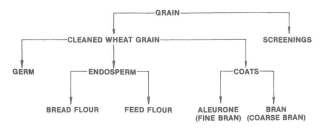

Fig. W-16. A simplified flow diagram of wheat milling.

Flour. Different grades and kinds of wheat are required for milling into the several grades and kinds of flours.

GRADES OF FLOUR; TEST BAKING.

The highest quality flour is known as *patent flour.*

The percentage of protein in a flour generally serves as one standard of quality. The protein of wheat and flour is composed of *gliadin* and *glutenin*, which, when mixed with liquid, form *gluten.* The elastic gluten developed by kneading the dough entraps the carbon dioxide formed by the fermentation of sugar and starches by yeast, or released by chemical leavening or air beaten into the mixture. The result is the unique *rising* or expansion characteristics of wheat flour doughs.

The mineral or ash content of flour is also used as an index of flour quality or grade; the lower the ash, or the less residue, the whiter or more refined the flour.

Small quantities of wheat are usually milled for test baking. Various flours are checked by different baking procedures according to end use. For hard wheat, test loaves are baked to determine the mixing time and tolerance of doughs, the degree to which the flour will absorb liquid, potential loaf volume, texture, crumb, color, and flavor. For soft wheat, similar determinations are made with test cakes, cookies, or crackers.

KINDS OF FLOURS.

There are many kinds of flours, each designed and best suited for a specific purpose, the main ones of which are discussed in the sections that follow:

• **Bakery flour**—Bakery bread wheat doughs are usually processed by machinery; therefore, the dough must have machine tolerance, or strength, in order to withstand the beating action of mechanical kneading and handling. This calls for *stronger* wheat—higher in the proteins, gliadin and glutenin, which form gluten—resulting in stronger doughs.

• **Family flour**—The decreasing production and sale of family flour reflects the decline in home baking.

Bread doughs worked by hand require strong arms, somewhat limiting the market to flour made from mellow gluten wheats that produce doughs that can be hand manipulated. There has been a trend away from the use of very strong wheats because the average homemaker does not have the strength or the patience to knead the dough long enough to develop the strong gluten.

The problems of the family flour requirements were partially solved with the making of *all-purpose* flours—mixed flours suited for both (1) good cakes, biscuits, and pastries, and (2) good bread.

Another innovation of family flour developed in the early 1960s was a granular or more dispersible type of flour. From the standpoint of the family, this flour offers the following advantages in comparison with regular flour: free-pouring—like salt; dust-free; alleviating the need for sifting; and dispersing in cold liquid, rather than balling or lumping.

Special soft wheat cake flours, and whole wheat flours are also available to homemakers. Graham flour is a whole wheat product made from 11.5% protein hard red winter wheat.

• **Specialty flours**—For the bakery trade, millers make a wide variety of different flours designed for different products, such as Italian and French breads and rolls.

Small retail shops keep a supply of several different kinds of flour, including: hard wheat flour for breads; soft wheat or cake flour for cookies, cakes, doughnuts, and cakelike products; and in-between flour for certain rolls, coffee cakes, and other pastries.

Special soft wheat flours are milled for cookie, cracker, and pretzel bakers; and other special flours are prepared for canners who make soups, gravies, desserts requiring thickening, and flour-based sauces.

• **Durum wheat flour**—The milling of durum wheat flour for making macaroni requires additional purifiers to separate the branny material from the desired semolina, a coarse granulation of the endosperm. By federal definition, semolina cannot contain more than 3% flour.

Durum millers also make *granulars,* or a coarse product with greater amounts of flour; or they grind the wheat into flour for special use in macaroni products, particularly in noodles.

• **Bleached flour**—At one time, it was necessary to age flour by storing it in a warehouse for a considerable time and turning it so as to expose it to the oxidizing action of the air. The stored flour lost some of its creamy color and baked better. But the process necessitated storage facilities, took time, and required labor; and the results were variable. Today, the same end result is achieved, more quickly and better, by the use of additives. In either case, the proteins are slightly oxidized. In storage, they're oxidized from the air; in bleaching, they're oxidized from the maturing agent. The oxidation of flour makes gluten stronger, or more elastic, and produces better baking results. Some oxidizing agents, such as those in which a calcium or phosphorus compound is used as a carrier, also improve the nutritive value of the flour. Only cookie, pie crust, and cracker flours are more satisfactory without bleach.

Today, flour must be plainly marked "bleached" if one of several government approved oxidizing-maturing agents has been added.

• **Phosphated flour**—Phosphated flour is popular in certain areas, particularly where biscuits are commonly made with sour milk and buttermilk. Since the acid content of sour milk varies, phosphated flour provides a greater assurance of leavening action from baking soda. The amount of calcium phosphate permitted in the flour is very small, not less than 0.25% and not more than 0.75%. Nevertheless, this low level of phosphation constitutes a significant nutritional enhancement, because a cup *(115 g)* of sifted phosphated flour supplies from 68 mg to 165 mg of calcium, and from 176 to 328 mg of phosphorus—compared to only 18 mg of calcium and 100 mg of phosphorus in unphosphated flour.

• **Self-rising flour**—Calcium phosphate is an ingredient of baking powder as well as in the leavening agents of self-rising flour, which also contains salt and soda. A typical self-rising flour contains 305 mg of calcium and 536 mg of phosphorus per cup *(115 g).*

In chemically-leavened flour, in the presence of liquid, the acid of the phosphate reacts with baking soda to produce a leavening gas, carbon dioxide—the same as the yeast organisms produce in yeast-raised products. Additionally, the calcium and phosphorus improve the nutritive qualities of the flour.

• **Whole wheat, cracked wheat, farina**—Whole wheat (or graham flour) is made from the entire wheat kernel.

Cracked or crushed wheat refers to wheat that has been broken into fragments.

Farina is purified middlings; it is the glandular endosperm that contains no bran and no more than 3% flour. Farina is most commonly used as a cooked breakfast cereal.

(Also see FLOUR, Table F-12, Major Flours.)

Enriched Flour. Since 1941, millers and bakers have followed the practice of enriching their products. Earlier surveys had indicated that the levels of certain nutrients were inadequate in the national diet; among them, the B vitamins, iron, iodine, and vitamin D. Since public need was demonstrated, the American Medical Association, leading nutritionists, government agencies, millers, and bakers worked together to develop an enrichment program with widely used, available, and inexpensive foods, like bread, to serve as carriers.

Originally, the FDA differentiated between enrichment and fortification, but now the two programs are used interchangeably. Nevertheless, the following vocabulary for and definition of food improvement persist:

• **Enriched**[8]—Refers to the addition of specific nutrients to a food *as established in a federal standard of identity and quality* (for example: enriched bread). The amounts added generally are moderate and include those commonly present at even lower levels.

• **Fortified**[9]—Refers to the addition to foods of specific nutrients. The amounts added are usually in excess of those normally found in the food because of the importance of providing additional amounts of the nutrients to the diet. Some foods are selected for fortification because they are an appropriate carrier for the nutrient. *Example*: Milk is frequently fortified with vitamin D.

---

[8]"Nutrition Labeling—Terms You Should Know," *FDA Consumer Memo,* DHEW Publication No. 74-2010, 1974, pp. 2-3.
[9]*Ibid.*

• **Restored**—The replacement of nutrients lost in processing food.

Although the principle followed with white flour is basically enrichment, rather than restoration, and although the quantity of the nutrients in the enrichment formula were established on the basis of dietary need, they also happen to restore to white flour the approximate amounts of those same nutrients as were present in whole wheat (see Table W-10).

Tables W-11, W-12, W-13, and W-14 show the standards for enriching flour, self-rising flour, macaroni products, and enriched bread, buns, and rolls. Calcium and vitamin D are permitted as optional additions to flour and bread, in the amounts specified. At about the same time that the flour enrichment program got underway (1941), iodine was added to table salt to prevent goiter, and vitamin D was added to milk to prevent rickets (Since most milk is now fortified with vitamin D, it is seldom added to bakery foods).

Since the start of the enrichment program in 1941, the available supplies of the B vitamins—thiamin, riboflavin, and niacin—and of iron in the national diet have increased. The enrichment program has played a significant role in the practical elimination of the vitamin deficiency diseases ariboflavinosis, beriberi, and pellagra, and of simple iron-deficiency anemia.

Today, enrichment is required in 34 states and Puerto Rico. In practice, however, all family flour is enriched, and about 90% of all commercially baked standard white bread is enriched.

TABLE W-10
COMPARISON OF NUTRIENTS IN WHEAT FLOUR AND WHEAT BREAD[1]
(Minimum and maximum enrichment levels for the three B vitamins—thiamin, riboflavin, and niacin—and iron are given in the table below)

One pound[2]	Protein	Fat	Calcium	Iron	Thiamin	Riboflavin	Niacin
	(g)	(g)	(mg)	(mg)	(mg)	(mg)	(mg)
Flour							
Whole Wheat	60.3	9.1	186	15.0	2.49	.54	19.7
Unenriched (all purpose)	47.6	4.5	73	3.6	.28	.21	4.1
Enriched (all purpose)	47.6	4.5	73-960[3]	13-16.5[3]	2.90[3]	1.80[3]	24.0[3]
Bread							
Whole Wheat (2% nonfat dry milk)	47.6	13.6	449	10.4	1.17	.56	12.9
Unenriched (white), (3 to 4% nonfat dry milk)	39.5	14.5	381	3.2	.31	.39	5.0
Enriched (white), (3 to 4% nonfat dry milk)	39.5	14.5	381-600[4]	8-12.5[4]	1.80[4]	1.10[4]	15.04[4]

[1]The source of this data is *From Wheat to Flour*, published by Wheat Flour Institute, Chicago, Ill., 1976, p. 68, except where noted otherwise.
[2]Multiply by 2.2 to obtain the amount per kg.
[3]*Code of Federal Regulations*, Title 21, Revised 4/1/78, Section 137.165 provides that enriched flour must contain the specified amounts of iron, thiamin, riboflavin, and niacin; but that the addition of calcium is optional. However, no claim for calcium on the label may be made if the flour contains less than 960 mg per lb.
[4]*Code of Federal Regulations*, Title 21, Revised 4/1/78, Section 136.115 provides that enriched bread, buns, and rolls must contain the specified amounts of iron, thiamin, riboflavin, and niacin; but that the addition of calcium is optional. However, no claim for calcium on the label may be made if the bread contains less than 600 mg per lb.

TABLE W-11
STANDARD FOR ENRICHED FLOUR[1]

Required Ingredients per Pound of Flour[2]	
Thiamin	2.9 mg
Riboflavin	1.8 mg
Niacin	24.0 mg
Iron	13.0-16.5 mg
Optional Ingredient	
Calcium	960 mg

[1]*Code of Federal Regulations*, Title 21, Revised 4/1/78, Section 137.165.
[2]Multiply by 2.2 to obtain the amount per kg.

TABLE W-12
STANDARD FOR ENRICHED SELF-RISING FLOUR[1]

Required Ingredients per Pound of Flour[2]	
Thiamin	2.9 mg
Riboflavin	1.8 mg
Niacin	24.0 mg
Iron	13.0-16.5 mg
Optional Ingredient	
Calcium	960 mg

[1]*Code of Federal Regulations*, Title 21, Revised 4/1/78, Section 137.185.
[2]Multiply by 2.2 to obtain the amount per kg.

TABLE W-13
STANDARD FOR
ENRICHED MACARONI PRODUCTS[1]

Required Ingredients per Pound of Flour[2]		
	(minimum)	(maximum)
Thiamin	4.0 mg	5.0 mg
Riboflavin	1.7 mg	2.2 mg
Niacin	27.0 mg	34.0 mg
Iron	13.0 mg	16.5 mg
Optional Ingredient		
Calcium	500.0 mg	625.0 mg
Vitamin D	250 U.S.P. units	1000 U.S.P. units

[1]*Code of Federal Regulations*, Title 21, Revised 4/1/78, Section 139.115.

[2]Multiply by 2.2 to obtain the amount per kg.

TABLE W-14
STANDARD FOR
ENRICHED BREAD, BUNS, AND ROLLS[1]

Required Ingredients per Pound of Flour[2]	
Thiamin	1.8 mg
Riboflavin	1.1 mg
Niacin	15.0 mg
Iron	8.0-12.5 mg
Optional Ingredient	
Calcium	600.0 mg

[1]*Code of Federal Regulations,* Revised 4/1/78, Section 136.115.

[2]Multiply by 2.2 to obtain the amount per kg.

It is noteworthy that England, Canada, and a few other countries have enrichment somewhat similar to the United States, but that France forbids enrichment.

NUTRITIONAL VALUE OF WHEAT. Whole wheat flour is made by grinding the entire kernel of wheat. Thus, it contains nutrients found in all three parts—the bran, endosperm, and germ. To produce white flour which has better keeping qualities than whole wheat flour, millers remove the bran and germ and use only the endosperm. Some of the B vitamins, calcium, and iron are lost. The processing of white flour removes additional nutrients.

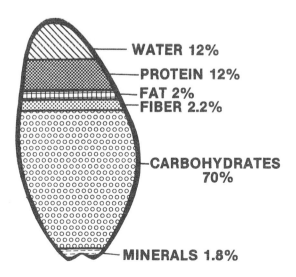

WATER 12%

PROTEIN 12%

FAT 2%
FIBER 2.2%

CARBOHYDRATES 70%

MINERALS 1.8%

Fig. W-17. Proximate analysis of whole wheat.

Wheat is high in carbohydrates; hence, it is a high energy source. Its contribution to the available food energy of the following countries is noteworthy: Israel, 35%; France, 30%; Australia, 25%; England, 20%; and United States, Canada, and West Germany, 18%. As a source of energy in the United States, wheat ranks on a par with three other major sources of energy: meats, processed fats, and processed sugar. Together, these four foods comprise three-fourths of the food energy available in the United States.

Wheat has the highest protein value of any of the cereals. Still, the protein is of poor quality. Table W-15 compares the essential amino acids in whole wheat, white flour, and wheat germ to a high quality protein product—cow's milk. Like rice and corn and their products, wheat is particularly deficient in lysine; hence, it should be eaten with a complementary protein source such as a legume. Although not indicated in Table W-15, wheat germ contains about 23% protein, whole wheat contains about 12% protein, and white flour contains about 11% protein. Interestingly, the protein in wheat germ is extra rich in lysine, the amino acid that is deficient in the rest of the wheat grain.

TABLE W-15
PROFILE OF ESSENTIAL AMINO ACIDS IN
WHOLE WHEAT, WHITE FLOUR, AND WHEAT GERM
COMPARED TO MILK—A HIGH-QUALITY PROTEIN[1,2]

Amino Acid	Whole Wheat	White Flour	Wheat Germ	Cow's Milk
		(mg/g protein)		
Histidine	19	20	27	27
Isoleucine	32	46	36	47
Leucine	58	77	64	95
Lysine	25	22	63	78
Methionine and cystine	33	33	37	33
Phenylalanine and tyrosine	68	89	68	102
Threonine	26	29	40	44
Tryptophan	11	12	12	14
Valine	38	43	48	64

[1]Additional information about the essential amino acid content of wheat and flour products may be found in Table P-16, Proteins and Amino Acids in Selected Foods.

[2]*Recommended Dietary Allowances*, 10th ed., 1989, National Academy of Sciences, p. 67, Table 6–5. The essential amino acid requirements of infants can be met by supplying 0.79 g of a high-quality protein per pound of body weight (1.73 g/kg body weight) per day.

The nutritive value of the many different forms of wheat and wheat products is given in Food Composition Table F-21.

(Also see BREADS AND BAKING, sections headed "Nutritive Values of Breads and Related Items," and "Improving the Nutritive Values of Breads and Baked Goods"; MINERAL[S]; PROTEIN[S]; and VITAMIN[S].)

For some individuals, the consumption of wheat creates health problems. If supplementary foods are inadequate, excessive amounts of wheat products can lead to kwashiorkor or beriberi. However, persons to whom wheat products are available are usually able to obtain a wider variety of foods, and deficiency diseases have not been directly attributed to overdependence on wheat products. Celiac disease, however, is apparently due to a genetically-determined intolerance to wheat gluten or an acquired defect in the intestinal lining. Also, the use of refined flour in wheat products, especially those of a sticky nature such as biscuits, is believed to be a major factor predisposing to dental caries.

WHEAT PRODUCTS AND USES.

The most common use of wheat is for food made from flour. Breakfast foods and beer or other alcoholic fermentations also take considerable quantities of wheat.

Breads date back thousands of years, in war and in peace, in famine and in plenty, to the discovery of wheat, the invention of milling, and the development of baking.

Wheat is not normally fed to animals because of its high cost and greater value for human food. However, surplus or damaged wheat and the by-products of milling find important use as livestock feeds and, eventually, enter the human diet as meat, milk, and eggs. Some wheat products are also used for industrial purposes.

Table W-16 presents in summary form the story of wheat products and uses for human food.

(Also see BREADS AND BAKING; CEREAL GRAINS; and BREAKFAST CEREALS.)

Fig. W-18. Home-made wheat bread. (Courtesy, University of Minnesota, St. Paul, Minn.)

Fig. W-19. There are an unlimited number of recipes for wheat flour production. (Courtesy, Washington State University, Pullman, Wash.)

TABLE W-16
WHEAT PRODUCTS AND USES FOR HUMAN FOOD

Product	Description	Uses	Comments
Flour	White flour is finely ground endosperm of wheat. Whole wheat flour contains the finely ground endosperm, bran, and germ.	Wheat flours are used for many purposes, such as bread, pastries, cakes, cookies, crackers, macaroni, puddings, and soups. (Also see BREADS AND BAKING, section headed "Types of Breads and Baked Doughs.")	For each purpose, a flour of particular properties is required. Additives, such as maturing agents, bleaching agents, and self-rising ingredients, are frequently blended into wheat flours at the mill. Family flour is enriched at the mill, while flour being shipped to the baker is enriched at the bakery.
Bulgur	A partly debranned, parboiled wheat product, used either whole or cracked.	Bulgur is used as a substitute for rice; e.g., in pilaf, an eastern European dish consisting of wheat, meat, oil, and herbs cooked together.	Bulgur is exported in large quantities, especially to the Far East. Bulgur provides a cheap food that is acceptable to rice-eating people because it can be cooked in the same way as rice, and it resembles rice superficially.
Fermented minchin	A product made by putting wheat gluten in a tightly covered container for 2 to 3 weeks and allowing it to be overgrown with both molds and bacteria. Then 10% salt is added and the mixture allowed to age for 2 weeks.	Aged fermented minchin is cut into thin strips and used as a condiment with other foods.	
Wheat germ	The germ is the embryo or sprouting part of the wheat seed.	As an ingredient of speciality breads.	Wheat germ is available for human food, but is usually added to animal feeds.
Wheat gluten	Wheat gluten is a mixture of two proteins, gliadin and glutenin. When mixed with water, they impart the characteristic stickiness to dough.	For making monosodium glutamate which imparts a meatlike flavor to foods. For adding to white flour to make a high-gluten bread.	Some people may develop an intolerance to gluten.
Wheat starch	A white, odorless, tasteless granular or powdery material.	Wheat starch is used as a thickener in soup where its lower gelatinization temperature and shorter gel temperature range is an advantage. Syrup and sugar.	
Wheat tempeh	Fermented cake of cooked wheat grains.	Vegetarian main dish when sliced and baked or fried, and/or an ingredient of salads, sandwiches, soups, sauces, casseroles, and spreads.	An experimental food developed by USDA scientists who utilized the mold *Rhizopus oligosporus* in a procedure similar to that used by Asians to make soybean tempeh. The nutritional value of the original grain is enhanced by the mold fermentation in that (1) proteins are partially digested, and (2) certain B vitamins are synthesized. It is noteworthy that the mold synthesizes vitamin B-12, which makes tempeh one of the few vegetable foods that contain the vitamin.
Breakfast foods, ready-to-eat: Bran flakes	Tan colored, paper thin flakes.	Ready-to-eat breakfast food	
Puffed wheat	Whole wheat depericarped (the outer covering removed), cooked, puffed by extruding through a machine, then dried-toasted.	Ready-to-eat breakfast food	(Also see BREAKFAST CEREALS.)
Shredded wheat	Whole wheat is cooked with water to gelatinize starch, cooled, shredded by machine, cut to shape, baked, then dried and cooled.	Ready-to-eat breakfast food	Shredded wheat has a protein content considerably lower than puffed wheat.
Wheat flakes	Whole wheat, that has been rolled, cooked, flavored, then dried.	Ready-to-eat breakfast food	
Breakfast foods, to be cooked: Whole wheat grains	Unground whole wheat	Cooked cereals, starchy vegetable, puddings such as frumenty.	Cook 40-50 minutes in an open pot or a pressure cooker after soaking overnight in water. Frumenty is cooked grains which have been reheated with milk and served with cream, butter, sugar, and/or jam.
Cracked wheat	Whole grains broken into small pieces	Cooked cereals, breads, soups, casseroles.	Cracked wheat has a nutlike flavor and texture, and cooks much faster than whole kernels of the grain. Only limited amounts can be used in breads because the sharp edges on the particles cut the strands of gluten which make the dough elastic.

(Continued)

TABLE W-16 (Continued)

Product	Description	Uses	Comments
Farina (Cream of Wheat)	This is granulated endosperm that contains no bran and no more than 3% flour; it's similar to semolina except that the wheat is not a durum variety.	A hot breakfast cereal.	Enriched farina is popularly known as *Cream of Wheat*, a product made famous by Nabisco.
Malted cereals	Cracked wheat and malted barley, mixed	Cooked cereals, baked items	May also be used to make muffins and other quick breads.
Rolled wheat	Whole grains rolled into flat pieces (flakes) after steaming	Cooked cereals, cookies, filler in meatloaf	Can be substituted for rolled oats in most recipes.
Wheat meal	Ground wheat from which the bran has been removed	Cooked wheat breakfast foods	
Fortified foods for developing countries: Bal Ahar	Coarsely ground mixture of bulgur wheat, peanut flour, skim milk powder, minerals, and vitamins. Contains 22% protein.	Infant cereal, suitable for mixing with water	Produced by the government of India. The protein quality is on a par with that of milk.
Bread fortified with low-fat cottonseed flour	Bread made from a mixture of 90% wheat flour and 10% cottonseed flour	High protein bread	Developed at Universidad Agraria La Molina in Lima, Peru. The special cottonseed flour (Protal) raises the protein level of wheat bread from 9% to 11%.
Cereal products fortified with fish protein concentrate (FPC)	Products made from wheat flour fortified with dried, defatted, whole fish powder	Breads, crackers, and noodles	Developed at Universidad Agraria La Molina in Lima, Peru.
Fortified atta	Whole wheat flour mixed with peanut flour, minerals, and vitamins	Baked goods commonly used by people in India	Produced by flour mills that are subsidized by the Indian government. Breads resemble unleavened breads in consistency, and contain about 13.5% protein.
Fortified couscous	Coarsely ground wheat mixed with high protein wheat fractions.	Cereals, casseroles, soups, and other traditional dishes prepared by people in Algeria and neighboring countries.	Developed by International Milling for U.S. Agency for International Development (AID). High protein wheat fractions are obtained by special milling procedures. Fortified couscous has higher quantity and quality of protein than ordinary couscous.
Laubina	Finely ground mixture of cooked bulgur wheat, cooked chickpeas, skim milk powder, sugar, minerals and vitamins.	Infant cereal	Developed at the American University of Beirut in Lebanon. Promoted prompt recovery of hospitalized infants suffering from severe protein-energy malnutrition, but was not effective in curing anemia.
Leche alim	Mixture of toasted wheat flour, fish protein concentrate (FPC), sunflower meal, and skim milk powder. Contains 27% protein.	High protein infant cereal suitable for mixing with water.	Produced by the Pediatrics Laboratory of the University of Chile in Santiago.
Rolled wheat and soy flake mixture	Mixture of 85% rolled wheat and 15% soy flakes.	Cereal	Developed by USDA and test marketed by AID.
Superamine	Finely ground mixture of wheat flour, chickpea flour, lentil flour, skim milk powder, sugar, minerals, vitamins, and flavoring. Contains 21% protein.	High protein powder which may be mixed with water and cooked briefly (2 to 4 minutes)	Developed as a cooperative project between FAO, WHO, UNICEF, and the Algerian government.
Vitalia macaroni	Pasta made from semolina, wheat, soy, corn, and rice derivitives. Contains 18% protein.	High protein macaroni product	Produced by Instituto de Investigaciones Tecnoligicas, in Bogota, Colombia.
Wheat protein beverage powder	A finely ground mixture of wheat protein concentrate, skim milk powder, sugar, vegetable oil, minerals, vitamins, and flavoring	Powder suitable for mixing with water to make a high protein beverage	Experimental products developed by USDA scientists, who devised means of producing a partially digested, low starch, high protein concentrate from wheat flour. Protein quality is intermediate between milk and egg proteins.
Fermented beverages: Wheat grain	Cleaned, whole wheat	Canadian whiskey	Canadian whiskey is produced from any one of the cereal grains; hence, corn, rye, or barley may also be used.
Wheat starch	Malt converts wheat starch to sugar, which is fermented by yeast to ethyl alcohol and carbon dioxide.	1. Ethyl alcohol, used largely for alcoholic beverages. 2. Denatured alcohol for industrial purposes. 3. Gasohol (10% anhydrous ethanol, 90% gasoline)—a motor fuel.	(Also see CORN, section headed "Distilling and Fermentation.")

WHEAT GERM

Wheat germ is the embryo or sprouting part of the wheat seed. A kernel of wheat is composed of three principal parts: The starchy endosperm, which constitutes about 83% of the kernel and is the source of white flour; the coat, which constitutes about 14.5% of the kernel and is the source of bran; and the germ, which constitutes 2.5% of the kernel and is the source of wheat germ meal and wheat germ oil. Wheat germ meal is used chiefly as a livestock feed, although some of it is now used as a human nutritional supplement. Wheat germ oil is used primarily as a rich vitamin E supplement for man and animals.

Wheat germ meal consists chiefly of wheat germ together with some bran and middlings or shorts. It is low in fiber, good in protein (minimum 25% crude protein), high in fat (7%), and rich in vitamin E. Defatted wheat germ meal is the product that remains after removal of part of the oil or fat from wheat germ meal. It must not contain less than 30% crude protein; and it is much lower than wheat germ meal in vitamin E.

Flour millers use the endosperm to make white flour and discard the bran and germ. In discarding the bran and germ, they are losing 77% of the thiamin, 80% of the riboflavin, 81% of the niacin, 72% of the vitamin B-6, 50% of the pantothenic acid, 67% of the folacin, 86% of the alpha-tocopherol, and 29% of the choline.[10]

Thus, wheat germ is a rich source of the nutrients that are discarded from the flour in milling. However, the oil of wheat germ decomposes readily after the germ is removed from the kernel, causing short shelf life and problems in marketing raw wheat germ. Thus, the product is usually defatted when offered for sale. The label will show whether the germ is defatted.

Wheat germ is added to many kinds of foods, such as bread, cookies, cereals, and milkshakes. Wheat germ oil may be taken to restore nutrients lost in defatting the germs.

(Also see WHEAT.)

WHEATMEAL, NATIONAL

The name given to 85% extraction wheat flour when introduced into the United Kingdom in 1941 (as distinct from whole wheat meal, which is 100% extraction). Later, the name was changed to *national flour*.

WHIPPLE'S DISEASE (INTESTINAL LIPODYSTROPHY)

A relatively rare, and previously considered fatal, disease of predominantly middle-age men. It is characterized by adominal pain, diarrhea progressing to steatorrhea (fatty diarrhea), marked weight loss, anemia, arthritis, fever, and impaired intestinal absorption—a malabsorption syndrome. Whipple's disease produces structural changes in the lymphatics and intestinal mucosa which seem responsible for the malabsorption. Usually there is evidence of the malabsorption of fat with the loss of fat-soluble vitamins, vitamin B-12, and xylose. Furthermore, the victim demonstrates low blood levels of calcium, magnesium, and potassium—hypocalcemia, hypomagnesemia, and hypokalemia, respectively. Excessive loss of albumin into the intestine results in low circulating levels of albumin in the blood—hypoalbuminemia. Evidence from intestinal biopsies suggests that it is due to a microorganism. Moreover, antibiotic therapy brings about dramatic improvement, but it is required on a long-term (10 to 12 months) basis.

Diet and replacement therapy are based on the nutritional deficiencies evident, on-going needs, and specific absorptive defects.

(Also see MODIFIED DIETS; and MALABSORPTION SYNDROME.)

WHITE BLOOD CELLS (LEUKOCYTES)

The colorless blood cells, generally called white blood cells or corpuscles.

WHOLE WHEAT MEAL

Meal that is made by grinding the entire wheat kernel.

WILD EDIBLE PLANTS

Contents   *Page*
Types of Wild Foods Available......................1127
Stalking Wild Edible Plants.........................1128
Edible Wild Plants of North America................1128

In the beginning, all plants were wild. Then, through the ages man collected, selected, and cultivated those plants which fitted his needs, while the others remained in the wild. As man started to travel more widely, he carried plants with him to introduce to new locations. Occasionally, these would escape his cultivation and fare so well in their new environment that they eventually became a wild plant—sometimes a weed. Still others came as weeds, but they came from other countries with crop seeds or on animals or birds.

Often what is called a wild plant, and an unfamiliar food source to most, was however, very familiar to the natives of a land. When colonists arrived in the New World, very few Indians had gardens or cultivated crops. Instead, most of them supplemented their diets by collecting edible wild plants with which they were familiar. Instead of the potatoes, carrots, radishes, parsnips, beets, and turnips known today, Indians relied on wild roots and tubers. They also collected various nuts, fruits, greens, and seeds. In some areas, natives still depend upon edible wild plants for food, and at times a distinction between wild and domesticated is difficult.

Perhaps, the future of edible wild plants lies ahead, as some wild plants may represent forgotten or undiscovered food sources. Wild edible indigenous plants, if exploited, could offer alternative food sources in some areas of the world. In addition, there are other plants which could be exploited for fiber, energy, and medicine.

TYPES OF WILD FOODS AVAILABLE. In the wild, a knowledgeable person may find the selection of foods as varied as that in the supermarket. Wild greens for use in salads or for cooking as a potherb include dandelions, chickweed, clover, chicory, miner's lettuce, mint, purslane, lamb's quarter, and watercress. Roots and tubers, or potato-like wild foods include the arrowhead, bulrush, cattail, and water lily. Shoots and stems that can be used as a vegetable include asparagus, burdock, bracken fern, and thistle. There is an endless array of wild nuts and berries available. Many plants produce seeds which can be ground and used like meal or flour. Nature even provides substitutes for coffee and tea. Many leaves can be brewed green or dried for tealike beverages; for example, wintergreen, catnip, colt's foot, Labrador tea, and cassina (holly). Caffeineless coffee substitutes can be made from roasted roots of chicory, chufa, dandelions, or salsify. Such examples of edible wild plants continue on and on.

---

[10]Schroeder, Henry A., M.D., "Losses of Vitamins and Trace Minerals Resulting from Processing and Preserving Foods," *The American Journal of Clinical Nutrition*, Vol. 24, May 1971, pp. 562-573, Table 3.

STALKING WILD EDIBLE PLANTS.

In the United States, people just buy their plants in the supermarkets in cans, or frozen, or fresh, so why should anyone be interested in what grows wild and what may have an unfamiliar flavor? There are several reasons:

• **Outdoor groups**—Groups such as the Boy Scouts, Girl Scouts, garden clubs, nature study groups, and campers and hikers always have a natural curiosity about the plants they observe, and often wish to know or need to know the identity of edible plants.

• **Hobby**—For many people, identifying and collecting wild edible plants represents a satisfying hobby. By starting with a few familiar plants, they gradually branch out to identifying more and more plants as they acquire new knowledge and sharpen their techniques of identification. To these individuals, the *weeds* in the garden, in the lawn, and along the roads take on new meaning. Collecting and eating wild greens, potherbs, berries, nuts, roots, and tubers, is an enjoyable adventure. Also, foods of the wild are tasty and can be preserved for future use by drying, freezing, canning, or use in jams. It is doubtful, however, that collecting wild edible plants would ever result in saving money. Gathering wild plants requires much time and often travel, and some must be hulled, or soaked, or prepared in some time-consuming way.

• **Survival**—Some individuals have experienced a time when life depended on being able to identify wild edible plants. Experienced outdoorsmen have stated: "If you starve to death in the wilderness it is because you are just plain tired of living." Indeed, manuals for training members of the United States Armed Forces contain sections on the identification, collection, and preparation of edible wild plants in areas throughout the world. Each year, campers, fishermen, and hunters become lost and many suffer needlessly because of their lack of knowledge of edible wild plants.

There are many more examples of the role of wild plants in the survival of people. When the Forty-Niners stampeded California's streams, deserts, and mountains in search of gold, the scarcity of fresh food brought scurvy to the camps. The Indians and Spanish introduced the Forty-Niners to what is now called miner's lettuce. To combat scurvy in the northern portions of this continent, frontiersmen, explorers, and gold diggers learned to identify and use a plant now called scurvy grass. When rations were low on the Lewis and Clark expedition, the sweet yellow fruit of the papaw was eaten. During World War II, the people of England and Scandinavia collected vitamin C-rich rose hips from wild roses.

While stalking the wild plants can spell fun, adventure, and survival, it may also spell danger. Before eating a wild plant it should be positively identified. For most plants, this requires knowledge, skill, and experience. Some poisonous plants are easily confused with edible plants. Some edible plants are poisonous at certain stages of development, or certain parts of the plant are poisonous. However, in times of *survival emergencies* the following advice is often given:

1. Never eat large quantities of a strange plant food without first testing it. (A disagreeable taste in a food item, which is otherise safe to eat, may sometimes be removed. If cooking is possible, boiling in one or more changes of water may remove the unpleasant taste.)

2. Take a teaspoonful of the plant food, prepared in the way it will be used (raw, boiled, baked, etc.), hold it in your mouth for about 5 minutes. If, by this time no burning sensation, or other unpleasant effect, has occurred, swallow it.

3. Wait 8 hours. If no ill effects such as nausea, cramps, or diarrhea occur, eat a handful and wait 8 hours. If no ill effects are noted at the end of this time, the plant may be considered edible.

4. If serious ill effects develop, induce vomiting.

5. Even foods deemed safe should be eaten with restraint until you become used to them.

6. Remember that olives are bitter and grapefruit is sour, so an unpleasant taste does not, in itself, mean poison. But a burning, nauseating, or bitter taste is a warning of danger. A small quantity of even poisonous food is not likely to prove fatal or even dangerous, whereas a large quantity may be. This, however, does not apply to mushrooms, which are best avoided in any case.

7. In general, it is safe to try foods that you observe being eaten by birds and mammals, but there are some exceptions.

Enthusiasts who are unable to escape completely into the wilds, and must do their foraging near civilization, should be somewhat cautious. Vacant lots and roadsides where wild plants may be found may get sprayed with herbicides. Also, plants along busy roads have been known to contain high levels of lead, due to the lead in some gasolines burned in cars. Plants should be gathered at least 100 ft (*31 m*) away from a busy road.

EDIBLE WILD PLANTS OF NORTH AMERICA.

In North America, there are hundreds of edible wild plants, and worldwide there are hundreds more.

Fig. W-20. Watercress, which requires an abundance of water, often grows wild. (Courtesy, Dept. of Food Science and Human Nutrition, University of Hawaii)

Foragers of wild plants should completely familiarize themselves with the identification of edible and poisonous wild plants in an area. Moreover, beginners would do well to take lessons from persons with experience in stalking the wild life. An alphabetical listing of a few of the edible wild plants of North America follows:

Acorn (oak), *Quercus* spp
Amaranth (pigweed, redroot, wild beet), *Amaranthus spp*
Arrowhead (wapatoo), *Sagittaria latifolia*
Asparagus, *Asparagus officinalis*
Beechnuts (American beech), *Fagus grandiflora*
Bitterroot, *Lewisia rediviva*
Bulrush, *Scirpus* spp
Burdock, *Arctium minus* and *Arctium lappa*
Cattail, *Typha* spp
Chokecherry, *Prunus virginiana*
Chufa (earth almond, nut grass, zula nuts), *Cyperus esculentus* and *Cyperus rotundus*
Chickweed, *Stellaria media*
Cranberry, *Vaccinium* spp
Currants and Gooseberries, *Ribes* spp
Dandelion, *Taraxacum officinale*
Dock, *Rumex* spp
Elderberry, *Sambusus* spp
Evening primrose, *Oenothera* spp
Glasswort (Samphire), *Salicornia* spp
Ground Cherry, *Physalis* spp
Groundnut (Indian Potatoes), *Apios americana*
Hazelnut (filbert), *Corylus* spp
Hickory, *Carya* spp
Jerusalem Artichoke, *Helianthus tuberosus*
Jojoba, *Simmondsia chinensis*
Kinnikinic, *Arctostaphylos* spp
Lamb's Quarter (Goosefoot), *Chenopodium* spp
Maple (Sugar maple), *Acer saccharum*
Miner's Lettuce, *Montia* spp
Mountain Sorrell (Scurvy grass), *Oxyria digyna*
Mulberry (Red Mulberry), *Morus rubra*
Nettles (Stinging Nettle), *Urtica* spp
Pinon Pine, *Pinus monophylla* and *Pinus edulis*
Plantain, *Plantago* spp
Pokeweed (Poke), *Phytolacca* spp
Poplar, *Populus* spp
Prairie Turnip (Bread root), *Psoralea* spp
Prickly Pear (Indian fig), *Opunita* spp
Purslane, *Portulaca oleracea*
Salsify (Oyster plant), *Tragopogon* spp
Tepary Bean, *Phaseolus acutifolius*
Thistle, *Cirsium* spp
Water Cress, *Nasturtium officianale*
Water Lily (Yellow water lily), *Nuphar* spp
Wild Rose, *Rosa* spp

NOTE WELL: For a more complete summary of edible wild plants of North America including their description, geographical distribution, and preparation and uses, see the two-volume *Foods & Nutrition Encyclopedia*, a work by the same authors as *Food For Health.*

WILD RICE *Zizania aquatica*

Wild rice (*Indian rice, water oats*), which is native to the Great Lakes Region—especially in what is now Minnesota, Wisconsin, and Manitoba, is the only native cereal crop to be domesticated in the United States; all other cereal grains were cultivated elsewhere and brought to America. Wild rice is not closely related to true rice, *Oryza;* it belongs to a different genera and even a different tribe of the grass family.

Taming wild rice was not easy. It ripened at different stages, and, as it ripened, the grain fell off the stalk, making harvest difficult. As a result, it was formerly harvested by Indians by *ricing,* which to the Indians meant harvesting from canoes by bending the stems over the boat and beating the heads with a stick. Following harvesting, the moist seeds were dried by parching in a heat-rotated drum or in an open kettle with constant stirring to prevent burning. However, in the 1960s, the University of Minnesota developed an improved variety of wild rice that ripened more uniformly without shattering, adapted to harvest with combines—just like wheat or barley. Today, 75% of the annual U.S. wild rice crop of 3 to 4 million lb (*1.4 to 1.8 mil kg*) is grown in the marshy areas of northern Minnesota (the remaining 25% of the supply comes from Canada, Wisconsin, Michigan, and California); and, today, 90% of Minnesota's wild rice is grown in paddies and harvested by combine, and only 10% is grown in lakes and harvested by canoes.

Fig. W-21. Harvesting wild rice in the Chippewa National Forest, Minnesota. One person *poles* the boat, and the other person harvests the rice by means of two sticks—he pulls a batch of ripe rice stalks over the boat with one stick, then knocks the rice into the bottom of the boat with the other stick. (Courtesy, USDA)

Wild rice can be cooked by the same methods as regular rice. Cooking times vary, depending on the size, color, and processing; generally, the larger the kernel and the darker the color, the longer it takes to cook. When properly cooked, the slender, round, purplish black, starchy grain is excellent food, highly esteemed by gourmets and others familiar with the taste. Wild rice is usually high in protein content and low in fat compared to other cereals. (For the composition of wild rice, see Food Composition Table F-21.) Normally, it sells at 2 to 3 times the price of cultivated rice.

(Also see CEREAL GRAINS; and RICE, section headed "Wild Rice.")

WINDBERRY (BILBERRY) *Vaccinium myrtillus*

A fruit of a low shrub (of the family *Ericaceae*) that is native to Europe and northern Asia. It is closely related to blueberries and huckleberries. Windberries are used fresh or in confections, jams, and pies.

(Also see FRUIT[S], Table F-23, Fruits Of The World—"Blueberry".)

WINE

Contents   Page
History of Wine Making............................1130
Basic Principles of Wine Production................1131
  Fermentation....................................1131
  Clarification 1131
  Aging..1131
Types of Wines and Related Products..............1132
Medicinal and Nutritional Effects..................1133
Dangers of Drinking Too Much Wine...............1133

Fig. W-22. Wine enhances a meal. (Courtesy, Cryovac Divison, W. R. Grace & Co.)

The term *wine* refers to the fermented juice of the grape. If other fruit juices are used to make wine, the name of the fruit must precede the word *wine*. Wine is believed to have been one of the oldest medicines in the world and has long been used as a base for many tonics. However, most people drink wine for enjoyment. Recently, Americans increased their consumption of this beverage at the expense of other alcoholic drinks, as evidenced by the fact that the per capita sales of all types of wine in the United States doubled in the period between 1956 and 1976.[11] Since 1976, the per capita consumption of wine has not changed much; it has gone from 2.7 gal to 3.0 gal.

By comparison, the per capita wine consumption in Portugal is 23 gal; in Italy, 21 gal; and in France, 20 gal.

HISTORY OF WINE MAKING. Some historians suspect that certain cave dwelling, primitive peoples of the Eurasian continent may have enjoyed fermented juices from spoiled wild fruit long before grapes were first cultivated some 7,000 years ago in the region around the Caspian Sea. (Grapes naturally contain the vital ingredients for fermentation because they are rich in sugar and have a waxy outer coating that collects wild airborne yeast cells.)

[11]Folwell, R. J., and J. L. Baritelle, *The U.S. Wine Market*, Agri. Economic Report No. 417, USDA, 1978.

Archaeologists have found 6,000 to 7,000 year old pottery wine jars that suggest that the first wines were made in the Middle East from grapes, figs, and dates. Apparently the art of grape growing (viticulture) was spread westward by migratory peoples, such as the Jews, since the *Bible* mentions that Noah raised grapes and made wine (Genesis 9:20-21). It is noteworthy that Egyptian accounts of wine making date back to about 2500 B.C. They also stored jars of wine in special wine cellars dug out of the earth, because they apparently realized that this highly esteemed drink kept better when it was stored in a cool environment. The ancient Minoan, Greek, and Etruscan civilizations also produced wine for their own use and trading, as evidenced by the airtight clay pots they used as containers. Later, the Greeks originated symposiums, which were gatherings of people who drank wine together, engaged in intellectual discussions, and played simple games. (The symposia may have represented a refinement of the earlier dionysian orgies that often ended in drunkenness and violence.)

Roman wine festivals, or bacchanalia (named after Bacchus, the god of wine), were characterized by drinking to intoxication and resembled the dionysian orgies, except that the Romans did not allow women to drink wine. The Romans were the first to retard the spoilage of wine by storing it in their smoke houses. They also became experts in vine growing and helped to spread this art throughout their European colonies.

By the early Christian era, the Gauls in France rivaled the Romans in their wine-making skills. When Rome fell, each region of Europe made its own types of wines from the fruits of the vineyards that were planted by the Romans. Because wine was used sacramentally in the celebration of the Mass, the monasteries developed special recipes for making local wines. Later, some groups of monks (such as the Benedictines) took winemaking a step further and made special distilled liqueurs by processes that were kept secret to insure their livelihood.

The European colonists of North America found fields of wild grapes when they arrived. However, the first wines made from the American species (*Vitis labrusca*) were inferior to those made from Old World grapes (*V. vinifera*). Nevertheless, the wild species were much hardier in the climate

Fig. W-23. Wine grapes. (Courtesy, USDA)

of eastern N. America than the Old World species. (There is some evidence that the Vikings who landed in eastern Canada found grapes there, also). Finally, in 1769 the Old World varieties were successfully introduced into California by the Spanish priest Father Junipero Serra who founded a string of missions with vineyards.

By the 19th century, many European varieties of wine grapes were brought to western United States, where they grew well in the mild climate. A better native American grape was developed after the Concord grape was first grown by E. W. Bull of Concord, Massachusetts in 1852. The hardiness of the American vines led to their introduction into European vineyards to which they carried the plant louse, *phylloxera*. This parasite caused massive destruction of vines throughout Europe until it was found that resistant vineyards could be produced by grafting European vines onto the louse-resistant American rootstocks. Eventually, most of the vineyards of Europe were reestablished, but many years were required for the production of the mature vines that yield the best wine grapes. Another important development of the 19th century was the discovery by the French microbiologist Pasteur that the heating of wines destroyed the undesirable bacteria that caused spoilage.

The growth of the American wine industry during the 20th century was interrupted by Prohibition, which lasted from 1919 to 1933. However, the consumption of wine has risen steadily since 1933; and there have been many innovations and improvements in the production of wines as a result of research programs in viticulture and enology. What were formerly the trade secrets jealously guarded by a few winemakers have now been formulated into scientific practices of the entire industry. As a result, many high quality American wines are available at very reasonable prices.

BASIC PRINCIPLES OF WINE PRODUCTION.
Actual production practices vary somewhat from winery to winery, since the grapes used and the wines produced are distinctively different for each growing region. For example, the high acid and low to moderate alcoholic content of the European-type table wines are produced from the grapes grown in the cooler areas of northern California, Oregon, and Washington, whereas the dessert wines are made from the high sugar content grapes grown in the hot Central Valley of California. (A long, hot growing season is needed to produce grapes that are very rich in sugars.) Only the basic principles that are utilized in most wineries will be presented here.

Fermentation. The juice from the crushed grapes is usually fermented by pure cultures of certain strains of yeasts, by processing steps like those that follow:

1. **Preparation of the must.** The must, or juice, to be fermented is extracted by passing the fresh grapes through a *stemmer-crusher*, after which the juice is treated with sulfur dioxide to kill any undesirable wild yeasts. **NOTE WELL:** Because sulfites trigger severe reactions among some people, since June 1978, the Bureau of Alcohol, Tobacco, and Firearms has required that all wines containing more than 10 parts per million of sulfites carry the phrase "Contains sulfites" on the neck, back, or side panels of containers.

2. **Fermentation.** A culture of the most suitable strain of yeast is added to the must in fermentation tanks. In most cases, the juice contains sufficient sugar to produce the desirable alcoholic content. However, some of the grapes grown in cool climates are too low in sugar and require additional sugar for complete fermentation. Large fermenta-

tion tanks may be equipped with cooling coils to prevent the temperature from rising too high, since the yeast cells may be killed by too much heat. The dark-colored wines are produced by fermenting the grape pulp with the skins present, whereas the lighter-colored wines are fermented with only minimal contact with the skins.

Clarification. Newly fermented wines contain considerable sediment which is usually removed by (1) fining, in which an agent such as gelatin or egg white is added to bring certain materials out of solution or suspension; (2) settling out of suspended materials; (3) filtration of the wine through special filter pads; and (4) centrifugation, when the wines are not fully clarified by the preceding processes.

Aging. Some wines are intended for consumption right after their production but others must be mellowed by aging for a few years to remove the harsh flavors and allow the more desirable ones to develop. This process has long been carried out in wooden barrels that contribute flavors of their own, while absorbing some of the astringent constituents. However, many of the less expensive wines are held in redwood, concrete, or lined wine tanks for periods ranging from a few months to 2 years or so.

After aging, the wines are filtered, treated with sulfur dioxide to prevent spoilage; and bottled.

A little more aging may occur in the bottle, as the sulfur dioxide content decreases gradually. Corked bottles of wine are stored with the cork end inclined downward so that the cork is kept wet and the admission of air is prevented. **NOTE WELL:** In 1991, the Food and Drug Administration issued a temporary standard for wine prohibiting the sale of wine with more than 300 parts per billion of lead as a temporary measure until a permanent standard can be set. Also, FDA has said that it will soon propose a ban on lead foil wrappers to cover corks in wine bottles.

Fig. W-24. These great wooden tanks, filled with aging wines, are in a cool winery in a California valley. (Courtesy, Wine Institute, San Francisco, Calif.)

TYPES OF WINES AND RELATED PRODUCTS.

The names listed on the labels of European wines are confusing to all but the experts in the field because they give vineyards and regions of production that are unfamiliar to most consumers. Furthermore, the quality of these wines varies from year to year because the climates of the grape producing regions are quite variable. Therefore, high class restaurants allow the customers to taste imported wines before they are served.

Fig. W-25. The opening and tasting of a bottle of wine is a pleasant ritual that is performed in many high class restaurants. (Courtesy, Wine Institute, San Francisco, Calif.)

The years in which good wines were produced in a region are called *vintage years*. However, wines produced in the more predictable climate of California vary little from year to year. Some of the common types of wines, along with brief descriptions, follow:

Apple wine (hard cider). Apple cider or juice that has been allowed to ferment.

Aromatic wine. A fortified wine flavored with one or more aromatic plant parts such as bark, flowers, leaves, roots, etc.

Bordeaux. A wine produced in the Bordeaux region of France. May be red, white, or rose.

Brandy. Distillate from a wine (Hence, the characteristics of each product stem from those of the original wine, the type of distillation, and the aging process).

Burgundy. A wine produced in the Burgundy region of France. May be red, white, or sparkling.

Cabernet. Wine made from the Cabernet Sauvignon grape, which was brought from Bordeaux, France to California.

Chablis. An excellent dry white wine (with a green-gold tint) from the French town of Chablis. However, the name is sometimes applied to similar dry, white wines made elsewhere.

Champagne. A sparkling wine that is made by allowing wine from Pinot grapes to undergo a second fermentation after a small amount of sugar has been added to the bottle.

Chianti. Red wine from the Tuscany region of Italy that is often sold in a round-bottom flask placed in a straw basket. However, the best wine comes in tall bottles that can be binned for aging.

Claret. A dry, red Bordeaux wine made from Cabernet Sauvignon grapes.

Cognac. Brandy that is double distilled from wine made in the Charente district of France.

Cold Duck. A sparkling wine that is similar to champagne.

Concord wine. A strong-flavored, dark red wine made from Concord grapes (a native American variety).

Crackling wines. Wines that are less carbonated than sparkling wines.

Cream sherry. A heavy, dark-colored, sweetened sherry that is made by a process similar to the one developed in Jerez de la Frontera, Spain.

Dessert wines. Fortified (with additional alcohol in the form of brandy), wine that contains from 15 to 20% alcohol by volume.

Dry wines. A wine that is *not* sweet or sweetened. (In other words, all or most of the natural sugar content has been converted to alcohol.)

Fortified wines. Wines that have had their natural alcohol content increased by the addition of a brandy.

Honey wine (mead). An ancient type of wine that was made from fermented honey flavored with herbs.

Light wines. A wine that has a low alcoholic content.

Madeira. One of the wines made on the island of Madeira, which is located 500 mi (800 km) southeast of the coast of Portugal. (The wines range from light and dry to heavy and rich.)

May wine. A light, white Rhine wine that is flavored with the herb woodruff.

Moselle wines. Light wines (the alcohol content is usually about 10% or less) made in the valley of the Moselle River in Germany which lies to the west of the Rhine.

Mulled wine. Heated, sweetened, spiced wine served in a cup.

Muscatel. A sweet fortified wine made from Muscat grapes.

Perry (pear wine). Light wine made from pear juice.

Pinot. Wine made from Pinot grapes.

Port. The type of fortified wine that originated in the town of Oporto in Portugal.

Pulque. Fermented juice of the agave plant that grows in Mexico and in southwestern United States.

Red wines. Wines produced from dark-colored grapes that are fermented together with their skins (which contains most of the color pigments).

Resinated (Greek) wines. Greek wines that contain a resin which imparts a pinelike flavor.

Rice wine (sake). A Japanese wine made from fermented white rice.

Riesling. White wine made from the Riesling grape, which is considered to be the finest wine grape grown in Germany.

Rhine wines. Wines from grapes grown in the Rhine River Valley of Germany. (The wines range from dry and light to rich and sweet.)

Rose wines. Rose-colored wines produced by fermenting dark-colored grapes without the skins present, or from lighter grapes in the presence of their skins.

Sauternes. Wines made in the Sauternes district of Bordeaux, France from grapes withered somewhat by a *Botrytis* mold that is also called *noble rot*.

Sherry. A fortified wine made by a process similar to the one developed in Jerez de la Frontera, Spain. (Sherries range from pale-colored dry wines to rich, sweet ones.)

Sparkling wines. Wines that are bubbly with carbon dioxide gas by virtue of having undergone a second fermentation initiated by the addition of a small amount of sugar.

Sweet wines. Fortified wines that contain considerable amounts of unfermented sugars. The addition of extra alcohol prevents the fermentation of the sugars which are present.

Table wines. Unfortified wines of low to moderate alcoholic content. (They usually contain 14% or less of alcohol.)

Tokay. A rich white dessert wine made in Hungary that comes in dry and sweet varieties.

Vermouth. A fortified wine that is flavored with a variety of aromatic herbs and comes in dry and sweet varieties.

White wines. Made by fermenting grapes separated from their skins in order to keep the content of colored pigments low.

Zinfandel. A red wine made from Zinfandel grapes grown in California.

NOTE WELL: For a more complete summary of types of wines and related products, including their uses, see the two-volume *Foods & Nutrition Encyclopedia,* a work by the same authors as *Food For Health.*

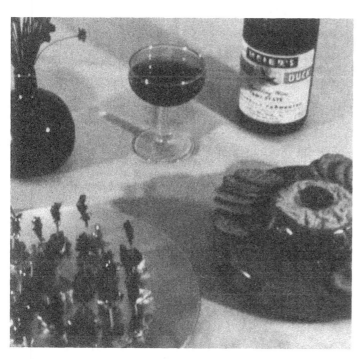

Fig. W-26. Wine and hors d'oeuvres are a favorite pre-dinner snack. (Courtesy, New Mexico State University, Las Cruces, N.M.)

MEDICINAL AND NUTRITIONAL EFFECTS. Many curative powers have been attributed to wine, since it has been used medicinally from the time of the ancient Greek physician, *(460 to 370 B.C.)* Hippocrates. However, the observations made in the early days of medicine were often clouded by unidentified factors that may have helped to alleviate or worsen diseases. Therefore, some of the beneficial effects ascribed to wine in a recent article are noteworthy:[12]

1. Wine contains many constituents other than alcohol that tend to slow the rate at which the alcoholic content is absorbed. (Earlier studies showed that wine consumed with a meal produced a peak blood alcohol level that was only about one-quarter of that resulting from the same

amount of alcohol taken in the form of gin or vodka.) Hence, intoxication is less likely to occur when moderate amounts of wine are consumed.

2. The anthocyanins (colored pigments) and tannins (astringent substances) in wines have greater antiviral effects than unfermented grape juice. However, these experiments were conducted in laboratory vessels, rather than in animal or human subjects.

3. Two 4 oz *(120 ml)* glasses of wine taken daily alter the blood patterns of cholesterol and other fats so that the likelihood of atherosclerotic heart disease is reduced by a small, but significant, amount.

4. Wine has a relaxing effect that may be due in part to ingredients other than alcohol.

5. The aroma and taste of good wines stimulate the appetite.

6. Substances present in both normal wine and dealcoholized wine promote better absorption of the essential minerals calcium, phosphorus, magnesium, and zinc than pure alcohol or deionized water. (It has long been thought that the absorption of iron is improved by wine. Hence, an old time remedy for iron deficiency anemia was "beef, iron, and wine.")

DANGERS OF DRINKING TOO MUCH WINE. Intoxication and other serious threats to health may result from the overconsumption of any of the alcoholic beverages. Experiments have shown that the body of a medium size man can metabolize about ½ oz *(15 ml)* of pure alcohol per hour. This is equivalent to the amount of alcohol present in a 4 oz *(120 ml)* glass of most table wines. It is noteworthy that four 8 oz *(250 ml)* portions of California Zinfandel wine were consumed with meals at 4 hour intervals (between 9 A.M. and 9 P.M.) by young male volunteers who showed no signs of intoxication.[13] However, the consumption of similar amounts of wine without food has been shown to produce blood alcohol levels that are twice as high as those resulting from the consumption of wine with food. Therefore, it seems best to restrict the consumption of table wines to not more than 1 pt *(480 ml)* per day in order to allow for differences in the rates of alcohol absorption under various conditions of drinking.

(Also see ALCOHOLISM.)

WINEBERRY *Rubus phoenicolasius*

The fruit of an oriental type of raspberry (fruit of the family *Rosaceae*).

Wineberries are orange-colored and originated in northern China and Japan. They may be eaten fresh or made into jam, jelly, juice, pies, and wine.

(Also see FRUIT[S], Table F-23, Fruits Of The World—"Raspberry".)

WINTERIZATION

Some oils solidify or crystallize and become cloudy at refrigerator temperatures, due to the presence of triglycerides containing saturated fatty acids which have a higher melting point. The process of winterization filters out these triglycerides from a chilled oil thus improving consumer acceptability by providing an oil that remains clear. The higher melting triglycerides removed from the oil may be used in margarines and shortening.

(Also see OILS, VEGETABLE, section on "Refining.")

---

[12]McDonald, J. B., "Not by Alcohol Alone," *Nutrition Today,* Vol. 14, January/February, 1979, pp. 14-19.

[13]*Ibid,* p. 18.

WITCHES' MILK

Secretion from the mammary glands of the newborn of both sexes thought to be due to placental permeability to the lactation-producing hormones of the mother.

WOOD ALCOHOL (METHANOL; METHYL ALCOHOL; CH₃OH)

Originally, this product was obtained from the destructive distillation of wood; hence, it was termed wood alcohol. It is extremely poisonous (causes blindness). Poisoning due to methyl alcohol is almost always a case of ingestion as a substitute for ethyl alcohol—the alcohol of liquors. Rather ironically, part of the treatment for methyl alcohol poisoning consists of administering ethyl alcohol. Aside from its adverse effects when ingested, methyl alcohol is an important industrial chemical utilized for numerous processes and products.

(Also see METHYL ALCOHOL.)

WOODCOCK

Many hunters consider the woodcock to be the best winged game bird; and many gourmets consider it one of the most succulent morsels. It is full of glory when roasted and served to the hunter who shot it.

WORK, MUSCULAR; ENERGY REQUIREMENTS

Work is accomplished when a force acts to move an object some distance. Muscles are capable of creating a force and of performing work. However, for muscles to create a force they require chemical energy. Thus, any time muscular work is performed the energy demands of the body increase. In fact, muscular activity is the most powerful stimulus for increasing the metabolic rate—the rate at which the body utilizes energy. Short bursts of strenuous exercise can increase the metabolic rate forty times that of the resting state. From day to day, a person engaged in sedentary work may require only 2,500 Calories (kcal) or less per day, while a person doing hard manual labor may require 5,000 Calories (kcal) or more per day.

(Also see CALORIC [ENERGY] EXPENDITURE; and METABOLIC RATE.)

WORLD FOOD

Contents	Page
Population, Production, and Wealth Around the World	1135
Type of Diet	1137
Outlook and Options Relative to World Food	1137
Who Shall Eat?	1137
Favoring Bread Alone	1137
Favoring Animals	1138
Meeting the Feeds Vs Foods Dilemma	1140
Improving the World Food Situation	1141
Why Be Your Brother's Keeper?	1143

In 1798, an English clergyman named Thomas Robert Malthus predicted part of the world food problem. He stated, "The power of population is infinitely greater than the power in the earth to provide subsistence for man." Malthus, however, failed to see the many changes of the future. For almost 200 years, we proved Malthus wrong because, as the population increased, new land was brought under cultivation; and machinery, chemicals, new crops and varieties, and irrigation were added to step up the yields.

Fig. W-27. The geography of world food problems. The darkened areas are countries where the undernourished population exceeded 15%.

Now, science has given us the miracle of better health and longer life; and world population is increasing at the rate of about 240,000 people a day. At this rate, world population will double by the year 2045. To meet the needs of a more sophisticated and demanding world population of this size, world food production needs to increase at an average rate of about 2.5% per year.

World food problems are more complex than just too many people and too little food. Thus, during the 25-year period 1961–63 to 1983–85, people as a whole were better fed than previously. On the average, the food available per capita rose from 2,320 calories to 2,660 calories. But the exceptions were many! In the low income countries as a group, apart from China and India, per capita food supplies in 1983–85 were no higher than 15 years earlier. More

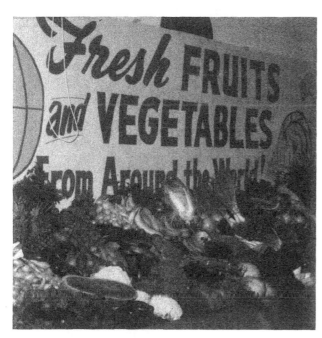

Fig. W-28. Fresh fruits and vegetables from around the world. (Courtesy, University of Tennessee, Knoxville)

disturbing yet, about 75% of the people in the world live in the developing countries where only 40% of the world's food is produced. Also, this is where most of the world's increase in population is occurring. Their rapid population growth causes severe economic strains on food production, processing and distribution. Finally, there is the problem of money. Countries able to produce sufficient food for their needs could supply food to the developing countries where production does not meet the needs. However, these developing countries generally lack the money necessary to purchase food from other countries. Therefore, the world food problems are related to (1) population, (2) production and distribution, and (3) wealth.

(Also see MALTHUS.)

POPULATION, PRODUCTION, AND WEALTH AROUND THE WORLD.
Around the world, population, food production, and wealth are unequally distributed, and will remain so in the future.

• **Population**—The world population is about 5.3 billion and by the year 2000 it will be 6.2 billion. Assuming all 6.2 billion people stood shoulder to shoulder and occupied an average width of 2 ft, 6.2 billion people would circle the world 91 times, or reach to the moon and back nearly five times!

Numbers are, however, only part of the problem. As Table W-17 shows, there is and will be dramatic unequal distribution of the people of the world. Only 21% of the population will be in developed regions of the world by the year 2000. Africa, Asia, and Oceania, will have 70% of the population. Among the individual nations the United States will have only 4% of the world's population, compared to 21% in the People's Republic of China, or 16% in India.

TABLE W-17
POPULATION PROJECTIONS FOR WORLD
MAJOR REGIONS, AND SELECTED COUNTRIES[1]

Area	Year 1975	Year 2000	Percent of World Population in 2000
	(millions)	(millions)	(%)
World	4,090	6,351	100
More developed regions	1,131	1,323	21
Less developed regions	2,959	5,028	79
Major regions			
Africa	399	814	13
Asia and Oceania	2,274	3,630	57
Latin America	325	637	10
U.S.S.R. and Eastern Europe	384	460	7
North America, Western Europe, Japan, Australia, and New Zealand	708	809	13
Selected countries and regions			
People's Republic of China	935	1,329	21
India	618	1,021	16
Indonesia	135	226	4
Bangladesh	79	159	2
Pakistan	71	149	2
Philippines	43	73	1
Thailand	42	75	1
South Korea	37	57	1
Egypt	37	65	1
Nigeria	63	135	2
Brazil	109	226	4
Mexico	60	131	2
United States	214	248	4
U.S.S.R.	254	309	5
Japan	112	133	2
Eastern Europe	130	152	2
Western Europe	344	378	6

[1]*Global 2000 Report to the President, Entering the Twenty-First Century*, p. 9, Table 1.

• **Production**—Food production depends upon a number of factors, among which are (1) the availability and use of land, (2) farming methods, (3) soil fertility, (4) water, (5) weather and climatic conditions, (6) civil order, and (7) incentives to producers. Therefore, the ability of countries to produce sufficient food for their need varies as shown in Table W-18.

TABLE W-18
PER CAPITA GRAIN PRODUCTION, CONSUMPTION, AND TRADE, ACTUAL AND PROJECTED, AND PERCENT INCREASE IN PER CAPITA TOTAL FOOD PRODUCTION AND CONSUMPTION[1]

Area	Grain			Food (Percent increase over the 1970-2000 period)
	1969-71	1973-75	2000	
	(kilograms per capita)			(%)
United States				
Production	1,018.6	1,079.3	1,640.3	51.1
Consumption	824.9	748.0	1,111.5	28.3
Trade[2]	+194.7	+344.0	+528.8	
Western Europe				
Production	364.9	388.4	394.0	1.0
Consumption	432.4	443.3	548.8	15.5
Trade	-65.4	-57.6	-154.8	
Japan				
Production	121.7	108.5	135.4	6.1
Consumption	267.5	274.4	452.3	54.2
Trade	-138.1	-175.9	-316.7	
U.S.S.R.				
Production	697.6	711.2	903.2	28.1
Consumption	663.1	796.1	949.9	41.4
Trade	+16.1	-42.0	-46.7	
People's Republic of China				
Production	216.3	217.6	259.0	17.4
Consumption	220.2	222.4	267.8	19.1
Trade	-4.0	-4.8	-8.8	
Latin America				
Production	236.1	241.0	311.4	33.7
Consumption	226.5	238.3	278.1	25.1
Trade	+11.8	+2.7	+33.3	
North Africa/Middle East				
Production	217.1	214.6	222.5	1.8
Consumption	276.2	273.8	292.8	2.2
Trade	-50.8	-69.8	-70.3	
South Asia				
Production	161.6	162.4	170.0	4.6
Consumption	170.0	171.8	181.0	5.8
Trade	-8.4	-11.8	-11.0	
Southeast Asia				
Production	244.7	214.5	316.5	35.9
Consumption	207.2	182.6	228.5	14.6
Trade	+37.5	+31.9	+87.5	
East Asia				
Production	137.3	136.0	163.5	22.8
Consumption	176.2	171.5	217.3	27.3
Trade	-40.4	-38.8	-53.8	

[1]*Global 2000 Report to the President, Entering the Twenty-First Century*, pp. 20 and 21, Table 6.
[2]In trade figures, a plus indicates export, while a minus sign indicates import.

Several important observations can be made from Table W-18. As far as past and future grain production is concerned, only three areas listed produce more grain than is consumed; thus leaving grain for export to other countries. Of the three—United States, Latin America, and Southeast Asia—the United States has the greatest excess for export. In terms of the percent increase in production and consumption of food between now and the year 2000, only the United States, Latin America, and Southeast Asia will significantly increase the percentage of food produced over

that consumed. Most areas of the world will consume more food than they produce.

● **Wealth**—Living standards have risen during the 20th century, but wealth remains unequally distributed. In many countries the economic development lags; hence, they are unable to purchase food or apply technology to increase food production within their country. By comparing the per capita income, some idea may be gained of the economic well-being of the countries and of the citizens' ability to purchase a nutritious diet. Table W-19 presents some of the countries with the highest incomes, and some of the lowest incomes.

TABLE W-19
PER CAPITA INCOME FOR SELECTED COUNTRIES[1]

Country	Per Capita Income
Highest:	(U.S. $)
Qatar	27,000
Switzerland	26,309
Iceland	21,660
Denmark and West Germany	19,750
U.S.A.	16,490
Japan	15,030
Australia	14,458
Norway	13,790
Luxembourg	13,380
United Kingdom	13,329
Netherlands	13,065
France	13,046
Austria	12,521
United Arab Emirates and Finland	11,900
Lowest:	
Niger	310
Zambia	304
India	300
Somalia	290
Zimbabwe	275
Tanzania and China	258
Togo	240
Uganda and Afghanistan	220
Myanmar (Burma)	210
Vietnam	180
Nepal	160
Ethiopia	121
Bangladesh	113

[1]*The World Almanac and Book of Facts 1991*, pp. 684–771.

Several crucial observations can be made from the information in Table W-19. There is a wide gap between the highest per capita income countries and the lowest per capita income countries, and the gap will widen by 2000. In general, areas with the greatest population have the lowest per capita income.

In most low income countries, consumers spend ½ or more of their income for food; whereas, in the higher income countries, the proportion drops to less than ⅕. For example, consumers in different countries spend the following proportion of their income for food: U.S.A., 10.3%; Canada, 11.3%; United Kingdom, 12.8%; Netherlands, 14.2%; Australia, 15%; Denmark, 15.7%; France, 16.4%; West Germany, 16.6%; Sweden, 17%; and Austria, 17.4%. Moreover, among the people of all countries, food takes a larger chunk of the income of the poor than it does of the rich, e.g., India where they spend 53% on food; the Philippines, 51%; and South Korea, 36%. The great disparities of wealth within countries can also be expected to widen by the year 2000. (Also see INCOME, PROPORTION SPENT FOR FOOD.)

TYPE OF DIET. A country's staple food is its primary source of carbohydrate, and for much of the poor population it must also serve as the principle source of protein. The degree to which this generally low-quality protein is supplemented by a more valuable or complementary source of protein (animal or legume protein) depends on income. The developing countries where wheat and rice are the principal crops have been the most successful in meeting food demand. In countries where corn is the chief crop, about 15% of the population is undernourished. In countries where the people subsist on millet and sorghum or root and tubers (much of Africa), some degree of malnutrition is virtually universal.

OUTLOOK AND OPTIONS RELATIVE TO WORLD FOOD. Increasingly, the charge will be made that much of the world goes hungry because of the substitution of meat, milk, and eggs for direct grain consumption. The goal is to feed the world a nutritious diet by the most efficient methods. Before deciding, individuals in both camps—(1) feed the grain to people, or (2) feed the grain to animals—should possess a sound knowledge of the options rather than moral indignation. When this is done, the outlook for the future may be brighter. To this end the important sections that follow are presented.

Who Shall Eat? Cereal grain is the most important single component of the world's food supply, accounting for between 30 and 70% of the food produced in all world regions. It is the major, and sometimes almost exclusive, source of food for many of the world's poorest people, supplying 60 to 75% of the total calories many of them consume. However, in many developed countries, more grain is fed to animals than is consumed directly by humans. Under such circumstances, sporadic food shortages and famine in different parts of the world give rise to the following recurring questions:

1. Who should eat grain—people or animals? Shall we have food or feed?

2. Can we have both food and feed?

FAVORING BREAD ALONE. Among the arguments sometimes advanced by those who favor bread alone—the direct human consumption of grain—are the following:

1. **More people can be fed.** About 2,000 lb (*907 kg*) of grain must be supplied to livestock in order to produce enough meat and other livestock products to support a man for a year, whereas 400 lb (*181 kg*) of grain (corn, wheat, rice, soybeans, etc.) eaten directly will support a man for the same period of time. Thus, a given quantity of grain eaten directly will feed five times as many people as it will if it is first fed to livestock and then is eaten indirectly by humans in the form of livestock products. This is precisely the reason why the people of the Orient have become vegetarians.

2. **On a feed, calorie, or protein conversion basis, it is not efficient to feed grain to animals and then to consume the livestock products.** This fact is pointed up in Figs. W-29 through W-31.

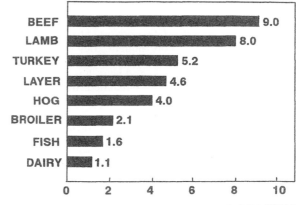

POUNDS OF FEED REQUIRED TO PRODUCE ONE POUND OF PRODUCT

Fig. W-29. Pounds of feed required to produce 1 lb of product. This shows that it takes 9 lb of feed to produce 1 lb of on-foot beef, whereas it takes only 1.1 lb of feed to produce 1 lb of milk. (One lb equals *0.45 kg.*)

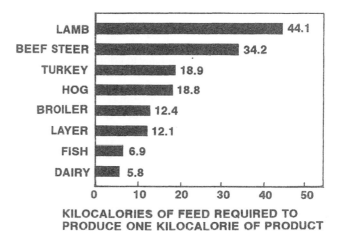

KILOCALORIES OF FEED REQUIRED TO PRODUCE ONE KILOCALORIE OF PRODUCT

Fig. W-30. Kilocalories in feed required to produce 1 kcal of product. This shows that it takes 44.1 kcal in feed to produce 1 kcal in lamb, whereas only 5.8 kcal in feed will produce 1 kcal in milk.

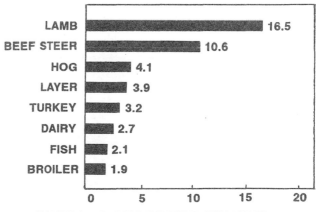

POUNDS OF FEED PROTEIN REQUIRED TO PRODUCE ONE POUND OF PRODUCT PROTEIN

Fig. W-31. Pounds of feed protein required to produce 1.0 lb of product protein. This shows that it takes 16.5 lb of feed protein to produce 1 lb of lamb protein, whereas only 1.9 lb of feed protein will produce 1 lb of broiler protein. (One lb equals *0.45 kg.*)

Thus, in the developing countries, where the population explosion is greatest, virtually all grain is eaten directly by people; precious little of it is converted to animal products.

As people become more affluent, they actually use more grain, but most of it is converted into animal products, for they consume more meat, milk, and eggs. It is noteworthy, too, that no nation appears to have reached such a level of affluency that its per capita grain requirement has stopped rising.

FAVORING ANIMALS. Practicality dictates that a hungry world should consider the following facts in favor of sharing grain with animals, then consuming the animal products:

1. **Animals provide needed power.** In the developing nations, cattle, water buffalo, donkeys, and horses still provide much of the agricultural power. Such draft animals are a part of the agricultural scene of Asia, Africa, the Near East, Latin America, and parts of Europe; areas characterized by small farms, low incomes, abundance of manpower, and lack of capital. They can be fueled on roughages to produce power, a most important consideration in times of energy shortage; and both cattle and water buffalo may be used for work, milk, and meat.

Although the general trend in the world is toward more and more mechanization, animals will continue to provide most of the agricultural power for the small farm food/crop agriculture in many of the developing countries.

If the entire world were suddenly to adopt American farming and food processing methods, increasing the diets of all people to the American level, the energy consumed would exhaust the world's known petroleum reserves in 13 years.

Modern mechanized feed and food production requires an extra input of fuel, which is mostly of fossil origin. This auxiliary energy is expended in endless ways to improve agricultural productivity; it is used for drainage and irrigation, clearing of forest land, seedbed preparation, weed and pest control, fertilization, and efficient harvesting. In addition to production as such, there are two other important steps in the feed-food line as it moves from the producer to the consumer; namely, processing and marketing, both of which require higher energy inputs than to produce the food on the farm. In 1990, U.S. farms expended an average of 2.8 Calories (kcal) on the farm per calorie of food grown. By contrast, the Chinese wet rice peasant, using animal power (water buffalo), expends only 1 Calorie (kcal) of energy to produce each 50 Calories (kcal) of food.

(Also see ENERGY REQUIRED FOR FOOD PRODUCTION.)

Fig. W-32. An Oriental wet rice peasant, using animal power (water buffalo), expends only 1 Calorie (kcal) of energy to produce each 50 calories (kcal) of food. By comparison, the average U.S. farmer, using mechanical power (tractors), expends 2.5 Calories (kcal) of fuel energy to produce 1 Calorie (kcal) of food. (Courtesy, International Bank for Reconstruction and Development, Washington, D.C.)

2. **Animals provide needed nutrients.** Animal products provide all the essential amino acids (including lysine and methionine in which vegetable sources are deficient), plus minerals and vitamins, along with palatability and digestibility.

Foods of animal origin (meat, milk, and their various by-products) are especially important in the American diet; they provide ⅔ of the total protein, about ⅓ of the total energy, ⅘ of the calcium, ⅔ of the phosphorus, and significant amounts of the other minerals and vitamins needed in the human diet.

In addition, meat, dairy products, and eggs are a rich source of vitamin B-12, which does not occur in plant foods—only in animal sources and fermentation products. Also, it is noteworthy that the availability of iron in beef is twice as high as in plants.

About ⅔ of the world's protein supply is provided from plant sources, ⅓ from animal sources. Since the Food and Agriculture Organization of the United Nations reports that the world's diet needs animal protein in amounts equivalent to ⅓ of the total protein requirements, there should be ample animal protein, *provided* it were equally distributed. But it isn't. The people in the developed countries have five times as much high-quality animal protein per person as the people living in the developing countries (Table W-20). The gap between total protein (animal and plant combined) is not as wide (106.2 g vs 58.3 g per person per day in the developed and developing countries, respectively).

TABLE W-20
WORLD PROTEIN PER PERSON PER DAY
IN GEOGRAPHIC AREAS OR COUNTRY[1]

Area or Country	Total Protein	Source	
		Animal	Plant
	(g/day)		
U.S.S.R.	106.2	56.1	50.1
Europe	102.0	59.6	42.4
North-Central America	95.6	56.7	38.9
Oceania	88.9	56.8	32.1
South America	65.0	29.5	35.5
Asia	61.0	13.2	47.8
Africa	58.3	12.5	45.8
World Total	70.4	24.7	45.7

[1]*FAO Production Yearbook 1991,* FAO/UN, Rome, Italy, p. 239.

The most important role of animal protein is to correct the amino acid deficiencies of the cereal proteins, which supply about two-thirds of the total protein intake, and which are notably deficient in the amino acid, lysine. The latter deficiency can also be filled by soybean meal, fish, protein concentrates and isolates, synthetic lysine, or high-lysine corn. But such products have neither the natural balance in amino acids nor the appetite appeal of animal protein.

(Also see PROTEIN[S], section headed "Protein for the World.")

3. **Much of the world's land is not cultivated.** Vast acreages throughout the world—including arid and semiarid grazing lands; and brush, forest, cutover, and swamplands—are unsuited to the production of bread grains or any other type of farming; their highest and best use is, and will remain, for grazing and forest.

Fig. W-33. Vast areas throughout the world, such as this rough terrain, are not suited to cultivation. Hence, their only use is for grazing or forest.

In the United States, only 21% of the land area of the 50 states is cultivated. About 900 million acres, or 46.8%, of the land area, exclusive of Alaska and Hawaii, is pasture and grazing land.

In China, only 10% of the land is cultivated. North of China's Great Wall, life centers on pastoral areas; large flocks and herds of cattle, sheep, and horses roam these vast grasslands.

4. **Forages provide most of the feed for livestock.** Pastures and other roughages—feeds not suitable for human consumption—provide most of the feed for livestock, especially for ruminants (four-stomached animals such as cattle, sheep, goats, buffalo, and certain wild species including deer, antelope, and elk), throughout the world. Fortunately, the uniqueness of the ruminant's stomach permits it to consume forages, and, through bacterial synthesis, to convert such inedible (to humans) roughages into high-quality proteins—meat and milk. Hence, cattle and sheep manufacture human food from nonedible forage crops. Additionally, they serve as the primary means of storing (on the hoof, without refrigeration) such forage from one season to the next.

Despite grains being relatively plentiful in the United States, forages provide the bulk of animal feeds; pastures and other roughages account for 94% of the total feed of sheep, 85% of the feed of beef cattle, 59% of the feed of dairy cattle, and 62% of the feed of all livestock.

5. **Food and feed grains are not synonymous.** Animals do not compete to any appreciable extent with the hungry people of the world for food grains, such as rice or wheat. Instead, they eat feed grains and by-product feeds—like field corn, grain sorghum, barley, oats, milling by-products, distillery wastes, and fruit and vegetable wastes—for which there is little or no demand for human use in most countries, plus forages and grasses—fibrous stuff that people cannot eat. For example, in the United States only 3% of the corn—the major animal feed grain—is used for human food. Also, it is noteworthy that the feed grains which the United States ships overseas are used almost entirely for livestock and poultry production abroad.

6. **Ruminants utilize low-quality roughages.** Cattle, sheep, and goats efficiently utilize large quantities of coarse, high-cellulose roughages, including crop residues, straw, and coarse low-grade hays. Such products are indigestible to humans, but from 30 to 80% of the cellulose material is digested by ruminants.

Of all U.S. crop residues, the residue of corn (cornstalks and husklage) is produced in the greatest abundance and offers the greatest potential for expansion in cow numbers. In 1990, 66,952,000 acres of corn, yielding 118.5 bushels per acre, were harvested in this country.

Normally, over and above the grain, corn produces over 200 million tons of corn residue each year. That's more than 200 million tons of potential cow feed, enough to winter more than 150 million pregnant cows. Mature cows are physiologically well adapted to utilizing such roughage. Moreover, when corn residue is used to the maximum as cow feed, acreage which would otherwise be used to pasture the herd is liberated to produce more corn and other crops. Also, there are many other crop residues which, if properly utilized, could increase the 150 million head figure given above.

Fig. W-34. Cattle can utilize efficiently large quantities of coarse, humanly inedible roughages, like cornstalks. This shows cows feeding on corn residue which had been harvested by mechanical means. (Courtesy, Iowa State University)

7. **Animals utilize by-products.** Animals provide a practical outlet for a host of by-product feeds derived from plants and animals, which are not suited for human consumption. Some of these residues (or wastes) have been used for animal feeds for so long, and so extensively, that they are commonly classed as feed ingredients, along with such things as the cereal grains, without reference to their by-product origin. Most of these processing residues have little or no value as a source of nutrients for human consumption. Among such by-products are corncobs, cottonseed hulls, gin trash, oilseed meals, beet pulp, citrus pulp, molasses (cane, beet, citrus, and wood), wood by-products, rice bran and hulls, wheat milling by-products, and fruit, nut, and vegetable refuse. It is estimated that each year ruminants convert more than 9 million tons of by-products into human food.

8. **Animals provide elasticity and stability to grain production.** Livestock feeding provides a large and flexible outlet for the year-to-year changes in grain supplies. When there is a large production of grain, more can be fed to livestock, with the animals carried to heavier weights and higher finish. On the other hand, when grain supplies are low, herds and flocks can be maintained by reducing the grain that is fed and by increasing the grasses and roughages

Fig. W-35. Chinese hogs in Kwang Tung Province, China. Their ration consisted of two by-products—rice millfeed and bagasse (the pith of sugarcane), along with water hyacinth—all of which the pigs ate with relish. In China, swine utilize millions of tons of otherwise wasted crop residues and by-products. (Photo by A. H. Ensminger)

in the ration. Thus, when grains are in short supply, fewer slaughter cattle are grain fed—more are grass finished. In the years ahead, depending on future grain supplies and prices, it is predicted that less than two-thirds of the U.S. domestic beef supply will come from feedlot cattle, in comparison with the 77% of U.S. slaughter cattle that were grain fed in 1973. Also, during periods of high-priced grains, heavier feeder cattle will go into feedlots, and they will be fed for a shorter period on less grain and more roughage than when grains are more abundant and cheaper.

In the future, animals will increasingly be *roughage burners,* with the proportion of grain to roughage determined by grain supplies and prices.

Beef cattlemen, dairymen, and sheepmen will more and more rely upon the ability of the ruminant to convert coarse forage, grass, and by-product feeds, along with a minimum of grain, into palatable and nutritious food for human consumption, thereby competing less for humanly edible grains. The longtime trend in animal feeding will be back to roughages; increasingly, all flesh will be grass.

9. **Animals step up the protein content and quality of foods.** Grains, such as corn, are much lower in protein content in cereal form than after conversion into meat, milk, or eggs. On a dry basis, the protein contents of selected products are corn, 10.45%; beef (Choice grade, total edible, trimmed to retail level, raw), 30.7%; milk, 26.4%; and eggs, 47.0%. Also, animals increase the quality (e.g., biological value) of the protein—a higher proportion of the protein is assimilated by the body.
(Also see PROTEIN[S].)

10. **Ruminants convert nonprotein nitrogen to protein.** Ruminant animals (cattle, sheep, and goats) can use nonprotein nitrogen, like urea, to produce protein for humans in the form of meat and milk.

11. **Animals provide medicinal and other products.** Animals are not processed for meat alone. They are the source of hundreds of important by-products, including some 100 medicines such as insulin, epinephrine, and heparin, without which the lifestyle and health of people would be altered.

Besides medicines, many familiar products are derived

from animals, including leather, shoe polish, photographic film, soap, lubricants, candles, glue, buttons, and bone china, to name a few.

12. **Animals maintain soil fertility.** Animals provide manure for the fields, a fact which was often forgotten during the era when chemical fertilizers were relatively abundant and cheap. One ton *(907 kg)* of average manure contains 18 lb *(8.2 kg)* of nitrogen, 9 lb *(4.1 kg)* of phosphorus, and 13 lb *(5.9 kg)* of potassium.

The energy crisis prompted concern that farmers would not have sufficient chemical fertilizers at reasonable prices in the years ahead. Since nitrogenous fertilizers are oil- and petroleum-based, there is cause for concern. As a result, a growing number of American farmers are returning to organic farming; they are using more manure—the unwanted barnyard centerpiece of the past, and they are discovering that they are just as good reapers of the land and far better stewards of the soil.
(Also see ORGANICALLY GROWN FOOD.)

Meeting the Feeds Vs Foods Dilemma. Life on earth is dependent upon *photosynthesis.* Without it, there would be no oxygen, no plants, no feed, no food, no animals, and no people.

Fig. W-36. Ruminants—cattle, sheep, and goats—convert the photosynthetic energy derived from solar energy and stored in grass into food for humans. (Courtesy, *The Progressive Farmer,* Birmingham, Ala.)

As fossil fuels (coal, oil, shale, and petroleum)—the stored photosynthates of previous millennia—become exhausted, the biblical statement, "all flesh is grass" (Isaiah 40:6), comes alive again. The focus is on photosynthesis. Plants, using solar energy, are by far the most important, and the only renewable, energy-producing method; the only basic food-manufacturing process in the world; and the only major source of oxygen in the earth's atmosphere. Even the chemical and electrical energy used in the brain cells of man is the product of sunlight and the chlorophyll of green plants. Thus, in an era of world food shortages, it is inevitable that the entrapment of solar energy through photosynthesis will, in the long run, prove more valuable than all the underground fossil fuels—for when the latter are gone, they are gone forever.
(Also see PHOTOSYNTHESIS.)

Practicality dictates that a hungry world should, and will, proceed in about the following order in meeting the feeds vs foods dilemma:

1. Consume a higher proportion of humanly edible grains and seeds, and their by-products, directly—without putting them through animals, simply because approximately five times more people can be fed by doing it this way.

2. Utilize a higher proportion of roughages to concentrates in animal rations as increasing quantities of cereal grains are needed for human consumption.

3. Retain more of those species that can utilize a maximum of humanly inedible feeds and a minimum of products suitable for human consumption. This would favor cows, sheep, and goats, provided they are fed a maximum of pasture and other roughages. Both poultry and swine may compete with man for grains. Nevertheless, it is expected that further increases in poultry will come, primarily because of their efficiency as converters of protein from feed to food, and their adaptability to small-scale production. Also, it is expected that there will be further increases in swine, especially in China, where pigs are scavengers and manure producers par excellence.

4. Propagate the most efficient feed to food species converters (see Figs. W-29, W-30, and W-31). This means dairy cows, fish, and poultry. Because beef cattle and sheep are at the bottom of the totem pole when it comes to feed efficiency, the pressure will be to eliminate them, except as roughage consumers. Although not mentioned yet, rabbits are also efficient converters of feed to food, and can utilize roughages to some extent. Moreover, rabbits are easily adapted to small-scale agriculture.

5. Increase the within-species efficiency of all animals and eliminate the inefficient ones. This calls for more careful selection and more rigid culling than ever before.

6. Improve pastures and ranges. Good pasture will produce 200 to 400 lb (91 to 181 kg) of beef or lamb per acre annually (in weight of young weaned, or in added weight of older animals); superior pastures will do much better.

Improving The World Food Situation. The world food situation can be improved provided major problems are solved, many of which are not self-correcting. Among the most pressing are curbing population growth, transferring food from the developed food-exporting countries to the developing food-deficit countries, providing for emergency disaster and famine relief, achieving an acceptable degree of stability of world food prices, and finding the proper combination of techniques and policies to bring about a substantial improvement in food production in developing countries. On a longtime basis, the world food situation can best be improved by a massive infusion of education, science, and technology—by self-help programs —so that they can produce more of their own food.

• **Population control**—Members of the animal kingdom other than man have their numbers held in check by the many factors encompassed in the term *balance of nature.* Man is different! His strong propensity is to overpopulate the earth and to create conditions which threaten his very existence—his food supply, the water he drinks, the environment in which he lives, and the very air he breathes.

Without doubt, people will continue to live longer. Hence, curbing population growth will be required to maintain the balance between production and demand for food. It will be necessary to bring the number of people and their supply of food into proper balance. Alternate methods of population control are starvation, disease, and/or war.

• **Fair prices and profits**—People do those things which

are most profitable to them; and farmers are people. The American farmer, and farmers in certain other countries of the world, can produce more, but higher prices than have existed in the past will be necessary to assure this.

Farmers, like any businessmen, have always demonstrated their willingness to respond to incentives—prices and profits.

• **Increase cultivable land**—Ever since man stopped living a nomadic life, he has been hunting for arable (cultivable) land. Fortunately, there is still much of it to be had. Studies show that about twice as much of the world's land is suitable for crops as is presently used. More than half of the potential, but presently unused, arable land is in the tropics, and about a sixth of it is in the humid tropics—the largest areas being in Africa and South America.

• **More irrigation**—The value of irrigation for increasing crop yields is generally known. Yet, a study of the world's 20 major irrigating countries, in areas irrigated, showed that only 15% of their total cultivated area is irrigated. Thus, the potential to increase crop yields through irrigation is very great.

• **Improve crop yields**—While the amount of land that could be brought into production is perhaps double that currently used, all recent studies of world food production conclude that, outside of Africa and Latin America, yield-increasing techniques—irrigation, fertilizer, new seeds, and improved technology—will be the primary source of future food increases.

In addition to irrigation, fertilizer is a key factor in yield increases, although it must be combined with improved varieties of seeds and improved cultural practices if it is to have much impact on yields. As evidence of the soundness of the fertilizer approach, it is noteworthy that almost half of the 50% gain in crop output per acre in the United States since 1940 is attributed to the increased use of fertilizers. From this, it may be concluded that increased use of fertilizer could increase world food output by 50% in the years ahead.

• **Full use of pastures and ranges**—Some sparsely populated areas of the world, such as Australia, New Zealand, Argentina's pampas area, and the western range areas of the United States and Canada, are now important sources of livestock products. But there are still vast areas of sparsely settled grasslands where the production of livestock products is small; among them, large portions of Africa, the highlands of central Asia, some portions of the Andean area of South America, and the nomadic grazing areas of the Near East. In these areas, subsistence is the goal and animal numbers are generally regarded as being more important than the yield of salable products. Nevertheless, the potential for increased production in these areas is considerable. Also, and most important, the only practical way of harvesting human food from many of these areas is through livestock. So, grass—the world's largest crop—should no longer be taken for granted. In an era of world food shortages, the contribution of properly managed grazing lands in terms of food and fiber production needs to be pursued. No other program offers so much potential to increase the world's food production capacity quickly and at so little cost; this is especially true of the grasslands in the tropics and subtropics.

• **Produce leaner beef**—Leaner beef is higher in protein content than fat beef. On a carcass basis, trimmed to retail level, Standard grade runs 19.4% protein vs 17.4% for Choice grade—that's 2% higher. Besides, leaner beef can be produced with much less grain.

Consumer preferences and costs of production underlie

the relative prices of fat and lean beef, but changes in U.S. grading standards help consumers adjust their consumption patterns. For this reason, when grain prices are high, producers exert pressure to have the beef-grading system changed so as to reduce the amount of grain fed.

• **Select efficient animals**—Improved genetics, along with improved feeding and management, have made for more meat, milk, and eggs. Yet, further improvements are possible and needed, especially in the developing countries. Although 60% of the animals of the world are raised in the developing countries, primarily in Africa and Asia, these nations produce less than 30% of the world's meat, milk, and eggs. This low productivity is largely due to the failure to utilize the scientific principles of husbandry and disease control.

• **Feed roughage and by-products**—In the future, cattle and sheep will increasingly be *roughage burners.* Stockmen will rely upon the ability of the ruminant to convert coarse forage, grass, and by-product feeds, along with a minimum of concentrate, into palatable and nutritious food for human consumption, thereby competing less for humanly edible grains.

Ruminants can make the transition to more roughage with ease. For them, it is merely a *return to nature,* for they evolved as consumers of forage.

• **Control disease and parasites**—Diseases in farm animals reduce the supply of meat and other products by a large quantity, and add substantially to the cost of food and fiber. The cost of animal diseases, parasites, and pests of livestock and poultry to U.S. producers and consumers is estimated to aggregate $10 to $12 billion.

Deaths of animals take a tremendous toll. Even greater economic losses—hidden losses—result from failure to reproduce living young, and from losses due to retarded growth and poor feed efficiency, carcass condemnations and decreases in meat quality, and labor and drug costs. Also, considerable cost is involved in keeping out diseases that do not exist in a country, such as keeping foot-and-mouth disease out of the United States. Quarantine of a diseased area may cause depreciation of land values or even restrict whole agricultural programs. Additionally, and most importantly, it is recognized that some 200 different types of infectious and parasitic diseases can be transmitted from animals to human beings; among them, such dreaded diseases as brucellosis (undulant fever), leptospirosis, anthrax, Q fever, rabies, trichinosis, tuberculosis, and tularemia. Thus, rigid meat and milk inspection is necessary for the protection of human health. This is added expense which the producer, processor, and consumer must share.

Thus, the potential throughout the world of providing more food through animal disease and parasite control is very great. The level of animal health attained in the advanced countries shows that tremendous scope exists for improvement in most developing areas. It is estimated that if the tsetse fly of the high rainfall belt of tropical Africa were brought under control, the savannah pastures could carry a cattle population of 120 million head, equal to the total cattle population of the United States.

• **Create new and improved protein sources**—It is generally recognized that diet customs are somewhat emotional in character—that many people will put synthetic clothes on their backs long before they will put synthetic food in their stomachs. Yet, when people are hungry or suffering from malnutrition, they are not finicky about the *pedigree* of their food.

Researchers are attempting to bolster traditional protein sources and to develop entirely new proteins, with their efforts centered around the following approaches and protein sources: (1) improvement of traditional sources through genetic manipulation, (2) fortification with synthetic nutrients, or addition of protein concentrates from fish, oilseeds, and other foods, (3) use of the versatile soybean, and (4) development of single cell protein technology.
(Also see SINGLE-CELL PROTEIN; and SOYBEAN.)

• **Improve small-scale farming methods**—The United States and other developed countries are well known for agriculture that is large in scale and high in capital investments, substituting machinery for human labor. Although it has made the United States the world's largest exporter of food, the technology for large-scale agriculture has been difficult to implement abroad.

Subsistence farming is the principal agricultural practice internationally, especially in the developing countries. Unlike the large U.S. operations, subsistence farms are small in scale and low in capital investments, requiring intensive labor of the approximately one billion farmers who operate them. About 40% of the world's farmers have fewer than 11 acres (4.5 ha) of land; 35% have fewer than 2.5 acres (1 ha). These farms feed a majority of the world population. To provide effective technical assistance to these farmers, scientists must understand the problems of small-scale agriculture.

Although leading scientists have recently identified small-scale agriculture as a priority for research and several international agricultural research centers have initiated mixed cropping and homestead farming studies, research is still lacking on integrated small-scale agriculture, including crop production, methodology, vegetable crops, small animal production, and energy generation.

History shows that world food production has never been immune from shortfalls. These food shortages become more likely as world population increases. As food prices increase because of higher energy costs, the needy will depend on diets made up of the cheapest, most-accessible food without regard for nutrition.

One way to increase the food base and income is for farmers in densely populated agrarian regions to increase the agricultural production on their small land holdings. These farmers would then have the incentive to remain on the farm, avoiding the social problems that result when farmers migrate to the city. Appropriate small-scale agriculture could integrate crops and small animals (poultry, rabbits, milking goats, and sheep) into a system that would meet the nutritional requirements of the family, improving the quality as well as the quantity of food. This integrated approach could make a significant contribution to national health, wealth, and stability.

• **Farm the sea**—The ocean, which covers 70.73% of the earth's surface and, therefore, receives a proportionate amount of all the solar energy reaching this planet, is one of the most promising potentials for providing added food for the world's spiraling human population. It is an immense reservoir of food which man has only lightly tapped. Hence, there is growing interest in the sea as a source of food supply, including both fish and vegetables.

• **Eliminate waste**—Waste of food supplies will increasingly nag the consciences and pocketbooks of all people—producers and consumers alike.

Pests cause an estimated 30% annual loss in the world-wide potential production of crops, livestock, and forests.

Every part of our food, feed, and fiber supply—including marine life, wild and domestic animals, field crops, horticultural crops, and wild plants—is vulnerable to pest attack. Obviously, if these losses could be prevented, or reduced, world food supplies would be increased by nearly one-third. The problems are complex, but the stakes are high.

This worldwide annual loss of 30% potential food productivity occurs despite the use of advanced farming technology and mechanized agriculture. Furthermore, in many of the developing countries losses greatly exceed this figure.

Pests of many kinds attack plants during all stages of their growth, and they attack food and food products after harvest—in storage, during transportation to market, in warehouses, in elevators, in ships, in supermarkets, and in homes after purchase. A few notable pest losses include: plant diseases, insects, weeds, rats, and birds.

• **Increase scientific and cultural exchange between countries**—There will always be international boundaries. But scientists and agriculturalists the world over can, and will, work together through the *tie that binds*—their desire to help mankind. Since no nation has a corner on all the brains, scientific and cultural exchange between countries offers our best hope for survival.

• **Increase teaching, research, and extension**—The best long-run solution to the problem of world food and nutrition shortages is to develop teaching, research, and extension programs that provide farmers in the developing countries with the necessary tools and techniques to increase their production, augmented by fortification of foods and education of consumers.

Each of the developing countries needs: (1) an agricultural teaching-research-extension program patterned after the Land Grant Colleges of America, but adapted to their respective countries; (2) seminars and agricultural short courses for adults; and (3) to send more of their best and brightest students abroad for graduate training. These programs need to be implemented by imaginative and bold approaches because, on a longtime basis, they afford the most logical way in which to provide adequate food and clothing for the world's expanding population.

Why Be Your Brother's Keeper? Famine and starvation are not new to the people of the world, but what is new are the electronic means of communication which are shrinking the size of the world and increasing our awareness. People can sit in their living rooms and watch others starve in some other area, or people in a remote village can become aware that there is another world where people have enough of life's essentials. There are numerous reasons to worry enough to take action, but the three major ones follow:

1. **Humanitarian.** Developing countries need help and many developed countries have the knowledge and resources to help. The challenge of a difficult task and the moral uplift that comes only from doing for others can serve to temper and balance the affluence of American life as exemplified in the late Albert Schweitzer's dictum: "It is only giving that stimulates."

2. **Security.** By the year 2000, there will be four times as many people in the developing countries as in the developed countries. Developed countries cannot afford to be too little and too late with their assistance. The idea that security is more than military might is not new. Seneca, nearly 2,000 years ago, warned the Roman Senate: "A hungry people listens not to reason nor is its demand turned aside by prayers."

The expectations of the poor are demanding fulfillment. Hopefully, some measure of their ambitions can be realized by peaceful means.

3. **Economic.** An important way to expand our own economy in the future will be the creation of additional markets for U.S. goods and products. This aim is not entirely self-serving, because achievement of sustained economic growth by the hungry countries will depend upon their participation in world markets on a competitive basis.

While people of the world do not share a common history, they do share a common future. By sharing and applying know-how, the whole world will have a brighter tomorrow. Dreams will come true—faster and more abundantly, with more food and animals in the future.

(Also see CEREAL GRAINS, sections headed "Feeds for Livestock" and "Future Prospects for the Cereal Grains"; ENERGY REQUIRED FOR FOOD PRODUCTION; GREEN REVOLUTION; HUNGER, WORLD; MALNUTRITION, PROTEIN-ENERGY; MALTHUS; NUCLEIC ACIDS, section headed "Genetic Engineering"; PHOTOSYNTHESIS; POPULATION, WORLD; and PROTEIN, WORLD PER CAPITA.)

WORLD HEALTH ORGANIZATION (WHO)

A specialized agency of the United Nations founded in 1948 to further international cooperation for improved health conditions—physical, mental, and social well-being. The main offices are located in Geneva, Switzerland; and there are six regional offices around the world. Primarily, member governments, on the basis of their relative ability, finance the WHO with yearly contributions. About 130 countries are members. The World Health Assembly, the policy-making body; the Executive Board, a board of health specialists; and the Secretariat, the regional offices and field staff, are the principle organs through which the WHO operates.

The WHO is involved in three distinct health-related areas:

1. It provides a research service and sets standards of international sanitary regulations. The WHO keeps member countries informed on the most recent developments in the use of vaccines, control of drug addiction, nutritional discoveries, cancer research, and health hazards of nuclear radiation. It also standardizes quarantine measures with minimal interference of international trade and air travel.

2. It encourages member nations to enlarge and strengthen their health programs. To do this, the WHO, on request, provides (a) technical advice to governments, (b) sends out international teams of experts, (c) helps set up health centers, and (d) offers aid for the training of medical and nursing personnel.

3. It helps control epidemic and endemic diseases by promoting mass campaigns involving nationwide vaccination programs, clinics for early diagnosis and prevention of diseases, pure water supplies, good sanitation systems, health education, and antibiotic and insecticide use. It has been effective with campaigns against tuberculosis, malaria, and smallpox.

Besides its own programs, the WHO often works closely with the FAO (Food and Agriculture Organization).

(Also see FOOD AND AGRICULTURE ORGANIZATION OF THE UNITED NATIONS; and DISEASES, section headed "International and Voluntary Organizations Engaged in Health and/or Nutrition Activities.")

Fig. W-37. Improving the world food situation involves both short-run and long-run arrangements. The short-run: grain reserves for droughts, floods, and similar emergencies. The long-run: population control and self-help programs through the application of science and technology.

The above picture shows U.S. sorghum grain unloaded at the docks in Dakar (Africa). (Courtesy, Agency for International Development, Washington, D.C.)

Fig. W-38. PRODUCING THE NATION'S FOOD

BEEF ON THE HOOF
(Courtesy, Union Pacific Railroad Co., Omaha, Neb.)

GROWING CORN
(Courtesy, USDA)

HARVESTING SUGAR BEETS
(Courtesy, J. C. Allen & Son, West Lafayette, Ind.)

PACKING LETTUCE
(Courtesy, California Iceberg Lettuce Commission, San Rafael, Calif.)

PIGS TO PORK
(Courtesy, National Hog Farmer, St. Paul, Minn.)

CUTTING WHEAT
(Courtesy, Henry Fisher, Oakesdale, Wash.)

XANTHINE

An intermediate in the metabolism of purines. It was first isolated from gallstones. It occurs in animal organs, yeast, potatoes, coffee, and tea.

XANTHOMA

A benign flat yellow tumor, containing a deposit of fatty substance, commonly located on the inner side of the lower eyelid.

XANTHOPHYLL

While it is one of the most widespread naturally occurring carotenoid alcohols, it does not possess any vitamin A activity. Xanthophyll is a yellow pigment, which can be (1) isolated from certain natural products, and (2) produced synthetically.

Feeds that contain large amounts of xanthophylls produce a deep yellow color in the beak, skin, and shank of yellow-skinned breeds of chickens. The consumer associates this pigmentation with quality and, in many cases, is willing to pay a premium price for a bird of this type. Also, processors of egg yolks are frequently interested in producing dark-colored yolks to maximize coloration of egg noodles and other food products. The latter can be accomplished by adding about 60 mg of xanthophyll per kilogram of diet. In recognition of these consumer preferences, many producers add ingredients that contain xanthophylls to poultry rations.

Rich natural sources of xanthophyll follow:

Feedstuff	Xanthophyll Content
	(mg/kg)
Marigold petal meal..................	7,000
Algae, common, dried...............	2,000
Alfalfa juice protein, 40% protein.......	800
Alfalfa meal, 20% protein	240

(Also see ZEAXANTHIN.)

XEROPHTHALMIA

The term describing keratinization and cloudiness of the cornea caused by a vitamin A deficiency. Xerophthalmia occurs most frequently in undernourished infants and children in India, the Middle East, Southeast Asia, and parts of Africa and South America, in areas where the diet is lacking in whole milk, butter, and green or yellow vegetables. Vitamin A should be administered immediately to persons with xerophthalmia in order to arrest the disease and prevent the loss of sight.

(Also see BLINDNESS DUE TO VITAMIN A DEFICIENCY; DEFICIENCY DISEASES, Table D-1 Major Dietary Deficiency Diseases; and VITAMIN A, section headed ''Deficiency Symptoms.'')

XEROSIS

Abnormal dryness of the skin or front of the eye.

XYLITOL ($C_5H_{12}O_5$)

Xylitol is a sugar alcohol. The aldehyde group (C = O) of the pentose sugar—5-carbon sugar—xylose is replaced by a hydroxyl group (OH). It can be made from birchwood chips, berries, leaves and mushrooms. Xylitol is commercially produced in Finland from birchwood chips hydrolyzed by acid to xylose. In the body, xylitol is formed as an intermediate during the formation of xylulose. Ingested xylitol is also converted to xylulose. Xylulose then enters the pentose-phosphate cycle of carbohydrate metabolism.

Xylitol tastes almost as sweet as sucrose, table sugar; additionally, it has a cool taste when dissolved in the mouth due to its negative heat of solution. Furthermore, xylitol cannot be metabolized by the acid producing bacteria of the mouth which cause tooth decay. Indeed, some clinical trials have demonstrated a marked reduction in tooth decay through the use of xylitol. Xylitol is approved by the FDA for use in special dietary foods, and some chewing gums are sweetened, in part, with it. However, a British study suggested that xylitol causes cancer in laboratory animals. This places the future use of xylitol in foods in question.

Xylitol is absorbed slowly from the intestine. Hence, the consumption of large quantities may cause an osmotic diarrhea.

(Also see SWEETENING AGENTS.)

XYLOSE (WOOD SUGAR; $C_5H_{10}O_5$)

A 5-carbon—pentose—sugar widely distributed in plant material such as maple wood, cherry wood, straw, cottonseed hulls, corncobs, and peanut shells. It is one of the most abundant plant sugars in the world. However, it is of little or no importance as an energy source for the body. Xylose is not found in its free state in nature but rather as xylan—a polysaccharide built from numerous units of xylose. The sweet-tasting alcohol of xylose, known as xylitol, has been approved by the FDA for use in special dietary foods. Also xylose is employed as a diagnostic aid for the detection of malabsorption. Urinary excretion of less than 4.5 g in 5 hours following ingestion of a 25 g load suggests decreased absorptive capacity.

(Also see CARBOHYDRATE[S], Table C-6, Classification of Carbohydrates and section headed ''Monosaccharides.'')

Fig. X-1. Roast turkey, a good source of amino acids. Poultry, which embraces a wide variety of birds of several species, appears more frequently in the diet of people throughout the world than any other type of meat. (Courtesy, National Turkey Federation, Reston, Va.)

YAM BEAN (MANIOC BEAN; POTATO BEAN; TURNIP BEAN) *Pachyrrhizus erosus; P. tuberosus*

These climbing vines, which are tropical legumes, are grown mainly for their edible tubers. Hence, the various common names contain both the word bean and the names of other more popular tuberous plants. The major characteristics of *P. erosus* and *P. tuberosus* are very similar, except that the bean pods of the latter are not eaten because they have irritating hairs. Therefore, the name yam bean will be the sole designation used in the rest of this article. Fig. Y-1 shows a typical yam bean.

Fig. Y-1. The yam bean, a leguminous plant which has an edible tuber.

ORIGIN AND HISTORY. Various species of yam beans are native to a wide area extending from northern South America to Mexico. The Spanish took these plants to the Philippines during the 16th century. Since then, they have been adopted as crops throughout the tropics of the world.

PRODUCTION. Most of the crop comes from the humid tropical areas of western Pacific and Southeast Asia. Statistics on production are not available.

Yam beans, unlike many other tropical tubers, are grown from seeds. In the Orient, the young plants are trained to climb bamboo trellises. About 10 months are required for the production of mature tubers, which are harvested by plowing or pulling them from the ground. The average yield of tubers is about 8,000 lb per acre (8,960 kg/ha).

PROCESSING. The tubers of the plant are rarely processed, although starch is sometimes extracted by washing the grated tubers on a screen.

PREPARATION. Raw yam bean tubers have a crisp texture and are good when served diced or sliced in a salad. The tubers may also be boiled, fried, or roasted. The immature pods of *P. erosus* may be cooked and eaten like snap beans.

CAUTION: The mature seeds and leaves of yam beans should not be consumed, because they contain a harmful agent that is not known to be destroyed by cooking.

NUTRITIONAL VALUE. The nutrient compositions of various forms of yam bean are given in Food Composition Table F-21.

Some noteworthy observations regarding the nutrient composition of yam bean follow:

1. Compared to Irish potatoes, yam bean tubers have a higher water content and are lower in calories, protein and vitamin C. However, the protein-to-calorie ratios are about the same (2.8 g per 100 kcal) for both tubers. This means that yam bean tubers may be a fair source of protein for adults, providing that sufficient quantities are consumed.

2. Yam bean tubers are low in minerals and vitamins. Hence, they should be eaten with foods that supply these essential nutrients.

YAUPON (CASSINA)

The leaves of the yaupon are used as a tea substitute. It is a holly (*Ilex cassine*) which grows in the southern United States. The Indians attributed many virtues to the tea, and allowed only men to drink it. Yaupon tea contains caffeine.

YEAST

Fungi, of which three types are used commercially in the food industry: *Brewers' yeast*, a by-product from the brewing of beer and ale; *dried yeast*, used in leavened breads; and *torula yeast*, cultured as a foodstuff for man and animals, as a source of protein, minerals, B vitamins, and unidentified factors.

(Also see BREADS AND BAKING, section headed "Breads Leavened with Yeast and Other Microorganisms"; BREWERS' YEAST; and TORULA YEAST.)

YEAST EXTRACT

A preparation of the water-soluble fraction of autolysed (self-digested) yeast, valuable both as a rich source of the B vitamins and for its strong savory flavor. Yeast is allowed to autolyse, extracted with hot water, and concentrated by evaporation.

YEAST FERMENTATION, BOTTOM

This refers to fermentation during the manufacture of beer with a yeast that sinks to the bottom of the tank. Most beers are produced this way. Ale, porter, and stout are the principal beers produced by top fermentation.

YOGURT

A fermented milk product prepared from lowfat milk, skim milk, or whole milk. After fermentation, the yogurt may be mixed with other ingredients such as nonfat dry milk solids, vegetable gums, flavoring, or fruit preserves.

(Also see MILK AND MILK PRODUCTS.)

YOLK INDEX

This is an expression of egg quality—freshness—in terms of the spherical nature of the yolk. It is derived by measuring the height and width of the yolk. As the egg deteriorates, the index decreases.

(Also see EGGS, section headed, ''Physical Characteristics of the Egg and Grading.'')

YORKSHIRE PUDDING

This is a quick bread made with flour, milk and eggs, and either cooked in the pan underneath the roast of beef, or in separate muffin tins, with the meat drippings poured into the cups first. It is a first cousin to the popover.

Fig. Y-2. Baking-day, 1870. (Courtesy, The Bettmann Archive, New York, N.Y.)

ZANTE CURRANT

This fruit is not a currant, but is actually a raisin; it is made from black Corinth grapes that are only about one-quarter of the size of an ordinary raisin and look like a dried blackcurrant. The name *Zante* was given to the raisin because it was first produced on a large scale on the island of Zante in the southeastern part of Greece. Now, they are also produced in Australia.

Zante currants have a tart, tangy flavor and are commonly used in baked products such as coffee cakes and hot cross buns.

(Also see RAISIN[S].)

ZEAXANTHIN

It is a carotenoid alcohol, and the pigment of yellow corn. However, it possesses no vitamin A activity. Zeaxanthin is widespread in nature and occurs together with xanthophyll.

(Also see XANTHOPHYLL.)

ZEDOARY ROOT

An aromatic root from one of two plants of the East Indies which belongs to the ginger family. Zedoary is a GRAS (generally recognized as safe) natural flavor additive.

ZEIN

A protein derived from corn. It lacks the essential amino acids lysine and tryptophan. Commercially, zein is extracted from corn gluten meal with alcohol. In the food industry, its prime use is as an edible coating for foodstuffs such as nut meats and candy. In manufacturing, it has a variety of uses: plastics, paper coatings, adhesives, shellac substitute, printing, laminated board, and microencapsulation.

(Also see CORN; and PROTEIN[S].)

ZEN MACROBIOTIC DIET

This phrase encompasses both a diet and a philosophy of life. Zen refers to meditation, and macrobiotic suggests a tendency to prolong life. The Zen philosophy is hundreds of years old; its beginnings may be traced back to India around 470 A.D. From India, the Zen philosophy moved to China in about 520 A.D., but it was not introduced into Japan until the 1100s and 1200s. Zen seeks to discipline the mind so that an individual comes into touch with the inner workings of his body—a ''larger awareness'' that cannot be taught. In the United States, the Zen philosophy and the Zen cooking—claimed traditional for the ancient Zen Japanese—was popularized by a Japanese named Georges Ohsawa, who coined the term macrobiotic.

According to the macrobiotic plan, there are ten diets or stages of the same diet. In order to live a happy, harmonious life, the follower progresses from the lowest dietary stage, −3, to the highest or +7. As the macrobiotic follower progresses, desserts, fruits and salads, animal foods, soup, and vegetables, in that order, are eliminated and replaced by increased amounts of cereal grains in the diet. Therefore, only certain stages are pure vegetarian diets, though meat is considered undesirable. All dietary stages encourage the restriction of fluid intake. Furthermore, fluid restriction is encouraged if an individual perspires. Table Z-1 lists the ten stages of the Zen macrobiotic diet and the percentage of food from the different sources.

TABLE Z-1
TEN STAGES OF THE ZEN MACROBIOTIC DIET,[1]
FROM LOWEST TO HIGHEST

Stage No.	Cereal Grains	Vege-tables	Soup	Animal	Fruits and Salads	Desserts
			(%)			
−3	10	30	10	30	15	5
−2	20	30	10	25	10	5
−1	30	30	10	20	10	
+1	40	30	10	20		
+2	50	30	10	10		
+3	60	30	10			
+4	70	20	10			
+5	80	20				
+6	90	10				
+7	100					

[1]All ten stages discourage the drinking of liquids.

The foods used at the different stages seek to establish a balance between foods classified as yang (the male principle) foods, and yin (the female principle) foods. The proper balance is 5 parts yin to 1 part yang. Such things as color, direction of growth, sodium and potassium level, water content, climate, taste, season, source (plant or animal), weight, and vitamin content are said to classify a food as more yang or more yin. For example, progressing from yang foods to yin foods, the following order would be observed: meat, eggs, fish, grains, vegetables, fruits, dairy products, sugar, alcohol, drugs, and chemicals. Brown rice is considered to contain a perfect balance of yang and yin. Hence, brown rice is the principal food of the diet, and the ultimate diet consists solely of brown rice. Brown rice contains no vitamin A, C, or B-12; low levels of other vitamins; low levels of calcium, iron, and other minerals; and a low-quality protein. Moreover, 2.2 lb (1 kg) of cooked brown rice would provide only 1,190 Calories (kcal) of energy—filling, but hardly adequate.

It is the overzealous individuals who persist in following the more rigid diet—high level of brown rice—who are in danger of developing serious nutritional deficiencies. Infants, children, and nursing or pregnant mothers are in particular peril. There have been reports of scurvy, anemia, hypoproteinemia (low blood protein), hypocalcemia (low blood calcium), slowed growth, rickets, loss of kidney function due to low fluid intake, and a few deaths attributed to the followers of the Zen macrobiotic diet.

(Also see VEGETARIAN DIETS.)

ZEST

The colored, oily outer layer of the peel on citrus fruits, which is also called the flavedo. It may be green due to the predominance of the pigment chlorophyll, or yellow to orange when ample amounts of carotene and xanthophyll pigments are present. The color of the zest is *not* always a good indicator of ripeness because some species of citrus fruits require cool nights for the development of the yellow and orange pigments in the peel. However, the color change from green to yellow, or from green to orange, may be induced artificially by exposing the fruit to ethylene gas in a warm room. It is noteworthy that the zest is the part of the peel that is grated and added to enhance various dishes.

(Also see FLAVEDO.)

ZINC (Zn)

Contents	Page
History	1150
Absorption, Metabolism, Excretion	1150
Functions of Zinc	1150
Deficiency Symptoms	1150
Interrelationships	1151
Recommended Daily Allowance of Zinc	1151
Toxicity	1151
Sources of Zinc	1151

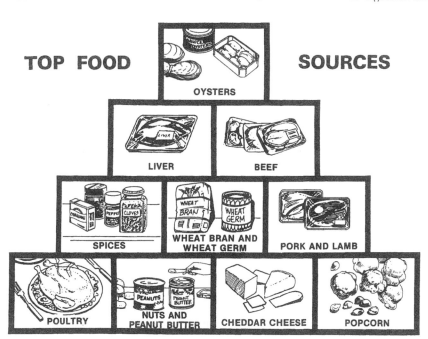

TOP FOOD SOURCES

OYSTERS
LIVER
BEEF
SPICES
WHEAT BRAN AND WHEAT GERM
PORK AND LAMB
POULTRY
NUTS AND PEANUT BUTTER
CHEDDAR CHEESE
POPCORN

Fig. Z-1. Top food sources of zinc.

Contrary to what many people think, zinc is needed for more than covering pipes, coating wire fences, and galvanizing buckets. It is an essential element for man.

Zinc is widely distributed throughout the body, but the highest concentrations are found in the skin (the skin contains 20% of the total body zinc), hair, nails, eyes, and prostate gland. Traces occur in the liver, bones, and blood. Also, it is a constituent of enzymes involved in most major pathways. In total, the human body contains about 2.2 g of zinc —more than any other trace element except iron.

HISTORY. Zinc was first shown to be biologically important more than 100 years ago when it was found to be needed for the growth of certain bacteria. In the 1920s, it was demonstrated to be required for the growth of experimental rats. But, it wasn't until the early 1960s that zinc was shown to be an essential nutrient of man.

ABSORPTION, METABOLISM, EXCRETION. Zinc is poorly absorbed; less than 10% of dietary zinc is taken into the body, primarily in the duodenum. It appears that metallic zinc and zinc in its carbonate, sulfate, and oxide forms are all absorbed equally well. Large amounts of calcium, phytic acid, or copper inhibit zinc absorption. Cadmium appears to be a zinc antimetabolite.

After zinc is absorbed in the small intestine, it combines with plasma proteins for transport to the tissues. Relatively large amounts of zinc are deposited in bones, but these stores do not move into rapid equilibrium with the rest of the organism. The body pool of biologically available zinc appears to be small and to have a rapid turnover, as evidenced by the prompt appearance of deficiency signs in experimental animals.

Most of the zinc derived from metabolic processes is excreted in the intestine—in pancreatic, intestinal, and bile secretions. Only small amounts are excreted in the urine.

FUNCTIONS OF ZINC. Zinc is needed for normal skin, bones, and hair. It imparts *bloom* to the hair. It is a component of several different enzyme systems which are involved in digestion and respiration. Also, zinc is required for the transfer of carbon dioxide in red blood cells; for proper calcification of bones; for the synthesis and metabolism of proteins and nucleic acids; for the development and functioning of reproductive organs; for wound and burn healing; for the functioning of insulin; and for normal taste acuity (the ability to taste accurately).

DEFICIENCY SYMPTOMS. The most common cause of zinc deficiency is an unbalanced diet, although other factors may be responsible. For example, the consumption of alcohol may precipitate a zinc deficiency by flushing stored zinc out of the liver and into the urine.

Lack of zinc in the human diet has been studied in detail in Egypt and Iran, where the major constituent of the diet is an unleavened bread prepared from low extraction wheat flour. The phytate present in the flour limits the availability of zinc in these diets, with the result that the requirements for the element are not satisfied. Zinc-response has also been observed in young children from middle-class homes in the United States who consume less than an ounce (28 g) of meat per day.

Zinc deficiency is characterized by loss of appetite, stunted growth in children, skin changes, small sex glands in boys, loss of taste sensitivity, lightened pigment in hair (dull hair), white spots on the fingernails, and delayed healing of wounds. In the Middle East, pronounced zinc deficiency in man has resulted in hypogonadism and dwarfism. In pregnant animals, experimental zinc deficiency has resulted in malformation and behavioral disturbances in the offspring—a finding which suggests that the same thing may happen to human fetuses.

INTERRELATIONSHIPS. Zinc is involved in many relationships: in the metabolism of carbohydrates, fats, proteins, and nucleic acids; in interference with the utilization of copper, iron, and other trace minerals, when there are excess dietary levels of zinc; in protection against the toxic effects of cadmium, when there is ample dietary zinc; in reduced absorption, when there are high dietary levels of calcium, phosphorus, and copper.

RECOMMENDED DAILY ALLOWANCE OF ZINC.
Studies have shown that, in healthy adults, equilibrium or positive balance is obtained with intakes of 12.5 mg of zinc per day when this intake is derived from a mixed diet. This has been accepted as a minimum requirement, as the balance studies did not take into account sweat and skin losses.

The National Academy of Sciences, National Research Council, recommended daily zinc allowances are given in the section on MINERAL(S), Table M-25, Mineral Table.

The recommended daily allowances are predicated on the consumption of a mixed diet containing animal products. Diets that supply sufficient animal protein usually also furnish enough zinc, but vegetarian diets may be somewhat low.

The RDA recommended zinc allowances of the National Academy of Sciences, National Research Council, 10th edition, 1989, are as follows:

Infants, birth to 1 year, 5 mg; children, 1 to 10 years, 10 mg; males, 11 to 51+ years, 15 mg; females, 11 to 51+ years, 12 mg; pregnant females, 15 mg; lactating females, first 6 months, 19 mg; lactating females, second 6 months, 16 mg.

Fig. Z-2. Beef tostada. Both beef and spices are good sources of zinc. (Courtesy, California Iceberg Lettuce Commission, Monterey, Calif.)

• **Zinc intake in average U.S. diet**—The average zinc content of mixed diets consumed by the American adult has been reported to be 12.7 mg/day, but vegetarian and low-protein diets may provide less.

TOXICITY. Toxicity of zinc in man occurs with the ingestion of 2 g or more. Zinc sulfate, taken in these amounts, produces acute gastrointestinal irritation and vomiting. However, zinc has been administered to patients in tenfold excess of the dietary allowances for months and years without adverse reactions. But there is evidence that excessive intakes of zinc may aggravate marginal copper deficiency. For the latter reason, the continuous taking of zinc supplements of more than 15 mg/day, in addition to the dietary intake, should not be done without medical supervision.

Toxicity of zinc is characterized by anemia, depressed growth, stiffness, hemorrhages in bone joints, bone resorption, depraved appetite, and in severe cases, death. The anemia appears to result from an interference with iron and copper utilization because addition of these two elements can overcome the anemia caused by excessive zinc.

Zinc poisoning may result from eating foods that have been stored in galvanized containers.

SOURCES OF ZINC. Human colostrum (the first secretion of a woman after childbirth) is a good source of zinc.

The zinc content of most municipal drinking water is negligible.

Groupings by rank of common food sources of zinc are given in the section on MINERAL(S), Table M-25, Mineral Table.

For additional sources and more precise values of zinc, see Food Composition Table F-21.

NOTE WELL: The biological availability of zinc in different foods varies widely; meats and seafoods are much better sources of available zinc than vegetables. Zinc availability is adversely affected by phytates (found in whole grains and beans), high calcium, oxalates (in rhubarb and spinach), high fiber, copper (from drinking water conveyed in copper piping), and EDTA (an additive used in certain canned foods).

(Also see MINERAL[S], Table M-25.

ZOLLINGER-ELLISON SYNDROME

Frequently, this disease produces malabsorption. It is caused by the development of a gastrin-secreting tumor in the pancreas or duodenal wall, which stimulates excessive, continued gastric secretion. This oversecretion by the stomach acidifies and dilutes the intestinal contents, leading to major disturbances in fat digestion and absorption, primarily due to the inactivation of the enzyme, pancreatic lipase. Other factors contributing to malabsorption include (1) alteration in the chemical nature of bile salts, reducing their effectiveness; (2) structural and function changes in the intestinal lining; and (3) hypermotility. Eventually, multiple peptic ulcers develop due to the excessive secretion of the hormone, gastrin. Total gastrectomy seems to be the therapy of choice.

ZWITTERION

The term used to describe the property of amino acids, when ionized in solution, to behave either as an acid or a base depending on the need of the solution in which they are present. This capacity makes amino acids good buffer substances.

ZYMASE

The name which is sometimes applied to the mixture of enzymes in yeast that change sugar to alcohol during fermentation.

ZYMOGENS

These are inactive forms of enzymes—proenzymes. Zymogens convert to active enzymes under the influence of various agents such a pH changes or other enzymes. Examples of zymogens are some of the enzymes involved in the digestion of protein. Trypsinogen and pepsinogen are both zymogens secreted into the intestine by the pancreas where they are converted to the active enzymes trypsin and pepsin, respectively.

ZYMOMETER

An instrument which is employed to measure the degree of fermentation of a fermenting liquid.

(Courtesy, Carnation, Los Angeles, Calif.)

(Courtesy, Phil Siegel/Ketchum, San Francisco, Calif.)

(Courtesy, University of Minnesota, St. Paul, Minn.)

(Courtesy, Hershey Foods Corp., Hershey, Pa.)

Fig. Z-3. A balanced diet, along with an adventure in good eating, is best achieved by consuming a variety of foods.

Abernethy biscuit 1
Abetalipoproteinemia 567
Absorption 1
Absorption meter 1
Acclimatize 1
Acerola 482, 528
Acetate 1
Acetic acid 1
Acetoacetic acid 1
Acetobacter 1
Aceto-glycerides 1
Acetoin 1
Acetone 1
Acetonemia 1
Acetonuria 1
Acetylcholine 1
Acetylcoenzyme A 1
Achalasia 2
ACH Index 2
Achlorhydria 2
Acholic 2
Achromotrichia 2
Acid 2
Acid-ash residue 2
Acid-ash residue diets 2
Acid-base balance 2, 1113
    abnormalities 3
    regulation 2
Acid-base reaction of food 4
Acid detergent fiber 4
Acid foods 6
Acid & alkaline foods 4
    dietary uses 6
Acid-forming foods 6
Acidophilus 528
    milk 528
Acidosis 6
Acid residue foods 5
Acid tartrate 243
Acid tide 6
Acini 6
Acne 6
Aconitase 6
Acrodynia 6
Acrolein 6
Acromegaly 6
ACTH 6
    hormone 308
Activate 6
Activation of enzymes 7
Activator 7
Active transport 7
Acute condition 7
ADA 27
Addison's disease 7, 9
Additives 7
    benefits 7
    consumption 8
    controls 8
    definition 7
    food 7, 154, 723, 887
    history 7
    kinds of 7
    need for 7
Adenine 8

Adenohyphysis 8
Adenosine 8
ADF 4
Adipose tissue 8
Adiposity 9
Adipsia 9
Adlay 182
Ad lib 9
Ad libitum 9
Adolescent nutrition 185
Adrenal cortex insufficiency 9
Adrenal gland 9
Adrenal hyperplasia 568
Adrenalin 9
Adrenaline 9
Adrenergic 9
Adrenocortical hormones 9
Adulteration 9
Adult nutrition 9
    body changes 10
    degenerative diseases 10
    dietary guidelines 12
    health status 10
    meal planning 12
    nutritional allowances 12
Adzuki bean 528, 611
Aequum 13
Aerobic 13
Aerophagia 13
Aesculin 13
Aflatoxin 13, 857
African sprouted grain beer 92
Agar-agar 13, 346, 519, 528
    history 13
    production 13
    properties 13
    uses 13
Ageusia 13
Aging 13
Aging meat 14
AIN 28
Akee 482
Alanine 14
Albedo 14
Albinism 14, 564
Albumen 14
Albumin 14
    index 14
Albuminoids 14, 346
Albuminuria 14
Alcaptonuria 14
Alcohol 14
    denatured 15
    ethyl 15
    grain 15
Alcoholic beverages 14
    beers 15
    distilled liquors 15
    history 14
    wine 15
Alcoholics Anonymous 19
Alcoholism 11, 15
    causes 17
    effects of 15
    signs of 17
    treatment 17
Aldehyde 19
Aldohexose 19
Ale 92
Aleurone 19
Alfalfa/Lucerne 611
    flour 346, 528
    seeds 528
    sprouts 528

tablets 528
Algae 19
Algin 519
Alginates 19
Alginic acid 19
Alimentary canal 280
Alimentary tract 19
Alkali 19
    reserve 19
Alkaline
    foods 6
    reserve 19
    residue foods 5
    tide 19
Alkaline-ash residue 19
Alkaloids 19
Alkalosis 19
Alkanet 20
Alkaptonuria 14, 20, 564
Allergen 20
Allergies 20, 96, 281
    causes 20-21
    corn 25
    cost 21
    definition 20
    diagnosis 22
    drugs 23
    egg 24, 302
    elimination diets 23
    food, infants 579
    gluten 24
    hereditary 21
    inhalants 21
    insects 21
    kinds 21
    legume 25
    milk 24, 698
    nut 25
    patch test 22
    prevalence 21
    protein 24
    Rowe elimination diets 23
    seafood 25
    skin test 22
    symptoms 21-22
    treatment 23
    wheat 24
Allergy-free
    products list 25
    recipes 25
Allicin 25
Alligator
    pear 64
    skin 25
Allopriophagy 25
Almond 25
    history 26
    nutritional value 27
    oil 26
    preparation 26
    production 26
Aloe vera 528
Alopecia 27, 252
Alpha-cholesterol 27
Alpha-ketoglutaric acid 27
Alpha-lipoproteins 27
Aluminum poisoning 857
Alveograph 27
Alzheimer's disease 27
Amebiasis 27
Ameloblasts 27
American
    Dietetic Association 27
    Institute of Nutrition 28

Society for Clinical Nutrition ... 28
Amidase 28
Amination 28
Amine 28
Amino acid 28, 893, 895, 898
    analyses 31, 34, 91, 906
    antagonism 30
    essential 28
    history 28
    nonessential 28
    peptide bonds 28
    reference pattern 30
    requirements 29
Aminoaciduria 30
Amino-peptidase 30
Aminopterin 30
Amino sugar 30, 161
Ammonia 30
Ammonium 30
Amoebic dysentery 286
Amphetamines, heart disease & .. 30
Amphoteric 30
Amydon 30
Amygdalin 601, 1067
Amylase 30
Amylodyspepsia 30
Amylopectin 30
Amylopsin 30
Amylose 30
Anabolism 31, 686, 688
Anacidity 31
Anaerobic 31
Analog 31
Analysis of foods 31
    amino acid 34
    biological 34
    bomb calorimetry 32
    chemical 31
    chromatography 32
    colorimetry 33
    digestibility trials 34
    history 31
    microbiological 35
    microscopic 35
    physical method 35
    procedures 31
    proximate 31
    spectrophotometry 33
    use in meal planning 35
    Van Soest 32
Anaphylaxis 35
Anchovy pear 35
Androgens 35
Androsterone 35
Anemia 36
    aplastic 38
    causes 37
    diagnosis 40
    food enrichment 42
    hemolytic 39, 249
    infant 580
    iron 37, 41
    iron-deficiency 250
    pernicious 38, 42, 1081
    prevention 40
    susceptibility 37
    symptoms 40
    tests for 40
    treatment 40
    types 37-39
Anemia, megaloblastic 42
Anemia, pernicious 42
Aneurine 43
Aneurysm 43

Angina pectoris 43, 535, 544
Angiogram 43
Angiotensin 43
Angiotensinogen 43
Ang-khak 932
Angular blepharo-conjunctivitis . 252
Angular fissures 253
Angular stomatitis 43
Animal
    energy step up 312
    fats 43
    products 43
    protein factor 43
    starch 43
Anion 43
Anorectic drugs 43
Anorexia nervosa 43
Anorexigenic drugs 43, 44
Anosmia 44
Anoxia 44
Antabuse 44
Antacids 44
    dangers 44
Antagonist 44
Anthocyanins 44
Anthropometric measurements .. 44
Anti- 44
Antibiotics 44, 74
    allergic reactions 45
    bacterial resistance to 45
    conjugation 46
    dangers 45
    foods, in 45
    future 46
    mode of action 45
    resistance to bacteria 45-46
    side effects 45
    transduction 46
    transformation 46
    uses in
        agriculture 45
        foods 45
Antibodies 46
Anticoagulant 47, 1103
Antidiuretic hormone 47
Antidote 47
Antifoaming agent 47
Antigen 47
Anti-grey hair vitamin 47
Antihemorrhagic 47
Antihistamines 47
Antiketogenesis 47
Antimetabolite 47
Antimycotics 47
Antineuritic 47
Antioxidant 47
Antisialagogues 47
Antistaling agents 47
Antitoxin 47
Antivitamin 47, 1057
Anuria 47
Anus 47
AOAC 60
Aorta 47
Apastia 48
Apathy 48
APF 43
Aphagosis 48
Aphrodisiacs 48
Aplastic anemia 38
Apoenzyme 48
Apoferritin 48
Apollinaris water 48
Apoplexy 48

Aposia 48
Apositia 48
Appendicitis 48
Appendix 48
Appestat 48
Appetite 48, 273
    controls 49
    factors affecting 49
    gastric influence on 49
    hypothalamic control 49
Appetizer 49
Apple 49
    brandy 52
    butter 51
    cider 51-52
    dried 52
    history 50
    jack 52
    nutritional value 51
    preparation 51
    production/world/U.S. 50
    uses 51
Apricot 52
    history 52
    kernels 528
    leather 53
    nutritional value 53
    preparation 52
    production/world/U.S. 52
Aquavit 292
Aqueous emulsion 53
Arabinose 53
Arachidonic acid 53
Arachis oil 53
Arbutus berry 53
Arginase 53
Arginine 28, 53
Argininosuccinic aciduria 564
Ariboflavinosis 53
Arrowroot 54
    flour 361, 364
    starch 528
Arryhthmia 54
Arsenic poisoning 857
Arteriole 54
Arteriosclerosis 54
Artery 54
Arthritis 54
    atropic 55
    causes 55
    gout 56
    hypertrophic 54
    osteoarthritis 54
    prevention 55
    rheumatoid 55
    symptoms 55
    treatment 55
Artichoke, globe 57
    history 57
    noodles 528
    nutritional value 58
    preparation 58
    production/world/U.S. 57
Artificial sweeteners 58
Artificially dried 58
Ascarids 59
Ascites 59
ASCN 28
Ascorbic acid 59, 1065, 1084
Ascorbin stearate 59
Ascorbyl palmitate 59
-ase 59
Ash 59
Asparagine 59

Asparagus **59**, 1036
  bean 614
  nutritional value 59
Aspartame **59**, 1000
Aspartic acid **60**
Aspergillus **60**
Assay **60**
Assimilation **60**
Association of Official Agricultural
  Chemists **60**
Asthma **60**
Asymptomatic **60**
Ataxia **60**
Atheroma **60**
Atherosclerosis **60**, 536
Athletics **60**, 844
  benefits 61
  energy expenditures 62
  history 60
  meals 62
  metabolism during 61
  nutrient requirements 61, 63
  salt requirements 62
  steroid alternatives 63
  water requirements 62
Atonic constipation **64**
Atony **64**
Atria **64**
Atrophy **64**
Atropic arthritis **55**
Atropine **64**
Atwater factors **64**
  values 64
Atwater, Wilbur **64**, 143, 892
Aubergine **304**
Aureomycin **191**
Auricles **64**
Autoimmunity **64**
Available nutrient **64**
Avidin **64**
Avitaminosis **64**
Avocado(s) **64**
  history 64
  nutritional value 65
  oil 528
  preparation 65
  production/world/U.S. 64
Avoirdupois **65**
A-V shunt **103**
Azarole **65**
Azotemia **65**
Azotobacter **65**

Babassu oil **67**
Babcock test **67**
Baby foods **67**
  guidelines for introducing 68
  history 67
  supplementing 67
Bacillary dysentery **70**
Bacillus cereus **70**, 71
Bacon **70**
  flitch of 70
Bacteria **70**
  control 74
  count 75
  food, in 70

food infections 71–73
  harmful 71
  helpful 70
  preventing 74
  toxins 73
Bacterial endotoxin **75**
  exotoxin 75
  spoilage of food 75
Bactericide **75**
Bacteriophage **75**
Baking **75**
  mixes, dietetic 528
  powder 75–76
    low-sodium 528
  soda 75–76
Balahar **1126**
Balanced diet **77**
Balance study **77**
Baldness **77**
Balsam
  apple 77
  pear 77
Bambarra groundnut **611**
Banana **77**
  figs 78
  flakes 528
  history 77
  nutritional value 78
  preparation 78
  production/world/U.S. 77
Banting, Frederick G. **263**
Baobab **78**
Barbados
  cherry 482
  gooseberry 78
  molasses 529
Barbecue **78**
  pit 78
Barbiturates **79**
Barding **79**
Barfoed's test **79**
Barium meal **79**
Barley
  fermented products 81
  flakes 81
  flour 81, 360, 363, 529
  groats 81
  history 79
  infant cereals 81
  malted 81
  malting 80
  milling 79
  nutritional value 80
  pearl 81
  pot 81
  processsing 79
  production/world/U.S. 79
  products 80–82
  sprouting 80
  uses 80–82
Barm **82**
Basal diet **82**
Basal metabolism **82**
  metabolism rate 147
Base **82**
  bicarbonate 82
Base-forming foods **2**, 5
Bataan experiment **83**, 316
Bean(s) **83**
  history 83
  kinds of 83
  production/world/U.S. 83
  utilization 84
Bean, common **84**

gas from 85
kinds 84
nutritional value 86
preparation 84
processisng 84
  dried 84
  snap 84
protein 87
utilization 84
Bean flour **87**
Beaten biscuit **87**
Beechwood sugar **87**
Beef **87**
  bread 90
  consumption, per capita ... 88
  cuts & cooking 88
  extract 90
  fat 43
  federal grades 88
  food, as 88
  measles 287
  nutritional value 88
  production, U.S. 88
  tapeworm 287
  tea 90
Beer(s) **81**, 90
  beneficial effect 94
  by-products 91
  harmful effect 94
  history 90
  hopping 91
  malt for 90
  mashing 91
  nutritional value 93
  production 92
  types 91–93
Beet **1036**
  coloring foods 220
  sugar 94, **990**
Beeturia **94**
Behavior & diet **94**
  causes 95
  cures 95
  problems 95
Belching **97**
Bel Fruit **97**
Belladonna drugs **97**
Benedict's test **97**
Bengal quince **97**
Benign disorders **97**
Benzedrine **97**
Bergamot orange **97**
Beriberi **97**
  Bataan experiment 316
  causes 98
  dry 98
  forms 98
  history 98
  juvenile 98
  prevention 98
  treatment 98
  wet 98
Berries **99**
Best, C. H. **263**
Beta **99**
Beta-carotene **99**
Beta cells of the pancreas **99**
Beta-cholesterol **99**
Beta-hydroxybutyric acid **99**
Betaine **99**
Beta-lipoproteins **99**
Beta-oxidation **99**
Betel
  leaves 100

B

nuts 100
Beverages 100
    alcoholic 14
Bezoar 100
BHA 134
BHT 134
Bibulous 100
Bicarbonate 100
    ion 100
    of soda 100
Bicycle ergometer 100
Bifidus factor 100
Bilberry 100
Bile 100
    acids 100
    ducts 100
    pigments 101
    salts 101
Bilharziasis 101
Biliary 101
    calculi 101
    cirrhosis 101
    colic 101
    dyskinesia 101
    obstruction 101
Bilimbe 101
Biliousness 101
Bilirubin 101
Bilirubinuria 101
Biliverdin 101
Bio- 101
Bioassay 102
Biochemistry 102
Bioflavonoids 102, 1066
    cooking losses 104
    deficiencies 104
    functions 103
    history 102
    RDA 104
    sources 104
Biological
    analyses 34
    assay for vitamins 1057
    value of proteins 105
Biology 105
Biopsy 105
Bios 105
Biotin 105, 1061
    cooking losses 107
    deficiency sysmptoms 107
    functions 106
    history 105
    RDA 107, 1061
    sources 107
Birch beer 108
Bitter
    beer 92
    melon 77
    orange 957
Bitters 108
Biuret
    assay 34
    test 108
Black
    bread 108
    tongue 758, 760
    walnut 1107
Black-eyed pea 612
Blackstrap molasses 108, 529
Black tongue 108
Bladder 108
Bland diets 108, 739
Bleeding
    gums 252

time 108
Blended 108
Bletting 108
Blindloop syndrome 108
Blindness
    night 252
    vitamin A deficiency 109
Bloaters 109
Blood
    cells 109
    cholesterol 109
    fats 109
    platelets 109
    pressure
        high 109
        low 109
    sugar 109
        blood sugar test 109
    vascular system 109
Blood sausage 109
Bloom 109
Blue
    cheese 109
    value (in vitamin A & starch) ... 110
Blueberry(s) 482
BMR 82
Bobolink 110
Bobwhite 110
Body
    build 110, 782
    composition 110
    fluid 110
    temperature 110, 314
    tissue 111
    types 111
    water 111
    weight 1027
Boiling point 111
Bologna 111
Bolting of food 111
Bolus 111
Bomb calorimetry 32
    calorimeter 111
Bone 112
    charcoal 112
    compact 112
    composition 112
    disorders 112
    growth 112
    meal 113, 529
    metabolism—vitamin D 112
    misshapened 253
Bonito 113
Books 113
Borax 113
Borden, Gail 572
Borlaug, Norman 517
Boron 113
Boston brown bread 113
Botulism 73
Bout 113
Bowel 113
Bradycardia 113
Bradyphagia 113
Brain transmitters and diet 96
Bran 113, 175, 346, 529
    wheat 117
Brandies 292
Brawn 113
Brazilian
    cherry 114
    nut 777
Breadfruit 482
Breads and baking 114–122

chemical composition 116–118, 121
    history 115
    nutritive values 117, 121
    principles 115
    special bakery products 119
    spreads 122
    types 115, 121
        Cornell bread 121
        leavened 115, 176
        quick 116, 176, 911
        sourdough 176
        sticks 116
        unleavened 115, 176
        white bread 121
        whole wheat bread 121
        yeast 115, 176
Breakfast cereals 122
    history 123
    hot 123
    infant 124, 126
    nutritional value 124
    quick 911
    ready-to-eat 124
    sugared 127
Breast feeding 127, 883
    advantages 128
    colostrum 129
    disadvantages 128
    milk composition 68
    mother's diet 129, 883
        drugs 883
        supplements 883
    technique 129
    weaning 130
Breathing rate 130
Brewers'
    rice 932
    yeast 81, 91, 130, 529
Brewing 90, 175
    history 90
Bright's disease 130
Brisket 130
British
    gums 130
    thermal unit 130
Brix 130
Broad bean 611
Broccoli 130
    history 130
    nutritional value 131
    preparation 131
    production/world/U.S. 130
Broilers 131, 869
Bromatology 131
Bromelin 131
Brominated vegetable oils 131
Bronchial asthma 131
Bronchiectasis 131
Bronchitis 132
Bronze diabetes 132
Brown
    rice 132
    sugar 132
Brucella 132
Brucellosis 71
Brunner's glands 132
Brush border 132
Brussels sprouts 1036
BTU 130
Buckwheat 132
    flour 360, 363, 529
    groats 529
    history 132
    milling 132

nutritional value 132
production/world/U.S. 132
products 132
uses 132
Buddhism dietary beliefs 920
Buerger's disease 133
Buffalo
  berry 133
  hump 133
Buffer 2, 3, 133
  acid-base 5
Bulgur 1125
  wheat 529
Bulimarexia 133
Bulimia 133
  nervosa 133
Bulk 133
Bulking agent 133
Burgoo 133
Burn 133
  burning feet syndrome 133
Burp 133
Burr, George Oswald 328
Burr, Mildred M. 328
Bushel 134
Butt (sirloin) 134
Butter 705
  bean 620
  black 134
  mowrah 134
Buttermilk 134
Butter stools 134
Butyric acid 134
Butyromel 134
By-product 134

C

Cabbage 135
  antiulcer factors 135
  history 135
  nutritional value 135
  preparation 135
  production/world/U.S. 135
Cachexia 136
Cadmium 136
Caffeine 136
  consumption 136
  content in foods and drugs ... 136
  effects 136
  health concerns 137
  metabolism 136
  sources 136
Calamondin 137
Calcemia 137
Calciferol 137
Calcification 137
Calcitonin 137, 309
Calcium 137, 714
  acetate 142
  acid phosphate 142
  deficiency symptoms 140
  diseases 142
  functions 140
  gluconate 142
  history 138
  losses 142
  metabolism 139
  phosphorus/vit. D ratio ... 140, 143

RDA 140
restricted diet 744
sources 142
Calculus 143
California quail 911
Caliper 143
Calorie/caloric 143, 151
  balancing energy 150
  deficiencies 145
  dietary energy 144
  diseases 150
  empty 151
  excesses 145
  expenditures 143
  intake 147
    adults 148
    children 148
  lactation 150
  pregnancy 149
  production/acre 518
  stresses 150
  value of foods 151
Calorimeter 151
Calorimetry 151
  direct 146, 151
  history 143
  indirect 146, 151
    basal metabolic rate 147
    respiratory quotient 147
Camomile 151
Camu-camu 151
Canadian bacon 151
Cancer 151
  carcinogens 153
  causes 153
  detection 155
  future of 158
  leading sites 152
  mortality 153
  prevention 157
  treatment 156
    chemotherapy 156
    nutrition & diet 156
    radiation 156
    surgery 156
  types 152
  vitamin A 1072
  vitamin C 1088
Cancrum oris 158
Candy 158
Cane sugar 158
Canistel 304
Canker 158
Canning foods 887
Canola (rape) 158, 917
Cantaloupe 158, 679
  history 159
  nutritional value 159
  preparation 159
  production/world/U.S. 159
Cape gooseberry 159
Caper 159
Capillaries 159
  exchange mechanism 159
  resistance test 159
Capric acid 159
Caproic acid 159
Caprylic acid 159
Carambola 159
Carbohydrases 160
Carbohydrate(s) 160, 684
  absorption 162
  amino sugars 161
  available 165
  classification 160

digestion 162
digestion/absorption 277
disaccharides 161
fermentation 164
functions 162
hexoses/hexosans 161–162
history 160
intolerance 165, 565
metabolism 162
monosaccharides 160
oligosaccharides 161
pentosans/pentoses 160–161
polysaccharides 161–162
protein-sparing action 162
required intake 163
restricted 740
sources 163
sugar
  acids 161
  alcohols 161
tetrasaccharides 161
trisaccharides 161
unavailable 165
Carbonated beverages 965
Carbon dioxide 165
Carbonic
  acid 165
  anhydrase 165
Carboxylase 165
Carboxymethyl cellulose 165
Carboxypeptidases 165
Carcass 165
Carcinogenic 165
Carcinogens 153, 165
Cardia 165
Cardiac 165
Cardiospasm 165
Cardiovascular 165
Cardiovascular disease 165
Cardon 166
Cardoon 166
Cariogenic agent 166
Carminative 166
Carnitine 166, 1066
  deficiencies 166
  functions 166
  history 166
  losses 167
  metabolism 166
  RDA 166
  sources 167
Carcinomas 152
Carob 611
  flour 361, 364
  powder 529
Carotene 167, 1068
Carotenemia 167
Carotenoid pigments 167
Carrageenan 166, 519, 529
  food uses 168
  properties 168
  types 168
Carrier 168
Carrot(s) 168
  history 169
  nutritional value 169
  preparation 169
  production/world/U.S. 169
Cartier, Jacques 1085
Cartilage 169
Casaba melon 679
Casal's necklace 169
Caseinates 706
Cashew nut 777

Cassava 169
  history 170
  nutritional value 170
  preparation 170
  production/world/U.S. 170
Castle, W. B. 1079
Castor oil 170
Catabolism 170, 684, **686**, 687
Catalase 170
Catalyst 170
Catalyze 170
Cataracts 170
Catecholamines 171
Cathartic 171
Cauliflower 171
  history 171
  nutritional value 171
  preparation 171
  production/world/U.S. 171
Cavy 171
Cecum (caecum) 171
Celeriac 1036
  history 172
Celery 172
  history 172
  nutritional value 172
  preparation 172
  production/world/U.S. 172
Celiac 172
  disease 24
  sprue 24
Cell 683
  membrane 173
Cellobiose 173
Cellulase 173
Cellulose 341
  derivatives 519
Cell wall 173
Cementum 173
Centigrade 173
Central intravenous nutrition .. 1017
Central nervous system 173
Centrifuge 173
Cephalin 173
Cereal(s) 173
  baking 175
  beverage 174, 935
  brewing 175
  by-products 175
  distilling 175
  enriched 177
  feeds from 174
  food 174
  future of 181
  history 174
  improved varieties 198
  malting 174
  milling 175
  nutritive value 177, 180
  processing 176, 890
  production/world/U.S. 173
  protein quality 177
  sprouting 174, 178
  uses 178
  world, of 174
Cerebral 182
  hemorrhage 182
  infarct 182
Cerebrospinal fluid 183
Ceriman 183
Ceroid 183
Ceruloplasmin 183
Cervelat 183
Chaff 183
Charcoal, activated 183

Charqui 183
Chayote 1038
Cheddaring 183
Cheese 705
Cheilitis 183
Cheilosis 183, 253
Chelate 183
Chelation 183
Chemical
  analysis 31
  assay 1057
  bonding 183
  nutritional value of leavening ... 76
  reaction 183
Chemotherapy 156
Cherimoya 183, 484
Cherries 183
  candied 184
  history 183
  maraschino 184
  nutritional value 184
  preparation/uses 184
  production/world/U.S. 183
Chestnut 777
Chia seeds 529
Chickens 869
Chickpea 529, **611**
Chicory 212, **1038**
  root powder 529
Chief cells 185
Child Care Food Program 512
Childhood and Adolescent
    Nutrition 184
  dietary guidelines 187
  food preferences 188
  growth 185
  height 185
  history 185
  nutrient allowances 187
  nutrition problems 188
    acne 189
    obesity 189
    skin blemishes 189
    underweight 189
  prepubescent 187
  school lunch & breakfast 188
  weights 186
Chinese cabbages 1038
Chinese
  date 485
  gooseberry 485
  restaurant syndrome 189
Chipolata 190
Chipped beef 190
Chitin 190, 519
Chitterling 190
Chives 1038
Chlorine/chloride 190, 715
  deficiencies 190
  dioxide 190
  functions 190
  history 190
  metabolism 190
  RDA 191
  shift 190
  sources 191
Chlorophyll 191, **530**
Chlorophyllase 191
Chlorophyllide 191
Chlorophyllin copper 191
Chlorophyllins 191
Chlorosis 191
Chlortetracycline 191
Chocolate 206
Choking 191

Cholagogue 191
Cholangitis 191
Cholecalciferol 191
Cholecystitis 192
Cholelithiasis 491
Cholera 71
Cholertic agent 192
Cholesterol 11, **192**, 302, 538
  blood levels 192
  chemistry 192
  food content 194
  heart disease 193
  history 192
  metabolism 192
    functions 193
    synthesis 192
  restricted diet 741
  sources 193
Cholestyramine 194
Cholic acid 194
Choline 96, 194, 1062
  deficiency symptoms 195
  functions 195
  history 194
  losses 195
  metabolism 194
  RDA 195
  sources 195
Cholinergic 196
Cholinesterase 196
Chondroitin sulfate 196
Chop suey 196
Chowchow 196
Chowder 196
Christianity dietary beliefs 920
Chromatin 196
Chromatography 32, 196
Chromium 196, 716
  deficiency symptoms 199
  functions 198
  history 196
  metabolism 197
  prevention of deficiency 201
  RDA 201
  sources 201
  treatment 201
Chronic conditions 202
Chutney 202
Chyle 202
Chylomicron 202
Chyme 202
Chymotrypsin 202
Cibophobia 202
Cider 203
Cieddu 203
Circulatory system 203
Cirrhosis 203
Cissa 203
Citral 203
Citric acid 203
  cycle 203
Citrin 203
Citron 484
  melon 203
Citrovorum factor 203
Citrulline 203
Citrullinemia 564
Citrus fruits 203
  nutritional value 204
  types 203
Clarification 204
Clarifixation 204
Claudication 204

intermittent 204
Clearing factor 204
Climacteric 204
Climacterium 204
Clod 204
Clofibrate 204
Clostridium perfringens 71
Cloudberry 204
CM Cellulose 165
Coagulated 205
Coarse-bolted 205
Coarse sifted 205
Cob 205
Cobalamins 1065, 1079
Cobalt 205, 716
  deficiencies 205
  functions 205
  history 205
  metabolism 205
  RDA 205
  sources 205
  vitamin B-12 205
Cocarboxylase 206
Cocci 206
Cockle 206
Cocoa & chocolate 206
  history 206
  nutritional value 207
  production 207
  uses 207
Cocoa butter 207
Coconut 207
  history 207
  milk 209
  nutritional value 208
  oil 208, 800
  palm 207
  preparation 208
  processing 208
  production/world/U.S. 207
Cocoyam 1004
Coddle 209
Cod-liver oil 209
Coefficient of digestibility .. 209, 280
Coenzyme 209
  I 209
  II 209
  A 209
  Q 209, 1066
  R 105, 210
Cofactor 210
Coffee 210
  chicory 212
  consumption/world/U.S. 212
  decaffeinated 212
  essence 214
  history 210
  instant 212
  processing 211
  production/world/U.S. 211
  substitute 530
  whiteners 706
Cognac 214
Cohune nut 214
Cold processed oils 530
Cold sterilization 214, 915
"Cold store" bacteria 214
Colds & flu 1087
Colic 214
  infant 579
Coliform bacteria 214
Colitis 214
  causes 215
  diagnosis 215

diets for 215-218
  prevention 215
  surgery 218
  treatment 215
Collagen 218
Collards 1038
Colloid 219
Colloidal osmotic pressure 219
Collop 219
Colon 219
Color additives 7
  coloring agents 7
Colorimetry analysis 33
Coloring of foods 219, 381
  natural in foods 220
  regulation of 219
  synthetic 219
  uses of 220
Colostomy 221
Colostrum 128, 129, 221
Coma 221
Comfit 221
Commodity Food Program 512
Common bean 84
Compote 221
Compound 221
Conalbumin 221
Condensed milk 702
Condiments 356
Confucianism dietary beliefs .. 920
Congenital 221
  lack/intrinsic factor 568
  nonhemolytic jaundice 568
Conjuntiva 221
Conjunctival hemorrhages 221
Connective tissue 221
Constipation 221
  infant 579
Consumer advocates 222
Consumer's food dollar,
  where it goes 664
Consumption, beef 88
Convenience & fast foods .. 222, 323
  nutritional analysis 224-229
  types 222
Convulsion 230
Cook, Captain James 1085
Cooking
  methods 382
  reasons for 382
Copper 230, 716
  deficiency symptoms 230
  functions 230
  history 230
  metabolism 230
  RDA 231
  sources 231
Copra 208
Coprophagy 231
Cordial 292
Cori cycle 231
Corn 231, 237, 1039
  fermented 238
  flakes 237, 530
  flour 237, 360, 363
    whole ground 530
  fortified 238
  fritters 116
  grits 237
  high-lysine 235
  history 232
  hominy 237
  kinds 232
  meal 237, 530

  nutritional value 234
  oil 237, 800
  opaque-2 235
  processing 233
    distilling 234
    fermentation 234
    milling 233
     dry 234
     wet 233
  production/world/U.S. 232
  products 236-238
  starch 237
  steepliquor 238
  steepwater 238
  sugar 237
  supplementing 236
  syrup 237, 993
  waxy 239
Cornea 238
Corneal vascularization 252
Corned beef 238
Corn-hull gum 519
Coronary 239
Coronary heart disease 239
Cortex 239
Corticosteroids 239
Corticosterone 239
Corticotropin 239
Cortisol 239
Cortisone 310
Coryza 239
Costiveness 239
Cotton (seed) 239
  flours 241, 361, 364, 530
  glandless (seed) 241
  history 240
  linters 241
  nutritional value 240
  oil 241, 800
  processing 240
  production/world/U.S. 240
  products 241
  seed, glandless 241
  uses 241
Coturnix quail 593
Couscous, fortified 530
Cowpea 612
Crabapple 484
Crackers
  cheese 116
  Graham 116
  saltine 116
Cranberry 242
  history 242
  nutritional value 243
  preparation 242
  production/world/U.S. 242
Cream 704
  clotted 243
  cornish 243
  line 243
    index 243
  plastic 243
  puffs 116
  tartar, of 243
Creatine 243
Creatinine 243
  coefficient 243
Creme 244
Crenshaw melon 679
Cretinism 249, 509
Crohn's disease 244
Crops, energy efficiency 312
Croquettes 244

Croutons 244
Crowdies 244
Cruciferous vegetables 244
Cryptoxanthin 244
Cucumber/gherkin 244
   history 244
   nutritional value 245
   preparation 245
   production/world/U.S. 244
   tree 101
Curd 246
   tension 246
Currant 484
Cushing's syndrome 133, 246
Custard(s)
   apple 484
   ice cream 705
Cyclamates 246, 1000
Cystathioninuria 564
Cysteine 28, 246
Cystic duct 246
Cystic fibrosis 246
Cystine 246
Cystinuria 564
Cystolithiasis 246
-cyte 246
Cytochrome 246
Cytology 246
Cytoplasm 246
Cytosine 246

Dadhi 247
Dairy products 247
Dam, Carl Peter Hendrik 1102
Dandelion 1039
Dark adaptation 247
Dasheen 1046
Date 247
   history 247
   nutritional value 247
   preparation 247
   production/world/U.S. 247
   syrup 530
Davis, Marguerite 1068
Davy, Humphrey 635, 713, 865
Deaeration 247
Deamination 247
Debility 248
Decalcification 248
Decarboxylation 248
Deciduous teeth 248
Decortication 248
Defibrillation 248
Defibrillator 248
Deficiency diseases 248
   causes 248
   prevention 248
   treatment 248
   types 248
   world, of 249–254
Defoaming agents 47
Degeneration 254
Degenerative
   diseases, risk factors 10
   joint disease 54
Dehydrate 254

Dehydration 254
   foods 887
Dehydroascorbic acid ... 255, 1084
Dehydrocholesterol 255
Dehydrogenases 255
Dehydroretinol 255
Delaney clause 255
Delirium tremens 255
Dementia 255
Demineralization 255
Demography 255
Demulcent 255
Demyelination 255
Denaturation 255
Dendritic salt 255
Density 255
Dental 256
   care 259
   caries 255, 257
   decay 257
   diet 257, 259
   disease 258
   fluoride 259
   fluorosis 256
   health 256
   nutrition 256–257, 259
   teeth 256
      permanent, ... 256
      primary 256
Deoxypyridoxine 260
Deoxyribonucleic acid 260
Depot fat 260
Dermatitis 260
Dermatology 260
Dermix 260
Desiccate 260
Desserts 260
   types 261
Detoxication 261
Development 261
Devonshire cream 243
Dewberry 261
Dextran 261, 519
Dextrimalose 262
Dextrin 262
Dextrose 262, 505
   equivalent value 262
Diabetes 262
   adult onset 262
   alloxan 262
   coma 266
   complications 264
   diagnosis 264–265
   diets, types of 266–267
   exercise 269
   factors 263
   foot care 269
   glucose tolerance test 265
   history 263
   hypoglycemia 266
   insipidus 262
   insulin 268
      drugs 268
      shock 266
   juvenile 262
   mellitus 262
   metabolic abnormalities 263
   prevention 265
   renal 269
   signs/symptoms 265
   treatment 265
Diabinase 269
Diacetyl 270
Dialysis 270

Diaphysis 270
Diarrhea 270
   infant 579
   newborn 286
Diatase 270
Diastole 270
Diastolic blood pressure 109
Dicoumarol 242, 270
Diet(s) 270
   alcoholic beverages 537
   caffeine beverages 537
   cholesterol 538
   diabetic 267
   obesity 787–789
   Rowe elimination 23
   saturated fats 538
   U.S. 270
   world 270
Dietary guidelines/adults 12
Dietetic foods 270
Dietitian 270
Diffuse 273
Diffusion 273
Digestibility 273
   apparent 273
   studies 342
Digestible nutrient 273
Digestion/absorption 273
   anatomy 274
   appetite 273
   carbohydrates 277
   chemical 276
   coefficient 280
   control
      hormonal 279
      neurological 279
   dysfunctions 280
   enzymatic 277
   food factors 279
   gastrointestinal hormones 279
   hunger 273
   minerals 278
   pancreatic secretions 275
   physiology 274
   processes 276
   proteins 278
   vitamins 278
   water 278
Digestive
   juices 280
   system 280
   tract 19
Diglycerides 280
Dipeptide 280
Dipsogen 280
Dipsomania 280
Dipsosis 280
Diptheria 286
Disaccharidase 280
Disaccharides 161
Diseases 280
   causes 280
   deficiency 248
   energy requirements 150
   food infections, bacterial .. 71–73
      Bacillus cereus 71
      brucellosis 71
      cholera 71
      Clostridium perfringens 71
      Escherichia coli 72
      leptospirosis 72
      salmonellosis 72
      shigellosis 72
      tuberculosis 72

tularemia 72
Versinis entero colitica 73
Vibrio parahaemolyticus 73
food poisonings, bacterial toxins
    botulism 73
    staphylococcal 73
genetic 496
Government agencies & 290
immunization 288
international agencies & 290
prevention 288
sanitation 289
signs/symptoms 285
spread of 285
vaccination 288
Disinfection 292
Disorder 292
Distal 292
Distillation 292
Distilled liquors 292
    medicinal effect 293
    nutritional effect 293
    proof 292
    types 292
Distilling 175
Diuresis 293
Diuretics 293, 1112
Diverticula 293
Diverticulitis 293
DNA 260, 293, 765, 893
Dock 293, 1045
Dolomite 293, 530
Donation of food to needy ... 512
Dopamine 294
Doughnuts 117
    cake 117
    yeast 117
Douglas bag 294
Dove 294
DPN 294
Dressings 294
Dried milk 294, 703
Drippings 294
Drooling 294
Dropsy 294
Drupe 294
Dry
    frying 294
    ice 294
    matter 294
Drying foods 887
Ducks 869
Duct 294
Ductless gland 294
Dumping syndrome 294
Duodenal ulcers 1023
Duodenum 295
Durian 484
Dutch
    cocoa 295
    oven 295
Dwarfism 295
Dwarf tapeworm 287
Dymelor 295
Dys– 295
Dysentery 295
    amoebic 286, 295
Dysgeusia 295
Dysinsulinism 295
Dysosmia 295
Dyspepsia 295
Dysphagia 296
Dyspnea 296
Dyssebacia 296

Dystrophy 296

E

Earth eating 297
Eating habits, trends 669
Eber's Papyrus 297
Eclampsia 297, 882, 1019
Ectomorph 111, 782
Eczema 254
Eddoe 1046
Edema 297
Edible portion 297
EDTA 297, 321
Effervescent 297
Egg(s) 297
    abnormalities 299
    allergies 24, 302
    amino acid pattern 29–30
    cholesterol 302
    color 299
    consumption 302
    dehydrated 303
    dried 303
    fertile 530
    grading 298
    history 298
    nutritional value 302
    production/world/U.S. 298
    proteins 304
    raw 530
    replacers 530
    shape 299
    size 299
    structure 299
    uses 300
    weight classes 299
    white 305
        injury 305
    whites, dried 530
Egg fruit 304
Eggplant 304, 1039
    history 304
    nutritional value 304
    preparation 304
    production/world/U.S. 304
Eijkman, Christiaan . 1009, 1055, 1083
EKG 305
Elastase 305
Elastin 305
Elderberry 484
Electrocardiograms 305
    abnormal 11
Electroencephalogram 305
Electrolyte(s) 305, 1110
Electron 305
    microscope 305
    transport chain 305
Electropure process 305
Elemental diets 305
    administration 306
    contraindications 306
    preparation 306
    use of 306
Elephant skin 254
Elimination diets 23
Elvehjem, Conrad 758, 835
Emaciated 306

Embden-Meyerhof pathway ... 306
Emblic 306
Embolism 307
Embryo 307
Emetic 307
–emia 307
Emmer 182
Emphysema 307
Empty calories 307
Emulsification 326
Emulsifiers 8
Emulsify 307
Emulsifying agents 307
    salts 307
Emulsion 307
Enamel, teeth 307
Encephalomalacia 307
Encephalopathy 307
Endemic 307
Endergonic 307
Endo- 307
Endocrine 307
    changes in aging 308
    glands 307
    hormones 308–310
    location 308
Endocrinology 311
Endogenous 311
Endomorph 782
Endopeptidase 311
Endoplasmic reticulum 311
Endosperm 311
    oil 311
Endotoxin 75, 311
Enema 311
Energy
    athletic activities 62
    balance 311
    biological chain 313
    body utilization 313, 314
        active transport 314
        maintenance of temp. 314
        muscle contraction 315
        synthesis 314
    digestible 144
    fecal 144
    food production 311
        animal efficiency 312
        crop efficiency 312
    gross 144
    heat of combustion 144
    heat increment 145
    metabolizable 144
    net 145
    urinary 144
English walnut 1107
Enrichment 315
    cereal products 891
    flours 362
    foods 317–318
    guidelines 318
    history 316
        Bataan exp. 316
        Beriberi/Philippines 316
    levels 318
    rice 930
    standards 318
Enteric 320
Enteritis 320
Entero- 320
Enterohepatic circulation 320
Enterokinase 320
Enterostomy 320
Environment 320

pollutants 283
Enzymatic 320
Enzyme 320
  activation 7
  digestion 277
  foods, in 380
Epidemic 320
Epidemiology 320
Epidermis 320
Epigastric 320
Epiglottis 320
Epilepsy 320
Epinephrine 310, 320
Epiphysis 320
Episome 45
Epithelial 320
Epsom salt 321
Ergocalciferol 321
Ergometer 321
Ergosterol 321
Ergot 857
Erucic acid 321
Erythorbic acid 321
Erythrocytes 321
Erythropoiesis 36
Erythropoietin 321
Escherichia coli 72
Esculin 13
Esophageal reflux 535
Esophagitis 535
Esophagus 321
Essential
  fatty acids 321
  oils 321
Ester 321
Estrogen 310
Ethanol 15, 321
Ether extract 321
Ethnic 321
Ethyl alcohol 15
Ethylene 321
Ethylenediamine tetra-acetic
  acid 321
Ethyl formate 321
Etiology 322
Evaporated milk 702
Evaporation, flash 322
Exacerbation 322
Exchange list 322, 746-748
Excipient 322
Excretion 322, 689
Exercise, caloric
  expenditures 150, 546
Exergonic 322
Exogenous 322
Exopeptidase 322
Exophthalmic goiter 82
Exotoxin 75
Experiment 322
Extensometer 322
Extracellular fluid 322
Extract 322
Extraction ratio 322
Extremities 322
Extrinsic factor 322
Exudate 322
Exudative diathesis 322

Facoldan 519
Fagot 323
Fahrenheit 323
Familial 323
  cholesterol ester def. 567
  high-density lipoprotein
    deficiency 567
  hyperlipoproteinemias 5, 567
Fanconi's syndrome 564
FAO 91, 372
Farina 323
Farmer's share/food prices 664
Fast foods 222, 323
  nutritional analysis 224-229
Fasting 324
Fats/other lipids 324, 686
  absorption 327
  back 324
  blood 324
  classification 325
  digestion 327
  fatty acids ...... (See Fatty Acids)
  functions 327
  history 325
  hydrogenated 339
  medium-chain triglycerides .. 332
  metabolism 327
  rancidity 326
  requirement 328
  soluble vitamins 339
  sources 329
Fat-soluble vitamins 339
Fattening 339
Fatty acids 325, 339
  carbon chain length 326
  essential 328-329, 538, 798
  foods 332
  hydrogenation 326
  iodine number 326
  polyunsaturated 330, 798
  saturated 330, 798
  saturation 325, 330
  sources 329
Fava bean 611
FDA 339, 372
Febrile 339
Federal Food, Drug and
  Cosmetic Act 339
Feed adjuvants 339
Feed animal efficiency 1137
Feedback mechanism 339
Feeds vs foods 1140
Feingold diet 95
Fennel, Florence 339, 1031
Fenugreek 612
Fermentation 339
  beer 91
  foods 888
  product 340
Fermented 340
  milk 203
  minchin 1125
Ferritin 340
Fetal alcohol syndrome 16
Fiber 340, 538
  beneficial effects 342
  constipation 342
  determination of 342
  diabetes 344
  dietary 344
  digestibility studies 342
  harmful effects 344
  high/diet 741
  history 340

  kinds 341
  lipid lowering 344
  preparation/foods 344
  proximate analysis 342
  sources 345-346
  Van Soest analysis 32, 342
  weight control 342
Fibrin 346
Fibrinogen 346
Fibrous 346
  proteins 346
Fig(s) 346
  dried 347
  history 347
  nutritional value 347
  preparation 347
  processing 347
  production/world/U.S. 347
  syrup 530
Filbert 777
Finely ground 347
Fines 347
Fingernails 755
Finish 347
Fireless cooker 347
Fish/seafoods 347
  buying 351
  catch/U.S. 349
  consumption/capita 354
  cooking 353
  farming 350
  flour 361, 364
  history 348
  industry 349
    U.S. 349
  nutritive value 353
  processing 350
  protein concentrate 354
Fistula 354
Flaky paint dermatosis 254
Flash-pasteurization 354
Flat sours 355
Flatulence 355
Flavedo 355
Flavin adenine dinucleotide ... 355
Flavin mononucleotide 355
Flavoprotein 355
Flavor
  agents 7
  compositions 7
  enhancers 7
  potentiator 358
  synthetic 358
Flavorings/seasonings 355
  definitions/terms 355-356
  history 356
  liquors 358
  popular 356
  quantity 357
  when 357
  why 357
  wines 358
Flax 358
Flax seed 346, 519, 530
Florence fennel 1041
Floridian starch 359
Flour(s) 175, 359
  by-products of 364
  enrichment 362, 422
  fortification 362
  history 359
  nutritive value 361
  processing 360
  recent developments 364

F

types 360–361
uses 363–364
wheat 1121
Fluid balance **365**
body 365
Flummery **365**
Fluoridation 365, 367
Fluoride/fluorine 365, 717
deficiencies 356
history 365
metabolism 356
sources 367
Fluorosis **367**
Foie gras **367**
Folacin 367, 1062
anemias & 369
cancer & 369
cooking losses 370
deficiency 370
functions 369
history 368
leukemia & 369
metabolism 368
RDA 370, 1062
sources 372
Folate **367**
Folic acid 367, 1062
deficiency in anemia 42
Folinic acid 203, **372**
Follicle stimulating hormone ... **308**
Follicular hyperkeratosis **254**
Food
acid 4, 6
acid-base reaction 4
acid residue 5
additives 7, 154
alkali 4, 6
alkaline residue 5
analysis of 31
& Agriculture Organization of
the United Nations .. 291, 372
& Drug Administration 372
buying 373
caloric values 151
colors 381
compositions 383–469
cooking 373, 381
data use 35
energy required for 311
exchange lists 322
fads 473
faddist 474
folk medicine 474
groups 470
calorie counting 472
five 471
four 470–471
meal planning 471
seven 471
six 470
snacking 472
strength 472
weakness 472
Guide Pyramid 470
infections, bacterial 71–73
low-energy 475
misinformation 473
myths 473
neutral 5
nutrients from animal products 694
poisonings 73, 475
preparing 373
properties 375
colors 381
enzymes 380
quack 474

safety of 939
sanitation 289
science 475
serving 373
Stamp Program 475, 511
benefits 476
concerns 475
costs 475
participants/number 476
value of stamps 476
storing 374
technology 477
Fool **477**
Forbidden fruit **477**
Formiminoglutamic acid **477**
Formula **477**
Fortification 315, **477**, 891
flour 362
Framingham study **537**
Frangipane **477**
Frankfurter **477**
Frappe **477**
Free fatty acid **477**
Freeze drying **887**
Freezing foods **888**
French
bean 84
toast 477
Frogs' legs **477**
Frostbite **102**
Fructose 478, 530, **1000**
intolerance 565
Fruit(s) **478**
beverages 489
butters 122
canning 480, 489
domestication 479
drying 480, 489
federal standards 480
freezing 480
history 478
losses in preparation 480
nutritive value 481
origin 479
processing 480
production/world/U.S. 479
sugar 489
syrups 489
types 478
world, of the 481–488
Frumenty **489**
Fryers **131**
Fuel factor **489**
Fungicide **489**
Funk, Casimir 1009, **1055**, 1083
Fustic **489**

G

Galactose 491, 1000
Galactosemia 491, 565
Gallates **491**
Gallbladder **491**
disease 491
Gallon **491**
Gallstones **491**
calcium 492
causes 492
cholesterol 492
diagnosis 493

diet 493
location of 491
symptoms 492
treatment 493
Game meat **493**
Gammon **493**
Garlic **494**, 530
history 494
nutritional value 494
preparation 494
production/world/U.S. 494
Gas storage **495**
Gastrectomy **495**
Gastric **495**
juice 495
mucosa 495
ulcer 1023
Gastritis **495**
acute 495
chronic 495
toxic 495
Gastroenteritis 286, **496**
Gastrointestinal **496**
disease 496
tract 496
Gastroscopy **496**
Gavage **496**
Geese **869**
Gel **496**
Gelatin **496**
Gelometer **496**
Gene **496**
Generally recognized as safe
(GRAS) **8**
Genetic
diseases 496
engineering 766
-genic **496**
Genipapo **496**
Geophagia **496**
Gerber, Dan **572**
Gerber test **498**
Geriatrics ...... (See Gerontology)
Germ **498**
Gerontology **498**
causes of aging 500
body changes 499
diets 501–503
factors/aging 499
history 499
menus 502
nutrition 498
nutritional allowances 499
Ghatti **519**
Gherkin **244**
Gigot **504**
Gin 82, 292
Ginger
beer 504
root 530
Gingivitis **504**
Ginkgo **504**
Ginseng **530**
Gland **504**
Gliadin **504**
Globulins **504**
Glomerulonephritis **505**
Glossitis 253, **505**
Glucoascorbic acid **505**
Glucose 262, **505**, 1000
metabolism 505
oxidase 506
TM 506
tolerance

factor 506
test 265, 560
transport maximum 506
Glucosuria **506**
Glucosyl ceramide lipidosis **567**
Glucuronic acid **506**
Glutamate, sodium **506**
Glutamic acid 28, **506**
Glutathione **507**
Gluten 175, **507**
allergy 24
flour 531
free
breads 119
diet 507, 743
foods 507
Glutenin **507**
Glyceride, esters, partial **1**
Glycerin **507**
Glycerine **507**
Glycerol **507**
lacto-stearate 507
Glycine 28, **507**
Glycogen **507**
storage, diseases 10, 565
Glycogenesis **507**
Glycogenic **507**
Glycogenolysis **507**
Glycolysis **507**
Glycoprotein **507**
Glycosides **508**
Glycyrrhiza **508**
Goat's milk 531, **692**
Goblet cells **508**
Gobley, N.T. **194**
Goiter **508**
causes 508
cretinism 509
history 508
iodine deficiency 508
myxedema 509
prevention 510
toxic 509
treatment 510
Goitrogenic
agents 509
broccoli 131
Goitrogens **510**
Goldberger, Joseph 958, **1076**
Golden
apple 97
gram 613
Gooseberry **485**
Gorge/purge **133**
Gossypol **510**
Gourds **511**
Gout 56, **568**
causes 56
prevention 57
symptoms 56
treatment 57
Government agencies **290**
Government food programs **511**
child care 512
commodity donation
to programs 513
costs 513
history 511
milk 512
school
breakfast 512
lunch 511

stamps 475, 511
WIC 512
Governor's plum **513**
Grades
beef 88
lamb 603
pork 862
veal 88
Graham
flour 513, 531
Sylvester 1048
Grain(s) **513**
alcohol 15
Gram **514**
Granadilla, giant 487, **514**
Granola cereal **531**
Grape **514**
history 514
juice 517
nutritional value 516
preparation 515
processing 514
canned 515
jam 515
jelly 515
juice 515
raisins 514
table grapes 515
wines 514
production/world/U.S. 514
sugar 505, 517
Grapefruit **516**
history 516
nutritional value 516
processing 516
production/world/U.S. 516
selection 516
Grapefruit, naringen **517**
GRAS 8, **517**
Graves' disease **82**
Green
bean 84, 517
gram 613
sickness 191
Green revolution 181, **517**
future improvement
of world food 517
Grijns, G. **1009**, 1083
Grind **518**
Griskin **518**
Grits 175, 237
Groats **175**
Ground beef **518**
Groundnut **829**
Grouse **518**
Growth **518**
hormone 308-309
Gruel **518**
GTF **506**
Guanbana **488**
Guanine **518**
Guar 519, **612**
Guarana **531**
Guargum **519**
Guava **485**
Guinea
corn 518
fowl 869
pigs 171, 519

Gum(s) 341, **519**
Arabic 519
sources 519
Gyorgy, Paul **1076**

H

Hackberry **521**
Haff disease **521**
Haggis **521**
Hair analysis 252, **521**
discoloration 252
loss 252
Half-life **521**
biological 521
physical 521
Halibut liver oil **522**
Hallucinogen **522**
Halva **522**
Ham **522**
Hamburger **522**
Hard cider **522**
Hardtack **522**
Hare **522**
Harelip **623**
Harris, biotin synthesizer **106**
Hart, E. B. **230**
Hartnup disease **564**
Haslet **522**
Haws **522**
Hawthorne **522**
Hay **522**
fever 21-22
Hazelnuts **777**
Headcheese **522**
Health **522**
athletics and 61
foods 524
common 527-534
history 525
popularity 526
rules of 524
signs of 523
status of adults 10
Heart 193, **535**
disease 535
disorders 535
angina pectoris 535
atherosclerosis 536
coronary occlusion 536
failure 536
irregular beats 536
myocardial infarction 536
factors affecting heart diseases . 536
dietary 537
ethnic groups 536
Framingham study 537
signs and symptoms of
heart disease 543
treatment and prevention of
heart disease 543

Heartburn 534
  causes 534
Heat
  combustion 144, 547
  exhaustion 282, 547
    salt 547
    water 547
  increment 145, 547
  labile 547
Height
  increments, children 185, 187
Heimlich choking maneuver 191
Hem-, hema-, hemo- 548
Hemagglutinins 548
  beans, in 86
Hematocrit 548
Hematopoiesis 548
Hematuria 548
Hemeralopia 548
Hemicellulose 341
Hemiplegia 548
Hemochromatosis 568
Hemodialysis 548
Hemoglobin 548
  synthesis 548
Hemolysis 548
Hemolytic 548
  anemia 38-39, 249, 566
Hemorrhage 548
Hemorrhagic disease of newborn 250
Hemorrhoids 549
Hemosiderin 549
Heparin 549
Hepatitis 286, 549
Hepatomegaly 252, 549
Herbs 356, 549
Hereditary spherocytosis 568
Hesperidin 102
Hess, A. 1092
Hexanoic acid 159
Hexosans 162
Hexosemonophosphate shunt .. 549
Hexoses 161
Hiatus hernia 535, 549
  causes 549
  prevention 549
  treatment 549
High blood pressure 11, 550
  age 550
  causes 550
  dangers 550
  emotions 550
  heredity 550
  high salt diet 550
  hormones 550
  obesity 550
  prevention 551
  smoking 550
  treatment 550
  types 550
High-density lipoproteins 551
High-nutrient density 551
High-ratio flour 551
Hinduism dietary beliefs 920
Hippocrates 585, 713
Histidenemia 564
Histidine 28, 551
Histology 551
History of nutrition 551
Hives 20, 552
Home economists 552
Homeostasis 552
Hominy 237

Homocystinuria 564
Homogenization 552
Honey 531, 552, 992
  nutritional value 553
  vinegar & 531
Honeydew melon 679
Hookworm 39
Hopkins, Frederick G. 1021
Hopping, beer 91
Hormone(s) 554
  endocrine glands 308-310
  gastrointestinal 279
Horse bean 612
Horsemeat 554
Horseradish 1041
Hospital diet 554
Hot dog 477
Hothouse lambs 554
Hot pot 554
Huckleberry 554
  garden 555
Hulls 555
Human assay for vitamins ... 1057
Human milk composition 693
Humble pie 555
Humectants 555
Humulone 555
Hungarian partridge 555
Hunger 273, 555
  ways to lessen 555
  world 555
Husks 556
Hyacinth bean 612
Hydraulic process 556
Hydrochloric acid 556
Hydrocooling 556
Hydrogenation 556
Hydrolysate 556
Hydrolysis 556
Hydroponics 556
Hydrostatic pressure 556
  sterilizer 556
Hydroxylation 556
Hydroxyprolinemia 564
Hyper- 556
Hyperactivity 95
Hyperalimentation, intravenous . 556
Hyperammonemia 564
Hypercalcemia 137, 557
Hypercalciuria 557
Hyperchlorhydria 557
Hypercholesteremia 557
Hypercholesterolemia 557
Hyperesthesia 557
Hyperkalemia 557
Hyperlipoproteinemia 557
Hyperlysinemia 564
Hyperoxaluria 564
Hyperphosphatemia 558
Hyperplasia 558
Hyperprolinemia 564
Hypertension 11, 550, 558
Hyperthyroidism 558
Hypertonic dehydration 558
Hypertriglyceridemia 558
Hypertrophic arthritis 54
Hypertrophied 558
Hyperuricemia 558
Hypervalinemia 564
Hypervitaminosis 558
Hypo- 558
Hypoalbuminemia 558
Hypocalcemia 558
Hypochloremic alkalosis 558

Hypochlorhydria 558
Hypochromic 558
Hypocupremia 558
Hypogeusia 558
Hypoglycemia 266, 558
  causes 559
    alcoholism 559
    coffee & 559
    diabetes 559
    drugs 559
    emotional stress 559
  diagnosis 560
  disorders 560
  factors controlling 559
  glucose tolerance test 560
  obesity 560
  prevention 561
  symptoms 560
  treatment 561
Hypoglycemic agents 562
Hypokalemia 562
Hypomagnesemia 562
Hypoproteinemia 562
Hypothalamus 562
Hypothyroidism 562
Hypoton dehydration 562
Hypovitaminosis 562
Hyssop 562

I

Ice
  cream 705
    custard 705
    milk 705
Iceland moss 563
Idiopathic 563
Ileal resection 563
Ileitis 563
Ileum 563
Ileus 563
Illipe butter 563
Immunization 288
Immunoglobulins 563
Impermeable 563
Inborn errors of metabolism . 563, 690
  amino acid 564
  carbohydrate 565
  lipid 567
  misc. 568
  nucleic acid 568
Incaparina 238, 569
Income, proportion
  for food 569, 570, 666
  time reqd to purchase food .. 570
Indian
  gooseberry 306
  ricegrass 182
Indigestion 570
Indispensable amino acid 571
Inert 571
Infant
  allergy 571
  anemia 571, 580
  cereal 81
  diet/nutrition 571
    breast vs bottle 575
    infant formulas 575-576

foods 575
growth curves 573-574
guidelines 516
history 572
modifications 580
nutritional requirements ... 575
problems 579
feeding new foods 68, 576
height curves 573-574
weight curves 573-574
Infantile
beriberi 98
paralysis 286
Infarction 580
Infection 580
Inflammation 580
Ingest 580
Inorganic 580
Inositol 580, 1067
deficiencies 581
functions 581
history 580
metabolism 580
RDA 581
sources 581
Insensible perspiration 581
Insidious 581
Institute of Food Technology .. 581
Insulin 268, 310
Interesterification 581
Intermediate hydrogen carrier .. 581
International Unit 581
Interstitial fluid 110, 581
Intestinal
flora 580
juice 581
Intestine
large 582
small 582
Intolerance 582
Intra- 582
Intracellular 582
Intramural nerve plexus 582
Intravenous feeding 582
Intravenous nutrition 582
benefits 583
contraindications 583
history 583
indications 583
precautions 583
types 583
Intrinsic factor 38, **584**
Inulin **584**
Inversion **584**
Invertase **584**
(sucrase)-isomaltase deficiency . 584
Invert sugar 584, **1000**
Iodine **584**, 717
deficiency 587
functions 585
goiter 508
history 585
metabolism 585
number 326, 587
RDA 586
sources 586
Iodine number 326, **587**
Iodine 131 587
Iodized salt 587
Iodopsin 587
Ion 587
Ionic iron 587
Ionization 587
Ionized calcium 587

Irish potato 866
Iron 40, **587**, 717
deficiency anemia 38
deficiency symptoms 589
enrichment 42
food sources 41
fortification 42
functions 589
history 588
injections 41
losses 590
metabolism 588
parenteral administration 41
RDA 590
roll in red blood cells 37
sources 590-591
utilization 41
Iron-binding capacity 592
Iron-deficiency anemia 250
Irradiated yeast 592
Irradiation 592
of foods 915
Ischemia 592
Ischemic heart disease 592
Isinglass 592
Islam dietary beliefs 920
Isles of Langerhans 592
Isoascorbic acid 592
Isocaloric 592
Isoenzyme 592
Isoleucine 592
Isomaltose 592
Isomer 592
Isotonic 592
Isotope 592
Isotopic label 592
Isovaleric acidema 564
-itis 592

Jaboticaba 593
Jack bean 612
Jackfruit 485
Jaggery 593
Jainism dietary beliefs 921
Jambalaya 593
Japanese
medlar 486
quail 593
Jaundice 593
Jejunum 593
Jellies 593
Jerusalem artichoke 1041
flour 122
Jicama 1041
Job's tears 182
Joule 593
Judaism dietary beliefs 921
Juices 593
Jujube, Indian 485, 593
Julep 593
June berry 593
Jungle rot 254
Juniper berry 594
Junk foods 594
Juvenile
beriberi 98

delinquency and diet 95

Kaffir 965
Kale 1042
Kangaroo meat 595
Karaya gum 519
Karell diet 595
Karrer, P. 924, 1069
KCAL 595
Kefir 531
Kei-apple 595
Kellogg, John 123, 1048
Kelp 531, 595, 948
Kempner diet 595
Keratin 595
Keratinization 595
Keratomalacia 250, 1073
Kernel 595
Keto- 595
Keto-acid 595
Ketogenesis 595
Ketogenic 595
amino acid 595
diets 595, 743
Ketone 595
Ketone bodies 596
Ketonic rancidity 596
Ketosis 596
Keys, Ancel 144
Kibbled 596
Kidney bean **84**
Kidney disease **596**
stones 596, 1029
Kidneys (animal), cooking 596
Kilocalorie 596
Kilogram 596
Kilojoule 597
Kinetic energy 597
King Charles II 87
King, Charles Glen 1085
Kiwi fruit 485
Kjeldahl 597
Kohlrabi 597, 1042
Koilonychia 253
Koji 598
Korsakoff psychoses and
syndrome 98, 598
K-P diet 95
Krebs 818
Krebs, E. 601
Krebs-Henseleit cycle ... 203, 1028
Krehl, Willard 758
Krick, Francis 892
Kuban 598
Kuhn 924
Kumquat 485
Kwashiorkor ... 251, 598, 642, 899

Labels
    code dating 600
    food 599
    grades 600
    guidelines 599
    imitation 599
    ingredients 599
    names 599
    nutritional 599
    open dating 600
Labile 600
    protein 600
Labrador tea 600
Lactalbumin 600
Lactase 600
Lactation 600, 875
    energy reqts 150
Lactic acid 601
Lactobacillus acidophilus ... 531, 601
Lactoflavin 601
Lactose 601
    free diet 743
Laetrile 601, 1067
    deficiency 602
    functions 602
    history 601
    metabolism 601
    sources 602
Lager beer 92
Lamb/mutton 603
    consumption/capita 603
    cuts 603–604
    grades 603
    production 603
    qualities 603
Laminaran 519
Larch gum 519
Lard 43, 605
Larding 605
Lark 605
Lathyrism 605
Lathyrus pea 612
Latin Am. chewed corn beer ... 93
Laubina 1126
Lavoisier, Antoine Laurent 143
Laxative 605
Lead 605
    poisoning 858
Leaven 605
Leavened breads 115
Leavening agents 75
    ingredients 76
    nutritional value 76
    principles of 76
    quick breads 116
    uses 76
    yeast breads 115
Leben 605
Lebkuchen 605
Leche alim 1126
Lecithin 531, 538, 605
Leek 1042
Legumes 606
    allergy 25
    antinutritional/toxic factors ... 608
    history 606
    nutritional value 609
        protein quantity/quality 609
    preparation 607
    production/world/U.S. 606
        contributions/legume crops .. 607
    world, of the 610
Lemon 614
    history 614

nutritional value 615
    preparation 615
    production/world/U.S. 614
Lenhartz diet 615
Lentil 615
    history 615
    nutritional value 616
    preparation 616
    production/world/U.S. 615
Leptospirosis 72
Lethargy 616
Lettuce 616
    history 616
    nutritional value 618
    preparation 617
    production/world/U.S. 616
Leucine 618
    induced hypoglycemia 566
Leukemia 39, 152, 618
Levulose 618
Licorice 508, 618
    root 531
Lieberkuhn's crypt 618
Life expectancy/nutrition 618
    food selections 620
    food selections/U.S. 618
    longevity 619
        Bulgarians 619
        Ecuadorians 619
        Georgians 619
        Hunzas 619
    physical activity 620
Light off-flavor 620
Lignin 341, 620
Lima bean 620
    antinutritional/toxic factors ... 621
    flour 361, 364
    gas-producing 621
    history 620
    nutritional value 621
    protein quality/quantity 622
    preparation 621
    production/world/U.S. 621
Lime 622
    history 622
    nutritional value 623
    preparation 621
    production/world/U.S. 621
Limiting amino acid 623
Limonin 623
Lind, James 1055, 1085
Lingonberry 623
Linoleic acid 623
Linolenic acid 623
Linseed 358
Lipase 623
Lip, cleft 623
Lipid(s) 623
    blood 109
    digestion/absorption 278
    hypothesis 332
Lipogenesis 624
Lipoic acid 624, 1067
    functions 624
    history 624
    RDA 624
    sources 624
Lipolysis 624
Lipolytic rancidity 624
Lipoproteins 11, 624
Lipotropic factor 624
Liqueurs 292, 358
Liquid diet 624
    clear 739
    types 624–626

Litchi nut 486
Lithiasis 627
Liver
    enlargement 252
    flukes 287
    powder 531
Livetins 627
Locust bean 519, 611
Lofenalac 627
Loganberry 627
Logarithmic phase 627
Longan 627
Longevity 627
Loquat 486
Low blood sugar 558
Low-density lipoproteins 627
Lowry assay 34
Lucerne/alfalfa 627
Lunin, Nicholas 1055
Lupines 613
Lupus erythematosus 627
Luteinizing hormone 308, 309
Lycopene 628
Lymph 628
Lymphatic system 628
Lymphoma 152
Lyophilization 628, 887
Lysergic acid 628
Lysine 28, 628
    availability 629
    herpes 629
    high-lysine grains 181
    requirement 628
    sources 628
    supplementation 629
Lysosomes 630
Lysozyme 630

M

Macadamia nut 778
Macaroni products 631
    enrichment 634
    fortification 634
    history 631
    nutritive value 633
    preparation 634
        production 632
        commercial 632
Macedoine 634
Macrobiotics, Zen 634
Macrocyte 634
Macrominerals 634
Macronutrient 634
Magma 634
Magnesium 635, 715, 718
    deficiencies 635
    functions 635
    history 635
    metabolism 635
    RDA 635
    sources 636
Maidenhair tree 504
Maillard reaction 636
Maize 231
    waxy 239
Malabsorption syndromes 636
Malaise 636
Malanga 1004

Malaria 636
  cycle 637
  types 637
Malay apple 637
Malformation 637
Malignant 637
Malnutrition 637
  behavior 96
  causes 639
  cereals for 126
  consequences 639
  energy-protein 642
    prevention 645
    signs 643-644
    treatment 645
    types 642
      adults 643
      kwashiorkor 643
      marasmus 643
      nutritional 643
    world supplies of 642
  prevention 641
  signs of 640-641
  treatment 641
  types 639
    energy-protein 639
    mineral 639
    vitamin 639
  world 638
Malt(ing) 174, **646**
  beer 82, 92
  derivatives 647
  flour 361, 364
  history 647
  liquor 92
  nutritive value 647
  uses 647
Maltase **648**
Malthus, Thomas Robert .. **648**, 861
Malting barley **80, 90**
Maltol **648**
  ethyl 648
  formation 648
  uses 648
Maltose **648**, 1000
Mamey **648**
Manatee **648**
Mandarin orange 1023
Manganese 648
  deficiency symptoms 649
  functions 649
  history 649
  metabolism 649
  RDA 650
  sources 650
Mango 650
  history 651
  nutritional value 651
  preparation 651
  production/world/U.S. 651
Mango melon 651
Mangosteen 486
Mangrove 651
Manioc 169, 1147
Manna 651
Mannitol 651
Mannose 651
Mao-tai 967
Maple
  sugar 531, 991
  syrup 531, 651
    nutritional value 652
    processing 652
    uses 652

  urine disease 564
Maraschino cherry 184
Marasmus 251, 643, **652**
Margarine 652
  consumption/capita 653
  history 652
  kosher 654
  nutritional value 653
  production 653
Marinade 654
Marmite 654
Marron 654
Marrow 112
Mate 654
Matoke 654
Matrix 654
May apple 654
Mayonnaise 654
Maypop 654
McCollum, E. V...865, 1009, **1068**, 1083
McKay, Clive **618**
Meal
  guide 655
  planning 654
Meal
  guide 655
  planning 654
Meat(s) **656**
  aging 14
  analogs 531, 676
  buying 664
  by-products 660, 676
  carving 668
  consumer's food dollar 666
  consumption/U.S. 668
  cooking 666
  eating trends 669
  extract 90, 676
  fabricated/packaged 663
  factor 676
  facts and myths 674
  grades 659
  history 657
  inspection 659
  kinds 661
  nutritional value 670-674
  packing 657
    kosher 660
    tenderizing 657
  preservation 661
  price determination 664
  production/U.S. 668
  qualities 663
  selecting 666
Mechanically extracted 676
Medicinal plants 676
  common 678
  history 676
  plant remedies 677
Mediterranean anemia 39
Medium-chain triglycerides ... 679
Medlar 486
Megadose 679
Megajoule 679
Megaloblast 679
Megaloblastic anemia 38
Megavitamin therapy & behavior . 96
Melanin 679
Melba toast 679
Mellorine 705
Melon(s) 679
  kinds 679
  pear 681
Melting point (fat) 798

Mendel, L.B. 892, **897**, 1068
Meniere's syndrome **681**
Mental
  development 681
  illness 681
  retardation 682
Mercury poisoning 858
Mertz, W. **197**
Mesomorph 111, **782**
Mesquite **613**
Metabolic
  pool 682
  rate 682
  reactions 682
  water 682
Metabolism **683**
  cells/functional units 683
  controlling factors 689
  excretion 689
  inborn errors of 690
Metabolite 690
Metalloenzyme 690
Methionine 28, **690**
  malabsorption 565
Methyl alcohol 690
Methylated spirits 690
Meulengracht diet 690
Mevalonic acid 690
Micellar bile-fat complex 690
Micelles 690
Microbiological assay 35, 1057
Microcyte 690
Microgram 690
Micronutrients 690
Microscopic analysis 35
Microvilli 690
Microwave cooking 690
  baking breads 122
Milk-alkali syndrome 691
Milk-free breads **120**
Milk/products **691**
  allergy 24, 698
  and cream diet 739
  anemia 699
  consumption per capita 709
  cow to table 693
  cream 704
  evaporated milk 702
  exports 708
  fermented 710
  freezing point test 710
  frozen 710
  galactose disease 699
  grades 708
  health & consumption 697
    benefits 698
    problems 698
  history 692
  imports 708
  intolerance 698
    lactose intolerance 698
  kinds 701
  marketing 706
    demand 707
    federal milk marketing 707
    state milk controls 707
    supply 706
  myths 699
  nutritional value 694-697
  problems 698
  processing 700
  production/U.S. 699
    dairy producers 699
    hauling 700

milking 699
program for children 512
raw 531
shake 711
species from 672
standards 708
substitutes 532
turbidity test 711
Milk program, special **512**
Millet 532, **711**
flour 361, 364
history 711
milling 711
nutritional value 711
production/world/U.S. 711
products 711
Milliequivalent **712**
Milligram **712**
Milling cereal grains **175**
barley 71
Milo **965**
Mince **712**
Mineral(s) 688, **712**
deficiencies
causes 719–728
diagnosis 729
dietary history 729
lab tests 729
symptoms 729
prevention 730
processing to maximize
minerals 730
treatment 730
history 712
macro 712
micro 712
supplements 736
table 713–719
Minot **1079**
Miscarriages **104**
Miscible **736**
Miso 532, **973**
Modified diets **736**
common 737–745
dietetic special foods 747
exchange lists 746–748
history 737
obesity 787–789
special bakery products 747
Moiety **749**
Moisture **749**
Moisture-free **749**
Molasses **749**, 993
blackstrap 108
food 749
production 749
uses 749
Mold(s) **750**
inhibitors 750
Molecule **750**
Molybdenum 718, **750**
deficiency symptoms 750
functions 750
history 750
metabolism 750
RDA 750
sources 751
Monkey bread **78**
Monoglycerides **751**
Monophagia **751**
Monosaccharides 160, **751**
Monosodium glutamate **506**
Monounsaturated **751**
Monovalent **751**

Morbidity **751**
Morning sickness **882**
Mortadella (sausage) **751**
Motility **751**
Mouth, dry **253**
MSG **506**
Mucillages **341**
Mucin **751**
Mucopolysaccharide **751**
Mucopolysaccharidoses **569**
Mucoprotein **751**
Mucosa **751**
Mucous membrane **751**
Muffins
blueberry 116
bran 116
corn 116
Mulberry, black **486**
Multiple sclerosis **751**
Mung bean **613**
sprouts 532
Murphy **1079**
Muscles **752**
circumference 1028
contraction 315
wasting 253
Muscovado **752**
Muscular
dystrophy 752
work 752
Mush **752**
Mushrooms 752, 1042
dried 532
nutritional value 753
types 752
Muskrat **754**
Mustard
greens 1043
seed 754
nutritional value 754
preparation 754
processing 754
Mutagens **754**
Myasthenia gravis **754**
Mycotoxins **858**
Myelin **754**
Myo- **754**
Myocardium **754**
Myofibrils **754**
Myogen **754**
Myoglobin **754**
Myosin **754**
Mysore flour **754**
Myxedema **509**

NADP **762**
Nails (finger) **755**
spoon 253
Naphthoquinone **755**
Naranjilla **755**
Naso-labial seborrhea **254**
Natal plum **755**
National flour **755**
National Institutes of Health .. **755**
Natto **974**
Natural foods **755**

Nausea **756**
Necrosis **756**
Nectar **756**
Nectarine **826**
Neonate **756**
Neoplasm **756**
Neopolitan medlar **65**
Nephritis 605, **756**
Nephrolithiasis **596**
Nephron **756**
Nephrosclerosis **756**
Nephrosis **756**
Neroli oil **756**
Nervous system **756**
disorders 757
Neuritic **757**
Neuritis **757**
Neuropathy **757**
Neutropenia **757**
Niacin 757, **1063**
deficiency symptoms 760
equivalent 761–762
functions 759
history 758
losses 761
metabolism 758
RDA 760, 1063
sources 761
Niacinamide **762**
Nickel **762**
Nicotinamide 757, **762**, 1063
dinucleotide 762
phosphate/NADP 762
Nicotinic acid 757, **762**
Niemann-Pick disease **567**
Night blindness 252, **1072**
Nitrates/nitrites **762**
cancer 763
dangers 763
exposure 762
FDA/Delaney Clause 763
food additives 762
naturally occurring 762
nitrosamines 763
sources 763
toxicity 763
Nitrilosides **1067**
Nitrites **762**
Nitrogen **763**
balance 764
negative 764
positive 764
-free extract 764
metabolic 764
Nitrosamines 763, **764**
Nitrosomyoglobin **764**
Nitrous oxide **764**
Nocturia **764**
Nonesterified fatty acids **764**
Nonheme **764**
Nonpareils **764**
Nonprotein nitrogen **764**
Nonvegan **764**
Norepinephrine **96**
Nourish **764**
NRC **755**
Nucleic acids **764**
DNA structure 765
functions 766
genetic engineering 766
metabolism 766
requirements 766
sources 766
synthesis 755

Nucleolus 766
Nucleoprotein 766
Nucleotides 766
Nucleus 766
Nut(s) 774
  history 774
  nutritional value 775
  processing 775
  production/world/U.S. 775
  world, of the 777
Nutrient(s) 766–771
  allowances 766–767
  deficient animals 34
  functions 766–767
  metabolism 684
  recommended daily dietary
    allowances 768–771
  requirements 766–767
    athletes 61
  sources 766–767
Nutrification 315
Nutrition 770
  adult 9
  education 771
  channels 773
    education institutions 772
    food industry 772
    history of 551
    government agencies 772
    prof. organizations 772
    sources 773
    voluntary agencies 772
Nutritional
  allowances, adults 12
  deficiency diseases 770
  status 770
  supplements 7, 770
Nutrition Foundation, Inc. 774
Nutritionist 774
Nyctalopia 778

Oat(s) 779
  flakes 780
  flour 361, 364, 532, 780
  history 779
  nutritional value 780
  processing 779
  production/world/U.S. 779
  products 780
  steel cut 532
Oatcakes 781
Oatmeal 780
Obese 781
Obesity 781
  adult 10, 12
  body builds 782
  causes 782
  children 786
  diets 786
  disorders 784
  drugs 791, 793
  eating behavior 790
  exercise 791
  formulas to detect 785
  height-weight tables 785
  pinch test 786
  slenderness index 786

surgery 793
treatment 786
types 781
weighing under water 786
Octacosanol 793
Octanoic acid 159
Offal 793
Oil(s) 793
  crops 793
  palm 793
    history 794
    nutritional value 794
    preparation 794
    processing 794
    production/world/U.S. 794
    uses 794
  process 804
  soybean, processing of 970
  vegetable 794
    composition 797
    consumption/capita 795
    extraction of 795
    fatty acids, essential 798
    history 794
    hydrogenation 796
    iodine number 797
    kinds 799
    melting point 798
    nutritive value 798
    processing 795
    production/world/U.S. 795
    refining 796
    saponification value 797
    toxic factors 799
    triglycerides, medium-chain . 798
    vitamin E 798
Oilseed 794
Okolehao 293
Okra 1043
Oleate 507
Olegasaccharide 161
Oleic acid 804
Oleomargarine 652
Olfactory 804
Oligosaccharide 804
Oliguria 804
Olive 804
  history 805
  nutritional value 806
  oil 801
  preparation 805
  production/world/U.S. 805
-ology 806
Omophagia 806
Onion 806
  dry 807
  flakes, dehydrated 346
  green 807
  history 806
  nutritional value 807
  preparation/uses 807
  production/world/U.S. 807
On-the-hoof 809
On-the-rail 809
Opthalmia 809
Opossum 809
Opsin 809
Opsomania 809
Oral 809
Orange
  blood 809
  butter 809
  juice 809
  pekoe 809

sweet 809
  history 810
  nutritional value 810
  preparation/uses 810
  production/world/U.S. 810
Organic 811
  acid 811
Organically grown food 811
  cost 812
  definition 811
  quality 811
  quantity 812
Organ meat 812
Organoleptic 812
Orgeat 812
Ornithine-Arginine Cycle 812
Orogenital syndrome 254
Orotic acid 1068, 1082
Orotic aciduria 568
Ortanique 812
Osborne, Thomas B. .. 892, 897, 1068
Osmophillic yeast 812
Osmosis 813
Osmotic pressure 813
Ossein 813
Ossification 813
Osteitis 813
Osteo– 813
Osteoarthritis 54
  causes 55
  prevention 55
    bone fractures 814
  symptoms 55
  treatment 55
Osteoblast 813
Osteodystrophy 813
Osteomalacia .. 112, 813, 933, 1096
Osteoporosis 112, 813
  causes 813
  prevention 814
  signs 814
  treatment 814
Otaheite
  apple 816
  gooseberry 816
Ovalbumin 816
Overweight 816
Ovomucoid 816
Oxalic acid 139, 816
Oxaloacetic acid 816
Oxaluria 816
Oxidation 816
Oxidative phosphorylation 816
Oxytocin 309
Oysterplant 1044

Pachyderma.................. 254
Palatability.................. 817
Palm........................ 817
  oil........................ 801
  sugar.................... 992
Palmitate 507
Palmitic acid 817
Pancreatin 817
Pancreatitis 817
Pandanus fruit.............. 817

O

P

Pangamic acid **817**, 1067
  chemistry 818
  deficiency symptoms 819
  disorders 818
  functions 818
  history 818
  metabolism 818
  properties 818
  sources 819
Pantothenic acid **820**, 1063
  chemistry 820
  deficiency symptoms 822
  functions 821
  history 820
  losses 822
  metabolism 820
  RDA 822, 1063
  sources 822
Papain 532, **823**
Papaw **486**
Papaya **823**
  history 823
  nutritional value 823
  preparation/uses 823
  production/world/U.S. 823
Para-aminobenzoic acid . **823**, 1037
  chemistry 824
  deficiency symptoms 824
  functions 824
  history 824
  metabolism 824
  properties 824
  RDA 824
  sources 824
Parabens 825
Paraplegia 825
Parasites 825
  infections 825
Parathyroid
  gland 825
  hormone 309
Parboil 825
Parenteral 825
Paresthesia 825
Parietal cells 825
Parkin 825
Parotid enlargement 252
Parsley 1043
Parsnip 1043
Parts
  per billion 825
  per million 825
Parturition 825
Passion fruit 487
Pasta 825
Pasteurization 825, 888
Pasteur, Louis 572
Patch test, allergy 22
Pâtés 825
Path-, patho-, -pathy 825
Pathogens 825
Pathology 826
Pavlov pouch 826
PBI test 826
Pea
  field 613
  flour 361, 364
  garden 827
    history 827
    nutritional value 828
    preparation/uses 828
    production/world/U.S. 827
Peach/nectarine 826
  history 826
  nutritional value 827

palm 827
  preparation/uses 827
  production/world/U.S. 826
Peacock 827
Peafowl 827
Peanut 829
  butter 831
    natural 532
  flakes 831
  flour 361, 364, 532, 831
  history 829
  nutritional value 830
  oil 802, 831
  processing 829
  production/world/U.S. 829
  products/uses 830-832
  roasted 832
Pear 832
  history 832
  nutritional value 833
  preparation/uses 833
  production/world/U.S. 832
Pearl barley 81
Pearled 833
Pecan 833
  history 833
  nutritional value 834
  preparation/uses 834
  production/world/U.S. 833
Pectin 341, 346, 519, 532, **834**
Pectinases 834
Pediatrics 835
Pekoe (tea) 836
Pellagra 835
  causes 835
  history 835
  prevention 835
  signs of 835
  treatment 835
Penicillin 45
Pentosans 161
Pentose 160
Pentosuria 566
Pepperoni 836
Peppers 836
  history 836
  nutritional value 837
  preparation/uses 836
  production/world/U.S. 836
Pepsin 837
Pepsinogen 837
Peptide 28
  assay 34
  bond 28
Peptone 837
Per capita consumption
  additives 8
  beef 88
Perinatal 837
Peripheral neuropathies 253
Periodontal disease 252
Peristalsis 837
Pernicious 837
  anemia 38, 837
Per oral 837
Per os 837
Peroxidation 837
Peroxide value 837
Persian
  berry 838
  melon 679
Persimmon 487
Pesticides 838, 859
Petigrain oil 838
pH 838

Phages 74
Phagocyte 838
Phagomania 838
Pheasant 838
Phenol oxidases 838
Phenylalanine 28, 838
Phenylketonuria (PKU) 565
Phenylpyruvic acid 838
Pheophytin 838
Phosphate baking powder 75
Phospholipids 838
Phosphoprotein 839
Phosphorus 714, 839
  calcium ratio 840
  deficiency symptoms 840
  functions 840
  history 839
  metabolism 840
  RDA 840
  related diseases 840
Phosphorylate 842
Phosphorylation 842
Photosynthesis 312, 318, 842
  diagrammatic summary 843
  ruminant/step-up energy 844
Phrynoderma 254
Phylloquinone 844
Physical
  activities 844, 847
  analysis methods 35
  assay 1057
  fitness & nutrition 844
  dietary guidelines 846
  importance 845
Physiological saline 847
Phytic acid 139, 847
Pica 25, 38, 847
Pickling 000
Pieplant 923
Pigeon pea 613
Pigeons 869
Pigmentation 847
  of the conjunctiva 252
Pig souse 847
Pilsner beer 92
Pinch test for fat 786
Pineapple 848
  history 848
  nutritional value 849
  preparation/uses 849
  processing 848
  production/world/U.S. 848
Pineapple guava 849
Pine nut 778
Pink disease 6
Pinocytosis 849
Pintailed duck 849
Pinworms 287
Pistachio nut 778
Pitanga 114
Pith 849
Pits 850
Pituitary gland 850
Pizza 115, 850
Placebos 850
Placenta 850
Plantain 77, 850
  history 850
  nutritional value 850
  preparation/uses 850
  processing 850
  production/world/U.S. 850
Plant proteins 850
Plaque 850

Plasma 850
Plate count 75
Platelets 109, 850
Pluck 850
Plum/prune 851
  history 851
  nutritional value 852
  preparation/uses 852
  processing 852
  production/world/U.S. 851
  types 851
Poach 852
Pocket bread 115
Poi 852
Poison Control Centers 854
Poisoning 852
Poison ivy 852
Poisoning, food 475
Poisonous plants 284, 852
  common 854
  history 853
  poison aid 853
  poison control centers 854
  precautions/cautions 853
Poisons 284, 854
  chemical 860
  control centers 855
  emergency care 855
  environmental 856
  prevention 855
  substances 856-860
  treatment 855
Poliomyelitis 286
Pollen 532
Pollutants, environment 283
Pollution 154
  control 860
Poly- 860
Polyarteritis nodosa 860
Polybrominated biphenyls 859
Polychlorinated biphenyls 859
Polycythemia 860
Polydipsia 860
Polyneuritis 860
Polyneuropathy 861
Polypeptides 861
Polyphagia 861
Polyphosphates 861
Polysaccharide 161, 162
Polyunsaturated fatty acids ... 538
Polyuria 861
Pomace 861
Pome 861
Pomegranate 487
Ponderal index 785
Popovers 116
Population
  growth 861
  Malthus 861
  world 861
Pork 862
  consumption/capita 864
  cuts 862-863
  grades 862
  production/U.S. 864
  qualities 862
Porphyria 569
Porphyrin 864
Portal 864
Porter (beer) 92
Post, C. W. 123, 1048
Postoperative nutrition 864
Potassium 539, 715, 865
  bitartrate 243

deficiency symptoms 865
high/diet 744
history 865
metabolism 865
RDA 866
sources 866
Potato 866
  bean 1147
  flour 361, 364, 532
  history 867
  Irish 866
  nutritional value 867
  preparation/uses 867
  production/world/U.S. 867
  syrup 869
Pot barley 81
Poultry 869
  buying 872
  classes 871
  consumption
    world 870-871
    U.S./capita 874
  cooking 872
  federal inspection 871
  grades 871
  kinds 869
  marketing 871
  nutritive qualities 874
  production/world 870-871
  storing 872
Prairie chicken 518
Precursor 875
Preeclampsia 882, 1019
Pregnancy/lactation 875
  anemia 38, 881, 886
  body changes 876-877
  breast feeding 883
  constipation 881
  counseling 883-885
  cravings 881
  diabetes 881
  diet 880
  energy required 149
  fluid retention 882
  heartburn 882
  high blood pressure 882
  history 875
  morning sickness 882
  nutrient allowances 877
  nutrition problems 881-882
  weight gain 881-882
Prenatal 886
Preservation of food 886
  methods 887-889
Preservatives 7
Preserves 122
Pressed 890
Pressure cooking 890
Pretzels 116, 890
Prickly pear 890
Processed foods, highly 890
  cereals 890
  dairy products 890
  eggs 890
  enrichment limitation 881
  fish 890
  fruits 891
  meal replacements 891
  meats 890
  seafoods 890
  trends 890
  vegetables 891
Processsing foods 891
Proenzyme 891

Progesterone 310
Prognosis 891
Prolactin 309
  hormone 308
Prophylaxis 891
Propyl gallate 891
Prostaglandins 891
Prosthetic group 891
Protease 891
Proteins 687, 891
  allergy 24
  allowance 895
  amino acids 893
    essential 893
    factors/utilization 899
    nonessential 893
    requirement 895
  analyses 34
  animal 906
  biological value 105
  bound iodine 906
  classification 894
  complementary 898
  complete 907
  consumption/capita 901
  controlled diet 742
  crude 907
  digestion/absorption 278, 894
  efficiency ratio 907
  egg 304
  energy malnutrition 251
  energy source, as 906
  factor 907
  food compositions 901-906
  functions 894
  high/diet 742
  high grains 181
  history 892
  hydrolysate 907
  incomplete 907
  malnutrition 899
  metabolism 894
  milk 907
  partially complete 907
  per acre crop 518
  quality 896
  quantity 896
  required intake 895
  score 907
  single-cell 907
  sources 896
  sparing action of
    carbohydrates 162
  supplements 532
  synthesis 893
  utilization 907
  world/capita 907
Proteinuria 907
Proteolysis 908
Proteose 908
Prothrombin 908
  activator 908
Proton 908
Protoplasm 908
Provitamin(s) 1057
  A 167, 908
Proximal 908
Proximate analysis 31, 342
Prune 851, 908
  juice 908
Psoriasis 908
Psychrophillic bacteria 908
Psyllium 519, 908
  seed 346, 532

Ptomaines 908
Puberty 908
Puerperium 909
Pulp 909
Pulse(s) 909
Pummelo 909
Pumpernickel bread 909
Pumpkin(s) 909
   history 909
   nutritional value 910
   preparation/uses 909
   production/world/U.S. 909
   seeds 532
Purgative 910
Purine(s) 56, 910
   restricted breads 120
   restricted diet 742
Puris 910
Purothionin 910
Putrefaction 910
Putrescine 910
Pyridoxine 1064, 1075
Pyrimidines 910
Pyruvic acid 910

Q

Quadriplegia 911
Quail 110, 911
Quick breads 116, 911
   history 911
   nutritional value 912
   trends 914
   types 912
Quillaja 914
Quince
   pineapple 487
   seed 519

R

Rabbit 915
   fever 72
Raccoon 915
Rad 915
Radiation 154, 915
   for cancer 156
   foods 889
   injury 282
   preservation of foods 915
      uses 917
   therapy 916
Radioactive
   fallout 916
   isotopes 916
   I¹³¹ test 916
Radish 1044
Raffinose 916, 1000
Ragout 916
Raisin(s) 514, 916
   oil 916
Raisine 916

Rambutan 487
Ramp 916
Rampion 916
Rancidity 326, 916
Rangpur 917
Rape (canola) 158, 917
   history 917
   nutritional value 917
   oil 802
   processing 917
   production/world/U.S. 917
   products/uses 917
Rasher (bacon) 917
Raspberry 487
Ratafia 917
Ravioli 917
Raw Sugar 918
RDA 918
   countries selected 918
Recommended daily
   allowances (RDA) 918
Recommended dietary
   allowances 918
Rectum 918
Red blood cells 36, 109
   functions 37
   iron role 37
   production 36
Red meat 919
Reducing diets 919
Reedbird 110
Reference
   man 919
   protein 919
   woman 919
Refrigeration of foods 889
Refsum's disease 567
Regional enteritis 244
Regurgitation 919
Reindeer 919
Relapse 919
Relaxin 310
Religion(s)
   11 living religions 920–921
   diets & 919
Relish/chutney 922
Remission 922
Renal
   calculi 596
   glucosuria 566, 922
   threshold 922
   tubular acidosis 569
Rendering 922
Rennet 922
Rennet, vegetable 922
Rennin 922
Repletion 922
Resection 922
Reserpine 922
Residue 922
Resorption 922
Respiration 922
Respiratory
   chain 922
   quotient 147
   system 922
Restaurant 922
Resting Energy Expenditure (REE) 149
Restoration 315
Reticulocyte 922
Reticuloendothelial system 922
Retinal 922
Retinoic acid 922
Retinol 922
Retinopathy 923

Retrogradation 923
Rhamnose 923
Rheumatic fever 923
Rheumatoid arthritis 55
   causes 55
   prevention 56
   symptoms 55
   treatment 56
Rh factor 884
Rhodopsin 923
Rhubarb 923, 1044
Riboflavin 923, 1064
   chemistry 924
   deficiency symptoms 53, 925
   functions 924
   history 924
   losses 926
   metabolism 924
   RDA 925, 1064
   sources 926
Ribonucleic acid 927
Ribose 927
Ribosomes 927
Rice 927
   Bataan experiment 930
   beer 93
   brown 532
   enriched 930, 932
   flour 360, 363, 532, 932
   glutinous 532
   history 927
   milled 928
   nutritional value 929
   paper 933
   parboiled 928, 932
   polish 533
   polishings 932
   processing 928
   production/world/U.S. 927
   products/uses 931–933
   white 933
   wild 932
   wine, Japanese 93
Ricebird 110
Rickets/osteomalacia 112, 933, 1095
   causes 933
   history 933
   prevention 934
   signs/symptoms 934
   treatment 934
Rigor mortis 935
Rissoles 935
RNA 765, 893
Roasted grain drinks 935
Roasters (pigs) 935
Rocambole 935
Rolled oats 780
   wheat 1126
Rolls 935
Rope (breads) 935
Rose
   apple 935
   hips 533, 936
Roselle 936
Rose, William C. 28, 892, 897
Roundworms 287
Roux 936
Rowe elimination diet 23, 936
Royal jelly 533, 936
Rum 292
Rumen 936
Ruminants step-up energy 844
Rusks 936
Russian rye bread beer 93

Rutabaga 1044
Rye 936
  bread 937
  flour 360, 363, 533, 937
  grain 533
  history 937
  nutritional value 937
  processing 937
  production/world/U.S. 937
  products/uses 937
  wafers 116

S

Saccharimeter 939
Saccharin 939, 999
Saccharometer 939
Safety of food 939
Safflower 939
  history 939
  nutritional value 939
  oil 533, 802
  processing 939
  production/world/U.S. 939
  products/uses 940
Saint Anthony's fire 940
Saint John's bread 611
Sake 93
Salami 940
Salep 940
Saline 940
Saliva 940
Sally Lunn 940
Salmonella 940
Salmonellosis 72, 302
Salsify 1044
Salt 940
  commercial uses 941
  curing 889
  deficiency 941
  diet, low 942
  excess 941
  foods, low 942
  functions 941
  history 940
  poisoning 860
  requirement for athletes 62
  salt-free diet 533, 942
  sea 533
  sources 942
  toxicity 941
Saltine 942
Saltpeter 942
Salt-rising bread 115
Sambal 942
Samp 942
Sanitation, food 289
Sapodilla 487
Saponification 326, 797, 942
Saponins 942
Sapote, green 488
Sarcoma 152, 942
Satiety 942
Saturated 942
  fat 942
Sauces 942
Sauerkraut 135, 943
Sausage 943
  casings 942
Scarlet fever 286

Scarlet runner bean 613
Schelle, Carl 750
Schistomiasis 101
School
  breakfast 188, 512
  lunch program 188, 511, 944
    benefits 944
    goals 949
    guidelines 945
    history 944
    provision of Act 944
Schwarz, K. 713
Scleroderma 946
Scleroproteins 346
Scotch barley 81
Scrapple 946
Screened 946
Screenings 946
Scurvy 946, 1089
  causes 947
  history 946
  prevention 947
  signs/symptoms 947
  treatment 947
Sea kale 948
  water 533
Seafood(s) 948
  allergy 25
Seasonings 356
Seaweed 13, 948
  composition 948
  uses 948
  varieties 948
Seborrheic dermatitis 107
Seeds 949
Selective permeability 949
Seleniferous 949
Selenium 718, 949
  areas 950
  deficiency symptoms 951
  functions 951
  metabolism 950
  prevention 954
  RDA 954
  soils, in 949
  sources 955
  treatment 954
  vitamin E & 951, 1100
Semidispensable amino acid .. 955
Semipermeable membrane 955
Senior citizen 9
  nutrition 498
Septicemia 955
Serendipity berry 956
Serotonin 96, 956
Serum 956
  cholesterol 956
  triglycerides 956
Sesame 956
  history 956
  nutritional value 956
  oil 803
  processsing 957
  production/world/U.S. 956
  products/uses 957
  seeds 533
  tahini 533
Seven food groups food plan .. 957
Seville orange 957
  history 958
  nutritional value 958
  production/world/U.S. 958
  uses 958
Shallot 1045

Shandy (beer) 92
Shea butter 958
Sherbets 705
Shigellosis 70, 72
Shinto dietary beliefs 921
Shortening 958
  types 958
Shredded 959
Sickle cell anemia 39
Siderotic anemia 38
Sifted 959
Sikhism dietary belief 921
Silicon 719, 959
  deficiency symptoms 959
  functions 959
  history 959
  metabolism 959
  RDA 959
  sources 959
Simple sugars 959
Single-cell protein 901, 959
Sippy diet 960
Sirloin butt 134
Skimmed 960
Skin 960
  atrophy 253
  bruising 254
  discoloration 253
  dryness 254
  eczema 254
  test 22
Skinfold thickness 1028
Slenderness index 785
Sloe 960
Smoking 12, 153
Smoking foods 889
Snack foods 960
  caloric values 62
Snipe 960
Soap 960
Soda water 961
Sodium 539, 714, 961
  bicarbonate 965
  carboxymethyl cellulose 165
  chloride 940
  deficiency symptoms 961
  diacetate 965
  functions 961
  history 961
  metabolism 961
  phytate 965
  potassium ratio 965
  RDA 961
  restricted diet 745
  sources 962
  tetraborate 113
Soft diet 738
Soft drinks 963
  consumption 963
  history 963
  ingredients 964
  nutritional value 965
Soils, minerals in 726
Solanine 965
Solids, milk 965
  non-fat 965
Solutes 965
Solvent-extracted 965, 970
Sorbic acid 965
Sorghum 965
  flour 967
  history 965
  kinds 966
  milling 966

dry 966
wet 966
nutritional value 966
oil 803, 967
production/world/U.S. 965
products/uses 966
starch 967
syrup 533, 967
Sorrels 293, **1045**
Soups **968**
kinds 968
nutritional value 969
Sour grasses **293**
Sour orange **957**
Soursop **488**
Southern beaten biscuit **87**
Soy
flour 533
protein powder 533
Soybean **969**
allergy 25
flour 361, 364, 974
history 969
lecithin 974
milk 974
milk powder 533
nutritional value 971
oil 803, 974
extraction 970
expeller 970
hydraulic 970
refining 970
processing 970
production/world/U.S. 969
products/uses 972–975
protein 974
shakes 975
Spanish
goat 975
lime 975
Specific
dynamic action 975
gravity 975
heat 975
Spectrophotometry analysis **33**
Spelt **182**
Sphincter **976**
Sphingomyelin **976**
Spices **356**
Spinach **1045**
Spirullina **533**
Spoilage of food **976**
Sponge cakes **116**
Spore **976**
Spray-dehydrated **976**
Sprouted grains **178**
Sprouting barley 80, **174**
Sprouts **533**
Spruce beer **976**
Sprue **976**
nontropical 976
tropical 977
Squab **977**
Squashes **977**
history 977
nutritional value 978
preparation/uses 977
production/world/U.S. 977
types 977
Stabilized **979**
Stabilizer 8, **979**
Stachyose **979**
Stalling **979**
Staphylococcal food poisoning . **73**

Star-apple **979**
Starch **979**
characteristics 980
derivatives 519
gelatinization 980
nutritional value 981
pregelatinized 982
production 980
uses 981
waxy 982
Starvation **982**
adaption to 982
causes 982
effects 983
prevention 983
signs 983
treatment 983
Stasis **984**
Stearic acid **984**
Steatorrhea **984**
Steenbock, Harry ... 933, 1069, **1092**
Stenosis **984**
Sterilization **984**
Steroids **984**
Sterol **984**
Stillage **984**
St. John's bread **611**
Stockmelon **203**
Stomatitis 253, **984**
Stool **984**
Strawberry **984**
history 985
nutritional value 986
pear 986
preparation/uses 985
production/world/U.S. 985
tomato 986
tree 53
Strep throat **286**
Streptococcal infections,
foodborne **986**
Streptomycin **45**
Stress 12, 282, **986**
diet 988
energy requirements 150
infections 987
injuries 987
nutritional demand 987
physical 987
response 987
weather 988
Stroke **988**
Strontium-90 .... **988**
Sub– **989**
Subcutaneous **989**
Substrate **989**
Succinic acid **989**
Succotash **989**
Sucrase **989**
Sucrose 989, 1000
esters 989
Sufu **975**
Sugar **989**
acids 161
alcohols 161
beets 990
behavior, undesirable 95
brown 993
by-products 990–991
cane 989
consumption/use 992
curing 889
disease & 994
doctor 994

granular 993
history 990–991
icing 994
invert 993
liquid 993
nutritional value 991
powdered 993
processing 990–991
production 990–991
raw 533, 993
sources 989, 991
turbinado 993
Sugaring of dried fruits **994**
Sugaring off **994**
Sulfa drug **994**
Sulfhydryl group **994**
Sulfur 716, **994**
amino acids 996
deficiencies 995
dioxide 996
functions 995
history 995
metabolism 995
RDA 995
sources 995
Sultana raisins **996**
Summer food program for
children **512**
Sun-cured **996**
Sunflower **996**
flour 361, 364, 998
history 996
nutritional value 997
oil 804
processing 996
production/world/U.S. 996
products/uses 997
seeds 998
Superamine **1126**
Superglycinerated fats **998**
Superoxide dismutase **998**
Supplements **999**
Surface area **999**
Surfactant **999**
Surinam cherry **114**
Swede **1044**
Sweet
corn 999
potato 1000
history 1001
nutritional value 1002
preparation/uses 1001
production/world/U.S. 1001
Sweetbreads **999**
Sweeteners, artificial **58**
aspartame 59
Sweetening agents **999**
comparison 999
Sweetsop 484, **1002**
Swiss chard **1045**
Symptom **1002**
Syn– **1002**
Syndrome **1002**
Synergism **1002**
Synthesis **1002**
Synthetic **1002**
Syrups **1002**
Systemic **1002**
Systolic blood pressure **109**
Szent-Gyorgy, Albert **1085**

Tachycardia 253, **1003**
Takaki, Kanehiro . 1009, 1055, 1083
Tallow 43, **1003**
Tamari 534
Tamarind 488, 519
    tree 614
Tangelo **1003**
Tangerine **1003**
    history 1003
    nutritional value 1004
    preparation/uses 1003
    production/world/U.S. 1003
Tangor **1003**
Tannia **1004**
    nutritional value 1004
    preparation 1004
Tannis **1004**
Tao-cho 974
Taoism dietary beliefs 921
Tao-si 975
Taotjo 975
Tapeworm 39
    beef 287
    dwarf 287
    fish 287
    pork 287
Tapioca 169, 534, **1004**
    flour 361, 364, 534
Tapioca-macaroni **1005**
Tardive dyskinesia 96
Taro 1046
Tartar **1005**
    emeric 1005
Tartaric acid **1005**
Tartazine **1005**
Taste **1005**
    buds 1005
Taurine **1005**
Tay-Sachs disease 567
Tea **1005**
    black 1007
    grades 1007
    green 1007
    history 1005
    nutritional value 1007
    oolong 1007
    processing 1007
    production/world/U.S. 1005
Teaseed oil **1007**
Teeth 256
    baby 256
    care of 259
    caries 257
    decay 257
    development 257
    diet effect 257, 259
    fluoride 259
    nutrition effect 257, 259
    periodontal disease 258
    permanent 256
    primary 256
    structure 256–257
Teff .... '**1007**
    faffa 1007
    use 1007
Tempeh 975
    wheat 1125

Template **1008**
Temple orange 1003
Tenderloin 1008
Teosinte 182
Tepary bean **614**
Tequila 293
Terramycin as preservative ... 1008
Testosterone 310
Tetany 1008
Tetrasaccharides 161
Texture 1008
Textured soybean proteins 975
    vegetable product 534
Theine 1008
Therapeutic 1008
    diets 1008
Therapy 1008
Thermal 1008
Thermoduric 1008
Thermogenesis 1008
Thermophiles 1008
Thi- (Thio-) 1008
Thiamin **1008**, 1064
    anti-factors 1012
    cardiovascular 1011
    chemistry 1009
    deficiency symptoms 1011
    diphosphate 1013
    functions 1010
    gastrointestinal 1011
    history 1009
    losses 1012
    metabolism 1009
    RDA 1011, 1064
    sources 1012
Thickeners 8
Thirst 1013
Threonine 1013
Thrombin 1013
Thrombocytes 109
Thrombosis 1013
Thymine 1013
Thyrocalcitonin 137, 1013
Thyroid
    gland 1013
    stimulating hormone ... 309, 1013
Thyroiditis **1013**
Thyrotoxicosis **1014**
Thyroprotein hormone **308**
Thyroxin 309, **1014**
TI **519**
Timbale **1014**
Tin 860
Tissue **1014**
    body 1014
Toadskin 254
Toasted 1014
Tocopherols 1014, 1060, 1098
Tocoterienol **1014**
Tofu 534, **975**
Tolbutamide 1014
Tomato 1014
    history 1014
    juice 1016
    nutritional value 1016
    preparation/uses 1015
    production/world/U.S. 1014
-tomy 1016
Tonic 1017
Tooth structure 256
Tophus 1017
Tortillas 115
Torularodin 1017
Torula yeast 534, **1017**

Total nutrition
    crude protein 1017
    milk solids 1017
    parenteral nutrition 1017
        benefits 1018
        composition 1018
        contraindications 1019
        home use 1018
        precautions 1019
Tox- 1019
Toxemia 1019
    decline 882
    pregnancy 1019
Toxicants 282
Toxins 1020
    bacterial, foodborne 1020
Trabeculae 112
Trace elements 712
Tragacanth gum 519
Transamination 1020
Trans fatty acid(s) 799, **1020**
Transferase 1020
Transferrin 1020
Transketolase 1020
Trauma 1020
Tree tomato 1020
Trichinosis 288
Triglycerides 1020
    medium-chain 798
Tripe **1020**
Tripeptides 1020
Tristearin 1020
Triticale 1020
Tropical ulcer 254
Truffles **1021**
Trypsin **1021**
    beans 86
Trypsinogen 1021
Tryptophan 28, **1021**
Tube feeding 1021
Tuber 1022
Tuberculosis 72
Tuberin **1022**
Tularemia 72
Tun 1022
Turbinado sugar 1022
Turkeys 869
Turnip 1046
    bean 1147
Turtle 1022
Twaddell 1022
Typhoid fever 287
Tyramine 1022
Tyrosinase 1022
Tyrosine 28
Tyrosinemia 565
Tyrosinosis 565

Ubiquinone 209, **1066**
Udo **1023**
UHT sterilization **1024**
Ulcers, peptic **1023**
    causes 1023
    diagnosis 1023
    diet 1024
    duodenal 44, 1023

gastric 44, 1023
occurrence 1023
prevention 1024
symptoms 1023
treatment 1024
Ultrahigh temperature
   sterilization **1024**
Ultrasonic homogenizer **1025**
Ultraviolet light **1025**
uses 1025
Umbles **1025**
Underweight **1025**
arm circumference 1028
causes 1026
effects 1025
   beneficial 1025
   harmful 1026
skinfold thickness 1028
weight status 1026
UNESCO **1028**
Unidentified factors **1028**
United States Pharmacopia .. **1029**
Unleavened breads **115**
Unsaturated
fat 1028
fatty acid 1028
Urea **1028**
cycle 1028
Urease **1029**
Uremia **1029**
Uric acid **1029**
kidney stones 1029
Uricase **1029**
Urinary
calculi 596, 1029
system 1029
Urine **1029**
Urolithiasis 596, **1029**
USDA 290, **1029**
U.S. Department of
   Agriculture 290, **1029**
U.S. Department of Health and
   Human Services 290, **1029**
USDH & HS **290**
USP **1029**
USPHS **1029**
U.S. Public Health Service ... **1029**
U.S. RDA **1030**
U.S. Recommended Daily
   Allowances **1030**
Uterus **1030**

V

Vaccination **288**
Vacuum cooling **1031**
Vacuum-dehydrated **1031**
Vacuumization **1031**
Valence **1031**
Valine **1031**
Vanadium **1031**
Vanaspati **1031**
Van Lent 1009, **1083**
Van Soest analysis 32, **342**
Vascular **1031**
Vasopressin **309**
Veal **87**
consumption 88

cooking 88
cuts 88
grades 88
nutritional value 88
production 88
Vegans **1031**
Vegetable(s) **1031**
butter 134, 1048
history 1032
nutritive value 1033
oils 1048
production/world/U.S. 1033
salt 534
types 1032
world, of the 1036–1047
Vegetarian
diets 1048
   benefits 1052
   meat substitutes 1051
   problems 1049
   types 1048
gelatin 534
Venison **1052**
Verjuice **1052**
Vermicelli **1052**
Vermouth **1052**
Versinia enterocolitica 73
Vetch **1052**
Vibrio parahalmolyticus 73
Vicilin **1053**
Vienna bread **1053**
Villi **1053**
Vinegar **1053**
definition 1053
uses 1053
Viral infections, foodborne .. **1053**
Virus(es) **1053**
Viscosity **1053**
Visual purple **1053**
Vitalia macaroni **1126**
Vitamers **1053**
Vitamin(s) 688, **1054**
biological assay 1057
chemical assay 1057
classification 1055
definition 1055
fat-soluble 1055
history/discovery 1054
microbiological assay 1057
physical assay 1057
supplementation 1058
supplements 1105
table 1059 1068
water-soluble 1055
Vitamin A 1060, **1068**
blindness 109
cancer 1072
chemistry 1069
deficiency 1072, 1075
functions 1070
history 1068
metabolism 1069
losses 1074
RDA 1060, 1073
sources 1074
Vitamin antagonist **1075**
Vitamin B complex 1056, **1083**
history 1083
nine fractions 1084
Vitamin B-1 .... 1008, **1064**
Vitamin B-2 .... 923, **1064**
Vitamin B-3 .... 820, 1063
Vitamin B-6 .... 1064, **1075**
chemistry 1076

deficiency symptoms 1077
functions 1076
history 1076
losses 1078
metabolism 1076
RDA 1064, 1077
sources 1078
Vitamin B-12 .... 1065, **1079**
anemia 38, 42
chemistry 1080
deficiency symptoms 1081
functions 1081
history 1079
losses 1082
metabolism 1080
pernicious anemia 1081
RDA 1065, 1081
sources 1082
sprue 1081
Vitamin B-13 .... 1068, **1082**
chemistry 1082
deficiency symptoms 1083
functions 1083
history 1082
RDA 1083
sources 1083
Vitamin B-15 .... 817, **1067**
Vitamin B-17 .... **1067**
Vitamin B-T 166, **1066**
Vitamin C 59, 1065, **1084**
body levels 1088
cancer & 1087
chemistry 1085
colds 1088
deficiency symptoms 1088
functions 1086
history 1084
losses 1090
metabolism 1085
RDA 1065, 1089
scurvy 1089
sources 1091
Vitamin D 191, 1060, **1091**
chemistry 1093
deficiency symptoms 1095
functions 1094
history 1092
losses 1097
metabolism 1093
osteomalacia & 1096
RDA 1060, 1096
rickets & 1095
sources 1097
Vitamin E 542, 1060, **1098**
chemistry 1098
deficiency symptoms 1100
functions 1099
history 1098
losses 1101
metabolism 1098
RDA 1060
selenium & 951, 1100
sources 1101
vitamin D & 1100
Vitamin H **106**
Vitamin K 1061, **1102**
antagonist 1103
chemistry 1102
deficiency symptoms ... 45, 1104
functions 1103
history 1102
losses 1104
metabolism 1102
RDA 1061, 1104

sources 1104
Vitamin L 1105
Vitamin P 102, 1066
Vodka 292
Voit, C. 892
Volatile oils 1105
Vomiting 1105
von Liebig, Justus 892
Von, Voit, Carl 892

W

Walnut, black, English, white 1107
  history 1107
  nutritional value 1107
  preparation/uses 1107
  processsing 1107
  production/world/U.S. 1107
Warburg-Christian assay 34
Warburg's yellow enzyme ... 1108
Water 689, 1108
  aerated 1110
  balance 1108, 1113
  belly 1029
  demineralized 1114
  electrolytes, & 1110
    acid-base balance 1113
    imbalances 1111
    regulation 1110
  extracellular 1114
  fluoridation of 367
  glass 1114
  hardness 1114
  importance 1108
  intracellular 1114
  lemon 1114
  requirements for 1109
    athletes 62
  sources 1109
Waterborne diseases 1113
Waterchestnut, Chinese 1046
Watercress 1046
Watermelon 679, 1114
  history 1114
  nutritional value 1115
  preparation/uses 1115
  production/world/U.S. 1114
Water-soluble vitamins 1115
Watson, James 892
Waugh, W. A. 1085
Wax
  apple 1115
  gourd 1115
Weaning 130
Weighing under water 786
Weight(s)
  average, (ages 25-35) 1027
  children 186-187
  control groups 791
  normal body 1027
Weight-height-age percentile
  standards 1115
Weights & measures 1116
  avoirdupois 65
  conversions 1117-1118
  English 1117
  foods 1116
  metric 1117
  U.S. customary 1117

Weiner 477
Weisse (beer) 92
Wernicke's disease 251
Whale oil 1118
Wheat 1118
  allergy 24
  cracked 534
  enriched 1122
  flour 360, 363, 1121
  germ 534, 1125, 1127
  germ flour 534
  germ oil 534
  gluten 1125
  grades, Federal 1121
  gum 519
  history 1119
  kinds of 1119, 1121
  milling 1120
  nutritional value 1123
  production/world/U.S. 1119
  products/uses 1124-1126
  protein beverage 1126
  starch 1125-1126
  whole grain berries 534
Wheatmeal, National 1127
Whey 706
  powder 534
  soy drink 975
Whipple, George Hoyt 1079
Whipple's disease 1127
Whiskey 82, 292
White blood cells 1127
White, Ellen 1048
WHO 291, 1143
Whole wheat meal 1127
WIC 512
Wiener 477
Wild edible plants 1127
  types 1127
Wild rice 1129
Williams, R. J. 368, 820
Williams, Robert R. 1009
Wills, Lucy 368
Wilson's disease 569
Windberry 1129
Wine 514, 1130
  aging 1131
  clarification 1131
  fermentation 1131
  history 1130
  nutritional value 1133
  production 1131
  types 1132
Wineberry 1133
Winterization 1133
Witches' milk 1134
Wolman's disease 567
Womb 1030
Women/infants/children 512
Wood
  alcohol 1134
  sugar 1145
Woodcock 1134
Woodward, R. B. 1093
Woolley, D. W. 580
Work, muscular 1134
World food 1134
  diet, type of 1137
  favoring bread 1137
  favoring animals 1138
  feeds vs foods 1140
  geography of 1134
  outlook 1137
  population 861, 1135

  production 1135
  wealth 1135
World Health
  Organization 291, 1143

X

Xanthan 519
Xanthine 1145
Xanthinuria 568
Xanthoma 1145
Xanthophyll 1145
Xeroderma 254
Xerophthalmia 109, 251, 1072, 1145
Xerosis 1072, 1145
  of the conjunctiva 252
Xerostoma 253
Xylitol 1145
Xylose 87, 1145

Y

Yam 1047
Yam bean 1041, 1147
  history 1147
  nutritional value 1147
  production 1147
Yard-long bean 614
Yaupon 1147
Yautia 1004
Yeast 1147
  extract 1147
  instant 534
  nutritional 534
Yogurt 534, 1148
Yolk, index 1148
Yorkshire pudding 1148
Yucca 169

Z

Zante currants 1149
Zeaxanthin 1149
Zedoary root 1149
Zein 1149
Zen macrobiotic diets 1149
Zest 1150
Zinc 719, 1150
  history 1150
  interrelationships 1151
  metabolism 1151
  RDA 1151
  sources 1151
  symptoms 1150
Zollinger-Ellison syndrome ... 1151
Zoroastrianism dietary beliefs . 921
Zwitterion 1151
Zymase 1151
Zymogen 1152
Zymometer 1152

T - #0534 - 071024 - C1200 - 279/216/56 - PB - 9780367401764 - Gloss Lamination